ABELIAN FUNCTIONS

ABELIAN FUNCTIONS
Abel's theorem and the allied theory of theta functions

H. F Baker
St John's College, Cambridge

CAMBRIDGE UNIVERSITY PRESS
Cambridge, New York, Melbourne, Madrid, Cape Town, Singapore, São Paulo

Cambridge University Press
The Edinburgh Building, Cambridge CB2 2RU, UK

Published in the United States of America by Cambridge University Press, New York

www.cambridge.org
Information on this title: www.cambridge.org/9780521498777

First published 1897
Reissued in the Cambridge Mathematical Library series 1995

A catalogue record for this publication is available from the British Library

ISBN-13 978-0-521-49877-7 paperback
ISBN-10 0-521-49877-5 paperback

Transferred to digital printing 2006

CONTENTS.

CHAPTER I.

THE SUBJECT OF INVESTIGATION.

CHAPTER II.

THE FUNDAMENTAL FUNCTIONS ON A RIEMANN SURFACE.

CHAPTER III.

THE INFINITIES OF RATIONAL FUNCTIONS.

CHAPTER IV.

Specification of a general form of Riemann's integrals.

CHAPTER V.

Certain forms of the fundamental equation of the Riemann surface.

CHAPTER VI.

Geometrical investigations.

CHAPTER VII.

Coordination of simple elements. Transcendental uniform functions.

CHAPTER VIII.

ABEL'S THEOREM. ABEL'S DIFFERENTIAL EQUATIONS.

CHAPTER IX.

JACOBI'S INVERSION PROBLEM.

CHAPTER X.

RIEMANN'S THETA FUNCTIONS. GENERAL THEORY.

CHAPTER XI.

THE HYPERELLIPTIC CASE OF RIEMANN'S THETA FUNCTIONS.

CHAPTER XII.

A PARTICULAR FORM OF FUNDAMENTAL SURFACE.

CHAPTER XIII.

RADICAL FUNCTIONS.

CHAPTER XIV.

FACTORIAL FUNCTIONS.

CHAPTER XV.

RELATIONS CONNECTING PRODUCTS OF THETA FUNCTIONS—INTRODUCTORY.

CHAPTER XVI.

A DIRECT METHOD OF OBTAINING THE EQUATIONS CONNECTING THETA PRODUCTS.

CHAPTER XVII.

THETA RELATIONS ASSOCIATED WITH CERTAIN GROUPS OF CHARACTERISTICS.

CHAPTER XVIII.

TRANSFORMATION OF PERIODS, ESPECIALLY LINEAR TRANSFORMATION.

CHAPTER XIX.

ON SYSTEMS OF PERIODS AND ON GENERAL JACOBIAN FUNCTIONS.

CHAPTER XX.

TRANSFORMATION OF THETA FUNCTIONS.

CHAPTER XXI.

COMPLEX MULTIPLICATION OF THETA FUNCTIONS. CORRESPONDENCE OF POINTS ON A RIEMANN SURFACE.

CHAPTER XXII.

DEGENERATE ABELIAN INTEGRALS.

APPENDIX I.

On algebraic curves in space.

APPENDIX II.

On matrices.

Foreword

Classical algebraic geometry, inseparably connected with the names of Abel, Riemann, Weierstrass, Poincaré, Clebsch, Jacobi and other outstanding mathematicians of the last century, has mainly been an analytical theory. In our century it has been enriched by the methods and ideas of topology and commutative algebra and has the authority of one of the most fundamental mathematical disciplines.

The traditional eclecticism (in the best sense of the word) of algebraic geometry has always been a source of its numerous applications to other branches of mathematics. The role of algebraic geometry as "an applied science" has grown immensely in the last 15–20 years, when its new applications to the problems of non-linear equations and quantum field theory were found.

Mechanics, mathematical and theoretical physics can be called "new" spheres of the application of algebraic geometry. These areas are non-traditional only for the algebraic geometry of the second third of our century, the period when the abstract language of Grothendieck's schemes seemed to replace once and for all the somewhat naive language of classical algebraic geometry.

The results of recent years, amongst which we should specially mention the solution of the Riemann–Schottky problem and the applications of topological gravity to the intersection theory of moduli spaces of algebraic curves, show that now, as in the last century, the relationship between algebraic geometry and physics is by no means a one-way street.

The sudden growth in the number of scholars for whom algebraic geometry has become a working tool has highlighted the lack of relevant mathematical literature. Practically all the books on this subject that are available to the reader of today have been written in abstract algebraic language. The idea of maximum generality inherent in the theory of schemes hampers the reader willing to be quickly introduced to the subject, especially if the reader is a physicist.

It would be an exaggeration to state that suitable literature is completely lacking. The neo-classical style is quite evident in some books of recent decades. Amongst them are *Principles of algebraic geometry* by Griffiths and Harris, *Tata lectures on Theta* by Mumford, and *Theta-functions* by Fay.

Undoubtedly, Baker's book assumes a special place in the list, one guaranteed by the mere fact that its first edition is dated at the end of the last century. But that is not its only special feature. It is surprisingly up-to-date and moreover contains some results which have until now remained beyond the scope of present-day textbooks. It is noteworthy that these particular results are closely related to the above-mentioned applications of algebraic geometry to modern mathematical physics.

With all the variety of results obtained within the framework of classical algebraic geometry, its core consists of relatively few basic definitions and theorems. The list includes the Riemann–Roch theorem, the notion of Jacobian variety of an algebraic curve, Abel's theorem and Jacobi's solution of the inversion problem with the help of theta-functions.

Though the author of the book put in its title only Abel's theorem and the theory of theta-functions, the modest words "and the allied theory" mean "all the rest". This "all the rest" includes, besides the theorems above, elements of uniformization theory of algebraic curves and related theory of automorphic forms and the Schottky model of algebraic curves. Of special importance for modern applications are the sections devoted to the "factorial functions".

It is necessary to emphasize another, and possibly the principal, merit of this book. It exhibits the characteristic feature of classical algebraic geometry — the wish to express the final results in exact analytical formulae. This presupposes the definition of a minimal set of new transcendental functions — the "bricks" of which the whole building can be constructed.

In order to demonstrate it we shall briefly present the key points of the so-called finite-gap (or algebraic–geometrical) integration theory of non-linear equations. At the same time this will allow us to do justice to the author of this book, whose name is perpetuated in the "Baker–Akhiezer" function, the concept of which plays a crucial role in various modern applications of algebraic geometry to non-linear physics.

The algebraic–geometrical integration scheme of non-linear equations

is applicable to all the equations that are considered in the framework of the inverse problem method. Among them are:
the Korteweg–de Vries (KdV) equation

$$u_t - \frac{3}{2}uu_x + \frac{1}{4}u_{xxx} = 0, \tag{0.1}$$

its two-dimensional generalization *the Kadomtsev–Petviashvili (KP) equation*

$$\frac{3}{4}u_{yy} = (u_t - \frac{3}{2}uu_x + \frac{1}{4}u_{xxx})_x, \tag{0.2}$$

the non-linear Schrödinger equation

$$i\psi_t = \psi_{xx} + |\psi|^2\psi = 0, \tag{0.3}$$

the sine–gordon equation

$$u_{tt} - u_{xx} = \sin u \tag{0.4}$$

and many other fundamental equations of modern mathematical physics.

All equations that are considered in the framework of the inverse problem can be represented as compatibility conditions for an over-determined system of auxiliary linear problems.

For example, for the KdV equation (0.1), this system has the form

$$L\psi = 0, \quad \partial_t\psi = A\psi, \tag{0.5}$$

where L and A are

$$L = -\partial_x^2 + u(x,t), \quad A = \partial_x^3 - \frac{3}{2}u\partial_x - \frac{3}{4}u_x. \tag{0.6}$$

The compatibility of (0.5) implies

$$[\partial_t - A, L] = 0 \Longleftrightarrow L_t = [A, L]. \tag{0.7}$$

The operator equation (0.7) is called *the Lax equation*. A wide class of non-linear equations can be represented in the form (0.7), where L and A are ordinary differential equations in the variable x with matrix or scalar coefficients that are functions of the variables x, t:

$$L = \sum_{i=1}^{n} u_i(x,t)\partial_x^i, \quad A = \sum_{i=1}^{m} v_i(x,t)\partial_x^i. \tag{0.8}$$

Each Lax equation is an infinite-dimensional analogue of the completely integrable systems. In particular, it can be included in a hierarchy of an infinite set of commuting flows. For the KdV equation they have the form

$$\partial_n u = f_n(u, u_x, \dots, u^{(2n+1)}), \quad u = u(x, t, t_3, t_4, \dots), \quad \partial_n = \frac{\partial}{\partial t_n}, \qquad (0.9)$$

and are equivalent to the operator equation

$$\partial_n L = [A_{2n+1}, L], \qquad (0.10)$$

where L is the Schrödinger operator and A_{2n+1} is a differential operator of order $2n+1$.

The initial definition of n-gap solutions of the KdV equation was proposed by Novikov who considered the restriction of the KdV equation to the space of stationary solutions of (0.10)

$$f_n(u, u_x, \dots, u^{(2n+1)}) = 0 \Longleftrightarrow [L, A_{2n+1}] = 0. \qquad (0.11)$$

The operator equation (0.11) is a particular case of the more general problem of the classification of commuting ordinary differential operators L_n and L_m of orders n and m, respectively. As a purely algebraic problem it was considered and partly solved in the remarkable works of Burchnall and Chaundy [1], [2] in the 1920s. They proved that for any pair of such operators there exists a polynomial $R(\lambda, \mu)$ in two variables such that

$$R(L_n, L_m) = 0. \qquad (0.12)$$

If the orders n and m of these operators are coprime, $(n, m) = 1$, then for each point $Q = (\lambda, \mu)$ of the curve Γ, that is defined in C^2 by the equation $R(\lambda, \mu) = 0$, there corresponds a unique (up to a constant factor) common eigenfunction $\psi(x, Q)$ of L_n and L_m:

$$L_n \psi(x, Q) = \lambda \psi(x, Q); \quad L_m \psi(x, Q) = \mu \psi(x, Q). \qquad (0.13)$$

The logarithmic derivative $\psi_x \psi^{-1}$ is a meromorphic function on Γ. In the general position (when Γ is smooth) it has g poles $\gamma_1(x), \dots, \gamma_g(x)$ in the affine part of the curve, where g is the genus of Γ. The commuting operators L_n, L_m (in this case of coprime orders) are uniquely defined by the polynomial R and by a set of g points $\gamma_1(x_0), \dots, \gamma_g(x_0)$ on Γ.

In such a form, the solution of the problem is one of pure classification: one set is equivalent to the other. Even the attempt to obtain exact formulae for the coefficients of commuting operators had not been made. Baker proposed making the programme effective by pointing out that the eigenfunction ψ has analytical properties that were introduced by Clebsch, Gordan and himself as a proper generalization of the notion of exponential functions on Riemann surfaces.

The Baker program was rejected by the authors of [1], [2] consciously (see the postscript of Baker's paper [3]) and all these results were forgotten for a long time.

Briefly, the key points of a proof of the above results are the following. The commutativity of L_n and L_m implies that the space $\mathcal{L}(\lambda)$ of solutions of the equation

$$L_n y(x) = \lambda y(x) \tag{0.14}$$

is invariant with respect to the operator L_m. The matrix elements L_m^{ij}, $i, j = 0, \ldots, n-1$, of the corresponding finite-dimensional operator $L_m(\lambda)$,

$$L_m|_{\mathcal{L}(\lambda)} = L_m(\lambda) : \ \mathcal{L}(\lambda) \longmapsto \mathcal{L}(\lambda) \tag{0.15}$$

in the canonical basis

$$c_i(x, \lambda, x_0) \in \mathcal{L}(\lambda), \quad c_i(x, \lambda, x_0)|_{x=x_0} = \delta_{ij}, \tag{0.16}$$

are polynomial functions in the variable λ. They depend on the choice of the normalization point $x = x_0$, i.e. $L_m^{ij} = L_m^{ij}(\lambda, x_0)$. The characteristic polynomial

$$R(\lambda, \mu) = \det(\mu - L_m^{ij}(\lambda, x_0)) \tag{0.17}$$

is a polynomial in both variables λ and μ and does not depend on x_0.

According to the property of characteristic polynomials we have

$$R(L_n, L_m) y(x, \lambda) = 0. \tag{0.18}$$

$R(L_n, L_m)$ is an ordinary differential operator. Therefore, if it is not equal to zero then its kernel is finite-dimensional. Hence, (0.18) implies (0.12), and the first statement of [1],[2] is proved.

The equation

$$R(\lambda, \mu) = 0 \tag{0.19}$$

defines the affine part of the algebraic curve Γ in C^2.

Surprisingly, the presentation of the contents in the present book is "parallel" to the solution of this problem. In the first lines we read: *This book is concerned with a particular development of the theory of the algebraic irrationality when a quantity y is defined in terms of a quantity x by mean of an equation of the form*

$$a_0 y^n + a_1 y^{n-1} + \cdots + a_{n-1} y + a_n = 0. \tag{0.20}$$

Possibly, this formulation and the consequent detailed discussion of "what is a *place*" of a Riemann surface, looks too naive for a modern reader, but it has its own advantages, because it allows us to touch the main subjects of the theory right from the start. The general structure of the book is from the particular definitions to a general structure and back to specific problems. For example, in the first chapter of the book just after the definition of algebraic irrationality (0.20) the notion of their rational equivalence is introduced. The invariance of the genus (deficiency) of irrationalities with respect to rational transformations is proved. At the end of this chapter it is proved that "the greatest number of irremovable parameters" for algebraic irrationalities of genus g is equal to $3g-3$ (in modern language this number is the dimension of the moduli space of genus g algebraic curves). And all this in only 13 pages!

Let us return to the classification problem of commuting ordinary differential operators. The consideration of the asymptotic behaviour of the algebraic equation (0.19) in a neighbourhood of "infinity" ($\lambda \to \infty$) allows one to prove that if the orders n and m of L_n and L_m are coprime, then the affine curve (0.19) is compactified by one smooth point, in a neighbourhood of which $\lambda^{-1/n}$ is a local co-ordinate, i.e. infinity is an n-fold branching point of Γ. Hence, equation (0.19) for a generic value of λ has n distinct roots and for each point $Q = (\lambda, \mu)$ of Γ there exists a unique eigenvector $h(Q) = (h_1(Q), \ldots, h_n(Q))$ of the operator $L_m(\lambda)$,

$$L_m(\lambda)h(Q) = \mu h(Q), \qquad (0.21)$$

normalized by the condition $h_0(Q) = 1$. All the other components h_i of this vector are rational functions in λ and μ, i.e. meromorphic functions on Γ. They depend on the choice of normalizing point x_0, $h_i = h_i(Q, x_0)$. In the affine part of Γ the poles of h coincide with zeros on the curve Γ of the minor $L_m^{ij}(\lambda)$, $i, j = 1, \ldots, n-1$. If the curve is smooth then the number of poles is equal to the genus of Γ. The poles $\gamma_1(x_0), \ldots, \gamma_g(x_0)$ depend on x_0.

The common eigenfunction $\psi(x, Q)$ of L_n and L_m is defined up to a constant factor, therefore its logarithmic derivative $\psi_x \psi^{-1}$ is defined uniquely. It follows from the definition of the canonical basis (0.16) that

$$\psi_x(x, Q)\psi^{-1}(x, Q)|_{x=x_0} = h_1(Q, x_0). \qquad (0.22)$$

That proves the second statement of [1],[2]. In order to prove the final statement that the coefficients of R and the divisor $\gamma_s(x_0)$ on the corresponding curve uniquely define the commuting operators, let us consider the analytical properties of $h_1(Q, x_0)$ on Γ, including the infinite point P_0. It turns out that besides the poles $\gamma_s(x_0)$ in the affine part of Γ it has a simple pole of the form

$$h_1(Q,x_0) = k + O(k^{-1}), \quad k^n = \lambda, \quad Q = (\lambda,\mu), \tag{0.23}$$

at infinity.

Let Γ be a smooth genus g algebraic curve with a local co-ordinate $k^{-1}(Q)$ in a neighbourhood of a puncture P_0. Then according to the Riemann–Roch theorem for a generic set of g points γ_s there exists a unique function h_1 that has at most simple poles at these points and has the form (0.23) in the neighbourhood of P_0.

The proof of this particular case of the Riemann–Roch theorem can be found in Chapter IV of the book. In the book the corresponding function is sometimes called a "Weierstrass function". It is one of the fundamental rational functions through which all the other rational functions can be expressed.

Let $z_s(Q)$ be a local co-ordinate near the point γ_s, then the corresponding function has an expansion

$$h_1(Q) = \frac{a_s}{z_s(Q) - z_s(\gamma_s)} + O(1). \tag{0.24}$$

The coefficients a_s of the expansion (0.24) are uniquely defined by the set $\gamma_1,\ldots,\gamma_g$, i.e. $a_s = a_s(\gamma_1,\ldots,\gamma_g)$.

The common eigenfunction $\psi(x,Q,x_0)$ of the operators L_n, L_m which is normalized by the condition $\psi(x=x_0,Q,x_0) = 1$ is equal to

$$\psi(x,Q,x_0) = \sum_{i=0}^{n-1} h_i(Q,x_0)c_i(x,\lambda,x_0). \tag{0.25}$$

The functions c_i are entire functions of the variable λ. Hence, ψ is a meromorphic function on Γ except for infinity. It has poles at $\gamma_s(x_0)$ and g zeros $\gamma_s(x)$ that are poles of its logarithmic derivative. In their vicinity we have

$$\psi_x\psi^{-1} = \frac{\partial_x z_s(\gamma_s(x))}{z_s(Q) - z_s(\gamma_s)} + O(1). \tag{0.26}$$

The comparison of (0.24) and (0.26) implies that

$$\partial_x z_s(\gamma_s(x)) = a_s(\gamma_1(x),\ldots,\gamma_g(x)). \tag{0.27}$$

Equations (0.27) consist of a well-defined system of differential equations of the first order. A solution of this system is defined by the initial data $\gamma_1(x_0),\ldots,\gamma_g(x_0)$. That proves the final statement of Burchnall and Chaundy.

In [3] it was proposed to consider the analytical properties of the common eigenfunction ψ on the compactified curve Γ. From the purely algebraic point of view it is a "forbidden" function, because it has an essential singularity at the infinite point P_0. But this essential singularity is of a very special form — it is of exponential type. It follows from (0.23) that

$$\begin{aligned}\psi(x, Q, x_0) &= \exp(\int_{x_0}^{x} h_1(Q, x)dx) \\ &= e^{k(x-x_0)}(1 + \sum_{s=1}^{\infty} \xi_s(x, x_0)k^{-s}), \ \lambda = k^n(Q) \to \infty.\end{aligned} \tag{0.28}$$

The theory of such functions, considered as a natural generalization of the exponential function to Riemann surfaces, is very deeply connected with the theory of the so-called *factorial* functions that can be found in Chapter XIV of the book. These functions are single-valued on the surface dissected along cycles and their values on different sides of the cuts satisfy special boundary conditions. In modern language they are solutions of the Riemann–Hilbert problem on a Riemann surface. The expression of such functions in terms of Riemann theta-functions is one of the main goals of the chapter.

Baker pointed out that these results should make it possible to find the exact formulae for the coefficients of the commuting operators of coprime orders. It turned out that this program was realized only in [4],[5] (though at that time the author was not aware of the remarkable results of Burchnall, Chaundy and Baker) where the commuting pairs of ordinary differential operators were considered in connection with the problem of constructing solutions to the KP equation.

The common eigenfunction of commuting operators is a particular case of the general definition of scalar the *multi-point* and *multi-variable* Clebsch–Gordan–Baker–Akhiezer function (or more simply the Baker–Akhiezer function). Let Γ be a non-singular algebraic curve of genus g with N punctures P_α and fixed local parameters $k_\alpha^{-1}(Q)$ in neighbourhoods of these punctures. For any set of points $\gamma_1, \ldots, \gamma_g$ in general position, there exists a unique (up to constant factor, $c(t_{\alpha,i})$) function $\psi(t, Q)$, $t = (t_{\alpha,i})$, $\alpha = 1, \ldots, N$, $i = 1, \ldots,$ such that:

(i) the function ψ (as a function of the variable $Q \in \Gamma$) is meromorphic everywhere except for the points P_α and it has at most simple poles at the points $\gamma_1, \ldots, \gamma_g$ (if all of them are distinct);

(ii) in a neighbourhood of the point P_α the function ψ has the form

$$\psi(t,Q) = \exp\Big(\sum_{i=1}^{\infty} t_{\alpha,i}k_\alpha^i\Big)\Big(\sum_{s=0}^{\infty} \xi_{s,\alpha}(t)k_\alpha^{-s}\Big), k_\alpha = k_\alpha(Q). \tag{0.29}$$

We see that the Baker–Akhiezer function ψ depends on the variables $t = \{t_{1,i},\dots,t_{n,i}\}$ and on external parameters.

From the uniqueness of the Baker–Akhiezer function it follows that for each pair (α,n) there exists a unique operator $L_{\alpha,n}$ of the form

$$L_{\alpha,n} = \partial_{\alpha,1}^n + \sum_{j=1}^{n-1} u_j^{(\alpha,n)}(t)\partial_{\alpha,1}^j \tag{0.30}$$

(where $\partial_{\alpha,i} = \partial/\partial t_{\alpha,i}$) such that

$$(\partial_{\alpha,i} - L_{\alpha,n})\psi(t,Q) = 0. \tag{0.31}$$

The idea of the proof of theorems of this type which was proposed in [4] is universal.

For any formal series of the form (0.29) there exists a unique operator $L_{\alpha,n}$ of the form (0.30) such that

$$(\partial_{\alpha,i} - L_{\alpha,n})\psi(t,Q) = O(k^{-1})\exp\Big(\sum_{i=1}^{\infty} t_{\alpha,i}k_\alpha^i\Big). \tag{0.32}$$

The coefficients of $L_{\alpha,n}$ are differential polynomials with respect to $\xi_{s,\alpha}$. They can be found after substitution of the series (0.29) into (0.32).

It turns out that, if the series (0.29) is not formal but is an expansion of the Baker–Akhiezer function in the neighbourhood of P_α, then the congruence (0.32) becomes an equality. Indeed, let us consider the function

$$\psi_1 = (\partial_{\alpha,n} - L_{\alpha,n})\psi(t,Q). \tag{0.33}$$

It has the same analytical properties as ψ, except for one. The expansion of this function in a neighbourhood of P_α starts from $O(k^{-1})$. From the uniqueness of the Baker–Akhiezer function it follows that $\psi_1 = 0$ and the equality (0.31) is proved.

A corollary is that the operators $L_{\alpha,n}$ satisfy the compatibility conditions

$$\big[\partial_{\alpha,n} - L_{\alpha,n}, \partial_{\alpha,m} - L_{\alpha,m}\big] = 0. \tag{0.34}$$

The equations (0.34) are gauge invariant. For any function $g(t)$ operators

$$\tilde{L}_{\alpha,n} = gL_{\alpha,n}g^{-1} + (\partial_{\alpha,n}g)g^{(-1)} \tag{0.35}$$

have the same form (0.30) and satisfy the same operator equations (0.34). The gauge transformation (0.35) corresponds to the gauge transformation of the Baker–Akhiezer function

$$\psi_1(t,Q) = g(t)\psi(t,Q) \tag{0.36}$$

In the one-point case the Baker–Akhiezer function has an exponential singularity at a single point P_1 and depends on a single set of variables. Let us choose the normalisation of the Baker–Akhiezer function with the help of the condition $\xi_{1,0} = 1$, i.e. an expansion of ψ in the neighbourhood of P_1 equals

$$\psi(t_1, t_2, \ldots, Q) = \exp\Big(\sum_{i=1}^{\infty} t_i k^i\Big)\Big(\sum_{s=0}^{\infty} \xi_s(t) k^{-s}\Big). \tag{0.37}$$

In this case the operator L_n has the form

$$L_n = \partial_1^n + \sum_{i=0}^{n-2} u_i^{(n)} \partial_1^i. \tag{0.38}$$

If we denote t_1, t_2, t_3 by x, y, t, respectively, then from (0.34) it follows (for $n = 2$, $m = 3$) that $u(x, y, t, t_4, \ldots)$ satisfies the KP equation (0.2). The exact formula for these solutions in terms of the Riemann theta-function is based on the exact formula for the Baker–Akhiezer function.

Let us fix the basis of cycles a_i, b_i, $i = 1, \ldots, g$ on Γ with the canonical matrix of intersections: $a_i \circ a_j = b_i \circ b_j = 0$, $a_i \circ b_j = \delta_{ij}$. The basis of normalized holomorphic differentials $\omega_j(Q)$, $j = 1, \ldots, g$ is defined by conditions

$$\oint_{a_i} \omega_j = \delta_{ij}. \tag{0.39}$$

The b-periods of these differentials define the so-called Riemann matrix

$$B_{kj} = \oint_{b_j} \omega_k. \tag{0.40}$$

The basic vectors e_k of C^g and the vectors B_k, which are the columns of matrix (0.40), generate a lattice $\mathcal{B}$ in C^g. The g-dimensional complex torus

$$J(\Gamma) = C^g/\mathcal{B}, \quad \mathcal{B} = \sum n_k e_k + m_k B_k, \quad n_k, m_k \in Z, \tag{0.41}$$

is called the Jacobian variety of Γ. A vector with co-ordinates

$$A_k(Q) = \int_{q_0}^{Q} \omega_k \tag{0.42}$$

defines the Abel map

$$A: \quad \Gamma \longmapsto J(\Gamma) \tag{0.43}$$

which depends on the choice of the initial point q_0.

The Riemann matrix has a positive-definite imaginary part. The entire function of g variables

$$\theta(z) = \theta(z|B) = \sum_{m \in Z^g} e^{2\pi i(z,m) + \pi i(Bm,m)}, \tag{0.44}$$

$$z = (z_1, \ldots, z_n), m = (m_1, \ldots, m_n), (z, m) = z_1 m_1 + \ldots + z_n m_n,$$

is called the Riemann theta-function. It has the following monodromy properties;

$$\theta(z + e_k) = \theta(z), \quad \theta(z + B_k) = e^{-2\pi i z_k - \pi i B_{kk}} \theta(z). \tag{0.45}$$

The function $\theta(A(Q) - Z)$ is a multi-valued function of Q. But according to (0.45), the zeros of this function are well-defined. For Z in a general position the equation

$$\theta(A(Q) - Z) = 0 \tag{0.46}$$

has g zeros $\gamma_1, \ldots, \gamma_g$. The vector Z and the divisor of these zeros are connected by the relation

$$Z_k = \sum_{i=1}^{g} A(\gamma_s) + \mathcal{K}, \tag{0.47}$$

where $\mathcal{K}$ is the vector of Riemann constants.

Let us introduce the normalized Abelian differentials $d\Omega_{\alpha,i}$ of the second kind. The differential $d\Omega_{\alpha,i}$ is holomorphic on Γ except for the puncture P_α. In the neighbourhood of this point it has the form

$$d\Omega_{\alpha,i} = d(k_\alpha^i + O(1)). \tag{0.48}$$

"Normalized" means that it has zero a-periods

$$\oint_{a_j} d\Omega_{\alpha,i} = 0. \tag{0.49}$$

Consider the function

$$\mathcal{E}(t, Q) = \exp\left(\sum_{\alpha,j} t_{\alpha,j} \int_{q_0}^{Q} d\Omega_{\alpha,j}\right). \tag{0.50}$$

It has the same exponential singularities of the form (0.29) at the punctures as the Baker–Akhiezer function, but it is a single-valued function on Γ dissected along a-cycles only. Its values on two sides of the cycle a_i differ by the factor

$$e^{2\pi i U_i} = \exp\big(\sum_{\alpha,j} t_{\alpha,j} U^i_{\alpha,j}\big), \tag{0.51}$$

where

$$U^i_{\alpha,j} = \frac{1}{2\pi i}\oint_{b_i} d\Omega_{\alpha,j}. \tag{0.52}$$

Consider the function

$$\phi(V,Q) = \frac{\theta(A(Q)+V-Z)}{\theta(A(Q)-Z)}, \tag{0.53}$$

where V is a vector with co-ordinates $V_1, \ldots, V_g$. This function is meromorphic on Γ dissected along a-cycles and has g poles (depending on Z). It follows from the monodromy properties (0.45), that the boundary values of ϕ on two sides of the a_j cycles satisfy the relation

$$\phi^+ = e^{-2\pi i V_j}\phi^-. \tag{0.54}$$

Such multi-valued functions are called "factorial" functions in the book.

Equalities (0.50-0.54) imply that the function

$$\psi(t,Q) = \mathcal{E}(t,Q)\frac{\theta(A(Q)+\sum_{\alpha,j} t_{\alpha,j}U_{\alpha,j} - Z)}{\theta(A(Q)-Z)} \tag{0.55}$$

is a single-valued function on Γ and has all the other properties of the desired function. Therefore, the existence of the Baker–Akhiezer function is proved. Let $\tilde{\psi}$ be any function with the same analytical properties. The ratio $\tilde{\psi}/\psi$ is a meromorphic function with at most g poles. The Riemann–Roch theorem implies that such a function is equal to a constant. Hence, the uniqueness of the Baker–Akhiezer function (up to a constant factor) is also proved.

The coefficients of the operators $L_{\alpha,j}$ which are defined by the equations (0.31) are differential polynomials in the coefficients of the expansions of the second factor in (0.55) near the punctures. Hence, they can be expressed as differential polynomials in terms of Riemann theta-functions. For example, the algebraic–geometrical solutions of the KP hierarchy have the form

$$u(x,y,t,t_4,\ldots) = 2\partial_x^2 \ln\theta(xU_1 + yU_2 + tU_3 + \cdots + Z) + const. \tag{0.56}$$

The common eigenfunction of commuting operators of coprime orders is the particular case of a one-point Baker–Akhiezer function corresponding to $t_1 = x, t_2 = 0, t_3 = 0, \ldots$. Therefore, the coefficients of such operators (in general position) are differential polynomials in terms of the Riemann theta-functions. This has an important corollary. The coefficients of commuting differential operators of coprime orders are meromorphic functions of the variable x. Moreover, in general position they are quasi-periodic functions of x. The last statement presents evidence that the theory of commuting operators is connected with the spectral Floquet theory of periodic differential operators. These connections were missing in [1], [2], [3].

The origin of Riemann surfaces in the spectral theory of ordinary periodic differential operators appears now to be self-evident. Indeed, for such an operator the space $\mathcal{L}(\lambda)$ of solutions of equation (0.14) is invariant with respect to the monodromy operator

$$\hat{T} : y(x) \longmapsto y(x+T). \tag{0.57}$$

Let $T(\lambda)$ be the corresponding finite-dimensional operator. The characteristic equation

$$R(w, \lambda) = \det(w - T(\lambda)) = 0 \tag{0.58}$$

defines the Riemann surface of Bloch solutions; the common eigenfunctions for the operator L and monodromy operator, i.e.

$$L\psi(x, Q) = \lambda\psi(x, Q), \quad \psi(x+T, Q) = w\psi(x, Q), \quad Q = (w, \lambda). \tag{0.59}$$

For the general periodic operator the Riemann surface of Bloch solutions has an infinite genus. Periodic operators for which this surface has a finite genus are operators that commute with some other ordinary differential operators. In this case the Riemann surface of Bloch solutions and the algebraic curve of common eigenfunctions of commuting ordinary differential operators are isomorphic.

Originally, the classification problem of commuting ordinary differential operators was posed for operators of arbitrary orders. In [1],[2] it was mentioned that there are no approaches to the solution of this problem if the orders of these operators are not coprime. The complete solution of the problem was obtained in [6]. It turned out that such operators are defined uniquely by the polynomial $R(\lambda, \mu)$ (0.17), by a vector bundle over Γ of rank r and degree rg and by a set of $r-1$ arbitrary functions $w_0(x), \ldots, w_{r-2}(x)$. Here r is a common divisor of the orders n, m. It equals the number of linear independent solutions of the equations (0.13). The problem of reconstruction of the coefficients of commuting differential operators of the rank $r > 1$ is reduced to the system of linear integral equations and is beyond the framework of the book.

To conclude let us give the list of reviews of finite-gap theory [7]-[11]. We would like to specially mention the work [12], where it was proved that the function that is given by formula (0.56) is a solution of the KP equation if and only if the matrix B that defines the theta-function is the Riemann matrix of some algebraic curve. This statement solves the Schottky problem and was conjectured by Novikov.

It would be fair to say that the most exciting results of recent years in algebraic geometry and in mathematical physics are connected with the application of so-called topological field theories, matrix integrals to the intersection theory on the moduli spaces of algebraic curves with punctures [13]-[14].

There is no doubt that this book will provide an excellent introduction to the algebraic–geometrical techniques that are necessary for those who are interested in this field.

We wish success to those who now begin to turn the pages which follow.

References

[1] Burchnall J.L., Chaundy T.W. *Commutative ordinary differential operators.I*, Proc. London Math Soc. **21** (1922), 420–440.

[2] Burchnall J.L., Chaundy T.W. *Commutative ordinary differential operators.II*, Proc. Royal Soc. London **118** (1928), 557–583.

[3] Baker H.F. *Note on the foregoing paper "Commutative ordinary differential operators"*, Proc. Royal Soc. London **118** (1928), 584–593.

[4] Krichever I.M. *The algebraic-geometrical construction of Zakharov-Shabat equations and their periodic solutions*, Doklady Akad. Nauk USSR **227** (1976), n 2, 291–294.

[5] Krichever I.M. *The integration of non-linear equations with the help of algebraic-geometrical methods*, Funk. anal. i pril. **11** (1977), n 1, 15–31.

[6] Krichever I.M. *The commutative rings of ordinary differential operators*, Funk. anal. i pril. **12** (1978), n 3, 20–31.

[7] Dubrovin B.A., Matveev V.B., Novikov S.P. *Non-lincar equations of the Korteweg-de Vries type, finite-gap operators and Abelian varieties*, Uspekhi Mat. Nauk **31** (1976), n 1, 55–136.

[8] Krichever I.M., Novikov S.P. *Holomorphic bundles over algebraic curves and non-linear equations*, Uspekhi Mat. Nauk **35** (1980), n 6.

[9] Dubrovin B.A. *Theta-functions and non-linear equations*, Uspekhi Mat. Nauk **36** (1981), n 2, 11–80.

[10] Dubrovin B.A., Krichever I.M., Novikov S.P. *Integrable systems*, Itogi Nauki i Tekhniki-Moscow, VINITI AN USSR, 1985, v.4, 179–285.

[11] Krichever I.M. *Spectral theory of two-dimensional periodic operators and its applications*, Uspekhi Mat. Nauk **44** (1989), n 2, 121–184.

[12] Shiota T. *Characterization of Jacobian varieties in terms of soliton equations*, Inv. Math. **83** (1986), 333–382.

[13] Witten E., *Two-dimensional gravity and intersection theory on moduli space*, Surveys in Diff. Geom. **1** (1991), 243–310.

[14] Kontsevich M. *Intersection theory on the moduli space of curves*, Funk. Anal. i Pril **25** (1991), 123–128.

[15] Kontsevich M. *Intersection theory on the moduli space of curves and matrix Airy function*, Comm. Math. Phys. **143**, n 2, (1992), 1–23.

Igor Krichever

PREFACE.

It may perhaps be fairly stated that no better guide can be found to the analytical developments of Pure Mathematics during the last seventy years than a study of the problems presented by the subject whereof this volume treats. This book is published in the hope that it may be found worthy to form the basis for such study. It is also hoped that the book may be serviceable to those who use it for a first introduction to the subject. And an endeavour has been made to point out what are conceived to be the most artistic ways of formally developing the theory regarded as complete.

The matter is arranged primarily with a view to obtaining perfectly general, and not merely illustrative, theorems, in an order in which they can be immediately utilised for the subsequent theory; particular results, however interesting, or important in special applications, which are not an integral portion of the continuous argument of the book, are introduced only so far as they appeared necessary to explain the general results, mainly in the examples, or are postponed, or are excluded altogether. The sequence and scope of ideas to which this has led will be clear from an examination of the table of Contents.

The methods of Riemann, as far as they are explained in books on the general theory of functions, are provisionally regarded as fundamental; but precise references are given for all results assumed, and great pains have been taken, in the theory of algebraic functions and their integrals, and in the analytic theory of theta functions, to provide for alternative developments of the theory. If it is desired to dispense with Riemann's existence theorems, the theory of algebraic functions may be founded either on the arithmetical ideas introduced by Kronecker and by Dedekind and Weber; or on the quasi-geometrical ideas associated with the theory of adjoint polynomials; while in any case it does not appear to be convenient to avoid reference to either class of ideas. It is believed that, save for some points in the periodicity of Abelian integrals, all that is necessary to the former elementary development will be found in Chapters IV. and VII., in connection with which the reader may consult the recent paper of Hensel, *Acta Mathematica*, XVIII. (1894), and also the papers of Kronecker and of

Dedekind and Weber, *Crelle's Journal*, XCI., XCII. (1882). And it is hoped that what is necessary for the development of the theory from the elementary geometrical point of view will be understood from Chapter VI., in connection with which the reader may consult the *Abel'sche Functionen* of Clebsch and Gordan (Leipzig, 1866) and the paper of Noether, *Mathematische Annalen*, VII. (1873). In the theory of Riemann's theta functions, the formulae which are given relatively to the ζ and $\wp$ functions, and the general formulae given near the end of Chapter XIV., will provide sufficient indications of how the theta functions can be algebraically defined; the reader may consult Noether, *Mathematische Annalen*, XXXVII. (1890), and Klein and Burkhardt, *ibid.* XXXII.—XXXVI. In Chapters XV., XVII., and XIX., and in Chapters XVIII. and XX., are given the beginnings of that analytical theory of theta functions from which, in conjunction with the general theory of functions of several independent variables, so much is to be hoped; the latter theory is however excluded from this volume.

To the reader who does not desire to follow the development of this volume consecutively through, the following course may perhaps be suggested; Chapters I., II., III. (in part), IV., VI. (to § 98), VIII., IX., X., XI. (in part), XVIII. (in part), XII., XV. (in part); it is also possible to begin with the analytical theory of theta functions, reading in order Chapters XV., XVI., XVII., XIX., XX.

The footnotes throughout the volume are intended to contain the mention of all authorities used in its preparation; occasionally the hazardous plan of adding to the lists of references during the passage of the sheets through the press, has been adopted; for references omitted, and for references improperly placed, only mistake can be pleaded. Complete lists of papers are given in the valuable report of Brill and Noether, "Die Entwicklung der Theorie der algebraischen Functionen in älterer und neuerer Zeit," *Jahresbericht der Deutschen Mathematiker-Vereinigung*, Dritter Band, 1892—3 (Berlin, Reimer, 1894); this report unfortunately appeared only after the first seventeen chapters of this volume, with the exception of Chapter XI., and parts of VII., were in manuscript; its plan is somewhat different from that of this volume, and it will be of advantage to the reader to consult it. Other books which have appeared during the progress of this volume, too late to effect large modifications, have not been consulted. The examples throughout the volume are intended to serve several different purposes; to provide practice in the ideas involved in the general theory; to suggest the steps of alternative developments without interrupting the line of reasoning in the text; and to place important consequences which are not utilised, if at all, till much subsequently, in their proper connection.

For my first interest in the subject of this volume, I desire to acknowledge my obligations to the generous help given to me during Göttingen vacations, on two occasions, by Professor Felix Klein. In the preparation of the book I have been largely indebted to his printed publications; the reader is recommended to consult also his lithographed lectures, especially the one dealing with Riemann surfaces. In the final revision of the sheets in their passage through the press, I have received help from several friends. Mr A. E. H. Love, Fellow and Lecturer of St John's College, has read the proofs of the volume; in the removal of obscurities of expression and in the correction of press, his untiring assistance has been of great value to me. Mr J. Harkness, Professor of Mathematics at Bryn Mawr College, Pennsylvania, has read the proofs from Chapter XV. onwards; many faults, undetected by Mr Love or myself, have yielded to his perusal; and I have been greatly helped by his sympathy in the subject-matter of the volume. To both these friends I am under obligations not easy to discharge. My gratitude is also due to Professor Forsyth for the generous interest he has taken in the book from its commencement. While, it should be added, the task carried through by the Staff of the University Press deserves more than the usual word of acknowledgment.

This book has a somewhat ambitious aim; and it has been written under the constant pressure of other work. It cannot but be that many defects will be found in it. But the author hopes it will be sufficient to shew that the subject offers for exploration a country of which the vastness is equalled by the fascination.

ST JOHN'S COLLEGE, CAMBRIDGE.
April 26, 1897.

In this volume no account is given of the differential equations satisfied by the theta functions, or of their expansion in integral powers of the arguments. The following references may be useful: Wiltheiss, *Crelle*, XCIX., *Math. Annal.* XXIX., XXXI., XXXIII., *Götting. Nachr.*, 1889, p. 381; Pascal, *Götting. Nachr.*, 1889, pp. 416, 547, *Ann. di Mat.*, Ser. 2ª, t. XVII.; Burkhardt (and Klein), *Math. Annal.* XXXII. The case $p=2$ is considered in Krause, *Transf. Hyperellip. Functionen.*

The following books of recent appearance, not referred to in the text, may be named here. (1) The completion of Picard, *Traité d'Analyse*, (2) Jordan, *Cours d'Analyse*, t. II. (1894), (3) Appell and Goursat, *Théorie des Fonctions algébriques et de leurs intégrales* (1895), (4) Stahl, *Theorie der Abel'schen Functionen* (1896).

CHAPTER I.

THE SUBJECT OF INVESTIGATION.

1. THIS book is concerned with a particular development of the theory of the algebraic irrationality arising when a quantity y is defined in terms of a quantity x by means of an equation of the form

$$a_0y^n + a_1y^{n-1} + \dots + a_{n-1}y + a_n = 0,$$

wherein $a_0, a_1, \dots, a_n$ are rational integral polynomials in x. The equation is supposed to be irreducible; that is, the left-hand side cannot be written as the product of other expressions of the same rational form.

2. Of the various means by which this dependence may be represented, that invented by Riemann, the so-called Riemann surface, is throughout regarded as fundamental. Of this it is not necessary to give an account here*. But the sense in which we speak of a *place* of a Riemann surface must be explained. To a value of the independent variable x there will in general correspond n distinct values of the dependent variable y—represented by as many *places*, lying in distinct sheets of the surface. For some values of x two of these n values of y may happen to be equal: in that case the corresponding sheets of the surface may behave in one of two ways. Either they may just touch at one point without having any further connexion in the immediate neighbourhood of the point†: in which case we shall regard the point where the sheets touch as constituting two places, one in each sheet. Or the sheets may wind into one another: in which case we shall regard this winding point (or branch point) as constituting one place: this place belongs then indifferently to either sheet; the sheets here merge into one another. In the first case, if a be the value of x for which the sheets just touch, supposed for convenience of statement to be finite, and x a value

* For references see Chap. II. § 12, note.

† Such a point is called by Riemann "ein sich aufhebender Verzweigungspunkt": *Gesammelte Werke* (1876), p. 105.

very near to a, and if b be the value of y at each of the two places, also supposed finite, and y_1, y_2 be values of y very near to b, represented by points in the two sheets very near to the point of contact of the two sheets, each of $y_1 - b$, $y_2 - b$ can be expressed as a power-series in $x-a$ with integral exponents. In the second case with a similar notation each of $y_1 - b$, $y_2 - b$ can be expressed as a power-series in $(x-a)^{\frac{1}{2}}$ with integral exponents. In the first case a small closed curve can be drawn on either of the two sheets considered, to enclose the point at which the sheets touch: and the value of the integral $\frac{1}{2\pi i}\int d \log (x-a)$ taken round this closed curve will be 1; hence, adopting a definition given by Riemann*, we shall say that $x-a$ is an infinitesimal of the first order at each of the places. In the second case the attempt to enclose the place by a curve leads to a curve lying partly in one sheet and partly in the other; in fact, in order that the curve may be closed it must pass twice round the branch place. In this case the integral $\frac{1}{2\pi i}\int d \log [(x-a)^{\frac{1}{2}}]$ taken round the closed curve will be 1: and we speak of $(x-a)^{\frac{1}{2}}$ as an infinitesimal of the first order at the place. In either case, if t denote the infinitesimal, x and y are uniform functions of t in the immediate neighbourhood of the place; conversely, to each point on the surface in the immediate neighbourhood of the place there corresponds uniformly a certain value of t†. The quantity t effects therefore a conformal representation of this neighbourhood upon a small simple area in the plane of t, surrounding $t=0$.

3. This description of a simple case will make the general case clear. In general for any finite value of x, $x=a$, there may be several, say k, branch points‡; the number of sheets that wind at these branch points may be denoted by w_1+1, w_2+1, ..., w_k+1 respectively, where

$$(w_1+1)+(w_2+1)+\dots+(w_k+1)=n,$$

so that the case of no branch point is characterised by a zero value of the corresponding w. For instance in the first case above, notwithstanding that two of the n values of y are the same, each of w_1, w_2, ..., w_k is zero and k is equal to n: and in the second case above, the values are $k=n-1$, $w_1=1$, $w_2=0$, $w_3=0$, ..., $w_k=0$. *In the general case each of these* k *branch points is called a place*, and at these respective places the quantities $(x-a)^{\frac{1}{w_1+1}}$, ..., $(x-a)^{\frac{1}{w_k+1}}$

* *Gesammelte Werke* (1876), p. 96.

† The limitation to the immediate neighbourhood involves that t is not necessarily a rational function of x, y.

It may be remarked that a rational function of x and y *can* be found whose behaviour in the neighbourhood of the place is the same as that of t. See for example Hamburger, *Zeitschrift f. Math. und Phys.* Bd. 16, 1871; Stolz, *Math. Ann.* 8, 1874; Harkness and Morley, *Theory of Functions*, p. 141.

‡ Cf. Forsyth, *Theory of Functions*, p. 171. Prym, *Crelle*, Bd. 70.

are infinitesimals of the first order. For the infinite value of x we shall similarly have n or a less number of places and as many infinitesimals, say $\left(\frac{1}{x}\right)^{\frac{1}{w_1+1}}, \ldots, \left(\frac{1}{x}\right)^{\frac{1}{w_r+1}}$, where $(w_1+1)+\ldots+(w_r+1)=n$. And as in the particular cases discussed above, the infinitesimal t thus defined for every place of the surface has the two characteristics that for the immediate neighbourhood of the place x and y are uniquely expressible thereby (in series of integral powers), and conversely t is a uniform function of position on the surface in this neighbourhood. Both these are expressed by saying that t effects a reversible conformal representation of this neighbourhood upon a simple area enclosing $t=0$. It is obvious of course that quantities other than t have the same property.

A place of the Riemann surface will generally be denoted by a single letter. And in fact a place (x, y) will generally be called the place x. When we have occasion to speak of the (n or less) places where the independent variable x has the same value, a different notation will be used.

4. We have said that the subject of enquiry in this book is a certain algebraic irrationality. We may expect therefore that the theory is practically unaltered by a rational transformation of the variables x, y which is of a reversible character. Without entering here into the theory of such transformations, which comes more properly later, in connexion with the theory of correspondence, it is necessary to give sufficient explanations to make it clear that the functions to be considered belong to a whole class of Riemann surfaces and are not the exclusive outcome of that one which we adopt initially.

Let ξ be any one of those uniform functions of position on the fundamental (undissected) Riemann surface whose infinities are all of finite order. Such functions can be expressed rationally by x and y*. For that reason we shall speak of them shortly as the rational functions of the surface. The order of infinity of such a function at any place of the surface where the function becomes infinite is the same as that of a certain *integral* power of the inverse $\frac{1}{t}$ of the infinitesimal at that place. The sum of these orders of infinity for all the infinities of the function is called the order of the function. The number of places at which the function ξ assumes any other value α is the same as this order: it being understood that a place at which $\xi-\alpha$ is zero in a finite ratio to the rth order of t is counted as r places at which ξ is equal to α†. Let ν be the order of ξ. Let η be another rational function of

* Forsyth, *Theory of Functions*, p. 370.

† For the integral $\frac{1}{2\pi i}\int d\log(\xi-\alpha)$, taken round an infinity of $\log(\xi-\alpha)$, is equal to the order of zero of $\xi-\alpha$ at the place, or to the negative of the order of infinity of ξ, as the case may be. And the sum of the integrals for all such places is equal to the value round the boundary of the surface—which is zero. Cf. Forsyth, *Theory of Functions*, p. 372.

order μ. Take a plane whose real points represent all the possible values of ξ in the ordinary way. To any value of ξ, say $\xi=\alpha$, will correspond ν positions $X_1, \ldots, X_\nu$ on the original Riemann surface, those namely where ξ is equal to α: it is quite possible that they lie at less than ν places of the surface. The values of η at $X_1, \ldots, X_\nu$ may or may not be different. Let H denote any definite rational symmetrical function of these ν values of η. Then to each position of α in the ξ plane will correspond a perfectly unique value of H, namely, H is a one-valued function of ξ. Moreover, since η and ξ are rational functions on the original surface, the character of H for values of ξ in the immediate neighbourhood of a value α, for which H is infinite, is clearly the same as that of a *finite* power of $\xi-\alpha$. Hence H is a rational function of ξ. Hence, if H_r denote the sum of the products of the values of η at $X_1, \ldots, X_\nu$, r together, η satisfies an equation

$$\eta^\nu - \eta^{\nu-1}H_1 + \eta^{\nu-2}H_2 - \ldots + (-)^\nu H_\nu = 0,$$

whose coefficients are rational functions of ξ.

It is conceivable that the left side of this equation can be written as the product of several factors each rational in ξ and η. If possible let this be done. Construct over the ξ plane the Riemann surfaces corresponding to these irreducible factors, η being the dependent variable and the various surfaces lying above one another in some order. It is a known fact, already used in defining the order of a rational function on a Riemann surface, that the values of η represented by *any* one of these superimposed surfaces include all possible values—each value in fact occurring the same number of times on each surface. To any place of the original surface, where ξ, η have definite values, and to the neighbourhood of this place, will correspond therefore a definite place (ξ, η) (and its neighbourhood) on each of these superimposed surfaces. Let $\eta_1, \ldots, \eta_r$ be the values of η belonging, on one of these surfaces, to a value of ξ: and $\eta_1', \ldots, \eta_s'$ the values belonging to the same value of ξ on another of these surfaces. Since for each of these surfaces there are only a finite number of values of ξ at which the values of η are not all different, we may suppose that all these r values on the one surface are different from one another, and likewise the s values on the other surface. Since each of the pairs of values $(\xi, \eta_1), \ldots, (\xi, \eta_r)$ must arise on both these surfaces, it follows that the values $\eta_1, \ldots, \eta_r$ are included among $\eta_1', \ldots, \eta_s'$. Similarly the values $\eta_1', \ldots, \eta_s'$ are included among $\eta_1, \ldots, \eta_r$. Hence these two sets are the same and $r=s$. Since this is true for an infinite number of values of ξ, it follows that these two surfaces are merely repetitions of one another. The same is true for every such two surfaces. Hence r is a divisor of ν and the equation

$$\eta^\nu - H_1\eta^{\nu-1} + \ldots + (-)^\nu H_\nu = 0,$$

when reducible, is the ν/rth power of a rational equation of order r in η. It will be sufficient to confine our attention to one of the factors and the (ξ, η)

surface represented thereby. Let now $X_1, \ldots, X_\nu$ be the places on the original surface where ξ has a certain value. Then the values of η at $X_1, \ldots, X_\nu$ will consist of ν/r repetitions of r values, these r values being different from one another except for a finite number of values of ξ. Thus to any place (ξ, η) on one of the ν/r derived surfaces will correspond ν/r places on the original surface, those namely where the pair (ξ, η) take the supposed values. Denote these by $P_1, P_2, \ldots$. Let Y be any rational symmetrical function of the ν/r pairs of values $(x_1, y_1), (x_2, y_2), \ldots$, which the fundamental variables x, y of the original surface assume at $P_1, P_2, \ldots$. Then to any pair of values (ξ, η) will correspond only one value of Y—namely, Y is a one-valued function on the (ξ, η) surface. It has clearly also only finite orders of infinity. Hence Y is a rational function of ξ, η. In particular $x_1, x_2, \ldots$ are the roots of an equation whose coefficients are rational in ξ, η—as also are $y_1, y_2, \ldots$.

There exists therefore a correspondence between the (ξ, η) and (x, y) surfaces—of the kind which we call a $\left(1, \frac{\nu}{r}\right)$ correspondence: to every place of the (x, y) surface corresponds one place of the (ξ, η) surface; to every place of this surface correspond $\frac{\nu}{r}$ places of the (x, y) surface.

The case which most commonly arises is that in which the rational irreducible equation satisfied by η is of the νth degree in η: then only one place of the original surface is associated with any place of the new surface. In that case, as will appear, the new surface is as general as the original surface. Many advantages may be expected to accrue from the utilization of that fact. We may compare the case of the reduction of the general equation of a conic to an equation referred to the principal axes of the conic.

5. The following method* is theoretically effective for the expression of x, y in terms of ξ, η.

Let the rational expression of ξ, η in terms of x, y be given by

$$\phi(x, y) - \xi\psi(x, y) = 0, \quad \vartheta(x, y) - \eta\chi(x, y) = 0,$$

and let the rational result of eliminating x, y between these equations and the initial equation connecting x, y be denoted by $F(\xi, \eta) = 0$, each of $\phi, \ldots, \chi$, F denoting integral polynomials. Let two terms of the expression $\phi(x, y) - \xi\psi(x, y) = 0$ be $ax^r y^s - \xi b x^{r'} y^{s'}$. This expression and therefore all others involved will be unaltered if a, b be replaced by such quantities $a+h$, $b+k$, that $hx^r y^s = \xi k x^{r'} y^{s'}$. In a formal sense this changes $F(\xi, \eta)$ into

$$F + \frac{1}{\underline{|\lambda}}\left[\frac{\partial^\lambda F}{\partial a^\lambda} h^\lambda + \binom{\lambda}{1}\frac{\partial^\lambda F}{\partial a^{\lambda-1}\partial b} h^{\lambda-1}k + \binom{\lambda}{2}\frac{\partial^\lambda F}{\partial a^{\lambda-2}\partial b^2} h^{\lambda-2}k^2 + \ldots + \frac{\partial^\lambda F}{\partial b^\lambda} k^\lambda\right] + \ldots\ldots$$

where $\lambda \geqq 1$, and F is such that all differential coefficients of it in regard to a and b of order less than λ are identically zero.

Hence the term within the square brackets in this expression must be zero. If it is possible, choose now $r = r' + 1$ and $s = s'$, so that $k = hx/\xi$.

* Salmon's *Higher Algebra* (1885), p. 97, § 103.

Then we obtain the equation

$$x^\lambda \frac{\partial^\lambda F}{\partial b^\lambda} - \binom{\lambda}{1} x^{\lambda-1} \xi \frac{\partial^\lambda F}{\partial b^{\lambda-1}\partial a} + \ldots\ldots + (-)^\lambda \xi^\lambda \frac{\partial^\lambda F}{\partial a^\lambda} = 0.$$

This is an equation of the form above referred to, by which x is determinate from ξ and η. And y is similarly determinate.

It will be noticed that the rational expression of x, y by ξ, η, when it is possible from the equations

$$\phi(x, y) - \xi\psi(x, y) = 0, \quad \vartheta(x, y) - \eta\chi(x, y) = 0, \quad f(x, y) = 0,$$

will not be possible, in general, from the first two equations: it is only the places x, y satisfying the equation $f(x, y) = 0$ which are rationally obtainable from the places ξ, η satisfying the equation $F(\xi, \eta) = 0$. There do exist transformations, rationally reversible, subject to no such restriction. They are those known as Cremona-transformations*. They can be compounded by reapplication of the transformation $x : y : 1 = \eta : \xi : \xi\eta$.

We may give an example of both of these transformations—

For the surface

$$y^5 - 5y^3(x^2+x+1) + 5y(x^2+x+1)^2 - 2x(x^2+x+1)^2 = 0$$

the function $\xi = y^2/(x^2+x+1)$ is of order 2, being infinite at the places where $x^2+x+1=0$, in each case like $(x-a)^{-\frac{1}{5}}$, and the function $\eta = x/y$ is of order 4, being infinite at the places $x^2+x+1=0$, in each case like $(x-a)^{-\frac{2}{5}}$, a being the value of x at the place.

From the given equation we immediately find, as the relation connecting ξ and η,

$$2\eta - \xi^2 + 5\xi - 5 = 0,$$

and infer, since the equation formed as in the general statement above should be of order 2 in η, that this general equation will be

$$(2\eta - \xi^2 + 5\xi - 5)^2 = 0.$$

Thence in accordance with that general statement we infer that to each place (ξ, η) on the new surface should correspond two places of the original surface: and in fact these are obviously given by the equations

$$\eta^2\xi = x^2/(x^2+x+1), \quad y = x/\eta.$$

If however we take

$$\xi = y^2/(x^2+x+1), \quad \eta = y/(x-\omega^2),$$

where ω is an imaginary cube root of unity, so that η is a function of order 3, these equations are reversible independently of the original equation, giving in fact

$$x = (\omega\xi - \omega^2\eta^2)/(\xi - \eta^2), \quad y = (\omega - \omega^2)\,\xi\eta/(\xi - \eta^2),$$

and we obtain the surface

$$\eta^2 - \tfrac{1}{2}(1-\omega^2)\,\eta\xi\,(\xi^2 - 5\xi + 5) - \omega^2\xi = 0,$$

having a (1, 1) correspondence with the original one.

It ought however to be remarked that it is generally possible to obtain reversible transformations which are not Cremona-transformations.

6. When a surface (x, y) is (1, 1) related to a (ξ, η) surface, the deficiencies of the surfaces, as defined by Riemann by means of the connectivity, must clearly be the same.

* See Salmon, *Higher Plane Curves* (1879), § 362, p. 322.

It is instructive to verify this from another point of view*.—Consider at how many places on the original surface the function $\frac{d\xi}{dx}$ is zero. It is infinite at the places where ξ is infinite: suppose for simplicity that these are separated places on the original surface or in other words are infinities of the first order, and are not at the branch points of the original surface. At a pole of ξ, $\frac{d\xi}{dx}$ is infinite twice. It is infinite like $\frac{1}{t^w}$ at a branch place (α) where $x-\alpha=t^{w+1}$: namely it is infinite $\Sigma w = 2n+2p-2$ times† at the branch places of the original surface. It is zero $2n$ times at the infinite places of the original surface. There remain therefore $2\nu+2n+2p-2-2n=2\nu+2p-2$ places where $\frac{d\xi}{dx}$ is zero. If a branch place of the original surface be a pole of ξ, and ξ be there infinite like $\frac{1}{t}$, $\frac{d\xi}{dx}$ is infinite like $\frac{1}{t^2 \cdot t^w}$, namely $2+w$ times: the total number of infinities of $\frac{d\xi}{dx}$ will therefore be the same as before. Now at a finite place of the original surface where $\frac{d\xi}{dx}=0$, there are two consecutive places for which ξ has the same value. Since $\frac{\nu}{r}=1$ they can only arise from consecutive places of the new surface for which ξ has the same value. The only consecutive places of a surface for which this is the case are the branch places. Hence† there are $2\nu+2p-2$ branch places of the new surface. This shews that the new surface is of deficiency p.

When ν/r is not equal to 1, the case is different. The consecutive places of the old surface, for which ξ has the same value, may either be those arising from consecutive places of the new surface—or may be what we may call accidental coincidences among the ν/r places which correspond to *one* place of the new surface. Conversely, to a branch place of the new surface, characterised by the same value for ξ for consecutive places‡, will correspond ν/r places on the old surface where ξ has the same value for consecutive places. In fact to two very near places of the new surface will correspond ν/r pairs each of very near places on the old surface. If then C denote the number of places on the old surface at which two of the ν/r places corresponding to a place on the new surface happen to coincide, and w' the number of branch points of the new surface, we have the equation

$$w'\frac{\nu}{r}+C=2\nu+2p-2,$$

* Compare the interesting geometrical account, Salmon, *Higher Plane Curves* (1879), p. 326, § 364, and the references there given.

† Forsyth, *Theory of Functions*, p. 348.

‡ Namely, near such a branch place $\xi=a$, $\xi-a$ is zero of higher order than the first.

and if p' be the deficiency of the new surface (of r sheets), this leads to the equation

$$(2r + 2p' - 2)\frac{\nu}{r} + C = 2\nu + 2p - 2,$$

from which

$$C = 2p - 2 - (2p' - 2)\frac{\nu}{r}.$$

*Corollary**. If $p = p'$, then $C = (2p - 2)\left(1 - \frac{\nu}{r}\right)$. Thus $\frac{\nu}{r} \not> 1$, so that $C = 0$, and the correspondence is reversible.

We have, herein, excluded the case when some of the poles of ξ are of higher than the first order. In that case the new surface has branch places at infinity. The number of finite branch places is correspondingly less. The reader can verify that the general result is unaffected.

Ex. In the example previously given (§ 5) shew that the function ξ takes any given value at *two* points of the original surface (other than the branch places where it is infinite), η having the same value for these two points, and that there are *six* places at which these two places coincide. (These are the place ($x=0$, $y=0$) and the five places where $x = -2$.)

There is one remark of considerable importance which follows from the theory here given. We have shewn that the number of places of the (x, y) surface which correspond to one place of the (ξ, η) surface is $\frac{\nu}{r}$, where ν is the order of ξ and r is not greater than ν, being the number of sheets of the (ξ, η) surface; hence, *if there were a function ξ of order* 1 *the correspondence would be reversible and therefore the original surface would be of deficiency* 0.

7. This notion of the transformation of a Riemann surface suggests an inference of a fundamental character.

The original equation contains only a finite number of terms: the original surface depends therefore upon a finite number of constants, namely, the coefficients in the equation. But conversely it is not necessary, in order that the equation be reversibly transformable into another given one, that the equation of the new surface contain as many constants as that of the original surface. For we may hope to be able to choose a transformation whose coefficients so depend on the coefficients of the original equation as to reduce this number. If we speak of all surfaces of which any two are connected by a rational reversible transformation as belonging to the same class†, it becomes a question whether there is any limit to the reduction obtainable, by rational reversible transformation, in the number of constants in the equation of a surface of the class.

* See Weber, *Crelle*, 76, 345.

† So that surfaces of the same class will be of the same deficiency.

It will appear in the course of the book* that there is a limit, and that the various classes of surfaces of given deficiency are of essentially different character according to the least number of constants upon which they depend. Further it will appear, that the most general class of deficiency p is characterised by $3p-3$ constants when $p>1$—the number for $p=1$ being one, and for $p=0$ none.

For the explanatory purposes of the present Chapter we shall content ourselves with the proof of the following statement—When a surface is reversibly transformed as explained in this Chapter, we cannot, even though we choose the new independent variable ξ to contain a very large number of disposeable constants, prescribe the position of all the branch points of the new surface; there will be $3p-3$ of them whose position is settled by the position of the others. Since the correspondence is reversible we may regard the new surface as fundamental, equally with the original surface. We infer therefore that the original surface depends on $3p-3$ parameters—*or on less*, for the $3p-3$ undetermined branch points of the new surface may have mutually dependent positions.

In order to prove this statement we recall the fact that a function of order Q contains† $Q-p+1$ linearly entering constants when its poles are prescribed: it may contain more for values of $Q<2p-1$, but we shall not thereby obtain as many constants as if we suppose $Q>2p-2$ and large enough. Also the Q infinities are at our disposal. We can then presumably dispose of $2Q-p+1$ of the branch points of the new surface. But these are, in number, $2Q+2p-2$ when the correspondence is reversible. Hence we can dispose of all but $2Q+2p-2-(2Q-p+1)=3p-3$ of the branch points of the new surface‡.

Ex. 1. The surface associated with the equation

$$y^2=x(1-x)(1-\kappa^2x)(1-\lambda^2x)(1-\mu^2x)(1-\nu^2x)(1-\rho^2x)$$

is of deficiency 3. It depends on $5=2p-1$ parameters, κ^2, λ^2, μ^2, ν^2, ρ^2.

Ex. 2. The surface associated with the equation

$$y^3+y^2(x, 1)_1+y(x, 1)_2+(x, 1)_4=0,$$

wherein the coefficients are integral polynomials of the orders specified by the suffixes, is of deficiency 3. Shew that it can be transformed to a form containing only $5=2p-1$ parametric constants.

* See the Chapters on the geometrical theory and on the inversion of Abelian Integrals. The reason for the exception in case $p=0$ or 1 will appear most clearly in the Chapter on the self-correspondence of a Riemann surface. But it is a familiar fact that the elliptic functions which can be constructed for a surface of deficiency 1 depend upon one parameter, commonly called the modulus: and the trigonometrical functions involve no such parameter.

† Forsyth, p. 459. The theorems here quoted are considered in detail in Chapter III. of the present book.

‡ Cf. Riemann, *Ges. Werke* (1876), p. 113. Klein, *Ueber Riemann's Theorie* (Leipzig, Teubner, 1882), p. 65.

8. But there is a case in which this argument fails. If it be possible to transform the original surface into itself by a rational reversible transformation involving r parameters, any r places on the surface are effectively equivalent with, as being transformable into, any other r places. Then the Q poles of the function ξ do not effectively supply Q but only $Q-r$ disposeable constants with which to fix the new surface. So that there are $3p-3+r$ branch points of the new surface which remain beyond our control. In this case we may say that all the surfaces of the class contain $3p-3$ disposeable parameters *beside* r parameters which remain indeterminate and serve to represent the possibility of the self-transformation of the surface. It will be shewn in the chapter on self-transformation that the possibility only arises for $p=0$ or $p=1$, and that the values of r are, in these cases, respectively 3 and 1. We remark as to the case $p=0$ that when the fundamental surface has only one sheet it can clearly be transformed into itself by a transformation involving three constants $x=\dfrac{a\xi+b}{c\xi+d}$: and in regard to $p=1$, the case of elliptic functions, that effectively a point represented by the elliptic argument u is equivalent to any other point represented by an argument $u+\gamma$. For instance a function of two poles is

$$F_{\alpha,\beta}=\frac{A}{\wp\left(u-\frac{\alpha+\beta}{2}\right)-\wp\frac{\alpha-\beta}{2}}+B,$$

and clearly $F_{\alpha,\beta}$ has the same value at u as has $F_{\alpha+\gamma,\beta+\gamma}$ at $u+\gamma$: so that the poles (α,β) are not, so far as absolute determinations are concerned, effective for the determination of more than one point.

9. The fundamental equation

$$a_0y^n+a_1y^{n-1}+\ldots+a_n=0,$$

so far considered as associated with a Riemann surface, may also be regarded as the equation of a plane curve: and it is possible to base our theory on the geometrical notions thus suggested. Without doing this we shall in the following pages make frequent use of them for purposes of illustration. It is therefore proper to remind the reader of some fundamental properties*.

The branch points of the surface correspond to those points of the curve where a line $x=$ constant meets the curve in two or more *consecutive* points: as for instance when it touches the curve, or passes through a cusp. On the other hand a double point of the curve corresponds to a point on the surface where two sheets just touch without further connexion. Thus the branch place of the surface which corresponds to a cusp is really a different singularity to that which corresponds to a place where the curve is touched by a

* Cf. Forsyth, *Theory of Functions*, p. 355 etc. Harkness and Morley, *Theory of Functions*, p. 273 etc.

line $x=$ constant, being obtained by the coincidence of an ordinary branch place with such a place of the Riemann surface as corresponds to a double point of the curve.

Properties of either the Riemann surface or a plane curve are, in the simpler cases, immediately transformed. For instance, by Plücker's formulae for a curve, since the number of tangents from any point is

$$t=(n-1)n-2\delta-3\kappa,$$

where n is the aggregate order in x and y, it follows that the number of branch places of the corresponding surface is

$$\begin{aligned} w &= t+\kappa=(n-1)n-2(\delta+\kappa) \\ &= 2n-2+2\left\{\tfrac{1}{2}(n-1)(n-2)-\delta-\kappa\right\}. \end{aligned}$$

Thus since $w=2n-2+2\ p$, the deficiency of the surface is

$$\tfrac{1}{2}(n-1)(n-2)-\delta-\kappa,$$

namely the number which is ordinarily called the deficiency of the curve.

To the theory of the birational transformation of the surface corresponds a theory of the birational transformation of plane curves. For example, the branch places of the new surface obtained from the surface $f(x, y)=0$ by means of equations of the form $\phi(x, y)-\xi\psi(x, y)=0$, $\vartheta(x, y)-\eta\chi(x, y)=0$ will arise for those values of ξ for which the curve $\phi(x, y)-\xi\psi(x, y)=0$ touches $f(x, y)=0$. The condition this should be so, called the tact invariant, is known to involve the coefficients of $\phi(x, y)-\xi\psi(x, y)=0$, and therefore in particular to involve ξ, to a degree* $n(n-3)-2\delta-3\kappa+2nn'$, where n' is the order of $\phi(x, y)-\xi\psi(x, y)=0$. Branch places of the new surface also arise corresponding to the cusps of the original curve. The total number is therefore $n(n-3)-2\delta-2\kappa+2nn'=2p-2+2nn'$. Now nn' is the number of intersections of the curves $f(x, y)=0$ and $\phi(x, y)-\xi\psi(x, y)=0$, namely it is the number of values of η arising for any value of ξ, and is thus the number of sheets of the new surface, which we have previously denoted by ν: so that the result is as before.

In these remarks we have assumed that the dependent variable occurs to the order which is the highest aggregate order in x and y together—and we have spoken of this as the order of the curve. And in regarding two curves as intersecting in a number of points equal to the product of their orders we have allowed count of branches of the curve which are entirely at infinity. Some care is necessary in this regard. In speaking of the Riemann surface represented by a given equation it is intended, unless the contrary be stated, that such infinite branches are unrepresented. As an example the curve $y^2=(x, 1)_6$ may be cited.

Ex. Prove that if from any point of a curve, ordinary or multiple, or from a point not on the curve, t be the number of tangents which can be drawn other than those touching

* See Salmon, *Higher Plane Curves* (1879), p. 81.

at the point, and κ be the number of cusps of the curve—and if ν be the number of points other than the point itself in which the curve is intersected by an arbitrary line through the point—then $t+\kappa-2\nu$ is independent of the position of the point. If the equation of the variable lines through the point be written $u-\xi v=0$, interpret the result by regarding the curve as giving rise to a Riemann surface whose independent variable is ξ*.

10. The geometrical considerations here referred to may however be stated with advantage in a very general manner.

In space of any (k) dimensions let there be a curve—(a one-dimensionality). Let points on this curve be given by the ratios of the $k+1$ homogeneous variables $x_1, \ldots, x_{k+1}$. Let u, v be any two rational integral homogeneous functions of these variables of the same order. The locus $u-\xi v=0$ will intersect the curve in a certain number, say ν, points—*we assume the curve to be such that this is the same for all values of ξ, and is finite.* Let all the possible values of ξ be represented by the real points of an infinite plane in the ordinary way. Let w, t be any two other integral functions of the coordinates of the same order. The values of $\eta=\frac{w}{t}$ at the points where $u-\xi v=0$ cuts the curve for any specified value of ξ will be ν in number. As before it follows thence that η satisfies an algebraic equation of order ν whose coefficients are one-valued functions of ξ. Since η can only be infinite to a finite order it follows that these coefficients are rational functions of ξ. Thence we can construct a Riemann surface, associated with this algebraic equation connecting ξ and η, such that every point of the curve gives rise to a place of the surface. In all cases in which the converse is true we may regard the curve as a representation of the surface, or conversely.

Thus such curves in space are divisible into sets according to their deficiency. And in connexion with such curves we can construct all the functions with which we deal upon a Riemann surface.

Of these principles sufficient account will be given below (Chapter VI.): familiar examples are the space cubic, of deficiency zero, and the most general space quartic of deficiency 1 which is representable by elliptic functions.

11. In this chapter we have spoken primarily of the algebraic equation —and of the curve or the Riemann surface as determined thereby. But this is by no means the necessary order. If the Riemann surface be given, the algebraic equation can be determined from it—and in many forms, according to the function selected as dependent variable (y). It is necessary to keep this in view in order fully to appreciate the generality of Riemann's methods. For instance, we may start with a surface in space whose shape is that of an

* The reader who desires to study the geometrical theory referred to may consult:— Cayley, *Quart. Journal*, VII.; H. J. S. Smith, *Proc. Lond. Math. Soc.* VI.; Noether, *Math. Annal.* 9; Brill, *Math. Annal.* 16; Brill u. Noether, *Math. Annal.* 7.

anchor ring*, and construct upon this surface a set of elliptic functions. Or we may start with the surface on a plane which is exterior to two circles drawn upon the plane, and construct for this surface a set of elliptic functions. Much light is thrown upon the functions occurring in the theory by thus considering them in terms of what are in fact different independent variables. And further gain arises by going a step further. The infinite plane upon which uniform functions of a single variable are represented may be regarded as an infinite sphere; and such surfaces as that of which the anchor ring above is an example may be regarded as generalizations of that simple case. Now we can treat of *branches* of a multiform function without the use of a Riemann surface, by supposing the branch points of the function marked on a single infinite plane and suitably connected by barriers, or cuts, across which the independent variable is supposed not to pass. In the same way, for any general Riemann surface, we may consider branches of functions which are not uniform upon that surface, the branches being separated by drawing barriers upon the surface. The properties obtained will obviously generalize the properties of the functions which are uniform upon the surface.

* Forsyth, p. 318; Riemann, *Ges. Werke* (1876), pp. 89, 415.

CHAPTER II.

THE FUNDAMENTAL FUNCTIONS ON A RIEMANN SURFACE.

12. IN the present chapter the theory of the fundamental functions is based upon certain *a priori* existence theorems*, originally given by Riemann. At least two other methods might be followed: in Chapters IV. and VI. sufficient indications are given to enable the reader to establish the theory independently upon purely algebraical considerations: from Chapter VI. it will be seen that still another basis is found in a preliminary theory of plane curves. In both these cases the ideas primarily involved are of a very elementary character. Nevertheless it appears that Riemann's descriptive theory is of more than equal power with any other; and that it offers a generality of conception to which no other theory can lay claim. It is therefore regarded as fundamental throughout the book.

It is assumed that the *Theory of Functions* of Forsyth will be accessible to readers of the present book; the aim in the present chapter has been to exclude all matter already contained there. References are given also to the treatise of Harkness and Morley*.

13. Let t be the infinitesimal† at any place of a Riemann surface: if it is a finite place, namely, a place at which the independent variable x is finite, the values of x for all points in the immediate neighbourhood of the place are expressible in the form $x = a + t^{w+1}$: if an infinite place, $x = t^{-(w+1)}$. There exists a function which save for certain additive moduli is one-valued on the whole surface and everywhere finite and continuous, save at the place in question, in the neighbourhood of which it can be expressed in the form

$$\frac{A}{t^r} + \frac{A_1}{t^{r-1}} + \dots + \frac{A_{r-1}}{t} + C + P(t).$$

* See for instance: Forsyth, *Theory of Functions of a Complex Variable*, 1893; Harkness and Morley, *Treatise on the Theory of Functions*, 1893; Schwarz, *Gesam. math. Abhandlungen*, 1890. The best of the early systematic expositions of many of the ideas involved is found in C. Neumann, *Vorlesungen über Riemann's Theorie*, 1884, which the reader is recommended to study. See also Picard, *Traité d'Analyse*, Tom. II. pp. 273, 42 and 77.

† For the notation see Chapter I. §§ 2, 3.

Herein, as throughout, $P(t)$ denotes a series of positive integral powers of t vanishing when $t=0$, C, A, ..., A_{r-1}, are constants whose values can be arbitrarily assigned beforehand, and r is a positive integer whose value can be assigned beforehand.

We shall speak of all such functions as integrals of the second kind: but the name will be generally restricted to that *particular function whose behaviour near the place is that of

$$-\frac{1}{t}+C+P(t).$$

This function is not entirely unique. We suppose the surface dissected by $2p$ cuts†, which we shall call period loops; they subserve the purpose of rendering the function one-valued over the whole of the dissected surface. We impose the further condition that the periods of the function for transit across the p loops of the first kind‡ shall be zero; then the function is unique save for an additive constant. It can therefore be made to vanish at an arbitrary place. The special function§ so obtained whose infinity is that of $-\frac{1}{t}$ is then denoted by $\Gamma_a^{x,\,c}$, c denoting the place where the function vanishes and x the current place. When the infinity is an ordinary place, at which either $x=a$ or $x=\infty$, the function is infinite either like $-\frac{1}{x-a}$ or $-x$. The periods of $\Gamma_a^{x,\,c}$ for transit of the period loops of the second kind will be denoted by $\Omega_1, \ldots, \Omega_p$.

14. Let (x_1y_1), (x_2y_2) be any two places of the surface: and let the infinitesimals be respectively denoted by t_1, t_2, so that in the neighbourhood of these places we have the equations $x-x_1=t_1^{w_1+1}$, $x-x_2=t_2^{w_2+1}$. Let a cut be made between the places (x_1y_1), (x_2y_2). There exists a function, here denoted by $\Pi_{x_1,\,x_2}^{x,\,c}$, which (α) is one-valued over the whole dissected surface, (β) has p periods arising for transit of the period loops of the second kind and has no periods at the period loop of the first kind, (γ) is everywhere continuous and finite save near (x_1y_1) and (x_2y_2), where it is infinite respectively like $\log t_1$ and $-\log t_2$, and, (δ), vanishes when the current place denoted by x is the place denoted by c. This function is unique. If the cut between (x_1y_1), (x_2y_2) be not made, the function is only definite apart from an additive integral multiple of $2\pi i$, whose value depends on the

* This particular function is also called an *elementary* integral of the second kind.

† Those ordinarily called the a, b curves; see Forsyth, p. 354. Harkness and Morley, p. 242, etc.

‡ Those called the a cuts.

§ The fact that the function has no periods at the period loops of the first kind is generally denoted by calling the function a *normal* integral of the second kind.

path by which the variable is supposed to pass from c. It will be called* the integral of the third kind whose infinity is like that of $\log(t_1/t_2)$.

15. Beside these functions there exist also certain integrals of the first kind—in number p. They are everywhere continuous and finite and one-valued on the dissected surface. For transit of the period loops of the first kind, one of them, say v_i, has no periods except for transit of the ith loop, a_i. This period is here taken to be 1. The periods of v_i for transit of the period loops of the second kind are here denoted by $\tau_{i1}, \ldots, \tau_{ip}$. We may therefore form the scheme of periods

	a_1	a_2	....	a_p	b_1	....	b_p
v_1	1	0		0	τ_{11}		τ_{1p}
v_2	0	1		0	τ_{21}		τ_{2p}
⋮							
v_p	0	0		1	τ_{p1}		τ_{pp}

Each of these functions v_i is unique when a zero is given. They will therefore be denoted by $v_1^{x,c}, \ldots, v_p^{x,c}$, the zero denoted by c being at our disposal.

The periods τ_{ij} have certain properties which will be referred to in their proper place: in particular $\tau_{ij} = \tau_{ji}$, so that they are certainly not equivalent to more than $\frac{1}{2}p(p+1)$ algebraically independent constants. As a fact, in accordance with the previous chapter, when $p > 1$ they are subject to $\frac{1}{2}p(p+1) - (3p-3) = \frac{1}{2}(p-2)(p-3)$ relations.

16. In regard to these enunciations, the reader will notice that the word period here used for that additive constant arising for transit of a period loop —namely, in consequence of a path leading from one edge of the period loop to the opposite edge—would be more properly applied to the period for circuit of this path than the period for transit of the loop.

The integrals here specified are more precisely called the *normal elementary* integrals of their kinds. The general integral of the first kind is a linear function of $v_1, \ldots, v_p$ with constant coefficients; its periods at the first p loops will not have the same simple forms as have those of $v_1 \ldots v_p$. The general integral of the third kind, infinite like $C\log(t_1/t_2)$, C being a constant, is obtained by adding a general integral of the first kind to $C\Pi_{x_1, x_2}^{x, c}$; similarly for the general integral of the second kind.

The function $\Pi_{x_1, x_2}^{x, c}$ has† the property expressed by the equation

$$\Pi_{x_1, x_2}^{x, c} = \Pi_{x, c}^{x_1, x_2}.$$

* More precisely, the *normal elementary* integral of the third kind.

† Forsyth, p. 453. Harkness and Morley, p. 445.

A more general integral of the third kind having the same property is

$$\Pi_{x_1, x_2}^{x, c} + \sum_{i, j}^{1 \ldots p} A_{ij}\, v_i^{x, c}\, v_j^{x_1, x_2},$$

wherein the arbitrary coefficients satisfy the equations $A_{ij} = A_{ji}$. The property is usually referred to as the theorem of the interchange of argument (x) and parameter (x_1).

The property allows the consideration of

$$\Pi_{x_1', x_2}^{x, c}$$

as *a function of* x_1' for fixed positions of x, c, x_2. In this regard a remark should be made:

For an ordinary position of x, the function

$$\Pi_{x_1', x_2}^{x, c} - \log(x_1' - x) = \Pi_{x, c}^{x_1', x_2} - \log(x_1' - x)$$

is a finite continuous function of x_1' when x_1' is in the neighbourhood of x. But if x_1 be a branch place where $w+1$ sheets wind, and x_1', x be two positions in its neighbourhood, the functions of x

$$\Pi_{x_1', x_2}^{x, c} - \log(x_1' - x), \quad \Pi_{x_1, x_2}^{x, c} - \frac{1}{w+1} \log(x_1 - x)$$

are respectively finite as x approaches x_1' and x_1, so that

$$\Pi_{x, c}^{x_1', x_2} - \log(x_1' - x)$$

is not a finite and continuous function *of* x_1' for positions of x_1' up to and including the branch place x_1.

In this case, let the neighbourhood of the branch place be conformally represented upon a simple plane closed area and let ξ_1, ξ_1', ξ be the representatives thereon of the places x_1, x_1', x. Then the correct statement is that

$$\Pi_{x, c}^{x_1', x_2} - \log(\xi_1' - \xi)$$

is a continuous function of x_1' or ξ_1' up to and including the branch place x_1. This is in fact the form in which the function $\Pi_{x, c}^{x_1', x_2}$ arises in the proof of its existence upon which our account is based*.

In a similar way the function

$$\Gamma_{x_1'}^{x, c},$$

regarded as a function of x_1', is such that

$$\Gamma_{x_1'}^{x, c} + \frac{1}{\xi - \xi_1'}$$

is a finite continuous function of ξ_1' in the immediate neighbourhood of x.

* The reader may consult Neumann, p. 220.

17. It may be desirable to give some simple examples of these integrals.

(a) For the surface represented by

$$y^2 = x(x - a_1)\dots(x - a_{2p+1}),$$

wherein $a_1, \dots, a_{2p+1}$ are all finite and different from zero and each other, consider the integral

$$\tfrac{1}{2}\int \frac{dx}{y}\left(\frac{y+\eta}{x-\xi} - \frac{y+\eta_1}{x-\xi_1}\right),$$

(ξ, η), (ξ_1, η_1) being places of the surface other than the branch places, which are

$$(0, 0), (a_1, 0), \dots, (a_{2p+1}, 0).$$

It is clearly infinite at these places respectively like $\log(x-\xi)$, $-\log(x-\xi_1)$.

It is not infinite at $(\xi, -\eta)$, $(\xi_1, -\eta_1)$; for $(y+\eta)/(x-\xi)$, $(y+\eta_1)/(x-\xi_1)$ are finite at these places respectively.

At a place $x=\infty$, where $x=t^{-1}$, $y=\epsilon t^{-p-1}(1+P_1(t))$, ϵ being ± 1, and $P_1(t)$ a series of positive integral powers of t vanishing for $t=0$, we have

$$\frac{y+\eta}{y} = 1 + \eta\epsilon t^{p+1}(1+P_1(t)),\quad \frac{dx}{x-\xi} = -\frac{dt}{t}(1-\xi t)^{-1},$$

and the integral has the form

$$-\tfrac{1}{2}\int \frac{dt}{t}\, t\,[A + P_2(t)],$$

A being a constant. It is therefore finite.

At a place $y=0$, for instance where

$$x = a_1 + t^2,\quad y = Bt\,[1 + P_3(t)],$$

B being a constant, the integral has the form

$$C\int dt\,[1 + P_4(t)],$$

C being a constant, and is finite.

Thus it is an elementary integral of the third kind with infinities at (ξ, η), (ξ_1, η_1).

It may be similarly shewn that the integral

$$\tfrac{1}{2}\int \frac{dx}{y}\left(\frac{y}{x} - \frac{y+\eta_1}{x-\xi_1}\right)$$

is infinite at (ξ_1, η_1) like $-\log(x-\xi_1)$ and is not elsewhere infinite except at $(0, 0)$.

Near $(0, 0)$, we have $x=t^2$, $y=Dt\,[1+P_5(t^2)]$ and this integral is infinite like

$$\int \frac{dt}{t} = \log t.$$

It is therefore an elementary integral of the third kind with one infinity at the branch place $(0, 0)$ and the other at (ξ_1, η_1).

Consider next the integral

$$\tfrac{1}{2}\int \frac{dx}{y}\frac{d}{d\xi}\left(\frac{y+\eta}{x-\xi}\right) = \tfrac{1}{2}\int \frac{dx}{y}\frac{y+\eta+(x-\xi)\eta'}{(x-\xi)^2},$$

where $\eta' = \dfrac{d\eta}{d\xi}$. It can easily be seen that it is not infinite save at (ξ, η). Writing for the neighbourhood of this place, which is supposed not to be a branch place,

$$y = \eta + (x-\xi)\eta' + \tfrac{1}{2}(x-\xi)^2\eta'' + \dots\dots,$$

the integral becomes

$$\int \frac{dx}{(x-\xi)^2} \frac{\eta+(x-\xi)\eta'+\frac{1}{4}(x-\xi)^2\eta''+\dots}{\eta+(x-\xi)\eta'+\frac{1}{2}(x-\xi)^2\eta''+\dots},$$

which is equal to

$$\int \frac{dx}{(x-\xi)^2}\left[1-\tfrac{1}{4}\frac{\eta''}{\eta}(x-\xi)^2+\dots\right].$$

Thus the integral is there infinite like $-\frac{1}{x-\xi}$, and is thus an elementary integral of the second kind.

The elementary integral of the second kind for a branch place, say $(0, 0)$, is a multiple of

$$\tfrac{1}{2}\int \frac{dx}{xy}.$$

In fact near $x=0$, writing $x=t^2$, $y=Dt[1+P(t^2)]$, this integral becomes

$$\frac{1}{D}\int \frac{dt}{t^2}[1+P(t^2)]^{-1}$$

or

$$\frac{1}{D}\int \frac{dt}{t^2}[1+Et^2+Ft^4+\dots]$$

which is equal to

$$\frac{1}{D}\left[-\frac{1}{t}+Et+\dots\right]$$

as desired.

The integral is clearly not infinite elsewhere.

Example 1. Verify that the integral last considered is the limit of

$$\frac{1}{2D}\int \frac{dx}{y}\left[\frac{y+\eta}{x-\xi}-\frac{y}{x}\right]$$

as the place (ξ, η) approaches indefinitely near to $(0, 0)$.

Example 2. Shew that the general integral of the first kind for the surface is

$$\int \frac{dx}{y}(A_1+A_2x+\dots+A_{p-1}x^{p-1}).$$

(β) We have in the first chapter §§ 2, 3 spoken of a circumstance that can arise, that two sheets of the surface just touch at a point and have no further connexion, and we have said that we regard the points of the sheets as distinct places. Accordingly we may have an integral of the third kind which has its infinities at these two places, or an integral of the third kind having one of its infinities at one of these places. For example, on the surface

$$f(x, y)=(y-m_1x)(y-m_2x)+(x, y)_3+(x, y)_4=0$$

where $(x, y)_3$, $(x, y)_4$ are integral homogeneous polynomials of the degrees indicated by the suffixes, with quite general coefficients, and m_1, m_2 are finite constants, there are at $x=0$ two such places, at both of which $y=0$.

In this case

$$\int \frac{dx}{f'(y)},$$

where $f'(y)=\frac{\partial f}{\partial y}$, is a constant multiple of an integral of the third kind with infinities at these two places $(0, 0)$; and

$$\int \frac{y-m_1x+Ax^2+Bxy+Cy^2}{Lx+My}\frac{dx}{f'(y)}$$

is a constant multiple of an integral of the third kind, provided A, B, C be so chosen that $y-m_1x+Ax^2+Bxy+Cy^2$ vanishes at one of the two places other than (0, 0) at which $Lx+My$ is zero. Its infinities are at (i) the uncompensated zero of $Lx+My$ which is not at (0, 0), (ii) the place (0, 0) at which the expression of y in terms of x is of the form

$$y=m_2x+Px^2+Qx^3+\dots$$

In fact, at a branch place of the surface where $x=a+t^2$, $f'(y)$ is zero of the first order, and $dx=2t\,dt$; thus $\int\frac{dx}{f'(y)}$ is finite at the branch places. At each of the places (0, 0), $f'(y)$ is zero of the first order, $Lx+My$ is zero of the first order and $y-m_1x+Ax^2+Bxy+Cy^2$ is zero at these places to the first and second order respectively. These statements are easy to verify; they lead immediately to the proof that the integrals have the character enunciated.

The condition given for the choice of A, B, C will not determine them uniquely—the integral will be determined save for an additive term of the form

$$\int(Px+Qy)\frac{dx}{f'(y)},$$

where P, Q are undetermined constants. The reader may prove that this is a general integral of the first kind. The constants P, Q may be determined so that the integral of the third kind has no periods at the period loops of the first kind, whose number in this case is two. The reasons that suggest the general form written down will appear in the explanation of the geometrical theory.

(γ) The reader may verify that for the respective cases

$$\begin{aligned} y^6&=(x-a)(x-b)^2(x-c)^3,\\ y^4&=(x-a)(x-b)\ (x-c)^2,\\ y^6&=(x-a)(x-b)\ (x-c)^4,\\ y^7&=(x-a)(x-b)\ (x-c)^6,\end{aligned}$$

the general integrals of the first kind are

$$\int\frac{dx}{y^5}(x-b)(x-c)^2,$$

$$\int\frac{dx}{y^3}(x-c),$$

$$\int\frac{dx}{y^5}(x-c)^2[Ay+B(x-c)],$$

$$\int\frac{dx}{y^6}(x-c)^2[Ay^2+By(x-c)+C(x-c)^2],$$

where A, B, C are arbitrary constants.

See an interesting dissertation "de Transformatione aequationis $y^n=R(x)$..." Eugen. Netto (Berlin, Gust. Schade, 1870).

(δ) *Ex.* Prove that if F denote any function everywhere one valued on the Riemann surface and expressible in the neighbourhood of every place in the form

$$\frac{A_1}{t}+\frac{A_2}{t^2}+\dots+B+B_1t+B_2t^2+\dots$$

the sum of the coefficients of the logarithmic terms $\log t$ of the integral $\int^x F dx$, for all places where such a term occurs, is zero.

It is supposed that the number of places where negative powers of t occur in the expansion of F is finite, but it is not necessary that the number of negative powers be finite. The theorem may be obtained by contour integration of $\int Fdx$, and clearly generalizes a property of the integral of the third kind.

18. The value of the integral* $\int \Gamma_a^{x,c}\, dv_i^{x,c}$ taken round the p closed curves formed by the two sides of the pairs of period loops $(a_1, b_1), \ldots, (a_p, b_p)$, in such a direction that the interior of the surface is always on the left hand, is equal to the value taken round the sole infinity, namely the place a, in a counter-clockwise direction. Round the pair a_r, b_r the value obtained is

$$\Omega_r \int dv_i^{x,c},$$

taken once positively in the direction of the arrow head round what in the figure is the outer side of b_r. This value is $\Omega_r(-\omega_{ir})$, where ω_{ir} denotes the period of v_i for transit of a_r, namely, from what in the figure is the inside of the oval a_r to the outside.

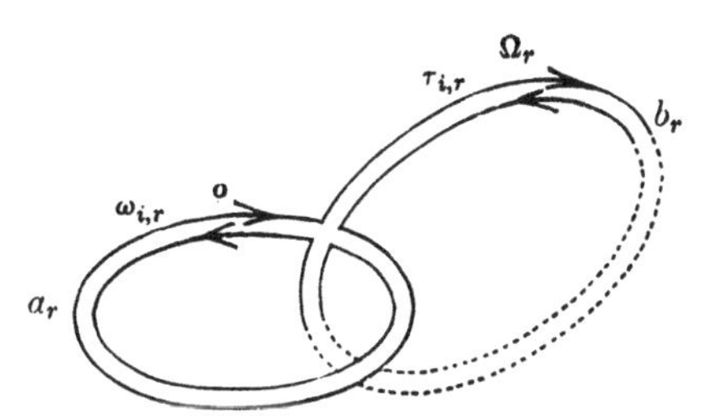

The relations indicated by the figure for the signs adopted for ω_{ir}, τ_{ir} and the periods of $\Gamma_a^{x,c}$ will be preserved throughout the book.

Since ω_{ir} is zero except when $r = i$, the sum of these p contour integrals is $-\omega_{i,i}\,\Omega_i$. Taken in a counter-clockwise direction, round the pole of $\Gamma_a^{x,c}$, where

$$\Gamma_a^{x,c} = -\frac{1}{t} + A + Bt + Ct^2 + \ldots,$$

the integral gives

$$\int \left[-\frac{1}{t} + A + Bt + Ct^2 + \ldots\right]\left[Dv_i^{a,c} + tD^2v_i^{a,c} + \ldots\right] dt,$$

where D denotes $\frac{d}{dt}$. Hence, as $\omega_{i,i} = 1$,

$$\Omega_i = 2\pi i \left(Dv_i^{x,c}\right)_a.$$

* Cf. Forsyth, pp. 448, 451. Harkness and Morley, p. 439.

This is true whether a be a branch place or a place at infinity (for which, if not a branch place, $x = t^{-1}$) or an ordinary finite place. In the latter case

$$\Omega_i = 2\pi i \frac{d}{dx}\left(v_i^{x,\, c}\right).$$

Similarly the reader may prove that the periods of $\Pi_{x_1,\, x_2}^{x,\, c}$ are

$$0, \ldots\ldots\ 0,\ 2\pi i v_i^{x_1,\, x_2}, \ldots\ldots\ 2\pi i v_p^{x_1,\, x_2}.$$

In this case it is necessary to enclose x_1 and x_2 in a curve winding $w_1 + 1$ times at x_1, $w_2 + 1$ times at x_2, in order that this curve may be closed.

19. From these results we can shew that the integral of the second kind is derivable by differentiation from the integral of the third kind. Apart from the simplicity thus obtained, the fact is interesting because, as will appear, the analytical expression of an integral of the third kind is of the same general form whether its infinities be branch places or not; this is not the case for integrals of the second kind.

We can in fact prove the equation

$$D_{t_{x_1}} \Pi_{x_1,\, x_2}^{x,\, c} = \Gamma_{x_1}^{x,\, c},$$

namely, if, to take the most general case, x_1 be a winding place and x_1' a place in its neighbourhood such that $x_1' = x_1 + t_{x_1}^{w+1}$, the equation,

$$\lim_{t_{x_1}=0} \frac{1}{t_{x_1}} \left[\Pi_{x_1',\, x_2}^{x,\, c} - \Pi_{x_1,\, x_2}^{x,\, c}\right] = \Gamma_{x_1}^{x,\, c}.$$

For, let the neighbourhood of the branch place x_1 be conformally represented upon a simple closed area without branch place, by means of the infinitesimal of x, as explained in the previous chapter. Let ξ_1', ξ_1 be the representatives of the places x_1', x_1, and ξ the representative of a place x which is very near to x_1, but is so situate that we may regard x_1' as ultimately infinitely closer to x_1 than x is.

Then
$$x - x_1 = (\xi - \xi_1)^{w+1},$$
$$x - x_1' = (\xi - \xi_1')[C + P(\xi - \xi_1')],$$

where C does not vanish for $x_1' = x$,

and
$$\Pi_{x_1',\, x_2}^{x,\, c} = \log(x - x_1') + \Phi' = \log(\xi - \xi_1') + \phi',$$

where ϕ' is finite for the specified positions of the places *and remains finite when* ξ_1' *is taken infinitely near to* ξ_1 (§ 16).

Also
$$\Pi_{x_1,\, x_2}^{x,\, c} = \frac{1}{w+1}\log(x - x_1) + \phi = \log(\xi - \xi_1) + \phi,$$

where ϕ is also finite. Therefore

$$\Pi^{x,\,c}_{x_1',\,x_2} - \Pi^{x,\,c}_{x_1,\,x_2} = \log\left(1 - \frac{\xi_1' - \xi_1}{\xi - \xi_1}\right) + \phi' - \phi$$

$$= -(\xi_1' - \xi_1)\left[\frac{1}{\xi - \xi_1} + \frac{1}{2}\frac{\xi_1' - \xi_1}{(\xi - \xi_1)^2} + \dots\right] + \phi' - \phi,$$

and thus

$$\lim_{\xi_1' = \xi_1} \left[\frac{\Pi^{x,\,c}_{x_1',\,x_2} - \Pi^{x,\,c}_{x_1,\,x_2}}{\xi_1' - \xi_1}\right] = -\frac{1}{\xi - \xi_1} + \psi,$$

where ψ is finite.

Now as ξ_1' moves up to ξ_1, for a fixed position of ξ, we have

$$\xi_1' - \xi_1 = (x_1' - x_1)^{\frac{1}{w+1}} = t_{x_1},$$

and $$\Gamma^{x,\,c}_{x_1} = \Gamma^{\xi,\,c}_{\xi_1} = -\frac{1}{\xi - \xi_1} + \vartheta,$$

where ϑ is finite.

Hence $$D_{t_{x_1}} \Pi^{x,\,c}_{x_1,\,x_2} - \Gamma^{x,\,c}_{x_1}$$
is finite when x is near to x_1.

Moreover it does not depend on x_2. For from the equation

$$\Pi^{x,\,c}_{x_1,\,x_2} = \Pi^{x_1,\,x_2}_{x,\,c},$$

we may regard $\Pi^{x,\,c}_{x_1,\,x_2}$ as a function of x_1, which is determinate save for an additive constant by the specification of x and c only. This additive constant, which is determined by the condition that the function vanishes when $x_1 = x_2$, is the only part of the function which depends on x_2. It disappears in the differentiation.

Finally, by the determination of the periods previously given, it follows that

$$D_{t_{x_1}} \Pi^{x,\,c}_{x_1,\,x_2} - \Gamma^{x,\,c}_{x_1}$$

has no periods at the $2p$ period loops. Hence it is a constant, and therefore zero since it vanishes when $x = c$.

Corollary i.

Hence $$D_{t_x} \Gamma^{x,\,c}_{x_1} = D_{t_x} D_{t_{x_1}} \Pi^{x,\,c}_{x_1,\,x_2} = D_{t_{x_1}} D_{t_x} \Pi^{x_1,\,x_2}_{x,\,c} = D_{t_{x_1}} \Gamma^{x_1,\,c}_{x},$$

of which neither depends on the constant position c.

Corollary ii.

The functions

$$D_{t_{x_1}} \Gamma^{x,\,c}_{x_1},\ D^2_{t_{x_1}} \Gamma^{x,\,c}_{x_1},\ D^3_{t_{x_1}} \Gamma^{x,\,c}_{x_1},\ \dots$$

are respectively infinite like

$$-\frac{1}{t_{x_1}{}^2},\quad -\frac{\lfloor 2}{t_{x_1}{}^3},\quad -\frac{\lfloor 3}{t_{x_1}{}^4},\ \ldots.$$

We shall generally write D_{x_1}, $D^2_{x_1}$, ... instead of $D_{t_{x_1}}$, $D^2_{t_{x_1}}$, ... When x_1 is an ordinary place D_{x_1} will therefore mean $\frac{d}{dx_1}$, etc.

Corollary iii.

By means of the example (δ) of § 17 it can now be shewn that the infinite parts of the integral

$$\int F dx,$$

in which F is any uniform function of position on the undissected surface having only infinities of finite order, are those of a sum of terms consisting of proper constant multiples of integrals of the third kind and differential coefficients of these in regard to the parametric place.

20. One particular case of Cor. iii. of the last Article should be stated. A function which is everywhere one-valued on the undissected surface must be somewhere infinite. As in the case of uniform functions on a single infinite plane (which is the particular case of a Riemann surface for which the deficiency is zero), such functions can be divided into rational and transcendental, according as all their infinities are of finite order and of finite number or not. Transcendental functions which are uniform on the surface will be more particularly considered later. A rational uniform function can be expressed rationally in terms of x and y*. But since the function can be expressed in the neighbourhood of any of its poles in the form

$$C + \frac{A_1}{t} + \frac{A_2}{t^2} + \ldots + \frac{A_m}{t^m} + P(t),$$

we can, by subtracting from the function a series of terms of the form

$$-\left[A_1 \Gamma_a^{x,c} + A_2 D_a \Gamma_a^{x,c} + \ldots + \frac{A_m}{\lfloor m-1} D_a^{m-1} \Gamma_a^{x,c}\right],$$

obtain a function nowhere infinite on the surface and having no periods at the first p period loops. Such a function is a constant†. Hence F can also be expressed by means of normal integrals of the second kind only. Since F has no periods at the period loops of the second kind there are for all rational functions certain necessary relations among the coefficients $A_1, \ldots, A_m$. These are considered in the next Chapter.

* Forsyth, p. 369. Harkness and Morley, p. 262.

† Forsyth, p. 439.

21. Of all rational functions there are p whose importance justifies a special mention here; namely, the functions

$$\frac{dv_1}{dx}, \frac{dv_2}{dx}, \dots \frac{dv_p}{dx}.$$

In the first place, these cannot be all zero for any ordinary finite place a of the surface. For they are, save for a factor $2\pi i$, the periods of the normal integral $\Gamma_a^{x, c}$. If the periods of this integral were zero, it would be a rational uniform function of the first order; in that case the surface would be representable conformally upon another surface of one sheet*, $\xi = \Gamma_a^{x, c}$ being the new independent variable; and the transformation would be reversible (Chap. I. § 6). Hence the original surface would be of deficiency zero; in which case the only integral of the first kind is a constant. The functions are all infinite at a branch place a. But it can be shewn as here that the quantities to which they are there proportional, namely $D_a v_1, \dots, D_a v_p$, cannot be all zero. The functions are all zero at infinity, but similarly it can be shewn that the quantities, $Dv_1, \dots, Dv_p$, cannot be all zero there.

Thus p linearly independent linear aggregates of these quantities cannot all vanish at the same place. We remark, in connexion with this property, that surfaces exist of all deficiencies such that $p-1$ linearly independent linear aggregates of these quantities vanish in an infinite number of sets of *two* places. Such surfaces are however special, and their equation can be put† into the form

$$y^2 = (x, 1)_{2p+2}.$$

We have seen that the statement of the property requires modification at the branch places, and at infinity; this particularity is however due to the behaviour of the independent variable x. We shall therefore state the property by saying: there is no place at which all the differentials $dv_1, \dots, dv_p$ vanish. A similar phraseology will be adopted in similar cases. For instance, we shall say that each of $dv_1, dv_2, \dots, dv_p$ has‡ $2p-2$ zeros, some of which may occur at infinity.

In the next place, since any general integral of the first kind

$$\lambda_1 v_1^x + \dots + \lambda_p v_p^x$$

must necessarily be finite all over any other surface upon which the original surface is conformally and reversibly represented and therefore must be an integral of the first kind thereon, it follows that the rational function

$$\lambda_1 \frac{dv_1}{dx} + \dots + \lambda_p \frac{dv_p}{dx}$$

* I owe this argument to Prof. Klein.

† See below, Chap. V.

‡ See Forsyth, p. 461. Harkness and Morley, p. 450.

is necessarily transformed with the surface into

$$M\left(\lambda_1 \frac{dV_1}{d\xi} + \ldots + \lambda_p \frac{dV_p}{d\xi}\right),$$

where $V_i = v_i$ is an integral of the first kind, not necessarily normal, on the new surface, ξ being the new independent variable, and $M = \frac{d\xi}{dx}$.

Thus, the ratios of the integrands of the first kind are transformed into ratios of integrands of the first kind; they may be said to be invariant for birational transformation.

This point may be made clearer by an example. The general integral of the first kind for the surface

$$y^2 = (x, 1)_8$$

can be shewn to be

$$\int \frac{dx}{y}(A + Bx + Cx^2),$$

A, B, C being arbitrary constants.

If then $\phi_1 : \phi_2 : \phi_3$ denote the ratios of any three linearly independent integrands of the first kind for this surface, we have

$$1 : x : x^2 = a_1\phi_1 + b_1\phi_2 + c_1\phi_3 : a_2\phi_1 + b_2\phi_2 + c_2\phi_3 : a_3\phi_1 + b_3\phi_2 + c_3\phi_3$$

for proper values of the constants $a_1, b_1, \ldots, c_3$,

and hence

$$(a_1\phi_1 + b_1\phi_2 + c_1\phi_3)(a_3\phi_1 + b_3\phi_2 + c_3\phi_3) = (a_2\phi_1 + b_2\phi_2 + c_2\phi_3)^2.$$

Such a relation will therefore hold for all the surfaces into which the given one can be birationally transformed.

22. It must be remarked that the determination of the normal integrals here described depends upon the way in which the fundamental period loops are drawn. An integral of the first kind which is normal for one set of period loops will be a linear function of the integrals of the first kind which are normal for another set; and an integral of the second or third kind, which is normal for one set of period loops, will for another set differ from a normal integral by an additive linear function of integrals of the first kind.

CHAPTER III.

THE INFINITIES OF RATIONAL UNIFORM FUNCTIONS.

23. IN this chapter and in general we shall use the term rational function to denote a uniform function of position on the surface of which all the infinities are of finite order, their number being finite. We deal first of all with the case in which these infinities are all of the first order.

If k places of the surface, say $a_1, a_2 \dots a_k$, be arbitrarily assigned we can always specify a function with p periods having these places as poles, of the first order, and otherwise continuous and uniform; namely, the function is of the form

$$\mu_0 + \mu_1 \Gamma^x_{a_1} + \dots + \mu_k \Gamma^x_{a_k},$$

where the coefficients $\mu_0, \mu_1 \dots \mu_k$ are constants, the zeros of the functions Γ being left undetermined. Conversely, as remarked in the previous chapter (§ 20), a rational function having $a_1, \dots, a_k$ as its poles must be of this form. In order that the expression may represent a rational function the periods must all be zero. Writing the periods of Γ^x_a in the form $\Omega_1(a), \dots, \Omega_p(a)$, this requires the equations

$$\mu_1 \Omega_i(a_1) + \mu_2 \Omega_i(a_2) + \dots + \mu_k \Omega_i(a_k) = 0,$$

for all the p values, $i = 1, 2, \dots, p$, of i. In what follows we shall for the sake of brevity say that a place c *depends upon* r places $c_1, c_2, \dots, c_r$ when for all values of i, the equations

$$\Omega_i(c) = f_1 \Omega_i(c_1) + \dots + f_r \Omega_i(c_r)$$

hold for finite values of the coefficients $f_1, \dots, f_r$, these coefficients being *independent of* i. Hence we may also say:

In order that a rational function should exist having k *assigned places as its poles, each simple, one at least of these places must depend upon the others.*

24. Taking the k places $a_1, a_2, \dots, a_k$ in the order of their suffixes, it may of course happen that several of them depend upon the others, say $a_{s+1}, \dots, a_k$

upon $a_1, \ldots, a_s$, the latter set $a_1, \ldots, a_s$ being independent: then we have equations of the form

$$\Omega_i(a_{s+1}) = n_{s+1,1}\,\Omega_i(a_1) + \ldots + n_{s+1,s}\,\Omega_i(a_s)$$
$$\ldots\ldots\ldots\ldots\ldots\ldots\ldots\ldots\ldots\ldots$$
$$\Omega_i(a_k) \;\; = n_{k,1}\,\Omega_i(a_1) \;\; + \ldots + n_{k,s}\,\Omega_i(a_s),$$

the coefficients in any of the rows here being the same for all the p values of i. In particular, if s be as great as p and $a_1, \ldots, a_s$ be independent, equations of this form will hold for all positions of $a_{s+1}, \ldots, a_k$. For then we have enough disposeable coefficients to satisfy the necessary p equations.

When it does so happen, that $a_{s+1}, \ldots, a_k$ depend upon $a_1 \ldots a_s$, there exist rational functions, of the form

$$R_{s+1} = \sigma_{s+1} + \lambda_{s+1}\left[\Gamma^x_{a_{s+1}} - n_{s+1,1}\,\Gamma^x_{a_1} - \ldots\ldots - n_{s+1,s}\,\Gamma^x_{a_s}\right],$$
$$\ldots\ldots\ldots\ldots\ldots\ldots\ldots\ldots\ldots\ldots$$
$$R_k \;\; = \sigma_k \;\; + \lambda_k \;\; \left[\Gamma^x_{a_k} \;\; - n_{k,1} \;\; \Gamma^x_{a_1} - \ldots\ldots - n_{k,s} \;\; \Gamma^x_{a_s}\right],$$

wherein $\sigma_{s+1} \ldots \sigma_k$, $\lambda_{s+1} \ldots \lambda_k$ are constants, which are all infinite once in $a_1 \ldots a_s$ and are, beside, infinite respectively at $a_{s+1}, \ldots, a_k$; and the most general function uniform on the dissected surface, which is infinite to the first order at $a_1, \ldots, a_k$, being, as remarked, of the form

$$\mu_0 + \mu_1\,\Gamma^x_{a_1} + \ldots\ldots + \mu_k\,\Gamma^x_{a_k},$$

can be written in the form

$$\mu_0 + \mu_1\,\Gamma^x_{a_1} + \ldots\ldots + \mu_s\,\Gamma^x_{a_s}$$
$$+ \mu_{s+1}\left[\frac{1}{\lambda_{s+1}}\,R_{s+1} + n_{s+1,1}\,\Gamma^x_{a_1} + \ldots\ldots + n_{s+1,s}\,\Gamma^x_{a_s} - \frac{\sigma_{s+1}}{\lambda_{s+1}}\right]$$
$$+ \ldots\ldots\ldots\ldots\ldots\ldots\ldots\ldots\ldots\ldots$$
$$+ \mu_k\left[\frac{1}{\lambda_k}\;\; R_k \;\; + n_{k,1} \;\; \Gamma^x_{a_1} + \ldots\ldots + n_{k,s} \;\; \Gamma^x_{a_s} - \frac{\sigma_k}{\lambda_k}\right],$$

namely, in the form

$$\nu_0 + \nu_1\,\Gamma^x_{a_1} + \ldots\ldots + \nu_s\,\Gamma^x_{a_s} + \nu_{s+1} R_{s+1} + \ldots\ldots + \nu_k R_k.$$

If this function is to have no periods, the equations

$$\nu_1\Omega_i(a_1) + \ldots\ldots + \nu_s\Omega_i(a_s) = 0, \quad (i = 1, 2, \ldots, p),$$

must hold. Since $a_1, \ldots, a_s$ are independent, such equations can only hold when $\nu_1 = 0 = \ldots = \nu_s$. Thus the *most general* rational function having k poles of the first order, at $a_1, \ldots, a_k$, is of the form

$$\nu_0 + \nu_{s+1} R_{s+1} + \ldots\ldots + \nu_k R_k,$$

and involves $k - s + 1$ linearly entering constants, s being the number of places among $a_1, \ldots, a_k$ which are independent. These constants will generally be called arbitrary: they are so only under the convention that a function which has all its poles *among* $a_1, \ldots, a_k$ be reckoned a particular case of a function having each of these as poles; for it is clear that, for instance, R_k is only infinite at $a_1, \ldots, a_s, a_k$. The proposition with a slightly altered enunciation, given below in § 27 and more particularly dealt with in § 37, is called the Riemann-Roch Theorem, having been first enunciated by Riemann*, and afterwards particularized by Roch†.

25. Take now other places $a_{k+1}, a_{k+2}, \ldots$ upon the surface *in a definite order*, and consider the possibility of forming a rational function, which beside simple infinities at $a_1, \ldots, a_k$ has other simple poles at, say, $a_{k+1}, a_{k+2}, \ldots, a_h$. By the first Article of the present chapter it follows that the least value of h for which this will be possible will be that for which a_h depends on $a_1 \ldots a_k a_{k+1} \ldots a_{h-1}$, that is, depends on $a_1 \ldots a_s a_{k+1} \ldots a_{h-1}$. This will certainly arise at latest when the number of these places $a_1 \ldots a_s a_{k+1} \ldots a_{h-1}$ is as great as p, namely $h-1=k+p-s$, and if none of the places $a_{k+1} \ldots a_{h-1}$ depend upon the preceding places $a_1 \ldots a_s$, it will not arise before; in that case there will be no rational function having for poles the places

$$a_1 \ldots\ldots a_k a_{k+1} \ldots\ldots a_{k+j}$$

for any value of j from 1 to $p-s$.

But in order to state the general case, suppose there is a value of j less than or equal to $p-s$, such that each of the places

$$a_{k+j+1} \ldots\ldots a_h$$

depends upon the places

$$a_1 \ldots\ldots a_s \; a_{k+1} \ldots\ldots a_{k+j},$$

the smallest value of j for which this occurs being taken, so that no one of $a_{k+1} \ldots a_{k+j}$ depends on the places which precede it in the series

$$a_1 \ldots\ldots a_s \; a_{k+1} \ldots\ldots a_{k+j}.$$

Then there exists no rational function with its poles at $a_1 \ldots a_k \, a_{k+1} \ldots a_{k+j}$, but there exist functions

$$R_{k+j+1} = \sigma_{k+j+1} + \lambda_{k+j+1} \Big[\Gamma^x_{a_{k+j+1}} - n_{k+j+1,1} \Gamma^x_{a_1} - \ldots\ldots - n_{k+j+1,s} \Gamma^x_{a_s} - n_{k+j+1,k+1} \Gamma^x_{a_{k+1}} - \ldots\ldots - n_{k+j+1,k+j} \Gamma^x_{a_{k+j}} \Big],$$

....

$$R_{k+j+i} = \sigma_{k+j+i} + \lambda_{k+j+i} \Big[\Gamma^x_{a_{k+j+i}} - n_{k+j+i,1} \Gamma^x_{a_1} - \ldots\ldots - n_{k+j+i,s} \Gamma^x_{a_s} - n_{k+j+i,k+1} \Gamma^x_{a_{k+1}} - \ldots\ldots - n_{k+j+i,k+j} \Gamma^x_{a_{k+j}} \Big],$$

whose poles are respectively at

$$a_1 \ldots\ldots a_s, \; a_{k+1} \ldots\ldots a_{k+j}, \; a_{k+j+i}$$

for all values of i from 1 to $h-k-j$.

* Riemann, *Ges. Werke*, 1876, p. 101 (§ 5) and p. 118 (§ 14) and p. 120 (§ 16).

† Crelle, 64. Cf. also Forsyth, pp. 459, 464. The geometrical significance of the theorem has been much extended by Brill and Noether. (*Math. Ann.* vii.)

Then the most general rational function with poles at

$$a_1 \ldots\ldots a_s a_{s+1} \ldots\ldots a_k a_{k+1} \ldots\ldots a_{k+j} a_{k+j+1} \ldots\ldots a_{k+j+i}$$

is in fact

$$\nu_0 + \nu_{s+1} R_{s+1} + \ldots\ldots + \nu_k R_k + \nu_{k+j+1} R_{k+j+1} + \ldots\ldots + \nu_{k+j+i} R_{k+j+i}$$

and involves $k - s + i + 1$ arbitrary constants, namely the same number as that of the places of the set

$$a_1 \ldots\ldots a_s a_{s+1} \ldots\ldots a_k a_{k+1} \ldots\ldots a_{k+j} a_{k+j+1} \ldots\ldots a_{k+j+i}$$

which depend upon the places that precede them.

For such a function must have the form

$$\mu_0 + \mu_1 \Gamma^x_{a_1} + \ldots\ldots + \mu_s \Gamma^x_{a_s} + \mu_{s+1} \Gamma^x_{a_{s+1}} + \ldots\ldots + \mu_k \Gamma^x_{a_k} + \mu_{k+1} \Gamma^x_{a_{k+1}} + \ldots\ldots$$
$$+ \mu_{k+j} \Gamma^x_{a_{k+j}} + \mu_{k+j+1} \Gamma^x_{a_{k+j+1}} + \ldots\ldots + \mu_{k+j+i} \Gamma^x_{a_{k+j+i}},$$

namely,

$$\mu_0 + \mu_1 \Gamma^x_{a_1} + \ldots\ldots + \mu_s \Gamma^x_{a_s} + \mu_{k+1} \Gamma^x_{a_{k+1}} + \ldots\ldots + \mu_{k+j} \Gamma^x_{a_{k+j}}$$
$$+ \sum_{r=1}^{k-s} \mu_{s+r} \left[\frac{1}{\lambda_{s+r}} R_{s+r} + n_{s+r,1} \Gamma^x_{a_1} + \ldots\ldots + n_{s+r,s} \Gamma^x_{a_s} - \frac{\sigma_{s+r}}{\lambda_{s+r}}\right]$$
$$+ \sum_{t=1}^{t=i} \mu_{k+j+t} \left[\frac{1}{\lambda_{k+j+t}} R_{k+j+t} + n_{k+j+t,1} \Gamma^x_{a_1} + \ldots\ldots \right.$$
$$\left. + n_{k+j+t,s} \Gamma^x_{a_s} + n_{k+j+t,k+1} \Gamma^x_{a_{k+1}} + \ldots\ldots + n_{k+j+t,k+j} \Gamma^x_{a_{k+j}} - \frac{\sigma_{k+j+t}}{\lambda_{k+j+t}}\right],$$

which is of the form

$$\nu_0 + \nu_1 \Gamma^x_{a_1} + \ldots\ldots + \nu_s \Gamma^x_{a_s} + \nu_{s+1} R_{s+1} + \ldots\ldots + \nu_k R_k$$
$$+ \nu_{k+1} \Gamma^x_{a_{k+1}} + \ldots\ldots + \nu_{k+j} \Gamma^x_{a_{k+j}} + \nu_{k+j+1} R_{k+i+1} + \ldots\ldots + \nu_{k+j+i} R_{k+j+i};$$

and the p periods of this, each of the form

$$\nu_1 \Omega(a_1) + \ldots\ldots + \nu_s \Omega(a_s) + \nu_{k+1} \Omega(a_{k+1}) + \ldots\ldots + \nu_{k+j} \Omega(a_{k+j}),$$

cannot be zero unless each of $\nu_1 \ldots \nu_s \nu_{k+1} \ldots \nu_{k+j}$ be zero, for it is part of the hypothesis that none of $a_{k+1} \ldots a_{k+j}$ depend upon preceding places.

26. Proceeding in this way we shall clearly be able to state the following result—

Let there be taken upon the surface, in a definite order, an unlimited number of places $a_1, a_2, \ldots.$ Suppose that each of $a_1 \ldots a_{Q_1 - q_1}$ is independent of those preceding it, but each of $a_{Q_1 - q_1 + 1} \ldots a_{Q_1}$ depends on $a_1 \ldots a_{Q_1 - q_1}$. Suppose that each of $a_{Q_1+1} a_{Q_1+2} \ldots a_{Q_2 - q_2}$ is independent of those that precede it in the series $a_1 \ldots a_{Q_1 - q_1} a_{Q_1+1} \ldots a_{Q_2 - q_2}$, but each of $a_{Q_2 - q_2 + 1} \ldots a_{Q_2}$ depends upon $a_1 \ldots a_{Q_1 - q_1} a_{Q_1+1} \ldots a_{Q_2 - q_2}$. This requires that

$$Q_1 - q_1 + [Q_2 - q_2 - Q_1] \not> p,$$

Suppose that each of $a_{Q_2+1} \dots a_{Q_3-q_3}$ is independent of those that precede it in the series $a_1 \dots a_{Q_1-q_1} a_{Q_1+1} \dots a_{Q_2-q_2} a_{Q_2+1} \dots a_{Q_3-q_3}$, but each of $a_{Q_3-q_3+1} \dots a_{Q_3}$ depends upon the places of this series. This requires that

$$Q_1 - q_1 + [Q_2 - q_2 - Q_1] + [Q_3 - q_3 - Q_2] \not> p.$$

Let this enumeration be continued. We shall eventually come to places $a_{Q_{h-1}+1}, a_{Q_{h-1}+2}, \dots a_{Q_h-q_h}$, each independent of the places preceding, for which the total number of independent places included, that is, of places which do not depend upon those of our series which precede them, is p—so that the equation

$$\begin{aligned} p &= (Q_h - q_h - Q_{h-1}) + \dots\dots + (Q_2 - q_2 - Q_1) + (Q_1 - q_1) \\ &= Q_h - q_1 - q_2 - \dots\dots - q_h \end{aligned}$$

will hold. Then every additional place of our series, those, namely, chosen in order from $a_{Q_h-q_h+1}, a_{Q_h-q_h+2}, \dots$ will depend on the preceding places of the whole series.

This being the case, it follows, using R_f as a notation for a rational function having its poles among $a_1 \dots a_f$, that *rational functions*

$$R_1 \dots R_{Q_1-q_1};\ R_{Q_1+1} \dots R_{Q_2-q_2};\ R_{Q_2+1} \dots R_{Q_3-q_3};\ \dots\dots;\ R_{Q_{h-1}+1} \dots R_{Q_h-q_h}$$

do not exist.

The number of these non-existent functions is p.

For all other values of f, a rational function R_f *exists.*

To exhibit the general form of these existing rational functions in the present notation, let m be one of the numbers $1, 2, \dots, h$; i be one of the numbers $1, 2, \dots q_m$, and let the dependence of $a_{Q_m-q_m+i}$ upon the preceding places arise by p equations of the form

$$\begin{aligned} \Omega(a_{Q_m-q_m+i}) = \left[\rho_1 \Omega(a_1) + \dots + \rho_{Q_1-q_1} \Omega(a_{Q_1-q_1})\right] + \dots \\ + \left[\rho_{Q_{m-1}+1} \Omega(a_{Q_{m-1}+1}) + \dots + \rho_{Q_m-q_m} \Omega(a_{Q_m-q_m})\right]; \end{aligned}$$

then, denoting $\Gamma^x_{a_r}$ by Γ_r, there is a rational function

$$\begin{aligned} R_{Q_m-q_m+i} = A + B\{\Gamma_{Q_m-q_m+i} - \left[\rho_1\Gamma_1 + \dots + \rho_{Q_1-q_1}\Gamma_{Q_1-q_1}\right] - \dots \\ - \left[\rho_{Q_{m-1}+1}\Gamma_{Q_{m-1}+1} + \dots + \rho_{Q_m-q_m}\Gamma_{Q_m-q_m}\right]\}, \end{aligned}$$

which has its poles at

$$a_1 \dots a_{Q_1-q_1}, a_{Q_1+1} \dots a_{Q_2-q_2}, \dots, a_{Q_{m-1}+1} \dots a_{Q_m-q_m}, a_{Q_m-q_m+i},$$

and the general rational function having its poles at

$$a_1 \dots a_{Q_1} a_{Q_1+1} \dots a_{Q_2} a_{Q_2+1} \dots a_{Q_m-q_m+i}$$

is of the form

$$\nu_0+\left[\nu_{Q-q_1+1}R_{Q_1-q_1+1}+\ldots+\nu_{Q_1}R_{Q_1}\right]+\left[\nu_{Q_2-q_2+1}R_{Q_2-q_2+1}+\ldots+\nu_{Q_2}R_{Q_2}\right]$$
$$+\ldots+\left[\nu_{Q_m-q_m+1}R_{Q_m-q_m+1}+\ldots+\nu_{Q_m-q_m+i}R_{Q_m-q_m+i}\right],$$

and involves $q_1+q_2+\ldots+q_{m-1}+i+1$ arbitrary coefficients.

The result may be summarised by putting down the line of symbols

$$1,\,2,\ldots(Q_1-q_1),\,\overline{(Q_1-q_1+1),\,\ldots,\,Q_1},\,Q_1+1,\ldots(Q_2-q_2),$$
$$\overline{(Q_2-q_2+1),\ldots,Q_2},\,Q_2+1,\ldots,\,Q_{h-1}+1,\,\ldots,(Q_h-q_h),\,\overline{(Q_h-q_h+1),\ldots}$$

with a bar drawn above the indices corresponding to the places which depend upon those preceding them in the series. The bar beginning over Q_h-q_h+1 is then continuous to any length. The total number of indices over which no bar is drawn is p. There exists a rational function R_f, in the notation above, for every index which is beneath a bar.

The proposition here obtained is of a very fundamental character. Suppose that for our initial algebraic equation or our initial surface, we were able only to shew, algebraically or otherwise, that for an arbitrary place a there exists a function K_a^x, discontinuous at a only and there infinite to the first order, this function being one valued save for additive multiples of k periods, and these periods finite and uniquely dependent upon a, then, taking arbitrary places $a_1, a_2, \ldots$ upon the surface, in a definite order, and considering functions of the form

$$\lambda_1 K_{a_1}^x+\ldots\ldots+\lambda_N K_{a_N}^x,$$

that is, functions having simple poles at $a_1, \ldots, a_N$, we could prove, just as above, that there are k values of N for which such functions cannot be one valued; and obtain the number of arbitrary coefficients in uniform functions of given poles. Namely, the proposition would furnish a definition of the characteristic number k—which is the deficiency, here denoted by p—based upon the properties of *the uniform rational functions*.

We shall sometimes refer to the proposition as *Weierstrass's gap theorem**.

27. When a place a is, in the sense here described, dependent upon places $b_1, b_2, \ldots, b_r$, it is clear that of the equations

* "Lückensatz." The proposition has been used by Weierstrass, I believe primarily under the form considered below, in which the places $a_1, a_2, \ldots$ are consecutive at one place of the surface, as the definition of p. Weierstrass's theory of algebraic functions, preliminary to a theory of Abelian functions, is not considered in the present volume. His lectures are in course of publication. The theorem here referred to is published by Schottky: Conforme Abbildung mehrfach zusammenhängender ebener Flächen, Crelle Bd. 83. A proof, with full reference to Schottky, is given by Noether, Crelle Bd. 97, p. 224.

$$A_1\Omega_1(b_1)+\ldots+A_p\Omega_p(b_1)=0$$
$$\ldots\ldots\ldots\ldots\ldots\ldots\ldots\ldots\ldots$$
$$A_1\Omega_1(b_r)+\ldots+A_p\Omega_p(b_r)=0$$
$$A_1\Omega_1(a)+\ldots+A_p\Omega_p(a)=0$$

the last is a consequence of those preceding—and conversely that when the last equation is a consequence of the preceding equations the place a depends upon the places $b_1, b_2, \ldots, b_r$.

Hence the conditions that the linear aggregate

$$\Omega(x)=A_1\Omega_1(x)+\ldots+A_p\Omega_p(x)$$

should vanish at the places

$$a_1 \ldots a_{Q_1} a_{Q_1+1} \ldots a_{Q_2} a_{Q_2+1} \ldots a_{Q_m-q_m+i},$$

wherein $i \not> q_m$, are equivalent to only

$$(Q_1-q_1)+(Q_2-q_2-Q_1)+\ldots+(Q_m-q_m-Q_{m-1})$$

or

$$Q_m-q_1-\ldots-q_m$$

linearly independent equations.

If then $\tau+1$ be the number of linearly independent linear aggregates of the form $\Omega(x)$, which vanish in the Q_m-q_m+i specified places, we have

$$\tau+1=p-(Q_m-q_1-\ldots-q_m).$$

Denoting Q_m-q_m+i by Q, and the number of constants in the general rational function with poles at the Q specified places, of which constants one is merely additive, by $q+1$,

$$q+1=q_1+q_2+\ldots+q_{m-1}+i+1.$$

We therefore have

$$Q-q=p-(\tau+1).$$

Recalling the values of $\Omega_1(x)\ldots\Omega_p(x)$ and the fact (Chapter II. § 21) that every linear aggregate of them vanishes in just $2p-2$ places, we see that when Q is greater than $2p-2$, $\tau+1$ is necessarily zero.

In the case under consideration in the preceding article the number $\tau+1$ for the function $R_{Q_{h-1}}$, namely the number of linearly independent linear aggregates $\Omega(x)$ which vanish in the places

$$a_1 a_2 \ldots a_{Q_1} a_{Q_1+1} \ldots a_{Q_{h-1}},$$

is given, by taking $m=h-1$ and $i=q_{h-1}$ in the formula of the present article, by the equation

$$\tau+1=p-(Q_{h-1}-q_1-\ldots-q_{h-1})$$
$$=Q_h-q_h-Q_{h-1}.$$

Hence one such linear aggregate vanishes in the places

$$a_1 a_2 \ldots a_{Q_{h-1}} a_{Q_{h-1}+1} \ldots a_{Q_h - q_h - 1}$$

and therefore

$$Q_h - q_h - 1 \not> 2p - 2$$

or, *the index associated with the last place* $a_{Q_h - q_h}$ *of our series, corresponding to which a rational function* $R_{Q_h - q_h}$ *does not exist, is not greater than* $2p - 1$. A case in which this limit is reached, which also furnishes an example of the theory, is given below § 37, Ex. 2.

28. A limiting case of the problem just discussed is that in which the series of points $a_1, a_2, \ldots$ are all consecutive at one place of the surface.

A rational function which becomes infinite only at a place, a, of the surface, and there like

$$\frac{C_1}{t} + \frac{C_2}{t^2} + \ldots + \frac{C_r}{t^r},$$

where any of the constants $C_1, C_2, \ldots C_{r-1}$, but *not* C_r, may be zero, t being the infinitesimal, is said to be there infinite to the rth order. If $-\lambda_i = C_i/(i-1)!$, such a function can be expressed in a form

$$\lambda + \lambda_1 \Gamma_a^x + \lambda_2 D_a \Gamma_a^x + \ldots + \lambda_r D_a^{r-1} \Gamma_a^x$$

where, in order that the function be one valued on the undissected surface, the p equations

$$\lambda_1 \Omega_i(a) + \lambda_2 D_a \Omega_i(a) + \ldots + \lambda_r D_a^{r-1} \Omega_i(a) = 0$$

must be satisfied: and conversely these equations give sufficient conditions for the coefficients $\lambda_1, \lambda_2, \ldots, \lambda_r$.

In other words, since λ_r cannot be zero because the function is infinite to the rth order, the p differential coefficients $D_a^{r-1}\Omega_i(a)$, each of the $\overline{r-1}$th order, must be expressible linearly in terms of those of lower order,

$$\Omega_i(a),\ D\Omega_i(a), \ldots, D^{r-2}\Omega_i(a),$$

with coefficients which are independent of i. We imagine the p quantities $D_a^{r-1}\Omega_i(a)$, for $i = 1, 2, \ldots, p$, written in a column, which we call the rth column; and for the moment we say that the necessary and sufficient condition for the existence of a rational function, infinite of the rth order at a, and not elsewhere infinite, is that the rth column be a linear function of the preceding columns.

Then as before, considering the columns in succession, they will divide themselves into two categories, those which are linear functions of the preceding ones and those which are not so expressible. And, since the number of elements in a column is p, the number of these latter independent columns will be just p. Let them be in succession the k_1th, k_2th, ..., k_pth. Then there exists no rational function infinite only at a, and there to these orders $k_1, k_2, \ldots, k_p$, though there are integrals of the second kind infinite to these orders. But if Q be a number different from $k_1, \ldots, k_p$, there does exist such a rational function of the Qth order, its most general expression being of the form

$$\lambda_Q D_a^{Q-1}\Gamma_a^x + \lambda_{Q-1} D_a^{Q-2}\Gamma_a^x + \ldots + \lambda_1 \Gamma_a^x + \lambda,$$

namely, the integral of the second kind whose infinity is of order Q is expressible linearly by integrals of the second kind of lower order of infinity, with the addition of a rational function.

If $q+1$ be the number of linearly independent coefficients in this function, one being additive, we have an equation

$$Q - q = p - (\tau + 1),$$

where $p - (\tau + 1)$ is the number of the linearly independent equations of the form

$$\lambda_1 \Omega_i(a) + \lambda_2 D\Omega_i(a) + \ldots + \lambda_Q D^{Q-1}\Omega_i(a) = 0, \quad (i = 1, 2, \ldots, p),$$

from which the others may be linearly derived. As before, $\tau + 1$ is the number of linearly independent linear aggregates of the form

$$A_1\Omega_1(x) + \ldots + A_p\Omega_p(x)$$

which satisfy the Q conditions

$$A_1 D^r \Omega_1(a) + \ldots + A_p D^r \Omega_p(a) = 0$$

for $r = 0, 1, 2, \ldots, Q-1$.

29. In regard to the numbers $k_1 \ldots k_p$ we remark firstly that, unless $p = 0$, $k_1 = 1$—for if there existed a rational function with only one infinity of the first order, the positive integral powers of this function would furnish rational functions of all other orders with their infinity at this one place, and there would be no gaps (compare the argument Chapter II. § 21); and further that in general they are the numbers 1, 2, 3 ... p, that is to say, there is only a finite number of places on the surface for which a rational function can be formed infinite there to an order less than $p+1$ and not otherwise infinite. We shall prove this immediately by finding an upper and a lower limit to the number of such places (§ 31).

30. Some detailed algebraic consequences of this theory will be given in Chapter V. It may be* here remarked, what will be proved in Chapter VI. in considering the geometrical theory, that the zeros of the linear aggregate

$$A_1\Omega_1(x) + \ldots + A_p\Omega_p(x)$$

* It is possible that the reader may find it more convenient to postpone the complete discussion of § 30 until after reading Chapter vi.

can be interpreted in general as the intersections of a certain *curve*, of the form

$$\phi = A_1\phi_1(x) + \ldots + A_p\phi_p(x) = 0,$$

wherein $\phi_1 \ldots \phi_p$ are integral polynomials in x and y, with the curve represented by the fundamental equation of our Riemann surface. In such interpretation, the condition for the existence of a rational function of order Q with poles only at the place a, is that the fundamental curve be of such character at this place that every curve ϕ, obtained by giving different values to $A_1 \ldots A_p$, which there cuts it in $Q-1$ consecutive points, necessarily cuts it in Q consecutive points. As an instance of such property, which seems likely also to make the general theory clearer, we may consider a Riemann surface associated with an equation of the form

$$f(x, y) = K + (x, y)_1 + (x, y)_2 + (x, y)_3 + (x, y)_4 = 0,$$

wherein $(x, y)_r$ is a homogeneous integral polynomial of the rth degree, with quite general coefficients, and K is a constant. Interpreted as a curve, this equation represents a general curve of the fourth degree; it will appear subsequently that the general integral of the first kind is

$$\int \frac{dx}{f'(y)}(A + Bx + Cy),$$

where $f'(y) = \partial f/\partial y$, and A, B, C are arbitrary constants; and thence, if we recall the fact that $\Omega_1(x), \ldots, \Omega_p(x)$ are differential coefficients of integrals of the first kind, that the zeros of the aggregate

$$A_1\Omega_1(x) + \ldots + A_p\Omega_p(x)$$

may be interpreted as the intersections of the quartic with a variable straight line.

Take now a point of inflexion of the quartic as the place a. Not every straight line there intersecting the curve in one point will intersect it in any other consecutive point; *but* every straight line there intersecting the curve in two consecutive points will necessarily intersect it there in three consecutive points. Hence it is possible to form a rational function of the third order whose only infinities are at the place of inflexion; in fact, if

$$A_0x + B_0y + 1 = 0$$

be the equation of the inflexional tangent, and

$$\lambda(A_0x + B_0y + 1) + \mu(Ax + By + 1) = 0$$

be the equation of any line through the fourth point of intersection of the inflexional tangent with the curve, the ratio of the expressions on the left hand side of these equations, namely

$$\lambda + \mu\frac{Ax + By + 1}{A_0x + B_0y + 1},$$

is a general rational function of the desired kind, as is immediately obvious on consideration of the places where it can possibly be infinite. Thus for the inflexional place the orders of two non-existent rational functions are 1, 2. It can be proved that in general there is no function of the fourth order—the gaps at the orders 1, 2, 4 are those indicated by Weierstrass' theorem.

In verification of a result previously enunciated we notice that since $Ax+By+1=0$ may be taken to be *any definite* line through the fourth intersection of the inflexional tangent with the curve, the function contains $q+1=2$ arbitrary constants. From the form of the integrals of the first kind which we have quoted, it follows that $p=3$; thus the formula

$$Q-q=p-(\tau+1),$$

wherein $Q=3$, requires $\tau+1=1$; now by § 28 $\tau+1$ should be the number of straight lines which can be drawn to have contact of the second order with the curve at the point: this is the case.

If the quartic possess also a point of osculation, a straight line passing through two consecutive points of the curve there will necessarily pass through three consecutive points and also necessarily through four. Hence, for such a place, we can form a rational function of the third order *and* one of the fourth. In fact, if $A_0x+B_0y+1=0$ be the tangent at the point of osculation and $A_1x+B_1y+1=0$ be any other line through this point, while $\lambda x+\mu y+\nu=0$ is any other line whatever, these functions are respectively, in their most general forms,

$$\lambda+\mu\frac{A_1x+B_1y+1}{A_0x+B_0y+1},\quad \frac{\lambda x+\mu y+\nu}{A_0x+B_0y+1},$$

wherein λ, μ, ν are arbitrary constants.

It can be shewn that in general we cannot form a rational function of the fifth order whose only infinity is at the place of osculation. Thus the gaps indicated by Weierstrass's theorem occur at the orders 1, 2, 5. (Cf. the concluding remark of § 34.)

In case, however, the place a is an ordinary point of the quartic, the lowest order of function, whose only infinity is there, is $p+1=4$: it will subsequently become clear that a general form of such a function in S'/S, where $S=0$ is *any* conic drawn to intersect the quartic in four consecutive points at a, and $S'=0$ is the most general conic drawn through the other four intersections of S with the quartic. S' will in fact be of the form $\lambda S+\mu T$, where T is any definite conic satisfying the conditions for S', and λ, μ are arbitrary constants; the equation $Q-q=p-(\tau+1)$ is clearly satisfied by $Q=4$, $q=1$, $p=3$, $\tau+1=0$.

The present article is intended only by way of illustration; the examples given appear to find their proper place here. The reader will possibly find it desirable to read them in connexion with the geometrical account given in Chapter VI.

31. Consider now what places of the surface are such that we can form a rational function infinite, only there, to an order as low as p.

For such a place, as follows from § 28, the determinant

$$\Delta = \begin{vmatrix} \Omega_1(x) & , \Omega_2(x) & ,\dots\dots, & \Omega_p(x) \\ D\Omega_1(x) & , D\Omega_2(x) & ,\dots\dots, & D\Omega_p(x) \\ \dots & \dots & \dots & \dots \\ \dots & \dots & \dots & \dots \\ \dots & \dots & \dots & \dots \\ D^{p-1}\Omega_1(x), & D^{p-1}\Omega_2(x), & \dots\dots, & D^{p-1}\Omega_p(x) \end{vmatrix}$$

must vanish. Assume for the present that none of the minors of Δ vanish at that place. It is clear by § 28 that Δ only vanishes at such places as we are considering.

Let v be any integral of the first kind. We can write

$$\Omega_i(x) = \frac{dv_i}{dt} \text{ in the form } \frac{dv}{dt}\frac{dv_i}{dv},$$

and similarly put

$$D\Omega_i(x) = \frac{d^2v}{dt^2}\frac{dv_i}{dv} + \left(\frac{dv}{dt}\right)^2 \frac{d^2v_i}{dv^2},$$

$$\dots\dots\dots\dots\dots\dots\dots\dots\dots\dots$$

$$D^{p-1}\Omega_i(x) = \dots\dots\dots + \left(\frac{dv}{dt}\right)^p \frac{d^pv_i}{dv^p},$$

and so write

$$\Delta = \left(\frac{dv}{dt}\right)^{\frac{1}{2}p(p+1)} D,$$

where D is the determinant whose rth row is formed with the quantities

$$\frac{d^rv_1}{dv^r}, \dots\dots, \frac{d^rv_p}{dv^r}.$$

Now $\frac{dv_i}{dv}$ is a rational function; and it is infinite only at the zeros of dv, whose aggregate number is $2p-2$; and $\frac{d^2v_i}{dv^2}$ is a rational function of the $(4p-4)$th order, its poles being also at the zeros of dv; and a similar statement can be made in regard to the other rows of D.

Hence D is a rational function whose infinities are of aggregate number

$$(2p-2)(1+2+\dots+p) = (p-1)p(p+1),$$

and this is therefore the number of zeros of D.

Now Δ can vanish either by the vanishing of the factor D or by the vanishing of the factor $\left(\frac{dv}{dt}\right)^{\frac{1}{2}p(p+1)}$. The zeros of the last factor are, however, the poles of D. Hence *the aggregate number of zeros of* Δ *is* $(p-1)p(p+1)$. We shall see immediately that these zeros do not necessarily occur at as many as $(p-1)p(p+1)$ distinct places of the surface.

In order that a rational function should exist of order less than p, its infinity being entirely at one place, say of order $p-r$, it would be necessary that the r determinants formed from the matrix obtained by omitting the last r rows of Δ should all vanish at that place. We can, as in the case of Δ, shew that each of these minors will vanish only at a finite number of places. It is therefore to be expected that in general these minors will not have common zeros; that is, that the surface will need to be one whose $3p-3$ moduli are connected in some special way.

Moreover it is not in general true that a rational function of order $p+1$ exists for a place for which a function of order p exists, these functions not being elsewhere infinite. For then we could simultaneously satisfy the two sets of p equations

$$\lambda_1\Omega_i(a)+\lambda_2 D\Omega_i(a)+\ldots\ldots+\lambda_{p-1}D^{p-2}\Omega_i(a)+\lambda_p D^{p-1}\Omega_i(a)=0,$$
$$\mu_1\Omega_i(a)+\mu_2 D\Omega_i(a)+\ldots\ldots+\mu_{p-1}D^{p-2}\Omega_i(a)+\mu_{p+1}D^{p}\Omega_i(a)=0,$$

namely, Δ and $\frac{d\Delta}{dt}$ would both be zero at such a place. The condition that this be so would require that a certain function of the moduli of the surface—what we may call an absolute invariant—should be zero.

Therefore when of the p gaps required by Weierstrass's theorem, $p-1$ occur for the orders $1, 2, \ldots, p-1$, the other will in general occur for the order $p+1$. The reader will see that there is no such reason why, when a function of order p exists, a function of order $p+2$ or higher order should not exist.

32. The reader who has followed the example of § 30 will recall that the number of inflexions of a non-singular plane quartic* is 24 which is equal to the value of $(p-1)p(p+1)$ when $p=3$. The condition that the quartic possess a point of osculation is that a certain invariant should vanish†.

When the curve has a double point, there are only two integrals of the first kind‡, and p is equal to two. Thus in accordance with the theory above, there should be $(p-1)p(p+1)=6$ places for which we can form functions

* Salmon, *Higher Plane Curves* (1879), p. 213.

† The equation can be written so as to involve only $5=3p-3-1$ parametric constants (Chap. V. p. 98, Exs. 1, 2).

‡ Their forms are given Chapter II. § 17 β. Reasons are given in Chapter VI. The reader may compare Forsyth, p. 395.

of the second order infinite only at one of these places. In fact six tangents can be drawn to the curve from the double point: if $A_0x + B_0y = 0$ be the equation of one of these and $\lambda(Ax + By) + \mu(A_0x + B_0y) = 0$ be the equation of any line through the double point, the ratio

$$\xi = \lambda \frac{Ax + By}{A_0x + B_0y} + \mu$$

represents a function of second order infinite only at the point of contact of $A_0x + B_0y = 0$*.

For the point of contact of one of these tangents the p gaps occur for the orders 1 and 3.

The quartic with a double point can be birationally related to a surface expressed by an equation of the form

$$\eta^2 = (\xi, 1)_6,$$

ξ being the function above. The reader should compare the theory in Chapter I. and the section on the hyperelliptic case, Chapter V. below.

33. *Ex.* For the surface represented by the equation

$$f(x, y) = x^2y^2\{x, y\}_1 + xy\{x, y\}_2 + (x, y)_3 + (x, y)_2 + (x, y)_1 = 0$$

where the brackets indicate general integral polynomials of the order of the suffixes, p is equal to 4, and the general integral of the first kind is

$$\int dx\,(Axy + Bx + Cy + D)/f'(y)$$

where $f'(y) = \dfrac{\partial f}{\partial y}$. Prove that at the $(p-1)p(p+1) = 60$ places for which rational functions of the 4th order exist, infinite only at these places, the following equations are satisfied

$$2y'''/y' - 3(y''/y')^2 = 0,$$

$$2f_xf_y\left[\frac{\partial^3 f}{\partial x^3}f_y^3 - 3\frac{\partial^3 f}{\partial x^2\partial y}f_y^2f_x + 3\frac{\partial^3 f}{\partial x\partial y^2}f_yf_x^2 - \frac{\partial^3 f}{\partial y^3}f_x^3\right]$$
$$-3\left[\frac{\partial^2 f}{\partial x^2}f_y^2 - \frac{\partial^2 f}{\partial y^2}f_x^2\right]\left[\frac{\partial^2 f}{\partial x^2}f_y^2 - 2\frac{\partial^2 f}{\partial x\partial y}f_yf_x + \frac{\partial^2 f}{\partial x^2}f_x^2\right] = 0,$$

where $y' = \dfrac{dy}{dx}$, etc., $f_x = \dfrac{\partial f}{\partial x}$, etc.

Explain how to express these functions of the fourth order.

Enumerate all the zeros of the second differential expression here given.

Ex. 2. In general, the corresponding places are obtained by forming the differential equation of the pth order of all adjoint ϕ curves. In a certain sense Δ is a differential invariant, for all reversible rational transformations. (See Chapter VI.)

* Here the number of integrands of the integrals of the first kind, which are of the form $(Lx + My)/f'(y)$ (cf. Chapter III. § 28), which vanish in two consecutive points at the point of contact of $A_0x + B_0y = 0$, is clearly 1, or $\tau + 1 = 1$: hence the formula $Q - q = p - (\tau + 1)$ is verified by $Q = 2$, $q = 1$, $p = 2$, so that the form of function of the second order given in the text is the most general possible.

34. We pass now to consider whether the $(p-1)p(p+1)$ zeros of Δ will in general fall at separate places*.

Consider the determinant

$$\nabla = \begin{vmatrix} 0, & \Omega_1(x) & , \dots\dots, & \Omega_p(x) \\ \lambda_1, & \Omega_1^{(k_1-1)}(\xi), & \dots\dots, & \Omega_p^{(k_1-1)}(\xi) \\ \dots & \dots & \dots & \dots \\ \dots & \dots & \dots & \dots \\ \lambda_p, & \Omega_1^{(k_p-1)}(\xi), & \dots\dots, & \Omega_p^{(k_p-1)}(\xi) \end{vmatrix},$$

wherein $\Omega_i^{(\mu)}(\xi) = D_\xi^\mu \Omega_i(\xi)$, and $k_1, \dots, k_p$ are the orders of non-existent rational functions for a place ξ, in ascending order of magnitude, $(k_1 = 1)$; and let its value be denoted by

$$\lambda_1 \omega_1(x) + \dots + \lambda_p \omega_p(x),$$

so that $u_r = \int \omega_r(x)\, dt_x$ is an integral of the first kind.

Then $\omega_r(x)$ *vanishes at* ξ *to the* $(k_r - 1)$*th order.*

For $\omega_r(x)$ is the determinant

$$\nabla_r = (-)^r \begin{vmatrix} \Omega_1(x) & , \dots\dots, & \Omega_p(x) \\ \Omega_1^{(k_1-1)}(\xi) & , \dots\dots, & \Omega_p^{(k_1-1)}(\xi) \\ \dots & \dots & \dots \\ \Omega_1^{(k_{r-1}-1)}(\xi), & \dots\dots, & \Omega_p^{(k_{r-1}-1)}(\xi) \\ \Omega_1^{(k_{r+1}-1)}(\xi), & \dots\dots, & \Omega_p^{(k_{r+1}-1)}(\xi) \\ \dots & \dots & \dots \\ \Omega_1^{(k_p-1)}(\xi) & , \dots\dots, & \Omega_p^{(k_p-1)}(\xi) \end{vmatrix};$$

now the $(k_r - 1)$th differential coefficient of this determinant (in regard to the infinitesimal at x) has at ξ a value which is in fact the minor of the element (1, 1) of ∇, save for sign. That this minor does not vanish is part of the definition of the numbers $k_1, k_2, \dots, k_p$. But all differential coefficients of ∇_r of lower than the $(k_r - 1)$th order do vanish at ξ: some, because for $x = \xi$ they are determinants having the first row identical with one of the following rows, this being the case for the differential coefficients of orders $k_1 - 1, k_2 - 1, \dots$; others, because when μ is not one of the numbers $k_1, k_2, \dots, k_p$, $D^{\mu-1}\Omega_i(\xi)$ is a linear function of those of $D^{k_1-1}\Omega_i(\xi)$, $D^{k_2-1}\Omega_i(\xi), \dots$ for which μ is greater than $k_1, k_2, \dots$, the coefficients of the linear functions being independent of i. This proves the proposition.

It is clear that the k_rth differential coefficient of ∇_r may also vanish at ξ. In particular $\omega_1(x)$ does not vanish at ξ: a result in accordance with a remark previously made (Chapter II. § 21), that there is no place at which the differentials of all the integrals of the first kind can vanish.

* The results in §§ 34, 35, 36 are given by Hurwitz, *Math. Annal.* 41, p. 409. They will be useful subsequently.

An important corollary is that *the highest order for which no rational function exists, infinite only at the place ξ, is less than $2p$.* For $\omega_p(x)$ vanishes only $2p-2$ times, namely, $k_p - 1 \overline{\gtrless} 2p-2$.

35. We can now prove that *if $k_2 > 2$, the sum of the orders $k_1, k_2, \ldots, k_p$ is less than p^2.* For if there be a rational function of order m, infinite only at ξ, and r be one of the non-existent orders* $k_1 \ldots k_p$, $r-m$ is also one of these non-existent orders—otherwise the product of the existent rational function of order $r-m$ with the function of order m would be an existent function of order r. The powers of the function of order m are existent functions, hence none of $k_1 \ldots k_p$ are divisible by m.

Let r_i be the greatest of the non-existent orders $k_1 \ldots k_p$ which is congruent to $i\,(< m)$ for the modulus m: then, by the remark just made,

$$r_i,\ r_i - m,\ r_i - 2m,\ \ldots,\ m+i,\ i$$

are all non-existent orders—and all congruent to i for the modulus m. Since r_i occurs among $k_1 \ldots k_p$, all these also occur. Take i in turn equal to $1, 2, \ldots m-1$.

Then, the number of non-existent orders being p,

$$p = \left(1 + \frac{r_1 - 1}{m}\right) + \left(1 + \frac{r_2 - 2}{m}\right) + \ldots + \left(1 + \frac{r_{m-1} - (m-1)}{m}\right),$$

so that

$$\begin{aligned} r_1 + r_2 + \ldots + r_{m-1} &= mp - \tfrac{1}{2} m(m-1) \\ &= \tfrac{1}{2} m(2p - m + 1). \end{aligned}$$

Now the sum of the non-existent orders is

$$\sum_{i=1}^{m-1} [r_i + (r_i - m) + (r_i - 2m) + \ldots + i],$$

which is equal to

$$\frac{1}{2m} \sum_{i=1}^{m-1} (r_i + m - i)(r_i + i)$$

$$= \frac{1}{2m} \sum_i r_i [r_i - (2p-1)] + \frac{1}{2m} \sum_i r_i [2p + m - 1] + \tfrac{1}{4} m(m-1) - \tfrac{1}{12}(m-1)(2m-1),$$

and, since $\Sigma r_i = \frac{1}{2} m(2p - m + 1)$, this is equal to

$$\frac{1}{2m} \sum_i r_i [r_i - (2p-1)] + \tfrac{1}{4}[4p^2 - (m-1)^2] + \tfrac{1}{12}(m-1)(m+1),$$

or

$$p^2 - \frac{1}{2m} \sum_i r_i (2p - 1 - r_i) - \tfrac{1}{6}(m-1)(m-2).$$

* i.e. orders of rational functions, infinite only at ξ, which do not exist: and similarly in what follows.

Since, by the corollary of the preceding article, $2p-1$ is not less than r_i, this is less than p^2 unless m is 1 or 2. Now m cannot be equal to 1; and if it is 2 then also $k_2 > 2$. Hence the statement made at the beginning of the present Article is justified.

When there *is* a rational function of order 2, it is easy to prove that there are places for which $k_1 \dots k_p$ are the numbers $1, 3, 5, \dots, 2p-1$, whose sum* is p^2. An example is furnished by § 32 above.

Ex. For the surface

$$y^3 + y^2 (x, 1)_1 + y (x, 1)_2 + (x, 1)_4 = 0,$$

for which $p=3$, there is, at $x=\infty$, only one place, and the non-existent orders are 1, 2, 5: whose sum is p^2-1.

36. We have in § 34 defined p integrals of the first kind

$$\int \omega_1(x)\, dt_x, \ \dots, \ \int \omega_p(x)\, dt_x$$

by means of a place ξ. Since the differential coefficients of these vanish at ξ to essentially *different* orders, these integrals cannot be connected by a homogeneous linear equation with constant coefficients. Hence a linear function of them with parametric constant coefficients is a general integral of the first kind. Therefore each of $\Omega_1(x) \dots \Omega_p(x)$ is expressible linearly in terms of $\omega_1(x) \dots \omega_p(x)$ in a form

$$\Omega_i(x) = c_{i1}\, \omega_1(x) + \dots + c_{ip}\, \omega_p(x),$$

where the coefficients are independent of x. Thus the determinant Δ (§ 31), which vanishes at places for which functions of order less than $p+1$ exist, is equal to

$$C \begin{vmatrix} \omega_1(x) & ,\dots\dots, & \omega_p(x) \\ D_x \omega_1(x) & ,\dots\dots, & D_x \omega_p(x) \\ \dots\dots & \dots\dots & \dots\dots \\ \dots\dots & \dots\dots & \dots\dots \\ D_x^{p-1} \omega_1(x), & \dots\dots, & D_x^{p-1} \omega_p(x) \end{vmatrix},$$

where C is the determinant of the coefficients c_{ij}. It follows from the result of § 34 that the determinant here multiplied by C vanishes at ξ to the order

$$(k_1 - 1) + (k_2 - 2) + \dots + (k_p - p) = k_1 + \dots + k_p - \tfrac{1}{2} p (p+1).$$

Thus, *the determinant Δ vanishes at any one of its zeros to an order equal to the sum of the non-existent orders for the place diminished by $\frac{1}{2}p(p+1)$.*

For example, it vanishes at a place where the non-existent orders are $1, 2, \dots, p-1, p+1$ to an order $\frac{1}{2}p(p-1) + p + 1 - \frac{1}{2}p(p+1)$ or to the first order. We have already remarked that such places are those which most usually occur.

* Cf. Burkhardt, *Math. Annal.* 32, p. 388, and the section in Chapter V., below, on the hyperelliptic case.

Hence, since $k_1 + \dots + k_p \leqq p^2$, Δ *vanishes at one of its zeros to an order* $\leqq \frac{1}{2}p(p-1)$.

Further, if r be the number of distinct places where Δ vanishes, and $m_1, m_2, \dots, m_r$ be the orders of multiplicity of zero at these places, it follows, from

$$m_1 + \dots + m_r = (p-1)\,p\,(p+1),$$

and

$$m_1 + \dots + m_r \leqq r\,\tfrac{1}{2}p\,(p-1),$$

that $r > 2p+2$, or

there are at least $2p+2$ *distinct places for which functions of less order than* $p+1$, *infinite only thereat, exist*; this lower limit to the number of distinct places is only reached when there are places for which functions of the second order exist.

Ex. For the surface given by

$$x^4 + y^4 + (ax+by+c)^4 = 0,$$

p is equal to 3; there are $12 = 2p+6$ distinct places where Δ vanishes.

37. We have called attention to the number of arbitrary constants contained in *the most general* rational function having simple poles in distinct places (§ 27) and to the number in the most general function infinite at a single place to prescribed order (§ 28): in this enumeration some of the constants may be multipliers of functions not actually becoming infinite in the most general way allowed them, that is, either of functions which are not really infinite at all the distinct places or of functions whose order of infinity is not so high as the prescribed order.

It will be convenient to state here the general result, the deduction of which follows immediately from the expression of the function in terms of integrals of the second kind:—

Let $a_1, a_2, \dots$ be any finite number of places on the surface, the infinitesimals at these places being denoted by $t_1, t_2, \dots$. The most general rational function whose expansion at the place a_i involves the terms

$$\frac{1}{t_i^{\lambda_i}},\ \frac{1}{t_i^{\mu_i}},\ \frac{1}{t_i^{\nu_i}},\ \dots$$

—whose number is finite, $= Q_i$ say,—and no other negative powers, involves $q+1$ linearly entering arbitrary constants, of which one is additive, q being given by the formula

$$Q - q = p - (\tau + 1),$$

where Q is the sum of the numbers Q_i, and $\tau + 1$ is the number of linearly independent linear aggregates of the form

$$\Omega(x) = A_1\Omega_1(x) + \dots + A_p\Omega_p(x),$$

which satisfy the sets of Q_i relations, whose total number is Q, given by

$$A_1 D^{\lambda_i - 1}\Omega_1(a_i) + A_2 D^{\lambda_i - 1}\Omega_2(a_i) + \dots + A_p D^{\lambda_i - 1}\Omega_p(a_i) = 0,$$
$$A_1 D^{\mu_i - 1}\Omega_1(a_i) + A_2 D^{\mu_i - 1}\Omega_2(a_i) + \dots + A_p D^{\mu_i - 1}\Omega_p(a_i) = 0,$$
$$\dots\dots\dots\dots\dots\dots\dots\dots$$

As before, this general function will as a rule be an aggregate of functions of which not every one is as fully infinite as is allowed, *and it is clear from the present chapter that in the absence of further information in regard to the places* $a_1, a_2, \dots$ *it may quite well happen that not one of these functions is as fully infinite as desired, the conditions analogous to those stated in §§ 23, 28 not being satisfied. See Example 2 below.*

The equation $Q - q = p - (\tau + 1)$ will be referred to as the Riemann-Roch Theorem.

Ex. 1. For a rational function having only simple poles or, more generally, such that the numbers $\lambda_i, \mu_i, \nu_i, \dots$ for any pole are the numbers $1, 2, 3, \dots Q_i$,

if $Q > 2p - 2$, $\tau + 1$ is zero, since $\Omega(x)$ has only an aggregate number $2p - 2$ of zeros: the function involves $Q - p + 1$ constants,

if $Q = 2p - 2$, $\tau + 1$ cannot be greater than 1; for the ratio of two of the aggregates $\Omega(x)$ then vanishing at the poles, being expressible in a form $\dfrac{dV}{dW}$, where V, W are integrals of the first kind, would be a rational function without poles, namely a constant; then the linear aggregates $\Omega(x)$ would be identical: thus the function involves $Q - p + 1$ or $Q - p + 2$ constants, namely $p - 1$ or p constants,

if $Q = 2p - 3$, $\tau + 1$ cannot be greater than 1, since the ratio of two of the aggregates $\Omega(x)$ then vanishing at the poles would be a rational function of the first order and therefore p be equal to unity—in which case $2p - 3$ is negative: thus the function involves $p - 2$ or $p - 1$ constants,

if $Q = 2p - 4$, and $\tau + 1$ be greater than unity, the ratio of two of the vanishing aggregates $\Omega(x)$ would be a rational function of the second order: we have already several times referred to this possibility as indicative that the surface is of a special character—called hyperelliptic—and depends in fact only on $2p - 1$ independent moduli. In general such a function would involve $p - 3$ constants.

Ex. 2. Let V be an integral of the first kind and a be an arbitrary definite place which is not among the $2p - 2$ zeros of dV. We can form a rational function infinite to the first order at the $2p - 2$ zeros of dV and to the second order at a; the general form of such a function would contain $2p - 2 + 2 - p + 1 = p + 1$ arbitrary constants. *But there exists no rational function infinite to the first order at the zeros of dV and to the first order at* *the place a.* Such a function would indeed by the Riemann-Roch theorem here stated, contain $2p-2+1-p+1=p$ arbitrary constants: *but the coefficients of these constants are in fact infinite only at the zeros of dV.* For when the places $a_1, \ldots, a_{2p-2}$ are all zeros of an aggregate of the form

$$A_1\Omega_1(x)+\ldots+A_p\Omega_p(x),$$

the conditions that the periods of an expression

$$\lambda+\lambda_1\Gamma_{a_1}^x+\ldots+\lambda_{2p-2}\Gamma_{a_{2p-2}}^x+\mu\Gamma_a^x$$

be all zero, namely the equations

$$\lambda_1\Omega_i(a_1)+\ldots+\lambda_{2p-2}\Omega_i(a_{2p-2})+\mu\Omega_i(a)=0, \quad (i=1, 2, \ldots, p),$$

lead to

$$\mu[A_1\Omega_1(a)+\ldots+A_p\Omega_p(a)]=0,$$

and therefore to $$\mu=0.$$

Thus the function in question will be a linear aggregate of p functions whose poles are among the places $a_1, \ldots, a_{2p-2}$. As a matter of fact, if W be a general integral of the first kind, expressible therefore in the form

$$\lambda V+\lambda_2V_2+\ldots+\lambda_pV_p,$$

wherein $V_2, \ldots, V_p$ are integrals of the first kind, $\dfrac{dW}{dV}$ involves the right number of constants and is the function sought.

In this case the place a does not, in the sense of § 23, depend upon the places $a_1, \ldots, a_{2p-2}$; the symbol suggested in § 26 for the places $a_1, \ldots, a_{2p-2}, a, \ldots$ is

$$1, 2, 3, \ldots, p-1, \overline{p, p+1, \ldots, 2p-2}, 2p-1, \overline{2p, 2p+1}, \ldots.$$

It may be shewn quite similarly that there is no rational function having simple poles in $a_1, a_2, \ldots, a_{2p-2}$ and infinite besides at a like the single term $\dfrac{1}{t^r}$, t being the infinitesimal at the place a.

Ex. 3. The most general rational function R which has the value c at each of Q given distinct places, $R-c$ being zero of the first order at each of these places, is obviously derivable by the remark that $1/(R-c)$ is infinite at these places.

CHAPTER IV.

Specification of a General Form of Riemann's Integrals.

38. In the present chapter the problem of expressing the Riemann integrals is reduced to the determination of certain fundamental rational functions, called integral functions. The existence of these functions, and their principal properties, is obtained from the descriptive point of view natural to the Riemann theory.

It appears that these integral functions are intimately related to certain functions, the differential-coefficients of the integrals of the first kind, of which the ratios have been shewn (Chapter II. § 21) to be invariant for birational transformations of the surface. It will appear, further, in the next chapter, that when these integral functions are given, or, more precisely, when the equations which express their products, of pairs of them, in terms of themselves, are given, we can deduce a form of equation to represent the Riemann surface; thus these functions may be regarded as anterior to any special form of fundamental equation.

Conversely, when the surface is given by a particular form of fundamental equation, the calculation of the algebraic forms of the integral functions may be a problem of some length. A method by which it can be carried out is given in Chapter V. (§§ 72 ff.). Compare § 50 of the present chapter.

It is convenient to explain beforehand the nature of the difficulty from which the theory contained in §§ 38—44 of this chapter has arisen. Let the equation associated with a given Riemann surface be written

$$Ay^n+A_1y^{n-1}+\dots+A_n=0,$$

wherein $A, A_1, \dots, A_n$ are integral polynomials in x. An *integral* function is one whose poles all lie at the places $x=\infty$ of the surface; in this chapter the integral functions considered are all rational functions. If y be an integral function, the rational symmetric functions of the n values of y corresponding to any value of x, whose values, given by the equation, are $-A_1/A$, A_2/A, $-A_3/A$, etc., will not become infinite for any finite value of x, and will, therefore, be integral polynomials in x. Thus when y is an integral function, the polynomial A divides all the other polynomials A_1, $A_2, \ldots\ldots, A_n$. Conversely, when A divides these other polynomials, the form of the equation shews that y cannot become infinite for any finite value of x, and is therefore an integral function.

When y is not an integral function, we can always find an integral polynomial in x, say β, vanishing to such an order at each of the finite poles of y, that βy is an integral function. Then also, of course, $\beta^2y^2, \beta^3y^3, \ldots$ are integral functions: though it often happens that there is a polynomial β_2 of less order than β^2, such that $\beta_2 y^2$ is an integral function, and similarly an integral polynomial β_3 of less order than β^3, such that $\beta_3 y^3$ is an integral function; and similarly for higher powers of y.

In particular, if in the equation given we put $Ay=\eta$, the equation becomes

$$\eta^n + A_1\eta^{n-1} + A_2A\eta^{n-2} + \ldots + A_nA^{n-1} = 0,$$

and η is an integral function.

Suppose that y *is* an integral function. Then any rational integral polynomial in x and y is, clearly, also an integral function. But it does not follow, conversely, though it is sometimes true, that every integral rational function can be written as an integral polynomial in x and y. For instance on the surface associated with the equation

$$y^3 + By^2x + Cyx^2 + Dx^3 - E(y^2 - x^2) = 0,$$

the three values of y at the places $x=0$ may be expressed by series of positive integral powers of x of the respective forms

$$y = x + \lambda x^2 + \ldots, \qquad y = -x + \mu x^2 + \ldots, \qquad y = E + \nu x + \ldots.$$

Thus, the rational function $(y^2 - Ey)/x$ is not infinite when $x=0$. Since y is an integral function, the function cannot be infinite for any other finite value of x. Hence $(y^2 - Ey)/x$ is an integral function. And it is not possible, with the help of the equation of the surface, to write the function as an integral polynomial in x and y. For such a polynomial could, by the equation of the surface, be reduced to the form of an integral polynomial in x and y of the second order in y; and, in order that such a polynomial should be equal to $(y^2 - Ey)/x$, the original equation would need to be reducible.

Ex. Find the rational relation connecting x with the function $\eta = (y^2 - Ey)/x$; and thus shew that η is an integral function.

39. We concern ourselves first of all with a method of expressing all rational functions whose poles are only at the places where x has the same finite value. For this value, say a, of x there may be several branch places: the most general case is when there are k places specified by such equations as

$$x - a = t_1^{w_1+1}, \ \ldots\ , \ x - a = t_k^{w_k+1}.$$

The orders of infinity, in these places, of the functions considered, will be specified by integral negative powers of $t_1, \ldots, t_k$ respectively. Let F be such a function. Let $\sigma + 1$ be the least positive integer such that $(x-a)^{\sigma+1}F$ is finite at every place $x=a$. We call $\sigma+1$ the *dimension* of F. Let $f(x, y) = 0$ be the equation of the surface. In order that there may be any branch places at $x=a$, it is necessary that $\partial f/\partial y$ should be zero for this value of x. Since this is only true for a finite number of values of x, we shall suppose that the value of x considered is one for which there are no branch places.

We prove that there are rational functions $h_1, \ldots, h_{n-1}$ infinite only at the n places $x=a$, such that every rational function whose infinities occur only at these n places can be expressed in the form

$$\left(\frac{1}{x-a}, 1\right)_\lambda + \left(\frac{1}{x-a}, 1\right)_{\lambda_1} h_1 + \ldots + \left(\frac{1}{x-a}, 1\right)_{\lambda_{n-1}} h_{n-1} \quad \ldots\ldots\ldots \text{(A)},$$

in such a way that no term occurs in this expression which is of higher dimension than the function to be expressed: namely, if $\sigma+1$ be the dimension of the function to be expressed and σ_i+1 the dimension of h_i, the function can be expressed in such a way that no one of the integers

$$\lambda,\ \lambda_1+\sigma_1+1,\ \ldots,\ \lambda_{n-1}+\sigma_{n-1}+1$$

is greater than $\sigma+1$. We may refer to this characteristic as the *condition of dimensions*. It is clear conversely that every expression of the form (A) will be a rational function infinite only for $x=a$.

Let the sheets of the surface at $x=a$ be considered in some definite order. A rational function which is infinite only at these n places may be denoted by a symbol $(R_1, R_2, \ldots, R_n)$, where $R_1, R_2, \ldots, R_n$ are the orders of infinity in the various sheets. We may call $R_1, R_2, \ldots, R_n$ the *indices* of the function. Since the surface is unbranched at $x=a$, it is possible to find a certain polynomial in $\frac{1}{x-a}$, involving only positive integral powers of this quantity, the highest power being $\left(\frac{1}{x-a}\right)^{R_n}$, such that the function

$$(R_1, R_2, \ldots, R_n) - \left(\frac{1}{x-a}, 1\right)_{R_n}, = (S_1, S_2, \ldots, S_{n-1}, 0) \text{ say} \quad \ldots\ldots\ldots \text{(i)},$$

is not infinite in the nth sheet at $x=a$.

Consider then all rational functions, infinite only at $x=a$, of which the nth index is zero. It is in general possible to construct a rational function having prescribed values for the $(n-1)$ other indices, provided their sum be $p+1$. When this is not possible a function can be constructed* whose indices have a less sum than $p+1$, none of them being greater than the prescribed values. Starting with a set of indices $(p+1, 0, \ldots, 0)$, consider how far the first index can be reduced by increasing the 2nd, 3rd, ..., $(n-1)$th indices. In constructing the successive functions with smaller first index, it will be necessary, in the most general case, to increase some of the 2nd, 3rd, ..., $(n-1)$th indices, and there will be a certain arbitrariness as to the way in which this shall be done. But if we consider only those functions of which the sum of the indices is less than $p+2$, there will be only a finite number

* The proof is given in the preceding Chapter, (§§ 24, 28).

possible for which the first index has a given value. There will therefore only be a finite number of functions of the kind considered*, for which the further condition is satisfied that *the first index is the least possible such that it is not less than any of the others.* Let this least value be r_1, and suppose there are k_1 functions satisfying this condition. Call them the reduced functions of the first class—and in general let any function whose nth index is zero be said to be of the first class when its first index is *greater or not less* than its other indices. In the same way reckon as functions of the second class all those (with nth index zero) whose second index is *greater than* the first index and *greater than or equal to* the following indices. Let the functions whose second index has the least value consistently with this condition be called the reduced functions of the second class; let their number be k_2 and their second index be r_2. In general, reckon to the ith class ($i < n$) all those functions, with nth index zero, whose ith index is *greater than* the preceding indices and *not less than* the succeeding indices. Let there be k_i reduced functions of this class, with ith index equal to r_i. Clearly none of the integers $r_1, \ldots, r_{n-1}$ are zero.

Let now $$(s_1 \ldots s_{i-1}\, r_i\, s_{i+1} \ldots s_{n-1}\, 0),$$

where $$r_i > s_1, \ldots, r_i > s_{i-1}, r_i \geqq s_{i+1}, \ldots, r_i \geqq s_{n-1},$$

be *any* definite one of the k_i reduced functions of the ith class. Make a similar selection from the reduced functions of every class. And let

$$(S_1 \ldots S_{i-1}\, R_i\, S_{i+1} \ldots S_{n-1}\, 0)$$

be any function of the ith class other than a reduced function, so that

$$R_i > S_1, \ldots, R_i > S_{i-1}, R_i \geqq S_{i+1}, \ldots, R_i \geqq S_{n-1}.$$

Then by choice of a proper constant coefficient λ we can write

$$(S_1 \ldots S_{i-1}\, R_i\, S_{i+1} \ldots S_{n-1}\, 0) - \lambda\,(x-a)^{-(R_i - r_i)}(s_1 \ldots s_{i-1}\, r_i\, s_{i+1} \ldots s_{n-1}\, 0)$$

in the form

$$(T_1 \ldots T_{i-1}\, R_i'\, T_{i+1} \ldots T_{n-1},\, R_i - r_i) \qquad \text{(ii)},$$

where $R_i' < R_i$; T_1 may be as great as the greater of S_1, $R_i - (r_i - s_1)$, but is certainly less than R_i; and similarly $T_2, \ldots, T_{i-1}$ are certainly less than R_i; while T_{i+1} may be as great as the greater of S_{i+1}, $R_i - (r_i - s_{i+1})$, and is therefore not greater than R_i; and similarly $T_{i+2}, \ldots, T_{n-1}$ are certainly not greater than R_i.

* Functions which have the same indices are here regarded as identical. Of course the general function with given indices may involve a certain number of arbitrary constants. By *the* function of given indices is here meant *any one* such, chosen at pleasure, which really becomes infinite in the specified way.

Further, if $\left(\frac{1}{x-a}, 1\right)_{R_i - r_i}$ be a suitable polynomial of order $R_i - r_i$ in $(x-a)^{-1}$, we can write

$$(T_1 \dots T_{i-1} R_i' \dots T_{n-1}, R_i - r_i) - \left(\frac{1}{x-a}, 1\right)_{R_i - r_i}$$
$$= (S'_1 \dots S'_{i-1} R''_i S'_{i+1} \dots S'_{n-1} 0) \text{ .... (iii)},$$

where R''_i may be as great as the greater of R'_i, $R_i - r_i$, but is certainly less than R_i; S'_1 may be as great as the greater of T_1, $R_i - r_i$, but is certainly less than R_i; and similarly $S'_2, \dots, S'_{i-1}$ are certainly less than R_i; while S'_{i+1} may be as great as the greater of T_{i+1}, $R_i - r_i$, and is certainly not greater than R_i; and similarly $S'_{i+2}, \dots, S'_{n-1}$ are certainly not greater than R_i.

Hence there are two possibilities.

(1) Either $(S'_1 \dots S'_{i-1} R''_i S'_{i+1} \dots S'_{n-1} 0)$ is still of the ith class, namely, $\quad R''_i > S_1, \dots, R''_i > S'_{i-1}, R''_i \geqq S'_{i+1}, \dots, R''_i \geqq S'_{n-1}$,

and in this case the greatest value occurring among its indices (R''_i) is less than the greatest value occurring in the indices of $(S_1 \dots S_{i-1} R_i S_{i+1} \dots S_{n-1} 0)$.

(2) Or it is a function of another class, for which the greatest value occurring among its indices may be smaller than or as great as R_i (though not greater); but when this greatest value is R_i, it is not reached by any of the first i indices.

If then, using a term already employed, the greatest value occurring among the indices of any function $(R_1, \dots, R_n)$ be called the dimension of the function, we can group the possibilities differently and say, either $(S'_1 \dots S'_{i-1} R''_i S'_{i+1} \dots S'_{n-1} 0)$ is of lower dimension than

$$(S_1 \dots S_{i-1} R_i S_{i+1} \dots S_{n-1} 0),$$

or it is of the same dimension and then belongs to a more advanced class, that is, to an $(i+k)$th class where $k > 0$.

In the same way if $(t_1 \dots t_{i-1} r_i t_{i+1} \dots t_{n-1} 0)$ be any reduced function of the ith class other than $(s_1 \dots s_{i-1} r_i s_{i+1} \dots s_{n-1} 0)$, we can, by choice of a suitable constant coefficient μ, write

$$(t_1 \dots t_{i-1} r_i t_{i+1} \dots t_{n-1} 0) - \mu (s_1 \dots s_{i-1} r_i s_{i+1} \dots s_{n-1} 0)$$
$$= (t'_1 \dots t'_{i-1} r'_i t'_{i+1} \dots t'_{n-1} 0) \text{ .... (iv)},$$

where $r'_i < r_i$, $t'_1 \dots t'_{i-1}$ may be respectively as great as the greater of the pairs $(t_1, s_1) \dots (t_{i-1}, s_{i-1})$ but are each certainly less than r_i, while similarly no one of $t'_{i+1}, \dots, t'_{n-1}$ is greater than r_i.

The function $(t'_1 \dots t'_{i-1} r'_i t'_{i+1} \dots t'_{n-1} 0)$ cannot be of the ith class, since no function of the ith class has its ith index less than r_i: and though the greatest value reached among its indices may be as great as r_i (and not greater), the number of indices reaching this value will be at least one less

4—2

than for $(s_1 \dots s_{i-1} r_i s_{i+1} \dots s_{n-1} 0)$. Namely $(t'_1 \dots t'_{i-1} r'_i t'_{i+1} \dots t'_{n-1} 0)$ is certainly of more advanced class than $(s_1 \dots s_{i-1} r_i s_{i+1} \dots s_{n-1} 0)$, and not of higher dimension than this.

Denote now by $h_1, \dots, h_{n-1}$ the selected reduced functions of the 1st, 2nd, ..., $(n-1)$th classes. Then, having regard to the equations given by (ii), (iii), (iv), we can make the statement,

Any function $(S_1 \dots S_{i-1} R_i S_{i+1} \dots S_{n-1} 0)$ can be expressed as a sum of (1) an integral polynomial in $(x-a)^{-1}$, (2) one of $h_1, \dots, h_{n-1}$ multiplied by such a polynomial, (3) a function F which is either of lower dimension than the function to be expressed or is of more advanced class.

In particular when the function to be expressed is of the $(n-1)$th class the new function F will necessarily be of lower dimension than the function to be expressed.

Hence by continuing the process as far as may be needful, every function

$$f = (S_1 \dots S_{i-1} R_i S_{i+1} \dots S_{n-1} 0)$$

can be expressed in the form

$$\left(\frac{1}{x-a}, 1\right)_\lambda + \left(\frac{1}{x-a}, 1\right)_{\lambda_1} h_1 + \dots + \left(\frac{1}{x-a}, 1\right)_{\lambda_{n-1}} h_{n-1} + F_1, \quad \dots\dots(\text{v})$$

where F_1 is of lower dimension than f.

Applying this statement and recalling that there are lower limits to the dimensions of existent functions of the various classes, namely, those of the $k_1 + \dots + k_{n-1}$ reduced functions, and noticing that the reduction formula (v) can be applied to these reduced functions, we can, therefore, put every function $f = (S_1 \dots S_{i-1} R_i S_{i+1} \dots S_{n-1} 0)$ into a form

$$\left(\frac{1}{x-a}, 1\right)_\lambda + \left(\frac{1}{x-a}, 1\right)_{\lambda_1} h_1 + \dots + \left(\frac{1}{x-a}, 1\right)_{\lambda_{n-1}} h_{n-1}.$$

Now it is to be noticed that in the equations (ii), (iii), (iv), upon which this result is based, no terms are introduced which are of higher dimension than the function which it is desired to express: and that the same remark is applicable to equation (i).

Hence every function $(R_1, \dots, R_n)$ can be written in the form (A) in such a way that the condition of dimensions is satisfied.

40. In order to give an immediate example of the theory we may take the case of a surface of four sheets, and assume that the places $x = a$ are such that no rational function exists, infinite only there, whose aggregate order of infinity is less than $p+1$. In that case the specification of the reduced functions is an easy arithmetical problem. The reduced functions of the first class are $(m_1, m_2, m_3, 0)$, where m_1 is to be as small as possible without being smaller than m_2 or m_3: by the hypothesis we may take

$$m_1 + m_2 + m_3 = p + 1.$$

Those of the second class require m_2 as small as possible subject to

$$m_1 + m_2 + m_3 = p + 1, \ m_2 > m_1, \ m_2 \overline{\overline{>}} m_3:$$

those of the third class require m_3 greater than m_1 and m_2 but otherwise as small as possible subject to $m_1 + m_2 + m_3 = p + 1$. We therefore immediately obtain the reduced functions given in the 2nd, 3rd and 4th columns of the following table The dimension of any function of the ith class being denoted by $\sigma_i + 1$, the values of σ_i are given in the fifth column, and the sum $\sigma_1 + \sigma_2 + \sigma_3$ in the sixth. The reason for the insertion of this value will appear in the next Article.

p	Reduced functions of the first class	Reduced functions of the second class	Reduced functions of the third class	$\sigma_1, \sigma_2, \sigma_3$	$\sigma_1+\sigma_2+\sigma_3$
$=3M-1$	$(M, M, M, 0)$	$(M-2, M+1, M+1, 0)$ $(M-1, M+1, M, 0)$ $(M, M+1, M-1, 0)$	$(M-1, M, M+1, 0)$	$M-1, M, M$	$3M-1$
$=3N-2$	$(N, N, N-1, 0)$ $(N, N-1, N, 0)$	$(N-1, N, N, 0)$	$(N-1, N-1, N+1, 0)$	$N-1, N-1, N$	$3N-2$
$=3P$	$(P+1, P, P, 0)$ $(P+1, P+1, P-1, 0)$ $(P+1, P-1, P+1, 0)$	$(P-1, P+1, P+1, 0)$ $(P, P+1, P, 0)$	$(P, P, P+1, 0)$	P, P, P	$3P$

Here the reduced functions of the various classes are written down in random order. Denoting those first written by h_1, h_2, h_3, we may exemplify the way in which the others are expressible by them in two cases.

(*a*) When $p = 3M - 1$, we have, μ being such a constant as in equation (iv) above (§ 39),

$$(M, M+1, M-1, 0) - \mu(M-2, M+1, M+1, 0) = \{M, M, M+1, 0\},$$

the right hand denoting a function whose orders of infinity in the various sheets are not higher than the indices given. If the order in the third sheet be less than $M + 1$, the right hand must be a function of the first class and therefore the order in the third sheet must be M. In that case, since a general function of aggregate order $p + 1$ contains two arbitrary constants, we have an expression of the form

$$(M, M+1, M-1, 0) = \mu h_2 + Ah_1 + B,$$

for suitable values of the constants A, B.

If however there be no such reduction, we can choose a constant λ so that

$$\{M, M, M+1, 0\} - \lambda(M-1, M, M+1, 0) = \{M, M, M, 0\} = A'h_1 + B',$$

and thus obtain on the whole

$$(M, M+1, M-1, 0) = \mu h_2 + \lambda h_3 + A' h_1 + B',$$

for suitable values of the constants A', B'.

(*b*) When $p = 3P$ we obtain

$$\begin{aligned}(P+1, P+1, P-1, 0) &= \lambda h_1 + A\,(P, P+1, P, 0) + B \\ &= \lambda h_1 + A\,\{\mu h_2 + C h_3 + D\} + B.\end{aligned}$$

Ex. 1. Shew for a surface of three sheets that we have the table

p	$h_1,\ h_2$	$\sigma_1,\ \sigma_2$	$\sigma_1+\sigma_2$
odd	$\left(\frac{p+1}{2}, \frac{p+1}{2}, 0\right)\left(\frac{p-1}{2}, \frac{p+3}{2}, 0\right)$	$\frac{p-1}{2}, \frac{p+1}{2}$	p
even	$\left(\frac{p+2}{2}, \frac{p}{2}, 0\right)\left(\frac{p}{2}, \frac{p+2}{2}, 0\right)$	$\frac{p}{2}, \frac{p}{2}$	p

Ex. 2. Shew, for a surface of n sheets, that if the places $x=a$ be such that it is impossible to construct a rational function, infinite only there, whose aggregate order of infinity is less than $p+1$, a set of reduced functions is given by

$$h_1 \dots h_{r+1} = (k, \dots k, k-1, \dots, k-1, 0), (k-1, k, \dots k, k-1, \dots, k-1, 0) \dots\dots (k-1, \dots, k-1, k, \dots k, 0)$$

$$h_{r+2} \dots h_{n-1} = (k-1, \dots, k-1, k+1, k, \dots k, 0)\ (k-1, \dots, k-1, k, k+1, k, \dots k, 0) \dots\dots$$
$$(k-1, \dots, k-1, k, \dots k, k+1, 0)$$

wherein $p+1=(n-1)k-r$ $(r<n-1)$ and, in the first row, there are r numbers $k-1$ in each symbol, and, in the second row, there are $r+1$ numbers $k-1$ in each symbol. In each case $k, \dots k$ denotes a set of numbers all equal to k and $k-1, \dots, k-1$ denotes a set of numbers all equal to $k-1$.

The values of $\sigma_1, \dots, \sigma_{r+1}$ are each $k-1$, those of $\sigma_{r+2}, \dots, \sigma_{n-1}$ are each k. Hence

$$\sigma_1 + \dots + \sigma_{r+1} + \sigma_{r+2} + \dots + \sigma_{n-1} = (r+1)(k-1) + (n-r-2)k = (n-1)k - r - 1 = p.$$

Ex. 3. Shew that the resulting set of reduced functions is effectively independent of the order in which the sheets are supposed to be arranged at $x=a$.

41. For the case where rational functions exist, infinite only at the places $x=a$, whose aggregate order of infinity is less than $p+1$, the specification of their indices is a matter of greater complexity.

But we can at once prove that the property already exemplified and expressed by the equation $\sigma_1 + \dots + \sigma_{n-1} = p$, or by the statement that the sum of the dimensions of the reduced functions is $p+n-1$, is true in all cases.

For consider a rational function which is infinite to the rth order in each sheet at $x=a$ and not elsewhere: if r be taken great enough, such a function necessarily exists and is an aggregate of $nr-p+1$ terms, one of these being an additive constant (Chapter III. § 37). By what has been proved, such a function can be expressed in the form

$$\left(\frac{1}{x-a}, 1\right)_\lambda + \left(\frac{1}{x-a}, 1\right)_{\lambda_1} h_1 + \dots + \left(\frac{1}{x-a}, 1\right)_{\lambda_{n-1}} h_{n-1},$$

where the dimensions of the several terms, namely the numbers

$$\lambda, \lambda_1 + \sigma_1 + 1, \dots, \lambda_{n-1} + \sigma_{n-1} + 1,$$

are not greater than the dimension, r, of the function.

Conversely, the most general expression of this form in which $\lambda_1, \lambda_2, \dots, \lambda_{n-1}$ attain the upper limits prescribed by these conditions, is a function of the desired kind.*

But such general expression contains

$$(\lambda + 1) + (\lambda_1 + 1) + \dots + (\lambda_{n-1} + 1),$$

that is
$$(r + 1) + (r - \sigma_1) + \dots + (r - \sigma_{n-1}),$$

or
$$nr - (\sigma_1 + \dots + \sigma_{n-1}) + 1$$

arbitrary constants.

Since this must be equal to $nr - p + 1$ the result enunciated is proved.

The result is of considerable interest—when the forms of the functions $h_1 \dots h_{n-1}$ are determined algebraically, we obtain the deficiency of the surface by finding the sum of the dimensions of $h_1 \dots h_{n-1}$. It is clear that a proof of the value of this sum can be obtained by considerations already adopted to prove Weierstrass's gap theorem. That theorem and the present result are in fact, here, both deduced from the same fact, namely, that the number of periods of a normal integral of the second kind is p.

42. Consider now the places $x = \infty$: let the character of the surface be specified by k equations

$$\frac{1}{x} = t_1^{w_1+1}, \dots, \frac{1}{x} = t_k^{w_k+1},$$

there being k branch places. A rational function g which is infinite only at these places will be called an integral function. If its orders of infinity at these places be respectively $r_1, r_2, \dots, r_k$ and $G[r_i/(w_i+1)]$ be the least positive integer greater than or equal to $r_i/(w_i+1)$, and $\rho + 1$ denote the greatest of the k integers thus obtained, then it is clear that $\rho + 1$ is the least positive integer such that $x^{-(\rho+1)} g$ is finite at every place $x = \infty$. We shall call $\rho + 1$ the dimension of g.

Of such integral functions there are $n - 1$ which we consider particularly, namely, using the notation of the previous paragraph, the functions

$$(x - a)^{\sigma_1+1} h_1, \dots\dots, (x - a)^{\sigma_{n-1}+1} h_{n-1},$$

which by the definitions of $\sigma_1, \dots\dots, \sigma_{n-1}$ are all finite at the places $x = a$, and are therefore infinite only for $x = \infty$. Denote $(x - a)^{\sigma_i+1} h_i$ by g_i. If h_i do not vanish at every place $x = \infty$, it is clear that the dimension of g_i is

* It is clear that this statement could not be made if any of the indices of the function to be expressed were less than the dimension of the function. For instance in the final equation of § 40 (a), unless μ, λ, A' be specially chosen, the right hand represents a function with its third index equal to $M+1$.

σ_i+1. If however h_i do so vanish, the dimension of g_i may conceivably be less than σ_i+1; denote it by ρ_i+1, so that $\rho_i \overline{\overline{<}} \sigma_i$. Then $x^{-(\rho_i+1)} g_i$, and therefore also $(x-a)^{-(\rho_i+1)} g_i, = (x-a)^{\sigma_i-\rho_i} h_i$, is finite at all places $x=\infty$: hence $(x-a)^{\sigma_i-\rho_i} h_i$ is a function which only becomes infinite at the places $x=a$. But, in the phraseology of § 39, it is clearly a function of the same class as h_i, it does not become infinite in the nth sheet at $x=a$, and is of less dimension than h_i if $\sigma_i > \rho_i$. That such a function should exist is contrary to the definition of h_i. Hence, in fact, $\sigma_i = \rho_i$. The reader will see that the same result is proved independently in the course of the present paragraph.

Let now F denote any integral function of dimension $\rho+1$. Then $x^{-(\rho+1)} F$ is finite at all places $x=\infty$: and therefore so also is $(x-a)^{-(\rho+1)} F$. This latter function is one of those which are infinite only at places $x=a$; if F do not vanish at all places $x=a$, the dimension $\sigma+1$ of $(x-a)^{-(\rho+1)} F$ will be $\rho+1$: in general we shall have $\sigma \overline{\overline{<}} \rho$.

By § 39 we can write

$$(x-a)^{-(\rho+1)} F = \left(\frac{1}{x-a}, 1\right)_\lambda + \left(\frac{1}{x-a}, 1\right)_{\lambda_1} h_1 + \dots\dots + \left(\frac{1}{x-a}, 1\right)_{\lambda_{n-1}} h_{n-1},$$

where $$\sigma + 1 \overline{\overline{>}} \lambda_i + \sigma_i + 1,$$

and therefore, *a fortiori*,

$$\rho + 1 \overline{\overline{>}} \lambda_i + \sigma_i + 1 \overline{\overline{>}} \lambda_i + \rho_i + 1.$$

Hence we can also write

$$F = (1, x-a)_\lambda (x-a)^{\rho-\lambda+1} + (1, x-a)_{\lambda_1} (x-a)^{\rho-\lambda_1-\sigma_1} g_1 + \dots\dots + (1, x-a)_{\lambda_{n-1}} (x-a)^{\rho-\lambda_{n-1}-\sigma_{n-1}} g_{n-1},$$

or say

$$F = (1, x)_\mu + (1, x)_{\mu_1} g_1 + \dots\dots + (1, x)_{\mu_{n-1}} g_{n-1}, \quad \dots\dots\dots\dots\text{(B)}$$

where $$\mu_i + \rho_i + 1 = \rho - \sigma_i + \rho_i + 1 = \rho + 1 - (\sigma_i - \rho_i) \overline{\overline{<}} \rho + 1,$$

namely, there is no term on the right whose dimension is greater than that of F (and each of $\mu, \mu_1, \dots\dots, \mu_{n-1}$ is a positive integer).

Hence the equation (B) is entirely analogous to the equation (A) obtained previously for the expression of functions which are infinite only at places $x=a$. The set $(1, g_1, \dots\dots, g_{n-1})$ will be called a *fundamental set* for the expression of rational integral functions*.

It can be proved precisely as in the previous Article that $\rho_1 + \rho_2 + \dots\dots + \rho_{n-1} = p$. For this purpose it is only necessary to consider a function

* The idea, derived from arithmetic, of making the integral functions the basis of the theory of all algebraic functions has been utilised by Dedekind and Weber, *Theor. d. alg. Funct. e. Veränd.* Crelle, t. 92. Kronecker, *U. die Discrim. alg. Fctnen.* Crelle, t. 91. Kronecker, *Grundzüge e. arith. Theor. d. algebr. Grössen*, Crelle, t. 92 (1882).

which is infinite at the places $x=\infty$ respectively to orders $r(w_1+1), \ldots, r(w_k+1)$. And the equations $\Sigma\rho = \Sigma\sigma = p$, taken with $\sigma_i \overline{\overline{>}} \rho_i$, suffice to shew that $\sigma_i = \rho_i$. It can also be shewn that from the set $g_1 \ldots g_{n-1}$ we can conversely deduce a fundamental set $1, (x-b)^{-(\rho_1+1)} g_1, \ldots, (x-b)^{-(\rho_{n-1}-1)} g_{n-1}$ for the expression of functions infinite only at places $x=b$; these have the same dimensions as $1, g_1, \ldots, g_{n-1}$*.

43. Having thus established the existence of fundamental systems for integral rational functions, it is proper to refer to some characteristic properties of all such systems.

(*a*) If $G_1 \ldots G_{n-1}$ be any set of rational integral functions such that every rational integral function can be expressed in the form

$$(x, 1)_\lambda + (x, 1)_{\lambda_1} G_1 + \ldots\ldots + (x, 1)_{\lambda_{n-1}} G_{n-1} \ldots\ldots\ldots\ldots\ldots (C),$$

there can exist no relations of the form

$$(x, 1)_\mu + (x, 1)_{\mu_1} G_1 + \ldots\ldots + (x, 1)_{\mu_{n-1}} G_{n-1} = 0.$$

For if k such relations hold, independent of one another, k of the functions $G_1 \ldots G_{n-1}$ can be expressed linearly, with coefficients which are rational in x, in terms of the other $n-1-k$. Hence also $\beta_1 y, \beta_2 y^2, \ldots, \beta_{n-1-k} y^{n-1-k}$, $\beta_{n-k} y^{n-k}$, which are integral functions when $\beta_1, \ldots, \beta_{n-k}$ are proper polynomials in x, can be expressed linearly in terms of the $n-1-k$ linearly independent functions occurring among $G_1 \ldots G_{n-1}$, with coefficients which are rational in x. By elimination of these $n-1-k$ functions we therefore obtain an equation

$$A + A_1 y + \ldots\ldots + A_{n-k} y^{n-k} = 0,$$

whose coefficients $A, A_1, \ldots\ldots, A_{n-k}$ are rational in x. Such an equation is inconsistent with the hypothesis that the fundamental equation of the surface is irreducible.

(*b*) Consider two places of the Riemann surface at which the independent variable, x, has the same value: suppose, first of all, that there are no branch places for this value of x. Let $\lambda, \lambda_1, \ldots\ldots, \lambda_{n-1}$ be constants. Then the linear function

$$\lambda + \lambda_1 G_1 + \ldots\ldots + \lambda_{n-1} G_{n-1}$$

cannot have the same value at these two places *for all values of* λ, $\lambda_1, \ldots\ldots, \lambda_{n-1}$.

For this would require that each of $G_1, \ldots\ldots, G_{n-1}$ has the same value at these two places. Denote these values by $a_1, \ldots\ldots, a_{n-1}$ respectively. We can choose coefficients $\mu_1, \ldots\ldots, \mu_{n-1}$ such that the function

$$\mu_1(G_1 - a_1) + \ldots\ldots + \mu_{n-1}(G_{n-1} - a_{n-1}),$$

* The dimension of an integral function is employed by Hensel, Crelle, t. 105, 109, 111; *Acta Math.* t. 18. The account here given is mainly suggested by Hensel's papers. For surfaces of three sheets see also Baur, *Math. Annal.* t. 43 and *Math. Annal.* t. 46.

which clearly vanishes at each of the two places in question, vanishes also at the other $n-2$ places arising for the same value of x. Denoting the value of x by c, it follows, since there are no branch places for $x=c$, that the function

$$[\mu_1(G_1-a_1)+\ldots\ldots+\mu_{n-1}(G_{n-1}-a_{n-1})]/(x-c)$$

is not infinite at any of the places $x=c$. It is therefore an integral rational function.

Now this is impossible. For then the function could be expressed in the form

$$(x, 1)_\lambda+(x, 1)_\mu\, G_1+\ldots\ldots+(x, 1)_\nu\, G_{n-1},$$

and it is contrary to what is proved under (a) that two expressions of these forms should be equal to one another.

Hence the hypothesis that the function

$$\lambda+\lambda_1 G_1+\ldots\ldots+\lambda_{n-1} G_{n-1}$$

can have the same value in each of two places at which x has the same value, is disproved.

If there be a branch place at $x=c$, at which two sheets wind, and no other branch place for this value of x, it can be proved in a similar way, that a linear function of the form

$$\lambda_1 G_1+\ldots\ldots+\lambda_{n-1} G_{n-1}$$

cannot vanish to the second order at the branch place, *for all values of* $\lambda_1,\ldots\ldots,\lambda_{n-1}$ namely, not all of $G_1,\ldots\ldots, G_{n-1}$ can vanish to the second order at the branch place. For then we could similarly find an integral function expressible in the form

$$(\mu_1 G_1+\ldots\ldots+\mu_{n-1} G_{n-1})/(x-c).$$

More generally, whatever be the order of the branch place considered, at $x=c$, and whatever other branch places may be present for $x=c$, it is always true that, *if all of* $G_1,\ldots\ldots, G_{n-1}$ *vanish at the same place* A *of the Riemann surface, they cannot all vanish at another place for which* x *has the same value; and if* A *be a branch place, they cannot all vanish at* A *to the second order.*

Ex. 1. Denoting the function

$$\lambda+\lambda_1 G_1+\ldots+\lambda_{n-1} G_{n-1}$$

by K, and its values in the n sheets for the same value of x by $K^{(1)}, K^{(2)},\ldots, K^{(n)}$, we have shewn that, for a particular value of x, we can always choose $\lambda, \lambda_1,\ldots, \lambda_{n-1}$, so that the equation $K^{(1)}=K^{(2)}$ is not verified. Prove, similarly, that we can always choose $\lambda, \lambda_1,\ldots, \lambda_{n-1}$ so that an equation of the form

$$m_1 K^{(1)}+m_2 K^{(2)}+\ldots+m_k K^{(k)}=0,$$

where $m_1,\ldots, m_{k-1}, m_k$ are given constants whose sum is zero, is not verified.

Ex. 2. Let $x=\gamma_1, \ldots, \gamma_k$ be k distinct given values of x: then it is possible to choose coefficients $\lambda, \lambda_1, \ldots, \mu, \mu_1, \ldots$, finite in number, such that the values of the function

$$(\lambda+\mu x+\nu x^2+\ldots)+(\lambda_1+\mu_1 x+\nu_1 x^2+\ldots)G_1+\ldots+(\lambda_{n-1}+\mu_{n-1}x+\nu_{n-1}x^2+\ldots)G_{n-1},$$

at the places $x=\gamma_1$, shall be all different, and also the values of the function, at the places $x=\gamma_2$, shall be all different, and, also, the values of the function, for each of the places $x=\gamma_3, \ldots, \gamma_k$, shall be all different.

(*c*) If $1, H_1, H_2, \ldots\ldots, H_{n-1}$ be another fundamental set of integral functions, with the same property as $1, G_1, \ldots\ldots, G_{n-1}$, we shall have linear equations of the form

$$1=1$$
$$H_i = \alpha_i + \alpha_{i,1}\, G_1 + \ldots\ldots + \alpha_{i,\,n-1}\, G_{n-1} \ldots\ldots\ldots\ldots\ldots\ldots \text{(D)},$$

where $\alpha_{i,\,j}$ is an integral polynomial in x.

Now in fact the determinant $|\,\alpha_{i,j}\,|$ is a constant $(i=1, 2, \ldots, \overline{n-1}; j=1, 2, \ldots, \overline{n-1})$. For if $H_i^{(r)}$ denote the value of H_i, for a general value of x, in the rth sheet of the surface, we clearly have the identity

$$\begin{vmatrix} 1, & 1, \ldots\ldots\ldots\ldots, & 1 \\ H_1^{(1)}, & H_1^{(2)}, \ldots\ldots\ldots, & H_1^{(n)} \\ \ldots & \ldots & \ldots \\ \ldots & \ldots & \ldots \\ \ldots & \ldots & \ldots \\ H_{n-1}^{(1)}, & H_{n-1}^{(2)}, \ldots\ldots, & H_{n-1}^{(n)} \end{vmatrix} = \begin{vmatrix} 1, & 0, \ldots\ldots\ldots, & 0 \\ \alpha_1, & \alpha_{1,1}, \ldots\ldots\ldots, & \alpha_{1,\,n-1} \\ \ldots & \ldots & \ldots \\ \ldots & \ldots & \ldots \\ \ldots & \ldots & \ldots \\ \alpha_{n-1}, & \alpha_{n-1,1}, \ldots\ldots, & \alpha_{n-1,\,n-1} \end{vmatrix} \begin{vmatrix} 1, & 1, \ldots\ldots\ldots\ldots, & 1 \\ G_1^{(1)}, & G_1^{(2)}, \ldots\ldots\ldots, & G_1^{(n)} \\ \ldots & \ldots & \ldots \\ \ldots & \ldots & \ldots \\ \ldots & \ldots & \ldots \\ G_{n-1}^{(1)}, & G_{n-1}^{(2)}, \ldots, & G_{n-1}^{(n)} \end{vmatrix}$$

If we form the square of this equation, the general term of the square of the left hand determinant, being of the form $H_i^{(1)}H_j^{(1)}+\ldots\ldots+H_i^{(n)}H_j^{(n)}$, will be a rational function of x which is infinite only for infinite values of x; it is therefore an integral polynomial in x. We shall therefore have a result which we write in the form

$$\Delta\,(1, H_1, \ldots\ldots, H_{n-1}) = \nabla^2 \,.\, \Delta\,(1, G_1, G_2, \ldots\ldots, G_{n-1}),$$

where ∇ is the determinant $|\,\alpha_{i,j}\,|$. $\Delta\,(1, H_1, \ldots\ldots, H_{n-1})$ may be called the discriminant of $1, H_1, \ldots\ldots, H_{n-1}$.

If β be such an integral polynomial in x that $\beta y, =\eta$, say, is an integral function, an equation of similar form exists when $1, \eta, \eta^2, \ldots\ldots, \eta^{n-1}$ are written instead of $1, H_1, \ldots\ldots, H_{n-1}$. Since then $\Delta\,(1, \eta, \eta^2, \ldots\ldots, \eta^{n-1})$ does not vanish for all values of x it follows that $\Delta\,(1, G_1, G_2, \ldots\ldots, G_{n-1})$ does not vanish for all values of x. (Cf. (*a*), of this Article.)

But because $1, H_1, H_2, \ldots\ldots, H_{n-1}$ are equally a set in terms of which all integral functions are similarly expressible, it follows that $\Delta\,(1, H_1, \ldots\ldots, H_{n-1})$ does not vanish for all values of x, and that

$$\Delta\,(1, G_1, \ldots\ldots, G_{n-1}) = \nabla_1^2\, \Delta\,(1, H_1, \ldots\ldots, H_{n-1}),$$

where ∇_1 is an integral function rationally expressible by x only.

Hence $\nabla^2 . \nabla_1^2 = 1$: *thus each of* ∇ *and* ∇_1 *is an absolute constant.*

Hence also the discriminants $\Delta(1, G_1, \ldots\ldots, G_{n-1})$ of all sets in terms of which integral functions are thus integrally expressible, are identical, save for a constant factor.

Let Δ denote their common value and $\eta_1, \ldots, \eta_n$ denote any n integral functions whatever; then if $\Delta(\eta_1, \eta_2, \ldots, \eta_n)$ denote the determinant which is the square of the determinant whose (s, r)th element is $\eta_s^{(r)}$, we can prove, as here, that there exists an equation of the form

$$\Delta(\eta_1, \eta_2, \ldots, \eta_n) = M^2\Delta,$$

wherein M is an integral polynomial in x. The function $\Delta(\eta_1, \eta_2, \ldots, \eta_n)$ is called the discriminant of the set $\eta_1, \eta_2, \ldots, \eta_n$. Since this is divisible by Δ, it follows, if, for shortness, we speak of $1, H_1, \ldots, H_{n-1}$, equally with $\eta_1, \eta_2, \ldots, \eta_n$, as a set of n integral functions, that Δ *is the highest divisor common to the discriminants of all sets of* n *integral functions.*

(*d*) The sets $(1, G_1, \ldots\ldots, G_{n-1})$, $(1, H_1, \ldots\ldots, H_{n-1})$ are not supposed subject to the condition that, in the expression of an integral function in terms of them, no term shall occur of higher dimension than the function to be expressed. If $(1, g_1, \ldots\ldots, g_{n-1})$ be a fundamental system for which this condition is satisfied, the equation which expresses G_i in terms of $1, g_1, g_2, \ldots\ldots, g_{n-1}$ will not contain any of these latter which are of higher dimension than that of G_i. Let the sets $G_1, \ldots\ldots, G_{n-1}$, $g_1, \ldots\ldots, g_{n-1}$ be each arranged in the ascending order of their dimensions. Then the equations which express $G_1, G_2, \ldots\ldots, G_k$ in terms of $g_1, \ldots\ldots, g_{n-1}$ must contain at least k of the latter functions; for if they contained any less number it would be possible, by eliminating those of the latter functions which occur, to obtain an equation connecting $G_1, \ldots\ldots, G_k$ of the form

$$(x, 1)_\lambda + (x, 1)_{\lambda_1} G_1 + \ldots\ldots + (x, 1)_{\lambda_k} G_k = 0;$$

this is contrary to what is proved under (*a*).

Hence the dimension of g_k is not greater than the dimension of G_k: hence the sum of the dimensions of $G_1, G_2, \ldots\ldots, G_{n-1}$ is not less than the sum of the dimensions of $g_1, g_2, \ldots\ldots, g_{n-1}$. Hence, *the least value which is possible for the sum of the dimensions of a fundamental set* $(1, G_1, \ldots\ldots, G_{n-1})$ *is that which is the sum of the dimensions for the set* $(1, g_1, \ldots\ldots, g_{n-1})$, *namely, the least value is* $p + n - 1$.

We have given in the last Chapter a definition of p founded on Weierstrass's gap theorem: in the property that the sum of the dimensions of $g_1, \ldots, g_{n-1}$ is $p + n - 1$ we have, as already remarked, another definition, founded on the properties of integral rational functions.

Ex. 1. Prove that if $(1, g_1, \ldots, g_{n-1})$, $(1, h_1, \ldots, h_{n-1})$ be two fundamental sets both having the property that, in the expression of integral functions in terms of them, no terms occur of higher dimension than the function to be expressed, the dimensions of the individual functions of one set are the same as those of the individual functions of the other set, taken in proper order.

Ex. 2. Prove, for the surface

$$y^3 - by^2 + a_1 cy - a_1^2 a_2 = 0,$$

that the function

$$\eta = (y^2 - by + a_1 c)/a_1$$

satisfies the equation

$$\eta^3 - c\eta^2 + a_2 b\eta - a_2^2 a_1 = 0;$$

and that

$$\Delta(1, y, \eta) = b^2c^2 + 18a_1a_2bc - 27a_1^2a_2^2 - 4a_1c^3 - 4a_2b^3,$$

$$\Delta(1, y, y^2) = a_1^2 \Delta(1, y, \eta) \quad \Delta(1, \eta, \eta^2) = a_2^2 \Delta(1, y, \eta) \quad \Delta(y, y^2, \eta) = a_1^2c^2 \Delta(1, y, \eta).$$

In general $1, y, \eta$ are a fundamental set for integral functions, in this case.

44. Let now $(1, g_1, g_2, \ldots\ldots, g_{n-1})$ be any set of integral functions in terms of which any integral function can be expressed in the form

$$(x, 1)_\mu + (x, 1)_{\mu_1} g_1 + \ldots\ldots + (x, 1)_{\mu_{n-1}} g_{n-1},$$

and let the sum of the dimensions of $g_1, \ldots\ldots, g_{n-1}$ be $p + n - 1$.

There will exist integral polynomials in $x, \beta_1, \beta_2, \ldots\ldots, \beta_{n-1}$, such that $\beta_i y^i$ is an integral function: expressing this by $g_1, \ldots\ldots, g_{n-1}$ in the form above and solving for $g_1, \ldots\ldots, g_{n-1}$ we obtain* expressions of which the most general form is

$$g_i = \frac{\mu_{i,\,n-1}\, y^{n-1} + \ldots\ldots + \mu_{i,\,1}\, y + \mu_i}{D_i},$$

where $\mu_{i,\,n-1}, \ldots\ldots, \mu_{i,1}, \mu_i, D_i$ are integral polynomials in x. Denote this expression by $g_i(y, x)$ or $g_i(x, y)$.

Let the equation of the surface, arranged so as to be an integral polynomial in x and y, be written

$$f(y, x) = Q_0 y^n + Q_1 y^{n-1} + \ldots\ldots + Q_{n-1} y + Q_n = 0,$$

and let $\chi_i(y, x)$ denote the polynomial

$$Q_0 y^i + Q_1 y^{i-1} + \ldots\ldots + Q_{i-1} y + Q_i,$$

so that $\chi_0(y, x)$ is Q_0.

Let $\phi_0', \phi_1', \ldots\ldots, \phi'_{n-1}$ be quantities determined by equating powers of y in the *identity*

$$\begin{aligned}\phi_0' + \phi_1' \cdot g_1(y, x) + \phi_2' \cdot g_2(y, x) + \ldots\ldots + \phi'_{n-1} \cdot g_{n-1}(y, x) \\ = \chi_0 y^{n-1} + y^{n-2} \chi_1(y', x) + \ldots\ldots + y\, \chi_{n-2}(y', x) + \chi_{n-1}(y', x):\end{aligned}$$

* Since $g_1, \ldots, g_{n-1}$ are linearly independent.

in other words, if the equations expressing $1, y, y^2, \ldots\ldots, y^{n-1}$ in terms of $g_1(y, x), \ldots\ldots, g_{n-1}(y, x)$ be

$$\begin{aligned} 1 &= 1, \\ y &= a_1 + a_{1,1} g_1 + \ldots\ldots + a_{1,\,n-1} g_{n-1}, \\ &\ldots\ldots\ldots\ldots\ldots\ldots\ldots\ldots \\ y^{n-1} &= a_{n-1} + a_{n-1,\,1} g_1 + \ldots\ldots + a_{n-1,\,n-1} g_{n-1}, \end{aligned}$$

where the coefficient $a_{i,j}$ is an integral polynomial in x divided by β_i, then

$$\begin{aligned} \phi_0' &= \chi_{n-1}(y', x) + a_1 \chi_{n-2}(y', x) + \ldots\ldots + a_{n-1}\chi_0 \\ \phi_1' &= a_{1,1} \chi_{n-2}(y', x) + \ldots\ldots + a_{n-1,1}\chi_0 \\ &\ldots\ldots\ldots\ldots\ldots\ldots\ldots\ldots\ldots\ldots \\ \phi'_{n-1} &= a_{1,\,n-1}\chi_{n-2}(y', x) + \ldots\ldots + a_{n-1,\,n-1}\chi_0. \end{aligned}$$

So that if we write

$$(1, y, y^2, \ldots\ldots, y^{n-1}) = \Omega\,(1, g, \ldots\ldots, g_{n-1}),$$

Ω being the matrix of the transformation, we have

$$(\phi_0', \phi_1', \ldots\ldots, \phi'_{n-1}) = \overline{\Omega}\,(\chi'_{n-1}, \chi'_{n-2}, \ldots\ldots\, \chi_1', \chi_0),$$

where $\chi_i' = \chi_i(y', x)$, and $\overline{\Omega}$ represents a transformation whose rows are the columns of Ω, its columns being the rows of Ω.

But if (Q) denote the substitution

$$\left|\begin{array}{llll} Q_{n-1}, & Q_{n-2}, \ldots\ldots, & Q_1, & Q_0 \\ Q_{n-2}, & Q_{n-3}, \ldots\ldots, & Q_0, & 0 \\ \ldots\ldots & \ldots\ldots & \ldots\ldots & \ldots \\ \ldots\ldots & \ldots\ldots & \ldots\ldots & \ldots \\ Q_1, & Q_0, & 0, & \ldots\ldots \\ Q_0, & 0, & \ldots\ldots & \ldots \end{array}\right|$$

we have

$$(\chi_{n-1}, \chi_{n-2}, \ldots\ldots, \chi_1, \chi_0) = (Q)\,(1, y, y^2, \ldots\ldots, y^{n-1}).$$

Hence, changing y' to y in ϕ_i' and writing therefore ϕ_i for ϕ_i', we may write

$$(\phi_0, \phi_1, \ldots\ldots, \phi_{n-1}) = \overline{\Omega}\,(Q)\,\Omega\,(1, g_1, g_2, \ldots\ldots, g_{n-1}) \;\ldots\ldots\ldots\ldots(E).$$

Either this, or the original definition, which is equivalent to

$$\begin{aligned} &\phi_0(y', x) + \phi_1(y', x)\, g_1(y, x) + \ldots\ldots + \phi_{n-1}(y', x)\, g_{n-1}(y, x) \\ &= \frac{f(y', x) - f(y, x)}{y' - y} \\ &= \chi_0\, y'^{n-1} + y'^{n-2}\chi_1(y, x) + \ldots\ldots + y'\,\chi_{n-2}(y, x) + \chi_{n-1}(y, x) \\ &= \chi_0\, y^{n-1} + y^{n-2}\chi_1(y', x) + \ldots\ldots + y\,\chi_{n-2}(y', x) + \chi_{n-1}(y', x) \ldots\ldots(F), \end{aligned}$$

may be used as the definition of the forms $\phi_0, \phi_1, \ldots\ldots, \phi_{n-1}$.

The latter form will now be further changed for the purposes of an immediate application: let $y_1, \ldots\ldots, y_n$ denote the values of y corresponding to any general value of x for which the values of y are distinct. Denote $\phi_i(y_r, x)$, $g_i(y_r, x)$, by $\phi_i^{(r)}$, $g_i^{(r)}$, etc.

Then putting in (F) in turn $y = y' = y_1$ and $y' = y_1$, $y = y_s$, we obtain

$$\phi_0^{(1)} + \phi_1^{(1)} g_1^{(1)} + \dots\dots + \phi_{n-1}^{(1)} g_{n-1}^{(1)} = \left(\frac{\partial f}{\partial y}\right)_{y_1} = f'(y_1) \text{ say,}$$

$$\phi_0^{(1)} + \phi_1^{(1)} g_1^{(s)} + \dots\dots + \phi_{n-1}^{(1)} g_{n-1}^{(s)} = 0, \qquad (s = 2, 3, \dots\dots, n).$$

Hence if, with arbitrary constant coefficients $c_0, c_1, \dots\dots, c_{n-1}$, we write

$$c_0\phi_0^{(1)} + c_1\phi_1^{(1)} + \dots\dots + c_{n-1}\phi_{n-1}^{(1)} = \phi^{(1)},$$

we have

$$\begin{vmatrix} c_0 & c_1 & \dots\dots & c_{n-1} & \phi^{(1)} \\ 1 & g_1^{(1)} & \dots\dots & g_{n-1}^{(1)} & f'(y_1) \\ \dots & \dots & \dots & \dots & \dots \\ 1 & g_1^{(r)} & \dots\dots & g_{n-1}^{(r)} & 0 \\ \dots & \dots & \dots & \dots & \dots \\ 1 & g_1^{(n)} & \dots\dots & g_{n-1}^{(n)} & 0 \end{vmatrix} = 0,$$

or

$$\frac{\phi^{(1)}}{f'(y_1)} \begin{vmatrix} 1 & g_1^{(1)} & \dots\dots & g_{n-1}^{(1)} \\ \dots & \dots & \dots & \dots \\ \dots & \dots & \dots & \dots \\ 1 & g_1^{(n)} & \dots\dots & g_{n-1}^{(n)} \end{vmatrix} = \begin{vmatrix} c_0 & c_1 & \dots\dots & c_{n-1} \\ 1 & g_1^{(2)} & \dots\dots & g_{n-1}^{(2)} \\ \dots & \dots & \dots & \dots \\ 1 & g_1^{(n)} & \dots\dots & g_{n-1}^{(n)} \end{vmatrix} \quad \dots\dots\dots(G);$$

and we shall find this form very convenient: it clearly takes an indeterminate form for some values of x.

If we put all of $c_1, \dots\dots, c_{n-1}, = 0$ except c_r, and put $c_r = 1$, and multiply both sides of this equation by the determinant which occurs on the left hand, the right hand becomes

$$S_r + S_{r,1} g_1^{(1)} + \dots\dots + S_{r,n-1} g_{n-1}^{(1)},$$

where, if $s_{i,j} = g_i^{(1)} g_j^{(1)} + g_i^{(2)} g_j^{(2)} + \dots\dots + g_i^{(n)} g_j^{(n)}$, $S_{i,j}$ means the minor of $s_{i,j}$ in the determinant

$$\Delta(1, g_1, g_2, \dots\dots, g_{n-1}) = \begin{vmatrix} n & s_1 & s_2 & \dots\dots & s_{n-1} \\ s_1 & s_{1,1} & s_{1,2} & \dots\dots & s_{1,n-1} \\ \dots & \dots & \dots & \dots & \dots \\ \dots & \dots & \dots & \dots & \dots \\ s_{n-1} & s_{n-1,1} & s_{n-1,2} & \dots\dots & s_{n-1,n-1} \end{vmatrix}$$

Since this is true for every sheet, we therefore have

$$\frac{\phi_r}{f'(y)} = \frac{S_r + S_{r,1} g_1 + \dots\dots + S_{r,n-1} g_{n-1}}{\Delta(1, g_1, \dots\dots, g_{n-1})}$$

$$= \frac{1}{\Delta}\frac{\partial \Delta}{\partial s_r} + \frac{1}{\Delta}\frac{\partial \Delta}{\partial s_{r,1}} g_1 + \dots\dots + \frac{1}{\Delta}\frac{\partial \Delta}{\partial s_{r,n-1}} g_{n-1} \dots\dots\dots\dots\dots (H),$$

and therefore, also

$$f'(y)\, g_r = s_r\, \phi_0 + s_{r,1}\, \phi_1 + \ldots\ldots + s_{r,n-1}\, \phi_{n-1} \ldots\ldots\ldots\ldots\ldots (\text{H}').$$

The equation (H) has the remarkable property that it determines the functions $\frac{\phi_r}{f'(y)}$ from the functions g_i *with a knowledge of these latter only.*

But we can also express $g_1, \ldots\ldots, g_{n-1}$ so that they are determined from $\frac{\phi_0}{f'(y)}, \frac{\phi_1}{f'(y)}, \ldots\ldots, \frac{\phi_{n-1}}{f'(y)}$, *with a knowledge of these only.*

For let these latter be denoted by $\gamma_0, \gamma_1, \ldots\ldots, \gamma_{n-1}$: and, in analogy with the definition of $s_{r,i}$, let $\sigma_{r,i} = \sum_{s=1}^{n} \gamma_r^{(s)}\, \gamma_i^{(s)}$.

Then from equation (H)

$$\sum_{s=1}^{n} \gamma_r^{(s)}\, g_i^{(s)} = \frac{1}{\Delta}\left[S_r s_i + S_{r,1}\, s_{i,1} + \ldots\ldots + S_{r,n-1}\, s_{i,n-1}\right]$$
$$= 0 \text{ or } 1 \text{ according as } i \neq r \text{ or } i = r.$$

Therefore, also, by equation (H),

$$\sigma_{r,i} = \sum_{s=1}^{n} \gamma_r^{(s)}\, \gamma_i^{(s)} = \frac{1}{\Delta}\left[S_r \sum_{s=1}^{n} \gamma_i^{(s)} + S_{r,1} \sum_{s=1}^{n} g_1^{(s)}\, \gamma_i^{(s)} + \ldots\ldots + S_{r,n-1} \sum_{s=1}^{n} g_{n-1}^{(s)}\, \gamma_i^{(s)}\right]$$
$$= \frac{1}{\Delta} S_{r,i},$$

so that equation (H) may be written

$$\gamma_r = \sigma_{r,0} + \sigma_{r,1}\, g_1 + \ldots\ldots + \sigma_{r,n-1}\, g_{n-1}.$$

If then $\Sigma_{r,i}$ denote the minor of $\sigma_{r,i}$ in the determinant of the quantities $\sigma_{r,i}$—which determinant we may call $\nabla(\gamma_0, \gamma_1, \ldots\ldots, \gamma_{n-1})$—we have, in analogy with (H),

$$g_r = \frac{1}{\nabla}(\Sigma_r\, \gamma_0 + \Sigma_{r,1}\, \gamma_1 + \ldots\ldots + \Sigma_{r,n-1}\, \gamma_{n-1}) \ldots\ldots\ldots\ldots\ldots (\text{K})^*.$$

Of course $\nabla = \frac{1}{\Delta}$ and $\Sigma_{r,i} = \frac{1}{\Delta} s_{r,i}$, and equation (K) is the same as (H′).

Ex. 1. Verify that if the integral functions $g_1, \ldots, g_{n-1}$ have the forms

$$g_1(x, y) = \frac{\chi_1(x, y)}{D_1},\ g_2(x, y) = \frac{\chi_2(x, y)}{D_2},\ \ldots,\ g_{n-1}(x, y) = \frac{\chi_{n-1}(x, y)}{D_{n-1}},$$

wherein $D_1, \ldots, D_{n-1}$ are integral polynomials in x, then $\phi_0, \ldots, \phi_{n-1}$ are given by

$$\phi_0(x, y) = y^{n-1},\ \phi_1(x, y) = D_1 y^{n-2},\ \ldots,\ \phi_{n-1}(x, y) = D_{n-1}.$$

* The equations (H) and (K) are given by Hensel. In his papers they arise immediately from the method whereby the forms of $\gamma_1, \gamma_2, \ldots\ldots$ are found.

Ex. 2. Prove from the expressions here obtained that

$$\sum_{s=1}^{n} [\phi_i/f'(y)]_s = 0, \qquad (i=1, 2, \dots, n-1),$$

and infer that
$$\sum_{s=1}^{n} (dv/dx)_s = 0,$$

v being any integral of the first kind.

45. We are now in a position to express the Riemann integrals.

Let $P_{x_1, x_2}^{x, c}$ be a general integral of the third kind, infinite only at the places x_1, x_2. Writing, in the neighbourhood of x_1, $x - x_1 = t_1^{w_1+1}$, dP/dx will (§§ 14, 16) be infinite like

$$\frac{1}{(w_1+1)\, t_1^{w_1}} \frac{d}{dt_1} \left[\log t_1 + A + A_1 t_1 + A_2 t_1^2 + \dots\dots \right],$$

namely, like

$$\frac{1}{w_1+1} \left[\frac{1}{x - x_1} + \frac{A_1}{t_1^{w_1}} + \frac{2A_2}{t_1^{w_1-1}} + \dots\dots \right];$$

thus $(x - x_1) \dfrac{dP}{dx}$ is finite at the place x_1 and is there equal to $\dfrac{1}{w_1+1}$.

Similarly $(x - x_2) \dfrac{dP}{dx}$ is finite at x_2 and there equal to $-\dfrac{1}{w_2+1}$.

Assume now, first of all, for the sake of simplicity, that at neither $x = x_1$ nor $x = x_2$ are there any branch places; let the finite branch places be at $x = a_1$, $x = a_2$, ...

At any one of these where, say, $x = a + t^{w+1}$, dP/dx is infinite like

$$\frac{1}{(w+1)\, t^w} \frac{d}{dt} [B + B_1 t + B_2 t^2 + \dots],$$

and therefore $(x - a) \dfrac{dP}{dx}$ is zero to the first order at the place.

Hence, if
$$\alpha = (x - a_1)(x - a_2)\dots$$
be the integral polynomial which vanishes at all the finite branch places of the surface, and g be any integral function whatever, the function

$$K = \alpha \, . \, g \, . \, (x - x_1)(x - x_2) \frac{dP}{dx}$$

is a rational function which is finite for all finite values of x and vanishes at every finite branch place.

Therefore the sum of the values of K in the n sheets, for any value of x, being a symmetrical function of the values of K belonging to that value of x, is a rational function of x only, which is finite for finite values of x and is therefore an integral polynomial in x. Since it vanishes for all the values of

x which make the polynomial α zero, it is divisible by α, and may be written in the form αJ.

Let the polynomial J be written in the form

$$\lambda_1 (x - x_2) - \lambda_2 (x - x_1) + (x - x_1)(x - x_2) H,$$

wherein λ_1 and λ_2 are constants and H is an integral polynomial in x. This is uniquely possible. Let H be of degree $\mu - 1$ in x; denote it by $(x, 1)^{\mu-1}$.

Then, on the whole,

$$\left(g\frac{dP}{dx}\right)_1 + \dots + \left(g\frac{dP}{dx}\right)_n = \frac{\lambda_1}{x - x_1} - \frac{\lambda_2}{x - x_2} + (x, 1)^{\mu-1}.$$

Multiply this equation by $x - x_1$ and consider the case when $x = x_1$, there being by hypothesis no branch place at $x = x_1$. Thus we obtain the value of λ_1; namely, it is the value of g at the place x_1. This we denote by $g(x_1, y_1)$. Similarly λ_2 is $g(x_2, y_2)$. Further, at an infinite place where $x = t^{-(w+1)}$,

$$\frac{dP}{dx} = -\frac{t^{w+2}}{w+1}\frac{dP}{dt},$$

so that $x^2 dP/dx$ is finite at all places $x = \infty$. Hence if $\rho + 1$ be the dimension of the integral function g, and we write

$$\left(gx^{-\rho-1} \,.\, x^2\frac{dP}{dx}\right)_1 + \dots + \left(gx^{-\rho-1} \,.\, x^2\frac{dP}{dx}\right)_n$$

$$= \frac{g(x_1, y_1)}{x^{\rho-1}(x - x_1)} - \frac{g(x_2, y_2)}{x^{\rho-1}(x - x_2)} + \frac{(x, 1)^{\mu-1}}{x^{\rho-1}},$$

we can infer, since ρ cannot be negative, that μ *is at most equal to* ρ.

Hence, taking g in turn equal to $1, g_1, \dots, g_{n-1}$, the dimensions of these functions being denoted by $0, \tau_1 + 1, \dots, \tau_{n-1} + 1$, we have the equations

$$\left(\frac{dP}{dx}\right)_1 + \dots\dots\dots + \left(\frac{dP}{dx}\right)_n = \frac{1}{x - x_1} - \frac{1}{x - x_2},$$

$$g_1^{(1)}\left(\frac{dP}{dx}\right)_1 + \dots + g_1^{(n)}\left(\frac{dP}{dx}\right)_n = \frac{g_1(x_1, y_1)}{x - x_1} - \frac{g_1(x_2, y_2)}{x - x_2} + (x, 1)^{\tau'_1 - 1},$$

$$\dots\dots\dots\dots\dots\dots\dots\dots\dots\dots\dots\dots\dots\dots\dots\dots$$

$$g_{n-1}^{(1)}\left(\frac{dP}{dx}\right)_1 + \dots + g_{n-1}^{(n)}\left(\frac{dP}{dx}\right)_n = \frac{g_{n-1}(x_1, y_1)}{x - x_1} - \frac{g_{n-1}(x_2, y_2)}{x - x_2} + (x, 1)^{\tau'_{n-1} - 1},$$

where $\tau'_1, \dots, \tau'_{n-1}$ are *positive integers not greater than* $\tau_1, \dots, \tau_{n-1}$ *respectively*.

Let these equations be solved for $\left(\frac{dP}{dx}\right)_1$: then in accordance with equations (G) on page 63 we have, after removal of the suffix,

$$f'(y)\frac{dP}{dx} = (x, 1)^{\tau'_1 - 1}\phi_1 + (x, 1)^{\tau'_2 - 1}\phi_2 + \dots + (x, 1)^{\tau'_{n-1} - 1}\phi_{n-1}$$
$$+ \frac{\phi_0 + \phi_1 g_1(x_1, y_1) + \dots + \phi_{n-1} g_{n-1}(x_1, y_1)}{x - x_1}$$
$$- \frac{\phi_0 + \phi_1 g_1(x_2, y_2) + \dots + \phi_{n-1} g_{n-1}(x_2, y_2)}{x - x_2},$$

where ϕ_i stands for $\phi_i(x, y)$.

This, by the method of deduction, is the most general form which dP/dx can have; the coefficients in the polynomials $(x, 1)^{\tau'_i - 1}$ are in number, *at most*,

$$\tau_1 + \tau_2 + \dots + \tau_{n-1},$$

or p; and no other element of the expression is undetermined. Now the most general form of dP/dx is known to be

$$\lambda_1 \frac{dv_1}{dx} + \dots + \lambda_p \frac{dv_p}{dx} + \left(\frac{dP}{dx}\right),$$

wherein $\left(\frac{dP}{dx}\right)$ is any special form of $\frac{dP}{dx}$ having the necessary character, and $\lambda_1, \dots, \lambda_p$ are arbitrary constants. Hence, by comparison of these forms, we can infer the two results—

(i) The most general form of integral of the first kind is

$$\int^x \frac{dx}{f'(y)} [(x, 1)^{\tau'_1 - 1}\phi_1(x, y) + \dots + (x, 1)^{\tau'_{n-1} - 1}\phi_{n-1}(x, y)],$$

wherein $\tau'_i \overline{\overline{<}} \tau_i$ and the coefficients in $(x, 1)^{\tau'_i - 1}$ are arbitrary:

(ii) A special and actual form of integral of the third kind logarithmically infinite at the two finite, ordinary, places (x_1, y_1), (x_2, y_2), namely like $\log[(x - x_1)/(x - x_2)]$, and elsewhere finite, is

$$\int^x \frac{dx}{f'(y)} \left[\frac{\phi_0(x, y) + \phi_1(x, y) g_1(x_1, y_1) + \dots + \phi_{n-1}(x, y) g_{n-1}(x_1, y_1)}{x - x_1}\right.$$
$$\left. - \frac{\phi_0(x, y) + \phi_1(x, y) g_1(x_2, y_2) + \dots + \phi_{n-1}(x, y) g_{n-1}(x_2, y_2)}{x - x_2}\right],$$

or

$$\int^x \frac{dx}{f'(y)} \int_{x_2}^{x_1} d\xi \frac{d}{d\xi} \left[\frac{\phi_0(x, y) + \phi_1(x, y) g_1(\xi, \eta) + \dots + \phi_{n-1}(x, y) g_{n-1}(\xi, \eta)}{x - \xi}\right].$$

In the actual way in which we have arranged the algebraic proof of this result we have only considered values of the current variable x for which the n sheets of the surface are distinct: the reader may verify that the result is valid for all values of x, and can be deduced by means of the definitions of the forms $\phi_0, \dots, \phi_{n-1}$, which have been given, other than the equation (G).

Ex. Apply the method to obtain the form of the general integral of the first kind only.

We shall find it convenient sometimes to use a single symbol for the expression

$$\frac{\phi_0(x, y) + \phi_1(x, y) g_1(\xi, \eta) + \ldots + \phi_{n-1}(x, y) g_{n-1}(\xi, \eta)}{(x - \xi) f'(y)},$$

and may denote it by (x, ξ). Then the result proved is that an elementary integral of the third kind is given by

$$P_{x_1, x_2}^{x, c} = \int_c^x dx \left[(x, x_1) - (x, x_2)\right].$$

This integral can be rendered normal, that is, chosen so that its periods at the p period loops of the first kind are zero, by the addition of a suitable linear aggregate of the p integrals of the first kind.

Now it can be shewn, as in Chapter II. § 19, that if $E_\xi^{x, c}$ denote an elementary integral of the second kind, the function of (x, y) given by the difference

$$D_\xi P_{\xi, x_2}^{x, c} - E_\xi^{x, c},$$

wherein D_ξ denotes a differentiation, is not infinite at (ξ, η). It follows from the form of $P_{\xi, x_2}^{x, c}$, here, that this function does not depend upon (x_2, y_2). Hence it is nowhere infinite, as a function of (x, y). Therefore, if not independent of (x, y), it is an aggregate of integrals of the first kind. Thus we infer that one form of an elementary integral of the second kind, which is once algebraically infinite at an *ordinary* place (ξ, η), like $-(x - \xi)^{-1}$, is given by

$$\int^x \frac{dx}{f'(y)} \frac{d}{d\xi} \left[\frac{\phi_0(x, y) + \phi_1(x, y) g_1(\xi, \eta) + \ldots + \phi_{n-1}(x, y) g_{n-1}(\xi, \eta)}{x - \xi}\right].$$

The direct deduction of the integral of the second kind when the infinity is at a branch place, which is given below, § 47, will furnish another proof of this result.

46. We proceed to obtain the form of an integral of the third kind when one or both of its infinities (x_1, y_1), (x_2, y_2) are at finite branch places; and when there may be other branch places for $x = x_1$ or $x = x_2$.

As before, let α be the integral polynomial vanishing at all the finite branch places. The function

$$g\alpha (x - x_1)(x - x_2)\, dP/dx$$

will vanish at all the places $x = x_1$: and though it may vanish at some of these to more than the first order, it will vanish at (x_1, y_1) only to as high order as $(x - x_1)$. Hence the sum of the values of this function in the several sheets for the same value of x is of the form αJ, where J is a polynomial in x which does not vanish, in general, for $x = x_1$ or $x = x_2$.

Hence as before (§ 45) we can write

$$\left(g\frac{dP}{dx}\right)_1 + \ldots + \left(g\frac{dP}{dx}\right)_n = \frac{\lambda_1}{x-x_1} - \frac{\lambda_2}{x-x_2} + (x, 1)^{\mu-1}.$$

Multiply this equation by $x - x_1$ and consider the limiting form of the resulting equation as (x, y) approaches to (x_1, y_1): let $w+1$ be the number of sheets which wind at this place. Recalling that the limiting value of $(x - x_1)\,dP/dx$ is $1/(w+1)$, we see that $w+1$ terms of the left hand, corresponding to the $w+1$ sheets at the discontinuity of the integral, will take a form

$$\frac{1}{w+1}\left[1 + A_1 t\epsilon + 2A_2 t^2\epsilon^2 + \ldots\right]\left[g(x_1, y_1) + Ct + Dt^2 + \ldots\right],$$

where ϵ is a $(w+1)$th root of unity. The limit of this when $t=0$ is $g(x_1, y_1)/(w+1)$; the corresponding terms of the left will therefore have $g(x_1, y_1)$ as limit. The other terms of the left hand will vanish.

Hence $\lambda_1 = g(x_1, y_1)$, $\lambda_2 = g(x_2, y_2)$. The determination of the upper limit for μ and the rest of the deduction proceed exactly as before. Thus,

The expression already given for an integral of the third kind holds whether (x_1, y_1), (x_2, y_2) *be branch places or ordinary places.*

If we denote the form of integral of the third kind thus determined by $P^{x,\,c}_{x_1,\,x_2}$, the zero c being assigned arbitrarily, it follows, as in § 45, above, that an elementary integral of the second kind, which is infinite at a branch place x_1, is given by

$$\lim._{x_1'=x_1}\left[P^{x,\,c}_{x_1',\,x_2} - P^{x,\,c}_{x_1,\,x_2}\right] t_{x_1}^{-1} = \lim._{t_{x_1}=0}\left[\int_c^x dx\,\{(x, x_1') - (x, x_1)\}\right] t_{x_1}^{-1}$$

$$- \lim.\, P^{x,\,c}_{x_1',\,x_1}\,.\,t_{x_1}^{-1}$$

Now if we write t for t_{x_1} and $x_1' = x_1 + t^{w+1}$, the coefficient of $dx/f'(y)$ in the integrand of the form here given for $P^{x,\,c}_{x_1',\,x_1}$ is

$$\frac{\phi_0 + \phi_1\,.\,(g_1 + tg_1' + \ldots) + \ldots + \phi_{n-1}\,.\,(g_{n-1} + tg'_{n-1} + \ldots)}{x - x_1 - t^{w+1}}$$

$$- \frac{\phi_0 + \phi_1\,.\,g_1 + \ldots + \phi_{n-1}\,.\,g_{n-1}}{x - x_1},$$

wherein $\phi_0, \ldots, \phi_{n-1}$ are functions of x, y, and $g_1, \ldots, g_{n-1}, g_1', g_2', \ldots$ are written for $g_1(x_1, y_1), \ldots, g_{n-1}(x_1, y_1)$, $Dg_1(x_1, y_1)$, $Dg_2(x_1, y_1), \ldots$, respectively, D denoting a differentiation in regard to t. Hence the ultimate form is

$$t\,.\,\frac{\phi_1 g_1' + \ldots + \phi_{n-1} g'_{n-1}}{x - x_1}.$$

That is, introducing ξ, η, instead of x_1, y_1, an elementary integral of the second kind, infinite at a finite branch place (ξ, η), is given by

$$\int^x \frac{dx}{f'(y)} \frac{\phi_1(x, y)\, g_1'(\xi, \eta) + \ldots + \phi_{n-1}(x, y)\, g'_{n-1}(\xi, \eta)}{x - \xi},$$

where $g_1'(\xi, \eta)$, ... are the differential coefficients in regard to the infinitesimal at the place. It has been shewn in (*b*) § 43 that these differential coefficients cannot be all zero.

Sufficient indications for forming the integrals when the infinities are at infinite places of the surface are given in the examples below (1, 2, 3, ...); in fact, by a linear transformation of the independent variable of the surface we are able to treat places at infinity as finite places.

Ex. 1. Shew that an integral of the third kind with infinities at (x_1, y_1), (x_2, y_2) can also be written in the form

$$\int \frac{dx}{f'(y)} \left[\frac{\lambda_1^{-1}\phi_0(x, y) + \Sigma \lambda_1^{\tau_r} \phi_r(x, y)\, g_r(x_1, y_1)}{x - x_1} - \frac{\lambda_2^{-1}\phi_0(x, y) + \Sigma \lambda_2^{\tau_r} \phi_r(x, y)\, g_r(x_2, y_2)}{x - x_2} \right],$$

wherein $\lambda_1 = (x-a)/(x_1 - a)$, $\lambda_2 = (x-a)/(x_2 - a)$, $\tau_r + 1$ is the dimension of g_r, and a is any arbitrary finite quantity.

It can in fact be immediately verified that the difference between this form and that previously given is an integral of the first kind. Or the result may be obtained by considering the surface with an independent variable $\xi = (x-a)^{-1}$ and using the forms of § 39 of this chapter for the fundamental set for functions infinite only at places $x = a$. The corresponding forms of the functions ϕ are then obtainable by equations (H) § 44.

Ex. 2. Obtain, as in the previous and present Articles, corresponding forms for integrals of the second kind.

Ex. 3. Obtain the forms for integrals of the third and second kinds which have an infinity at a place $x = \infty$.

It is only necessary to find the limits of the results in Examples 1 and 2 as (x_1, y_1) approaches the prescribed place at infinity. It is clearly convenient to take $a = 0$.

Ex. 4. For a surface of the form

$$y^2 = x(x - a_1) \ldots\ldots (x - a_{2p+1}),$$

wherein a_1, ..., a_{2p+1} are finite and different from zero and from each other, we may* take the fundamental set $(1, g_1)$ to be $(1, y)$, and so obtain $(\phi_0, \phi_1) = (y, 1)$. Assuming this, obtain the forms of all the integrals, for infinite and for finite positions of the infinities.

Ex. 5. In the case of Example 4 for which $p = 1$, the integral of Example 1, when a is taken 0, is

$$\tfrac{1}{2} \int \frac{dx}{y} \left[\frac{x_1}{x} \frac{y + x^2 x_1^{-2} y_1}{x - x_1} - \frac{x_2}{x} \frac{y + x^2 x_2^{-2} y_2}{x - x_2} \right].$$

Putting $x_1 = \infty$ and $y_1 = m x_1^2 + n x_1 + A + B x_1^{-1} + \ldots$, this takes the form

$$-\tfrac{1}{2} \int \frac{dx}{y} \left[\frac{y + m x^2}{x} + \frac{x_2 y}{(x - x_2)\, x} + \frac{x y_2}{x_2 (x - x_2)} \right]$$

or

$$-\tfrac{1}{2} \int \frac{dx}{y} \left[m x + \frac{y + y_2}{x - x_2} + \frac{y_2}{x_2} \right].$$

* Chap. V. § 56.

Prove that this integral is infinite at one place $x=\infty$ like $\log\left(\frac{1}{x}\right)$ and is otherwise infinite only at (x_2, y_2), namely like $-\log(x-x_2)$, if (x_2, y_2) be not a branch place.

Ex. 6. Prove in Example 5 that the limit of

$$\tfrac{1}{2}\int\frac{dx}{y}\left[\frac{x_1}{x}\,\frac{y+x^2{x_1}^{-2}y_1}{x-x_1}+\frac{y+mx^2}{x}\right]$$

as (x_1, y_1) approaches that place (∞, ∞) where $y=mx^2+nx+A+B/x+\dots$, is

$$-\tfrac{1}{2}\int\frac{dx}{y}(y+mx^2+nx),$$

and that the expansion of this integral in the neighbourhood of this place is

$$-x-\frac{A}{2m}\,\frac{1}{x}+\dots\dots,$$

and that it is otherwise finite. It is therefore an integral of the second kind with this place as its infinity. The process by which the integral is obtained is an example of the method followed in the present and the last Articles, for obtaining an elementary integral of the second kind from an elementary integral of the third kind.

47. We give now a direct deduction of the integral of the second kind whose infinity is at a finite place (ξ, η): we suppose that $(w+1)$ sheets of the surface wind at this place, and find the integral which is there infinite like an expression of the form

$$\frac{A_1}{t}+\frac{A_2}{t^2}+\dots+\frac{A_w}{t^w}+\frac{A_{w+1}}{x-\xi},$$

t being the infinitesimal at the place.

Firstly, let F be an integral which is infinite like the single term $(x-\xi)^{-1}$, so that in the neighbourhood of the infinity its expansion has a form

$$F=\frac{1}{x-\xi}+A+Bt+Ct^2+\dots.$$

Forming as before the sum of the values of the functions $g\,.\,(x-\xi)^2\,dF/dx$ in the n sheets of the surface, g being any integral function, we obtain an expression

$$\sum_{i=1}^{i=n}\left[g\,(x-\xi)^2\frac{dF}{dx}\right]_i=\lambda+\mu\,(x-\xi)+(x-\xi)^2\,.\,(x, 1)^{\mu-1}.$$

Putting $x=\xi$ we infer, since all terms on the left except those belonging to the place (ξ, η) vanish, that

$$\lambda=-(w+1)\,g\,(\xi, \eta).$$

Differentiating, and then putting $x=\xi$, we obtain, from the terms on the left belonging to the infinity,

$$\mu\,\underline{|w+1}=\lim.\,\Sigma\left\{\frac{d^{w+1}g}{dt^{w+1}}\,.\,(x-\xi)^2\frac{dF}{dx}+g\,.\,\frac{d^{w+1}}{dt^{w+1}}\left[-1+(x-\xi)^2\frac{d}{dx}(A+Bt+\dots)\right]\right\},$$

the summation extending to $(w+1)$ terms.

Now

$$\frac{d}{dx}\left[(x-\xi)^2 \frac{d}{dx}(A+Bt+\dots)\right] = \frac{1}{(w+1)^2 t^w}\frac{d}{dt}[t^{w+2}(B+2Ct+\dots)]$$

vanishes when t is zero: hence

$$\mu = -\frac{1}{\underline{|w}} D^{w+1} g(\xi, \eta).$$

Hence we can prove as before that, save for additive terms which are integrals of the first kind, the integral which is infinite like $(x-\xi)^{-1}$ is given by

$$F = -(w+1)\int \frac{dx}{f'(y)} \frac{\phi_0 + \phi_1 g_1(\xi,\eta) + \dots + \phi_{n-1} g_{n-1}(\xi,\eta)}{(x-\xi)^2}$$
$$-\frac{1}{\underline{|w}}\int \frac{dx}{f'(y)} \frac{D^{w+1}[\phi_0 + \phi_1 g_1(\xi,\eta) + \dots + \phi_{n-1} g_{n-1}(\xi,\eta)]}{x-\xi}.$$

This result is true whether (ξ, η) be a branch place or an ordinary place.

Consider now the integral, say E, which is infinite at (ξ, η) like t^{-m}, m being a positive integer less than $w+1$. At this place, therefore, $(x-\xi)\,dE/dx$ is infinite like $-\frac{m}{w+1}\cdot\frac{1}{t^m}$. If, as before, we consider the sum of the n values of the expression $\alpha \,.\, g \,.\, (x-\xi)\,dE/dx$, wherein g is any integral function and α is the integral polynomial before used, which vanishes at all the finite branch points of the surface, we shall obtain

$$\sum_{i=1}^{n}\left[g\,.\,(x-\xi)\frac{dE}{dx}\right]_i = \lambda + (x-\xi)(x,1)^{\mu-1}.$$

To find λ, let x approach to ξ. Then all the terms on the left, except those for the $w+1$ sheets which wind at the infinity of E, vanish: for such a non-vanishing term we have an expansion of the form

$$\left[g + tDg + \frac{t^2}{\underline{2}}D^2 g + \dots\right]\left[-\frac{m}{w+1}\frac{1}{t^m} + A + Bt + Ct^2 + \dots\right],$$

where D denotes, as usual, a differentiation in regard to the infinitesimal of the surface at (ξ, η), and g is written for $g(\xi, \eta)$. The sum of these $w+1$ expansions is

$$-\left[\frac{1}{\underline{|m-1}} D^m g + \frac{m}{w+1} g \Sigma \frac{1}{t^m} + \frac{m}{w+1} g' . \Sigma \frac{1}{t^{m-1}} + \dots + \frac{m}{w+1}\frac{1}{\underline{|m-1}} D^{m-1} g \,.\, \Sigma \frac{1}{t}\right]$$
$$+ (w+1)Ag + (Ag' + Bg)\Sigma t + \dots.$$

Now in fact every summation Σt^r, being a sum of terms of the form

$$t^r + \epsilon^r t^r + \dots + \epsilon^{(w+1)r} t^r,$$

wherein ϵ is a primitive $(w+1)$th root of unity, will be zero unless r be a multiple of $w+1$. Thus the terms involving negative powers of t in the sum will vanish: those involving positive powers of t will vanish ultimately when $t=0$; and in fact A is zero, otherwise E would contain the logarithmic term $A\log(x-\xi)$ when (x, y) is near to (ξ, η). Hence on the whole

$$\lambda = -\frac{1}{\lfloor m-1}D^m g(\xi, \eta).$$

Then, proceeding as before, we obtain an expression of the integral in the form,

$$-\frac{1}{\lfloor m-1}\int^x \frac{dx}{f'(y)}\cdot\frac{1}{x-\xi}D^m[\phi_0(x, y)+\dots+\phi_{n-1}(x, y)\,g_{n-1}(\xi, \eta)].$$

Thus, denoting the expression

$$\phi_0(x, y)+\sum_1^{n-1}\phi_r(x, y)\,g_r(\xi, \eta)$$

by Φ, an integral which is infinite like an expression

$$\frac{A_1}{t}+\dots+\frac{A_w}{t^w}+\frac{A_{w+1}}{x-\xi},$$

is given by

$$-(w+1)A_{w+1}\int^x \frac{dx}{f'(y)}\frac{\Phi}{(x-\xi)^2}$$

$$-\int^x \frac{dx}{f'(y)}\cdot\frac{1}{x-\xi}\left[A_1 D+\frac{A_2}{\lfloor 1}D^2+\frac{A_3}{\lfloor 2}D^3+\dots+\frac{A_w}{\lfloor w-1}D^w+\frac{A_{w+1}}{\lfloor w}D^{w+1}\right]\Phi.$$

Of course the differentiations at the place (ξ, η) must be understood in the sense in which they arise in the work. If $\phi(\xi, \eta)$ be any function of ξ, η, $D\phi(\xi, \eta)$ means that we substitute in $\phi(x, y)$, for x, $\xi+t^{w+1}$, and for y, an expression of the form $\eta+P(t)$, that we then differentiate this function of t in regard to t, and afterwards regard t as evanescent.

Ex. 1. Obtain this result by repeated differentiation of the integral $P_{\xi, a}^{x, c}$.

Ex. 2. Obtain by the formula the integral which is infinite like $A/t+B/t^2$ in the neighbourhood of (0, 0), the surface being $y^2=x(x, 1)_3$. Verify that the integral obtained actually has the property required.

48. The determinant $\Delta(1, g_1, \dots, g_{n-1})$, of which the general element is

$$s_{ij}=g_i^{(1)}g_j^{(1)}+\dots+g_i^{(n)}g_j^{(n)},$$

can be written in the form

$$\begin{vmatrix} n, & x^{-\tau_1-1}s_1, & \dots\dots, & x^{-\tau_{n-1}-1}s_{n-1} \\ x^{-\tau_1-1}s_1, & x^{-2\tau_1-2}s_{1,1}, & \dots\dots, & x^{-\tau_1-\tau_{n-1}-2}s_{1,n-1} \\ \dots & \dots & \dots & \dots \\ x^{-\tau_{n-1}-1}s_{n-1}, & x^{-\tau_{n-1}-\tau_1-2}s_{n-1,1}, & \dots\dots, & x^{-2\tau_{n-1}-2}s_{n-1,n-1} \end{vmatrix} x^{2n-2+2p}.$$

In this form the determinant factor is finite at every place $x=\infty$: hence also $x^{-(2p-2+2n)}\Delta(1, g_1, \dots, g_{n-1})$ is finite (including zero) at infinity. Thus

$\Delta(1, g_1, \dots, g_{n-1})$, which is an integral polynomial in x, is of not higher order than $2n-2+2p$ in x.

But when the sheets of the surface for $x=\infty$ are separate, it is not of less order; it is in fact easy to shew that if *for any value of x, $x=a$, there be several branch places, at which respectively w_1+1, w_2+1, ... sheets wind, then $\Delta(1, g_1, \dots, g_{n-1})$ contains the factor $(x-a)^{w_1+w_2+\dots}$.*

For, writing, in the neighbourhood of these places respectively,

$$x-a=t_1^{w_1+1},\ x-a=t_2^{w_2+1},\ \dots,$$

the determinant (§ 43)

$$\begin{vmatrix} 1, & g_1^{(1)}, & . & . & , & g_{n-1}^{(1)} \\ 1, & g_1^{(2)}, & . & . & , & g_{n-1}^{(2)} \\ . & . & . & . & & . \\ 1, & g_1^{(n)}, & . & . & , & g_{n-1}^{(n)} \end{vmatrix},$$

of which $\Delta(1, g_1, \dots, g_{n-1})$ is the square, can, for values of x very near to $x=a$, be written in a form in which one row divides by t_1, another row by $t_1^2, \dots$, another row by $t_1^{w_1}$, in which also another row divides by t_2, another row by $t_2^2, \dots$, and another row by $t_2^{w_2}$, and so on.

Thus this determinant has the factor $t_1^{\frac{1}{2}w_1(w_1+1)}\, t_2^{\frac{1}{2}w_2(w_2+1)} \dots$, and hence the square of this determinant has the factor $(x-a)^{w_1}(x-a)^{w_2} \dots$.

Therefore, when there are no branch places at infinity, $\Delta(1, g_1, \dots, g_{n-1})$ has at least an order Σw, $=2n+2p-2$ (§ 6).

In that case then $\Delta(1, g_1, \dots, g_{n-1})$ *is exactly of order $2n+2p-2$: and, when all the branch places occur for different values of x, its zeros are the branch places of the surface, each entering to its appropriate order.*

When the surface is branched at infinity, choose a value $x=a$ where all the sheets are separate: and let $g_i=(x-a)^{r_i+1} h_i$. Then by putting $\xi=(x-a)^{-1}$ we can similarly prove that $\Delta(1, h_1, \dots, h_{n-1})$ is an integral polynomial in ξ of precisely the order $2n+2p-2$. But it is immediately obvious that

$$\Delta(1, g_1, \dots, g_{n-1})=(x-a)^{2n+2p-2}\,\Delta(1, h_1, \dots, h_{n-1}).$$

Hence if the lowest power of ξ in $\Delta(1, h_1, \dots, h_{n-1})$ be ξ^s, $\Delta(1, g_1, \dots, g_{n-1})$ is an integral polynomial of order $2n+2p-2-s$. In this case the zeros of $\Delta(1, g_1, \dots, g_{n-1})$, which arise for finite values of x, are the branch places, each occurring to its appropriate order, provided all the branch places occur for different values of x: and $\Delta(1, h_1, \dots, h_{n-1})$ vanishes for $x=\infty$ to an order expressing the number of branch places there.

Ex. 1. For the surface $y^4=x^2(x-1)(x-a)$ there are two branch places at $x=0$, and a branch place at each of the places $x=1$, $x=a$, where all the sheets wind. Thus

$$2n+2p-2=w=2\,.\,1+3+3=8.$$

* Chap. II. § 21.

For this surface fundamental integral functions are given by $g_1=y$, $g_2=y^2/x$, $g_3=y^3/x$. With these values, prove that $\Delta(1, g_1, g_2, g_3)=-256x^2(x-1)^3(x-a)^3$, there being a factor x^2 corresponding to the superimposed branch places at $x=0$, while the other factors are of the same orders as the branch places corresponding to them.

Ex. 2. The surface $y^4=x^2(x-1)$ is similar to that in the last example, but there is a branch place at infinity at which the four sheets wind, so that, in the notation of this Article, $s=3$. As in the last example $2n+2p-2=8$, and $1, y, y^2/x, y^3/x$ are a fundamental system of integral functions. Prove that, now, $\Delta(1, g_1, g_2, g_3)$ is equal to $-256x^2(x-1)^3$, its order in x being $2n+2p-2-s=8-3=5$.

49. In accordance with the previous Chapter* the most general rational function having poles at $p+1$ independent places, is of the form $AF+B$, where F is a special function of this kind and A, B are arbitrary constants. The function will therefore become quite definite if we prescribe the coefficient of the infinite term at one of the $p+1$ poles—the so-called residue there—and also prescribe a zero of the function.

Limiting ourselves to the case where the $p+1$ poles are finite ordinary places of the surface, we proceed, now, to shew that the unique function thus determined can be completely expressed in terms of the functions introduced in this chapter. It will then be seen that we are in a position to express any rational function whatever.

If the general integral of the third kind here obtained with unassigned zero be denoted by $P^z_{x,a}$, the current variables being now (z, s), instead of (x, y), the infinities of the function being at x and a, the function

$$f'(s)\frac{dP^z_{x,a}}{dz}=\frac{\phi_0(z,s)+\phi_1(z,s)\,g_1(x,y)+\dots\dots+\phi_{n-1}(z,s)\,g_{n-1}(x,y)}{z-x}$$

$$-\frac{\phi_0(z,s)+\phi_1(z,s)\,g_1+\dots\dots+\phi_{n-1}(z,s)\,g_{n-1}}{z-a}$$

$$+\phi_1(z,s)(z,1)^{\tau_1-1}+\dots\dots+\phi_{n-1}(z,s)(z,1)^{\tau_{n-1}-1},$$

wherein $g_1, \dots, g_{n-1}$ are written for the values of the functions $g_1(z,s), \dots, g_{n-1}(z,s)$ at the place denoted by a, contains p disposeable coefficients, namely, those in the polynomials $(z,1)^{\tau_1-1}, \dots\dots, (z,1)^{\tau_{n-1}-1}$.

Let now $c_1, \dots\dots, c_p$ denote p finite, ordinary places of the surface, the values of z at these places being actually $c_1, \dots, c_p$, which are so situated that the determinant

$$\Delta=\begin{vmatrix}\phi_1^{(1)}, \phi_1^{(1)}c_1, \dots\dots, \phi_1^{(1)}c_1^{\tau_1-1}, \dots\dots, \phi^{(1)}_{n-1}, \phi^{(1)}_{n-1}c_1, \dots\dots, \phi^{(1)}_{n-1}c_1^{\tau_{n-1}-1}\\ \dots\dots\dots\dots\dots\dots\dots\dots\dots\dots\dots\dots\dots\dots\dots\dots\dots\dots\\ \phi_1^{(p)}, \phi_1^{(p)}c_p, \dots\dots, \phi_1^{(p)}c_p^{\tau_1-1}, \dots\dots, \phi^{(p)}_{n-1}, \phi^{(p)}_{n-1}c_p, \dots\dots, \phi^{(p)}_{n-1}c_p^{\tau_{n-1}-1}\end{vmatrix}$$

wherein $\phi_i^{(r)}$ is the value of $\phi_i(z,s)$ at the place c_r, does not vanish. That it is always possible to choose such p places is clear: for if $v_1, \dots\dots, v_p$ denote a

* Chap. III. § 37.

set of independent integrals of the first kind, the vanishing of Δ expresses the condition that a rational function of the form

$$f'(s)\left[\lambda_1 \frac{dv_1}{dz} + \dots\dots + \lambda_p \frac{dv_p}{dz}\right],$$

involving only $p-1$ disposeable ratios $\lambda_1 : \lambda_2 : \dots\dots : \lambda_p$, vanishes at each of the places $c_1, \dots\dots, c_p$.

Choose the p coefficients in the function $f'(s)\, dP/dz$, so that this function vanishes at $c_1, \dots\dots, c_p$: and denote the function dP/dz, with these coefficients, by $\psi(x, a; z, c_1, \dots\dots, c_p)$, so that $\Delta f'(s)\, \psi(x, a; z, c_1 \dots\dots c_p)$ is equal to the determinant

$$\left|\begin{matrix} [z, x]-[z, a], & \phi_1(z, s), & z\phi_1(z, s), & \dots, & z^{\tau_1-1}\phi_1(z, s), & \dots, & z^{\tau_{n-1}-1}\phi_{n-1}(z, s) \\ [c_1, x]-[c_1, a], & \phi_1^{(1)}, & c_1\phi_1^{(1)}, & \dots, & c_1^{\tau_1-1}\phi_1^{(1)}, & \dots, & c_1^{\tau_{n-1}-1}\phi_{n-1}^{(1)} \\ \dots & \dots & \dots & \dots & \dots & \dots & \dots \\ [c_p, x]-[c_p, a], & \phi_1^{(p)}, & c_p\phi_1^{(p)}, & \dots, & c_p^{\tau_1-1}\phi_1^{(p)}, & \dots, & c_p^{\tau_{n-1}-1}\phi_{n-1}^{(p)} \end{matrix}\right|,$$

where $[z, x]$ denotes the expression

$$\frac{\phi_0(z, s) + \phi_1(z, s)\, g_1(x, y) + \dots + \phi_{n-1}(z, s)\, g_1(x, y)}{z - x}.$$

Suppose now that (z, s) is a finite place, not a branch place, such that none of the minors of the elements of the first row of this determinant vanish. Consider $\psi(x, a; z, c_1, \dots\dots, c_p)$ *as a function of* (x, y). It is clearly a rational function; *and is in fact rationally expressed in terms of all the quantities involved.* It is infinite at each of the places $z, c_1, c_2, \dots\dots, c_p$— and in fact as x approaches z, the limit of $(z-x)\,\psi(x, a; z, c_1, \dots\dots, c_p)$ is the same as that of

$$\frac{\phi_0(z, s) + \Sigma\, \phi_r(z, s)\, g_r(x, y)}{f'(s)},$$

namely, unity (§ 44, F): so that at $x = z$, ψ is infinite like $-(x-z)^{-1}$. And at $c_1, \dots, c_p$ it is similarly seen to be infinite to the first order.

To obtain its behaviour when x is at infinity, we notice that, by the definition of the dimension of $g_i(x, y)$, the expression

$$\frac{g_i(x, y)}{z - x} + g_i(x, y)\left[\frac{1}{x} + \frac{z}{x^2} + \dots + \frac{z^{\tau_i - 1}}{x^{\tau_i}}\right],$$

which is of the form

$$-x^{-(\tau_i+1)}\, g_i(x, y)\left[z^{\tau_i} + \frac{z^{\tau_i+1}}{x} + \frac{z^{\tau_i+2}}{x^2} + \dots\right]$$

is finite for infinite values of x. If then we add to the first column of the determinant which expresses the value of $\Delta f'(s)\, \psi(x, a; z, c_1, \dots, c_p)$, the following multiples of the succeeding p columns

$$\frac{g_1(x, y)}{x^{\tau_1'}} - \frac{g_1(a, b)}{a^{\tau_1'}}, \quad \frac{g_2(x, y)}{x^{\tau_2'}} - \frac{g_2(a, b)}{a^{\tau_2'}}, \dots (\tau_1' = 1, 2, \dots, \tau_1;\ \tau_2' = 1, 2, \dots, \tau_2;\ \dots),$$

the determinant will contain only quantities which remain finite for infinite values of x.

On the whole then, as the reader can now immediately see, we can summarise the result as follows.

$\psi(x, a; z, c_1, \ldots\ldots, c_p)$ *is a rational function of* x, *having only* $p+1$ *poles, each of the first order, namely* $z, c_1, \ldots\ldots, c_p$. *It is infinite at* z *like* $-(x-z)^{-1}$ *and it vanishes at* $x=a$.

It is immediately seen that if a function of x of the form

$$\lambda_1 \frac{dv_1}{dx} + \ldots\ldots + \lambda_p \frac{dv_p}{dx},$$

which is so chosen that it is zero at all of $c_1, \ldots, c_p$ except c_i and is unity at c_i, be denoted by $\omega_i(x)$, then $\psi(x, a; z, c_1 \ldots c_p)$ is infinite at c_i like $\dfrac{\omega_i(z)}{x-c_i}$.

Let now $R(x, y)$ be a rational function of (x, y) with poles at the finite ordinary places $z_1, z_2, \ldots, z_Q$: let its manner of infinity at z_i be the same as that of $-\lambda_i(x-z_i)^{-1}$. Then the function

$$R(x, y) - \lambda_1 \psi(x, a; z_1, c_1, \ldots, c_p) - \ldots - \lambda_Q \psi(x, a; z_Q, c_1, \ldots, c_p)$$

is a rational function of (x, y) which is only infinite at $c_1, \ldots, c_p$. Since however these latter places are independent*, no such function exists—nor does there exist a rational function infinite only in places falling *among* $c_1, \ldots, c_p$. Hence the function just formed is a constant; thus

$$R(x, y) = \lambda_1 \psi(x, a; z_1, c_1, \ldots, c_p) + \ldots + \lambda_Q \psi(x, a; z_Q, c_1, \ldots, c_p) + \lambda.$$

Conversely an expression such as that on the right hand here will represent a rational function having $z_1, \ldots, z_Q$ for poles, for *all* values of the coefficients $\lambda_1, \ldots, \lambda_Q, \lambda$, which satisfy the conditions necessary that this expression be finite at each of $c_1, \ldots, c_p$; these conditions are expressed by the p equations

$$\lambda_1 \omega_i(z_1) + \lambda_2 \omega_i(z_2) + \ldots + \lambda_Q \omega_i(z_Q) = 0,$$

where $i = 1, 2, \ldots, p$.

When these conditions are independent the function contains therefore

$$Q - p + 1$$

arbitrary constants—in accordance with the result previously enunciated (Chapter III. § 37). The excess arising when these conditions are not independent is immediately seen to be also expressible in the same way as before.

We thus obtain the Riemann-Roch Theorem for the case under consideration.

The function $\psi(x, a; z, c_1, \ldots, c_p)$ will sometimes be called Weierstrass's function. The modification in the expression of it which is necessary when

* In the sense employed Chapter III. § 23.

some of its poles are branch points, will appear in a subsequent utilization of the function (Chapter VII.*). The modification necessary when some of these poles are at infinity is to be obtained, conformably with § 39 of the present chapter by means of the transformation $x=(\xi-m)^{-1}$, whereby the place $x=\infty$ becomes a finite place $\xi=m$.

50. The theory contained in this Chapter can be developed in a different order, on an algebraical basis.

Let the equation of the surface be put into such a form as

$$y^n + y^{n-1} a_1 + \ldots + y a_{n-1} + a_n = 0,$$

wherein $a_1, \ldots, a_n$ are *integral* polynomials in x: so that y is an integral function of x.

By algebraical methods only it can be shewn that a set of integral functions $g_1, \ldots, g_{n-1}$ exists having the property that every integral function can be expressed by them in a form

$$(x, 1)_\lambda + (x, 1)_{\lambda_1} g_1 + \ldots + (x, 1)_{\lambda_{n-1}} g_{n-1},$$

in such a way that no term occurs in the expression which is of higher dimension than the function to be expressed; and that the sum of the dimensions of $g_1, \ldots, g_{n-1}$ is not less than $n-1$ but is less than that of any other set $(1, h_1, \ldots, h_{n-1})$, in terms of which all integral functions can be expressed in such a form as

$$[(x, 1)_\lambda + (x, 1)_{\lambda_1} h_1 + \ldots + (x, 1)_{\lambda_{n-1}} h_{n-1}]/(x, 1)_m.$$

If the sum of the dimensions of $g_1, \ldots, g_{n-1}$ be then written in the form $p+n-1$, p is called the deficiency of the fundamental algebraic equation.

The expressions of the functions $g_1, g_2, \ldots, g_{n-1}$ being once obtained, and the forms $\phi_0, \phi_1, \ldots, \phi_{n-1}$ thence deduced as in this Chapter, the integrals of the first kind can be shewn, as in this Chapter or otherwise†, to have the form

$$\int \frac{dx}{f'(y)} [(x, 1)^{\tau'_1 - 1} \phi_1 + \ldots\ldots + (x, 1)^{\tau'_{n-1} - 1} \phi_{n-1}],$$

wherein $\tau'_1 \leqq \tau_1$, etc., $\tau_i + 1$ being the dimension of g_i. Thus the number of terms which enter is at most $\tau_1 + \ldots\ldots + \tau_{n-1}$ or p. But it can in fact be shewn algebraically that every one of these terms is an integral of the first kind, namely, that an integral of the form

$$\int \frac{dx}{f'(y)} x^r \phi_i \qquad (i = 1, 2, \ldots\ldots, n-1)$$

is everywhere finite‡ provided $0 \not> r \not> \tau_i - 1$.

* The reader may, with advantage, consult the early parts (e.g. §§ 122, 130) of that chapter at the present stage.

† Hensel, *Crelle*, 109.

‡ For this we may use the definition (G) or the definition (H) (§ 44). The reader may refer to Hensel, *Crelle*, 105, p. 336.

Then the forms of the integrals of the second and third kind will follow as in this Chapter: and an algebraic theory of the expression of rational functions of given poles can be built up on the lines indicated in the previous article (§ 49) of this Chapter. In this respect Chapter VII. may be regarded as a continuation of the present Chapter.

A method for realising the expressions of $g_1, \dots, g_{n-1}$ for a given form of fundamental equation is explained in Chapter V. (§ 73).

For Kronecker's determination of a fundamental set of integral functions, for which however the sum of the dimensions is not necessarily so small as $p+n-1$, the reader may refer to the account given in Harkness and Morley, *Theory of Functions*, p. 262. It is one of the points of interest of the system here adopted that the method of obtaining them furnishes an algebraic determination of the deficiency of the surface.

CHAPTER V.

On certain forms of the Fundamental Equation of the Riemann Surface.

51. We have already noticed that the Riemann surface can be expressed in many different ways, according to the rational functions used as variables. In the present chapter we deal with three cases: the first, the hyperelliptic case (§§ 51—59), is a special case, and is characterised by the existence of a rational function of the second order; the second, which we shall often describe as that of Weierstrass's canonical surface (§§ 60—68), is a general case obtained by choosing, as independent variables, two rational functions whose poles are at one place of the surface: the third case referred to (§§ 69—71) is also a general case, which may be regarded as a generalization of the second case. It will be seen that both the second and third cases involve ideas which are in close connexion with those of the previous chapter. The chapter concludes with an account of a method for obtaining the fundamental integral functions for any fundamental algebraic equation whatever (§§ 73—79).

It may be stated for the guidance of the reader that the results obtained for the second and third cases (§§ 60—71) are not a necessary preliminary to the theory of the remainder of the book; but they will be found to furnish useful examples of the actual application of the theory.

52. We have seen that when p is greater than zero, no rational function of the first order exists. We consider now the consequences of the hypothesis of the existence of a rational function of the second order. Let ξ denote such a function; let c be any constant and α, β denote the two places where $\xi = c$, so that $(\xi - c)^{-1}$ is a rational function of the second order with poles at α, β. The places α, β cannot coincide for all values of c, because the rational function $d\xi/dx$ has only a finite number of zeros. We may therefore regard α, β as distinct places, in general. The most general rational function which has simple poles at α, β cannot contain more than two linearly entering arbitrary constants. For if such a function be $\lambda + \lambda_1 f_1 + \lambda_2 f_2 + \ldots$, λ, λ_1, ... being arbitrary constants, each of the functions f_1, f_2, ... must be of the second order at most and therefore actually of the second order: by choosing the constants so that the sum of the residues at α is zero, we can therefore

obtain a function infinite only at β, which is impossible*. Thus the most general rational function having simple poles at α, β is of the form $A(\xi - c)^{-1} + B$. Therefore, from the Riemann-Roch Theorem (Chapter III., § 37), $Q - q = p - (\tau + 1)$, putting $Q = 2$, $q = 1$, we obtain $p - (\tau + 1) = 1$; namely, the number of linearly independent linear aggregates ($p > 1$)

$$\Omega(x) = \lambda_1 \Omega_1(x) + \ldots + \lambda_p \Omega_p(x),$$

which vanish in the two places α, β is $p - 1$. Since α may be taken arbitrarily and c determined from it, and $p - 1$ is the number of these linear aggregates which vanish in an arbitrary place, we have therefore the result—*When there exists a function of the second order, every place α of the surface determines another place β: and the determination may be expressed by the statement that every linearly independent linear aggregate $\Omega(x)$ which vanishes in one of these places vanishes necessarily in the other.*

53. Conversely when there are two places α, β in which $p - 1$ linearly independent $\Omega(x)$ aggregates vanish, there exists a rational function having these two places for simple poles. To see this we may employ the formula of § 37, putting $Q = 2$, $\tau + 1 = p - 1$, and obtaining $q = 1$. Or we may repeat the argument upon which that result is founded, thus—Not every one of $\Omega_1(x), \ldots, \Omega_p(x)$ can vanish at α; let $\Omega_1(\alpha)$ be other than zero. Since $p - 1$ linearly independent $\Omega(x)$ aggregates vanish in α, and, by hypothesis, $p - 1$ linearly independent $\Omega(x)$ aggregates vanish in both α and β, it follows that every $\Omega(x)$ aggregate which vanishes in α vanishes also in β. Hence each of the $p - 1$ aggregates

$$\Omega_2(\alpha)\,\Omega_1(x) - \Omega_1(\alpha)\,\Omega_2(x), \ldots\ldots, \Omega_p(\alpha)\,\Omega_1(x) - \Omega_1(\alpha)\,\Omega_p(x),$$

vanishes in β, namely, we have the $p - 1$ equations

$$\Omega_i(\alpha)\,\Omega_1(\beta) - \Omega_1(\alpha)\,\Omega_i(\beta) = 0, \quad (i = 2, 3, \ldots, p).$$

Therefore the function

$$\Omega_1(\beta)\,\Gamma_\alpha^x - \Omega_1(\alpha)\,\Gamma_\beta^x$$

has each of its periods zero. Thus it is a rational function whose poles are at α and β: and $\Omega_1(\beta)$ cannot be zero since otherwise the function would be of the first order.

Hence when there are two places at which $p - 1$ linearly independent $\Omega(x)$ aggregates vanish, there is an infinite number of pairs of places having the same character. For any pair of places the relation is reciprocal, namely, if the place α determine the place β, α is the place which is similarly determined by β: in other words, the surface has a reciprocal (1, 1) correspondence with itself. It can be shewn by such reasoning as is employed in

* By the equation $Q - q = p - (\tau + 1)$, if q were 2, $\tau + 1$ would be p, or all linear aggregates $\Omega(x)$ would vanish in the same places, which is impossible (Chap. II. § 21).

Chap. I. (p. 5), that if (x_1, y_1), (x_2, y_2) be the values of the fundamental variables of the surface at such a pair of places, each of x_1, y_1 is a rational function of x_2 and y_2, and that conversely x_2, y_2 are the same rational functions of x_1 and y_1.

54. We proceed to obtain other consequences of the existence of a rational function, ξ, of the second order. If the poles of ξ do not fall at finite distinct ordinary places of the surface, choose a function of the form $(\xi - c)^{-1}$, in accordance with the explanation given, for which the poles are so situated. Denote this function by z. Then* the function dz/dx has $2.2+2p-2=2p+2$ zeros at each of which z is finite. Denote their positions by $x_1, x_2, \dots, x_{2p+2}$. If these are not all finite places we may, if we wish, suppose that, instead of x, such a linear function of x is taken that each of $x_1, \dots, x_{2p+2}$ becomes a finite place. They are distinct places. For if the value of z at x_i be c_i, $z - c_i$ is there zero to the second order: that another place x_j should fall at x_i would mean that $z - c_i$ is there zero to higher than the second order, which is impossible because z is only of the second order. By the explanations previously given it follows that a linear aggregate $\Omega(x)$, which vanishes at any one of these places $x_1, \dots, x_{2p+2}$, vanishes to the second order there. Hence there is no linear aggregate $\Omega(x)$ vanishing at p or any greater number of these places, for $\Omega(x)$ has only $2p-2$ zeros. The general rational function which has infinities of the first order at the places $x_1, \dots, x_{p+r}$ will therefore† contain a number of $q+1$ of constants given by $p+r-q=p$, namely, will contain $r+1$ constants. Such a function will therefore not exist when $r=0$. In order to prove that a function actually infinite in the prescribed way does exist for all values of r greater than zero, it is sufficient, in accordance with §§ 23—27 (Chap. III.), to shew that there exists no rational function having $x_1, x_2, \dots, x_i$ for poles of the first order for any value of i less than $p+1$. Without stopping to prove this fact, which will appear *a posteriori*, we shall suppose r chosen so that a function of the prescribed character actually exists. For this it is certainly sufficient that r be as great as p‡. Denote the function by h, so that h has the form

$$h = \lambda + \lambda_1 \Sigma_1 + \dots + \lambda_r \Sigma_r,$$

$\lambda, \lambda_1, \dots, \lambda_r$ being arbitrary constants.

Let h, h' denote the values of h at the two places (x, y), (x', y'), where z has the same value. Then to each value of z corresponds one and only one value of $h+h'$, or $h+h'$ may be regarded as an uniform function of z: the infinities of $h+h'$ are clearly of finite order, so that $h+h'$ is a rational function of z. Consider now the function $(z-c_1)(z-c_2)\dots(z-c_{p+r})(h+h')$.

* Chap. I. § 6.

† Chap. III. § 37.

‡ Chap. III. § 27. For the need of the considerations here introduced compare § 37 of Chap. III.

Since h and h' are only infinite at places of the original surface at which z is equal to one or other of $c_1, \dots, c_{p+r}$, this function is only infinite for infinite values of z. As it is a rational function of z, it must therefore be a polynomial in z of order not greater than $p+r$. Hence we may write

$$h + h' = (z,\ 1)_{p+r}/(z - c_1) \dots (z - c_{p+r}).$$

But here the left hand is only infinite to the first order, at most, at any one of $c_1, \dots, c_{p+r}$—and the denominator of the right hand is zero to the second order at such a place. Hence the numerator of the right hand must be zero at each of these places, and must therefore be divisible by the denominator. Thus $h + h'$ is an absolute constant, $= 2C$ say. From the equations

$$h = \lambda + \lambda_1 \Sigma_1 + \dots + \lambda_r \Sigma_r,$$
$$h' = \lambda + \lambda_1 \Sigma'_1 + \dots + \lambda_r \Sigma'_r,$$

we infer then that $\Sigma_i + \Sigma'_i$ is also a constant, $= 2C_i$ say: for h was chosen to be the most general function of its assigned character and the coefficients $\lambda, \dots, \lambda_r$ are arbitrary. Thence we obtain

$$C = \lambda + \lambda_1 C_1 + \dots + \lambda_r C_r.$$

We can therefore put

$$s = h - C = -s' = -(h' - C) = \lambda_1 (\Sigma_1 - C_1) + \dots + \lambda_r (\Sigma_r - C_r),$$

so that s will be a function of the same general character as h, such however that $s + s' = 0$: in its expression the constants $\lambda_1, \dots, \lambda_r$ are arbitrary, while the constants $C_1, \dots, C_r$ depend on the choice made for the functions $\Sigma_1, \dots, \Sigma_r$.

55. Consider now the two places α, α' at which z is infinite. Choose the ratios $\lambda_1 : \lambda_2 : \dots : \lambda_r$ so that s is zero to the $(r-1)$th order at α. This can always be done, and will define s precisely save for a constant multiplier, unless it is the case that when s is made to vanish to the $(r-1)$th order at α, it vanishes, of itself, to a higher order. In order to provide for this possibility, let us assume that s vanishes to the $(r-1+k)$th order at α. Since $s' = -s$, s will also vanish to the $(r-1+k)$th order at α'. There will then be other $p + r - 2(r - 1 + k)$, or $p - r + 2 - k$, zeros of s. From the manner of formation this number is certainly not negative. Consider now the function

$$f = (z - c_1) \dots (z - c_{p+r})\, s^2.$$

At the places where z is infinite f is infinite of order $p + r - 2(r - 1 + k)$, or $p - r + 2 - 2k$ times. At the places, $x_1, \dots, x_{p+r}$ where s is infinite, it is finite; each of the factors $z - c_1, \dots, z - c_{p+r}$ is zero to the second order at the place where it vanishes. Since $s^2 = -ss'$, f is a symmetrical function of the values which s takes at the places where z has any prescribed value. Hence, by such reasoning as is previously employed, it follows that the function f is a rational integral polynomial in z of order $p-r+2-2k$. Denote this polynomial by H. By consideration of the zeros of f it follows that the $2(p-r+2-2k)$ zeros of the polynomial H are the zeros of s^2 which do not fall at α or α'. But since the sum of the values of s at the two places where z has any prescribed value is zero, it follows that s is zero at each of the places $x_{p+r+1}, \ldots, x_{2p+2}$. For each of these is formed by a coalescence of two places where z has the same value, and at each of them s is not infinite. Hence the polynomial H must be divisible by $(z-c_{p+r+1}) \ldots (z-c_{2p+2})$. Thus, as H is a polynomial of order $p-r+2-2k$ in z, $p-r+2-2k$ must be at least equal to $2p+2-(p+r)$ or to $p-r+2$. Hence k is zero, and the value of H is determinate save for a constant multiplier. Supposing this multiplier absorbed in s we may therefore write

$$(z-c_1) \ldots (z-c_{p+r}) s^2 = (z-c_{p+r+1}) \ldots (z-c_{2p+2}) \qquad \text{(A)};$$

and s is determined uniquely by the conditions, (1) of being once infinite at $x_1, \ldots, x_{p+r}$, (2) of being $(r-1)$ times zero at each of the places α, α' where z is infinite. Denote s, now, by s_{p+r}, and denote the function h from which we started, which was defined by the condition of being once infinite at each of $x_1, \ldots, x_{p+r}$, by h_{p+r}, and consider the function $(z-c_{p+r}) s_{p+r}$. *This* function is once infinite at each of $x_1, \ldots, x_{p+r-1}$, it is zero to the first order at x_{p+r}, and it is $r-1-1, =r-2$ times zero at each of the places α, α' where z is infinite. Hence the function

$$(z-c_{p+r}) s_{p+r} (A + A_1 z + \ldots + A_{r-2} z^{r-2}) + B,$$

wherein $B, A, A_1, \ldots, A_{r-2}$ are arbitrary constants, has the property of being once infinite at each of $x_1, \ldots, x_{p+r-1}$, and not elsewhere. It is then exactly such a function as would be denoted, in the notation suggested, by h_{p+r-1}, and it contains the appropriate number of arbitrary constants—and we can from it obtain a function s_{p+r-1}, having the property of being once infinite at each of $x_1, \ldots, x_{p+r-1}$ and vanishing $(r-2)$ times at each of the places α, α' where z is infinite.

Ex. 1. Determine s_{p+r-1} in accordance with this suggestion.

Ex. 2. Prove that h_{p+r} is of the form $s_{p+r}(A + A_1 z + \ldots + A_{r-1} z^{r-1}) + B$.

Ex. 3. Prove that h_{p+r+t} is of the form $\dfrac{s_{p+r}(A + A_1 z + \ldots + A_{r+t-1} z^{r+t-1})}{(z-c_{p+r+1}) \ldots (z-c_{p+r+t})} + B$.

Ex. 4. Shew that the square root $\sqrt{\dfrac{(z-c_{p+r+1}) \ldots (z-c_{2p+2})}{(z-c_1) \ldots (z-c_{p+r})}}$ can be interpreted as an one-valued function on the original surface.

56. The functions, z, s_{p+r} are defined as rational functions of the x, y of the original surface. Conversely x, y are rational functions of z, s_{p+r}. For* we have found a rational irreducible equation (A) connecting z and

* See Chap. I. § 4.

s_{p+r} wherein the highest power of s_{p+r} is the same as the order of z. *Hence this equation (A) gives rise to a new surface, of two sheets, with branch places at* $z = c_1, \ldots, c_{2p+2}$, *whereon the original surface is rationally and reversibly represented.*

It is therefore of interest to obtain the forms of the fundamental integral functions and the forms of the various Riemann integrals for this new surface. It is clear that the function

$$(z - c_1) \ldots (z - c_{p+r})\, s_{p+r}\, (z, 1)_{k-1},$$

where k is a positive integer, and $(z, 1)_{k-1}$ denotes any polynomial of order $k-1$, is infinite only at the places α, α' where z is infinite, and in fact to order $p + r - (r - 1) + k - 1, = p + k$: and that, therefore, by suitable choice of the coefficients in another polynomial $(z, 1)_{p+k}$, we can find a rational function

$$(z - c_1) \ldots (z - c_{p+r})\, s_{p+r}\, (z, 1)_{k-1} + (z, 1)_{p+k},$$

which is not infinite at α', and is infinite at α to any order, $p + k$, greater than p. Now, of rational functions which are infinite only at α, there are p orders for which the function does not exist*. Hence these must be the orders $1, 2, \ldots, p$.

Hence, of functions infinite only in one sheet at $z = \infty$, on the surface

$$(z - c_1) \ldots (z - c_{p+r})\, s^2_{p+r} = (z - c_{p+r+1}) \ldots (z - c_{2p+2}),$$

that of lowest order is a function of the form

$$\eta = (z - c_1) \ldots (z - c_{p+r})\, s_{p+r} + (z, 1)_{p+1},$$

which becomes infinite to the $(p + 1)$th order. Hence by Chapter IV. § 39, every rational function which becomes infinite only at the places $z = \infty$, can be expressed in the form

$$(z, 1)_\lambda + (z, 1)_\mu\, \eta,$$

and if the dimension of the function, namely, the number which is the order of its higher infinity at these places, be $\rho + 1$, λ and μ are such that

$$\rho + 1 \overline{\overline{>}} \lambda, \quad \rho + 1 \overline{\overline{>}} \mu + p + 1.$$

Therefore also, if $\sigma = (z - c_1) \ldots (z - c_{p+r})\, s_{p+r} = \eta - (z, 1)_{p+1}$, in which case equation (A) may be replaced by the equation

$$\sigma^2 = (z - c_1)(z - c_2) \ldots (z - c_{2p+2}),$$

we have the result that all such functions can be also expressed in the form

$$(z, 1)_{\lambda'} + (z, 1)_{\mu'}\, \sigma,$$

with

$$\rho + 1 \overline{\overline{>}} \lambda', \quad \rho + 1 \overline{\overline{>}} \mu' + p + 1.$$

* Chap. III. § 28.

By means of this result, hitherto assumed, the forms for the various integrals given Chapter II., § 17, Chapter IV., § 46, are immediately obtainable by the methods of Chapter IV.

57. Or we can obtain the forms of the integrals of the first kind thus— Let v be such an integral. Consider the rational function

$$s_{p+r}(z-c_1)\dots(z-c_{p+r})\frac{dv}{dz}.$$

It can only be infinite (1) where z is infinite (2) where $dz=0$, that is at the branch places of the (s_{p+r}, z) surface. It is immediately seen that the latter possibility does not arise. Where z is infinite the function is infinite to the order $p+1-2$, or $p-1$. Hence it is an integral polynomial in z of order $p-1$. Namely, the general integral of the first kind* is

$$\int\frac{(z,\ 1)_{p-1}\,dz}{(z-c_1)\dots(z-c_{p+r})\,s_{p+r}}.$$

58. *Ex.* 1. A rational function h_{p-k}, infinite only at the places where $z=c_1, \dots, c_{p-k}$, contains $p-k-p+\tau+1+1=\tau+2-k$ arbitrary constants, where $\tau+1$ is the number of coefficients in a general polynomial $(z,\ 1)_{p-1}$ which remain arbitrary after the prescription that $(z,\ 1)_{p-1}$ shall vanish at $c_1, \dots, c_{p-k}$. Prove this: and infer that $h_p, h_{p-1}, \dots$ do not exist.

Ex. 2. It can be shewn as in § 57 that at any ordinary place of the surface

$$\sigma^2=(z-c_1)\dots(z-c_{2p+2}),$$

rational functions exist, infinite only there, of orders $p+1, p+2, \dots$: the gaps indicated by Weierstrass's theorem (Chapter III. § 28) come therefore at the orders $1, 2, \dots, p$. At a branch place, say at $z=c$, the gaps occur for the orders $1, 3, 5, \dots, (2p-1)$. For, all other possible orders, which a rational function, infinite only there, can have, are expressible in one of the forms $2(p-k)$, $2p+2r+1$, $2p+2r$, where k is a positive integer less than p, or zero, and r is a positive integer: and we can immediately put down rational functions infinite to these orders at the branch place $z=c$ and nowhere else infinite. Prove in fact that the following functions have the respective characters

$$\frac{(z,\ 1)_{p-k}}{(z-c)^{p-k}},\quad \frac{(z,\ 1)_r\,\sigma+(z-c)\,(z,\ 1)_{p+r}}{(z-c)^{p+r+1}},\quad \frac{(z,\ 1)_{p+r}}{(z-c)^{p+r}},$$

wherein $(z,\ 1)_{p-k}$, $(z,\ 1)_r$, $(z,\ 1)_{p+r}$ are polynomials of the orders indicated by their suffixes with arbitrary coefficients.

Shew further that the most general $\Omega(x)$ aggregate which vanishes $2p-2k$ times at the branch place contains k arbitrary coefficients: and infer that the expressions given represent the most general functions of the prescribed character (see Chapter III. § 37).

Ex. 3. Prove for the surface

$$Ax^2+Bxy+Cy^2+Px^3+Qx^2y+Rxy^2+Sy^3+a_0x^4+a_1x^3y+a_2x^2y^2+a_3xy^3+a_4y^4=0$$

that the function

$$z=\mu+\lambda x/y,$$

* Cf. the forms quoted from Weierstrass. Forsyth, *Theory of Functions*, p. 456.

wherein λ and μ are arbitrary constants, is of the second order. And that there are six values of z for which the pairs of places at which z takes the same value, coincide, these places of coincidence being zeros of the function

$$2(Ax^2+Bxy+Cy^2)+Px^3+Qx^2y+Rxy^2+Sy^3.$$

Prove further that a rational function which is infinite at these six places is given by

$$h=\frac{2(Ax^2+Bxy+Cy^2)+P'x^3+Q'x^2y+R'xy^2+S'y^3}{2(Ax^2+Bxy+Cy^2)+Px^3+Qx^2y+Rxy^2+Sy^3}$$

for arbitrary values of the constants P', Q', R', S'.

This function is, therefore, such a function as has been here called h_{p+r}: and since there are six places at which dz is zero, p is equal to 2 and r equal to 4.

Prove that the sum of the values of h at the two places other than (0, 0) at which z has the same value is constant and equal to 2.

We may then proceed as in the text and obtain the transformed surface in the simple hyperelliptic form. But a simpler process in practice is to form the equation connecting z and h. Writing $k=h-1$ and $Z=x/y$, prove that

$$k^2\{(PZ^3+QZ^2+RZ+S)^2-4(AZ^2+BZ+C)(a_0Z^4+a_1Z^3+a_2Z^2+a_3Z+a_4)\}$$
$$=\{(P'-P)Z^3+(Q'-Q)Z^2+(R'-R)Z+(S'-S)\}^2.$$

Hence, if the coefficient of k^2 on the left be written $(Z, 1)_6$, and we write

$$Y=[(P'-P)Z^3+(Q'-Q)Z^2+(R'-R)Z+(S'-S)]/k$$
$$=[2(Ax^2+Bxy+Cy^2)+Px^3+Qx^2y+Rxy^2+Sy^3]/y^3,$$

we have

$$Y^2=(Z, 1)_6,$$

which is the equation of the transformed surface. And, as remarked in the text, the transformation is reversible; verify in fact that x, y are given by

$$x=2Z(AZ^2+BZ+C)/[Y-(PZ^3+QZ^2+RZ+S)],$$
$$y=2(AZ^2+BZ+C)/[Y-(PZ^3+QZ^2+RZ+S)].$$

Hence any theorem referred to one form of equation can be immediately transformed so as to refer to the other form.

59. The equation

$$\sigma^2=(z-c_1)(z-c_2)\dots(z-c_{2p+2})$$

by which, as we have shewn, any hyperelliptic surface can be represented, contains $2p+2$ constants, namely $c_1, c_2, \dots, c_{2p+2}$. If we write $z=(ax+b)/(x+c)$ we introduce three new disposable constants; by suitable choice of these the equation of the surface can be reduced to a form in which there are only $2p-1$ parametric constants. For instance if we put

$$(z-c_1)(c_3-c_2)/(z-c_2)(c_3-c_1)=x/(x-1)$$

and then, further,

$$s=A\sigma(z-c_3)^{-p-1},$$

where the constant A is given by

$$A=(c_3-c_1)^p(c_3-c_2)^p/(c_1-c_2)^{p+\frac{1}{2}}(c_3-c_4)^{\frac{1}{2}}(c_3-c_5)^{\frac{1}{2}}\dots(c_3-c_{2p+2})^{\frac{1}{2}},$$

the equation becomes

$$s^2 = x(x-1)(x-a_4)(x-a_5)\dots(x-a_{2p+2}),$$

wherein

$$a_r = (c_2 - c_3)(c_r - c_1)/(c_1 - c_2)(c_3 - c_r),$$

and the right-hand side of the equation is now a polynomial of order $2p+1$ only. Of its branch places three are now at $x=0$, $x=1$, $x=\infty$, and the values of x for the others are the parametric constants upon which the equation depends. It is quite clear that the transformation used gives s, x as rational function of σ, z. Thus

The hyperelliptic surface depends on $2p-1$ moduli only. Among the positions of the $3p-3$ branch places upon which a general surface depends (Chapter I. § 7), *there are, in this case, $3p-3-(2p-1)=p-2$ relations.*

Thus a surface for which $p=2$ is hyperelliptic in all cases. There are in fact $(p-1)p(p+1)=6$ places* for which we can construct a rational function of order 2 infinite only at the place.

A surface for which $p=1$ is also hyperelliptic—but it is more than this (Chapter I. § 8), being susceptible of a reversible transformation into itself in which an *arbitrary* parameter enters.

Ex. 1. On the surface of six sheets associated with the equation

$$y^6 = x(x-a)(x-b)^4$$

there are four branch places, one at (0, 0) where six sheets wind, and at $(a, 0)$ where six sheets wind, two at $(b, 0)$ at each of which three sheets wind. These count† in all as

$$w = 6-1+6-1+2(3-1) = 14.$$

Hence, by the formula

$$w = 2n + 2p - 2,$$

putting $n=6$, we obtain $p=2$.

Thus there exists a rational function ξ of the second order, and the surface can be reversibly transformed into the form $\eta^2 = (\xi, 1)_6$. In fact the function

$$\xi = \frac{x-b}{y}$$

is infinite to the first order at each of the branch places $(b, 0)$, $(a, 0)$ and is not elsewhere infinite.

To obtain the values of ξ at the branch places of the new surface, we may express either x or y in terms of ξ. Since there are two places at which ξ takes any value, each of x and y will be determined from ξ by a quadratic equation—which may reduce to a simple equation in particular cases. When ξ has a value such that the corresponding two places coincide, each of these quadratic equations will have a repeated root.

Now we have

$$\xi^6 = \frac{(x-b)^2}{x(x-a)} = \frac{y^2\xi^2}{(b+y\xi)(b-a+y\xi)}.$$

* Chap. III. § 31.

† Forsyth, *Theory of Functions*, p. 349.

Hence

$$y^2(\xi^6-1)-y\xi^5(a-2b)-b(a-b)\xi^4=0.$$

The condition then is

$$\xi^{10}(a-2b)^2+4b(a-b)\xi^4(\xi^6-1)=0, \text{ or } \xi^4[a^2(\xi^6-1)+(a-2b)^2]=0.$$

The factor

$$a^2(\xi^6-1)+(a-2b)^2,$$

is equal to

$$[a^2\{(x-b)^2-x(x-a)\}+(a-2b)^2x(x-a)]/x(x-a),$$

which is immediately seen to be the same as

$$[x(a-2b)+ab]/x(x-a)$$

or

$$\{[x(a-2b)+ab][x-b]^2/y^3\}^2.$$

Thus this factor gives rise to the six places at which $x=-ab/(a-2b)$. And if we put

$$\eta=[x(a-2b)+ab][x-b]^2/y^3,$$

we obtain

$$\eta^2=a^2(\xi^6-1)+(a-2b)^2,$$

which is then the equation associated with the transformed surface.

Then, from the equation

$$\eta\xi^{-3}=[x(a-2b)+ab]/[x-b],$$

we obtain

$$x=[b\eta+ab\xi^3]/[\eta-\xi^3(a-2b)],$$
$$y=[2b(a-b)\xi^2]/[\eta-\xi^3(a-2b)],$$

which give the reverse transformation.

Ex. 2. Prove for the surface

$$y^3=x(x-a)(x-b)^2(x-c)^2$$

that $p=2$ and that the function

$$\xi=(x-b)(x-c)/y$$

is of the second order. Prove further that

$$[a\xi^3-b-c]^2+4bc(\xi^3-1)=\{[a-b-c)x^2+2bcx-abc]/x(x-a)\}^2$$

Hence shew that the surface can be transformed to

$$\eta^2=[a\xi^3-b-c]^2+4bc(\xi^3-1)$$

and that

$$x=[a^2\xi^3+a\eta+2bc-ab-ac]/[a\xi^3+\eta+b+c-2a],$$
$$y=2\xi^2[bc+a^2-ab-ac][a^2\xi^3+a\eta+2bc-ab-ac]/[a\xi^3+\eta+b+c-2a]^2.$$

Ex. 3. In the following five cases shew that $p=2$, that ξ is a function of the second order, that in each case η^2 is either a quintic or a sextic polynomial in ξ, and obtain each of x and y as rational functions of ξ and η;

(α) $y^{10}=x(x-a)^4(x-b)^5$, $\quad\xi=(x-a)(x-b)/y^2$, $\quad\eta=\sqrt{a}\,.\,(x-a)^2(x-b)^3$

(β) $y^8=x(x-a)^3(x-b)^4$, $\quad\xi=(x-a)(x-b)/y^2$, $\quad\eta=\sqrt{a}\,.\,(x-a)^2(x-b)^3/y^5$

(γ) $y^5=x(x-a)(x-b)^3$, $\quad\xi=(x-b)/y$, $\quad\eta=[x(a-2b)+ab][x-b]^2/y^3$

(δ) $y^6=x^2(x-a)^3(x-b)^3(x-c)^4$, $\quad\xi=x(x-a)(x-b)(x-c)/y^2$, $\quad\eta=cx(x-a)^2(x-b)^2(x-c)/y^3$

(ϵ) $y^4=x(x-a)^2(x-b)^2(x-c)^3$, $\quad\xi=(x-a)(x-b)(x-c)^2/y^2$, $\quad\eta=c(x-a)(x-b)(x-c)/xy$.

Ex. 4. Shew that the surface

$$y^n=(x-a_1)^{n_1}\dots(x-a_r)^{n_r}$$

can always be transformed to such form that $n_1, \dots, n_r$ are positive integers whose sum is divisible by n: and in that form determine the deficiency of the surface. Shew also that, in that form, the only cases in which the deficiency is 2 are those given in Exs. 1, 2, 3. Prove that the cases in which $p=1$ are*

$$y^6=x(x-a)^2(x-b)^3,\quad y^3=x(x-a)(x-b),$$
$$y^4=x(x-a)(x-b)^2,\quad y^2=x(x-a)(x-b)(x-c).$$

The results here given have been derived, with alterations, from the dissertation, E. Netto, *De Transformatione Aequationis* $y^n=R(x)$ (Berlin, 1870, G. Schade).

The equation

$$y^n=(x-a_1)^{n_1}\dots(x-a_r)^{n_r}$$

is considered by Abel, *Œuvres Complètes* (Christiania, 1881), vol. I., pp. 188, etc.

It is to be noticed that in virtue of Chapter IV. we are now in a position, immediately to put down the fundamental integrals for the surfaces considered in Examples 1, 2, 3.

60. Passing from the hyperelliptic case we resume now the consideration of the circumstances considered in Chapter III. §§ 28, 31—36.

Consider any place, c, of a Riemann surface: and consider rational functions which are infinite only at this place: all such functions will be denoted by symbols of the form g_N, the suffix N denoting the order of infinity of the function at the place.

Let g_a be the function of the lowest existing order. The suffixes of all other existing functions g_N can be written in the form $N=\mu a+i$, where $i<a$. Since there are only p orders for which functions of the prescribed character do not exist, all the values $i=0, 1, \dots, (a-1)$ will arise. Let $\mu_i a+i$ be the suffix of the function of lowest order whose order is congruent to i for modulus a. We obtain thus a functions

$$g_a,\ g_{\mu_1 a+1},\ g_{\mu_2 a+2},\ \dots,\ g_{\mu_{a-1}a+a-1}.$$

Then, if g_{ma+i} be any other function that occurs, m cannot be less than μ_i, and a constant λ can be chosen so that $g_{ma+i}-\lambda g_a^{m-\mu_i} g_{\mu_i a+i}$, which is clearly a rational function infinite only at c, is not infinite to the order $\mu_i a+i$. Thus we have an equation of the form

$$g_{ma+i}=\lambda g_a^{m-\mu_i} g_{\mu_i a+i}+g_{\mu a+j},$$

wherein $\mu a+j$ is less than $ma+i$. Proceeding then similarly with $g_{\mu a+j}$, we clearly reach an equation of the form

$$g_{ma+i}=A+Bg_{\mu_1 a+1}+Cg_{\mu_2 a+2}+\dots+Kg_{\mu_{a-1}a+a-1} \qquad \text{(i)}$$

wherein the coefficients $A, B, \dots, K$, whose number is a, are rational integral polynomials in g_a.

* Cf. Forsyth, p. 486. Briot and Bouquet, *Théorie des Fonct. Ellipt.* (Paris, 1875), p. 390.

In particular, if g_r be any rational function whatever of the g_N functions, we have equations

$$\begin{aligned} g_r &= A_1 + B_1 g_{\mu_1 a+1} + \dots\dots + K_1 g_{\mu_{a-1} a+a-1} \\ g_r^2 &= A_2 + B_2 g_{\mu_1 a+1} + \dots\dots + K_2 g_{\mu_{a-1} a+a-1} \qquad \text{(ii)}. \\ &\dots\dots\dots\dots\dots\dots\dots\dots \\ g_r^{a-1} &= A_{a-1} + B_{a-1} g_{\mu_1 a+1} + \dots\dots + K_{a-1} g_{\mu_{a-1} a+a-1}. \end{aligned}$$

61. If these equations, regarded as equations for obtaining $g_{\mu_1 a+1}, \dots, g_{\mu_{a-1} a+a-1}$ in terms of g_a and g_r, be linearly independent, we can obtain, by solving, such results as

$$g_{\mu_i a+i} = Q_{i,1}(g_r - A_1) + Q_{i,2}(g_r^2 - A_2) + \dots + Q_{i,a-1}(g_r^{a-1} - A_{a-1}),$$

wherein $Q_{i,1}, \dots, Q_{i,a-1}$ are rational functions of g_a, which are not necessarily of integral form.

If however the equations be not linearly independent, there exist equations of the form

$$P_1(g_r - A_1) + P_2(g_r^2 - A_2) + \dots + P_{a-1}(g_r^{a-1} - A_{a-1}) = 0$$

or say

$$P_{a-1} g_r^{a-1} + P_{a-2} g_r^{a-2} + \dots + P_1 g_r + P = 0 \qquad \text{(iii)},$$

wherein $P_1, P_2, \dots, P_{a-1}, P$ are integral rational polynomials in g_a. Denote the orders of these in g_a by $\lambda_1, \lambda_2, \dots, \lambda_{a-1}, \lambda$ respectively; here P denotes the expression

$$P_1 A_1 + P_2 A_2 + \dots + P_{a-1} A_{a-1}.$$

Then $P_k g^k$ is of order $a\lambda_k + rk$ at the place c of the surface. In order that such an equation as (iii) may exist, the terms of highest infinity at the place c must destroy one another: hence there must be such an equation as

$$a\lambda_k + rk = a\lambda_{k'} + rk',$$

and therefore

$$r/a = (\lambda_{k'} - \lambda_k)/(k - k').$$

Now k and k' are both less than a: this equation requires therefore that r and a have a common divisor.

62. Take now r *prime to* a; then it follows that the equations (ii) must be linearly independent. And in that case each of $g_{\mu_1 a+1}, \dots, g_{\mu_{a-1} a+a-1}$ can be expressed rationally in terms of g_a and g_r, the expression being integral in g_r but not necessarily so in g_a.

Also by equation (i) it follows that every function infinite only at c is rationally expressible by g_a and g_r: and in particular that there is an equation of the form

$$L g_r^a + L_1 g_r^{a-1} + \dots + L_{a-1} g_r + L_a = 0 \qquad \text{(iv)},$$

wherein $L, L_1, \ldots, L_a$ are integral rational polynomials in g_a, of which however, since g_r is only infinite when g_a is infinite, L is an absolute constant. It follows from the reasoning given that the equation (iv) is irreducible, and therefore belongs to a new Riemann surface, wherein g_a and g_r are independent and dependent variables. Further, any rational function whatever on the original surface can be modified into a rational function which is infinite only at the place c, by multiplication by an integral polynomial in g_a of the form $(g_a - E_1)^{r_1}(g_a - E_2)^{r_2}\ldots\ldots$ Hence any rational function on the surface is expressible rationally by g_a and g_r. Hence the surface represented by (iv) is a surface upon which the original surface can be rationally and reversibly represented.

Since g_a^{-1} is zero to order a at the place where g_a is infinite, it is clear that the new surface is one for which there is a branch place at infinity at which all the sheets wind.

To every value of g_r there belong r places of the old surface, at which g_r takes this value, and therefore also, in general*, r values of g_a. Hence the highest power of g_a in equation (iv) is the rth, and this term does actually enter. While, because g_a only becomes infinite when g_r is infinite, the coefficient of the term g_a^r is a constant (and not an integral polynomial in g_r).

The equation (iv) is the generalization of that which is used in introducing what are called Weierstrass's elliptic functions, namely of the equation

$$g_3{}^2 - (4g_2{}^3 - a_2 g_2 - a_3) = 0.$$

This equation is satisfied by writing $g_2 = \wp(u)$, $g_3 = \wp'(u)$: it is a known fact that the poles of $\wp(u)$ are at one place (where $u = 0$). This is not true of the Jacobian function sn u.

63. It follows from equation (i) that the functions

$$(1, g_{\mu_1 a+1}, \ldots, g_{\mu_{a-1} a+a-1})$$

form a fundamental set for the expression of rational functions infinite only at the place c of the surface, that is, a fundamental set for the expression of the integral rational functions of the surface (iv). And, defining the dimension D of such an integral function F as the lowest positive integer such that $g_a^{-D}F$ is finite at infinity on the surface (iv), in accordance with Chap. IV., § 39, it is clear that in the expression of an integral function by this fundamental system there arise no terms of higher dimension than the function to be expressed: this fundamental set is therefore entirely such an one as that used in Chapter IV. If k be the order of infinity of an integral function F, at the *single* infinite place of the surface (iv), it is obvious that the dimension of F is the least integer equal to or greater than $\frac{k}{a}$.

* That is, for an infinite number of values of g_r.

64. We shall generally call the equation (iv) Weierstrass's canonical form; a certain interest attaches to the tabulation of the possible forms which the equation can have for different values of the deficiency p. It will be sufficient here to obtain these forms for some of the lowest values of p; it will be seen that the method is an interesting application of Weierstrass's gap theorem.

Take the case $p=4$, and consider rational functions which are only infinite at a single place c of a surface which is of deficiency 4. Such functions do not exist of all orders—there are four orders for which such functions do not exist; these four orders may be 1, 2, 3, 4, and this is the commonest case*, or they may fall otherwise. We desire to specify all the possibilities: their number is limited by the considerations—

(i) If functions of orders $k_1, k_2, \ldots$ exist, say $F_1, F_2, \ldots$, then there exists a function of order $n_1k_1 + n_2k_2 + \ldots$, where $n_1, n_2, \ldots$ are any positive integers. In fact $F_1^{n_1} F_2^{n_2} \ldots$ is such a function.

(ii) The number of non-existent functions must be 4.

(iii) The highest order of non-existent function cannot be† greater than $2p-1$ or 7.

It follows that a function of order 1 does not exist, and if a function of order 2 exists then a function of order 3 does not exist; for every positive integer can be written as a sum of integral multiples of 2 and 3.

Consider then first the case when a function of order 2 exists. Write down all positive integers up to $2p$ or 8. Draw‡ a bar at the top of the numbers 2, 4, 6, 8 to indicate that all functions of these orders exist—

$$1 \ \bar{2} \ 3 \ \bar{4} \ 5 \ \bar{6} \ 7 \ \bar{8} \qquad (\alpha).$$

If then the functions of orders 5 or 7 existed there would need to be a gap beyond 8, which is contrary to the consideration (iii) above. Hence the non-existent orders are 1, 3, 5, 7. We have thus a verification of the results obtained earlier in this chapter (§ 58, Ex. 2).

Consider next the possibility that a function of order 3 exists, there being no function of order 2. If then a function of order 4 exists, the symbol will be

$$1 \ 2 \ \bar{3} \ \bar{4} \ 5 \ \bar{6} \ \bar{7} \ \bar{8},$$

a function of order 6 being formed by the square of the function of order 3, that of order 7 by the product of the functions of orders 3 and 4, and the function of order 8 by the square of the function of order 4. Thus there would need to be a gap beyond 8. Hence when a function of order 3 exists

* Chap. III. 31.

† Chap. III. § 34. Also Chap. III. § 27.

‡ Cf. Chap. III. § 26.

there cannot be one of order 4. If however functions of orders 3 and 5 exist the symbol would be

$$1\ 2\ \bar{3}\ 4\ \bar{5}\ \bar{6}\ 7\ \bar{8} \qquad (\beta),$$

the function of order 8 being formed by the product of the functions of orders 3 and 5. So far then as our conditions are concerned this symbol represents a possibility. Another is represented by the symbol

$$1\ 2\ \bar{3}\ 4\ 5\ \overline{6\ 7\ 8} \qquad (\gamma).$$

In this case however the existent integral function of order 8 is not expressible as an integral polynomial in the existent functions of orders 3 and 7.

When a function of order 3 exists there are no other possibilities; otherwise more than 4 gaps would arise.

Consider next the possibility that the lowest order of existent function is 4. Then possibilities are expressed by

$$1\ 2\ 3\ \overline{4\ 5\ 6}\ 7\ \bar{8} \qquad (\delta),$$

$$1\ 2\ 3\ \overline{4\ 5}\ 6\ \overline{7\ 8} \qquad (\epsilon),$$

$$1\ 2\ 3\ \bar{4}\ 5\ \overline{6\ 7\ 8} \qquad (\zeta),$$

as is to be seen just as before.

Finally, there is the ordinary case when no function of order less than 5 exists, given by

$$1\ 2\ 3\ 4\ \overline{5\ 6\ 7\ 8} \qquad (\eta).$$

For these various cases let a denote the lowest order of existent function and r the lowest next existent order prime to a. Then the results can be summarised in the table

$p=4$	a	r	Gaps at orders	Fundamental system of orders	Dimensions of functions of fundamental system	Sum of these dimensions	$p+a-1$	$\frac{1}{2}(a-1)(r-1)-p$
α	2	9	1, 3, 5, 7	0, 9	0, 5	5	5	0
β	3	5	1, 2, 4, 7	0, 5, 10	0, 2, 4	6	6	0
γ	3	7	1, 2, 4, 5	0, 7, 8	0, 3, 3	6	6	2
δ	4	5	1, 2, 3, 7	0, 5, 6, 11	0, 2, 2, 3	7	7	2
ϵ	4	5	1, 2, 3, 6	0, 5, 7, 10	0, 2, 2, 3	7	7	2
ζ	4	7	1, 2, 3, 5	0, 6, 7, 9	0, 2, 2, 3	7	7	5
η	5	6	1, 2, 3, 4	0, 6, 7, 8, 9	0, 2, 2, 2	8	8	6

That the seventh and eighth columns of this table should agree is in accordance with Chapter IV., § 41. The significance of the last column is explained in § 68 of this Chapter.

Similar tables can easily be constructed in the same way for the cases $p=1, 2, 3$.

Ex. 1. Prove that for $p=3$ the results are given by

$p=3$	a	r	Gaps at orders	Fundamental system of orders	Dimensions of functions of fundamental system	Sum of these dimensions	$p+a-1$
α	2	7	1, 3, 5	0, 7	0, 4	4	4
β	3	4	1, 2, 5	0, 4, 8	0, 2, 3	5	5
γ	3	5	1, 2, 4	0, 5, 7	0, 2, 3	5	5
δ	4	5	1, 2, 3	0, 5, 6, 7	0, 2, 2, 2	6	6

Ex. 2. Prove that for $p=5$, 6, 7, 8, the possible cases in which the lowest existing function is of the third order are those denoted by the symbols

$$p=5\begin{cases}1\ 2\ \overline{3}\ 4\ 5\ \overline{6}\ 7\ \overline{8}\ \overline{9}\ \overline{10}\\ 1\ 2\ \overline{3}\ 4\ 5\ \overline{6}\ \overline{7}\ 8\ \overline{9}\ \overline{10}\end{cases}$$

$$p=6\begin{cases}1\ 2\ \overline{3}\ 4\ 5\ \overline{6}\ 7\ 8\ \overline{9}\ \overline{10}\ \overline{11}\ \overline{12}\\ 1\ 2\ \overline{3}\ 4\ 5\ \overline{6}\ 7\ \overline{8}\ \overline{9}\ 10\ \overline{11}\ \overline{12}\end{cases}$$

$$p=7\begin{cases}1\ 2\ \overline{3}\ 4\ 5\ \overline{6}\ 7\ 8\ \overline{9}\ 10\ \overline{11}\ \overline{12}\ \overline{13}\ \overline{14}\\ 1\ 2\ \overline{3}\ 4\ 5\ \overline{6}\ 7\ 8\ \overline{9}\ \overline{10}\ 11\ \overline{12}\ \overline{13}\ \overline{14}\\ 1\ 2\ \overline{3}\ 4\ 5\ \overline{6}\ 7\ \overline{8}\ \overline{9}\ 10\ \overline{11}\ \overline{12}\ 13\ \overline{14}\end{cases}$$

$$p=8\begin{cases}1\ 2\ \overline{3}\ 4\ 5\ \overline{6}\ 7\ 8\ \overline{9}\ 10\ 11\ \overline{12}\ \overline{13}\ \overline{14}\ \overline{15}\ \overline{16}\\ 1\ 2\ \overline{3}\ 4\ 5\ \overline{6}\ 7\ 8\ \overline{9}\ 10\ \overline{11}\ \overline{12}\ 13\ \overline{14}\ \overline{15}\ \overline{16}\\ 1\ 2\ \overline{3}\ 4\ 5\ \overline{6}\ 7\ 8\ \overline{9}\ \overline{10}\ 11\ \overline{12}\ \overline{13}\ 14\ \overline{15}\ \overline{16}\end{cases}$$

65. We have already stated (Chap. IV. § 38) that when the fundamental set of integral functions are so far given that we know the relations expressing their products in terms of themselves, the form of an equation to represent the surface can be deduced. We give now two examples of how this may be done: these examples will be sufficient to explain the general method.

Take first the case $p=4$, $a=3$, $r=7$. Denote the corresponding functions by g_3, g_7. In accordance with § 60 preceding, all integral functions can be expressed by means of g_3 and two functions g_7, g_8 whose orders are respectively $\equiv 1$ and 2 for modulus 3: in particular there are equations of the form

$$g_7^2 = g_8 (g_3, 1)_2 + g_7 (g_3, 1)_2 + (g_3, 1)_4$$

$$g_7 g_8 = g_8 (g_3, 1)_2 + g_7 (g_3, 1)_2 + (g_3, 1)_5$$

$$g_8^2 = g_8 (g_3, 1)_2 + g_7 (g_3, 1)_3 + (g_3, 1)_5$$

wherein $(g_3, 1)_2$ denotes an integral polynomial in g_3 of order 2 at most, the upper limit for the suffix being determined by the condition that no terms shall occur on the right of higher dimension than those on the left. Similarly for the other polynomials occurring here on the right.

Instead of g_7, g_8 we may clearly use any functions $g_7 - (g_3, 1)_2$, $g_8 - (g_3, 1)_2$. Choosing these polynomials to be those occurring on the right in the value of g_7g_8, we may write our equations

$$g_7^2 = \alpha_2 g_8 + \beta_2 g_7 + \alpha_4, \quad g_8^2 = \gamma_2 g_8 + \alpha_3 g_7 + \alpha_5, \quad g_7 g_8 = \beta_5 \qquad \text{(A)},$$

where the Greek letters denote polynomials in g_3 of the orders given by their suffixes.

Multiplying the first and last equations by g_8 and g_7 respectively, and subtracting, we obtain

$$\begin{aligned} g_7\beta_5 &= g_8(\alpha_2 g_8 + \beta_2 g_7 + \alpha_4) \\ &= \alpha_2(\gamma_2 g_8 + \alpha_3 g_7 + \alpha_5) + \beta_2\beta_5 + g_8\alpha_4, \end{aligned}$$

and thence, since* $1, g_7, g_8$ cannot be connected by an integral equation of such form,

$$\alpha_2\gamma_2 + \alpha_4 = 0, \quad \alpha_2\alpha_3 - \beta_5 = 0, \quad \alpha_2\alpha_5 + \beta_2\beta_5 = 0,$$

from which, as α_2 is not identically zero,—for then g_7 would satisfy a quadratic equation with rational functions of g_3 as coefficients—we infer

$$\alpha_5 + \beta_2\alpha_3 = 0 \qquad \text{(B)}.$$

Similarly from the last two equations (A) we have

$$\begin{aligned} g_8\beta_5 &= g_7(\gamma_2 g_8 + \alpha_3 g_7 + \alpha_5) \\ &= \gamma_2\beta_5 + \alpha_3(\alpha_2 g_8 + \beta_2 g_7 + \alpha_4) + \alpha_5 g_7, \end{aligned}$$

and thence

$$\beta_5 - \alpha_2\alpha_3 = 0, \quad \alpha_3\beta_2 + \alpha_5 = 0, \quad \gamma_2\beta_5 + \alpha_3\alpha_4 = 0,$$

so that, since α_3 cannot be zero—as follows from the second of equations (A)—we have

$$\gamma_2\alpha_2 + \alpha_4 = 0 \qquad \text{(C)}.$$

The equations (B) and (C) have been formed by the condition that the equations (A) should lead to the same values for $g_7^2 g_8$ and $g_8^2 g_7$, however these latter products be formed from equations (A). We desire to shew that, conversely, these equations (B) and (C) are sufficient to ensure that any integral polynomial in g_7 and g_8 should have an unique value however it be formed from the equations (A). Now any product of powers of g_7 and g_8 is of one of the three forms g_7, g_8^μ, $g_7 g_8 K$. In the first two cases it can be formed from equations (A) in one way only. In the third case let us suppose it proved that K has an unique value however it be derived from the equations (A);

* Chap. IV. § 43.

then to prove that $g_7g_8 K$ has an unique value we require only to prove that $g_7 . g_8 K = g_8 . g_7 K$. Let K be written in the form $g_8L + g_7M + N$. Then the condition is that $g_7(Lg_8^2 + Mg_7g_8 + Ng_8)$ shall be equal to $g_8(Lg_7g_8 + Mg_7^2 + Ng_7)$. This requires only $g_7 . g_8^2 = g_8 . g_7g_8$ and $g_7 . g_7g_8 = g_8 . g_7^2$: and it is by these conditions that we have derived equations (B) and (C). Hence also $g_7g_8 K$ has an unique value.

Thus every rational integral polynomial in g_7 and g_8 will, when the conditions (B), (C) are satisfied, have an unique value however it be formed from equations (A).

The equations (B) and (C) are equivalent to $\alpha_4 = -\alpha_2\gamma_2$, $\beta_5 = \alpha_2\alpha_3$, $\alpha_5 = -\alpha_3\beta_2$, and lead to

$$g_7^2 = \alpha_2 g_8 + \beta_2 g_7 - \alpha_2\gamma_2, \quad g_8^2 = \gamma_2 g_8 + \alpha_3 g_7 - \alpha_3\beta_2, \quad g_7g_8 = \alpha_2\alpha_3.$$

Thence

$$g_7^2 = \alpha_2 \frac{\alpha_2\alpha_3}{g_7} + \beta_2 g_7 - \alpha_2\gamma_2,$$

or

$$g_7^3 - \beta_2 g_7^2 + \alpha_2\gamma_2 g_7 - \alpha_2^2\alpha_3 = 0,$$

which is the form of equation (iv) which belongs to the possibility under consideration.

The expression of the fundamental set of integral functions 1, g_7, g_8 in terms of g_3 and g_7 is therefore

$$1, \; g_7, \; \frac{g_7^2 - \beta_2 g_7 + a_2\gamma_2}{a_2}.$$

66. Take as another example the possibility ϵ, § 64 above, where $a = 4$, $r = 5$, the orders of non-existent functions being 1, 2, 3, 6. For a fundamental system of integral functions we may take 1, g_5, g_5^2, g_7.

We have then such an equation as

$$g_5g_7 = g_7(g_4, \; 1)_1 + cg_5^2 + g_5(g_4, \; 1)_1 + (g_4, \; 1)_3$$

where c is a constant: let this be written in the form

$$g_5g_7 = \alpha_1 g_7 + g_5^2 + \beta_1 g_5 + \alpha_3,$$

the constant c being supposed absorbed in g_5^2.

Write h_5 for $g_5 - \alpha_1$ and h_7 for $g_7 - h_5 - \beta_1 - 2\alpha_1$.

Then

$$h_5h_7 = \alpha_1^2 + \alpha_1\beta_1 + \alpha_3.$$

Replacing now h_5, h_7 by the notation g_5, g_7 and $\alpha_3 + \alpha_1\beta_1 + \alpha_1^2$ by α_3 we may write

$$g_5g_7 = \alpha_3, \quad g_7^2 = \beta_3 + \alpha_2 g_5 + \alpha_1 g_5^2 + \beta_1 g_7, \quad g_5^3 = \gamma_3 + \beta_2 g_5 + \gamma_1 g_5^2 + \gamma_2 g_7.$$

Hence the condition $g_5 \cdot g_7^2 = g_5 g_7 \cdot g_7$ requires

$$\alpha_3 g_7 = \beta_3 g_5 + \alpha_2 g_5^2 + \alpha_1 [\gamma_3 + \beta_2 g_5 + \gamma_1 g_5^2 + \gamma_2 g_7] + \beta_1 \alpha_3,$$

from which

$$\alpha_3 = \alpha_1 \gamma_2, \quad \beta_3 + \alpha_1 \beta_2 = 0, \quad \alpha_2 + \alpha_1 \gamma_1 = 0, \quad \alpha_1 \gamma_3 = -\beta_1 \alpha_3,$$

and thence

$$\alpha_1 \gamma_3 = -\beta_1 \alpha_1 \gamma_2, \text{ or if } \alpha_1 \text{ is not zero, } \gamma_3 = -\beta_1 \gamma_2.$$

Substituting this value for γ_3 and the value $g_7 = \alpha_3/g_5 = \alpha_1\gamma_2/g_5$ in the expression for g_5^3 we obtain

$$g_5^3 = -\beta_1 \gamma_2 + \beta_2 g_5 + \gamma_1 g_5^2 + \alpha_1 \gamma_2^2/g_5$$

or

$$g_5^4 - \gamma_1 g_5^3 - \beta_2 g_5^2 + \beta_1 \gamma_2 g_5 - \alpha_1 \gamma_2^2 = 0,$$

which is then a form of the equation (iv) corresponding to the possibility (ϵ).

In this case the fundamental integral functions may be taken to be

$$1, \; g_5, \; g_5^2, \; (g_5^3 - \gamma_1 g_5^2 - \beta_2 g_5 + \beta_1 \gamma_2)/\gamma_2.$$

It is true in general, as in these examples, that the terms of highest order of infinity in the equation (iv) are the terms $\overset{r}{g_a}, \overset{a}{g_r}$. For there must be two terms (at least) of the highest order of infinity which occurs; and since r is prime to a, two such terms as $\overset{\lambda}{g_a}\overset{\mu}{g_r}, \overset{\lambda'}{g_a}\overset{\mu'}{g_r}$ cannot be of the same order of infinity.

Ex. 1. Prove that for $p=3$ the form of the equation of the surface in the case where $a=3$, $r=4$ is

$$g_4^3 + g_4^2 (g_3, 1)_1 + g_4 (g_3, 1)_3 + (g_3, 1)_4 = 0,$$

and shew that this is reducible to the form

$$y^3 + yx(x+a) + x^4 + a_1 x^3 + a_2 x^2 + a_3 x + a_4 = 0,$$

x being of the form $Ag_3 + B$, y of the form $Cg_4 + Dg_3 + E$, A, B, C, D, E being constants.

Thus the surface depends on $3p-4$ or 5 constants, at most.

Ex. 2. The reader who is acquainted with the theory of plane curves may prove that the homogeneous equation of a quartic curve which has a point of osculation, can be put into the form

$$\omega^3 \xi + \omega \xi \eta (\xi, \eta)_1 + (\xi, \eta)_4 = 0.$$

By putting $x = \eta/\xi$, $y = \omega/\xi$, this takes the form of the final equation of Example 1. Compare Chapter III. § 32.

Ex. 3. Prove that for $p=3$, the form of the equation of the surface in the case where $a=3$, $r=5$ is

$$g_5^3 + g_5^2 (g_3, 1)_1 + g_5 g_3 (g_3, 1)_2 + g_3^2 (g_3, 1)_3 = 0.$$

Ex. 4. Denoting the left hand of equation (iv) by $f(g_r, g_a)$, $\partial f/\partial g_r$ by $f'(g_r)$ and the operator

$$f'(g_r) \frac{d}{dg_a}$$

by D, prove that if g_m be any rational function which is infinite only where g_a and g_r are infinite, there exists an equation

$$X_0 D^{a-1} g_m + X_1 D^{a-2} g_m + \ldots\ldots + X_{a-1} g_m = 0,$$

where $X_0, \ldots\ldots, X_{a-1}$ are polynomials in g_a.

67. We have already in Chapter IV. referred to the fact that an integral function is not necessarily expressible integrally in terms of the coordinates x, y by which the equation of the surface is expressed, even though y be an integral function. The consideration of the Weierstrass canonical surface suggests interesting examples of integral functions which are not expressible integrally.

In order that an integral function g whose order is μ should be expressible as an integral polynomial in the coordinates g_a, g_r of the surface, in the form

$$g = \overset{m}{g_a} \overset{n}{g_r} + \ldots\ldots$$

it is necessary that there should be a term on the right hand whose order of infinity is the same as that of the function; we must therefore have an equation of the form

$$\mu = ma + nr$$

wherein m, n are positive integers. Since a polynomial in g_a and g_r can be reduced by the equation of the surface until the highest power of g_r which enters is less than a, we may suppose n less than a.

This equation is impossible for any value of μ of the form $nr - ka$. And since herein k may be taken equal to any positive integer less than nr/a, the number of integers of this form, with any value of n, is $E(nr/a)$, or the greatest integer contained in the fraction nr/a. Hence on the whole there are

$$\sum_{n=1}^{a-1} E(nr/a)$$

orders of integral functions which are not expressible integrally by g_a and g_r.

Corresponding to any order which is not expressible in the form $nr - ka$, which is therefore of the form $nr + ma$, we can assign an integrally expressible integral function* namely $\overset{n}{g_r} \overset{m}{g_a}$: hence the p orders corresponding to which, according to Weierstrass's gap theorem, no integral functions whatever exist, must be among the excepted orders whose number we have proved to be

$$\sum_{n=1}^{a-1} E(nr/a) \text{ or}\dagger\ \tfrac{1}{2}(a-1)(r-1).$$

* Though it does not follow that every integral function whose order is of the form $nr + ma$ can be expressed wholly in integral form.

† If a right-angled triangle be constructed whose sides containing the right angle are respectively a and r, and the interior of the triangle be ruled by lines parallel to the sides

7—2

Hence the number of orders of actually existing integral functions which are not expressible integrally is

$$\tfrac{1}{2}(a-1)(r-1)-p.$$

In the table which we have given for $p=4$ (§ 64) the existing integral functions which are not expressible integrally are, for the case (γ), of orders 8 and 11; for the case (δ) of orders 6 and 11; for the case (ϵ) of orders 7 and 11; for the case (ζ) of orders 6, 9, 10, 13, 17; for case (η) of orders 7, 8, 9, 13, 14, 19. The reader can easily assign the numbers for the cases in which $p=3$.

Ex. 1. Prove that for the surface

$$g_5^3+g_5^2(g_3-c)+g_5g_3(g_3, 1)_2+g_3^2(g_3, 1)_3=0,$$

the function

$$g_7=g_5(g_5-c)/g_3$$

is an integral function which is not expressible as an integral polynomial in g_3 and g_5.

Ex. 2. Prove that for the surface

$$g_7^3+g_7^2\beta_2+g_7a_2\gamma_2+a_2^2a_3=0,$$

where

$$a_2=c(g_3-k_1)(g_3-k_2),$$
$$\beta_2=(g_3-k_1)f_1+b_1,$$

f_1 being of the first order in g_3, and c, b_1, k_1, k_2 being constants, the two following functions are integral functions not integrally expressible—

$$g_8=g_7(g_7+\beta_2)/a_2, \quad g_{11}=g_7(g_7+b_1)/(g_3-k_1).$$

68. The number $\tfrac{1}{2}(a-1)(r-1)-p$ is susceptible of another interpretation which is in close connexion with the last. Let the set of fundamental integral functions for the Weierstrass canonical surface be denoted by $1, G_1, G_2, \dots, G_{a-1}$. From the equations whereby $1, g_r, g_r^2, \dots, g_r^{a-1}$ are expressed in terms of them we are able (Chapter IV., § 43) to deduce an equation

$$\Delta(1, g_r, \dots, g_r^{a-1})=\nabla^2.\,\Delta(1, G_1, G_2, \dots, G_{a-1}),$$

wherein $\Delta(1, g_r, \dots, g_r^{a-1})$ is formed as a determinant whose (i, j)th element is the sum of the values of g_r^{i+j-2} at the a places of the surface where g_a has the same value, and is therefore an integral polynomial in g_a, $\Delta(1, G_1, \dots, G_{a-1})$ is formed as a determinant whose (i, j)th element is the sum of the values of $G_{i-1}G_{j-1}$ for the same value of g_a, which also is an integral polynomial in

containing the right angle, and at unit distances from these sides and each other, so describing squares interior to the triangle, the number of angular points interior to the triangle is easily seen to be $\sum_{n=1}^{a-1} E(nr/a)$. On the other hand if the right-angled triangle be regarded as the half of a rectangle whose diagonal is the hypotenuse of the right-angled triangle, and the ruled lines be continued into the other half, it is easily seen that the total number of angular points of the squares interior to the whole rectangle is $(a-1)(r-1)$.

g_a, and ∇ is a determinant whose elements are those integral polynomials in g_a which arise in the expressions of $1, g_r, \ldots, g_r^{a-1}$ in terms of $1, G_1, \ldots, G_{a-1}$.

The determinant $\Delta(1, g_r, \ldots, g_r^{a-1})$ is the square of the product of all the differences of the values of g_r which correspond to any value of g_a. It therefore vanishes, for finite values of g_a, when and only when two of these are equal. If the form of the equation of the surface be denoted by $f(g_r, g_a) = 0$, this happens when, and only when, $\partial f/\partial g_r = 0$. Now $\partial f/\partial g_r$ is an integral polynomial in g_a and g_r, of order $a-1$ in the latter. Regarded as a rational function on the surface it is only infinite when g_a and g_r are infinite. It follows from the fact (§ 66), that g_r^a is a term of the highest order of infinity which enters in the polynomial $f(g_r, g_a)$, that $\partial f/\partial g_r$ is infinite, at $g_a = \infty$, to an order $r(a-1)$. This is therefore the number of finite places on the surface at which $\partial f/\partial g_r$ vanishes. Hence we infer that the polynomial $\Delta(1, g_r, \ldots, g_r^{a-1})$ is of degree $r(a-1)$ in g_a.

Since there is a branch place at infinity counting for $(a-1)$ branch places, the polynomial $\Delta(1, G_1, \ldots, G_{a-1})$ is of order $2a + 2p - 2 - (a-1) = a - 1 + 2p$ in g_a (§§ 48, 61).

Thus ∇ is of order

$$\tfrac{1}{2}\left[r(a-1) - (a - 1 + 2p)\right],$$

that is, of order

$$\tfrac{1}{2}(r-1)(a-1) - p,$$

in g_a.

This interpretation of the degree of ∇ is of interest when taken in connexion with the theorem—Every integral function can be written in the form

$$(g_a, g_r)/(g_a, 1),$$

the numerator being an integral polynomial in g_a and g_r, and the denominator being an integral polynomial in g_a. *All the polynomials* $(g_a, 1)$ *thus occurring are divisors of the polynomial* ∇. See § 48 and § 88 Exx. ii, iii*.

When the factors of ∇ are all simple we may therefore expect to be able to associate each of them, as denominator, with an integral function which is not integrally expressible. In this connexion some indications are given in a paper, *Camb. Phil. Trans.* xv. pp. 430, 436. For Weierstrass's canonical surface see also a dissertation, De aequatione algebraica...in quandam formam canonicam transformata. G. Valentin. Berlin, 1879. (A. Haack.) Also Schottky, *Crelle*, 83. Conforme Abbildung...ebener Flächen.

69. The method which has been exemplified in §§ 65, 66 for the formation of the general form of the equation of a surface when the fundamental set of integral functions is given, is not limited to Weierstrass's canonical surface.

Take for instance any surface of three sheets, and let $1, g_1, g_2$ be any set

* Cf. Harkness and Morley, *Theory of Functions*, p. 268, § 186.

of fundamental integral functions with the properties assigned in Chapter IV. § 42. Then there exist equations of the form

$$g_1 g_2 = \gamma + \beta\, g_1 + \alpha\, g_2$$
$$g_1^2 = \gamma_1 + \beta_1 g_1 + \alpha_1 g_2$$
$$g_2^2 = \gamma_2 + \alpha_2 g_1 + \beta_2 g_2$$

wherein the Greek letters denote polynomials in the independent variable of the surface, x, whose degrees are limited by the condition that no terms occur on the right of higher dimensions than those on the left.

Thus the dimension of β is not greater than that of g_2 and the dimension of α is not greater than that of g_1. Hence we may use $g_1 - \alpha$, $g_2 - \beta$ instead of g_1 and g_2 respectively, and so take the first equation in the form $g_1 g_2 = \gamma$, the form of the other equations being unaltered. As before, there are conditions that these equations should lead to unique values for every integral polynomial in g_1 and g_2, namely

$$g_2(\gamma_1 + \beta_1 g_1 + \alpha_1 g_2) = g_1 \gamma, \quad g_1(\gamma_2 + \alpha_2 g_1 + \beta_2 g_2) = g_2 \gamma.$$

These lead to the equations

$$\gamma = \alpha_1 \alpha_2, \quad \gamma_1 = -\alpha_1 \beta_2, \quad \gamma_2 = -\alpha_2 \beta_1,$$

and thence to

$$g_1^3 - \beta_1 g_1^2 + \alpha_1 \beta_2 g_1 - \alpha_1^2 \alpha_2 = 0$$
$$g_2^3 - \beta_2 g_2^2 + \alpha_2 \beta_1 g_2 - \alpha_2^2 \alpha_1 = 0. \qquad \text{(v)}$$

Since every rational function can be represented rationally by x and g_1 and $g_2 = \alpha_1 \alpha_2 / g_1$, it follows that every rational function can be represented rationally by x and g_1. Hence the surface represented by the first of these two final equations is one upon which the original surface is rationally and reversibly represented. So also is the surface represented by the second of these equations.

The fundamental integral functions are derived immediately from the equation, being

$$1,\ g_1,\ (g_1^2 - \beta_1 g_1 + \alpha_1 \beta_2)/\alpha_1.$$

Ex. 1. Prove that the integrals of the first kind for the surface

$$f(g_1, x) = g_1^3 - \beta_1 g_1^2 + a_1 \beta_2 g_1 - a_1^2 a_2 = 0$$

are given by

$$\int \frac{dx}{f'(g_1)} [(x, 1)^{\tau_1 - 1} g_1 + (x, 1)^{\tau_2 - 1} a_1],$$

where $\tau_1 + 1$, $\tau_2 + 1$ are the dimensions of g_1 and g_2 and $f'(g_1) = \partial f / \partial g_1$.

Ex. 2. Prove that for the case quoted in Ex. i, § 40, Chapter IV, the form of the equation is, (i) when p is odd $= 2n - 1$, say,

$$g_n^3 - a_n g_n^2 + a_{n-1} a_{n+1} g_n - a^2_{n-1} a_{n+2} = 0,$$

where a_{n-1}, a_n, a_{n+1}, a_{n+2} are polynomials in x of the orders indicated by their suffixes, (ii) when p is even $=2n-2$, say,

$$g_n^3 - a_n g_n^2 + \beta_n \delta_n g_n - \beta_n^2 \gamma_n = 0,$$

where a_n, β_n, γ_n, δ_n are polynomials in x of the nth order.

Ex. 3. Writing $g_1 = a_1 y$, the first of the equations (v) becomes

$$a_1 y^3 - \beta_1 y^2 + \beta_2 y - a_2 = 0. \qquad \text{(A)}$$

If the dimensions of g_1 and g_2 be τ_1+1, τ_2+1, find the degrees of the polynomials a_1, β_1, a_2, β_2. And prove that if the positive quadrant of a plane of rectangular coordinates (x, y) be divided into squares whose sides are each 1 unit in length, and a convex polygon be constructed whose angular points are determined from this equation (A), by the rule that a term $x^r y^s$ in the equation determines the point (r, s) of the plane, then the number of angular points of the squares which lie within this polygon is p.

70. In obtaining the equation

$$g_1^3 - \beta_1 g_1^2 + \alpha_1 \beta_2 g_1 - \alpha_1^2 \alpha_2 = 0 \qquad \text{(E)}$$

we have spoken as if the original surface were of three sheets. *It is important to notice that this is not necessary.*

Suppose our given surface to be any surface for which a rational function of the third order, ξ, exists. Take c so that the poles of the function $(\xi - c)^{-1}$, which is also a function of the third order, are distinct ordinary places of the surface. So determined denote the function by x. Let a_1, a_2, a_3 denote these poles. Then just as in § 39 of Chapter IV. it can be shewn that there exist two rational functions g_1 and g_2, only infinite in a_1 and a_2, such that every rational function which is infinite only in a_1, a_2, a_3 can be expressed in the form

$$\gamma + \alpha g_1 + \beta g_2,$$

wherein γ, α, β are integral polynomials in x whose degrees have certain upper limits determined by the condition of dimensions.

And as before we can obtain the equation (E). Further, if F be any rational function whatever and A_1, A_2, ... be the values of x at the places other than a_1, a_2, a_3 at which F becomes infinite, it is clearly possible to find a polynomial K of the form $(x - A_1)^{n_1} (x - A_2)^{n_2} \ldots$ such that KF only becomes infinite at a_1, a_2, a_3. Hence every rational function of the original surface can be expressed rationally by x and g_1.

Thus as x, g_1 are rational functions on the original surface, (E) represents a new surface upon which our canonical surface is rationally and reversibly represented. *And it is as much the proper normal form for surfaces upon which a rational function of the third order exists as is the equation*

$\sigma^2=(z, 1)_{2p+2}$, *previously derived, for the hyperelliptic surfaces upon which a function of the second order exists.*

Ex. Obtain the hyperelliptic equation in this way.

71. In the same way we can obtain a canonical form for surfaces upon which a function of the fourth order exists. We can shew that there exist three functions g_1, g_2, g_3 satisfying such equations as

$$g_3^2 = a_1g_1 + b_1g_2 + c_1g_3 + k_1$$
$$g_2g_3 = a_2g_1 \qquad\qquad\qquad + k_2$$
$$g_1g_3 = a_3g_1 + b_3g_2 \qquad\quad + k_3,$$

wherein the nine coefficients are integral polynomials in a rational function x, which is of the fourth order; and that the surface is rationally and reversibly representable upon a surface given by the equation

$$g_3^4 - g_3^3(a_3+c_1) - g_3^2(a_2b_3+k_1-a_3c_1) + g_3(a_1k_3 - b_1k_2 + a_2b_3c_1 + a_3k_1) + a_1b_3k_2 + a_3b_1k_2 + a_2b_3k_1 = 0.$$

Ex. These coefficients $a_1, \dots, k_3$ satisfy certain relations; prove that the conditions that $g_2\cdot g_3^2=g_2g_3\cdot g_3$, $g_1\cdot g_3^2=g_1g_3\cdot g_3$, $g_1g_3\cdot g_2=g_2g_3\cdot g_1$ are that the following nine polynomials should be divisible by a polynomial Δ, whose value is $a_1^2b_3-a_3a_1b_1-a_2b_1^2$;

$$a_2n_1(a_1b_3-a_3b_1)-b_1(a_2h_1-a_1k_2),\quad a_1b_3h_1-b_1(a_2b_3n_1-a_1k_3),\quad a_1b_3k_2-a_2b_1k_3$$
$$-a_2^2n_1b_1+a_1a_2h_1+a_1^2k_2,\quad -h_1(a_1a_3+a_2b_1)+a_1(a_2b_3n_1-a_1k_3),\quad -k_2(a_1a_3+a_2b_1)+a_1a_2k_3$$
$$(h_1+a_3n_1)(a_1b_3-a_3b_1)-b_1(a_2b_3n_1-b_1k_3),\quad n_1b_3(a_1b_3-a_3b_1)-b_3(h_1b_3+b_1k_3),$$
$$k_3(a_1b_3-a_3b_1)-b_1b_3k_2.$$

Herein $n_1=a_3-c_1$, $h_1=a_2b_3-k_1$.

In fact if

$$g_1g_2=a_5g_1+b_5g_2+c_5g_3+k_5,\quad g_2^2=a_4g_1+b_4g_2+c_4g_3+k_4,\quad g_1^2=a_6g_1+b_6g_2+c_6g_3+k_6,$$

the results of the division of these nine polynomials by Δ are respectively

$$a_5,\ b_5,\ c_5,\ a_4,\ b_4,\ c_4,\ a_6,\ b_6,\ c_6,$$

while

$$k_4=a_2c_5-c_1c_4,\quad k_5=n_1c_5+b_3c_4,\quad k_6=n_1c_6+b_3c_5.$$

72. When the order of the independent function, denoted in §§ 69—71 by x, is known, and the dimensions of the fundamental integral functions in regard thereto, the general forms of the polynomial coefficients in the equations, whereby the products of pairs of these integral functions are expressed as linear functions of themselves, can be written down. And thence, if the necessary algebra (such as that indicated in the example of § 71), which serves to limit the forms of these polynomial coefficients, can be carried out, a canonical form of the equation of the surface can be deduced.

But the converse process may arise: when we are given a form of the fundamental equation associated with the surface, we may require to replace the given equation by one in which the dependent variable is one of the set of fundamental integral functions. More generally we may replace it by an equation in which the dependent variable is an integral function of the form

$$\eta=(x, 1)_\lambda+(x, 1)_{\lambda_1}g_1+\dots+(x, 1)_{\lambda_{n-1}}g_{n-1}.$$

This replacement possesses a high degree of interest (§ 88. Ex. iii). In either case it is necessary to be able to calculate the fundamental integral functions.

73. We give now sufficient explanation to enable the reader to calculate the expression of the fundamental integral functions for any given form of the fundamental equation associated with the Riemann surface. This equation may* be taken in the form

$$y^n + y^{n-1} a_1 + \dots + y a_{n-1} + a_n = 0, \qquad \text{(A)}$$

$a_1, \dots, a_n$ being integral polynomials in x; thus y is an integral function of x (§ 38).

The n values of any rational function, η, which arise for the same value of x, will be denoted by $\eta^{(1)}, \dots, \eta^{(n)}$ and called conjugate values; their sum will be denoted by $\Sigma\eta$. If any of the possible rational expressions of η be $\phi(x, y)/\psi(x, y)$, ϕ and ψ being integral polynomials in x and y, and if in the expression of $\eta^{(i)}$,

$$\eta^{(i)} = \phi(x, y^{(i)})/\psi(x, y^{(i)}),$$

we multiply numerator and denominator by the product of the $n-1$ values conjugate to $\psi(x, y^{(i)})$, the denominator will become an integral symmetric function of $y^{(1)}, \dots, y^{(n)}$, and can therefore be expressed by means of the equation (A), as an integral polynomial in x; and the numerator will take a form which can be expressed as an integral polynomial in x and $y^{(i)}$. Hence the value of any rational function, on the surface associated with the equation (A), can be expressed in the form

$$\eta = \frac{A + A_1 y + \dots + A_{n-1} y^{n-1}}{D}, \qquad \text{(B)}$$

$A, \dots, A_{n-1}$, D denoting integral polynomials in x, with no common divisor.

Thus, to determine the expression of the fundamental integral functions, we may enquire what modification this general form undergoes when η is an integral function.

74. In the first place the denominator D must be such that D^2 is a factor of the integral polynomial† $\Delta(1, y, \dots, y^{n-1})$; so that D is capable only of a limited number of forms. For let $x-a$ be a factor of D, repeated r times, and write

$$A_i = (x-a)^r B_i + C_i, \qquad (i = 0, 1, \dots, (n-1))$$

wherein C_i is a polynomial of order less than r; since $A, \dots, A_{n-1}$ have no common divisor which divides D, not all of $C, C_1, \dots, C_{n-1}$ can be divisible by $x-a$. Then the function

$$\eta D/(x-a)^r - (B + B_1 y + \dots + B_{n-1} y^{n-1}), = (C + C_1 y + \dots + C_{n-1} y^{n-1})/(x-a)^r,$$

is an integral function, when η is an integral function, as appears from its first form of expression. Denote it by ζ.

Suppose C_i not divisible by $x-a$. From the equation†

$$\Delta(1, y, \dots, y^{i-1}, \zeta, y^{i+1}, \dots, y^{n-1}) = \nabla^2_i \Delta(1, g_1, \dots, g_{n-1}),$$

recalling the form of the determinant which is the square root of the left hand side, we infer

$$\frac{C_i^2}{(x-a)^{2r}} \Delta(1, y, \dots, y^{i-1}, y^i, y^{i+1}, \dots, y^{n-1}) = \nabla_i^2 \Delta(1, g_1, \dots, g_{n-1}).$$

Hence, save for sign,

$$\nabla/\nabla_i = (x-a)^r/C_i,$$

so that $(x-a)^r$ divides ∇.

Thus the first step in the determination of the integral functions is to put $\Delta(1, y, \dots, y^{n-1})$ into the form $u_1^{k_1} \dots u_r^{k_r}$, wherein $u_1, \dots, u_r$ are polynomials having only simple

* Chap. IV. § 38. † Chap. IV. § 43.

factors. This can always be done by the rational process of finding the highest divisor common to $\Delta(1, y, \ldots, y^{n-1})$ and its differential coefficients in regard to x. It will include most cases of practical application if we further suppose all the linear factors of $\Delta(1, y, \ldots, y^{n-1})$ to be known*.

75. Suppose then that $x-a$ is a factor which occurs to at least the second order in $\Delta(1, y, \ldots, y^{n-1})$. Denote $x-a$ by u. By the solution of a system of linear equations, we can (below, § 78) find all the existing linearly independent expressions of the form

$$(a+a_1 y+\ldots+a_{n-1} y^{n-1})/u,$$

wherein $a, a_1, \ldots, a_{n-1}$ are constants, which represent integral functions. If the highest power of y actually entering be the same in two of these integral functions, say in ζ and ζ', we can use instead of ζ' a function of the form $\zeta'-\mu\zeta$, where μ is a certain constant. By continued application of this method of reduction we obtain, suppose, k integral functions, of the form

$$\zeta_r=(a'+a'_1 y+\ldots+a'_r y^r)/u, \qquad (C)$$

wherein, since these functions are linearly independent, k is less than n, and the values of r that occur are all different. These values of r that occur are among the sequence $1, 2, \ldots, (n-1)$; let s denote in turn all the $n-1-k$ other integers in this sequence. Put ζ_s for y^s. Consider now the set of integral functions

$$1, \zeta_1, \ldots, \zeta_{n-1}.$$

As before we can determine by the solution of a system of linear equations all the linearly independent functions of the form

$$(\beta+\beta_1 \zeta_1+\ldots+\beta_{n-1} \zeta_{n-1})/u,$$

wherein $\beta, \beta_1, \ldots, \beta_{n-1}$ are constants, which are integral functions; and, as before, we can choose them so that the ζ's of highest suffix which occur shall not be the same in any two of these integral functions. Then in place of $1, \zeta_1, \ldots, \zeta_{n-1}$ we obtain a set $1, \xi_1, \ldots, \xi_{n-1}$, wherein ξ_r is ζ_r unless there be an integral function of the form

$$(\beta'+\beta'_1 \zeta_1+\ldots+\beta'_r \zeta_r)/u, \qquad (D)$$

wherein the ζ of highest suffix occurring is ζ_r, in which case ξ_r denotes this function.

Then we enquire whether there are any integral functions of the form

$$(\gamma+\gamma_1 \xi_1+\ldots+\gamma_{n-1} \xi_{n-1})/u,$$

$\gamma, \ldots, \gamma_{n-1}$ being constants. If there are, the process is to be continued†. If there are none, let v denote any other linear factor occurring in $\Delta(1, y, \ldots, y^{n-1})$ to at least the second order. Then, as for the set $1, y, \ldots, y^{n-1}$, we investigate what linearly independent integral functions exist of the form

$$(a+a_1 \xi_1+\ldots+a_{n-1} \xi_{n-1})/v,$$

and continue the process for v as for u: and afterwards for all other repeated factors of $\Delta(1, y, \ldots, y^{n-1})$.

76. When these processes are completed, we shall obtain a set of integral functions

$$1, \eta_1, \ldots, \eta_{n-1},$$

such that there exists no integral function of the form

$$(a+a_1 \eta_1+\ldots+a_{n-1} \eta_{n-1})/(x-c),$$

* In the work below, if u be a polynomial of order r, it is necessary to suppose $a, a_1, \ldots, a_k$ to be polynomials of order $r-1$.

† The number of steps is finite, by § 74.

wherein $a, \dots, a_{n-1}$ are constants, for *any* value of c. It is obvious now from the successive definitions (C), (D), ... of the sets $(1, \zeta_1, \dots, \zeta_{n-1})$, $(1, \xi_1, \dots, \xi_{n-1})$, ..., $(1, \eta_1, \dots, \eta_{n-1})$, that every power of y can be represented in the form

$$y^i = v + v_1\eta_1 + \dots + v_{n-1}\eta_{n-1},$$

wherein $v, v_1, \dots, v_{n-1}$ are integral polynomials in x. Hence every integral function can be written in the form

$$\eta = (E + E_1\eta_1 + \dots + E_{n-1}\eta_{n-1})/F,$$

wherein $E, \dots, E_{n-1}, F$ are integral polynomials in x without common divisor. If now $x-c$ be a factor of F and we write

$$E_i = (x-c)\,G_i + a_i, \quad i = 0, 1, 2, \dots, (n-1),$$

a_i being a constant, the function

$$\eta F/(x-c) - [G + G_1\eta_1 + \dots + G_{n-1}\eta_{n-1}] = (a + a_1\eta_1 + \dots + a_{n-1}\eta_{n-1})/(x-c)$$

is an integral function, as appears from the form of the left-hand side. By the property of the set $1, \eta_1, \dots, \eta_{n-1}$ there is no integral function having the form of the right-hand side, unless each of $a, a_1, \dots, a_{n-1}$ be zero.

Hence each of $E, \dots, E_{n-1}$ are divisible by $x-c$. By successive steps of this kind it can be shewn that every integral function can be written in the form

$$\eta = H + H_1\eta_1 + \dots + H_{n-1}\eta_{n-1}, \tag{E}$$

wherein $H, H_1, \dots, H_{n-1}$ are *integral* polynomials in x.

77. But in order that the set $1, \eta_1, \dots, \eta_{n-1}$ should be such a fundamental set as $1, g_1, \dots, g_{n-1}$, used in Chap. IV., there must be no terms occurring on the right-hand side here, which are of higher dimension than η. We prove now that this requires a further reduction in the forms of $1, \eta_1, \dots, \eta_{n-1}$, which is of a kind precisely analogous to the reductions already described.

Let $\sigma+1$ be the dimension of η, ρ_i the order, and therefore also the dimension of the polynomial H_i (§ 76) and σ_i+1 the dimension of η_i; we suppose $\sigma_1 \not> \sigma_2 \not> \dots \not> \sigma_{n-1}$; then

$$\eta/x^{\sigma+1} = \dots + (H_i x^{-\rho_i})(\eta_i/x^{\sigma_i+1})\,x^{\rho_i+\sigma_i-\sigma} + \dots.$$

Putting $x = 1/\xi$, $h = \eta/x^{\sigma+1}$, $h_i = \eta_i/x^{\sigma+1}$, $H_i x^{-\rho_i} = (1, \xi)_{\rho_i}$, an integral polynomial in ξ, this equation is

$$h = \dots + (1, \xi)_{\rho_i} h_i/\xi^{\rho_i+\sigma_i-\sigma} + \dots.$$

If now in equation (E) a term arises of higher dimension than η, one of the integers

$$\rho - (\sigma+1), \dots, \rho_i + \sigma_i - \sigma, \dots$$

is greater than zero. In that case let $r+1$ be the greatest of these integers. Then we can write

$$\xi^r h = (\dots + (1, \xi)_{m_i} h_i + \dots)/\xi,$$

wherein the symbols $(1, \xi)_{m_i}$ denote integral polynomials in ξ. Putting

$$(1, \xi)_{m_i} = \xi K_i + a_i, \quad (i = 0, 1, 2, \dots, n-1),$$

wherein a_i is a constant, we have

$$\xi^r h - (K + K_1 h_1 + \dots + K_{n-1} h_{n-1}) = (a + a_1 h_1 + \dots + a_{n-1} h_{n-1})/\xi.$$

Herein the left hand is a function which is not infinite when x is infinite. Hence, when the set $1, \eta_1, \ldots, \eta_{n-1}$ are such that the condition of dimensions* is not satisfied, there exist functions of the form

$$(a+a_1h_1+\ldots+a_{n-1}h_{n-1})/\xi,$$

i.e. of the form

$$x\left[a+a_1\eta_1/x^{\sigma_1+1}+\ldots+a_{n-1}\eta_{n-1}/x^{\sigma_{n-1}+1}\right],$$

wherein $a, \ldots, a_{n-1}$ are constants which are not infinite when ξ is zero or x is infinite.

In virtue of their definition the functions $h_1, \ldots, h_{n-1}$ are not infinite when x is infinite, and are therefore infinite only when x is zero or ξ infinite. We may therefore regard them as integral functions of ξ. And since there exists no integral function of the form η_i/x, the dimensions of $h_1, \ldots, h_{n-1}$ as functions of ξ are $\sigma_1+1, \ldots, \sigma_{n-1}+1$.

As before determine a set of linearly independent functions of the form

$$(a+a_1h_1+\ldots+a_{n-1}h_{n-1})/\xi,$$

$a, \ldots, a_{n-1}$ being constants, which are not infinite when $\xi=0$, choosing them so that the h of highest suffix which occurs is not the same in any two of the functions. Let the function wherein the h of highest suffix is h_r be denoted by k_r, so that k_r is of the form

$$k_r=(\mu+\mu_1h_1+\ldots+\mu_rh_r)/\xi.$$

Then

$$k_rx^{\sigma_r}=x^{\sigma_r+1}\left(\mu+\mu_1\eta_1/x^{\sigma_1+1}+\ldots+\mu_r\eta_r/x^{\sigma_r+1}\right)$$

is a function which is not infinite when $x=0$, as appears from the form of the right-hand side; it is therefore an integral function of x, and since k_r is not infinite when x is infinite it is an integral function of x whose dimension is only σ_r. Denote it by G_r. Then η_r can be expressed in the form

$$\eta_r=-\frac{1}{\mu_r}\left[\mu x^{\sigma_r+1}+\mu_1\eta_1x^{\sigma_r-\sigma_1}+\ldots+\mu_{r-1}\eta_{r-1}x^{\sigma_r-\sigma_{r-1}}-G_r\right], \qquad \text{(F)}$$

and in the right hand no term occurs of higher dimension than that of η_r, while G_r is of less dimension than η_r. If then there be m functions such as k_r, m of the functions $\eta_1, \ldots, \eta_{n-1}$ can be expressed in the form (F) in terms of the remaining $n-1-m$ functions of $\eta_1, \ldots, \eta_{n-1}$ and m functions G_r; the sum of the dimensions of these m functions G_r is less by m than that of the dimensions of the functions η_r which they replace. Denoting the functions among $\eta_1, \ldots, \eta_{n-1}$ which are not thus replaced by functions G, also by the symbol G, for the sake of uniformity, every integral function is expressible in the form

$$(x, 1)_\lambda+(x, 1)_{\lambda_1}G_1+\ldots+(x, 1)_{\lambda_{n-1}}G_{n-1},$$

and the sum of the dimensions of $G_1, \ldots, G_{n-1}$ is less by m than the sum of the dimensions of $\eta_1, \ldots, \eta_{n-1}$.

If now in this expression of integral functions by $G_1, \ldots, G_{n-1}$ any terms can arise which are of higher dimension than the functions to be expressed, we can similarly replace the set $G_1, \ldots, G_{n-1}$ by another set whose dimensions have a still less sum.

Since no integral function can have a less dimension than 1, the sum of the dimensions of the functions whereby integral functions are expressed, cannot be diminished below $n-1$. We shall therefore arrive at length at a set $g_1, \ldots, g_{n-1}$ of integral functions, in terms of which all integral functions can be expressed so that the condition of dimensions is satisfied.

It is this system which it was our aim to deduce.

* Chap. IV. § 39.

Ex. For the surface associated with the equation $y^2=(x, 1)_{2p+2}$ all integral functions can in fact be represented in the form $(x, 1)_\lambda+(x, 1)_{\lambda_1}\eta_1$, where $\eta_1=y+x^m$. If $m>p+1$ the dimension of η_1 is m. In order to ascertain whether the condition of dimensions is satisfied we enquire whether there exist any functions of the form $x[a+a_1(y+x^m)/x^m]$, wherein a, a_1 are constants, which are finite for $x=\infty$, namely whether $[a+a_1(y\xi^m+1)]/\xi$ can be an integral function of ξ.

Shew that this can only be the case when $a+a_1=0$. Putting $k_r=[-a+a_1(y\xi^m+1)]/\xi$ it is clear that $k_r x^{m-1}=a_1 y$. Thus all integral functions can be represented in the form $(x, 1)_\lambda+(x, 1)_{\lambda_1}y$. Shew that the condition of dimensions is now satisfied.

78. There is one part of the process given here which has not been explained. Let $\eta_1, \dots, \eta_{n-1}$ be integral functions, and let u denote a linear function of the form $x-c$. It is required to find all possible functions of the form

$$(a+a_1\eta_1+\dots+a_{n-1}\eta_{n-1})/u,$$

wherein $a, \dots, a_{n-1}$ are constants, which are not infinite when $u=0$. We suppose $\eta_1, \dots, \eta_{n-1}$ to be such that the product of every two of them is expressible in the form $v+v_1\eta_1+\dots+v_{n-1}\eta_{n-1}$, $v, \dots, v_{n-1}$ being integral polynomials in x; this condition is always satisfied in the actual case under consideration.

The integral function $H=a+a_1\eta_1+\dots+a_{n-1}\eta_{n-1}$ will satisfy an equation of the form

$$(H-H^{(1)})\dots(H-H^{(n)})=H^n+K_1H^{n-1}+\dots+K_{n-1}H+K_n=0,$$

wherein K_i is an integral polynomial in $a, \dots, a_{n-1}$ of the ith order; K_i is also an integral polynomial in x. In order that H/u be an integral function it is sufficient that K_i be divisible by u^i, and when H/u is an integral function these n conditions will always be satisfied. And it is easy to see that if S_i denote the sum of the ith powers of the n values of H which arise for any value of x, these conditions may be replaced by the conditions that S_i be divisible by u_i. It is clear that it may not be an easy matter to obtain the values of $a, \dots, a_{n-1}$, which satisfy the conditions thus expressed.

But in fact these conditions can be reduced to a set of linear congruences, and eventually to a set of linear equations for $a, \dots, a_{n-1}$. We shall not give here the proof of this reduction*, but give the resulting equations. For in many practical cases we can obtain the results, geometrically or otherwise, in a much shorter way.

Let

$$\frac{1}{n}, \frac{1}{n-1}, \dots, \epsilon, \epsilon', \dots, 1$$

denote in order of magnitude all the positive rational numerical fractions not greater than unity, whose denominators are not greater than n; each being in its lowest terms. Let $\eta_1, \dots, \eta_r$ denote any linearly independent integral functions. Let Σ denote the sum of the n values of a function which arise for any value of x. Determine all the possible sets of values of the constants $a, a_1, \dots, a_r$ such that the congruence

$$\Sigma(a+a_1\eta_1+\dots+a_r\eta_r)(c+c_1\eta_1+\dots+c_r\eta_r)\equiv 0 \pmod{u}$$

is satisfied for all values of the quantities $c, c_1, \dots, c_r$. Substituting in the left hand the value of x for which $u=0$ and equating separately to zero the coefficients of $c, c_1, \dots, c_r$, we obtain $r+1$ linear *equations* for the constants $a, a_1, \dots, a_r$. By these equations we can

* Which is given by Hensel, *Acta Math.* 18, pp. 284—292. His use of homogeneous variables is explained below Chap. VI. § 85. But it is unessential to the theory of the reduction referred to.

express a certain number* of $a, a_1, \ldots, a_r$ in terms of the others; denoting these others by $\beta_1, \ldots, \beta_s$ the function $a+a_1\eta_1+\ldots+a_r\eta_r$ takes the form $\beta_1\zeta_1+\ldots+\beta_s\zeta_s$, wherein $\zeta_1, \ldots, \zeta_s$ are *definite* linear functions of $1, \eta_1, \ldots, \eta_r$ with constant coefficients, and the equations in question are then satisfied for *all constant values* of $\beta_1, \ldots, \beta_s$. We *associate*† the functions $\zeta_1, \ldots, \zeta_s$ with the first term $\frac{1}{n}$ of the series of fractions specified above. We proceed thence to deduce a set of integral functions associated with the next term of the series, $\frac{1}{n-1}$. But in order to be able to describe the successive processes in as few words as possible, let us assume we have obtained a set of integral functions $\xi_1, \ldots, \xi_m$ which in the sense employed are *associated with*‡ the fraction ϵ of the series, and wish to deduce a set of functions associated with the next following fraction of the series, ϵ'. Put down the congruence

$$\Sigma(\gamma_1\xi_1+\ldots+\gamma_m\xi_m)(e_1\xi_1+\ldots+e_m\xi_m)^{i-1}\equiv 0 \quad (\text{mod. } u^{|i\epsilon'|}).$$

Herein $\gamma_1, \ldots, \gamma_m$ denote constants, i denotes in turn all positive integers not greater than n which are exact multiples of the denominator of the fraction ϵ, so that $i\epsilon$ is an integer, $|i\epsilon'|$ denotes the least integer which is not less than $i\epsilon'$, and, for any proper value of i, the congruence is to be satisfied for *all values* of the quantities $e_1, \ldots, e_m$. It will be found in practice that the left-hand side divides by $u^{|i\epsilon'|-1}$ for all values of $\gamma_1, \ldots, \gamma_m$, $e_1, \ldots, e_m$. If we carry out the division, then, in the result, substitute the value of x which makes $u=0$, and equate separately to zero the coefficients of the $\binom{m}{i-1}$ products of $e_1, \ldots, e_m$ which enter on the left, we shall have this number of linear equations for $\gamma_1, \ldots, \gamma_m$. Solving these, and thereby expressing as many as possible of $\gamma_1, \ldots, \gamma_m$ in terms of the remaining, which we may denote by $\gamma_1', \ldots, \gamma'_{m'}$, $\gamma_1\xi_1+\ldots+\gamma_m\xi_m$ will take a form $\gamma_1'\xi_1'+\ldots+\gamma'_{m'}\xi'_{m'}$, wherein $\gamma_1', \ldots, \gamma'_{m'}$ are arbitrary constants, and $\xi_1', \ldots, \xi'_{m'}$ are definite linear functions of $\xi_1, \ldots, \xi_m$. We say that $\xi_1', \ldots, \xi'_{m'}$ are associated with the fraction ϵ'.

This process is to be continued beginning with the case when $\epsilon=\frac{1}{n}$ and ending with the case when $\epsilon'=1$. The functions associated with the last term, 1, of the series of fractions, say $G_1, \ldots, G_k$, are all the functions of the form $a+a_1\eta_1+\ldots+a_{n-1}\eta_{n-1}$, wherein $a, a_1, \ldots, a_{n-1}$ are constants, which are such that $G_1/u, \ldots, G_k/u$ are finite when $u=0$.

For the case $n=3$, of a surface of three sheets, the series is $\frac{1}{3}, \frac{1}{2}, \frac{2}{3}, 1$. The successive congruences may therefore be denoted by

$$(S_2)\equiv 0 \text{ (mod. } u), \ (S_3)\equiv 0 \text{ (mod. } u^2), \ (S_2)\equiv 0 \text{ (mod. } u^2), \ (S_3)\equiv 0 \text{ (mod. } u^3),$$

wherein (S_i) denotes such an expression as $\Sigma(\gamma_1\xi_1+\ldots+\gamma_m\xi_m)(e_1\xi_1+\ldots+e_m\xi_m)^{i-1}$.

In fact 3 is the only integer not greater than 3 such that $3\,.\,\frac{1}{3}$ is integral and $|3\,.\,\frac{1}{2}|=2$. And 2 is the only integer not greater than 3 such that $2\,.\,\frac{1}{2}$ is integral and $|2\,.\,\frac{2}{3}|=2$; finally 3 is the only integer such that $3\,.\,\frac{2}{3}$ is integral, and $|3\,.\,1|=3$.

For a surface of four sheets the fractions are

$$\tfrac{1}{4}, \tfrac{1}{3}, \tfrac{1}{2}, \tfrac{2}{3}, \tfrac{3}{4}, 1.$$

* At most, and in general, equal to r.

† In a certain sense the functions $\zeta_1, \ldots, \zeta_s$ are all divisible by $u^{\frac{1}{n}}$.

‡ Divisible by x^{ϵ}, in a sense.

We therefore have

ϵ	ϵ'	i such that $i\epsilon$ = integral	$\lvert i\epsilon' \rvert$	congruence
0	$\frac{1}{4}$	$i=2$	1	$(S_2)\equiv 0$ (mod. u)
$\frac{1}{4}$	$\frac{1}{3}$	$i=4$	$\lvert\frac{4}{3}\rvert=2$	$(S_4)\equiv 0$ (mod. u^2)
$\frac{1}{3}$	$\frac{1}{2}$	$i=3$	$\lvert\frac{3}{2}\rvert=2$	$(S_3)\equiv 0$ (mod. u^2)
$\frac{1}{2}$	$\frac{2}{3}$	$i=2$	$\lvert\frac{4}{3}\rvert=2$	$(S_2)\equiv 0$ (mod. u^2)
		$i=4$	$\lvert\frac{8}{3}\rvert=3$	$(S_4)\equiv 0$ (mod. u^3)
$\frac{2}{3}$	$\frac{3}{4}$	$i=3$	$\lvert\frac{9}{4}\rvert=3$	$(S_3)\equiv 0$ (mod. u^3)
$\frac{3}{4}$	1	$i=4$	$\lvert 4\rvert=4$	$(S_4)\equiv 0$ (mod. u^4)

It must be borne in mind that the results of the solution of each of the seven congruences of the sequence in the right-hand column, are here supposed to be substituted in the next one: so that, for instance, the fourth congruence here may be quite other than a slightly harder case of the first congruence.

Ex. Prove that for a surface of five sheets the congruences are, in order,

(1) $(S_2)\equiv 0\ (\,,\ u)$; (2) $(S_5)\equiv 0\ (\,,\ u^2)$; (3) $(S_4)\equiv 0\ (\,,\ u^2)$; (4) $(S_3)\equiv 0\ (\,,\ u^2)$; (5) $(S_5)\equiv 0\ (\,,\ u^3)$; (6) $(S_2)\equiv 0\ (\,,\ u^2)$; (7) $(S_4)\equiv 0\ (\,,u^3)$; (8) $(S_5)\equiv 0\ (\,,\ u^4)$; (9) $(S_3)\equiv 0\ (\,,\ u^3)$; (10) $(S_4)\equiv 0\ (\,,\ u^4)$; (11) $(S_5)\equiv 0\ (\,,\ u^5)$.

79. *Ex.* i. Prove for the equation $y^4=x^2(x-1)$ that $\Delta\,(1, y, y^2, y^3)=-256\,x^6(x-1)^3$.

Shew that the equations

$$\Sigma\,(a+a_1y+a_2y^2+a_3y^3)^i\equiv 0 \text{ (mod. } (x-1)^i),$$

where a, a_1, a_2, a_3 are constants, and i is in turn equal to 1, 2, 3, 4, are only satisfied by $a=a_1=a_2=a_3=0$.

Shew that the equations

$$\Sigma\,(\beta+\beta_1y+\beta_2y^2+\beta_3y^3)^i\equiv 0 \text{ (mod. } x^i),$$

where β, β_1, β_2, β_3 are constants, and i is in turn equal to 1, 2, 3, 4, require $\beta=\beta_1=0$ and leave β_2 and β_3 arbitrary. Hence $\frac{y^2}{x}$, $\frac{y^3}{x}$ are the only integral functions of the form

$$(a+a_1y+a_2y^2+a_3y^3)/x.$$

Shew that the equations

$$\Sigma\left(\gamma+\gamma_1y+\gamma_2\frac{y^2}{x}+\gamma_3\frac{y^3}{x}\right)^i\equiv 0 \text{ (mod. } x^i)$$

require $\gamma=\gamma_1=\gamma_2=\gamma_3=0$.

Prove that the dimensions of $1, y, \frac{y^2}{x}, \frac{y^3}{x}$ are 0, 1, 1, 2. Prove then that there is no function of the form

$$\left(\delta+\delta_1\frac{y}{x}+\delta_2\frac{y^2}{x^2}+\delta_3\frac{y^3}{x^3}\right)x,$$

which is finite for x infinite.

Hence 1, y, $\frac{y^2}{x}$, $\frac{y^3}{x}$ are a fundamental system such as 1, g_1, g_2, g_3 in Chap. IV.; and the deficiency of the surface is $1+1+2-(4-1)=1$.

Ex. ii. In partial illustration of Hensel's method of reduction consider the case of the equation

$$y^3-3xy^2+3yx(x-1)+x^2(x-1)^2(9x^3+7x^2+5x+3)=0,$$

for which the sums of the powers of y are given by

$$s_1=3x,\quad s_2=3x^2+6x,\quad s_3=-27x^7+33x^6+3x^3+18x^2,$$
$$s_4=-108x^8+132x^7+3x^4+36x^3+18x^2.$$

The determinant $\Delta(1, y, y^2)$ is divisible by x^3 and by $(x-1)^2$, as appears on calculation. By forming the equation satisfied by y^2/x it appears that y^2/x is an integral function. Denote it by η. We consider now what functions exist of the form

$$(a+a_1y+a_2\eta)/(x-1),$$

wherein a, a_1, a_2 are constants, which are integral functions.

The congruence $(S_2)=\Sigma(a+a_1y+a_2\eta)(c+c_1y+c_2\eta)\equiv 0 \pmod{x-1}$ leads, considering the coefficients of c, c_1, c_2 separately, to the congruences

$$3a+a_1s_1+a_2\frac{s_2}{x}\equiv 0\,(\ ,x-1),\ as_1+a_1s_2+a_2s_3/x\equiv 0\,(\ ,x-1),\ a\frac{s_2}{x}+a_1\frac{s_3}{x}+a_2\frac{s_4}{x^2}\equiv 0\,(\ ,x-1),$$

and therefore to the equations

$$3a+3a_1+9a_2=0,\quad 3a+9a_1+27a_2=0,\quad 9a+27a_1+81a_2=0,$$

which give $a=0$, $a_1=-3a_2$, and shew that the only function of the kind required is, save for a constant multiplier,

$$(\eta-3y)/(x-1).$$

The other three congruences reduce then to conditions for this function; for example, the congruence $(S_3)\equiv 0\,(\ ,x^2)$ becomes

$$\Sigma\left[\frac{y^2}{x(x-1)}-\frac{3y}{x-1}\right]^3\equiv 0\,(\ ,x^2).$$

But in fact, if we write $g=(y^2-3xy)/x(x-1)$, $A=9x^3+7x^2+5x+3$, we immediately find from the original equation that

$$g^3+6g^2-3g(Ax-3)+A^2x(x-1)+9Ax=0,$$

so that g is an integral function.

Apply the method to shew that y^2/x is the only integral function of the form $(a+a_1y+a_2y^2)/x$.

Prove that the dimensions of the functions

$$1,\ y,\ (y^2-3xy)/x(x-1)$$

are respectively 0, 3, 3.

Putting $x=1/\xi$, $y/x^3=h$, examine whether there exists any integral function of ξ of the form

$$[a+a_1h+3a_2(h^2-3\xi^2h)/\xi(1-\xi)]/\xi,$$

and deduce the fundamental integral functions.

The deficiency of the surface is $3+3-(3-1)=4$.

CHAPTER VI.

Geometrical Investigations.

80. It has already been pointed out (§ 9) that the algebraical equation, associated with a Riemann surface, may be regarded as the equation of a plane curve; for the sake of distinctness we may call this curve the fundamental curve. The most general form of a rational function on the Riemann surface is a quotient of two expressions which are integral polynomials in the variables (x, y) in terms of which the equation associated with the surface is expressed. Either of these polynomials, equated to zero, may be regarded as representing a curve intersecting the fundamental curve. Thus we may expect that a comparison of the theory of rational functions on the Riemann surface with the theory of the intersection of a fundamental curve with other variable curves, will give greater clearness to both theories.

In the present chapter we shall make full use of the results obtainable from Riemann's theory and seek to deduce the geometrical results as consequences of that theory.

81. The converse order of development, though of more elementary character, requires much detailed preliminary investigation, if it is to be quite complete, especially in regard to the theory of the multiple points of curves. But the following account of this order of development may be given here with advantage (§§ 81—83). Let the term of highest aggregate degree in the equation of the fundamental curve $f(y, x) = 0$ be of degree n; and, in the usual way, regard the equation as having its most general form when it consists of all terms whose aggregate degree, in x and y, is not greater than n; this general form contains therefore $\frac{1}{2}(n+1)(n+2)$ terms. Suppose, further, that the curve has no multiple points other than ordinary double points and cusps, δ being the number of double points and κ of cusps. Consider now another curve, $\psi(x, y) = 0$, of order m, whose coefficients are at our disposal. By proper choice of these coefficients in ψ we can determine ψ to pass through any given points of f, whose number is not greater than the number of disposeable coefficients in ψ. Let k be the number of the prescribed points, and interpret the infinite intersections of f and ψ, in the usual way, so that their total number of intersections is mn. Then there

remain $mn-k$ intersections of f and ψ which are determined by the others already prescribed. We proceed to prove that if $m > n-3$, and if we utilise all the coefficients of ψ to prescribe as many of the intersections of ψ and f as possible, and introduce further the condition that ψ shall pass once through each cusp and double point of f, then the number of remaining intersections which are determined by the others will be $p=\frac{1}{2}(n-1)(n-2)-\delta-\kappa$*, for all values of m. For, if $m \geqq n$, the intersections of ψ with f are the same as those of a curve

$$\psi + U_{m-n}f = 0,$$

wherein U_{m-n} is any integral polynomial in the coordinates x and y, in which no term of higher aggregate dimension than $m-n$ occurs. By suitable choice of the $\frac{1}{2}(m-n+1)(m-n+2)$ coefficients which occur in the general form of U_{m-n} we can reduce $\frac{1}{2}(m-n+1)(m-n+2)$ coefficients in $\psi + U_{m-n}f$ to zero†. It will therefore contain, in its new form,

$$M+1 = 1+\tfrac{1}{2}m(m+3)-\tfrac{1}{2}(m-n+1)(m-n+2)$$

arbitrary coefficients. M is therefore the number of the intersections of ψ with f which we can dispose of at will, by choosing the coefficients in ψ suitably. Of these intersections, by hypothesis, $2(\delta+\kappa)$ are to be taken at the double points and cusps of the curve f. This can be effected by the disposal of $\delta+\kappa$ of the arbitrary coefficients. There remain then

$$1+\tfrac{1}{2}m(m+3)-\tfrac{1}{2}(m-n+1)(m-n+2)-\delta-\kappa$$

disposeable coefficients and $mn-2(\delta+\kappa)$ intersections. Of these, therefore,

$$mn-2(\delta+\kappa)-[\tfrac{1}{2}m(m+3)-\tfrac{1}{2}(m-n+1)(m-n+2)-\delta-\kappa]$$

is the number of intersections determined by the others which are at our disposal; and this number is

$$\tfrac{1}{2}(n-1)(n-2)-(\delta+\kappa).$$

In case $m < n$, of the $mn-2(\delta+\kappa)$ intersections of ψ with f, which are not at the double points or cusps of f, we can, by means of the $\frac{1}{2}m(m+3)-\delta-\kappa$ coefficients of ψ which remain arbitrary when ψ is prescribed to vanish at each double point and cusp, dispose of all except

$$mn-2(\delta+\kappa)-[\tfrac{1}{2}m(m+3)-(\delta+\kappa)];$$

when $m=n-1$ or $n-2$ it is easily seen that this is the same as before.

82. Let us assume now that the polynomials which occur, as the numerator and denominator, in the expression of a rational function, have the

* Reasons are given, Forsyth, *Theory of Functions*, p. 356, § 182, for the conclusion that this number is the deficiency of the Riemann surface having $f(y, x)=0$ as an associated equation. We shall assume this result here.

† As, for instance, the coefficients of y^m, y^{m-1}, $y^{m-1}x$, ..., y^n, $y^n x$, ..., $y^n x^{m-n}$, in which case the highest power of y, in $\psi + U_{m-n}f$, that remains, is y^{n-1}.

property here assigned to ψ, of vanishing once at each double point and cusp of f. Without attempting to justify this assumption completely, we remark that if it is not verified at any particular double point, the rational function will clearly take the same value at the double point by whichever of the two branches of the curve f the double point be approached. As a matter of fact this is not generally the case. Suppose then we wish to obtain a general form of rational function which has Q given finite points of f, $A_1, \dots, A_Q$, as poles of the first order. Draw through these poles, $A_1, \dots, A_Q$, any curve ψ whatever, of degree greater than $n-3$, which passes once through each double point and cusp of f. Then ψ will intersect f in

$$mn - 2(\delta + \kappa) - Q$$

other points $B_1, B_2, \dots$. Through these other points $B_1, B_2, \dots$ of f, and through the double points, draw another curve, ϑ, of the same degree as ψ. The curve ϑ will in general not be entirely determined by the prescription of the $mn - 2(\delta+\kappa) - Q$ points $B_1, B_2, \dots$. Let the number of its coefficients which still remain arbitrary be denoted by $q+1$. Then it would be possible by the prescription of, in all,

$$mn - 2(\delta + \kappa) - Q + q$$

points of ϑ, to determine ϑ completely. But by what has just been proved, ϑ is determined completely when all but p of its intersections are prescribed. Wherefore

$$mn - 2(\delta+\kappa) - Q + q = mn - 2(\delta + \kappa) - p.$$

Hence $Q - q = p$, and ϑ has the form

$$\lambda\psi + \lambda_1\vartheta_1 + \dots + \lambda_q\vartheta_q,$$

where $\lambda, \lambda_1, \dots, \lambda_q$ are arbitrary constants and $\psi, \vartheta_1, \dots, \vartheta_q$ are $q+1$ linearly independent curves, all passing through the $mn - 2(\delta+\kappa) - Q$ points $B_1, B_2, \dots$, as well as through the double points and cusps; and the general rational function with the Q prescribed poles will have the form

$$\lambda + \lambda_1 R_1 + \dots + \lambda_q R_q,$$

where $R_i = \vartheta_i/\psi$; and this function contains $q+1$ arbitrary coefficients.

83. In this investigation, which is given only for purposes of illustration, we have assumed that the prescription of a point of a curve determines one of its coefficients in terms of the remaining coefficients, and that the prescription of this one point does not of itself necessitate that the curve pass through other points; and we have obtained not the exact form of the Riemann-Roch Theorem (Chap. III. § 37), but the first approximation to that theorem which is expressed by $Q - q = p$; this result is true for all cases only when $Q > n(n-3) - 2(\delta+\kappa)$.

We may illustrate the need of the hypothesis that the curves ψ and ϑ pass through the double points and cusps, by considering the more particular case when the fundamental curve

$$f = (x, y)_2 + (x, y)_3 + (x, y)_4 = 0,$$

wherein $(x, y)_2$ is an integral homogeneous polynomial in x and y of the second degree, etc., is a quartic with a double point at the origin $x=0$, $y=0$. Since here $n=4$ and $\delta+\kappa=1$, we have

$$p=\tfrac{1}{2}(n-1)(n-2)-\delta-\kappa=\tfrac{1}{2}\,.\,3\,.\,2-1=2,$$

and therefore (in accordance with Chap. III. §§ 23, 24, etc.) there exists a rational function having any three prescribed points as poles of the first order. Let us attempt to express this function in the form ϑ/ψ, wherein ϑ, ψ are curves, of degree m, $(m>1)$, which do not vanish at the double point. Beside the three prescribed poles A_1, A_2, A_3 of the function, ψ will intersect f in $4m-3$ points B_1, B_2, ... The intersections with f of the general curve ϑ of degree m, are the same as those of a curve

$$\vartheta-U_{m-4}f=0,$$

provided $m \not< 4$, and are therefore determined by $\tfrac{1}{2}m(m+3)-\tfrac{1}{2}(m-4+1)(m-4+2)$, or $4m-3$ of them. And it is easily seen that the same result follows when $m=3$ or 2. Hence no curve ϑ can be drawn through the points B_1, B_2, ... other than the curve ψ, which already passes through them; and the rational function cannot be determined in the way desired. It will be found moreover that this is still true when the hypothesis, here made, that ψ and ϑ shall be of the same degree, is allowed to lapse. As in the general case, this hypothesis is made in order that the function obtained may be finite for infinite values of x and y.

A curve which passes through each double point and cusp of the fundamental curve f is said to be *adjoint*. When f has singularities of more complicated kind there is a corresponding condition, of greater complexity. For example in the case of the curve

$$f=y^2-(1-x^2)(1-k^2x^2)=0,$$

which, in the present point of view, we regard as a quartic, there is a singularity at the infinite end of the axis of y. If, in the usual way, we introduce the variable z to make the equation homogeneous, and then* put $y=1$, whereby the equation becomes

$$z^2=(z^2-x^2)(z^2-k^2x^2),$$

we see that the branches are, approximately, given by $z=\pm kx^2$, namely there is a point of self contact, the common tangent being $z=0$. If we assume that it is legitimate to regard this self contact as the limit of two coincident double points, we shall infer that the condition of adjointness for a curve ψ is that it shall touch the two branches of f at the point. For example this condition is satisfied by the parabola

$$y=ax^2+bx+c,$$

which, by the same transformation as that above, reduces to

$$z=ax^2+bxz+cz^2,$$

and it is obvious that the four intersections with f of this parabola, other than those at the singular point, are determined by all but p of them, p being in this case equal to 1.

We shall see in this chapter that we can obtain these results in a somewhat different way: the equation $y^2=(1-x^2)(1-k^2x^2)$ is a good example of those in which it is *not* convenient to regard the equation as a particular case of a curve of degree equal to the highest degree which occurs. Though this method, of regarding any given curve as a particular case of one whose degree is the degree of the highest term which occurs in the given equation of the curve, is always allowable, it is often cumbersome.

Ex. 1. Prove that the theorem, that the intersections with f of a variable curve ψ are determined by all but p of them, may be extended to the case where f has multiple points

* This process is equivalent to projecting the axis $y=0$ to infinity.

of order k, with separated tangents, by assuming that the condition of adjointness is that ψ should have a multiple point of order $k-1$ at every such multiple point of f, whose tangents are distinct from each other and from those of f. (In this case any such multiple point of f furnishes a contribution $\frac{1}{2}k(k-1)$ to the number $\delta+\kappa$ of f.)

Ex. 2. The curve $y^2=(x, 1)_6$ may be regarded as a sextic. Shew that the singular point at infinity may be regarded as the limit of eight double points, and that a general adjoint curve is

$$(x, 1)^{4+\mu}+y(x, 1)^{\mu+1}=0.$$

Ex. 3. Shew that for the curve $y^2=(x, 1)_{2p+2}$ a general adjoint curve is

$$(x, 1)^{\mu+2p}+y(x, 1)^{\mu+p-1}=0.$$

For further information on this subject consult Salmon, *Higher Plane Curves* (Dublin, 1879), pp. 42—48, and the references given in this volume, § 9 note, § 93, § 97, § 112 note, § 119.

84. In the remaining analytical developments of this chapter we suppose* the equation associated with the Riemann surface to be given in the form

$$f(y, x)=y^n+y^{n-1}(x, 1)_{\lambda_1}+\dots+y(x, 1)_{\lambda_{n-1}}+(x, 1)_{\lambda_n}=0,$$

so that y is an integral function of x. Let $\sigma+1$ be the dimension of y; then $\sigma+1$ is the least positive integer such that $y/x^{\sigma+1}$ is finite when x is infinite; thus if we put $x=1/\xi$ and $y=\eta/\xi^{\sigma+1}$, $\sigma+1$ is the least positive integer, such that η is an integral function of ξ. This substitution gives $f(y, x)=\xi^{-n(\sigma+1)}F(\eta, \xi)$, where

$$F(\eta, \xi)=\eta^n+\eta^{n-1}\xi^{\sigma+1-\lambda_1}(1, \xi)_{\lambda_1}+\dots+\eta\xi^{(n-1)(\sigma+1)-\lambda_{n-1}}(1, \xi)_{\lambda_{n-1}}$$
$$+\xi^{n(\sigma+1)-\lambda_n}(1, \xi)_{\lambda_n},$$

so that $\sigma+1$ is the least positive integer which is not less than any of the quantities

$$\lambda_1, \lambda_2/2, \dots, \lambda_{n-1}/(n-1), \lambda_n/n.$$

Ex. 1. For the case

$$y^4+y^2x^2(x, 1)_3+yx^3(x, 1)_4+x^4(x, 1)_5=0$$

the dimension of y as an integral function of x is 3. Writing $y=\eta/\xi^3$, where $x=1/\xi$, the equation becomes

$$\eta^4+\eta^2\xi(1, \xi)_3+\eta\xi^2(1, \xi)_4+\xi^3(1, \xi)_5=0$$

and η is an integral function of ξ of dimension 2. In fact $y_1=\eta/\xi^2=y/x$ satisfies the equation

$$y_1^4+y_1^2(x, 1)_3+y_1(x, 1)_4+(x, 1)_5=0$$

and is finite when $\xi=\infty$, or $x=0$.

Ex. 2. Shew that in the case in which the equation associated with the Riemann surface contains y to a degree equal to the highest aggregate degree which occurs, $\sigma=0$.

* Chap. IV. § 38.

Whenever we are considering the places of the surface for which $x = \infty$, we shall consider the surface in association with the equation $F(\eta, \xi) = 0$; and shall speak of the infinite places as given by $\xi = 0$. The original equation is practically unaffected by writing $x - c$ for x, c being a constant. We may therefore suppose the equation so written that at $x = 0$, the n sheets of the surface are distinct; and may speak of the places $x = 0$ as the places $\xi = \infty$.

85. By the simultaneous use of the equations $f(y, x) = 0$, $F(\eta, \xi) = 0$, we shall be better able to formulate our results in accordance with the view, hitherto always adopted, whereby the places $x = \infty$ are regarded as exactly like any finite places. But it should be noticed that both these equations may be regarded as particular cases of another in which homogeneous variables, of a particular kind*, are used. For put $x = \omega/z$, $y = u/z^{\sigma+1}$; we obtain $f(y, z) = z^{-n(\sigma+1)} U(u; \omega, z)$, where

$$U(u; \omega, z) = u^n + u^{n-1} z^{\sigma+1-\lambda_1} (\omega, z)_{\lambda_1} + \dots + u z^{(n-1)(\sigma+1)-\lambda_{n-1}} (\omega, z)_{\lambda_{n-1}} + z^{n(\sigma+1)-\lambda_n} (\omega, z)_{\lambda_n},$$

and it is clear that $U(u; \omega, z)$ is changed into $f(y, x)$ by writing $u = y$, $\omega = x$, $z = 1$, and is changed into $F(\eta, \xi)$ by writing $u = \eta$, $\omega = 1$, $z = \xi$. We may speak of ω, z as *forms*, of degree 1, and suppose that they do not become infinite, the values $x = \infty$ being replaced by the values $z = 0$. When ω, z are replaced by $t\omega, tz$, t being any quantity whatever, u is replaced by $t^{\sigma+1}u$, y and x remaining unaltered. We may therefore speak of u as a *form* of degree $\sigma + 1$.

Similarly $U(u; \omega, z)$ is a form of degree $n(\sigma+1)$, being multiplied by $t^{n(\sigma+1)}$ when u, ω, z are replaced by $t^{\sigma+1}u$, $t\omega$, tz respectively. That there is some advantage in using such homogeneous forms to express the results of our theory will sufficiently appear; but it seems proper that the results should first be obtained independently, in order that the implications of the notation may be made clear. We shall adopt this course.

Some examples of the change which our expressions will undergo when the results are expressed by homogeneous forms, may be fitly given here:— Instead of $f(y, x)$ we shall have $U(u; \omega, z)$ which is equal to $z^{n(\sigma+1)} f(y, x)$; instead of $f'(y)$ we shall have $U'(u) = z^{(n-1)(\sigma+1)} f'(y)$; instead of the integral function† g_i, of dimension $\tau_i + 1$, an integral form $\bar{g}_i$ of degree $\tau_i + 1$, equal to $z^{\tau_i+1} g_i$, will arise; since $\Sigma(\tau_i + 1) = n + p - 1$, it is easy to see that the determinant‡ $\Delta(1, \bar{g}_1, \dots, \bar{g}_{n-1})$ is equal to $z^{2n+2p-2} \Delta(1, g_1, \dots, g_{n-1})$. In accordance with § 48, Chap. IV. the former determinant will have a factor

* This homogeneous equation is used by Hensel. See the references given in Chap. IV. (§ 42). It may be regarded as a generalization of the familiar case when $\sigma = 0$.

† Chap. IV. § 42.

‡ Chap. IV. § 43.

$(\omega - cz)^r$ corresponding to a finite branch place of order r where $x = c$, and a factor z^s corresponding to a branch place of order s at $x = \infty$. Further, if, by the formula (H) of page 63, we calculate the form $\bar{\phi}_i(u, \omega, z)$ from $\bar{g}_1, \ldots, \bar{g}_{n-1}$, as $\phi_i(x, y)$ is there calculated from $g_1, \ldots, g_{n-1}$, it is easy to see that we obtain a form, $\bar{\phi}_i(u, \omega, z)$, which is equal to $z^{(n-1)(\sigma+1)-(\tau_i+1)}\phi_i(x, y)$. Hence also, if u_1, ω_1, z_1 denote special values of u, ω, z, the integral

$$\int \frac{z\,d\omega - \omega\,dz}{U'(u)} \frac{\mu^{-1}\bar{\phi}_0(u, \omega, z) + \Sigma\mu^{\tau_r}\bar{\phi}_r(u, \omega, z)\bar{g}_r(u_1, \omega_1, z_1)}{\omega z_1 - \omega_1 z},$$

wherein $\mu = (b\omega - az)/(b\omega_1 - az_1)$, a and b being arbitrary constants, is equal to

$$\int \frac{z^2dx \,.\, z^{(n-1)(\sigma+1)}}{z^{(n-1)(\sigma+1)}f'(y)} \cdot \frac{\mu^{-1}\phi_0(x, y) + \Sigma\mu^{\tau_r}(z_1/z)^{\tau_r+1}\phi_r(x, y)g_r(x_1, y_1)}{zz_1(x - x_1)},$$

and is thus equal to

$$\int \frac{dx}{f'(y)} \cdot \frac{\lambda^{-1}\phi_0(x, y) + \Sigma\lambda^{\tau_r}\phi_r(x, y)g_r(x_1, y_1)}{x - x_1},$$

where $\lambda = \mu z_1/z = (bx - a)/(bx_1 - a)$.

If in this we put $b = 0$, we obtain the form which we have already shewn to be part of the expression of an integral of the third kind (Chap. IV. p. 67). But if we put $b = 1$, the integral is exactly what we have already deduced (Chap. IV. p. 70, Ex. 1) by the ordinary process of putting $x = 1/(\xi - a)$ and regarding ξ as the independent variable.

We may, if we please, further specialise the quantities ω, z, of which hitherto only the ratio has been used, supposing* them defined by $\omega = x/(x - c)$, $z = 1/(x - c)$, where c is a constant. Then $\omega - cz = 1$.

Ex. 1. The integral of the first kind obtained in Chap. IV. § 45, p. 67, can similarly be written

$$\int \frac{z\,d\omega - \omega\,dz}{U'(u)}\left[(\omega, z)^{\tau_1 - 1}\bar{\phi}_1(u, \omega, z) + \ldots\ldots + (\omega, z)^{\tau_{n-1}-1}\bar{\phi}_{n-1}(u, \omega, z)\right].$$

Ex. 2. In the case $y^2 = (x, 1)_{2p+2}$, wherein y is of dimension $p+1$, the equation $U(u;\ \omega, z) = 0$ is

$$u^2 = (\omega, z)_{2p+2}$$

obtained by putting $y = u/z^{p+1}$, $x = \omega/z$.

86. We shall be largely concerned here with rational polynomials which are integral in x and y. The values of such a polynomial here considered are only those which it has for values of y and x satisfying the fundamental equation. We shall therefore suppose every integral polynomial in x and y reduced, by means of the fundamental equation, to a form in which the highest power of y which enters is y^{n-1}, say to a form

$$\psi(y, x) = y^{n-1}(x, 1)_{\mu_0} + \ldots + y^{n-1-i}(x, 1)_{\mu_i} + \ldots + (x, 1)_{\mu_{n-1}}.$$

* In this view ω and z are functions. If we regard c as throughout undetermined, we may regard these functions as having no *definite* infinities.

If herein we write $y = \eta/\xi^{\sigma+1}$, $x = 1/\xi$, $\sigma + 1$ being, as before, the dimension of y as an integral function of x, we shall obtain $\psi(y, x) = \xi^{-G}\,\Psi(\eta, \xi)$, where $\Psi(\eta, \xi)$ is an integral polynomial in η and ξ of which a representative term is

$$\eta^{n-1-i}\,\xi^{G-(n-1-i)(\sigma+1)-\mu_i}(1, \xi)_{\mu_i}, \qquad i=0, 1, \ldots\ldots, (n-1)$$

and G is the positive integer equal to the greatest of the quantities

$$(n-1-i)(\sigma+1)+\mu_i.$$

Thus G is the highest dimension occurring for the terms of $\psi(y, x)$, and $\Psi(\eta, \xi)$ is not identically divisible by ξ. The dimension of the integral function $\psi(y, x)$ may be G; but if $\Psi(\eta, \xi)$ vanish in every sheet at $\xi = 0$, the dimension of $\psi(y, x)$ will be less than G. For this reason we shall speak of G as the *grade* of $\psi(y, x)$. It is clear that if all the values of η for $\xi = 0$ be distinct, that is, if $F'(\eta)$ do not vanish for any place $\xi = 0$, the polynomial $\Psi(\eta, \xi)$, of order $n-1$ in η, cannot vanish for all the n places $\xi = 0$. In that case the grade and the dimension of $\psi(y, x)$ are necessarily the same. Further, by the vanishing of one of the coefficients, a polynomial of grade G may reduce to one of lower grade. In this sense a polynomial of low grade may be regarded as a particular case of one of higher grade.

In what follows we shall consider all polynomials whose grade is lower than $(n-1)\sigma + n - 3$ or $(n-1)(\sigma+1) - 2$, as particular cases of polynomials of grade $(n-1)\sigma + n - 3$: the general expression of the grade will therefore* be $(n-1)\sigma + n - 3 + r$, or $(n-1)(\sigma+1) + r - 2$, where r is zero or a positive integer. The most general form of a polynomial of grade $(n-1)(\sigma+1)+r-2$ is easily seen to be

$$\begin{aligned}\psi(y, x) = y^{n-1}(x, 1)_{r-2} + y^{n-2}(x, 1)_{r-1} + \ldots + y^{n-1-i}(x, 1)_{r-1} + \ldots + (x, 1)_{r-1}\\ + x^r\{y^{n-2}(x, 1)_{\sigma+1-2} + \ldots\ldots + y^{n-1-i}(x, 1)_{i(\sigma+1)-2} + \ldots\ldots + (x, 1)_{(n-1)(\sigma+1)-2}\},\end{aligned}$$

wherein the first line is to be entirely absent if $r = 0$, the first term of the first line is to be absent if $r = 1$, and the first term of the second line is to be absent if $\sigma = 0$.

Hence when $r > 0$, the general polynomial of grade $(n-1)\sigma + n - 3 + r$ contains

$$nr - 1 + \tfrac{1}{2}(n-1)(n-2+n\sigma)$$

terms, this being still true if $\sigma = 0$; but when $r = 0$, the general polynomial of grade $(n-1)\sigma + n - 3$ contains

$$\tfrac{1}{2}(n-1)(n-2+n\sigma)$$

terms. This is not the number obtained by putting $r = 0$ in the number obtained for $r > 0$.

* The number is written in the former way to point out the numbers for the common case when $\sigma = 0$.

Further, putting

$$\psi(y, x) = \xi^{-(n-1)\sigma-(n-3)-r}\,\Psi(\eta, \xi),$$

and denoting the aggregate number of zeros of $\Psi(\eta, \xi)$ at $\xi = 0$ by μ, it is clear that the aggregate number of infinities of $\psi(y, x)$ at $x = \infty$ is $[(n-1)\sigma + n - 3 + r]\,n - \mu$. Since $\psi(y, x)$ is only infinite for $x = \infty$, this is also the total number of zeros of $\psi(y, x)$. We shall find it extremely convenient to introduce a certain artificiality of expression, and to speak of the sum of the number of zeros of $\psi(y, x)$ and the number of zeros of $\Psi(\eta, \xi)$ at $\xi = 0$ as the number of *generalized* zeros of $\psi(y, x)$. This number is then $n(n-1)(\sigma+1) + n(r-2)$.

If by a change in the values of the coefficients in $\psi(y, x)$, $\Psi(\eta, \xi)$ should take the form $\xi\Psi_1(\eta, \xi)$ where $\Psi_1(\eta, \xi)$ is an integral polynomial in η and ξ, so that $\psi(y, x)$ is equal to $\xi^{-(n-1)\sigma-(n-3)-(r-1)}\,\Psi_1(\eta, \xi)$, the sum of the number of finite zeros of $\psi(y, x)$ and the number of zeros of $\Psi_1(\eta, \xi)$ is $n(n-1)(\sigma+1) + n(r-3)$. But, since $\Psi(\eta, \xi)$ is equal to $\xi\Psi_1(\eta, \xi)$, the number of zeros of $\Psi(\eta, \xi)$ at $\xi = 0$ is n more than the number of zeros of $\Psi_1(\eta, \xi)$ at $\xi = 0$. Hence the sum of the number of finite zeros of $\psi(y, x)$ and the number of zeros of $\Psi(\eta, \xi)$ at $\xi = 0$, is still equal to

$$n(n-1)(\sigma+1) + n(r-2).$$

Ex. i. The number $n(n-1)(\sigma+1)+n(r-2)$ is clearly the number of zeros of the *integral form*

$$z^{(n-1)\sigma+n-3+r}\,\psi(uz^{-\sigma-1}, \omega z^{-1}).$$

Ex. ii. The generalized number of zeros of $f'(y)$, for which $r=2$, is $n(n-1)(\sigma+1)$.

Ex. iii. The general polynomial of grade d, $<(n-1)\sigma+n-3$, contains

$$\left[1+E\left(\frac{d}{\sigma+1}\right)\right]\left[1+d-\tfrac{1}{2}(\sigma+1)E\left(\frac{d}{\sigma+1}\right)\right] \text{ terms,}$$

$E(x)$ being the greatest integer in x. Its generalized number of zeros is nd.

87. We introduce now a certain speciality in the integral polynomials under consideration, that known as *adjointness*.

An integral polynomial $\psi(y, x)$ is said to be adjoint at a finite place $(x = a,\ y = b)$ when the integral

$$\int^x \frac{\psi(y, x)}{f'(y)}\,dx$$

is finite at this place. If t be the infinitesimal at the place (Chap. I. §§ 2, 3) the condition is equivalent to postulating that the expression

$$\frac{\psi(y, x)}{f'(y)}\,\frac{dx}{dt}$$

shall be finite at the place; or again equivalent to postulating that the expression

$$\frac{(x-a)\,\psi(y, x)}{f'(y)}$$

shall be zero at the place, to the first order at least.

As a limitation for the polynomial $\psi(y, x)$, the condition is therefore ineffective at all places where $f'(y)$ is not zero. And if at a finite place where $f'(y)$ vanishes, $i+w$ denote the order of zero of $f'(y)$, $w+1$ being the number of sheets that wind at this place*, the condition is that $\psi(y, x)$ vanish to at least order i at the place. We shall call $\frac{1}{2}i$ the index of the place; the condition of adjointness is therefore ineffective at all places of zero index.

If $\psi(y, x)$ be of grade $(n-1)\sigma+n-3+r$, and

$$\psi(y, x) = \xi^{-(n-1)\sigma-(n-3)-r}\,\Psi(\eta, \xi),$$

the condition of adjointness of $\psi(y, x)$ for infinite places, is that, at all places $\xi=0$ where $F'(\eta)=0$, the function

$$\frac{\xi\Psi(\eta, \xi)}{F'(\eta)}$$

should be zero, to the first order at least. It is easily seen that this is the same as the condition that the integral

$$\int^x \frac{1}{x^r}\frac{\psi}{f'(y)}\,dx$$

should be finite at the place considered.

When the condition of adjointness is satisfied at all finite and infinite places where $f'(y)=0$ or $F'(\eta)=0$, the polynomial $\psi(y, x)$ is said to be *adjoint*. If $\Pi(x-a)$ denote the integral polynomial which contains a simple factor corresponding to every finite value of x for which $f'(y)$ vanishes, and if N denote the number of these factors, it is immediately seen that the polynomial $\psi(y, x)$ is adjoint provided the function

$$\frac{\Pi(x-a)}{x^{N+r-1}f'(y)}\,\psi(y, x)$$

is zero, to the first order at least, at all the places where $f'(y)=0$ or $F'(\eta)=0$.

Ex. i. For the surface associated with the equation

$$f(y, x) = (x, y)_2 + (x, y)_3 + (x, y)_4 = 0$$

there are two places at $x=0$, at each of which $y=0$. At each of these places $f'(y)$ vanishes to the first order, and $w=0$. Hence the condition of adjointness is that $\psi(y, x)$ vanishes

* It is easy to see that i is not a *negative* integer. Cf. Forsyth, *Theory of Functions*, p. 169.

to the first order at each of these places. The general adjoint polynomial will therefore not contain any term independent of x and y.

Ex. ii. For the surface

$$y^4 - y^2[(1+k^2)x^2+1] + k^2x^4 = 0$$

there are two places at $x=0$, at each of which y is zero of the second order: they are not branch places. At each of these $f'(y)$ vanishes to the second order.

The dimension of y is 1, and the general polynomial of grade $(n-1)\sigma+n-3+1$ or 2, is*

$$Ay^2+By+C+x[Dy+Ex+F].$$

In order that this may vanish to the second order at the places in question, it is sufficient that $C=0$ and $F=0$. Then the polynomial takes the form

$$By+Ay^2+Dxy+Ex^2,$$

and if we put x/η for x and $1/\eta$ for y this becomes, save for a factor η^{-2},

$$B\eta+A+Dx+Ex^2,$$

which is therefore an adjoint polynomial for the surface

$$1-(1+k^2)x^2-\eta^2+k^2x^4=0.$$

Compare § 83.

Ex. iii. Prove that the general adjoint polynomial for the surface

$$y^2=(x-a)^3,$$

is

$$y(x, 1)_{r-2}+(x-a)(x, 1)_{r-1}=0.$$

(The index of the place at $x=a$ is 1.)

88. Since the number of generalized zeros of $f'(y)$ is $n(n-1)(\sigma+1)$, (§ 86, Ex. ii), we have, in the notation here adopted,

$$\Sigma(i+w)=n(n-1)(\sigma+1),$$

or if I denote Σi and W denote Σw, the summation extending to all finite and infinite places of the surface

$$I+W=n(n-1)(\sigma+1).$$

Hence, as†

$$W=2n+2p-2,$$

we can infer

$$p=\tfrac{1}{2}(n-1)(n-2+n\sigma)-\tfrac{1}{2}I,$$

shewing that I is an even integer.

Further if X denote the number of zeros of an adjoint polynomial $\psi(y, x)$, of grade $(n-1)\sigma+n-3+r$, exclusive of those occurring at places where $f'(y)=0$ or $F'(\eta)=0$, and calculated on the hypothesis that the adjoint polynomial vanishes, at a place where $f'(y)$ or $F'(\eta)$ vanishes, to an order equal to twice the index of the place‡, we have the equation

$$X+I=n(n-1)(\sigma+1)+n(r-2).$$

* § 86 preceding.

† Forsyth, *Theory of Functions*, p. 349.

‡ So that a place of index $\tfrac{1}{2}i$ where $\psi(y, x)$, or $\Psi(\eta, \xi)$, vanishes to order $i+\lambda$, will furnish a contribution λ to the number X.

Thus, as

$$I = n(n-1)(\sigma+1) - 2(n-1) - 2p,$$

we have

$$X = nr + 2p - 2;$$

and this is true when $r = 0$.

These important results may be regarded as a generalization of some of Plücker's equations* for the case $\sigma = 0$.

Ex. i. The number of terms in the general polynomial of grade $(n-1)\sigma+n-3+r$ was proved to be $\frac{1}{2}(n-1)(n-2+n\sigma)+nr-1$ or $\frac{1}{2}(n-1)(n-2+n\sigma)$, according as $r>0$ or $r=0$. This number may therefore be expressed as $p+\frac{1}{2}I+nr-1$ or $p+\frac{1}{2}I$ in these two cases.

Ex. ii. It is easy to see, in the notation explained in § 85, that the homogeneous form $\Delta(1, u, u^2, \dots, u^{n-1})$ is of degree $n(n-1)(\sigma+1)$ in ω and z, and the form $\Delta(1, \bar{g}_1, \dots, \bar{g}_{n-1})$ of degree W. The quotient $\Delta(1, u, \dots, u^{n-1})/\Delta(1, \bar{g}_1, \dots, \bar{g}_{n-1})$ is (§ 43) an integral form in ω, z, which, by an equation proved here, is of degree I. It is the square of an integral homogeneous form ∇ whose degree in ω, z together is $\frac{1}{2}I$.

Ex. iii. It can be proved (compare § 43 *b*, Exx. 1, 2, and § 48; also Harkness and Morley, *Theory of Functions*, pp. 269, 270, 272, or Kronecker's original paper, *Crelle*, t. 91) that if for y we take the function

$$\lambda+\lambda_1 g_1+\dots+\lambda_{n-1} g_{n-1},$$

wherein $\lambda, \lambda_1, \dots, \lambda_{n-1}$ are integral polynomials in x, of sufficient (but finite) order, the polynomial ∇ occurring in the equation,

$$\Delta(1, y, \dots, y^{n-1}) = \nabla^2 \Delta(1, g_1, \dots, g_{n-1}),$$

cannot, for general values of the coefficients in $\lambda, \lambda_1, \dots, \lambda_{n-1}$, have any repeated factor, or have any factor which is also a factor of $\Delta(1, g_1, \dots, g_{n-1})$. And the inference can be made† that for this dependent variable y, there is no place at which the index is greater than $\frac{1}{2}$, and no value of x for which two places occur at which $f'(y)$, or $F'(\eta)$, is zero.

89. We proceed, now, to shew the utility of the notion of adjoint polynomials for the solution of the problem of finding the expression of a rational function of given poles.

Let R be any rational function, and suppose, first, that none of the finite poles of R are at places where $f'(y) = 0$. Let ψ be any integral polynomial, chosen so as to be zero at every finite pole of R, to an order at least as high as the order of the pole of R, and to be adjoint at every finite place where $f'(y)$ vanishes. Denote the integral polynomial $\Pi(x-a)$, which contains a linear factor corresponding to every finite value of x for which $f'(y)$ vanishes, by μ. Then the rational function

$$\mu\Lambda(y, x) = \mu R\psi/f'(y)$$

* Salmon, *Higher Plane Curves* (Dublin, 1879), p. 65.

† See also Noether, *Math. Annal.* t. xxiii. p. 311 (Rationale Ausführung, u. s. w.), and Halphen, *Comptes Rendus*, t. 80 (1875), where a proof is given that every algebraic plane curve may be regarded as the projection of a space curve having only one multiple point at which all the tangents are distinct. But see Valentiner, *Acta Math.*, ii. p. 137.

is finite at all finite places where R is infinite, and is finite, being zero, at every finite place at which $f'(y)=0$. If $y_1, \ldots, y_n$ denote the n values of y which belong to any value of x, and c be an arbitrary constant, the function

$$\sum_{i=1}^{n} \frac{(c-y_1)(c-y_2)\ldots(c-y_n)}{c-y_i}\, \mu\Lambda(y_i, x),$$

is a symmetrical function of $y_1, \ldots, y_n$ and, therefore, expressible as a rational function in x only; moreover the function is finite for all finite values of x and, therefore, expressible as an integral polynomial in x. Since this polynomial vanishes for every finite value of x which reduces the product μ to zero, it must divide by μ. Finally, the function is an integral polynomial in c, of degree $n-1$. Hence we have an equation of the form

$$\sum_{i=1}^{n} \frac{(c-y_1)\ldots(c-y_n)}{c-y_i}\, \Lambda(y_i, x) = c^{n-1}A_0 + c^{n-2}A_1 + \ldots + cA_{n-2} + A_{n-1},$$

wherein $A_0, A_1, \ldots, A_{n-1}$ are integral polynomials in x.

Therefore, putting $c=y_i$, recalling the form of the function $\Lambda(y, x)$, and replacing y_i by y, we have the result

$$R\psi = y^{n-1}A_0 + y^{n-2}A_1 + \ldots + yA_{n-2} + A_{n-1},$$

which we may write in the form

$$R = \vartheta/\psi,$$

ϑ being an integral polynomial in x and y.

Since

$$\frac{(x-a)\vartheta}{f'(y)} = R\frac{(x-a)\psi}{f'(y)},$$

ϑ, like ψ, is adjoint at every finite place where $f'(y)$ vanishes.

Suppose, next, that the function R has finite poles at places where $f'(y)$ vanishes. Then the polynomial ψ is to be chosen so that $R(x-a)\psi/f'(y)$ is zero at such a place, a being the value of x at the place. This may be stated by saying that ψ is adjoint at such a place and, *besides*, satisfies the condition of being zero at the place to as high order as R is infinite.

Corollary. Suppose R to be an integral function; and for a finite place, $x=a$, $y=b$, where $f'(y)$ vanishes, suppose $t+1$ to be the least positive integer such that $(x-a)^{t+1}/f'(y)$ has limit zero at the place. Then the polynomial ψ of the preceding investigation may be replaced by the product $\Pi(x-a)^t$, extended to all the finite values of x for which $f'(y)$ is zero. Hence, any integral function is expressible in the form

$$\vartheta/\Pi(x-a)^t,$$

where ϑ is an integral polynomial in x and y, which is adjoint at every finite place where $f'(y)$ vanishes.

If the order of a zero of $f'(y)$ be represented as before by $i+w$, it is clear that the corresponding value of $t+1$ is the least positive integer for which $(t+1)(w+1) > i+w$, or, for which $t > (i-1)/(w+1)$. Hence the denominator $\Pi(x-a)^t$ only contains factors corresponding to places at which the index $\frac{1}{2}i$ is greater than zero; if the index be zero at all the finite places at which $f'(y)$ vanishes, every integral function is expressible integrally.

It does not follow that when the index is zero at all finite places, the functions $1, y, \dots, y^{n-1}$, form a fundamental system of integral functions for which the condition of dimensions is satisfied. For the sum of the dimensions of $1, y, \dots, y^{n-1}$ is greater than $p+n-1$ by the sum of the indices at all the places $x=\infty$.

It is clear that if R be any rational function whatever, it is possible to find an integral polynomial in x only, say λ, such that λR is an integral function. To this integral function we may apply the present Corollary. The reader who recalls Chapter IV. will compare the results there obtained.

90. Let the polynomial ψ be of grade $(n-1)\sigma+n-3+r$, and the polynomial ϑ of grade $(n-1)\sigma+n-3+s$, so that

$$\psi = \xi^{-(n-1)\sigma-(n-3)-r}\Psi, \quad \vartheta = \xi^{-(n-1)\sigma-(n-3)-s}\Theta,$$

and

$$R = \xi^{r-s}\Theta/\Psi,$$

Θ, Ψ being integral polynomials in η and ξ.

If R have poles for $\xi=0$, it will generally be convenient to choose the polynomial ψ so that $R\Psi$ is finite at all places $\xi=0$; if $F'(\eta)$ vanish for any places $\xi=0$, it is also convenient, as a rule, to choose ψ so that $\xi\Psi/F'(\eta)$ vanishes at every place $\xi=0$ where $F'(\eta)$ vanishes, namely, so that ψ is adjoint at infinity. When both R is infinite and $F'(\eta)$ vanishes at a place where $\xi=0$, we may suppose ψ so chosen that $\xi R\Psi/F'(\eta)$ is zero at the place. Let ψ be chosen to satisfy these conditions. Then, since $R\Psi$, $=R\psi\,.\,\xi^{(n-1)\sigma+n-3+r}$, is finite at every place, except $\xi=\infty$, and $(1-a\xi)\Psi/F'(\eta)$, $=\xi^{r-1}(x-a)\psi/f'(y)$, vanishes at every place $x=a$, $y=b$, where x is finite, at which $f'(y)$ vanishes, except $\xi=\infty$, it follows, as here, that R can be written in a form

$$R = \Theta_1/\Psi,$$

wherein Θ_1 is an integral polynomial in η and ξ.

Hence $\Theta_1 = \xi^{r-s}\Theta$, and therefore $r-s$ is not negative: namely, the polynomial ϑ which occurs in the expression of a rational function in the form $R=\vartheta/\psi$, is not of higher grade than the denominator ψ, provided ψ be chosen to be adjoint at infinity, and, at the same time, to compensate the poles of R which occur for $x=\infty$. Since a polynomial of low grade is a particular case of one of higher grade we may regard ϑ and ψ as of the same grade.

Hence we can formulate a rule for the expression of a rational function of assigned poles as follows—*Choose any integral polynomial ψ which is adjoint at all finite places and is adjoint at infinity, which, moreover, vanishes at every finite place and at every infinite place* where R is infinite, to as high order as that of the infinity of R. If a pole of R fall at a place where $f'(y)$, or $F'(\eta)$, vanishes, these two conditions may be replaced by a single one in accordance with the indications of the text. Then, choose an integral polynomial ϑ, of the same grade as ψ, also adjoint at all finite and infinite places, which, moreover, vanishes at every zero of the polynomial ψ other than the poles of R, to as high order as the zero of ψ at that place. Then the function can be expressed in the form ϑ/ψ.*

91. We may apply the rule just given to determine the form of the integrals of the first kind.

If v be any integral of the first kind, dv/dx is a rational function having no poles, for finite values of x, except at the branch places of the surface. If a be the value of x at one of these branch places, the product $(x-a)\,dv/dx$ vanishes at the place. Hence we may apply to dv/dx the same reasoning as was applied to the function $\Lambda(y, x)$ in § 89, and obtain the result, that dv/dx can be expressed in the form

$$\frac{dv}{dx}=\frac{y^{n-1}A_0+y^{n-2}A_1+\dots+yA_{n-2}+A_{n-1}}{f'(y)},$$

wherein $A_0, \dots, A_{n-1}$ are integral polynomials in x. Denote the numerator by ϕ, and let its grade be denoted by $(n-1)\sigma+n-3+r$; then

$$-\frac{dv}{d\xi}=\xi^{-2}\frac{dv}{dx}=\xi^{-2}\frac{\xi^{-(n-1)\sigma-(n-3)-r}\Phi}{\xi^{-(n-1)\sigma-(n-1)}F'(\eta)}=\frac{\xi^{-r}\Phi}{F'(\eta)}.$$

But, as a function of ξ, $dv/d\xi$ has exactly the same character as has dv/dx as a function of x. Thus by a repetition of the argument $F'(\eta)\,dv/d\xi$ is expressible as an integral function of η and ξ. Thus r is either zero or negative.

Wherefore, $f'(y)\dfrac{dv}{dx}$ is an integral polynomial in x and y, of grade $(n-1)\sigma+n-3$ or less. It is clearly adjoint at all finite places, and, reckoned as a particular case of a polynomial of grade $(n-1)\sigma+n-3$, it is clearly also adjoint at infinity.

Conversely, it is immediately seen, that if ϕ be any integral polynomial of

* That is, if the polynomial be ψ, of grade $(n-1)\sigma+n-3+r$ and $\psi=\Psi\xi^{-(n-1)r-(n-3)-r}$, Ψ vanishes at $\xi=0$ to the order stated. A similar abbreviated phraseology is constantly employed.

grade $(n-1)\sigma+n-3$, which is adjoint at all finite and infinite places, the integral

$$\int \frac{\phi}{f'(y)}\,dx,$$

is an integral of the first kind.

Corollary. We have seen that the general adjoint polynomial of grade $(n-1)\sigma+n-3$ contains $p+\frac{1}{2}I$ terms, and we know that there are just p linearly independent integrals of the first kind. We can therefore make the inference

The condition of adjointness, for a polynomial of grade $(n-1)\sigma+n-3$, is equivalent to $\frac{1}{2}I$ linearly independent conditions for the coefficients of the polynomial, and reduces the number of terms in the polynomial to p.

92. We have shewn that a general polynomial of grade $(n-1)\sigma+n-3+r$ is of the form

$$\psi_{n-3+r} = y^{n-1}(x, 1)_{r-2} + y^{n-2}(x, 1)_{r-1} + \ldots + y(x, 1)_{r-1} + (x, 1)_{r-1} + x^r \psi_{n-3}.$$

We shall assume in the rest of this chapter that the condition of adjointness for a general polynomial of grade $(n-1)\sigma+n-3+r$ is equivalent to as many independent linear conditions as for a general polynomial of grade $(n-1)\sigma+n-3$. Thence, *the general adjoint polynomial of grade $(n-1)\sigma+n-3+r$ contains $nr-1+p$ terms.*

Further we shewed that the adjoint polynomial of grade $(n-1)\sigma+n-3$ has $2p-2$ zeros exclusive of those falling at places where $f'(y)=0$, or $F'(\eta)=0$.

Hence, the $2p-2$ zeros of the differential dv (Chap. II. § 21) are the zeros of the polynomial $f'(y)\,dv/dx$, exclusive of those where $f'(y)=0$, or $F'(\eta)=0$.

It is in fact an obvious corollary from the condition of adjointness that

$$dv/dt = [\phi/f'(y)]\frac{dx}{dt}$$

only vanishes when ϕ vanishes. For, at a place where $f'(y)=0$, ϕ vanishes i times, $\dfrac{dx}{dt}$ vanishes w times, and $f'(y)$ vanishes $i+w$ times.

Ex. i. For the surface associated with the equation

$$f(y, x) = y^4 + y^3(x, 1)_1 + y^2(x, 1)_2 + y(x, 1)_3 + (x, 1)_4 = 0,$$

where $(x, 1)_1, \ldots$ are integral polynomials in x of the degrees indicated by their suffixes, $\sigma=0$; and the general polynomial of grade $(n-1)\sigma+n-3$ or 1, is of the form (§ 86)

$$Ay + Bx + C.$$

The indices of the places where $f'(y)=0$ are easily seen to be everywhere zero—there are no places, beside branch places, at which $f'(y)$ vanishes. Hence p is equal to the number of terms in this polynomial, or $p=3$. And this polynomial vanishes in $2p-2=4$ places. These results may be modified when the coefficients in the equation have special values.

Ex. ii. For the more particular case when the equation is

$$f(y, x)=y^4+y^3(x, 1)_1+y^2(x, 1)_2+yx(x, 1)_2+x^2(x, 1)_2=0$$

there are two places at $x=0$ at which $y=0$. For general values of the coefficients in the equation these are not branch places and $f'(y)$ vanishes to the first order at each; the index at each place is therefore $\frac{1}{2}i$ where $i=1$, and the condition for adjointness of the general polynomial of grade 1, is that it shall vanish once at each of these places. These conditions are equivalent to one condition only, that $C=0$. Hence, as there are no other places where the index is greater than zero, the general integral of the first kind is

$$\int (Ay+Bx)\,dx/f'(y)$$

and $p=2$; the polynomial $Ay+Bx$ vanishes in $2p-2$ or 2 places other than the places $x=0$, $y=0$ at which $f'(y)=0$.

Ex. iii. In general when the equation of the surface represents a plane curve with a double point, the condition of adjointness at the places which correspond to this double point, is the one condition that the adjoint polynomial vanish at the double point*.

Ex. iv. Prove that for each of the surfaces

$$y^3+y^2(x, 1)_1+y(x, 1)_2+(x, 1)_4=0,$$
$$y^3+y^2(x, 1)_2+y(x, 1)_4+(x, 1)_7=0,$$

there is only one place at infinity and the index there, in both cases, is 1.

Shew that the index at the infinite place of Weierstrass's canonical surface† is in all cases

$$\tfrac{1}{2}(a-1)\left(a\left|\frac{r}{a}\right|-r-1\right),$$

where $\left|\frac{r}{a}\right|$ means the least integer greater than r/a, and that the deficiency is given by

$$p-\tfrac{1}{2}(r-1)(a-1)-I',$$

where I' denotes the sum of the indices at all finite places of the surface.

Cf. *Camb. Phil. Trans.* xv. iv. p. 430. The practical method of obtaining adjoint polynomials of grade $(n-1)\sigma+n-3$ which is explained in that paper (pp. 414—416) is often of great use.

Ex. v. In the notation of Chap. IV. the polynomial

$$(x, 1)^{\tau_1-1}\phi_1+\dots+(x, 1)^{\tau_{n-1}-1}\phi_{n-1}$$

is an adjoint polynomial of grade $(n-1)\sigma+n-3$.

Ex. vi. We can prove in exactly the same way as in the text that an integral of the third kind infinite only at the ordinary finite places (x_1, y_1), (x_1', y_1'), at the former like $C\log(x-x_1)$ and at the latter like $-C\log(x-x_1')$, C being a constant, can be written in the form

$$P=\int\frac{\psi}{(x-x_1)(x-x_1')}\frac{dx}{f'(y)},$$

where ψ is an adjoint integral polynomial in x and y, of grade $(n-1)\sigma+n-1$, which

* The sum of the indices at the k places of the surface corresponding to an ordinary k-ple point of the curve is $\frac{1}{2}k(k-1)$; the index at each of the places is in fact $\frac{1}{2}(k-1)$. Cf. § 83, Ex. i.

† Chap. V. § 64.

vanishes at the $(n-1)$ places $x=x_1$ where y is not equal to y_1 and at the $(n-1)$ places $x=x_1'$ where y is not equal to y_1'. Putting ψ in the form

$$\psi=(x-x_1')(C_0y^{n-1}+C_1y^{n-2}+\dots+C_{n-1})-(x-x_1)(C_0'y^{n-1}+C_1'y^{n-2}+\dots+C'_{n-1})$$
$$+(x-x_1)(x-x_1')(R_0y^{n-1}+R_1y^{n-2}+\dots+R_{n-1}),$$

where $C_0, \dots, C_{n-1}, C_0', \dots, C'_{n-1}$ are constants, it follows, since $(x-x_1')y^{n-1}$ is of grade $(n-1)\sigma+n$, and $(R_0y^{n-1}+R_1y^{n-2}+\dots+R_{n-1})(x-x_1)(x-x_1')$ is of grade $(n-1)\sigma+n+1$ at least, that R_0 is zero and $C_0=C_0'$. Further, if the equation associated with the surface be written

$$f(y, x)=y^n+Q_1y^{n-1}+Q_2y^{n-2}+\dots+Q_{n-1}=0,$$

and $\chi_i(x)$ denote

$$y^i+Q_1y^{i-1}+\dots+Q_i,$$

it follows, from the condition for ψ which ensures that the integral P is not infinite at all the n places $x=x_1$, that the factors of the polynomial

$$C_0y^{n-1}+C_1y^{n-2}+\dots+C_{n-1}$$

are the same as those of $f(y, x)/(y-y_1)$, or of

$$y^{n-1}+\chi_1(x_1)\cdot y^{n-2}+\chi_2(x_1)\cdot y^{n-3}+\dots+\chi_{n-1}(x_1).$$

Hence, save for a constant multiplier, P has the form

$$P=\int\frac{dx}{f'(y)}\left[(x, x_1)-(x, x_1')+y^{n-2}(x, 1)_{\sigma-1}+y^{n-3}(x, 1)_{2\sigma}+\dots+(x, 1)_{(n-1)\sigma+n-3}\right],$$

where (x, x_1) denotes

$$[y^{n-1}+y^{n-2}\chi_1(x_1)+\dots+\chi_{n-1}(x_1)]/(x-x_1),$$

so that $(x, x_1)=(x_1, x)$, and (x, x_1') denotes a similar expression.

A general polynomial ψ of grade $(n-1)\sigma+n-1$ contains $2n-1$ more terms than a general polynomial of grade $(n-1)\sigma+n-3$. In accordance with the assumption made in § 92 the general adjoint polynomial ψ of grade $(n-1)\sigma+n-1$ will contain $2n-1+p$ terms. The condition that ψ vanishes in the $2n-2$ places $x=x_1$, $x=x_1'$ other than those where $y=y_1$, $y=y_1'$ respectively, will reduce the number of terms to $p+1$. This is exactly the proper number of terms for a general integral of the third kind (cf. § 45, p. 67). The assumption of § 92 is therefore verified in this instance.

The practical determination of an integral of the third kind here sketched is often very useful. In the hyperelliptic case it gives the integral immediately.

Ex. vii. Prove that if the matrix of substitution Ω occurring on p. 62, in the equation

$$(1, y, y^2, \dots, y^{n-1})=\Omega(1, g, \dots, g_{n-1}),$$

be denoted by Ω_x, and the general element of the product-matrix $\Omega_x\Omega_{x_1}^{-1}$ be denoted by $c_{r,s}$, and if, for distinctness of expression, we denote the elements

$$\chi_{n-1}(x), \chi_{n-2}(x), \dots, \chi_1(x), 1, 1, y_1, y_1^2, \dots, y_1^{n-1},$$

respectively by

$$u_1, u_2, \dots, u_{n-1}, u_n, k_1, k_2, k_3, \dots, k_n,$$

then the function

$$\phi_0(x)+\phi_1(x)g_1(x_1)+\dots+\phi_{n-1}(x)g_{n-1}(x_1),$$

which occurs in the expression of an integral of the third kind given in § 45, is equal to

$$c_{11}u_1k_1+\dots+c_{ii}u_ik_i+\dots+c_{rs}u_rk_s+c_{sr}u_sk_r+\dots.$$

This takes the form $u_1k_1+\dots+u_nk_n$ obtained in Ex. vi. when $c_{rs}=0$ and $c_{ii}=1$, namely when Ω is a constant. This condition will be satisfied when the index is zero at all finite and infinite places.

Ex. viii. Prove for the surface associated with the equation

$$y^3+y^2(x,1)_1+y(x,1)_2+(x,1)_4=0,$$

that the condition of adjointness for any polynomial is that it vanish to the second order at the place $\xi=0$.

Thence shew that the polynomial

$$(x-x_1')[y^2+y\chi_1(x_1)+\chi_2(x_1)]-(x-x_1)[y^2+y\chi_1(x_1')+\chi_2(x_1')]$$
$$+(Ay+Bx^2+Cx+D)(x-x_1)(x-x_1')$$

is adjoint provided $B=0$; and thence that the integral of the third kind is

$$\int\frac{dx}{f'(y)}\left[\frac{y^2+y\chi_1(x_1)+\chi_2(x_1)}{x-x_1}-\frac{y^2+y\chi_1(x_1')+\chi_2(x_1')}{x-x_1'}+Ay+Cx+D\right].$$

Ex. ix. There is a very important generalization* of the method of Ex. vi. for forming an integral of the third kind. Let μ be any positive integer. Let a general non-adjoint polynomial of grade μ be chosen so as to vanish in the two infinities of the integral, which we suppose, first of all, to be ordinary finite places. Denote this polynomial by L. It will vanish† in $n\mu-2$ other places $B_1, B_2, \ldots$. Take an adjoint polynomial ψ, of grade $(n-1)\sigma+n-3+\mu$, chosen so as to vanish in the places $B_1, B_2, \ldots$. The polynomial will presumably contain (§ 92) $n\mu-1+p-(n\mu-2)$ or $p+1$ homogeneously entering arbitrary coefficients, and will vanish (§ 88) in $n\mu+2p-2-(n\mu-2)$ or $2p$ places other than the places $B_1, B_2, \ldots$ and places where $f'(y)$, or $F'(\eta)$, vanishes. Then the integral

$$P=\int\frac{\psi}{L}\frac{dx}{f'(y)}$$

is a constant multiple of an elementary integral of the third kind.

The proof is to be carried out exactly on the lines of the proof of the form of an integral of the first kind in § 91, with reference to the investigation in § 89.

Further as we know (§ 16) that dP/dx is of the form

$$C(dP/dx)_0+\lambda_1(dv_1/dx)+\ldots+\lambda_p(dv_p/dx),$$

where $C, \lambda_1, \ldots, \lambda_p$ are arbitrary constants, $(dP/dx)_0$ is a special form of dP/dx with the proper behaviour at the infinities, and $v_1, \ldots, v_p$ are integrals of the first kind, it follows that the polynomial ψ, which is an adjoint polynomial of grade $(n-1)\sigma+n-3+\mu$, prescribed to vanish at all but two of the zeros of a non-adjoint polynomial L of grade μ, is of the form

$$\psi=\psi_0+L\phi,$$

where ψ_0 is a particular form of ψ satisfying the conditions, and ϕ is any adjoint polynomial of grade $(n-1)\sigma+n-3$; for this is the only form of ψ which will reduce dP/dx to the form specified.

Ex. x. Shew that if in Ex. ix. one or both of the infinities of the integral be places where $f'(y)=0$, the condition for L is that it vanish to the first order in each place.

Ex. xi. For the case of the surface associated with the equation

$$(y,x)_4+(y,x)_3+(y,x)_2=0,$$

* Given, for $\sigma=0$, $\mu=1$, in Clebsch and Gordan, *Abel. Functionen* (Leipzig, 1866), p. 22, and Noether, "Abel. Differentialausdrücke," *Math. Annal.* t. 37, p. 432.

† Counting zeros which occur for $x=\infty$, or supposing all the zeros to be at finite places. Zeros which occur at $x=\infty$ are to be obtained by considering $\xi^\mu L$, which is an integral polynomial in ξ and η (§ 86).

for which the dimension of y is 1, let us form the integral of the third kind with its infinities at the two places $x=0$, $y=0$ by the rules of Exs. ix. and x.; taking $\mu=1$, the general polynomial of grade 1 which vanishes at the two places in question is $\lambda x+\mu y$. The general polynomial of grade $n-3+\mu$, or 2, is of the form $ax^2+by^2+2hxy+2gx+2fy+c$. In order that this may be adjoint, c must vanish; in order that it may vanish at the two points, other than (0, 0) at which $\lambda x+\mu y$ vanishes, it must reduce to the form

$$(\lambda x+\mu y)(Ax+By+C).$$

Hence the integral of the third kind is $\int (Ax+By+C)\,dx/f'(y)$. (Cf. § 6 β, p. 19.)

Ex. xii. Obtain the other result of § 6 β, p. 19 in a similar way.

Ex. xiii. It will be instructive to compare the method of expressing rational functions which is explained here, with a method founded on the use of the integral functions obtained in Chap. IV. We consider, as example, the case of a rational function which has simple poles at k_1 places where $x=a_1$, k_2 places where $x=a_2$, ..., k_r places at $x=a_r$, and for simplicity we suppose all these values of x to be finite, and assume that the sheets of the surface are all distinct for each of these values of x. If R be the rational function, the function $(x-a_1)...(x-a_r)\,R$ is an integral function of dimension r, and is expressible in the form

$$(x,\,1)_r+(x,\,1)_{r-\tau_1-1}\,g_1+\ldots+(x,\,1)_{r-\tau_{n-1}-1}\,g_{n-1};$$

this form contains $(r+1)+(r-\tau_1)+\ldots+(r-\tau_{n-1})$ or $nr-p+1$ coefficients; these coefficients are not arbitrary, for the function $(x-a_1)...(x-a_r)\,R$ must vanish at each of the $n-k_1$ places $x=a_1$ where R is not infinite, and must vanish at each of the places $x=a_2$ where R is not infinite, and so on. The number of linear conditions thus imposed is $rn-(k_1+k_2+\ldots+k_r)$ or $rn-Q$, if Q be the total number of poles of the function R. Hence the number of coefficients left arbitrary is $nr-p+1-(nr-Q)$ or $Q-p+1$; this is in accordance with results already obtained.

Ex. xiv. If the differential coefficients of $\tau+1$ linearly independent integrals of the first kind vanish in the Q poles, in Ex. xiii., the conditions for the coefficients are equivalent to only $nr-Q-(\tau+1)$ independent conditions.

93. Let $A_1, \ldots, A_Q$ be Q arbitrary places of the Riemann surface. We shall suppose these places so situated that a rational function exists of which they are the poles, each being of the first order*. This is a condition which is always satisfied† when $Q>p$. The general rational function in question is of the form

$$\lambda+\lambda_1 Z_1+\ldots+\lambda_q Z_q,$$

wherein $\lambda, \lambda_1, \ldots, \lambda_q$ are arbitrary constants and $Z_1, \ldots, Z_q$ are definite rational functions whose poles, together, are the places $A_1, \ldots, A_Q$.

The number q is connected with Q by an equation

$$Q-q=p-\tau-1,$$

where $\tau+1$ is‡ the number of linearly independent linear aggregates of the form

$$\mu_1\Omega(x)+\ldots\ldots+\mu_p\Omega_p(x),$$

* We speak as if the poles were distinct. This is unimportant.

† Cf. Chap. III.

‡ Chap. III. §§ 27, 37.

which vanish in $A_1, \dots, A_Q$. This aggregate is the differential coefficient, in regard to the infinitesimal at the place x, of the general integral of the first kind. We have seen* that this differential coefficient only vanishes at a zero of the integral polynomial of grade $(n-1)\sigma+n-3$, which occurs in the expression of the integral of the first kind. Hence $\tau+1$ *is the number of linearly independent adjoint polynomials of grade* $(n-1)\sigma+n-3$ *which vanish in the places* $A_1, \dots, A_Q$; in other words, $\tau+1$ is the number of coefficients in the general adjoint polynomial of grade $(n-1)\sigma+n-3$ which are left arbitrary after the prescription that the polynomial shall vanish in $A_1, \dots, A_Q$.

Now we have proved that if any adjoint polynomial ψ, of grade $(n-1)\sigma+n-3+r$ be taken to vanish at the places $A_1, \dots, A_Q$†, its other zeros being $B_1, \dots, B_R$, where‡ $R = nr+2p-2-Q$, and ϑ be a proper general adjoint polynomial of grade $(n-1)\sigma+n-3+r$ vanishing at $B_1, \dots, B_R$, any rational function having $A_1, \dots, A_Q$ as poles, is of the form ϑ/ψ. Hence the rational functions $Z_1, \dots, Z_q$ are of the forms $\vartheta_1/\psi, \dots, \vartheta_q/\psi$, and the general form of an adjoint polynomial of grade $(n-1)\sigma+n-3+r$ vanishing at $B_1, \dots, B_R$ must be

$$\vartheta = \lambda\psi + \lambda_1\vartheta_1 + \dots\dots + \lambda_q\vartheta_q,$$

wherein $\lambda, \lambda_1, \dots, \lambda_q$ are arbitrary constants, and $\psi, \vartheta_1, \dots, \vartheta_q$ are special adjoint polynomials of grade $(n-1)\sigma+n-3+r$ which vanish in $B_1, \dots, B_R$, some of them possibly vanishing also in some of $A_1, \dots, A_Q$.

Since the general adjoint polynomial ϑ of grade $(n-1)\sigma+n-3+r$ contains $nr-1+p$ arbitrary coefficients, and these, in this case, by the prescription of the zeros $B_1, \dots, B_R$ for ϑ, reduce to $q+1$, we may say that the places $B_1, \dots, B_R$, as determinators of adjoint polynomials of grade $(n-1)\sigma+n-3+r$, have the strength $nr-1+p-q-1$, or $R-(p-1)+Q-q-1$, or $R-(\tau+1)$. And, calling these places $B_1, \dots, B_R$ the residual of the places $A_1, \dots, A_Q$, because they are the remaining zeros of the adjoint polynomial ψ of grade $(n-1)\sigma+n-3+r$ which vanishes in $A_1, \dots, A_Q$, we have the result:—

When Q places $A_1, \dots, A_Q$ have the strength $p-(\tau+1)$ or $Q-q$ as determinators of adjoint polynomials of grade $(n-1)\sigma+n-3$, their residual of $R=nr+2p-2-Q$ places, which are the other zeros of any adjoint polynomial of grade $(n-1)\sigma+n-3+r$ prescribed to vanish in the places $A_1, \dots, A_Q$, have the strength $R-(\tau+1)$ as determinators of adjoint polynomials of grade $(n-1)\sigma+n-3+r$.

Particular cases are, (i), when no adjoint polynomial of grade $(n-1)\sigma+n-3$ vanishes in $A_1, \dots, A_Q$; then the places $B_1, \dots, B_R$ have a strength equal to their number; (ii), when one adjoint polynomial of grade $(n-1)\sigma+n-3$ vanishes in $A_1, \dots, A_Q$; then

* § 92. † A condition requiring in general $Q < nr-1+p$. ‡ § 88.

there are $R-1$ of the places $B_1, \dots, B_R$ such that every adjoint polynomial of grade $(n-1)\sigma+n-3+r$, vanishing at these places, vanishes at the remaining place. For an example of this case we may cite the theorem: If a cubic curve be drawn through three collinear points A_1, A_2, A_3 of a plane quartic curve, the remaining nine intersections $B_1, \dots, B_9$ are such that every cubic through a proper set of eight of them necessarily passes through the ninth. In general any set of eight of them may be chosen.

When $\tau+1$ is greater than zero we may take the polynomial ψ itself to be of grade $(n-1)\sigma+n-3$. Since then a general polynomial ϑ of grade $(n-1)\sigma+n-3$ contains p arbitrary coefficients, we can similarly prove that

When $\tau+1$ adjoint polynomials of grade $(n-1)\sigma+n-3$ vanish in Q places $A_1, \dots, A_Q$, so that the Q places have the strength $Q-q$ as determinators of adjoint polynomials of grade $(n-1)\sigma+n-3$, their residual $B_1, \dots, B_R$, of $R=2p-2-Q$ places, have the strength $p-q-1$, or $R-\tau$, as determinators of adjoint polynomials of grade $(n-1)\sigma+n-3$. In this case the numbers are connected by the equations

$$Q+R=2p-2, \quad Q-R=2(q-\tau),$$

and the characters of the sets $A_1, \dots, A_Q$, $B_1, \dots, B_R$ are perfectly reciprocal.*

Ex. When the strength of a set $A_1, \dots, A_Q$, wherein $Q<p$, as determinators of adjoint polynomials of grade $(n-1)\sigma+n-3$, is equal to their number, so that the number of linearly independent adjoint polynomials of grade $(n-1)\sigma+n-3$ which vanish in the places of the set is given by $\tau+1=p-Q$, it follows that $q=0$. Thus if $B_1, \dots, B_R$ be the residual zeros of an adjoint polynomial, ϕ, of grade $(n-1)\sigma+n-3$, which vanishes in $A_1, \dots, A_Q$, so that $R+Q=2p-2$, only one adjoint polynomial of grade $(n-1)\sigma+n-3$ vanishes in $B_1, \dots, B_R$, namely ϕ.

94. It is known that the number of places† of the Riemann surface at which a rational function takes an arbitrary value c, is the same as the number of places at which the function is infinite. The sets of places at which c has its different values, may be called *equivalent* sets of places for the function under consideration. For such sets we can prove the result:— *if a set of places $A_1', \dots, A'_Q$ be equivalent to a set $A_1, \dots, A_Q$, in the sense that a rational function g takes the value c' at each place of the former set and at no other places, and takes the value c at each of $A_1, \dots, A_Q$ and at no other places of the Riemann surface, then the general rational function with simple poles at $A_1', \dots, A'_Q$ contains as many linearly entering arbitrary constants as the general rational function whose poles are at $A_1, \dots, A_Q$.*

* For the theory of such reciprocal sets from the point of view of the algebraical theory of curves, see the classical paper, Brill u. Noether, "Ueber die algebraischen Functionen u.s.w.", *Math. Annal.* vii. p. 283 (1873).

† In this Article, when a rational function g is said to have the value c at a place, it is intended that $g-c$ is zero of the first order at the place. A place where $g-c$ is zero of the k-th order is regarded as arising by the coalescence of k places where g is equal to c.

For let the general rational function with poles at $A_1, \ldots, A_Q$ be denoted by G, and be given by

$$G = \nu_0 + \nu_1 G_1 + \ldots\ldots + \nu_q G_q,$$

where $\nu_0, \ldots, \nu_q$ are arbitrary constants, and $G_1, \ldots, G_q$ are particular functions whose poles are among $A_1, \ldots, A_Q$—of which one, say G_1, may be taken to be the function $(g-c')/(g-c)$. Then if G' denote any function whatever having poles $A_1', \ldots, A'_Q$, and not elsewhere infinite, the function $G'(g-c')/(g-c)$ is one whose poles are at $A_1, \ldots, A_Q$; thus $G'(g-c')/(g-c)$ can be expressed in the form

$$G'(g-c')/(g-c) = \nu_0 + \nu_1 G_1 + \ldots\ldots + \nu_q G_q,$$

for proper values of $\nu_0, \ldots, \nu_q$. Therefore G' can be expressed in the form

$$G' = \nu_0 \frac{g-c}{g-c'} + \nu_1 + \nu_2 G_2 \frac{g-c}{g-c'} + \ldots\ldots + \nu_q G_q \frac{g-c}{g-c'}.$$

Since this is true of every function whose poles are at $A_1', \ldots, A'_Q$, and that the functions $G_i(g-c)/(g-c')$ are functions whose poles are at $A_1', \ldots, A'_Q$, the result is obvious.

95. If the symbol ∞ be used to denote the number of values of an arbitrary (real or complex) constant, the general adjoint polynomial ϑ, of grade $(n-1)\sigma + n - 3 + r$, of the form

$$\vartheta = \lambda\psi + \lambda_1\vartheta_1 + \ldots\ldots + \lambda_q\vartheta_q,$$

which vanishes in the places $B_1, \ldots, B_R$, gives rise to ∞^q sets of places, constituted by the zeros of ϑ other than $B_1, \ldots, B_R$, each set consisting of, say, Q places. Let $A_1, \ldots, A_Q$ be one of these sets.

We shall say that these sets are a *lot* of sets; that each *set* is a residual of $B_1, \ldots, B_R$, and that they are co-residual with one another; in particular they are all co-residual with the set $A_1, \ldots, A_Q$. Further we shall say that the multiplicity of the sets, or of the lot, is q, and that each set has the *sequence* $Q-q$; in fact an individual set is determined by q independent linear conditions, namely, of the Q places of a set, q can be prescribed and the remaining $Q-q$ are *sequent*.

It is clear then that any set, $A_1', \ldots, A'_Q$, which is co-residual with $A_1, \ldots, A_Q$, is equivalent with $A_1, \ldots, A_Q$, in the sense of the last article; for these two sets are respectively the zeros and poles of the same rational function; in fact if ψ be the polynomial vanishing in $B_1, \ldots, B_R, A_1, \ldots, A_Q$, and ϑ the polynomial vanishing in $B_1, \ldots, B_R, A_1', \ldots, A'_Q$, the rational function ϑ/ψ has $A_1', \ldots, A'_Q$ for zeros and $A_1, \ldots, A_Q$ for poles. Hence by the preceding article it follows that the number $q+1$ of linear, arbitrary, coefficients in a general rational function prescribed to have its poles at $A_1, \ldots, A_Q$, is the same as the number in the general function prescribed to have its poles at the co-residual set $A_1', \ldots, A'_Q$. In other words, co-residual sets of places have the same multiplicity, this being determined by the number of constants in the general rational function having one of these sets as poles; they have therefore also the same strength $Q-q$, or $p-(\tau+1)$, as determinators of adjoint polynomials of grade $(n-1)\sigma+n-3$.

96. In the determination of the sets co-residual to a given one, $A_1, \ldots, A_Q$, we have made use of a particular residual, $B_1, \ldots, B_R$. It can however be shewn that this is unnecessary—*and that, if two sets be co-residual for any one common residual, they are co-residual for any residual of one of them.* In other words, let an adjoint polynomial ψ, of grade $(n-1)\sigma+n-3+r$, be taken to vanish in a set $A_1, \ldots, A_Q$, its other zeros (besides those where $f'(y)=0$, or $F'(\eta)=0$), being $B_1, \ldots, B_R$, and an adjoint polynomial ϑ, of grade $(n-1)\sigma+n-3+r$, be taken to vanish in $B_1, \ldots, B_R$, its other zeros being the set $A_1', \ldots, A'_Q$, co-residual with $A_1, \ldots, A_Q$; then if an adjoint polynomial, ψ', of grade $(n-1)\sigma+n-3+r'$, which vanishes in $A_1, \ldots, A_Q$, have $B_1', \ldots, B'_{R'}$ for its residual zeros, R' being equal to $nr'+2p-2-Q$, it is possible to find an adjoint polynomial ϑ', of grade $(n-1)\sigma+n-3+r'$, whose zeros are the places $B_1', \ldots, B'_{R'}, A_1', \ldots, A'_Q$.

For we have shewn that any rational function having $A_1, \ldots, A_Q$ as its poles can be written as the quotient of two adjoint polynomials, of which the denominator is arbitrary save that it must vanish in the poles of the function, and be of sufficiently high grade to allow this. In particular therefore the function ϑ/ψ, whose zeros are $A_1', \ldots, A'_Q$, can be written as the quotient of two polynomials of which ψ' is the denominator, namely in the form ϑ'/ψ'. The polynomial ϑ' will therefore vanish in the places $B_1', \ldots, B'_{R'}, A_1', \ldots, A'_Q$, as stated.

Hence, not only are equivalent sets necessarily co-residual, but co-residual sets are necessarily equivalent, independently of their residual*.

97. The equivalence of the representations ϑ/ψ, ϑ'/ψ', here obtained, of the same function, has place algebraically in virtue of an identity of the form

$$\vartheta\psi'=\vartheta'\psi+Kf,$$

where $f=0$ is the equation associated with the Riemann surface and K is an integral polynomial in x and y. Reverting to the phraseology of the theory of plane curves, it can in fact be shewn that if three curves $f=0$, $\psi=0$, $H=0$ be so related that, at every common point of f and ψ, which is a multiple point of order k for f and of order l for ψ, whereat f and ψ intersect in $kl+\beta$ points, the curve H have a multiple point of order $k+l-1+\beta$, so that in particular H passes through every simple intersection of f and ψ, then there exist curves $\vartheta'=0$, $K=0$, such that, identically,

$$H=\vartheta'\psi+Kf.$$

Now in the case under consideration in the text, if the only multiple points of f be multiple points at which all the tangents are distinct, the adjointness of ψ ensures that ψ

* For the theory of co-residual sets for a plane cubic curve see Salmon, *Higher Plane Curves* (Dublin, 1879), p. 137. That theory is ascribed to Sylvester; cf. *Math. Annal.*, t. vii., p. 272 note.

has a multiple point of order $k-1$ at every multiple point of f of order k. The adjointness of the polynomials ϑ, ψ' ensures that the compound curve $\vartheta\psi'$ has a multiple point of order $2(k-1)$ or $k+k-1-1$ at every multiple point of f of order k. Further, the curve $\vartheta\psi'$ passes through the simple intersections of f and ψ, which consist of the sets $A_1, \dots, A_Q$, $B_1, \dots, B_R$; for ϑ passes through $B_1, \dots, B_R$, and ψ' is drawn through $A_1, \dots, A_Q$. Hence the conditions are fully satisfied in this case by taking $H=\vartheta\psi'$; thus there is an equation of the form

$$\vartheta\psi'=\vartheta'\psi+Kf,$$

from which it follows that the curve ϑ' is adjoint at the multiple points of f and passes through the remaining intersections of f and $\vartheta\psi'$, namely through $A'_1, \dots, A'_Q$ and $B'_1, \dots, B'_{R'}$. This is the result of the text.

In case of greater complication in the multiple points of f, there is need for more care in the application of the theorem here quoted from the algebraic theory of plane curves. But this theorem is of great importance. For further information in regard to it the reader may consult Cayley, *Collected Works*, Vol. I. p. 26; Noether, *Math. Annal.* vi. p. 351; Noether, *Math. Annal.* xxiii. p. 311; Noether, *Math. Annal.* xl. p. 140; Brill and Noether, *Math. Annal.* vii. p. 269. Also papers by Noether, Voss, Bertini, Brill, Baker in the *Math. Annal.* xvii, xxvii, xxxiv, xxxix, xlii respectively. See also Grassmann, *Die Ausdehnungslehre von* 1844 (Leipzig, 1878), p. 225. Chasles, *Compt. Rendus*, xli. (1853). de Jonquières, *Mém. par divers savants*, xvi. (1858).

98. From the theorem, that a lot of co-residual sets, of Q places, may be regarded as the residual of any residual of one set, S_Q, of the lot, it follows, that every lot wherein the sequence of a set is less than p, may be determined as the residual zeros of a lot of adjoint polynomials of grade $(n-1)\sigma+n-3$, which have $R=2p-2-Q$ common zeros. For the sequence $Q-q$ is equal to $p-(\tau+1)$, and when $\tau+1>0$ an adjoint polynomial (involving $\tau+1$ arbitrary coefficients) can be determined which is zero in any one set, S_Q, of the lot, and in R other places.

Hence also, when Q places are such that the most general rational function, of which they are the poles, contains more than $Q-p+1$ arbitrary constants, this general rational function can be expressed as the quotient of two adjoint polynomials of grade $(n-1)\sigma+n-3$; the same is true when the Q places are known to be zeros of an adjoint polynomial of grade $(n-1)\sigma+n-3$.

It follows from what was shewn in Chap. III. §§ 23, 27, that if p places be the poles of a rational function, an adjoint polynomial of grade $(n-1)\sigma+n-3$ vanishes in these places; and an adjoint polynomial of that grade can always be chosen to vanish in $p-1$, or a less number, of arbitrary places. Hence, *every rational function of order less than* $p+1$, *is expressible as the quotient of two adjoint polynomials of grade* $(n-1)\sigma+n-3$.

Ex. i. *A rational function of order* $2p-2$ *which contains* p, *or more, arbitrary constants (one being additive) is expressible as the quotient of two adjoint polynomials of grade* $(n-1)\sigma+n-3$.

Ex. ii. For a general quartic curve, co-residual sets of 4 places with multiplicity 1 are determined by variable conics having 4 given zeros; but co-residual sets of 4 places with multiplicity 2 are determined as the zeros of variable polynomials of degree 1, i.e. by straight lines.

Ex. iii. The equation of a plane quintic curve with two double points, can be written in the form $\vartheta S' - \vartheta' S = 0$, where ϑ, ϑ' are cubics passing through the double points and seven other common points, and S, S' are conics passing through the double points and two other common points.

Ex. iv. When $\tau+1$ adjoint polynomials of grade $(n-1)\sigma+n-3$ vanish in a set, S_Q, of Q places, there must be $p-\tau-1$ independent places $A_1, \ldots, A_{p-\tau-1}$, in S_Q, such that every adjoint polynomial of grade $(n-1)\sigma+n-3$ which vanishes in them vanishes of itself in the remaining q places $A_{p-\tau}, \ldots, A_Q$. Let S_R be a residual of S_Q, R being equal to $2p-2-Q$. Then, regarding S_R and $A_{p-\tau}, \ldots, A_Q$, together, as forming a residual of $A_1, \ldots, A_{p-\tau-1}$, it follows (§ 93) that there is only one adjoint polynomial of grade $(n-1)\sigma+n-3$ which vanishes in S_R and in $A_{p-\tau}, \ldots, A_Q$. Hence there exists no rational function having poles only at the places $A_1, \ldots, A_{p-\tau-1}$. For such a function could be expressed as the quotient of two adjoint polynomials of grade $(n-1)\sigma+n-3$ having S_R and $A_{p-\tau}, \ldots, A_Q$ as common zeros. Compare § 26, Chap. III.

It can also be shewn, in agreement with the theory given in Chapter III., that if $B_1, \ldots, B_{\tau'+1}$ be any $\tau'+1$ independent places, τ' being less than τ, there exists no rational function having poles in S_Q and $B_1, \ldots, B_{\tau'+1}$. In fact $\tau+1-(\tau'+1)$ linearly independent adjoint polynomials of grade $(n-1)\sigma+n-3$ vanish in S_Q and $B_1, \ldots, B_{\tau'+1}$. Let $S_{R'}$, where $R'=2p-2-(Q+\tau'+1)$, be the residual zeros of one of these polynomials. Then the strength of $S_{R'}$, as determinators of adjoint polynomials of grade $(n-1)\sigma+n-3$ is (§ 93) $R'-(\tau-\tau')+1=R-\tau$, where $R=2p-2-Q$, namely the strength of $S_{R'}$ is the same as the strength of $S_{R'}$ and $B_1, \ldots, B_{\tau'+1}$ together; hence every adjoint polynomial of grade $(n-1)\sigma+n-3$ which vanishes in $S_{R'}$, vanishes also in $B_1, \ldots, B_{\tau'+1}$. Now every rational function having S_Q and $B_1, \ldots, B_{\tau'+1}$ as poles, could be expressed as the quotient of two adjoint polynomials of grade $(n-1)\sigma+n-3$ having $S_{R'}$ as common zeros; since each of these polynomials will also have $B_1, \ldots, B_{\tau'+1}$ as zeros, the result is clear.

99. The remaining Articles of this Chapter are devoted to developments more intimately connected with the algebraical theory of curves.

We have seen that an individual set of a lot of co-residual sets of Q places is determined by the prescription of a certain number, q, of the places; this number q being less than* Q and not greater than $Q-p$.

But it does not follow that *any* q places of a set are effective for this purpose; it may happen that q places, chosen at random, are ineffective to give q independent conditions.

We give an example of this which leads (§ 100) to a result of some interest.

Suppose that a set of Q places, S_Q, is given, in which no adjoint polynomial of grade $(n-1)\sigma+n-3$ vanishes; then $\tau+1$ is zero, and co-residual sets are determined by $Q-p$ places. Suppose that among the Q places there are $p+s-1$ places, forming a set which we shall denote by σ_{p+s-1}, which are common zeros of $\tau'+1$ adjoint polynomials of grade $(n-1)\sigma+n-3$; denote the other $Q-p-s+1$ or $q-s+1$ places by σ_{q-s+1}.

* The formula is $Q-q=p-(\tau+1)$; if q were Q and therefore $\tau+1=p$, all adjoint polynomials of grade $(n-1)\sigma+n-3$ would vanish in the same Q places, contrary to what is proved in § 21, Chap. II.

Take an adjoint polynomial of grade $(n-1)\sigma+n-3+r$ which vanishes in the places of the set S_Q, and let S_R denote its remaining zeros, so that $R+Q=nr+2p-2$. If we now regard the sets S_R, σ_{q-s+1} together as the residual of the set σ_{p+s-1}, it follows (§ 93) that S_R, σ_{q-s+1} together have only the strength $R+q-s+1-(\tau'+1)$, or $nr+p-2-(\tau'+s)$, as determinators of polynomials of grade $(n-1)\sigma+n-3+r$; and if we choose $s-1$ places $A_1, \dots, A_{s-1}$ from σ_{p+s-1}, the polynomial of grade $(n-1)\sigma+n-3+r$ with zeros in S_R, which vanishes in the q places constituted by σ_{q-s+1} and $A_1, \dots, A_{s-1}$ together, will not be entirely determined, but will contain* $\tau'+2$ arbitrary coefficients, at least†: thus $\tau'+1$ further zeros must be prescribed to make the polynomial determinate.

A particular case of this result is as follows:—Consider a lot of co-residual sets of Q, $=q+p$, places, in which no adjoint polynomial of grade $(n-1)\sigma+n-3$ vanishes. If p of the places of a set be zeros of $\tau'+1$ adjoint polynomials of grade $(n-1)\sigma+n-3$, then the other q places are not sufficient to individualise the set; $\tau'+1$ additional places are necessary.

For instance a particular set from the double infinity of sets of 5 places, on a plane quartic curve, determined by variable cubic curves having seven fixed zeros, is generally determined by prescribing 2 places of the set. But if there be one of the sets for which 3 of the five places are collinear, then the other two places do not determine this set; we require also to specify one of the three collinear places. It is easy to verify this result in an elementary way.

100. Consider now two sets S_R, S_{Q_1}, which are residual zeros of an adjoint polynomial, ψ_1, of grade $(n-1)\sigma+n-3+r_1$, so that

$$Q_1+R=nr_1+2p-2.$$

Let $X_{r-r_1}+1$ be the number of terms in the general non-adjoint polynomial of grade $r-r_1$ and N_{r-r_1} be the total number of zeros of such a non-adjoint polynomial of grade $r-r_1$. Take X_{r-r_1} independent places on the Riemann surface, forming a set which we shall denote by T_{r-r_1}, and determine a non-adjoint polynomial, χ, of grade $r-r_1$, to vanish in T_{r-r_1}. It will vanish in $N_{r-r_1}-X_{r-r_1}$ other places, U_{r-r_1}. Suppose that no adjoint polynomials of grade $(n-1)\sigma+n-3$ vanish in all the places of S_{Q_1} and T_{r-r_1}. The product of the polynomials ψ_1 and χ is an adjoint polynomial of grade $(n-1)\sigma+n-3+r$. A general adjoint polynomial of grade $(n-1)\sigma+n-3+r$ which vanishes in S_R will vanish in all the places forming S_{Q_1}, T_{r-r_1}, U_{r-r_1} together, provided we choose the polynomial to have a sufficient number of these places as zeros. Divide the set S_{Q_1} into two parts, one, $\overline{T}$, consisting of $Q_1-p+(N_{r-r_1}-X_{r-r_1})$ places, the other $\overline{U}$ consisting of $p-(N_{r-r_1}-X_{r-r_1})$

* For $nr+p-2$ is the number of independent zeros necessary to determine an adjoint polynomial of grade $(n-1)\sigma+n-3+r$.

† More if the $s-1$ places $A_1, \dots, A_{s-1}$ be not independent of the others already chosen.

places. The sets $\overline{T}$ and T_{r-r_1} together consist of $Q_1 - p + N_{r-r_1}$, or $Q - p$, places, where

$$Q = Q_1 + N_{r-r_1}, \ = nr + 2p - 2 - R,$$

for $N_{r-r_1} = n(r - r_1)$, (§ 86, Ex. iii.); if then the sets $\overline{U}$ and U_{r-r_1} together are not zeros of any adjoint polynomial of grade $(n-1)\sigma + n - 3$, the general adjoint polynomial, of grade $(n-1)\sigma + n - 3 + r$, which vanishes in S_R, will be entirely determined by the condition of vanishing also in the places of $\overline{T}$ and T_{r-r_1}, and will of itself vanish in the remaining places $\overline{U}$ and U_{r-r_1}. If, however, $\tau' + 1$ adjoint polynomials of grade $(n-1)\sigma + n - 3 - (r - r_1)$ vanish in the places U, the products of these with the non-adjoint polynomial χ give $\tau' + 1$ adjoint polynomials of grade $(n-1)\sigma + n - 3$ vanishing in $\overline{U}$ and U_{r-r_1}. In that case, assuming that no adjoint polynomials of grade $(n-1)\sigma + n - 3$ vanish in the p places $\overline{U}$, U_{r-r_1}, other than those containing χ as a factor, the adjoint polynomial of grade $(n-1)\sigma + n - 3 + r$ which vanishes in S_R, $\overline{T}$ and T_{r-r_1}, will require $\tau' + 1$ further zeros for its complete determination (§ 99).

Since now the set T_{r-r_1} entirely determines the set U_{r-r_1}, we may drop the consideration of it, and obtain the result—

The adjoint polynomial, of grade $(n-1)\sigma + n - 3 + r$, which vanishes in all but $p - (N_{r-r_1} - X_{r-r_1})$ of the zeros of an adjoint polynomial of grade $(n-1)\sigma + n - 3 + r_1$, will have a multiplicity $\tau' + 1 + X_{r-r_1}$, where $\tau' + 1$ is the number of adjoint polynomials of grade $(n-1)\sigma + n - 3 - (r - r_1)$ which vanish in these other $p - N_{r-r_1} + X_{r-r_1}$ zeros. When $\tau' + 1$ is zero the adjoint polynomial of grade $(n-1)\sigma + n - 3 + r$ vanishes of itself in the remaining $p - N_{r-r_1} + X_{r-r_1}$ zeros of the adjoint polynomial of grade $(n-1)\sigma + n - 3 + r_1$. When $\tau' + 1$ is not zero it is necessary, for this, to prescribe $\tau' + 1$ further places of these $p - N_{r-r_1} + X_{r-r_1}$ zeros (provided $\tau' + 1 \overline{\overline{<}} p - N_{r-r_1} + X_{r-r_1}$).

We have noticed (§ 86, Ex. iii.) that

$$N_{r-r_1} = n(r - r_1),$$

$$X_{r-r_1} = \left[E\left(\frac{r - r_1}{\sigma + 1}\right) + 1\right]\left[r - r_1 + 1 - \tfrac{1}{2}(\sigma + 1) E\left(\frac{r - r_1}{\sigma + 1}\right)\right] - 1,$$

where $E(x)$ denotes the greatest integer in x.

For $\sigma = 0$, therefore, the number $p - N_{r-r_1} + X_{r-r_1}$ is immediately seen to be equal to

$$\tfrac{1}{2}(\gamma - 1)(\gamma - 2) - \tfrac{1}{2}I,$$

where $\gamma = n - (r - r_1)$, and $\frac{1}{2}I$ is the sum of the indices, of the surface, for finite and infinite places (§ 88).

Thus the result, for $\sigma = 0$,—*an adjoint polynomial of degree $n - 3 + r$ which vanishes in all but $\frac{1}{2}(\gamma - 1)(\gamma - 2) - \frac{1}{2}I$ of the zeros of an adjoint polynomial of degree $n - 3 + r_1$ ($r > r_1$, $\gamma = n - (r - r_1) \not< 3$) will have a* *multiplicity* $\tau'+1+\frac{1}{2}(n-\gamma)(n-\gamma+3)$, *where* $\tau'+1$ *is the number of adjoint polynomials of degree* $\gamma-3$ *which vanish in the* $\frac{1}{2}(\gamma-1)(\gamma-2)-\frac{1}{2}I$ *unassigned zeros; if* $\tau'+1$ *is zero this polynomial of degree* $n-3+r$ *will of itself vanish in these unassigned zeros: if* $\tau'+1>0$ *it is necessary, for this, to prescribe* $\tau'+1$ *or, if* $\tau'+1>\frac{1}{2}(\gamma-1)(\gamma-2)-\frac{1}{2}I$, *to prescribe all the unassigned zeros.*

For example let $n=5$; take as the fundamental curve a plane quintic with 2 double points ($p=4$); let the remaining point of intersection with the quintic, of the straight line drawn through these double points, be denoted by A.

(i) Take $r=2$, $r_1=1$. Then $\gamma=5-1=4$, $\gamma-3=1$; thus, an adjoint quartic curve vanishing in all but $\frac{1}{2}(\gamma-1)(\gamma-2)-2$, or 1, of the zeros of an adjoint cubic, that is, vanishing in 10 of these zeros, beside vanishing at the double points, will have a multiplicity $\tau'+1+\frac{1}{2}4$, or $\tau'+1+2$, where $\tau'+1$ is zero if the non-assigned zero be not the point A: and this quartic will then, of itself, pass through the unassigned zero. In this case, in fact, the prescription of the $10+2$ zeros of the quartic on the cubic, is a prescription of more than $4.3-p_1$, where p_1 is the deficiency of the cubic. Hence the quartic will contain the cubic wholly, as part of itself. (In general, the condition to provide against this can be seen to be $r \overline{\gtrless} 3$.)

(ii) Take the same fundamental quintic, with $r=4$, $r_1=3$. Then an adjoint sextic curve, ψ, passing through all but $\frac{1}{2}3.2-2$, or 1, of the zeros of an adjoint quintic, ϑ, that is through 20 of them, will have multiplicity $\tau'+1+2$, where $\tau'+1$ is zero unless the other zero of the quintic, ϑ, be the point A.

If however the unassigned zero of the quintic, ϑ, be the point A, the 20 points are not sufficient; the sextic, ψ, has multiplicity 3 and the 20 points *plus* A are necessary to make ψ go through the remaining 7 points.

It should be noticed that an adjoint curve of degree $\gamma-3$ can always be made to pass through $\frac{1}{2}(\gamma-1)(\gamma-2)-\frac{1}{2}I-1$ places. The peculiarity in the case considered is that such curves pass through one more place.

The theorem here proved was first given by Cayley in 1843 (*Collected Works*, Vol. I. p. 25) without special reference to adjoint curves. A further restriction was added by Bacharach (*Math. Annal.* t. 26, p. 275 (1886)).

101. In the following articles of this chapter we shall speak of an adjoint polynomial of grade $(n-1)\sigma+n-3$ as a ϕ-polynomial. In chapter III. (§ 23) we have seen that the set of places constituted by the poles of a rational function, is such that one of them 'depends' upon the others; thus (§ 27) there is one place of the set such that every ϕ-polynomial vanishing in the other places, vanishes also in this. Conversely when a set of places is such that every ϕ-polynomial vanishing in all but one of the places, vanishes of necessity also in the remaining place, this remaining place depends upon the others*. When a set S is such that every ϕ-polynomial

* Or on *some* of them. For instance, if in a two-sheeted hyperelliptic surface, associated with the equation $y^2=(x, 1)_{2p+2}$, we take three places (x_1, y_1), (x_2, y_2), $(x_2, -y_2)$, every ϕ-polynomial, $(x-x_1)(x-x_2)(x, 1)_{p-3}$, of order $p-1$ in x, which vanishes in (x_1, y_1), (x_2, y_2), vanishes also in $(x_2, -y_2)$. But this last place does not, strictly, 'depend' on (x_1, y_1) and (x_2, y_2); it depends on (x_2, y_2) only.

vanishing in S, vanishes also in places $A, B, \ldots$, it will be convenient, here, to say that these places are determined by S.

Take now any $p-3$ places of the surface, which we suppose chosen in order in such a way that no one of them is determined by those preceding. Then the general ϕ-polynomial vanishing in them will be of the form $\lambda\phi + \mu\vartheta + \nu\psi$, wherein λ, μ, ν are arbitrary constants and ϕ, ϑ, ψ are ϕ-polynomials vanishing in the $p-3$ places. We desire now to find a place (x_1) such that all ϕ-polynomials vanishing in the $p-3$ given places and in x_1, shall vanish in another place x_2. For this it is sufficient that the ratios $\phi(x_1) : \vartheta(x_1) : \psi(x_1)$ be equal to the ratios $\phi(x_2) : \vartheta(x_2) : \psi(x_2)$. From the two equations thus expressed, with help of the fundamental equation of the surface, we can eliminate x_2, and obtain an equation for x_1, so that the problem is in general a determinate one and has a finite number of solutions: as a matter of fact (§ 102, p. 144, § 107) the number of positions for x_1 is $\frac{1}{2}p(p-3)$*, and each determines the corresponding position of x_2. Hence there exist on the Riemann surface ∞^{p-3} sets of $p-1$ places such that a single infinity of ϕ-polynomials vanish in them; such a set can be determined from $p-3$ quite arbitrarily chosen places, and, from them, in $\frac{1}{2}p(p-3)$ ways. Putting $Q=p-1$, $\tau+1=2$, we obtain, by the Riemann-Roch Theorem $q=1$. Hence to each set once obtained there corresponds a single infinity of co-residual sets.

102. The reasoning employed in the last article, to prove that there are a finite number of positions possible for x_1, and the reasoning subsequently to be given to determine the number of these positions, is of a kind that may be fallacious for special forms of the fundamental equation associated with the Riemann surface. An extreme case is when the surface is hyperelliptic, in which case all the ϕ-polynomials vanishing in any given place have another common zero (Chap. V. § 52). In what follows we consider only surfaces which are of perfectly general character for the deficiency assigned.

In particular we assume, what is in accordance with the reasoning of the last article, that not every set of $p-2$ places is such that the two (or more) linearly independent ϕ-polynomials vanishing in them, have another common zero†.

* This result is given in Clebsch and Gordan, *Theorie der Abel. Funct.* (Leipzig, 1866) p. 213.

† Noether (*Math. Annal.* xvii.) gives a proof that this is true for every surface which is not hyperelliptic. Take a set of $p-2$ *independent* places, denoted, say, by S, and, if every $p-2$ places determine another place, let A be the place determined by the set S. Take a further quite arbitrary place, B. When the surface is not hyperelliptic, B will not determine another place. Each of the $\frac{1}{2}(p-1)(p-2)$ sets, of $p-3$ places, which can be selected from the $p-1$ places formed by S and A, constitutes, with B, a set of $p-2$ places, and, in accordance with the hypothesis allowed, each of these sets determines another place. *It is assumed that the $p-2$ places S, and the place B, can be so chosen that the $\frac{1}{2}(p-1)(p-2)$ other places, thus determined, are different from each other and from the p places constituted by S, A and B together.* Since the places S are independent, the ϕ-polynomial vanishing in S and B is unique; and, by what we have proved,

Then it will be possible to choose $p-3$ independent places, S, as in the last article, such that there is a finite number of solutions of the problem of finding a place (x_1) such that the ϕ-polynomials vanishing in S and (x_1), have another common zero; let $p-3$ places, forming a set denoted by S, be so chosen. Let A be a place not coinciding with any of the positions possible for x_1, and not determined by S. Let ϕ, ϑ be two linearly independent ϕ-polynomials vanishing in S and A. Then the general ϕ-polynomial vanishing in S and A is of the form $\lambda\phi + \mu\vartheta$, λ and μ being arbitrary constants, and the general ϕ-polynomial vanishing in the places S only can be written in a form $\lambda\phi + \mu\vartheta + \nu\psi$, wherein ν is an arbitrary constant and ψ is a ϕ-polynomial so chosen as not to vanish at the place A.

Consider now the rational functions* $z = \phi/\psi$, $s = \vartheta/\psi$, each of the $(p+1)$th order. They both vanish at the place A.

These functions will be connected by a rational algebraic equation, $(s, z) = 0$, obtained by eliminating (x, y) between the fundamental equation and the equations $z\psi = \phi$, $s\psi = \vartheta$; associated with the equation $(s, z) = 0$ will be a new Riemann surface; to every place (x, y) of the old surface will belong a definite place $z = \phi/\psi$, $s = \vartheta/\psi$, of the new surface; to every place of the new surface will belong one or more places of the original surface, the number being the same for every place of the new surface†; since there is only one place of the old surface at which both z and s are zero, namely the place which was denoted by A, it follows that there is only one place of the old surface corresponding to any place of the new surface. Hence each of x, y can be expressed as rational functions of s, z, the expression being obtained from the equations $z\psi = \phi$, $s\psi = \vartheta$, $(s, z) = 0$‡.

Since a linear function, $\lambda z + \mu s + \nu$, equal to $(\lambda\phi + \mu\vartheta + \nu\psi)/\psi$, vanishes* at the variable zeros of the polynomial $\lambda\phi + \mu\vartheta + \nu\psi$, namely in $p+1$ places, it follows that the equation $(s, z) = 0$ may be interpreted as the equation of a plane curve of order $p+1$; the number

it vanishes in $p + \frac{1}{2}(p-1)(p-2)$ places. This number, however, is greater than $2p-2$ when $p > 3$. Hence the hypothesis, that every $p-2$ places determine another is invalid. In case $p=3$ the surface is clearly hyperelliptic when every $p-2$ places determine another. In case $p=2$ or 1 the surface is always hyperelliptic. It may be remarked that when we are once assured of the existence of a rational function of p poles, we can infer the existence of a set of $p-2$ places which do not determine another (cf. § 103). We have already shewn (Chap. III. § 31) that in general a rational function of order p does exist. The reader may prove that for a hyperelliptic surface whose deficiency is an odd number there does not exist any rational function of order p.

* It must be borne in mind that, in dealing with a rational function expressed as a ratio of two adjoint polynomials, we speak of its poles as all given by the zeros of the denominator; some of these may be at $x=\infty$ (cf. § 86), and in that case their existence is to be shewn by considering (§ 84), instead of the polynomial, ψ, of grade μ, the polynomial in η and ξ, given by $\xi^{\mu}\psi$. Or we may use homogeneous variables (§ 85). For instance, for $p=3$, we may, in the text, have (§ 92, Ex. i.) $\phi=x$, $\vartheta=y$, $\psi=1$. Then $\phi : \vartheta : \psi = 1 : \eta : \xi = \omega : u : z$; and ψ has a zero at $z=\infty$.

† Chap. I. § 4.

‡ Or by the direct process of § 5, Chap. I.

of its double points will, therefore*, be $\frac{1}{2}p(p-1)-p$, or $\frac{1}{2}p(p-3)$, though it is not shewn here that they occur as simple double points. These double points are the transformations of the pairs of places, (x_1), (x_2), on the old surface, which were such that every ϕ-polynomial, vanishing in the $p-3$ fixed places S, and in x_1, also vanished in x_2.

Since a double point of a curve requires one condition among its coefficients, and the number of coefficients that can be introduced or destroyed, in the equation of a curve, by general linear transformation of the coordinates is 8, it follows that a curve of order m has

$$\tfrac{1}{2}m(m+3)-(\delta+\kappa)-8, \text{ or } \tfrac{1}{2}m(m+3)-\tfrac{1}{2}(m-1)(m-2)+p-8, \text{ or } 3m+p-9$$

constants which are not removeable by linear transformation. In the case under consideration here, there are $p-3$ places, S, of each of which an infinite number of positions is possible, independently of the others, and the most general linear transformation of s and z is equivalent only to adopting three new linear functions of ϕ, ϑ, ψ, instead of ϕ, ϑ, ψ, in order to express the general ϕ-polynomial through the places S. Hence there are, in the new surface (s, z) effectively

$$3(p+1)-9+p-(p-3),$$

that is, $3p-3$ intrinsic constants: this is in agreement with a result previously obtained (Chap. I. § 7).

103. The $p-3$ places S may be defined in a particular way, thus:— In general there are (Chap. III. § 31) $(p-1)p(p+1)$ places of the original surface, for each of which a rational function can be found, infinite only at such place and infinite to the pth order. Every rational function, whose order is less than $p+1$, can be expressed as the quotient of two ϕ-polynomials (§ 98). The ϕ-polynomial, ϕ, occurring in the denominator of the function, will† vanish p times at the place where the function has a pole of order p‡, and will vanish in $p-2$ other places forming a set T. The general ϕ-polynomial § through these $p-2$ places T will not have another fixed zero, or it would be impossible to form a rational function of order p with ϕ as denominator. Let now A denote any place of the set T, the remaining $p-3$ places being denoted by S. Then we may continue the process exactly as in the last Article.

The p variable zeros of the ϕ-polynomials, of the form $\lambda\phi+\mu\vartheta$, which vanish in the $p-2$ places T will, for the transformed curve, become the variable intersections of it with the straight lines, $\lambda z+\mu s=0$, which pass through the place $s=0$, $z=0$. We enquire now how many of these straight lines will touch the new curve. This number may be found either by the ordinary methods of analytical geometry ‖ or as the number of places where

* By the formula $p=\frac{1}{2}(n-1)(n\sigma+n-2)-\frac{1}{2}\Sigma i$, for it is clear that s is an integral function of z of dimension 1, so that $\sigma=0$. And we have remarked that i is 1 at each of the places corresponding to a double point of the curve, so that $\delta+\kappa=\frac{1}{2}\Sigma i$; cf. Forsyth, *Theory of Functions*, § 182.

† See the note (*) of § 102.

‡ This is the fact expressed by the vanishing of the determinant Δ in § 31, Chap. III.

§ Which we assume to be of the form $\lambda\phi+\mu\vartheta$, involving $q+1=2$ arbitrary coefficients. If q were greater than unity, it would be possible to construct a function of lower than the pth order. This possibility is considered below (§ 105 ff.).

‖ See for example Salmon's *Higher Plane Curves*.

the differential of the function ϑ/ϕ, of order p, vanishes to the second order, namely* $2p+2p-2$. Among these tangents, however, there is one which touches the transformed curve in p points, counting as $p-1$ tangents. There are, therefore, $3p-1$ other tangents. Of the $3p$ distinct tangent lines thus obtained, there are $3p-3$ distinct cross ratios, formed from the $3p-3$ distinct sets of four of them, and these cross ratios are independent of any linear transformation of the coordinates s and z.

There are thus $3p-3$ quantities obtainable for the transformed curve. We prove, now†, that they entirely determine this curve, and may, therefore, since the transformation is reversible, be regarded as the absolute constants of the original curve. For take any arbitrary point O; draw through it 3 arbitrary straight lines and draw $3p-3$ other straight lines which form with the 3 straight lines first drawn pencils of given cross ratios. Then the coefficients of a curve of order $p+1$, which passes through O, has $\frac{1}{2}p(p-3)$ double points, and touches $3p$ straight lines through O, one of them in p consecutive points, are subject to $1+\frac{1}{2}p(p-3)+3p-1+p-1$ or $\frac{1}{2}p^2+\frac{5}{2}p-1$ linear conditions. The number of these coefficients is $\frac{1}{2}(p+1)(p+4)$ or $\frac{1}{2}p^2+\frac{5}{2}p+2$. Hence there are three coefficients left arbitrary; besides these there are five other constants in the equation of the curve, namely, those which settle the position of O and the three arbitrary straight lines through O. The eight constants thus involved in the curve can be disposed of by a linear transformation.

The reader will recognise here a verification of the argument sketched in § 7, Chap. I.; the present argument is in fact only a particular case of that, obtained by specialising the dependent variable of the new surface, and the order of the independent variable g. The restriction that the p poles of g shall be in one place can be removed, with a certain loss of definiteness and conviction.

The argument employed clearly fails for the hyperelliptic case, since then the $p-2$ fixed zeros of the polynomials ϕ and ϑ determine other places, and the function ϑ/ϕ is not of the pth order.

For $p=3$ we have the result:—If an inflexional tangent of a plane quartic curve meet the curve again in O, eight other tangents to the curve can be drawn from O. The cross ratios of the six independent sets of four tangents, which can be formed from these nine tangents, determine the curve completely—save for constants which can be altered by projection.

More generally, from any point O of the quartic, ten tangents to the curve can be drawn. The seven cross ratios of these tangents leave, by elimination of the coordinates of O, six quantities from which the curve is determinate, save for quantities altered by projection.

* Chap. I. § 6.

† Cayley, *Collected Works*, vol. VI. p. 6. Brill u. Noether, *Math. Annal.* t. VII. p. 303.

104. It is a very slight step from the process of the last Article to take the independent variable to be $g = \vartheta/\phi$, where ϑ, ϕ are ϕ-polynomials, having $p-2$ common zeros forming a set such that a single infinity of ϕ-polynomials vanish in the places of the set. And it may be convenient to take another dependent variable.

In the process of Article 102, the fixed zeros of the polynomials used are $p-3$ in number, and a double infinity of ϕ-polynomials vanish in the places of the set.

These two processes are capable of extension. If we can find a set S_Q, of Q places, in which just $(\tau + 1 =) 3$ ϕ-polynomials vanish, and if the places S_Q be such that these three ϕ-polynomials have no other common zero, while the problem of finding a further place x_1, such that the two ϕ-polynomials vanishing in S_Q and x_1 have another common zero x_2, is capable of only a finite number of solutions, then we can extend the process of Article 102; we can then, in fact, transform the surface into one of $2p-2-Q$ sheets. The dependent variable in the new equation will be of dimension unity, and the equation such as represents a curve of order $2p-2-Q$. If, therefore, we can find sets S_Q in which $Q > p-3$, the new surface will have a less number of sheets, and therefore, in general, a simpler form of equation, than the surface obtained in § 102.

Similarly, if we can find a set, S_Q, which are the common zeros of $(\tau + 1 =) 2$ ϕ-polynomials, say ϑ and ϕ, we can use the function $g = \vartheta/\phi$, with a suitable other function, as independent and dependent variables respectively, to obtain a new form of equation for which there are $2p-2-Q$ sheets: and if we can get $Q > p-2$ the new surface will be simpler than that obtained in § 103.

105. We are thus led to enquire what are the conditions that $\tau + 1$ linearly independent ϕ-polynomials should vanish in any Q places $a_1, \dots, a_Q$.

If the general ϕ-polynomial be written in the form $\lambda_1\phi_1(x) + \dots + \lambda_p\phi_p(x)$, where $\lambda_1, \dots, \lambda_p$ are arbitrary constants, the conditions are that the Q equations

$$\lambda_1\phi_1(a_i) + \dots + \lambda_p\phi_p(a_i) = 0, \qquad (i = 1, 2, \dots, Q)$$

should be equivalent to only $p-\tau-1$ equations, for the determination of the ratios $\lambda_1 : \dots : \lambda_p$; we suppose $Q > p-\tau-1$, and further that the notation is so chosen that the independent equations are the first $p-\tau-1$ of them. Then there exist $Q-(p-\tau-1)$ sets, each of p equations, of the form

$$\phi_j(a_{p-\tau-1+\sigma}) = m_1\phi_j(a_1) + \dots + m_{p-\tau-1}\,\phi_j(a_{p-\tau-1}), \qquad (j = 1, 2, \dots, p)$$

for each value of σ from 1 to $Q-(p-\tau-1)$, the values of $m_1, \dots, m_{p-\tau-1}$ being, for any value of σ, the same for every value of j. The set, of p, of these equations, for which σ has any definite value, lead to $\tau+1$ equations, of the form

$$\left|\begin{array}{llll} \phi_1(a_1), & \ldots, & \phi_1(a_{p-\tau-1}) & , \ \phi_1(a_{p-\tau-1+\sigma}) \\ \cdots & \cdots & \cdots & \cdots \\ \phi_{p-\tau-1}(a_1), & \ldots, & \phi_{p-\tau-1}(a_{p-\tau-1}) & , \ \phi_{p-\tau-1}(a_{p-\tau-1+\sigma}) \\ \phi_{p-\tau-1+k}(a_1), & \ldots, & \phi_{p-\tau-1+k}(a_{p-\tau-1}), & \ \phi_{p-\tau-1+k}(a_{p-\tau-1+\sigma}) \end{array}\right| = 0,$$

arising for $k = 1, 2, \ldots, \tau+1$.

Putting $q = Q-(p-\tau-1)$, we have therefore $q(\tau+1)$ such equations* connecting the Q places $a_1, \ldots, a_Q$.

It is obvious from the method of formation that these $q(\tau+1)$ equations are in general independent; in what follows we consider only the cases in which they are independent and determinate. Then, taking $Q-q(\tau+1)$ quite arbitrary places, it is possible to determine $q(\tau+1)$ other places, such that there are $\tau+1$ linearly independent ϕ-polynomials vanishing in the total Q places.

The determination of the $q(\tau+1)$ places, from the arbitrary $Q-q(\tau+1)$ places, may be conceived of as the problem of finding $p-\tau-1-[Q-q(\tau+1)]$, or $q\tau$, places, T, to add to the $Q-q(\tau+1)$ arbitrary places, S, such that all ϕ-polynomials vanishing in the resulting $p-\tau-1$ places S, T, may have $Q-(p-\tau-1)$, or q, other common zeros. The $p-\tau-1$ places S, T are independent determinators of ϕ-polynomials.

For instance, when $Q=p-1$, $\tau+1=2$, it follows that $q=1$ and $Q-q(\tau+1)=p-3$, and hence, from the theory here given, it follows that we can determine $p-1$ places in which two ϕ-polynomials vanish, and, of these, $p-3$ places are arbitrary. The problem of determining the other two places may be conceived of as the problem of determining $p-\tau-1-[Q-q(\tau+1)]$, or *one*, other place, to add to the $p-3$ places, such that all ϕ-polynomials vanishing in the resulting $p-2$ places, which are independent determinators of ϕ-polynomials, may have $q=1$ other common zero. We have already seen reason for believing that, when the $p-3$ places are given, the other two places can be determined in $\frac{1}{2}p(p-3)$ ways.

To every set of Q places thus determined, there corresponds a co-residual lot of sets of Q places, the multiplicity of the lot being q; and every co-residual set will have the same character as the original set. The number, q, of places of a co-residual set which are arbitrary, cannot, obviously, be greater than the number, $Q-q(\tau+1)$, of the original set, which are arbitrary. Hence, the self-consistence of the theory clearly requires that $Q-q(\tau+1) \geqq q$. From this, by means of the relation $Q-q=p-\tau-1$, we can deduce the two important results

$$p \geqq (q+1)(\tau+1), \quad Q \geqq q+p\frac{q}{q+1}.$$

* These equations are necessary in order that $a_1, \ldots, a_Q$ should be the poles of a rational function.

Putting $Q - q(\tau+1) = q + \alpha$, we obtain

$$p = (\tau+1)(q+1) + \alpha, \quad Q = q + p\frac{q}{q+1} + \frac{\alpha}{q+1}.$$

From each such set S_Q we can deduce, as its residuals, sets, S_R, of R, $= 2p - 2 - Q$, places, in which $q+1$ ϕ-polynomials vanish, and it is immediately seen that

$$Q - q(\tau+1) - q = \alpha = R - \tau(q+1) - \tau.$$

106. If now we determine, in accordance with this theory, a set S_Q in which $\tau + 1 = 3$ ϕ-polynomials vanish, it being assumed that these three ϕ-polynomials have no other common zero, and determine ϕ, ϑ to be two ϕ-polynomials vanishing in S_Q and in one other place O, ψ being another ϕ-polynomial vanishing in S_Q but not in O, then the equations $z = \phi/\psi$, $s = \vartheta/\psi$, determine, as before, a reversible transformation of the surface, to a new surface of which the number of sheets is $R = 2p - 2 - Q$, and in which s is of dimension 1 in regard to z.

Since $R \overline{\overline{>}} \tau + p\tau/(\tau+1)$, the value of R is $\overline{\overline{>}}\, 2 + \frac{2}{3}p$. Thus writing $p = 3\pi$, or $3\pi + 1$, or $3\pi + 2$, according as it is a multiple of 3 or not, R is $p - \pi + 2$ in all cases.

From $R = p - \pi + 2$ follows $Q = p - 4 + \pi$; thus $q = Q - p + 3 = \pi - 1$, and $Q - q(\tau+1) = p + \pi - 4 - 3\pi + 3 = p - 2\pi - 1$. This is the number of places of the set S_Q which may be taken arbitrarily. If this number be equal to $q = \pi - 1$, it follows that, by taking two different sets of $Q - q(\tau+1)$, $= p - 2\pi - 1$, places, we get only two co-residual sets, and for the purposes of forming the functions ϕ/ψ, ϑ/ψ, one is as good as the other. If however $Q - q(\tau+1) > q$, we do not get co-residual sets by taking different arbitrary sets of $Q - q(\tau+1)$ places :—and there is a disposeableness which is expressed by the number of the arbitrary places, $Q - q(\tau+1)$, which is in excess of the number, q, which determines the sets co-residual to any given one.

Now $Q - q(\tau+1) - q = p - 2\pi - 1 - \pi + 1 = p - 3\pi$. And, in a surface of m sheets and deficiency p, the number of constants independent of linear transformations is $3m + p - 9$ (§ 102). Hence the number of unassignable quantities in the equation of the surface is

$$3(p - \pi + 2) + p - 9 - (p - 3\pi) \text{ or } 3p - 3;$$

and this is in accordance with a result previously obtained (§ 7, Chap. I.).

Ex. i. The values of π for $p=4$, 5 are 1, 1 respectively, and $p - \pi + 2$, in these cases, $=5$, 6 respectively.

Hence a quintic curve with two double points ($p=4$), can be transformed into a quintic; this will also have two double points, in general, since the deficiency must be unaltered. We determine a set consisting of Q, $=1$, quite arbitrary place. Let the general conic through this place, and the two double points, be $\lambda\phi+\mu\vartheta+\nu\psi=0$. Then the formulae of transformation are $z=\phi/\psi$, $s=\vartheta/\psi$. As in the text, we may suppose ϕ, ϑ to have another common point, in which ψ does not vanish.

Ex. ii. A quintic with one double point ($p=5$) can be transformed into a sextic with, in general, $\frac{1}{2}(6-1)(6-2)-5=5$ double points. For this we take $p-2\pi-1=2$ arbitrary points; if $\lambda\phi+\mu\vartheta+\nu\psi$ be the general conic through the two points and the double point, the equations of transformation are $z=\phi/\psi$, $s=\vartheta/\psi$.

Ex. iii. Shew that the orders $p-\pi+2$ of the curves obtainable by this method to represent curves of deficiencies

$$p=6,\ 7,\ 8,\ 9$$

are respectively

$$R=6,\ 7,\ 8,\ 8.$$

107. But, as remarked (§ 104), we can also make use of sets of R places for which $\tau+1=2$, to obtain transformations of our original surface.

We can obtain such a set by taking $R-\tau(q+1)$, or $R-q-1$, arbitrary places, and determining the remaining $q+1$ such that $q+1$ ϕ-polynomials vanish in the whole set of R places.

It is proved by Brill* that the number of sets of $q+1$ thus obtainable from $R-q-1$ arbitrary places, is

$$\sum_{0}^{\mu}(-1)^{\lambda}\binom{p}{\lambda}\binom{2p-1-R-2\lambda}{2p-1-R-q-1},$$

where $\mu=\frac{1}{2}q$ or $\frac{1}{2}(q+1)$, according as q is even or odd, and $\binom{\lambda}{\nu}$ denotes $\lambda(\lambda-1)\dots(\lambda-\nu+1)/\nu!$.

For instance with $R=p$, $q=0$, the series reduces to one term, whose value is $p-1$, which is clearly right; while, when $R=p-1$, $q=1$, the series reduces to

$$\binom{p}{p-2}-p\binom{p-2}{p-2},$$

or $\frac{1}{2}p(p-3)$, as in § 101, § 102, p. 144.

When p is even and $R=\frac{1}{2}p+1$, $q=\frac{1}{2}p-1$, this series can be summed, and is equal to

$$2\,\underline{|p-1}/\underline{|\tfrac{1}{2}p-1}\ \underline{|\tfrac{1}{2}p+1}.$$

When p is odd and $R=\frac{1}{2}(p+1)+1$, $q=\frac{1}{2}(p-1)-1$, the series can be summed, and is equal to

$$4p\,\underline{|p-2}/\underline{|\tfrac{1}{2}(p-3)}\ \underline{|\tfrac{1}{2}(p+3)}.$$

Now let $\lambda\phi+\mu\vartheta$ be the general ϕ-polynomial vanishing in a set which is residual to one of these sets of R places, λ and μ being arbitrary constants; we may transform the surface with $z=\vartheta/\phi$ as the new independent variable. The new surface obtained will have R sheets. The new dependent variable may be chosen at will, provided only the transformation be reversible.

* *Math. Annal.* xxxvi, pp. 354, 358, 369. See also Brill and Noether, *Math. Annal.* VII. p. 296.

The function $\mu z+\lambda$, $=\mu\vartheta/\phi+\lambda$, depends on $2+R-q-1$ arbitrary quantities, namely the constants λ, μ and the position of the $R-q-1$ arbitrarily taken places. There are $2R+2p-2$ places where dz is zero to the second order, namely, $2R+2p-2$ places where the curve $a\vartheta+b\phi=0$ touches the fundamental curve ; there remain then

$$2R+2p-2-(R-q+1),\ =R-1-p+q+1+3p-3,\ =3p-3$$

of the $2R+2p-2$ values which z has when dz vanishes to the second order, which are quite arbitrary. Compare § 7, Chap. I.

The least possible value of R is given by the formula $R \overline{\overline{>}} \tau+p\tau/(\tau+1)$. If then p be written equal to 2π, or $2\pi+1$, according as p is even or odd, we may take* $R=p-\pi+1$, that is $\frac{1}{2}p+1$ or $\frac{1}{2}(p+1)+1$, according as p is even or odd.

Hence, when p is even, we can determine a single infinity of co-residual sets of $\frac{1}{2}p+1$ places, these sets being the zeros of ϕ-polynomials, $\lambda\phi+\mu\vartheta$, which have $\frac{3}{2}p-3$ common zeros. To determine one of these sets of $\frac{1}{2}p+1$ places, we may take one place, A, arbitrarily. The other $\frac{1}{2}p$ places can then be determined in $2\,\underline{|p-1}/\underline{|\tfrac{1}{2}p-1}\ \underline{|\tfrac{1}{2}p+1}$ ways. Let two of these ways be adopted, corresponding to one arbitrary place A; the resulting sets of $\frac{1}{2}p+1$ places will not be co-residual; for the sets co-residual with a given set have a multiplicity 1, and therefore no two of these sets can have a place common without coinciding altogether. Let the sets co-residual to these two sets be given by $\lambda\phi+\mu\vartheta=0$, $\lambda'\phi'+\mu'\vartheta'=0$, ϕ and ϕ' being chosen so as to vanish in A: we assume that ϕ, ϕ' have no other common zero.

Then the equations $z=\phi/\vartheta$, $s=\phi'/\vartheta'$ will determine a reversible transformation, as is immediately seen in a way analogous to those already adopted. In the new equation z and s enter to a degree $\frac{1}{2}p+1$, and, since there exists* no rational function of lower order than $\frac{1}{2}p+1$, no further reduction of the degree to which z and s enter, is possible.

The new equation may be interpreted as the equation of a curve of order $p+2$: it will have the form

$$(z,\ 1)^m s^m+(z,\ 1)^m s^{m-1}+\ldots+(z,\ 1)^m=0,$$

wherein $m=\frac{1}{2}p+1$.

By putting $z=1/z_1$, $s=1/s_1$, it is reduced to the equation of a curve of order p. The form possesses the interest that it was employed by Riemann.

Ex. Obtain the 2 sets of $\frac{1}{2}p+1$ places corresponding to a given arbitrary point for a quintic curve with two double points, and transform the equation.

108. If we have a set of R places†, for which $\tau+1=4$, the co-residual places being given by the variable zeros of ϕ-polynomials of the form $\lambda\phi_1+\mu\phi_2+\nu\phi_3+\psi$, we can, by writing

$$X=\phi_1/\psi,\quad Y=\phi_2/\psi,\quad Z=\phi_3/\psi,$$

* Thus, for perfectly general surfaces of deficiency p, no rational function exists of order less than $1+\frac{1}{2}p$. Cf. Forsyth, *Theory of Functions*, p. 460. Riemann, *Gesam. Werke* (1876), p. 101.

† Wherein $R-\tau<p$, or $R<p+3$.

and eliminating x, y from these three equations and the fundamental equation associated with the Riemann surface, obtain two rational algebraic equations connecting X, Y, Z; these equations determine a curve in space, of order R; for this is the number of variable zeros of the function $\lambda X + \mu Y + \nu Z + 1$. To a point $X = X_1$, $Y = Y_1$, $Z = Z_1$ of the curve in space, will correspond the places of the surface, other than the fixed zeros of ϕ_1, ϕ_2, ϕ_3, ψ, at which

$$X_1\psi - \phi_1 = 0, \quad Y_1\psi - \phi_2 = 0, \quad Z_1\psi - \phi_3 = 0,$$

and it is generally possible to choose ϕ_1, ϕ_2, ϕ_3, ψ so that these equations have only one solution.

The lowest order possible for the space curve is given by

$$R \gtreqless \tau + \tau p/(\tau + 1) \gtreqless 3 + 3p/4.$$

If then $p = 4\pi$, or $4\pi + 1$, or $4\pi + 2$, or $4\pi + 3$, R may be taken equal to $p - \pi + 3$.

For instance with* $p=4$, $R=6$, taking a plane curve with double points at the places $x=\infty$, $y=0$ and $x=0$, $y=\infty$, given by

$$x^2y^2(x, y)_1 + xy(x, y)_2 + (x, y)_3 + (x, y)_2 + (x, y)_1 + A = 0,$$

we may† take $\lambda\phi_1 + \mu\phi_2 + \nu\phi_3 + \psi = \lambda xy + \mu x + \nu y + 1$; the places residual to the variable set of R places are, in number, $2p - 2 - 6$, $=0$. Then the equations of transformation are

$$X = xy, \quad Y = x, \quad Z = y,$$

and these give points (X, Y, Z) lying on the surfaces,

$$X = YZ,$$
$$X^2(Y, Z)_1 + X(Y, Z)_2 + (Y, Z)_3 + (Y, Z)_2 + (Y, Z)_1 + A = 0,$$

of which the first is a quadric and the second a cubic.

A set of R places with multiplicity $\tau = 3$ may of course also be used to obtain a transformation to another Riemann surface. With the same notation we may put $z = \phi_1/\psi$, $s = \phi_2/\psi$. It is clear that the resulting equation, regarded as that of a plane curve, is the orthogonal projection, on to the plane $Z = 0$, of the space curve just obtained.

A set of R places with multiplicity $\tau > 3$ may be used similarly to obtain a curve of order R in space of τ dimensions. Some considerations in this connexion will be found in the concluding articles of this chapter.

109. It has already been explained that the methods of transformation given in §§ 101—108 of this chapter are not intended to apply to surfaces which are not of general character for their deficiency, and that, in particular, hyperelliptic surfaces are excluded from consideration. We may give here a practical method of obtaining the canonical form of a hyperelliptic surface,

* Since p must be $\gtreqless (\tau+1)(q+1)$, this is the first case to which the theory applies.

† It is easy to shew that this is the general adjoint polynomial of degree $n-3$. We may also shew that the integrals, $\int xydx/f'(y)$, etc., are finite, or use the method given *Camb. Phil. Trans.* xv. iv. p. 413, there being no finite multiple points.

whose existence has already been demonstrated (Chap. V. § 54). Suppose first that $p>1$. In the hyperelliptic case every ϕ-polynomial vanishing in any place A will vanish, of itself, in another place A'. Any one of these ϕ-polynomials will have $2p-4$ other zeros, forming a set which we shall denote by S. Putting $Q=2$ and $\tau+1=p-1$ in the formula $Q-q=p-\tau-1$, we find $q=1$, so that the general ϕ-polynomial vanishing in the places S will be of the form $\lambda_1\phi_1-\lambda_2\phi_2$, wherein λ_1, λ_2 are arbitrary constants; in fact these $2p-4$ places S consist of $p-2$ independent places and the other $p-2$ places determined by them, one by each. Thus a function of the second order is given by $z=\phi_1/\phi_2$. A general adjoint polynomial of grade $(n-1)\sigma+n-2$ will contain $n+p-1$ terms and vanish, in all, in $n+2p-2$ places; thus the general adjoint polynomial, of this grade, which is prescribed to vanish in a set T of $n+p-3$ arbitrary places, will be of the form $\mu_1\psi_1+\mu_2\psi_2$, μ_1, μ_2 being arbitrary constants, and will vanish in $p+1$ other places. We may suppose ψ_1 so chosen that it vanishes in one of the two zeros of ϕ_1 which are not among the set S, and we shall assume that ψ_2 does not vanish in this place, and that ψ_1 does not vanish in the other of these two zeros of ϕ_1. Then the functions $z=\phi_1/\phi_2$, $s=\psi_1/\psi_2$, are connected by a rational equation, $(s, z)=0$, with which a new Riemann surface may be associated; to any place of the old surface there corresponds only one place $z=\phi_1/\phi_2$, $s=\psi_1/\psi_2$, of the new surface; to the place $z=0$, $s=0$ of the new surface corresponds only one place of the original surface, and the same is therefore true of every place of the new surface. Thus the equation $(s, z)=0$ is of degree 2 in s and degree $p+1$ in z. The highest aggregate degree in s and z together, in the equation $(s, z)=0$, is the same as the number of zeros of functions of the form $\lambda z+\mu s+\nu$, for arbitrary values of λ, μ, ν, and therefore if the poles s be different from the poles of z, namely, if the zeros of ψ_2 other than T, be different from the zeros of ϕ_2 other than S, the aggregate degree of (s, z) in s and z together will be $p+3$; thus the equation will be included in the form

$$s^2\alpha+s\beta+\gamma=0,$$

where α, β, γ are integral polynomials in z of degree $p+1$.

If we put $\sigma=s\alpha+\frac{1}{2}\beta$, this takes the form

$$\sigma^2=\tfrac{1}{4}\alpha^2-\alpha\gamma,$$

which is of the canonical form in question.

Ex. A plane quartic curve with a double point ($p=2$) may be regarded as generated by the common variable zero A of (i) straight lines through the double point, vanishing also in variable points A and A', (ii) conics through the double point and three fixed points, vanishing also in variable points A, B, C.

When p is 1 or 0, the method given here does not apply, since then adjoint ϕ-polynomials (which in general vanish in $2p-2$ variable places) have no variable zeros. In case $p=1$ or $p=0$, if $\mu_1\psi_1+\mu_2\psi_2+\mu_3\psi_3$, with μ_1, μ_2, μ_3 arbitrary, be the general adjoint polynomial of grade $(n-1)\sigma+n-2$ which vanishes in $n+p-4$ fixed places, ψ_1, ψ_3 being chosen to have one other common zero beside these $n+p-4$ fixed places, we may use the transformation $z=\psi_1/\psi_3$, $s=\psi_2/\psi_3$, z being a function of order $p+1$, and s being a function of order $p+2$. Then, since the function $\lambda z+\mu s+\nu$ vanishes in $p+2$ places, we obtain an equation of the form*

$$s^2(z, 1)_p + s(z, 1)_{p+1} + (z, 1)_{p+2} = 0,$$

of which the further reduction is immediate.

Ex. For a plane quartic curve with two double points ($p=1$) let $\mu_1\psi_1+\mu_2\psi_2+\mu_3\psi_3$ be the general conic through the double points and a further point A, ψ_1 and ψ_3 being chosen also to vanish at any point B. Then we may use the transformation $z=\psi_1/\psi_3$, $s=\psi_2/\psi_3$.

110. In the transformations which have been given we have made frequent use of the polynomials which we have called ϕ-polynomials, namely adjoint polynomials of grade $(n-1)\sigma+n-3$. For this there is the special reason, already referred to†, that, in any reversible transformation of the surface, their ratios are changed into ratios of ϕ-polynomials belonging to the transformed surface; thus any property, or function, which can be expressed by these ϕ-polynomials only, is invariant for all birational transformations. We give now some important examples of such properties.

Let the general ϕ-polynomial be always supposed expressed in the form $\lambda_1\phi_1+\ldots+\lambda_p\phi_p$, $\lambda_1, \ldots, \lambda_p$ being arbitrary constants. Instead of $\phi_1, \ldots, \phi_p$ we may use any p linearly independent linear functions of $\phi_1, \ldots, \phi_p$, agreed upon beforehand. A convenient method is to take p independent places $c_1, \ldots, c_p$ and define ϕ_i as the ϕ-polynomial vanishing in all of $c_1, \ldots, c_p$ except c_i; but we shall not adhere to that convention in this place. Let any general integral homogeneous polynomial in $\phi_1, \ldots, \phi_p$, of degree μ, be denoted by $\Phi^{(\mu)}$ or $\Phi'^{(\mu)}$. This polynomial contains $p(p+1)\ldots(p+\mu-1)/\mu!$ terms.

In a polynomial $\Phi^{(2)}$ there are $\frac{1}{2}p(p+1)$ products of two of $\phi_1, \ldots, \phi_p$. But these $\frac{1}{2}p(p+1)$ products of pairs are not linearly independent. For example in a hyperelliptic case, we can choose a function of the second order, z, such that the ratios of p independent ϕ-polynomials are given by

$$\phi_1 : \phi_2 : \ldots : \phi_p = 1 : z : z^2 : \ldots : z^{p-1};$$

then there will be $p-2$ identities of the form

$$\phi_2/\phi_1 = \phi_3/\phi_2 = \ldots = \phi_p/\phi_{p-1},$$

* Further developments are given by Clebsch, *Crelle*, t. 64, pp. 43, 210. For this subject and for many other matters dealt with in this Chapter, the reader may also consult Clebsch-Lindemann-Benoist, *Leçons sur la Géométrie* (Paris 1883), t. III.

† Chap. II. § 21.

whereby the number of linearly independent products of pairs of $\phi_1, \ldots, \phi_p$ is reduced to $\frac{1}{2}p(p+1)-(p-2)$, at most. But we can in fact shew, whether the surface be hyperelliptic or not, that there are not more than $3(p-1)$ linearly independent products of pairs of $\phi_1, \ldots, \phi_p$. For consider the $4(p-2)$ places in which any general quadratic polynomial, $\Phi^{(2)}$, vanishes. If $\phi_i\phi_j$ be any product of two of the polynomials $\phi_1, \ldots, \phi_p$, the quotient $\phi_i\phi_j/\Phi^{(2)}$ represents a rational function having no poles except such as occur among the zeros* of $\Phi^{(2)}$; there are therefore at least as many linearly independent rational functions, with poles among the zeros of $\Phi^{(2)}$, as there are linearly independent products of pairs of $\phi_1, \ldots, \phi_p$. But the general rational function having its poles among the $4(p-1)$ zeros of $\Phi^{(2)}$, contains only $4(p-1)-p+1$, $=3(p-1)$, arbitrary constants. Hence there are not more than this number of linearly independent pairs of $\phi_1, \ldots, \phi_p$. In precisely the same way it follows that there are not more than $(2\mu-1)(p-1)$ linearly independent products of μ of the polynomials $\phi_1, \ldots, \phi_p$.

111. But it can be further shewn that in general† there are just $(2\mu-1)(p-1)$ linearly independent products of μ of the polynomials $\phi_1, \ldots, \phi_p$; so that there are

$$\frac{p(p+1)\ldots(p+\mu-1)}{\underline{|\mu}} - (2\mu-1)(p-1)$$

identical relations connecting the products of μ of the polynomials $\phi_1, \ldots, \phi_p$.

Consider the case $\mu=2$. Take $p-2$ places such that the general ϕ-polynomial vanishing in them is of the form $\lambda\phi_1+\mu\phi_2$, λ and μ being arbitrary, and ϕ_1, ϕ_2 having no zero common beside these $p-2$ places. Let $\Phi^{(1)}$, $\Phi'^{(1)}$ denote two general linear functions of $\phi_1, \ldots, \phi_p$. The polynomial

$$\phi_1\Phi^{(1)} + \phi_2\Phi'^{(1)}$$

is quadratic in $\phi_1, \ldots, \phi_p$. It contains $2p$ terms. But clearly these terms are not linearly independent, for the term $\phi_2\phi_1$ occurs both in $\phi_1\Phi^{(1)}$ and in $\phi_2\Phi'^{(1)}$. Suppose, then, that there are terms, $\phi_2\Psi'^{(1)}$, occurring in $\phi_2\Phi'^{(1)}$, which are equal to terms, $\phi_1\Psi^{(1)}$, occurring in $\phi_1\Phi^{(1)}$. The necessary equation for this,

$$-\frac{\Psi'^{(1)}}{\Psi^{(1)}} = \frac{\phi_1}{\phi_2},$$

shews that $\Psi^{(1)}$ vanishes in the p zeros of ϕ_2 which are not zeros of ϕ_1. But since these p zeros form a set which is a residual of a set (of $p-2$ places)

* Here, as in all similar cases, the zeros of the polynomial are its generalised zeros when it is regarded as of its specified grade.

† Precisely, the theorem is true when the surface is sufficiently general to allow the existence of $p-2$ places such that the general ϕ-polynomial, vanishing in them, is of the form $\lambda\phi_1+\mu\phi_2$, λ and μ being arbitrary constants, and ϕ_1, ϕ_2 having no common zero other than the $p-2$ places. We have already given a proof that this is always the case when the surface is not hyperelliptic (§ 102).

in which two ϕ-polynomials vanish, it follows* that only one ϕ-polynomial vanishes in these p places; and such an one is ϕ_2. Hence $\Psi^{(1)}$ must be a multiple of ϕ_2, and therefore $\Psi'^{(1)}$ a multiple of ϕ_1. Thus the polynomial

$$\phi_1\Phi^{(1)} + \phi_2\Phi'^{(1)}$$

contains $2p-1$ linearly independent products of pairs of $\phi_1, \dots, \phi_p$.

Let now ϕ_3 be a ϕ-polynomial not vanishing in the common zeros of ϕ_1, ϕ_2, and let $\phi_4, \dots, \phi_p$ be chosen so that $\phi_1, \phi_2, \phi_3, \dots, \phi_p$ are linearly independent. Consider the polynomial

$$\Phi = \phi_1\Phi^{(1)} + \phi_2\Phi'^{(1)} + \phi_3[\lambda_3\phi_3 + \dots + \lambda_p\phi_p],$$

wherein $\lambda_3, \dots, \lambda_p$ are arbitrary constants. Herein $\phi_3(\lambda_3\phi_3 + \dots + \lambda_p\phi_p)$ cannot contain any terms $\phi_3(\lambda_3'\phi_3 + \dots + \lambda_p'\phi_p)$ which are equal to terms already occurring in the part $\phi_1\Phi^{(1)} + \phi_2\Phi'^{(1)}$, or else $\lambda_3'\phi_3 + \dots + \lambda_p'\phi_p$ would vanish in the $p-2$ common zeros of ϕ_1 and ϕ_2; and this is contrary to the hypothesis that $\lambda\phi_1 + \mu\phi_2$ is the most general ϕ-polynomial vanishing in these $p-2$ places. Hence the polynomial Φ contains $2p-1+p-2$, or $3p-3$, independent products of twos of the polynomials $\phi_1, \dots, \phi_p$. As we have proved that a greater number does not exist, $3p-3$ is the number of such products of pairs.

Consider next the case $\mu = 3$. Since co-residual sets of $2p-1$ places have † a multiplicity $p-1$, it follows that the general polynomial, $\Psi^{(2)}$, of the second degree in $\phi_1, \dots, \phi_p$, which vanishes in $2p-3$ fixed places, and therefore in $2p-1$ variable places, contains p arbitrary coefficients. If then the $2p-3$ fixed zeros of $\Psi^{(2)}$ be zeros of a definite polynomial, ϕ_2, it follows that $\Psi^{(2)}$ is of the form $\phi_2\Psi^{(1)}$, $\Psi^{(1)}$ being of the first degree in $\phi_1, \phi_2, \dots, \phi_p$. Hence, as in the case $\mu = 2$, it can be proved that if ϕ_1, ϕ_2 be ϕ-polynomials with one common zero, the reduction in the number, $2(3p-3)$, of terms in a polynomial $\phi_1\Phi^{(2)} + \phi_2\Phi'^{(2)}$, which arises in consequence of the occurrence of terms, $\phi_2\Psi'^{(2)}$, in $\phi_2\Phi'^{(2)}$, which are equal to terms, $-\phi_1\Psi^{(2)}$, occurring in $\phi_1\Phi^{(2)}$, is at most equal to p. Hence the polynomial $\phi_1\Phi^{(2)} + \phi_2\Phi'^{(2)}$ contains at least $5p-6$ linearly independent products of threes of $\phi_1, \dots, \phi_p$. Hence taking ϕ_3, and a quadratic polynomial $\Phi''^{(2)}$, such as do not vanish in the common zero of ϕ_1, ϕ_2, it follows that a cubic polynomial with at least $5p-5$ linearly independent products, is given by

$$\phi_1\Phi^{(2)} + \phi_2\Phi'^{(2)} + \phi_3\Phi''^{(2)}.$$

We have thus proved that in the cases $\mu = 2$, $\mu = 3$, the polynomial $\Phi^{(\mu)}$ contains $(2\mu-1)(p-1)$ linearly independent products. Assume now that $\Phi^{(\mu-1)}$ contains $(2\mu-3)(p-1)$ independent terms, and that $\Phi^{(\mu-2)}$

* From the formula (Chap. VI. § 93)

$$Q - R = 2(q - r),$$

putting $Q = p-2$, $R = p$, $r = 1$, we obtain $q = 0$.

† From $Q - q = p - (r+1)$, putting $r+1=0$ (because $2p-1 > 2p-2$) $Q = 2p-1$, $q = p-1$.

contains $(2\mu-5)(p-1)$ independent terms. A general polynomial $\Psi^{(\mu-1)}$ vanishing in the zeros of a definite ϕ-polynomial, ϕ_2, will have $2(\mu-2)(p-1)$ variable zeros; and the multiplicity of co-residual sets of $2(\mu-2)(p-1)$ places, when $\mu>3$, is $(2\mu-5)(p-1)-1$, which by hypothesis is the same as the multiplicity of the sets of zeros of a polynomial $\phi_2\Psi^{(\mu-2)}$, in which $\Psi^{(\mu-2)}$ has its most general form possible. Hence the general polynomial $\Psi^{(\mu-1)}$ vanishing in the zeros of ϕ_2, is of the form $\phi_2\Psi^{(\mu-2)}$. If then, in a polynomial, $\phi_1\Phi^{(\mu-1)}+\phi_2\Phi'^{(\mu-1)}$, of the μth degree in $\phi_1, \ldots, \phi_p$, wherein ϕ_1, ϕ_2 have no common zeros, there be terms, $\phi_2\Psi'^{(\mu-1)}$, occurring in $\phi_2\Phi'^{(\mu-1)}$, which are equal to terms, $-\phi_1\Psi^{(\mu-1)}$, occurring in $\phi_1\Phi^{(\mu-1)}$, then $\Psi^{(\mu-1)}$ must be of the form $\phi_2\Psi^{(\mu-2)}$, and $\Psi'^{(\mu-1)}$ of the form $\phi_1\Psi'^{(\mu-2)}$, and the resulting reduction in the number, $2(2\mu-3)(p-1)$, of terms in $\phi_1\Phi^{(\mu-1)}+\phi_2\Phi'^{(\mu-1)}$, is at most equal to the number, $(2\mu-5)(p-1)$, of terms in a polynomial $\Psi^{(\mu-2)}$. Thus, there are at least

$$2(2\mu-3)(p-1)-(2\mu-5)(p-1), \;=(2\mu-1)(p-1),$$

linearly independent terms in the polynomial $\phi_1\Phi^{(\mu-1)}+\phi_2\Phi'^{(\mu-1)}$; as we have proved that no greater number exists, it follows that $(2\mu-1)(p-1)$ is the number of linearly independent products of μ of the polynomials $\phi_1, \ldots, \phi_p$.

112. Another most important theorem follows from the results just obtained: *Every rational function whose poles are among the zeros of a polynomial $\Psi^{(\mu)}$ can be expressed in a form $\Phi^{(\mu)}/\Psi^{(\mu)}$.* For the most general function having poles in these $2\mu(p-1)$ places contains $2\mu(p-1)-p+1$ arbitrary constants*, and we have shewn that a polynomial $\Phi^{(\mu)}$ contains just this number of terms; thus the quotient $\Phi^{(\mu)}/\Psi^{(\mu)}$, which clearly has its poles in the assigned places, is of sufficiently general character to represent any such function.

For further information on the matter here discussed the reader may consult Noether, *Math. Annal.* t. XVII. p. 263, "Ueber die invariante Darstellung algebraischer Functionen." And† *ibid.* t. XXVI. p. 143, "Ueber die Normalcurven für $p=5$, 6, 7."

In order to explain the need for the theorem just obtained, we may consider the simple case where the fundamental equation is that of a general plane quartic curve, $f(x, y, z)=0$, homogeneous coordinates being used. If we take the four polynomials,

$$\psi_1=x^2, \;\psi_2=y^2, \;\psi_3=xy, \;\psi_4=xz,$$

which are not ϕ-polynomials, from which we obtain

$$x:y:z=\psi_1:\psi_3:\psi_4,$$

* When $\mu>1$. The theorem has already been proved for $\mu=1$ (§ 98, Chap. VI.).

† In the present chapter all the polynomials considered in connexion with the fundamental equation have been adjoint; there is also a geometrical theory for polynomials of any grade in extension of the theory here given, in which the associated polynomials are not adjoint. For its connexion with the theory here, the reader may compare Klein, "Abel. Functionen," *Math. Annal.* t. 36, p. 60, Clebsch-Lindemann-Benoist, *Leçons sur la Géométrie*, Paris 1883, t. III., also Lindemann, *Untersuchungen über den Riemann-Roch'schen Satz* (Teubner 1879), pp. 10, 30 etc., Noether, *Math. Annal.* t. 15, p. 507, "Ueber die Schnittpunktssysteme einer algebraischen Curve mit nicht adjungirten Curven."

then the general rational function with poles at the sixteen zeros of a polynomial, $\Psi^{(2)}$, of the second order in $\psi_1, \psi_2, \psi_3, \psi_4$, contains 14 homogeneously entering arbitrary constants. Now there are only ten terms in the general polynomial $\Phi^{(2)}$, of the second order in $\psi_1, \ldots, \psi_4$; and these are equivalent to only nine linearly independent terms, because of the relation $\psi_1\psi_2 = \psi_3^2$. Hence the rational function in question cannot be expressed in the form $\Phi^{(2)}/\Psi^{(2)}$.

113. The investigations in regard to the ϕ-polynomials $\phi_1, \ldots, \phi_p$, which have been referred to in §§ 110—112, find their proper place in the consideration of the theory of algebraic curves in space of higher than two dimensions.

Let $\phi_1, \ldots, \phi_p$ be linearly independent adjoint polynomials of grade $(n-1)\sigma + n - 3$, defined, suppose, by the invariant condition that if $c_1, \ldots, c_p$ be p independent places on the Riemann surface, ϕ_i vanishes in all of $c_1, \ldots, c_p$ except c_i. Let $x_1, \ldots, x_p$ be quantities whose ratios are defined by the equations

$$x_1 : x_2 : \ldots : x_p = \phi_1 : \phi_2 : \ldots : \phi_p.$$

We may suppose * that there is no place of the original surface at which all of $x_1, \ldots, x_p$ are zero, and, since only the ratios of these quantities are defined, we may suppose that none of them become infinite.

Hence we may interpret $x_1, \ldots, x_p$ as the homogeneous coordinates of a point in space of $p-1$ dimensions; we may call this the point x. Corresponding then to the one-dimensionality constituted by the original Riemann surface, we shall have a curve, in space of $p-1$ dimensions. Its order, measured by the number of zeros of a general linear function $\lambda_1 x_1 + \ldots + \lambda_p x_p$, will be $2p-2$. To any place x of this curve there cannot correspond two places c, c' of the original surface, unless

$$\phi_1(c) : \phi_2(c) : \ldots : \phi_p(c) = \phi_1(c') : \phi_2(c') : \ldots : \phi_p(c').$$

Now, from these equations we can infer that the ϕ polynomials corresponding to the normal integrals of the first kind, have the same mutual ratios at c as at c'; such a possibility, however, necessitates the existence of a rational function of *the second order*, expressible in the form

$$\lambda \Gamma_c^{x,} - \mu \Gamma_{c'}^{x,},$$

where λ, μ are constants whose ratio is definite, and $\Gamma_c^{x,}$, $\Gamma_{c'}^{x,}$ are normal elementary integrals of the second kind with unassigned zeros. Hence the correspondence between the original Riemann surface and the space curve, C_{2p-2}, is reversible except in the hyperelliptic case.

In the hyperelliptic case the equations of transformation are reducible to a form

$$x_1 : x_2 : \ldots : x_p = 1 : z : z^2 : \ldots : z^{p-1}.$$

* Chap. II. § 21.

To any point x of the space curve corresponds, therefore, not only the place (s, z) of the Riemann surface, but equally the place $(-s, z)$. The space curve may be regarded as a doubled curve of order $p-1$. (Cf. Klein, *Vorles. üb. d. Theorie der ellip. Modulfunctionen*, Leipzig, 1890, t. I. p. 569.)

For the general case in which $p=3$, the curve, C_{2p-2}, is the ordinary plane quartic curve. For the general case, $p=4$, the curve C_{2p-2} is a sextic curve in space of three dimensions, lying* on $\frac{1}{2}p(p+1)-(3p-3)$, $=1$, surface of the second order and $\frac{1}{6}p(p+1)(p+2)-(5p-5)$, $=5$, linearly independent surfaces of the third order.

Ex. If, for the case $p=4$, we suppose the original surface to be associated with the equation†

$$f(x, y)=x^2y^2(Lx+My)+xy(ax^2+2hxy+by^2)+Px^3+Qx^2y+Rxy^2$$
$$+Sy^3+Ax^2+2Hxy+By^2+Cx+Dy+1=0,$$

and put $Z=xy$, $X=x$, $Y=y$, as the non-homogeneous coordinates of the points of the curve C_{2p-2}, the single quadric surface containing the curve is clearly given by

$$U_2=Z-XY=0,$$

and one cubic surface, containing the curve, is given by

$$U_3=Z^2(LX+MY)+Z(aX^2+2hXY+bY^2)+PX^3+QX^2Y+RXY^2$$
$$+SY^3+AX^2+2HXY+BY^2+CX+DY+1=0.$$

Four other cubic surfaces, $V_1=0$, $V_2=0$, $V_3=0$, $V_4=0$, can be obtained from $U_3=0$ by replacing XY by Z, respectively in, (i) the coefficient of h, (ii) the coefficient of Q, (iii) the coefficient of R, (iv) the coefficient of H; these are linearly independent of $U_3=0$, and of one another. Other cubic surfaces can be obtained from $U_3=0$ by replacing XY by Z in two of its terms simultaneously; for instance, if we replace XY by Z in the coefficients of h and H, we obtain a surface of which the equation is $V_1-U_3+V_4=0$. Similarly all others than $U_3=0$, $V_1=0$, ..., $V_4=0$, are linearly deducible from these.

114. As an example of more general investigations, consider now the correspondence between the space curve C_{2p-2}, for $p=4$, and the original Riemann surface. Let us seek to form a rational function having $p+1=5$ given poles on the sextic curve. A surface of order μ can be drawn through 5 arbitrary points of the curve when μ is great enough; we may denote its equation by $\Psi^{(\mu)}=0$, in accordance with § 110. It was proved that the rational function can be written in the form $\Phi^{(\mu)}/\Psi^{(\mu)}$, $\Phi^{(\mu)}$ being another polynomial, of order μ in the space coordinates, which vanishes in the $6\mu-5$ zeros of $\Psi^{(\mu)}$ other than the 5 given points. Since a general surface of order μ contains $(\mu+3, 3)$‡ terms, the most general form possible for $\Phi^{(\mu)}$, when subject to the conditions enunciated, will contain

$$(\mu+3, 3)-(6\mu-5)$$

arbitrary, homogeneously entering, coefficients; the polynomials which multiply these coefficients, represent, equated to zero, all the linearly inde-

* § 111 preceding.

† Cf. § 108.

‡ Where (μ, ν) is used for the number $\mu(\mu-1)...(\mu-\nu+1)/\nu!$.

pendent surfaces of order μ which vanish in the $6\mu-5$ points spoken of; they will therefore include the

$$\frac{4.5\ldots\ldots(p+\mu-1)}{\lfloor\mu}-(2\mu-1)(p-1), \text{ or } (\mu-3,3)-(6\mu-3),$$

surfaces of the μth order which* contain the sextic curve. Denote the number of these surfaces by r and their equations by $U_1=0, \ldots, U_r=0$. Then the general form of the equation of a surface, $\Phi^{(\mu)}=0$, vanishing in the $6\mu-5$ given points will be

$$\Phi^{(\mu)}=\lambda_1 U_1+\ldots\ldots+\lambda_r U_r+\lambda\Psi^{(\mu)}+\mu U=0,$$

wherein $\lambda_1, \ldots, \lambda_r, \lambda, \mu$ are arbitrary constants, and U is a surface of order μ, other than $\Psi^{(\mu)}$, which vanishes in the $6\mu-5$ points, and does not wholly contain the curve. The intersections of the surface $\Phi^{(\mu)}=0$ with the sextic are the same as those of the surface $\lambda\Psi^{(\mu)}+\mu U=0$; and the general form of the rational function having the $p+1=5$ given points as poles is

$$\lambda+\mu U/\Psi^{(\mu)},$$

involving the right number $(q+1=Q-p+1=5-4+1)$ of arbitrary constants.

Ex. i. There are sixteen of the surfaces $\lambda\Psi^{(\mu)}+\mu U=0$ which *touch* the sextic (in points other than the $6\mu-5$ fixed points).

For there are $2.5+2.4-2$, $=16$, places at which the differential, dz, of the rational function $z=U/\Psi^{(\mu)}$, is zero to the second order.

Ex. ii. In the example of the previous Article, prove that

$$f'(y)=\frac{\partial U_2}{\partial Y}\cdot\frac{\partial U_3}{\partial Z}-\frac{\partial U_3}{\partial Y}\cdot\frac{\partial U_2}{\partial Z}, \ =\Delta \text{ say},$$

and that the integrals of the first kind, expressed in terms of X, Y, Z, are given by

$$\int(\lambda_1 X+\lambda_2 Y+\lambda_3 Z+\lambda_4)\,dX/\Delta,$$

for arbitrary values of the constants $\lambda_1, \lambda_2, \lambda_3, \lambda_4$†.

115. We abstain from entering on the theory of curves in space in this place. But some general considerations on the same elementary lines as those referred to in §§ 81—83, as applicable to plane curves, may fitly conclude the present chapter‡. The general theorem considered is, that of the intersections of a curve, in space of k dimensions, which is defined as the complete locus satisfying $k-1$ algebraic equations, with a surface

* § 111.

† The canonical curve discussed by Klein, *Math. Annal.* t. 36, p. 24, is an immediate generalisation of the curve C_{2p-2} here explained. But it includes other cases also.

‡ See the note in Salmon, *Higher Plane Curves* (Dublin 1879), p. 22, "on an apparent contradiction in the Theory of Curves" and the references there given, which include a reference to a paper by Euler of date 1748. For further consideration of curves in space see Appendix I. to the present volume.

of sufficiently high order, r, there are a certain number, P, which are determined by prescribing the others, *P being independent of r.*

We take first the case of the curve in three dimensions, defined as the complete intersection of two surfaces of orders m and n, say $U_m=0$, $U_n=0$. The curve is here supposed to be of the most general kind possible, having only such singularities as those considered in Salmon, *Solid Geometry* (Dublin, 1882, p. 291). For instance the surfaces $U_m=0$, $U_n=0$ are not supposed to touch; for at such a place the curve would have a double point. We prove that if $r>m+n-4$, all but $\frac{1}{2}mn(m+n-4)+1$ of the intersections of the curve $U_m=0$, $U_n=0$ with a surface of order r, $U_r=0$, are determined by prescribing the others, whose number is

$$rmn-\tfrac{1}{2}mn(m+n-4)-1.$$

For when, firstly, $r>m+n-1$, the intersections of $U_r=0$ with the curve are the same as those of a surface

$$U_r-U_mV_{r-m}-U_nV_{r-n}-U_mU_nV_{r-m-n}=0,$$

wherein V_{r-m}, V_{r-n}, V_{r-m-n} are general polynomials whose highest aggregate order in the coordinates is that given by their suffixes. Hence, in analogy with the argument given in § 81, it may at first sight appear that, of the $(r+3, 3)$ coefficients in U_r, we can reduce a certain number, K, given by

$$K=(r-m+3,3)+(r-n+3,3)+(r-m-n+3,3),$$

to zero, by using the arbitrary coefficients in V_{r-m}, V_{r-n}, V_{r-m-n}. This however is not the case. For if W_{r-m-n}, T_{r-m-n} denote general polynomials, of the orders of their suffixes, we can write the modified equation of the surface of order r in the form

$$U_r-U_m(V_{r-m}-U_nW_{r-m-n})-U_n(V_{r-n}-U_mT_{r-m-n})$$
$$-U_mU_n(V_{r-m-n}-W_{r-m-n}-T_{r-m-n})=0.$$

Now, whatever be the values assigned to the coefficients in W_{r-m-n}, T_{r-m-n}, the coefficients in $V_{r-m-n}-W_{r-m-n}-T_{r-m-n}$ are just as arbitrary as those of V_{r-m-n}. And we may use the coefficients in W_{r-m-n}, T_{r-m-n} to reduce $(r-m-n+3, 3)$ of the coefficients in each of the polynomials

$$V_{r-m}-U_nW_{r-m-n},\quad V_{r-n}-U_mT_{r-m-n},$$

to zero.

Hence the K equations by which we should reduce the number of effective coefficients in U_r to $(r+3, 3)-K$, are really unaltered when $2(r-m-n+3, 3)$ of the disposeable quantities entering therein, are put equal to zero. Thus we may conclude, that so far as the intersections of U_r with the curve are concerned, its coefficients are effectively

$$(r+3,3)-(r-m+3,3)-(r-n+3,3)+(r-m-n+3,3)$$

in number. Provided the linear equations reducing the others to zero are independent, what we prove is that the number of effective coefficients is certainly not more than this.

This number can immediately be seen to be equal to

$$rmn - \tfrac{1}{2}mn(m+n-4).$$

Hence, we cannot arbitrarily prescribe more than $rmn - \tfrac{1}{2}mn(m+n-4) - 1$ of the intersections of $U_r = 0$ with the curve.

This result is obtained on the condition that $r > m+n-1$. If $r = m+n-1$, $m+n-2$ or $m+n-3$, the number of effective coefficients in U_r cannot be more than in the polynomial

$$U_r - U_m V_{r-m} - U_n V_{r-n},$$

namely, than

$$(r+3, 3) - (r-m+3, 3) - (r-n+3, 3).$$

By the previous result this number is equal to

$$rmn - \tfrac{1}{2}mn(m+n-4) - (r-m-n+3, 3),$$

and $(r-m-n+3, 3)$, $=(r-m-n+1)(r-m-n+2)(r-m-n-3)/3!$, vanishes when $r = m+n-1$, $m+n-2$, or $m+n-3$. Hence the result obtained holds provided $r > m+n-4$.

If we denote the number $\tfrac{1}{2}mn(m+n-4)+1$ by P, the result is, that when $r > m+n-4$, we cannot prescribe more than $mnr - P$ of the intersections of the curve $U_m = 0$, $U_n = 0$ with a surface of order r; the prescription of this number of independent points determines the remaining intersections.

Corollary. Hence it follows, when $(r+3, 3) - 1 \geqq rmn - P + 1$, that a surface of order r described through $rmn - P + 1$ quite general points of the curve, will entirely contain the curve. Hence, in general, the curve lies upon $(r+3, 3) - rmn + P - 1$ linearly independent surfaces of order r, r being greater than $m+n-4$.

Ex. i. For the curve of intersection of two quadric surfaces, $P=1$; every surface of order r drawn through $4r$ quite arbitrary points of the curve entirely contains the curve; the $4r$ intersections of a surface of order r, which does not contain the curve, are determined by $4r-1$ of them. When $r=2$, the number $(r+3, 3) - rmn + P - 1$ is equal to 2. This is the number of linearly independent quadric surfaces containing the curve.

Ex. ii. For the curve of intersection of a quadric surface with a cubic surface, $P=4$; of the $6r$ intersections of the curve with a surface whose order r is >1, $6r-4$ determine the others. The number $(r+3, 3) - rmn + P - 1$ is equal to 1 when $r=2$, and equal to 5 when $r=3$; thus, as previously found, the curve lies on one quadric surface and on five linearly independent cubic surfaces; the number, for any value of r, is in agreement with the result of § 111.

116. In regard to the intersections, with the curve, of a surface of order $m+n-4$, such a surface has effectively not more coefficients than are contained in the polynomial

$$U_{m+n-4} - U_m V_{n-4} - U_n V_{m-4},$$

for arbitrary values of the coefficients in V_{n-4} and V_{m-4}. Here we firstly suppose $m > 3$, $n > 3$.

Now we can prove, as before, that

$$(m+n-1, 3)-(n-1, 3)-(m-1, 3)=\tfrac{1}{2}mn(m+n-4)+1, \ =P.$$

Hence, also when $m > 3$ and $n=3$, 2 or 1,

$$(m+n-1, 3)-(m-1, 3), \ =\tfrac{1}{2}mn(m+n-4)+1+(n-1)(n-2)(n-3)/6,$$

is equal to P, and the number of effective coefficients in a polynomial $U_{m+n-4} - U_n V_{m-4}$, wherein the coefficients in V_{m-4} are arbitrary, is as before equal to P. Similarly for other cases.

Hence P is the number of coefficients in a polynomial U_{m+n-4}, which are effective so far as the intersections of the curve with the surface $U_{m+n-4}=0$ are concerned; in other words, $P-1$ of the intersections determine the others. The total number of intersections is $mn(m+n-4)$, $=2P-2$.

The analogy of these polynomials of order $m+n-4$ with the ϕ-polynomials in the case of a plane curve is obvious.

117. If now, the homogeneous coordinates of the points of the curve in space being denoted by X_1, X_2, X_3, X_4, the symbol $[i, j]$ denote the Jacobian $\partial(U_m, U_n)/\partial(X_i, X_j)$, and $(X_1+dX_1, X_2+dX_2, X_3+dX_3, X_4+dX_4)$ denote a point of the curve consecutive to (X_1, X_2, X_3, X_4), it follows from the equations

$$\frac{\partial U_m}{\partial X_1}dX_1+\frac{\partial U_m}{\partial X_2}dX_2+\frac{\partial U_m}{\partial X_3}dX_3+\frac{\partial U_m}{\partial X_4}dX_4=0$$
$$=X_1\frac{\partial U_m}{\partial X_1}+X_2\frac{\partial U_m}{\partial X_2}+X_3\frac{\partial U_m}{\partial X_3}+X_4\frac{\partial U_m}{\partial X_4},$$

and the similar equations holding for U_n, that the ratios

$$X_2dX_3-X_3dX_2 : X_3dX_1-X_1dX_3 : X_1dX_2-X_2dX_1 : X_1dX_4$$
$$-X_4dX_1 : X_2dX_4-X_4dX_2 : X_3dX_4-X_4-dX_3,$$

are the same as the ratios

$$[1, 4] : [2, 4] : [3, 4] : [2, 3] : [3, 1] : [1, 2];$$

each of these rows is in fact constituted by the coordinates of the tangent line of the curve. If then $u_1, u_2, u_3, u_4, v_1, v_2, v_3, v_4$ denote any quantities whatever, and, in each of these rows, we multiply the elements respectively by

$$u_2v_3-u_3v_2,\ u_3v_1-u_1v_3,\ u_1v_2-u_2v_1,\ u_1v_4-u_4v_1,\ u_2v_4-u_4v_2,\ u_3v_4-u_5v_3,$$

and add the results, we shall obtain for the first row

$$\Sigma(u_2v_3-u_3v_2)(X_2dX_3-X_3dX_2)=udv-vdu,$$

where

$$u=u_1X_1+u_2X_2+u_3X_3+u_4X_4,\ du=u_1dX_1+u_2dX_2+u_3dX_3+u_4dX_4, \text{ etc.},$$

and, for the second row we shall obtain the determinant

$$\begin{vmatrix} u_1, & u_2, & u_3, & u_4 \\ v_1, & v_2, & v_3, & v_4 \\ \dfrac{\partial U_m}{\partial X_1}, & \dfrac{\partial U_m}{\partial X_2}, & \dfrac{\partial U_m}{\partial X_3}, & \dfrac{\partial U_m}{\partial X_4} \\ \dfrac{\partial U_n}{\partial X_1}, & \dfrac{\partial U_n}{\partial X_2}, & \dfrac{\partial U_n}{\partial X_3}, & \dfrac{\partial U_n}{\partial X_4} \end{vmatrix},$$

which we may denote by (uvU_mU_n).

From the proportionality of the elements of the two rows considered, it follows, therefore, that the ratio $(udv - vdu)/(uvU_mU_n)$ is independent of the values of the quantities $u_1, \ldots, v_4$. This ratio is of degree

$$-(m-1+n-2-2) = -(m+n-4)$$

in the homogeneous coordinates; namely, if X_1, X_2, X_3, X_4 be replaced by ρX_1, ρX_2, ρX_3, ρX_4, the ratio will be multiplied by $\rho^{-(m+n-4)}$. Hence, if U_{m+n-4} be any polynomial of degree $m+n-4$, the product

$$U_{m+n-4}(udv - vdu)/(uvU_mU_n)$$

is a functional differential, independent of the arbitrary factor of the homogeneous coordinates.

The integral,

$$\int U_{m+n-4} \frac{udv - vdu}{(uvU_mU_n)},$$

can only be infinite at the places where the curve is intersected by the surface $(uvU_mU_n) = 0$: if $u = 0$, $v = 0$ be regarded as the equations of planes, this equation expresses that the straight line $u = 0$, $v = 0$, is intersected by the tangent line of the curve at the point (X_1, X_2, X_3, X_4). The differential

$$udv - vdu, \; = \Sigma\,(u_2v_3 - u_3v_2)(X_2dX_3 - X_3dX_2),$$

is zero, to the second order, when the line $u = 0 = v$ is intersected by the tangent line, whose coordinates are $X_2dX_3 - X_3dX_2$, etc. Hence the ratio $(udv - vdu)/(uvU_mU_n)$ is never infinite, and the integral above is finite for all points of the curve.

Hence*, since U_{m+n-4} contains P terms, *we can obtain P everywhere-finite algebraical integrals.*

The same result is obtained if $u_1, \ldots, v_4$ be polynomials in the coordinates, $u_1, \ldots, u_4$ being of the same degree, and $v_1, \ldots, v_4$ of the same degree.

* As stated, we are considering a curve without singular points. If the curve had a double point, the polynomial (uvU_mU_n) would vanish at that point, for all values of $u_1, \ldots, v_4$. We could then prescribe $U_{m+n-4}=0$ to pass through the double point, thus obtaining a reduction of one in the number of finite integrals. Etc.

Ex. i. For a plane curve of order n, without multiple points, prove similarly that we can obtain p finite algebraical integrals in the form

$$\int \phi_{n-3}\,(udv - vdu)/(uvf),$$

where $f(x_1, x_2, x_3)=0$ is the homogeneous equation of the curve, $u=u_1x_1+u_2x_2+u_3x_3$, etc., and (uvf) denotes a determinant of three rows.

Ex. ii. Shew that a surface of order $m+n-4+\mu$ which vanishes in all but two of the intersections of the curve in space with a surface of order μ, $U_\mu=0$, is of the form

$$\psi=\lambda U+(\lambda_1 V_1+\ldots+\lambda_P V_P)\, U_\mu=0,$$

where $\lambda, \lambda_1, \ldots, \lambda_P$ are arbitrary; and that an integral of the third kind is of the form

$$\int \frac{\psi}{U_\mu} \frac{udv - vdu}{(uvU_m U_n)}.$$

118. Retaining still the convention that $u = 0$, $v = 0$ are the equations of planes, let $u' = 0$, $v' = 0$ be the equations of other planes whose line of intersection does not coincide with the line $u = 0 = v$.

From the equations

$$zu - v = 0,\ su' - v' = 0,\ U_m = 0,\ U_n = 0,$$

wherein z, s have any values, we can eliminate the coordinates of the points of the curve in space, and obtain a rational equation, $(s, z) = 0$, with which we may associate a Riemann surface*. To any point of the curve corresponds a single point, $z = v/u$, $s = v'/u'$, of the Riemann surface; to any point of the Riemann surface will in general correspond conversely only one point of the curve in space. Hence the Riemann surface will have mn sheets, the places, at which z has any value, being those which correspond to the places, on the curve in space, at which the plane $zu-v=0$ intersects this curve. Thus the Riemann surface will have $2mn+2p-2$ branch places, p being the deficiency of the surface. These are the places where dz is zero of the second order. Thus they correspond to the places, on the curve in space, where $udv - vdu$ is zero to the second order. We have seen that these are given as the intersections of this curve with the surface $(uvU_mU_n) = 0$, of order $m+n-2$; their number is therefore $mn(m+n-2) = 2mn+2P-2$. Hence the number P, obtained for the curve in space, is equal to the deficiency p of the Riemann surface with which it is reversibly related. The same result can be proved when u, v are polynomials of any, the same, order, and u', v' are polynomials of any, the same, order.

And from the reversibility of this transformation it follows that the everywhere-finite integrals for the Riemann surface are the same as those here obtained for the curve in space.

* We may of course interpret the equation as that of a plane curve; a particular case is that in which this curve is a central projection of the space curve.

Ex. Prove that if e_1, e_2, e_3 be such that $e_1+e_2+e_3=0$,

$$(b-c)(c-a)(a-b)=(b-c)(a-d)/(e_2-e_3)=(c-a)(b-d)/(e_3-e_1)=(a-b)(c-d)/(e_1-e_2),$$

the points of the curve $aX^2+bY^2+cZ^2+dT^2=0$, $X^2+Y^2+Z^2+T^2=0$ can be expressed in terms of two quantities, x, y, satisfying the equation $y^2=4(x-e_1)(x-e_2)(x-e_3)$, in the form $T:X:Y:Z$

$$=y:\sqrt{b-c}\,[(x-e_1)^2-(e_1-e_2)(e_1-e_3)]:\sqrt{c-a}\,[(x-e_2)^2-(e_2-e_3)(e_2-e_1)]$$
$$:\sqrt{a-b}\,[(x-e_3)^2-(e_3-e_1)(e_3-e_2)].$$

Find x, y in terms of X, Y, Z, T in the form

$$[e_1(e_2-e_3)X/\sqrt{b-c}+e_2(e_3-e_1)Y/\sqrt{c-a}+e_3(e_1-e_2)Z/\sqrt{a-b}]/x$$
$$=(e_2-e_3)X/\sqrt{b-c}+(e_3-e_1)Y/\sqrt{c-a}+(e_1-e_2)Z/\sqrt{a-b}=2(e_2-e_3)(e_3-e_1)(e_1-e_2)T/y.$$

See Mathews, *London Math. Soc.* t. xix. p. 507.

119. As already remarked we have considered here only the case of a non-singular curve in space which is completely defined as the intersection of two algebraical surfaces. For this case the reader may consult Jacobi, *Crelle*, t. 15 (1836), p. 298; Plücker, Crelle, t. 16, p. 47; Clebsch, *Crelle*, t. 63, p. 229; Clebsch, *Crelle*, t. 64, p. 43; Salmon, *Solid Geometry* (Dublin, 1882), p. 308; White, *Math. Annal.* t. 36, p. 597; Cayley, *Collected Works, passim.* For the more general case, in connexion however with an extension of the theory of this volume to the case of two *independent* variables, the following, inter alia, may be consulted: Noether, *Math. Annal.* t. 8 (1873), p. 510; Clebsch, *Comptes Rendus de l'Acad. des Sciences*, t. 67, July—December, 1868, p. 1238; Noether, *Math. Annal.* t. 2, p. 293, and t. 29, p. 339 (1887); Valentiner, *Acta Math.* t. ii. p. 136 (1883); Halphen, *Journal de l'École Polyt.* t. lii. (1882), p. 1; Noether, *Abh. der Akad. zu Berlin* (1882); Cayley, *Collected Works*, Vol. v. p. 613, etc.; and Picard, *Liouv. Journ. de Math.* 1885, 1886 and 1889.

Ex. i. Prove that

$$(r+k,k)-\sum_1(r+k-m_i,k)+\sum_2(r+k-m_i-m_j,k)-\dots+(-)^{k-1}(r+k-m_1-\dots-m_{k-1},k)$$
$$=rm_1m_2\dots m_{k-1}-\tfrac{1}{2}m_1m_2\dots m_{k-1}(m_1+m_2+\dots+m_{k-1}-k-1),$$

where (r,μ) denotes $r(r-1)\dots(r-\mu+1)/\mu!$, $m_1,\dots,m_{k-1}$, k are any positive integers, r is a positive integer greater than $m_1+m_2+\dots+m_{k-1}-k-1$, $\sum_1$ denotes a summation extending to all the values $i=1,2,\dots,(k-1)$, $\sum_2$ denotes a summation extending to every pair of two unequal numbers chosen from the series $m_1,m_2,\dots,m_{k-1}$, and so on. Hence infer that of the intersections of a general curve in space of k dimensions, which is determined as the complete locus common to $k-1$ algebraic surfaces of orders $m_1,m_2,\dots,m_{k-1}$, with a surface of order r, all but

$$\tfrac{1}{2}m_1m_2\dots m_{k-1}(m_1+m_2+\dots+m_{k-1}-k-1)+1$$

are determined by the others. The result is known to hold for $k=2$. We have here been considering the case $k=3$.

Ex. ii. With the notation and hypotheses employed in Salmon's *Solid Geometry* (1882), Chap. XII. (p. 291) (see also a note by Cayley, *Quarterly Journal*, t. VII., or *Collected Works*, Vol. v. p. 517), where m is the degree of a curve in space, n is its class, namely the number of its osculating planes which pass through an arbitrary point, r is its rank, namely the number of its tangents which intersect an arbitrary line, α is the number of osculating planes containing four consecutive points of the curve, β the number of points through which four consecutive planes pass, x the number of points of intersections of non-consecutive tangents which lie in an arbitrary plane, y the number of planes containing two non-consecutive tangents which pass through an arbitrary point, h the number of chords of the curve which can be drawn through an arbitrary point, g the number of lines of intersection of two non-consecutive osculating planes which lie in an arbitrary plane, ϑ the number of tangent lines of the curve which contain three consecutive points, prove, by using Plücker's equations (Salmon, *Higher Plane Curves*, 1879, p. 65) for the plane curve traced on any plane by the intersections, with this plane, of the tangent lines of the curve in space, that the equations hold,

$$(1)\quad n = r(r-1) - 2x - 3m - 3\vartheta, \qquad (3)\quad r = n(n-1) - 2g - 3a,$$

$$(2)\quad a = 3r(r-2) - 6x - 8(m+\vartheta), \qquad (4)\quad m+\vartheta = 3n(n-2) - 6g - 8a,$$

$$p_1 - 1 = \tfrac{1}{2} r(r-3) - x - m - \vartheta = \tfrac{1}{2} n(n-3) - g - a \quad\ldots\ldots\ldots\ldots\ldots\ldots\text{(A)},$$

p_1 being the deficiency of this plane curve.

Prove further, by projecting the curve in space from an arbitrary point, and using Plücker's equations for the plane curve in which the cone of projection is cut by an arbitrary plane, the equations

$$(5)\quad r = m(m-1) - 2h - 3\beta, \qquad (7)\quad m = r(r-1) - 2y - 3(\vartheta+n),$$

$$(6)\quad \vartheta + n = 3m(m-2) - 6h - 8\beta, \qquad (8)\quad \beta = 3r(r-2) - 6y - 8(\vartheta+n),$$

$$p_2 - 1 = \tfrac{1}{2} m(m-3) - h - \beta = \tfrac{1}{2} r(r-3) - y - n - \vartheta \quad\ldots\ldots\ldots\ldots\ldots\ldots\text{(B)},$$

p_2 being the deficiency of this plane curve.

From the equations (1) and (7) we can infer $n - m = 3n - 3m - 2(x-y)$, and therefore

$$y + n = x + m.$$

Hence $p_1 = p_2$.

Ex. iii. For the non-singular curve which is the complete intersection of two algebraic surfaces of orders μ, ν, prove (cf. Salmon, *Solid Geometry*, pp. 308, 309) that in the notation of Ex. ii. here,

$$\beta = 0,\quad m = \mu\nu,\quad r = \mu\nu(\mu+\nu-2),\quad h = \tfrac{1}{2}\mu\nu(\mu-1)(\nu-1).$$

Hence, by the equations (B) of Ex. ii. prove that, now,

$$p_1 = p_2 = \tfrac{1}{2}\mu\nu(\mu+\nu-4) + 1.$$

This is the number we have denoted by P.

Ex. iv. Denoting the number $p_1 = p_2$, in Ex. ii., by p, prove from equations (5) and (B) that

$$6(p-1) = m(m-7) - 2h + 2r = 3(r + \beta - 2m).$$

Hence shew that if, through a curve C of order m, lying on a surface S of order μ, we draw a surface of order ν, cutting the surface S again in a curve C' of order m', and if p, p' denote the values of p for these curves C, C' respectively, then

$$m'(\mu+\nu-4) - (2p'-2) = m(\mu+\nu-4) - (2p-2)$$

(see Salmon, pp. 311, 312). Shew that each of these numbers is equal to the number, i, of points in which the curves C, C' intersect, and interpret geometrically the relation

$$i + r + \beta = m(\mu+\nu-2).$$

Ex. v. If in Ex. iv. a surface ϕ of order $\mu+\nu-4$ be drawn through $(\mu+\nu-4)m' - p' + 1$, or $i - 1 + p'$, of the points of the curve C', prove that, so far as its intersections with the curve C are concerned, the surface ϕ contains effectively p terms. Prove further that ϕ contains the curve C' entirely.

Ex. vi. Prove that a surface of order $\mu+\nu-4$ passing through $i-1$ of the intersections of the curves C, C', in Ex. iv., will pass through the other intersection.

Ex. vii. An example of the case in Ex. iv. is that in which $\mu=2$, $\nu=2$, $m=3$, $m'=1$. Then C' is a straight line and $p'=0$: hence p is given by $-2=2p-2$. Hence, for the cubic curve of intersection of two quadrics having a common generator, $p=0$. And in fact coordinate planes can be chosen so that the homogeneous coordinates of the points of the cubic can be expressed in the form

$$X : Y : Z : T=1 : \theta : \theta^2 : \theta^3,$$

θ being a variable parameter. For instance (using Cartesian coordinates) the polar planes of a fixed point $(X'Y'Z')$ in regard to quadrics confocal with $X^2/a+Y^2/b+Z^2/c=1$ are the osculating planes of such a cubic curve, the coordinates of whose points are expressible in the form

$$XX'=(a+\lambda)^3/(a-b)(a-c),\quad YY'=(b+\lambda)^3/(b-c)(b-a),\quad ZZ'=(c+\lambda)^3/(c-a)(c-b),$$

λ being a variable parameter.

Ex. viii. For the quintic curve of intersection of a quadric and a cubic surface having a common generator we obtain, from Ex. iv., putting $m'=1$, $p'=0$, $m=5$, that $p=2$; the results of Exx. iv., v., vi. can be immediately verified for this curve; further, if the surfaces be taken to be $yU-zV=0$, $yS-zT=0$, where U, V are of the first degree in x, y, z and S, T of the second degree, and we put $y=z\xi$, $x=z\eta$, we obtain

$$z(\eta a_1+a_2)=\lambda_1,\quad z^2(\eta^2\beta_1+\eta\beta_2+\beta_3)+z(\eta\gamma_1+\gamma_2)+\delta_1=0,$$

where the Greek letters a_1, a_2,... denote polynomials in ξ of the degrees of their suffixes. Hence, if σ be defined by the equation,

$$\lambda_1\sigma=2\eta(\lambda_1^2\beta_1+\lambda_1 a_1\gamma_1+\delta_1 a_1^2)+\lambda_1^2\beta_2+\lambda_1(a_1\gamma_2+a_2\gamma_1)+2\delta_1 a_1 a_2,$$

we obtain $\sigma^2=(\xi, 1)_6$; ξ, σ are rational functions of x, y, z and x, y, z are rational functions of ξ, σ.

Ex. ix. Prove that if the sextic intersection of a cubic surface and a quadric surface, break up into a quartic curve and a curve of the second order, the numbers p, p' for these curves are $p=1$, $p'=0$ or $p=0$, $p'=-1$ according as the curve of the second order is a plane curve or is two non-intersecting straight lines.

Ex. x. In analogy with Ex. iv., shew that the deficiencies of two non-singular plane curves of orders m, m' are connected by the equation

$$m(m+m'-3)-(2p-2)=mm'=m'(m+m'-3)-(2p'-2),$$

and further in analogy with Ex. v. that if a plane curve, of order $m+m'-3$, be drawn through $(m+m'-3)m'-p'+1$ independent points of the curve of order m', only $p-1$ of its intersections with the curve of order m can be prescribed.

Further indications of the connexion of the theory of curves in space with the subject of this chapter will be found in Appendix I.

CHAPTER VII.

COORDINATION OF SIMPLE ELEMENTS. TRANSCENDENTAL UNIFORM FUNCTIONS.

120. WE have shewn in Chapter II. (§§ 18, 19, 20), that all the fundamental functions are obtainable from the normal elementary integral of the third kind. The actual expression of this integral for any given form of fundamental equation, is of course impracticable without precise conventions as to the form of the period loops, and for numerical results it may be more convenient to use an integral which is defined algebraically. Of such integrals we have given two forms, one expressed by the fundamental integral functions (Chap. IV. §§ 45, 46), the other expressed in the terms of the theory of plane curves (Chap. VI. § 92, Ex. ix.). In the present Chapter we shew how from the integral $P_{z,c}^{x,a}$, obtained in Chap. IV.*, to determine algebraically an integral $Q_{z,c}^{x,a}$ for which the equation $Q_{z,c}^{x,a} = Q_{x,a}^{z,c}$ has place; incidentally the character of $P_{z,c}^{x,a}$, as a function of z, becomes plain; and therefore also the character of the integral of the second kind, $E_z^{x,a}$, which was found in Chap. IV. (§§ 45, 47).

This determination arises in close connexion with the investigation of the algebraic expression of the rational function of x which was obtained in § 49 and denoted by $\psi(x, a; z, c_1, \dots c_p)$. It was there shewn that every rational function of x can be expressed in terms of this function. It is shewn in this Chapter that any uniform function whatever, which has a finite number of distinct infinities, which may be essential singularities, can be expressed by such a function.

Further, it is here shewn how to obtain an uniform function of x having only one zero, at which it vanishes to the first order, and one infinity; and that any uniform function can be expressed in factors by means of this function.

* For the integral of the third kind obtained in Chap. VI. the reader may compare Clebsch and Gordan, *Theorie der Abel. Functionen* (Leipzig, 1866), p. 117, and, for other important results, Noether, *Math. Annal.* XXXVII. (1890), pp. 442, 448; also Cayley, *Amer. Journal*, V. (1882), p. 173.

121. Let $u_1^{x,a}, \ldots, u_p^{x,a}$ denote any p linearly independent integrals of the first kind, vanishing at the arbitrary place a. Let t denote the infinitesimal at x, and let $Du_1^x, \ldots\ldots, Du_p^x$ denote the differential coefficients of the integrals in regard to t, all of which are everywhere finite. Let $c_1, \ldots, c_p$ denote any p fixed places of the Riemann surface, so chosen that no linear aggregate of the form

$$\lambda_1 Du_1^x + \ldots\ldots + \lambda_p Du_p^x,$$

where $\lambda_1, \ldots, \lambda_p$ are constants, vanishes in all the places $c_1, \ldots, c_p$, but such that one linear aggregate of this form vanishes in every set of $p-1$ of these places*; and let $\omega_i(x)$ denote the linear aggregate, of this form, which vanishes in all of $c_1, \ldots, c_p$ except c_i, and is equal to 1 at the place c_i.

Then $\omega_i(x)$ is expressible as the quotient of two determinants; the denominator has $Du_s^{c_r}$ for its (r, s)th element, the numerator differs from the denominator only in the i-th row, which consists of the quantities $Du_1^x, \ldots, Du_p^x$; thus $\omega_1(x), \ldots, \omega_p(x)$ are determinable algebraically when $u_1^x, \ldots, u_p^x$ are given. Conversely the differential coefficients of the normal integrals of the first kind (§§ 18, 23) are clearly expressible by $\omega_1(x), \ldots, \omega_p(x)$, in the form

$$\Omega_i(x) = \omega_1(x)\,\Omega_i(c_1) + \ldots\ldots + \omega_p(x)\,\Omega_i(c_p).$$

We have already used $v_i^{x,a}$ as a notation for the normal integral $\frac{1}{2\pi i}\int_a^x \Omega_i(x)\,dt_x$. In this chapter we shall use the notation $V_i^{x,a} = \int_a^x \omega_i(x)\,dt_x$.

If the period of the integral $u_i^{x,a}$ at the j-th period loop of the first kind† be denoted by $C_{i,j}$, we can express $v_i^{x,a}$ as the quotient of two determinants, the denominator having $C_{j,i}$ for its (i, j)th element, and the numerator being different from the denominator only in the ith row which consists of the elements $u_1^{x,a}, \ldots, u_p^{x,a}$.

122. Consider now the function of x expressed‡ by

$$\Gamma_z^{x,a} - \sum_{r=1}^{p} \omega_r(z)\,\Gamma_{c_r}^{x,a},$$

z being any place whatever. The function is clearly infinite to the first order at the place z, like $-t_z^{-1}$, t_z being the infinitesimal at z; it is also infinite at each of the places $c_1, \ldots, c_p$, and, at c_i, like $\omega_i(z)\,t_{c_i}^{-1}$, t_{c_i} being the infinitesimal at c_i. The function has no periods at the period loops of the

* Thus there exists no rational function infinite only to the first order at each of $c_1, \ldots, c_p$. Cf. §§ 23, 26.

† $C_{i,j}$ is the quantity by which the value of $u_i^{x,a}$ on the left side of this period loop exceeds the value on the right side. See the figure, § 18, Chap. II.

‡ Klein, *Math. Annal.* xxxvi. p. 9 (1890), Neumann, *loc. cit.* p. 14, p. 259.

first kind. At the ith period loop of the second kind the function has the period

$$\Omega_i(z) - \sum_{r=1}^{p} \omega_r(z)\,\Omega_i(c_r),$$

which, as remarked (§ 121), is also zero. Hence *the function is a rational function of x.* It vanishes at the place a. We shall denote the function by $\psi(x, a; z, c_1, \dots, c_p)$. It is easy to see that it entirely agrees, in character, with the function given in § 49.

For the places $c_1, \dots, c_p$ have been chosen so that no aggregate of the form

$$\lambda_1\Omega_1(x) + \dots\dots + \lambda_p\Omega_p(x)$$

vanishes in all of them. Hence (Chap. III. § 37) the general rational function having poles of the first order at the places $z, c_1, \dots, c_p$ is of the form $Ag + B$, where g is such a function, and A, B are constants. These constants can be uniquely determined so that the residue at the pole, z, is -1, and so that the function vanishes at the place a.

Ex. For the case $p=1$, if we use Weierstrass's elliptic functions, the places x, a, z, c, being represented by the arguments u, a, v, γ_1, and put $x = \wp u$, $y = \wp'(u)$ etc., we may take, supposing v not to be a half period,

$$\Gamma_z^{x, a} = -\frac{1}{\wp' v}\left[\zeta(u-v) - \zeta(a-v) - \frac{\eta}{\omega}(u-a)\right], \quad \omega_1(z) = \frac{\wp'(\gamma_1)}{\wp'(v)},$$

$$\Gamma_{c_1}^{x, a} = -\frac{1}{\wp'(\gamma_1)}\left[\zeta(u-\gamma_1) - \zeta(a-\gamma_1) - \frac{\eta}{\omega}(u-a)\right],$$

and obtain

$$\psi(x, a; z, c_1) = -\frac{1}{\wp' v}\{\zeta(u-v) - \zeta(u-\gamma_1) - \zeta(a-v) + \zeta(a-\gamma_1)\},$$

or

$$\psi(x, a; z, c_1) = \frac{1}{2\wp'(v)}\left[\frac{\wp'(u-v) + \wp'(u-\gamma_1)}{\wp(u-v) - \wp(u-\gamma_1)} - \frac{\wp'(a-v) + \wp'(a-\gamma_1)}{\wp(a-v) - \wp(a-\gamma_1)}\right];$$

and any doubly periodic function can be expressed linearly by functions of this form, in which the same value occurs for γ_1 and different values for v. (Cf. § 49, Chap. IV.)

123. Since $\omega_i(z), = \dfrac{d}{dt_z} V_i^{z, c}$, is a linear function of $\Omega_1(z), \dots, \Omega_p(z)$, it follows that $\omega_i(z)\Big/\dfrac{dz}{dt}$ is a rational function of z; and $\Gamma_z^{x, a}, = \dfrac{d}{dt_z}\Pi_{z, c}^{x, a}$, $= \left(\dfrac{d}{dz}\Pi_{z, c}^{x, a}\right)\dfrac{dz}{dt_z}$, is such that* $\Gamma_z^{x, a}\Big/\dfrac{dz}{dt}$ is a rational function of z; hence

* Throughout this chapter such an expression as $f(z)\dfrac{dz}{dt}$ is used to denote the limit, when a variable place ξ approaches the place z, of the expression $f(\xi)\dfrac{d\xi}{dt}$, t being the infinitesimal for

$\psi(x, a; z, c_1, \dots, c_p) \Big/ \frac{dz}{dt}$ is a rational function of z. It is easy also to see, from the determinant expression of $\omega_i(z)$, that $\omega_i(z)\frac{dc_i}{dt}$ is a rational function of $c_1, \dots, c_p$.

Hence $\psi(x, a; z, c_1, \dots, c_p) \Big/ \frac{dz}{dt}$ *is a rational function of the variables of all the places* $x, a, z, c_1, \dots, c_p$.

Further, as depending upon z, $\psi(x, a; z, c_1, \dots, c_p)$ is infinite only when $\Gamma_z^{x,a}$ is infinite; and $\Gamma_z^{x,a}, = \frac{d}{dt_z}\Pi_{x,a}^{z,a}$, is infinite only when z is at x or at a. At the place x, $\Gamma_z^{x,a}$ is infinite like $\frac{d}{dt_x}\log t_x$, namely like the inverse of the infinitesimal at the place x.

Hence $\psi(x, a; z, c_1, \dots, c_p)$, *regarded as depending upon* z, *is infinite only when* z *is in the neighbourhood of the place* x, *or in the neighbourhood of the place* a. *At the place* x, $\psi(x, a; z, c_1, \dots, c_p)$ *is infinite like the positive inverse of the infinitesimal, at the place* a *it is infinite like the negative inverse of the infinitesimal.* The rational function of z denoted by

$$\psi(x, a; z, c_1, \dots, c_p) \Big/ \frac{dz}{dt}$$

will therefore be infinite at the place x like $\frac{1}{w_1+1}\frac{1}{z-x}$ and at the place a like $-\frac{1}{w_2+1}\frac{1}{z-a}$, where w_1+1, w_2+1 denote the number of sheets that wind at the places x, a respectively; and will be infinite at every branch place, like $\frac{A}{(w+1)t^w}$, t being the infinitesimal at the place, $w+1$ the number of sheets that wind there, and A the value of $\psi(x, a; z, c_1, \dots, c_p)$ when z is at the branch place.

The actual expression of the function $\psi(x, a; z, c_1, \dots, c_p)$ is given below (§ 130).

124. From the function $\psi(x, a; z, c_1, \dots, c_p)$ we obtain a function,

$$\bar{E}(x, z) = e^{\int_c^z \psi(x, a;\, z, c_1, \dots, c_p)\, dt_z}, = e^{\Pi_{z,c}^{x,a} - \sum_{r=1}^{p} V_r^{z,c}\Gamma_{c_r}^{x,a}},$$

wherein c is an arbitrary place, which has the following properties, as a function of x.

the neighbourhood of the place z. When z is not a branch place $\frac{d\xi}{dt}=1$; when $w+1$ sheets wind at z, $\frac{d\xi}{dt}=(w+1)t^w$ (cf. §§ 2, 3; Chap. I.). Ample practice in the notation is furnished by the examples of this chapter.

(i) It is an uniform function of x. For the exponent has no periods at the period loops of the first kind, and at the ith period loop of the second kind it has the period

$$2\pi i v_i^{z, c} - \sum_{r=1}^{p} V_r^{z, c} \Omega_i(c_r)$$

which, as follows from the equation

$$\Omega_i(z) = \omega_1(z)\Omega_i(c_1) + \dots\dots + \omega_p(z)\Omega_i(c_p),$$

is equal to zero. Further the integral multiples of $2\pi i$, which may accrue to $\Pi_{z, c}^{x, a}$ when x describes a contour enclosing one of the places z, c, do not alter the value of the function.

(ii) The function vanishes only at the place z, and to the first order.

(iii) The function has a pole of the first order at the place c.

(iv) The function is infinite at the place c_i, like $e^{V_i^{z, c} t_{c_i}^{-1}}$, t_{c_i} being the infinitesimal at the place. We may therefore speak of $c_1, \dots, c_p$ as essential singularities of the function.

125. In order to call attention to the importance of such a function as this, we give an application. Let $R(x)$ denote a rational function, having simple poles at $\alpha_1, \dots, \alpha_m$, and simple zeros at $\beta_1, \dots, \beta_m$. We suppose these places different from the fixed places c, a, $c_1, \dots, c_p$. Then the product

$$F(x) = R(x) \frac{\bar{E}(x, \alpha_1) \dots\dots \bar{E}(x, \alpha_m)}{\bar{E}(x, \beta_1) \dots\dots \bar{E}(x, \beta_m)},$$

is an uniform function of x, which becomes infinite only at the places $c_1, \dots c_p$; at c_i it is infinite like a constant multiple of

$$e^{-\sum_{r=1}^{m} V_i^{\alpha_r, \beta_r} \Gamma_{c_i}^{x, a}}.$$

Now, in fact, $\log F(x)$ is also an uniform function of x: for it is only infinite at the places $c_1, \dots, c_p$, and, at the place c_i, like $-\left(\sum_{r=1}^{m} V_i^{\alpha_r, \beta_r}\right)\Gamma_{c_i}^{x, a}$. Hence the integral $\int d \log F(x)$, $= \int \frac{F'(x)}{F(x)} dx$, taken round any closed area on the Riemann surface which does not enclose any of the places $c_1, \dots, c_p$, is certainly zero, and taken round the place c_i is equal to $-\sum_{r=1}^{m} V_i^{\alpha_r, \beta_r} \int \frac{dt_{c_i}}{t_{c_i}^2}$, taken round c_i, and is, therefore, also zero.

But an uniform function of x which is infinite only to the first order at each of $c_1, \dots, c_p$ does not exist. For the places $c_1, \dots, c_p$ were chosen so that the conditions that the periods of a function, of the form

$$\lambda_1 \Gamma_{c_1}^{x, a} + \dots\dots + \lambda_p \Gamma_{c_p}^{x, a},$$

wherein $\lambda_1, \ldots, \lambda_p$ are constants, should be zero, namely the conditions

$$\lambda_1 \Omega_r (c_1) + \ldots\ldots + \lambda_p \Omega_r (c_p) = 0, \qquad r = 1, 2, \ldots\ldots, p,$$

are impossible unless each of $\lambda_1, \ldots, \lambda_p$ be zero.

Hence we can infer that $\sum_{r=1}^{m} V_i^{\alpha_r, \beta_r} = 0$, for $i = 1, 2, \ldots, p$, and that $F(x)$ is a constant; this constant is clearly equal to $F(a)$, for $\bar{E}(a, z) = 1$ for all values of z.

Hence, any rational function can be expressed as a product of uniform functions of x, in the form

$$R(x) = R(a) \frac{\bar{E}(x, \beta_1) \ldots\ldots \bar{E}(x, \beta_m)}{\bar{E}(x, \alpha_1) \ldots\ldots \bar{E}(x, \alpha_m)},$$

where $\alpha_1, \ldots, \alpha_m$ are the poles and $\beta_1, \ldots, \beta_m$ the zeros of the function. We have given the proof in the case in which the poles and zeros are of the first order. But this is clearly not important.

Further, the zeros and poles of a rational function are such that

$$\sum_{r=1}^{m} V_i^{\alpha_r, c} = \sum_{r=1}^{m} V_i^{\beta_r, c}, \qquad i = 1, 2, \ldots, p,$$

c being an arbitrary place. This is a case of Abel's Theorem, which is to be considered in the next Chapter. We remark that in the definition of the function $\bar{E}(x, z)$ by means of Riemann integrals, the ordinary conventions as to the paths joining the lower and upper limits of the integrals are to be regarded; these paths must not intersect the period loops.

Ex. i. For the case $p = 0$, $\Pi_{z, c}^{x, a} = \log \left(\frac{x - z}{x - c} \frac{a - c}{a - z} \right)$ and $\bar{E}(x, z) = \frac{(x - z)(a - c)}{(x - c)(a - z)}$.

Ex. ii. For the case $p = 1$, supposing the place c represented by the argument γ, we have

$$\psi(x, a; z, c_1, \ldots, c_p) = -\frac{1}{\wp'(v)} \{\zeta(u - v) - \zeta(u - \gamma_1) - \zeta(a - v) + \zeta(a - \gamma_1)\}$$

$$\log E(x, z) = \int_c^z \psi(x, a; z, c_1, \ldots, c_p)\, dz = -\int_\gamma^v dv \{\zeta(u - v) - \zeta(u - \gamma_1) - \zeta(a - v) + \zeta(a - \gamma_1)\}$$

$$= \log \frac{\sigma(u - v)}{\sigma(u - \gamma)} \frac{\sigma(a - \gamma)}{\sigma(a - v)} + (v - \gamma)[\zeta(u - \gamma_1) - \zeta(a - \gamma_1)],$$

and therefore

$$\bar{E}(x, z) = \frac{\sigma(u - v)\, \sigma(a - \gamma)}{\sigma(u - \gamma)\, \sigma(a - v)} e^{(v - \gamma)[\zeta(u - \gamma_1) - \zeta(a - \gamma_1)]}.$$

Ex. iii. Prove, if a', c' denote any places whatever, that

$$\frac{\bar{E}(x, z)\, \bar{E}(a', c')}{\bar{E}(x, c')\, \bar{E}(a', z)} = e^{\Pi_{z, c'}^{x, a'} - \sum_{i=1}^{p} V_i^{z, c'} \Gamma_{c_i}^{x, a'}}.$$

Ex. iv. The rational function of x, $\psi(x, \zeta; z, c_1, \ldots, c_p)$, will, beside ζ, have p zeros, say $\gamma_1, \ldots, \gamma_p$, such that the set $\zeta, \gamma_1, \ldots, \gamma_p$ is equivalent with or coresidual with the set $z, c_1, \ldots, c_p$ (§§ 94, 96, Chap. VI.). Hence, in the product

$$\psi(x, \zeta; z, c_1, \ldots, c_p)\, \psi(x, z; \zeta, \gamma_1, \ldots, \gamma_p),$$

the zeros of either factor are the poles of the other, and the product is therefore a constant. To find the value of this constant, let x approach to the place z. Then the product becomes equal to

$$-t_x^{-1} \cdot t_x \left[D_x \psi (x, z; \zeta, \gamma_1, \ldots, \gamma_p)\right]_{x=z}.$$

It is clear from the expression of $\psi(x, a; z, c_1, \ldots, c_p)$ which has been given, that $D_x \psi(x, a; z, c_1, \ldots, c_p)$ does not depend upon the place a. Thus, by the symmetry, we have the result

$$\psi(x, \zeta; z, c_1, \ldots, c_p)\, \psi(x, z; \zeta, \gamma_1, \ldots, \gamma_p) = -D_z \psi(z, a; \zeta, \gamma_1, \ldots, \gamma_p)$$
$$= -D_\zeta \psi(\zeta, a; z, c_1, \ldots, c_p),$$

where a is a perfectly arbitrary place, and the sets $z, c_1, \ldots, c_p, \zeta, \gamma_1, \ldots, \gamma_p$ are subject to the condition of being coresidual.

Hence also if $W(x; z, c_1, \ldots, c_p)$ denote the expression

$$D_x \left[\psi(x, a; z, c_1, \ldots, c_p) - \Gamma_z^{x, a}\right],$$

we have

$$W(z; \zeta, \gamma_1, \ldots, \gamma_p) = W(\zeta; z, c_1, \ldots, c_p),$$

provided only the set $z, c_1, \ldots, c_p$ be coresidual with the set $\zeta, \gamma_1, \ldots, \gamma_p$.

Ex. v. Prove, with the notation of Ex. iv., that

$$\psi(x, a; z, c_1, \ldots, c_p)\, \psi(z, a; \zeta, \gamma_1, \ldots, \gamma_p) = \psi(x, \zeta; z, c_1, \ldots, c_p)\, \psi(x, a; \zeta, \gamma_1, \ldots, \gamma_p).$$

126. These investigations can be usefully modified*; we can obtain a rational function $\psi(x, a; z, c)$, having the same general character as $\psi(x, a; z, c_1, \ldots, c_p)$ but simpler in that its poles occur only at two distinct places z, c, of the Riemann surface, and we can obtain an uniform function $E(x, z)$ having only one zero, of the first order, at the place z, which is infinite at only one place, c, of the surface.

The limit, when the place x approaches the place c, of the rth differential coefficient of $\Omega_i(x)$ in regard to the infinitesimal at the place c, will be denoted by $\Omega_i^{(r)}(c)$, or simply by $\Omega_i^{(r)}$. We have shewn (Chap. III. § 28) that there are certain numbers $k_1, \ldots, k_p$, such that no rational function exists, infinite only at the place c, to the orders $k_1, \ldots, k_p$. The periods of a function of the form

$$D_c^{k-1}\, \Gamma_c^{x, a} - \lambda_1 D_c^{k_1 - 1}\, \Gamma_c^{x, a} - \ldots\ldots - \lambda_p\, D_c^{k_p - 1}\, \Gamma_c^{x, a},$$

wherein $\lambda_1, \ldots, \lambda_p$ are constants, and $D_c^{k-1}\, \Gamma_c^{x, a}$ denotes† the limit, when z approaches c, of the kth differential coefficient of the function $\Pi_{z, \mu}^{x, a}$ in regard to the infinitesimal at c, μ being an arbitrary place, are all of the form

$$\Omega_i^{(k-1)} - \lambda_1 \Omega_i^{(k_1 - 1)} - \ldots\ldots - \lambda_p\, \Omega_p^{(k_p - 1)}, \qquad (i = 1, 2, \ldots, p).$$

These periods cannot all vanish when k is any one of the numbers $k_1, \ldots, k_p$; thus the determinant formed with the p^2 quantities $\Omega_i^{(k_r - 1)}$ does

* Günther, *Crelle*, cix. p. 199 (1892).

† For purposes of calculation, when c is a branch place, it is necessary to have care as to the definition.

not vanish; but $\lambda_1, \dots, \lambda_p$ can be chosen to make all these periods vanish when k is not one of the numbers $k_1, \dots, k_p$.

127. Consider now the function

$$\psi(x, a; z, c) = \begin{vmatrix} \Gamma_z^{x, a} & , \Omega_1(z) , & . & , \Omega_p(z) \\ . & . & . & . \\ D_c^{k_r - 1}\Gamma_c^{x, a}, & \Omega_1^{(k_r-1)}, & . & , \Omega_p^{(k_r-1)} \\ . & . & . & . \end{vmatrix} \div \begin{vmatrix} \Omega_1^{(k_1-1)}, & . & , \Omega_p^{(k_1-1)} \\ . & . & . \\ \Omega_1^{(k_p-1)}, & . & , \Omega_p^{(k_p-1)} \end{vmatrix}$$

wherein $r = 1, 2, \dots, p$.

Since the period of $\Gamma_z^{x, a}$, at the ith period loop of the second kind, is $\Omega_i(z)$, the periods of the elements of the first column of the first determinant are the elements of the various other columns of that determinant. Thus the function is a rational function of x.

We shall denote the minors of the elements of the first column of the first determinant, divided by the second determinant, by $1, -\omega_1(z), \dots, -\omega_p(z)$, although that notation has already (§ 121) been used in a different sense. Before, $\omega_i(z)$ was such that $\omega_i(c_r) = 0$ unless $r = i$ in which case $\omega_i(c_i) = 1$; now, as is easy to see, $\left[D_z^{k_r-1}\omega_i(z)\right]_{z=c}$ is 0 or 1 according as r is not equal or is equal to i. The integrals $\int^z \omega_i(z)\, dt_z$ are linearly independent integrals of the first kind (cf. Chap. III. § 36).

Then the function can be written

$$\psi(x, a; z, c) = \Gamma_z^{x, a} - \sum_{i=1}^{p} \omega_i(z)\, D_c^{k_i - 1}\,\Gamma_c^{x, a};$$

the function is infinite at z like $-t_z^{-1}$, t_z being the infinitesimal at the place z, and is infinite at c like*

$$\underline{|k_1 - 1}\,\omega_1(z)\, t_c^{-k_1} + \dots\dots + \underline{|k_p - 1}\,.\,\omega_p(z)\, t_c^{-k_p},$$

t_c being the infinitesimal at the place c. It is not elsewhere infinite. The function vanishes when x approaches the place a. As before (§ 123) $\psi(x, a; z, c) \Big/ \dfrac{dz}{dt}$ is a rational function of all the quantities involved; and $\psi(x, a; z, c)$, as depending upon z, is infinite only at the places x, a, in each case to the first order.

* This is clear when c is not a branch place, since then, when x is near to c, $\Gamma_c^{x, a}$ is infinite like $-\dfrac{1}{x - c}$; and the $(k-1)$th differential coefficient of this in regard to c is $-\underline{|k-1}\,(x-c)^{-k}$. When c is a branch place, exactly similar reasoning applies if we first make a conformal representation of the neighbourhood of the place, as explained in Chap. II. §§ 16, 19.

128. If now $R(x)$ be a rational function with poles of the first order at the places $z_1, \dots, z_m$, it is possible to choose the constants $\lambda_1, \dots, \lambda_p$ so that the difference

$$R(x) - \lambda_1 \psi(x, a; z_1, c) - \lambda_2 \psi(x, a; z_2, c) - \dots\dots - \lambda_m \psi(x, a; z_m, c)$$

is not infinite at any of the places $z_1, \dots, z_m$; this difference is therefore infinite only at the place c, and is infinite at c like

$$-(A_1 \underline{|k_1 - 1}\, t_c^{-k_1} + \dots\dots + A_p \underline{|k_p - 1}\, t_c^{-k_p}),$$

where

$$A_i = \lambda_1 \omega_i(z_1) + \dots\dots + \lambda_m \omega_i(z_m), \qquad (i = 1, 2, \dots, p).$$

But, a rational function whose only infinity is that given by this expression, can be taken to have a form

$$A + A_1 D_c^{k_1 - 1} \Gamma_c^{x, a} + \dots\dots + A_p D_c^{k_p - 1} \Gamma_c^{x, a},$$

wherein A is a constant; and we have already remarked (§ 126) that the periods of this function cannot all be zero unless each of $A_1, \dots, A_p$ be zero. Hence this is the case, and we have the equation

$$R(x) = A + \lambda_1 \psi(x, a; z_1, c) + \dots\dots + \lambda_m \psi(x, a; z_m, c),$$

whereby any rational function with poles of the first order is expressed by means of the function $\psi(x, a; z, c)$. It is immediately seen that the equations $A_1 = 0 = \dots = A_p$ enable us to reduce the constants $\lambda_1, \dots, \lambda_m$ to the number given by the Riemann-Roch Theorem (Chap. III. § 37).

When some of the poles of the function $R(x)$ are multiple, the necessary modification consists in the introduction of the functions

$$D_z \psi(x, a; z, c),\ D_z^2 \psi(x, a; z, c), \dots\dots.$$

Ex. If $\bar{\omega}_1(x), \dots, \bar{\omega}_p(x)$ denote what are called $\omega_1(x), \dots, \omega_p(x)$ in § 121, and the notation of § 127 be preserved, prove that

$$\bar{\omega}_r(z) = \sum_{i=1}^{p} \omega_i(z)\, D_c^{k_i - 1} \bar{\omega}_r(c),$$

and that

$$\psi(x, a; z, c) = \psi(x, a; z, c_1, \dots, c_p) - \Sigma \omega_i(z)\, D_c^{k_i - 1} \psi(x, a; c, c_1, \dots, c_p)$$

$$\psi(x, a; z, c_1, \dots, c_p) = \psi(x, a; z, c) - \Sigma \bar{\omega}_i(z)\, \psi(x, a; c_i, c).$$

129. From the function $\psi(x, a; z, c)$ we derive a function of x, given by

$$E(x, z) = e^{\int_c^z \psi(x, a; z, c)\, dt_z}, \quad = e^{\Pi_{z, c}^{x, a} - \sum_{r=1}^{p} V_r^{z, c} D^{k_r - 1} \Gamma_c^{x, a}},$$

where, in the notation of § 127, $V_r^{z, c} = \int_c^z \omega_r(z)\, dt_z$, which has the following properties:

(i) It is an uniform function of x; there exists in fact an equation

$$2\pi i v_i^{z, c} = \sum_{r=1}^{p} V_r^{z, c}\, \Omega_i^{(k_r - 1)}.$$

(ii) The function vanishes to the first order when the place x approaches the place z; and is equal to unity at the place a.

(iii) The function is infinite only at the place c, and there like

$$t_c^{-1}\, e^{\sum_{r=1}^{p} V_r^{z,c} \lfloor k_r - 1\, t_c^{-k_r}} .$$

As before we can shew that any rational function $R(x)$, with poles at $a_1, \ldots, a_m$, and zeros at $\beta_1, \ldots, \beta_m$, can be written in the form

$$R(a)\,\frac{E(x, \beta_1)\, \ldots\, E(x, \beta_m)}{E(x, a_1)\, \ldots\, E(x, a_m)},$$

this being still true when some of the places $a_1, \ldots, a_m$, or some of the places $\beta_1, \ldots, \beta_m$ are coincident.

130. We pass now to the algebraical expression of the functions which have been described here*. We have already (Chap. IV. § 49) given the expression of the function $\psi(x, a; z, c_1, \ldots, c_p)$ in the case when all the places $a, z, c_1, \ldots, c_p$ are ordinary finite places. In what follows we shall still suppose these places to be finite places; the necessary modifications when this is not so can be immediately obtained by a transformation of the form $x = (\xi - k)^{-1}$, or by the use of homogeneous variables (cf. § 46, Chap. IV., § 85, Chap. VI.).

If, s being the value of y when $x = z$, we denote the expression

$$\frac{\phi_0(s, z) + \sum_{r=1}^{n-1} \phi_r(s, z)\, g_r(y, x)}{(z - x) f'(s)}$$

by† (z, x), and use the integrands $\omega_1(x), \ldots, \omega_p(x)$ defined in § 121, the rational expression of $\psi(x, a; z, c_1, \ldots, c_p)$, which was given in § 49, can be put into the form

$$\psi(x, a; z, c_1, \ldots, c_p) = (z, x) - (z, a) - \sum_{r=1}^{p} \omega_r(z)\, [(c_i, x) - (c_i, a)].$$

In case z be a branch place, the expression (z, x) is identically infinite in virtue of the factor $f'(s)$ in the denominator, and this expression can no longer be valid. But, then, the limit, as ζ approaches z, of the expression

* It is known (Klein, *Math. Annal.* xxxvi. p. 9 (1890); Günther, *Crelle*, cix. p. 199 (1892)) that the actual expressions of functions having the character of the functions $\psi(x, a; z, c_1, \ldots, c_p)$, $E(x, z)$, $Q_{z,c}^{x,a}$, have been given by Weierstrass, in lectures. Unfortunately these expressions have not yet (August, 1895) been published, so far as the writer is aware. Indications of some value are given by Hettner, *Götting. Nachr.* 1880, p. 386; Bolza, *Götting. Nachr.* 1894, p. 268; Weierstrass, *Gesamm. Werke*, Bd. ii. p. 235 (1895), and in the *Jahresbericht der Deuts. Math.-Vereinigung*, Bd. iii. (Nov. 1894), pp. 403—436. But it does not appear how far the last of these is to be regarded as authoritative; and it has not been used here. The reader is recommended to consult the later volumes of Weierstrass's works.

† This notation has already been used (§ 45). It will be adhered to.

$(\zeta, x)\dfrac{d\zeta}{dt}$, wherein t is the infinitesimal at the place z, is finite*; if we denote this limit by $(z, x)\dfrac{dz}{dt}$, and introduce a similar notation for the places $c_1, \dots, c_p$, we obtain the expression

$$\psi(x, a; z, c_1, \dots, c_p) = [(z, x) - (z, a)]\frac{dz}{dt} - \sum_{r=1}^{p} \omega_r(z) . [(c_i, x) - (c_i, a)]\frac{dc_i}{dt},$$

which, as in § 49, has the necessary behaviour, for all finite positions of $z, a, c_1, \dots, c_p$.

From this expression we immediately obtain (§ 45)

$$\overline{E}(x, z), = e^{\int_c^z \psi(x, a; z, c_1, \dots, c_p)\, dt_z}, = e^{P_{x,a}^{z,c} - \sum\limits_{r=1}^{p} V_r^{z,c}[(c_i, x) - (c_i, a)]\frac{dc_i}{dt}}.$$

131. In a precisely similar way it can be seen (see § 127) that

$$\psi(x, a; z, c) = [(z, x) - (z, a)]\frac{dz}{dt} - \sum_{r=1}^{p} \omega_r(z) D_c^{k_r-1}\left\{[(c, x) - (c, a)]\frac{dc}{dt}\right\},$$

wherein $D_c^{k_r-1}\left\{[(c, x) - (c, a)]\dfrac{dc}{dt}\right\} = \text{limit}_{\zeta=c}\left[\left(\dfrac{d}{dt_c}\right)^{k_r-1}\left\{[(\zeta, x) - (\zeta, a)]\dfrac{d\zeta}{dt_c}\right\}\right]$; for this expression can be written as the quotient of two determinants, in the manner of § 49, and the integrands $\Omega_1(z), \dots, \Omega_p(z)$ are linear functions of the p integrands

$$\frac{\phi_1(z)}{f'(s)}\frac{dz}{dt}, \frac{z\phi_1(z)}{f'(s)}\frac{dz}{dt}, \dots, \frac{z^{\tau_1-1}\phi_1(z)}{f'(s)}\frac{dz}{dt}, \frac{\phi_2(z)}{f'(s)}\frac{dz}{dt}, \dots\dots;$$

these latter quantities can therefore be introduced in the determinants in place of $\Omega_1(z), \dots, \Omega_p(z)$, the same change being made, at the same time, for the quantities $\Omega_1(c), \dots, \Omega_p(c)$, throughout. Then it can be shewn precisely as in § 49 that the expression is not infinite when x is at infinity. In regard to finite places, it is clear that the expression

$$D_c^{k_r-1}\left\{[(c, x) - (c, a)]\frac{dc}{dt}\right\}, = D_c^{k_r} P_{x,a}^{c,y},$$

regarded as a function of x, has the same character, when x is near to c, as the function $D_c^{k_r-1}\Gamma_c^{x,a}$.

Hence, also, it follows that $E(x, z)$ has the form

$$E(x, z) = e^{P_{x,a}^{z,c} - \sum\limits_{r=1}^{p} V_r^{z,c} D_c^{k_r-1}\left\{[(c, x) - (c, a)]\frac{dc}{dt}\right\}}.$$

* $f'(\eta)$, when η is very nearly s, vanishes to order $i+w$, and $d\zeta/dt$ to order w (see Chap. VI. § 87). Or the result may be seen from the formula

$$(z, x) - (z, a) = \frac{d}{dz}P_{x,a}^{z,c}$$

(Chap. IV. § 45).

132. *Ex.* i. For the case $(p=1)$ where the surface is associated with the equation

$$y^2=(x,\ 1)_4,$$

if the values of the variables x, y at the place a be respectively a, b, and the values at the place c_1 be c_1, d_1 respectively, then

(a) when $(c_1,\ d_1)$ is not a branch place $\omega_1(z)=\frac{d_1}{s}\frac{dz}{dt}$, $(z_1,\ x)=\frac{s+y}{2s(z-x)}$

and

$$\psi(x,\ a;\ z,\ c_1)=\left[\frac{s+y}{2s(z-x)}-\frac{s+b}{2s(z-a)}\right]\frac{dz}{dt}-\frac{d_1}{s}\frac{dz}{dt}\left[\frac{d_1+y}{2d_1(c_1-x)}-\frac{d_1+b}{2d_1(c_1-a)}\right]$$
$$=\frac{1}{2s}\frac{dz}{dt}\left[\frac{s+y}{z-x}-\frac{s+b}{z-a}-\frac{d_1+y}{c_1-x}+\frac{d_1+b}{c_1-a}\right];$$

(β) when $(c_1,\ d_1)$ is a branch place, in the neighbourhood of which

$$x=c_1+t^2,\ y=At+\dots,\ \omega_1(z)=\frac{A}{2s}\frac{dz}{dt},\ (c_1,\ x)\frac{dc_1}{dt}=\text{limit of }\frac{At+y}{2At(c_1-x)}\cdot 2t=\frac{y}{A(c_1-x)},$$

and

$$\psi(x,\ a;\ z,\ c_1)=\left[\frac{s+y}{2s(z-x)}-\frac{s+b}{2s(z-a)}\right]\frac{dz}{dt}-\frac{A}{2s}\frac{dz}{dt}\left\{\frac{y}{A(c_1-x)}-\frac{b}{A(c_1-a)}\right\}$$
$$=\frac{1}{2s}\frac{dz}{dt}\left\{\frac{s+y}{z-x}-\frac{s+b}{z-a}-\frac{y}{c_1-x}+\frac{b}{c_1-a}\right\}.$$

If $(s,\ z)$ be not a branch place, $\frac{1}{2s}\frac{dz}{dt}=\frac{1}{2s}$; if $(s,\ z)$ be a branch place, in the neighbourhood of which $x=z+t^2$, $y=Bt+\dots$, $\frac{1}{2s}\frac{dz}{dt}$, $=$limit of $\frac{1}{2Bt}2t$, $=\frac{1}{B}$.

Ex. ii. For the case $(p=2)$ where the surface is associated with the equation $y^2=f(x)$, where $f(x)$ is an integral function of x of the sixth order, we shall form the function $\psi(x,\ a;\ z,\ c_1,\ c_2)$ for the case where c_1, c_2 are branch places, so that $f(c_1)=f(c_2)=0$, and shall form the function $\psi(x,\ a;\ z,\ c)$ for the case when c is a branch place, so that $f(c)=0$.

When c_1, c_2 are branch places, in the neighbourhood of which, respectively, $x=c_1+t_1^2$, $y=A_1t_1+\dots$, and $x=c_2+t_2^2$, $y=A_2t_2+\dots$, so that $A_1^2=f'(c_1)$, $A_2^2=f'(c_2)$, we have

$$\omega_1(z)=\frac{z-c_2}{c_1-c_2}\frac{A_1}{2s}\frac{dz}{dt},\quad \omega_2(z)=\frac{z-c_1}{c_2-c_1}\frac{A_2}{2s}\frac{dz}{dt},\quad [(c_1,\ x)-(c_1,\ a)]\frac{dc_1}{dt}=\frac{1}{A_1}\left(\frac{y}{c_1-x}-\frac{b}{c_1-a}\right),$$

and

$$\psi(x,\ a;\ z,\ c_1,\ c_2)=\left[\frac{s+y}{2s(z-x)}-\frac{s+b}{2s(z-a)}\right]\frac{dz}{dt}-\frac{1}{2s}\frac{dz}{dt}\left\{\frac{z-c_2}{c_1-c_2}\left(\frac{y}{c_1-x}-\frac{b}{c_1-a}\right)\right.$$
$$\left.+\frac{z-c_1}{c_2-c_1}\left(\frac{y}{c_2-x}-\frac{b}{c_2-a}\right)\right\}.$$

When c is a branch place, in the neighbourhood of which $x=c+t^2$, $y=At+Bt^3+\dots$, so that $A^2=f'(c)$, the numbers k_1, k_2 are 1, 3 respectively (Chap. V. § 58, Ex. ii.). In the definition of the forms $\omega_1(z)$, $\omega_2(z)$ (§ 127) we may, by linear transformation of the 2nd, 3rd, ..., $(p+1)$th columns of the numerator determinant, and the same linear transformation of the columns of the denominator determinant, replace $\Omega_1(z)$, ..., $\Omega_p(z)$ by the differential coefficients of any linearly independent integrals of the first kind. In the case now under consideration we may replace them by the differential coefficients of the integrals $\int\frac{dz}{2s}$, $\int\frac{zdz}{2s}$. Hence the denominator determinant becomes

$$\text{limit}_{x=c}\begin{vmatrix}\dfrac{1}{2y}\dfrac{dx}{dt}, & \dfrac{x}{2y}\dfrac{dx}{dt}\\ D_x^2\left(\dfrac{1}{2y}\dfrac{dx}{dt}\right), & D_x^2\left(\dfrac{x}{2y}\dfrac{dx}{dt}\right)\end{vmatrix}=\text{limit}_{x=c}\begin{vmatrix}\dfrac{2t}{2At}, & \dfrac{c+t^2}{2At}2t\\ \left(\dfrac{d}{dt}\right)^2\left(\dfrac{t}{At+Bt^3+\dots}\right), & \left(\dfrac{d}{dt}\right)^2\left(\dfrac{t(c+t^2)}{At+Bt^3+\dots}\right)\end{vmatrix}$$

$$=\begin{vmatrix}\dfrac{1}{A}, & \dfrac{c}{A}\\ -\dfrac{2B}{A^2}, & \dfrac{2}{A}\left(1-\dfrac{cB}{A}\right)\end{vmatrix}, =2/A^2.$$

Hence $$\omega_1(z)\frac{2}{A^2}=\text{limit}_{x=c}\begin{vmatrix}\dfrac{1}{2s}\dfrac{dz}{dt}, & \dfrac{z}{2s}\dfrac{dz}{dt}\\ D_x^2\left(\dfrac{1}{2y}\dfrac{dx}{dt}\right), & D_x^2\left(\dfrac{x}{2y}\dfrac{dx}{dt}\right)\end{vmatrix}=\left\{\frac{2B}{A^2}z+\frac{2}{A}\left(1-\frac{cB}{A}\right)\right\}\frac{1}{2s}\frac{dz}{dt}$$

$$=\frac{1}{A^2}[A+B(z-c)]\frac{1}{s}\frac{dz}{dt},$$

and

$$\omega_2(z)\frac{2}{A^2}=-\text{limit}_{x=c}\begin{vmatrix}\dfrac{1}{2s}\dfrac{dz}{dt}, & \dfrac{z}{2s}\dfrac{dz}{dt}\\ \dfrac{1}{2y}\dfrac{dx}{dt}, & \dfrac{x}{2y}\dfrac{dx}{dt}\end{vmatrix}=-\frac{1}{2s}\frac{dz}{dt}\frac{c-z}{A}.$$

Hence

$$\omega_1(z)=[A+B(z-c)]\frac{1}{2s}\frac{dz}{dt},\qquad \omega_2(z)=\tfrac{1}{2}A(z-c)\frac{1}{2s}\frac{dz}{dt}.$$

Further $$[(c,x)-(c,a)]\frac{dc}{dt}=\frac{1}{A}\left(\frac{y}{c-x}-\frac{b}{c-a}\right),$$ as in Example i.,

but

$$D_c^2\left\{[(c,x)-(c,a)]\frac{dc}{dt}\right\}=\text{limit}_{t=0}\left[\left(\frac{d}{dt}\right)^2\left\{\left[\frac{At+Bt^3+y}{2(At+Bt^3)(c-x+t^2)}\right.\right.\right.$$
$$\left.\left.\left.-\frac{At+Bt^3+b}{2(At+Bt^3)(c-a+t^2)}\right]2t\right\}\right]=-\frac{2y}{A^2(x-c)^2}[A-B(x-c)]+\frac{2b}{A^2(a-c)^2}[A-B(a-c)].$$

Hence the function $\psi(x, a;\ z, c)$ is given by the expression

$$\left[\frac{s+y}{2s(z-x)}-\frac{s+b}{2s(z-a)}\right]\frac{dz}{dt}-\frac{A+B(z-c)}{A}\frac{1}{2s}\frac{dz}{dt}\left(\frac{y}{c-x}-\frac{b}{c-a}\right)$$
$$+\frac{z-c}{A}\frac{1}{2s}\frac{dz}{dt}\left(\frac{A-B(x-c)}{(x-c)^2}y-\frac{A-B(a-c)}{(a-c)^2}b\right).$$

Ex. iii. Apart from the algebraical determination of the function $\psi(x, a;\ z, c_1, \dots, c_p)$ which is here explained, it will in many cases be very easy* to determine the function by the methods of Chapter VI. It is therefore of interest to remark that, when the function $\psi(x, a;\ z, c_1, \dots, c_p)$ is once obtained the forms of independent integrals of the first and second kinds can be immediately obtained as the coefficients in the first few terms of the expansion of the function in the neighbourhood of its poles, in terms of the infinitesimals at these poles.

* An adjoint polynomial Ψ of grade $(n-1)\sigma+n-2$ which vanishes in the $p+1$ places $z, c_1, \dots, c_p$ will vanish in $n+p-3$ other places. The general adjoint polynomial of grade $(n-1)\sigma+n-2$ which vanishes in these $n+p-3$ places will be of the form $\lambda\Psi+\mu\Theta$, where λ and μ are constants. The function $\psi(x, a;\ z, c_1, \dots, c_p)$ is obtained from $\lambda+\mu\Theta/\Psi$, by determining λ and μ properly. Cf. Noether (*loc. cit.*) Math. Annal. xxxvii.

In fact, if t_i be the infinitesimal in the neighbourhood of the place c_i, and $M_{r,i}$ denote

$$D_{c_i}\left[(c_r, c_i)\frac{dc_r}{dt}\right], \; M_{i,i} \text{ denoting } \left\{D_x\left[(c_i, x)\frac{dc_i}{dt}+\frac{1}{t_i}\right]\right\}_{x=c_i},$$

the expansion of $\psi(x, a; z, c_1, \ldots, c_p)$, as a function of x, in the neighbourhood of the place c_i, has, as the coefficient of t_i^{-1}, the expression $\omega_i(z)$, which is one of a set of linearly independent integrands of the first kind, while the coefficient of t_i is

$$D_{c_i}\left[(z, c_i)\frac{dz}{dt}\right] - \sum_{r=1}^{p} \omega_r(z) M_{r,i}.$$

Now the elementary integral of the second kind obtained in Chap. IV. (§§ 45, 47) with its pole at a place c, when z is the current place, is $E_c^{z,a} = \int_a^z dz\, D_c(z, c)$, whether c be a branch place or not, and when z is near a branch place this must be taken in the form

$$E_c^{z,a} = \int_a^z dt_z D_c\left[(z, c)\frac{dz}{dt}\right].$$

Hence the coefficient of t_i in the expansion of $\psi(x, a; z, c_1, \ldots, c_p)$, when x is near to c_i, is equal to

$$D_z E_{c_i}^{z,a} - \sum_{r=1}^{p} \omega_r(z) M_{r,i}.$$

This is the differential coefficient of an integral of the second kind, with its pole at c_i, the current place being z. We shall see that the integral of the second kind with its pole at any place z can be expressed by means of the functions $E_{c_1}, \ldots, E_{c_p}$ (§ 135, Equation x.).

Ex. iv. Similar results hold for the expansion of the function $\psi(x, a; z, c)$, as a function of x, when x is in the neighbourhood of the place c. If t_c be the infinitesimal at this place, the terms involving negative powers are

$$\frac{\lfloor k_1 - 1}{t_c^{k_1}}\omega_1(z) + \ldots + \frac{\lfloor k_p - 1}{t_c^{k_p}}\omega_p(z),$$

of which the coefficients of the various powers of t_c are differential coefficients of linearly independent integrals of the first kind; the terms involving positive powers are

$$\sum_{k=1} \frac{t_c^k}{\lfloor k}\left\{D_c^k\left((z, c)\frac{dz}{dt}\right) - \sum_{i=1}^{p} \omega_i(z) P_{i,k}\right\},$$

where $P_{i,k}$ is the limit, when the place x approaches the place c of the expression

$$D_x^k\left\{D_c^{k_i-1}\left[(c, x)\frac{dc}{dt}\right] + \frac{\lfloor k_i - 1}{t_c^{k_i}}\right\}.$$

Among the coefficients of these positive powers of t_c, only those are important for which k is one of the numbers $k_1, \ldots, k_p$. This follows from the fact that $D_c^{k-1}\Gamma_c^{x,a}$, when k is not one of the numbers $k_1, \ldots, k_p$, is expressible by those of

$$D_c^{k_1-1}\Gamma_c^{x,a}, \ldots, D_c^{k_p-1}\Gamma_c^{x,a},$$

of which the indices $k_1 - 1, k_2 - 1, \ldots$, are less than $k - 1$, together with a rational function of x (Chap. III. § 28).

Ex. v. In the expansion of the function $\psi(x, a; z, c)$ whose expression is given in Example ii., the terms involving negative powers are

$$\frac{A+B(z-c)}{2s}\frac{dz}{dt}\cdot\frac{1}{t_c}+\frac{z-c}{2s}A\frac{dz}{dt}\cdot\frac{1}{t_c^3};$$

and the terms involving positive powers are

$$\frac{1}{2s}\frac{dz}{dt}\cdot t\left[\frac{A}{z-c}+B+C(z-c)\right]+\tfrac{1}{2}\frac{dz}{dt}t^2\frac{1}{(z-c)^2}+\frac{1}{2s}\frac{dz}{dt}t^3\left[\frac{A}{(z-c)^2}+\frac{B}{z-c}+C+D(z-c)\right]$$
$$+\tfrac{1}{2}\frac{dz}{dt}t^4\frac{1}{(z-c)^3}+\frac{1}{2s}\frac{dz}{dt}t^5\left[\frac{A}{(z-c)^3}+\frac{B}{(z-c)^2}+\frac{C}{z-c}+D+E(z-c)\right]+\dots,$$

where the quantities $A, B, \dots, E$ are those occurring in the expansion of y in the neighbourhood of the place c; this expansion is of the form $y=At+Bt^3+Ct^5+Dt^7+Et^9+\dots$.

Ex. vi. If in Ex. v. the integrals of the coefficients of t, t^3 and t^5 be denoted by F_1^s, F_3^s, F_5^s, find the equation of the form

$$F_5^s=\lambda F_1^s+\mu F_3^s+\text{integrals of the first kind}+\text{rational function of }(s, z)$$

which is known to exist (Chap. III. §§ 28, 26; Chap. V. § 57, Ex. ii.), λ and μ being constants.

Prove, in fact, if the surface be associated with the equation

$$y^2=(x-c)^6+p_1(x-c)^5+p_2(x-c)^4+p_3(x-c)^3+p_4(x-c)^2+p_5(x-c)$$

that

$$\int^s\frac{dz}{2s}\left[\frac{3p_3(z-c)^2+4p_4(z-c)+5p_5}{(z-c)^3}+2p_2+p_1(z-c)\right]=-\frac{s}{(z-c)^3}+\text{constant}.$$

133. We pass now to a comparison of the two forms we have obtained for each of the rational functions $\psi(x, a; z, c_1, \dots, c_p)$, $\psi(x, a; z, c)$, one of which was expressed by the Riemann integrals, the other in explicit algebraical form.

The cases of the two functions are so far similar that it will be sufficient to give the work only for one case $\psi(x, a; z, c_1, \dots, c_p)$, and the results for the other case.

From the two equations (§§ 122, 130)

$$\psi(x, a; z, c_1, \dots, c_p)=\Gamma_z^{x, a}-\sum_{i=1}^{p}\omega_i(z)\Gamma_{c_i}^{x, a},$$

$$\psi(x, a; z, c_1, \dots, c_p)=[(z, x)-(z, a)]\frac{dz}{dt}-\sum_{i=1}^{p}\omega_i(z)[(c_i, x)-(c_i, a)]\frac{dc_i}{dt},$$

we infer, denoting the function

$$\Gamma_z^{x, a}-[(z, x)-(z, a)]\frac{dz}{dt} \qquad \text{(i)}$$

by $H_z^{x, a}$, that

$$H_z^{x, a}=\sum_{i=1}^{p}\omega_i(z)H_{c_i}^{x, a} \qquad \text{(ii)}.$$

The function $H_z^{x,a}$ is not infinite at the place z, but is algebraically infinite at infinity; it has the same periods as $\Gamma_z^{x,a}$. The equation (ii) shews that $H_z^{x,a} \Big/ \frac{dz}{dt}$ is a rational function of z, while the equation

$$\Gamma_z^{x,a} = [(z, x) - (z, a)] \frac{dz}{dt} + \sum_{i=1}^{p} \omega_i(z) H_{c_i}^{x,a} \qquad \text{(iii)}$$

gives the form of $\Gamma_z^{x,a} \Big/ \frac{dz}{dt}$ as a rational function of z.

Integrating the equation (iii) in regard to z, we obtain

$$\Pi_{z,c}^{x,a} = P_{x,a}^{z,c} + \sum_{i=1}^{p} V_i^{z,c} H_{c_i}^{x,a} \qquad \text{(iv)},$$

where c is an arbitrary place, and $P_{x,a}^{z,c}$ is the integral of the third kind, as a function of z, which was determined in Chap. IV. (§§ 45, 46).

Since the integral of the second kind $E_z^{x,a}$, obtained in Chap. IV. (§§ 45, 46), is equal to $D_z P_{z,c}^{x,a}$, we deduce from the last equation, interchanging x and z, and also a and c, and then differentiating in regard to z,

$$E_z^{x,a} + \sum_{i=1}^{p} V_i^{x,a} D_z H_{c_i}^{z,c} = D_z \Pi_{x,a}^{z,c}, = D_z \Pi_{z,c}^{x,a}, = \Gamma_z^{x,a} \qquad \text{(v)},$$

and thence, using equation (iii) to express $\Gamma_z^{x,a}$,

$$E_z^{x,a} = [(z, x) - (z, a)] \frac{dz}{dt} + \sum_{i=1}^{p} [\omega_i(z) H_{c_i}^{x,a} - V_i^{x,a} D_z H_{c_i}^{z,c}] \qquad \text{(vi)},$$

which* gives the form of $E_z^{x,a} \Big/ \frac{dz}{dt}$ as a rational function of z.

The difference of two elementary integrals of the second kind must needs be a function which is everywhere finite, and therefore an aggregate of integrals of the first kind. The equation (v) expresses the difference of $E_z^{x,a}$ and $\Gamma_z^{x,a}$ in this way. But it should be noticed that the coefficients of the integrals of the first kind in this equation, which depend upon z, become infinite for infinite values of z. They are the quantities

$$D_z H_{c_i}^{z,c}.$$

From the equation (iv) we have

$$P_{z,c}^{x,a} = \Pi_{z,c}^{x,a} - \sum_{i=1}^{p} V_i^{x,a} H_{c_i}^{z,c},$$

wherein the coefficients of $V_i^{x,a}$ on the right may be characterised as integrals of the second kind. From this equation also, if the periods of $V_i^{x,a}$ at the jth period loops of the

* An equation of this form is given by Clebsch and Gordan, *Abel. Functnen.* (Leipzig, 1866), p. 120.

first and second kind be denoted by $C_{i,j}$ and $C'_{i,j}$ respectively, we obtain, as the corresponding periods of $P^{x,a}_{z,c}$

$$\left(P^{x,a}_{z,c}\right)_j = \qquad - \sum_{i=1}^{p} C_{i,j} H^{z,c}_{c_i}$$

$$\left(P^{x,a}_{z,c}\right)'_j = 2\pi i v_j^{z,c} - \sum_{i=1}^{p} C'_{i,j} H^{z,c}_{c_i};$$

from these equations the periods of $E^{x,a}_z$ are immediately obtainable. These equations may be used to express the integrals $H^{z,c}_{c_i}$ in terms of the periods of $P^{x,a}_{z,c}$ at the period loops of the first kind.

134. But all these equations are in the nature of transition equations; they connect functions which are algebraically derivable with functions whose definition depends upon the form of the period loops. We proceed further to eliminate these latter functions as far as is possible, replacing them by certain constants, which, in the nature of the case, are not determinable algebraically.

The function of x expressed by $H^{x,a}_z$ is not infinite at the place z. Hence we may define p^2 finite constants $A_{i,r}$ by the equation

$$A_{i,r} = D_{c_r} H^{c_r,c}_{c_i},$$

where c is an arbitrary place. And if, as in § 132, Ex. iii., we use the algebraically determinable quantities given by

$$M_{i,r} = D_{c_r}\left[(c_i, c_r)\frac{dc_i}{dt}\right], \quad M_{i,i} = \left\{D_x\left[(c_i, x)\frac{dc_i}{dt} + \frac{1}{t_i}\right]\right\}_{x=c_i},$$

we have

$$M_{i,r} + A_{i,r} = D_{c_r}\Gamma^{c_r,c}_{c_i} = D_{c_i}\Gamma^{c_i,c}_{c_r} = M_{r,i} + A_{r,i},$$

and

$$M_{i,i} + A_{i,i} = \left[D_x\left(\Gamma^{x,c}_{c_i} + \frac{1}{t_i}\right)\right]_{x=c_i}.$$

Then, from equation (v), putting therein c_r for z,

$$H^{x,a}_{c_r} = \Gamma^{x,a}_{c_r} - [(c_r,x)-(c_r,a)]\frac{dc_r}{dt} = E^{x,a}_{c_r} - [(c_r,x)-(c_r,a)]\frac{dc_r}{dt} + \sum_{i=1}^{p} A_{i,r} V_i^{x,a} \quad \text{(vii)}$$

and thence, since $E^{x,a}_{c_r} = \int_a^x dx\, D_{c_r}(x, c_r)$

$$D_x H^{x,a}_{c_r} = D_{c_r}\left[(x, c_r)\frac{dx}{dt}\right] - D_x\left[(c_r, x)\frac{dc_r}{dt}\right] + \sum_{i=1}^{p} A_{i,r}\,\omega_i(x).$$

If in this equation we replace x by z and i by r and then substitute in equation (v), we obtain

$$\Gamma^{x,a}_z = E^{x,a}_z + \sum_{i=1}^{p} V_i^{x,a}\left\{D_{c_i}\left[(z, c_i)\frac{dz}{dt}\right] - D_z\left[(c_i, z)\frac{dc_i}{dt}\right] + \sum_{r=1}^{p} A_{r,i}\,\omega_r(z)\right\};$$

and thus, if we define an, algebraically determinable, integral by the equation

$$G_z^{x,a}=E_z^{x,a}+\sum_{i=1}^{p}V_i^{x,a}\left\{D_{c_i}\left[(z,c_i)\frac{dz}{dt}\right]-D_z\left[(c_i,z)\frac{dc_i}{dt}\right]\right.$$
$$\left.-\tfrac{1}{2}\sum_{r=1}^{p}(M_{r,i}-M_{i,r})\,\omega_r(z)\right\} \quad \text{(viii)},$$

we have

$$\Gamma_z^{x,a}=G_z^{x,a}+\sum_{i=1}^{p}V_i^{x,a}\sum_{r=1}^{p}(A_{r,i}+\tfrac{1}{2}M_{r,i}-\tfrac{1}{2}M_{i,r})\,\omega_r(z),$$

or

$$\Gamma_z^{x,a}=G_z^{x,a}+\tfrac{1}{2}\sum_{i=1}^{p}V_i^{x,a}\sum_{r=1}^{p}(A_{r,i}+A_{i,r})\,\omega_r(z), \quad \text{(viii)}',$$

from which, by integration in regard to z, we obtain an equation

$$Q_{z,c}^{x,a}=\int_c^z G_z^{x,a}\,dt_z=\Pi_{z,c}^{x,a}-\tfrac{1}{2}\sum_{i=1\ldots p}^{r=1\ldots p}\sum(A_{r,i}+A_{i,r})\,V_i^{x,a}\,V_r^{z,c} \quad \text{(ix)},$$

either of these expressions being, by equation (viii), also equal to

$$P_{z,c}^{x,a}+\sum_{i=1}^{p}V_i^{x,a}\left[E_{c_i}^{z,c}-[(c_i,z)-(c_i,c)]\frac{dc_i}{dt}\right]$$
$$+\tfrac{1}{2}\sum_{i=1}^{p}\sum_{r=1}^{p}(V_i^{x,a}V_r^{z,c}-V_r^{x,a}V_i^{z,c})(M_{i,r}-M_{r,i}) \quad \text{(ix)}'.$$

The equation (ix) shews that the integral $Q_{z,c}^{x,a}$ is such that

$$Q_{z,c}^{x,a}=Q_{x,a}^{z,c}$$

while every term of (ix)′ is capable of algebraic determination.

135. From the equation (ix), when none of the places $x, z, c_1, \ldots, c_p$ are branch places, we obtain

$$\frac{\partial^2 Q_{z,c}^{x,a}}{\partial x\partial z}=\frac{\partial}{\partial z}(x,z)+\sum_{i=1}^{p}\omega_i(x)\left[\frac{\partial}{\partial c_i}(z,c_i)-\frac{\partial}{\partial z}(c_i,z)\right]$$
$$+\tfrac{1}{2}\sum_{i=1}^{p}\sum_{r=1}^{p}[\omega_i(x)\,\omega_r(z)-\omega_r(x)\,\omega_i(z)]\,[M_{i,r}-M_{r,i}] \quad \text{(x)},$$

and hence, from the characteristic property $\frac{\partial^2}{\partial x\partial z}Q_{z,c}^{x,a}=\frac{\partial^2}{\partial x\partial z}Q_{x,a}^{z,c}$, we infer

$$\frac{\partial}{\partial z}(x,z)-\frac{\partial}{\partial x}(z,x)+\sum_{i=1}^{p}\left\{\omega_i(x)\left[\frac{\partial}{\partial c_i}(z,c_i)-\frac{\partial}{\partial z}(c_i,z)\right]-\omega_i(z)\left[\frac{\partial}{\partial c_i}(x,c_i)-\frac{\partial}{\partial x}(c_i,x)\right]\right\}$$
$$+\tfrac{1}{2}\sum_{i=1}^{p}\sum_{r=1}^{p}[\omega_i(x)\,\omega_r(z)-\omega_r(x)\,\omega_i(z)]\,[M_{i,r}-M_{r,i}]=0 \quad \text{(xi)},$$

wherein every quantity which occurs is defined algebraically. The form when some of the places are branch places is obtainable by slight modifications. *This is then the general algebraic relation underlying the fundamental property of the interchange of argument and parameter, which was originally denoted, in this volume, by the equation* $\Pi_{z,c}^{x,a} = \Pi_{x,a}^{z,c}$.

The relation is of course independent of the places $c_1, \ldots, c_p$. For an expression in which these places do not enter, see § 138, Equation 17.

The equation (xi) can be obtained in an algebraic manner (§ 137, Ex. vi.). The method followed here gives the relations connecting the Riemann normal integrals and the particular integrals obtained in Chap. IV., with the canonical integrals $G_z^{x,a}$, $Q_{z,c}^{x,a}$.

It should be noticed, in equation (xi), that in the last summation each term occurs twice. By a slight change of notation the factor $\frac{1}{2}$ can be omitted.

The interchange of argument and parameter was considered by Abel; some of his formulae, with references, are given in the examples in § 147.

136. From the equation (viii)′ we have

$$\Gamma_{c_s}^{x,a} = G_{c_s}^{x,a} + \tfrac{1}{2}\sum_{i=1}^{p}(A_{s,i} + A_{i,s})\, V_i^{x,a}.$$

From this equation, and the equation (viii)′, we infer that

$$G_z^{x,a} - \sum_{s=1}^{p}\omega_s(z)\, G_{c_s}^{x,a} = \Gamma_z^{x,a} - \sum_{s=1}^{p}\omega_s(z)\, \Gamma_{c_s}^{x,a}$$

$$= \psi(x, a;\ z, c_1, \ldots, c_p) \qquad \text{(xii)},$$

which result may be regarded as giving an expression of the function $\psi(x, a;\ z, c_1, \ldots, c_p)$ in terms of the integrals G; but, written in the form

$$G_z^{x,a} = \sum_{s=1}^{p}\omega_s(z)\,\Gamma_{c_s}^{x,a} + [(z, x) - (z, a)]\frac{dz}{dt} - \sum_{i=1}^{p}\omega_i(z)\,[(c_i, x) - (c_i, a)]\frac{dc_i}{dt},$$

the equation (xii) has another importance; if we call $Q_{z,c}^{x,a}$ an elementary *canonical* integral of the third kind, and $G_z^{x,a}$, $= D_z Q_{z,c}^{x,a}$, an elementary canonical integral of the second kind, we may express the result in words thus—*The elementary canonical integral of the second kind with its pole at any place z is expressible in the form*

$$\sum_{s=1}^{p}\omega_s(z)\, G_{c_s}^{x,a} + (\text{rational function of } x, z, c_1, \ldots, c_p)\Big/\frac{dz}{dt},$$

wherein the elementary canonical integrals occurring, have their poles at p arbitrary independent places $c_1, \ldots, c_p$.

Further, by equation (xii) the function $\bar{E}(x, z)$, of § 124, can be written in the form

$$\bar{E}(x, z) = e^{Q_{z,c}^{x,a} - \sum_{s=1}^{p} V_s^{z,c}\, G_{c_s}^{x,a}} \qquad \text{(xiii)}.$$

If we put

$$K_z^{x,a} = G_z^{x,a} - [(z, x) - (z, a)] \frac{dz}{dt} \qquad \text{(xiv)},$$

the equation following equation (xii) gives

$$K_z^{x,a} = \sum_{i=1}^{p} \omega_i(z) K_{c_i}^{x,a} \qquad \text{(xv)},$$

and therefore, also

$$Q_{z,c}^{x,a} = P_{z,c}^{x,a} + \sum_{i=1}^{p} V_i^{x,a} K_{c_i}^{z,c} \qquad \text{(xvi)},$$

and

$$D_x\left((z, x)\frac{dz}{dt}\right) - D_z\left((x, z)\frac{dx}{dt}\right) = \sum_{i=1}^{p} [\omega_i(x) D_z K_{c_i}^{z,c} - \omega_i(z) D_x K_{c_i}^{x,a}] \quad \text{(xvii)}$$

which is another form of equation (xi).

It is easy to see that

$$G_{c_s}^{x,a} = E_{c_s}^{x,a} - \tfrac{1}{2}\sum_{i=1}^{p} (M_{i,s} - M_{s,i}) V_i^{x,a}.$$

137. *Ex.* i. Prove that the most general elementary integral of the third kind, with its infinities at the places z and c, and vanishing at the place a, which is unaltered when x, z are interchanged and also a and c, is of the form

$$\Pi_{z,c}^{x,a} - \sum_{i=1}^{p}\sum_{r=1}^{p} a_{i,r} V_i^{x,a} V_r^{z,c},$$

wherein $a_{i,r}$ are constants satisfying the equations $a_{i,r} = a_{r,i}$.

Ex. ii. If the integral of Ex. i. be denoted by $\overline{Q}_{z,c}^{x,a}$, and $D_z \overline{Q}_{z,c}^{x,a}$ be denoted by $\overline{G}_z^{x,a}$, prove that

$$\psi(x, a; z, c_1, \dots, c_p) = \overline{G}_z^{x,a} - \sum_{s=1}^{p} \omega_s(z) \overline{G}_{c_s}^{x,a}.$$

Ex. iii. If, in particular, $\overline{Q}_{z,c}^{x,a}$ be given by

$$\overline{Q}_{z,c}^{x,a} = Q_{z,c}^{x,a} - \tfrac{1}{2}\sum_{i=1}^{p}\sum_{r=1}^{p} (M_{r,i} + M_{i,r}) V_i^{x,a} V_r^{z,c},$$

prove that

$$\overline{G}_{c_i}^{z,a} = E_{c_i}^{z,a} - \sum_{r=1}^{p} M_{r,i} V_r^{z,a}.$$

This is the integral, in regard to z, of the coefficient of t_i in the expansion of $\psi(x, a; z, c_1, \dots, c_p)$, as a function of x, in the neighbourhood of the place c_i (§ 132, Ex. iii.).

The integral $Q_{z,c}^{x,a}$ is algebraically simpler than the integral $\overline{Q}_{z,c}^{x,a}$, of this example, in that its calculation does not require the determination of the limits denoted by $M_{i,i}$.

Ex. iv. For the case $p=1$, when the fundamental equation is of the form

$$y^2 = (x, 1)_4,$$

if the variables at the place c_1 be denoted by $x=c_1$, $y=d_1$, the place not being a branch place, prove that

$$\frac{\partial}{\partial z}(x, z)-\frac{\partial}{\partial x}(z, x)=\frac{1}{24ys}[f^{(\text{III})}(z).(x-z)+\tfrac{1}{2}f^{(\text{IV})}(z).(x-z)^2],$$

and calculate $Q_{z,\,c}^{x,\,a}$, from the equation xi, in the form

$$Q_{z,\,c}^{x,\,a}=\int_a^x\int_c^z\frac{ys+f(x,z)}{2(x-z)^2}\frac{dx}{y}\frac{dz}{s}-\frac{1}{24}f''(c_1)\int_a^x\frac{dx}{y}\int_c^z\frac{dz}{s},$$

where, if $y^2=f(x)=a_0x^4+4a_1x^3+6a_2x^2+4a_3x+a_4$, the symbol $f(x, z)$ denotes the symmetrical expression

$$x^2(a_0z^2+2a_1z+a_2)+2x(a_1z^2+2a_2z+a_3)+(a_2z^2+2a_3z+a_4).$$

Prove also that in this case $M_{1,\,1}=-f'(c_1)/4f(c_1)$.

Calculate the integral $Q_{z,\,c}^{x,\,a}$ when the place c_1 is a branch place, and prove that in that case $M_{1,\,1}$, $=\text{limit}_{t=0}\left(\frac{1}{A}\frac{y}{c_1-x}+\frac{1}{t}\right)$, wherein $x=c_1+t^2$, $y=At+Bt^3+\dots$, vanishes.

Ex. v. For the case $(p=2)$ in which the fundamental equation is

$$y^2=f(x),$$

where $f(x)$ is a sextic polynomial, taking c_1, c_2 to be the branch places $(c_1, 0)$, $(c_2, 0)$, in the neighbourhood of which, respectively, $x=c_1+t_1^2$, $y=A_1t_1+B_1t_1^3+\dots$, and $x=c_2+t_2^2$, $y=A_2t_2+B_2t_2^3+\dots$, prove that

$$E_{c_1}^{z,\,c}=\int_c^z\frac{dz}{2s}\frac{A_1}{z-c_1};\ \omega_1(z)=A_1\frac{z-c_2}{c_1-c_2}\frac{1}{2s}\frac{dz}{dt};\ (c_1, z)\frac{dc_1}{dt}=\frac{1}{A_1}\frac{s}{c_1-z};\ M_{1,\,2}=\frac{A_2}{A_1(c_1-c_2)}$$

and infer that

$$[\omega_1(x)\,\omega_2(z)-\omega_2(x)\,\omega_1(z)][M_{2,\,1}-M_{1,\,2}]=-\frac{A_1^2+A_2^2}{(c_1-c_2)^2}(x-z)\frac{1}{2s}\frac{dz}{dt}\frac{1}{2y}\frac{dx}{dt}.$$

Supposing x and z have general positions, deduce from equation (ix) that

$$4ys(x-z)^2\frac{\partial^2 Q}{\partial x\partial z}-2ys=+\tfrac{1}{2}\frac{f'(c_1)+f'(c_2)}{(c_1-c_2)^2}(x-z)^3+f'(z)(x-z)+2f(z)$$
$$+(x-z)^2\left\{\frac{[f'(c_1)+f'(z)](z-c_1)-2f(z)}{(z-c_1)^2}\frac{x-c_2}{c_1-c_2}-\frac{[f'(c_2)+f'(z)][z-c_2]-2f(z)}{(z-c_2)^2}\frac{x-c_1}{c_1-c_2}\right\},$$

where A_1^2, A_2^2 have been replaced by $f'(c_1)$, $f'(c_2)$ respectively.

Prove that this form leads to

$$Q_{z,\,c}^{x,\,a}=\int_c^z\int_a^x\frac{sy+f(x,z)}{2(x-z)^2}\frac{dx}{y}\frac{dz}{s}+\int_c^z\int_a^x\frac{dx}{2y}\frac{dz}{2s}[Lxz+M(x+z)+N],$$

where, if $f(x)$ be $a_0x^6+6a_1x^5+15a_2x^4+20a_3x^3+15a_4x^2+6a_5x+a_6$, $f(x, z)$ denotes the expression

$$x^3(a_0z^3+3a_1z^2+3a_2z+a_3)+3x^2(a_1z^3+3a_2z^2+3a_3z+a_4)$$
$$+3x(a_2z^3+3a_3z^2+3a_4z+a_5)+(a_3z^3+3a_4z^2+3a_5z+a_6),$$

and L, M, N are certain constants depending upon c_1 and c_2.

Ex. vi. Let $R(x)$ be any rational function. By expressing the fact that the value of the integral $\int R(x)\,dx$ taken round the complete boundary of the Riemann surface, is equal to the sum of its value taken round all the places of the surface at which the integral is infinite, we shall (cf. also p. 232) obtain the theorem

$$\Sigma \left[R(x) \frac{dx}{dt} \right]_{t^{-1}} = 0,$$

where the summation extends to all places at which the expansion of $R(x)\frac{dx}{dt}$, in terms of the infinitesimal, contains negative powers of t, and $\left[R(x) \frac{dx}{dt} \right]_{t^{-1}}$ means the coefficient of t^{-1} in the expansion. If all the poles of $R(x)$ occur for finite values of x, this summation will contain terms arising from the fact that $\frac{dx}{dt}$ contains negative powers of t when x is infinite, as well as terms arising at the finite poles of $R(x)$. If however $R(x)$ be of the form $U(x)\frac{d}{dx}V(x)$, wherein $U(x)$, $V(x)$ are rational functions of x, whose poles are at finite places of the surface, there will be no terms arising from the infinite places of the surface.

Now let ξ denote the current variable, and x, z denote fixed finite places: prove, by applying the theorem to the case* when

$$R(\xi) = \psi(\xi, a;\ z, c_1, \dots, c_p)\frac{d}{d\xi}\psi(\xi, a;\ x, c_1, \dots, c_p),$$

that

$$D_x\psi(x, z) - D_z\psi(z, x) = \sum_{i=1}^{p} \{\omega_i(x)\,[\psi(x, z)]^x_{t_{c_i}} - \omega_i(z)\,[\psi(z, x)]^z_{t_{c_i}}\},$$

where $\psi(x, z)$ is written for shortness for $\psi(x, a;\ z, c_1, \dots, c_p)$, and $[\psi(x, z)]^x_{t_{c_i}}$ denotes the coefficient of t_{c_i} in the expansion of $\psi(x, z)$, regarded as a function of x, in the neighbourhood of the place c_i.

Shew, when all the involved places are ordinary places, that this equation is the same as equation (xi) obtained in the text.

Prove also that

$$D_x D_z Q^{x, a}_{z, c} - \tfrac{1}{2}\sum_{i=1}^{p}\sum_{r=1}^{p}\omega_i(x)\,\omega_r(z)\,(M_{r, i} + M_{i, r}) = D_z\psi(z, x) + \sum_{i=1}^{p}\omega_i(x)\,[\psi(x, z)]^x_{t_{c_i}}.$$

Hence, as the forms $\omega_i(x)$ are also obtainable by expansion of the function $\psi(z, x)$, *every term on the right hand is immediately calculable when the form of the function* $\psi(x, z)$ *is known;* then by integrating the right hand in regard to x and z we obtain an integral of the third kind for which the property of the interchange of argument and parameter holds. (Cf. Ex. iii. p. 180.)

Ex. vii. By comparison of the two forms given for the function $\psi(x, a;\ z, c)$ (§§ 126, 131), we can obtain results analogous to those obtained in §§ 133—136 for the function $\psi(x, a;\ z, c_1, \dots, c_p)$.

Putting, as before, $H^{x, a}_z = \Gamma^{x, a}_z - [(z, x) - (z, a)]\frac{dz}{dt}$, and, when z is a branch place, understanding by $D^{k-1}_z H^{x, a}_z$ the expression $D^k_z(\Pi^{x, a}_{z, c} - P^{z, c}_{x, a})$, and, further, putting

$$B_{i, r} = (D^{k_r}_z D^{k_i - 1}_c H^{z, m}_c)_{z=c},\quad N_{i, r} = \left[D^{k_r}_z D^{k-1}_c\left((c, z)\frac{dc}{dt} + \frac{1}{t_c}\right)\right]_{z=c},$$

* Günther, *Crelle*, cix. p. 206.

wherein m is an arbitrary place and t_c the infinitesimal at the place c, so that

$$K_{i,r}=N_{i,r}-N_{r,i}=B_{r,i}-B_{i,r}=\left\{D_x^{k_i-1}D_z^{k_r-1}\left[D_z\left((x,z)\frac{dx}{dt}\right)-D_x\left((z,x)\frac{dz}{dt}\right)\right]\right\}_{\substack{x=c\\z=c}},$$

prove, in order, the following equations, which are numbered as the corresponding equations in §§ 133—136;

$$H_z^{x,a}=\sum_{i=1}^{p}\omega_i(z)\,D_c^{k_i-1}\,H_c^{x,a} \qquad \text{(ii)},$$

$$P_{z,m}^{x,a}=\Pi_{z,m}^{x,a}-\sum_{i=1}^{p}V_i^{x,a}\,D_c^{k_i-1}\,H_c^{z,m} \qquad \text{(iv)},$$

$$E_z^{x,a}=[(z,x)-(z,a)]\frac{dz}{dt}+\sum_{i=1}^{p}\left[\omega_i(z)\,D_c^{k_i-1}\,H_c^{x,a}-V_i^{x,a}\,D_z D_c^{k_i-1}\,H_c^{z,m}\right] \qquad \text{(vi)},$$

$$D_c^{k_r-1}\,H_c^{x,a}=D_c^{k_r-1}\left\{E_c^{x,a}-[(c,x)-(c,a)]\frac{dc}{dt}\right\}+\sum_{i=1}^{p}B_{i,r}\,V_i^{x,a} \qquad \text{(vii)},$$

wherein, when c is a branch place, the first term of the right hand is to be interpreted as

$$D_c^{k_r}\left(P_{c,m}^{x,a}-P_{x,a}^{c,m}\right);$$

also the equations

$$D_z D_c^{k_i-1}\,H_c^{z,m}=D_c^{k_i-1}\left[D_c\left((z,c)\frac{dz}{dt}\right)-D_z\left((c,z)\frac{dc}{dt}\right)\right]+\sum_{r=1}^{p}B_{r,i}\,\omega_r(z),$$

$$\Gamma_z^{x,a}=E_z^{x,a}+\sum_{i=1}^{p}V_i^{x,a}\,D_c^{k_i-1}\left[D_c\left((z,c)\frac{dz}{dt}\right)-D_z\left((c,z)\frac{dc}{dt}\right)\right]$$
$$+\sum_{i=1}^{p}\sum_{r=1}^{p}A_{r,i}\,V_i^{x,a}\,\omega_r(z),$$

and thence, that the algebraically determinable integral

$$G_z^{x,a}=E_z^{x,a}+\sum_{i=1}^{p}V_i^{x,a}\,D_c^{k_i-1}\left[D_c\left((z,c)\frac{dz}{dt}\right)-D_z\left((c,z)\frac{dc}{dt}\right)\right]$$
$$-\tfrac{1}{2}\sum_{i=1}^{p}\sum_{r=1}^{p}(N_{r,i}-N_{i,r})\,V_i^{x,a}\,\omega_r(z),$$

is equal to

$$\Gamma_z^{x,a}-\tfrac{1}{2}\sum_{i=1}^{p}\sum_{r=1}^{p}V_i^{x,a}\,\omega_r(z)\,(B_{r,i}+B_{i,r}) \qquad \text{(viii)};$$

and, finally, that the integral

$$Q_{z,m}^{x,a}=\Pi_{z,m}^{x,a}-\tfrac{1}{2}\sum_{i=1}^{p}\sum_{r=1}^{p}V_i^{x,a}\,V_r^{z,m}\,(B_{r,i}+B_{i,r}) \qquad \text{(ix)},$$

which, clearly, is such that $Q_{z,m}^{x,a}=Q_{x,a}^{z,m}$, can be algebraically defined by the equation

$$Q_{z,m}^{x,a}=P_{z,m}^{x,a}+\sum_{i=1}^{p}V_i^{x,a}\,D_c^{k_i-1}\left\{E_c^{z,m}-[(c,z)-(c,m)]\frac{dc}{dt}\right\}$$
$$-\tfrac{1}{2}\sum_i\sum_r\left(V_i^{x,a}\,V_r^{z,m}-V_r^{x,a}\,V_i^{z,m}\right)K_{r,i} \qquad \text{(ix)}'.$$

Further shew that the function $\psi(x,a;\,z,c)$ can be written in the form

$$\psi(x,a;\,z,c)=G_z^{x,a}-\sum_{s=1}^{p}\omega_s(z)\,D_c^{k_s-1}\,G_c^{x,a} \qquad \text{(xii)}.$$

The algebraical formula expressing the property of interchange of argument and parameter is to be obtained from the equation

$$D_x D_z Q_{z,m}^{x,a} = D_z\left((x,z)\frac{dx}{dt}\right) + \sum_{i=1}^{p} \omega_i(x) D_c^{k_i-1}\left\{D_c\left((z,c)\frac{dz}{dt}\right) - D_z\left((c,z)\frac{dc}{dt}\right)\right\}$$
$$+\tfrac{1}{2}\Sigma\Sigma\left[\omega_i(x)\omega_r(z) - \omega_r(x)\omega_i(z)\right]K_{i,r} \qquad \text{(x)}.$$

Lastly, if $L_k(z)$ denote the coefficient of $t^k/\underline{|k}$ (k positive) in the expansion of the function $\psi(x, a; z, c)$ as a function of x in the neighbourhood of the place c, so that (Ex. iv. § 132)

$$L_k(z) = D_c^k\left((z,c)\frac{dz}{dt}\right) - \sum_{i=1}^{p}\omega_i(z) P_{i,k},$$

where $P_{i,k}$ denotes a certain constant such that P_{i,k_r} is $N_{i,r}$, prove, by equating to zero the sum of the coefficients of the first negative powers of the infinitesimals in the expansions of the function of ξ, $\psi(\xi, a; z, c) D_\xi \psi(\xi, a; x, c)$, at all places where negative powers occur, that

$$D_x\psi(x, a; z, c) - D_z\psi(z, a; x, c) = \sum_{i=1}^{p}\left[\omega_i(x) L_{k_i}(z) - \omega_i(z) L_{k_i}(x)\right] \qquad \text{(A)},$$

wherein, on the right, only functions $L_k(z)$ occur for which k is one of the p numbers $k_1, k_2, \dots, k_p$, and that

$$D_x D_z Q_{z,m}^{x,a} - \tfrac{1}{2}\sum_{i=1}^{p}\sum_{r=1}^{p}\omega_i(x)\omega_r(z)(N_{r,i}+N_{i,r}) = D_z\psi(z, a; x, c) + \sum_{i=1}^{p}\omega_i(x) L_{k_i}(z) \qquad \text{(B)};$$

thus an elementary integral of the third kind, permitting interchange of argument and parameter, is obtained immediately from the function $\psi(x, a; z, c)$ by integrating the right hand of equation (B) in regard to x and z.

Prove also, that if

$$K_z^{x,a} = G_z^{x,a} - [(z,x) - (z,a)]\frac{dz}{dt},$$

we have the formulae

$$K_z^{x,a} = \sum_{i=1}^{p}\omega_i(z) D^{k_i-1} K_c^{x,a} \qquad \text{(xv)}$$

$$Q_{z,m}^{x,a} = P_{z,m}^{x,a} + \sum_{i=1}^{p} V_i^{x,a} D_c^{k_i-1} K_c^{z,m} \qquad \text{(xvi)}$$

$$D_x\left((z,x)\frac{dz}{dt}\right) - D_z\left((x,z)\frac{dx}{dt}\right) = \sum_{i=1}^{p}\left[\omega_i(x) D_z D_c^{k_i-1} K_c^{z,m} - \omega_i(z) D_x D_c^{k_i-1} K_c^{x,a}\right] \qquad \text{(xvii)}.$$

Ex. viii. To calculate the integral $Q_{z,m}^{x,a}$ for the case ($p=2$) where the fundamental equation is

$$y^2 = f(x),$$

wherein $f(x)$ is a sextic polynomial divisible by $x-c$, which is expansible in the form

$$f(x) = A^2(x-c) + Q(x-c)^2 + R(x-c)^3 + \dots,$$

we may use the equation (xi) of Ex. vii. When x, z are near the place c, putting

$$x = c + t_1^2,\quad z = c + t_2^2,\quad y = At_1 + \frac{Q}{2A}t_1^3 + \dots,\quad s = At_2 + \frac{Q}{2A}t_2^3 + \dots,$$

prove that

$$D_z\left((x,z)\frac{dx}{dt}\right) - D_x\left((z,x)\frac{dz}{dt}\right) = \frac{R}{A^2}(t_1^2 - t_2^2) + \text{cubes and higher powers of } t_1 \text{ and } t_2,$$

and thence (see Ex. ii. § 132) that

$$K_{12}[\omega_1(x)\,\omega_2(z)-\omega_2(x)\,\omega_1(z)]=\frac{R(x-z)}{4yz}\frac{dx}{dt}\frac{dz}{dt}.$$

Also, when z is not a branch place, if c_1 be a place near to c, and the expansion of the function $\left[\frac{\partial}{\partial c_1}(z,c_1)-\frac{\partial}{\partial z}(c_1,z)\right]\frac{dc_1}{dt}$ in powers of the infinitesimal at c, contain the terms $M+\,.\,+Nt^2+\dots$, so that

$$M=\left[\frac{\partial}{\partial c}(z,c)-\frac{\partial}{\partial z}(c,z)\right]\frac{dc}{dt},\qquad 2N=D_c^2\left\{\left[\frac{\partial}{\partial c}(z,c)-\frac{\partial}{\partial z}(c,z)\right]\frac{dc}{dt}\right\},$$

prove that

$$M=\frac{[A^2+f'(z)][z-c]-2f(z)}{2As(z-c)^2},$$

$$N=\frac{3A^2(z-c)[A^2+\tfrac{1}{2}Q(z-c)]+f'(z)\,.\,(z-c)[A^2-\tfrac{1}{2}Q(z-c)]-2f(z)[2A^2-\tfrac{1}{2}Q(z-c)]}{2A^3s(z-c)^3};$$

substituting these results in the formula (xi) of Ex. vii., prove that

$$\frac{\partial^2 Q}{\partial x\partial z}=\frac{ys+f(x,z)}{2ys(x-z)^2}-\frac{1}{240}\left\{(x-c)(z-c)\frac{\partial^4 f}{\partial c^4}+6(x+z-2c)\frac{\partial^3 f}{\partial c^3}+12\frac{\partial^2 f}{\partial c^2}\right\}\Big/ys,$$

where $f(x,z)$ has the same signification as in Example v. The part within the brackets $\{\ \}$ is of the form $ys\,\Sigma\Sigma a_{i,r}\,\omega_i(x)\,\omega_r(z)$, where $a_{i,r}=a_{r,i}$.

Obtain the same result by the formula (B) of Ex. vii., using the form of $\psi(x,a;z,c)$ found in Ex. ii. § 132.

138. The formulæ in §§ 133—136 enable us to express the form of a canonical integral of the third kind, in the most general case; and to calculate the integral for any fundamental algebraic equation, when the integral functions are known. But they have the disadvantage of presenting the result in a form in which there enter p arbitrary places $c_1,\dots,c_p$. We proceed now to shew how to formulate the theory in a more general way; though the results obtained are not so explicit as those previously given, they are in some cases more suitable for purposes of calculation.

Let $u_1^{x,a},\dots,u_p^{x,a}$ denote any p linearly independent integrals of the first kind; denote $D_x u_i^{x,a}$ by $\mu_i(x)$. Let the matrix whose (i,j)th element is $\mu_j(c_i)$ be denoted by μ, $c_1,\dots,c_p$ being the places used (§ 121) to define the quantities $\omega_1(x),\dots,\omega_p(x)$. Let $\nu_{i,j}$ denote the minor of the (i,j)th element in the determinant of the matrix μ, divided by the determinant of μ; so that the matrix inverse * to μ is that whose (i,j)th element is $\nu_{j,i}$. Then we clearly have

$$\omega_i(x)=\nu_{i,1}\mu_1(x)+\dots\dots+\nu_{i,p}\mu_p(x)\qquad (i=1,2,\dots,p).$$

* Since $u_1^{x,a},\dots,u_p^{x,a}$ are linearly independent, and the places $c_1,\dots,c_p$ are independent (see §§ 23, 121), the matrix μ^{-1} can always be formed.

Let a denote any symmetrical matrix of p^2 quantities, $a_{i,j}$, in which $a_{i,j} = a_{j,i}$. Then we define p quantities by the p equations

$$L_i^{x,a} = \nu_{1,i} H_{c_1}^{x,a} + \nu_{2,i} H_{c_2}^{x,a} + \dots + \nu_{p,i} H_{c_p}^{x,a} - 2\,(a_{i,1} u_1^{x,a} + \dots + a_{i,p} u_p^{x,a}),$$

and call them fundamental integrals of the second kind *associated* with the integrals $u_1^{x,a}, \dots, u_p^{x,a}$. For instance when $\mu_i(x) = \omega_i(x)$, $\nu_{i,j} = 0$ unless $i = j$, in which case $\nu_{i,i} = 1$. Thus by taking $a_{i,j} = \frac{1}{4}(A_{i,j} + A_{j,i})$, the integrals $K_{c_1}^{x,a}, \dots, K_{c_p}^{x,a}$ (p. 187. xiv.) are a fundamental system associated with the set $V_1^{x,a}, \dots, V_p^{x,a}$.

It will be convenient in what follows to employ the notation of matrices to express the determinant relations of which we avail ourselves *. We shall therefore write the definition given above in the form

$$L^{x,a} = \bar{\nu} H^{x,a} - 2a u^{x,a},$$

wherein $L^{x,a}$ stands for the row of p quantities $L_1^{x,a}, \dots, L_p^{x,a}$, $H^{x,a}$ stands for the row of p quantities $H_{c_1}^{x,a}, \dots, H_{c_p}^{x,a}$, and $\bar{\nu}$ denotes the matrix obtained by changing the rows of ν into its columns, and is in fact equal to the matrix denoted by μ^{-1}, so that we may also write

$$L^{x,a} = \mu^{-1} H^{x,a} - 2a u^{x,a}, \; = \mu^{-1} K^{x,a} - 2a' u^{x,a},$$

where (§ 137)

$$H_{c_i}^{x,a} = K_{c_i}^{x,a} + \tfrac{1}{2} \sum_{r=1}^{p} (A_{r,i} + A_{i,r}) V_r^{x,a}.$$

Explicit forms of the integrals $K_{c_i}^{x,a}$ have been given (§§ 134, 136).

Then, from the equations defining the integrals $L_i^{x,a}$, we have

$$\sum_{i=1}^{p} \mu_i(z) L_i^{x,a} = \sum_{j=1}^{p} H_{c_j}^{x,a} \sum_{i=1}^{p} \nu_{j,i} \mu_i(z) - 2 \sum_{r=1}^{p} \sum_{s=1}^{p} a_{r,s} u_r^{x,a} \mu_s(z),$$

$$= \sum_{j=1}^{p} \omega_j(z) H_{c_j}^{x,a} - 2 \sum_{r=1}^{p} \sum_{s=1}^{p} a_{r,s} u_r^{x,a} \mu_s(z),$$

$$= H_z^{x,a} - 2 \sum_{r=1}^{p} \sum_{s=1}^{p} a_{r,s} u_r^{x,a} \mu_s(z);$$

and this is an important result. For, putting for z in turn any p independent places, the p functions $L_i^{x,a}$ are determined by this equation. Thus the functions $L_1^{x,a}, \dots, L_p^{x,a}$ *do not depend upon the places* $c_1, c_2, \dots, c_p$.

* See for instance Cayley, *Collected Works*, vol. ii. p. 475, and the Appendix II. to the present volume, where other references are given.

Also, from this equation we infer

$$D_x\left[(z, x)\frac{dz}{dt}\right] - D_z\left[(x, z)\frac{dx}{dt}\right] = D_z H_x^{z, c} - D_x H_z^{x, a}$$

$$= \sum_{i=1}^{p} [\mu_i(x) D_z L_i^{z, c} - \mu_i(z) D_x L_i^{x, a}] \quad (17),$$

c being any arbitrary place. Now it is immediately seen that if $R_1(x), \dots, R_p(x)$ be any rational functions of x such that

$$\sum_{i=1}^{p} [\mu_i(x) D_z L_i^{z, c} - \mu_i(z) D_x L_i^{x, a}] = \sum_{i=1}^{p} [\mu_i(x) R_i(z) - \mu_i(z) R_i(x)],$$

then $R_i(x)$ can only be a form of $D_x L_i^{x, a}$, obtained from $D_x L_i^{x, a}$ by altering the values of the constant elements of the symmetrical matrix a. Hence the equation (17) furnishes a method of calculating the integrals $L_i^{x, a}$, whenever it is possible to put the left-hand side into the form of the right-hand side.

The equation (17) shews that the expression

$$D_z\left((x, z)\frac{dx}{dt}\right) + \sum_{i=1}^{p} \mu_i(x) D_z L_i^{z, c},$$

is unaltered by the interchange of x and z. This expression is also equal to

$$D_z\left((x, z)\frac{dx}{dt}\right) + D_z H_x^{z, c} - 2\sum_{r=1}^{p}\sum_{s=1}^{p} a_{r, s}\mu_r(x)\mu_s(z)$$

and, therefore, to

$$D_z \Gamma_x^{z, c} - 2\sum_{r=1}^{p}\sum_{s=1}^{p} a_{r, s}\mu_r(x)\mu_s(z).$$

Hence, the formula (§ 134, ix.)

$$R_{z, c}^{x, a} = P_{z, c}^{x, a} + \sum_{i=1}^{p} u_i^{x, a} L_i^{z, c} = \Pi_{z, c}^{x, a} - 2\sum_{r=1}^{p}\sum_{s=1}^{p} a_{r, s} u_r^{x, a} u_s^{z, c}$$

$$= Q_{z, c}^{x, a} + \tfrac{1}{2}\sum_{r=1}^{p}\sum_{s=1}^{p}(A_{r, s} + A_{s, r}) V_r^{x, a} V_s^{z, c} - 2\sum_{r=1}^{p}\sum_{s=1}^{p} a_{r, s} u_r^{x, a} u_s^{z, c}$$

gives us a form of canonical integral of the third kind not depending upon the places $c_1, \dots, c_p$, and immediately calculable when the forms of the functions $L_i^{x, a}$ are found.

The formula

$$\Gamma_z^{x, a} = [(z, x) - (z, a)]\frac{dz}{dt} + \sum_{\mu=1}^{p} \mu_i(z) L_i^{x, a} + 2\sum_{r=1}^{p}\sum_{s=1}^{p} a_{r, s} u_r^{x, a}\mu_s(z)$$

serves to express any integral of the second kind in terms of the integrals $L_1, \dots, L_p$

Ex. i. For the surface $y^2=f(x)$, where $f(x)$ is a rational polynomial of order $2p+2$, the function

$$R(\xi)=\frac{\eta}{s(\xi-z)}\cdot\frac{d}{d\xi}\left(\frac{\eta}{y(\xi-x)}\right), =\frac{1}{2ys}\left\{\frac{f'(\xi)}{(\xi-x)(\xi-z)}-\frac{2f(\xi)}{(\xi-x)^2(\xi-z)}\right\},$$

wherein $s^2=f(z)$, $\eta^2=f(\xi)$, is a rational function of ξ (without η). Prove by applying the theorem, $\Sigma\left[R(\xi)\frac{d\xi}{dt}\right]_{t^{-1}}=0$, (Ex. vi, § 137) that

$$\frac{\partial}{\partial x}(z, x)-\frac{\partial}{\partial z}(x, z)=\sum_k\sum_{k'}(k'-k)\lambda_{k+k'+2}\left(\frac{x^k}{2y}\frac{z^{k'}}{2s}-\frac{x^{k'}}{2y}\frac{z^k}{2s}\right),$$

where k, k' represent in turn every pair of unequal numbers from 0, 1, 2, ..., $2p$, whose sum is not greater than $2p$, k' being greater than k, and the coefficients λ are given by the fact that

$$y^2=f(x)=\lambda+\lambda_1x+\lambda_2x^2+\ldots+\lambda_{2p+1}x^{2p+1}+\lambda_{2p+2}x^{2p+2}.$$

Hence, a set of integrals of the second kind associated with the integrals of the first kind

$$\int_a^x\frac{dx}{y},\quad \int_a^x\frac{xdx}{y},\quad \ldots\ldots,\quad \int_a^x\frac{x^{p-1}dx}{y},$$

is given by

$$L_i^{x,a}=\int_a^x\frac{dx}{4y}\sum_{k=i}^{k=2p+1-i}\lambda_{k+1+i}(k+1-i)x^k,\qquad (i=1, 2, \ldots, p);$$

and a canonical integral of the third kind is given by

$$\int_a^x\int_c^z\frac{dz}{2s}\frac{dx}{2y}\left[\frac{2ys+2f(z)+f'(z)(x-z)}{(x-z)^2}+\sum_{i=1}^p x^{i-1}\sum_{k=1}^{2p+1-i}(k+1-i)\lambda_{k+1+i}z^k\right].$$

This is equal to

$$\int_a^x\int_c^z\frac{dz}{2s}\frac{dx}{2y}\frac{2ys+\sum_{i=0}^{p+1}x^iz^i\left[2\lambda_{2i}+\lambda_{2i+1}(x+z)\right]}{(x-z)^2},$$

which is clearly symmetric in x and z.

The value of $\frac{\partial}{\partial x}(z, x)-\frac{\partial}{\partial z}(x, z)$ used in this example is given by Abel, *Œuvres Complètes* (Christiania, 1881), Vol. i. p. 49.

Ex. ii. Shew in Ex. i., for $p=1$, that the integral associated with $\int_a^x\frac{dx}{y}$ is $\int_a^x\frac{\lambda_3x+2\lambda_4x^2}{4y}dx$; and express these in the notation of Weierstrass's elliptic functions when the fundamental equation is $y^2=4x^3-g_2x-g_3$.

139. Suppose now that the integrals $u_1^{x,a}, \ldots, u_p^{x,a}$ are connected with the normal integrals $v_1^{x,a}, \ldots, v_p^{x,a}$ by means of the equations

$$\pi i\mu_r(x)=\lambda_{r,1}\Omega_1(x)+\ldots\ldots+\lambda_{r,p}\Omega_p(x),$$

which, since $\Omega_i(x)=2\pi iDv_i^{x,a}$, are equivalent to

$$u_r^{x,a}=2\left(\lambda_{r,1}v_1^{x,a}+\ldots\ldots+\lambda_{r,p}v_p^{x,a}\right).$$

Then the periods of the integral $u_r^{x,a}$, at the first p period loops, form the rth row of a matrix, 2λ, and the periods of the integral $u_r^{x,a}$ at the second

p period loops form the rth row of a matrix $2\lambda\tau$; we shall write $\omega = \lambda$ and $\omega' = \lambda\tau$, so that $\omega_{i,j} = \lambda_{i,j}$. The two suffixes of the quantities $\omega_{i,j}$ will prevent confusion between them and the differential coefficients $\omega_i(x)$.

Let the periods of $L_i^{x,a}$ at the jth period loops of the first and second kind be denoted by $-2\eta_{i,j}$ and $-2\eta'_{i,j}$ respectively. The matrix whose ith row consists of the quantities $\eta_{i,1}, \dots, \eta_{i,p}$ will be denoted by η; similarly the matrix of the quantities $\eta'_{i,j}$ will be denoted by η'. The matrix of the periods of the integrals $H_{c_1}^{x,a}, \dots, H_{c_p}^{x,a}$ at the first period loops is zero; the (i, j)th element of the matrix at the second period loops is the jth period of $H_{c_i}^{x,a}$, namely $\Omega_j(c_i)$. We shall denote this matrix by Δ.

By the definitions of the integrals $L_i^{x,a}$ we therefore have

$$2\eta_{i,j} = 4(a_{i,1}\omega_{1,j} + \dots + a_{i,p}\omega_{p,j}), \qquad (i, j = 1, 2, \dots, p)$$
$$2\eta'_{i,j} = 4(a_{i,1}\omega_1{}',_j + \dots + a_{i,p}\omega'_{p,j}) - \big(\nu_{1,i}\Omega_j(c_1) + \dots + \nu_{p,i}\Omega_j(c_p)\big),$$

and all these equations are contained in the equations

$$\eta = 2a\omega,$$
$$\eta' = 2a\omega' - \tfrac{1}{2}\bar{\nu}\Delta = 2a\omega' - \tfrac{1}{2}\mu^{-1}\Delta.$$

Now from the equations connecting $\mu_r(x)$ and $\Omega_s(x)$, we obtain

$$\pi i\mu_r(c_i) = \lambda_{r,1}\Omega_1(c_i) + \dots\dots + \lambda_{r,p}\Omega_p(c_i),$$

wherein $\mu_r(c_i)$ is the (i, r)th element of the matrix μ, and the right hand is the (i, r)th element of the matrix $\Delta\bar{\lambda}$; hence we may put

$$\pi i\mu = \Delta\bar{\lambda}.$$

If then we denote the matrix $\tfrac{1}{2}\mu^{-1}\Delta$ by $\bar{h}$, we have

$$2\Delta\overline{\lambda h} = 2\pi i\mu\bar{h} = \pi i\Delta = \Delta\pi i,$$

and infer that $2\overline{\lambda h} = \pi i$, and thence that $2h\lambda = \pi i$. Thus $2h\omega = \pi i$, $2h\omega' = \pi i\tau$. Also the integrals $u_1^{x,a}, \dots, u_p^{x,a}, \dots, v_1^{x,a}, \dots, v_p^{x,a}$ are connected by the equation $hu^{x,a} = 2h\lambda v^{x,a} = \pi iv^{x,a}$.

140. The four equations

$$2h\omega = \pi i, \quad 2h\omega' = \pi i\tau, \quad \eta = 2a\omega, \quad \eta' = 2a\omega' - \bar{h} \qquad \text{(A)}$$

will prove to be of fundamental importance in the theory of the theta functions. They express the periods η, η' independently of the places $c_1, \dots, c_p$, used in defining $L_i^{x,a}$.

If beside the symmetrical matrix τ, and the arbitrary symmetrical matrix a, we suppose the matrix h, which is in general unsymmetrical, to be arbitrarily given, the integrals $u_1^{x,a}, \dots, u_p^{x,a}$ being then determined by the equation $hu^{x,a} = \pi v i^{x,a}$, the first equation, $2h\omega = \pi i$, gives rise to p^2 equations whereby the p^2 quantities $\omega_{i,j}$ are to be found, and similarly the other equations give rise each to p^2 equations determining respectively the quantities $\omega'_{i,j}$, $\eta_{i,j}$, $\eta'_{i,j}$. But, thereby, the $4p^2$ quantities thus involved are determined in terms of less than $4p^2$ given quantities. For the symmetrical matrices a, τ involve each only $\frac{1}{2}p(p+1)$ quantities, and the number of given quantities is thus only $p(p+1)+p^2$. There are therefore, presumably,

$$4p^2 - [p^2 + p(p+1)], \; = 2p^2 - p,$$

relations connecting the $4p^2$ quantities $\omega_{i,j}$, $\omega'_{i,j}$, $\eta_{i,j}$, $\eta'_{i,j}$; we can in fact express these relations in various forms.

One of these forms is

$$\bar{\omega}\eta = \bar{\eta}\omega, \quad \bar{\omega}'\eta' = \bar{\eta}'\omega', \quad \bar{\eta}\omega' - \bar{\omega}\eta' = \tfrac{1}{2}\pi i = \bar{\omega}'\eta - \bar{\eta}'\omega, \qquad \text{(B)}$$

of which, for instance, the first equation is equivalent to the $\frac{1}{2}p(p-1)$ equations

$$\sum_{r=1}^{p} (\omega_{r,i}\,\eta_{r,j} - \eta_{r,i}\,\omega_{r,j}) = 0,$$

in which $i = 1, 2, \dots, p$, $j = 1, 2, \dots, p$, and i is not equal to j. The second equation is similarly equivalent to $\frac{1}{2}p(p-1)$ equations, and the third to p^2 equations. The total number of relations thus obtained is therefore the right number $p^2 + p(p-1)$, In this form the equations are known as Weierstrass's equations.

Another form in which the $2p^2 - p$ relations can be expressed is

$$\omega\bar{\omega}' = \omega'\bar{\omega}, \quad \eta\bar{\eta}' = \eta'\bar{\eta}, \quad \omega'\bar{\eta} - \omega\bar{\eta}' = \tfrac{1}{2}\pi i = \eta\bar{\omega}' - \eta'\bar{\omega} \qquad \text{(C)}$$

These equations are distinguished from the equations (B) as Riemann's equations.

141. The equations (B) and (C) are entirely equivalent; either set can be deduced from the equations (A) or from the other set. A natural way of obtaining the set (B) is to use the equation (17). A natural way of obtaining the set (C) is to make use of the Riemann method of contour integration.

The equations (A) give, recalling that $\bar{a} = a$, $\omega' = \omega\tau$, $\bar{\tau} = \tau$,

$$\bar{\omega}\eta = 2\bar{\omega}a\omega \;, = \beta, \text{ say, a symmetrical matrix,}$$

$$\bar{\omega}\eta' = 2\bar{\omega}a\omega' - \bar{\omega}\bar{h} = 2\bar{\omega}a\omega\tau - \overline{h\omega} = \beta\tau - \tfrac{1}{2}\pi i.$$

Hence $\qquad \bar{\eta}\omega' = \bar{\eta}\omega\tau = \bar{\beta}\tau = \beta\tau,$

and because $\bar{\omega}' = \tau\bar{\omega}$,

$$\bar{\omega}'\eta' = \tau\bar{\omega}\eta' = \tau\beta\tau - \tfrac{1}{2}\pi i\tau,$$

and thus, as $\overline{\tau\beta\tau} = \tau\beta\tau$, we have

$$\bar{\omega}\eta = \bar{\eta}\omega, \quad \bar{\omega}'\eta' = \bar{\eta}'\omega', \quad \bar{\eta}\omega' - \bar{\omega}\eta' = \tfrac{1}{2}\pi i = \bar{\omega}'\eta - \bar{\eta}'\omega,$$

which are the equations (B). And it should be noticed that these results are all derived from the three $\omega'=\omega\tau$, $\bar{\omega}\eta=\beta$, $\bar{\omega}\eta'=\beta\tau-\frac{1}{2}\pi i$, assuming only that β and τ are symmetrical.

From the equations (B), putting $\bar{\omega}\eta=\beta$, $\bar{\omega}'\eta'=\gamma$, so that β and γ are symmetrical matrices, we obtain*

$$\eta=(\bar{\omega})^{-1}\beta,\quad \eta'=\gamma(\omega')^{-1},\ \text{and thence}\ \bar{\omega}'(\bar{\omega})^{-1}\beta-\gamma(\omega')^{-1}\omega=\tfrac{1}{2}\pi i.$$

Hence, if $\omega^{-1}\omega'=\kappa$, so that $\omega\kappa=\omega'$, $\bar{\omega}'=\bar{\kappa}\bar{\omega}$, $\bar{\omega}'(\bar{\omega})^{-1}=\bar{\kappa}$, and $\kappa^{-1}=(\omega')^{-1}\omega$, we have

$$\bar{\kappa}\beta-\gamma\kappa^{-1}=\tfrac{1}{2}\pi i,\ \text{or}\ \bar{\kappa}\beta\kappa-\gamma=\tfrac{1}{2}\pi i\kappa,$$

and therefore, as the matrices $\bar{\kappa}\beta\kappa$ and γ are symmetrical, so also is the matrix κ; and thus

$$\omega^{-1}\omega'=\bar{\omega}'(\bar{\omega})^{-1},\ \text{and therefore}\ \omega\bar{\omega}'=\omega'\bar{\omega},$$

which is one of the equations (C).

Further

$$\bar{\omega}\eta'=\eta\omega'-\tfrac{1}{2}\pi i=\eta\omega\kappa-\tfrac{1}{2}\pi i=\beta\kappa-\tfrac{1}{2}\pi i,$$

and therefore

$$\bar{\eta}'\omega=\bar{\kappa}\beta-\tfrac{1}{2}\pi i=\kappa\beta-\tfrac{1}{2}\pi i,$$

leading to

$$\bar{\omega}\eta\bar{\eta}'\omega=\beta\kappa\beta-\tfrac{1}{2}\pi i\beta,$$

and the right hand is a symmetrical matrix, and therefore equal to $\bar{\omega}\eta'\bar{\eta}\omega$; thus also

$$\eta\bar{\eta}'=\eta'\bar{\eta},$$

which is the second of the equations (C).

Finally $(\omega'\bar{\eta}-\omega\bar{\eta}')\omega=\omega'\bar{\eta}\omega-\omega(\bar{\omega}'\eta-\tfrac{1}{2}\pi i)-\omega'\bar{\omega}\eta-\omega\bar{\omega}'\eta+\tfrac{1}{2}\pi i\omega=(\omega'\bar{\omega}-\omega\bar{\omega}')\eta+\tfrac{1}{2}\pi i\omega$

$$=\tfrac{1}{2}\pi i\omega,$$

and thus

$$\omega'\bar{\eta}-\omega\bar{\eta}'=\tfrac{1}{2}\pi i,\ =\ ,\ \text{therefore},\ \eta\bar{\omega}'-\eta'\bar{\omega},$$

which is the third of equations (C).

We have deduced both the equations (B) and (C) from the equations (A). A similar method can be used to deduce the equations (B) from the equations (C).

Other methods of obtaining the equations (B) and (C) are explained in the Examples which follow (§ 142, Exx. ii—v).

142. *Ex.* i. Shew that the p integrals given by the equation

$$\Lambda_i^{x,a}=t_{1,i}H_{c_1}^{x,a}+\dots+t_{p,i}H_{c_p}^{x,a},$$

where $t_{i,j}$ is the minor of $\Omega_j(c_i)$ in the determinant of the matrix Δ (§ 139), divided by the determinant of Δ, namely by the equation

$$\Lambda^{x,a}=\Delta^{-1}H^{x,a},$$

are a set of fundamental integrals of the second kind associated with the set of integrals of the first kind $2\pi i v_1^{x,a},\dots,2\pi i v_p^{x,a}$, and are such that

$$D_x\left((z,x)\frac{dz}{dt}\right)-D_z\left((x,z)\frac{dx}{dt}\right)$$

$$=\sum_{i=1}^{p}\left(\omega_i(x)\,D_z H_{c_i}^{z,c}-\omega_i(z)\,D_x H_{c_i}^{x,a}\right)=\sum_{i=1}^{p}\left(\omega_i(x)\,D_z K_{c_i}^{z,c}-\omega_i(z)\,D_x K_{c_i}^{x,a}\right)$$

$$=\sum_{i=1}^{p}\left(\Omega_i(x)\,D_z\Lambda_i^{z,c}-\Omega_i(z)\,D_x\Lambda_i^{x,a}\right)=\sum_{i=1}^{p}\left(\mu_i(x)\,D_z L_i^{z,c}-\mu_i(z)\,D_x L_i^{x,a}\right).$$

* The determinant of the matrix ω, $=\lambda$, cannot vanish, because $u_1^{x,a},\dots,u_p^{x,a}$ are linearly independent. The determinant of the matrix τ does not vanish, since otherwise we could determine an integral of the first kind with no periods at the period loops of the second kind (cf. Forsyth, *Theory of Functions*, § 231, p. 440).

Prove that the function $\Lambda_i^{x, a}$ has only one period, namely at the ith period loop of the second kind, and that this period is equal to 1. For the sets

$$2\pi i v_1^{x, a}, \ldots, 2\pi i v_p^{x, a}, \Lambda_1^{x, a}, \ldots, \Lambda_p^{x, a},$$

we have in fact $\omega = \pi i,\ \omega' = \pi i \tau,\ \eta = 0,\ \eta' = -\tfrac{1}{2}$.

Shew that these values satisfy the equations (B) and (C).

Ex. ii. From Ex. i. we deduce

$$2\pi i \sum_{i=1}^{p} (v_i^{x, a} \Lambda_i^{z, c} - v_i^{z, c} \Lambda_i^{x, a}) = \sum_{i=1}^{p} (u_i^{x, a} L_i^{z, c} - u_i^{z, c} L_i^{x, a}).$$

Hence, supposing x and z separately to pass, on the dissected Riemann surface, respectively from one side to the other* of the rth period loop of the first kind, and from one side to the other of the sth period loop of the first kind, we obtain, for the increment of the right-hand side

$$-4 \sum_{i=1}^{p} (\omega_{i, r} \eta_{i, s} - \eta_{i, r} \omega_{i, s}),$$

which is the (r, s)th element of the matrix $-4(\bar{\omega}\eta - \bar{\eta}\omega)$. For the functions on the left-hand side the matrix $\bar{\omega}\eta - \bar{\eta}\omega$ vanishes (Ex. i.). Hence the same is true for those on the right hand.

Supposing x to pass from one side to the other of the rth period loops of the first kind, and z from one side to the other of the sth period loop of the second kind, we similarly prove that $\bar{\omega}\eta' - \bar{\eta}\omega'$ has the same value for the functions on the two sides of the equation, and therefore, as we see by considering the functions on the left hand, has the value $-\tfrac{1}{2}\pi i$.

While, if both x and z pass from one side to the other of period loops of the second kind we are able to infer $\bar{\omega}'\eta' = \bar{\eta}'\omega'$.

We thus obtain Weierstrass's equations (B).

Ex. iii. If $U_1^{x, a}, \ldots, U_p^{x, a}$ be any integrals, the periods of $U_i^{x, a}$ at the jth period loops of the first and second kind be respectively $\zeta_{i, j}$, $\zeta'_{i, j}$, and the matrices of these elements be respectively denoted by ζ, ζ'; and $W_1^{x, a}, \ldots, W_p^{x, a}$ be other integrals for which the corresponding matrices are ξ and ξ', prove that the integral $\int U_i^{x, a}\, dW_j^{x, a}$, taken positively round all the period-loop-pairs has the value

$$\sum_{r=1}^{p} (\zeta_{i, r} \xi'_{j, r} - \zeta'_{i, r} \xi_{j, r}),$$

which is the (i, j)th element of the matrix $\zeta\bar{\xi}' - \zeta'\bar{\xi}$.

Ex. iv. If $R_i(x)$ denote the rational function of x given by

$$R_i(x) = \sum_{r=1}^{p} \nu_{r, i} [(c_r, x) - (c_r, a)] \frac{dc_r}{dt},$$

the function $L_i^{x, a} + R_i(x)$ is infinite only at $c_1, \ldots, c_p$, and has the same periods $L_i^{x, a}$, Denote this function by $Y_i^{a, a}$.

* To that side for which the periods count positively (see the diagram, § 18).

Prove that if the expansion of the integral $Y_i^{x,a}$ in the neighbourhood of the place c_s be written in the form

$$Y_i^{x,a} = -\frac{\nu_{s,i}}{t} + f_{i,s} + g_{i,s}t + \dots,$$

then

$$g_{i,s} = \nu_{1,i}(A_{1,s} + M_{1,s}) + \dots + \nu_{p,i}(A_{p,s} + M_{p,s}),$$

where $A_{i,s}$, $M_{i,s}$ are as defined in § 134, and are such that $A_{i,s} + M_{i,s} = A_{s,i} + M_{s,i}$.

Hence shew that the sum of the values of the integral $\int Y_i^{x,a}\, dY_j^{x,a}$ taken round all the places $c_1, \dots, c_p$ is zero.

Ex. v. Infer from Exs. iii. and iv., by taking

(α) $U_i^{x,a} = u_i^{x,a} = W_i^{x,a}$, that $\omega\bar{\omega}' = \omega'\bar{\omega}$,

(β) $U_i^{x,a} = Y_i^{x,a}$, $W_i^{x,a} = u_i^{x,a}$, that $\eta\bar{\omega}' - \eta'\omega = \frac{1}{2}\pi i$,

(γ) $U_i^{x,a} = Y_i^{x,a} = W_i^{x,a}$ that $\eta\bar{\eta}' = \eta'\bar{\eta}$.

These are Riemann's equations.

Ex. vi. If instead of the places $c_1, \dots, c_p$ and the matrix μ, we use a matrix depending only on one place c, the ith row being formed with the elements $D_c^{k_i-1}\mu_1(c), \dots, D_c^{k_i-1}\mu_p(c)$, we can similarly obtain a set $L_1^{x,a}, \dots, L_p^{x,a}$ associated with the set $u_1^{x,a}, \dots, u_p^{x,a}$.

Shew that the periods of $L_1^{x,a}, \dots, L_p^{x,a}$ thus determined are independent of the position of the place c.

Ex. vii. If the differential coefficients $\mu_1(x), \dots, \mu_p(x)$, be those derived from a set of p independent places $b_1, b_2, \dots, b_p$, just as $\omega_1(x), \dots, \omega_p(x)$ are derived from $c_1, \dots, c_p$, so that $\mu_i(b_i) = 1$, $\mu_i(b_r) = 0$, prove that $\nu_{r,i} = \omega_r(b_i)$ and that

$$L_i^{x,a} = H_{b_i}^{x,a} - 2(a_{i1}u_1^{x,a} + \dots + a_{ip}u_p^{x,a}).$$

143. We conclude this chapter with some applications* of the functions $\psi(x, a; z, c)$, $E(x, z)$ to the expression of functions which are single-valued on the (undissected) Riemann surface. Such functions include, but are more general than, rational functions, in that they may possess essential singularities.

Consider first a single-valued function which is infinite only at one place; denote the place by m, and the function by $F(x)$.

Since $\psi(x, a; z, c) \Big/ \frac{dz}{dt}$ is a rational function of z, the integral

$$\int F(z)\left[\psi(x, a; z, c) \Big/ \frac{dz}{dt}\right] dz, \text{ or } \int F(z)\,\psi(x, a; z, c)\, dt_z,$$

taken round the edges of the period-pair-loops, has zero for its value. But this integral is also equal to the sum of its values taken round the place m,

* Appell, *Acta Math.* i. pp. 109, 132 (1882), Günther, *Crelle* cix. p. 199 (1892).

where $F(z)$ is infinite, and the places x and a at which $\psi(x, a; z, c)$ is infinite.

Now, when z is in the neighbourhood of the place m, since $\psi(x, a; z, c) \Big/ \frac{dz}{dt}$ is a rational function of z, we can put

$$\psi(x, a; z, c) = \sum_{r=0}^{\infty} \frac{t_m^r}{\underline{|r}} D_m^r \psi(x, a; m, c),$$

where t_m is the infinitesimal at the place m.

Thus the integral $\int F(z) \psi(x, a; z, c)\, dt_z$, taken round the place m, gives

$$2\pi i \sum_{r=0}^{\infty} \frac{A_r}{\underline{|r}} D_m^r \psi(x, a; m, c),$$

where A_r is the value of the integral $\frac{1}{2\pi i}\int t_m^r F(z)\, dt_z$ taken round the place m.

When z is in the neighbourhood of the place x, $\psi(x, a; z, c)$ is infinite like t_x^{-1}, t_x being the infinitesimal at the place x, and therefore, taken round the place x, the integral

$$\int F(x) \psi(x, a; z, c)\, dt_z$$

gives

$$2\pi i F(x).$$

Similarly round the place a, the integral gives $-2\pi i F(a)$.

Hence the function $F(x)$ can be expressed in the form

$$F(x) = F(a) - \sum_{r=0}^{\infty} \frac{A_r}{\underline{|r}} D_m^r \psi(x, a; m, c),$$

the places a and c being arbitrary (but not in the neighbourhood of the place m).

For example, when $p=0$, $\psi(x, a; z, c) = -\left(\frac{1}{x-z} - \frac{1}{a-z}\right)$, and

$$F(x) - F(a) = \sum_{r=0}^{\infty} A_r \left[\frac{1}{(x-m)^{r+1}} - \frac{1}{(a-m)^{r+1}}\right],$$

wherein

$A_r = \frac{1}{2\pi i}\int (z-m)^r F(z)\, dz$, the integral being taken round the place m.

A similar result can be obtained for the case of a single valued function with only a finite number of essential singularities. When one of these singularities is only a pole, say of order μ, the integral $\int t_m^r F(z)\, dz$, taken round this pole, will vanish when $r \geqq \mu$, and the corresponding series of functions $D_m^r \psi(x, a; m, c)$ will terminate.

144. We can also obtain a generalization of Mittag Leffler's Theorem. If $c_1, c_2, \ldots$ be a series of distinct places, of infinite number, which converge* to one place c, and $f_1(x), f_2(x), \ldots$ be a corresponding series of rational functions, of which $f_i(x)$ is infinite only at the place c_i, then we can find a single valued function $F(x)$, with one essential singularity (at the place c), which is otherwise infinite only at the places $c_1, c_2, \ldots$, and in such a way that the difference $F(x) - f_i(x)$ is finite in the neighbourhood of the place c_i.

Since $f_i(x)$ is a rational function, infinite only at the place c_i, and $\psi(x, a; z, c)$ does not become infinite when z comes to c, we can put

$$f_i(x) = f_i(a) - \sum_{r=0}^{\lambda_i} \frac{A_r}{\underline{|r}} D_{c_i}^r \psi(x, a; c_i, c), \qquad \text{(A)}$$

wherein a is an arbitrary place not in the neighbourhood of any of the places $c_1, c_2, \ldots, c$, and λ_i is a finite positive integer, and A_r a constant.

Also, when z is sufficiently near to c, and x is not near to c, we can put

$$\psi(x, a; z, c) = \sum_{k=0}^{\infty} \frac{t_c^k}{\underline{|k}} [D_z^k \psi(x, a; z, c)]_{z=c},$$

wherein t_c is the infinitesimal at the place c. Thus also, when z is near to c,

$$D_z^r \psi(x, a; z, c) = \sum_{k=0}^{\infty} t_c^k R_k(x), \qquad \text{(B)},$$

wherein $R_k(x)$ is a rational function, which is only infinite at the place c. There are p values of k which do not enter on the right hand; for it can easily be seen that if $k_1, \ldots, k_p$ denote the orders of non-existent rational functions infinite only at the place c, each of the functions

$$[D_z^{k_1-1} \psi(x, a; z, c)]_{z=c}, \ldots\ldots, [D_z^{k_p-1} \psi(x, a; z, c]_{z=c}$$

vanishes identically. Let the neighbourhood of the place c, within which z must lie in order that the expansions (B) may be valid, be denoted by M.

Of the places $c_1, c_2, \ldots$, an infinite number will be within the region M; let these be the places $c_{s+1}, c_{s+2}, \ldots$; then s will be finite and, when $i > s$, we have

$$D_{c_i}^r \psi(x, a; c_i, c) = \sum_{k=0}^{\infty} t_i^k R_{i,k}(x),$$

wherein t_i is the value of t_c, in the equation (B), when z is at c_i. Hence also, from the equation (A), wherein there are only a finite number of terms on the right hand, we can put

$$f_i(x) - f_i(a) = \sum_{k=0}^{\infty} t_i^k S_{i,k}(x), \qquad \text{(C)},$$

wherein $S_{i,k}$ is a rational function, $i > s$, and x is not near to the place c.

* so that c is what we may call the *focus* of the series $c_1, c_2, \ldots$ (Häufungsstelle).

It is the equation (C) which is the purpose of the utilisation of the function $\psi(x, a; z, c)$ in the investigation. The functions $S_{i,k}(x)$ will be infinite only at the place c. The series (C) are valid so long as x is outside a certain neighbourhood of c. We may call this the region M'.

Let now $\epsilon_{s+1}, \epsilon_{s+2}, \dots$ be any infinite series of real positive quantities, such that the series

$$\epsilon_{s+1} + \epsilon_{s+2} + \epsilon_{s+3} + \dots$$

is convergent; let μ_i be the smallest positive integer such that, for $i > s$, the terms

$$\sum_{k=\mu_i+1}^{\infty} t_i^k S_{i,k}(x),$$

taken from the end of the convergent series (C), are, in modulus, less than ϵ_i, for all the positions of x outside M'; then, defining a function $g_i(x)$, when $i > s$, by the equation

$$g_i(x) = f_i(x) - f_i(a) - \sum_{k=0}^{\mu_i} t_i^k S_{i,k}(x),$$

we have, for $i > s$,

$$|g_i(x)| < \epsilon_i.$$

Thus the series

$$\sum_{i=1}^{s} [f_i(x) - f_i(a)] + \sum_{i=s+1}^{\infty} g_i(x)$$

is absolutely and uniformly convergent for all positions of x not in the neighbourhood of the places $c, c_1, c_2, \dots$, and represents a continuous single valued function of x. When x is near to c_i, the function represented by the series is infinite like $f_i(x)$.

The function is not unique; if $\psi(x)$ denote any single-valued function which is infinite only at the place c, the addition of $\psi(x)$ to the function obtained will result in a function also having the general character required in the enunciation of the theorem. As here determined the function vanishes at the arbitrary place a; but that is an immaterial condition.

For instance when $p=0$, and the place m is at infinity, the places $m_1, m_2, m_3, \dots$, being $0, 1, \omega, 1+\omega, \dots, p+q\omega, \dots$, wherein ω is a complex quantity and p, q are any rational integers, let the functions $f_1(x), f_2(x), \dots$ be $x^{-1}, (x-1)^{-1}, (x-\omega)^{-1}, \dots, (x-p-q\omega)^{-1}, \dots$.

Here $$\psi(x, a; z, c) = -\left(\frac{1}{x-z} - \frac{1}{a-z}\right) = \frac{x-a}{z^2} + \frac{x^2-a^2}{z^3} + \frac{x^3-a^3}{z^4} + \dots$$

when z is great enough and $|x| < |z|$, $|a| < |z|$.

Also

$$\frac{1}{x-m_i} = \frac{1}{a-m_i} - \psi(x, a; m_i, c)$$

$$= \frac{1}{a-m_i} - \left(\frac{x-a}{m_i^2} + \frac{x^2-a^2}{m_i^3} + \dots\right),$$

when m_i is great enough, and $|x| < |m_i|$, $|a| < |m_i|$.

Now the series

$$\Sigma \left| \frac{1}{m_i^3} \right| = \Sigma \Sigma \left| \frac{1}{(p+q\omega)^3} \right|$$

is convergent. Hence when x and a are not too great

$$\left| \frac{x^2-a^2}{m_i^3} + \frac{x^3-a^3}{m_i^4} + \dots \right| < \epsilon_i,$$

where ϵ_i is a term of a convergent series of positive quantities. This equation holds for all values of i except $i=1$, in which case $m_i=0$.

Hence we may write

$$g_i(x) = \frac{1}{x-m_i} + \frac{1}{a-m_i} + \frac{x-a}{m_1^2}$$

and obtain the function

$$\frac{1}{x} - \frac{1}{a} + \sum_{p=-\infty}^{\infty} \sum_{q=-\infty}^{\infty} \left[\frac{1}{x-p-q\omega} - \frac{1}{a-p-q\omega} + \frac{x-a}{(p+q\omega)^2} \right],$$

which has the property required. This function is in fact equal, in the notation of Weierstrass's elliptic functions, to $\zeta(x \mid 1, \omega) - \zeta(a \mid 1, \omega)$.

145. We can always specify a rational function of x which, beside being infinite at the place c, is infinite at a place c_i like an expression of the form

$$\frac{A_0}{t_{c_i}} + \frac{A_1}{t_{c_i}^2} + \dots\dots + \frac{A_{\lambda_i}}{t_{c_i}^{\lambda_i+1}},$$

namely, such a function is

$$-\sum_{r=0}^{\lambda_i} \frac{A_r}{\lfloor r} D_{c_i}^r \psi(x, a; c_i, c),$$

and this may be used in the investigation instead of the function $f_i(x) - f_i(a)$.

Hence, in the enunciation of the theorem of § 144, it is not necessary that the expressions of the rational functions $f_i(x)$ be known, or even that there should exist rational functions infinite only at the places c_i in the assigned way. All that is necessary is that the character of the infinity of the function F, at the pole c_i, should be assigned.

Conversely, any single-valued function F whose singularities consist of one essential singularity and an infinite number of distinct poles which converge to the place of the essential singularity, can be represented by a series of rational functions of x, which beside the essential singularity have each only one pole.

146. Let the places $c_1, c_2, \dots, c$ be as in § 144. We can construct a single-valued function, having the places $c_1, c_2, \dots$, as zeros, of assigned positive integral orders $\lambda_1, \lambda_2, \dots$, which is infinite only at the place c, where it has an essential singularity.

For the function $$E(x, z) = e^{\int_c^z \psi(x, a; z, c)\, dt_z}$$

is zero at the place z and infinite only at the place c. When z is near to c we can put

$$D_z \log E(x, z) = \sum_{r=0}^{\infty} \frac{t_c^r}{\underline{|r}} [D_z^r \psi(x, a; z, c)]_{z=c},$$

and therefore, when c_i is near to c, and x is not near to the place c, we can put

$$\lambda_i \log E(x, c_i) = \sum_{k=0}^{\infty} t_i^k R_{i,k}(x),$$

wherein $R_{i,k}(x)$ is a rational function of x which is infinite only at the place c, and t_i has the same significance as in § 144.

Let the least value of i for which this equation is valid be denoted by $s+1$, and, taking $\epsilon_{s+1}, \epsilon_{s+2}, \ldots$ any positive quantities such that the series

$$\epsilon_{s+1} + \epsilon_{s+2} + \ldots,$$

is convergent, let μ_i be the least number such that, for $i > s$,

$$\left| \sum_{k=\mu_i+1}^{\infty} t_i^k R_{i,k}(x) \right| < \epsilon_i.$$

Then the series

$$\sum_{i=1}^{s} \lambda_i \log E(x, c_i) + \sum_{i=s+1}^{\infty} \left(\lambda_i \log E(x, m_i) - \sum_{k=0}^{\mu_i} t_i^k R_{i,k}(x) \right)$$

consists of single-valued finite functions provided x is not near to any of $c_1, c_2, \ldots, c$, and, by the condition as to the numbers μ_i, is absolutely and uniformly convergent.

Hence the product

$$\prod_{i=1}^{s} [E(x, c_i)]^{\lambda_i} \prod_{i=s+1}^{\infty} \left\{ [E(x, c_i)]^{\lambda_i} e^{-\sum_{k=0}^{\mu_i} t_i^k R_{i,k}(x)} \right\}$$

represents a single-valued function, which is infinite only at c where it has an essential singularity, which is moreover zero only at the places $c_1, c_2, \ldots$ respectively to the orders $\lambda_1, \lambda_2, \ldots$.

With the results obtained in §§ 144—146, the reader will compare the well-known results for single-valued functions of one variable (Weierstrass, *Abhandlungen aus der Functionenlehre*, Berlin, 1886, pp. 1—66, or *Mathem. Werke*, Bd. ii. pp. 77, 189).

147. The following results possess the interest that they are given by Abel; they are related to the problems of this chapter. (Abel, *Œuvres Complètes*, Christiania, 1881, vol. i. p. 46 and vol. ii. p. 46.)

Ex. i. If $\phi(x)$ be a rational polynomial in x, $=\Pi(x+a_k)^{\beta_k}$,

and $f(x)$ be a rational function of x, $=\Sigma\gamma_k x^k+\Sigma\dfrac{\delta_k}{(x+\epsilon_k)^{\mu_k}}$,

then

$$\int\frac{e^{f(x)-f(z)}\phi(x)}{(x-z)\phi(z)}dx-\int\frac{e^{f(x)-f(z)}\phi(x)}{(z-x)\phi(z)}dz=\sum_k\sum_{k'}k\gamma_k\int\frac{e^{-f(z)}z^{k'}}{\phi(z)}dz\int e^{f(x)}\phi(x)\,.\,x^{k-k'-2}\,dx$$

$$-\sum_k\beta_k\int\frac{e^{-f(z)}}{(z+a_k)\phi(z)}dz\int\frac{e^{f(x)}\phi(x)}{x+a_k}dx+\sum_{k}\sum_{k'}\mu_k\delta_k\int\frac{e^{-f(z)}dz}{(z+\epsilon_k)^{\mu_k-k'+2}\phi(z)}\int\frac{e^{f(x)}\phi(x)}{(x+\epsilon_k)^{k'}}dx.$$

The theorem can be obtained most directly by noticing that if $\phi(x, z)=\dfrac{e^{f(x)-f(z)}\phi(x)}{\phi(z)(x-z)}$ then

$$\phi(X, z)\frac{d}{dX}\phi(x, X)=\frac{e^{f(x)-f(z)}\phi(x)}{\phi(z)}\left\{\frac{f'(X)+\dfrac{\phi(X)}{\phi(X)}}{(X-x)(X-z)}+\frac{1}{(X-z)(X-x)^2}\right\}$$

is a rational function of X. Denoting it by $R(X)$ and applying the theorem

$$\Sigma\left[R(X)\frac{dX}{dt}\right]_{t^{-1}}=0,$$

we obtain Abel's result.

Ex. ii. With the same notation, but supposing $f(x)$ to be an integral polynomial, prove that

$$\int\phi(x, z)\,dx+\int\frac{\psi(x)}{\psi(z)}\phi(x, z)\,dz=\Sigma\Sigma A_{k,k'}\int\frac{e^{-f(z)}z^{k'}\,dz}{\phi(z)\psi(z)}\int e^{f(x)}\phi(x)\,x^k\,dx,$$

wherein $A_{k,k'}$, is a certain constant, and $\psi(x)$ is the product of all the simple factors of $\phi(x)$.

This result may be obtained from the rational function

$$R(X)=\frac{\psi(X)}{\psi(z)}\phi(X, z)\frac{d}{dX}\phi(x, X)$$

as in the last example.

Ex. iii. Obtain the theorem of Ex. ii. when $f(x)=0$, and $\phi(x)=[\psi(x)]^m$. In the result put $m=-\frac{1}{2}$, and obtain the result of the example in § 138.

These results are extended by Abel to the case of linear differential equations. Further development is given by Jacobi, *Crelle* xxxii. p. 194, and by Fuchs, *Crelle* lxxvi. p. 177.

CHAPTER VIII.

ABEL'S THEOREM; ABEL'S DIFFERENTIAL EQUATIONS.

148. THE present chapter is mainly concerned with that theorem with which the subject of the present volume may be said to have begun. It will be seen that with the ideas which have been analysed in the earlier part of the book, the statement and proof of that theorem is a matter of great simplicity.

The problem of the integration of a rational algebraical function (of a single variable) leads to the introduction of a transcendental function, the logarithm; and the integral of any such rational function can be expressed as a sum of rational functions and logarithms of rational functions. More generally, an integral of the form

$$\int dx\, R(x, y, y_1, \ldots, y_k),$$

wherein $x, y, y_1, y_2, \ldots$ are capable of rational expression in terms of a single parameter, and R denotes any rational algebraic function, can be expressed as a sum of rational functions of this parameter, and logarithms of rational functions of the same. This includes the case of an integral of the form

$$\int dx\, R(x, \sqrt{ax^2+bx+c}).$$

But an integral of the form

$$\int dx\, R(x, \sqrt{ax^4+bx^3+cx^2+dx+e})$$

cannot, in general, be expressed by means of rational or logarithmic functions; such integrals lead in fact to the introduction of other transcendental functions than the logarithm, namely to elliptic functions; and it appears that the nearest approach to the simplicity of the case, in which the subject of integration is a rational function, is to be sought in the relations which exist for the *sums* of like elliptic integrals. For instance, we have the equation

$$\int_0^{x_1} \frac{dx}{\sqrt{(1-x^2)(1-k^2x^2)}} + \int_0^{x_2} \frac{dx}{\sqrt{(1-x^2)(1-k^2x^2)}} - \int_0^{x_3} \frac{dx}{\sqrt{(1-x^2)(1-k^2x^2)}} = 0,$$

provided

$$x_3(1-k^2x_1^2x_2^2)=x_1\sqrt{(1-x_2^2)(1-k^2x_2^2)}+x_2\sqrt{(1-x_1^2)(1-k^2x_1^2)}.$$

On further consideration, however, it is clear that this is not a complete statement; and it is proper, beside the quantity x, to introduce a quantity y, such that

$$y^2-(1-x^2)(1-k^2x^2)=0,$$

and to regard y, for any value of x, as equally capable either of the positive or negative sign; in fact by varying x continuously from any value, through one of the values $x=\pm 1$, $x=\pm\frac{1}{k}$, and back to its original value, we can suppose that y varies continuously from one sign to the other. Then the theorem in question can be written thus;

$$\int_{(0,\,1)}^{(x_1,\,y_1)}\frac{dx_1}{y_1}+\int_{(0,\,1)}^{(x_2,\,y_2)}\frac{dx_2}{y_2}+\int_{(0,\,1)}^{(x_3,\,y_3)}\frac{dx_3}{y_3}=0,$$

where the limits specify the value of y as well as the value of x. The theorem holds when, in the first two integrals the variables (x, y) are taken through any continuous succession of simultaneous values, from the lower to the upper limits, the variables in the last integral being, at every stage of the integration, defined by the equations

$$-x_3(1-k^2x_1^2x_2^2)=x_1y_2+x_2y_1,$$
$$y_3(1-k^2x_1^2x_2^2)^2=y_1y_2(1+k^2x_1^2x_2^2)-x_1x_2(1-k^2x_1^2x_2^2)(1-k^2).$$

The quantity y is called an algebraical function of x; and the notion thus introduced is a fundamental one in the theorems to be considered; its complete establishment has been associated, in this volume, with a Riemann surface.

In the case where $y^2=(1-x^2)(1-k^2x^2)$ we have the general theorem that, if $R(x, y)$ be any rational function of x, y, the sum of any number, m, of similar integrals

$$\int_{(a_1,\,b_1)}^{(x_1,\,y_1)}R(x,y)\,dx+\dots\dots+\int_{(a_m,\,b_m)}^{(x_m,\,y_m)}R(x,y)\,dx$$

can be expressed by rational functions of $(x_1, y_1), \dots, (x_m, y_m)$, and logarithms of such rational functions, with the addition of an integral

$$-\int_{(a_{m+1},\,b_{m+1})}^{(x_{m+1},\,y_{m+1})}R(x,y)\,dx.$$

Herein the lower limits $(a_1, b_1), \dots, (a_m, b_m)$ represent arbitrary pairs of corresponding values of x and y, and the succession of values for the pairs $(x_1, y_1), \dots, (x_m, y_m)$ is quite arbitrary; but in the last integral x_{m+1}, y_{m+1} are each rational functions of $(x_1, y_1), \dots, (x_m, y_m)$, which must be properly determined, and it is understood that the relations are preserved at all stages of the integration, so that for example a_{m+1}, b_{m+1} are respectively taken to be the same rational functions of $(a_1, b_1), \ldots, (a_m, b_m)$. The question of what alteration is necessary in the enunciation when this convention is not observed, is the question of the change in the value of an integral

$$\int_{(a_{m+1}, b_{m+1})}^{(x_{m+1}, y_{m+1})} R(x, y)\, dx$$

when the path of integration is altered. This question is fully treated in the consideration of the Riemann surface, with the help of what have been called period loops.

149. Abel's theorem may be regarded as a generalization of the theorem just stated, and may be enunciated as follows: Let y be the algebraical function of x defined by an equation of the form

$$f(y, x) = y^n + A_1 y^{n-1} + \ldots\ldots + A_n = 0,$$

wherein $A_1, \ldots, A_n$ are rational polynomials in x, and the left-hand side of the equation is supposed incapable of resolution into the product of factors of the same rational form; let $R(x, y)$ be any rational function of x and y; then the sum of any number, m, of similar integrals

$$\int^{(x_1, y_1)} R(x, y)\, dx + \ldots\ldots + \int^{(x_m, y_m)} R(x, y)\, dx,$$

with arbitrary lower limits, is expressible by rational functions of $(x_1, y_1), \ldots, (x_m, y_m)$, and logarithms of such rational functions, with the addition of the sum of a certain number, k, of integrals,

$$-\int^{(z_1, s_1)} R(x, y)\, dx - \ldots\ldots - \int^{(z_k, s_k)} R(x, y)\, dx,$$

wherein $z_1, \ldots, z_k$ are values of x, determinable from $x_1, y_1, \ldots, x_m, y_m$ as the roots of an algebraical equation whose coefficients are rational functions of $x_1, y_1, \ldots, x_m, y_m$, and $s_1, \ldots, s_k$ are the corresponding values of y, of which any one, say s_i, is determinable as a rational function of z_i, and $x_1, y_1, \ldots, x_m, y_m$. The relations thus determining $(z_1, s_1), \ldots, (z_k, s_k)$ from $(x_1, y_1), \ldots, (x_m, y_m)$ may be supposed to hold at all stages of the integration; in particular they determine the lower limits of the last k integrals from the arbitrary lower limits of the first m integrals. The number k does not depend upon m, nor upon the form of the rational function $R(x, y)$; and in general it does not depend upon the values of $(x_1, y_1), \ldots, (x_m, y_m)$, but only upon the fundamental equation which determines y in terms of x.

150. In this enunciation there is no indication of the way in which the equations determining $z_1, s_1, \ldots, z_k, s_k$ from $x_1, y_1, \ldots, x_m, y_m$ are to be found. Let $\theta(y, x)$ be an integral polynomial in x and y, wherein some or all of the coefficients are regarded as variable. By continuous variation of these

coefficients the set of corresponding values of x and y which satisfy both the equations $f(y, x) = 0$, $\theta(y, x) = 0$, will also vary continuously. Then, if m be the number of variable coefficients of $\theta(y, x)$, and $m + k$ the total number of variable pairs (x, y) which satisfy both the equations $f(y, x) = 0$, $\theta(y, x) = 0$, the necessary relations between $(x_1, y_1), \ldots, (x_m, y_m), (z_1, s_1), \ldots, (z_k, s_k)$ are expressed by the fact that these pairs are the common solutions of the equations $f(y, x) = 0$, $\theta(y, x) = 0$. The polynomial $\theta(y, x)$ may have any form in which there enter m variable coefficients; by substitution, in $\theta(y, x)$, of the m pairs of values $(x_1, y_1), \ldots, (x_m, y_m)$, we can determine these variable coefficients as rational functions of $x_1, y_1, \ldots, x_m, y_m$; by elimination of y between the equations $\theta(y, x) = 0$, $f(y, x) = 0$, we obtain an algebraic equation for x, breaking into two factors, $P_0(x) P(x) = 0$, one factor, $P_0(x)$, not depending on $x_1, y_1, \ldots, x_m, y_m$, and vanishing for the values of x at the fixed solutions of $f(y, x) = 0$, $\theta(y, x) = 0$, which do not depend on $x_1, y_1, \ldots, x_m, y_m$, the other factor, $P(x)$, having the form

$$(x - x_1) \ldots (x - x_m)(x^k + R_1 x^{k-1} + \ldots + R_k),$$

where $R_1, \ldots, R_k$ are rational functions of $x_1, y_1, \ldots, x_m, y_m$. Finally, from the equations $f(s_i, z_i) = 0$, $\theta(s_i, z_i) = 0$ we can determine s_i rationally in terms of $z_i, x_1, y_1, \ldots, x_m, y_m$. As a matter of fact the rational functions of $x_1, y_1, \ldots, x_m, y_m$, which appear on the right-hand side of the equation which expresses Abel's theorem, are rational functions of the variable coefficients in $\theta(y, x)$.

151. When $\theta(y, x)$ is quite general save for the condition of having certain fixed zeros satisfying $f(y, x) = 0$, the forms of $(z_1, s_1), \ldots, (z_k, s_k)$ as functions of $(x_1, y_1), \ldots, (x_m, y_m)$ are independent of the form of $\theta(y, x)$. This appears from the following enunciation of the theorem, which introduces ideas that have been elaborated since Abel's time, and which we regard as the final form—Let $(a_1, b_1), \ldots, (a_Q, b_Q)$ be any places of the Riemann surface whatever, such that sets coresidual therewith have a multiplicity q, and a sequence $Q - q = p - \tau - 1$, where $\tau + 1$ is the number of ϕ polynomials vanishing in the places $(a_1, b_1), \ldots, (a_Q, b_Q)$; let $(x_1, y_1), \ldots, (x_q, y_q)$ be q arbitrary places determining a set coresidual with $(a_1, b_1), \ldots, (a_Q, b_Q)$, and $(z_1, s_1), \ldots, (z_{p-\tau-1}, s_{p-\tau-1})$ be the sequent places of this set*; then, $R(x, y)$ being any rational function of (x, y), the sum

$$\int_{(a_1, b_1)}^{(x_1, y_1)} R(x, y)\, dx + \ldots\ldots + \int_{(a_q, b_q)}^{(x_q, y_q)} R(x, y)\, dx$$

is expressible by rational functions of $(x_1, y_1), \ldots, (x_q, y_q)$, and logarithms of such rational functions, with the addition of a sum

$$-\int_{(a_{q+1}, b_{q+1})}^{(z_1, s_1)} R(x, y)\, dx - \ldots\ldots - \int_{(a_Q, b_Q)}^{(z_{p-\tau-1}, s_{p-\tau-1})} R(x, y)\, dx$$

* See Chap. VI. § 95.

herein it is understood that the paths of integration are such that at every stage the variables form a set coresidual with $(a_1, b_1), \ldots, (a_Q, b_Q)$.

The places $(a_1, b_1), \ldots, (a_Q, b_Q)$ may therefore be regarded as the poles, and $(x_1, y_1), \ldots, (x_q, y_q), (z_1, s_1), \ldots, (z_{p-\tau-1}, s_{p-\tau-1})$ as the zeros, of the same rational function $Z(x)$; if $\theta_1(y, x)$ denote the form of the polynomial $\theta(y, x)$ when it vanishes in $(a_1, b_1), \ldots, (a_Q, b_Q)$, and $\theta_2(y, x)$ denote its form when its zeros are $(x_1, y_1), \ldots, (z_1, s_1), \ldots$, the function $Z(x)$ may be expressed in the form $\theta_2(y, x)/\theta_1(y, x)$. If the polynomials $\theta_1(y, x)$, $\theta_2(y, x)$ are not adjoint, the function will be of the kind, hitherto regarded as special, which takes the same value at all the places of the Riemann surface which correspond to a multiple point of the plane curve represented by the equation $f(y, x) = 0$; this fact does not affect the application of Abel's theorem to the case.

152. To prove the theorem thus enunciated, with the greatest possible definiteness, we shew first that it may be reduced to two simple cases.

In the neighbourhood of any place of the Riemann surface, at which t is the infinitesimal, we can express $R(x, y)\dfrac{dx}{dt}$ in a series of positive and negative powers of t, in which the number of negative powers is finite. Let the expression at some place, ξ, where negative powers actually enter, be denoted by

$$\underline{|m-1}\,\frac{A_m}{t^m} + \underline{|m-2}\,\frac{A_{m-1}}{t^{m-1}} + \ldots\ldots + \frac{A_2}{t^2} + \frac{A_1}{t} + B + B_1 t + B_2 t^2 + \ldots\ldots;$$

then, if $P_{\xi, \gamma}^{x, c}$ denote any elementary integral of the third kind, with infinities at ξ, γ, and $E_{\xi}^{x, c}$ denote the differential coefficient of $P_{\xi, \gamma}^{x, c}$ in regard to the infinitesimal at ξ, the places γ, c being arbitrary, the difference

$$\int_{(a, b)}^{(x, y)} R(x, y)\, dx - A_1 P_{\xi, \gamma}^{x, c} - A_2 E_{\xi}^{x, c} - A_3 D_{\xi} E_{\xi}^{x, c} - \ldots\ldots - A_m D_{\xi}^{m-2} E_{\xi}^{x, c},$$

wherein D_ξ denotes differentiation in regard to the infinitesimal at ξ, is finite at the place ξ. The number of places, ξ, at which negative powers of t enter in the expansion of $R(x, y)\dfrac{dx}{dt}$, is finite; dealing with each in turn we obtain an expression of the form

$$\int_{(a, b)}^{(x, y)} R(x, y)\, dx - \sum_{\xi} \left[A_1 P_{\xi, \gamma}^{x, c} + A_2 E_{\xi}^{x, c} + A_3 D_{\xi} E_{\xi}^{x, c} + \ldots\ldots + A_m D_{\xi}^{m-2} E_{\xi}^{x, c}\right],$$

wherein γ, c are taken the same for every place ξ; this is finite at all places of the Riemann surface, except possibly the place γ. If t_γ be the infinitesimal at this place the function is there infinite like $(\Sigma A_1)\log t_\gamma$. But in fact ΣA_1 is zero (Chap. II. § 17, Ex. (δ); Chap. VII. § 137, Ex. vi.). Hence the function under consideration is nowhere infinite, and is therefore necessarily* a linear aggregate of integrals of the first kind, plus a constant. Hence if $u_1^{x,a}, \ldots, u_p^{x,a}$ be a set of linearly independent integrals of the first kind, a denoting the place (a, b), and $C_1, \ldots, C_p$ be proper constants, we have

$$\int_a^x R(x, y)\,dx = \sum_\xi (A_1 + A_2 D_\xi + \ldots\ldots + A_m D_\xi^{m-2}) P_{\xi,\gamma}^{x,a} + C_1 u_1^{x,a} + \ldots\ldots + C_p u_p^{x,a}.$$

The consideration of the sum

$$\int_{a_1}^{x_1} R(x, y)\,dx + \ldots\ldots + \int_{a_Q}^{x_Q} R(x, y)\,dx,$$

wherein $a_1, \ldots, a_Q$ denote the places $(a_1, b_1), \ldots, (a_Q, b_Q)$, and $x_1, \ldots, x_Q$ denote the places $(x_1, y_1), \ldots, (x_q, y_q), (z_1, s_1), \ldots, (z_{p-\tau-1}, s_{p-\tau-1})$, is thus reduced to the consideration of the two sums

$$u_i^{x_1, a_1} + \ldots\ldots + u_i^{x_Q, a_Q}, \qquad (i = 1, 2, \ldots, p.)$$

$$P_{\xi,\gamma}^{x_1, a_1} + \ldots\ldots + P_{\xi,\gamma}^{x_Q, a_Q}.$$

Ex. i. By the proposition here repeated from § 20, Chap. II., it follows that any rational function can be written in the form

$$R(x, y) = \sum_\xi \{A_1[(x, \xi) - (x, \gamma)] + A_2 D_\xi (x, \xi) + \ldots + A_m D_\xi^{m-2}(x, \xi)\}$$

$$+ [(x, 1)^{\tau_1' - 1} \phi_1(x, y) + \ldots + (x, 1)^{\tau'_{n-1} - 1} \phi_{n-1}(x, y)]/f'(y)$$

where (cf. § 45, Chap. IV.)

$$(x, \xi) = [\phi_0(x, y) + \sum_1^{n-1} \phi_r(x, y)\, g_r(\xi, \eta)]/(x - \xi) f'(y),$$

η being the value of y at the place ξ.

Ex. ii. Prove also that any rational function with simple poles at $\xi_1, \xi_2, \ldots$ can be written in the form

$$\lambda_1[(\xi_1, x) - (\xi_1, a)] + \lambda_2[(\xi_2, x) - (\xi_2, a)] + \ldots,$$

$\lambda_1, \lambda_2, \ldots$ being constants, and a denoting an arbitrary place (cf. § 130, Chap. VII.).

153. We shall prove, now, in regard to these two sums, under the conventions that the upper limits are coresidual with the lower limits, and that the Q paths of integration are such that at every stage the variables are at places also coresidual with the lower limits, a convention under which the paths of integration may quite well cross the period loops on the Riemann surface, that the first sum is zero for all values of i, and the second equal to $\log Z(\xi)/Z(\gamma)$, $Z(x)$ being the† rational function which has $a_1, \ldots, a_Q$ as poles and $x_1, \ldots, x_Q$ as zeros. The sense in which the logarithm is to be understood will appear from the proof of the theorem. If we suppose the lower limits arbitrarily assigned, the general function $Z(x)$, of which these

* Forsyth, *Theory of Functions*, § 234.

† If two rational functions have the same poles and the same zeros their ratio is necessarily a constant.

places $a_1, \ldots, a_Q$ are the poles, will contain $q+1$ arbitrary linear coefficients, entering homogeneously, and the assignation of q of the zeros, say $x_1, \ldots, x_q$, will determine the others, as explained.—The equations giving the determination will be such functions of $a_1, \ldots, a_Q$ as are identically satisfied by these places, $a_1, \ldots, a_Q$. Hence the general form of Abel's theorem is

$$\sum_{i=1}^{Q} \int_{a}^{x_i} R(x, y)\, dx = \sum_{\xi} \left[A_1 \log \frac{Z(\xi)}{Z(\gamma)} + A_2 \frac{Z'(\xi)}{Z(\xi)} + \ldots\ldots \right]$$
$$= \sum_{\xi} \left[A_1 \log Z(\xi) + A_2 \frac{Z'(\xi)}{Z(\xi)} + \ldots\ldots \right]$$

where $Z'(\xi) = D_\xi Z(\xi)$; the term $\sum_{\xi} A_1 \log Z(\gamma) = \log Z(\gamma) \Sigma A_1$ can be omitted because $\Sigma A_1 = 0$ (Chap. II. p. 20 (δ)). Herein $Z(\xi)$ is a rational function of $a_1, \ldots, a_Q$ and $x_1, \ldots, x_q$.

154. In carrying out the proof we make at first a simplification—Let $Z(x)$, or Z, be the rational function having $a_1, \ldots, a_Q$ as simple poles and $x_1, \ldots, x_Q$ as simple zeros, these places being supposed to be all different; trace on the Riemann surface an arbitrary path joining a_1 to x_1, chosen so as to avoid all places where dZ is zero to higher than the first order, and let μ be the value of Z at any place of this path; then there will be $Q-1$ other places at which Z has the same value μ; the paths traced by these $Q-1$ places as μ varies from ∞ to 0 are the paths we assign for the $Q-1$ integrals following the first. The simultaneous positions thus defined for the variables in the Q integrals are, for $q > 1$, not so general* as those allowed by the convention that the simultaneous positions are coresidual with $a_1, \ldots, a_Q$; but it will be seen that the more general case is immediately deducible from the particular one.

Consider now, for any value of μ, the rational function

$$\frac{1}{Z-\mu} \frac{dI}{dx},$$

$I, = \int R(x, y)\, dx$, being any Abelian integral whatever. In accordance with a theorem previously used (Chap. II. p. 20 (δ); Chap. VII. § 137, Ex. vi.) the sum of the coefficients of t^{-1} in the expansions of $(Z-\mu)^{-1} dI/dt$, in terms of the infinitesimal t, at all places where negative powers of t occur, is equal to zero. Of such places there are first the Q places where Z is equal to μ. We shall suppose that dI/dt is finite at all these places; then the sum of the coefficients of t^{-1} at these places is

$$\Sigma \frac{1}{d\mu/dt} \left(\frac{dI}{dt}\right), \quad = \left(\frac{dI}{d\mu}\right)_1 + \ldots\ldots + \left(\frac{dI}{d\mu}\right)_Q,$$

* Sets coresidual with two given coresidual sets have a multiplicity q; but sets equivalent with two given coresidual sets have a variability expressible by one parameter only (cf. Chap. VI. §§ 94—96).

provided $Z-\mu$ be not zero to the second order at any of the places, that is, provided dZ be not zero to higher than the first order. In accordance with the convention made as to the paths of the variables in the integrals, we suppose this condition to be satisfied.

Hence this sum is equal to the sum of the coefficients of t^{-1} in the expansions of the function $-(Z-\mu)^{-1}\,dI/dt$ at all places, only, where dI/dt is infinite; this result we may write in the form

$$\left(\frac{dI}{d\mu}\right)_1+\ldots\ldots+\left(\frac{dI}{d\mu}\right)_Q=-\left(\frac{dI}{dt}\,\frac{1}{Z-\mu}\right)_{t^{-1}};$$

we may regard this equation as a convenient way of stating Abel's theorem for many purposes; and may suppose the case, in which an infinity of dI/dt coincides with a place at which $Z=\mu$, to be included in this equation, the left hand being restricted to all places at which $Z=\mu$ and dI/dt is not infinite.

In this equation, in case $I, = u_i^{x, a}$, be any integral of the first kind, the right hand vanishes; then, integrating in regard to μ from ∞ to 0, we obtain

$$u_i^{x_1, a_1}+\ldots\ldots+u_i^{x_Q, a_Q}=0. \tag{A}$$

In case I be an integral of the third kind, $=P_{\xi,\gamma}^{x, c}$ say, and Z be not equal to μ either at ξ or γ, the right hand is equal to

$$-\frac{1}{Z(\xi)-\mu}+\frac{1}{Z(\gamma)-\mu};$$

hence, integrating,

$$P_{\xi,\gamma}^{x_1, a_1}+\ldots\ldots+P_{\xi,\gamma}^{x_Q, a_Q},\ =\int_\infty^0 d\mu\left(-\frac{1}{Z(\xi)-\mu}+\frac{1}{Z(\gamma)-\mu}\right),\ =\log\frac{Z(\xi)}{Z(\gamma)}, \tag{B}$$

while, if the places at which the rational function $Z(x)$ has the values μ, ν be respectively denoted by

$$x_1', \ldots\ldots, x'_Q,$$

and

$$a_1', \ldots\ldots, a'_Q$$

we have

$$P_{\xi,\gamma}^{x_1', a_1'}+\ldots\ldots+P_{\xi,\gamma}^{x'_Q, a'_Q},\ =\int_\nu^\mu d\mu\left(-\frac{1}{Z(\xi)-\mu}+\frac{1}{Z(\gamma)-\mu}\right),$$

$$=\log\left[\frac{Z(\xi)-\mu}{Z(\xi)-\nu}\Big/\frac{Z(\gamma)-\mu}{Z(\gamma)-\nu}\right].$$

For any Abelian integral we similarly have

$$I^{x_1', a_1'}+\ldots\ldots+I^{x'_Q, a'_Q}=\left[\frac{dI}{dt}\log\frac{Z(x)-\mu}{Z(x)-\nu}\right]_{t^{-1}},$$

which is a complete statement of Abel's theorem.

155. In the equation (B), and in the equation which follows it, the significance of the logarithm is determined by the path of μ in the integral expression which defines the logarithm; we may also define the logarithm by considering the two sides of the equation as functions of ξ.

There is no need to extend the equation (B) to the case where one of the paths of integration on the left passes through either ξ or γ, since in that case a corresponding infinite term enters on both sides of the equation.

But it is clear that the condition that no two of the upper limits $x_1, \dots, x_Q$ should be coincident is immaterial, and may be removed. And if two (or more) of the places at which Z takes any value, μ, should coincide, the equations (A) and (B) can be formed each as the sum of two equations in which the course of integration is respectively from $Z=\infty$ to $Z=\mu$ and from $Z=\mu$ to $Z=0$, and the final outcome can only be that the order in which the upper limits $x_1, \dots, x_Q$ are associated with the lower limits $a_1, \dots, a_Q$ may undergo a change. But in the general case we may equally put, for example, in equations (A), (B),

$$\int_{a_1}^{x_1} dI + \int_{a_2}^{x_2} dI, = \int_{a_1}^{x_2} dI + \int_{x_2}^{x_1} dI + \int_{x_1}^{x_2} dI + \int_{a_2}^{x_1} dI, = \int_{a_1}^{x_2} dI + \int_{a_2}^{x_1} dI,$$

with proper conventions as to the paths; hence the condition that dZ shall not be zero to higher than the first order at any stage of the integration may be discarded also, with a certain loss of definiteness. The most general form of equation (A), when each of the Q paths of integration are arbitrary, is of course

$$u_i^{x_1, a_1} + \dots\dots + u_i^{x_Q, a_Q} = M_1 \omega_{i,1} + \dots\dots + M_p \omega_{i,p} + M_1' \omega'_{i,1} + \dots\dots + M_p' \omega'_{i,p}, \quad \text{(C)}$$

where $\omega_{i,1}, \dots, \omega'_{i,p}$ are the periods of $u_i^{x,a}$ and $M_1, \dots, M_p'$ are rational integers, independent of i. We shall subsequently see that this equation is sufficient to prove that the places $x_1, \dots, x_Q$ are coresidual with the set $a_1, \dots, a_Q$.

If, in equation (B), we substitute for $Z(x)$ any one of its rational expressions, say* $\theta_2(x)/\theta_1(x)$, we shall obtain

$$P_{\xi,\gamma}^{x_1, a_1} + \dots\dots + P_{\xi,\gamma}^{x_Q, a_Q} = \log \frac{\theta_2(\xi)}{\theta_1(\xi)} \Big/ \frac{\theta_2(\gamma)}{\theta_1(\gamma)},$$

where, now, $\theta_2(x)$, $\theta_1(x)$ are any two polynomials, integral in x and y, of which, beside common zeros, $\theta_2(x)$ has $x_1, \dots, x_Q$ for zeros, and $\theta_1(x)$ has $a_1, \dots, a_Q$ for zeros. If in this equation we suppose any of the coefficients in $\theta_2(x)$ to vary infinitesimally in any way, such that the common zeros of $\theta_2(x)$

* $\theta(x)$ is, for shortness, put for what would more properly be denoted by $\theta(y, x)$.

and $\theta_1(x)$ remain fixed, $\theta_2(x)$ changing thereby into $\theta_2(x)+\delta\theta_2(x)$, the places $x_1, \dots, x_Q$ changing thereby to $x_1+dx_1, \dots, x_Q+dx_Q$, we shall obtain

$$\frac{dP_{\xi,\gamma}^{x_1,c}}{dx_1}dx_1+\dots\dots+\frac{dP_{\xi,\gamma}^{x_Q,c}}{dx_Q}dx_Q=\delta\log\frac{\theta_2(\xi)}{\theta_2(\gamma)},$$

which is slightly more general than any equation before given, in that the places $x_1+dx_1, \dots, x_Q+dx_Q$, though coresidual with $x_1, \dots, x_Q$, are not necessarily such that the function $\theta_2(x)/\theta_1(x)$ has the same value at all of them. This general equation is obtained by Abel in the course of his proof of his theorem.

For any Abelian integral we have, similarly, the equation

$$\frac{dI}{dx_1}dx_1+\dots\dots+\frac{dI}{dx_Q}dx_Q=\left[\frac{dI}{dt}\delta\log\theta(x)\right]_{t^{-1}},$$

which, also, may be regarded as a complete statement of Abel's theorem.

156. In equation (B) the logarithm of the right hand will disappear if $Z(\xi)=Z(\gamma)$, namely if the infinities of the integral be places at which the function $Z(x)$ has the same value.

One case of this may be noticed; if $\psi(y, x)$ be an integral polynomial of grade $(n-1)\sigma+n-3$ (cf. Chap. VI. §§ 86, 91), which is adjoint at all places except those two, say A, A', which correspond to an ordinary double point of the curve represented by the equation $f(y, x)=0$, the integral

$$V^{x,a}, =\int_a^x \frac{\psi(y, x)}{f'(y)}dx,$$

will be an integral of the third kind having A, A' as its infinities. Hence, if in forming the function $Z(x)$, $=\theta_2(x)/\theta_1(x)$, the places A, A' have been disregarded, so that the polynomials $\theta_1(x)$, $\theta_2(x)$ do not vanish in these places, the function $Z(x)$ will take the same value at A as at A', and we shall obtain

$$V^{x_1,a_1}+\dots\dots+V^{x_Q,a_Q}=0.$$

Hence we obtain the result: if, in the formation of the integrals of the first kind for a given fundamental curve, we overlook the existence of a certain number, say δ, of double points, we shall obtain $p+\delta$ integrals, where p is the true deficiency of the curve; and these integrals will be linear aggregates of the actual integrals of the first kind and of δ integrals of the third kind. If in the formation of the rational functions also we overlook the existence of these double points, Abel's theorem will have the same form of equation for the $p+\delta$ integrals as if they were integrals of the first kind (cf. §§ 83, 90, and Abel, *Œuvres Comp.*, Christiania, 1881, Vol. I. p. 167).

For example, let $a_1, \dots, a_Q$ be arbitrary places in which $\tau+1$ ϕ-polynomials vanish (Chap. VI. §§ 101, 93). Take $q\,(=Q-p+\tau+1)$ arbitrary places $c_1, \ldots, c_q$, and so determine the set $c_1, \ldots, c_Q$ coresidual with $a_1, \ldots, a_Q$. A rational function, $\zeta(x)$, which has the places $a_1, \ldots, a_Q$ for poles and the places $c_1, \ldots, c_Q$ for zeros is quite determinate save for a constant multiplier. Let $x_1, \ldots, x_Q$ be any set of places at which $\zeta(x)$ has the same value, A say, so that $x_1, \ldots, x_Q$ are the zeros of $\zeta(x)-A$; then, as $a_1, \ldots, a_Q$ are the poles of $\zeta(x)-A$, we have

$$P_{c_1, c_2}^{x_1, a_1} + \ldots\ldots + P_{c_1, c_2}^{x_Q, a_Q} = \log \frac{\zeta(c_1)-A}{\zeta(c_2)-A},$$

and as $\zeta(c_1)=\zeta(c_2)=0$, the right hand is zero.

Hence, calling the places where a definite rational function has the same value a set of *level points* for the function, we can make the statement—the level points of a definite function satisfy the equations

$$\frac{dP_{c_1, c_2}^{x_1}}{dx_1} dx_1 + \ldots\ldots + \frac{dP_{c_1, c_2}^{x_Q}}{dx_Q} dx_Q = 0,$$

c_1, c_2 being any two of the zeros of the function.

In particular, when $q=1$, the sets of level points are the most general sets coresidual with the poles or zeros of the function. Hence, if $x_1, \ldots, x_{p+1}$ be any set of places coresidual with a fixed set $c_1, c_2, \ldots, c_{p+1}$, in which no ϕ-polynomials vanish, we have the equations

$$\frac{dP_{c_1, c_2}^{x_1}}{dx_1} dx_1 + \ldots\ldots + \frac{dP_{c_1, c_2}^{x_{p+1}}}{dx_{p+1}} dx_{p+1} = 0.$$

157. *Ex.* i. We give an example of the application of Abel's theorem.

For the surface associated with the equation

$$y^2 = 4x^{2p+1} - g_1 x^{2p-1} - g_2 x^{2p-2} - \ldots - g_{2p}$$

the integral

$$I = \int \frac{x^p + c_1 x^{p-1} + \ldots + c_p}{y} dx$$

is of the second kind, becoming infinite only at the (single) place $x=\infty$. Consider the rational function

$$Z = \frac{y + Ax^p + Bx^{p-1} + \ldots + Kx + L}{y + A_0 x^p + B_0 x^{p-1} + \ldots + K_0 x + L_0},$$

which, for general values of $A, \ldots, L_0$, is of the $(2p+1)$th order, its zeros, for instance, being given by

$$4x^{2p+1} - g_1 x^{2p-1} - \ldots - g_{2p} - (Ax^p + \ldots + L)^2 = 0.$$

To evaluate the expression

$$\left(\frac{dI}{dt} \frac{1}{Z-\mu}\right)_{t^{-1}},$$

the place $x=\infty$ being the only one to be considered, we put $x=t^{-2}$ and obtain

$$y = \frac{2}{t^{2p+1}} \left(1 - \tfrac{1}{8} g_1 t^4 - \tfrac{1}{8} g_2 t^6 - \ldots\ldots\right),$$

$$Z=\frac{1+\frac{1}{2}At+\dots\dots}{1+\frac{1}{2}A_0t+\dots\dots}=1+\tfrac{1}{2}(A-A_0)\,t+\dots\dots,$$

$$\frac{1}{Z-\mu}=\frac{1}{1-\mu}-\tfrac{1}{2}\frac{A-A_0}{(1-\mu)^2}t+\dots\dots,$$

$$\frac{dI}{dt}=\frac{\dfrac{1}{t^{2p}}+\dfrac{c_1}{t^{2p-2}}+\dots\dots}{\dfrac{2}{t^{2p+1}}(1-\frac{1}{8}g_1t^4-\dots\dots)}\,\frac{-2}{t^3},$$

$$=-\frac{1}{p^2}(1+c_1t^2+c_2t^4+\dots)(1+\tfrac{1}{8}g_1t^4+\dots),$$

$$=-\frac{1}{t^2}-c_1-(c_2+\tfrac{1}{8}g_1)\,t^2-\dots\dots,$$

and therefore

$$\frac{dI}{dt}\frac{1}{Z-\mu}=-\frac{1}{1-\mu}\frac{1}{t^2}+\tfrac{1}{2}\frac{A-A_0}{(1-\mu)^2}\frac{1}{t}-\frac{c_1}{1-\mu}-\dots\dots,$$

wherein the coefficient of t^{-1} is $\frac{1}{2}(A-A_0)(1-\mu)^{-2}$.

Hence, if $x_1,\dots,x_{2p+1}$ be the zeros, and $a_1,\dots,a_{2p+1}$ be the poles of Z, we have

$$I^{x_1,\,a_1}+\dots+I^{x_{2p+1},\,a_{2p+1}}=-\tfrac{1}{2}(A-A_0)\int_\infty^0\frac{d\mu}{(1-\mu)^2}=-\tfrac{1}{2}(A-A_0).$$

Now the zeros of Z are zeros of the polynomial

$$y+U(x)=y+Ax^p+Bx^{p-1}+\dots\dots+Kx+L=0\,;$$

denoting the values of y by $y_1,\dots,y_{2p+1}$, and using $F(x)$ for $(x-x_1)\dots\dots(x-x_{p+1})$, where $(x_1,y_1),\dots,(x_{p+1},y_{p+1})$ are any $p+1$ of the places $(x_1,y_1),\dots,(x_{2p+1},y_{2p+1})$, we have, from the $p+1$ equations

$$y_i+Ax_i^p+Bx_i^{p-1}+\dots\dots+Kx_i+L=0,\qquad (i=1,2,\dots\dots,(p+1)),$$

$$\sum_{i=1}^{p+1}\frac{y_i}{F'(x_i)}=\left[\sum_{i=1}^{p+1}\frac{xy_i}{(x-x_i)F'(x_i)}\right]_{x=\infty}=-\left[\sum_{i=1}^{p+1}\frac{x\,U(x_i)}{(x-x_i)F'(x_i)}\right]_{x=\infty}=-\left[x\frac{U(x)}{F(x)}\right]_{x=\infty}=-A,$$

and hence, if $b_1,b_2,\dots$ be the values of y when $x=a_1,a_2,\dots$, and $F_0(x)=(x-a_1)\dots(x-a_{p+1})$, we have

$$I^{x_1,\,a_1}+\dots\dots+I^{x_{2p+1},\,a_{2p+1}}=\tfrac{1}{2}\sum_{i=1}^{p+1}\frac{y_i}{F'(x_i)}-\tfrac{1}{2}\sum_{i=1}^{p+1}\frac{b_i}{F_0'(a_i)}.$$

If in the integral I the term x^p be absent, the value obtained for the sum

$$I^{x_1,\,a_1}+\dots\dots+I^{x_{2p+1},\,a_{2p+1}}$$

will be zero.

The reader will notice that for $p=1$, we obtain an equation from which the equation

$$-\zeta(u_1)-\zeta(u_2)-\zeta(u_3)=\tfrac{1}{2}\frac{\wp'u_1-\wp'u_2}{\wp u_1-\wp u_2}$$

can be deduced, u_1,u_2,u_3 being arguments whose sum is zero; and that the algebraic equation whose roots are $x_1,\dots,x_{2p+1}$ gives

$$x_1+x_2+\dots\dots+x_{2p+1}=\tfrac{1}{4}A^2=\tfrac{1}{4}\left(\sum_{i=1}^{p+1}\frac{y_i}{F'(x_i)}\right)^2,$$

which for $p=1$ becomes

$$\wp(u_1)+\wp(u_2)+\wp(u_3)=\tfrac{1}{4}\left(\frac{\wp'u_1-\wp'u_2}{\wp u_1-\wp u_2}\right)^2.$$

Ex. ii. If Y, Z be any two rational functions, and u any integral of the first kind, prove by the theorem

$$\left(\frac{1}{(Y-b)(Z-c)}\frac{du}{dx}\frac{dx}{dt}\right)_{t^{-1}}=0$$

that the sum of the values of $(Y-b)^{-1}du/dZ$, at all places where $Z=c$, added to the sum of the values of $(Z-c)^{-1}du/dY$ at all places where $Y=b$, is zero.

It is assumed that all the zeros of the functions $Y-b$, $Z-c$ are of the first order.

Hence prove the equation

$$\sum_{r=1}^{Q}\int_{a_r}^{x_r}\frac{du}{x-b}=\sum_{i=1}^{n}\left(\frac{du}{dx}\log\frac{Z(x)-\mu}{Z(x)-\nu}\right)_i,$$

where $a_1, \ldots, a_Q$ are the places at which $Z(x)=\nu$, $x_1, \ldots, x_Q$ the places at which $Z(x)=\mu$, and the suffix on the right hand indicates that the values of the expression in the brackets are to be taken for the n places of the surface at which $x=b$.

It is assumed that there are no branch places for $x=b$.

Ex. iii. If $\phi(x)$ be any integral polynomial in x, $y^2=(x, 1)_{2p+2}$, $=f(x)$ say, and $M(x)$, $N(x)$ be any two integral polynomials in x of which some coefficients are variable, and

$$f(x) . M^2(x)-N^2(x)=K(x-x_1)\ldots\ldots(x-x_Q),$$

where K is a constant or an integral polynomial whose coefficients do not depend upon the variable coefficients in $M(x)$, $N(x)$, and $y_1, \ldots, y_Q$ be determined by the equations $y_i M(x_i)+N(x_i)=0$, then, on the hypothesis that z is not one of the quantities $x_1, \ldots, x_Q$, and is not a root of $f(x)=0$, prove that

$$\int^{x_1}\frac{\phi(x)\,dx}{(x-z)y}+\ldots\ldots+\int^{x_Q}\frac{\phi(x)\,dx}{(x-z)y}=\frac{\phi(z)}{\sqrt{f(z)}}\log\frac{N(z)+M(z)\sqrt{f(z)}}{N(z)-M(z)\sqrt{f(z)}}-R+C,$$

where C is a constant, and R is the coefficient of $\frac{1}{x}$ in the development of the function

$$\frac{\phi(x)}{(x-z)\sqrt{f(x)}}\log\frac{N(x)+M(x)\sqrt{f(x)}}{N(x)-M(x)\sqrt{f(x)}}$$

in descending powers of x; herein the signs of $\sqrt{f(x)}$, $\sqrt{f(z)}$ are arbitrary, but must be used consistently.

Shew that the statement remains valid when $f(x)$ is of order $2p+1$ (in which case the development from which r is chosen is to be regarded as a development in powers of $\sqrt{x}$); prove that r is zero when $\phi(x)$ is of order p, or of less order. Obtain the corresponding theorem when z is a root of $f(x)=0$.

Ex. iv. The result of Ex. iii. is given by Abel (*Œuvres Compl.*, Vol. i. p. 445), with a direct proof. We explain now the nature of this proof, in the general case. Let $f(y, x)=0$ be the fundamental equation, and let $\theta(y, x)$ be a polynomial of which some of the coefficients are variable; if $y_1, \ldots, y_n$ be the n conjugate roots of $f(y, x)=0$ corresponding to any general value of x, the equation

$$r(x)=\theta(y_1, x)\,\theta(y_2, x)\ldots\ldots\theta(y_n, x)=0,$$

gives the values of x at the finite zeros of the polynomial $\theta(y, x)$. Suppose that the left-hand side breaks into two factors $F_0(x)$ and $F(x)$, of which the former does not contain any of the variable coefficients of $\theta(y, x)$. Let ξ be a root of $F(x)=0$, and $\eta_1, \ldots, \eta_n$ be the corresponding values of y; then one or more of the places $(\xi, \eta_1), \ldots\ldots,$

(ξ, η_n) are zeros of $\theta(y, x)$; fix attention upon one of these, and denote it by (ξ, η). Then if, by a slight change in the variable coefficients of $\theta(y, x)$, whereby it becomes changed into $\theta(y, x)+\delta\theta(y, x)$, $F(x)$ become $F(x)+\delta F(x)$, the symbol δ referring only to the coefficients of $\theta(y, x)$, and ξ become $\xi+d\xi$, we have the equations

$$\delta F(\xi)+F'(\xi)\,d\xi=0,$$

$$F_0(\xi)\,\delta F(\xi)=\delta r(\xi)=\sum_{i=1}^{n}\theta(\eta_1, \xi)\ldots\ldots\theta(\eta_{i-1}, \xi)\,\theta(\eta_{i+1}, \xi)\ldots\ldots\theta(\eta_n, \xi)\,\delta\theta(\eta_i, \xi),$$

where $F'(\xi)=dF(\xi)/d\xi$. Denote now by $U(x)$ the rational function of x, given by

$$U(x)=\sum_{i=1}^{n}\theta(y_1, x)\ldots\ldots\theta(y_{i-1}, x)\,\theta(y_{i+1}, x)\ldots\ldots\theta(y_n, x)\,\delta\theta(y_i, x);$$

then if $R(x, y)$ be any rational function of x and y, we have

$$R(\xi, \eta)\,d\xi=-R(\xi, \eta)\frac{U(\xi)}{F_0(\xi)\,F'(\xi)},$$

where, on account of $\theta(\eta, \xi)=0$ we can write

$$U(\xi)=\frac{r(\xi)}{\theta(\eta, \xi)}\,\delta\theta(\eta, \xi)$$

and

$$R(\xi, \eta)\,U(\xi)=\sum_{i=1}^{n}R(\xi, \eta_i)\,\theta(\eta_1, \xi)\ldots\ldots\theta(\eta_{i-1}, \xi)\,\theta(\eta_{i+1}, \xi)\ldots\ldots\theta(\eta_n, \xi)\,\delta\theta(\eta_i, \xi)$$
$$=\phi(\xi), \text{ say,}$$

$\phi(\xi)$ being a rational function of ξ only. Taking the sum of the equations of this form, for all the zeros of $\theta(y, x)$, we have

$$\Sigma R(\xi, \eta)\,d\xi=-\Sigma\frac{\phi(\xi)}{F_0(\xi)\,F'(\xi)};$$

herein the summation on the right hand can be carried out, and the result written as the perfect differential of a function of the variable coefficients of $\theta(y, x)$, in fact in the form

$$\left[R(x, y)\frac{dx}{dt}\,\delta\log\theta(y, x)\right]_{t^{-1}},$$

as we have shewn.

For example, when

$$f(y, x)=y^3+x^3-3ayx-1,\quad \theta(y, x)=y-mx-n, \text{ we have } F_0(x)=1,$$
$$F(x)=x^3+(mx+n)^3-3ax\,(mx+n)-1,$$

and

$$\frac{\xi\eta d\xi}{\eta^2-a\xi}=-\frac{3\xi\eta\delta F(\xi)}{f'(\eta)\,F'(\xi)}=-\frac{3\xi\eta f'(\eta)\,(\xi\delta m+\delta n)}{f'(\eta)\,F(\xi)}=-\frac{3\xi\,(m\xi+n)\,(\xi\delta m+\delta n)}{F'(\xi)},\ =\frac{\psi(\xi)}{F'(\xi)}, \text{ say.}$$

Now
$$\frac{\psi(x)}{F(x)}=-\frac{3m\delta m}{1+m^3}+\sum_{i=1}^{3}\frac{\psi(\xi_i)}{(x-\xi_i)\,F'(\xi_i)},$$

and hence
$$\Sigma\frac{\xi\eta d\xi}{\eta^2-a\xi}=\left[\frac{x\psi(x)}{F(x)}+\frac{3xm\delta m}{1+m^3}\right]_{x=\infty},\ =-3\delta\left(\frac{mn-a}{1+m^3}\right),$$

as is easily seen. From this we infer

$$\sum_{i=1}^{3}\int_0^{x_i}\frac{xy\,dx}{y^2-ax}=-3\,\frac{mn-a}{1+m^3}+3\left(\frac{mn-a}{1+m^3}\right)_{\substack{m=a\\ n=1}},\ =(x_1+x_2+x_3)\frac{x_1-x_2}{y_1-y_2}.$$

In this example it is easily seen that the integral is only infinite when x is infinite; putting $x=t^{-1}$, the equation $f(y, x)=0$ gives $y=-\omega t^{-1}-a\omega^2+At+Bt^2+\dots\dots$, where $\omega=1$, or $(-1\pm\sqrt{-3})/2$; then $\log\theta(y, x)\, dI/dt$, $=\log(y-mx-n)\,[xy/(y^2-ax)]\,dx/dt$, has $(a\omega^2+n)\,\omega^2/(\omega+m)$ for coefficient of t^{-1}, and we easily find

$$\frac{a+n}{m+1}+\frac{a\omega^2+n}{m+\omega}\omega^2+\frac{a\omega+n}{m+\omega^2}\omega=\frac{3(a-mn)}{m^3+1}.$$

Ex. v. If Y, Z denote any two rational functions (in x and y), such that there is no finite value of x for which both have infinities, and $\Sigma(YZ)$ denote the sum of the n conjugate values of YZ for any value of x, and $[\Sigma(\bar{Y}Z)]_{(x-a)^{-1}}$ denote the sum of the coefficients of $(x-a)^{-1}$ in the expansions of the rational function of x, $\Sigma(YZ)$, for all finite values of x for which Y is infinite, and $[\Sigma(YZ)]_{x^{-1}}$ denote the coefficient of x^{-1} in the expansion of $\Sigma(YZ)$ in descending powers of x, it is easy (cf. § 162 below) to prove that

$$\left(Y\frac{dx}{dt}\bar{Z}\right)'_{t^{-1}}=[\Sigma(YZ)]_{x^{-1}}-[\Sigma(\bar{Y}Z)]_{(x-a)^{-1}},$$

wherein, on the left hand, the dash indicates that the sum is to be taken only for the *finite* places at which Z is infinite. Hence if I be any Abelian integral, $=\int R(x, y)\,dx$, we have

$$\left(\frac{dI}{dt}\delta\log\theta(y, x)\right)'_{t^{-1}}=\left[\Sigma\left(\frac{dI}{dx}\delta\log\theta(y, x)\right)\right]_{x^{-1}}-\left[\Sigma\left(\frac{\bar{dI}}{dx}\delta\log\theta(y, x)\right)\right]_{(x-a)^{-1}}.$$

Hence, if we assume that $\theta(y, x)$ has no variable zeros at infinity, we can obtain Abel's theorem in the form

$$\Sigma\frac{dI}{dx}dx=-\left[\Sigma\left(\frac{dI}{dx}\delta\log\theta(y, x)\right)\right]_{x^{-1}}+\left[\Sigma\left(\frac{\overline{dI}}{dx}\delta\log\theta(y, x)\right)\right]_{(x-a)^{-1}},$$

wherein the summation on the left refers to all the zeros of $\theta(y, x)$.

This is the form in which the result is given by Abel (*Œuvres Compl.*, Christiania, 1881, Vol. i. p. 159, and notes, Vol. ii. p. 296), the right hand being obtained by actual evaluation of the summation which we have written, in the last example, in the form

$$-\Sigma\frac{\phi(\xi)}{F_0(\xi)\,F'(\xi)}.$$

The reader is recommended to study Abel's paper*, which, beside the theorem above, contains two important enquiries; first, as to the form necessary for the rational function dI/dx, in order that the right-hand side of the equation of Abel's theorem may reduce to a constant, next, as to the least number of the integrals in the equation of Abel's theorem, of which the upper limits may not be taken arbitrarily but must be taken as functions of the other upper limits. Though the results have been incorporated in the theory here given (§§ 156, 151, 95), Abel's investigation must ever have the deepest interest.

Ex. vi. Obtain the result of Ex. i. (§ 157) by the method explained in Ex. iv.

* Which was presented to the Academy of Sciences of Paris in Oct. 1826, and published by the Academy in 1841 (*Mémoires par divers savants*, t. vii.). During this period many papers were published in *Crelle's Journal* on Abel's theorem, by Abel, Minding, Jürgensen, Broch, Richelot, Jacobi and Rosenhain. (See *Crelle*, i—xxx. I have not examined all these papers with care. Jürgensen uses a method of fractional differentiation.)

Ex. vii. Prove that the sum of the values of the expression

$$\frac{U \cdot v}{J},$$

wherein v is any linear expression in the homogeneous coordinates x, y, z, U is any integral polynomial of degree $m+n-3$, J is the Jacobian of any two curves $f=0$, $\phi=0$, of degrees n and m, and the line $v=0$, and the sum extends to all the common points of $f=0$ and $\phi=0$, vanishes, multiple points of $f=0$, $\phi=0$ being disregarded.

Hence deduce Abel's theorem for integrals of the first kind.

(See Harnack, Alg. Diff. *Math. Annal.* t. ix.; Cayley, *Amer. Journ.* Vol. v. p. 158; Jacobi, theoremata nova algebraica, *Crelle*, t. xiv. The theorem is due to Jacobi; for geometrical applications, see also Humbert, *Liouville's Journal* (1885) Ser. iv. t. i. p. 347)*.

Ex. viii. For the surface

$$y^2 = \phi(x)\,\psi(x), \quad = f(x),$$

wherein $\phi(x)$, $\psi(x)$ are cubic polynomials in x, prove the equation

$$P_{\xi,\gamma}^{x_1, m_1} + P_{\xi,\gamma}^{x_2, m_2} + P_{\xi,\gamma}^{\bar{\xi}, \bar{\gamma}} + 2\log\{[\sqrt{\phi(\xi)\,\psi(\gamma)} + \sqrt{\phi(\gamma)\,\psi(\xi)}]/2\sqrt[4]{f(\xi)\,f(\gamma)}\} = 0,$$

wherein $x_1, x_2, \bar{\xi}$ and $m_1, m_2, \bar{\gamma}$ are coresidual with the roots of $\phi(x)=0$, and $\bar{\xi}, \bar{\gamma}$ are the places conjugate to ξ and γ; conjugate places being those for which the values of x are the same.

158. When the places $x_1, \dots, x_Q$ are determined as coresidual with the fixed places $a_1, \dots, a_Q$, $p-\tau-1$ of the places $x_1, \dots, x_Q$ are fixed by the assignation of the others. Hence the $p+1$ relations, which are given by Abel's theorem,

$$u_i^{x_1, a_1} + \dots\dots + u_i^{x_Q, a_Q} = 0,$$

$$P_{\xi,\gamma}^{x_1, a_1} + \dots\dots + P_{\xi,\gamma}^{x_Q, a_Q} = \log[Z(\xi)/Z(\gamma)],$$

cannot be independent. We prove now first of all that the last may be regarded as a consequence of the other p equations. *In fact, if $x_1, \dots, x_Q$ and $a_1, \dots, a_Q$ be any two sets of places, such that, for any paths of integration,*

$$u_i^{x_1, a_1} + \dots\dots + u_i^{x_Q, a_Q} = M_1\omega_{i,1} + \dots\dots + M_p\omega_{i,p} + M_1'\omega'_{i,1} + \dots\dots + M'_p\omega'_{i,p},$$

$(i = 1, 2, \dots, p)$, wherein $u_1^{x,a}, \dots, u_p^{x,a}$ are any set of linearly independent integrals of the first kind, $\omega_{i,1}, \dots, \omega'_{i,p}$ are the periods of the integral $u_i^{x,a}$, and $M_1, \dots, M'_p$ are rational integers independent of i, then there exists a rational function having the places $a_1, \dots, a_Q$ for poles and the places $x_1, \dots, x_Q$ for zeros.

For if $v_1^{x,a}, \dots, v_p^{x,a}$ be the normal integrals of the first kind, so that we have equations of the form,

$$v_i^{x,a} = C_{i,1}\,u_1^{x,a} + \dots\dots + C_{i,p}\,u_p^{x,a},$$

* Further algebraical consideration of Abel's theorem may be found in Clebsch-Lindemann-Benoist, *Leçons sur la Géométrie* (Paris 1883) Vol. iii. Geometrical applications are given by Humbert, *Liouville's Journal*, 1887, 1889, 1890 (Ser. iv. t. iii. v. vi.).

wherein $C_{i,1}, \ldots, C_{i,p}$ are constants, and therefore, also,

$$C_{i,1}\,\omega_{1,j} + \ldots\ldots + C_{i,p}\,\omega_{p,j} = 0 \text{ or } 1, \text{ according as } i \neq j, \text{ or } i = j,$$

and

$$C_{i,1}\,\omega'_{1,j} + \ldots\ldots + C_{i,p}\,\omega'_{p,j} = \tau_{i,j},$$

we can deduce

$$v_i^{x_1, a_1} + \ldots\ldots + v_i^{x_Q, a_Q} = M_i + M_1'\tau_{i,1} + \ldots\ldots + M'_p\,\tau_{i,p}.$$

Consider now the function

$$Z(x) = e^{\Pi_{x_1, a_1}^{x, c} + \ldots\ldots + \Pi_{x_Q, a_Q}^{x, c} - 2\pi i (M'_1 v_1^{x, c} + \ldots\ldots + M'_p v_p^{x, c})},$$

c being an arbitrary place.

Herein an integral, $\Pi_{x_1, a_1}^{x, c}$, suffers an increment $2\pi i$ when x makes a circuit about the place x_1; but this does not alter the value of $Z(x)$. And in fact $Z(x)$ is a single-valued function of x; for the functions $\Pi_{x_i, a_i}^{x, a}$ have no periods at the first p period loops, while, if x describe a circuit equivalent to crossing the i-th period loop of the second kind, the function $Z(x)$ is only multiplied by the factor

$$e^{2\pi i (v_i^{x_1, a_1} + \ldots\ldots + v_i^{x_Q, a_Q}) - 2\pi i (M'_1\tau_{i,1} + \ldots\ldots + M'_p\tau_{i,p})},$$

or $e^{2\pi i M_i}$, whose value is unity.

Further the function $Z(x)$ has no essential singularities; for it has poles at the places $a_1, \ldots, a_Q$, and is elsewhere finite.

Since the function has zeros at $x_1, \ldots, x_Q$ and not elsewhere, the statement made above is justified.

Ex. i. It is impossible to find two places γ, ξ, such that each of the p integrals $u_i^{\xi, \gamma}$ is zero. For then there would exist a rational function, given by

$$e^{\Pi_{\xi, \gamma}^{x, a}},$$

having only one pole, at the place γ. (Cf. § 6, Chap. I.) It is also impossible that the equations

$$v_i^{\xi, \gamma} = M_1 + M'_1\,\tau_{i,1} + \ldots\ldots + M'_p\,\tau_{i,p},$$

wherein $M_1, \ldots, M_p, M'_1, \ldots, M'_p$ are rational integers independent of i, should be simultaneously true.

Ex. ii. If p equations, of the form

$$v_i^{\xi_1, \gamma_1} + v_i^{\xi_2, \gamma_2} = M_i + M'_1\,\tau_{i,1} + \ldots\ldots + M'_p\,\tau_{i,p}$$

exist, γ_1 and γ_2 are the poles of a rational function of the second order, and the surface is hyperelliptic. (Chap. V. § 52.)

159. In regard now to the equations

$$u_i^{x_1, a_1} + \ldots\ldots + u_i^{x_Q, a_Q} = 0,$$

which express that the places $x_1, \ldots, x_Q$ are coresidual with the places $a_1, \ldots, a_Q$, if $\tau + 1$ be the number of ϕ-polynomials which vanish in the places $a_1, \ldots, a_Q$ (Chap. VI. § 93), or (Chap. III. §§ 27, 37) the number of linearly independent linear aggregates of the form

$$C_1\Omega_1(x) + \ldots\ldots + C_p\Omega_p(x),$$

wherein $C_1, \ldots, C_p$ are constants, which vanish in these places, then, $Q - p + \tau + 1$ of the places $x_1, \ldots, x_Q$ can be assumed arbitrarily, and the equations are therefore equivalent to only $p - \tau - 1$ equations, determining the other places of $x_1, \ldots, x_Q$ in terms of those assumed. This can be stated also in another way: the p differential equations

$$\frac{du_i}{dx_1} dx_1 + \ldots\ldots + \frac{du_i}{dx_Q} dx_Q = 0, \quad (i = 1, 2, \ldots, p),$$

express that the places $x_1, \ldots, x_Q$ are coresidual with the places $x_1 + dx_1, \ldots, x_Q + dx_Q$; if the places $x_1, \ldots, x_Q$ have quite general positions these equations are independent; if however $\tau + 1$ linearly independent linear aggregates, of the form,

$$C_1 \frac{du_1}{dx} + \ldots\ldots + C_p \frac{du_p}{dx} = 0,$$

wherein $C_1, \ldots, C_p$ are constants, vanish in the places $x_1, \ldots, x_Q$, then the p differential equations are linearly determinable from $p - \tau - 1$ of them.

Ex. i. A rational function having $x_1, \ldots, x_Q$ as poles of the first order, and such that $\lambda_1, \ldots, \lambda_p$ are the coefficients of the inverses of the infinitesimals in the expansion of the function in the neighbourhood of these places, can be written in the form

$$-\lambda_1 \Gamma_{x_1}^{x, c} - \ldots\ldots - \lambda_Q \Gamma_{x_Q}^{x, c};$$

the conditions that the periods be zero are then the p equations

$$\lambda_1 \Omega_i(x_1) + \ldots\ldots + \lambda_Q \Omega_i(x_Q) = 0, \quad (i = 1, 2, \ldots, p).$$

But, if we take consecutive places coresidual with $x_1, \ldots, x_Q$, and $t_1, \ldots, t_Q$ be the corresponding values of the infinitesimals at $x_1, \ldots, x_Q$, we also have

$$\Omega_i(x_1) t_1 + \ldots\ldots + \Omega_i(x_Q) t_Q = 0;$$

thus, if the first q $(= Q - p + \tau + 1)$ of $t_1, \ldots, t_Q$ be taken proportional to $\lambda_1, \ldots, \lambda_q$, we shall have the equations

$$t_1/\lambda_1 = \ldots\ldots = t_Q/\lambda_Q.$$

Ex. ii. When the set $x_1, \ldots, x_Q$, beside being coresidual with $a_1, \ldots, a_Q$, has other specialities of position, Abel's theorem may be incompetent to express them. For instance, in the case of a Riemann surface whose equation represents a plane quartic curve with two double points, there is one finite integral; if $a_1, \ldots, a_4$ represent any 4 collinear points, and $x_1, \ldots, x_4$ represent any other 4 collinear points, the equation of Abel's theorem is

$$u^{x_1, a_1} + \ldots + u^{x_4, a_4} = 0;$$

but this equation does not express the *two* relations which are necessary to ensure that $x_1, \ldots, x_4$ are collinear; it expresses only that x_1, x_2, x_3, x_4 are on a conic, S, passing through the double points, or that x_1, x_2, x_3, x_4 are the zeros, and $a_1, \ldots, a_4$ are the poles of the rational function S/LL_0, where $L=0$ is the line containing $a_1, \ldots, a_4$ and $L_0=0$ is the line joining the double points.

160. From these results there follows the interesting conclusion that the p simultaneous differential equations

$$\frac{du_i}{dx_1} dx_1 + \ldots\ldots + \frac{du_i}{dx_Q} dx_Q = 0, \qquad (i = 1, 2, \ldots, p),$$

have algebraical integrals, Q being $> p$, and $u_1, \ldots, u_p$ being a set of p linearly independent integrals of the first kind. The problem of determining these integrals consists only in the expression of the fact that $x_1, \ldots, x_Q$ constitute a set belonging to a lot of coresidual sets of places.

The most general lot will consist of the sets coresidual with Q arbitrary fixed places $a_1, \ldots, a_Q$, in which no ϕ-polynomials vanish. But the lot does not therefore depend on Q arbitrary constants; for in place of the set $a_1, \ldots, a_Q$ we can equally well use a set $A_1, \ldots, A_Q$, whereof $q, = Q-p$, places have positions arbitrarily assigned beforehand; in other words, all possible lots of sets of Q places with multiplicity q can be regarded as derived from fundamental sets of Q places in which q places are the same for all. *A lot depends therefore on $Q-q, =p$, arbitrary constants,* and this number of arbitrary constants should appear in the integrals of the equations (Chap. VI. § 96).

We may denote the Q arbitrary places, with which $x_1, \ldots, x_Q$ are coresidual, by $A_1, \ldots, A_q, a_1, \ldots, a_p$, so that $A_1, \ldots, A_q$ are arbitrarily assigned beforehand, in any way that is convenient, and the positions of $a_1, \ldots, a_p$ are the arbitrary constants of the integration.

Then one way in which we can express the integrals of the equations is as follows: form the rational function with poles, of the first order, in the places $x_1, \ldots, x_Q$, and determine the ratios of the $q+1$ homogeneous arbitrary coefficients entering therein, so that the function vanishes in $A_1, \ldots, A_q$. Then the function is determined save for an arbitrary multiplier, and must vanish also in $a_1, \ldots, a_p$. The expression of the fact that it does so gives p equations, each containing one of $a_1, \ldots, a_p$ as an arbitrary constant.

From these p equations we may suppose p of the places $x_1, \ldots, x_Q$, say $x_1, \ldots, x_p$, to be expressed in terms of $a_1, \ldots, a_p$ and $x_{p+1}, \ldots, x_Q$ (and $A_1, \ldots, A_q$). The resulting equations may be derived also by forming the general rational function with its poles in $a_1, \ldots, a_p, A_1, \ldots, A_q$ and eliminating the arbitrary constants by the condition that this function vanishes in $x_i, x_{p+1}, x_{p+2}, \ldots, x_Q$, i being in turn taken equal to $1, 2, \ldots, p$.

For example, for $Q=p+1$, if $\psi(x, a; z, c_1, \ldots, c_p)$ denote the definite rational function which has poles of the first order in the places $z, c_1, \ldots, c_p$, the coefficient of the inverse of the infinitesimal at the place z being taken $=-1$, which function also vanishes at the place a (Chap. VII. § 122), then a complete set of integrals is given by

$$\psi(a_1, A; x_{p+1}, x_1, \ldots, x_p)=0=\ldots\ldots=\psi(a_p, A; x_{p+1}, x_1, \ldots, x_p),$$

and a complete set is also given by

$$\psi(x_1, x_{p+1}; A, a_1, \ldots, a_p)=0=\ldots\ldots=\psi(x_p, x_{p+1}; A, a_1, \ldots, a_p).$$

The first of these integrals is in fact the equation

$$\begin{vmatrix} \frac{du_1}{dx_1}, & \frac{du_1}{dx_2}, & \cdot & \cdot, & \frac{du_1}{dx_{p+1}} \\ \cdot & \cdot & \cdot & & \cdot \\ \frac{du_p}{dx_1}, & \frac{du_p}{dx_2}, & \cdot & \cdot, & \frac{du_p}{dx_{p+1}} \\ \frac{dP}{dx_1}, & \frac{dP}{dx_2}, & \cdot & \cdot, & \frac{dP}{dx_{p+1}} \end{vmatrix}=0,$$

wherein $P=P_{a_1, A}^{x_1, c}$, and may be regarded as derived by elimination of $dx_1, \ldots, dx_{p+1}$ from the p given differential equations and the differential of the equation (§ 156)

$$P_{a_1, A}^{x_1, c_1}+\ldots\ldots+P_{a_1, A}^{x_{p+1}, c_{p+1}}=0,$$

which holds when $(x_1, \ldots, x_{p+1})$, $(c_1, \ldots, c_{p+1})$, and $(A, a_1, \ldots, a_p)$ are coresidual sets.

Ex. i. For $p=1$, the fundamental equation being $y^2=(x, 1)_4=\lambda^2x^4+\ldots$, shew that the differential equation

$$\frac{dx_1}{y_1}+\frac{dx_2}{y_2}=0$$

has the integral

$$\frac{y_1+b}{x_1-a}+\lambda x_1=\frac{y_2+b}{x_2-a}+\lambda x_2,$$

where $b^2=(a, 1)_4$. (Here the place A has been taken at infinity.)

Shew also that this integral expresses that the places (x_1, y_1), (x_2, y_2), $(a, -b)$, are the variable zeros of the polynomial $-y+p+qx-\lambda x^2$, when p and q are varied.

Ex. ii. For $p=2$, the fundamental equation being $y^2=(x, 1)_6=\lambda^2x^6+\ldots$, using the form of the function $\psi(x, a; z, c_1, \ldots, c_p)$ given in Ex. ii. § 132, Chap. VII., and putting the place A at infinity, obtain, for the differential equations

$$\frac{dx_1}{y_1}+\frac{dx_2}{y_2}+\frac{dx_3}{y_3}=0, \qquad \frac{x_1dx_1}{y_1}+\frac{x_2dx_2}{y_2}+\frac{x_3dx_3}{y_3}=0,$$

the integral

$$\frac{y_1}{(x_1-a)F'(x_1)}+\frac{y_2}{(x_2-a)F'(x_2)}+\frac{y_3}{(x_3-a)F'(x_3)}+\frac{b}{F(a)}=-\lambda,$$

wherein $F(x)=(x-x_1)(x-x_2)(x-x_3)$, $b^2=(a, 1)_6$, and the position of the place (a, b) is the arbitrary constant of integration. By taking three positions of (a, b) we obtain a system of complete integrals.

Shew that this integral is obtained by eliminating p, q, r from the equations which express that the places (x_1, y_1), (x_2, y_2), (x_3, y_3), (a, b) are zeros of the polynomial $-y-\lambda x^3+px^2+qx+r$.

Ex. iii. For the case $(p=3)$ in which the fundamental equation is of the form

$$f(y, x)=(x, y)_4+(x, y)_3+(x, y)_2+(x, y)_1=0,$$

$(x, y)_4$ being a homogeneous polynomial of the fourth degree with general coefficients, etc., prove that an integral of the equations

$$\frac{dx_1}{f'(y_1)}+\frac{dx_2}{f'(y_2)}+\frac{dx_3}{f'(y_3)}+\frac{dx_4}{f'(y_4)}=0, \quad \frac{x_1dx_1}{f'(y_1)}+\text{etc.}=0, \quad \frac{y_1dx_1}{f'(y_1)}+\text{etc.}=0,$$

is given by

$$(2, 3, 4)\, U_1+(3, 1, 4)\, U_2+(1, 2, 4)\, U_3-(1, 2, 3)\, U_4=0,$$

where
$$(2, 3, 4)=\begin{vmatrix} x_2 & x_3 & x_4 \\ y_2 & y_3 & y_4 \\ 1 & 1 & 1 \end{vmatrix} \text{ etc.,}$$

and
$$U_i=\frac{f\left(\frac{b}{a}x_i, x_i\right)}{x_i(x_i-a)\left(y_i-\frac{b}{a}x_i\right)},$$

$f(b, a)$ being $=0$, and the position of (a, b) being the arbitrary constant of integration. A complete system of integrals is obtained by giving (a, b) any three arbitrary positions. To obtain these equations the place A has been put at $x=0$, $y=0$.

Ex. iv. When the fundamental equation is $x^4+y^4=1$, shew, putting the place A at $x=1$, $y=0$, that, as in Ex. iii., we have integrals of the form

$$(2, 3, 4)\, U_1+(3, 1, 4)\, U_2+(1, 2, 4)\, U_3-(1, 2, 3)\, U_4=0,$$

wherein
$$U_i=\frac{x_i^2(2a^2-a+1)-x_i(a+1)^2+a^2-a+2}{(a-1)y_i-(x_i-1)b},$$

and $a^4+b^4=1$.

161. The method of forming the integrals of the differential equations which is explained in the last article may also be stated thus: take any adjoint polynomial ψ which vanishes in the Q places $A_1, \dots, A_q, a_1, \dots, a_p$; let $C_1, \dots, C_R$ be the other zeros* of ψ; let the general adjoint polynomial of the same grade as ψ, which vanishes in $C_1, \dots, C_R$, be denoted by

$$\lambda\psi+\lambda_1\psi_1+\dots\dots+\lambda_q\psi_q,$$

$\lambda, \lambda_1, \dots, \lambda_q$ being arbitrary constants. By expressing that the places $x_i, x_{p+1}, x_{p+2}, \dots, x_Q$ are zeros of this polynomial we obtain a relation whereby x_i is determined from $x_{p+1}, \dots, x_Q$ in terms of the arbitrary positions

* Beside those where $f'(y)$ or $F'(\eta)$ vanishes (cf. Chap. VI. § 86).

$a_1, \dots, a_p$ (and $A_1, \dots, A_q$). By taking $i = 1, 2, \dots, p$ we obtain a complete system* of integrals.

Now instead of regarding the set $A_1, \dots, A_q, a_1, \dots, a_p$ as the arbitrary quantities of the integration, we may regard the set $C_1, \dots, C_R$ as the arbitrary quantities, or, more accurately, we may regard the p quantities upon which the lot of sets coresidual with $C_1, \dots, C_R$ depends, as the arbitrary quantities. To this end, and under the hypothesis that no ϕ-polynomials vanish in the places $C_1, \dots, C_R$, imagine a set of places $B_1, \dots, B_{R-p}, b_1, \dots, b_p$ determined coresidual with $C_1, \dots, C_R$, in which $B_1, \dots, B_{R-p}$ have any convenient positions assigned beforehand, so that the lot of sets coresidual with $C_1, \dots, C_R$ depends upon the positions of $b_1, \dots, b_p$. Let a general adjoint polynomial with $Q + R$ variable zeros be of the form

$$\Theta = \mu\vartheta + \mu_1\vartheta_1 + \dots\dots + \mu_k\vartheta_k,$$

wherein $\mu, \dots, \mu_k$ are arbitrary constants, and k is for shortness written for $Q + R - p$. Then an integral of the differential equations under consideration is obtained by expressing that the places

$$B_1, \dots, B_{R-p}, b_1, \dots, b_p, x_i, x_{p+1}, x_{p+2}, \dots, x_Q$$

are zeros of the polynomial Θ; and a complete system of integrals is obtained by putting i in turn equal to $1, 2, \dots, p$.

Similarly a complete set of integrals is obtained by expressing that the places

$$x_1, \dots, x_p, x_{p+1}, \dots, x_Q, b_i, B_1, \dots, B_{R-p}$$

are zeros of the polynomial Θ, i being taken in turn equal to $1, 2, \dots, p$.

In this enunciation there is no restriction as to the value of R, save that it must not be less than p.

Ex. i. For the general surface of the form

$$f(y, x) = (x, y)_4 + (x, y)_3 + (x, y)_2 + (x, y)_1 + \text{constant} = 0,$$

a set of integrals of the equations

$$\sum_1^4 \frac{dx_i}{f'(y_i)} = 0, \quad \sum_1^4 \frac{x_i dx_i}{f'(y_i)} = 0, \quad \sum_1^4 \frac{y_i dx_i}{f'(y_i)} = 0,$$

is given by

$$\begin{vmatrix} x_1^2 & x_1y_1 & y_1^2 & x_1 & y_1 & 1 \\ x_2^2 & x_2y_2 & y_2^2 & x_2 & y_2 & 1 \\ x_3^2 & x_3y_3 & y_3^2 & x_3 & y_3 & 1 \\ x_4^2 & x_4y_4 & y_4^2 & x_4 & y_4 & 1 \\ a_i^2 & a_ib_i & b_i^2 & a_i & b_i & 1 \\ A^2 & AB & B^2 & A & B & 1 \end{vmatrix} = 0,$$

* And we can of course obtain quite similarly a set of p integrals, each connecting $x_1, \dots, x_Q, A_1, \dots, A_q$, and one of the arbitrary positions $a_1, \dots, a_p$.

where $f(b_i, a_i)=0$, $f(B, A)=0$, $i=1, 2, 3$, and the place (A, B) may be taken at any convenient position.

Ex. ii. Taking as before $Q=p+1$, and considering the hyperelliptic case, the fundamental equation being

$$y^2=(x, 1)_{2p+2}=\lambda^2 x^{2p+2}+\mu u^{2p+1}+\dots\dots,$$

we require a polynomial having $R+p+1$ variable zeros: such an one is

$$\Theta=-y+\lambda x^{p+1}+Fx^p+Gx^{p-1}+\dots\dots+H,$$

R being equal to p, and we have

$$(\lambda^2 x^{2p+2}+\mu x^{2p+1}+\dots)-(\lambda x^{p+1}+Fx^p+\dots+H)^2=(\mu-2\lambda F)\,F(x)\,\phi(x),$$

where $F(x)=(x-x_1)\dots\dots(x-x_{p+1})$, $\phi(x)=(x-b_1)\dots\dots(x-b_p)$.

An integral of the differential equations may be obtained by eliminating $F, G, \dots, H$ from the equations expressing that the places

$$b_1, \dots, b_p, x_i, x_{p+1}$$

are zeros of the polynomial Θ, or from the equations expressing that

$$x_1, \dots, x_p, x_{p+1}, b_i$$

are zeros of this polynomial, and a complete system of integrals, in either case, by taking i in turn equal to $1, 2, \dots, p$.

Or a complete system of p integrals may be obtained by eliminating $F, G, \dots, H$ from the $2p+1$ equations obtained by equating the coefficients of the same powers of x on the two sides of the equation.

We may of course also take Θ in the form

$$-y+Ex^{p+1}+Fx^p+\dots\dots+H;$$

then $R=p+1$, and the places $B_1, \dots, B_{R-p}$ are not evanescent; putting the place B_1 at infinity we obtain $E=\lambda$, as above.

Ex. iii. The integration in the previous example may be carried out in various ways. By introducing again a set of fixed places $a_1, \dots, a_p, A$, coresidual with $x_1, \dots, x_p, x_{p+1}$, we can draw a particular inference as to the forms of the coefficients $F, G, \dots, H$. For if $U(x)$ denote $\lambda x^{p+1}+Fx^p+\dots+G$, and $U_0(x)$ denote what $U(x)$ becomes when $x_1, \dots, x_{p+1}$ take the positions $a_1, \dots, a_p, A$, the coefficients $F, G, \dots, H$ being then $F_0, G_0, \dots, H_0$, and also $F_0(x)=(x-a_1)\dots\dots(x-a_p)(x-A)$, then, because each of the polynomials $-y+U(x)$, $-y+U_0(x)$ vanishes in the places $b_1, \dots, b_p$, the polynomial $U(x)-U_0(x)$ must divide by $\phi(x)$, namely $U(x)=U_0(x)+t\,\phi(x)$, where t is a variable parameter; or, if we write $\phi(x)=x^p+t_1 x^{p-1}+\dots\dots+t_p$, $t_1, \dots, t_p$ being then regarded, instead of $b_1, \dots, b_p$, as the arbitrary constants of the integration, we have

$$F=F_0+t, \quad G=G_0+tt_1, \dots\dots\dots, \quad H=H_0+tt_p,$$

and the quantities $G-t_1 F, \dots, H-t_p F$ are constants in the integration, being unaltered when the places $x_1, \dots, x_{p+1}$ come to $a_1, \dots, a_p, A$. Hence we can formulate the following result: let the $p+1$ quantities $F_0, G_0, \dots, H_0$ be determined so that the polynomial $-y+U_0(x)$ vanishes in the fixed places $a_1, \dots, a_p, A$. Then denoting $(x-a_1)\dots(x-a_p)(x-A)$ by $F_0(x)$, the fraction

$$[y^2-U_0^2(x)]/F_0(x)$$

is an integral polynomial; denote it by $(\mu-2F_0\lambda)(x^p+t_1 x^{p-1}+\dots\dots+t_p)$, so that

$\phi_0, t_1, \dots, t_p$ are uniquely determined in terms of the places $a_1, \dots, a_p, A$, and put $F(x)$ for $x^p + t_1 x^{p-1} + \dots\dots + t_p$. Then $x_1, \dots, x_{p+1}$ are the roots of the equation

$$\frac{y^2 - [U_0(x) + t\,\phi(x)]^2}{\phi(x)} = (\mu - 2F_0\lambda)\,F_0(x) - 2t\,U_0(x) - t^2\phi(x) = 0;$$

and the set $x_1, \dots, x_{p+1}$ varies with the value of t, which is the only variable quantity in this equation. By equating the coefficients of the various powers of x in the polynomial on the left-hand side of this equation to the coefficients in the polynomial $(\mu - 2F_0\lambda)\,F(x)$, we can express each of the symmetric functions

$$h_1 = x_1 + \dots\dots + x_{p+1}$$
$$h_2 = x_1x_2 + x_1x_3 + \dots\dots + x_p\,x_{p+1}$$
$$\dots\dots\dots\dots\dots\dots\dots\dots\dots\dots$$

as rational quadratic functions of a variable parameter t, containing definite rational functions of the variables at the places $a_1, \dots, a_p, A$; the place A may be given any fixed position that is convenient; the positions of the places $a_1, \dots, a_p$ are the arbitrary constants of the integration.

Ex. iv. By eliminating t between the $p+1$ equations obtained at the end of Ex. iii. we obtain the complete system of p integrals. In particular any two of the quantities $h_1, h_2, \dots$ are connected by a quadratic relation, and any three of them are connected by a linear relation (Jacobi, *Crelle*, t. 32, p. 220).

Ex. v. From the equation

$$\frac{U(x)}{F(x)} = \lambda + \sum_{r=1}^{p+1} \frac{y_r}{(x - x_r)\,F'(x_r)}$$

we infer

$$F + \lambda h_1 = \sum_{r=1}^{p+1} \frac{y_r}{F'(x_r)},\quad \mu - 2F\lambda = \mu + 2\lambda^2 h_1 - 2\lambda \sum_{r=1}^{p+1} \frac{y_r}{F'(x_r)},$$

where $h_1 = x_1 + \dots + x_{p+1}$; hence if a be the value of x at a branch place of the surface, we have from Ex. ii.

$$-F(a)\left[\lambda + \sum_{r=1}^{p+1} \frac{y_r}{(a - x_r)\,F'(x_r)}\right]^2 = \phi(a)\left[\mu + 2\lambda^2 h_1 - 2\lambda \sum_{r=1}^{p+1} \frac{y_r}{F'(x_r)}\right],$$

and if, herein, a be put in turn at any p of the branch places of the surface, the resulting values of $\phi(a)$ may be regarded as the arbitrary constants of the integration, and the resulting equations as a complete set of integrals; and if $\lambda = 0$, as we may always suppose without loss of generality (Chap. V.), we thus obtain the p integrals

$$(a_i - x_1)\dots(a_i - x_{p+1})\left[\sum_{r=1}^{p+1} \frac{y_r}{(a_i - x_r)\,F'(x_r)}\right]^2 = C_i, \qquad (i = 1, 2, \dots\dots, p)$$

$C_1, \dots, C_p$ being the constants of integration (Richelot, *Crelle*, xxiii. (1842), p. 369. In this paper is also shewn how to obtain integrals by extension of Lagrange's method for the case $p=1$. See Lagrange, *Theory of Functions*, Chap. II., and Cayley, *Elliptic Functions*, 1876, p. 337).

Ex. vi. By comparing coefficients of x^{2p} in the equation of Ex. ii., we obtain

$$\nu - (2\lambda G + F^2) = (\mu - 2\lambda F)\,(t_1 - h_1),$$

where $h_1 = x_1 + \dots + x_{p+1}$; hence prove that

$$\left\{\sum_{r=1}^{p+1} \frac{y_r}{F'(x_r)}\right\}^2 - \mu\,(x_1 + \dots + x_{p+1}) - \lambda^2\,(x_1 + \dots + x_{p+1})^2 = \nu - t_1\,\mu - 2\lambda\,(G - Ft_1);$$

by Ex. ii. the right-hand side is a constant in the integration; hence this equation is an integral of the differential equations; in particular if $\lambda=0$, $\mu=4$, which is not a loss of generality, we have the integral

$$x_1+\dots+x_{p+1}+C=\tfrac{1}{4}\left[\sum_{r=1}^{p+1}\frac{y_r}{F'(x_r)}\right]^2,$$

where C is a constant; this is a generalization of the equation, for $p=1$,

$$\wp u+\wp v+\wp(u+v)=\tfrac{1}{4}\left(\frac{\wp' u-\wp' v}{\wp u-\wp v}\right)^2$$

(cf. Ex. i. § 157).

Ex. vii. Shew that if the fundamental equation be

$$y^2=(x,\ 1)^{2p+2}=\lambda^2 x^{2p+2}+\mu x^{2p+1}+\dots\dots+Lx+M,$$

then another integral is

$$x_1^2\dots x_{p+1}^2\left[\sum_{r=1}^{p+1}\frac{y_r}{x_r^2 F'(x_r)}\right]^2-L\left(\frac{1}{x_1}+\dots+\frac{1}{x_{p+1}}\right)-M\left(\frac{1}{x_1}+\dots+\frac{1}{x_{p+1}}\right)^2=\text{Const.}$$

(Richelot, *loc. cit.*)

Ex. viii. If a_0, a_i be the values of x at two branch places of the surface, obtain the equations

$$\frac{(a_i-x_1)\dots\dots(a_i-x_{p+1})}{(a_i-A)\dots\dots(a_i-a_p)}\Bigg/\frac{(a_0-x_1)\dots\dots(a_0-x_{p+1})}{(a_0-A)\dots\dots(a_0-a_p)}=(1+\mu\rho_i)^2,$$

wherein the quantities $A, \dots, a_p$ are the values of x at fixed places coresidual with $x_1, \dots, x_{p+1}$, ρ_i is an absolute constant, and μ is a parameter varying with the places $x_1, \dots, x_{p+1}$. Take i in turn equal to $1, 2, \dots, (p+1)$, and, eliminating μ, we obtain a complete set of integrals. In particular if the left-hand side of this equation be denoted by G_i we have such equations as

$$(G_i-1)\,\rho_j\,\rho_k\,(\rho_j-\rho_k)+(G_j-1)\,\rho_k\,\rho_i\,(\rho_k-\rho_i)+(G_k-1)\,\rho_i\,\rho_j\,(\rho_i-\rho_j)=0.$$

(Weierstrass, *Collected Works*, Vol. I. p. 267.)

162. The proof of Abel's theorem which has been given in this chapter can be extended to the case of an algebraical curve in space. Taking the case of three dimensions, and denoting the coordinates by x, y, z, we shall assume that for any finite value of x, say $x=a$, the curve is completely given by a series of equations of the form

$$\begin{array}{llll} x=a+t_1^{w_1+1}, & x=a+t_2^{w_2+1}, & \dots\dots\dots, & x=a+t_k^{w_k+1}, \\ y=P_1(t_1)\quad, & y=P_2(t_2)\quad, & \dots\dots\dots, & y=P_k(t_k)\quad, \\ z=Q_1(t_1)\quad, & z=Q_2(t_2)\quad, & \dots\dots\dots, & z=Q_k(t_k)\quad, \end{array} \qquad \text{(D)}$$

wherein $w_1+1, \dots, w_k+1$ are positive integers, $t_1, \dots, t_k$ are infinitesimals, and $P_1, Q_1, \dots, P_k, Q_k$, denote power series of integral powers of the variable, with only a finite number of negative powers, which have a finite radius of convergence. The values represented by any of these k columns, for all values of the infinitesimal within the radius of convergence involved, are the coordinates of all points of the curve which lie within the neighbourhood of a single *place* (cf. § 3, Chap. I.); the sum

$$(w_1+1)+(w_2+1)+\dots\dots+(w_k+1)$$

is the same for all values of x, and equal to n, the order of the curve. A similar result holds for infinite values of x; we have only to write $\frac{1}{x}$ for $x-a$.

We assume further that any rational symmetric function of the n sets of values for the pair (y, z), which are represented by the equations (D), is a rational function of x.

Then we can prove that if $R(x, y, z)$ be any rational function of x, y, z, the sum of the coefficients of t^{-1} in the expression $R(x, y, z)\frac{dx}{dt}$, at all the k places of the curve represented by the equations (D), is equal to the coefficient of $\frac{1}{x-a}$ in the rational function of x,

$$U(x) = R(x, y_1, z_1) + R(x, y_2, z_2) + \dots\dots + R(x, y_n, z_n).$$

And further that the sum of the coefficients of t^{-1} in $R(x, y, z)\frac{dx}{dt}$ at all the places arising for $x=\infty$ is equal to the coefficient of $-\frac{1}{x}$ in the expansion of the same rational function of x, namely, equal to the coefficient of t^{-1} in $U(x)\frac{dx}{dt}$, when $x=\frac{1}{t}$.

Hence, the theorem

$$\left[U(x)\frac{dx}{dt}\right]_{t^{-1}} = 0,$$

which holds for any rational function, $U(x)$, of a single variable (as may be immediately proved by expressing the function in partial fractions in the ordinary way), enables us to infer, in the case of the curve considered, that also

$$\left[R(x, y, z)\frac{dx}{dt}\right]_{t^{-1}} = 0.$$

By this theorem, applied to the case

$$\left[\frac{1}{R(x, y, z)}\frac{d}{dx}R(x, y, z)\frac{dx}{dt}\right]_{t^{-1}} = 0,$$

we can prove that the number of poles of $R(x, y, z)$ is equal to the number of its zeros, and therefore also equal to the number of places where $R(x, y, z)$ has any assigned value μ, a place being counted as r coincident zeros when the expression, in $R(x, y, z)$, of the appropriate values for x, y, z, in terms of the infinitesimal, leads to a series in which the lowest power of t is t^r; similarly for the poles.

Hence, if I be any integral of the form $\int R(x, y, z)\,dx$, we can apply this theorem in the form

$$\left(\frac{dI}{dt}\frac{1}{Z-\mu}\right)_{t^{-1}}=0,$$

Z being any rational function of x, y, z, and so obtain, as before (§§ 154, 155), the theorem

$$I^{x_1, a_1}+\dots\dots+I^{x_k, a_k}=\left(\frac{dI}{dt}\int_{\mu=\infty}^{0}\delta\mu\frac{\partial}{\partial\mu}\log(Z-\mu)\right)_{t^{-1}},$$

and if Z is of the form $\theta_2(x, y, z)/\theta_1(x, y, z)$, where θ_2, θ_1 are integral polynomials, we can put the right-hand side

$$=\left[\frac{dI}{dt}\log\frac{\theta_2(x, y, z)}{\theta_1(x, y, z)}\right]_{t^{-1}},$$

wherein $x_1, \dots, x_k$ are the places at which $Z=0$, or $\theta_2(x, y, z)=0$, and $a_1, \dots, a_k$ are the places where $Z=\infty$ or $\theta_1(x, y, z)=0$, and the places to be considered on the right hand are the infinities of dI/dt.

The reader may also consult the investigation given by Forsyth, *Phil. Trans.*, 1883, Part i. p. 337.

Take for example the curve which is the complete intersection of the cylinders

$$y^2=x(1-x)$$
$$z^2=x.$$

For any finite value of x, except $x=0$ or $x=1$, we have 4 places given by

$$y=\pm\sqrt{x(1-x)},\quad z=\pm\sqrt{x}.$$

For infinite values of x, putting $x=\frac{1}{t^2}$, we have two places given by

$$y=i\frac{1}{t^2}+\dots\quad,\qquad y=-i\frac{1}{t^2}+\quad,$$
$$z=\frac{1}{t}\quad,\qquad z=\frac{1}{t}\quad.$$

For $x=1$, putting $x=1+t^2$, we have two places given by

$$y=it+\dots\quad,\qquad y=it+\dots\quad,$$
$$z=+(1+\tfrac{1}{2}t^2+\dots),\qquad z=-(1+\tfrac{1}{2}t^2+\dots)\;.$$

For $x=0$, putting $x=t^2$, we have two places given by

$$y=t(1-\tfrac{1}{2}t^2-\dots),\qquad y=-t(1-\tfrac{1}{2}t^2-\dots),$$
$$z=t\quad,\qquad z=t\quad,$$

and, at $x=0$, $y=0$, $z=0$, $dx:dy:dz=2t:1:1$ or $=2t:-1:1=0:1:1$ or $=0:-1:1$ so that there is a double point with $x=0$, $y=\pm z$ for tangents.

Consider now $\Sigma\int\frac{dx}{yz}$, from the intersections of $z+ax+by=0$ to those of $z+a'x+b'y=0$.

Put $I=\int \frac{dx}{yz}$; then $\frac{dI}{dt}, =\frac{1}{yz}\frac{dx}{dt}$, when x is near to 0, has, for one value,

$$\frac{1}{t^2(1-\frac{1}{2}t^2 \dots)}2t=\frac{2}{t}(1+\frac{1}{2}t^2+\dots),$$

while
$$\log\frac{z+a'x+b'y}{z+ax+by}=\log\frac{t+a't^2+b't(1-\frac{1}{2}t^2\dots)}{t+at^2+bt(1-\frac{1}{2}t^2\dots)}=\log\frac{1+b'}{1+b}\frac{1+\frac{a'}{1+b'}t+\dots}{1+\frac{a}{1+b}t+\dots}$$

$$=\log\frac{1+b'}{1+b}+\left(\frac{a'}{1+b'}-\frac{a}{1+b}\right)t+\dots\dots,$$

and the contribution to the sum $\left(\frac{dI}{dt}\log\frac{z+a'x+b'y}{z+ax+by}\right)_{t^{-1}}$ is $2\log\frac{1+b'}{1+b}$.

If we take the other place at $x=0$ we shall get, as the contribution to

$$\left(\frac{dI}{dt}\log\frac{z+a'x+b'y}{z+ax+by}\right)_{t^{-1}},$$

the quantity $-2\log\frac{1-b'}{1-b}$.

Thus, on the whole we get, at $x=0$,

$$2\log\left(\frac{1+b'}{1-b'}\Big/\frac{1+b}{1-b}\right).$$

It is similarly seen that no contribution arises at the places $x=1$, $x=\infty$.

Thus on the whole

$$\int\frac{dx_1}{x_1\sqrt{1-x_1}}+\int\frac{dx_2}{x_2\sqrt{1-x_2}}=2\log\left(\frac{1+b'}{1-b'}\Big/\frac{1+b}{1-b}\right).$$

Now from the equations $z_1+ax_1+by_1=0$, $z_2+ax_2+by_2=0$, we find

$$b=\frac{z_1x_2-z_2x_1}{x_1y_2-x_2y_1},$$

and thus

$$\int^{x_1}\frac{dx}{x\sqrt{1-x}}+\int^{x_2}\frac{dx}{x\sqrt{1-x}}=2\log\frac{\sqrt{x_1(1-x_2)}-\sqrt{x_2(1-x_1)}+\sqrt{x_2}-\sqrt{x_1}}{\sqrt{x_1(1-x_2)}-\sqrt{x_2(1-x_1)}-\sqrt{x_2}+\sqrt{x_1}}+\text{constant}$$

which is a result that can be directly verified.

CHAPTER IX.

JACOBI'S INVERSION PROBLEM.

163. It is known what advance was made in the theory of elliptic functions by the adoption of the idea, of Abel and Jacobi, that the value of the integral of the first kind should be taken as independent variable, the variables, x and y, belonging to the upper limit of this integral being regarded as dependent. The question naturally arises whether it may not be equally advantageous, if possible, to introduce a similar change of independent variable in the higher cases. We have seen in the previous chapter that, if $u_1^{x, a}, \dots, u_p^{x, a}$ be any p linearly independent integrals of the first kind, the p equations

$$u_i^{x_1, a_1} + \dots\dots + u_i^{x_p, a_p} = - u_i^{x_{p+1}, a_{p+1}} - \dots\dots - u_i^{x_Q, a_Q}, \quad (i = 1, 2, \dots, p),$$

justify us in regarding the places $x_1, \dots, x_p$ as rationally determinable from the arbitrary places $a_1, \dots, a_Q, x_{p+1}, \dots, x_Q$; hence is suggested the problem, known as Jacobi's inversion problem*, which may be stated thus: *if $U_1, \dots, U_p$ be arbitrary quantities, regarded as variable, and $a_1, \dots, a_p$ be arbitrary fixed places, required to determine the nature and the expression of the dependence of the places $x_1, \dots, x_p$, which satisfy the p equations*

$$u_i^{x_1, a_1} + \dots\dots + u_i^{x_p, a_p} = U_i, \qquad (i = 1, 2, \dots, p),$$

upon the quantities $U_1, \dots, U_p$. It is understood that the path of integration from a_r to x_r is to be taken the same in each of the p equations, and is not restricted from crossing the period loops.

164. It is obvious first of all that if for any set of values $U_1, \dots, U_p$ there be one set of corresponding places $x_1, \dots, x_p$ of such general positions that no ϕ-polynomial (§ 101) vanishes in them, there cannot be another set of places, $x_1', \dots, x_p'$, belonging to the same values of $U_1, \dots, U_p$. For then we should have

$$u_i^{x_1', x_1} + \dots\dots + u_i^{x_p', x_p} = 0, \qquad (i = 1, 2, \dots, p),$$

* Jacobi, *Crelle* XIII. (1835), p. 55.

and therefore (§ 158, Chap. VIII.) there would exist a rational function having $x_1, \ldots, x_p$ as poles and $x_1', \ldots, x_p'$ as zeros, which is contrary (§ 37, Chap. III.) to the hypothesis that no ϕ-polynomial vanishes in $x_1, \ldots, x_p$.

But a further result follows from the § referred to (§ 158, Chap. VIII.). Let $2\omega_{i,1}, \ldots, 2\omega_{i,p}, 2\omega_{i,1}', \ldots, 2\omega_{i,p}'$ denote the periods of $u_i^{x,a}$, and $m_1, \ldots, m_p, m_1', \ldots, m_p'$ denote any rational integers which are the same for all values of i. On the hypothesis that the inversion problem is capable of solution for all values of the quantities $U_1, \ldots, U_p$, suppose these quantities to vary continuously from the values $U_1, \ldots, U_p$ to the values $V_1, \ldots, V_p$, where

$$V_i = U_i + 2m_1\omega_{i,1} + \ldots\ldots + 2m_p\omega_{i,p} + 2m_1'\omega_{i,1}' + \ldots\ldots + 2m_p'\omega_{i,p}', \qquad (i = 1, 2, \ldots, p),$$

$$= U_i + 2\Omega_i, \text{ say,}$$

and let $z_1, \ldots, z_p$ be the places such that

$$u_i^{z_1, a_1} + \ldots + u_i^{z_p, a_p} = V_i;$$

then it follows from § 158, that the places $z_1, \ldots, z_p$ are, in some order, the same as the places $x_1, \ldots, x_p$. For this reason it is proper to write the equations of the inversion problem in the form

$$u_i^{x_1, a_1} + \ldots\ldots + u_i^{x_p, a_p} \equiv U_i,$$

where the sign $\equiv$ indicates that the two sides of the congruence differ by a quantity of the form $2\Omega_i$. And further, if the set $x_1, \ldots, x_p$ be uniquely determined by the values $U_1, \ldots, U_p$, any symmetrical function of the values of x, y at the places of this set, must be a single-valued function of $U_1, \ldots, U_p$. Denoting such a function by $\phi(U_1, \ldots, U_p)$, we have, therefore,

$$\phi(U_1 + 2\Omega_1, U_2 + 2\Omega_2, \ldots, U_p + 2\Omega_p) = \phi(U_1, \ldots, U_p).$$

The functions that arise are therefore such as are unaltered when the p variables $U_1, \ldots, U_p$ are simultaneously increased by the same integral multiples of any one of the $2p$ sets of quantities denoted by

$$2\omega_{1,r},\ 2\omega_{2,r},\ \ldots,\ 2\omega_{p,r}$$
$$2\omega_{1,r}',\ 2\omega_{2,r}',\ \ldots,\ 2\omega_{p,r}'. \qquad (r = 1, 2, \ldots, p).$$

165. The sign $\equiv$ will often be employed in what follows, in the sense explained above. There is one case in which it is absolutely necessary. In what has preceded the paths of integration have not been restricted from crossing the period loops. But it is often convenient, for the sake of definiteness, to use only integrals for which this restriction is enforced. In such case the problem expressed by the equations

$$u_i^{x_1, a_1} + \ldots\ldots + u_i^{x_p, a_p} = U_i$$

may be incapable of solution for some values of $U_1, \ldots, U_p$. This can be seen as follows: if both the sets of equations

$$u_i^{x_1, a_1} + \ldots\ldots + u_i^{x_p, a_p} = U_i,$$

$$u_i^{z_1, a_1} + \ldots\ldots + u_i^{z_p, a_p} = U_i + 2\Omega_i,$$

were capable of solution, it would follow, by § 158, that the set $z_1, \ldots, z_p$ is the same as the set $x_1, \ldots, x_p$. And thence, as the paths are restricted not to cross the period loops, we should have

$$u_i^{x_1, a_1} + \ldots\ldots + u_i^{x_p, a_p} = u_i^{z_1, a_1} + \ldots\ldots + u_i^{z_p, a_p},$$

and thence

$$2\Omega_1 = 2m_1\omega_{i,1} + \ldots\ldots + 2m_p\omega_{i,p} + 2m_1'\omega_{i,1}' + \ldots\ldots + 2m_p'\omega_{i,p}' = 0;$$

but these equations are reducible to

$$m_i + m_1'\tau_{i,1} + \ldots\ldots + m_p'\tau_{i,p} = 0,$$

and, therefore, there would exist a function, expressed by

$$e^{2\pi i (m_1' v_1^{x, a} + \ldots\ldots + m_p' v_p^{x, a})},$$

(where $v_1^{x, a}, \ldots, v_p^{x, a}$ are Riemann's elementary integrals of the first kind), everywhere finite and without periods. Such a function must be a constant; thus the conclusion would involve that $v_1^{x, a}, \ldots, v_p^{x, a}$ are not linearly independent, which is untrue.

Hence when the paths of integration are restricted not to cross the period loops, the equations of the inversion problem must be written

$$u_i^{x_1, a_1} + \ldots\ldots + u_i^{x_p, a_p} \equiv U_i;$$

in this case the integral sum on the left-hand side is not capable of assuming all values; and the particular period which must be added to the right-hand side to make the two sides of the congruence equal is determined by the solution of the problem.

166. Before passing to the proof that Jacobi's inversion problem does admit of solution, another point should be referred to. It is not at first sight apparent why it is necessary to take p arguments, $U_1, \ldots, U_p$, and p dependent places $x_1, \ldots, x_p$. It may be thought, perhaps, that a single equation

$$u^{x, a} = U,$$

wherein $u^{x, a}$ is any definite integral of the first kind, suffices to determine the place x as a function of the argument U. We defer to a subsequent place the enquiry whether this is true when the path of integration on the left hand is not allowed to cross the period loops of the Riemann surface; it is obvious enough that in such a case all conceivable values of U would not arise, for instance $U=\infty$ would not arise, and the function of U obtained would only be defined for restricted values of the argument. But it is possible to see that when the path of integration is not limited, the place x cannot be definitely determinate from U. For, then, putting $x=f(U)$, we must have $f(U+2\Omega)=f(U)$, wherein

$$\Omega = m_1\omega_1 + \ldots\ldots + m_p\omega_p + m_1'\omega_1' + \ldots\ldots + m_p'\omega_p',$$

$m_1, \ldots, m_p'$ being arbitrary rational integers, and $2\omega_1, \ldots, 2\omega_p'$ being the periods of $u^{x,a}$; and it can be shewn, when $p>1$, that in general it is possible to choose the integers $m_1, \ldots, m_p'$ so that Ω shall be within assigned nearness of any prescribed arbitrary value whatever. Thus not only would the function $f(U)$ have infinitesimal periods, but any assigned value of this function would arise for values of the argument lying within assigned nearness of any value whatever. We shall deal later with the possibility of the existence of infinitesimal periods; for the present such functions are excluded from consideration.

The arithmetical theorem referred to* may be described thus; if a_1, a_2 be any real quantities, the values assumed by the expression $N_1a_1+N_2a_2$, when N_1, N_2 take all possible rational integer values independently of one another, are in general infinite in number; exception arises only in the case when the ratio a_1/a_2 is rational; and it is in general possible to find rational integer values of N_1 and N_2 to make $N_1a_1+N_2a_2$ approach within assigned nearness of any prescribed real quantity. Similarly if a_1, a_2, a_3, b_1, b_2, b_3 be real quantities, of the expressions $N_1a_1+N_2a_2+N_3a_3$, $N_1b_1+N_2b_2+N_3b_3$, where N_1, N_2, N_3 take all possible rational integer values independently of one another, there are, in general, values which lie within assigned nearness respectively to two arbitrarily assigned real quantities a, b. More generally, if $a_1, \ldots, a_k, b_1, \ldots, b_k, \ldots\ldots, c_1, \ldots, c_k$ be any $(k-1)$ sets each of k real quantities, and $a, b, \ldots, c$ be $(k-1)$ arbitrary real quantities, it is in general possible to find rational integers $N_1, \ldots, N_k$ such that the $(k-1)$ quantities

$$N_1a_1+\ldots\ldots+N_ka_k-a,\ N_1b_1+\ldots\ldots+N_kb_k-b,\ \ldots,\ N_1c_1+\ldots\ldots+N_kc_k-c,$$

are all within assigned nearness of zero.

Hence it follows, taking $k=2p$, that we can choose values of the integers $m_1, \ldots, m_p'$, to make $p-1$ of the quantities

$$\Omega_r = m_1\omega_{r,1} + \ldots\ldots + m_p\omega_{r,p} + m_1'\omega_{r,1}' + \ldots\ldots + m_p'\omega_{r,p}',$$

say $\Omega_1, \ldots, \Omega_{p-1}$, approach within assigned nearness of any $(p-1)$ prescribed values, and at the same time to make the real part of the remaining quantity Ω_p approach within assigned nearness of any prescribed value; but the imaginary part of Ω_p will thereby be determined. We cannot therefore

* Jacobi, *loc. cit.*; Hermite, *Crelle*, LXXXVIII. p. 10.

expect to obtain an intelligible inversion by taking less than p new variables $U_1, U_2, \ldots$; and it is manifest that we ought to use the same number of dependent places $x_1, x_2, \ldots$. On the other hand, the proof which has been given that there can in general only be one set of places $x_1, \ldots, x_p$ corresponding to given values of $U_1, \ldots, U_p$ would not remain valid in case the left-hand sides of the equations of the problem of inversion consisted of a sum of more than p integrals; for it is generally possible to construct a rational function with $p+1$ assigned poles.

167. It follows from the argument here that when $p > 1$ an integral of the first kind, $u^{x, a}$, is capable, for given positions of the extreme limits, x, a, of the integration, of assuming values within assigned nearness of any prescribed value whatever. Though not directly connected with the subject here dealt with it is worth remark that it does not thence follow that the integral is capable of assuming all possible values. For the values represented by an expression of the form

$$m_1\omega_1 + \ldots\ldots + m_p\omega_p + m_1'\omega_1' + \ldots\ldots + m_p'\omega_p',$$

for all values of the integers $m_1, \ldots, m_p, m_1', \ldots, m_p'$, form an enumerable aggregate—that is, they can be arranged in order and numbered $-\infty, \ldots, -3, -2, -1, 0, 1, 2, 3, \ldots, \infty$. To prove this we may begin by proving that all values of the form $m_1\omega_1 + m_2\omega_2$ form an enumerable aggregate; the proof is identical with the proof that all rational fractions form an enumerable aggregate; and may then proceed to shew that all values of the form $m_1\omega_1 + m_2\omega_2 + m_3\omega_3$ form an enumerable aggregate, and so on, step by step. Since then the aggregate of all conceivable complex values is not an enumerable aggregate, the statement made is justified.

The reader may consult Harkness and Morley, *Theory of Functions*, p. 280, Dini, *Theorie der Functionen einer reellen Grösse* (German edition by Lüroth and Schepp), pp. 27, 191, Cantor, *Acta Math.* II. pp. 363—371, Cantor, *Crelle*, LXXVII. p. 258, *Rendiconti del Circolo Mat. di Palermo*, 1888, pp. 197, 135, 150, where also will be found a theorem of Poincaré's to the effect that no multiform analytical function exists whose values are not enumerable.

168. Consider now* the equations

$$\text{(A)} \qquad u_i^{x_1, a_1} + \ldots\ldots + u_i^{x_p, a_p} = U_i, \qquad (i = 1, 2, \ldots, p)$$

wherein, denoting the differential coefficient of $u_i^{x, a}$ in regard to the infinitesimal at x by $\mu_i(x)$, the fixed places $a_1, \ldots, a_p$ are supposed to be such that the determinant of p rows and columns whose (i, j)th element is $\mu_j(a_i)$ does not vanish; wherein also the p paths of integration a_1 to $x_1, \ldots, a_p$ to x_p, are to be the same in all the p equations, and are not restricted from crossing the period loops.

When $x_1, \ldots, x_p$ are respectively in the neighbourhoods of $a_1, \ldots, a_p$ and $U_1, \ldots, U_p$ are small, these equations can be written

$$\left[t_1\mu_i(a_1) + \frac{t_1^2}{\underline{|2}}\mu_i'(a_1) + \ldots\ldots\right] + \ldots\ldots + \left[t_p\mu_i(a_p) + \frac{t_p^2}{\underline{|2}}\mu_i'(a_p) + \ldots\ldots\right] = U_i,$$

* The argument of this section is derived from Weierstrass; see the references given in connection with § 170.

wherein t_r is the infinitesimal in the neighbourhood of the place a_r, and $\mu_r'(x)$ is derived from $\mu_r(x)$ by differentiation. From these equations we obtain

$$t_r = \nu_{r,1} U_1 + \ldots\ldots + \nu_{r,p} U_p + U_r^{(2)} + U_r^{(3)} + \ldots\ldots, \qquad (r = 1, 2, \ldots, p),$$

where, if Δ denote the determinant whose (i, j)th element is $\mu_j(a_i)$, $\nu_{i,j}$ denotes the minor of this element divided by Δ, and $U_r^{(k)}$ denotes a homogeneous integral polynomial in $U_1, \ldots, U_p$ of the kth degree. These series will converge provided $U_1, \ldots, U_p$ be of sufficient, not unlimited, smallness. Hence also, so long as the place x_r lies within a certain finite neighbourhood of the place c_r, the values of the variables x_r, y_r associated with this place, which are expressible by convergent series of integral powers of t_r, are expressible by series of integral powers of $U_1, \ldots, U_p$ which are convergent for sufficiently small values of $U_1, \ldots, U_p$.

Suppose that the values of $U_1, \ldots, U_p$ are such that the places $x_1, \ldots, x_p$ thus obtained are not such that the determinant whose (i, j)th element is $\mu_j(x_i)$ is zero; then if $U_1', \ldots, U_p'$ be small quantities, it is similarly possible to obtain p places $x_1', \ldots, x_p'$, lying respectively in the neighbourhoods of $x_1, \ldots, x_p$, such that

$$u_i^{x_1', x_1} + \ldots\ldots + u_i^{x_p', x_p} = U_i', \qquad (i = 1, 2, \ldots, p);$$

by adding these equations to the former we therefore obtain

$$u_i^{x_1', a_1} + \ldots\ldots + u_i^{x_p', a_p} = U_i + U_i', \quad (i = 1, 2, \ldots, p).$$

Since all the series used have a finite range of convergence, we are thus able, step by step, to obtain places $x_1, \ldots, x_p$ to satisfy the p equations

$$u_i^{x_1, a_1} + \ldots\ldots + u_i^{x_p, a_p} = U_i, \qquad (i = 1, 2, \ldots, p),$$

for any finite values of the quantities $U_1, \ldots, U_p$ which can be reached from the values $0, 0, \ldots, 0$ without passing through any set of values for which the corresponding positions of $x_1, \ldots, x_p$ render a certain determinant zero.

169. The method of continuation thus sketched has a certain interest; but we can arrive at the required conclusion in a different way. Let $U_1, \ldots, U_p$ be any finite quantities; and let m be a positive integer. When m is large enough, the quantities $U_1/m, \ldots, U_p/m$ are, in absolute value, as small as we please. Hence there exist places $z_1, \ldots, z_p$, lying respectively in the neighbourhoods of the places $a_1, \ldots, a_p$, such that

$$u_i^{z_1, a_1} + \ldots\ldots + u_i^{z_p, a_p} = -U_i/m \qquad (i = 1, 2, \ldots, p).$$

In order then to obtain places $x_1, \ldots, x_p$, to satisfy the equations

$$u_i^{x_1, a_1} + \ldots\ldots + u_i^{x_p, a_p} = U_i, \qquad (i = 1, 2, \ldots, p),$$

it is only necessary to obtain places $x_1, \dots, x_p$, such that

$$u_i^{x_1, a_1} + \dots\dots + u_i^{x_p, a_p} + m u_i^{z_1, a_1} + \dots\dots + m u_i^{z_p, a_p} = 0, \quad (i = 1, 2, \dots, p);$$

and it has been shewn (Chap. VIII. § 158), that these equations express only that the set of $mp + p$ places formed of $z_1, \dots, z_p$, each m times repeated and the places $x_1, \dots, x_p$, are coresidual with the set of $(m+1)p$ places formed of $a_1, \dots, a_p$ each $(m+1)$ times repeated.

Now, when $(m+1)p$ places are not zeros of a ϕ-polynomial, we may (Chap. VI.) arbitrarily assign all but p of the places of a set of $(m+1)p$ places which are coresidual with them; and the other p places will be algebraically and rationally determinable from the mp assigned places.

Hence with the general positions assigned to the places $a_1, \dots, a_p$, it follows, if Z denote any rational function, that the values of Z at the places $x_1, \dots, x_p$ are the roots of an algebraical equation,

$$Z^p + Z^{p-1} R_1 + \dots\dots + R_p = 0,$$

whose coefficients $R_1, \dots, R_p$ are rationally determinable from the places $z_1, \dots, z_p$, and are therefore, by what has been shewn, expressible by series of integral powers of $U_1/m, \dots, U_p/m$, which converge for sufficiently large values of m. Thus the problem expressed by the equations

$$u_i^{x_1, a_1} + \dots\dots + u_i^{x_p, a_p} = U_i, \qquad (i = 1, 2, \dots, p),$$

is always capable of solution, for any finite values of $U_1, \dots, U_p$.

It has already been shewn (§ 164), that for general values of $U_1, \dots, U_p$ the set $x_1, \dots, x_p$ obtained is necessarily unique; the same result follows from the method of the present article. It is clear in § 164, in what way exception can arise; to see how a corresponding peculiarity may present itself in the present article the reader may refer to the concluding result of § 99 (Chap. VI.). (See also Chap. III. § 37, Ex. ii.)

In case the places $a_1, \dots, a_p$ in the equations (A) be such that the determinant denoted by Δ vanishes, we may take places $b_1, \dots, b_p$, for which the corresponding determinant is not zero, and follow the argument of the text for the equations

$$u_i^{x_1, b_1} + \dots\dots + u_i^{x_p, b_p} = V_i,$$

in which $V_i = U_i + u_i^{a_1, b_1} + \dots\dots + u_i^{a_p, b_p}$.

We do not enter into the difficulty arising as to the solution of the inversion problem expressed by the equations (A) in the case where $U_1, \dots, U_p$ have such values that $x_1, \dots, x_p$ are zeros of a ϕ-polynomial. This point is best cleared up by actual examination of the functions which are to be obtained to express the solution of the problem (cf.* § 171, and

* See also Clebsch and Gordan, *Abel. Functnen.*, pp. 184, 186.

Props. xiii. and xv., Cor. iii., of Chap. X.). But it should be noticed that the method of § 168 shews that a solution exists in all cases in which the fixed places $a_1, \ldots, a_p$ do not make the determinant Δ vanish; the peculiarity in the special case is that instead of an unique solution $x_1, \ldots, x_p$, all the $\infty^{\tau+1}$ sets coresidual with $x_1, \ldots, x_p$ are equally solutions, $\tau+1$ being the number of linearly independent ϕ-polynomials which vanish in $x_1, \ldots, x_p$. This follows from §§ 154, 158.

170. We consider now how to form functions with which to express the solution of the inversion problem.

Let $P_{\xi, \gamma}^{x, a}$ denote any elementary integral of the third kind, with infinities at the arbitrary fixed places ξ, γ. Then if $a_1, \ldots, a_p, x_1, \ldots, x_p$ denote the places occurring on the left hand in equation (A), it can be shewn that the function

$$T = P_{\xi, \gamma}^{x_1, a_1} + \ldots\ldots + P_{\xi, \gamma}^{x_p, a_p}$$

is the logarithm of a single valued function of $U_1, \ldots, U_p$, and that the solution of the inversion problem can be expressed by this function; and further that, if $I^{x, a}$ denote any Abelian integral, the sum

$$I^{x_1, a_1} + \ldots\ldots + I^{x_p, a_p}$$

can also* be expressed by the function T.

It is clear that in this statement it is immaterial what integral of the third kind is adopted. For the difference between two elementary integrals of the third kind with infinities at ξ, γ is of the form

$$\lambda_1 u_1^{x, a} + \ldots\ldots + \lambda_p u_p^{x, a} + \lambda,$$

where $\lambda_1, \ldots, \lambda_p, \lambda$ may depend on ξ, γ but are independent of x; hence the difference between the two corresponding values of T is of the form

$$\lambda_1 U_1 + \ldots\ldots + \lambda_p U_p + \lambda;$$

and this is a single-valued function of $U_1, \ldots, U_p$.

For definiteness we may therefore suppose that $P_{\xi, \gamma}^{x, a}$ denotes the integral of the third kind obtained in Chap. IV. (§ 45. Also Chap. VII. § 134).

Then, firstly, when $x_1, \ldots, x_p$ are very near to $a_1, \ldots, a_p$, and $U_1, \ldots, U_p$ are small, T is given by

$$\sum_{i=1}^{p} \left\{ t_i \left[(a_i, \xi) - (a_i, \gamma) \right] \frac{da_i}{dt} + \frac{t_i^2}{\underline{|2}} D_{a_i}^2 P_{\xi, \gamma}^{a_i, c} + \ldots\ldots \right\},$$

* The introduction of the function T is, I believe, due to Weierstrass. See *Crelle*, LII. p. 285 (1856) and *Mathem. Werke* (Berlin, 1894), I. p. 302. The other functions there used are considered below in Chaps. XI., XIII.

where t_i denotes the infinitesimal in the neighbourhood of the place a_i, c is an arbitrary place, and the notation is as in § 130, Chap. VII. It is intended of course that neither of the places ξ or γ is in the neighbourhood of any of the places $a_1, \ldots, a_p$. Now we have shewn that the infinitesimals $t_1, \ldots, t_p$ are expressible as convergent series in $U_1, \ldots, U_p$. Thus T is also expressible as a convergent series in $U_1, \ldots, U_p$ when $U_1, \ldots, U_p$ are sufficiently small.

Nextly, suppose the places $x_1, \ldots, x_p$ are not near to the places $a_1, \ldots, a_p$; determine, as in § 168, places to satisfy the equations

$$u_i^{z_1, a_1} + \ldots\ldots + u_i^{z_p, a_p} = -U_i/m,$$

$$u_i^{x_1, a_1} + \ldots\ldots + u_i^{x_p, a_p} = U_i,$$

m being a large positive integer; then we shall also have (§ 158, Chap. VIII.)

$$P_{\xi, \gamma}^{x_1, a_1} + \ldots\ldots + P_{\xi, \gamma}^{x_p, a_p} + m\,(P_{\xi, \gamma}^{z_1, a_1} + \ldots\ldots + P_{\xi, \gamma}^{z_p, a_p}) = \log \frac{Z(\xi)}{Z(\gamma)},$$

where $Z(x)$ denotes the rational function which has a pole of the $(m+1)$th order at each of the places $a_1, \ldots, a_p$, and has a zero of the mth order at each of the places $z_1, \ldots, z_p$. The function $Z(x)$ has also a simple zero at each of the places $x_1, \ldots, x_p$, but this fact is not part of the definition of the function.

This equation can be written in the form

$$e^T = e^{-mT_0} \frac{Z(\xi)}{Z(\gamma)},$$

wherein T_0 denotes the sum

$$P_{\xi, \gamma}^{z_1, a_1} + \ldots\ldots + P_{\xi, \gamma}^{z_p, a_p}.$$

It follows by the proof just given that T_0 is expressible as a series of integral powers of the variables $U_1/m, \ldots, U_p/m$, which converges for sufficiently great values of m; and it is easy to see that the expression $Z(\xi)/Z(\gamma)$ is also expressible by series of integral powers of $U_1/m, \ldots, U_p/m$. For let the most general rational function having a pole of the $(m+1)$th order in each of $a_1, \ldots, a_p$ be of the form

$$Z(x) = \lambda_1 Z_1(x) + \ldots\ldots + \lambda_{mp} Z_{mp}(x) + \lambda,$$

wherein $Z_1(x), \ldots, Z_{mp}(x)$ are definite functions, and $\lambda, \lambda_1, \ldots, \lambda_{mp}$ are arbitrary constants. Then the expression of the fact that this function vanishes to the mth order at each of the places $z_1, \ldots, z_p$ will consist of mp equations determining $\lambda_1, \ldots, \lambda_{mp}$ rationally and symmetrically in terms of the places $z_1, \ldots, z_p$. Hence (by § 168) $\lambda_1, \ldots, \lambda_{mp}$ are expressible as series of integral powers of $U_1/m, \ldots, U_p/m$. Hence $Z(\xi)/Z(\gamma)$ is expressible by series of integral powers of $U_1/m, \ldots, U_p/m$.

Hence, for any finite values of $U_1, \ldots, U_p$ the function e^T is expressible by series of integral powers of $U_1, \ldots, U_p$. It is also obvious, from the method of proof adopted, that the series obtained for any set of values of $U_1, \ldots, U_p$ are independent of the range of values for $U_1, \ldots, U_p$ by which the final values are reached from the initial set $0, 0, \ldots, 0$; so that the function e^T is a single valued function of $U_1, \ldots, U_p$. The function e^T reduces to unity for the initial set $0, 0, \ldots, 0$.

171. An actual expression of the function e^T, in terms of $U_1, \ldots, U_p$, will be obtained in the next chapter (§ 187, Prop. xiii.). We shew here that if that expression be known, the solution of the inversion problem can also be given in explicit terms. Let $\Pi_{\xi, \gamma}^{x, a}$ denote the normal elementary integral of the third kind (Chap. II., § 14). Then if K denote the sum

$$K = \Pi_{\xi, \gamma}^{x_1, a_1} + \ldots\ldots + \Pi_{\xi, \gamma}^{x_p, a_p},$$

it follows, as here, that e^K is a single valued function of $U_1, \ldots, U_p$, whose expression is known when that of e^T is known, and conversely. Denote e^K by $V(U_1, \ldots, U_p;\ \xi, \gamma)$. Let $Z(x)$ denote any rational function whatever, its poles being the places $\gamma_1, \ldots, \gamma_k$; and let the places at which $Z(x)$ takes an arbitrary value X be denoted by $\xi_1, \ldots, \xi_k$. Then, from the equation (Chap. VIII., § 154),

$$\Pi_{x_i, a_i}^{\xi_1, \gamma_1} + \ldots\ldots + \Pi_{x_i, a_i}^{\xi_k, \gamma_k} = \log \frac{Z(x_i) - X}{Z(a_i) - X}, \qquad (i = 1, 2, \ldots, p),$$

we obtain *

$$V(U_1, \ldots, U_p;\ \xi_1, \gamma_1) \ldots V(U_1, \ldots, U_p;\ \xi_k, \gamma_k) = \frac{[X - Z(x_1)] \ldots [X - Z(x_p)]}{[X - Z(a_1)] \ldots [X - Z(a_p)]};$$

the left-hand side of this equation has, we have said, a well ascertained expression, when the values of $U_1, \ldots, U_p$, the function $Z(x)$, and the value X, are all given; hence, substituting for X in turn any p independent values, we can calculate the expression of any symmetrical function of the quantities

$$Z(x_1), \ldots, Z(x_p),$$

and this will constitute the complete solution of the inversion problem.

It has been shewn in § 152, Chap. VIII. that any Abelian integral $I^{x,a}$ can be written as a sum of elementary integrals of the third kind and of differential coefficients of such integrals, together with integrals of the first kind. Hence, when the expression of $V(U_1, \ldots, U_p;\ \xi, \gamma)$ is obtained, that of the sum

$$I^{x_1, a_1} + \ldots\ldots + I^{x_p, a_p}$$

can also be obtained.

* Clebsch u. Gordan, *Abels. Functionen*, (1866), p. 175.

172. The consideration of the function

$$\Pi_{\xi, \gamma}^{x_1, a_1} + \ldots\ldots + \Pi_{\xi, \gamma}^{x_p, a_p},$$

which is contained in this chapter is to be regarded as of a preliminary character. It will appear in the next chapter that it is convenient to consider this function as expressed in terms of another function, the theta function. It is possible to build up the theta function in an *à priori* manner, which is a generalization of that, depending on the equation

$$\wp u = -\frac{d^2}{du^2} \log \sigma(u),$$

whereby, in the elliptic case, the σ-function may be supposed derived from the function $\wp(u)$. But this process is laborious, and furnishes only results which are more easily evident *à posteriori*. For this reason we proceed now immediately to the theta functions; formulae connecting these functions with the algebraical integrals so far considered are given in chapters X. XI. and XIV.

CHAPTER X.

RIEMANN'S THETA FUNCTIONS. GENERAL THEORY.

173. THE theta functions, which are, certainly, the most important elements of the theory of this volume, were first introduced by Jacobi in the case of elliptic functions.* They enabled him to express his functions sn u, cn u, dn u, in the form of fractions having the same denominator, the zeros of this denominator being the common poles of the functions sn u, cn u, dn u. The ratios of the theta functions, expressed as infinite products, were also used by Abel †. For the case $p = 2$, similar functions were found by Göpel ‡, who was led to his series by generalizing the form in which Hermite had written the general exponent of Jacobi's series, and by Rosenhain §, who first forms degenerate theta functions of two variables by multiplying together two theta functions of one variable, led thereto by the remark that two integrals of the first kind which exist for $p = 2$, become elliptic integrals respectively of the first and third kind, when two branch places of the surface for $p = 2$, coincide. Both Göpel and Rosenhain have in view the inversion problem enunciated by Jacobi; their memoirs contain a large number of the ideas that have since been applied to more general cases. In the form in which the theta functions are considered in this chapter they were first given, for any value of p, by Riemann ‖. Functions which are quotients of theta functions had been previously considered by Weierstrass, without any mention of the theta series, for any hyperelliptic case ¶. These functions occur in the memoir of Rosenhain, for the case $p = 2$. It will be seen that

* *Fundamenta Nova* (1829); *Ges. Werke* (Berlin, 1881), Bd. I. See in particular, Dirichlet, Gedächtnissrede auf Jacobi, *loc. cit.* Bd. I., p. 14, and Zur Geschichte der Abelschen Transcendenten, *loc. cit.*, Bd. II., p. 516.

† *Œuvres* (Christiania, 1881), t. I. p. 343 (1827). See also Eisenstein, *Crelle*, XXXV. (1847), p. 153, etc. The equation (*b*) p. 225, of Eisenstein's memoir, is effectively the equation

$$\wp'^2(u) = 4\wp^3(u) - g_2\wp(u) - g_3.$$

‡ *Crelle*, XXXV. (1847), p. 277.

§ *Mém. sav. étrang.* XI. (1851), p. 361. The paper is dated 1846.

‖ *Crelle*, LIV. (1857); *Ges. Werke*, p. 81.

¶ *Crelle*, XLVII. (1854); *Crelle*, LII. (1856); *Ges. Werke*, pp. 133, 297.

the Riemann theta functions are not the most general form possible. The subsequent development of the general theory is due largely to Weierstrass.

174. In the case $p=1$, the convergence of the series obtained by Jacobi depends upon the use of two periods 2ω, $2\omega'$, for the integral of the first kind, such that the ratio ω'/ω has its imaginary part positive. Then the quantity $q=e^{\pi i \frac{\omega'}{\omega}}$ is, in absolute value, less than unity.

Now it is proved by Riemann that if we choose normal integrals of the first kind $v_1^{x,a}, \dots, v_p^{x,a}$, so that $v_r^{x,a}$ has the periods $0\dots 0, 1, 0, \dots, \tau_{r,1}, \dots, \tau_{r,p}$, the imaginary part of the quadratic form

$$\phi = \tau_{11}n_1^2 + \dots\dots + \tau_{r,r}n_r^2 + \dots\dots + 2\tau_{1,2}n_1n_2 + \dots\dots + 2\tau_{r,s}n_rn_s + \dots\dots$$

is positive* for all real values of the p variables $n_1, \dots, n_p$. Hence for all rational integer values of $n_1, \dots, n_p$, positive or negative, the quantity $e^{i\pi\phi}$ has its modulus less than unity. Thus, if we write $\tau_{r,s} = \rho_{r,s} + i\kappa_{r,s}$, $\rho_{r,s}$ and $\kappa_{r,s}$ being real, and $a_1, = b_1 + ic_1, \dots, a_p, = b_p + ic_p$, be any p constant quantities, the modulus of the general term of the p-fold series

$$\sum_{n_1=-\infty}^{n_1=\infty} \sum_{n_2=-\infty}^{n_2=\infty} \dots\dots \sum_{n_p=-\infty}^{n_p=\infty} e^{a_1n_1+\dots\dots+a_pn_p+i\pi\phi},$$

wherein each of the indices $n_1, \dots, n_p$ takes every real integer value independently of the other indices, is e^{-L}, where

$$L = -(b_1n_1 + \dots\dots + b_pn_p) + \pi(\kappa_{11}n_1^2 + \dots\dots + 2\kappa_{1,2}n_1n_2 + \dots\dots),$$
$$= -(b_1n_1 + \dots\dots + b_pn_p) + \psi, \text{ say,}$$

where ψ is a real quadratic form in $n_1, \dots, n_p$, which is essentially positive for all the values of $n_1, \dots, n_p$ considered. When one (or more) of $n_1, \dots, n_p$ is large, L will have the same sign as ψ, and will be positive; and if μ be any positive integer $e^{L/\mu}$ is greater than $1 + L/\mu$, and therefore $e^{-L} < \left(1 + \frac{L}{\mu}\right)^{-\mu}$; now the series whose general term is $\left(1 + \frac{L}{\mu}\right)^{-\mu}$ will be convergent or not according as the series whose general term is $\psi^{-\mu}$ is convergent or not, for the ratio $1 + \frac{L}{\mu} : \psi$ has the finite limit $1/\mu$ for large values of $n_1, \dots, n_p$; and the series whose general term is $\psi^{-\mu}$ is convergent provided μ be taken

* The proof is given in Forsyth, *Theory of Functions*, § 235. If $w_1^{x,a}, \dots, w_p^{x,a}$ denote a set of integrals of the first kind such that $w_r^{x,a}$ has no periods at the b period loops except at b_r, and has there the period 1, and $\sigma_{r,1}, \dots, \sigma_{r,p}$ be the periods of $w_r^{x,a}$ at the a period loops, the quadratic function

$$\sigma_{11}n_1^2 + \dots\dots + 2\sigma_{12}n_1n_2 + \dots\dots$$

has its imaginary part negative. *Cf.* p. 531 note†.

$>\frac{1}{2}p$. (Jordan, *Cours d'Analyse*, Paris, 1893, vol. I., § 318.) Hence the series whose general term is

$$e^{a_1 n_1 + \dots\dots + a_p n_p + i\pi\phi},$$

is absolutely convergent.

In what follows we shall write $2\pi i u_r$ in place of a_r and speak of $u_1, \dots, u_p$ as the arguments; we shall denote by un the quantity $u_1 n_1 + \dots\dots + u_p n_p$, and by τn^2 the quadratic $\tau_{11} n_1^2 + \dots\dots + 2\tau_{12} n_1 n_2 + \dots\dots$. Then the Riemann theta function is defined by the equation

$$\Theta(u) = \Sigma e^{2\pi i u n + i\pi\tau n^2},$$

where the sign of summation indicates that each of the indices $n_1, \dots, n_p$ is to take all positive and negative integral values (including zero), independently of the others. By what has been proved it follows that $\Theta(u)$ is a single-valued, integral, analytical function of the arguments $u_1, \dots, u_p$.

The notation is borrowed from the theory of matrices (cf. Appendix ii.); τ is regarded as representing the symmetrical matrix whose (r, s)th element is $\tau_{r,s}$, n as representing a row, or column, letter, whose elements are $n_1, \dots, n_p$, and u, similarly, as representing such a letter with $u_1, \dots, u_p$ as its elements.

It is convenient, with $\Theta(u)$, to consider a slightly generalized function, given by

$$\Theta(u;\, q, q'), \text{ or } \Theta(u, q) = \Sigma e^{2\pi i u(n+q') + i\pi\tau(n+q')^2 + 2\pi i q(n+q')};$$

herein q denotes the set of p quantities $q_1, \dots, q_p$, and q' denotes the set of p quantities $q_1', \dots, q_p'$, and, for instance, $u(n+q')$ denotes the quantity $un + uq'$, namely

$$u_1 n_1 + \dots\dots + u_p n_p + u_1 q_1' + \dots\dots + u_p q_p',$$

and $\tau(n+q')^2$ denotes $\tau n^2 + 2\tau n q' + \tau q'^2$, namely

$$(\tau_{11} n_1^2 + \dots + 2\tau_{1,2} n_1 n_2 + \dots) + 2 \sum_{s=1}^{p} \sum_{r=1}^{p} \tau_{r,s} n_r q_s' + (\tau_{11} q_1'^2 + \dots + 2\tau_{1,2} q_1' q_2' + \dots).$$

The quantities $q_1, \dots, q_p, q_1', \dots, q_p'$ constitute, in their aggregate, the *characteristic* of the function $\Theta(u;\, q)$; they may have any constant values whatever; in the most common case they are each either 0 or $\frac{1}{2}$.

The quantities $\tau_{i,j}$ are the periods of the Riemann normal integrals of the first kind at the second set of period loops. It is clear however that any symmetrical matrix, σ, which is such that for real values of $k_1, \dots k_p$ the quadratic form σk^2 has its imaginary part positive, may be equally used instead of τ, to form a convergent series of the same form as the Θ series. And it is worth while to make this remark in order to point out that the Riemann theta functions are not of as general a character as possible. For such a symmetrical matrix σ contains $\frac{1}{2}p(p+1)$ different quantities, while the periods $\tau_{r,s}$ are (Chap. I., § 7), functions of only $3p-3$ independent quantities. The difference $\frac{1}{2}p(p+1)-(3p-3)=\frac{1}{2}(p-2)(p-3)$, vanishes for $p=2$ or $p=3$; for $p=4$ it is equal to 1, and for greater values of p is still greater. We shall afterwards be concerned with the more general theta-function here suggested.

The function $\Theta(u)$ is obviously a generalization of the theta functions used in the theory of elliptic functions. One of these, for instance, is given by

$$\vartheta_1(u;\ \tfrac{1}{2},\ \tfrac{1}{2}) = \frac{\vartheta_1'(0)}{2\omega} e^{-2\eta\omega u^2} \sigma(2\omega u) = -\Sigma e^{2\pi i u(n+\frac{1}{2}) + \pi i\tau(n+\frac{1}{2})^2 + \pi i(n+\frac{1}{2})};$$

and the four elliptic theta functions are in fact obtained by putting respectively $q, q' = 0, \tfrac{1}{2}$; $= \tfrac{1}{2}, \tfrac{1}{2}$; $= \tfrac{1}{2}, 0$; $= 0, 0$.

175. There are some general properties of the theta functions, immediately deducible from the definition given above, which it is desirable to put down at once for purposes of reference. Unless the contrary is stated it is always assumed in this chapter that the characteristic consists of half integers; we may denote it by $\tfrac{1}{2}\beta_1, \dots, \tfrac{1}{2}\beta_p, \tfrac{1}{2}\alpha_1, \dots, \tfrac{1}{2}\alpha_p$, or shortly, by $\tfrac{1}{2}\beta, \tfrac{1}{2}\alpha$, where $\beta_1, \dots, \beta_p, \alpha_1, \dots, \alpha_p$ are integers, in the most common case either 0 or 1. Further we use the abbreviation $\Omega_{m,m'}$, or sometimes only Ω_m, to denote the set of p quantities

$$m_i + \tau_{i,1} m_1' + \dots\dots + \tau_{i,p} m_p', \qquad (i = 1, 2, \dots, p),$$

wherein $m_1, \dots, m_p, m_1', \dots, m_p'$ are $2p$ constants. When these constants are integers, the p quantities denoted by Ω_m are the periods of the p Riemann normal integrals of the first kind when the upper limit of the integrals is taken round a closed curve which is reducible to m_i circuits of the period loop b_i (or m_i crossings of the period loop a_i) and to m_i' circuits of the period loop a_i, i being equal to $1, 2, \dots, p$. (Cf. the diagram Chap. II. p. 21.) The general element of the set of p quantities denoted by Ω_m, will also sometimes be denoted by $m_i + \tau_i m'$, τ_i denoting the row of quantities formed by the ith row of the matrix τ. When $m_1, \dots, m_p'$ are integers, the quantity $m_i + \tau_i m'$ is the period to be associated with the argument u_i.

Then we have the following formulae, (A), (B), (C), (D), (E):

$$\Theta(-u;\ \tfrac{1}{2}\beta,\ \tfrac{1}{2}\alpha) = e^{\pi i\beta\alpha}\, \Theta(u;\ \tfrac{1}{2}\beta,\ \tfrac{1}{2}\alpha), \qquad \text{(A)}.$$

Thus $\Theta(u;\ \tfrac{1}{2}\beta, \tfrac{1}{2}\alpha)$ is an odd or even function of the variables $u_1, \dots, u_p$ according as $\beta\alpha$, $= \beta_1\alpha_1 + \dots\dots + \beta_p\alpha_p$, is an odd or even integer; in the former case we say that the characteristic $\tfrac{1}{2}\beta, \tfrac{1}{2}\alpha$ is an odd characteristic, in the latter case that it is an even characteristic.

The behaviour of the function $\Theta(u)$ when proper simultaneous periods are added to the arguments, is given by the formulae immediately following, wherein r is any one of the numbers $1, 2, \dots, p$,

$$\Theta(u_1, \dots, u_r + 1, \dots, u_p;\ \tfrac{1}{2}\beta,\ \tfrac{1}{2}\alpha) = e^{\pi i\alpha_r}\, \Theta(u;\ \tfrac{1}{2}\beta,\ \tfrac{1}{2}\alpha),$$

$$\Theta(u_1 + \tau_{1,r}, u_2 + \tau_{2,r}, \dots, u_p + \tau_{p,r};\ \tfrac{1}{2}\beta,\ \tfrac{1}{2}\alpha) = e^{-2\pi i(u_r + \frac{1}{2}\tau_{r,r}) - \pi i\beta_r}\, \Theta(u;\ \tfrac{1}{2}\beta,\ \tfrac{1}{2}\alpha).$$

Both these are included in the equation

$$\Theta(u + \Omega_m;\ \tfrac{1}{2}\beta,\ \tfrac{1}{2}\alpha) = e^{-2\pi i m'(u + \frac{1}{2}\tau m') + \pi i(m\alpha - m'\beta)}\, \Theta(u;\ \tfrac{1}{2}\beta,\ \tfrac{1}{2}\alpha), \qquad \text{(B)};$$

herein the quantities $m_1, \ldots, m_p, m_1', \ldots, m_p'$ are integers, $u+\Omega_m$ stands for the p quantities such as $u_r + m_r + m_1'\tau_{r,1} + \ldots\ldots + m_p'\tau_{r,p}$, and the notation in the exponent on the right hand is that of the theory of matrices; thus for instance $m'\tau m'$ denotes the expression

$$\sum_{r=1}^{p} m_r'(\tau_{r,1} m_1' + \ldots\ldots + \tau_{r,p} m_p'),$$

and is the same as the expression denoted by $\tau m'^2$.

Equation (B) shews that the partial differential coefficients, of the second order, of the logarithm of $\Theta(u;\tfrac{1}{2}\beta,\tfrac{1}{2}\alpha)$, in regard to $u_1, \ldots, u_p$, are functions of $u_1, \ldots, u_p$, with $2p$ sets of simultaneous periods.

Equation (B) is included in another equation; if each of β', α' denotes a row of p integers, we have

$$\Theta(u+\tfrac{1}{2}\Omega_{\beta',\alpha'};\tfrac{1}{2}\beta,\tfrac{1}{2}\alpha) = e^{-\pi i\alpha'(u+\frac{1}{2}\beta+\frac{1}{2}\beta'+\frac{1}{4}\tau\alpha')}\,\Theta(u;\tfrac{1}{2}\beta+\tfrac{1}{2}\beta',\tfrac{1}{2}\alpha+\tfrac{1}{2}\alpha'),\quad (C);$$

to obtain equation (B) we have only to put $\beta_r' = 2m_r$, $\alpha_r' = 2m_r'$ in equation (C). If, in the same equation, we put $\beta' = -\beta$, $\alpha' = -\alpha$, we obtain

$$\Theta(u-\tfrac{1}{2}\Omega_{\beta,\alpha};\tfrac{1}{2}\beta,\tfrac{1}{2}\alpha) = e^{\pi i\alpha(u-\frac{1}{4}\tau\alpha)}\,\Theta(u;0,0) = e^{\pi i\alpha(u-\frac{1}{4}\tau\alpha)}\,\Theta(u);$$

from this we infer

$$\Theta(u;\tfrac{1}{2}\beta,\tfrac{1}{2}\alpha) = e^{\pi i\alpha(u+\frac{1}{2}\beta+\frac{1}{4}\tau\alpha)}\,\Theta(u+\tfrac{1}{2}\Omega_{\beta,\alpha}),\qquad (D);$$

this is an important equation because it reduces a theta function with any half-integer characteristic to the theta function of zero characteristic.

Finally, when each of m, m' denotes a set of p integers, we have the equation

$$\Theta(u;\tfrac{1}{2}\beta+m,\tfrac{1}{2}\alpha+m') = e^{\pi i m\alpha}\,\Theta(u;\tfrac{1}{2}\beta,\tfrac{1}{2}\alpha),\qquad (E);$$

thus the addition of integers to the quantities $\tfrac{1}{2}\alpha$ does not alter the theta function $\Theta(u;\tfrac{1}{2}\beta,\tfrac{1}{2}\alpha)$, and the addition of integers to the quantities $\tfrac{1}{2}\beta$ can at most change the sign of the function. Hence all the theta functions with half-integer characteristics are reducible to the 2^{2p} theta functions which arise when every element of the characteristic is either 0 or $\tfrac{1}{2}$.

176. We shall verify these equations in order in the most direct way. The method consists in transforming the exponent of the general term of the series, and arranging the terms in a new order. This process is legitimate, because, as we have proved, the series is absolutely convergent.

(A) If in the general term

$$e^{2\pi i u(n+\frac{1}{2}\alpha)+i\pi\tau(n+\frac{1}{2}\alpha)^2+\pi i\beta(n+\frac{1}{2}\alpha)},$$

we change the signs of $u_1, \ldots, u_p$, the exponent becomes

$$2\pi i u(-n-\alpha+\tfrac{1}{2}\alpha)+i\pi\tau(-n-\alpha+\tfrac{1}{2}\alpha)+\pi i\beta(-n-\alpha+\tfrac{1}{2}\alpha)+2\pi i\beta\tau i+\pi i\beta\alpha.$$

Since a consists of integers we may write m for $-n-a$, that is $m_r=-(n_r+a_r)$, for $r=1, 2, \ldots, p$; then, since β consists of integers, and therefore $e^{2\pi i\beta n}=1$, the general term becomes

$$e^{\pi i\beta a} \cdot e^{2\pi iu(m+\frac{1}{2}a)+i\pi\tau(m+\frac{1}{2}a)+\pi i\beta(m+\frac{1}{2}a)};$$

save for the factor $e^{\pi i\beta a}$, this is of the same form as the general term in the original series, the summation integers $m_1, \ldots, m_p$ replacing $n_1, \ldots, n_p$. Thus the result is obvious.

(B) The exponent

$$2\pi i(u+m+\tau m')(n+\tfrac{1}{2}a)+i\pi\tau(n+\tfrac{1}{2}a)^2+\pi i\beta(n+\tfrac{1}{2}a),$$

wherein $m+\tau m'$ stands for a row, or column, of p quantities of which the general one is

$$m_r+\tau_{r,1}m_1'+\ldots\ldots+\tau_{r,p}m_p',$$

is equal to

$$2\pi iu(n+\tfrac{1}{2}a)+i\pi\tau(n+\tfrac{1}{2}a)^2+\pi i\beta(n+\tfrac{1}{2}a)+2\pi imn+\pi ima+2\pi i\tau m'n+\pi i\tau m'a$$
$$=2\pi iu(n+m'+\tfrac{1}{2}a)+i\pi\tau(n+m'+\tfrac{1}{2}a)^2+\pi i\beta(n+m'+\tfrac{1}{2}a)-2\pi im'(u+\tfrac{1}{2}\tau m')$$
$$+\pi i(ma-m'\beta)+2\pi imn.$$

Replacing $e^{2\pi imn}$ by 1 and writing n for $n+m'$, the equation (B) is obtained.

(C) By the work in (B), replacing m, m' by $\frac{1}{2}\beta'$, $\frac{1}{2}a'$ respectively, we obtain

$$2\pi i(u+\tfrac{1}{2}\beta'+\tfrac{1}{2}\tau a')(n+\tfrac{1}{2}a)+i\pi\tau(n+\tfrac{1}{2}a)^2+\pi i\beta(n+\tfrac{1}{2}a)$$
$$=2\pi iu(n+\tfrac{1}{2}a'+\tfrac{1}{2}a)+i\pi\tau(n+\tfrac{1}{2}a'+\tfrac{1}{2}a)+\pi i\beta(n+\tfrac{1}{2}a'+\tfrac{1}{2}a)-\pi ia'(u+\tfrac{1}{4}\tau a')$$
$$+\tfrac{1}{2}\pi i(\beta'a-a'\beta)+\pi i\beta'n,$$

and this is immediately seen to be the same as

$$2\pi iu(n+\tfrac{1}{2}a'+\tfrac{1}{2}a)+i\pi\tau(n+\tfrac{1}{2}a'+\tfrac{1}{2}a)+\pi i(\beta+\beta')(n+\tfrac{1}{2}a'+\tfrac{1}{2}a)-\pi ia'(u+\tfrac{1}{2}\beta+\tfrac{1}{2}\beta'+\tfrac{1}{4}\tau a').$$

This proves the formula (C).

It is obvious that equations (D) are only particular cases of equation (C), and the equation (E) is immediately obvious.

It follows from the equation (A) that the number of odd theta functions contained in the formula $\Theta(u;\ \frac{1}{2}\beta,\ \frac{1}{2}a)$ is $2^{p-1}(2^p-1)$, and therefore that the number of even functions is $2^{2p}-2^{p-1}(2^p-1)$, or $2^{p-1}(2^p+1)$.

For the number of odd functions is the same as the number of sets of integers, $x_1, y_1, \ldots, x_p, y_p$, each either 0 or 1, for which

$$x_1y_1+\ldots\ldots+x_py_p=\text{an odd integer}.$$

These sets consist, (i), of the solutions of the equation

$$x_1y_1+\ldots\ldots+x_{p-1}y_{p-1}=\text{an odd integer},$$

in number, say, $f(p-1)$, each combined with each of the three sets

$$(x_p, y_p)=(0, 1),\ (1, 0),\ (0, 0),$$

together with, (ii), the solutions of the equation

$$x_1y_1+\ldots\ldots+x_{p-1}y_{p-1}=\text{an even integer},$$

in number $2^{2p-2}-f(p-1)$, each combined with the set

$$(x_p, y_p)=(1, 1).$$

Thus

$$f(p)=3f(p-1)+2^{2p-2}-f(p-1)=2^{2p-2}+2f(p-1)$$
$$=2^{2p-2}+2\{2^{2p-4}+2f(p-2)\}=\text{etc.}$$
$$=2^{2p-2}+2^{2p-3}+2^{2p-4}+\ldots\ldots+2^p+2^{p-1}f(1)$$
$$=2^{p-1}(2^p-1).$$

Hence the number of even half periods is $2^{p-1}(2^p+1)$.

177. Suppose now that $e_1, \ldots, e_p$ are definite constants, that m denotes a fixed place of the Riemann surface, and x denotes a variable place of the surface. We consider p arguments given by $u_r = v_r^{x, m} + e_r$, where $v_1^{x, m}, \ldots, v_p^{x, m}$ are the Riemann normal integrals of the first kind. Then the function $\Theta(u)$ is a function of x. By equation (B) it satisfies the conditions

$$\Theta(u + k) = \Theta(u), \quad \Theta(u_r + \tau_r k') = e^{-2\pi i k'(u + \frac{1}{2}\tau k')}\,\Theta(u),$$

wherein k denotes a row, or column, of integers $k_1, \ldots, k_p$ and k' denotes a row or column * of integers $k_1', \ldots, k_p'$. As a function of x, the function $\Theta(v^{x, m} + e)$ cannot, clearly, become infinite, for the arguments $v_r^{x, m} + e_r$ are always finite; but the function does vanish; we proceed in fact to prove the fundamental theorem—*the function* $\Theta(v^{x, m} + e)$ *has always* p *zeros of the first order or zeros whose aggregate multiplicity is* p.

For brevity we denote $v_r^{x, m} + e_r$ by u_r. When the arguments $u_1, \ldots, u_p$ are nearly equal to any finite values $U_1, \ldots, U_p$, the function $\Theta(u)$ can be represented by a series of positive integral powers of the differences $u_1 - U_1, \ldots, u_p - U_p$. Hence the zeros of the function $\Theta(u)$, $= \Theta(v^{x, m} + e)$, are all of positive integral order. The sum of these orders of zero is therefore equal to the value of the integral

$$\frac{1}{2\pi i}\int d\log\Theta(u) = \frac{1}{2\pi i}\int \sum_{s=1}^{p} du_s \Theta_s'(u)/\Theta(u) = \frac{1}{2\pi i}\int dx \sum_{s=1}^{p} (du_s/dx)(\Theta_s'(u)/\Theta(u)),$$

wherein the dash denotes a partial differentiation in regard to the argument u_s, and the integral is to be taken round the complete boundary of the p-ply connected surface on which the function is single-valued, namely round the p closed curves formed by the sides of the period-pair-loops. (Cf. the diagram, p. 21.)

Now the values of $\dfrac{\Theta_s'(u)}{\Theta(u)}\dfrac{du_s}{dx}$ at two points which are opposite points on a period-loop a_r are equal, and in the contour integration the corresponding values of dx are equal and opposite. Hence the portions of the integral arising from the two sides of a period-loop a_r destroy one another. The values of $\dfrac{\Theta_s'(u)}{\Theta(u)}$ at two points which are opposite points on a period-loop b_r differ by $-2\pi i$, or 0, according as $s = r$ or not.

Hence the part of the integral which arises from the period-loop-pair (a_r, b_r) is equal to $-\int du_r$, taken once positively round the left-hand side of the loop b_r, namely equal to $-(-1) = 1$.

The whole value of the integral is, therefore, p; this is then the sum of the orders of zero of the function $\Theta(v^{x, m} + e)$.

* The notation $u_r + \tau_r k'$ denotes the p arguments $u_1 + \tau_1 k', \ldots, u_p + \tau_p k'$.

178. In regard to the position of the zeros of this function we are able to make some statement. We consider first the case when there are p distinct zeros, each of the first order. It is convenient to dissect the Riemann surface in such a way that the function $\log \Theta (v^{x, m} + e)$ may be regarded as single-valued on the dissected surface. Denoting the p zeros of $\Theta (v^{x, m} + e)$ by $z_1, \ldots, z_p$, we may suppose the dissection made by p closed curves such as the one represented in Figure [2], so that a zero of $\Theta (v^{x, m} + e)$ is associated with every one of the period-loop-pairs. Then the surface is still p-ply connected, and $\log \Theta (u)$ is single-valued on the surface bounded by the

Fig. 2.

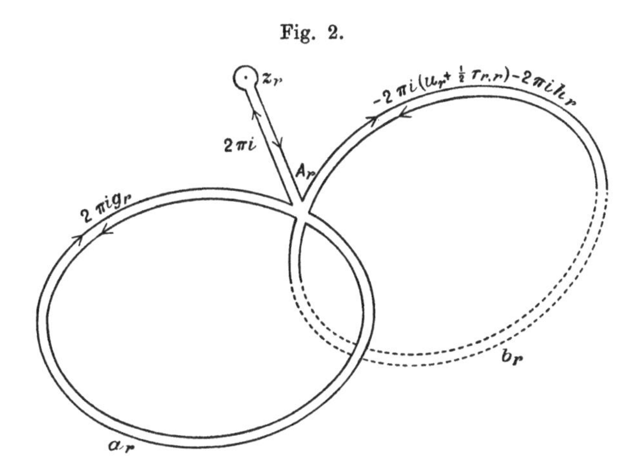

p closed curves such as the one in the figure. For we proved that a complete circuit of the closed curve formed by the sides of the (a_r, b_r) period-loop-pair, gives an increment of $2\pi i$ for the function $\log \Theta (u)$; when the surface is dissected as in the figure this increment of $2\pi i$ is again destroyed in the circuit of the loop which encloses the point z_r. Any closed circuit on the surface as now dissected is equivalent to an aggregate of repetitions of such circuits as that in the figure; thus if x be taken round any closed circuit the value of $\log \Theta (u)$ at the conclusion of that circuit will be the same as at the beginning. From the formulae

$$\Theta (u_1, \ldots, u_r + 1, \ldots, u_p) = \Theta (u),$$

$$\Theta (u_1 + \tau_{r,1}, \ldots, u_r + \tau_{r,r}, \ldots, u_p + \tau_{r,p}) = e^{-2\pi i (u_r + \frac{1}{2}\tau_{r,r})} \Theta (u),$$

which we express by the statement that $\Theta (u)$ has the factors unity and $e^{-2\pi i (u_r + \frac{1}{2}\tau_{r,r})}$ for the period loops a_r and b_r respectively, it follows that $\log \Theta (u)$ can, at most, have, for opposite points of a_r, b_r, respectively, differences of the form $2\pi i g_r$, $-2\pi i (u_r + \frac{1}{2}\tau_{r,r}) - 2\pi i h_r$, wherein g_r and h_r are integers. The sides of the loops for which these increments occur are marked in the figure, u_r denoting the value of $v_r^{x, m} + e_r$ at the side opposite to that where the increment is marked; thus $u_r + \frac{1}{2}\tau_{r,r}$ is the mean of the values, u_r and $u_r + \tau_{r,r}$, which the integral u_r takes at the two sides of the loop b_r.

Since $\log \Theta(u)$ is now single-valued, the integral $\frac{1}{2\pi i}\int \log \Theta(u) \,.\, du_s$, taken round all the p closed curves constituting the boundary of the surface, will have the value zero. Consider the value of this integral taken round the single boundary in the figure. Let A_r denote the point where the loops a_r, b_r, and that round z_r, meet together. The contribution to the integral arising from the two sides of a_r will be $\int g_r dv_s^{x,m}$, *this* integral being taken once positively round the left side of a_r, from A_r back to A_r. This contribution is equal to $g_r \tau_{r,s}$. The contribution to the integral $\frac{1}{2\pi i}\int \log \Theta(u)\, du_s$ which arises from the two sides of the loop b_r is equal to

$$-\int [v_r^{x,m} + e_r + \tfrac{1}{2}\tau_{r,r} + h_r]\, dv_s^{x,m},$$

taken once positively round the left side of the curve b_r, from A_r back to A_r; this is equal to

$$-\int (v_r^{x,m} + \tfrac{1}{2}\tau_{r,r})\, dv_s^{x,m} + (e_r + h_r) f_{r,s},$$

where $f_{r,s}$ is equal to 1 when $r = s$, and is otherwise zero. Finally the part of the integral $\frac{1}{2\pi i}\int \log \Theta(u)\, du_s$, which arises by the circuit of the loop enclosing the point z_r, from A_r back to A_r, in the direction indicated by the arrow head in the figure, is $\int_{A_r}^{z_r} dv_s^{x,m}$ where A_r denotes now a definite point on the boundary of the loop b_r. If we are careful to retain this signification we may denote this integral by $v_s^{z_r, A_r}$. When we add the results thus obtained, for the p boundary curves, taking r in turn equal to $1, 2, \ldots, p$, we obtain

$$h_s + g_1\tau_{1,s} + \ldots\ldots + g_p\tau_{p,s} + e_s = \sum_{r=1}^{p}\left[-v_s^{z_r, A_r} + \int_{b_r}(v_r^{x,m} + \tfrac{1}{2}\tau_{r,r})\, dv_s^{x,m}\right],$$

wherein, on the right hand, the b_r attached to the integral sign indicates a circuit once positively round the left side of b_r from A_r back to A_r; and if k_s denote the quantity defined by the equation

$$k_s = \sum_{r=1}^{p}\int_{b_r}(v_r^{x,m} + \tfrac{1}{2}\tau_{r,r})\, dv_s^{x,m},$$

which, beside the constants of the surface, depends only on the place m, we have the result

$$h_s + g_1\tau_{1,s} + \ldots + g_p\tau_{p,s} + e_s = -v_s^{z_1, A_1} - \ldots - v_s^{z_p, A_p} + k_s \qquad (s = 1, 2, \ldots, p).$$

179. Suppose now that places $m_1, \ldots, m_p$ are chosen to satisfy the congruences

$$v_s^{m_1, A_1} + \ldots\ldots + v_s^{m_p, A_p} \equiv k_s; \qquad (s = 1, 2, \ldots, p);$$

this is always possible (Chap. IX. §§ 168, 169); it is not necessary for our purpose, to prove that only one set* of places $m_1, \ldots, m_p$, satisfies the conditions; these places, beside the fixed constants of the surface, depend only on the place m. Then, by the equations just obtained, we have

$$e_s \equiv -(v_s^{z_1, m_1} + \ldots\ldots + v_s^{z_p, m_p}); \qquad (s = 1, 2, \ldots, p).$$

Thus if we express the zero in the function $\Theta(v^{x, m} + e)$, it takes the form

$$\Theta(v_s^{x, m} - v_s^{z_1, m_1} - \ldots\ldots - v_s^{z_p, m_p} - h_s' - \tau_s g'),$$

where $g_1', \ldots, g_p', h_1', \ldots, h_p'$ are certain integers, and this, by the fundamental equation (B), § 175, is equal to

$$\Theta(v_s^{x, m} - v_s^{z_1, m_1} - \ldots\ldots - v_s^{z_p, m_p}),$$

save for the factor $e^{-2\pi i g'(v^{x, m} - v^{z_1, m_1} - \ldots\ldots - v^{z_p, m_p} - \frac{1}{2}\tau g')}$. This factor does not vanish or become infinite. Hence we have the result: *It is possible, corresponding to any place m, to choose p places, $m_1, \ldots, m_p$, whose position depends only on the position of m, such that the zeros of the function,*

$$\Theta(v^{x, m} - v^{z_1, m_1} - \ldots\ldots - v^{z_p, m_p}),$$

regarded as a function of x, are the places $z_1, \ldots, z_p$. This is a very fundamental result†.

It is to be noticed that the arguments expressed by $v^{x, m} - v^{z_1, m_1} - \ldots - v^{z_p, m_p}$ do not in fact depend on the place m. For the equations for $m_1, \ldots, m_p$, corresponding to any arbitrary position of m, were

$$v_s^{m_1, A_1} + \ldots\ldots + v_s^{m_p, A_p} \equiv k_s, = \sum_{r=1}^{p} \int_{b_r} (v_r^{x, m} + \tfrac{1}{2}\tau_{r, r})\, dv_s^{x, a},$$

a being an arbitrary place. If, instead of m, we take another place μ, we shall, similarly, be required to determine places $\mu_1, \ldots, \mu_p$ by the equations

$$v_s^{\mu_1, A_1} + \ldots\ldots + v_s^{\mu_p, A_p} \equiv k_s, = \sum_{r=1}^{p} \int_{b_r} (v_r^{x, \mu} + \tfrac{1}{2}\tau_{r, r})\, dv_s^{x, a}, \qquad (s = 1, 2, \ldots, p);$$

* If two sets satisfy the conditions, these sets will be coresidual (Chap. VIII., § 158).

† Cf. Riemann, *Ges. Werke* (1876), p. 125, (§ 22). The places $m_1, \ldots, m_p$ are used by Clebsch u. Gordan (*Abel. Functionen*, 1866), p. 195. In Riemann's arrangement the existence of the solution of the inversion problem is not proved before the theta functions are introduced.

thus

$$v_s^{\mu_1, m_1} + \ldots\ldots + v_s^{\mu_p, m_p} \equiv \sum_{r=1}^{p} \int_{b_r} v_r^{m, \mu} \, dv_s^{x, a}, = \sum_{r=1}^{p} f_{s, r} v_r^{\mu, m}, \qquad (s = 1, 2, \ldots, p),$$

wherein $f_{s,r} = 1$ when $r = s$, and is otherwise zero, as we see by recalling the significance of the b_r attached to the integral sign. Thus (Chap. VIII., § 158), the places $\mu_1, \ldots, \mu_p, m$ are coresidual with the places $m_1, \ldots, m_p, \mu$, and the arguments

$$v_s^{x, m} - v_s^{z_1, m_1} - \ldots\ldots - v_s^{z_p, m_p},$$

are congruent to arguments of the form

$$v_s^{x, \mu} - v_s^{z_1, \mu_1} - \ldots\ldots - v_s^{z_p, \mu_p}.$$

The fact that the places $\mu_1, \ldots, \mu_p, m$ are coresidual with the places $m_1, \ldots, m_p, \mu$, which is expressed by the equations

$$v_s^{\mu_1, m_1} + \ldots\ldots + v_s^{\mu_p, m_p} + v_s^{m, \mu} \equiv 0, \qquad (s = 1, 2, \ldots, p),$$

will also, in future, be often represented in the form

$$(\mu_1, \ldots, \mu_p, m) \equiv (m_1, \ldots, m_p, \mu).$$

If the places $m_1, \ldots, m_p$ are not zeros of a ϕ-polynomial, this relation determines $\mu_1, \ldots, \mu_p$ uniquely from the place μ.

Ex. In case $p=1$, prove that the relation determining $m_1, \ldots, m_p$ leads to

$$v^{m_1, m} \equiv \tfrac{1}{2}(1 + \tau).$$

Hence the function $\Theta(v^{x, z} + \frac{1}{2} + \frac{1}{2}\tau)$ vanishes for $x = z$, as is otherwise obvious.

180. The deductions so far made, on the supposition that the p zeros of the function $\Theta(v^{z, m} + e)$ are distinct, are not essentially modified when this is not so. Suppose the zeros to consist of a p_1-tuple zero at z_1, a p_2-tuple zero at $z_2, \ldots$, and a p_k-tuple zero at z_k, so that $p_1 + \ldots\ldots + p_k = p$. The surface may be dissected into a simply connected surface as in Figure 3. The function $\log \Theta(v^{x, m} + e)$ becomes a single-valued function of x on the dissected surface; and its differences, for the two sides of the various cuts, are those given in the figure. To obtain these differences we remember that $\log \Theta(v^{x, m} + e)$ increases by $2\pi i$ when x is taken completely round the four sides of a pair of loops (a_r, b_r). The mode of dissection of Fig. 3, may of course also be used in the previous case when the zeros of $\Theta(v^{x, m} + e)$ are all of the first order.

The integral $\dfrac{1}{2\pi i}\displaystyle\int \log \Theta(v^{x, m} + e)\, dv_s^{x, m}$, taken along the single closed boundary constituted by the sides of all the cuts, has the value zero. Its value is, however, in the case of Figure 3,

$$p_1 v_s^{z_1, A_1} + \ldots\ldots + p_k v_s^{z_k, A_1}$$

$$+ g_1 \int_{a_1} dv_s^{x, m} - h_1 \int_{b_1} dv_s^{x, m} - \int_{b_1} (v_1^{x, m} + e_1 + \tfrac{1}{2}\tau_{1,1})\, dv_s^{x, m} - (p-1)\, v_s^{A_2, A_1}$$

$$+ g_2 \int_{a_2} dv_s^{x, m} - h_2 \int_{b_2} dv_s^{x, m} - \int_{b_2} (v_2^{x, m} + e_2 + \tfrac{1}{2}\tau_{2,2})\, dv_s^{x, m} - (p-2)\, v_s^{A_3, A_1}$$

$$+ \ldots\ldots\ldots\ldots\ldots\ldots\ldots\ldots\ldots\ldots\ldots\ldots\ldots\ldots\ldots\ldots$$

$$+ g_p \int_{a_p} dv_s^{x, m} - h_p \int_{b_p} dv_s^{x, m} - \int_{b_p} (v_p^{x, m} + e_p + \tfrac{1}{2}\tau_{p,p})\, dv_s^{x, m},$$

wherein the first row is that obtained by the sides of the cuts, from A_1, excluding the zeros $z_1, \ldots, z_k$, and the second row is that obtained from the cuts a_1, b_1, c_1, and so on. The suffix a_1 to the first integral sign in

Fig. 3.

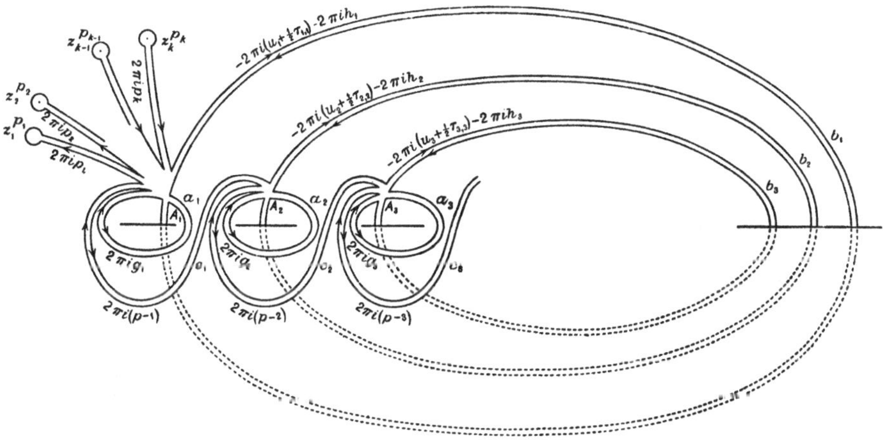

the second row indicates that the integral is to be taken once positively round the left side* of the cut a_1, the suffix b_1 indicates a similar path for the cut b_1, and so on. If, as before, we put k_s for the sum

$$k_s, = \sum_{r=1}^{p} \int_{b_r} (v_r^{x, m} + \tfrac{1}{2}\tau_{r, r})\, dv_s^{x, m},$$

we obtain, therefore, as the result of the integration, that the quantity

$$h_s + g_1 \tau_{s, 1} + \ldots\ldots + g_p \tau_{s, p} + e_s$$

* By the left side of a cut a_1, or b_1, is meant the side upon which the increments of $\log \Theta(u)$ are marked in the figure. The general question of the effect of variation in the period cuts is most conveniently postponed until the transformation of the theta functions has been considered.

is equal to

$$k_s - p_1 v_s^{z_1, A_1} - \ldots\ldots - p_k v_s^{z_k, A_1} + (p-1)\, v_s^{A_2, A_1} + (p-2)\, v_s^{A_3, A_2} + \ldots\ldots + v_s^{A_p, A_{p-1}}$$

and this is immediately seen to be the same as

$$k_s - v_s^{z_1, A_1} - \ldots\ldots - v_s^{z_1, A_{p_1}} - v_s^{z_2, A_{p_1+1}} - \ldots\ldots - v_s^{z_2, A_{p_1+p_2}} - \ldots\ldots - v_s^{z_k, A_p}.$$

We thus obtain, of course, the same equations as before (§ 179), save that z_1 is here repeated p_1 times, ..., and z_k is repeated p_k times. And we can draw the inference that $\Theta(v^{x, m} + e)$ can be written in the form $\Theta(v_s^{x, m} - v_s^{z_1, m_1} - \ldots\ldots - v_s^{z_p, m_p} - h_s - \tau_s g)$, which, save for a finite non-vanishing factor, is the same as $\Theta(v_s^{x, m} - v_s^{z_1, m_1} - \ldots\ldots - v_s^{z_p, m_p})$; the argument $v_s^{x, m} - v_s^{z_1, m_1} - \ldots\ldots - v_s^{z_p, m_p}$ does not depend on the place m.

181. From the results of §§ 179, 180, we can draw an inference which leads to most important developments in the theory of the theta functions.

For, from what is there obtained it follows that if $z_1, \ldots, z_p$ be any places whatever, the function $\Theta(v^{x, m} - v^{z_1, m_1} - \ldots\ldots - v^{z_p, m_p})$ has $z_1, \ldots, z_p$ for zeros. Hence, putting z_p for x we infer that *the function*

$$\Theta(v^{m_p, m} - v^{z_1, m_1} - \ldots\ldots - v^{z_{p-1}, m_{p-1}}) \qquad \text{(F)}$$

vanishes identically for all positions of $z_1, \ldots, z_{p-1}$. Putting

$$f_s = v_s^{z_1, m_1} + \ldots\ldots + v_s^{z_{p-2}, m_{p-2}} - v_s^{m_p, m},$$

for $s = 1, 2, \ldots, p$, this is the same as the statement that the function $\Theta(v^{x, m_{p-1}} + f)$ vanishes identically for all positions of x and for all values of $f_1, \ldots, f_p$ which can be expressed in the form arising here. When $f_1, \ldots, f_p$ are arbitrary quantities it is not in general possible to determine places $z_1, \ldots, z_{p-2}$ to express $f_1, \ldots, f_p$ in the form in question. Nevertheless the case which presents itself reminds us that in the investigation of the zeros of $\Theta(v^{x, m} + e)$ we have assumed that the function does not vanish identically, and it is essential to observe that this is so for general values of $e_1, \ldots, e_p$. If, for a given position of x, the function $\Theta(v^{x, m} + e)$ vanished identically for all values of $e_1, \ldots, e_p$, the function $\Theta(r)$ would vanish for all values of the arguments $r_1, \ldots, r_p$. We assume * from the original definition of the theta function, by means of a series, that this is not the case.

Further the function $\Theta(v^{x, m} + e)$ is by definition an analytical function of each of the quantities $e_1, \ldots, e_p$; and if an analytical function do not vanish

* The series is a series of integral powers of the quantities $e^{2\pi i r_1}, \ldots, e^{2\pi i r_p}$.

for all values of its argument, there must exist a continuum of values of the argument, of finite extent in two dimensions, within which the function does not vanish *. Hence, for each of the quantities $e_1, \ldots, e_p$ there is a continuum of values of two dimensions, within which the function $\Theta(v^{x,m}+e)$ does not vanish identically. And, by equation (B), § 175, this statement remains true when the quantities $e_1, \ldots, e_p$ are increased by any simultaneous periods. Restricting ourselves then, first of all, to values of $e_1, \ldots, e_p$ lying within these regions, there exist (Chap. IX. § 168) positions of $z_1, \ldots, z_p$ to satisfy the congruences

$$e_s \equiv v_s^{z_1, m_1} + \ldots\ldots + v_s^{z_p, m_p}, \qquad (s=1, 2, \ldots, p);$$

and, since to each set of positions of $z_1, \ldots, z_p$, there corresponds only one set of values for $e_1, \ldots, e_p$, the places $z_1, \ldots, z_p$ are also, each of them, variable within a certain two-dimensionality. Hence, within certain two-dimensional limits, there certainly exist arbitrary values of $z_1, \ldots, z_p$ such that the function $\Theta(v^{x,m} - v^{z_1, m_1} - \ldots\ldots - v^{z_p, m_p})$ does not vanish identically. For such values, and the corresponding values of $e_1, \ldots, e_p$, the investigation so far given holds good. And therefore, for such values, the function $\Theta(v^{m_p, m} - v^{z_1, m_1} - \ldots\ldots - v^{z_{p-1}, m_{p-1}})$ vanishes identically. Since this function is an analytical function of the places† $z_1, \ldots, z_{p-1}$, and vanishes identically for all positions of each of these places within a certain continuum of two dimensions, it must vanish identically for all positions of these places.

Hence the theorem (F) holds without limitation, notwithstanding the fact that for certain special forms of the quantities $e_1, \ldots, e_p$, the function $\Theta(v^{x,m}+e)$ vanishes identically. The important part played by the theorem (F) will be seen to justify this enquiry.

182. It is convenient now to deduce in order a series of propositions in regard to the theta functions (§§ 182—188); and for purposes of reference it is desirable to number them.

(I.) If $\zeta_1, \ldots, \zeta_p$ be p places which are zeros of one or more linearly independent ϕ-polynomials, that is, of linearly independent linear aggregates of the form $\lambda_1\Omega_1(x) + \ldots\ldots + \lambda_p\Omega_p(x)$ (Chap. II. § 18, Chap. VI. § 101), then the function

$$\Theta(v^{x,m} - v^{\zeta_1, m_1} - \ldots\ldots - v^{\zeta_p, m_p})$$

vanishes identically for all positions of x.

For then, if $\tau+1$ be the number of linearly independent ϕ-polynomials which vanish in the places $\zeta_1, \ldots, \zeta_p$, we can, taking $\tau+1$ arbitrary places

* E.g. a single-valued analytical function of an argument z, $=x+iy$, cannot vanish for all rational values of x and y without vanishing identically.

† By an analytical function of a place z on a Riemann surface, is meant a function whose values can be expressed by series of integral powers of the infinitesimal at the place.

$z_1, \ldots, z_{\tau+1}$, determine $p-\tau-1$ places $z_{\tau+2}, \ldots, z_p$, such that $(z_1, \ldots, z_p) \equiv (\zeta_1, \ldots, \zeta_p)$ (see Chap. VI. § 93, etc., and for the notation, § 179). Then the argument

$$v_s^{x, m} - v_s^{\zeta_1, m_1} - \ldots\ldots - v_s^{\zeta_p, m_p}, \qquad (s = 1, 2, \ldots, p),$$

can be put in the form

$$v_s^{x, m} - v_s^{z_1, m_1} - \ldots\ldots - v_s^{z_p, m_p},$$

save for integral multiples of the periods; thus (§§ 179, 180) the theta function vanishes when x is at any one of the perfectly arbitrary places $z_1, \ldots, z_{\tau+1}$. Thus, since by hypothesis $\tau+1$ is at least equal to 1, the theta function vanishes identically.

It follows from this proposition that if $z_2', \ldots, z_p'$ be the remaining zeros of a ϕ-polynomial determined to vanish in each of $z_2, \ldots, z_p$, and neither x nor z_1 be among $z_2', \ldots, z_p'$, then the zeros of the function

$$\Theta(v^{x, m} - v^{z_1, m_1} - \ldots\ldots - v^{z_p, m_p}),$$

regarded as a function of z_1, are the places $x, z_2', \ldots, z_p'$.

From this Proposition and the results previously obtained, we can infer that *the function* $\Theta(v^{x, m} - v^{z_1, m_1} - \ldots\ldots - v^{z_p, m_p})$ *vanishes only* (i) *when* x *coincides with one of the places* $z_1, \ldots, z_p$, or (ii) *when* $z_1, \ldots, z_p$ *are zeros of a* ϕ*-polynomial.*

(II.) Suppose a rational function exists, of order, Q, not greater than p, and let $\tau+1$ be the number of ϕ-polynomials vanishing in the poles of this function. Take $\tau+1$ arbitrary places

$$\zeta_1, \ldots, \zeta_q, x_1, \ldots, x_{\tau+1-q},$$

wherein $q = Q - p + \tau + 1$, and suppose $z_1, \ldots, z_Q$ to be a set of places coresidual with the poles of the rational function, of which, therefore, q are arbitrary. Then the function

$$\Theta\,(v^{m_p, m} + v^{\zeta_1, z_1} + \ldots\ldots + v^{\zeta_q, z_q} - v^{x_1, m_1} - \ldots\ldots$$
$$- v^{x_{\tau+1-q}, m_{\tau+1-q}} - v^{z_{q+1}, m_{\tau+2-q}} - \ldots\ldots - v^{z_Q, m_{p-q}})$$

vanishes identically.

For if we choose $\zeta_{q+1}, \ldots, \zeta_Q$ such that $(\zeta_1, \ldots, \zeta_Q) \equiv (z_1, \ldots, z_Q)$, the general argument of the theta function under consideration is congruent to the argument

$$v^{m_p, m} - v^{x_1, m_1} - \ldots\ldots - v^{x_{\tau+1-q}, m_{\tau+1-q}} - v^{\zeta_{q+1}, m_{\tau+2-q}} - \ldots\ldots - v^{\zeta_Q, m_{p-q}}.$$

This value of the argument is a particular case of that occurring in (F), § 181, the last $q-1$ of the upper limits in (F) being put equal to the lower limits. Hence the proposition follows from (F).

(III.) If r denote such a set of arguments $r_1, \ldots, r_p$ that $\Theta(r) = 0$, and, for the positions of z under consideration, the function $\Theta(v^{x, z} + r)$ does not vanish for all positions of x, then there are unique places $z_1, \ldots, z_{p-1}$, such that

$$r \equiv v^{m_p, m} - v^{z_1, m_1} - \ldots\ldots - v^{z_{p-1}, m_{p-1}}.$$

In this statement of the proposition a further abbreviation is introduced which will be constantly employed. The suffix indicating that the equation stands as the representative of p equations is omitted.

Before proceeding to the proof it may be remarked that if $m', m_1', \ldots, m_p'$ be places such that (cf. § 179)

$$(m', m_1, \ldots, m_p) \equiv (m, m_1', \ldots, m_p')$$

and therefore, also,

$$v^{m', m} - v^{m_1', m_1} - \ldots\ldots - v^{m_p', m_p} \equiv 0,$$

then the equation

$$r \equiv v^{m_p, m} - v^{z_1, m_1} - \ldots\ldots - v^{z_{p-1}, m_{p-1}}$$

is the same as the equation

$$r \equiv v^{m_p', m'} - v^{z_1, m_1'} - \ldots\ldots - v^{z_{p-1}, m'_{p-1}}.$$

This proposition (III.) is in the nature of a converse to equation (F). Since the function $\Theta(v^{x, z} + r)$ does not vanish identically, its zeros, $z_1, \ldots, z_p$, are such that

$$v^{x, z} + r \equiv v^{x, m} - v^{z_1, m_1} - \ldots\ldots - v^{z_p, m_p};$$

now we have

$$v^{z_1, m_1} + v^{z_p, m_p} \equiv v^{z_p, m_1} + v^{z_1, m_p},$$

so that the zeros $z_1, \ldots, z_p$ may be taken in any order; since $\Theta(r)$ vanishes, z is one of the zeros of $\Theta(v^{x, z} + r)$; hence, we may put $z_p = z$, and obtain

$$r \equiv v^{x, m} - v^{z_1, m_1} - \ldots\ldots - v^{z_p, m_p} - v^{x, z_p},$$

$$\equiv v^{m_p, m} - v^{z_1, m_1} - \ldots\ldots - v^{z_{p-1}, m_{p-1}},$$

which is the form in question.

If the places $z_1, \ldots, z_{p-1}$ in this equation are not unique, but, on the contrary, there exists also an equation of the form

$$r \equiv v^{m_p, m} - v^{z_1', m_1} - \ldots\ldots - v^{z'_{p-1}, m_{p-1}},$$

then, from the resulting equation

$$v^{z_1', z_1} + \ldots\ldots + v^{z'_{p-1}, z_{p-1}} \equiv 0,$$

we can (Chap. VIII. § 158) infer that there is an infinite number of sets of places $z_1', \ldots, z'_{p-1}$, all coresidual with the set $z_1, \ldots, z_{p-1}$; hence we can put

$$v^{x,z} + r \equiv v^{x,m} - v^{z_1', m_1} - \ldots\ldots - v^{z'_{p-1}, m_{p-1}} - v^{z, m_p},$$

wherein at least one of the places $z_1', \ldots, z'_{p-1}$ is entirely arbitrary. Then the function $\Theta(v^{x,z} + r)$ vanishes for an arbitrary position of x, that is, it vanishes identically; this is contrary to the hypothesis made.

It follows also that whenever it is possible to find places $z_1, \ldots, z_{p-1}$ to satisfy the inversion problem expressed by the p equations

$$v^{z_1, m_1} + \ldots\ldots + v^{z_{p-1}, m_{p-1}} = u,$$

the function $\Theta(v^{m_p, m} - u)$ vanishes; conversely, when u is such that this function vanishes we can solve the inversion problem referred to.

(IV.) When r is such that $\Theta(r)$ vanishes, and $\Theta(v^{x,z} + r)$ does not, for the values of z considered, vanish identically for all positions of x, the zeros of $\Theta(v^{x,z} + r)$, other than z, are independent of z and depend only on the argument r.

This is an immediate corollary from Proposition (III.); but it is of sufficient importance to be stated separately.

(V.) If $\Theta(r) = 0$, and $\Theta(v^{x,z} + r)$ vanish identically for all positions of x and z, but $\Theta(v^{x,z} + v^{\xi,\zeta} + r)$ do not vanish identically, in regard to x, for the positions of z, ξ, ζ considered, then it is possible to find places $z_1, \ldots, z_{p-2}$ such that

$$r \equiv v^{m_p, m} - v^{z_1, m_1} - \ldots\ldots - v^{z_{p-2}, m_{p-2}} - v^{\xi, m_{p-1}},$$

and these places $z_1, \ldots, z_{p-2}$ are definite.

Under the hypotheses made, we can put

$$v^{x,z} + v^{\xi,\zeta} + r \equiv v^{x,m} - v^{z_1, m_1} - \ldots\ldots - v^{z_p, m_p},$$

wherein $z_1, \ldots, z_p$ are the zeros of $\Theta(v^{x,z} + v^{\xi,\zeta} + r)$; now z is clearly a zero; for the function $\Theta(v^{\xi,\zeta} + r)$ is of the same form as $\Theta(v^{x,z} + r)$, and vanishes identically; and ζ is also a zero; for, putting ζ for x, the function $\Theta(v^{x,z} + v^{\xi,\zeta} + r)$ becomes $\Theta(v^{\xi,z} + r)$, which also vanishes identically. Putting, therefore, ζ, z for z_{p-1} and z_p respectively, the result enunciated is obtained, the uniqueness of the places $z_1, \ldots, z_{p-2}$ being inferred as in Proposition (III.).

We may state the theorem differently thus: If $\Theta(v^{x,z} + r)$ vanish for all positions of x and z, and $\Theta(v^{x,z} + v^{\xi,\zeta} + r)$ do not in general vanish identically, the equations

$$r \equiv v^{m_p, m} - v^{z_1, m_1} - \ldots\ldots - v^{z_{p-2}, m_{p-2}} - v^{z_{p-1}, m_{p-1}}$$

can be solved, and in the solution one of $z_1, \ldots, z_{p-1}$ may be taken arbitrarily, and the others are thereby determined. Hence also we can find places $z_1', \ldots, z'_{p-1}$, other than $z_1, \ldots, z_{p-1}$, such that

$$v^{z_1', z_1} + \ldots\ldots + v^{z'_{p-1}, z_{p-1}} = 0,$$

one of the places $z_1', \ldots, z'_{p-1}$ being arbitrary. Hence by the formula $Q - q = p - \tau - 1$, putting $Q = p - 1$, $q = 1$, we infer $\tau + 1 = 2$, so that a ϕ-polynomial vanishing in $z_1, \ldots, z_{p-1}$ can be made to vanish in the further arbitrary place z. Thus, when $\Theta(v^{x, z} + r)$ vanishes identically, we can write

$$v^{x, z} + r \equiv v^{x, m} - v^{z_1, m_1} - \ldots\ldots - v^{z_{p-1}, m_{p-1}} - v^{z, m_p},$$

wherein the places $z_1, \ldots, z_{p-1}, z$ are zeros of a ϕ-polynomial (cf. Prop. I.).

(VI.) The propositions (III.) and (V.) can be generalized thus: If $\Theta(v^{x_1, z_1} + \ldots\ldots + v^{x_q, z_q} + r)$ be identically zero for all positions of the places $x_1, z_1, \ldots, x_q, z_q$, and the function $\Theta(v^{x, z} + v^{x_1, z_1} + \ldots\ldots + v^{x_q, z_q} + r)$ do not vanish identically in regard to x, then places $\zeta_1, \ldots, \zeta_{p-1}$ can be found to satisfy the equations

$$r \equiv v^{m_p, m} - v^{\zeta_1, m_1} - \ldots\ldots - v^{\zeta_{p-1}, m_{p-1}},$$

and, of these places, q are arbitrary, the others being thereby determined.

These arbitrary places, $\zeta_1, \ldots, \zeta_q$, say, must be such that the function $\Theta(v^{x, z} + v^{\zeta_1, z_1} + \ldots\ldots + v^{\zeta_q, z_q} + r)$ does not vanish identically.

For as before we can put

$$v^{x, z} + v^{x_1, z_1} + \ldots\ldots + v^{x_q, z_q} + r \equiv v^{x, m} - v^{\zeta_1, m_1} - \ldots\ldots - v^{\zeta_p, m_p},$$

wherein $\zeta_1, \ldots, \zeta_p$ are the zeros of the function $\Theta(v^{x, z} + v^{x_1, z_1} + \ldots + v^{x_q, z_q} + r)$. It is clear that z is one zero of this function; also putting z_1 for x the function becomes $\Theta(v^{x_1, z} + v^{x_2, z_2} + \ldots\ldots + v^{x_q, z_q} + r)$, which vanishes, by the hypothesis. Thus the places $z, z_1, \ldots, z_q$ are all zeros of the function

$$\Theta(v^{x, z} + v^{x_1, z_1} + \ldots\ldots + v^{x_q, z_q} + r).$$

Putting then $z_1, \ldots, z_q, z$ respectively for $\zeta_1, \ldots, \zeta_q, \zeta_p$ in the congruence just written, it becomes

$$v^{x, z} + v^{x_1, z_1} + \ldots\ldots + v^{x_q, z_q} + v^{z_1, m_1} + \ldots\ldots + v^{z_q, m_q} + v^{\zeta_{q+1}, m_{q+1}} + \ldots\ldots$$
$$+ v^{\zeta_{p-1}, m_{p-1}} + v^{z, m_p} + r \equiv v^{x, m},$$

and this is the same as

$$r \equiv v^{m_p, m} - v^{x_1, m_1} - \ldots\ldots - v^{x_q, m_q} - v^{\zeta_{q+1}, m_{q+1}} - \ldots\ldots - v^{\zeta_{p-1}, m_{p-1}};$$

replacing $x_1, \ldots, x_q$ by $\zeta_1, \ldots, \zeta_q$ we have the result stated.

Hence also, we can find places $\zeta_1', \dots, \zeta'_{p-1}$, other than $\zeta_1, \dots, \zeta_{p-1}$, such that

$$v^{\zeta_1', \zeta_1} + \dots\dots + v^{\zeta'_{p-1}, \zeta_{p-1}} \equiv 0,$$

q of the places $\zeta_1', \dots, \zeta'_{p-1}$ being arbitrary. Therefore a ϕ-polynomial can be chosen to vanish in $\zeta_1, \dots, \zeta_{p-1}$ and in $q \, (= p - 1 - (Q - q)$, when $Q = p - 1)$ other arbitrary places. Thus the argument

$$v^{x, z} + v^{x_1, z_1} + \dots\dots + v^{x_{q-1}, z_{q-1}} + r,$$

for which the theta function vanishes identically, can be written in the form

$$v^{x, m} - v^{z_1, m_1} - \dots\dots - v^{z_{q-1}, m_{q-1}} - v^{\zeta_q, m_q} - \dots\dots - v^{\zeta_{p-1}, m_{p-1}} - v^{z, m_p},$$

wherein $z_1, \dots, z_{q-1}, \zeta_q, \dots, \zeta_{p-1}, z$ are zeros of $q+1$ linearly independent ϕ-polynomials.

(VII.) If the function $\Theta\,(v^{x_1, z_1} + \dots\dots + v^{x_q, z_q} + r)$ be identically zero for all positions of the places $x_1, z_1, x_2, z_2, \dots, x_q, z_q$, and, for general positions of $x_1, z_1, \dots, x_q, z_q$, the function $\Theta\,(v^{x, z} + v^{x_1, z_1} + \dots\dots + v^{x_q, z_q} + r)$ be not identically zero, as a function of x, for proper positions of z, and be not identically zero, as a function of z, for proper positions of x, then we can find places $\zeta_1, \dots, \zeta_{p-1}$, of which q places are arbitrary, such that

$$r \equiv v^{m_p, m} - v^{\zeta_1, m_1} - \dots\dots - v^{\zeta_{p-1}, m_{p-1}},$$

and can also find places $\xi_1, \dots, \xi_{p-1}$, of which q places are arbitrary, such that

$$-r \equiv v^{m_p, m} - v^{\xi_1, m_1} - \dots\dots - v^{\xi_{p-1}, m_{p-1}}.$$

This is obvious from the last proposition, if we notice that

$$\Theta\,(v^{z, x} + v^{z_1, x_1} + \dots\dots + v^{z_q, x_q} - r) = \Theta\,(v^{x, z} + v^{x_1, z_1} + \dots\dots + v^{x_q, z_q} + r).$$

We can hence infer that

$$2v^{m_p, m} + v^{m_1, \zeta_1} + v^{m_1, \xi_1} + \dots\dots + v^{m_{p-1}, \zeta_{p-1}} + v^{m_{p-1}, \xi_{p-1}} \equiv 0,$$

and this is the same (Chap. VIII. § 158) as the statement that the set of $2p$ places constituted by $\xi_1, \dots, \xi_{p-1}, \zeta_1, \dots, \zeta_{p-1}$ and the place m, repeated, is coresidual with the set of $2p$ places constituted by the places $m_1, \dots, m_p$, each repeated. This result we write (cf. § 179) in the form

$$(m^2, \xi_1, \dots, \xi_{p-1}, \zeta_1, \dots, \zeta_{p-1}) \equiv (m_1^2, m_2^2, \dots, m_p^2).$$

(VIII.) We can now prove that if $\zeta_1, \dots, \zeta_{p-1}$ be arbitrary places, places $\xi_1, \dots, \xi_{p-1}$ can be found such that

$$(m^2, \xi_1, \dots, \xi_{p-1}, \zeta_1, \dots, \zeta_{p-1}) \equiv (m_1^2, m_2^2, \dots, m_p^2).$$

Let r denote the set of p arguments given by

$$r \equiv v^{m_p, m} - v^{\zeta_1, m_1} - \dots\dots - v^{\zeta_{p-1}, m_{p-1}},$$

$\zeta_1, \ldots, \zeta_{p-1}$ being quite arbitrary. Then, by theorem (F), (§ 181), the function $\Theta(r)$ certainly vanishes. It may happen that also the function $\Theta(v^{x,z}+r)$ vanishes identically for all positions of x and z. It may further happen that also the function $\Theta(v^{x,z}+v^{x_1,z_1}+r)$ vanishes identically for all positions of x, z, x_1, z_1. We assume* however that there is a finite value of q such that the function $\Theta(v^{x,z}+v^{x_1,z_1}+\ldots\ldots+v^{x_q,z_q}+r)$ does not vanish identically for all positions of $x, z, x_1, z_1, \ldots, x_q, z_q$. Then by Proposition VII. it follows that we can find places $\xi_1, \ldots, \xi_{p-1}$, such that

$$-r \equiv v^{m_p,m} - v^{\xi_1,m_1} - \ldots\ldots - v^{\xi_{p-1},m_{p-1}};$$

comparing this with the equations defining the argument r, we can, as in Proposition (VII.) infer that the congruence stated at the beginning of this Proposition also holds.

(IX.) Hence follows a very important corollary. Taking any other arbitrary places $\zeta_1', \ldots, \zeta'_{p-1}$, we can find places $\xi_1', \ldots, \xi'_{p-1}$ such that

$$(m^2, \xi_1', \ldots, \xi'_{p-1}, \zeta_1', \ldots, \zeta'_{p-1}) \equiv (m_1^2, m_2^2, \ldots, m_p^2);$$

therefore the set $\xi_1, \ldots, \xi_{p-1}, \zeta_1, \ldots, \zeta_{p-1}$ is coresidual with the set $\xi_1', \ldots, \xi'_{p-1}, \zeta_1', \ldots, \zeta'_{p-1}$. Now, of a set of $2p-2$ places coresidual with a given set we can in general take only $p-2$ arbitrarily; when, as here, we can take $p-1$ arbitrarily, each of the sets must be the zeros of a ϕ-polynomial (Chap. VI. § 93). Thus the places $\xi_1, \ldots, \xi_{p-1}, \zeta_1, \ldots, \zeta_{p-1}$ are zeros of a ϕ-polynomial.

Therefore, if $a_1, \ldots, a_{2p-2}$ be the zeros of any ϕ-polynomial whatever, that is, the zeros of the differential of any integral of the first kind, *the places $m_1, \ldots, m_p$ are so derived from the place m that we have*

$$(m^2, a_1, \ldots, a_{2p-2}) \equiv (m_1^2, m_2^2, \ldots, m_p^2), \qquad (G);$$

in other words, if $c_1, \ldots, c_p$ denote any independent places, the places $m_1, \ldots, m_p$ satisfy the equations

$$2\left[v_s^{m_1,c_1}+\ldots\ldots+v_s^{m_p,c_p}\right] \equiv 2v_s^{m,c_p}+v_s^{a_1,c_1}+v_s^{a_2,c_1}+\ldots\ldots+v_s^{a_{2p-3},c_p}+v_s^{a_{2p-2},c_p},$$

for $s=1, 2, \ldots, p$. Denoting the right hand, whose value is perfectly definite, by A_s, and supposing $g_1, \ldots, g_p, h_1, \ldots, h_p$ to denote proper integers, these equations are the same as

$$v_s^{m_1,c_1}+\ldots\ldots+v_s^{m_p,c_p} \equiv \tfrac{1}{2}A_s + \tfrac{1}{2}(h_s + g_1\tau_{s,1}+\ldots\ldots+g_p\tau_{s,p}), \qquad (G'),$$

where $s=1, 2, \ldots, p$.

* It will be seen in Proposition XIV. that if $\Theta(v^{x,z}+v^{x_1,z_1}+\ldots\ldots+v^{x_q,z_q}+r)$ vanishes identically, then all the partial differential coefficients of $\Theta(u)$, in regard to $u_1, \ldots, u_p$, up to and including those of the $(q+1)$th order, also vanish for $u=r$.

There are however 2^{2p} sets of places $m_1, \ldots, m_p$, corresponding to any position of the place m, which satisfy the equation* (G). For in equations (G′) there are 2^{2p} values possible for the right-hand side in which each of $g_1, \ldots, g_p, h_1, \ldots, h_p$ is either 0 or 1, and any two sets of values $g_1, \ldots, g_p, h_1, \ldots, h_p$ and $g_1', \ldots, g_p', h_1', \ldots, h_p'$, such that g_i, g_i' differ by an even integer, and h_i, h_i' differ by an even integer, for $i = 1, 2, \ldots, p$, lead to the same positions for the places $m_1, \ldots, m_p$. (Chap. VIII. § 158.)

We have seen (§ 179) that the places $m_1, \ldots, m_p$ depend only on the place m and on the mode of dissection of the Riemann surface. We are to see, in what follows, that the 2^{2p} solutions of the equation (G) are to be associated, in an unique way, each with one of the 2^{2p} essentially distinct theta functions with half integer characteristics.

183. The equation (G) can be interpreted geometrically. Take a non-adjoint polynomial, Δ, of any grade μ, which has a zero of the second order at the place m; it will have $n\mu - 2$ other zeros. Take an adjoint polynomial ψ, of grade $(n-1)\sigma + n - 3 + \mu$, which vanishes in these other $n\mu - 2$ zeros of Δ. Then (Chap. VI. § 92, Ex. ix.) ψ will be of the form $\lambda\psi_0 + \Delta\phi$, where ψ_0 is a special form of ψ, λ is an arbitrary constant, and ϕ is a general ϕ-polynomial. The polynomial ψ will have $2p$ zeros other than those prescribed; denote them by $k_1, \ldots, k_{2p}$. If ϕ' be any ϕ-polynomial, with $a_1, \ldots, a_{2p-2}$ as zeros, we can form a rational function, given by $(\lambda\psi_0 + \Delta\phi)/\Delta\phi'$, whose poles are the places $a_1, \ldots, a_{2p-2}$, together with the place m repeated, its zeros being the places $k_1, \ldots, k_{2p}$. Hence (Chap. VI. § 96) we have

$$(m^2, a_1, \ldots, a_{2p-2}) \equiv (k_1, k_2, \ldots, k_{2p-1}, k_{2p}),$$

and therefore, by equation (G),

$$(m_1^2, \ldots, m_p^2) \equiv (k_1, k_2, \ldots, k_{2p-1}, k_{2p}) \qquad (G'');$$

hence (Chap. VI. § 90) it is possible to take the polynomial ψ so that its zeros $k_1, \ldots, k_{2p}$ consist of p zeros each of the second order, and *the places $m_1, \ldots, m_p$ are one of the sets of p places thus obtained.*

There are 2^{2p} possible polynomials ψ which have the necessary character, as we have already seen by considering the equation (G′); but, in fact, a certain number of these are composite polynomials formed by the product of the polynomial Δ and a ϕ-polynomial of which the $2p-2$ zeros consist of $p-1$ zeros each repeated. To prove this it is sufficient to prove that there exist such ϕ-polynomials having only $p-1$ zeros, each of the second order; for it is clear that if Φ denote such a polynomial, the product $\Delta\Phi$ is of grade

* If for any set of values for $g_1, \ldots, g_p, h_1, \ldots, h_p$ the equations (G′) are capable of an infinity of (coresidual) sets of solutions, the correct statement will be that there are 2^{2p} lots of coresidual sets, belonging to the place m, which satisfy the equation (G). The corresponding modification may be made in what follows.

$(n-1)\sigma + n - 3 + \mu$ and satisfies the conditions imposed on the polynomial ψ. That there are such ϕ-polynomials Φ is immediately obvious algebraically. If we form the equation giving the values of x at the zeros of the general ϕ-polynomial,

$$\lambda_1\phi_1 + \ldots\ldots + \lambda_p\phi_p,$$

the $p-1$ conditions that the left-hand side should be a perfect square, will determine the necessary ratios $\lambda_1 : \lambda_2 : \ldots : \lambda_p$, and, in general, in only a finite number of ways. (Cf. also Prop. XI. below.)

It is immediately seen, from equation (G″), that if $m_1, \ldots, m_p$ be the double zeros of one such polynomial ψ as described, and $m_1', \ldots, m_p'$ of another, both sets being derived from the same place m, then

$$v^{m_1', m_1} + \ldots\ldots + v^{m_p', m_p} = \tfrac{1}{2}\Omega_{\beta, \alpha}, \qquad \text{(H)}$$

where $\Omega_{\beta, \alpha}$ stands for p quantities such as

$$\beta_s + \alpha_1\tau_{s,1} + \ldots\ldots + \alpha_p\tau_{s,p},$$

$\alpha_1, \ldots, \alpha_p, \beta_1, \ldots, \beta_p$ being integers.

We may give an example of the geometrical relation thus introduced, which is of great importance. It will be sufficient to use only the usual geometrical phraseology.

Suppose the fundamental equation is of the form

$$C + (x, y)_1 + (x, y)_2 + (x, y)_3 + (x, y)_4 = 0,$$

representing a plane quartic curve ($p=3$). Then if a straight line be drawn touching the curve at a point m, it will intersect it again in 2 points A, B. Through these 2 points A, B, ∞^3 conics can be drawn; of these conics there are a certain number which touch the fundamental quartic in three points P, Q, R other than A and B. There are $2^{2p}=64$ sets of three such points P, Q, R; but of these some consist of the two points of contact of double tangents of the quartic taken with the point m itself.

In fact there are (Salmon, *Higher Plane Curves*, Dublin, 1879, p. 213) 28, $=2^{p-1}(2^p-1)$, double tangents; these do not depend at all on the point m; there are therefore 36, $=2^{p-1}(2^p+1)$, proper sets of three points P, Q, R in which conics passing through A and B touch the curve. One of these sets of three points is formed by the points m_1, m_2, m_3. It has been proved that the numbers $2^{p-1}(2^p-1)$, $2^{p-1}(2^p+1)$ are respectively the numbers of odd and even theta functions of half integer characteristics (§ 176).

184. (X.) We have seen in Proposition (VIII.) (§ 182) that the places $m_1, \ldots, m_p$ are one set from 2^{2p} sets of p places all satisfying the same equivalence (G). We are now to see the interpretation of the other $2^{2p}-1$ solutions of this equation.

Let $m_1', \ldots, m_p'$ be any set, other than $m_1, \ldots, m_p$, which satisfies the congruence (G). Then, by equations (G′), we have

$$2\left(v_s^{m_1', m_1} + \ldots\ldots + v_s^{m_p', m_p}\right) \equiv 0, \qquad (s = 1, 2, \ldots, p),$$

and therefore, if $\Omega_{\beta,\alpha}$ denote the set of p quantities of which a general one is given by

$$\beta_s + \alpha_1\tau_{s,1} + \dots\dots + \alpha_p\tau_{s,p}, \qquad (s=1, 2, \dots, p),$$

where $\alpha_1, \dots, \alpha_p, \beta_1, \dots, \beta_p$ are certain integers, we have

$$v_s^{m_1', m_1} + \dots\dots + v_s^{m_p', m_p} = \tfrac{1}{2}\Omega_{\beta,\alpha};$$

hence the function

$$\Theta(v^{x,m} - v^{z_1, m_1'} - \dots\dots - v^{z_p, m_p'};\ \tfrac{1}{2}\beta, \tfrac{1}{2}\alpha),$$

$$= e^{\pi i\beta\alpha}\,\Theta(v^{z_1, m_1'} + \dots\dots + v^{z_p, m_p'} - v^{x,m};\ \tfrac{1}{2}\beta, \tfrac{1}{2}\alpha),$$

$$= e^{\pi i\beta\alpha}\,\Theta(u - \tfrac{1}{2}\Omega_{\beta,\alpha};\ \tfrac{1}{2}\beta, \tfrac{1}{2}\alpha),$$

where

$$u_s = v_s^{z_1, m_1} + \dots\dots + v_s^{z_p, m_p} - v^{x,m}, \qquad (s=1, 2, \dots, p);$$

the function is therefore equal to

$$e^{\pi i\beta\alpha - \pi i\alpha(u - \frac{1}{4}\tau\alpha)}\,\Theta(u),$$

by equation (C), §175; thus *the function* $\Theta(v^{x,m} - v^{z_1, m_1'} - \dots\dots - v^{z_p, m_p'};\ \tfrac{1}{2}\beta, \tfrac{1}{2}\alpha)$ *vanishes when x is at either of the places $z_1, \dots, z_p$.*

We can similarly prove that

$$\Theta(v^{x,m} - v^{z_1, m_1'} - \dots\dots - v^{z_p, m_p'}) = e^{-\pi i\alpha(u + \frac{1}{2}\beta + \frac{1}{4}\tau\alpha)}\,\Theta(-u;\ \tfrac{1}{2}\beta, \tfrac{1}{2}\alpha).$$

It has been remarked (§ 175) that there are effectively 2^{2p} theta functions, corresponding to the 2^{2p} sets of values of the integers α, β in which each is either 0 or 1. The present proposition enables us to associate each of the functions with one of the solutions of the equivalence (G). When the function $\Theta(v^{x,m};\ \tfrac{1}{2}\beta, \tfrac{1}{2}\alpha)$ does not vanish identically in respect to x, its zeros are the places $m_1', \dots, m_p'$. Therefore, instead of the function $\Theta(u)$, we may regard the function $\Theta(u;\ \tfrac{1}{2}\beta, \tfrac{1}{2}\alpha)$ as fundamental, and shall only be led to the places $m_1', \dots, m_p'$, instead of $m_1, \dots, m_p$.

(XI.) The sets of places $m_1', \dots, m_p'$ which are connected with the places $m_1, \dots, m_p$ by means of the equations

$$v_s^{m_1', m_1} + \dots\dots + v_s^{m_p', m_p} \equiv \tfrac{1}{2}\Omega_{\beta,\alpha}, \qquad \text{(H)},$$

wherein $\alpha_1, \dots, \alpha_p, \beta_1, \dots, \beta_p$ denote in turn all the 2^{2p} sets of values in which each element is either 0 or 1, may be divided into two categories, according as the integer $\beta\alpha, = \beta_1\alpha_1 + \dots\dots + \beta_p\alpha_p$, is even or odd. We have remarked, in Proposition (IX.), that they may be divided into two categories according as they are the zeros, of the second order, of a proper polynomial $\lambda\psi_0 + \Delta\phi$, or consist of the $p-1$ zeros, each of the second order, of a ϕ-polynomial together with the place m. *When the fundamental Riemann surface is perfectly general these two methods of division of the 2^{2p} sets entirely agree. When $\beta\alpha$ is odd, $m_1', \dots, m_p'$ consist of the place m and the $p-1$ zeros, each of the second order, of a ϕ-polynomial. When $\beta\alpha$ is even, $m_1', \dots, m_p'$*

consist of the zeros, each of the second order, of a proper polynomial ψ. In the latter case we may speak of the places $m_1', \dots, m_p'$ as a set of *tangential derivatives* of the place m.

For by the equations (D), (A), (§ 175), we have

$$e^{\pi i a u}\,\Theta\left(\tfrac{1}{2}\Omega_{\beta,\,a}+u\right)/e^{-\pi i a u}\,\Theta\left(\tfrac{1}{2}\Omega_{\beta,\,a}-u\right)=e^{-\pi i\beta a};$$

hence, when $\beta\alpha$ is odd, $e^{\pi i a u}\,\Theta(\frac{1}{2}\Omega_{\beta,\,a}+u)$ is an odd function of u, and must vanish when u is zero; since then $\Theta(\frac{1}{2}\Omega_{\beta,\,a})$ vanishes, there exist, by Proposition (VII.), places $n_1, \dots, n_{p-1}$, such that

$$-\tfrac{1}{2}\Omega_{\beta,\,a}\equiv v^{m_p,\,m}-v^{n_1,\,m_1}-\dots\dots-v^{n_{p-1},\,m_{p-1}},\qquad\text{(K)},$$

or

$$2\left(v^{n_1,\,m_1}+\dots\dots+v^{n_{p-1},\,m_{p-1}}+v^{m,\,m_p}\right),\;=\Omega_{\beta,\,a},\;\equiv 0.$$

Hence (Chap. VIII. § 158) we have

$$(m^2, n_1^2, \dots, n^2_{p-1})\equiv(m_1^2, \dots, m_p^2),$$

so that, by equation (G), the places $n_1, \dots, n_{p-1}$ are the zeros of a ϕ-polynomial, each being of the second order.

When $\beta\alpha$ is even, the function $e^{\pi i a u}\,\Theta(\frac{1}{2}\Omega_{\beta,\,a}+u)$ is an even function, and it is to be expected that it will not vanish for $u=0$. This is generally the case, but exception may arise when the fundamental Riemann surface is of special character. We are thus led to make a distinction between the general case, which, noticing that $\Theta(\frac{1}{2}\Omega_{\beta,a}+u)$ is equal to $e^{-\pi i a(u+\frac{1}{4}\beta-\frac{1}{4}\tau a)}\,\Theta(u;\,\frac{1}{2}\beta,\frac{1}{2}\alpha)$, may be described as that in which no even theta function vanishes for zero values of the argument, and special cases in which one or more even theta functions do vanish for zero values of the argument.

Suppose then, firstly, that no even theta function vanishes for zero values of the argument. Then if $n_1', \dots, n'_{p-1}$ be places which, repeated, are the zeros of a ϕ-polynomial, we have

$$(m^2, n_1'^2, \dots, n'^2_{p-1})\equiv(m_1^2, m_2^2, \dots, m_p^2);$$

hence the argument

$$v^{m_p,\,m}-v^{n_1',\,m_1}-\dots\dots-v^{n'_{p-1},\,m_{p-1}}$$

is a half-period, $\equiv-\frac{1}{2}\Omega_{\beta',\,a'}$, say. Thus, by the result (F), $\Theta(\frac{1}{2}\Omega_{\beta',a'})$ is zero; therefore, by the hypothesis $\beta'\alpha'$ is an odd integer. So that, in this case, every odd half-period corresponds to a ϕ-polynomial of which all the zeros are of the second order, and conversely.

Further, in this case it is immediately obvious that the places $m_1, \dots, m_p$ do not consist of the place m and the zeros of a ϕ-polynomial whose zeros are of the second order; for if $m_1, \dots, m_p$ were the places $n_1, \dots, n_{p-1}, m$, then, by the result (F), the function $\Theta(v^{z_1,\,n_1}+\dots\dots+v^{z_{p-1},\,n_{p-1}})$ would vanish for all positions of $z_1, \dots, z_{p-1}$, and therefore $\Theta(0)$ would vanish.

185. If, however, nextly, there be even theta functions which vanish for zero values of the argument, it does not follow as above that every ϕ-polynomial with double zeros corresponds to an odd half-period; there will still be such ϕ-polynomials corresponding to the $2^{p-1}(2^p-1)$ odd half-periods, but there will also be such ϕ-polynomials corresponding to even half-periods.

For if $\alpha_1, \ldots, \alpha_p, \beta_1, \ldots, \beta_p$ be integers such that $\beta\alpha$ is even, and $\Theta(u+\frac{1}{2}\Omega_{\beta,\alpha})$ vanishes for $u=0$, the first differential coefficients, in regard to $u_1, \ldots, u_p$, of the even function $e^{\pi i a u}\Theta(u+\frac{1}{2}\Omega_{\beta,\alpha})$, being odd functions, will vanish for $u=0$. By an argument which, for convenience, is postponed to Prop. XIV., it follows that then the function $\Theta(v^{x,z}+\frac{1}{2}\Omega_{\beta,\alpha})$ vanishes identically for all positions of x and z. Therefore, by Prop. V., there is at least a single infinity of places $z_1, \ldots, z_{p-1}$ satisfying the equations

$$-\tfrac{1}{2}\Omega_{\beta,\alpha} \equiv v^{m_p, m} - v^{z_1, m_1} - \ldots\ldots - v^{z_{p-1}, m_{p-1}};$$

these equations are equivalent to

$$(m^2, z_1^2, \ldots, z^2_{p-1}) \equiv (m_1^2, m_2^2, \ldots, m_p^2);$$

hence there is a single infinity of ϕ-polynomials with double zeros corresponding to the even half-period $\frac{1}{2}\Omega_{\beta,\alpha}$, and their $p-1$ zeros form coresidual sets with multiplicity at least equal to 1.

By similar reasoning we can prove another result*; the argument is repeated in the example which follows; *if, for any set of values of the integers* $\beta_1, \ldots, \beta_p, \alpha_1, \ldots, \alpha_p$, *it is possible to obtain more than one set of places* $n_1, \ldots, n_{p-1}$ *to satisfy the equations*

$$-\tfrac{1}{2}\Omega_{\beta,\alpha} \equiv v^{m_p, m} - v^{n_1, m_1} - \ldots\ldots - v^{n_{p-1}, m_{p-1}},$$

then it is, of course, possible to obtain an infinite number of such sets. Let ∞^q *be the number of sets obtainable. Then* $\beta\alpha \equiv q+1 \pmod{2}$. *And this may be understood to include the general cases when* (i) *for an even value of* $\beta\alpha$, *no solution of the congruence is possible* ($q=-1$), (ii), *for an odd value of* $\beta\alpha$, *only a single solution is possible* ($q=0$).

As an example of the exceptional case here referred to, consider the hyperelliptic surface; and first suppose $p=3$, the equation associated with the surface being

$$y^2=(x-a_1)\ldots\ldots(x-a_8);$$

then we clearly have $\binom{8}{2}=28=2^{p-1}(2^p-1)$ ϕ-polynomials, each of the form $(x-a_i)(x-a_j)$, of which the zeros are both of the second order. We have, however, also, a ϕ-polynomial, of the form $(x-c)^2$, in which c is arbitrary, of which the zeros are both of the second order; denote these zeros by c and $\bar{c}$; then if $\frac{1}{2}\Omega_{\beta,\alpha}$ be a proper half-period

$$-\tfrac{1}{2}\Omega_{\beta,\alpha} \equiv v^{m_3, m} - v^{c, m_1} - v^{\bar{c}, m_2};$$

* Weber, *Math. Ann.* XIII. p. 42.

but, since, if e be any other place, the function $(x-c)/(x-e)$ is a rational function, it follows that $(c, \bar{c}) \equiv (e, \bar{e})$, and therefore that in the value just written for $\frac{1}{2}\Omega_{\beta, a}$, c may be replaced by e, and therefore, regarded as quite arbitrary. By the result (F), the function $\Theta(u)$ vanishes when u is replaced by $\frac{1}{2}\Omega_{\beta, a}$, and therefore $\Theta(v^{x, z} - \frac{1}{2}\Omega_{\beta, a})$, which is equal to $\Theta(v^{x, m} - v^{c, m_1} - v^{\bar{c}, m_2} - v^{z, m_3})$, vanishes when x is at c; since c is arbitrary the function $\Theta(v^{x, z} - \frac{1}{2}\Omega_{\beta, a})$ vanishes identically in regard to x, for all positions of z. If the function $\Theta(v^{x, z} + v^{x_1, z_1} - \frac{1}{2}\Omega_{\beta, a})$ vanished identically, it would, by Prop. VI., be possible, in the equation

$$-\tfrac{1}{2}\Omega_{\beta, a} \equiv v^{m_3, m} - v^{z_1, m_1} - v^{z_2, m_2}$$

to choose both z_1 and z_2 arbitrarily. As this is not the case, it follows, by Prop. XIV. below, that the function $\Theta(u + \frac{1}{2}\Omega_{\beta, a})$, and its first, but not its second differential coefficients, vanish for $u=0$. Hence $\frac{1}{2}\Omega_{\beta, a}$ is an even half-period. (See the tables for the hyperelliptic case, given in the next chapter, §§ 204, 205.)

There is therefore, in the hyperelliptic case in which $p=3$, one even theta function which vanishes for zero values of the argument.

In any hyperelliptic case in which p is odd, the equation associated with the surface being

$$y^2 = (x-a_1) \ldots\ldots (x - a_{2p+2})$$

ϕ-polynomials with double zeros are given by

(i) the $\binom{2p+2}{p-1}$ polynomials such as $(x-a_1) \ldots\ldots (x-a_{p-1})$. As there is no arbitrary place involved, the q of the theorem enunciated (§ 185) is zero, and the half-period given by the equation

$$-\tfrac{1}{2}\Omega_{\beta, a} \equiv v^{m_p, m} - v^{n_1, m_1} - \ldots\ldots - v^{n_{p-1}, m_{p-1}},$$

where $n_1^2, \ldots, n^2_{p-1}$ are the zeros of the ϕ-polynomial under consideration, is consequently odd.

(ii) the $\binom{2p+2}{p-3}$ polynomials such as $(x-a_1) \ldots\ldots (x-a_{p-3})(x-c)^2$, wherein c is arbitrary. Here $q=1$ and $\beta a \equiv 0$ (mod. 2).

(iii) the $\binom{2p+2}{p-5}$ polynomials such as $(x-a_1) \ldots\ldots (x-a_{p-5})(x-c)^2(x-e)^2$, for which $q=2$, $\beta a \equiv 1$ (mod. 2); and so on. And, finally,

the single polynomial of the form $(x-c_1)^2 \ldots\ldots (x-c_{\frac{p-1}{2}})^2$, in which all of $c_1, \ldots, c_{\frac{p-1}{2}}$ are arbitrary; in this case $q = \frac{p-1}{2}$, $\beta a \equiv \frac{p+1}{2}$ (mod. 2).

On the whole there arise

$$\binom{2p+2}{p-1} + \binom{2p+2}{p-5} + \ldots\ldots + 1, \text{ or } \binom{2p+2}{p-1} + \binom{2p+2}{p-5} + \ldots\ldots + \binom{2p+2}{2}$$

ϕ-polynomials corresponding to odd half-periods, according as $p \equiv 1$ or 3 (mod. 4).

Now in fact, when $p \equiv 1$ (mod. 4)

$$1 + \binom{2p+2}{4} + \ldots\ldots + \binom{2p+2}{p-1}, \quad = \tfrac{1}{8}\left[(1+x)^{2p+2} + (1-x)^{2p+2} + (1+ix)^{2p+2} + (1-ix)^{2p+2}\right]_{x=1},$$

is equal to

$$\tfrac{1}{8}\left(2^{2p+2}+2^{p+2}\cos\frac{p+1}{2}\pi\right) \text{ or } 2^{2p-1}-2^{p-1} \text{ or } 2^{p-1}(2^p-1),$$

while, when $p\equiv 3$ (mod. 4)

$$\binom{2p+2}{2}+\binom{2p+2}{6}+\ldots\ldots+\binom{2p+2}{p-1},$$

$$=\tfrac{1}{8}\left[(1+x)^{2p+2}+(1-x)^{2p+2}-(1+ix)^{2p+2}-(1-ix)^{2p+2}\right]_{x=1},$$

is equal to $\frac{1}{8}\left(2^{2p+2}-2^{p+2}\cos\frac{p+1}{2}\pi\right)$, and therefore, also to $2^{p-1}(2^p-1)$.

Thus all the odd half-periods are accounted for. And there are

$$\binom{2p+2}{p-3}+\binom{2p+2}{p-7}+\ldots\ldots$$

even half-periods which reduce the theta function to zero. This number is equal to

$$-\tfrac{1}{2}\binom{2p+2}{p+1}+\{2^{2p}-2^{p-1}(2^p-1)\},$$

namely to $2^{p-1}(2^p+1)-\binom{2p+1}{p}$. This is the number of even theta functions which vanish for zero values of the argument. It is easy to see that the same number is obtained when p is even. For instance when $p=4$, there are 10 even theta functions which vanish for zero values of the argument. They correspond to the 10 ϕ-polynomials of the form $(x-c)^2(x-a_1)$, wherein c is arbitrary, and a_1 is one of the 10 branch places. There are therefore $\binom{2p+1}{p}$ even theta functions which do not vanish for zero values of the argument.

In regard to the places $m_1, \ldots, m_p$ in the hyperelliptic case the following remark may conveniently be made here. Suppose the place m taken at the branch place a_{2p+2}; using the geometrical rule given in § 183, we may take for the polynomial Δ, of grade μ, the polynomial $x-a_{2p+2}$, of grade 1; its remaining $n\mu-2$, $=0$, zeros, give no conditions for the polynomial ψ of grade $(n-1)\sigma+n-3+\mu$, $=(2-1)p+2-3+1$, $=p$. Since $\sigma+1$, the dimension of y, is $p+1$, the only possible form for ψ is that of an integral polynomial in x of order p. This is to be chosen so that its $2p$ zeros consist of p repeated zeros. When $p=3$, for example, it must, therefore, be of one of the forms $(x-a_i)(x-a_j)(x-a_k)$, $(x-a_i)(x-c)^2$, where c is arbitrary. It will be seen in the next chapter that the former is the proper form.

186. Another matter* which connects the present theory with a subject afterwards (Chap. XIII.) dealt with may be referred to here. Let $\frac{1}{2}\Omega$ be a half-period such that the congruence

$$\tfrac{1}{2}\Omega\equiv v^{m_p,\,m}-v^{z_1,\,m_1}-\ldots\ldots-v^{z_{p-1},\,m_{p-1}}$$

can be satisfied by ∞^q coresidual sets of places $z_1, \ldots, z_{p-1}$ (as in Proposition VI.). Then we have

$$(m^2, z_1^2, \ldots, z_{p-1}^2)=(m_1^2, \ldots, m_p^2),$$

so that (Prop. IX.) $z_1, \ldots, z_{p-1}$, each repeated, are the zeros of a ϕ-polynomial; denote this polynomial by ϕ. If $z_1', \ldots, z_{p-1}'$ be another set, which, repeated, are the zeros of a ϕ-polynomial ϕ', and are such that

$$\tfrac{1}{2}\Omega\equiv v^{m_p,\,m}-v^{z_1',\,m_1}-\ldots\ldots-v^{z_{p-1}',\,m_{p-1}},$$

* Cf. Weber, *Math. Annal.* XIII. p. 35; Noether, *Math. Annal.* XVII. 263.

then we have

$$0 \equiv 2v^{m_p,\, m} - v^{z_1,\, m_1} - v^{z_1',\, m_1} - \dots\dots - v^{z_{p-1},\, m_{p-1}} - v^{z'_{p-1},\, m_{p-1}},$$

so that $z_1, \dots, z_{p-1}, z_1', \dots, z'_{p-1}$ are the zeros of a ϕ-polynomial; denote this polynomial by ψ.

The rational functions ψ/ϕ, ϕ'/ψ have the same poles, the places $z_1, \dots, z_{p-1}$, and the same zeros, the places $z_1', \dots, z'_{p-1}$. Therefore, absorbing a constant multiplier in ψ, we have

$$\psi^2 = \phi\phi', \text{ and } \phi'/\phi = (\psi/\phi)^2,$$

and thus the function $\sqrt{\phi'/\phi}$ may be regarded as a rational function if a proper sign be always attached. The function has $z_1, \dots, z_{p-1}$ for poles and $z_1', \dots, z'_{p-1}$ for zeros. Conversely any rational function having $z_1, \dots, z_{p-1}$ for poles can be written in this form. For if $z_1'', \dots, z''_{p-1}$ be the zeros of such a function, we have

$$v^{z_1'',\, z_1} + \dots\dots + v^{z''_{p-1},\, z_{p-1}} \equiv 0,$$

and therefore, by the first equation of this §, also

$$\tfrac{1}{2}\Omega \equiv v^{m_p,\, m} - v^{z_1'',\, m_1} - \dots\dots - v^{z''_{p-1},\, m_{p-1}};$$

thus q of the zeros can be taken arbitrarily; and if Φ be any ϕ-polynomial whose zeros $\zeta_1, \dots, \zeta_{p-1}$ are all of the second order, and such that

$$\tfrac{1}{2}\Omega \equiv v^{m_p,\, m} - v^{\zeta_1,\, m_1} - \dots\dots - v^{\zeta_{p-1},\, m_{p-1}},$$

we can put

$$\sqrt{\frac{\Phi}{\phi}} = \lambda + \lambda_1 \sqrt{\frac{\phi_1}{\phi}} + \dots\dots + \lambda_q \sqrt{\frac{\phi_q}{\phi}},$$

where $\phi_1, \dots, \phi_q$ are particular polynomials such as ϕ' or Φ, and $\lambda, \lambda_1, \dots, \lambda_q$ are constants. In other words, corresponding to the ∞^q sets of solutions of the original equation of this §, we have an equation of the form

$$\sqrt{\Phi} = \lambda\sqrt{\phi} + \lambda_1\sqrt{\phi_1} + \dots\dots + \lambda_q\sqrt{\phi_q},$$

wherein proper signs are to be attached to the ratios of any two of the square roots, and any two of the $q+1$ polynomials $\phi, \phi_1, \dots, \phi_q$, are such that their product is the square of a ϕ-polynomial. There are therefore $\tfrac{1}{2}q(q+1)$ linearly independent quadratic relations connecting the ϕ-polynomials. (Cf. Chap. VI. §§ 110—112.)

For example in the hyperelliptic case in which $p=3$, the vanishing of an even theta function corresponds to the existence of a ϕ-polynomial $\Phi = (x-c)^2$, such that

$$\sqrt{\Phi} = -c\sqrt{1} + \sqrt{x^2}, \;\; = -c\sqrt{\phi_1} + \sqrt{\phi_3},$$

where $\phi_1\phi_3, = (x)^2, = \phi_2^2$.

Ex. i. Prove, for $p=3$, that if an even theta function vanishes for zero values of the arguments the surface is necessarily hyperelliptic.

Ex. ii. Prove, for $p=4$, that if two even theta functions vanish for zero values of the arguments the surface is necessarily hyperelliptic; so that, then, eight other even theta functions also vanish for zero values of the arguments. The number, 2, of conditions thus necessary for the fundamental constants of the surface, in order that it be hyperelliptic, is the same as the difference, $9-7$, between the number, $3p-3$, of constants in the general surface of deficiency 4, and the number, $2p-1$, of constants in the general hyperelliptic surface of deficiency 4.

187. (XII.) If r denote any arguments such that $\Theta(r)=0$, and such that $\Theta(v^{x,z}+r)$ does not vanish identically for all positions of x and z, the Riemann normal integral of the third kind can be expressed in the form

$$\Pi_{a,\beta}^{x,z} = \log\left[\frac{\Theta(v^{x,a}+r)}{\Theta(v^{x,\beta}+r)} \Big/ \frac{\Theta(v^{z,a}+r)}{\Theta(v^{z,\beta}+r)}\right].$$

For consider the function of x given by

$$e^{-\Pi_{a,\beta}^{x,z}} \frac{\Theta(v^{x,a}+r)\,\Theta(v^{z,\beta}+r)}{\Theta(v^{x,\beta}+r)\,\Theta(v^{z,a}+r)};$$

(α) it is single-valued on the Riemann surface dissected by the a and b period loops;

(β) it does not vanish or become infinite, for the zeros of $\Theta(v^{x,z}+r)$, other than z, do not depend upon z (by Proposition IV.);

(γ) it is unaffected by a circuit of any one of the period loops. At a loop a_i it has clearly (Equation B, § 175) the factor unity; at a loop b_i it has the factor

$$e^{-2\pi i v_i^{a,\beta}} \,.\, e^{-2\pi i (v_i^{x,a}+r_i+\frac{1}{2}\tau_{i,i})} \,.\, e^{2\pi i (v_i^{x,\beta}+r_i+\frac{1}{2}\tau_{i,i})},$$

which is also unity. Thus the function is single-valued on the undissected surface;

(δ) thus the function is independent of x; and hence equal to the value it has when the place x is at z, namely 1.

A particular case is obtained by taking

$$r = v^{m_p, m} - v^{z_1, m_1} - \dots\dots - v^{z_{p-1}, m_{p-1}},$$

where $z_1, \dots, z_{p-1}$ are any places such that $\Theta(v^{x,z}+r)$ does not vanish identically. Then by the result (F) the function $\Theta(r)$ vanishes.

Hence we have

$$\Pi_{a,\beta}^{x,z} = \log\left[\frac{\Theta(v^{x,m} - v^{z_1,m_1} - \dots\dots - v^{z_{p-1},m_{p-1}} - v^{a,m_p})}{\Theta(v^{x,m} - v^{z_1,m_1} - \dots\dots - v^{z_{p-1},m_{p-1}} - v^{\beta,m_p})} \Big/ \frac{\Theta(v^{z,m} - v^{z_1,m_1} - \dots\dots - v^{z_{p-1},m_{p-1}} - v^{a,m_p})}{\Theta(v^{z,m} - v^{z_1,m_1} - \dots\dots - v^{z_{p-1},m_{p-1}} - v^{\beta,m_p})}\right].$$

Another particular case, of great importance, is obtained by taking $r=\frac{1}{2}\Omega_{k,k'}$, k, k' denoting respectively p integers $k_1, \dots, k_p, k_1', \dots, k_p'$, such that kk' is odd, the assumption being made that the equations

$$\tfrac{1}{2}\Omega_{k,k'} \equiv v^{m_p, m} - v^{\zeta_1, m_1} - \dots\dots - v^{\zeta_{p-1}, m_{p-1}}$$

are not satisfied by more than one set of places $\zeta_1, \ldots, \zeta_{p-1}$ (cf. Props. III., V.). Then the function $\Theta(v^{x,z}+\frac{1}{2}\Omega_{k,k'})$ does not vanish identically, and we have

$$\Pi_{\alpha,\beta}^{x,z} = \log \frac{\Theta(v^{x,\alpha}+\frac{1}{2}\Omega_{k,k'})\,\Theta(v^{z,\beta}+\frac{1}{2}\Omega_{k,k'})}{\Theta(v^{x,\beta}+\frac{1}{2}\Omega_{k,k'})\,\Theta(v^{z,\alpha}+\frac{1}{2}\Omega_{k,k'})}.$$

(XIII.) Suppose k equal to or less than p; consider the function given by the product of

$$e^{-\Pi_{\alpha_1,\beta_1}^{x,z} - \Pi_{\alpha_2,\beta_2}^{x,z} - \ldots\ldots - \Pi_{\alpha_k,\beta_k}^{x,z}}$$

and

$$\frac{\Theta(v^{x,m}-v^{\alpha_1,m_1}-\ldots\ldots-v^{\alpha_k,m_k}+r)}{\Theta(v^{x,m}-v^{\beta_1,m_1}-\ldots\ldots-v^{\beta_k,m_k}+r)} \Big/ \frac{\Theta(v^{z,m}-v^{\alpha_1,m_1}-\ldots\ldots-v^{\alpha_k,m_k}+r)}{\Theta(v^{z,m}-v^{\beta_1,m_1}-\ldots\ldots-v^{\beta_k,m_k}+r)},$$

wherein r denotes arguments given by

$$r = -(v^{\gamma_{k+1},m_{k+1}} + \ldots\ldots + v^{\gamma_p,m_p}),$$

and each of the sets $\alpha_1, \ldots, \alpha_k, \gamma_{k+1}, \ldots, \gamma_p, \beta_1, \ldots, \beta_k, \gamma_{k+1}, \ldots, \gamma_p$ is such that the functions involved do not vanish identically in regard to x.

This function is single-valued on the dissected Riemann surface, does not become infinite or zero, and, for example, at the period loop b_i it has the factor e^L, where

$$L, = -2\pi i(v^{\alpha_1,\beta_1}+\ldots\ldots+v^{\alpha_k,\beta_k}) - 2\pi i(v^{x,m}-v^{\alpha_1,m_1}-\ldots\ldots-v^{\alpha_k,m_k})$$
$$+ 2\pi i(v^{\alpha,m}-v^{\beta_1,m_1}-\ldots\ldots-v^{\beta_k,m_k}),$$

is zero. Thus the function has the constant value, unity, which it has when x is at z. Therefore

$$\Pi_{\alpha_1,\beta_1}^{x,z}+\ldots+\Pi_{\alpha_k,\beta_k}^{x,z} = \log\left[\frac{\Theta(v^{x,m}-v^{\alpha_1,m_1}-\ldots-v^{\alpha_k,m_k}-v^{\gamma_{k+1},m_{k+1}}-\ldots-v^{\gamma_p,m_p})}{\Theta(v^{x,m}-v^{\beta_1,m_1}-\ldots-v^{\beta_k,m_k}-v^{\gamma_{k+1},m_{k+1}}-\ldots-v^{\gamma_p,m_p})}\right.$$
$$\left.\Big/\frac{\Theta(v^{z,m}-v^{\alpha_1,m_1}-\ldots\ldots-v^{\alpha_k,m_k}-v^{\gamma_{k+1},m_{k+1}}-\ldots\ldots-v^{\gamma_p,m_p})}{\Theta(v^{z,m}-v^{\beta_1,m_1}-\ldots\ldots-v^{\beta_k,m_k}-v^{\gamma_{k+1},m_{k+1}}-\ldots\ldots-v^{\gamma_p,m_p})}\right],$$

the places $\gamma_{k+1}, \ldots, \gamma_p$ being arbitrarily chosen so that $\alpha_1, \ldots, \alpha_k, \gamma_{k+1}, \ldots, \gamma_p$ are not zeros of a ϕ-polynomial, and $\beta_1, \ldots, \beta_k, \gamma_{k+1}, \ldots, \gamma_p$ are not zeros of a ϕ-polynomial.

Thus, when $k=p$, we have the expression of the function considered in § 171, Chap. IX. in terms of theta functions. For the case where $\alpha_1, \ldots, \alpha_k$ are the zeros of a ϕ-polynomial, cf. Prop. XV. Cor. iii.

188. (XIV.) We return now to the consideration of the identical vanishing of the Θ function. We have proved (Prop. VII.), that if $\Theta(v^{x_1,z_1}+\ldots\ldots+v^{x_q,z_q}+r)$ be identically zero for all positions of $x_1, \ldots, x_q, z_1, \ldots, z_q$, but $\Theta(v^{x,z}+v^{x_1,z_1}+\ldots\ldots+v^{x_q,z_q}+r)$ be not identically zero for all positions of

x and z, then there exist ∞^q sets of places $\zeta_1, \ldots, \zeta_{p-1}$, and ∞^q sets of places $\xi_1, \ldots, \xi_{p-1}$, such that

$$r = v^{m_p, m} - v^{\zeta_1, m_1} - \ldots\ldots - v^{\zeta_{p-1}, m_{p-1}},$$

and

$$-r = v^{m_p, m} - v^{\xi_1, m_1} - \ldots\ldots - v^{\xi_{p-1}, m_{p-1}}.$$

Now, if in the equation $\Theta(v^{x_1, z_1} + \ldots\ldots + v^{x_q, z_q} + r) = 0$, we make x_q approach to and coincide with z_q, we obtain

$$\sum_{i=1}^{p} \Theta_i'(v^{x_1, z_1} + \ldots\ldots + v^{x_{q-1}, z_{q-1}} + r)\, \Omega_i(z_q) = 0,$$

wherein $\Theta_i'(u)$ is put for $\frac{\partial}{\partial u_i}\Theta(u)$, $\Omega_i(x)$ for $2\pi i\, D_x v_i^{x, a}$, a being arbitrary; and this equation holds for all positions of $x_1, z_1, \ldots, x_{q-1}, z_{q-1}$. Since, however, the quantities $\Omega_1(z_q), \ldots, \Omega_q(z_q)$ cannot be connected by any linear equation whose coefficients are independent of z_q, we can thence infer that the first differential coefficients of $\Theta(u)$ vanish identically when u is of the form $v^{x_1, z_1} + \ldots\ldots + v^{x_{q-1}, z_{q-1}} + r$. It follows then in the same way that the second differential coefficients of $\Theta(u)$ vanish identically when u has the form $v^{x_1, z_1} + \ldots\ldots + v^{x_{q-2}, z_{q-2}} + r$; in particular all the first and second differential coefficients vanish when $u = r$. Proceeding thus we finally infer that $\Theta(u)$ and all its differential coefficients up to and including those of the qth order vanish when $u = r$.

We proceed now to shew conversely that when $\Theta(u)$ and all its differential coefficients up to and including those of the qth order, vanish for $u = r$, then $\Theta(v^{x_1, z_1} + \ldots\ldots + v^{x_q, z_q} + r)$ vanishes identically for all positions of $x_1, z_1, x_2, z_2, \ldots, x_q, z_q$. By what has just been shewn $\Theta(v^{x, z} + v^{x_1, z_1} + \ldots\ldots + v^{x_q, z_q} + r)$ will not vanish identically unless the differential coefficients of the $(q+1)$th order also vanish.

We begin with the case $q = 1$. Suppose that $\Theta(u), \Theta_1'(u), \ldots, \Theta_p'(u)$, all vanish for $u = r$; we are to prove that $\Theta(v^{x, z} + r)$ vanishes identically for all positions of x and z.

Let e, f be such arguments that $\Theta(e) = 0$, $\Theta(f) = 0$, but such that $\Theta_i'(e)$ are not all zero and $\Theta_i'(f)$ are not all zero, and therefore $\Theta(v^{x, z} + e)$, $\Theta(v^{x, z} + f)$ do not vanish identically; consider the function

$$\frac{\Theta(e + v^{x, z})\,\Theta(e - v^{x, z})}{\Theta(f + v^{x, z})\,\Theta(f - v^{x, z})};$$

firstly, it is rational in x and z; for, considered as a function of x, it has, at the period loop b_r, (Equation B, § 175) the factor

$$e^{-2\pi i(v_r^{x, z} + e + v_r^{x, z} - e) - \pi i \tau_{r, r}} \Big/ e^{-2\pi i(v_r^{x, z} + f + v_r^{x, z} - f) - \pi i \tau_{r, r}}$$

whose value is unity; and a similar statement holds when the expression is considered as a function of z, for the expression is immediately seen to be symmetrical in x and z; secondly, regarded as a function of x, the expression has $2(p-1)$ zeros, and the same number of poles, and these (Prop. IV.) are independent of z. Similarly as a function of z it has $2(p-1)$ zeros and poles, independent of x; therefore the expression can be written in the form $F(x)F(z)$, where $F(x)$ denotes the definite rational function having the proper zeros and poles, multiplied by a suitable constant factor, and $F(z)$ is the same rational function of z.

Putting, then, x to coincide with z, and extracting a square root, we infer

$$F(x) = \pm \frac{\sum_{i=1}^{p} \Theta_i'(e)\,\Omega_i(x)}{\sum_{i=1}^{p} \Theta_i'(f)\,\Omega_i(x)},$$

where $\Omega_i(x) = 2\pi i\, D_x v_i^{x,\,a}$, for a arbitrary, is the differential coefficient of an integral of the first kind; thence we have

$$\frac{\Theta(v^{x,\,z}+e)\,\Theta(v^{x,\,z}-e)}{\Theta(v^{x,\,z}+f)\,\Theta(v^{x,\,z}-f)} = \frac{[\Sigma\Theta_i'(e)\,\Omega_i(x)]\,[\Sigma\Theta_i'(e)\,\Omega_i(z)]}{[\Sigma\Theta_i'(f)\,\Omega_i(x)]\,[\Sigma\Theta_i'(f)\,\Omega_i(z)]}.$$

In this equation suppose that e approaches indefinitely near to r, for which $\Theta(r)=0$, $\Theta_i'(r)=0$. Then the right hand becomes infinitesimal, independently of x and z. Therefore also the left hand becomes infinitesimal independently of x and z; and hence $\Theta(v^{x,\,z}+r)$ vanishes identically, for all positions of x and z.

We have thus proved the case of our general theorem in which $q=1$. The theorem is to be inferred for higher values of q by proving that if the function $\Theta(v^{x_1,\,z_1}+\ldots\ldots+v^{x_{m-1},\,z_{m-1}}+r)$ vanish identically for all positions of $x_1, z_1, \ldots, x_{m-1}, z_{m-1}$, and also the differential coefficients of $\Theta(u)$, of order m, vanish for $u=r$, then the function $\Theta(v^{x_1,\,z_1}+\ldots\ldots+v^{x_m,\,z_m}+r)$ vanishes identically. For instance if this were proved, it would follow, putting $m=2$, from what we have just proved, that also $\Theta(v^{x_1,\,z_1}+v^{x_2,\,z_2}+r)$ vanished identically, and so on.

As before let f be such that $\Theta(f)=0$, but all of $\Theta_i'(f)$ are not zero; so that $\Theta(v^{x,\,z}+f)$ does not vanish identically in regard to x and z. Let e be such that $\Theta(v^{x_1,\,z_1}+\ldots\ldots+v^{x_{m-1},\,z_{m-1}}+e)$ vanishes identically for all positions of $x_1, z_1, \ldots, x_{m-1}, z_{m-1}$, but such that the differential coefficients of $\Theta(u)$ of the first order do not vanish identically for $u=v^{x_1,\,z_1}+\ldots+v^{x_{m-1},\,z_{m-1}}+e$; so that the function $\Theta(v^{x_1,\,z_1}+\ldots\ldots+v^{x_m,\,z_m}+e)$ does not vanish identically. Consider the product of the expressions

$$\Theta(v^{x_1,\,z_1}+\ldots\ldots+v^{x_m,\,z_m}+e)\,\Theta(v^{x_1,\,z_1}+\ldots\ldots+v^{x_m,\,z_m}-e)$$

$$\frac{\Pi'\Theta(v^{x_h,\,x_k}+f)\,\Theta(v^{x_h,\,x_k}-f)\,\Pi'\Theta(v^{z_h,\,z_k}+f)\,\Theta(v^{z_h,\,z_k}-f)}{\Pi\Pi\Theta(v^{x_\lambda,\,z_\mu}+f)\,\Theta(v^{x_\lambda,\,z_\mu}-f)},$$

wherein h, k in the numerator denote in turn every pair of the numbers $1, 2, \ldots, m$, so that the numerator contains $4 \cdot \frac{1}{2}m(m-1)+2 = 2(m^2-m+1)$ theta functions, and λ, μ in the denominator are each to take all the values $1, 2, \ldots, m$, so that there are $2m^2$ theta functions in the denominator.

Firstly, this product is a rational function of each of the $2m$ places $x_1, z_1, \ldots, x_m, z_m$. Consider for instance x_1; it is clear that if the product be rational in x_1, it will be entirely rational. As a function of x_1, the product has at the period loop b_r a factor $e^{-2\pi i K}$ where

$$K = 2\left(v_r^{x_1, z_1} + \ldots\ldots + v_r^{x_m, z_m} + \tfrac{1}{2}\tau_{r,r}\right) + 2\sum_{k=2}^{m}\left(v_r^{x_1, x_k} + \tfrac{1}{2}\tau_{r,r}\right) - 2\sum_{\mu=1}^{m}\left(v_r^{x_1, z_\mu} + \tfrac{1}{2}\tau_{r,r}\right),$$

and this expression is identically zero.

Secondly, considering the product as a rational function of x_1, the denominator is zero to the second order when x_1 coincides with any one of the m places $z_1, \ldots, z_m$, and is otherwise zero at $2m(p-1)$ places depending on f only; of these latter places $2(m-1)(p-1)$ are also zeros of the factors $\Pi'\Theta(v^{x_h, x_k}+f)\,\Theta(v^{x_h, x_k}-f)$; there are then $2(p-1)$ poles of the function which depend on f only. The factors $\Pi'\Theta(v^{x_h, x_k}+f)\,\Theta(v^{x_h, x_k}-f)$ have also the zeros $x_2, \ldots, x_m$, each of the second order. The factors $\Theta(v^{x_1, z_1}+\ldots+v^{x_m, z_m}+e)\,\Theta(v^{x_1, z_1}+\ldots+v^{x_m, z_m}-e)$ have, by the hypothesis as to e, the zeros $z_1, z_2, \ldots, z_m$, each of the second order, as well as $2(p-m)$ other zeros depending on e only. On the whole then, regarded as a function of x_1, the product has

for zeros, $2(p-m)$ zeros depending on e, as well as the zeros $x_2, \ldots, x_m$, each of the second order,

for poles, $2(p-1)$ poles depending on f;

the function is thus of order $2(p-1)$; and it is determined, save for a factor independent of x_1, by the assignation of its zeros and poles. It is to be noticed that these do not depend on $z_1, z_2, \ldots, z_m$.

It is easy now to see that the product, regarded as a function of z_1, depends on $z_2, \ldots, z_m, e, f$ in just the same way as, regarded as a function of x_1, it depends on $x_2, \ldots, x_m, e, f$.

The expression is therefore of the form $F(x_1, x_2, \ldots, x_m)\,F(z_1, z_2, \ldots, z_m)$, wherein F denotes a rational function of all the variables involved.

The form of F can be determined by supposing $x_1, \ldots, x_m$ to approach indefinitely near to $z_1, \ldots, z_m$ respectively; then we obtain

$$\Theta(v^{x_1, z_1}+\ldots+v^{x_m, z_m}+e) = \frac{1}{2\pi i}\, t_m \sum_{i=1}^{p} \Theta_i'(v^{x_1, z_1}+\ldots+v^{x_{m-1}, z_{m-1}}+e)\,\Omega_i(z_m),$$

where t_m is the infinitesimal for the neighbourhood of the place z_m,

$$\Theta_i'(v^{x_1, z_1}+\ldots\ldots+v^{x_{m-1}, z_{m-1}}+e)$$
$$= \frac{1}{2\pi i}\, t_{m-1} \sum_{j=1}^{p} \Theta'_{i,j}(v^{x_1, z_1}+\ldots\ldots+v^{x_{m-1}, z_{m-1}}+e)\,\Omega_j(z_{m-1}),$$

where t_{m-1} is the infinitesimal for the neighbourhood of the place z_{m-1}, and so on, and eventually,

$$\Theta(v^{x_1, z_1} + \dots\dots + v^{x_m, z_m} + e)$$
$$= \frac{t_1 t_2 \dots t_m}{(2\pi i)^m} \sum_{i_m=1}^{p} \dots \sum_{i_1=1}^{p} \Theta'_{i_1, i_2, \dots i_m}(e)\, \Omega_{i_1}(z_1)\, \Omega_{i_2}(z_2) \dots \Omega_{i_m}(z_m).$$

Similarly

$$\Pi\Pi\Theta(v^{x_\lambda, z_\mu} + f) = \Pi'\Theta(v^{z_h, z_k} + f)\, \Theta(v^{z_h, z_k} - f) \left[\frac{t_1 t_2 \dots t_m}{(2\pi i)^m} \prod_{\mu=1}^{m} \sum_{i=1}^{p} \Theta_i'(f)\, \Omega_i(z_\mu)\right],$$

where h, k refers to all pairs of different numbers from among 1, 2, ..., m.

Therefore, dividing by a factor

$$(-)^m\, \Pi'\Theta^2(v^{z_h, z_k} + f)\, \Theta^2(v^{z_h, z_k} - f) \left[\frac{t_1 \dots t_m}{(2\pi i)^m}\right]^2,$$

which is common to numerator and denominator, and taking the square root, we have

$$F(z_1, \dots, z_m) = \frac{\sum\limits_{i_m=1}^{p} \dots \sum\limits_{i_1=1}^{p} \Theta'_{i_1, i_2, \dots, i_m}(e)\, \Omega_1(z_1)\, \Omega_2(z_2) \dots \Omega_m(z_m)}{\prod\limits_{\mu=1}^{m} \left[\sum\limits_{i=1}^{p} \Theta_i'(f)\, \Omega_i(z_\mu)\right]}.$$

On the whole therefore we have the equation

$$\Theta(v^{x_1, z_1} + \dots\dots + v^{x_m, z_m} + e)\, \Theta(v^{x_1, z_1} + \dots\dots + v^{x_m, z_m} - e)$$
$$\cdot \frac{\Pi'\Theta(v^{x_h, x_k} + f)\, \Theta(v^{x_h, x_k} - f)\, \Pi'\Theta(v^{z_h, z_k} + f)\, \Theta(v^{z_h, z_k} - f)}{\Pi\Pi\Theta(v^{x_\lambda, z_\mu} + f)\, \Theta(v^{x_\lambda, z_\mu} - f)}$$
$$= \frac{\Psi(x_1, \dots, x_m, e)\, \Psi(z_1, \dots, z_m, e)}{\prod\limits_{1}^{p} \Phi(x_\mu, f) \prod\limits_{1}^{p} \Phi(z_\mu, f)},$$

where

$$\Phi(x, f) = \sum_{i=1}^{p} \Theta_i'(f)\, \Omega_i(x),$$

$$\Psi(x_1, \dots, x_m, e) = \sum_{i_m=1}^{p} \dots \sum_{i_1=1}^{p} \Theta'_{i_1, i_2, \dots, i_m}(e)\, \Omega_{i_1}(x_1) \dots \Omega_{i_m}(x_m).$$

Suppose now that e_i is made to approach to r_i; then the conditions we have imposed for e are satisfied, and there is added the further condition that the differential coefficients of order m, $\Theta'_{i_1, i_2, \dots, i_m}$, also vanish. Hence it follows that $\Theta(v^{x_1, z_1} + \dots\dots + v^{x_m, z_m} + r)$ vanishes identically.

The whole theorem enunciated is thus demonstrated.

(XV.) The remarkable investigation of Prop. XIV. is due to Riemann; it is worth while to give a separate statement of one of the results obtained. Using q instead of $m-1$, we have proved that if the equations

$$e \equiv v^{m_p,\, m} - v^{\zeta_1,\, m_1} - \ldots\ldots - v^{\zeta_{p-1},\, m_{p-1}}$$

are satisfied by ∞^q sets of places $\zeta_1, \ldots, \zeta_{p-1}$, so that also the equations

$$-e \equiv v^{m_p,\, m} - v^{\xi_1,\, m_1} - \ldots\ldots - v^{\xi_{p-1},\, m_{p-1}}$$

are satisfied by ∞^q sets of places $\xi_1, \ldots, \xi_{p-1}$, then their exists a rational function, which has (i) for *poles*, the $2(p-1)$ places $t_1, \ldots, t_{p-1}, z_1, \ldots, z_{p-1}$, which satisfy the equations

$$f \equiv v^{m_p,\, m} - v^{t_1,\, m_1} - \ldots\ldots - v^{t_{p-1},\, m_{p-1}}$$

$$-f \equiv v^{m_p,\, m} - v^{z_1,\, m_1} - \ldots\ldots - v^{z_{p-1},\, m_{p-1}},$$

f being supposed such that these equations have one and only one set of solutions, and has (ii) for *zeros*, the arbitrary places $x_1, \ldots, x_q$, each of the second order, together with $2(p-1-q)$ places $\zeta_{q+1}, \ldots, \zeta_{p-1}, \xi_{q+1}, \ldots, \xi_{p-1}$, satisfying the equations

$$e \equiv v^{m_p,\, m} - v^{x_1,\, m_1} - \ldots\ldots - v^{x_q,\, m_q} - v^{\zeta_{q+1},\, m_{q+1}} - \ldots\ldots - v^{\zeta_{p-1},\, m_{p-1}},$$

$$-e \equiv v^{m_p,\, m} - v^{x_1,\, m_1} - \ldots\ldots - v^{x_q,\, m_q} - v^{\xi_{q+1},\, m_{q+1}} - \ldots\ldots - v^{\xi_{p-1},\, m_{p-1}},$$

and the function can be given in the form

$$\Psi(x_1, x_2, \ldots, x_q, x, e) \div \Phi(x, f),$$

the notation being that employed at the conclusion of Proposition (XIV.). The expressions Ψ, Φ occurring here have the zeros of certain ϕ-polynomials, to which they are proportional.

Corollary i. If we take $p-1$ places $\zeta_1, \ldots, \zeta_{p-1}$, so situated that only one ϕ-polynomial vanishes in all of them, and define e by the equations

$$e \equiv v^{m_p,\, m} - v^{\zeta_1,\, m_1} - \ldots\ldots - v^{\zeta_{p-1},\, m_{p-1}},$$

there will be no other set $\zeta_1, \ldots, \zeta_{p-1}$, satisfying these equations, or $q=0$. If $\xi_1, \ldots, \xi_{p-1}$ be the remaining zeros of the ϕ-polynomial which vanishes in $\zeta_1, \ldots, \zeta_{p-1}$, we have (Prop. IX.)

$$(m^2, \zeta_1, \ldots, \zeta_{p-1}, \xi_1, \ldots, \xi_{p-1}) \equiv (m_1^2, \ldots, m_p^2),$$

and therefore

$$-e \equiv v^{m_p,\, m} - v^{\xi_1,\, m_1} - \ldots\ldots - v^{\xi_{p-1},\, m_{p-1}}.$$

Similarly if $t_1, \ldots, t_{p-1}$ be arbitrary places which are the zeros of only one ϕ-polynomial, we can put

$$f \equiv v^{m_p,\, m} - v^{t_1,\, m_1} - \ldots\ldots - v^{t_{p-1},\, m_{p-1}}$$

$$-f \equiv v^{m_p,\, m} - v^{z_1,\, m_1} - \ldots\ldots - v^{z_{p-1},\, m_{p-1}}.$$

Then the rational function having $t_1, \ldots, t_{p-1}, z_1, \ldots, z_{p-1}$ for poles, and $\zeta_1, \ldots, \zeta_{p-1}, \xi_1, \ldots, \xi_{p-1}$ for zeros is given by $\Phi(x, e) \div \Phi(x, f)$. Thus the ϕ-polynomial which vanishes in $\zeta_1, \ldots, \zeta_{p-1}, \xi_1, \ldots, \xi_{p-1}$ is given by

$$\sum_{i=1}^{p} \Theta_i' \left(v^{m_p, m} - v^{\zeta_1, m_1} - \ldots\ldots - v^{\zeta_{p-1}, m_{p-1}}\right) \phi_i(x),$$

where $\phi_1(x), \ldots, \phi_p(x)$ are the ϕ-polynomials occurring in the differential coefficients of Riemann's normal integrals of the first kind.

Hence if $n_1, \ldots, n_{p-1}$ be places which, repeated, are all the zeros of a ϕ-polynomial, the form of this polynomial is known. Since, then, we have (Prop. XI. p. 269)

$$\tfrac{1}{2}\Omega \equiv v^{m_p, m} - v^{n_1, m_1} - \ldots\ldots - v^{n_{p-1}, m_{p-1}},$$

we can write this polynomial

$$\sum_{i=1}^{p} \Theta_i'(\tfrac{1}{2}\Omega)\, \phi_i(x),$$

$\tfrac{1}{2}\Omega$ being an odd half-period.

If another ϕ-polynomial than this one vanished in $n_1, \ldots, n_{p-1}$, there would be other places $n_1', \ldots, n'_{p-1}$, such that

$$\tfrac{1}{2}\Omega \equiv v^{m_p, m} - v^{n_1', m_1} - \ldots\ldots - v^{n'_{p-1}, m_{p-1}},$$

and therefore (Prop. VI.) the function $\Theta(v^{x, z} + \tfrac{1}{2}\Omega)$ would vanish identically; in that case (Prop. XIV. p. 276) the coefficients $\Theta_i'(\tfrac{1}{2}\Omega)$ would vanish.

We can express the ϕ-polynomial in terms of any integrals of the first kind; if $V_1^{x, m}, \ldots, V_p^{x, m}$ be any linearly independent integrals of the first kind, expressible in terms of the Riemann normal integrals $v_1^{x, m}, \ldots, v_p^{x, m}$ by linear equations of the form

$$v_i^{x, m} = \lambda_{i, 1} V_1^{x, m} + \ldots\ldots + \lambda_{i, p} V_p^{x, m}, \qquad (i = 1, 2, \ldots, p),$$

and the function $\Theta(u)$ be regarded as a function of $U_1, \ldots, U_p$ given by

$$u_i = \lambda_{i, 1} U_1 + \ldots\ldots + \lambda_{i, p} U_p, \qquad (i = 1, 2, \ldots, p),$$

and, so regarded, be written $\vartheta(U)$, the ϕ-polynomial which has zeros of the second order at $n_1, \ldots, n_{p-1}$ can be written

$$\sum_{i=1}^{p} \vartheta_i'(\tfrac{1}{2}\overline{\Omega})\, \psi_i(x),$$

where $\psi_1(x), \ldots, \psi_p(x)$ are the ϕ-polynomials corresponding to $V_1^{x, m}, \ldots, V_p^{x, m}$, and $\tfrac{1}{2}\overline{\Omega}$ denotes a set of simultaneous half-periods of the integrals $V_1^{x, m}, \ldots, V_p^{x, m}$. If $\tfrac{1}{2}\Omega$ stand for p quantities of which a general one is

$$\tfrac{1}{2}(k_i + k_1'\tau_{i, 1} + \ldots\ldots + k_p'\tau_{i, p}), \qquad (i = 1, 2, \ldots, p),$$

and $\omega_{r,s}$, $\omega'_{r,s}$ be $2p^2$ quantities given by

$$\left.\begin{matrix}1\\0\end{matrix}\right\} = 2\lambda_{i,1}\omega_{1,s} + 2\lambda_{i,2}\omega_{2,s} + \dots\dots + 2\lambda_{i,p}\omega_{p,s}, \quad (i, s = 1, 2, \dots, p),$$

$$\tau_{i,s} = 2\lambda_{i,1}\omega'_{1,s} + 2\lambda_{i,2}\omega'_{2,s} + \dots\dots + 2\lambda_{i,p}\omega'_{p,s},$$

where, in the first equation, we are to take 1 or 0 according as $i = s$ or $i \neq s$, then $\frac{1}{2}\overline{\Omega}$ will stand for p quantities of which one is

$$k_1\omega_{i,1} + \dots\dots + k_p\omega_{i,p} + k_1'\omega'_{i,1} + \dots\dots + k_p'\omega'_{i,p}, \quad (i = 1, 2, \dots, p).$$

For example when the fundamental Riemann surface is that whose equation may be interpreted as the equation of a plane quartic curve, every double tangent is associated with an odd half-period and its equation may be put into the form

$$x\vartheta_1'(\tfrac{1}{2}\overline{\Omega}) + y\vartheta_2'(\tfrac{1}{2}\overline{\Omega}) + \vartheta_3'(\tfrac{1}{2}\overline{\Omega}) = 0.$$

Corollary ii. If the equations

$$e \equiv v^{m_p, m} - v^{x_1, m_1} - v^{\zeta_2, m_2} - \dots\dots - v^{\zeta_{p-1}, m_{p-1}}$$

can be satisfied with an arbitrary position of x_1 and suitable positions of $\zeta_2, \dots, \zeta_{p-1}$, and therefore, also, the equations

$$-e \equiv v^{m_p, m} - v^{x_1, m_1} - v^{\xi_2, m_2} - \dots\dots - v^{\xi_{p-1}, m_{p-1}}$$

can be satisfied, then a ϕ-polynomial vanishing at x_1 to the second order, and otherwise vanishing in $\zeta_2, \dots, \zeta_{p-1}, \xi_2, \dots, \xi_{p-1}$, is given by

$$\sum_{i=1}^{p} \Omega_i(x) \sum_{j=1}^{p} \Theta'_{i,j}(e)\,\Omega_j(x_1) = 0.$$

Ex. In the case of a plane quintic curve having two double points, this gives us the equation of the straight lines joining these double points to an arbitrary point x_1, of the curve.

Corollary iii. We have seen (Chap. VI. § 98) that any rational function of which the multiplicity (q) is greater than the excess of the order of the function over the deficiency of the surface, say, $q = Q - p + \tau + 1$, can be expressed as the quotient of two ϕ-polynomials. If the function have $\zeta_1, \dots, \zeta_Q$ for zeros, and $\xi_1, \dots, \xi_Q$ for poles, and the common zeros of the ϕ-polynomials expressing the function be $z_1, \dots, z_R$, where $R = 2p - 2 - Q$, the function is in fact expressed by

$$\sum_{i=1}^{p} \Theta_i'(e)\,\Omega_i(x) \div \sum_{i=1}^{p} \Theta_i'(f)\,\Omega_i(x),$$

where (cf. § 93, Chap. VI.)

$$e \equiv v^{m_p, m} - v^{z_1, m_1} - \dots\dots - v^{z_{R-\tau}, m_{R-\tau}} - v^{\zeta_1, m_{R-\tau+1}} - \dots\dots - v^{\zeta_q, m_{p-1}},$$

$$f \equiv v^{m_p, m} - v^{z_1, m_1} - \dots\dots - v^{z_{R-\tau}, m_{R-\tau}} - v^{\xi_1, m_{R-\tau+1}} - \dots\dots - v^{\xi_q, m_{p-1}}.$$

189. Before concluding this chapter it is convenient to introduce a slightly more general function* than that so far considered; we denote by $\vartheta(u; q, q')$, or by $\vartheta(u, q)$, the function

$$\vartheta(u; q, q') = \Sigma e^{au^2+2hu(n+q')+b(n+q')^2+2i\pi q(n+q')},$$

wherein the summation extends to all positive and negative integer values of the p integers $n_1, \ldots, n_p$, a is any symmetrical matrix whatever of p rows and columns, h is any matrix whatever of p rows and columns, in general not symmetrical, b is any symmetrical matrix whatever of p rows and columns, such that the real part of the quadratic form bm^2 is necessarily negative for all real values of the quantities $m_1, \ldots, m_p$, other than zero, and q, q' denote two sets, each of p constant quantities, which constitute the characteristic of the function. In the most general case the matrix b depends on $\frac{1}{2}p(p+1)$ independent constants; if however we put $i\pi\tau$ for b, τ being the symmetrical matrix hitherto used, depending only on $3p-3$ constants, and denote the p quantities hu by U, we shall obtain

$$\vartheta(u; q, q') = e^{au^2}\Theta(U; q, q').$$

We make consistent use of the notation of matrices (see Appendix ii.). If u denote a row (or column) letter of p elements, and h denote any matrix of p rows and columns, then hu is a row letter; we shall generally write huv for $hu \,.\, v$; and we have $huv = \bar{h}vu$, where $\bar{h}$ is the matrix obtained from h by transposition of rows and columns. Further if k be any matrix of p rows and columns, $hu \,.\, kv = \bar{h}kvu = khuv$. For the present every matrix denoted by a single letter is a square matrix of p rows and columns.

Now let $\omega, \omega', \eta, \eta'$ be any such matrices, and P, P' be row letters of elements $P_1, \ldots, P_p, P_1', \ldots, P_p'$. Then, by the sum of the two row letters $\omega P + \omega' P'$ we denote a row letter consisting of p elements, each being the sum of an element of ωP with the corresponding element of $\omega' P'$. This row letter, with every element multiplied by 2, will be denoted by Ω_P, so that

$$\Omega_P = 2\omega P + 2\omega' P';$$

in a similar way we define a row letter of p elements by the equation

$$H_P = 2\eta P + 2\eta' P';$$

then $u + \Omega_P$ will denote a row letter of p elements, like u.

The equation we desire to prove, subject to proper relations connecting $\omega, \omega', \eta, \eta'$, is the following,

$$\vartheta(u + \Omega_P, q) = e^{H_P(u+\frac{1}{2}\Omega_P) - \pi i PP' + 2\pi i(Pq' - P'q)}\, e^{-2\pi i Pq'}\,\vartheta(u, P+q), \qquad \text{(L)},$$

which is a generalization of some of the fundamental equations given for $\Theta(u)$.

* Schottky, *Abriss einer Theorie der Abelschen Functionen von drei Variabeln*, Leipzig, 1880. The introduction of the matrix notation is suggested by Cayley, *Math. Annal.* (xvii.), p. 115.

In order that this equation may hold it is sufficient that the terms on the two sides of the equation, which contain the same values of the summation letters $n_1, \ldots, n_p$, should be equal; this will be so if

$$a(u+\Omega_P)^2 + 2h(u+\Omega_P)(n+q') + b(n+q')^2 + 2\pi i q(n+q')$$
$$= H_P(u+\tfrac{1}{2}\Omega_P) - \pi i P P' - 2\pi i P' q + au^2 + 2hu(n+q'+P') + b(n+q'+P')^2$$
$$+ 2\pi i (P+q)(n+q'+P');$$

picking out in this conditional equation respectively the terms involving squares, first powers, and zero powers of $n_1, \ldots, n_p$, we require

$$b = b,$$
$$h(u+\Omega_P) + \bar{b}q' + \pi i q = hu + \bar{b}(q'+P') + \pi i (P+q),$$

and

$$a(u+\Omega_P)^2 + 2h(u+\Omega_P)q' + bq'^2 + 2\pi i q q' = H_P(u+\tfrac{1}{2}\Omega_P) - \pi i P P' - 2\pi i P' q$$
$$+ au^2 + 2hu(q'+P') + b(q'+P')^2 + 2\pi i (P+q)(q'+P').$$

190. In working out these conditions it will be convenient at first to neglect the fact that a and b are symmetrical matrices, in order to see how far it is necessary.

The second of these conditions gives

$$h\Omega_P = \pi i P + \bar{b}P',$$

and therefore gives the two conditions $h\omega = \frac{1}{2}\pi i$, $h\omega' = \frac{1}{2}\bar{b}$, whereby ω, ω' are determined in terms of the matrices h, b. In particular when $h = \pi i$ and $b = i\pi\tau$, as in the case of the function $\Theta(u)$, we have $2\omega = 1$, $2\omega' = \tau$, namely 2ω, $2\omega'$ are the matrices of the periods of the Riemann normal integrals of the first kind, respectively at the first kind, and at the second kind of period loops.

The third condition gives

$$2au\Omega_P + a\Omega^2_P + 2h\Omega_P q' = H_P(u+\tfrac{1}{2}\Omega_P)$$
$$- \pi i P P' - 2\pi i P' q + 2huP' + b(2q'P' + P'^2) + 2\pi i (qP' + Pq' + PP'),$$

that is

$$(2\bar{a}\Omega_P - H_P - 2\bar{h}P')u + (a\Omega_P - \tfrac{1}{2}H_P)\Omega_P - \pi i P P' - bP'^2$$
$$+ 2(h\Omega_P - \pi i P - \bar{b}P')q' = 0;$$

in order that this may be satisfied for all values of $u_1, \ldots, u_p$, we must have, referring to the equation already obtained from the second condition,

$$H_P = 2\bar{a}\Omega_P - 2\bar{h}P',$$

and

$$(a\Omega_P - \tfrac{1}{2}H_P)\Omega_P = (\pi i P + bP')P';$$

from the first of these, by the equation already obtained, we have

$$(\bar{a}\Omega_P - \tfrac{1}{2}H_P)\Omega_P = \bar{h}P'\Omega_P = h\Omega_P P' = (\pi i P + \bar{b}P')P';$$

subtracting this from the second equation, there results

$$(\bar{a}-a)\,\Omega^2_P=(\bar{b}-b)\,P'^2,$$

and in order that this may hold independently of the values assigned to P, P' it is necessary that $\bar{a}=a$, $b=\bar{b}$; when this is so, these two equations give, in addition to the one already obtained, only the equation

$$H_P=2a\Omega_P-2\bar{h}P',$$

leading to

$$\eta=2a\omega,\;\; \eta'=2a\omega'-2\bar{h},$$

which express the matrices η and η' in terms of the matrices a and h. These equations, with

$$h\Omega_P=\pi iP+bP',$$

or

$$h\omega=\tfrac{1}{2}\pi i,\;\; h\omega'=\tfrac{1}{2}b,$$

are all the conditions necessary, and they are clearly sufficient. When they are satisfied we have

$$\vartheta(u+\Omega_P,q)=e^{\lambda_P(u)-2\pi iP'q}\,\vartheta(u\,;\,q+P), \qquad \text{(L)},$$

where

$$\lambda_P(u)=H_P(u+\tfrac{1}{2}\Omega_P)-\pi iPP'.$$

Ex. Weierstrass's function σu is given by

$$A\sigma u=\Sigma e^{\frac{\eta}{2\omega}u^2+\frac{2\pi iu}{2\omega}(n+\frac{1}{2})+i\pi\tau(n+\frac{1}{2})^2+\pi i(n+\frac{1}{2})}$$

where A is a certain constant.

The equations obtained express the $4p^2$ elements of the matrices $\omega,\omega',\eta,\eta'$ in terms of the $p^2+p(p+1)$ quantities occurring in the matrices a, h, b; there must therefore be $2p^2-p$ relations connecting the quantities in ω, ω', η, η'. The equations are in fact of precisely the same form as those already obtained in § 140, Chap. VII., equation (A), and precisely as in § 141 it follows that the necessary relations connecting ω, ω', η, η' may be expressed by either of the equations (B), (C) of § 140. Using the notation of matrices in greater detail we may express these relations in a still further way.

For

$$\begin{aligned}\tfrac{1}{2}(H_P\Omega_Q-H_Q\Omega_P)&=(a\Omega_P-\bar{h}P')\,\Omega_Q-(a\Omega_Q-\bar{h}P')\,\Omega_P\\&=-\bar{h}P'\Omega_Q+\bar{h}Q'\Omega_P\\&=h\Omega_P\,.\,Q'-h\Omega_Q\,.\,P'\\&=(\pi iP+bP')\,Q'-(\pi iQ+bQ')\,P',\end{aligned}$$

so that

$$H_P\Omega_Q-H_Q\Omega_P=2\pi i\,(PQ'-P'Q)\,;$$

this relation includes all the $2p^2-p$ necessary relations; for it gives

$$(\eta P+\eta'P')(\omega Q+\omega'Q')-(\eta Q+\eta'Q')(\omega P+\omega'P')=\tfrac{1}{2}\pi i\,(PQ'-P'Q),$$

or (using the matrix relation already quoted in the form $hu.kv = \overline{h}kvu = \overline{k}huv$)

$$(\overline{\omega}\eta - \overline{\eta}\omega) PQ + (\overline{\omega}\eta' - \overline{\eta}\omega') P'Q + (\overline{\omega}'\eta - \overline{\eta}'\omega) PQ' + (\overline{\omega}'\eta' - \overline{\eta}'\omega') P'Q'$$
$$= \tfrac{1}{2}\pi i (PQ' - P'Q),$$

and expressing that this equation holds for all values of P, Q, P', Q', we obtain the Weierstrassian equations ((B) § 140).

Similarly the Riemann equations ((C) § 140) are all expressed by

$$(2\overline{\omega}'P + 2\overline{\eta}'Q)(2\overline{\omega}P' + 2\overline{\eta}Q') - (2\overline{\omega}P + 2\overline{\eta}Q)(2\overline{\omega}'P' + 2\overline{\eta}'Q') = 2\pi i (PQ' - P'Q).$$

Ex. i. If we substitute for the variables u in the ϑ function linear functions of any p new variables v, with non-vanishing determinant of transformation, and L_P be formed from the new form of the ϑ function, regarded as a function of v, just as H_P was formed from the original function, prove that $L_P v = H_P u$, and that $\lambda_P(u)$ remains unaltered.

Ex. ii. Prove that

$$\lambda_P(u + \Omega_M) + \lambda_M(u) - 2\pi i M'P = \lambda_Q(u + \Omega_N) + \lambda_N(u) - 2\pi i N'Q,$$

provided

$$M + P = N + Q.$$

The equation (L) is simplified when P, P' both consist of integers. For if M, M' be rows of integers, it is easy (putting a new summation letter, m, for $n + M'$, in the exponent of the general term of $\vartheta(u;\ q + M,\ q' + M')$,) to verify that

$$\vartheta(u;\ q + M,\ q' + M') = e^{2\pi i M q'}\vartheta(u;\ q,\ q').$$

Therefore, if m, m' consist of integers, we find

$$\vartheta(u + \Omega_m,\ q) = e^{\lambda_m(u) + 2\pi i (mq' - m'q)}\vartheta(u,\ q),$$

and in particular

$$\vartheta(u + \Omega_m) = e^{\lambda_m(u)}\vartheta(u),$$

where $\vartheta(u)$ is written for $\vartheta(u;\ 0,\ 0)$. The reader will compare the equations obtained at the beginning of this chapter, where $a = 0$, $\eta = 0$, $\eta' = -2\pi i$, $\omega = \tfrac{1}{2}$, $\omega' = \tfrac{1}{2}\tau$, $\Omega_P = P + \tau P'$, $H_P = -2\pi i P'$, $\lambda_P(u) = -2\pi i P'(u + \tfrac{1}{2}P + \tfrac{1}{2}\tau P') - \pi i P P'$.

One equation, just used, deserves a separate statement; we have

$$\vartheta(u;\ q + M) = e^{2\pi i M q'}\vartheta(u;\ q),$$

where M stands for a row of integers $M_1, \ldots, M_p, M_1', \ldots, M_p'$.

191. Finally, to conclude these general explanations as to the function $\vartheta(u)$, we may enquire in what cases $\vartheta(u)$ can be an odd or even function.

When m, m' are rows of integers the general formula gives

$$\vartheta(-u + \Omega_m,\ q) = e^{\lambda_m(-u) + 2\pi i (mq' - m'q)}\vartheta(-u,\ q);$$

hence when $\vartheta(u, q)$ is odd, or is even, since $\lambda_m(-u) = \lambda_{-m}(u)$, we have

$$\vartheta(u - \Omega_m, q) = e^{\lambda_{-m}(u) + 2\pi i(mq' - m'q)}\,\vartheta(u, q);$$

therefore, by equation (L),

$$\vartheta(u + \Omega_m, q), \;= \vartheta(u - \Omega_m, q) \,.\, e^{\lambda_{2m}(u - \frac{1}{2}\Omega_m) + 4\pi i(mq' - m'q)},$$
$$= \vartheta(u, q)\, e^{\lambda_{-m}(u) + \lambda_{2m}(u - \frac{1}{2}\Omega_m) + 6\pi i(mq' - m'q)},$$

while also, by the same equation,

$$\vartheta(u + \Omega_m, q) = \vartheta(u, q)\, e^{\lambda_m(u) + 2\pi i(mq' - m'q)}.$$

Thus the expression

$$\lambda_{2m}(u - \tfrac{1}{2}\Omega_m) + \lambda_{-m}(u) - \lambda_m(u) + 4\pi i(mq' - m'q)$$

must be an integral multiple of $2\pi i$. This is immediately seen to require only that $2(mq' - m'q - mm')$ be integral for all integral values of m, m'. Hence the necessary and sufficient condition is that q and q' consist of half-integers. In that case we prove as before that $\vartheta(u, q)$ is odd or even according as $4qq'$ is an odd or even integer.

192. In what follows in the present chapter we consider only the case in which $b = i\pi\tau$, τ being the matrix of the periods of Riemann's normal integrals at the second kind of period loops. And if $u_1^{x,a}, \ldots, u_p^{x,a}$ denote any p linearly independent integrals of the first kind, such as used in §§ 138, 139, Chap. VII., the matrix h is here taken to be such that

$$2\pi i v_i^{x,a} = h_{i,1} u_1^{x,a} + \ldots\ldots + h_{i,p} u_p^{x,a}, \qquad (i = 1, 2, \ldots, p),$$

so that h is as in § 139, and

$$\vartheta(u^{x,a}, q) = e^{au^2}\,\Theta(v^{x,a}, q),$$

where $u = u^{x,a}$.

From the formula

$$\vartheta(u + \Omega_m) = e^{H_m(u + \frac{1}{2}\Omega_m) - \pi i m m'}\,\vartheta(u),$$

wherein m, m' denote rows of integers, we infer, using the abbreviation

$$\zeta_i(u) = \frac{\partial}{\partial u_i}\log\vartheta(u),$$

that

$$\zeta_i(u + \Omega_m) - \zeta_i(u) = 2(\eta_{i,1} m_1 + \ldots\ldots + \eta_{i,p} m_p + \eta'_{i,1} m_1' + \ldots\ldots + \eta'_{i,p} m_p');$$

particular cases of this formula are

$$\zeta_i(u_1 + 2\omega_{1,r}, \ldots, u_p + 2\omega_{p,r}) = \zeta_i(u) + 2\eta_{i,r},$$
$$\zeta_i(u_1 + 2\omega'_{1,r}, \ldots, u_p + 2\omega'_{p,r}) = \zeta_i(u) + 2\eta'_{i,r}.$$

Thus if u_s be the argument

$$u_s^{x,m} - u_s^{x_1,m_1} - \ldots\ldots - u_s^{x_p,m_p},$$

where $u_1^{x,a}, \ldots, u_p^{x,a}$ are any p linearly independent integrals of the first kind, and the matrix a here used in the definition of $\vartheta(u)$ be the same as that previously used (Chap. VII. § 138) in the definition of the integral $L_i^{x,a}$, so that the matrices η, η' will be the same in both cases, then it follows that the periods of the expression

$$\zeta_i(u) + L_i^{x,a},$$

regarded as a function of x, are zero.

193. And in fact, when the matrix a is thus chosen, there exists the equation

$$-\zeta_i(u^{x,m} - u^{x_1,m_1} - \ldots\ldots - u^{x_p,m_p}) + \zeta_i(u^{a,m} - u^{x_1,m_1} - \ldots\ldots - u^{x_p,m_p})$$
$$= L_i^{x,a} + \sum_{r=1}^{p} \tilde{\nu}_{r,i}\left[(x_r, x) - (x_r, a)\right]\frac{dx_r}{dt},$$

wherein $\tilde{\nu}_{r,i}$ denotes the minor of the element $\mu_i(x_r)$ in the determinant whose (r, i)th element is $\mu_i(x_r)$, divided by this determinant itself; thus $\tilde{\nu}_{r,i}$ depends on the places $x_1, \ldots, x_p$ exactly as the quantity $\nu_{r,i}$ (Chap. VII. § 138) depends on the places $c_1, \ldots, c_p$.

For we have just remarked that the two sides of this equation regarded as functions of x have the same periods; the left-hand side is only infinite at the places $x_1, \ldots, x_p$; if in $L_i^{x,a}$, which does not depend on the places $c_1, \ldots, c_p$ used in forming it (Chap. VII. § 138), we replace $c_1, \ldots, c_p$ by $x_1, \ldots, x_p$, it takes the form

$$\tilde{\nu}_{1,r}\Gamma_{x_1}^{x,a} + \ldots\ldots + \tilde{\nu}_{p,i}\Gamma_{x_p}^{x,a} - 2(a_{i,1}u_1^{x,a} + \ldots\ldots + a_{i,p}u_p^{x,a}),$$

and becomes infinite only at the places $x_1, \ldots, x_p$. Hence the difference of the two sides of the equation is a rational function with only p poles, $x_1, \ldots, x_p$, having arbitrary positions. Such a function is a constant (Chap. III. § 37, and Chap. VI.); and by putting $x = a$, we see that this constant is zero.

194. It will be seen in the next chapter that in the hyperelliptic case the equation of § 193 enables us to obtain a simple expression for $\zeta_i(u^{x,m} - u^{x_1,m_1} - \ldots\ldots - u^{x_p,m_p})$ in terms of algebraical integrals and rational functions only. In the general case we can also obtain such an expression*;

* See Clebsch und Gordan, *Abels. Functnen.* p. 171, Thomae, *Crelle*, LXXI. (1870), p. 214, Thomae, *Crelle*, CI. (1887), p. 326, Stahl, *Crelle*, CXI. (1893), p. 98, and, for a solution on different lines, see the latter part of chapter XIV. of the present volume.

though not of very simple character (§ 196). In the course of deriving that expression we give another proof of the equation of § 193.

The function of x given by $\vartheta(u^{x,m}; \tfrac{1}{2}\beta, \tfrac{1}{2}\alpha)$ will have p zeros, unless $\vartheta(u^{x,m} + \tfrac{1}{2}\Omega_{\beta,\alpha})$ vanish identically (§§ 179, 180); we suppose this is not the case. Denote these zeros by $m_1', \dots, m_p'$. Then (Prop. X. § 184) the function $\vartheta(u^{x,m} - u^{x_1, m_1'} - \dots\dots - u^{x_p, m_p'}; \tfrac{1}{2}\beta, \tfrac{1}{2}\alpha)$ will vanish when x coincides with $x_1, x_2, \dots$, or x_p. Determining $m_1, \dots, m_p$ so that

$$u^{m_1, m_1'} + \dots\dots + u^{m_p, m_p'} \equiv \tfrac{1}{2}\Omega_{\beta,\alpha},$$

and supposing the exact value of the left-hand side to be $\tfrac{1}{2}\Omega_{\beta,\alpha} + \Omega_{k,h}$, where k, h are integral, this function is equal to

$$\vartheta(u^{x,m} - u^{x_1, m_1} - \dots\dots - u^{x_p, m_p} - \tfrac{1}{2}\Omega_{\beta,\alpha} - \Omega_{k,h}; \tfrac{1}{2}\beta. \tfrac{1}{2}\alpha),$$

and this, by equation (L) is equal to

$$e^{-\frac{1}{2}H_{\beta,\alpha}(u - \frac{1}{4}\Omega_{\beta,\alpha}) + \frac{1}{4}\pi i\beta\alpha}\, \vartheta(u),$$

where $u = u^{x,m} - u^{x_1, m_1} - \dots\dots - u^{x_p, m_p} - \Omega_{k,h}$.

Therefore (§ 190) the expression

$$\Phi, = \frac{\vartheta(u^{x,m} - u^{x_1, m_1'} - \dots\dots - u^{x_p, m_p'}; \tfrac{1}{2}\beta, \tfrac{1}{2}\alpha)}{\vartheta(u^{\mu,m} - u^{x_1, m_1'} - \dots\dots - u^{x_p, m_p'}; \tfrac{1}{2}\beta, \tfrac{1}{2}\alpha)} \Bigg/ \frac{\vartheta(u^{x,m} - u^{\mu_1, m_1'} - \dots\dots - u^{\mu_p, m_p'}; \tfrac{1}{2}\beta, \tfrac{1}{2}\alpha)}{\vartheta(u^{\mu,m} - u^{\mu_1, m_1'} - \dots\dots - u^{\mu_p, m_p'}; \tfrac{1}{2}\beta, \tfrac{1}{2}\alpha)},$$

is equal to

$$\frac{\vartheta(u^{x,m} - u^{x_1, m_1} - \dots\dots - u^{x_p, m_p})}{\vartheta(u^{\mu,m} - u^{x_1, m_1} - \dots\dots - u^{x_p, m_p})} \Bigg/ \frac{\vartheta(u^{x,m} - u^{\mu_1, m_1} - \dots\dots - u^{\mu_p, m_p})}{\vartheta(u^{\mu,m} - u^{\mu_1, m_1} - \dots\dots - u^{\mu_p, m_p})};$$

we may write this in the form

$$\frac{\vartheta(U - r)}{\vartheta(V - r)} \Bigg/ \frac{\vartheta(U - s)}{\vartheta(V - s)};$$

the expression is therefore equal to

$$e^{L}\, \frac{\Theta(v^{x,m} - v^{x_1, m_1} - \dots\dots - v^{x_p, m_p})}{\Theta(v^{\mu,m} - v^{x_1, m_1} - \dots\dots - v^{x_p, m_p})} \Bigg/ \frac{\Theta(v^{x,m} - v^{\mu_1, m_1} - \dots\dots - v^{\mu_p, m_p})}{\Theta(v^{\mu,m} - v^{\mu_1, m_1} - \dots\dots - v^{\mu_p, m_p})},$$

where

$$L, = a(U - r)^2 - a(V - r)^2 - a(U - s)^2 + a(V - s)^2,$$

is equal to

$$-2aU(r - s) + 2aV(r - s),$$

or

$$-2a(U - V)(r - s),$$

that is

$$-2au^{x,\mu}(u^{x_1, \mu_1} + \dots\dots + u^{x_p, \mu_p}),$$

which denotes

$$-\sum_{r=1}^{p}\left(\sum_{i,j}\sum 2a_{i,j}\, u_j^{x,\mu}\, u_i^{x_r,\mu_r}\right).$$

Hence, by Prop. XIII. § 187, supposing that the matrix a, here used, is the same as that used in § 138, Chap. VII., and denoting the canonical integral

$$\Pi_{z,c}^{x,a} - 2\sum_{r=1}^{p}\sum_{s=1}^{p} a_{r,s}\, u_r^{x,a}\, u_s^{z,c},$$

which has already occurred (page 194), by $R_{z,c}^{x,a}$, we have

$$R_{x_1,\mu_1}^{x,\mu} + \dots\dots + R_{x_p,\mu_p}^{x,\mu} = \log \Phi.$$

195. From the formula

$$\sum_{r=1}^{p} R_{x,\mu}^{x_r,\mu_r} = \log \frac{\vartheta(u^{x,m} - u^{x_1,m_1} - \dots - u^{x_p,m_p})}{\vartheta(u^{x,m} - u^{\mu_1,m_1} - \dots - u^{\mu_p,m_p})} \Big/ \frac{\vartheta(u^{\mu,m} - u^{x_1,m_1} - \dots - u^{x_p,m_p})}{\vartheta(u^{\mu,m} - u^{\mu_1,m_1} - \dots - u^{\mu_p,m_p})},$$

since

$$R_{x,\mu}^{x_r,\mu_r} = P_{x,\mu}^{x_r,\mu_r} + \sum_{i=1}^{p} u_i^{x_r,\mu_r} L_i^{x,\mu},$$

we obtain

$$\sum_{r=1}^{p} P_{x,\mu}^{x_r,\mu_r} + \sum_{i=1}^{p}\sum_{r=1}^{p} u_i^{x_r,\mu_r} L_i^{x,\mu} = \log \frac{\vartheta(u^{x,m} - U)}{\vartheta(u^{x,m} - U_0)} \Big/ \frac{\vartheta(u^{\mu,m} - U)}{\vartheta(u^{\mu,m} - U_0)},$$

where

$$U = u^{x_1,m_1} + \dots\dots + u^{x_p,m_p},$$

$$U_0 = u^{\mu_1,m_1} + \dots\dots + u^{\mu_p,m_p},$$

and therefore

$$U - U_0 = \sum_{r=1}^{p} u^{x_r,\mu_r}.$$

Hence, differentiating,

$$\sum_{r=1}^{p} \frac{\partial x_r}{\partial U_i}[(x_r, x) - (x_r, \mu)] + L_i^{x,\mu} = -\zeta_i(u^{x,m} - U) + \zeta_i(u^{\mu,m} - U),$$

where

$$\zeta_i(u) = \frac{\partial}{\partial u_i}\log \vartheta(u);$$

but, from

$$dU_i = Du_i^{x_1,m_1}\,.\,dx_1 + \dots\dots + Du_i^{x_p,m_p}\,.\,dx_p,$$

where $dx_1, \dots, dx_p$ denote the infinitesimals at $x_1, \dots, x_p$, we obtain

$$\frac{\partial x_r}{\partial U_i} = \bar{\nu}_{r,i}\frac{dx_r}{dt};$$

thus

$$-\zeta_i(u^{x,m} - U) + \zeta_i(u^{\mu,m} - U) = L_i^{x,m} + \sum_{r=1}^{p} \bar{\nu}_{r,i}[(x_r, x) - (x_r, \mu)]\frac{dx_r}{dt},$$

which is the equation of § 193.

196. From the equation

$$R_{z_1, \mu_1}^{x, \mu} + \ldots\ldots + R_{z_p, \mu_p}^{x, \mu} = \log \Phi,$$

differentiating in regard to x, we obtain an equation which we write in the form

$$\sum_{r=1}^{p} F_x^{z_r, \mu_r} = \sum_{r=1}^{p} \mu_r(x) \left[\zeta_r(u^{x, m} - U) - \zeta_r(u^{x, m} - U_0)\right],$$

where $U = u^{z_1, m_1} + \ldots\ldots + u^{z_p, m_p}$, $U_0 = u^{\mu_1, m_1} + \ldots\ldots + u^{\mu_p, m_p}$.

Thus, if we take for $\mu_1, \ldots, \mu_p$ places determined from x just as $m_1, \ldots, m_p$ are determined from m, so that

$$(m, \mu_1, \ldots, \mu_p) \equiv (x, m_1, \ldots, m_p),$$

the arguments $u^{x, m} - U_0$ will be $\equiv 0$; as the odd function $\zeta_r(u)$ vanishes for zero values of the argument, we therefore have (§ 192), writing Ω_P for the exact value of $u^{x, m} - U_0$,

$$\begin{aligned} F_x^{z_1, \mu_1} + \ldots\ldots + F_x^{z_p, \mu_p} &= \sum_{r=1}^{p} \mu_r(x) \left[\zeta_r(u^{x, m} - u^{z_1, m_1} - \ldots\ldots - u^{z_p, m_p}) - (H_P)_r\right] \\ &= \sum_{r=1}^{p} \mu_r(x)\, \zeta_r(u^{x, m} - u^{z_1, m_1} - \ldots - u^{z_p, m_p} - \Omega_P) \\ &= -\sum_{r=1}^{p} \mu_r(x)\, \zeta_r(u^{z_1, \mu_1} + \ldots + u^{z_p, \mu_p}). \end{aligned}$$

If in this equation we put x at m we derive

$$F_m^{z_1, m_1} + \ldots\ldots + F_m^{z_p, m_p} = -\sum_{r=1}^{p} \mu_r(m)\, \zeta_r(u^{z_1, m_1} + \ldots\ldots + u^{z_p, m_p}), \qquad \text{(M)},$$

where $z_1, \ldots, z_p$ are arbitrary.

If however we put x in turn at p independent places $c_1, \ldots, c_p$, and denote the places determined from c_i, as $m_1, \ldots, m_p$ are determined from m, by $c_{i,1}, \ldots, c_{i,p}$, so that

$$(c_i, m_1, \ldots, m_p) \equiv (m, c_{i,1}, \ldots, c_{i,p}),$$

we obtain p equations of the form

$$F_{c_i}^{z_1, c_{i,1}} + \ldots\ldots + F_{c_i}^{z_p, c_{i,p}} = -\sum_{r=1}^{p} \mu_r(c_i)\, \zeta_r(u^{z_1, c_{i,1}} + \ldots\ldots + u^{z_p, c_{i,p}}).$$

Suppose then that $x, x_1, \ldots, x_p$ are arbitrary independent places; for $z_1, \ldots, z_p$ put the places $x_{i,1}, \ldots, x_{i,p}$ determined by the congruence

$$(x, x_{i,1}, \ldots, x_{i,p}) \equiv (c_i, x_1, \ldots, x_p);$$

then, if Ω_Q denote a certain period, $-u^{x_{i,1}, c_{i,1}} - \ldots - u^{x_{i,p}, c_{i,p}}$ is equal to $\Omega_Q + u^{x, m} - u^{x_1, m_1} - \ldots\ldots - u^{x_p, m_p}$, and we have

$$F_{c_i}^{x_{i,1}, c_{i,1}} + \ldots + F_{c_i}^{x_{i,p}, c_{i,p}} = \sum_{r=1}^{p} \mu_r(c_i)\, \zeta_r(\Omega_Q + u^{x, m} - u^{x_1, m_1} - \ldots - u^{x_p, m_p});$$

therefore

$$\zeta_i(\Omega_Q + u^{x,\,m} - u^{x_1,\,m_1} - \dots - u^{x_p,\,m_p}) = \sum_{r=1}^{p} \nu_{r,\,i}\,[F_{c_r}^{x_r,\,1,\,c_{r,\,1}} + \dots + F_{c_r}^{x_r,\,p,\,c_{r,\,p}}],$$

where $\nu_{r,\,i}$ is the minor of $\mu_i(c_r)$ in the determinant whose (r, s)th element is $\mu_s(c_r)$, divided by the determinant itself.

In particular, when the differential coefficients $\mu_1(x), \dots, \mu_p(x)$ are those already denoted (§ 121, Chap. VII.) by $\omega_1(x), \dots, \omega_p(x)$, and $V_i^{x,\,a} = \int_a^x \omega_i(x)dt_x$, and the paths of integration are properly taken, we have*

$$\frac{\partial}{\partial V_i}\log \vartheta(V^{x,\,m} - V^{x_1,\,m_1} - \dots\dots - V^{x_p,\,m_p}) = F_{c_i}^{x_{i,\,1},\,c_{i,\,1}} + \dots\dots + F_{c_i}^{x_{i,\,p},\,c_{i,\,p}}.$$

197. A further result should be given. Let $x, x_1, \dots, x_p$ be fixed places. Take a variable place z, and thereby determine places $z_1, \dots, z_p$, functions of z, such that

$$(x, z_1, \dots, z_p) \equiv (z, x_1, \dots, x_p).$$

Then from the formula

$$-\zeta_i(u^{z,\,m} - u^{z_1,\,m_1} - \dots\dots - u^{z_p,\,m_p}) + \zeta_i(u^{a,\,m} - u^{z_1,\,m_1} - \dots\dots - u^{z_p,\,m_p})$$

$$= L_i^{z,\,a} + \sum_{s=1}^{p} \nu_{s,\,i}\,[(z_s, z) - (z_s, a)]\,\frac{dz_s}{dt},$$

wherein $\nu_{s,\,i}$ is formed with $z_1, \dots, z_p$, we have, by differentiating in regard to z and denoting $-\dfrac{\partial}{\partial u_j}\zeta_i(u)$ by $\wp_{i,\,j}(u)$,

$$\sum_{j=1}^{p} \wp_{i,\,j}(U)\left[\mu_j(z) - \mu_j(z_1)\frac{dz_1}{dz} - \dots\dots - \mu_j(z_p)\frac{dz_p}{dz}\right]$$

$$- \sum_{j=1}^{p} \wp_{i,\,j}(\overline{U})\left[-\mu_j(z_1)\frac{dz_1}{dz} - \dots\dots - \mu_j(z_p)\frac{dz_p}{dz}\right]$$

$$= D_z L_i^{z,\,a} + \sum_{s=1}^{p} [(z_s, z) - (z_s, a)]\frac{dz_s}{dt}\sum_{r=1}^{p}\frac{dz_r}{dz}\frac{d}{dz_r}(\nu_{s,\,i})$$

$$+ \sum_{s=1}^{p} \nu_{s,\,i}\left[\frac{d}{dz_s}\left((z_s, z)\frac{dz_s}{dt}\right) - \frac{d}{dz_s}\left((z_s, a)\frac{dz_s}{dt}\right)\right]\frac{dz_s}{dz} + \sum_{s=1}^{p} \nu_{s,\,i}\,D_z\left((z_s, z)\frac{dz_s}{dt}\right),$$

where $U = u^{z,\,m} - u^{z_1,\,m_1} - \dots\dots - u^{z_p,\,m_p}$, $\overline{U} = u^{a,\,m} - u^{z_1,\,m_1} - \dots\dots - u^{z_p,\,m_p}$.

In this equation a is arbitrary. Let it now be put to coincide with z; hence

$$\sum_{j=1}^{p} \mu_j(z)\,\wp_{i,\,j}(U) = D_z L_i^{z,\,a} + \sum_{s=1}^{p} \nu_{s,\,i}\,D_z\left[(z_s, z)\frac{dz_s}{dt}\right].$$

* This form is used by Noether, *Math. Annal.* XXXVII. (1890), p. 488.

Therefore

$$\sum_{i=1}^{p}\sum_{j=1}^{p}\mu_i(k)\,\mu_j(z)\,\wp_{i,j}(U)$$
$$=\sum_{i=1}^{p}\mu_i(k)\,D_z L_i^{z,a}+\sum_{s=1}^{p}\sum_{i=1}^{p}\nu_{s,i}\,\mu_i(k)\,D_z\left[(z_s, z)\frac{dz_s}{dt}\right]$$
$$=\sum_{i=1}^{p}\mu_i(k)\,D_z L_i^{z,a}+\sum_{s=1}^{p}\omega_s(k)\,D_z\left[(z_s, z)\frac{dz_s}{dt}\right]$$
$$=D_z'\left\{\sum_{i=1}^{p}\mu_i(k)\,L_i^{z,a}+\sum_{s=1}^{p}\omega_s(k)\,[(z_s, z)-(z, a)]\frac{dz_s}{dt}\right\},$$

where D_z' means a differentiation taking no account of the fact that $z_1, \dots, z_p$ are functions of z,

$$=D_z'\left\{\sum_{i=1}^{p}\mu_i(k)\,L_i^{z,a}-\psi(z, a;\ k, z_1, \dots, z_p)+\left[(k, z)-(k, a)\frac{dk}{dt}\right]\right\},$$
$$=D_z'\left\{D_k R_{z,a}^{k,c}-\psi(z, a;\ k, z_1, \dots, z_p)\right\},$$

in which form the expression is algebraically calculable when the integrals $L_i^{x,a}$ are known (Chap. VII. § 138),

$$=D_z'\left\{\Gamma_k^{z,a}-\psi(z, a;\ k, z_1, \dots, z_p)-2\Sigma\Sigma a_{r,s}\,\mu_r(k)\,u_s^{z,c}\right\},$$

where c is an arbitrary place; and this (cf. Ex. iv. § 125)

$$=-W(z;\ k, z_1, \dots, z_p)-2\sum_{r=1}^{p}\sum_{s=1}^{p}a_{r,s}\,\mu_r(z)\,\mu_r(k).$$

If now

$$(k, z_1, \dots, z_p)\equiv(z, k_1, \dots, k_p),$$

so that

$$U\equiv u^{x,m}-u^{x_1,m_1}-\dots\dots-u^{x_p,m_p}\equiv u^{z,m}-u^{z_1,m_1}-\dots\dots-u^{z_p,m_p}$$
$$\equiv u^{k,m}-u^{k_1,m_1}-\dots\dots-u^{k_p,m_p},$$

and

$$(x, z_1, \dots, z_p)\equiv(z, x_1, \dots, x_p),$$
$$(x, k_1, \dots, k_p)\equiv(k, x_1, \dots, x_p),$$

then the formula is

$$-\sum_i\sum_j\wp_{i,j}(U)\,.\,\mu_i(k)\,\mu_j(z)=W(z;\ k, z_1, \dots, z_p)+2\sum_{r=1}^{p}\sum_{s=1}^{p}a_{r,s}\,\mu_r(z)\,\mu_s(k),$$
$$=W(k;\ z, k_1, \dots, k_p)+2\sum_{r=1}^{p}\sum_{s=1}^{p}a_{r,s}\,\mu_r(z)\,\mu_s(k),$$

by Ex. iv. § 125.

By the congruences

$$u^{z_1, x_1} + \ldots\ldots + u^{z_p, x_p} \equiv u^{z, x}$$

the places $z_1, \ldots, z_p$ are algebraically determinable from the places $x, x_1, \ldots, x_p, z$, and therefore the function $W(z; k, z_1, \ldots, z_p)$ can be expressed by $x, x_1, \ldots, x_p, k, z$ only. In fact we have

$$\psi(z_1, x; z, x_1, \ldots, x_p) = 0, \ldots\ldots, \psi(z_p, x; z, x_1, \ldots, x_p) = 0.$$

The interest of the formula lies in the fact that the left-hand side is a multiply periodic function of the arguments $U_1, \ldots, U_p$.

A particular way of expressing the right-hand side in terms of $x, x_1, \ldots, x_p, z, k$ is to put down $\frac{1}{2}p(p+1)$ linearly independent particular cases of this equation, in which the right-hand side contains only $x, x_1, \ldots, x_p, z, k$, and then to solve for the $\frac{1}{2}p(p+1)$ quantities $\wp_{i,j}$. Since $\psi(z, a; k, z_1, \ldots, z_p)$ vanishes when $k = z_p$, we clearly have, as one particular case,

$$\sum_i \sum_j \wp_{i,j}(u^{z, m} - u^{z_1, m_1} - \ldots\ldots - u^{z_p, m_p})\, \mu_i(z)\, \mu_j(z_p) = D_z D_{z_p} R^{z, a}_{z_p, c},$$

and therefore

$$\sum_i \sum_j \wp_{i,j}(u^{x, m} - u^{x_1, m_1} - \ldots\ldots - u^{x_p, m_p})\, \mu_i(x)\, \mu_j(x_r) = D_x D_{x_r} R^{x, a}_{x_r, c}, \qquad \text{(N)}$$

and there are p equations of this form, in which $x_1, \ldots, x_p$ occur instead of x_r.

If we determine $x_1', \ldots, x'_{p-1}$ by the congruences

$$u^{x, m} - u^{x_1, m_1} - \ldots\ldots - u^{x_p, m_p} \equiv -[u^{x_p, m} - u^{x_1', m_1} - \ldots\ldots - u^{x'_{p-1}, m_{p-1}} - u^{x, m_p}],$$

so that $x_1', \ldots, x'_{p-1}$ are the other zeros of a ϕ-polynomial vanishing in $x_1, \ldots, x_{p-1}$, we can infer $p-1$ other equations, of the form

$$\sum_i \sum_j \wp_{i,j}(u^{x, m} - u^{x_1, m_1} - \ldots\ldots - u^{x_p, m_p})\, \mu_i(x_p)\, \mu_j(x_r') = D_{x_p} D_{x'_r} R^{x_p, a}_{x_r', a},$$

where $r = 1, 2, \ldots, (p-1)$. Here the right hand side does not depend upon the place x. And we can obtain p such sets of equations.

We have then sufficient * equations. For the hyperelliptic case the final formula is given below (§ 217, Chap. XI.).

198. *Ex.* i. Verify the formula (N) for the case $p = 1$.

Ex. ii. Prove that

$$\zeta_i(u^{x, m} - u^{x_1, m_1} - \ldots\ldots - u^{x_p, m_p}) + L_i^{x, a} - L_i^{x_1, a} - \ldots\ldots - L_i^{x_p, a}$$

is a rational function of $x, x_1, \ldots, x_p$.

Ex. iii. Prove that if

$$(x, z_1, \ldots, z_p) \equiv (z, x_1, \ldots, x_p) \equiv (a, a_1, \ldots, a_p),$$

then

$$\psi(x, a; z, x_1, \ldots, x_p) = \Gamma_z^{x, a} + \Gamma_z^{z_1, a_1} + \ldots\ldots + \Gamma_z^{z_p, a_p}.$$

Deduce the first formula of § 193 from the final formula of § 196.

* The function $\wp_{i,j}(u)$, here employed, is remarked, for the hyperelliptic case, by Bolza, *Göttinger Nachrichten*, 1894, p. 268.

Ex. iv. Prove that if

$$Q_i = \Gamma_{c_i}^{x_{i,1}, a_1} + \ldots\ldots + \Gamma_{c_i}^{x_{i,p}, a_p},$$

where $a_1, \ldots, a_p$ are arbitrary places, and

$$V_r = V_r^{x, m} - V_r^{x_1, m_1} - \ldots\ldots - V_r^{x_p, m_p} = V_r^{c_i, m} - V_r^{x_{i,1}, m_1} - \ldots\ldots - V_r^{x_{i,p}, m_p},$$

then

$$\frac{\partial Q_i}{\partial V_r} = W(c_i;\ c_r, x_{i,1}, \ldots, x_{i,p}),$$

where W denotes the function used in Ex. iv. § 125 ; it follows therefore by that example, that $\frac{\partial Q_i}{\partial V_r} = \frac{\partial Q_r}{\partial V_i}$. Hence the function

$$Q_1 dV_1 + \ldots\ldots + Q_p dV_p$$

is a perfect differential ; it is in fact, by the final equation of § 196, practically equivalent to the differential of the function $\log \Theta\,(V^{x, m} - V^{x_1, m_1} - \ldots\ldots - V^{x_p, m_p})$. Thus the theory of the Riemann theta functions can be built up from the theory of algebraical integrals. Cf. Noether, *Math. Annal.* XXXVII. For the step to the expression of the function by the theta series, see Clebsch and Gordan, *Abelsche Functionen* (Leipzig, 1866), pp. 190—195.

Ex. v. Prove that if

$$(m^2, x_{i,1}, \ldots, x_{i,p}, z_1, \ldots, z_p) \equiv (c_i^2, m_1^2, \ldots, m_p^2)$$

then

$$\frac{\partial}{\partial V_i} \log \Theta\,(V^{x, m} - V^{x_1, m_1} - \ldots\ldots - V^{x_p, m_p}) = \tfrac{1}{2}\,(\Gamma_{c_i}^{x_{i,1}, z_1} + \ldots\ldots + \Gamma_{c_i}^{x_{i,p}, z_p}).$$

Ex. vi. Prove that

$$-\sum_{i=1}^{p} \mu_i(z)\,[\zeta_i(u^{x, m} - u^{x_1, m_1} - \ldots\ldots - u^{x_p, m_p}) - \zeta_i(u^{a, m} - u^{x_1, m_1} - \ldots\ldots - u^{x_p, m_p})]$$
$$= F_z^{x, a} - \psi(x, a;\ z, x_1, \ldots, x_p).$$

Ex. vii. If

$$T(x, a;\ x_1, \ldots, x_p) = [\psi(x, a;\ z, x_1, \ldots, x_p) - F_z^{x, a}]_{z=x},$$

prove that

$$\log \vartheta\,(u^{x, m} - u^{x_1, m_1} - \ldots\ldots - u^{x_p, m_p})$$
$$= A + A_1 u_1^{x, a} + \ldots\ldots + A_p u_p^{x, a} + \int^x dx\, T(x, a;\ x_1, \ldots, x_p),$$

where $A, A_1, \ldots, A_p$ are independent of x.

Ex. viii. Prove that

$$-\sum_{r=1}^{p} \mu_r(x)\,\wp_{i,r}(u^{x, m} - u^{x_1, m_1} - \ldots\ldots - u^{x_p, m_p}) = \sum_{r=1}^{p} \bar{\nu}_{r,i} D_x D_{x_r} R_{x_r, c}^{x, a},$$

where a, c are arbitrary places and the notation is as in § 193.

CHAPTER XI.

THE HYPERELLIPTIC CASE OF RIEMANN'S THETA FUNCTIONS.

199. WE have seen (Chap. V.) that the hyperelliptic case* is a special one, characterised by the existence of a rational function of the second order. In virtue of this circumstance we are able to associate the theory with a simple algebraical relation, which we may take to be of the form

$$y^2 = 4(x-a_1)\dots(x-a_p)(x-c_1)\dots(x-c_{p+1}).$$

We have seen moreover (Chap. X. § 185) that in the hyperelliptic case, when p is greater than 2, there are always even theta functions which vanish for zero values of the argument. We may expect, therefore, that the investigation of the relations connecting the Riemann theta functions with the algebraical functions will be comparatively simple, and furnish interesting suggestions for the general case. It is also the fact that the grouping of the characteristics of the theta functions, upon which much of the ultimate theory of these functions depends, has been built up directly from the hyperelliptic case.

It must be understood that the present chapter is mainly intended to illustrate the general theory. For fuller information the reader is referred to the papers quoted in the chapter, and to the subsequent chapters of the present volume.

* For the subject-matter of this chapter, beside the memoirs of Rosenhain, Göpel, and Weierstrass, referred to in § 173, Chap. X., which deal with the hyperelliptic case, and general memoirs on the theta functions, the reader may consult, Prym, *Zur Theorie der Functionen in einer zweiblättrigen Fläche* (Zürich, 1866); Prym, *Neue Theorie der ultraellip. Funct.* (zweite Aus., Berlin, 1885); Schottky, *Abriss einer Theorie der Abel. Functionen von drei Variabeln* (Leipzig, 1880), pp. 147—162; Neumann, *Vorles. über Riem. Theorie* (Leipzig, 1884); Thomae, *Sammlung von Formeln welche bei Anwendung der.. Rosenhain'schen Functionen gebraucht werden* (Halle, 1876); Brioschi, *Ann. d. Mat.* t. x. (1880), and t. xiv. (1886); Thomae, *Crelle*, LXXI. (1870), p. 201; Krause, *Die Transformation der hyperellip. Funct. erster Ordnung* (Leipzig, 1886); Forsyth, "Memoir on the theta functions," *Phil. Trans.*, 1882; Forsyth, "On Abel's theorem," *Phil. Trans.*, 1883; Cayley, "Memoir on the.. theta functions," *Phil. Trans.*, 1880, and *Crelle*, Bd. 83, 84, 85, 87, 88; Bolza, *Göttinger Nachrichten* 1894, p. 268. The addition equation is considered in a dissertation by Hancock, Berlin, 1894 (Bernstein). For further references see the later chapters of this volume which deal with theta functions.

200. Throughout this chapter we suppose the relative positions of the branch places and period loops to be as in the annexed figure (4), the branch place a being at infinity.

Fig. 4.

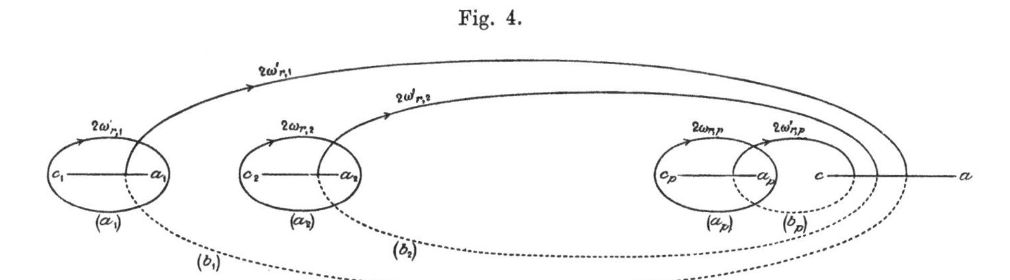

In the general case, in considering the zeros of the function $\vartheta\,(u^{x,\,m}-e)$, we were led to associate with the place m, other p places $m_1, \ldots, m_p$, such that $\vartheta\,(u^{x,\,m})$ has $m_1, \ldots, m_p$ for its zeros (Chap. X. § 179). In this case we shall always take m at the branch place a, that is at infinity. It can be shewn that if b, b' denote any two of the branch places, the p integrals $u_1^{b,\,b'}, \ldots, u_p^{b,\,b'}$ are the p simultaneous constituents of a half-period, so that

$$u_r^{b,\,b'} = m_1\omega_{r,\,1} + \ldots\ldots + m_p\omega_{r,\,p} + m_1'\omega'_{r,\,1} + \ldots\ldots + m_p'\omega'_{r,\,p}, \quad (r = 1, 2, \ldots, p),$$

wherein $m_1, \ldots, m_p, m_1', \ldots, m_p'$ are integers, independent of r; this fact we shall often denote by putting $u^{b,\,b'} = \tfrac{1}{2}\Omega$. It can further be shewn that if, b remaining any branch place, b' is taken to be each of the other $2p+1$ branch places in turn, the $2p+1$ half-periods, $u^{b,\,b'}$, thus obtained, consist of p odd half-periods, and $p+1$ even half-periods. Thus if the branch places, b', for which $u^{b,\,b'}$ is an odd half-period be denoted by $b_1, \ldots, b_p$, we have, necessarily, $\vartheta\,(u^{b,\,b_1}) = 0, \ldots, \vartheta\,(u^{b,\,b_p}) = 0$, and we may take, for the places $m, m_1, \ldots, m_p$, the places $b, b_1, \ldots, b_p$. In particular it can be shewn that, when for b the branch place a is taken, and the branch places are situated as in the figure (4), each of $u^{a,\,a_1}, \ldots, u^{a,\,a_p}$ is an odd half-period. We have therefore the statement, which is here fundamental, *the function* $\vartheta(u^{x,\,a} - u^{x_1,\,a_1} - \ldots - u^{x_p,\,a_p})$ *has the places* $x_1, \ldots, x_p$ *as its zeros. It is assumed that the function* $\vartheta\,(u^{x,\,a})$ *does not vanish identically. This assumption will be seen to be justified.*

For our present purpose it is sufficient to prove (i) that each of the integrals $u^{b,\,b'}$ is a half-period, (ii) that each of the integrals $u^{a,\,a_1}, \ldots, u^{a,\,a_p}$ is an odd half-period. In regard to (i) the general statement is as follows: Let the period loops of the Riemann surface be projected on to the plane upon which the Riemann surface is constructed, forming such a network as that represented in the figure (4); denote the projection of the loop (a_r) by (A_r), and that of (b_r) by (B_r), and suppose (A_r), (B_r) affected with arrow heads, as in the figure, whereby to define the left-hand side, and the right-hand side; finally let a continuous curve be drawn on the plane of projection, starting from the projection of the branch place b' and ending in the projection of the branch place b; then if this curve cross the loop (A_r) m_r times from right to left, so that m_r is either $+1$ or -1, or 0, and cross the loop (B_r) m_r' times from right to left, we have

$$u_r^{b, b'} = m_1\omega_{r, 1} + \ldots\ldots + m_p\omega_{r, p} + m_1'\omega'_{r, 1} + \ldots\ldots + m_p'\omega'_{r, p}.$$

Thus, for instance, in accordance with this statement we should have $u_r^{a_1, c_1} = -\omega'_{r, 1}$, and $u_r^{c_2, a_1} = \omega_{r, 1} - \omega_{r, 2}$, and it will be sufficient to prove the first of these results; the general proof is exactly similar. Now we can pass from c_1 to a_1, on the Riemann surface, by a curve lying in the upper

Fig. 5.

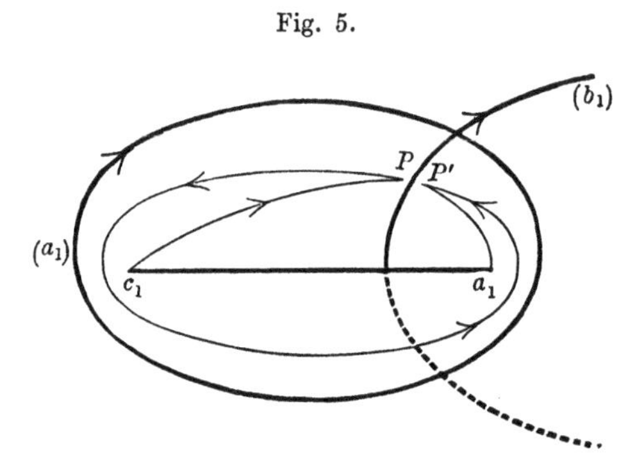

sheet which goes first to a point P on the left-hand side of the loop (b_1), and thence, following a course coinciding roughly with the right-hand side of the loop (a_1), goes to the point P', opposite to P on the right-hand side of (b_1), and thence, from P', goes to a_1. Thus we have

$$u_r^{a_1, c_1} = u_r^{P, c_1} - 2\omega'_{r, 1} + u_r^{a_1, P'}.$$

On the other hand we can pass from c_1 to a_1 by a path lying entirely in the lower sheet, and consisting of two portions, from c_1 to P, and from P' to a_1, lying just below the paths from c_1 to P and from P' to a_1, which are in the upper sheet. Thus we have a result which we may write in the form

$$u_r^{a_1, c_1} = (u_r^{P, c_1})' + (u_r^{a_1, P'})'.$$

But, in fact, as the integral $u_r^{x, a}$ is of the form $\int \frac{(x, 1)_{p-1}}{y} dx$, and y has different signs in the two sheets, we have

$$(u_r^{P, c_1})' = -u_r^{P, c_1}, \text{ and } (u_r^{a_1, P'})' = -u_r^{a_1, P'}.$$

Therefore, by addition of the equations we have

$$u_r^{a_1, c_1} = -\omega'_{r,1},$$

which proves the statement made.

In regard now to the proof that $u^{a, a_1}, \ldots, u^{a, a_p}$ are all odd half-periods, we clearly have, in accordance with the results just obtained,

$$u_r^{a, a_i} = \omega_{r,i} - (\omega_{r,i+1} + \omega'_{r,i+1}) - \ldots\ldots - (\omega_{r,p} + \omega'_{r,p}) + (\omega'_{r,1} + \ldots\ldots + \omega'_{r,p}),$$

which is equal to

$$(\omega'_{r,1} + \omega'_{r,2} + \ldots\ldots + \omega'_{r,i}) + (\omega_{r,i} - \omega_{r,i+1} - \ldots\ldots - \omega_{r,p}),$$

and if this be written in the form

$$m_1\omega_{r,1} + \ldots\ldots + m_p\omega_{r,p} + m_1'\omega'_{r,1} + \ldots\ldots + m'_p\omega'_{r,p}$$

we obviously have $m_1 m_1' + \ldots\ldots + m_p m_p' = 1$.

Ex. i. We have stated that if b be any branch place there are p other branch places $b_1, b_2, \ldots, b_p$, such that $u^{b, b_1}, u^{b, b_2}, \ldots, u^{b, b_p}$ are odd half-periods, and that, if b' be any branch place other than $b, b_1, \ldots, b_p$, $u^{b, b'}$ is an even half-period. Verify this statement in case $p=2$, by calculating all the fifteen, $=\frac{1}{2}6.5$, integrals of the form $u^{b, b'}$, and prove that when b is in turn taken at a, c, c_1, c_2, a_1, a_2 the corresponding pairs b_1, b_2 are respectively

$$(a_1, a_2), (c_1, c_2), (c_2, c), (c_1, c), (a_2, a), (a_1, a).$$

Prove also that

$$u_r^{c, a} + u_r^{c_1, a_1} + u_r^{c_2, a_2} = 0.$$

Ex. ii. The reader will find it an advantage at this stage to calculate some of the results of the second and fifth columns in the tables given below (§ 204).

201. Consider now the $2p+1$ half-periods $u^{b, a}$ wherein b is any of the branch places other than a. From these we can form $\binom{2p+1}{2}$ half-periods, of the form $u^{b, a} + u^{b', a}$, wherein b, b' are any two different branch places, other than a, and $\binom{2p+1}{3}$ half-periods of the form $u^{b, a} + u^{b', a} + u^{b'', a}$, where b, b', b'' are any three different branch places other than a, and so on, and finally we can form $\binom{2p+1}{p}$ half-periods by adding any p of the half-periods $u^{b, a}$. The number

$$\binom{2p+1}{1} + \binom{2p+1}{2} + \ldots\ldots + \binom{2p+1}{p}$$

is equal to $-1 + \frac{1}{2}[(x+1)^{2p+1}]_{x=0}$, or to $2^{2p} - 1$, and therefore equal to the whole number of existent half-periods of which no two differ by a period, with the exclusion of the identically zero half-period; we may say that this number is equal to the number of incongruent half-periods, omitting the identically zero half-period.

And in fact the $2^{2p}-1$ half-periods thus obtained are themselves incongruent. For otherwise we should have congruences of the form

$$u^{b_1, a} + u^{b_2, a} + \dots\dots + u^{b_r, a} \equiv u^{b_1', a} + u^{b_2', a} + \dots\dots + u^{b_s', a},$$

wherein any integral $u^{b_\kappa, a}$ that occurs on both sides of the congruence may be omitted. Since every one of these integrals is a half-period, and therefore $u^{b_\kappa, a} \equiv -u^{b_\kappa, a}$, we may put this congruence in the form

$$u^{b_1, a} + u^{b_2, a} + \dots\dots + u^{b_m, a} \equiv 0,$$

and here, since we are only considering the half-periods formed by sums of p, or less, different periods, m cannot be greater than $2p$. Now this congruence is equivalent with the statement that there exists a rational function having a for an m-fold pole and having $b_1, \dots, b_m$ for zeros of the first order (Chap. VIII. § 158). Since a is at infinity, such a function can be expressed in the form (Chap. V. § 56)

$$(x, 1)_r + y\,(x, 1)_s,$$

and the number of its zeros is the greater of the integers $2r$, $2p+1+s$. Thus the function under consideration would necessarily be expressible in the form $(x, 1)_r$. But such a function, if zero at a branch place, would be zero to the second order. Thus no such function exists.

On the other hand the rational function y is zero to the first order at each of the branch places $a_1, \dots, a_p, c_1, \dots, c_p, c$, and is infinite at a to the $(2p+1)$th order; hence we have the congruence

$$u^{a_1, a} + \dots\dots + u^{a_p, a} + u^{c_1, a} + \dots\dots + u^{c_p, a} + u^{c, a} \equiv 0.$$

202. With the half-period of which one element is expressed by

$$m_1 \omega_{r,1} + \dots\dots + m_p \omega_{r,p} + m_1' \omega'_{r,1} + \dots\dots + m_p' \omega'_{r,p},$$

we may associate the symbol

$$\begin{pmatrix} k_1', & k_2', & \dots, & k_p' \\ k_1, & k_2, & \dots, & k_p \end{pmatrix},$$

wherein k_s, equal to 0 or 1, is the remainder when m_s is divided by 2. The sum of two or more such symbols is then to be formed by adding the $2p$ elements separately, and replacing the sum by the remainder on division by 2. Thus for instance, when $p=2$, we should write $\binom{01}{11}+\binom{11}{01}=\binom{10}{10}$. If we call this symbol the characteristic-symbol, we have therefore proved, in the previous article, that *each of the* $2^{2p}-1$ *possible characteristic-symbols other than that one which has all its elements zero can be obtained as the sum of not more than* p *chosen from* $2p+1$ *fundamental characteristic-symbols, these* $2p+1$ *fundamental characteristic-symbols having as their sum the symbol of which all the elements are zero. In the method here adopted* p *of the fundamental symbols are associated with odd half-periods* (namely those given by $u^{a, a_1}, \ldots, u^{a, a_p}$), *and the other* $p+1$ *with even half-periods.* It is manifest that this theorem for characteristic-symbols, though derived by consideration of the hyperelliptic case, is true for all cases*. We may denote the fundamental symbols which correspond to the odd half-periods by the numbers $1, 3, 5, \ldots, 2p-1$, and those which correspond to the even half-periods by the numbers $0, 2, 4, 6, \ldots, 2p$, reserving the number $2p+1$ to represent the symbol of which all the elements are zero. Then a symbol which is formed by adding k of the fundamental symbols may be represented by placing their representative numbers in sequence.

Thus for instance, for $p=2$, Weierstrass has represented the symbols

$$\binom{10}{11}\ \binom{01}{01}\ \binom{00}{11}\ \binom{10}{01}\ \binom{01}{00}\ \binom{00}{00}$$

respectively by the numbers

$$1 \quad 3 \quad 0 \quad 2 \quad 4 \quad 5;$$

and, accordingly, represented the symbol $\binom{10}{10}$, which is equal to $\binom{00}{11}+\binom{10}{01}$, by the compound number 02. The $\binom{5}{2}=10$ combinations of the symbols 1, 3, 0, 2, 4 in pairs, represent the $2^{2p}-6$ symbols other than those here written. Further illustration is afforded by the table below (§ 204).

In case $p=3$, there will be seven fundamental symbols which may be represented by the numbers 0, 1, 2, 3, 4, 5, 6. All other symbols are represented either by a combination of two of these, or by a combination of three of them.

It may be mentioned that the fact that, for $p=3$, all the symbols are thus representable by seven fundamental symbols is in direct correlation with the fact that a plane quartic is determined when seven proper double tangents are given.

* The theorem is attributed to Weierstrass (Stahl, *Crelle*, LXXXVIII. pp. 119, 120). A further proof, and an extension of the theorem, are given in a subsequent chapter.

203. If in the half-period $\frac{1}{2}\Omega_{m, m'}$, of which an element is given by

$$\tfrac{1}{2}\Omega_{m, m'} = m_1\omega_{r, 1} + \ldots\ldots + m_p\omega_{r, p} + m_1'\omega'_{r, 1} + \ldots\ldots + m_p'\omega'_{r, p},$$

we write $\frac{1}{2}m_s = M_s + \frac{1}{2}k_s$, $\frac{1}{2}m_s' = M_s' + \frac{1}{2}k_s'$, where M_s, M_s' denote integers, and each of k_s, k_s' is either 0 or 1, we have (cf. the formulæ § 190, Chap. X.)

$$\vartheta(u + \tfrac{1}{2}\Omega_{m, m'}) = \vartheta(u\,;\ M + \tfrac{1}{2}k,\ M' + \tfrac{1}{2}k')\,e^{\lambda},$$

where

$$\lambda = [2\eta(M + \tfrac{1}{2}k) + 2\eta'(M' + \tfrac{1}{2}k')]\,[u + \omega(M + \tfrac{1}{2}k) + \omega'(M' + \tfrac{1}{2}k')] - \pi i(M + \tfrac{1}{2}k)(M' + \tfrac{1}{2}k'),$$

and therefore

$$\vartheta(u\,;\ \tfrac{1}{2}k,\ \tfrac{1}{2}k') = e^{-\lambda - \pi i Mk'}\,\vartheta(u + \tfrac{1}{2}\Omega_{m, m'}).$$

The function represented by either side of this equation will sometimes be represented by $\vartheta(u \mid \frac{1}{2}\Omega_{m, m'})$; or if $\frac{1}{2}\Omega_{m, m'} = u^{b_1, a} + u^{b_2, a} + \ldots\ldots + u^{b_s, a}$, the function will sometimes be represented by $\vartheta(u \mid u^{b_1, a} + \ldots\ldots + u^{b_s, a})$, or by $\vartheta_{b_1 b_2 \ldots b_s}(u)$.

We have proved in the last chapter (§§ 184, 185) that every odd half-period can be represented in the form

$$\tfrac{1}{2}\Omega \equiv u^{m_p, m} - u^{n_1, m_1} - \ldots\ldots - u^{n_{p-1}, m_{p-1}},$$

and, when there are no even theta functions which vanish for zero values of the argument, that every even half-period can be represented in the form

$$\tfrac{1}{2}\Omega' \equiv u^{m_1', m_1} + \ldots\ldots + u^{m_p', m_p}\,;$$

in the hyperelliptic case every odd half-period can be represented in the form

$$\tfrac{1}{2}\Omega \equiv u^{a_p, a} - u^{n_1, a_1} - \ldots\ldots - u^{n_{p-1}, a_{p-1}},$$

and every even half-period $\frac{1}{2}\Omega'$, for which $\vartheta(\frac{1}{2}\Omega')$ does not vanish, can be represented in the form

$$\tfrac{1}{2}\Omega' \equiv u^{b_1, a_1} + \ldots\ldots + u^{b_p, a_p},$$

and (§ 182, Chap. X.) the zeros of the function $\vartheta(u^{x, z} \mid \frac{1}{2}\Omega)$ consist of the place z and the places $n_1, \ldots, n_p$, while the zeros of the function $\vartheta(u^{x, a} \mid \frac{1}{2}\Omega')$ are the places $b_1, \ldots, b_p$. In case $p = 2$ there are no even theta functions vanishing for zero values of the argument; in case $p = 3$ there is one such function (§ 185, Chap. X.), and the corresponding even half-period $\frac{1}{2}\Omega''$ is such that we can put

$$\tfrac{1}{2}\Omega'' \equiv u^{a_3, a} - u^{x_1, a_1} - u^{x_2, a_2},$$

wherein x_1 is an arbitrary place and x_2 is the place conjugate to x_1. Since then $u^{x_2, a_2} \equiv -u^{x_1, a_2}$, this equation gives

$$\tfrac{1}{2}\Omega'' \equiv u^{a_3, a} - u^{a_2, a_1};$$

now, as in § 200, we easily find

$$u_r^{a_3, a} = -(\omega_{r,3} + \omega'_{r,1} + \omega'_{r,2} + \omega'_{r,3}), \quad u^{a_2, a_1} = \omega_{r,1} - \omega_{r,2} - \omega'_{r,2},$$

and therefore

$$\tfrac{1}{2}\Omega'' \equiv -\omega_{r,1} + \omega_{r,2} - \omega_{r,3} - (\omega'_{r,1} + \omega'_{r,3}).$$

Thus the even theta function which vanishes for zero values of the argument is that associated with the characteristic symbol $\begin{pmatrix}101\\111\end{pmatrix}$.

In the same way for $p=4$, the 10 even theta functions which vanish for zero values of the argument are (§ 185, Chap. X.) associated with even half-periods given by

$$\tfrac{1}{2}\Omega'' = u^{a_4, a} - u^{b, a_3} - u^{a_1, a_2},$$

where b is in turn each of the ten branch places.

204. The following table gives the results for $p=2$. The reader is recommended to verify the second and fifth columns. The set of p equations represented by the equation $(\tfrac{1}{2}\Omega)_r = m_1\omega_{r,1} + m_2\omega_{r,2} + m_1'\omega'_{r,1} + m_2'\omega'_{r,2}$ is denoted by putting $\tfrac{1}{2}\Omega = \tfrac{1}{2}\begin{pmatrix}m_1' m_2'\\ m_1 m_2\end{pmatrix}$.

I. *Six odd theta functions in the case $p=2$.*

Function	We have	Weierstrass's number associated with this symbol		Putting the corresponding half-period $\equiv u^{a_2, a} - u^{n_1, a_1}$, we have for n_1 respectively
$\vartheta_{aa_1}(u)$	$u^{a, a_1} = \tfrac{1}{2}\begin{pmatrix}10\\10\end{pmatrix}$	02	(1)	a_2
$\vartheta_{aa_2}(u)$	$u^{a, a_2} = \tfrac{1}{2}\begin{pmatrix}11\\01\end{pmatrix}$	24	(3)	a_1
$\vartheta_{a_1a_2}(u)$	$u^{a_1, a_2} = \tfrac{1}{2}\begin{pmatrix}01\\-11\end{pmatrix}$	04	(13)	a
$\vartheta_{c_1c_2}(u)$	$u^{c_1, c_2} = \tfrac{1}{2}\begin{pmatrix}10\\-11\end{pmatrix}$	1	(24)	c
$\vartheta_{cc_1}(u)$	$u^{c_1, c} = \tfrac{1}{2}\begin{pmatrix}11\\-10\end{pmatrix}$	13	(02)	c_2
$\vartheta_{cc_2}(u)$	$u^{c_2, c} = \tfrac{1}{2}\begin{pmatrix}0\ \ 1\\0-1\end{pmatrix}$	3	(04)	c_1

II. Ten even theta functions in the case $p=2$.

Function	We have	Weierstrass's number associated with this symbol		Putting the corresponding half-period $\equiv u^{b_1, a_1} + u^{b_2, a_2}$, we have for b_1, b_2
$\vartheta(u)$	$\frac{1}{2}\begin{pmatrix}00\\00\end{pmatrix}$	5		a_1, a_2
$\vartheta_{ac}(u)$	$u^{a, c} = \frac{1}{2}\begin{pmatrix}11\\00\end{pmatrix}$	23	(0)	c_1, c_2
$\vartheta_{ac_1}(u)$	$u^{a, c_1} = \frac{1}{2}\begin{pmatrix}00\\10\end{pmatrix}$	12	(2)	c, c_2
$\vartheta_{ac_2}(u)$	$u^{a, c_2} = \frac{1}{2}\begin{pmatrix}10\\01\end{pmatrix}$	2	(4)	c, c_1
$\vartheta_{a_1c_1}(u)$	$u^{c_1, a_1} = \frac{1}{2}\begin{pmatrix}10\\00\end{pmatrix}$	01	(12)	a_2, c_1
$\vartheta_{a_1c_2}(u)$	$u^{c_2, a_1} = \frac{1}{2}\begin{pmatrix}0 & 0\\1 & -1\end{pmatrix}$	0	(14)	a_2, c_2
$\vartheta_{a_2c_1}(u)$	$u^{c_1, a_2} = \frac{1}{2}\begin{pmatrix}1 & 1\\-1 & 1\end{pmatrix}$	14	(23)	a_1, c_1
$\vartheta_{a_2c_2}(u)$	$u^{c_2, a_2} = \frac{1}{2}\begin{pmatrix}01\\00\end{pmatrix}$	4	(34)	a_1, c_2
$\vartheta_{ca_2}(u)$	$u^{c, a_2} = \frac{1}{2}\begin{pmatrix}00\\01\end{pmatrix}$	34	(03)	a_1, c
$\vartheta_{ca_1}(u)$	$u^{c, a_1} = \frac{1}{2}\begin{pmatrix}0 & -1\\1 & 0\end{pmatrix}$	03	(01)	a_2, c

The numbers in brackets in the fourth column might be employed instead of the Weierstrass numbers; they are based on the branch places according to the correspondence

$$\begin{matrix}1 & 3 & 0 & 2 & 4\\ a_1 & a_2 & c & c_1 & c_2.\end{matrix}$$

But the Weierstrass notation is now so fully established that it will be employed here whenever any such notation is used.

It should be noticed that the letter notation for an odd function consists always of two a's or two c's; the letter notation for an even function contains one a and one c.

The expression of the half-period associated with any function as a sum of not more than two of the integrals $u^{b, a}$, which has been described in § 202, is of course immediately indicated by the letter notation employed for the functions.

Ex. Prove that if $\mathbf{a} = \frac{1}{2}\begin{pmatrix}01\\11\end{pmatrix}$

$$u^{a, a_1} + \mathbf{a} \equiv u^{a, a_2} \qquad u^{c_1, c_2} + \mathbf{a} \equiv u^{a, c} \qquad u^{c, c_1} + \mathbf{a} \equiv u^{a, c_2}$$

$$u^{a, a_2} + \mathbf{a} \equiv u^{a, a_1} \qquad u^{c, c_2} + \mathbf{a} \equiv u^{a, c_1}.$$

These equations effect a correspondence between five of the odd functions and the branch places.

205. Next we give the corresponding results for $p=3$. Each half-period can be formed as a sum of not more than 3 of the seven integrals $u^{b,\,a}$ (§ 202); the proper integrals are indicated by the suffix letters employed to represent the function. We may also associate the branch places with the numbers 0, 1, 2, 3, 4, 5, 6, say, in accordance with the scheme

$$\begin{array}{ccccccc} a_1, & a_2, & a_3, & c, & c_1, & c_2, & c_3 \\ 1, & 3, & 5, & 0, & 2, & 4, & 6\,; \end{array}$$

then the functions $\vartheta_1(u)$, $\vartheta_3(u)$, $\vartheta_5(u)$ will be odd, and the functions $\vartheta_0(u)$, $\vartheta_2(u)$, $\vartheta_4(u)$, $\vartheta_6(u)$ will be even; and every function will have a suffix formed of 1 or 2 or 3 of these numbers. There is however another way in which the 64 characteristics can be associated with the combinations of seven numbers, and one which has the advantage that all the seven numbers and their 21 combinations of two are associated with odd functions, while all the even functions except that in which the associated half-period is zero are associated with their 35 combinations of three. It will be seen in a later chapter in how many ways such a scheme is possible. One way is that in which the numbers

$$1, \quad 2, \quad 3, \quad 4, \quad 5, \quad 6, \quad 7$$

are associated respectively with the half-periods given by

$$u^{a_1,\,a},\quad u^{a_2,\,a},\quad u^{a_3,\,a},\quad u^{c,\,a}+u^{c_2,\,a}+u^{c_3,\,a},\quad u^{c,\,a}+u^{c_3,\,a}+u^{c_1,\,a},\quad u^{c,\,a}+u^{c_1,\,a}+u^{c_2,\,a},$$
$$u^{c_1,\,a}+u^{c_2,\,a}+u^{c_3,\,a}.$$

By § 201 the sum of these integrals is $\equiv 0$. The numbers thus obtained are given in the second column. Further every odd half-period can be represented by a sum $u^{a_3,\,a}-u^{n_1,\,a_1}-u^{n_2,\,a_2}$, and all the even half-periods except one as a sum $u^{b_1,\,a_1}+u^{b_2,\,a_2}+u^{b_3,\,a_3}$; the positions of n_1, n_2 or of b_1, b_2, b_3 are given in the fourth column.

I. 28 odd theta functions for $p=3$.

			$n_1, n_2=$
$\vartheta_{a_1}(u)$	1	$u^{a_1,\,a}\equiv\frac{1}{2}\begin{pmatrix}100\\100\end{pmatrix}$	$a_2,\ a_3$
$\vartheta_{a_2}(u)$	2	$u^{a_2,\,a}\equiv\frac{1}{2}\begin{pmatrix}110\\010\end{pmatrix}$	$a_3,\ a_1$
$\vartheta_{a_3}(u)$	3	$u^{a_3,\,a}\equiv\frac{1}{2}\begin{pmatrix}111\\001\end{pmatrix}$	$a_1,\ a_2$
$\vartheta_{a_1a_2}(u)$	12	$u^{a_1,\,a}+u^{a_2,\,a}\equiv\frac{1}{2}\begin{pmatrix}010\\110\end{pmatrix}$	$a\ ,\ a_3$
$\vartheta_{a_1a_3}(u)$	13	$u^{a_1,\,a}+u^{a_3,\,a}\equiv\frac{1}{2}\begin{pmatrix}011\\101\end{pmatrix}$	$a\ ,\ a_2$
$\vartheta_{a_2a_3}(u)$	23	$u^{a_2,\,a}+u^{a_3,\,a}\equiv\frac{1}{2}\begin{pmatrix}001\\011\end{pmatrix}$	$a\ ,\ a_1$
$\vartheta_{cc_1}(u)$	74	$u^{c_1,\,a}+u^{c,\,a}\equiv\frac{1}{2}\begin{pmatrix}111\\100\end{pmatrix}$	$c_2,\ c_3$
$\vartheta_{cc_2}(u)$	75	$u^{c_2,\,a}+u^{c,\,a}\equiv\frac{1}{2}\begin{pmatrix}011\\010\end{pmatrix}$	$c_3,\ c_1$

Table I. (continued.)

			$n_1, n_2 =$
$\vartheta_{cc_3}(u)$	76	$u^{c_3, a} + u^{c, a} \equiv \frac{1}{2}\begin{pmatrix}001\\001\end{pmatrix}$	c_1, c_2
$\vartheta_{c_2c_3}(u)$	56	$u^{c_2, a} + u^{c_3, a} \equiv \frac{1}{2}\begin{pmatrix}010\\011\end{pmatrix}$	c, c_1
$\vartheta_{c_3c_1}(u)$	64	$u^{c_1, a} + u^{c_3, a} \equiv \frac{1}{2}\begin{pmatrix}110\\101\end{pmatrix}$	c, c_2
$\vartheta_{c_1c_2}(u)$	45	$u^{c_1, a} + u^{c_2, a} \equiv \frac{1}{2}\begin{pmatrix}100\\110\end{pmatrix}$	c, c_3
$\vartheta_{ca_1a_2}(u)$	37	$u^{c, a} + u^{a_1, a} + u^{a_2, a} \equiv \frac{1}{2}\begin{pmatrix}101\\110\end{pmatrix}$	c, a_3
$\vartheta_{ca_1a_3}(u)$	27	$u^{c, a} + u^{a_1, a} + u^{a_3, a} \equiv \frac{1}{2}\begin{pmatrix}100\\101\end{pmatrix}$	c, a_2
$\vartheta_{ca_2a_3}(u)$	17	$u^{c, a} + u^{a_2, a} + u^{a_3, a} \equiv \frac{1}{2}\begin{pmatrix}110\\011\end{pmatrix}$	c, a_1
$\vartheta_{c_1a_2a_3}(u)$	14	$u^{c_1, a} + u^{a_2, a} + u^{a_3, a} \equiv \frac{1}{2}\begin{pmatrix}001\\111\end{pmatrix}$	c_1, a_1
$\vartheta_{c_1a_3a_1}(u)$	24	$u^{c_1, a} + u^{a_3, a} + u^{a_1, a} \equiv \frac{1}{2}\begin{pmatrix}011\\001\end{pmatrix}$	c_1, a_2
$\vartheta_{c_1a_1a_2}(u)$	34	$u^{c_1, a} + u^{a_1, a} + u^{a_2, a} \equiv \frac{1}{2}\begin{pmatrix}010\\010\end{pmatrix}$	c_1, a_3
$\vartheta_{c_2a_2a_3}(u)$	15	$u^{c_2, a} + u^{a_2, a} + u^{a_3, a} \equiv \frac{1}{2}\begin{pmatrix}101\\001\end{pmatrix}$	c_2, a_1
$\vartheta_{c_2a_3a_1}(u)$	25	$u^{c_2, a} + u^{a_3, a} + u^{a_1, a} \equiv \frac{1}{2}\begin{pmatrix}111\\111\end{pmatrix}$	c_2, a_2
$\vartheta_{c_2a_1a_2}(u)$	35	$u^{c_2, a} + u^{a_1, a} + u^{a_2, a} \equiv \frac{1}{2}\begin{pmatrix}110\\100\end{pmatrix}$	c_2, a_3
$\vartheta_{c_3a_2a_3}(u)$	16	$u^{c_3, a} + u^{a_2, a} + u^{a_3, a} \equiv \frac{1}{2}\begin{pmatrix}111\\010\end{pmatrix}$	c_3, a_1
$\vartheta_{c_3a_3a_1}(u)$	26	$u^{c_3, a} + u^{a_3, a} + u^{a_1, a} \equiv \frac{1}{2}\begin{pmatrix}101\\100\end{pmatrix}$	c_3, a_2
$\vartheta_{c_3a_1a_2}(u)$	36	$u^{c_3, a} + u^{a_1, a} + u^{a_2, a} \equiv \frac{1}{2}\begin{pmatrix}100\\111\end{pmatrix}$	c_3, a_3
$\vartheta_{cc_2c_3}(u)$	4	$u^{c, a} + u^{c_2, a} + u^{c_3, a} \equiv \frac{1}{2}\begin{pmatrix}101\\011\end{pmatrix}$	a, c_1
$\vartheta_{cc_3c_1}(u)$	5	$u^{c, a} + u^{c_3, a} + u^{c_1, a} \equiv \frac{1}{2}\begin{pmatrix}001\\101\end{pmatrix}$	a, c_2
$\vartheta_{cc_1c_2}(u)$	6	$u^{c, a} + u^{c_1, a} + u^{c_2, a} \equiv \frac{1}{2}\begin{pmatrix}011\\110\end{pmatrix}$	a, c_3
$\vartheta_{c_1c_2c_3}(u)$	7	$u^{c_1, a} + u^{c_2, a} + u^{c_3, a} \equiv \frac{1}{2}\begin{pmatrix}010\\111\end{pmatrix}$	a, c

II. 36 *even characteristics for* $p=3$.

			b_1	b_2	b_3
$\vartheta(u)$		$\frac{1}{2}\begin{pmatrix}000\\000\end{pmatrix}$	a_1	a_2	a_3
$\vartheta_{a_1a_2a_3}(u)$	123	$u^{a_1,a}+u^{a_2,a}+u^{a_3,a}\equiv\frac{1}{2}\begin{pmatrix}101\\111\end{pmatrix}$	*$a,$	$x,$	$\bar{x}$
$\vartheta_c(u)$	456	$u^{c,a} \equiv\frac{1}{2}\begin{pmatrix}111\\000\end{pmatrix}$	c_1	c_2	c_3
$\vartheta_{c_1}(u)$	567	$u^{c_1,a} \equiv\frac{1}{2}\begin{pmatrix}000\\100\end{pmatrix}$	c	c_2	c_3
$\vartheta_{c_2}(u)$	647	$u^{c_2,a} \equiv\frac{1}{2}\begin{pmatrix}100\\010\end{pmatrix}$	c	c_3	c_1
$\vartheta_{c_3}(u)$	457	$u^{c_3,a} \equiv\frac{1}{2}\begin{pmatrix}110\\001\end{pmatrix}$	c	c_1	c_2
$\vartheta_{ca_1}(u)$	237	$u^{c,a}+u^{a_1,a} \equiv\frac{1}{2}\begin{pmatrix}011\\100\end{pmatrix}$	c	a_2	a_3
$\vartheta_{ca_2}(u)$	317	$u^{c,a}+u^{a_2,a} \equiv\frac{1}{2}\begin{pmatrix}001\\010\end{pmatrix}$	c	a_3	a_1
$\vartheta_{ca_3}(u)$	127	$u^{c,a}+u^{a_3,a} \equiv\frac{1}{2}\begin{pmatrix}000\\001\end{pmatrix}$	c	a_1	a_2
$\vartheta_{c_1a_1}(u)$	234	$u^{c_1,a}+u^{a_1,a} \equiv\frac{1}{2}\begin{pmatrix}100\\000\end{pmatrix}$	c_1	a_2	a_3
$\vartheta_{c_1a_2}(u)$	314	$u^{c_1,a}+u^{a_2,a} \equiv\frac{1}{2}\begin{pmatrix}110\\110\end{pmatrix}$	c_1	a_3	a_1
$\vartheta_{c_1a_3}(u)$	124	$u^{c_1,a}+u^{a_3,a} \equiv\frac{1}{2}\begin{pmatrix}111\\101\end{pmatrix}$	c_1	a_1	a_2
$\vartheta_{c_2a_1}(u)$	235	$u^{c_2,a}+u^{a_1,a} \equiv\frac{1}{2}\begin{pmatrix}000\\110\end{pmatrix}$	c_2	a_2	a_3
$\vartheta_{c_2a_2}(u)$	315	$u^{c_2,a}+u^{a_2,a} \equiv\frac{1}{2}\begin{pmatrix}010\\000\end{pmatrix}$	c_2	a_3	a_1
$\vartheta_{c_2a_3}(u)$	125	$u^{c_2,a}+u^{a_3,a} \equiv\frac{1}{2}\begin{pmatrix}011\\011\end{pmatrix}$	c_2	a_1	a_2
$\vartheta_{c_3a_1}(u)$	236	$u^{c_3,a}+u^{a_1,a} \equiv\frac{1}{2}\begin{pmatrix}010\\101\end{pmatrix}$	c_3	a_2	a_3
$\vartheta_{c_3a_2}(u)$	316	$u^{c_3,a}+u^{a_2,a} \equiv\frac{1}{2}\begin{pmatrix}000\\011\end{pmatrix}$	c_3	a_3	a_1
$\vartheta_{c_3a_3}(u)$	126	$u^{c_3,a}+u^{a_3,a} \equiv\frac{1}{2}\begin{pmatrix}001\\000\end{pmatrix}$	c_3	a_1	a_2
$\vartheta_{a_1c_2c_3}(u)$	156	$u^{a_1,a}+u^{c_2,a}+u^{c_3,a} \equiv\frac{1}{2}\begin{pmatrix}110\\111\end{pmatrix}$	a_1	c_1	c
$\vartheta_{a_1c_3c_1}(u)$	164	$u^{a_1,a}+u^{c_3,a}+u^{c_1,a} \equiv\frac{1}{2}\begin{pmatrix}010\\001\end{pmatrix}$	a_1	c_2	c
$\vartheta_{a_1c_1c_2}(u)$	145	$u^{a_1,a}+u^{c_1,a}+u^{c_2,a} \equiv\frac{1}{2}\begin{pmatrix}000\\010\end{pmatrix}$	a_1	c_3	c
$\vartheta_{a_1cc_1}(u)$	147	$u^{a_1,a}+u^{c,a}+u^{c_1,a} \equiv\frac{1}{2}\begin{pmatrix}011\\000\end{pmatrix}$	a_1	c_2	c_3

Table II. (continued).

			b_1 b_2 b_3
$\vartheta_{a_1cc_2}(u)$	157	$u^{a_1,a} + u^{c,a} + u^{c_2,a} \equiv \frac{1}{2}\begin{pmatrix}111\\110\end{pmatrix}$	a_1 c_3 c_1
$\vartheta_{a_1cc_3}(u)$	167	$u^{a_1,a} + u^{c,a} + u^{c_3,a} \equiv \frac{1}{2}\begin{pmatrix}101\\101\end{pmatrix}$	a_1 c_1 c_2
$\vartheta_{a_2c_2c_3}(u)$	256	$u^{a_2,a} + u^{c_2,a} + u^{c_3,a} \equiv \frac{1}{2}\begin{pmatrix}100\\001\end{pmatrix}$	a_2 c_1 c
$\vartheta_{a_2c_3c_1}(u)$	264	$u^{a_2,a} + u^{c_3,a} + u^{c_1,a} \equiv \frac{1}{2}\begin{pmatrix}000\\111\end{pmatrix}$	a_2 c_2 c
$\vartheta_{a_2c_1c_2}(u)$	245	$u^{a_2,a} + u^{c_1,a} + u^{c_2,a} \equiv \frac{1}{2}\begin{pmatrix}010\\100\end{pmatrix}$	a_2 c_3 c
$\vartheta_{a_2cc_1}(u)$	247	$u^{a_2,a} + u^{c,a} + u^{c_1,a} \equiv \frac{1}{2}\begin{pmatrix}001\\110\end{pmatrix}$	a_2 c_2 c_3
$\vartheta_{a_2cc_2}(u)$	257	$u^{a_2,a} + u^{c,a} + u^{c_2,a} \equiv \frac{1}{2}\begin{pmatrix}101\\000\end{pmatrix}$	a_2 c_3 c_1
$\vartheta_{a_2cc_3}(u)$	267	$u^{a_2,a} + u^{c,a} + u^{c_3,a} \equiv \frac{1}{2}\begin{pmatrix}111\\011\end{pmatrix}$	a_2 c_1 c_2
$\vartheta_{a_3c_2c_3}(u)$	356	$u^{a_3,a} + u^{c_2,a} + u^{c_3,a} \equiv \frac{1}{2}\begin{pmatrix}101\\010\end{pmatrix}$	a_3 c_1 c
$\vartheta_{a_3c_3c_1}(u)$	364	$u^{a_3,a} + u^{c_3,a} + u^{c_1,a} \equiv \frac{1}{2}\begin{pmatrix}001\\100\end{pmatrix}$	a_3 c_2 c
$\vartheta_{a_3c_1c_2}(u)$	345	$u^{a_3,a} + u^{c_1,a} + u^{c_2,a} \equiv \frac{1}{2}\begin{pmatrix}011\\111\end{pmatrix}$	a_3 c_3 c
$\vartheta_{a_3cc_1}(u$	347	$u^{a_3,a} + u^{c,a} + u^{c_1,a} \equiv \frac{1}{2}\begin{pmatrix}000\\101\end{pmatrix}$	a_3 c_2 c_3
$\vartheta_{a_3cc_2}(u)$	357	$u^{a_3,a} + u^{c,a} + u^{c_2,a} \equiv \frac{1}{2}\begin{pmatrix}100\\011\end{pmatrix}$	a_3 c_3 c_1
$\vartheta_{a_3cc_3}(u)$	367	$u^{a_3,a} + u^{c,a} + u^{c_3,a} \equiv \frac{1}{2}\begin{pmatrix}110\\000\end{pmatrix}$	a_3 c_1 c_2

It is to be noticed that every odd theta function is associated with either (i) any single one of a_1, a_2, a_3 or (ii) any pair of a_1, a_2, a_3 or any pair of c, c_1, c_2, c_3, or (iii) a triplet consisting of one of c, c_1, c_2, c_3 and two of a_1, a_2, a_3 or consisting of three from c, c_1, c_2, c_3. This may be stated by saying that *odd* suffixes are of one of the forms a, a^2, c^2, a^2c, c^3. Similarly an *even* suffix is of one of the forms c, ac, ac^2, a^3.

In the tables just given the fundamental characteristic-symbols, denoted by the numbers 1, 2, 3, 4, 5, 6, 7, are those associated with sums of integrals which may be denoted by

$$a_1,\ a_2,\ a_3,\ cc_2c_3,\ cc_3c_1,\ cc_1c_2,\ c_1c_2c_3.$$

We can equally well choose seven fundamental odd characteristic-symbols, associated with the integrals denoted by any one of the following sets:

$$\begin{array}{ccccccc}
c\,c_1, & c\,c_2, & c\,c_3, & c\,a_2a_3, & c\,a_3a_1, & c\,a_1a_2, & c_1c_2c_3 \\
c_1c\,, & c_1\,c_2, & c_1\,c_3, & c_1a_2a_3, & c_1a_3a_1, & c_1a_1a_2, & c\,c_2c_3 \\
c_2c\,, & c_2\,c_1, & c_2\,c_3, & c_2a_2a_3, & c_2a_3a_1, & c_2a_1a_2, & c\,c_3c_1 \\
c_3c\,, & c_3\,c_1, & c_3\,c_2, & c_3a_2a_3, & c_3a_3a_1, & c_3a_1a_2, & c\,c_1c_2 \\
a_1, & a_1a_2, & a_1a_3, & c_1a_2a_3, & c_2a_2a_3, & c_3a_2a_3, & ca_2a_3 \\
a_2, & a_2a_3, & a_2a_1, & c_1a_3a_1, & c_2a_3a_1, & c_3a_3a_1, & ca_3a_1 \\
a_3, & a_3a_1, & a_3\,a_2, & c_1a_1a_2, & c_2a_1a_2, & c_3a_1a_2, & ca_1a_2
\end{array}$$

The general theorem is—it is possible, corresponding to every even characteristic ϵ, to determine, in 8 ways, 7 odd characteristics $\alpha, \beta, \gamma, \kappa, \lambda, \mu, \nu$, such that the combinations

$$\alpha,\ \beta,\ \gamma,\ \kappa,\ \lambda,\ \mu,\ \nu,\ \epsilon\alpha\beta,\ \epsilon\alpha\kappa,\ \epsilon\lambda\mu$$

constitute all the 28 odd characteristics, and the combinations

$$\epsilon,\ \alpha\beta\gamma,\ \alpha\kappa\lambda,\ \beta\gamma\kappa$$

constitute all the 36 even characteristics. In the cases above $\epsilon = 0$. The proof is given in a subsequent chapter.

206. Consider now what are the zeros of the functions

$$\vartheta(u),\ \vartheta(u \,|\, u^{b_1, a} + \dots\dots + u^{b_k, a}),$$

where $b_1, \dots, b_k$ denote any k of the branch places other than a $(k \ngtr p)$, and u is given by

$$u_r = u_r^{x_1, a_1} + \dots\dots + u_r^{x_p, a_p}, \qquad (r = 1, 2, \dots, p),$$

the functions being regarded as functions of x_1.

The zeros of $\vartheta(u)$ are the places $z_1, \dots, z_p$ determined by the congruence

$$u^{x_1, a_1} + \dots\dots + u^{x_p, a_p} \equiv u^{x_1, a} - u^{z_1, a_1} - \dots\dots - u^{z_p, a_p},$$

or, by*

$$u^{z_1, a} + u^{z_2, \bar{x}_2} + \dots\dots + u^{z_p, \bar{x}_p} \equiv 0.$$

Provided the places $a, \bar{x}_2, \dots, \bar{x}_p$ be not the zeros of a ϕ-polynomial, that is, provided none of the places $x_2, \dots, x_p$ be at a, and there be no coincidence expressible in the form $x_i = \bar{x}_j$, the places $z_1, z_2, \dots, z_p$ cannot be coresidual with any p other places (Chap. VI. § 98, and Chap. III.) and therefore (Chap. VIII. § 158) this congruence can only be satisfied when the places $z_1, \dots, z_p$ are the places

$$a,\ \bar{x}_2,\ \bar{x}_3,\ \dots,\ \bar{x}_p;$$

these are then the zeros of $\vartheta(u)$, regarded as a function of x_1.

* The two places for which x has the same value, and y has the same value with opposite signs, are frequently denoted by x and $\bar{x}$.

The zeros of $\vartheta(u|u^{b_1,a}+\ldots\ldots+u^{b_k,a})$ are to be determined by the congruence

$$u^{x_1,a_1}+\ldots\ldots+u^{x_p,a_p}+u^{b_1,a}+\ldots\ldots+u^{b_k,a}\equiv u^{x_1,a}-u^{z_1,a_1}-\ldots\ldots-u^{z_p,a_p},$$

or, by

$$u^{z_1,b_1}+u^{z_2,\bar{x}_2}+\ldots\ldots+u^{z_p,\bar{x}_p}+u^{b_2,a}+\ldots\ldots+u^{b_k,a}\equiv 0,$$

which we may write also

$$(z_1, z_2, \ldots, z_p, a^{k-1})\equiv(b_1, \ldots, b_k, \bar{x}_2, \ldots, \bar{x}_p);$$

in particular the zeros of $\vartheta(u|u^{b,a})$ are the places $b, \bar{x}_2, \ldots, \bar{x}_p$.

207. Now, in fact, if the sum of the characteristics $q_1, \ldots, q_n$ differs from the sum of the characteristics $r_1, \ldots, r_n$ by a characteristic consisting wholly of integers, n being an integer not less than 2, then the quotient

$$f(u)=\frac{\vartheta(u;\, q_1)\,\vartheta(u;\, q_2)\ldots\ldots\vartheta(u;\, q_n)}{\vartheta(u;\, r_1)\,\vartheta(u;\, r_2)\ldots\ldots\vartheta(u;\, r_n)}$$

is a periodic function of u.

For, by the formula (§ 190, Chap. X.)

$$\vartheta(u+\Omega_m;\, q)=e^{\lambda_m(u)+2\pi i(mq'-m'q)}\,\vartheta(u;\, q),$$

where m denotes a row of integers, we have

$$\frac{f(u+\Omega_m)}{f(u)}=e^{2\pi i[m(\Sigma q'-\Sigma r')-m'(\Sigma q-\Sigma r)]},$$

and if $\Sigma q'-\Sigma r'$, $\Sigma q-\Sigma r$, each consist of a row of integers the right-hand side is equal to 1.

Hence, when the arguments, u, are as in § 206, the function $f(u)$ is a rational function of the places $x_1, \ldots, x_p$.

208. It follows therefore that the function

$$\frac{\vartheta^2(u|u^{b,a})}{\vartheta^2(u)}$$

is a rational function of the places $x_1, \ldots, x_p$. By what has been proved in regard to the zeros of the numerator and denominator it has, as a function of x_1, the zero b, of the second order, and is infinite at a, that is, at infinity, also to the second order. Thus it is equal to $M(b-x_1)$, where M does not depend on x_1. As the function is symmetrical in $x_1, x_2, \ldots, x_p$, it must therefore be equal to $K(b-x_1)\ldots(b-x_p)$, where K is an absolute constant. Therefore the function

$$\sqrt{(b-x_1)(b-x_2)\ldots(b-x_p)}=\frac{1}{\sqrt{K}}\,\frac{\vartheta(u|u^{b,a})}{\vartheta(u)}$$

may be interpreted as a single valued function of the places $x_1, \ldots, x_p$, on the Riemann surface, dissected by the $2p$ period loops. The values of the function on the two sides of any period loop have a quotient which is constant along that loop, and equal to ± 1.

The function has been considered by Rosenhain*, Weierstrass†, Riemann‡ and Brioschi§. We shall denote the quotient $\vartheta(u|u^{b,a})/\vartheta(u)$ by $q_b(u)$. There are $2p+1$ such functions, according to the position of b. Of these $q_{a_1}(u), \ldots, q_{a_p}(u)$ are odd functions, and $q_c(u), q_{c_1}(u), \ldots, q_{c_p}(u)$ are even functions. The functions are clearly generalisations of the functions $\sqrt{x} = \text{sn}\, u$, $\sqrt{1-x} = \text{cn}\, u$, $\sqrt{1-k^2x} = \text{dn}\, u$, obtained from the consideration of the integral

$$u = \int_0^x \frac{dx}{\sqrt{4x(1-x)(1-k^2x)}}.$$

209. Consider next the function

$$F = \frac{\vartheta(u|u^{b_1,a} + \ldots\ldots + u^{b_k,a})\,\vartheta^{k-1}(u)}{\vartheta(u|u^{b_1,a}) \ldots\ldots \vartheta(u|u^{b_k,a})},$$

wherein $b_1, \ldots, b_k$ are any k branch places other than a. We consider only the cases $k < p+1$. By what has been shewn, the function is rational in x_1, and if $z_1, \ldots, z_p$ denote the zeros of $\vartheta(u|u^{b_1,a} + \ldots\ldots + {}^{b_k,a})$ the zeros of the numerator, as here written, consist of the places

$$z_1, \ldots, z_p, a^{k-1}, \bar{x}_2^{k-1}, \ldots, \bar{x}_p^{k-1}$$

and the zeros of the denominator consist of the places

$$b_1, b_2, \ldots, b_k, \bar{x}_2^k, \ldots, \bar{x}_p^k.$$

Thus the rational function of x_1 has for zeros the places $z_1, \ldots, z_p, a^{k-1}$, and, for poles, the places $b_1, \ldots, b_k, \bar{x}_2, \ldots, \bar{x}_p$. It has already been otherwise shewn that these two sets of $p+k-1$ places are coresidual. Now any rational function, of the place x, which has these poles, can (Chap. VI. § 89) be written in the form

$$\frac{uy + v(x-b_1)\ldots(x-b_k)}{(x-b_1)\ldots(x-b_k)(x-x_2)\ldots(x-x_p)},$$

wherein u, v are suitable integral polynomials in x, so chosen that the numerator vanishes at the places $x_2, \ldots, x_p$. The denominator, as here written, vanishes to the second order at each of $b_1, \ldots, b_k$, and also vanishes at the places $x_2, \bar{x}_2, \ldots, x_p, \bar{x}_p$.

Let λ, μ be the highest powers of x respectively in u and v. Then, in order that this function may be zero at the place a, that is, at infinity, to the order $k-1$, it is necessary that the greater of the two numbers

$$2\lambda + 2p + 1 - 2(p+k-1),\ 2\mu + 2k - 2(p+k-1)$$

* *Mémoires par divers savants*, t. XI. (1851), pp. 361—468.

† By Weierstrass the function is multiplied by a certain constant factor and denoted by $al(u)$.

‡ In the general form enunciated, as a quotient of products of theta functions, *Werke* (Leipzig, 1876), p. 134 (§ 27).

§ *Annali di Mat.* t. X. (1880), t. XIV. (1886).

(wherein $2(p+k-1)$ is the order of infinity, at infinity, of the denominator) should be equal to $-(k-1)$. Since one of these numbers is odd and the other even, they cannot be both equal to $-(k-1)$. Further in order that the ratios of the $\lambda+\mu+2$ coefficients in u, v may be capable of being chosen so that the numerator vanishes in the places $x_2, \ldots, x_p$, it is necessary that $\lambda+\mu+1$ should not be less than $p-1$. And, since a rational function is entirely determined when its poles and all but p of its zeros are given, these conditions should entirely determine the function.

In fact we easily find from these conditions that the case $2\lambda+2p+1>2(\mu+k)$ can only occur when k is even, and then $\lambda=\frac{1}{2}k-1$, $\mu=p-1-\frac{1}{2}k$, and that the case $2\lambda+2p+1<2\mu+2k$ can only occur when k is odd, and then $\lambda=\frac{1}{2}(k-3)$, $\mu=p-\frac{1}{2}(k+1)$. In both cases $\lambda+\mu+2=p$.

By introducing the condition that the polynomial $uy+v(x-b_1)\ldots(x-b_k)$ should vanish in the places $x_2, \ldots, x_p$ we are able, save for a factor not depending on x, y, to express this polynomial as the product of $(x-b_1)\ldots(x-b_k)$ by a determinant of p rows and columns of which, for $r>1$, the rth row is formed with the elements

$$\frac{x_r^{\lambda}y_r}{\phi(x_r)}, \frac{x_r^{\lambda-1}y_r}{\phi(x_r)}, \ldots, \frac{y_r}{\phi(x_r)}, x_r^{\mu}, x_r^{\mu-1}, \ldots, 1,$$

wherein $\phi(x)$ denotes $(x-b_1)\ldots(x-b_k)$, the first row being of the same form with the omission of the suffixes.

Therefore, noticing that F is symmetrical in the places $x_1, \ldots, x_p$, we infer, denoting the product of the differences of $x_1, \ldots, x_p$ by $\Delta(x_1, \ldots, x_p)$, that

$$\frac{\vartheta(u|u^{b_1, a}+\ldots\ldots+u^{b_k, a})\,\vartheta^{k-1}(u)}{\vartheta(u|u^{b_1, a})\ldots\ldots\vartheta(u|u^{b_k, a})}=C\frac{\left|\frac{x_r^{\lambda}y_r}{\phi(x_r)}, \frac{x_r^{\lambda-1}y_r}{\phi(x_r)}, \ldots, \frac{y_r}{\phi(x_r)}, x_r^{\mu}, x_r^{\mu-1}, \ldots, 1\right|}{\Delta(x_1, \ldots, x_p)},$$

where C is an absolute constant, and the numerator denotes a determinant in which the first, second, ... rows contain, respectively, $x_1, x_2, \ldots$; and here

when k is even, $\quad \lambda=\frac{1}{2}k-1, \quad \mu=p-1-\frac{1}{2}k$

and when k is odd, $\quad \lambda=\frac{1}{2}(k-3), \quad \mu=p-\frac{1}{2}(k+1)$.

210. By means of the algebraic expression which we have already obtained for the quotients $\vartheta(u|u^{b, a})/\vartheta(u)$, we are now able to deduce an algebraic expression for the quotients

$$\vartheta(u|u^{b_1, a}+\ldots\ldots+u^{b_k, a})/\vartheta(u);$$

since it has already been shewn that by taking k in turn equal to $1, 2, \ldots, p$, and taking all possible sets $b_1, \ldots, b_k$ corresponding to any value of k, the half-periods represented by $u^{b_1, a}+\ldots\ldots+u^{b_k, a}$ consist of all possible half-periods except that one which is identically zero, it follows that, *in the* *hyperelliptic case, if u denote $u^{x_1, a_1} + \ldots\ldots + u^{x_p, a_p}$, and q denote in turn all possible half-integer characteristics except the identically zero characteristic, all the $2^{2p}-1$ ratios $\vartheta(u\,;\,q)/\vartheta(u)$ can be expressed algebraically in terms of $x_1, \ldots, x_p$, by the formulae which have been given.*

The simplest case is when $k=2$; then we have $\lambda=0$, $\mu=p-2$, and

$$\frac{\vartheta(u|u^{b_1, a}+u^{b_2, a})\,\vartheta(u)}{\vartheta(u|u^{b_1, a})\,\vartheta(u|u^{b_2, a})}=C\sum_{r=1}^{p}\frac{y_r}{(x_r-b_1)(x_r-b_2)}\frac{1}{R'(x_r)},$$

where $R(x)=(x-x_1)(x-x_2)\ldots(x-x_p)$, and C is an absolute constant. Denoting the quotient $\vartheta(u|u^{b_1, a}+u^{b_2, a})/\vartheta(u)$ by q_{b_1, b_2}, we have

$$q_{b_1, b_2}=A_{1, 2}q_{b_1}q_{b_2}\sum_{r=1}^{p}\frac{y_r}{(x_r-b_1)(x_r-b_2)}\frac{1}{R'(x_r)},$$

where $A_{1,2}$ is an absolute constant; and there are $p(2p+1)$ such functions.

When $k=3$, we have $\lambda=0$, $\mu=p-2$, and, if q_{b_1, b_2, b_3} denote the quotient $\vartheta(u|u^{b_1, a}+u^{b_2, a}+u^{b_3, a})/\vartheta(u)$, we obtain

$$q_{b_1, b_2, b_3}=B_{1, 2, 3}q_{b_1}q_{b_2}q_{b_3}\sum_{r=1}^{p}\frac{y_r}{(x_r-b_1)(x_r-b_2)(x_r-b_3)}\frac{1}{R'(x_r)},$$

where $B_{1, 2, 3}$ is an absolute constant. It is however clear that

$$\frac{q_{b_1, b_2}}{A_{12}q_{b_1}q_{b_2}}-\frac{q_{b_1, b_3}}{A_{13}q_{b_1}q_{b_3}}=(b_2-b_3)\,\frac{q_{b_1, b_2, b_3}}{B_{123}q_{b_1}q_{b_2}q_{b_3}},$$

so that the functions with three suffixes are immediately expressible by those with one and those with two suffixes.

More generally, the $2^{2p}-1$ quotients $\vartheta(u\,;\,q)/\vartheta(u)$, depending only on the p places $x_1, \ldots, x_p$, must be connected by $2^{2p}-p-1$ algebraical relations; and since (Chap. IX.) any argument can be expressed in the form $u^{x_1, a_1}+\ldots\ldots+u^{x_p, a_p}$, it follows that these may be regarded as relations connecting Riemann theta functions of arbitrary argument. This statement is true whether the surface be hyperelliptic or not.

Of such relations one simple and obvious one for the hyperelliptic case under consideration may be mentioned at once. We clearly have

$$\frac{q_{b_2, b_3}}{A_{23}q_{b_2}q_{b_3}}(b_2-b_3)+\frac{q_{b_3, b_1}}{A_{31}q_{b_3}q_{b_1}}(b_3-b_1)+\frac{q_{b_1, b_2}}{A_{12}q_{b_1}q_{b_2}}(b_1-b_2)=0,$$

and therefore

$$\frac{b_2-b_3}{A_{23}}\vartheta_{b_2b_3}(u)\,\vartheta_{b_1}(u)+\frac{b_3-b_1}{A_{31}}\vartheta_{b_3b_1}(u)\,\vartheta_{b_2}(u)+\frac{b_1-b_2}{A_{12}}\vartheta_{b_1b_2}(u)\,\vartheta_{b_3}(u)=0.$$

It is proved below (§ 213) that $A^2_{23} : A^2_{31} : A^2_{12}=(b_2\sim b_3) : (b_3\sim b_1) : (b_1\sim b_2)$.

Other relations will be given for the cases $p=2$, $p=3$. A set of relations connecting the q's of single and double suffixes, for any value of p, is given by Weierstrass (*Crelle* LII. *Werke* I. p. 336).

211. *Ex.* i. Prove that the rational function having the places $\bar{x}_1, \ldots, \bar{x}_p, a$, as poles, and the branch place b as one zero, is given by

$$Z=(b-x)\ldots\ldots(b-x_p)\sum_0^p \frac{y_r}{x_r-b}\frac{1}{R'(x_r)},$$

where $R(\xi)=(\xi-x)(\xi-x_1)\ldots\ldots(\xi-x_p)$, and, in the summation, x_0, y_0 are to be replaced by x, y.

Prove that if u denote the argument

$$u=u^{x, a}+u^{x_1, a_1}+\ldots\ldots+u^{x_p, a_p},$$

then

$$\frac{\vartheta^2(u|u^{b, a})}{\vartheta^2(u)}=A\frac{Z^2}{(b-x)(b-x_1)\ldots\ldots(b-x_p)},$$

where A is an absolute constant.

Prove for example, in the elliptic case, with Weierstrass's notation, that

$$\frac{\sigma_i(u+v)}{\sigma(u+v)}=\sqrt{\wp(u+v)-e_i}=\tfrac{1}{2}\sqrt{(\wp u-e_i)(\wp v-e_i)}\left(\frac{\wp' u}{\wp u-e_i}-\frac{\wp' v}{\wp v-e_i}\right)\frac{1}{\wp u-\wp v}.$$

Ex. ii. If Z_r denote the function Z when the branch place b_r is put in place of b, and $R(b_r)$ denote $(b_r-x)(b_r-x_1)\ldots\ldots(b_r-x_p)$, and we put

$$\Phi=\frac{\vartheta(u|u^{b_1, a}+\ldots\ldots+u^{b_k, a})\,\vartheta^{k-1}(u)}{\vartheta(u|u^{b_1, a})\ldots\ldots\vartheta(u|u^{b_k, a})},$$

prove that

$$\Phi Z_1\ldots\ldots Z_k=BR(b_1)\ldots\ldots R(b_k)\left|\frac{y_r}{\phi(x_r)}x_r^{\lambda}, \frac{y_r}{\phi(x_r)}x_r^{\lambda-1}, \ldots, \frac{y_r}{\phi(x_r)}, x_r^{\mu}, x_r^{\mu-1}, \ldots, 1\right| \div\Delta(x, x_1, \ldots, x_p),$$

where B is an absolute constant, $\Delta(x, x_1, \ldots, x_p)$ denotes the product of all the differences of the $(p+1)$ quantities $x, x_1, \ldots, x_p$, $\phi(x_r)=(x_r-b_1)\ldots\ldots(x_r-b_k)$, and the determinant is one of $p+1$ rows and columns in which, in the first row, x_0, y_0 are to be replaced by x, y.

Prove that, when k is even, $\lambda=\frac{1}{2}(k-2)$, $\mu=p-\frac{1}{2}k$, and, when k is odd, $\lambda=\frac{1}{2}(k-1)$, $\mu=p-\frac{1}{2}(k+1)$.

Ex. iii. Hence prove that the function $\dfrac{\vartheta(u|u^{b_1, a}+\ldots\ldots+u^{b_k, a})}{\vartheta(u)}$ is a constant multiple of

$$\frac{\sqrt{R(b_1)\ldots\ldots R(b_k)}\left|\frac{y_r}{\phi(x_r)}x_r^{\lambda}, \frac{y_r}{\phi(x_r)}x_r^{\lambda-1}, \ldots, \frac{y_r}{\phi(x_r)}, x_r^{\mu}, x_r^{\mu-1}, \ldots, x_r, 1\right|}{\Delta(x, x_1, \ldots, x_p)}.$$

This formula is true when $k=1$.

Ex. iv. A particular case is when $k=2$. Then the function $\vartheta(u|u^{b_1, a}+u^{b_2, a})/\vartheta(u)$ is a constant multiple of

$$\sqrt{(b_1-x)(b_1-x_1)\ldots\ldots(b_1-x_p)}\sqrt{(b_2-x)(b_2-x_1)\ldots\ldots(b_2-x_p)}\sum_0^p\frac{y_r}{(x_r-b_1)(x_r-b_2)}\frac{1}{R'(x_r)},$$

wherein $R(\xi)=(\xi-x)(\xi-x_1)\ldots\ldots(\xi-x_p)$.

Ex. v. Verify that the formula of Ex. iii. includes the formulae of the text (§ 210); shew that when x is put at infinity the values of λ, μ in the determinant of § 209 are properly obtained.

Ex. vi. Verify that the expression $\psi(x, b; a, \bar{x}_1, \ldots, \bar{x}_p)$ of § 130, Chap. VII., takes the form given for the function Z of Ex. i. when a is the place infinity.

Ex. vii. If $f(x)$ denote the polynomial

$$\lambda+\lambda_1 x+\lambda_2 x^2+\ldots\ldots+\lambda_{2p+2}x^{2p+2},$$

prove that any rational integral polynomial, $F(x, z)$, which is symmetric in the two variables x, z and of order $p+1$ in each of them, and satisfies the conditions

$$F(z, z)=2f(z), \quad \left[\frac{d}{dx}F(x, z)\right]_{x=z}=\frac{d}{dz}f(z),$$

is of the form

$$F(x, z)=f(x, z)+(x-z)^2\psi(x, z),$$

where (cf. p. 195), with $\lambda_0=\lambda$, $\lambda_{2p+3}=0$,

$$f(x, z)=\sum_{i=0}^{p+1} x^i z^i\{2\lambda_{2i}+\lambda_{2i+1}(x+z)\},$$

and $\psi(x, z)$ is an integral polynomial, symmetric in x, z, of order $p-1$ in each*.

In case $p=2$, and $f(x)=(x-a_1)(x-a_2)(x-c)(x-c_1)(x-c_2)$, prove that a form of $F(x, z)$ is given by

$$F(x, z)=(x-a_1)(x-a_2)(z-c)(z-c_1)(z-c_2)+(z-a_1)(z-a_2)(x-c)(x-c_1)(x-c_2).$$

Ex. viii. If for purposes of operation we introduce homogeneous variables and write

$$f(x)=\lambda x_2^{2p+2}+\lambda_1 x_2^{2p+1}x_1+\ldots\ldots+\lambda_{2p+1}x_2 x_1^{2p+1}+\lambda_{2p+2}x_1^{2p+2},$$

prove that a form of $F(x, z)$ is given by

$$\bar{f}(x, z)=\frac{\lfloor p}{\lfloor 2p+1}\left(x_1\frac{\partial}{\partial z_1}+x_2\frac{\partial}{\partial z_2}\right)^{p+1}f(z),$$

where, after differentiation, x_1, x_2, z_1, z_2 are to be replaced by x, 1, z, 1 respectively.

This is the same as that which in the ordinary symbolical notation for binary forms is denoted by $\bar{f}(x, z)=2a_x^{p+1}a_z^{p+1}$, $f(x)$ being a_x^{2p+2}.

Ex. ix. Using the form of Ex. viii. for $F(x, z)$, prove that if e_1, e_2, x, x_1, $\ldots$, x_p be any values of x, we have

$$\sum_{r=0}^{p}\frac{f(x_r)}{[G'(x_r)]^2}+\Sigma\Sigma\frac{\bar{f}(x_r, x_s)}{G'(x_r)\,G'(x_s)}=\frac{f(e_1)}{[G'(e_1)]^2}+\frac{f(e_2)}{[G'(e_2)]^2}+\frac{\bar{f}(e_1, e_2)}{G'(e_1)\,G'(e_2)},$$

where $G(\xi)=(\xi-e_1)(\xi-e_2)(\xi-x)(\xi-x_1)\ldots\ldots(\xi-x_p)$, and the double summation on the left refers to every one of the $\frac{1}{2}p(p+1)$ pairs of quantities chosen from x, x_1, $\ldots$, x_p.

Ex. x. Hence it follows†, when $y^2=f(x)$, $y_r{}^2=f(x_r)$, etc., and $R(\xi)=(\xi-x)(\xi-x_1)\ldots(\xi-x_p)$, that

$$R(e_1)\,R(e_2)\left[\sum_0^p\frac{y_r}{(e_1-x_r)(e_2-x_r)\,R'(x_r)}\right]^2-\frac{f(e_1)\,R(e_2)}{(e_1-e_2)^2R(e_1)}-\frac{f(e_2)\,R(e_1)}{(e_1-e_2)^2R(e_2)}+\frac{\bar{f}(e_1, e_2)}{(e_1-e_2)^2}$$

is equal to

$$R(e_1)\,R(e_2)\,\Sigma\Sigma\frac{2y_r y_s-f(x_r, x_s)}{G'(x_r)\,G'(x_s)},$$

* It follows that the hyperelliptic canonical integral of the third kind obtained on page 195 can be changed into the most general canonical integral, $R_{z,c}^{x,a}$ (p. 194), in which the matrix a has any value, by taking, instead of $f(x, z)$, a suitable polynomial $F(x, z)$ satisfying the conditions of Ex. vii.

† The result of this Example is given by Bolza, *Götting. Nachrichten*, 1894, p. 268.

where the summation refers to every pair from the $p+1$ quantities $x, x_1, \ldots, x_p$, and $\bar{f}(x, z)$ denotes the special value of $F(x, z)$ obtained in Ex. viii.

Ex. xi. It follows therefore by Ex. iv. that when b_1, b_2 are any branch places of the surface associated with the equation $y^2-f(x)=0$, there exists an equation of the form

$$C\frac{\vartheta^2(u|u^{b_1, a}+u^{b_2, a})}{\vartheta^2(u)}=R(b_1)R(b_2)\Sigma\Sigma\frac{2y_ry_s-f(x_r, x_s)}{G'(x_r)G'(x_s)}-\frac{\bar{f}(b_1, b_2)}{(b_1-b_2)^2},$$

where C is an absolute constant, $G(\xi)=(\xi-b_1)(\xi-b_2)(\xi-x)(\xi-x_1)\ldots\ldots(\xi-x_p)$, and $u=u^{x, a}+u^{x_1, a_1}+\ldots\ldots+u^{x_p, a_p}$. The importance of this result will appear below.

212. The formulae of §§ 208, 210 furnish a solution of the inversion problem expressed by the p equations

$$u_i^{x_1, a_1}+\ldots\ldots+u_i^{x_p, a_p}\equiv u_i; \qquad (i=1, 2, \ldots, p).$$

For instance the solution is given by the $2p+1$ equations

$$\frac{\vartheta^2(u|u^{b, a})}{\vartheta^2(u)}=A(b-x_1)(b-x_2)\ldots(b-x_p);$$

from any p of these equations $x_1, \ldots, x_p$ can be expressed as single valued functions of the arbitrary arguments $u_1, \ldots, u_p$.

And it is easy to determine the value of A^2. For let $b_1, \ldots, b_p, b_1', \ldots, b_p'$ denote the finite branch places other than b. As already remarked (§ 201) we have

$$(c, c_1, \ldots, c_p)\equiv(a, a_1, \ldots, a_p)$$

and therefore

$$(b, b_1, \ldots, b_p)\equiv(a, b_1', \ldots, b_p').$$

Now we easily find by the formulae of § 190, Chap. X. that if P be a set of $2p$ integers, $P_1, \ldots, P_p, P_1', \ldots, P_p'$,

$$\frac{\vartheta^2(u+\tfrac{1}{2}\Omega_P, \tfrac{1}{2}\Omega_P)}{\vartheta^2(u+\tfrac{1}{2}\Omega_P)}=\frac{\vartheta^2(u)}{\vartheta^2(u;\tfrac{1}{2}P)}e^{-\pi iPP'};$$

hence, if $u^{b, a}=\tfrac{1}{2}\Omega_{P, P'}$, and $u_0=u^{b_1, a}+\ldots\ldots+u^{b_p, a}$, we have, by the formula under consideration, writing $b_1, \ldots, b_p$ in place of $x_1, \ldots, x_p$, the equation

$$\frac{\vartheta^2(u_0|u^{b, a})}{\vartheta^2(u)}=A(b-b_1)\ldots(b-b_p),$$

and, writing $b_1', \ldots, b_p'$ in place of $x_1, \ldots, x_p$, we have

$$\frac{\vartheta^2(u_0+u^{b, a}|u^{b, a})}{\vartheta^2(u_0+u^{b, a})}=A(b-b_1')\ldots(b-b_p');$$

thus, by multiplication

$$e^{-\pi iPP'}=A^2(b-b_1)\ldots(b-b_p)(b-b_1')\ldots(b-b_p'),$$

and hence

$$\frac{\vartheta^2(u|u^{b,a})}{\vartheta^2(u)} = \pm \frac{(b-x_1)(b-x_2)\dots(b-x_p)}{\sqrt{e^{\pi i PP'} f'(b)}},$$

where $f(x)$ denotes $(x-a_1)\dots(x-a_p)(x-c)(x-c_1)\dots(x-c_p)$, and $e^{\pi i PP'} = \pm 1$ according as $u^{b,a}$ is an odd or even half-period.

The reader should deduce this result from the equation (§ 171, Chap. IX.)

$$V(U_1, \dots, U_p;\ \xi_1, \gamma_1) \dots\dots V(U_1, \dots, U_p;\ \xi_k, \gamma_k) = \frac{(X - Z(x_1)) \dots\dots (X - Z(x_p))}{(X - Z(a_1)) \dots\dots (X - Z(a_p))}$$

by taking Z to be the rational function of the second order, x.

When $u = u^{x,a} + u^{x_1, a_1} + \dots\dots + u^{x_p, a_p}$, we deduce (see Ex. i. § 211)

$$\frac{\vartheta^2(u|u^{b,a})}{\vartheta^2(u)} = \pm \frac{(b-x)(b-x_1)\dots\dots(b-x_p)}{4\sqrt{e^{\pi i PP'}}\, f'(b)} \left[\sum_{r=0}^{p} \frac{y_r}{x_r - b} \frac{1}{R'(x_r)}\right]^2,$$

where $R(\xi) = (\xi - x)(\xi - x_1)\dots\dots(\xi - x_p)$.

If in particular we put b in turn at the places $a_1, \dots, a_p$, write $P(x) = (x-a_1)\dots(x-a_p)$ and $Q(x) = (x-c)(x-c_1)\dots(x-c_p)$, and use the equation

$$\frac{(x-x_1)\dots(x-x_p)}{P(x)} = 1 + \sum_{1}^{p} \frac{(a_i - x_1)\dots(a_i - x_p)}{(x-a_i)P'(a_i)},$$

we can infer that $x_1, \dots, x_p$ are the roots of the equation*

$$\sum_{i-1}^{p} \epsilon_i \sqrt{-\frac{Q(a_i)}{P'(a_i)}} \frac{\vartheta^2(u|u^{a_i,a})}{a_i - x} = \vartheta^2(u),$$

where ϵ_i is ± 1 and is such that we have

$$\frac{\vartheta^2(u|u^{a_i,a})}{\vartheta^2(u)} = \epsilon_i \frac{(a_i - x_1)\dots(a_i - x_p)}{\sqrt{P'(a_i)\,Q(a_i)}}.$$

Another form of this equation for $x_1, \dots, x_p$ is given below (§ 216), where the equation determining y_i from x_i is also given.

213. We can also obtain the constant factor in the algebraic expression of the function $\vartheta(u|u^{b_1,a} + u^{b_2,a})\,\vartheta(u) \div \vartheta(u|u^{b_1,a})\,\vartheta(u|u^{b_2,a})$.

Let b_1, b_2 denote any branch places, and choose $z_1, \dots, z_p$ so that

$$u^{x_1,a_1} + \dots\dots + u^{x_p,a_p} + u^{b_1,a} \equiv u^{z_1,a_1} + \dots\dots + u^{z_p,a_p};$$

then $z_1, \dots, z_p, a$ are the zeros of a rational function which vanishes in $x_1, \dots, x_p, b_1$. Such a function can be expressed in the form

$$\frac{y + (x-b_1)(x, 1)^{p-1}}{(x-x_1)\dots\dots(x-x_p)},$$

* Cf. Weierstrass, *Math. Werke* (Berlin, 1894), vol. I. p. 328.

where $(x, 1)^{p-1}$ is an integral polynomial in x whose coefficients are to be chosen to satisfy the p equations

$$-y_i+(x_i-b_1)(x_i, 1)^{p-1}=0, \qquad (i=1, 2, \ldots, p);$$

thus the function is

$$\frac{y}{F(x)}+(x-b_1)\sum_{i=1}^{p}\frac{y_i}{x_i-b_1}\frac{1}{(x-x_i)F'(x_i)},$$

where $F(x)=(x-x_1)\ldots(x-x_p)$; and, if the coefficient of x^{2p+1} in the equation associated with the Riemann surface be taken to be 4, we have

$$y^2-(x-b_1)^2[F(x)]^2\left[\sum_{i=1}^{p}\frac{y_i}{x_i-b_1}\frac{1}{(x-x_i)F'(x_i)}\right]^2=4(x-x_1)\ldots(x-x_p)(x-z_1)\ldots(x-z_p)(x-b_1),$$

and therefore, putting b_2 for x,

$$\frac{(b_2-z_1)\ldots\ldots(b_2-z_p)}{(b_2-x_1)\ldots\ldots(b_2-x_p)}=(b_1-b_2)\left[\tfrac{1}{2}\sum_{i=1}^{p}\frac{y_i}{(x_i-b_1)(x_i-b_2)}\frac{1}{F'(x_i)}\right]^2.$$

Now we have found, denoting $u^{x_1, a_1}+\ldots\ldots+u^{x_p, a_p}$ by u, and $u^{z_1, a_1}+\ldots\ldots+u^{z_p, a_p}$ by v, the results

$$\frac{\vartheta^2(u|u^{b_2, a})}{\vartheta^2(u)}=\pm\frac{(b_2-x_1)\ldots\ldots(b_2-x_p)}{\sqrt{e^{\pi iPP'}f'(b)}}, \qquad \frac{\vartheta^2(v|u^{b_2, a})}{\vartheta^2(v)}=\pm\frac{(b_2-z_1)\ldots\ldots(b_2-z_p)}{\sqrt{e^{\pi iPP'}f'(b)}},$$

where $u^{b_2, a}=\tfrac{1}{2}\Omega_{P, P'}$; hence we have

$$\frac{\vartheta^2(v|u^{b_2, a})\,\vartheta^2(u)}{\vartheta^2(v)\,\vartheta^2(u|u^{b_2, a})}=\pm(b_1-b_2)\left[\tfrac{1}{2}\sum_{i=1}^{p}\frac{y_i}{(x_i-b_1)(x_i-b_2)}\frac{1}{F'(x_i)}\right]^2,$$

which, by the formulae of § 190, is the same as

$$\frac{\vartheta(u|u^{b_1, a}+u^{b_2, a})\,\vartheta(u)}{\vartheta(u|u^{b_1, a})\,\vartheta(u|u^{b_2, a})}=\epsilon\sqrt{b_1-b_2}\sum_{i=1}^{p}\frac{y_i}{2(x_i-b_1)(x_i-b_2)F'(x_i)},$$

where ϵ is a certain fourth root of unity.

Thus the method of this § not only reproduces the result of § 210, but determines the constant factor.

Ex. Determine the constant factors in the formulae of §§ 208, 210, 211.

214. Beside such formulae as those so far developed, which express products of theta functions algebraically, there are formulae which express differential coefficients of theta functions algebraically; as the second differential coefficients of $\vartheta(u)$ in regard to the arguments $u_1, \ldots, u_p$ are periodic functions of these arguments, this was to be expected.

We have (§ 193, Chap. X.) obtained* the formula

$$-\zeta_i(u^{x, m}-u^{x_1, m_1}-\ldots\ldots-u^{x_p, m_p})+\zeta_i(u^{\mu, m}-u^{x_1, m_1}-\ldots\ldots-u^{x_p, m_p})$$

$$=L_i^{x, \mu}+\sum_{k=1}^{p}\tilde{\nu}_{k, i}\left[(x_k, x)-(x_k, \mu)\right]\frac{dx_k}{dt};$$

* Cf. also Thomae, *Crelle*, LXXI., XCIV.

we denote by h_r the sum of the homogeneous products of $x_1, \ldots, x_p$, r together, without repetitions, and use the abbreviation

$$\chi_{p-i}(x;\, x_1, \ldots, x_p) = x^{p-i} - h_1 x^{p-i-1} + h_2 x^{p-i-2} - \ldots\ldots + (-)^{p-i} h_{p-i};$$

further, for the p fundamental integrals $u_1^{x,\mu}, \ldots, u_p^{x,\mu}$, we take the integrals

$$\int_\mu^x \frac{dx}{y},\ \int_\mu^x \frac{x\,dx}{y},\ \ldots,\ \int_\mu^x \frac{x^{p-1}\,dx}{y};$$

then it is immediately verified that

$$\tilde{\nu}_{k,i} = y_k \frac{\chi_{p-i}(x_k;\, x_1, \ldots, x_p)}{F'(x_k)} \Big/ \frac{dx_k}{dt},$$

where $F(x)$ denotes $(x - x_1) \ldots (x - x_p)$.

Thus, if μ, ν denote the values of x and y at the place μ, we have, writing $a, a_1, \ldots, a_p$ for $m, m_1, \ldots, m_p$ (§ 200),

$$-\zeta_i(u^{x,a} - u^{x_1,a_1} - \ldots\ldots - u^{x_p,a_p}) + \zeta_i(u^{\mu,a} - u^{x_1,a_1} - \ldots\ldots - u^{x_p,a_p})$$

$$= L_i^{x,\mu} + \tfrac{1}{2}\sum_{k=1}^{p} \frac{\chi_{p-i}(x_k;\, x_1, \ldots, x_p)}{F'(x_k)} \left[\frac{y + y_k}{x_k - x} - \frac{y_k + \nu}{x_k - \mu}\right];$$

therefore, also, the function

$$\zeta_i(u^{x,a} + u^{x_1,a_1} + \ldots\ldots + u^{x_p,a_p}) + L_i^{x,\mu} - \tfrac{1}{2}\sum_{k=1}^{p} \frac{\chi_{p-i}(x_k;\, x_1, \ldots, x_p)}{F'(x_k)} \frac{y - y_k}{x - x_k}$$

is equal to

$$\zeta_i(u^{\mu,a} + u^{x_1,a_1} + \ldots\ldots + u^{x_p,a_p}) - \tfrac{1}{2}\sum_{k=1}^{p} \frac{\chi_{p-i}(x_k;\, x_1, \ldots, x_p)}{F'(x_k)} \frac{\nu - y_k}{\mu - x_k},$$

which is independent of the place x.

Now let $R(t)$ denote $(t - x)(t - x_1) \ldots (t - x_p)$, and use the abbreviation given by the equation

$$\frac{y\chi_{p-i}(x;\, x_1, \ldots, x_p)}{R'(x)} + \frac{y_1\chi_{p-i}(x_1;\, x, x_2, \ldots, x_p)}{R'(x_1)} + \ldots + \frac{y_p\chi_{p-i}(x_p;\, x, x_1, \ldots, x_{p-1})}{R'(x_p)}$$
$$= f_{p-i}(x, x_1, \ldots, x_p);$$

then also

$$\frac{y_1\chi_{p-i-1}(x_1;\, x_2, \ldots, x_p)}{F'(x_1)} + \ldots\ldots + \frac{y_p\chi_{p-i-1}(x_p;\, x_1, \ldots, x_{p-1})}{F'(x_p)} = f_{p-i-1}(x_1, \ldots, x_p).$$

Now $\qquad \chi_{p-i}(x_1;\, x, x_2, \ldots, x_p) - \chi_{p-i}(x_1;\, x_1, x_2, \ldots, x_p)$

is equal to

$$[x_1^{p-i} - x_1^{p-i-1}(x + k_1) + x_1^{p-i-2}(xk_1 + k_2) - \ldots\ldots + (-1)^{p-i} x k_{p-i-1}]$$
$$- [x_1^{p-i} - x_1^{p-i-1}(x_1 + k_1) + x_1^{p-i-2}(x_1 k_1 + k_2) - \ldots\ldots + (-1)^{p-i} x_1 k_{p-i-1}],$$

wherein k_r denotes the sum of the homogeneous products of $x_2, \ldots, x_p$, without repetitions, r together, and is therefore equal to

$$(x_1 - x)\left[x_1^{p-i-1} - x_1^{p-i-2}k_1 + \ldots\ldots + (-)^{p-i-1}k_{p-i-1}\right]$$

or to

$$(x_1 - x)\,\chi_{p-i-1}(x_1;\, x_2, \ldots, x_p).$$

Hence

$$\frac{\chi_{p-i}(x_1;\, x, x_2, \ldots, x_p)}{R'(x_1)} = \frac{\chi_{p-i}(x_1;\, x_1, x_2, \ldots, x_p) + (x_1 - x)\,\chi_{p-i-1}(x_1;\, x_2, \ldots, x_p)}{(x_1 - x)\,F'(x_1)}$$

$$= -\frac{\chi_{p-i}(x_1;\, x_1, x_2, \ldots, x_p)}{F'(x_1)}\,\frac{1}{x - x_1} + \frac{\chi_{p-i-1}(x_1;\, x_2, \ldots, x_p)}{F'(x_1)}.$$

While, also,

$$\frac{\chi_{p-i}(x;\, x_1, \ldots, x_p)}{R'(x)} = \sum_{k=1}^{p} \frac{\chi_{p-i}(x_k;\, x_1, \ldots, x_p)}{F'(x_k)}\,\frac{1}{x - x_k}.$$

Thus

$$f_{p-i}(x, x_1, \ldots, x_p) = \sum_{k=1}^{p} \frac{\chi_{p-i}(x_k;\, x_1, \ldots, x_p)}{F'(x_k)}\,\frac{y - y_k}{x - x_k} + f_{p-i-1}(x_1, \ldots, x_p).$$

Therefore the expression

$$\zeta_i(u^{x,a} + u^{x_1, a_1} + \ldots + u^{x_p, a_p}) + L_i^{x,\mu} + L_i^{x_1,\mu} + \ldots + L_i^{x_p,\mu} - \tfrac{1}{2}f_{p-i}(x, x_1, \ldots, x_p)$$

is equal to

$$\zeta_i(u^{x,a} + u^{x_1, a_1} + \ldots + u^{x_p, a_p}) + L_i^{x_1,\mu} + \ldots + L_i^{x_p,\mu} - \tfrac{1}{2}f_{p-i}(a, x_1, \ldots, x_p).$$

In this equation the left-hand side is symmetrical in $x, x_1, \ldots, x_p$, and the right-hand side does not contain x. Hence the left-hand side is a constant in regard to x, and, therefore, also in regard to $x_1, \ldots, x_p$. That is, the left-hand side is an absolute constant, depending on the place μ. Denoting this constant by $-C$ we have

$$-\zeta_i(u^{x,a} + u^{x_1, a_1} + \ldots\ldots + u^{x_p, a_p}) = L_i^{x,\mu} + L_i^{x_1,\mu} + \ldots\ldots + L_i^{x_p,\mu}$$

$$-\frac{y\chi_{p-i}(x;\, x_1, \ldots, x_p)}{2R'(x)} - \ldots\ldots - \frac{y_p\chi_{p-i}(x_p;\, x, x_1, \ldots, x_{p-1})}{2R'(x_p)} + C.$$

215. From this equation another important result can be deduced. It is clear that the function

$$-\zeta_i(u^{x,a} + u^{x_1, a_1} + \ldots\ldots + u^{x_p, a_p}) - L_i^{x_1, a_1} - \ldots\ldots - L_i^{x_p, a_p}$$

does not become infinite when x approaches the place a, that is, the place infinity. If we express the value of this function by the equation just obtained, it is immediately seen that the limit of

$$\frac{-y_k\chi_{p-i}(x_k;\, x, x_1, \ldots, x_p)}{2R'(x_k)} \text{ is } -\frac{y_k\chi_{p-i-1}(x_k;\, x_1, \ldots, x_p)}{2F'(x_k)},$$

and that the expression

$$\frac{y\chi_{p-i}(x\,;\ x_1,\ \dots,\ x_p)}{2R'(x)},$$

when expanded in powers of t by the substitutions $x=\frac{1}{t^2},\ y=\frac{2}{t^{2p+1}}(1+At^2+\dots)$, where A is a certain constant, contains *only odd powers of* t. Hence the limit when t is zero of the terms of the expansion of this expression other than those containing negative powers of t, is absolute zero, and therefore, *does not depend on the places* $x_1,\ \dots,\ x_p$. The terms of the expansion which contain negative powers of t are cancelled by terms arising from the integral $L_i^{x,\,\mu}$. Since this integral does not contain $x_1,\ \dots,\ x_p$ we infer that the difference

$$L_i^{x,\,\mu}-\frac{y\chi_{p-i}(x\,;\ x_1,\ \dots,\ x_p)}{2R'(x)}$$

has a limit independent of $x_1,\ \dots,\ x_p$, and, therefore, that

$$-\zeta_i(u^{x_1,\,a_1}+\dots+u^{x_p,\,a_p})=L_i^{x_1,\,a_1}+\dots+L_i^{x_p,\,a_p}-\sum_{k=1}^{p}\frac{y_k\chi_{p-i-1}(x_k;\ x_1,\ \dots,\ x_p)}{2F'(x_k)},$$

no additive constant being necessary because, as $\zeta_i(u)$ is an odd function, both sides of the equation vanish when $x_1,\ \dots,\ x_p$ are respectively at the places $a_1,\dots,a_p$. As any argument can be written, save for periods, in the form $u^{x_1,\,a_1}+\dots+u^{x_p,\,a_p}$, this equation is theoretically sufficient to enable us to express $\zeta_i(u)$ for any value of u.

Ex. i. It can easily be shewn (§ 200) that

$$u^{c,\,a}+u^{c_1,\,a_1}+\dots\dots+u^{c_p,\,a_p}=0.$$

Thus the final formula of § 214 immediately gives

$$-\zeta_i(u^{x_1,\,c_1}+\dots\dots+u^{x_p,\,c_p})=L_i^{x_1,\,c_1}+\dots\dots+L_i^{x_p,\,c_p}-\sum_{k=1}^{p}\frac{y_k\chi_{p-i}(x_k;\ c,\ x_1,\ \dots,\ x_p)}{2\,(x_k-c)\,F'(x_k)}.$$

Ex. ii. In case $p=1$ we infer from the formula just obtained, and from the final formula of § 214, respectively, the results

$$-\zeta_1(u^{x_1,\,a_1})=L_1^{x_1,\,a_1},\quad -\zeta_1(u^{x,\,a}+u^{x_1,\,a_1})=L_1^{x,\,a_1}+L_1^{x_1,\,a_1}-\tfrac{1}{2}\frac{y-y_1}{x-x_1}+D,$$

where D is an absolute constant. Thus

$$\zeta_1(u^{x,\,a}+u^{x_1,\,a_1})=\zeta_1(u^{x,\,a_1})+\zeta_1(u^{x_1,\,a_1})+\tfrac{1}{2}\frac{y-y_1}{x-x_1}-D.$$

This is practically equivalent with the well-known formula

$$\zeta(u+v)=\zeta(u)+\zeta(v)+\tfrac{1}{2}\frac{\wp'u-\wp'v}{\wp u-\wp v}.$$

The identification can be made complete by means of the facts (i) The Weierstrass argument u is equal to $u^{a,\,x}$, in our notation, so that $y=-\wp'(u)$, (ii) $u^{x,\,a_1}=\omega+\omega'-u$, so that $\zeta_1(u^{x,\,a_1})=\zeta_1(\omega+\omega'-u)=-L_1^{x,\,a_1}=-\int_{a_1}^{x}\frac{x\,dx}{y}$, as we easily find when $L_1^{x,\,\mu}$ is chosen as in § 138, Ex. i., (iii) $d\zeta u=\frac{x\,dx}{y}$, (iv) therefore $\zeta_1(u^{x,\,a_1})=-\zeta u$, (v) the branch places c_1, a_1, c are chosen by Weierstrass (in accordance with the formula $e_1+e_2+e_3=0$) so that the limit of $\wp u-\frac{1}{u^2}$, when $u=0$, is 0. The effect of this is that the constant D is zero.

Ex. iii. For $p=2$ we have

$$-\zeta_1(u^{x,\,a}+u^{x_1,\,a_1}+u^{x_2,\,a_2})=L_1^{x,\,\mu}+L_1^{x_1,\,\mu}+L_1^{x_2,\,\mu}$$
$$-\frac{y(x-x_1-x_2)}{2(x-x_1)(x-x_2)}-\frac{y_1(x_1-x-x_2)}{2(x_1-x)(x_1-x_2)}-\frac{y_2(x_2-x-x_1)}{2(x_2-x)(x_2-x_1)}+C_1$$

$$-\zeta_2(u^{x,\,a}+u^{x_1,\,a_1}+u^{x_2,\,a_2})=L_2^{x,\,\mu}+L_2^{x_1,\,\mu}+L_2^{x_2,\,\mu}$$
$$-\frac{y}{2(x-x_1)(x-x_2)}-\frac{y_1}{2(x_1-x)(x_1-x_2)}-\frac{y_2}{2(x_2-x)(x_2-x_1)}+C_2$$

and

$$-\zeta_1(u^{x_1,\,a_1}+u^{x_2,\,a_2})=L_1^{x_1,\,a_1}+L_2^{x_2,\,a_2}-\tfrac{1}{2}\frac{y_1-y_2}{x_1-x_2},\quad -\zeta_2(u^{x_1,\,a_1}+u^{x_2,\,a_2})=L_2^{x_1,\,a_1}+L_2^{x_2,\,a_2},$$

where with a suitable determination of the matrix a which occurs in the definition of the integrals $L_i^{x,\,\mu}$ and in the function $\vartheta(u)$, we may take (§ 138, Ex. i. Chap. VII.)

$$L_1^{x,\,\mu}=\int_\mu^x\frac{dy}{4y}(\lambda_3x+2\lambda_4x^2+3\lambda_5x^3),\quad L_2^{x,\,\mu}=\int_\mu^x\frac{dy}{4y}\lambda_5x^2.$$

For any values of p we obtain

$$-\zeta_p(u^{x_1,\,a_1}+\dots\dots+u^{x_p,\,a_p})=L_p^{x_1,\,a_1}+\dots\dots+L_p^{x_p,\,a_p}=\frac{\lambda_{2p+1}}{4}\sum_{k=1}^{p}\int_{a_k}^{x_k}\frac{x^p dx}{y}.$$

Ex. iv. We have (§ 210) obtained $2^{2p}-1$ formulae of the form

$$\frac{\vartheta(u\,|\,u^{b_1,\,a}+\dots\dots+u^{b_k,\,a})}{\vartheta(u)}=Z,$$

where Z is an algebraical function, and the arguments $u_1,\dots,u_p$ are given by

$$u=u^{x_1,\,a_1}+\dots\dots+u^{x_p,\,a_p};$$

the integrals being taken as in § 214, these equations lead to

$$\frac{\partial x_r}{\partial u_i}=\nu_{r,\,i}\frac{dx_r}{dt}=y_r\frac{\chi_{p-i}(x_r;\ x_1,\dots,x_p)}{F'(x_r)}.$$

Hence we have

$$\zeta_i(u\,|\,u^{b_1,\,a}+\dots\dots+u^{b_k,\,a})-\zeta_i(u)=\sum_{r=1}^{p}y_r\frac{\chi_{p-i}(x_r;\ x_1,\dots,x_p)}{F'(x_r)}\frac{1}{Z}\frac{\partial Z}{\partial x_r}.$$

For instance, when $k=1$, and Z is a constant multiple of $\sqrt{(b_1-x_1)\dots\dots(b_1-x_p)}$, we obtain

$$\zeta_i(u\,|\,u^{b_1,\,a})-\zeta_i(u)=\sum_{r=1}^{p}y_r\frac{\chi_{p-i}(x_r;\ x_1,\dots,x_p)}{2F'(x_r)}\frac{1}{x_r-b_1},$$

so that

$$-\zeta_i(u\,|\,u^{b,\,a})=L_i^{x_1,\,a_1}+\dots\dots+L_i^{x_p,\,a_p}-\sum_{r=1}^{p}\frac{y_r}{2F'(x_r)}\left[\chi_{p-i-1}(x_r;\ x_1,\dots,x_p)+\frac{\chi_{p-i}(x_r;\ x_1,\dots,x_p)}{x_r-b}\right]$$
$$=L_i^{x_1,\,a_1}+\dots\dots+L_i^{x_p,\,a_p}-\sum_{r=1}^{p}\frac{y_r}{2F'(x_r)}\frac{\chi_{p-i}(x_r;\ b,\ x_1,\dots,x_p)}{x_r-b}.$$

By means of the formula

$$\zeta_i(u+\tfrac{1}{2}\Omega_{P,P'})=\eta_{i,1}P_1+\dots\dots+\eta_{i,p}P_p+\eta'_{i,1}P_1'\dots\dots+\eta'_{i,p}P_p'+\zeta_i(u|\tfrac{1}{2}\Omega_{P,P'}),$$

which is easily obtained from the formulae of § 190, we can infer that the formula just obtained is in accordance with the final formula of § 214.

Ex. v. We have seen (§ 185, Chap. X.) that in the hyperelliptic case there are $\binom{2p+1}{p}$ even theta functions which do not vanish; and the corresponding half-periods are congruent to expressions of the form

$$u^{x_1, a_1}+\dots\dots+u^{x_p, a_p}.$$

It may be shewn in fact that these half-periods are obtained by taking for $x_1, \dots, x_p$ the $\binom{2p+1}{p}$ possible sets of p branch places that can be chosen from $a_1, \dots, a_p, c, c_1, \dots, c_p$. Hence it follows from the formula of the text (p. 321) that if $\tfrac{1}{2}\Omega_k$ be any even half-period corresponding to a non-vanishing theta function, we have

$$\zeta_i(\tfrac{1}{2}\Omega_k)=(\tfrac{1}{2}H_k)_i.$$

This formula generalises the well-known elliptic function formula expressed by $\zeta\omega=\eta$. To explain the notation a particular case may be given; we have

$$\zeta_i(\omega_{1,r}, \omega_{2,r}, \dots, \omega_{p,r})=\eta_{i,r}, \text{ or } \zeta_i(u^{c_{r+1}, a_r})=-L_i^{c_{r+1}, a_r}$$

and

$$\zeta_i(\omega'_{1,r}, \omega'_{2,r}, \dots, \omega'_{p,r})=\eta'_{i,r}, \text{ or } \zeta_i(u^{c_r, a_r}) \quad =-L_i^{c_r, a_r}.$$

Thus each of the $2p^2$ quantities $\eta_{i,r}, \eta'_{i,r}$ can be expressed as ζ-functions of half-periods.

Ex. vi. The formula of the text (p. 321) is equivalent to

$$-\zeta_i(u^{x_1, a_1}+\dots\dots+u^{x_p, a_p})=L_i^{x_1, a_1}+\dots\dots+L_i^{x_p, a_p}-\tfrac{1}{2}\sum_{k=1}^{p}\frac{\partial x_k}{\partial u_{i+1}},$$

where

$$u_r=u_r^{x_1, a_1}+\dots\dots+u_r^{x_p, a_p}.$$

For example when $p=2$

$$-\zeta_1(u)+\tfrac{1}{2}\frac{\partial}{\partial u_2}(x_1+x_2)=L_1^{x_1, a_1}+L_1^{x_2, a_2}$$

$$-\zeta_2(u) \qquad\qquad =L_2^{x_1, a_1}+L_2^{x_2, a_2}.$$

216. It is easy to prove, as remarked in Ex. iii. § 215, that if

$$u=u^{x_1, a_1}+\dots\dots+u^{x_p, a_p},$$

and the matrix a (§ 138, Chap. VII.) be determined so that the integrals $L_i^{x, \mu}$ have the value found in § 138, Ex. i., then

$$-\zeta_p(u)=\tfrac{1}{4}\lambda_{2p+1}\sum_{k=1}^{p}\int_{a_k}^{x_k}\frac{x^p\,dx}{y}.$$

Therefore, if $-\frac{\partial}{\partial u_i}\zeta_r(u)$ be denoted by $\wp_{r,i}(u)$, we have

$$\wp_{p,i}(u)=-\frac{\partial\zeta_p(u)}{\partial u_i}=\tfrac{1}{4}\lambda_{2p+1}\sum_{k=1}^{p}\frac{x_k^p}{y_k}\frac{\partial x_k}{\partial u_i},$$

and thus, as follows from the definition of the arguments u,

$$\wp_{p,i}(u) = \tfrac{1}{4}\lambda_{2p+1} \sum_{k=1}^{p} \frac{x_k^p \chi_{p-i}(x_k;\, x_1, \ldots, x_p)}{F'(x_k)},$$

where $F(x)$ denotes $(x - x_1) \ldots (x - x_p)$.

Whence, if x be any argument whatever,

$$\sum_{i=1}^{p} x^{i-1} \wp_{p,i}(u), = \tfrac{1}{4}\lambda_{2p+1} \sum_{k=1}^{p} \frac{x_k^p \sum_{i=1}^{p} x^{i-1} \chi_{p-i}(x_k;\, x_1, \ldots, x_p)}{F'(x_k)},$$

$$= \tfrac{1}{4}\lambda_{2p+1} \sum_{k=1}^{p} \frac{x_k^p F(x)}{(x - x_k) F'(x_k)};$$

but we have

$$\frac{\sum_{i=1}^{p} x^{i-1} \wp_{p,i}(u)}{F(x)} = \sum_{k=1}^{p} \frac{\sum_{i=1}^{p} x_k^{i-1} \wp_{p,i}(u)}{(x - x_k) F'(x_k)}.$$

Thus

$$\tfrac{1}{4}\lambda_{2p+1} x_k^p = \sum_{i=1}^{p} x_k^{i-1} \wp_{p,i}(u).$$

Thus, if we suppose $\lambda_{2p+1} = 4$, the values of $x_1, \ldots, x_p$ satisfying the inversion problem expressed by the equations

$$u \equiv u^{x_1, a_1} + \ldots\ldots + u^{x_p, a_p}$$

are the roots of the equation

$$F(x) = x^p - x^{p-1} \wp_{p,p}(u) - x^{p-2} \wp_{p,p-1}(u) - \ldots\ldots - \wp_{p,1}(u) = 0.$$

In other words, if the sum of the homogeneous products of r dimensions, without repetitions, of the quantities $x_1, \ldots, x_p$ be denoted by h_r, we have

$$h_r = (-)^{r-1} \wp_{p,p-r+1}(u).$$

Further, from the equation

$$\frac{\partial x_k}{\partial u_i} = \frac{y_k \chi_{p-i}(x_k;\, x_1, \ldots, x_p)}{F'(x_k)},$$

putting p for i, we infer that

$$y_k = F'(x_k) \frac{\partial x_k}{\partial u_p}, = - \left[\frac{\partial F(x)}{\partial u_p}\right]_{x = x_k},$$

because $F(x_k) = 0$. Thus, if we use the abbreviation

$$\psi(x) = -\frac{\partial F(x)}{\partial u_p} = x^{p-1} \wp_{p,p,p}(u) + x^{p-2} \wp_{p,p,p-1}(u) + \ldots\ldots + \wp_{p,p,1}(u)$$

we obtain

$$y_k = \psi(x_k).$$

These equations constitute a complete solution of the inversion problem. In the $\wp$-functions the matrix a is as in § 138, Ex. i., and the integrals of the first kind are as in § 214.

We have previously (§ 212) shewn that $x_1, \ldots, x_p$ are determinable from p such equations as

$$\frac{\vartheta^2(u|u^{a_i,a})}{\vartheta^2(u)} = \pm\frac{(a_i - x_1)\ldots(a_i - x_p)}{\sqrt{-P'(a_i)\,Q(a_i)}}, = \frac{(a_i - x_1)\ldots(a_i - x_p)}{\mu_i}, \text{ say.}$$

Thus we have p equations of the form

$$\mu_i\frac{\vartheta^2(u|u^{a_i,a})}{\vartheta^2(u)} = a_i^p - a_i^{p-1}\wp_{p,p}(u) - a_i^{p-2}\wp_{p,p-1}(u) - \ldots\ldots - \wp_{p,1}(u).$$

Ex. i. For $p=1$ we have

$$\mu_1\frac{\vartheta^2(u|u^{a_1,a})}{\vartheta^2(u)} = a_1 - \wp_{1,1}(u), \ = a_1 + \frac{\partial^2}{\partial u^2}\log\vartheta(u).$$

This is equivalent to the equation which is commonly written in the form

$$\wp u = e_3 + \frac{e_1 - e_3}{\operatorname{sn}^2(u\sqrt{e_1 - e_3})}.$$

Ex. ii. For $p=2$ we have

$$\mu_1\frac{\vartheta^2(u|u^{a_1,a})}{\vartheta^2(u)} = a_1^2 - a_1\wp_{2,2}(u) - \wp_{2,1}(u),$$

$$\mu_2\frac{\vartheta^2(u|u^{a_2,a})}{\vartheta^2(u)} = a_2^2 - a_2\wp_{2,2}(u) - \wp_{2,1}(u).$$

We may denote the left-hand sides of these equations respectively by $\mu_1 q_1^2$, $\mu_2 q_2^2$.

Ex. iii. Prove that, with $\mu_1 q_1^2 = a_1^2 - a_1\wp_{2,2}(u) - \wp_{1,2}(u)$, etc., $\mu_1 = \pm\sqrt{-f'(a_1)}$, we have

$$\frac{\mu_1\mu_2}{a_1 - a_2}(q_1^2 q_2'^2 - q_2^2 q_1'^2)$$
$$= \wp_{22}(u)\wp_{12}(u') - \wp_{12}(u)\wp_{22}(u') + (a_1 + a_2)[\wp_{12}(u) - \wp_{12}(u')] + a_1 a_2[\wp_{22}(u) - \wp_{22}(u')].$$

Ex. iv. Prove that

$$y_s = \frac{\partial x_s}{\partial u_1} + x_s\frac{\partial x_s}{\partial u_2} + \ldots\ldots + x_s^{p-1}\frac{\partial x_s}{\partial u_p}.$$

Ex. v. If, with $P(x)$ to denote $(x - a_1)\ldots\ldots(x - a_p)$, we put

$$V_r = \int_{a_1}^{x_1}\frac{P(x)}{x - a_r}\frac{dx}{2y} + \ldots\ldots + \int_{a_p}^{x_p}\frac{P(x)}{x - a_r}\frac{dx}{2y},$$

prove that

$$\frac{\partial}{\partial V_1} + \ldots\ldots + \frac{\partial}{\partial V_p} = 2\frac{\partial}{\partial u_p}.$$

Ex. vi. With the same notation, shew that if

$$G = \int_{a_1}^{x_1}P(x)\frac{dx}{2y} + \ldots\ldots + \int_{a_p}^{x_p}P(x)\frac{dx}{2y},$$

then

$$\frac{\partial G}{\partial V_i} = -\frac{(a_i - x_1)\ldots\ldots(a_i - x_p)}{P'(a_i)}.$$

The arguments $V_1, \ldots, V_p$ are those used by Weierstrass (*Math. Werke*, Bd. I. Berlin, 1894, p. 297). The result of Ex. iv. is necessary to compare his results with those here obtained. The equation $y_r = \psi(x_r)$ is given by Weierstrass. The relation of Ex. vi. is given by Hancock (*Eine Form des Additionstheorem u. s. w. Diss.* Berlin, 1894, Bernstein).

With these arguments we have

$$\mu_i \frac{\vartheta^2(u|u^{a_1, a})}{\vartheta^2(u)} = a_i^p - \tfrac{1}{2}P'(a_i)\frac{\partial}{\partial V_i}\zeta_p(u) = a_i^p - \tfrac{1}{4}P'(a_i)\frac{\partial}{\partial V_i}\left(\frac{\partial}{\partial V_1} + \ldots\ldots + \frac{\partial}{\partial V_p}\right)\log\vartheta(u).$$

Ex. vii. Prove from the formula

$$-\zeta_i(u^{x, a} + u) + \zeta_i(u^{\mu, a} + u) = L_i^{x, \mu} + \sum_{k=1}^{p} \tilde{\nu}_{k, i}\left[(x_k, x) - (x_k, \mu)\right]\frac{dx_k}{dt},$$

where

$$u = u^{x_1, a_1} + \ldots\ldots + u^{x_p, a_p},$$

that the function

$$\frac{\partial}{\partial u_i}\log\left[\frac{\vartheta^2(u^{x, a} + u)}{F(x)} e^{2\sum_{r=1}^{p} u_r L_r^{x, c}}\right] - \frac{y\chi_{p-i}(x;\, x_1, \ldots, x_p)}{F(x)}$$

is independent of the place x. Here c is an arbitrary place and $F(x) = (x - x_1)\ldots\ldots(x - x_p)$.

Ex. viii. If $R_{z, c}^{x, a}$ denote the integral $\Pi_{z, c}^{x, a} - 2\Sigma\Sigma a_{i, j} u_i^{z, c} u_j^{x, a}$, obtained in § 138, and $F_z^{x, a}$ denote $D_z R_{z, c}^{x, a}$, prove that in the hyperelliptic case, with the matrix a determined as in Ex. i. § 138, when the place a is at infinity,

$$F_a^{x, \mu} = -\frac{\sqrt{\lambda_{2p+1}}}{2}\int_\mu^x \frac{x^p dx}{y}.$$

Hence, when $\lambda_{2p+1} = 4$, shew that the equation obtained in § 215 (p. 321) is deducible from the equation (Chap. X. § 196)

$$F_m^{z_1, m_1} + \ldots\ldots + F_m^{z_p, m_p} = -\sum_{r=1}^{p} \mu_r(m)\,\zeta_r(u^{z_1, m_1} + \ldots\ldots + u^{z_p, m_p}).$$

Ex. ix. We can also express the function $\zeta_p(u+v) - \zeta_p(u) - \zeta_p(v)$, which is clearly a periodic function of the arguments u, v, in an algebraical form, and in a way which generalizes the formula of Jacobi's elliptic functions given by

$$Z(u) + Z(v) - Z(u+v) = k^2 \operatorname{sn} u \operatorname{sn} v \operatorname{sn}(u+v).$$

For if we take places $x_1, \ldots, \zeta_p$, such that

$$u \equiv u^{x_1, a_1} + \ldots\ldots + u^{x_p, a_p}$$

$$v \equiv u^{z_1, a_1} + \ldots\ldots + u^{z_p, a_p}$$

$$-u - v \equiv u^{\zeta_1, a_1} + \ldots\ldots + u^{\zeta_p, a_p},$$

these $3p$ places will be the zeros of a rational function which has $a_1, \ldots, a_p$ as poles, each to the third order. This function is expressible in the form $(My + NP)/P^2$, where P denotes $(x - a_1)\ldots\ldots(x - a_p)$, M is an integral polynomial in x of order $p-1$, and N is an integral polynomial in x of order p. Denoting this function by Z, we have

$$\zeta_p(u) + \zeta_p(v) - \zeta_p(u+v), = L_p^{x_1, a_1} + \ldots + L_p^{x_p, a_p} + L_p^{z_1, a_1} + \ldots + L_p^{z_p, a_p} + L_p^{\zeta_1, a_1} + \ldots + L_p^{\zeta_p, a_p},$$

$$= -\int_\infty^0 \left[\frac{dI}{dt}\frac{1}{Z-\mu}\right]_{t^{-1}} d\mu, \;= K \text{ say},$$

by § 154, Chap. VIII., where $I = L_p^{x,\mu} = \frac{1}{4}\lambda_{2p+1}\int_\mu^x \frac{x^p dx}{y}$. Writing Z in the form

$$\frac{(Ax^{p-1} + \ldots\ldots)\, y + (x^p + \ldots\ldots)\, P}{P^2},$$

and taking $\lambda_{2p+1} = 4$, we find the value of the integral K to be $-2A$.

But from the equation

$$N^2P - 4M^2Q = (x - x_1)\ldots\ldots(x - x_p)\,(x - z_1)\ldots\ldots(x - z_p)\,(x - \zeta_1)\ldots\ldots(x - \zeta_p),$$

where $Q = (x - c)(x - c_1)\ldots\ldots(x - c_p)$, we have, putting a_i for x,

$$p_i q_i \varpi_i = 2\sqrt{-Q(a_i)}\,(Aa_i^{p-1} + \ldots), \qquad (i = 1, 2, \ldots, p),$$

where $p_i = \sqrt{(a_i - x_1)\ldots\ldots(a_i - x_p)}$, $q_i = \sqrt{(a_i - z_1)\ldots\ldots(a_i - z_p)}$, $\varpi_i = \sqrt{(a_i - \zeta_1)\ldots\ldots(a_i - \zeta_p)}$; solving these equations for A we eventually have*

$$\zeta_p(u) + \zeta_p(v) - \zeta_p(u + v) = \sum_{i=1}^{p} \frac{p_i q_i \varpi_i}{P(a_i)\sqrt{-Q(a_i)}}.$$

Ex. x. Obtain, for $p = 2$, the corresponding expression for $\zeta_1(u) + \zeta_1(v) - \zeta_1(u + v)$.

Ex. xi. Denoting $\frac{1}{P(a_i)\sqrt{-Q(a_i)}}$ by C_i, the equation

$$\zeta_p(u) + \zeta_p(v) - \zeta_p(u + v) = \sum_{i=1}^{p} C_i p_i q_i \varpi_i$$

gives

$$-\wp_{p,r}(u) + \wp_{p,r}(v) = \sum_{i=1}^{p} C_i\,[p_i^{(r)} q_i - p_i q_i^{(r)}]\,\varpi_i, \qquad (r = 1, 2, \ldots, p),$$

where $p_i^{(r)}$ denotes $\frac{\partial}{\partial u_r}\sqrt{(a_i - x_1)\ldots\ldots(a_i - x_p)}$. It has been shewn that p_i is a single valued function of u and it may be denoted by $p_i(u)$. Similarly ϖ_i is a single valued function of $u + v$, being equal to $p_i(-u - v)$. The equation here obtained enables us therefore to express $p_i(u + v)$ in terms of $p_i(u)$, $p_i(v)$, and the differential coefficients of these; for we have obtained sufficient equations to express $\wp_{p,r}(u)$, $\wp_{p,r}(v)$ in terms of the functions $p_i(u)$, $p_i(v)$. A developed result is obtained below in the case $p = 2$, in a more elementary way.

217. We have obtained in the last chapter (§ 197) the equation

$$\sum_i \sum_j \wp_{i,j}(u^{x,m} - u^{x_1,m_1} - \ldots\ldots - u^{x_p,m_p})\,\mu_i(x)\,\mu_j(x_p) = D_x D_{x_p} R_{x_p,c}^{x,a}.$$

Hence, adopting that determination of the matrix a, occurring in the integrals $L_i^{x,\mu}$, and the function $\vartheta(u)$ (§ 192, Chap. X.), which gives the particular forms for $L_i^{x,\mu}$ obtained in § 138, Ex. i., we have in the hyperelliptic case

$$\sum_i \sum_j \wp_{i,j}(u^{x,a} + u^{x_1,a_1} + \ldots\ldots + u^{x_p,a_p})\,x^{i-1}x_r^{j-1} = \frac{f(x, x_r) - 2yy_r}{4(x - x_r)^2},$$

where $f(x, z) = \sum_{i=0}^{p+1} x^i z^i\,[2\lambda_{2i} + \lambda_{2i+1}(x + z)]$. This equation is, however, in-

* This equation, with the integrals $L_p^{x,a}$ on the left-hand side, is given by Forsyth, *Phil. Trans.* 1883, Part I.

dependent of the particular matrix a adopted. For suppose, instead of the particular integral

$$L_i^{x,\mu}, = \int_\mu^x \frac{dx}{y} \sum_{k=i}^{2p+1-i} \lambda_{k+1+i}(k+1-i)\,x^k,$$

we take

$$L_i^{x,\mu} - \sum_{k=1}^{p} C_{i,k} u_k^{x,\mu},$$

where $C_{i,k} = C_{k,i}$; then (§ 138) this is equivalent to replacing the particular matrix a by $a + \frac{1}{2}C$, where C is an arbitrary symmetrical matrix, and we have the following resulting changes (p. 315)

$R_{z,c}^{x,a}$ (p. 194) becomes changed to $R_{z,c}^{x,a} - \Sigma\Sigma C_{i,k} u_i^{x,a} u_k^{z,c}$, so that,

$f(x, z)$ (p. 195) becomes changed to $f(x, z) - 4(x-z)^2 \Sigma\Sigma C_{i,k} x^{i-1} z^{k-1}$,

$\vartheta(u)$ (§ 189) becomes multiplied by $e^{\frac{1}{2}Cu^2}$,

and thus $\zeta_i(u)$ is increased by $C_{i,1}u_1 + \ldots\ldots + C_{i,p}u_p$, and instead of $\wp_{i,j}(u)$ we have $\wp_{i,j}(u) - C_{i,j}$.

Since now $u^{x,a} + u^{x_s,a_s} = u^{x_s,a} + u^{x,a_s}$, we have $\frac{1}{2}p(p+1)$ equations of the form

$$\sum_i \sum_j \wp_{i,j}(u)\, x_r^{i-1} x_s^{j-1} = \frac{f(x_r, x_s) - 2y_r y_s}{4(x_r - x_s)^2},$$

where $u = u^{x,a} + u^{x_1,a_1} + \ldots\ldots + u^{x_p,a_p}$, $r = 0, 1, \ldots, p$, and $s = 0, 1, \ldots, p$. Hence, if e_1, e_2 denote any quantities we obtain by calculation

$$\sum_i \sum_j \wp_{i,j}(u)\, e_1^{i-1} e_2^{j-1} = R(e_1)\, R(e_2) \sum_r \sum_s \frac{2y_r y_s - f(x_r, x_s)}{4G'(x_r)\, G'(x_s)};$$

here the matrix a is arbitrary, the polynomial $f(x_r, x_s)$ being correspondingly chosen, and

$$G(\xi) = (\xi - e_1)(\xi - e_2)(\xi - x)(\xi - x_1)\ldots(\xi - x_p), \quad R(\xi) = (\xi - x)(\xi - x_1)\ldots(\xi - x_p).$$

Suppose now that $f(x, z) = \bar{f}(x, z) + 4(x-z)^2 \sum_i \sum_j A_{i,j} x_r^{i-1} x_s^{j-1}$, where $\bar{f}(x, z)$ is the form obtained in Ex. viii. § 211; then we obtain

$$\sum_i \sum_j [\wp_{i,j}(u) - A_{i,j}]\, e_1^{i-1} e_2^{j-1} = R(e_1)\, R(e_2) \sum_r \sum_s \frac{2y_r y_s - \bar{f}(x_r, x_s)}{4G'(x_r)\, G'(x_s)},$$

and by Ex. x. § 211 this is equal to

$$\frac{1}{4} R(e_1)\, R(e_2) \left[\sum_0^p \frac{y_r}{(e_1 - x_r)(e_2 - x_r)\, R'(x_r)} \right]^2 - \frac{f(e_1)\, R(e_2)}{4(e_1 - e_2)^2\, R(e_1)}$$

$$- \frac{f(e_2)\, R(e_1)}{4(e_1 - e_2)^2\, R(e_2)} + \frac{\bar{f}(e_1, e_2)}{4(e_1 - e_2)^2},$$

and therefore

$$\sum_i\sum_j \wp_{i,j}(u)\, e_1^{i-1} e_2^{j-1} = \tfrac{1}{4} R(e_1) R(e_2) \left[\sum_0^p \frac{y_r}{(e_1-x_r)(e_2-x_r) R'(x_r)}\right]^2$$
$$- \frac{f(e_1) R(e_2)}{4(e_1-e_2)^2 R(e_1)} - \frac{f(e_2) R(e_1)}{4(e_1-e_2)^2 R(e_2)} + \frac{f(e_1, e_2)}{4(e_1-e_2)^2}.$$

This is a very general formula*; in it the matrix a is arbitrary.

It follows from Ex. xi. § 211 that if b_1, b_2 be any branch places, we have

$$\sum_i\sum_j \wp_{i,j}(u)\, b_1^{i-1} b_2^{j-1} = \frac{f(b_1, b_2)}{4(b_1-b_2)^2} + E\frac{\vartheta^2(u|u^{b_1, a} + u^{b_2, a})}{\vartheta^2(u)},$$

where E is a certain constant (cf. §§ 213, 212). This equation is also independent of the determination of the matrix a.

By solving $\frac{1}{2}p(p+1)$ equations of this form, wherein b_1, b_2 are in turn taken to be every pair chosen from any $p+1$ branch places, we can express $\sum_i\sum_j \wp_{i,j}(u)\, e_1^{i-1} e_2^{j-1}$ as a linear function of $\frac{1}{2}p(p+1)$ squared theta quotients, e_1, e_2 *being any quantities whatever.*

By putting b_2 at a, that is at infinity (first dividing by b_2^{p-1}), and putting x also at a, this becomes the formula already obtained (§ 216)

$$\mu_i \frac{\vartheta^2(u|u^{a_i, a})}{\vartheta^2(u)} = a_i^p - a_i^{p-1}\wp_{p,p}(u) - \dots\dots - \wp_{p,1}(u).$$

Ex. i. When $p=1$, taking the fundamental equation to be

$$y^2 = 4x^3 - g_2 x - g_3,$$

the expression

$$f(x,z), \; = \sum_0^{p+1} x^i z^i [2\lambda_{2i} + \lambda_{2i+1}(x+z)], \; = -2g_3 - g_2(x+z) + 4xz(x+z),$$

and

$$\frac{2ys - f(x,z)}{4(x-z)^2} = \frac{2ys - (y^2+s^2) + 4(x^2-z^2)(x-z)}{4(x-z)^2} = x + z - \tfrac{1}{4}\left(\frac{y-s}{x-z}\right)^2,$$

if $s^2 = 4z^3 - g_2 z - g_3$.

Therefore, by the formula at the middle of page 328, taking the matrix a to have the particular determination of § 138, Ex. i.,

$$\wp_{1,1}(u^{x,a} + u^{x_1, a_1}) = -(e_1-x)(e_1-x_1)(e_2-x)(e_2-x_1)\frac{x + x_1 - \frac{1}{4}\left(\frac{y-y_1}{x-x_1}\right)^2}{(x-e_1)(x-e_2)(x_1-e_1)(x_1-e_2)}$$
$$= -x - x_1 + \tfrac{1}{4}\left(\frac{y-y_1}{x-x_1}\right)^2;$$

this is a well-known result.

Ex. ii. When $p=2$, we easily find

$$\frac{R(e_1) R(e_2)}{G'(x_r) G'(x_s)} = -\frac{(x-e_1)(x-e_2)}{(x-x_r)(x-x_s)} \frac{1}{(x_r-x_s)^2}$$

* It is given by Bolza, *Göttinger Nachrichten*, 1894, p. 268.

and thus the expression

$$\wp_{1,1}(u)+(e_1+e_2)\wp_{1,2}(u)+e_1e_2\wp_{2,2}(u)$$

is equal to

$$-\frac{(x-e_1)(x-e_2)}{(x-x_1)(x-x_2)}\frac{2y_1y_2-f(x_1,x_2)}{4(x_1-x_2)^2}-\frac{(x_1-e_1)(x_1-e_2)}{(x_1-x)(x_1-x_2)}\frac{2yy_2-f(x,x_2)}{4(x-x_2)^2}$$
$$-\frac{(x_2-e_1)(x_2-e_2)}{(x_2-x)(x_2-x_1)}\frac{2yy_1-f(x,x_1)}{4(x-x_1)^2}.$$

Herein the matrix a is perfectly general. Adopting the particular determination of § 138, Ex. i., we have, since the term in $f(x,z)$ of highest degree in x is $\lambda_{2p+1}x^{p+1}z^p$, $=4x^3z^2$, say, by putting the place x at a, that is at infinity, the result

$$\wp_{1,1}(u)+(e_1+e_2)\wp_{1,2}(u)+e_1e_2\wp_{2,2}(u)=-\frac{2y_1y_2-f(x_1,x_2)}{4(x_1-x_2)^2}-x_1x_2(e_1+e_2)+e_1e_2(x_1+x_2),$$

where $u=u^{x_1,a_1}+u^{x_2,a_2}$.

Ex. iii. Prove, for $p=2$, when the matrix a is as in § 138, Ex. i., that

$$\wp_{11}(u)+\wp_{12}(u).(e_1+e_2)+\wp_{22}(u).e_1e_2=\frac{(a_1-e_1)(a_1-e_2)}{a_1-a_2}\mu_2q_2^2-\frac{(a_2-e_1)(a_2-e_2)}{a_1-a_2}\mu_1q_1^2$$
$$+\frac{\mu_1\mu_2}{a_1-a_2}q_{12}^2+\frac{f(a_1,a_2)}{4(a_1-a_2)^2}+e_1e_2(a_1+a_2)-(e_1+e_2)a_1a_2,$$

where e_1, e_2 are any quantities, $u=u^{x_1,a_1}+u^{x_2,a_2}$, and μ_1, μ_2 are as in § 216 (cf. § 213).

Ex. iv. From the formula, for $p=2$ (§§ 217, 216, 213),

$$\wp_{11}(u)+\wp_{12}(u).(a_1+a_2)+\wp_{22}(u).a_1a_2=\frac{\mu_1\mu_2}{a_1-a_2}q_{12}^2+\frac{f(a_1,a_2)}{4(a_1-a_2)^2},$$

where a_1, a_2 are the branch places as before denoted, infer (§ 216, Ex. iii.) that

$$\wp_{11}(u)-\wp_{11}(u')+\wp_{12}(u)\wp_{22}(u')-\wp_{12}(u')\wp_{22}(u)=\frac{\mu_1\mu_2}{a_1-a_2}[q_{12}^2-q'^2_{12}-q_1^2q_2'^2+q_2^2q_1'^2].$$

Prove also that, for any value of u, and any position of x,

$$\wp_{11}(u^{x,a}+u)-\wp_{11}(u)+\wp_{12}(u^{x,a}+u)\wp_{22}(u)-\wp_{22}(u^{x,a}+u)\wp_{12}(u)=0.$$

Ex. v. If $b_1, \dots, b_{p+1}$ be any $(p+1)$ branch places, and e_1, e_2 any quantities whatever, and $L(x)=(x-b_1)\dots\dots(x-b_{p+1})$, $M(x)=(x-e_1)(x-e_2)(x-b_1)\dots\dots(x-b_{p+1})$, prove that

$$\Sigma\Sigma\wp_{i,j}(u)e_1^{i-1}e_2^{j-1}=-L(e_1)L(e_2)\sum_r\sum_s\frac{(b_r-b_s)^2}{M'(b_r)M'(b_s)}\left[\frac{f(b_r,b_s)}{4(b_r-b_s)^2}+E_{r,s}\frac{\vartheta^2(u|u^{b_r,a}+u^{b_s,a})}{\vartheta^2(u)}\right],$$

where the matrix a has a perfectly general value, r, s consist of every pair of different numbers from the numbers $1, 2, \dots, (p+1)$, and $E_{r,s}$ are constants.

218. We conclude this chapter with some further details in regard to the case $p=2$, which will furnish a useful introduction to the problems of future chapters of the present volume. We have in case $p=1$ such a formula as that expressed by the equation

$$\frac{\sigma(u+u')\sigma(u-u')}{\sigma^2(u)\sigma^2(u')}=\wp(u')-\wp(u);$$

we investigate now, in case $p=2$, corresponding formulae for the functions

$$\frac{\vartheta(u+u')\vartheta(u-u')}{\vartheta^2(u)\vartheta^2(u')},\qquad\frac{\vartheta(u+u'|u^{b,a})\vartheta(u-u')}{\vartheta^2(u)\vartheta^2(u')};$$

by division of the results we obtain a formula expressing the theta quotient $\vartheta(u+u'|u^{b,a})\div\vartheta(u+u')$ by theta quotients of the arguments u, u'; this formula may be called the addition equation for the theta quotient $\vartheta(u|u^{b,a})\div\vartheta(u)$. Though we shall in a future chapter obtain the result in another way, it will be found that a certain interest attaches to the mode of proof employed here.

Determine the places x_1, x_2, x_1', x_2' so that

$$u \equiv u^{x_1, a_1} + u^{x_2, a_2}, \quad u' \equiv u^{x_1', a_1} + u^{x_2', a_2};$$

then, in order to find where the function $\vartheta(u^{x_1, a_1} + u^{x_2, a_2} + u^{x_1', a_1} + u^{x_2', a_2})$ vanishes, regarded as a function of x_1, we are to put

$$u^{x_1, a_1} + u^{x_2, a_2} + u^{x_1', a_1} + u^{x_2', a_2} \equiv u^{x_1, a} - u^{z_1, a_1} - u^{z_2, a_2},$$

or

$$(a, x_2, x_1', x_2', z_1, z_2) \equiv (a_1^3, a_2^3);$$

thus the places z_1, z_2 are positions of x_1 for which the determinant

$$\nabla = \begin{vmatrix} \dfrac{x_1 y_1}{P(x_1)}, & \dfrac{y_1}{P(x_1)}, & x_1, & 1 \\ \dfrac{x_2 y_2}{P(x_2)}, & \dfrac{y_2}{P(x_2)}, & x_2, & 1 \\ \dfrac{x_1' y_1'}{P(x_1')}, & \dfrac{y_1'}{P(x_1')}, & x_1', & 1 \\ \dfrac{x_2' y_2'}{P(x_2')}, & \dfrac{y_2'}{P(x_2')}, & x_2', & 1 \end{vmatrix}$$

wherein $P(x)$ denotes $(x-a_1)(x-a_2)$, vanishes. By considerations analogous to those of § 209 we therefore find, $\bar{\nabla}$ denoting the determinant derived from ∇ by changing the sign of y_1', y_2',

$$\frac{\vartheta(u+u')\,\vartheta(u-u')}{\vartheta^2(u)\,\vartheta^2(u')} = A\,\frac{\nabla\bar{\nabla} P(x_1) P(x_2) P(x_1') P(x_2')}{(x_1-x_2)^2 (x_1'-x_2')^2 (x_1-x_1')(x_1-x_2')(x_2-x_1')(x_2-x_2')},$$

where A is an absolute constant.

Now, if $\eta_1 = y_1/P(x_1)$, etc., we find by expansion and multiplication,

$$\nabla\bar{\nabla} = (\eta_1\eta_2 + \eta_1'\eta_2')^2 (x_1-x_2)^2 (x_1'-x_2')^2 - [(\eta_1\eta_1' + \eta_2\eta_2')(x_1'-x_1)(x_2'-x_2) - (\eta_1\eta_2' + \eta_2\eta_1')(x_1'-x_2)(x_2'-x_1)]^2,$$

and, if $\alpha = (x_1'-x_1)(x_2'-x_2)$, $\beta = (x_1'-x_2)(x_2'-x_1)$, $\alpha-\beta = (x_1'-x_2')(x_1-x_2)$, this leads to

$$\frac{\nabla\bar{\nabla}}{\alpha-\beta} = (\eta_1^2-\eta_2'^2)(\eta_2^2-\eta_1'^2)\,\alpha - (\eta_1^2-\eta_1'^2)(\eta_2^2-\eta_2'^2)\,\beta - \frac{\alpha\beta}{\alpha-\beta}(\eta_1-\eta_2)^2(\eta_1'-\eta_2')^2;$$

but, putting $y^2 = 4P(x)\,Q(x)$, $= 4(x-a_1)(x-a_2)(x-c)(x-c_1)(x-c_2)$, we have

$$\frac{P(x_1)P(x_2)P(x_1')P(x_2')}{(\alpha-\beta)\,\alpha\beta}\,[(\eta_1^2-\eta_2'^2)(\eta_2^2-\eta_1'^2)\,\alpha - (\eta_1^2-\eta_1'^2)(\eta_2^2-\eta_2'^2)\,\beta]$$

$$= \frac{16}{(x_1-x_2)(x_1'-x_2')}\left[\frac{Qx_1Px_2' - Qx_2'Px_1}{x_2'-x_1}\cdot\frac{Qx_2Px_1' - Qx_1'Px_2}{x_1'-x_2} - \frac{Qx_1Px_1' - Qx_1'Px_1}{x_1'-x_1}\cdot\frac{Qx_2Px_2' - Qx_2'Px_2}{x_2'-x_2}\right],$$

and this expression is equal to

$$16\left[Qa_1\,.\,Qa_2+\frac{Qa_2}{P'a_2}(a_1-x_1)(a_1-x_2)(a_1-x_1')(a_1-x_2')\right.$$
$$\left.+\frac{Qa_1}{P'a_1}(a_2-x_1)(a_2-x_2)(a_2-x_1')(a_2-x_2')\right],$$

as may be proved in various ways; now we have proved (§§ 208, 212, 213) that

$$(a_1-x_1)(a_1-x_2)=\pm\sqrt{-P'(a_1)\,Q(a_1)}\,q_1^2,\quad (a_2-x_1)(a_2-x_2)=\pm\sqrt{-P'(a_2)\,Q(a_2)}\,q_2^2$$

and

$$\tfrac{1}{4}\left(\frac{\eta_1-\eta_2}{x_1-x_2}\right)^2=\pm\frac{1}{a_1-a_2}\,\frac{q_{12}^2}{q_1^2q_2^2},$$

where $q_1=\vartheta(u|u^{a_1,\,a})\div\vartheta(u)$, $q_2=\vartheta(u|u^{a_2,\,a})\div\vartheta(u)$, $q_{1,\,2}=\vartheta(u|u^{a_1,\,a}+u^{a_2,\,a})\div\vartheta(u)$; thus as $q_1^2q_2^2q_1'^2q_2'^2=\dfrac{P(x_1)\,P(x_2)\,P(x_1')\,P(x_2')}{P'(a_1)\,P'(a_2)\,Qa_1\,Qa_2}$, we have

$$\frac{1}{A}\,\frac{\vartheta(u+u')\,\vartheta(u-u')}{\vartheta^2(u)\,\vartheta^2(u')},\ =\frac{\nabla\overline{\nabla}P(x_1)\,P(x_2)\,P(x_1')\,P(x_2')}{\alpha\beta\,(\alpha-\beta)^2},$$
$$=16\,Qa_1Qa_2\left[1-\frac{P'a_1}{P'a_2}q_1^2q_1'^2-\frac{P'a_2}{P'a_1}q_2^2q_2'^2\right]-16\,\frac{P'(a_1)\,P'(a_2)\,Qa_1Qa_2}{(a_1-a_2)^2}\,q_{12}^2q_{12}'^2,$$

where however we have assumed that the sign to be attached to the quotient

$$(a_1-x_1)(a_1-x_2)\div\sqrt{-P'(a_1)\,Q(a_1)}\,q_1^2$$

is the same for the places x_1', x_2' as for the places x_1, x_2. The product $\sqrt{-P'(a_1)\,Q(a_1)}$ $\sqrt{-P'(a_1)\,Q(a_1)}$ is, of course, here equal to $-P'(a_1)\,Q(a_1)$. Now,

$$P'(a_1)=(a_1-a_2)=-P'(a_2);$$

thus we obtain

$$\frac{\vartheta(u+u')\,\vartheta(u-u')\,\vartheta^2}{\vartheta^2(u)\,\vartheta^2(u')}=1+q_1^2q_1'^2+q_2^2q_2'^2+q_{12}^2q_{12}'^2,$$

the value of the constant multiplier, ϑ^2, $=[\vartheta(0)]^2$, being determined by putting $u'=0$, in which case q_1', q_2', $q'_{1,\,2}$ all vanish.

If in this formula we write $v=u+u^{a_1,\,a}+u^{a_2,\,a}$ in place of u, we obtain, from the formulae

$$q_1^2(u+u^{a_1,\,a}+u^{a_2,\,a}),\ =q_1^2(v),\ =\frac{\vartheta^2(v|u^{a_1,\,a})}{\vartheta^2(v)},\ =\frac{\vartheta^2(u|u^{a_2,\,a})}{\vartheta^2(u|u^{a_1,\,a}+u^{a_2,\,a})}=\frac{q_2^2(u)}{q_{12}^2(u)},$$
$$q_2^2(u+u^{a_1,\,a}+u^{a_2,\,a})=-\frac{q_1^2(u)}{q_{12}^2(u)},\quad q_{12}^2(u+u^{a_1,\,a}+u^{a_2,\,a})=-\frac{1}{q_{12}^2(u)},$$

which are easy to verify from the formulae of § 190, Chap. X. and the table of characteristics given in this chapter, that

$$\frac{\vartheta^2\,.\,\vartheta(u+u'|u^{a_1,\,a}+u^{a_2,\,a})\,\vartheta(u-u'|u^{a_1,\,a}+u^{a_2,\,a})}{\vartheta^2(u|u^{a_1,\,a}+u^{a_2,\,a})\,\vartheta^2(u')}=1+\frac{q_2^2q_1'^2}{q_{12}^2}-\frac{q_2'^2q_1^2}{q_{12}^2}-\frac{q_{12}'^2}{q_{12}^2},$$

and therefore

$$\frac{\vartheta^2\,.\,\overline{\vartheta}(u+u')\,\overline{\vartheta}(u-u')}{\vartheta^2(u)\,\vartheta^2(u')}=q_{12}^2-q_{12}'^2-q_1^2q_2'^2+q_2^2q_1'^2,$$

where $\bar{\vartheta}(u)$ denotes $\vartheta(u \mid u^{a_1, a} + u^{a_2, a})$. But we can use the result of Ex. iv. § 217, to give the right-hand side a still further form, namely

$$\frac{a_1 - a_2}{\mu_1 \mu_2}[\wp_{11}(u) + \wp_{11}(u') + \wp_{12}(u)\wp_{22}(u') - \wp_{12}(u')\wp_{22}(u)].$$

Further if $u^{a_1, a} + u^{a_2, a} \equiv \frac{1}{2}\Omega_{m, m'}$, where m, m' consist of integers each either 0 or 1, we find, by adding $\frac{1}{2}\Omega_{m, m'}$ to u and u' and utilising the fact (§ 190) that

$$\lambda_m(u + u') = 2\lambda_{\frac{1}{2}m}(u) + 2\lambda_{\frac{1}{2}m}(u'),$$

that

$$\frac{\vartheta^2 \bar{\vartheta}(u + u')\bar{\vartheta}(u - u')}{\bar{\vartheta}^2(u)\bar{\vartheta}^2(u')}\frac{\mu_1 \mu_2}{a_1 - a_2} = \wp_{11}(v) - \wp_{11}(v') + \wp_{12}(v)\wp_{22}(v') - \wp_{12}(v')\wp_{22}(v),$$

where $v = u + \frac{1}{2}\Omega_{m, m'}$, $v' = u' + \frac{1}{2}\Omega_{m, m'}$. It should be noticed that

$$\wp_{i, j}(v) = -\frac{\partial^2}{\partial u_i \partial u_j}\log \vartheta(u\,;\ \tfrac{1}{2}m, \tfrac{1}{2}m')\,;\ \text{hence}$$

this formula can be expressed so as to involve only a single function in the form

$$\frac{\vartheta^2 \mu_1 \mu_2}{a_1 - a_2} \cdot \frac{\sigma(u + v)\sigma(u - v)}{\sigma^2(u)\sigma^2(v)} = \wp_{11}(u) - \wp_{11}(v) + \wp_{12}(u)\wp_{22}(v) - \wp_{12}(v)\wp_{22}(u)$$

where $\sigma(u)$ denotes $\vartheta\left(u \,\middle|\, \frac{1}{2}\begin{pmatrix}01\\11\end{pmatrix}\right)$, and $\wp_{i, j}(u) = -\frac{\partial^2}{\partial u_i \partial u_j}\log \sigma(u)$. In Weierstrass's corresponding formula for $p = 1$, the function $\sigma(u)$ is determined so that $\sigma(u)/u = 1$ when $u = 0$. To introduce the corresponding conditions here would carry us further into detail. (See §§ 212, 213.)

Ex. Prove that if a_3 denote any one of the branch places c, c_1, c_2, $\alpha = (a_2 - a_3)$, $\beta = (a_3 - a_1)$, $\gamma = (a_1 - a_2)$, $P_1 = (a_1 - x_1)(a_1 - x_2)$, etc., $P_1' = (a_1 - x_1')(a_1 - x_2')$, etc., and

$$A = \left[\frac{y_1}{(x_1 - a_1)(x_1 - a_3)} - \frac{y_2}{(x_2 - a_1)(x_2 - a_3)}\right]\frac{1}{x_2 - x_1},$$

$$B = \left[\frac{y_1}{(x_1 - a_2)(x_1 - a_3)} - \frac{y_2}{(x_2 - a_2)(x_2 - a_3)}\right]\frac{1}{x_2 - x_1},$$

with similar notation for A', B', then the determinant Δ can be expressed in the form

$$\Delta = \frac{y_1 y_2}{P(x_1)P(x_2)} + \frac{y_1' y_2'}{P(x_1')P(x_2')} - X,$$

where

$$\gamma^2 X = AA'(P_1 P_3' + P_3 P_1') + BB'(P_2 P_3' + P_3 P_2') - AB'(\gamma\alpha P_3 + \gamma\beta P_3' + P_1 P_3' + P_2' P_3)$$
$$- A'B(\gamma\beta P_3 + \gamma\alpha P_3' + P_2 P_3' + P_1' P_3).$$

In this form Δ can be immediately expressed in terms of theta quotients.

219. Consider, nextly, the function

$$\frac{\vartheta(u + u' \mid u^{a_1, a})\vartheta(u - u')}{\vartheta^2(u)\vartheta^2(u')}.$$

This is not a periodic function of u, u'. Thus we take in the first place the function

$$\frac{\vartheta(u+u'|u^{a_1,a})\,\vartheta(u-u')}{\vartheta(u)\,\vartheta(u|u^{a_1,a})\,\vartheta(u')\,\vartheta(u'|u^{a_1,a})}.$$

Put

$$u \equiv u^{x_1,a_1} + u^{x_2,a_2},\ u' \equiv u^{x_1',a_1} + u^{x_2',a_2};$$

then, as functions of x_1, the zeros of $\vartheta(u)$, $\vartheta(u|u^{a_1,a})$ respectively are a, $\bar{x}_2$ and a_1, $\bar{x}_2$, the zeros of $\vartheta(u+u'|u^{a_1,a})$ are found in the usual way to be zeros of a rational function of the fifth order having a_1^2, a_2^3 as poles, and x_2, x_1', x_2' as zeros; such a function of x_1 is $\Delta_1/P(x_1)$, where $P(x_1)=(x_1-a_1)(x_1-a_2)$ and

$$\Delta_1 = \begin{vmatrix} \eta_1(x_1-a_1), & x_1^2, & x_1, & 1 \\ \eta_2(x_2-a_1), & x_2^2, & x_2, & 1 \\ \eta_1'(x_1'-a_1), & x_1'^2, & x_1', & 1 \\ \eta_2'(x_2'-a_1), & x_2'^2, & x_2', & 1 \end{vmatrix},$$

wherein $\eta_1 = y_1/P(x_1)$, etc.; the zeros of $\vartheta(u-u')$, as a function of x_1, are similarly zeros of a function of the sixth order having a_1^3, a_2^3 as poles and a, x_2, $\bar{x}_1'$, $\bar{x}_2'$ for its other zeros; such a function of x_1 is $\bar{\Delta}/P(x_1)$, where

$$\bar{\Delta} = \begin{vmatrix} \eta_1 x_1, & \eta_1, & x_1, & 1 \\ \eta_2 x_2, & \eta_2, & x_2, & 1 \\ -\eta_1' x_1', & -\eta_1', & x_1', & 1 \\ -\eta_2' x_2', & -\eta_2', & x_2', & 1 \end{vmatrix};$$

hence we find

$$\frac{\vartheta(u+u'|u^{a_1,a})\,\vartheta(u-u')}{\vartheta(u)\,\vartheta(u|u^{a_1,a})\,\vartheta(u')\,\vartheta(u'|u^{a_1,a})}$$

$$= C\frac{\Delta_1\bar{\Delta}(x_1-a_2)(x_2-a_2)(x_1'-a_2)(x_2'-a_2)}{(x_1-x_2)^2(x_1-x_1')(x_1-x_2')(x_2-x_1')(x_2-x_2')(x_1'-x_2')^2},$$

wherein C is an absolute constant; for it is immediately seen that the two sides of this equation have the same poles and zeros.

We proceed to put the right-hand side into a particular form; for this purpose we introduce certain notations; denote the quantities c, c_1, c_2, which refer to the branch places other than a_1, a_2 by a_3, a_4, a_5 in any order; denote $(a_i-x_1)(a_i-x_2)$ by p_i, $(a_i-x_1')(a_i-x_2')$ by p_i'; denote by $\pi_{i,j}$ the expression

$$\tfrac{1}{2}\left[\frac{y_1}{(x_1-a_i)(x_1-a_j)} - \frac{y_2}{(x_2-a_i)(x_2-a_j)}\right]\frac{1}{x_2-x_1},$$

and write $p_{i,j}$ for $p_i p_j \pi_{i,j}$, with a similar notation $\pi'_{i,j}$, $p'_{i,j}$; also let $P(x)=(x-a_1)(x-a_2)$, $\eta_1 = y_1/P(x_1)$, etc.

Then, by regarding the expression

$$(x_1'-x_2')^2(x_2-a_3)\frac{(a_2-a_4)(a_2-a_5)}{(a_2-x_2)(a_2-x_1')(a_2-x_2')}$$

as a function of a_2, and putting it into partial fractions in the ordinary way, we find that it is equal to

$$\frac{1}{a_2-x_2}(x_1'-x_2')^2(x_2-a_3)\frac{(x_2-a_4)(x_2-a_5)}{(x_2-x_1')(x_2-x_2')}+\frac{1}{a_2-x_1'}(x_1'-x_2')^2(x_2-a_3)\frac{(x_1'-a_4)(x_1'-a_5)}{(x_1'-x_2)(x_1'-x_2')}$$
$$+\frac{1}{a_2-x_2'}(x_1'-x_2')^2(x_2-a_3)\frac{(x_2'-a_4)(x_2'-a_5)}{(x_2'-x_2)(x_2'-x_1')};$$

using then the identities

$$-(x_2-a_3)(x_1'-x_2')=(x_2'-x_2)(x_1'-a_3)-(x_1'-x_2)(x_2'-a_3),$$
$$(x_2-a_3)(x_1'-x_2')=(x_1'-x_2)(x_2'-a_3)-(x_2'-x_2)(x_1'-a_3),$$

we are able to give the same expression the form

$$\tfrac{1}{4}\eta_2^2(x_2-a_1)\frac{(x_1'-x_2')^2}{(x_1'-x_2)(x_2'-x_2)}-\tfrac{1}{4}\eta_1'^2(x_1'-a_1)\frac{x_2'-x_2}{x_1'-x_2}-\tfrac{1}{4}\eta_2'^2(x_2'-a_1)\frac{x_1'-x_2}{x_1'-x_2}$$
$$+\frac{x_2'-a_3}{x_1'-a_2}(x_1'-a_4)(x_1'-a_5)+\frac{x_1'-a_3}{x_2'-a_2}(x_2'-a_4)(x_2'-a_5),$$

where $\tfrac{1}{4}\eta_1^2=(x_1-a_3)(x_1-a_4)(x_1-a_5)$, etc.; thus

$$\tfrac{1}{4}\eta_1'^2(x_1'-a_1)(x_2'-x_2)^2+\tfrac{1}{4}\eta_2'^2(x_2'-a_1)(x_1'-x_2)^2-\tfrac{1}{4}\eta_2^2(x_2-a_1)(x_1'-x_2')^2$$
$$=-(a_2-a_4)(a_2-a_5)(x_1'-x_2')^2(x_1'-x_2)(x_2'-x_2)(x_1-a_2)(x_2-a_3)\frac{1}{p_2^2p_2'^2}$$
$$+\frac{1}{p_2'^2}(x_1'-x_2)(x_2'-x_2)\{(x_2'-a_2)(x_2'-a_3)(x_1'-a_4)(x_1'-a_5)$$
$$+(x_1'-a_2)(x_1'-a_3)(x_2'-a_4)(x_2'-a_5)\}.$$

Now we have, by expansion,

$$\bar{\Delta}=(\eta_1\eta_2+\eta_1'\eta_2')(x_1-x_2)(x_1'-x_2')+(\eta_1\eta_1'+\eta_2\eta_2')(x_1'-x_1)(x_2'-x_2)$$
$$-(\eta_1\eta_2'+\eta_2\eta_1')(x_1'-x_2)(x_2'-x_1),$$

$$\Delta_1=\eta_1(x_1-a_1)(x_1'-x_2)(x_2'-x_2)(x_1'-x_2')-\eta_2(x_2-a_1)(x_1'-x_1)(x_2'-x_1)(x_1'-x_2')$$
$$+\eta_1'(x_1'-a_1)(x_2'-x_1)(x_2'-x_2)(x_1-x_2)-\eta_2'(x_2'-a_1)(x_1'-x_1)(x_1'-x_2)(x_1-x_2),$$

and in the product $\bar{\Delta}\Delta$ there will be two kinds of terms

(i) $$-\eta_1'\eta_2'(\eta_1-\eta_2)\gamma(x_1-x_2)(x_1'+x_2'-2a_1),$$

where γ denotes $(x_1'-x_1)(x_1'-x_2)(x_2'-x_1)(x_2'-x_2)$, there being four terms of this kind obtainable from this by the interchange of the suffixes 1 and 2, and the interchange of dashed and undashed letters,

(ii) $$\eta_1(x_2'-x_1)(x_1'-x_1)(x_1-x_2)\{\eta_1'^2(x_1'-a_1)(x_2'-x_2)^2+\eta_2'^2(x_2'-a_1)(x_1'-x_2)^2$$
$$-\eta_2^2(x_2-a_1)(x_1'-x_2')^2\},$$

there being three other terms similarly derivable from this one.

Consider now the expression

$$(a_2-a_4)(a_2-a_5)(p_{13}p_3p_1'+p'_{13}p_3'p_1)+p_{12}p_2p'_{23}p'_{45}+p'_{12}p_2'p_{23}p_{45},$$

and, of this, consider only the terms

$$(a_2-a_4)(a_2-a_5)p_{13}p_3p_1'+p_{12}p_2p'_{23}p'_{45};$$

by substitution of the values for p_{13} etc., and arrangement, we immediately find that these terms are equal to

$$-\tfrac{1}{8}p_2'^2p_2^2p_1'p_1\frac{\eta_1'\eta_2'(\eta_1-\eta_2)}{(x_1-x_2)(x_1'-x_2')^2}(x_1'+x_2'-2a_1)$$

$$-\tfrac{1}{2}(a_2-a_4)(a_2-a_5)p_1p_1'\frac{\eta_1(x_1-a_2)(x_2-a_3)-\eta_2(x_2-a_2)(x_1-a_3)}{x_1-x_2}$$

$$+\tfrac{1}{2}p_1p_1'p_2^2\frac{\eta_1-\eta_2}{(x_1-x_2)(x_1'-x_2')^2}\{(x_1'-a_2)(x_1'-a_3)(x_2'-a_4)(x_2'-a_5)$$

$$+(x_1'-a_4)(x_1'-a_5)(x_2'-a_2)(x_2'-a_3)\};$$

this expression, as we see by utilising an identity which was developed at the commencement of the investigation, is equal to

$$-\tfrac{1}{8}p_1p_1'p_2^2p_2'^2\frac{\eta_1'\eta_2'(\eta_1-\eta_2)}{(x_1-x_2)(x_1'-x_2')^2}(x_1'+x_2'-2a_1)+\tfrac{1}{8}\frac{p_2^2p_2'^2p_1p_1'}{(x_1-x_2)(x_1'-x_2')^2(x_1'-x_2)(x_2'-x_2)}K,$$

where K denotes

$$\eta_1[\eta_1'^2(x_1'-a_1)(x_2'-x_2)^2+\eta_2'^2(x_2'-a_1)(x_1'-x_2)-\eta_2^2(x_2-a_1)(x_1'-x_2')^2]$$

$$-\eta_2[\eta_2'^2(x_2'-a_1)(x_1'-x_1)^2+\eta_1'^2(x_1'-a_1)(x_2'-x_1)-\eta_1^2(x_1-a_1)(x_1'-x_2')^2].$$

Comparing this form with the terms occurring in the expansion for $\bar{\Delta}\Delta_1$, we obtain the result

$$\tfrac{1}{8}\frac{p_1p_1'p_2^2p_2'^2\bar{\Delta}\Delta_1}{(x_1-x_2)^2(x_1'-x_2')^2(x_1'-x_1)(x_1'-x_2)(x_2'-x_1)(x_2'-x_2)}$$

$$=(a_2-a_4)(a_2-a_5)(p_{13}p_3p_1'+p'_{13}p_3'p_1)+p_{12}p_2p'_{23}p'_{45}+p'_{12}p_2'p_{23}p_{45}.$$

Now we have (§§ 216, 213, 212) the formulae $p_i^2=\mu_iq_i^2$, $\dfrac{q^2_{i,j}}{q_i^2q_j^2}=\pm(a_i-a_j)\dfrac{p^2_{i,j}}{p_i^2p_j^2}$; we shall therefore put $p_i=M_iq_i$, $p_{i,j}=N_{i,j}q_{i,j}$; hence by the formula (p. 334) the quotient

$$\frac{\vartheta(u+u'|u^{a_1,a})\,\vartheta(u-u')}{\vartheta^2(u)\,\vartheta^2(u')}$$

is a certain constant multiple of the function

$$(a_2-a_4)(a_2-a_5)M_1M_3N_{13}(q_{13}q_3q_1'+q'_{13}q_3'q_1)+N_{12}N_{23}N_{45}M_2(q_{12}q_2q'_{23}q'_{45}+q'_{12}q_2'q_{23}q_{45}).$$

Also we have $M_i^2=\mu_i$, $N^2_{i,j}=\pm\mu_i\mu_j/(a_i-a_j)$, where $\mu_i=\pm\sqrt{-f'(a_i)}$ when $i=1$ or 2, and $\mu_i=\pm\sqrt{f'(a_i)}$ when $i=3$, 4, 5. Hence it is easy to prove that the fourth powers of the quantities $(a_2-a_4)(a_2-a_5)M_1M_3N_{13}$, $N_{12}N_{23}N_{45}M_2$ are equal.

Hence we have

$$A\frac{\vartheta(u+u'|u^{a_1,a})\,\vartheta(u-u')}{\vartheta^2(u)\,\vartheta^2(u')}=\epsilon(q_{13}q_3q_1'+q'_{13}q_3'q_1)+q_{12}q_2q'_{23}q'_{45}+q'_{12}q_2'q_{23}q_{45},$$

where A is a certain constant, and ϵ a certain fourth root of unity. The value of ϵ is determined by a subsequent formula.

220. The equation just obtained (§ 219) taken with a previous formula gives the result

$$C\frac{\vartheta(u+u'|u^{a_1,a})}{\vartheta(u+u')}=\frac{\epsilon(q_{13}q_3q_1'+q'_{13}q_3'q_1)+q_{12}q_2q'_{23}q'_{45}+q'_{12}q_2'q_{23}q_{45}}{1+q_1^2q_1^{12}+q_2^2q_2^{12}+q_{12}^2q_{12}^{12}},$$

and limiting ourselves to one case, we may now take the places a_3, a_4, a_5 to be, respectively, c_1, c_2, c, and introduce Weierstrass's theta functions; *defining* the ten even functions* $\vartheta_5(u)$, $\vartheta_{23}(u), \dots, \vartheta_{03}(u)$ *to be respectively identical with the functions* $\vartheta(u)$, $\vartheta_{ac}(u), \dots, \vartheta_{ca_1}(u)$, *and the six odd functions* $\vartheta_{02}(u), \dots, \vartheta_3(u)$ *to be respectively the negatives of the functions* $\vartheta_{aa_1}(u), \dots, \vartheta_{cc_2}(u)$, the right-hand side of the equation is equivalent to

$$-\frac{\epsilon(\vartheta_5\vartheta_{02}\vartheta'_{01}\vartheta'_{12}+\vartheta_5'\vartheta'_{02}\vartheta_{01}\vartheta_{12})+\vartheta_{04}\vartheta_{24}\vartheta'_{14}\vartheta_3'+\vartheta'_{04}\vartheta'_{24}\vartheta_{14}\vartheta_3}{\vartheta_5^2\vartheta_5'^2+\vartheta_{02}^2\vartheta'^2_{02}+\vartheta_{24}^2\vartheta'^2_{24}+\vartheta_{04}^2\vartheta'^2_{04}};$$

here ϑ denotes $\vartheta(u)$, ϑ' denotes $\vartheta(u')$, and C is an absolute constant. This equation may be called the addition formula for the function q_1, and is one of a set which are the generalisation to the case $p=2$ of such formulae as that arising for $p=1$ in the form

$$\operatorname{sn}(u+u')=\frac{\operatorname{sn} u \operatorname{cn} u' \operatorname{dn} u'+\operatorname{sn} u' \operatorname{cn} u \operatorname{dn} u}{1-k^2\operatorname{sn}^2 u \operatorname{sn}^2 u'}.$$

By interchanging the suffixes 1 and 2 we obtain an analogous expression for $\vartheta(u+u'|u^{a_2, a}) \div \vartheta(u+u')$; if in this expression we add the half-period $u^{a_1, a}$ to u we obtain an expression for the function $\vartheta(u+u'|u^{a_1, a}+u^{a_2, a})$ $\div\vartheta(u+u'|u^{a_1, a})$; and if this be multiplied by the expression just developed for the function $\vartheta(u+u'|u^{a_1, a}) \div \vartheta(u+u')$ we obtain an expression for $\vartheta(u+u'|u^{a_1, a}+u^{a_2, a}) \div \vartheta(u+u')$, and it can be shewn that the form obtained can be reduced to have the same denominator as in the expression here developed at length. The formulae are however particular cases of results obtained in subsequent chapters, and will not be further developed here. For that development such results as those contained in the following examples are necessary; these results are generalisations of such formulae as $\operatorname{sn}(u+K)=\operatorname{cn} u/\operatorname{dn} u$ which occur in the case $p=1$.

Ex. Prove, if $q_i(u)=\vartheta(u|u^{a_i, a})\div\vartheta(u)$, $q_{i,j}(u)=\vartheta(u|u^{a_i, a}+u^{a_j, a})\div\vartheta(u)$, etc., that (see the table § 204, and the formulae Chap. X. § 190)

$$q_1(u+u^{a_1, a})=-e^{\frac{1}{2}\pi i}/q_1(u), \quad q_2(u+u^{a_1, a})=-\frac{q_{1,2}(u)}{q_1(u)},$$

$$q_2(u+u^{a_1, a}+u^{a_2, a})=e^{\frac{1}{2}\pi i}\frac{q_1(u)}{q_{12}(u)},$$

and obtain the complete set of formulae.

221. In case $p=2$ there are five quotients of the form $\vartheta(u|u^{b, a})\div\vartheta(u)$, and ten of the form $\vartheta(u|u^{b_1, a}+u^{b_2, a})\div\vartheta(u)$, wherein b, b_1, b_2 denote any finite branch places. Since the arguments u may be written in the form $u^{x_1, a_1}+u^{x_2, a_2}$, the fifteen quotients are connected by thirteen algebraic relations. In virtue of the algebraic expression of these fifteen quotients, they may be studied independently of the theta functions. We therefore give below some examples of the equations connecting them.

* Königsberger, *Crelle*, LXIV. (1865), p. 22. In the letter notation (§ 204) the reduced characteristic symbols are such (§ 203) that each of k_s, k'_s is positive, or zero, and less than 2. In Weierstrass's notation the reduced symbols have the elements k'_s positive, or zero, and the elements k_s negative, or zero.

Ex. i. There is one relation, known as Göpel's biquadratic relation, which is of importance in itself, in view of developments that have arisen from it, and is of some historical interest.

Let $$q_c = \frac{\vartheta(u|u^{c,\,a})}{\vartheta(u)},\quad q_{a_1,\,a_2} = \frac{\vartheta(u|u^{a_1,\,a} + u^{a_2,\,a})}{\vartheta(u)},\quad q_{c_1,\,c_2} = \frac{\vartheta(u|u^{c_1,\,a} + u^{c_2,\,a})}{\vartheta(u)},$$

be three functions whose suffixes, together, involve all the five finite branch places. Then these three functions satisfy a biquadratic relation, which, if the functions be regarded as Cartesian coordinates in a space of three dimensions, represents a quartic surface with sixteen nodal points.

In fact, if p_a denote $\sqrt{(a-x_1)(a-x_2)}$, and $p_{b_1,\,b_2}$ denote the function

$$\tfrac{1}{2}p_{b_1}p_{b_2}\left[\frac{y_1}{(x_1-b_1)(x_1-b_2)} - \frac{y_2}{(x_2-b_1)(x_2-b_2)}\right]\frac{1}{x_2-x_1},$$

we have

$$p^2_{b_1,\,b_2}$$
$$= \frac{4(x_2-b_1)(x_2-b_2)(x_1-e_1)(x_1-e_2)(x_1-e_3) + 4(x_1-b_1)(x_1-b_2)(x_2-e_1)(x_2-e_2)(x_2-e_3) - 2y_1y_2}{4(x_1-x_2)^2},$$

where b_1, b_2, e_1, e_2, e_3 are the finite branch places in any order; and if this be denoted by

$$\frac{\psi(x_1, x_2) - 2y_1y_2}{4(x_1-x_2)^2},$$

it is immediately obvious that $\psi(x, x) = 2y^2$, $= 2f(x)$, say, and $\left[\frac{\partial}{\partial z}\psi(x, z)\right]_{z=x} = \frac{\partial f(x)}{\partial x}$; thus there is (§ 211, Ex. vii.) an equation of the form

$$p^2_{b_1,\,b_2} = \frac{f(x_1, x_2) - 2y_1y_2}{4(x_1-x_2)^2} + Ax_1x_2 + B(x_1+x_2) + C,$$

where $f(x_1, x_2)$ is a certain symmetrical expression of frequent occurrence (cf. § 217), the same whatever branch places b_1, b_2 may be, and A, B, C are such that $\psi(x_1, x_2)$ vanishes when for x_1, x_2 are put any one of the four pairs of values (b_1, b_2), (e_2, e_3), (e_3, e_1), (e_1, e_2); therefore the difference between any two expressions such as $p^2_{b_1,\,b_2}$, formed for different pairs of finite branch places, is expressible in the form $Lx_1x_2 + M(x_1+x_2) + N$; thus there must be an equation of the form

$$p^2_{a_1,\,c_1} = \lambda p^2_{a_1,\,a_2} + \mu p^2_{c_1,\,c_2} + \nu p^2_c + \rho,$$

where λ, μ, ν, ρ are independent of the places x_1, x_2.

Similarly

$$p^2_{a_2,\,c_2} = \lambda' p^2_{a_1,\,a_2} + \mu' p^2_{c_1,\,c_2} + \nu' p^2_c + \rho'.$$

But also it can be verified that

$$p_{a_1,\,a_2}p_{c_1,\,c_2} - p_{a_1,\,c_1}p_{a_2,\,c_2} = -(a_2-c_1)(a_1-c_2)p_c, \; = \kappa p_c, \text{ say};$$

thus we have

$$[\lambda p^2_{a_1,\,a_2} + \mu p^2_{c_1,\,c_2} + \nu p^2_c + \rho][\lambda' p^2_{a_1,\,a_2} + \mu' p^2_{c_1,\,c_2} + \nu' p^2_c + \rho'] = [p_{a_1,\,a_2}p_{c_1,\,c_2} - \kappa p_c]^2,$$

and when the expressions $p_{a_1,\,a_2}$, etc., are replaced by the functions $q_{a_1,\,a_2}$, etc. (§ 210), this is the biquadratic relation in question. This proof is practically that given by Göpel (*Crelle*, xxxv. 1847, p. 291).

Ex. ii. Prove that

$$\frac{p^2_{a_1, a_2} - p^2_{a_1, c_1}}{a_2 - c_1} + p^2_{a_1} = (a_1 - a_2)(a_1 - c_1),$$

$$\frac{p^2_{c_1, c_2} - p^2_{a_1, c_1}}{c_2 - a_1} + p^2_{c_1} = (c_1 - c_2)(c_1 - a_1),$$

$$\frac{p^2_{a_1}}{(a_1 - c_1)(a_1 - c)} + \frac{p^2_{c_1}}{(c_1 - a_1)(c_1 - c)} + \frac{p^2_{c}}{(c - a_1)(c - c_1)} = 1,$$

and hence develop the method of *Ex.* i. in detail.

Ex. iii. For any value of p prove

(a) that the squares of any p of the theta quotients q_b, $= \vartheta(u \mid u^{b, a}) \div \vartheta(u)$, are connected by a linear relation,

(β) that the squares of any p of the theta quotients

$$q_b, \; q_{b, b_1}, \; q_{b, b_2}, \; q_{b, b_3}, \; \ldots\ldots$$

are connected by a linear relation. (Weierstrass, *Math. Werke*, vol. I. p. 332.) These equations generalise the relations of *Ex.* ii.

Ex. iv. Another method of obtaining the biquadratic relations is as follows; if

$$\vartheta_q(v) = \Sigma e^{2\pi i v (n+q') + i\pi\tau (n+q')^2 + 2\pi i q (n+q')}$$

$$\Theta_q(V) = \Sigma e^{2\pi i V (n+q') + \frac{1}{2} i\pi\tau (n+q')^2 + 2\pi i q (n+q')},$$

$V = \frac{1}{2} v$, and, in Weierstrass's notation,

$$x = \vartheta_5(v), \quad y = \vartheta_{01}(v), \quad z = \vartheta_4(v), \quad t = \vartheta_{23}(v),$$

so that $x : y : z : t = 1 : q_{a_1, c_1} : q_{a_2, c_2} : q_c$, and if a, b, c, d denote the values of x, y, z, t when $v = 0$, and the linear function $cx + dy - az - bt$ be denoted by $(c, d, -a, -b)$, etc., then it can be proved, by actual multiplication of the series, that

$$\Theta_3^2(V) = (c, d, -a, -b), \quad \Theta_{14}^2(V) = (d, -c, -b, a), \quad \Theta_{02}^2(V) = (b, -a, d, -c)$$

$$\Theta_5^2(V) = (a, b, c, d) \quad , \quad \Theta_1^2(V) = (b, -a, -d, c), \quad \Theta_{34}^2(V) = (a, b, -c, -d).$$

Relations of this character are actually obtained by Göpel, in this way. It will be sufficient, for the purpose of introducing the subject of a subsequent chapter, if the method of obtaining one of these relations be explained here. The general term of the series $\Theta_{02}(V)$ is (cf. the table § 204 and § 220)

$$-e^{\pi i v (n+q') + \frac{1}{2} i\pi\tau (n+q')^2 + 2\pi i q (n+q')},$$

where $q' = \frac{1}{2}(1, 0)$, $q = \frac{1}{2}(1, 0)$, namely is

$$-e^{\pi i [v_1 (n_1 + \frac{1}{2}) + v_2 n_2] + \frac{1}{2}\pi i [\tau_{11} (n_1 + \frac{1}{2})^2 + 2\tau_{12} (n_1 + \frac{1}{2})(n_2) + \tau_{22} n_2^2] + i\pi (n_1 + \frac{1}{2})};$$

thus the exponent of the general term in the product $\Theta_{02}^2(V)$ is $\pi i L$, where L is equal to

$$v_1 (n_1 + m_1 + 1) + v_2 (n_2 + m_2) + \tfrac{1}{2}\tau_{11} [(n_1 + \tfrac{1}{2})^2 + (m_1 + \tfrac{1}{2})^2] + \tau_{12} [(n_1 + \tfrac{1}{2}) n_2 + (m_1 + \tfrac{1}{2}) m_2]$$
$$\tfrac{1}{2}\tau_{22} (n_2^2 + m_2^2) + n_1 + m_1 + 1;$$

there are therefore four kinds of terms in the product according to the evenness or oddness of the two integers n_1+m_1, n_2+m_2. Consider only one kind, namely when n_1+m_1, n_2+m_2 are both even, respectively equal to $2N_1$, $2N_2$, say; then L is equal to

$$2v_1(N_1+\tfrac{1}{2})+2v_2N_2+\tau_{11}(N_1+\tfrac{1}{2})^2+2\tau_{12}(N_1+\tfrac{1}{2})N_2+\tau_{22}N_2^2$$
$$+\tau_{11}\left(\frac{n_1-m_1}{2}\right)^2+2\tau_{12}\left(\frac{n_1-m_1}{2}\right)\left(\frac{n_2-m_2}{2}\right)+\tau_{22}\left(\frac{n_2-m_2}{2}\right)^2$$
$$+2N_1+1;$$

if now we put $\frac{n_1-m_1}{2}=M_1$, $\frac{n_2-m_2}{2}=M_2$, we have

$$n_1=N_1+M_1,\quad m_1=N_1-M_1,\quad n_2=N_2+M_2,\quad m_2=N_2-M_2;$$

thus, to any assigned values of the integers N_1, N_2, M_1, M_2 there correspond integers n_1, n_2, m_1, m_2 such that n_1+m_1, n_2+m_2 are both even; therefore, as

$$e^{2\pi i v_1(N_1+\frac{1}{2})+2\pi i v_2N_2+i\pi\tau_{11}(N_1+\frac{1}{2})^2+2i\pi\tau_{12}(N_1+\frac{1}{2})N_2+i\pi\tau_{22}N_2^2}$$

is a term of the series $\vartheta\left(v;\tfrac{1}{2}\begin{pmatrix}10\\00\end{pmatrix}\right)$, that is, of $\vartheta_{01}(v)$, and

$$e^{i\pi\tau_{11}M_1^2+2\pi i\tau_{12}M_1M_2+\tau_{22}M_2^2}$$

is a term of the series $\vartheta\left(0;\tfrac{1}{2}\begin{pmatrix}00\\00\end{pmatrix}\right)$, that is, of $\vartheta_5(v)$, and $e^{i\pi(2N_1+1)}=-1$, it follows that the terms of $\Theta_{02}{}^2(V)$ which are of the kind under consideration consist of all the terms of the product $-\vartheta_5.\vartheta_{01}(v)$, or $-ay$. It can similarly be seen that the three other sorts of terms, when n_1+m_1 is even and n_2+m_2 odd, when n_1+m_1 is odd and n_2+m_2 odd or even, are, in their aggregate the terms of the sum $bx+dz-ct$.

We can also, in a similar way, prove the equations

$$\Theta_{03}\Theta_{23}\Theta_3(V)\Theta_{14}(V)+\Theta_0\Theta_2\Theta_{02}(V)\Theta_5(V)=\Theta_{12}\Theta_{01}\Theta_1(V)\Theta_{34}(V),$$
$$\Theta_{03}{}^2=2(ac-bd),\ \Theta_{23}{}^2=2(ad+bc),\ \Theta_2{}^2=2(ab-cd),\ \Theta_{01}{}^2=2(ab+cd),$$
$$\Theta_0{}^2=a^2-b^2-c^2+d^2,\ \Theta_{12}{}^2=a^2-b^2+c^2-d^2,$$

Θ_{03} denoting $\Theta_{03}(0)$, etc.

Hence the equation of the quartic surface is obtainable in the form

$$\sqrt{2(ac-bd)(ad+bc)(c,\ d,\ -a,\ -b)(d,\ -c,\ -b,\ a)}$$
$$+\sqrt{(a^2-b^2-c^2+d^2)(ab-cd)(b,\ -a,\ d,\ -c)(a,\ b,\ c,\ d)}$$
$$=\sqrt{(a^2-b^2+c^2-d^2)(ab+cd)(b,\ -a,\ -d,\ c)(a,\ b,\ -c,\ -d)}.$$

A relation of this form is rationalised by Cayley in *Crelle's Journal*, LXXXIII. (1877), p. 215. The form obtained is shewn by Borchardt, *Crelle*, LXXXIII. (1877), p. 239, to be the same as that obtained by Göpel. See also Kummer, *Berlin. Monats.* 1864, p. 246, and *Berlin. Abhand.* 1866, p. 64; Cayley, *Crelle*, LXXXIV., XCIV.; and Humbert, *Liouville*, 4me Sér., t. IX. (1893); Schottky, *Crelle*, CV. pp. 233, 269; Wirtinger, *Untersuchungen über Thetafunctionen* (Leipzig, 1895).

The rationalised form of the equation, from which the presence of the sixteen nodes is obvious, is obtained in chapter XV. of the present volume.

Ex. v. Obtain the following relations, connecting the ratios of the values of the even theta functions for zero values of the arguments when $p=2$. They may be obtained from the relations (§ 212)

$$(b-x_1)(b-x_2)=\pm\sqrt{e^{\pi i PP'} f'(b)}\,\vartheta^2(u|u^{b,a})\div\vartheta^2(u)$$

by substituting special values for x_1 and x_2.

$$\vartheta^4 : \vartheta^4_c : \vartheta^4_{c_1} : \vartheta^4_{c_2} : \vartheta^4_{a_1c_1} : \vartheta^4_{a_1c_2} : \vartheta^4_{a_2c_1} : \vartheta^4_{a_2c_2} : \vartheta^4_{ca_2} : \vartheta^4_{ca_1}$$

$$=(c_1-c_2)(c_2-c)(c-c_1).(a_1-a_2) : (a_1-a_2)(a_2-c)(c-a_1).(c_1-c_2)$$
$$: (a_1-a_2)(a_2-c_1)(c_1-a_1).(c_2-c) : (a_1-a_2)(a_2-c_2)(c_2-a_1).(c_1-c)$$
$$: (c_2-c)(c-a_1)(a_1-c_2).(c_1-a_2) : (c-c_1)(c_1-a_1)(a_1-c).(c_2-a_2)$$
$$: (c-c_2)(c_2-a_2)(a_2-c).(c_1-a_1) : (c-c_1)(c_1-a_2)(a_2-c).(c_2-a_1)$$
$$: (c_1-c_2)(c_2-a_2)(a_2-c_1).(a_1-c) : (c_1-c_2)(c_2-a_1)(a_1-c_1).(a_2-c).$$

Infer that

$$\vartheta^4_{ca_2}\vartheta^4_{c_2a_2} : \vartheta^4_{a_1c_2}\vartheta^4_{ca_1} : \vartheta^4_c\vartheta^4_{c_2}=(a_2-c_1)^2 : (c_1-a_1)^2 : (a_1-a_2)^2.$$

We have proved (§§ 210, 213) that

$$\sqrt{a_2-c_1}\,\vartheta_{a_1}(u)\,\vartheta_{a_2c_1}(u)+\sqrt{c_1-a_1}\,\vartheta_{a_2}(u)\,\vartheta_{a_1c_1}(u)+\sqrt{a_1-a_2}\,\vartheta_{c_1}(u)\,\vartheta_{a_1a_2}(u)=0$$

and we have in fact, as follows from formulae developed subsequently, the equation

$$\vartheta_{ca_2}\vartheta_{c_2a_2}\vartheta_{a_1}(u)\,\vartheta_{a_2c_1}(u)+\vartheta_{a_1c_2}\vartheta_{ca_1}\vartheta_{a_2}(u)\,\vartheta_{a_1c_1}(u)=\vartheta_c\vartheta_{c_2}\vartheta_{c_1}(u)\,\vartheta_{a_1a_2}(u).$$

Ex. vi. Obtain formulae to express the ratios of the differential coefficients of the odd theta functions for zero values of the arguments.

Ex. vii. Prove that

$$\vartheta(u)\frac{\partial}{\partial u_2}\vartheta(u|u^{b_1,a}+u^{b_2,a})-\vartheta(u|u^{b_1,a}+u^{b_2,a})\frac{\partial}{\partial u_2}\vartheta(u)=\epsilon\sqrt{b_1-b_2}\,\vartheta(u|u^{b_1,a})\,\vartheta(u|u^{b_2,a}),$$

wherein b_1, b_2 are any two finite branch places, and ϵ is a certain fourth root of unity.

This result can be obtained in various ways; one way is as follows: Writing $u=u^{x_1,a_1}+u^{x_2,a_2}$, $u+u^{b_1,a}=v$, and $v=u^{z_1,b_1}+u^{z_2,b_2}$, we find, by the formula $\vartheta(u+\Omega_P)=e^{\lambda_P(u)}\vartheta(u;P)$, that

$$\frac{\partial}{\partial u_2}\log\frac{\vartheta(u|u^{b_1,a}+u^{b_2,a})}{\vartheta(u)}=L_2^{b_1,b_2}+\zeta_2(v-u^{b_2,a})-\zeta_2(v-u^{b_1,a}),$$

and, by the formula expressing $\zeta_i(u^{x,m}-u^{x_1,m_1}-\dots\dots-u^{x_p,m_p})-\zeta_i(u^{\mu,m}-u^{x_1,m_1}-\dots\dots-u^{x_p,m_p})$ by integrals and rational functions, the right-hand side is equal to

$$-\tfrac{1}{2}\frac{b_1-b_2}{z_1-z_2}\left[\frac{s_1}{(z_1-b_1)(z_1-b_2)}-\frac{s_2}{(z_2-b_1)(z_2-b_2)}\right],$$

where s_1, z_1 are the values of y, x respectively at the place z_1, and s_2, z_2 at the place z_2. This rational function of z_1, z_2 is however (§ 210) a certain constant multiple of $\vartheta(v|u^{b_1,a}+u^{b_2,a})/\vartheta(v)$, and hence the result can immediately be deduced.

One case of the relation, when b_1, b_2 are the places a_1, a_2, is expressible by Weierstrass's notation in the form

$$\vartheta_5(u)\frac{\partial}{\partial u_2}\vartheta_{04}(u)-\vartheta_{04}(u)\frac{\partial}{\partial u_2}\vartheta_5(u)=\epsilon\sqrt{a_1-a_2}\,\vartheta_{02}(u)\,\vartheta_{24}(u),$$

and it is interesting, using results which belong to the later part of this volume, to compare this with other methods of proof. We have*

$$\vartheta_4\vartheta_0\bar{\vartheta}_{04}(u+v)\,\vartheta_5(u-v)=\vartheta_5(u)\,\bar{\vartheta}_{04}(u)\,\vartheta_4(v)\,\vartheta_0(v)+\vartheta_2(u)\,\bar{\vartheta}_{13}(u)\,\bar{\vartheta}_{02}(v)\,\bar{\vartheta}_{24}(v)$$
$$+\vartheta_5(v)\,\bar{\vartheta}_{04}(v)\,\vartheta_4(u)\,\vartheta_0(u)+\vartheta_2(v)\,\bar{\vartheta}_{13}(v)\,\bar{\vartheta}_{02}(u)\,\bar{\vartheta}_{24}(u),$$

where ϑ_4, ϑ_0 denote $\vartheta_4(0)$, $\vartheta_0(0)$, and the bar denotes an odd function; if, herein, the arguments v_1, v_2 be taken very small, we may write $\vartheta(u+v)=\vartheta(u)+\left(v_1\frac{\partial}{\partial u_1}+v_2\frac{\partial}{\partial u_2}\right)\vartheta(u)$. Thus we obtain, eventually, remembering that the odd functions, and the first differential coefficients of the even functions, vanish for zero values of the arguments,

$$\vartheta_5(u)\,\vartheta'_{04}(u)-\vartheta_{04}(u)\,\vartheta'_5(u)=\frac{\vartheta_5\vartheta'_{04}}{\vartheta_4\vartheta_0}\vartheta_4(u)\,\vartheta_0(u)+\frac{\vartheta_2\vartheta'_{13}}{\vartheta_4\vartheta_0}\vartheta_{02}(u)\,\vartheta_{24}(u),$$

where $\vartheta'(u)=\frac{\partial}{\partial u_2}\vartheta(u)$, $\vartheta=\vartheta(0)$, $\vartheta'=\vartheta'(0)$.

Thus, by the formula of this example, putting $u=0$, we infer that

$$\left[\frac{\partial}{\partial u_2}\vartheta(u\,|\,u^{a_1,\,a}+u^{a_2,\,a})\right]_{u=0}=0$$

or $\vartheta'_{04}=0$, and the result of the general formula agrees with the formula of this example.

In the cases $p>2$ we have even theta functions vanishing for zero values of the argument; here we have one of the differential coefficients of an odd function vanishing for zero values of the argument.

Note. Beside the references given in this chapter there is a paper by Bolza, *American Journal*, XVII. 11 (1895), "On the first and second derivatives of hyperelliptic σ-functions" (see *Acta Math.* XX. (Feb. 1896), p. 1: "Zur Lehre von den hyperelliptischen Integralen, von Paul Epstein"), which was overlooked till the chapter was completed. The fundamental formula of Klein, utilised by Bolza, is developed, in what appeared to be its proper place, in chapter XIV. of the present volume. See also Wiltheiss, *Crelle*, XCIX. p. 247, *Math. Annal.* XXXI. p. 417; Brioschi, *Rend. d. Acc. dei Lincei*, (Rome), 1886, p. 199; and further, Königsberger, *Crelle*, LXV. (1866), p. 342; Frobenius, *Crelle*, LXXXIX. (1880), p. 206.

To the note on p. 301 should be added the references; Prym, *Zur Theorie der Functnen. in einer zweiblätt. Fläche* (Zürich, 1866), p. 12; Königsberger, *Crelle*, LXIV. p. 20. To the note on p. 296 should be added; Harkness and Morley, *Theory of Functions*, chapter VIII., on double theta functions. In connection with § 205, notations for theta functions of three variables are given by Cayley and Borchardt, *Crelle*, LXXXVII. (1878).

* Krause, *Hyperelliptische Functionen*, p. 44; Königsberger, *Crelle*, LXIV. p. 28.

CHAPTER XII.

A PARTICULAR FORM OF FUNDAMENTAL SURFACE.

222. JACOBI'S inversion theorem, and the resulting theta functions, with which we have been concerned in the three preceding chapters, may be regarded as introducing a method for the change of the independent variables upon which the fundamental algebraic equation, and the functions associated therewith, depend. The theta functions, once obtained, may be considered independently of the fundamental algebraic equation, and as introductory to the general theory of multiply-periodic functions of several variables; the theory is resumed from this point of view in chapter XV., and the reader who wishes may pass at once to that chapter. But there are several further matters of which it is proper to give some account here. The present chapter deals with a particular case of a theory which is historically a development* of the theory of this volume; it is shewn that on a surface which is in many ways simpler than a Riemann surface, functions can be constructed entirely analogous to the functions existing on a Riemann surface. The suggestion is that there exists a conformal representation of a Riemann surface upon such a surface as that here considered, which would then furnish an effective change of the independent variables of the Riemann surface. We do not however at present undertake the justification of that suggestion, nor do we assume any familiarity with the general theory referred to. The present particular case has the historical interest that in it a function has arisen, which we may call the Schottky-Klein prime function, which is of great importance for any Riemann surface.

223. Let α, β, γ, δ be any quantities whatever, whereof three are definitely assigned, and the fourth thence determined by the relation $\alpha\delta - \beta\gamma = 1$. Let ζ, ζ' be two corresponding complex variables associated together by the relation $\zeta' = (\alpha\zeta + \beta)/(\gamma\zeta + \delta)$. This relation can be put into the form

$$\frac{\zeta' - B}{\zeta' - A} = \mu e^{i\kappa} \frac{\zeta - B}{\zeta - A},$$

* Referred to by Riemann himself, *Ges. Werke* (Leipzig, 1876), p. 413.

wherein μ is real, and B, A are the roots of the quadratic equation $\zeta = (\alpha\zeta + \beta)/(\gamma\zeta + \delta)$, distinguished from one another by the condition that μ shall be less than unity. In all the linear substitutions which occur in this chapter it is assumed that B, A are not equal, and that μ is not equal to unity. We introduce now the ordinary representation of complex quantities by the points of a plane. Let the points A, B be marked as in the figure (6),

Fig. 6.

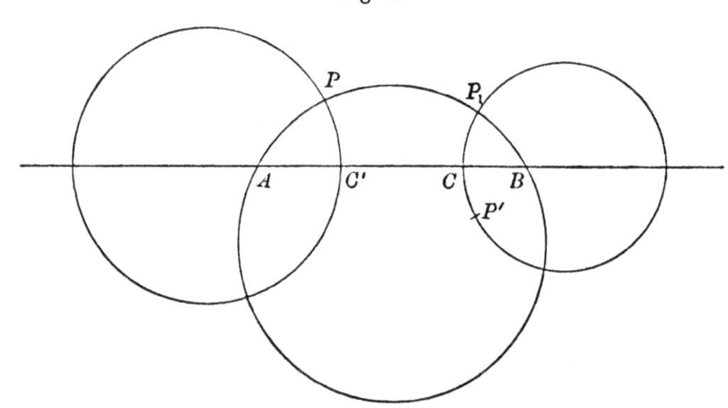

and a point C' be taken between A, B in such a way that $1 > AC'/C'B > \mu$, but otherwise arbitrarily; then the locus of a point P such that $AP/PB = AC'/C'B$ is a circle. Take now a point C also between A and B, such that $CB/AC = \mu C'B/AC'$, and mark the circle which is the locus of a point P' for which $P'B/AP' = CB/AC$; since $P'B/AP'$ is less than unity, this circle will lie entirely without the other circle. If now any circle through the points A, B cut the first circle, which we shall call the circle C', in the points P, Q, and cut the second circle, C, in P_1 and Q_1, P and P_1 being on the same side of AB, we have angle $AP_1B =$ angle APB, and $P_1B/AP_1 = \mu PB/AP$; therefore, if the point P be ζ, and the point P_1 be ζ_1, we have

$$\frac{\zeta_1 - B}{\zeta_1 - A} = \mu \frac{\zeta - B}{\zeta - A},$$

the argument of P vanishing when P is at the end of the diameter of the C' circle remote from C', and varying from 0 to 2π as P describes the circle C' in a clockwise direction; if then we pass along the circle C in a counter clockwise direction to a point P' such that the sum of the necessary positive rotation of the line BP_1 about B into the position BP', and the necessary negative rotation of the line AP_1 about A into the position AP', is κ, and ζ' be the point P', we have

$$\frac{\zeta' - B}{\zeta' - A} = e^{i\kappa} \frac{\zeta_1 - B}{\zeta_1 - A} = \mu e^{i\kappa} \frac{\zeta - B}{\zeta - A}.$$

Thus the transformation under consideration transforms any point ζ on the circle C' into a point on the circle C. If ζ denote any point within C'

the modulus of $(\zeta - B)/(\zeta - A)$ is greater than when ζ is on the circumference of C', and the transformed point ζ' is without the circle C, though not necessarily without the circle C'. If ζ denote any point without C' the transformed point is within the circle C.

224. Suppose * now we have given p such transformations as have been described, depending therefore on $3p$ given complex quantities, whereof 3 can be given arbitrary values by a suitable transformation $z' = (Pz + Q)/(Rz + S)$ applied to the whole plane; denote the general one by

$$\zeta' = \frac{\alpha_i \zeta + \beta_i}{\gamma_i \zeta + \delta_i}, \text{ wherein } \alpha_i \delta_i - \beta_i \gamma_i = 1, \qquad (i = 1, 2, \ldots, p),$$

or also by

$$\zeta' = \vartheta_i \zeta, \; \zeta = \vartheta_i^{-1} \zeta',$$

the quantities corresponding to A, B, μ, α being denoted by A_i, B_i, μ_i, α_i; construct as here a pair of circles corresponding to each substitution, and *assume that the constants are such that, of the $2p$ circles obtained, each is exterior to all the others;* let the region exterior to all the circles be denoted by S, and the region derivable therefrom by the substitution ϑ_i be denoted by $\vartheta_i S$.

If the whole plane exterior to the circle C_i be subjected to the transformation ϑ_i, the circle C_i' will be transformed into C_i, the circle C_i itself will be transformed into a circle interior to C_i, which we denote by $\vartheta_i C_i$, and the other $2p - 2$ circles which lie in a space bounded by C_i and C_i' will be transformed into circles lying in the region bounded by $\vartheta_i C_i$ and C_i, and, corresponding to the region S, exterior to all the $2p$ circles, we shall have a region $\vartheta_i S$ also bounded by $2p$ circles. But suppose that before we thus transform the whole plane by the transformation ϑ_i, we had transformed the whole plane by another transformation ϑ_j and so obtained, within C_j, a region $\vartheta_j S$ bounded by $2p$ circles, of which C_j is one. Then, in the subsequent transformation, ϑ_i, all the $2p - 1$ circles lying within C_j will be transformed, along with C_j, into $2p - 1$ other circles lying in a region, $\vartheta_i \vartheta_j S$, bounded by the circle $\vartheta_i C_j$. They will therefore be transformed into circles lying *within* $\vartheta_i C_j$—they cannot lie without this circle, namely in $\vartheta_i S$, because $\vartheta_i S$ is the picture of a space, S, whose only boundaries are the $2p$ fundamental circles $C_1, C_1', \ldots, C_p, C_p'$. Proceeding in the manner thus indicated we shall obtain by induction the result enunciated in the following statement, wherein ϑ_i^{-1} is the inverse transformation to ϑ_i, and transforms the circle C_i into C_i': *Let all possible multiples of powers of $\vartheta_1, \vartheta_1^{-1}, \ldots, \vartheta_p, \vartheta_p^{-1}$ be formed, and the corresponding regions, obtained by applying to S the transformations*

* The subject-matter of this section is given by Schottky, *Crelle*, CI. (1887), p. 227, and by Burnside, *Proc. London Math. Soc.* XXIII. (1891), p. 49.

Fig. 7.

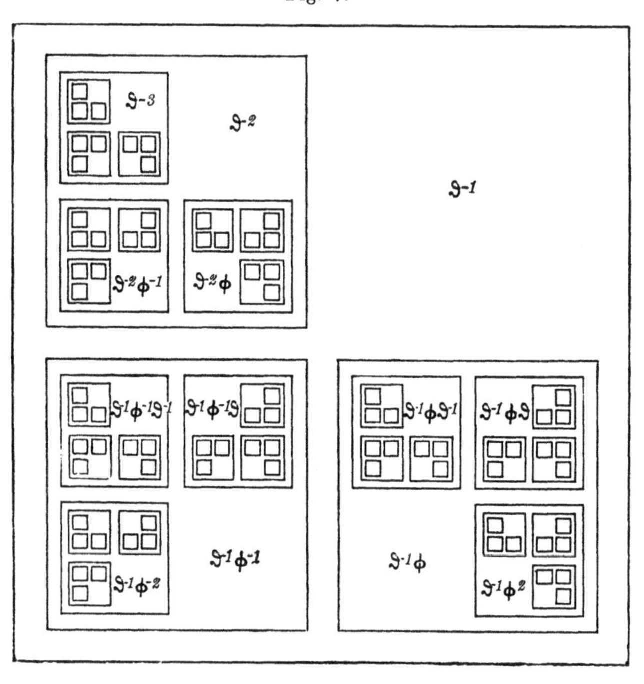
ϑ^{-3}
ϑ^{-2}
ϑ^{-1}
$\vartheta^{-2}\phi^{-1}$
$\vartheta^{-2}\phi$
$\vartheta^{-1}\phi^{-1}\vartheta^{-1}$
$\vartheta^{-1}\phi^{-1}\vartheta$
$\vartheta^{-1}\phi\vartheta^{-1}$
$\vartheta^{-1}\phi\vartheta$
$\vartheta^{-1}\phi^{-1}$
$\vartheta^{-1}\phi$
$\vartheta^{-1}\phi^{-2}$
$\vartheta^{-1}\phi^{2}$

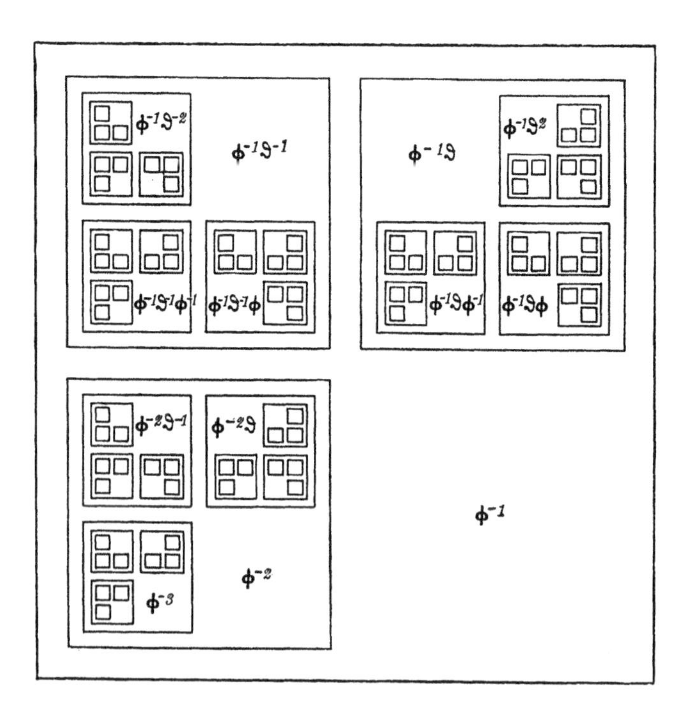
$\phi^{-1}\vartheta^{-2}$
$\phi^{-1}\vartheta^{-1}$
$\phi^{-1}\vartheta$
$\phi^{-1}\vartheta^{2}$
$\phi^{-1}\vartheta^{-1}\phi^{-1}$
$\phi^{-1}\vartheta^{-1}\phi$
$\phi^{-1}\vartheta\phi^{-1}$
$\phi^{-1}\vartheta\phi$
$\phi^{-2}\vartheta^{-1}$
$\phi^{-2}\vartheta$
ϕ^{-1}
ϕ^{-2}
ϕ^{-3}

Fig. 7.

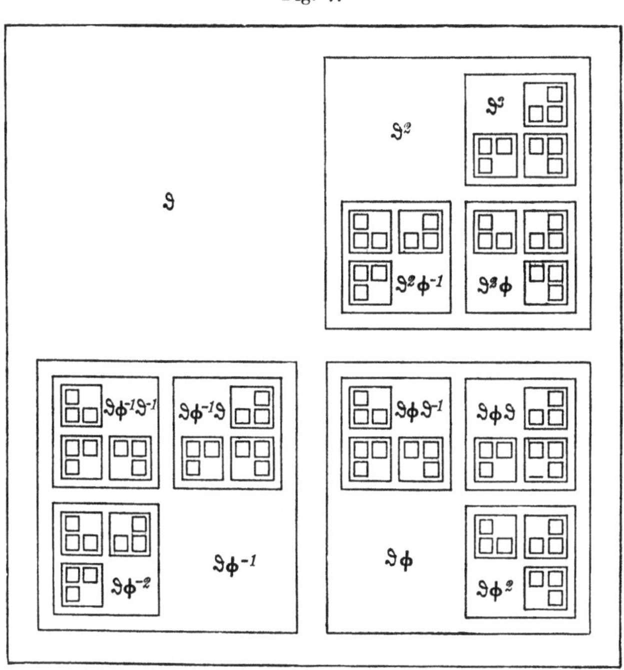
ϑ
ϑ^2
ϑ^3
$\vartheta^2\phi^{-1}$
$\vartheta^2\phi$
$\vartheta\phi^{-1}\vartheta^{-1}$
$\vartheta\phi^{-1}\vartheta$
$\vartheta\phi\vartheta^{-1}$
$\vartheta\phi\vartheta$
$\vartheta\phi^{-2}$
$\vartheta\phi^{-1}$
$\vartheta\phi$
$\vartheta\phi^2$

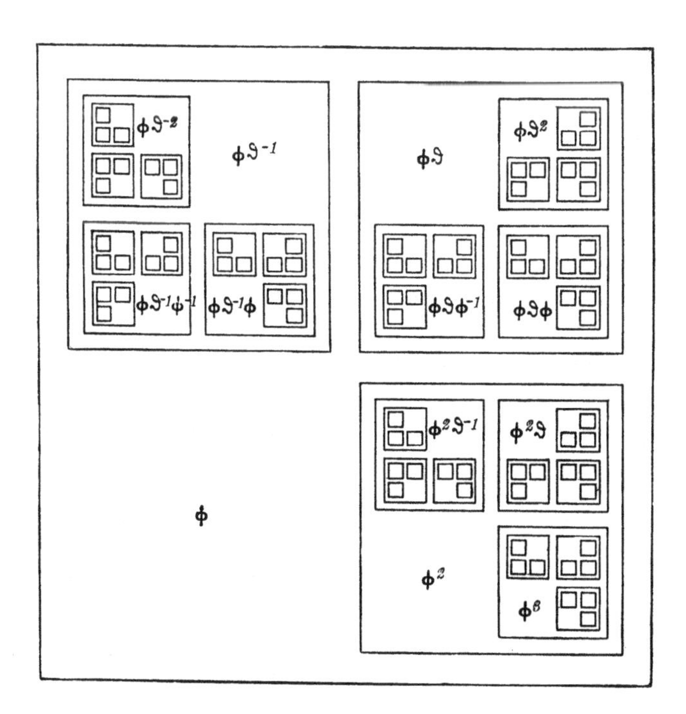
$\phi\vartheta^{-2}$
$\phi\vartheta^{-1}$
$\phi\vartheta$
$\phi\vartheta^2$
$\phi\vartheta^{-1}\phi^{-1}$
$\phi\vartheta^{-1}\phi$
$\phi\vartheta\phi^{-1}$
$\phi\vartheta\phi$
$\phi^2\vartheta^{-1}$
$\phi^2\vartheta$
ϕ
ϕ^2
ϕ^3

corresponding to all such products of powers, be marked out. In any such product the transformation first to be applied is that one which stands to the right. Let m be any one such product, of the form

$$m = \ldots\ldots \vartheta_i^{r_i} \vartheta_j^{r_j} \vartheta_k^{r_k},$$

formed by

$$\ldots\ldots + r_i + r_j + r_k, \ = h$$

factors, and let ϑ be any transformation other than the inverse of ϑ_k, so that $m\vartheta_k$ is formed by the product of $h + 1$, not $h - 1$, factors. Then the region mS entirely surrounds the region $m\vartheta S$.

Thus, the region $\vartheta_i S$ entirely surrounds the space $\vartheta_i \vartheta_j S$, and the latter surrounds $\vartheta_i \vartheta_j^2 S$, or $\vartheta_i \vartheta_j \vartheta_k S$; but $\vartheta_i S$ is surrounded by $\vartheta_i \vartheta_i^{-1} S$ or S. The reader may gain further clearness on this point by consulting the figure (7), wherein, for economy of space, rectangles are drawn in place of circles, and the case of only two fundamental substitutions, ϑ, ϕ, is taken.

The consequence of the previous result is—*The group of substitutions consisting of the products of positive and negative powers of $\vartheta_1, \ldots, \vartheta_p$ gives rise to a single covering of the whole plane, every point being as nearly reached as we desire, by taking a sufficient number of factors, and no point being reached by two substitutions.*

225. There are in fact certain points which are not reached as transformations of points of S, by taking the product of any *finite* number of substitutions. For instance the substitution ϑ_i^m is

$$\frac{\zeta' - B_i}{\zeta' - A_i} = \mu_i{}^m e^{im\kappa_i} \frac{\zeta - B_i}{\zeta - A_i},$$

and thus when m is increased indefinitely ζ' approaches indefinitely near to B_i, whatever be the position of ζ; but B_i is not reached for any finite value of m. In general the result of any infinite series of successive substitutions, $K = \alpha\beta\gamma\ldots$, applied to the region S, is, by what has been proved, a region lying within αS, in fact lying within $\alpha\beta S$, nay more, lying within $\alpha\beta\gamma S$, and so on—namely is a region which may be regarded as a point; denoting it by K, the substitution K transforms every point of the region S and in fact every other point of the plane into the same point K; and transforms the point K into itself. There will similarly be a point K' arising by the same infinite series of substitutions taken in the reverse order.

Such points are called the singular points of the group. There is an infinite number of them; but two of them for which the corresponding products of the symbols ϑ agree to a sufficient number of the left-hand factors are practically indistinguishable; none of them lie within regions that are obtained from S with a finite number of substitutions. The most important of these singular points are those for which the corresponding series of substitutions is periodic; of these the most obvious are those formed by indefinite repetition of one of the fundamental substitutions; we have already introduced the notation

$$\vartheta_i^{\infty} S = B_i, \quad \vartheta_i^{-\infty} S = A_i,$$

to represent the results of such substitutions.

226. If ϑ, ϕ be any two substitutions given respectively by

$$\zeta' = \frac{\alpha\zeta + \beta}{\gamma\zeta + \delta}, \quad \zeta' = \frac{A\zeta + B}{C\zeta + D},$$

wherein $\alpha\delta - \beta\gamma = 1 = AD - BC$, the compound substitution $\vartheta\phi$ is given by

$$\zeta' = \frac{\alpha(A\zeta + B) + \beta(C\zeta + D)}{\gamma(A\zeta + B) + \delta(C\zeta + D)} = \frac{(\alpha A + \beta C)\zeta + (\alpha B + \beta D)}{(\gamma A + \delta C)\zeta + (\gamma B + \delta D)},$$

and if this be represented by $\zeta' = (\alpha'\zeta + \beta')/(\gamma'\zeta + \delta')$, we have, in the ordinary notation of matrices

$$\begin{pmatrix} \alpha' & \beta' \\ \gamma' & \delta' \end{pmatrix} = \begin{pmatrix} \alpha & \beta \\ \gamma & \delta \end{pmatrix} \begin{pmatrix} A & B \\ C & D \end{pmatrix},$$

and $\alpha'\delta' - \beta'\gamma' = (\alpha\delta - \beta\gamma)(AD - BC) = 1$. We suppose all possible substitutions arising by products of positive and negative powers of the fundamental substitutions $\vartheta_1, \dots, \vartheta_p$ to be formed, and denote any general substitution by $\zeta' = (\alpha\zeta + \beta)/(\gamma\zeta + \delta)$, wherein, by the hypothesis in regard to the fundamental substitutions, $\alpha\delta - \beta\gamma = 1$. We may suppose all the substitutions thus arising to be arranged in order, there being first the identical substitution $\zeta' = (\zeta + 0)/(0 \,.\, \zeta + 1)$, then the $2p$ substitutions whose products contain one factor, ϑ_i or ϑ_i^{-1}, then the $2p(2p-1)$ substitutions whose products are of one of the forms $\vartheta_i\vartheta_j$, $\vartheta_i\vartheta_j^{-1}$, $\vartheta_i^{-1}\vartheta_j$, $\vartheta_i^{-1}\vartheta_j^{-1}$, in which the two substitutions must not be inverse, containing two factors, then the $2p(2p-1)^2$ substitutions whose products contain three factors, and so on. So arranged consider the series

$$\Sigma\,(\text{mod}\ \gamma)^{-k},$$

wherein k is a real positive quantity, and the series extends to every substitution of the group except the identical substitution. Since the inverse substitution to $\zeta' = (\alpha\zeta + \beta)/(\gamma\zeta + \delta)$ is $\zeta = (\delta\zeta' - \beta)/(-\gamma\zeta' + \alpha)$, each set of $2p(2p-1)^{n-1}$ terms corresponding to products of n substitutions will contain each of its terms twice over.

Let now Θ_n denote a substitution formed by the product of n factors, and $\Theta_{n+1} = \Theta_n\vartheta_i$, where ϑ_i denotes any one of the primary $2p$ substitutions $\vartheta_1, \vartheta_1^{-1}, \dots, \vartheta_p, \vartheta_p^{-1}$ other than the inverse of the substitution whose symbol stands at the right hand of the symbol Θ_n, so that Θ_{n+1} is formed with $n+1$

factors; then by the formula just set down $\gamma_{n+1} = \gamma_n \alpha_i + \delta_n \gamma_i$, where, if ϑ_i, or $\zeta' = (\alpha_i \zeta + \beta_i)/(\gamma_i \zeta + \delta_i)$, be put in the form $(\zeta' - B_i)/(\zeta' - A_i) = \rho_i (\zeta - B_i)/(\zeta - A_i)$, we have

$$\alpha_i, \qquad \beta_i, \qquad \gamma_i, \qquad \delta_i$$

respectively equal to

$$\frac{B_i \rho_i^{-\frac{1}{2}} - A_i \rho_i^{\frac{1}{2}}}{B_i - A_i}, \quad - \frac{A_i B_i (\rho_i^{-\frac{1}{2}} - \rho_i^{\frac{1}{2}})}{B_i - A_i}, \quad \frac{\rho_i^{-\frac{1}{2}} - \rho_i^{\frac{1}{2}}}{B_i - A_i}, \quad - \frac{A_i \rho_i^{-\frac{1}{2}} - B_i \rho_i^{\frac{1}{2}}}{B_i - A_i};$$

the signification of $\rho_i^{\frac{1}{2}}$ is not determined when the corresponding pair of circles is given; but we have supposed that the values of $\alpha_i, \beta_i, \gamma_i, \delta_i$ are given, and thereby the value of $\rho_i^{\frac{1}{2}}$. By these formulae we have

$$\frac{\gamma_{n+1}}{\gamma_n} = \rho_i^{-\frac{1}{2}} \frac{B_i + \delta_n/\gamma_n}{B_i - A_i} - \rho_i^{\frac{1}{2}} \frac{A_i + \delta_n/\gamma_n}{B_i - A_i}.$$

Herein the modulus of ρ_i may be either μ_i or μ_i^{-1}, according as ϑ_i is one of $\vartheta_1, \ldots, \vartheta_p$ or one of $\vartheta_1^{-1}, \ldots, \vartheta_p^{-1}$; the modulus of ρ_i is accordingly either less or greater than unity. If now $\Theta_n = \ldots \psi \phi \vartheta_r^{-1}$, where ϑ_r is one of the $2p$ fundamental substitutions $\vartheta_1, \ldots, \vartheta_p^{-1}$, and therefore $\Theta_n^{-1} = \vartheta_r \phi^{-1} \psi^{-1} \ldots$, the region $\Theta_n^{-1} S$ lies entirely within the region $\vartheta_r S$ (§ 224) or coincides with it; wherefore the point $\Theta_n^{-1}(\infty)$, or $-\delta_n/\gamma_n$, lies within the circle C_r when ϑ_r is one of $\vartheta_1, \ldots, \vartheta_p$, and lies within the circle C_r' when ϑ_r is one of $\vartheta_1^{-1}, \ldots, \vartheta_p^{-1}$; thus the points B_i and $-\delta_n/\gamma_n$ can only lie within the same one of the $2p$ fundamental circles $C_1, \ldots, C_p'$ when $r = i$ and ϑ_r is one of $\vartheta_1, \ldots, \vartheta_p$; and the points A_i and $-\delta_n/\gamma_n$ can only lie within the same one of the $2p$ fundamental circles $C_1, \ldots, C_p'$ when $r = i$ and ϑ_r is one of $\vartheta_1^{-1}, \ldots, \vartheta_p^{-1}$. Now, if the modulus of ρ_i be less than unity, and $r = i$, ϑ_r must be one of $\vartheta_1^{-1}, \ldots, \vartheta_p^{-1}$, namely must be ϑ_i^{-1}, since otherwise $\Theta_n \vartheta_i$ would consist of $n-1$ factors, and not $n+1$ factors; in that case therefore $B_i + \dfrac{\delta_n}{\gamma_n}$ is not of infinitely small modulus; if, however, the modulus of ρ_i be greater than unity, and $r = i$, ϑ_r must be ϑ_i, namely one of $\vartheta_1, \ldots, \vartheta_p$, and in that case the modulus of $A_i + \delta_n/\gamma_n$ is not infinitely small. Thus, according as $|\rho_i| \lessgtr 1$, we may put

$$| B_i + \delta_n/\gamma_n | > \lambda, \quad | A_i + \delta_n/\gamma_n | > \lambda,$$

where λ is a positive real quantity which is certainly not less than the distance of B_i, A_i, respectively, from the nearest point of the circle within which $-\delta_n/\gamma_n$ lies.

It follows from this that we have

$$\text{mod}\,(\gamma_{n+1}/\gamma_n) > \sigma, \text{ or mod }(\gamma_{n+1}^{-1}/\gamma_n^{-1}) < \frac{1}{\sigma},$$

where σ is a positive finite quantity, for which an arbitrary lower limit may be assigned independent of the substitutions of which Θ_n is compounded, and independent of n, *provided the moduli* $\mu_1, \ldots, \mu_p$ *be supposed sufficiently small, and the p pairs of circles be sufficiently distant from one another.*

Ex. Prove, in § 223, that if C' be chosen so that $C'C$ is as great as possible

$$\frac{1}{\sqrt{\mu}}\frac{C'C}{AB} = \frac{1-\sqrt{\mu}}{1+\sqrt{\mu}}\frac{1}{\sqrt{\mu}},$$

and the circles are both of radius $d\sqrt{\mu}/(1-\mu)$, where d is the length of AB.

We suppose the necessary conditions to be satisfied; then if γ_0 be the least of the p quantities mod $[(\mu_i^{-\frac{1}{2}}e^{-\frac{1}{2}i\kappa_i} - \mu_i^{\frac{1}{2}}e^{\frac{1}{2}i\kappa_i})/(B_i - A_i)]$, and k be positive, the series Σ mod γ^{-k} is less than

$$\gamma_0^{-k}\left[2p + \frac{2p(2p-1)}{\sigma^k} + \frac{2p(2p-1)^2}{\sigma^{2k}} + \ldots\ldots\right],$$

and therefore certainly convergent if $\sigma^k > 2p-1$, which, as shewn above, may be supposed, $\mu_1, \ldots, \mu_p$ being sufficiently small.

227. Hence we can draw the following inference: Let $\sigma_1, \ldots, \sigma_p$ be assigned quantities, called multipliers, each of modulus unity, associated respectively with the p fundamental substitutions $\vartheta_1, \ldots, \vartheta_p$; with any compound substitution $\vartheta_1^{r_1}\vartheta_2^{r_2}\ldots$, let the compound quantity $\sigma_1^{r_1}\sigma_2^{r_2}\ldots$ be associated: let $f(x)$ denote any uniform function of x with only a finite number of separated infinities; let $\zeta' = (\alpha\zeta+\beta)/(\gamma\zeta+\delta)$ denote any substitution of the group, and σ be the multiplier associated with this substitution: then the series, extending to all the substitutions of the group,

$$\Sigma\sigma f\left(\frac{\alpha\zeta+\beta}{\gamma\zeta+\delta}\right)(\gamma\zeta+\delta)^{-k}$$

converges absolutely and uniformly * for all positions of ζ other than (i) the singular points of the group, and the points $\zeta = -\delta/\gamma$, namely the points derivable from $\zeta = \infty$ by the substitutions of the group, including the point $\zeta = \infty$ itself, (ii) the infinities of $f(\zeta)$ and the points thence derived by the substitutions of the group. The series represents therefore a well-defined continuous function of ζ for all the values of ζ other than the excepted ones. The function will have poles at the poles of $f(\zeta)$ and the points thence derived by the substitutions of the group; it may have essential singularities at the singular points of the group and at the essential singularities of

$$f((\alpha\zeta+\beta)/(\gamma\zeta+\delta)).$$

* In regard to ζ; for the convergence was obtained independently of the value of ζ.

Denote this function by $F(\zeta)$; if ϑ_0 denote any assigned substitution of the group, and ϑ denote all the substitutions of the group in turn, it is clear that $\vartheta\vartheta_0$ denotes all the substitutions of the group in turn including the identical substitution; recognising this fact, and denoting the multiplier associated with ϑ_0 by σ_0, we immediately find

$$F(\vartheta_0(\zeta)) = \sigma_0^{-1}(\gamma_0\zeta + \delta_0)^k F(\zeta),$$

or, the function is multiplied by the factor $\sigma_0^{-1}(\gamma_0\zeta + \delta_0)^k$ when the variable ζ is transformed by the substitution, ϑ_0, of the group. Thence also, if $G(\zeta)$ denote a similar function to $F(\zeta)$, formed with the same value of k and a different function $f(\zeta)$, the ratio $F(\zeta)/G(\zeta)$ remains entirely unaltered when the variable is transformed by the substitutions of the group. In order to point out the significance of this result we introduce a representation whereof the full justification is subsequent to the present investigation. Let a Riemann surface be taken, on which the $2p$ period loops are cut; let the circumference of the circle C_i of the ζ plane be associated with one side of the period loop (b_i) of the second kind, and the circumference of the circle C_i' with the other side of this loop; let an arbitrary curve which we shall call the i-th barrier be drawn in the ζ plane from an arbitrary point P of the circle C_i' to the corresponding point P' of the circle C_i, and let the two sides of this curve be associated with the two sides of the period loop (a_i) of the Riemann surface. Then the function $F(\zeta)/G(\zeta)$, which has the same value at any two near points on opposite sides of the barrier, and has the same value at any point Q of the circle C_i' as at the corresponding point Q' of the circle C_i, will correspond to a function uniform on the undissected Riemann surface. In this representation the whole of the Riemann surface corresponds to the region S; any region $\vartheta_i S$ corresponds to a repetition of the Riemann surface; thus if the only essential singularities of $F(\zeta)/G(\zeta)$ be at the singular points of the group, none of which are within S, $F(\zeta)/G(\zeta)$ corresponds to a rational function on the Riemann surface. It will appear that the correspondence thus indicated extends to the integrals of rational functions; of such integrals not all the values can be represented on the dissected Riemann surface, while on the undissected surface they are not uniform; for instance, of an integral of the first kind, u_i, the values u_i, $u_i + 2\omega_{i,r}$, $u_i + 2\omega'_{i,r}$, $u_i + 2\omega_{i,r} + 2\omega'_{i,r}$ may be represented, but in that case not the value $u_i + 4\omega_{i,r}$; in view of this fact the repetition of the Riemann surface associated with the regions derived from S by the substitutions of the group is of especial interest—*we are able to represent more of the values of the integral in the ζ plane than on the Riemann surface.* These remarks will be clearer after what follows.

228. In what follows we consider only a simple case of the function $F(\zeta)$, that in which the multipliers $\sigma_1, \dots, \sigma_p$ are all unity, $k = 2$, and $f(\zeta) = 1/(\zeta - a)$, a being a point which, for the sake of definiteness, we

suppose to be in the region S. We denote by $\zeta_i = \vartheta_i(\zeta) = (\alpha_i\zeta + \beta_i)/(\gamma_i\zeta + \delta_i)$ all the substitutions of the group, in turn, and call ζ_i the *analogue* of ζ by the substitution in question. The function

$$\Phi(\zeta, a) = \Sigma \frac{(\gamma_i\zeta + \delta_i)^{-2}}{\zeta_i - a},$$

has essential singularities at the singular points of the group, and has poles at the places $\zeta = a$, $\zeta = \infty$ and at the analogues of these places. Let the points ∞, a be joined by an arbitrary barrier lying in S, and the analogues of this barrier be drawn in the other regions. Then the integral of this uniformly convergent series, from an arbitrary point ξ, namely, the series

$$\Sigma \log \frac{\zeta_i - a}{\xi_i - a}, \quad = \Pi_{a,\,\infty}^{\zeta,\,\xi}, \text{ say},$$

is competent to represent a function of ζ which can only deviate from uniformity when ζ describes a contour enclosing more of the points a and its analogues than of the points ∞ and its analogues; this is prevented by the barriers. Thus the function is uniform over the whole ζ plane; it is infinite at $\zeta = a$ like $\log(\zeta - a)$, and at $\zeta = \infty$ like $-\log\left(\frac{1}{\zeta}\right)$, as we see by considering the term of the series corresponding to the identical substitution; its value on one side of the barrier $a\infty$ is $2\pi i$ greater than on the other side; it has analogous properties in the analogues of the points a, ∞, and the barrier $a\infty$; further, if $\zeta_n = \vartheta_n(\zeta)$ be any of the fundamental substitutions $\vartheta_1, \dots, \vartheta_p$,

$$\Pi_{a,\,\infty}^{\zeta_n,\,\xi} - \Pi_{a,\,\infty}^{\zeta,\,\xi} = \sum_i \log \frac{\zeta_{in} - a}{\zeta_i - a} = \sum_i \log \frac{\zeta_{in} - a}{\xi_{in} - a} + \sum_i \log \frac{\xi_{in} - a}{\xi_i - a} - \sum_i \log \frac{\zeta_i - a}{\xi_i - a},$$

where ζ_{in} is obtained from ζ by the substitution $\vartheta_i\vartheta_n$; since the first and last of these sums contain the same terms, we have

$$\Pi_{a,\,\infty}^{\zeta_n,\,\xi} - \Pi_{a,\,\infty}^{\zeta,\,\xi} = \Pi_{a,\,\infty}^{\xi_n,\,\xi},$$

and the right-hand side is independent of ξ, being equal to $\Pi_{a,\,\infty}^{\zeta_n,\,\zeta}$; in order to prove this in another way, and obtain at the same time a result which will subsequently be useful, we introduce an abbreviated notation; denote the substitution ϑ_r simply by the letter r; then if j be in turn every substitution of the group whose product symbol has not a positive or negative power of the substitution n at its right-hand end, all the substitutions of the group have the symbol jn^h, h being in turn equal to all positive and negative integers (including zero); hence

$$\sum_i [\log(\xi_{in} - a) - \log(\xi_i - a)], \quad = \sum_j \sum_h [\log(\xi_{jn^{h+1}} - a) - \log(\xi_{jn^h} - a)],$$

is equal to

$$\sum_j \log \frac{\vartheta_j(\xi_N) - a}{\vartheta_j(\xi_M) - a},$$

where $N=n^{\infty}$, $M=n^{-\infty}$; but, in fact, ξ_N is B_n, and ξ_M is A_n; thus $\Pi_{a,\infty}^{\xi_n,\xi}$ is independent of ξ; and if we introduce the definition

$$v_n^{\zeta}=\frac{1}{2\pi i}\sum_j \log\frac{\zeta-\vartheta_j(B_n)}{\zeta-\vartheta_j(A_n)},$$

where ϑ_n is one of the p fundamental substitutions, and, as before, j denotes all the substitutions whose product symbols have not a power of n at the right-hand end, we have

$$\Pi_{a,\infty}^{\zeta_n,\xi}-\Pi_{a,\infty}^{\zeta,\xi}=\Pi_{a,\infty}^{\zeta_n,\zeta}=2\pi i v_n^a.$$

Ex. If for abbreviation we put

$$P_{a,\infty}^{\zeta,\xi}=\sum_i \sigma_i \log\frac{\zeta_i-a}{\xi_i-a},$$

prove that

$$P_{a,\infty}^{\zeta_n,\xi}-\frac{1}{\sigma_n}P_{a,\infty}^{\zeta,\xi}=P_{a,\infty}^{c_n,c}+\frac{1-\sigma_n}{\sigma_n}P_{a,\infty}^{\xi,c},$$

c being an arbitrary point.

229. Introduce now the function $\Pi_{a,b}^{\zeta,\xi}$ defined by the equation

$$\Pi_{a,b}^{\zeta,\xi}=\Pi_{a,\infty}^{\zeta,\xi}-\Pi_{b,\infty}^{\zeta,\xi}=\sum_i \log\left(\frac{\zeta_i-a}{\xi_i-a}\bigg/\frac{\zeta_i-b}{\xi_i-b}\right);$$

then, because a cross ratio of four quantities is unaltered by the same linear transformation applied to all the variables, we have also

$$\Pi_{a,b}^{\zeta,\xi}=\sum_i \log\left[\frac{\zeta-\vartheta_i^{-1}(a)}{\xi-\vartheta_i^{-1}(a)}\bigg/\frac{\zeta-\vartheta_i^{-1}(b)}{\xi-\vartheta_i^{-1}(b)}\right]=\sum_r \log\left(\frac{a_r-\zeta}{b_r-\zeta}\bigg/\frac{a_r-\xi}{b_r-\xi}\right),$$

where r, denoting ϑ_r, $=\vartheta_i^{-1}$, becomes in turn every substitution of the group. Thus we have

$$\Pi_{a,b}^{\zeta,\xi}=\Pi_{\zeta,\xi}^{a,b},\quad \Pi_{a,b}^{\zeta_n,\xi}-\Pi_{a,b}^{\zeta,\xi}=2\pi i v_n^{a,b},$$

where

$$v_n^{a,b},\ =v_n^a-v_n^b,\ =\frac{1}{2\pi i}\sum_j \log\left[\frac{a-\vartheta_j(B_n)}{a-\vartheta_j(A_n)}\bigg/\frac{b-\vartheta_j(B_n)}{b-\vartheta_j(A_n)}\right],\ =\frac{1}{2\pi i}\Pi_{a,b}^{\xi_n,\xi},$$

j denoting as before every substitution whose product symbol has not a positive or negative power of n at the right-hand end and ξ being arbitrary; hence also

$$v_n^{\zeta,a}=\frac{1}{2\pi i}\Pi_{\zeta,a}^{\xi_n,\xi}=\frac{1}{2\pi i}\sum_i \log\left(\frac{\xi_{in}-\zeta}{\xi_i-\zeta}\bigg/\frac{\xi_{in}-a}{\xi_i-a}\right)=\frac{1}{2\pi i}\sum_r \log\left(\frac{\zeta_r-\xi_n}{a_r-\xi_n}\bigg/\frac{\zeta_r-\xi}{a_r-\xi}\right),$$

where r, $=i^{-1}$, denotes every substitution of the group.

There are essentially only p such functions $v_n^{\zeta, a}$, according as ϑ_n denotes $\vartheta_1, \vartheta_2, \dots, \vartheta_p$; for, taking the expression given last but one, and putting $n = st$, that is, $\vartheta_n = \vartheta_s \vartheta_t$, we have

$$2\pi i v_{st}^{\zeta, a} = \Pi_{\zeta, a}^{\xi_{st}, \xi} = \Pi_{\zeta, a}^{\xi_{st}, \xi_t} + \Pi_{\zeta, a}^{\xi_t, \xi}$$

$$= \Pi_{\zeta, a}^{\eta_s, \eta} + \Pi_{\zeta, a}^{\xi_t, \xi},$$

where $\eta = \xi_t$, so that

$$v_{st}^{\zeta, a} = v_s^{\zeta, a} + v_t^{\zeta, a},$$

and in particular, when st is the identical substitution, as we see by the formula itself,

$$0 = v_s^{\zeta, a} + v_{s^{-1}}^{\zeta, a};$$

thus, if r denote $\vartheta_1^{\lambda_1} \vartheta_2^{\lambda_2} \dots \vartheta_p^{\lambda_p} \dots$, we obtain

$$v_r^{\zeta, a} = \lambda_1 v_1^{\zeta, a} + \dots\dots + \lambda_p v_p^{\zeta, a} + \dots\dots,$$

so that all the functions $v_r^{\zeta, a}$ are expressible as linear functions of $v_1^{\zeta, a}, \dots, v_p^{\zeta, a}$.

230. It follows from the formula

$$v_n^{\zeta, a} = \frac{1}{2\pi i} \sum_j \log \left(\frac{\zeta - \vartheta_j(B_n)}{\zeta - \vartheta_j(A_n)} \Big/ \frac{a - \vartheta_j(B_n)}{a - \vartheta_j(A_n)} \right),$$

that the function $v_n^{\zeta, a}$ is never infinite save at the singular points of the group. But it is not an uniform function of ζ; for let ζ describe the circumference of the circle C_n in a counter clockwise direction; then, by the factor $\zeta - B_n$, $v_n^{\zeta, a}$ increases by unity; and no other increase arises; for, when the region within the circle C_n, constituted by $\vartheta_n S$ and regions of the* form $\vartheta_n \phi S$, contains a point $\vartheta_j(B_n)$, the product representing the substitution j has a positive power of ϑ_n as its left-hand factor, and in that case the region contains also the point $\vartheta_j(A_n)$. Similarly if ζ describe the circle C_n' in a clockwise direction, $v_n^{\zeta, a}$ increases by unity. But if ζ describe the circumference of any other of the $2p$ circles, no increase arises in the value of $v_n^{\zeta, a}$, for the existence of a point $\vartheta_j(B_n)$ in such a circle involves the existence also of a point $\vartheta_j(A_n)$.

It follows therefore that the function can be made uniform in the region S by drawing the barrier, before described, from an arbitrary point P of C_n' to the corresponding point P' of C_n. Then $v_n^{\zeta, a}$ is greater by unity on one side of this barrier than on the other side. Further if m denote any one of the substitutions $\vartheta_1, \dots, \vartheta_p$, we have

$$v_n^{\zeta_m, b} - v_n^{\zeta, b} = v_n^{\zeta_m} - v_n^{\zeta} = v_n^{\zeta_m, \zeta} = \Pi_{\zeta_m, \zeta}^{\xi_n, \xi},$$

* Where ϕ denotes a product of substitutions in which ϑ_n^{-1} is not the left-hand factor.

where ξ is arbitrary; thus as $\Pi_{\xi_n, \zeta}^{\zeta_m, \zeta} = \Pi_{\zeta_m, \zeta}^{\xi_n, \xi}$, the difference is also independent of ζ, and we have, introducing a symbol for this constant difference,

$$v_n^{\zeta_m, b} - v_n^{\zeta, b} = \tau_{n, m} = \tau_{m, n}.$$

It follows therefore that if the p barriers, connecting the pairs of circles C_n', C_n, and their analogues for all the substitutions, be drawn in the interiors of the circles, the functions $v_1^{\zeta, a}, \dots, v_p^{\zeta, a}$ are uniform in the region S, and in all the regions derivable therefrom by the substitutions of the group. The behaviour of the functions $v_1^{\zeta, a}, \dots, v_p^{\zeta, a}$ in the region S is therefore entirely analogous to that of the Riemann normal integrals upon a Riemann surface, the correspondence of the pair of circumferences C_n, C_n' and the two sides of the barrier $P'P$, to the two sides of the period loops (b_n), (a_n), on the Riemann surface, being complete. And the regions within the circles $C_1, \dots, C_p'$ enable us to represent, in an uniform manner, all the values of the integrals which would arise on the Riemann surface if the period loops (b_n) were not present. Thus the ζ plane has greater powers of representation than the Riemann surface. Further it follows, by what has preceded, that the integral $\Pi_{a, b}^{\zeta, \xi}$ is entirely analogous to the Riemann normal elementary integral of the third kind which has been denoted by the same symbol in considering the Riemann surface. On the Riemann surface the period loops (a_n) are not wanted for this function, which appears as a particular case of a more general canonical integral having symmetrical behaviour in regard to the first and second kinds of period loops; but the loops (b_n) are necessary; they render the function uniform by preventing the introduction of all the values of which the function is capable. In the ζ plane, however*, the function is uniform for all values of ζ, and the regions interior to the circles enable us to represent all the values of which the function is susceptible. Thus the introduction of Riemann's normal integrals appears a more natural process in the case of the ζ plane than in the case of the Riemann surface itself.

231. We may obtain a product expression for $\tau_{n, m}$ directly from the formula

$$\tau_{n, m} = \frac{1}{2\pi i} \sum_j \log \left[\frac{\zeta_m - \vartheta_j(B_n)}{\zeta - \vartheta_j(B_n)} \Big/ \frac{\zeta_m - \vartheta_j(A_n)}{\zeta - \vartheta_j(A_n)} \right];$$

let k denote in turn every substitution whose product symbol neither has a power of ϑ_m at its left-hand end nor a power of ϑ_n at its right-hand end; thus we may write $\vartheta_j = \vartheta_m^{-h} \vartheta_k$, or, for abbreviation, $j = m^{-h}k$; and for every substitution k, the substitution j has all the forms derivable by giving to h all positive and negative integral values including zero, except that, when k

* Barriers being drawn to connect the infinities of the function.

is the identical substitution, if $m=n$, h can only have the one value zero; then applying ϑ_j^{-1} to every quantity of the cross ratio under the logarithm sign, we have

$$\tau_{n,m}=\frac{1}{2\pi i}\sum_j \log\left(\frac{\zeta_{j^{-1}m}-B_n}{\zeta_{j^{-1}}-B_n}\Big/\frac{\zeta_{j^{-1}m}-A_n}{\zeta_{j^{-1}}-A_n}\right)$$

$$=\frac{1}{2\pi i}\sum_{k,h}\log\left(\frac{\zeta_{k^{-1}m^{h+1}}-B_n}{\zeta_{k^{-1}m^h}-B_n}\Big/\frac{\zeta_{k^{-1}m^{h+1}}-A_n}{\zeta_{k^{-1}m^h}-A_n}\right),$$

and therefore, if m be not equal to n,

$$\tau_{n,m}=\frac{1}{2\pi i}\sum_k \log\left(\frac{\vartheta_k^{-1}(B_m)-B_n}{\vartheta_k^{-1}(A_m)-B_n}\Big/\frac{\vartheta_k^{-1}(B_m)-A_n}{\vartheta_k^{-1}(A_m)-A_n}\right),$$

while when $m=n$, separating away the term for which k is the identical substitution,

$$\tau_{n,n}=\frac{1}{2\pi i}\log\left(\frac{\zeta_n-B_n}{\zeta-B_n}\Big/\frac{\zeta_n-A_n}{\zeta-A_n}\right)$$

$$+\frac{1}{2\pi i}\sum_k{}'\log\left(\frac{\vartheta_k^{-1}(B_n)-B_n}{\vartheta_k^{-1}(A_n)-B_n}\Big/\frac{\vartheta_k^{-1}(B_n)-A_n}{\vartheta_k^{-1}(A_n)-A_n}\right),$$

where Σ' denotes that the identical substitution, $\vartheta_k=1$, is not included; thus

$$\tau_{n,n}=\frac{1}{2\pi i}\log(\mu_n e^{i\kappa_n})+\frac{1}{2\pi i}\sum_s{}'\log\left[\frac{B_n-\vartheta_s(B_n)}{A_n-\vartheta_s(B_n)}\Big/\frac{B_n-\vartheta_s(A_n)}{A_n-\vartheta_s(A_n)}\right]^2,$$

where s denotes every substitution of the group other than the identical substitution, not beginning or ending with a power of ϑ_n, and excluding every substitution of which the inverse has already occurred.

These formulæ, like that for $v_n^{\zeta,a}$, are not definite unless the barriers (§ 227) are drawn.

232. *Ex.* i. If $v_n^{\zeta,a}=u_n+iw_n$, u_n, iw_n being the real and imaginary parts of $v_n^{\zeta,a}$, prove, as in the case of a Riemann surface, by taking the integral $\int u\,dw$ round the p closed curves each formed by the circumferences of a pair of circles and the two sides of the barrier joining them, that the imaginary part of $N_1{}^2\tau_{11}+\dots\dots+2N_1N_2\tau_{12}+\dots\dots$ is positive, $N_1,\dots,N_p$ being any real quantities and $u+iw=N_1v_1^{\zeta,a}+\dots\dots+N_pv_p^{\zeta,a}$. Prove also the result $\tau_{m,n}=\tau_{n,m}$ by contour integration.

Ex. ii. Prove that the function of ζ expressed by

$$\Gamma_a^{\zeta,\xi}=\frac{d}{da}\Pi_{a,b}^{\zeta,\xi}=\sum_r(\gamma_r a+\delta_r)^{-2}\left[\frac{1}{a_r-\zeta}-\frac{1}{a_r-\xi}\right],$$

has analogous properties to Riemann's normal elementary integral of the second kind.

Ex. iii. Prove that

$$\Gamma_{a_i}^{\zeta,\xi}=(\gamma_i a+\delta_i)^2\,\Gamma_a^{\zeta,\xi},$$

where $a_i=(\alpha_i a+\beta_i)/(\gamma_i a+\delta_i)$.

Ex. iv. With the notation

$$\Phi(z, \zeta) = \sum_r \frac{(\gamma_r z + \delta_r)^{-2}}{z_r - \zeta},$$

prove that

$$\Phi(z, \zeta_n) - \Phi(z, \zeta) = 2\pi i \frac{d}{dz} v_n^{z, a} = \Phi(z, \xi_n) - \Phi(z, \xi),$$

where ξ is an arbitrary point, and hence prove that if $z, c_1, \dots, c_p, \xi$ be any arbitrary points, and $\xi_1 = \vartheta_1(\xi), \dots, \xi_p = \vartheta_p(\xi)$, the function of ζ expressed by

$$\begin{vmatrix} \Phi(z, \zeta), & \Phi(z, \xi), & \Phi(z, \xi_1), \dots, & \Phi(z, \xi_p) \\ \Phi(c_1, \zeta), & \Phi(c_1, \xi), & \Phi(c_1, \xi_1), \dots, & \Phi(c_1, \xi_p) \\ \dots & \dots & \dots & \dots \\ \Phi(c_p, \zeta), & \Phi(c_p, \xi), & \Phi(c_p, \xi_1), \dots, & \Phi(c_p, \xi_p) \\ 1, & 1, & 1, \dots, & 1 \end{vmatrix},$$

is unchanged by the substitutions of the group, and has simple poles at $z, c_1, \dots, c_p$, and their analogues, and a simple zero at ξ, and its analogues. Thus the function is similar to the function $\psi(x, a; z, c_1, \dots, c_p)$ of § 122, and every function which is unchanged by the substitutions of the group can be expressed by means of it.

As a function of z, the function is infinite at $z = \xi$, $z = \zeta$, beside being infinite at $z = \infty$, and its analogues; when $(a_i z + \beta_i)/(\gamma_i z + \delta_i)$ is put for z, the function becomes multiplied by $(\gamma_i z + \delta_i)^2$. This last circumstance clearly corresponds with the fact (§ 123) that $\psi(x, a; z, c_1, \dots, c_p)$ is not a rational function of z, but a rational function multiplied by $\frac{dz}{dt}$ (cf. Ex. iii.)

Ex. v. Prove that

$$\Gamma_a^{\zeta, \xi} = \sum_r \left(\frac{1}{a - \zeta_r} - \frac{1}{a - \xi_r} \right).$$

Ex. vi. In case $p = 1$, we have

$$v^{\zeta, a} = \frac{1}{2\pi i} \log\left(\frac{\zeta - B}{\zeta - A} \Big/ \frac{a - B}{a - A} \right), \quad \Pi_{a,b}^{\zeta, \xi} = \log \prod_{r=-\infty}^{\infty} \left(\frac{a_r - \zeta}{b_r - \zeta} \Big/ \frac{a_r - \xi}{b_r - \xi} \right), \quad \tau = \frac{1}{2\pi i} \log(\mu e^{i\kappa}),$$

where

$$(a_r - B)/(a_r - A) = (\mu e^{i\kappa})^r (a - B)/(a - A).$$

Putting, for abbreviation, $q = e^{i\pi\tau} = \sqrt{\mu e^{i\kappa}}$, and

$$\Theta(\zeta) = \sum_{n=-\infty}^{\infty} (-)^n q^{(n+\frac{1}{2})^2} \left(\frac{\zeta - B}{\zeta - A} \Big/ \frac{a - B}{a - A} \right)^{n+\frac{1}{2}},$$

prove, by applying the fundamental transformation once, that

$$\Theta(\zeta') = -\frac{1}{q} \frac{\zeta - A}{\zeta - B} \Big/ \frac{a - A}{a - B} \Theta(\zeta) = -e^{-2\pi i(v^{\zeta, a} + \frac{1}{2}\tau)} \Theta(\zeta),$$

and shew that $\Theta(\zeta)$ is a multiple of the Jacobian theta function $\Theta(v^{\zeta, a}, q; \frac{1}{2}, \frac{1}{2})$.

Ex. vii. Taking two circles as in figure 6 (§ 223), let $C'B/AC' = \sigma$ and $\frac{CB}{AC} \Big/ \frac{C'B}{AC'} = \mu$;

take an arbitrary real quantity ω, and a pure imaginary quantity $\omega' = \frac{\omega}{i\pi} \log \mu$, and let

$\wp(u)$ denote Weierstrass's elliptic function of u with 2ω, $2\omega'$ as periods. Then prove, if a, c denote points outside both the circles, a' denote the inverse point of a in regard to either one of the circles, and P, Q be arbitrary real quantities,

(α) that the function

$$\left\{\wp\left[\frac{\omega}{i\pi}\log\left(\frac{\zeta-B}{\zeta-A}\right)^2 \Big/ \frac{a-B}{a-A}\,\frac{c-B}{c-A}\right]-\wp\left[\frac{\omega}{i\pi}\log\frac{a-B}{a-A}\Big/\frac{c-B}{c-A}\right]\right\}^{-1},$$

is unaltered by the substitution $(\zeta'-B)/(\zeta'-A)=\mu(\zeta-B)/(\zeta-A)$, and has poles of the first order, outside both the circles, only at the points $\zeta=a$, $\zeta=c$.

(β) that the function,

$$\frac{P+iQ}{\wp\left[\frac{\omega}{i\pi}\log\frac{1}{\sigma}\frac{\zeta-B}{\zeta-A}\right]-\wp\left[\frac{\omega}{i\pi}\log\frac{1}{\sigma}\frac{a-B}{a-A}\right]}+\frac{P-iQ}{\wp\left[\frac{\omega}{i\pi}\log\frac{1}{\sigma}\frac{\zeta-B}{\zeta-A}\right]-\wp\left[\frac{\omega}{i\pi}\log\frac{1}{\sigma}\frac{a'-B}{a'-A}\right]}$$

is real on the circumference of each circle, and, outside both the circles, has a pole of the first order only at the point $\zeta=a$. The arbitraries P, Q can be used to prescribe the residue at this pole.

Ex. viii. Prove that any two uniform functions of ζ having no discontinuities except poles, which are unaltered by the substitutions of the group, are connected by an algebraic relation (cf. § 235); and that, if these two be properly chosen, any other uniform function of ζ having no discontinuities except poles, which is unaltered by the substitutions of the group, can be expressed rationally in terms of them. The development of the theory on these lines is identical with the theory of rational functions on a Riemann surface, but is simpler on account of the absence of branch places. Thus for instance we have a theory of fundamental integral functions, an integral function being one which is only infinite in the poles of an arbitrarily chosen function x. And we can form a function such as $\bar{E}(x, z)$ (§ 124, Chap. VII.); but the essential part of that function is much more simply provided by the function, $\varpi(\zeta, \gamma)$, investigated in the following article.

233. The preceding investigations are sufficient to explain the analogy between the present theory and that of a Riemann surface. We come now to the result which is the main purpose of this chapter. In the equation

$$\Pi_{\zeta,\gamma}^{z,c}=\sum_i \log\left(\frac{z_i-\zeta}{c_i-\zeta}\Big/\frac{z_i-\gamma}{c_i-\gamma}\right)=\sum_i \log\{\zeta,\gamma/z_i,c_i\},$$

where $\{\zeta,\gamma/z_i,c_i\}$ denotes a cross ratio, let the point z approach indefinitely near to ζ, and the point c approach indefinitely near to γ; then separating away the term belonging to the identical substitution, and associating with the term belonging to any other substitution that belonging to the inverse substitution, we have, after applying a linear transformation to every element of the cross ratio arising from the inverse substitution

$$\Pi_{\zeta,\gamma}^{z,c}=\log\frac{(z-\zeta)(c-\gamma)}{(z-\gamma)(c-\zeta)}+\sum_i{}'\log\frac{(z_i-\zeta)(c_i-\gamma)}{(z_i-\gamma)(c_i-\zeta)}\cdot\frac{(z-\zeta_i)(c-\gamma_i)}{(z-\gamma_i)(c-\zeta_i)},$$

where Σ' denotes that, in the summation, of terms arising by a substitution and its inverse, only one is to be taken, and the identical substitution is excluded. Thus we have*

$$\underset{z=\zeta,\, c=\gamma}{\text{limit}} \left[-(z-\zeta)(c-\gamma)e^{-\Pi_{\zeta,\gamma}^{z,c}}\right]^{\frac{1}{2}} = (\zeta-\gamma)\prod_i{}' \frac{(\zeta_i-\gamma)(\gamma_i-\zeta)}{(\zeta_i-\zeta)(\gamma_i-\gamma)},$$

$$= (\zeta-\gamma)\prod_i{}' \{\zeta, \gamma/\gamma_i, \zeta_i\},$$

where $\prod_i{}'$ has a similar signification to $\sum_i{}'$ and $\{\zeta, \gamma/\gamma_i, \zeta_i\}$ denotes a cross ratio. Consider now the expression

$$\varpi(\zeta, \gamma), \ = (\zeta-\gamma)\prod_i{}' \{\zeta, \gamma/\gamma_i, \zeta_i\};$$

it has clearly the following properties—it represents a perfectly definite function of ζ and γ, single-valued on the whole ζ-plane; it depends only on two variables, and $\varpi(\zeta, \gamma) = -\varpi(\gamma, \zeta)$; as a function of ζ it is infinite, save for the singular points of the group, only at $\zeta = \infty$, and not at the analogues of $\zeta = \infty$; it vanishes only at $\zeta = \gamma$ and the analogues of this point, and $\text{limit}_{\zeta=\gamma}\, \varpi(\zeta, \gamma)/(\zeta-\gamma) = 1$. Thus the function may be expected to generalise the irreducible factor of the form $x - a$, in the case of rational functions, and the factor $\sigma(u-a)$ in the case of elliptic functions, and to serve as a prime function for the functions of ζ now under consideration (cf. also Chap. VII. § 129 and Chaps. XIII. and XIV.). It should be noticed that the value of $\varpi(\zeta, \gamma)$ does not depend upon the choice we make in the product between any substitution and its inverse; this follows by applying the substitution ϑ_i^{-1} to every element of any factor.

234. We enquire now as to the behaviour of the function $\varpi(\zeta, \gamma)$ under the substitutions of the group. It will be proved that

$$\frac{\varpi(\zeta_n, \gamma)}{\varpi(\zeta, \gamma)} = (-1)^{g_n+h_n} \frac{e^{-2\pi i\left(v_n^{\zeta,\gamma} + \frac{1}{2}\tau_{n,n}\right)}}{\gamma_n\zeta + \delta_n},$$

where $(-1)^{g_n}$, $(-1)^{h_n}$ are certain $\pm$ signs to be explained.

This result can be obtained, save for a sign, from the definition of $\varpi(\zeta, \gamma)$, as a limit, from the function $\Pi_{z,c}^{\zeta,\gamma}$; but since, for our purpose, it is essential to avoid any such ambiguity, and because we wish to regard the function $\varpi(\zeta, \gamma)$ as fundamental, we adopt the longer method of dealing directly with the product $(\zeta-\gamma)\prod_i{}' \{\zeta, \gamma/\gamma_i, \zeta_i\}$. We imagine the barriers, each connecting a pair of circles, which are necessary to render the functions $v_1^{\zeta,a}, \ldots, v_p^{\zeta,a}$

* This function occurs in Schottky, *Crelle*, CI. (1887), p. 242 (at the top of the page). See also p. 253, at the top. The function is modified, for a Riemann surface, by Klein, *Math. Annal.* XXXVI. (1890), p. 13. The modified function occurs also, in particular cases, in a paper by Pick, *Math. Annal.* XXIX., and in Klein, *Math. Annal.* XXXII. (1888), p. 367. For $p=1$, the theta function was of course expressed in factors by Jacobi. The function employed by Ritter, *Math. Annal.* XLIV. (p. 291), has a somewhat different character.

uniform, to be drawn; then the quantities $\tau_{n,m}$, $\tau_{n,n}$ given in § 231, and defined by $v_n^{\zeta_m, \zeta}$, $v_n^{\zeta_n, \zeta}$ are definite; so therefore is also $e^{\pi i v_n^{\zeta, \gamma}}$ and the quantity $e^{\pi i \tau_{n,n}}$, which is equal to

$$\mu_n^{\frac{1}{2}} e^{\frac{1}{2} i\kappa_n} \prod_s{}' \left[\frac{B_n - (B_n)_s}{A_n - (B_n)_s} \Big/ \frac{B_n - (A_n)_s}{A_n - (A_n)_s} \right],$$

where s denotes a substitution, other than the identical substitution, not beginning or ending with a power of ϑ_n, and excluding the inverse of a substitution which has already occurred. This formula raises the question whether κ_n, which we take positive, is to be regarded as less than 2π or not, since otherwise the sign of $e^{\frac{1}{2}i\kappa_n}$ is not definite. But in fact, as it arises in this formula, from $v_n^{\zeta_n, \zeta}$, $\log \mu_n + i\kappa_n$ is the value of $\log \left(\frac{\zeta' - B_n}{\zeta' - A_n} \Big/ \frac{\zeta - B_n}{\zeta - A_n} \right)$ when ζ' has reached ζ_n from ζ *by a path which does not cross the barriers.* Thus κ_n is perfectly definite when the barriers are drawn, and the sign of the quantity

$$e^{-\pi i \tau_{n,n}} \mu_n^{\frac{1}{2}} e^{\frac{1}{2} i\kappa_n} \prod_s{}' \left[\frac{B_n - (B_n)_s}{A_n - (B_n)_s} \Big/ \frac{B_n - (A_n)_s}{A_n - (A_n)_s} \right]$$

is perfectly definite and independent of the barriers. We denote it by $(-1)^{g_n - 1}$. The annexed figure illustrates two ways of drawing a barrier PP'. In the first case κ_n is less than 2π. In the second case ζ' must pass

Fig. 8.

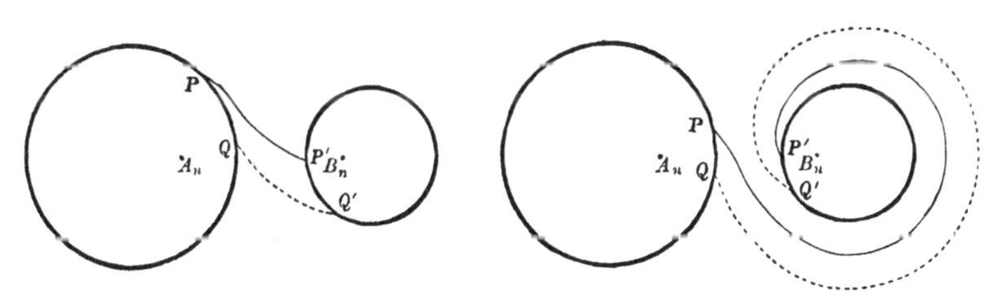

once round the point B, and κ_n is greater than 2π. When κ_n is thus determined, the expression by means of κ_n of the $\rho_n^{\frac{1}{2}}$ which occurs in the formulae connecting $\alpha_n, \beta_n, \gamma_n, \delta_n$ and A_n, B_n, ρ_n, for instance in the formula $\rho_n^{\frac{1}{2}} = (1 + \rho_n)/(\alpha_n + \delta_n)$, is also definite; it may be $\rho_n^{\frac{1}{2}} = \mu_n^{\frac{1}{2}} e^{\frac{1}{2}i\kappa_n}$ or $\rho_n^{\frac{1}{2}} = -\mu_n^{\frac{1}{2}} e^{\frac{1}{2}i\kappa_n}$. We shall put $\rho_n^{\frac{1}{2}} = (-1)^{h_n} \mu_n^{\frac{1}{2}} e^{\frac{1}{2}i\kappa_n}$. If the whole investigation had been commenced with a different sign for each of $\alpha_n, \beta_n, \gamma_n, \delta_n$, h_n would have become $h_n - 1$, but g_n, depending only on the circles and the barrier, would have the same value.

We have

$$\frac{\varpi(\zeta_n, \gamma)}{\varpi(\zeta, \gamma)} = \frac{\zeta_n - \gamma}{\zeta - \gamma} \prod_i{}' \frac{\zeta_{in} - \gamma}{\zeta_i - \gamma} \cdot \frac{\gamma_i - \zeta_n}{\gamma_i - \zeta} \cdot \frac{\zeta_i - \zeta}{\zeta_{in} - \zeta_n},$$

where i denotes in turn all substitutions which with their inverses give the whole group, except the identical substitution; thus i denotes all substitutions n^λ for $\lambda = 1, 2, 3, \ldots, \infty$, as well as all substitutions $n^h s n^k$, where s has the significance just explained and h, k take all positive and negative integer values including zero. Therefore

$$\frac{\varpi(\zeta_n, \gamma)}{\varpi(\zeta, \gamma)} = \frac{\zeta_n - \gamma}{\zeta - \gamma} \prod_\lambda \frac{\zeta_{n^{\lambda+1}} - \gamma}{\zeta_{n^\lambda} - \gamma} \cdot \frac{\gamma_{n^\lambda} - \zeta_n}{\gamma_{n^\lambda} - \zeta} \cdot \frac{\zeta_{n^\lambda} - \zeta}{\zeta_{n^{\lambda+1}} - \zeta_n}$$

$$\prod_{h,s,k} \frac{\zeta_{n^h s n^{k+1}} - \gamma}{\zeta_{n^h s n^k} - \gamma} \cdot \frac{\gamma_{n^h s n^k} - \zeta_n}{\gamma_{n^h s n^k} - \zeta} \cdot \frac{\zeta_{n^h s n^k} - \zeta}{\zeta_{n^h s n^{k+1}} - \zeta_n}$$

$$= \frac{\zeta_n - \gamma}{\zeta - \gamma} \prod_\lambda \frac{\zeta_{n^{\lambda+1}} - \gamma}{\zeta_{n^\lambda} - \gamma} \prod_\lambda \frac{\zeta_{n^\lambda} - \zeta}{\zeta_{n^{\lambda+1}} - \zeta} \prod_\lambda \frac{\gamma_{n^\lambda} - \zeta_n}{\gamma_{n^\lambda} - \zeta} \cdot \frac{\zeta_{n^{\lambda+1}} - \zeta}{\zeta_{n^{\lambda+1}} - \zeta_n}$$

$$\prod_{h,s} \frac{(B_n)_{n^h s} - \gamma}{(A_n)_{n^h s} - \gamma} \cdot \frac{(A_n)_{n^h s} - \zeta}{(B_n)_{n^h s} - \zeta} \prod_{h,s,k} \frac{\gamma_{n^h s n^k} - \zeta_n}{\gamma_{n^h s n^k} - \zeta} \cdot \frac{\zeta_{n^h s n^{k+1}} - \zeta}{\zeta_{n^h s n^{k+1}} - \zeta_n},$$

the transformation of the second part of the product being precisely as in the first part,

$$= \frac{\zeta_n - \gamma}{\zeta - \gamma} \cdot \frac{B_n - \gamma}{\zeta_n - \gamma} \cdot \frac{\zeta_n - \zeta}{B_n - \zeta} \prod_\lambda \frac{\gamma - \zeta_{n^{1-\lambda}}}{\gamma - \zeta_{n^{-\lambda}}} \cdot \frac{\zeta_n - \zeta_{n^{-\lambda}}}{\zeta_n - \zeta_{n^{1-\lambda}}}$$

$$\prod_{h,s} \frac{(B_n)_{n^h s} - \gamma}{(A_n)_{n^h s} - \gamma} \cdot \frac{(A_n)_{n^h s} - \zeta}{(B_n)_{n^h s} - \zeta} \prod_{h,s,k} \frac{\gamma - \zeta_{n^{-k} s^{-1} n^{1-h}}}{\gamma - \zeta_{n^{-k} s n^{-h}}} \cdot \frac{\zeta_n - \zeta_{n^{-k} s^{-1} n^{-h}}}{\zeta_n - \zeta_{n^{-k} s^{-1} n^{1-h}}}$$

$$= \frac{B_n - \gamma}{B_n - \zeta} \cdot \frac{\zeta_n - \zeta}{\zeta - \gamma} \cdot \frac{\gamma - \zeta}{\gamma - A_n} \cdot \frac{\zeta_n - A_n}{\zeta_n - \zeta} \prod_{h,s} \frac{(B_n)_{n^h s} - \gamma}{(A_n)_{n^h s} - \gamma} \cdot \frac{(A_n)_{n^h s} - \zeta}{(B_n)_{n^h s} - \zeta}$$

$$\prod_{s,k} \frac{\gamma - (B_n)_{n^{-k} s^{-1}}}{\gamma - (A_n)_{n^{-k} s^{-1}}} \cdot \frac{\zeta_n - (A_n)_{n^{-k} s^{-1}}}{\zeta_n - (B_n)_{n^{-k} s^{-1}}};$$

since h and $-k$ have the same range of signification we may replace $-k$ by h, in the last form, and obtain, by a rearrangement of the second product,

$$\frac{\varpi(\zeta_n, \gamma)}{\varpi(\zeta, \gamma)} = -\frac{B_n - \gamma}{B_n - \zeta} \cdot \frac{\zeta_n - A_n}{\gamma - A_n} \prod_{h,s} \frac{\zeta - (A_n)_{n^h s}}{\zeta - (B_n)_{n^h s}} \cdot \frac{\gamma - (B_n)_{n^h s}}{\gamma - (A_n)_{n^h s}}$$

$$\prod_{s,h} \frac{\gamma - (B_n)_{n^h s^{-1}}}{\gamma - (A_n)_{n^h s^{-1}}} \cdot \frac{\zeta_n - (A_n)_{n^h s^{-1}}}{\zeta_n - (B_n)_{n^h s^{-1}}};$$

but, from the formula

$$v_n^{\zeta, \gamma} = \frac{1}{2\pi i} \sum_j \log \frac{\zeta - \vartheta_j(B_n)}{\zeta - \vartheta_j(A_n)} \cdot \frac{\gamma - \vartheta_j(A_n)}{\gamma - \vartheta_j(B_n)},$$

where j can have the forms $n^h s$, $n^h s^{-1}$, or be the identical substitution, we have

$$e^{2\pi i v_n^{\zeta, \gamma}} = \frac{\zeta - B_n}{\zeta - A_n} \cdot \frac{\gamma - A_n}{\gamma - B_n} \prod_{h,s} \frac{\zeta - (B_n)_{n^h s}}{\zeta - (A_n)_{n^h s}} \cdot \frac{\gamma - (A_n)_{n^h s}}{\gamma - (B_n)_{n^h s}} \prod_{s,h} \frac{\zeta - (B_n)_{n^h s^{-1}}}{\zeta - (A_n)_{n^h s^{-1}}} \cdot \frac{\gamma - (A_n)_{n^h s^{-1}}}{\gamma - (B_n)_{n^h s^{-1}}};$$

therefore

$$\frac{\varpi(\zeta_n, \gamma)}{\varpi(\zeta, \gamma)} e^{2\pi i v_n^{\zeta, \gamma}} = -\frac{\zeta_n - A_n}{\zeta - A_n} \prod_{s, h} \frac{\zeta - (B_n)_{nh_s - 1}}{\zeta_n - (B_n)_{nh_s - 1}} \cdot \frac{\zeta_n - (A_n)_{nh_s - 1}}{\zeta - (A_n)_{nh_s - 1}}$$

$$= -\frac{\zeta_n - A_n}{\zeta - A_n} \prod_{s, h} \frac{\zeta_{sn-h} - B_n}{\zeta_{sn1-h} - B_n} \cdot \frac{\zeta_{sn1-h} - A_n}{\zeta_{sn-h} - A_n}$$

$$= -\frac{\zeta_n - A_n}{\zeta - A_n} \prod_{s} \frac{(A_n)_s - B_n}{(B_n)_s - B_n} \cdot \frac{(B_n)_s - A_n}{(A_n)_s - A_n},$$

and hence

$$\frac{\varpi(\zeta_n, \gamma)}{\varpi(\zeta, \gamma)} e^{2\pi i v_n^{\zeta, \gamma} + \pi i \tau_{n, n}} = \frac{\zeta_n - A_n}{\zeta - A_n} (-1)^{g_n} \mu_n^{\frac{1}{2}} e^{\frac{1}{2} i \kappa_n};$$

now from the formula $(\zeta_n - B_n)/(\zeta_n - A_n) = \rho_n (\zeta - B_n)/(\zeta - A_n)$, and the values of α_n, β_n, γ_n, δ_n given in § 226, we immediately find

$$(\zeta - A_n)/(\zeta_n - A_n) = [\zeta - A_n - \rho_n (\zeta - B_n)]/(B_n - A_n),$$

$$\gamma_n \zeta + \delta_n = [\rho_n^{-\frac{1}{2}} (\zeta - A_n) - \rho_n^{\frac{1}{2}} (\zeta - B_n)]/(B_n - A_n);$$

thus, as $\rho_n^{\frac{1}{2}} = (-1)^{h_n} \mu_n^{\frac{1}{2}} e^{\frac{1}{2} i \kappa_n}$, we have

$$(\zeta - A_n)/(\zeta_n - A_n) = (-1)^{h_n} \mu_n^{\frac{1}{2}} e^{\frac{1}{2} i \kappa_n} (\gamma_n \zeta + \delta_n);$$

hence, finally

$$\frac{\varpi(\zeta_n, \gamma)}{\varpi(\zeta, \gamma)} = (-1)^{g_n + h_n} \frac{e^{-2\pi i (v_n^{\zeta, \gamma} + \frac{1}{2} \tau_{n, n})}}{\gamma_n \zeta + \delta_n},$$

where $(-1)^{g_n} e^{-\pi i \tau_{n, n}} e^{\frac{1}{2} i \kappa_n}$ is independent of how the barriers are drawn, and $(-1)^{h_n} \gamma_n$, $(-1)^{h_n} \delta_n$ are independent of the signs attached to γ_n and δ_n.

235. The function $\varpi(\zeta, \gamma)$, whose properties have thus been deduced immediately from its expression as an infinite product, supposed to be convergent, may be regarded as fundamental. Thus, as can be immediately verified, the integral $\Pi_{\zeta, \gamma}^{z, c}$ is expressible by $\varpi(\zeta, \gamma)$, in the form

$$\Pi_{\zeta, \gamma}^{z, c} = \log \frac{\varpi(z, \zeta) \varpi(\gamma, c)}{\varpi(z, \gamma) \varpi(\zeta, c)},$$

and thence the integrals $v_n^{\zeta, \gamma}$ arise, by the definition $v_n^{\zeta, \gamma} = \frac{1}{2\pi i} \Pi_{\zeta, \gamma}^{z_n, z}$, and thence, also, integrals with algebraic infinities, by the definition

$$\Gamma_x^{\zeta, \gamma} = \frac{d}{dx} \Pi_{x, a}^{\zeta, \gamma}$$

(cf. Ex. ii, § 232). Further, if $F(\zeta)$ denote any uniform function of ζ whose value is unaltered by the substitutions of the group, which has no discontinuities except poles, it is easy to prove, by contour integration, as in the case of a Riemann surface, (i) That $F(\zeta)$ must be somewhere infinite in the region S, (ii) That $F(\zeta)$ takes any assigned value as many times within S as the sum of its orders of infinity within S, (iii) That if $\alpha_1, \ldots, \alpha_k$ be the poles and $\beta_1, \ldots, \beta_k$ the zeros of $F(\zeta)$ within S, and the barriers be supposed drawn,

$$v_i^{\beta_1, \alpha_1} + \ldots\ldots + v_i^{\beta_k, \alpha_k} = m_i + m_1' \tau_{i,1} + \ldots\ldots + m_p' \tau_{i,p}, \qquad (i = 1, \ldots, p),$$

where $m_1, \ldots, m_p, m_1', \ldots, m_p'$ are definite integers. Thence it is easy to shew that the ratio

$$F(\zeta) \Big/ \frac{\varpi(\zeta, \beta_1) \ldots\ldots \varpi(\zeta, \beta_k)}{\varpi(\zeta, \alpha_1) \ldots\ldots \varpi(\zeta, \alpha_k)} e^{-2\pi i (m_1' v_1^{\zeta, a} + \ldots + m'_p v_p^{\zeta, a})}$$

is a constant for all values of ζ. And replacing some of $\beta_1, \ldots, \alpha_k$ in this expression by suitable analogues, the exponential factor may be absorbed.

Ex. In the elliptic case where there is one fundamental substitution $(\zeta' - B)/(\zeta' - A) = \rho(\zeta - B)/(\zeta - A)$, we have $(\zeta_i - B)/(\zeta_i - A) = \rho^i (\zeta - B)/(\zeta - A)$, and thence putting u, v, respectively for the integrals v^ζ, v^γ, so that $e^{2\pi i u} = (\zeta - B)/(\zeta - A)$, $e^{2\pi i v} = (\gamma - B)/(\gamma - A)$, we immediately find

$$\frac{\zeta - \gamma_i}{\gamma - \gamma_i} \Big/ \frac{\zeta - \zeta_i}{\gamma - \zeta_i} = \frac{1 - 2\rho^i \cos 2\pi (u - v) + \rho^{2i}}{(1 - \rho^i)^2}, \quad \zeta - \gamma = \frac{B - A}{2i} \frac{\sin \pi (u - v)}{\sin \pi u \sin \pi v},$$

and hence

$$\varpi(\zeta, \gamma) = \frac{B - A}{2i} \frac{\sin \pi (u - v)}{\sin \pi u \sin \pi v} \prod_{i=1}^{\infty} \frac{1 - 2\rho^i \cos 2\pi (u - v) + \rho^{2i}}{(1 - \rho^i)^2},$$

which*, putting $e^{\pi i \tau} = \rho^{\frac{1}{2}}$, is equal to

$$\frac{(B - A)\pi}{4i\omega} e^{-2\eta\omega (u - v)^2} \sigma[2\omega(u - v);\ 2\omega,\ 2\omega\tau] \div \sin \pi u \sin \pi v,$$

where ω is an arbitrary quantity, and

$$\eta\omega = \frac{\pi^2}{12} - 2\pi^2 \sum_1^{\infty} \frac{\rho^n}{(1 - \rho^n)^2}.$$

236. The further development of the theory of functions in the ζ plane may be carried out on the lines already followed in the case of the Riemann surface. We limit ourselves to some indications in regard to matters bearing on the main object of this chapter.

The excess of the number of zeros over the number of poles, in any region, of a function of ζ, $f(\zeta)$, which is uniform and without essential singularities within that region, is of course equal to the integral

$$\frac{1}{2\pi i} \int d \log f(\zeta),$$

* See, for instance, Halphen, *Fonct. Ellipt.* (Paris, 1886), vol. I. p. 400.

taken round the boundary of the region. If we consider, for example, the function $\Omega_n(\zeta), = dv_n^{\zeta,\gamma}/d\zeta$, which is nowhere infinite, in the region S, the number of its zeros within the region S is

$$\frac{1}{2\pi i}\sum_{r=1}^{p}\int\left[\frac{\Omega_n'(\zeta_r)}{\Omega_n(\zeta_r)} - \frac{\Omega_n'(\zeta)}{\Omega_n(\zeta)}\right]d\zeta,$$

where the dash denotes a differentiation in regard to ζ, and the sign of summation means that the integral is taken round the circles $C_1', \dots, C_p'$, in a counter-clockwise direction. Since $\Omega_n(\zeta_r) = (\gamma_r\zeta + \delta_r)^2\,\Omega_n(\zeta)$, the value is

$$\frac{1}{2\pi i}\sum_{r=1}^{p}\int\frac{2d\zeta}{\zeta - \underset{r}{\vartheta}^{-1}(\infty)}$$

or $2p$; thus as $\Omega_n(\zeta)$ vanishes to the second order at $\zeta = \infty$ in virtue of the denominator $d\zeta$, we may say that $dv_n^{\zeta,\gamma}$ has $2p-2$ zeros in the region S, in general distinct from $\zeta = \infty$. The function $\Omega_n(\zeta)$ vanishes in every analogue of these $2p-2$ places, but does not vanish in the analogues of $\zeta = \infty$.

The theory of the theta functions, constructed from the integrals $v_n^{\zeta,\gamma}$, and their periods $\tau_{n,m}$, will subsist, and, as in the case of the Riemann surface there will, corresponding to an arbitrary point m, which we take in the region S, be points $m_1, \dots, m_p$ in the region S, such that the zeros of the function $\Theta(v^{\zeta,m} - v^{\zeta_1,m_1} - \dots\dots - v^{\zeta_p,m_p})$ are the places $\zeta_1, \dots, \zeta_p$. And corresponding to any odd half period, $\frac{1}{2}\Omega_{s,s'}$, there will be places $n_1, \dots, n_{p-1}$, in the region S, which, repeated, constitute the zero of a differential $dv^{\zeta,\gamma}$, and satisfy the equations typified by

$$\tfrac{1}{2}\Omega_{s,s'} = v^{m_p,m} - v^{n_1,m_1} - \dots\dots - v^{n_{p-1},m_{p-1}}.$$

The values of the quantities $e^{\pi i\tau_{n,n}}$ and the positions of $m_1, \dots, m_p$ may vary when the barriers which are necessary to define the periods $\tau_{n,m}$ are changed.

But it is one of the main results of the representation now under consideration that a particular theta function is derivable immediately from the function $\varpi(\zeta, \gamma)$; and hence, as is shewn in chapter XIV., that any theta function can be so derived. Let v denote the integral whose differential vanishes to the second order in each of the places $n_1, \dots, n_{p-1}$. Consider the expression $\sqrt{dv/d\zeta}$ in the region S. It has no infinities and it is single-valued in the neighbourhood of its zeros, as follows from the fact that the p zeros of $dv/d\zeta$ are all of the second order. Hence if the region S be made simply connected by drawing the p barriers, and joining the p pairs of circles by $p-1$ further barriers $(c_1), \dots, (c_{p-1})$, of which (c_r) joins the circumference C_r' to the circumference C_{r+1}, $\sqrt{dv/d\zeta}$ will be uniform in the region S so long as ζ does not cross any of the barriers. For the change in the value of $\sqrt{dv/d\zeta}$ when ζ is taken round any closed circuit may then be obtained by considering the equivalent circuits enclosing the zeros. But in fact the barriers $(c_1), \ldots, (c_{p-1})$ are unnecessary; to see this it is sufficient to see that any circuit in the region S which entirely surrounds a pair of circles, such as C_1', C_1, encloses an even number of the infinities of $dv/d\zeta$ which are at the singular points of the group. Since these infinities are among the logarithmic zeros and poles of $v_1^{\zeta, \gamma}, \ldots, v_p^{\zeta, \gamma}$, whereof v is a linear function, the proof required is included in the proof that any one of the functions $v_1^{\zeta, \gamma}, \ldots, v_p^{\zeta, \gamma}$ is unaltered when taken round a circuit entirely surrounding a pair of the circles, such as C_1', C_1. Thus when the barriers which render the functions $v_1^{\zeta, \gamma}, \ldots, v_p^{\zeta, \gamma}$ uniform are drawn, the function $\sqrt{dv/d\zeta}$ is entirely definite within the region S, save for an arbitrary constant multiplier, provided the sign of the function be given for some one point in the region S. And, this being done, if γ be any point, the function $\sqrt{\dfrac{dv}{d\zeta}}\sqrt{\dfrac{dv}{d\gamma}}$ is independent of this sign. This function, with a certain constant multiplier, which will be afterwards assigned, may be denoted by $\psi(\zeta)$.

237. We proceed now to prove the equation

$$\varpi(\zeta, \gamma) = A\,\frac{\Theta(v^{\zeta, \gamma} + \tfrac{1}{2}\Omega_{s, s'})e^{\pi i s' v^{\zeta, \gamma}}}{\psi(\zeta)} = Ae^{-\frac{1}{2}\pi i s'(s + \frac{1}{2}\tau s')}\,\frac{\Theta(v^{\zeta, \gamma};\ \tfrac{1}{2}s,\ \tfrac{1}{2}s')}{\psi(\zeta)},$$

where $s'v^{\zeta, \gamma} = s_1'v_1^{\zeta, \gamma} + \ldots\ldots + s_p'v_p^{\zeta, \gamma}$, and A is constant, independent of ζ and γ. It is clear first of all that the two sides of this equation have the same poles and zeros in the region S. For $\Theta(v^{\zeta, \gamma} + \tfrac{1}{2}\Omega_{s, s'})$ vanishes to the first order at the places $\gamma, n_1, \ldots, n_{p-1}$, and $\psi(\zeta)$ vanishes to the first order at $n_1, \ldots, n_{p-1}, \infty$, while $\varpi(\zeta, \gamma)$ vanishes to the first order at $\zeta = \gamma$, and is infinite to the first order at $\zeta = \infty$.* Thus the quotient of the two sides of the equation has no infinities within the region S. Further the square of this quotient is uniform within the region S, independently of the barriers; for this statement holds of each of the factors

$$\varpi(\zeta, \gamma), \quad \psi^2(\zeta), \quad \Theta(v^{\zeta, \gamma} + \tfrac{1}{2}\Omega_{s, s'}), \quad e^{2\pi i s' v^{\zeta, \gamma}}.$$

And, if ζ be replaced by ζ_n, the square of the quotient of the two sides of the equation becomes (cf. § 175, Chap. X.) multiplied by the factor

$$\left[(-1)^{g_n + h_n}\frac{\psi(\zeta_n)/\psi(\zeta)}{\gamma_n\zeta + \delta_n}\right]^2,$$

which is equal to unity. Now† a function of ζ, which is unaltered by the substitutions of the group, and is uniform within the region S, and has no

* At the analogues of $\zeta = \infty$ neither $\varpi(\zeta, \gamma)$ nor $1/\psi(\zeta)$ becomes infinite.

† If $U + iV$ be the function, the integral $\int U dV$, taken round the $2p$ fundamental circles is expressible as a surface integral over S whose elements are positive or zero. In the case considered the former integral vanishes.

infinities, must, like a rational function on a Riemann surface, be a constant. Since the square root of a constant is also a constant the proof of the equation is complete.

From it we infer (i) that

$$\psi(\zeta_n)/\psi(\zeta) = (-1)^{g_n+h_n}(\gamma_n\zeta + \delta_n)(-1)^{s_n},$$

and (ii) that the values of $\psi(\zeta)$ on the two sides of a barrier have a quotient of the form $(-1)^{s'_n}$. The constant factor to be attached to $\psi(\zeta)$ may be chosen so that $A = 1$. For this it is sufficient to take for the integral v the expression

$$v = \sum_{i=1}^{p} \Theta_i'(\tfrac{1}{2}\Omega_{s,s'})\, v_i^{\zeta,\gamma},$$

where $\Theta_i'(u) = \partial\Theta(u)/\partial u_i$. Then (cf. § 188, p. 281) the right-hand side, when ζ is near to γ, is equal to $A(\zeta - \gamma) + \ldots$, while the left-hand side has the value $(\zeta - \gamma) + \ldots$.

238. The developments of an equation analogous to that just obtained, which will be given in Chap. XIV. in connection with the functions there discussed, render it unnecessary for us to pursue the matter further here. The following forms an interesting example of theta functions, of another kind.

Suppose that the quantities $\mu_1, \ldots, \mu_p$ are small enough to ensure (cf. § 226) the convergence of the series

$$\lambda(\zeta, \mu) = \sum_i \frac{[\gamma_i\zeta + \delta_i]^{-1}}{\zeta_i - \mu},$$

wherein μ denotes an arbitrary place within the region S, and i denotes a summation extending to every substitution of the group. It will appear that this function is definite in all cases in which the function $\varpi(\zeta, \mu)$ is definite. The function is immediately seen to verify the equations

$$\lambda(\zeta_n, \mu) = (\gamma_n\zeta + \delta_n)\lambda(\zeta, \mu), \quad \lambda(\zeta, \mu_n) = (\gamma_n\mu + \delta_n)\lambda(\zeta, \mu),$$

and

$$\begin{aligned} \lambda(\mu, \zeta) &= \sum_i \frac{1}{\alpha_i\mu + \beta_i - \zeta(\gamma_i\mu + \delta_i)} \\ &= -\sum_i \frac{1}{\delta_i\zeta - \beta_i - \mu(-\gamma_i\zeta + \alpha_i)} \\ &= -\sum \frac{(\gamma_r\zeta + \delta_r)^{-1}}{\zeta_r - m}, \end{aligned}$$

where r denotes the substitution inverse to that denoted by i. Thus

$$\lambda(\zeta, \mu) = -\lambda(\mu, \zeta).$$

The function has one pole in the region S, namely at μ, and no other infinities, and if the series be uniformly convergent near $\zeta = \infty$, as we assume, the function vanishes to the first order at $\zeta = \infty$. The excess of the number of its zeros over the number of its poles in S, which is given by

$$\frac{1}{2\pi i} \sum_{n=1}^{p} \int \left[\frac{\lambda'(\zeta_n, \mu)}{\lambda(\zeta_n, \mu)} - \frac{\lambda'(\zeta, \mu)}{\lambda(\zeta, \mu)} \right] d\zeta,$$

where the dash denotes a differentiation in regard to ζ, and the integrals are taken counter-clockwise round the circles C_1', ..., C_p', namely by

$$\frac{1}{2\pi i} \sum_{n=1}^{p} \int \frac{d\zeta}{\zeta - \vartheta_n^{-1}\infty},$$

is equal to p. Thus the function has p zeros in S other than $\zeta = \infty$; denote these by $\mu_1, \ldots, \mu_p$. Within any region $\vartheta_n S$ the function has the analogue of μ for a pole, and the analogues of $\mu_1, \ldots, \mu_p$ for zeros; it does not vanish at the analogue of $\zeta = \infty$. This result may be verified also by investigating similarly the excess of the number of zeros over the number of poles in any such region; the result is found to be $p - 1$.

Consider the ratio

$$f(\zeta) = [\lambda(\zeta, \mu)]^2 \div \frac{dv}{d\zeta},$$

where v is any linear function of $v_1^{\zeta, \gamma}, \ldots, v_p^{\zeta, \gamma}$; let $\zeta_1, \ldots, \zeta_{2p-2}$ denote the zeros of dv. Then $f(\zeta)$ is uniform within the region S, and is unaltered by the substitutions of the group. It has poles $\mu^2, \zeta_1, \ldots, \zeta_{2p-2}$, and no other infinities in S, and has zeros $\mu_1^2, \ldots, \mu_p^2$, the square of a symbol being written to denote a zero or pole of the second order. Thus we have, precisely as for the case of rational functions on a Riemann surface,

$$2v_n^{\mu_p, \mu} + v_n^{\zeta_1, \mu_1} + v_n^{\zeta_2, \mu_1} + \ldots\ldots + v_n^{\zeta_{2p-3}, \mu_{p-1}} + v_n^{\zeta_{2p-2}, \mu_{p-1}} \equiv 0, \qquad (\mu = 1, 2, \ldots, p),$$

or (§ 179, p. 256),

$$(\mu^2, \zeta_1, \ldots, \zeta_{2p-2}) \equiv (\mu_1^2, \ldots, \mu_p^2),$$

and therefore, if $m_1, \ldots, m_p$ denote the points in S, derivable from μ (§ 236), such that $\Theta(v^{\zeta, \mu} - v^{x_1, m_1} - \ldots\ldots - v^{x_p, m_p})$ vanishes in $\zeta = x_1, \ldots, \zeta = x_p$, we have (§ 182, p. 265).

$$(\mu_1^2, \ldots, \mu_p^2) \equiv (m_1^2, \ldots, m_p^2).$$

When the barriers are drawn, let

$$v_n^{\mu_1, m_1} + \ldots\ldots + v_n^{\mu_p, m_p} = \tfrac{1}{2}(k_i + k_1'\tau_{2,1} + \ldots\ldots + k_p'\tau_{i,p}), \qquad (i = 1, 2, \ldots, p),$$

$k_1, \ldots, k_p, k_1', \ldots, k_p'$ being integers.

Now consider the product $\lambda(\zeta, \mu)\, \varpi(\zeta, \mu)$. It has no poles, in S, and its zeros are $\mu_1, \ldots, \mu_p$. It is an uniform function of ζ, and, subjected to one of the fundamental substitutions of the group it takes the factor

$$\frac{\lambda(\zeta_n, \mu)\, \varpi(\zeta_n, \mu)}{\lambda(\zeta, \mu)\, \varpi(\zeta, \mu)}, = (-1)^{g_n + h_n} e^{-2\pi i (v_n^{\zeta, \mu} + \frac{1}{2}\tau_{n,n})}.$$

Hence the function

$$F(\zeta) = \frac{\lambda(\zeta, \mu)\,\varpi(\zeta, \mu)}{\Theta(v^{\zeta, \mu} - \tfrac{1}{2}\Omega)} e^{\pi i k' v^{\zeta, \mu}},$$

wherein $k'v^{\zeta, \mu}$ denotes $k_1'v_1^{\zeta, \mu} + \ldots\ldots + k_p'v_p^{\zeta, \mu}$, and Ω denotes the p quantities $k_i + k_1'\tau_{i, 1} + \ldots\ldots + k_p'\tau_{i, p}$, has, within S, no zeros or poles, and is such that, for a fundamental substitution,

$$F(\zeta_n)/F(\zeta) = (-1)^{g_n + h_n - k_n}$$

(cf. § 175, Chap. X.); thus, as in the previous article, $F(\zeta)$ is a constant thus, also, $g_n + h_n - k_n$ is an even integer, $= 2H_n$, say, and we have

$$\lambda(\zeta, \mu)\,\varpi(\zeta, \mu) = Ae^{-\pi i k' v^{\zeta, \mu}}\,\Theta(v^{\zeta, \mu} - \tfrac{1}{2}P),$$

where P denotes the p quantities $g_i + h_i + k_1'\tau_{i, 1} + \ldots\ldots + k_p'\tau_{i, p}$, and A is independent of ζ. But, if ζ describe the circumference C_n, the left-hand side is unchanged, and the right-hand side obtains the factor $e^{-\pi i k'_n}$. Thus the integers $k_1', \ldots, k_p'$ are all even; put $k_r' = 2H_r'$; then, as

$$\Theta\left(v^{\zeta, \mu} - \frac{g+h}{2} - \tau H'\right) = e^{2\pi i H'\left(v^{\zeta, \mu} - \frac{g+h}{2}\right) - \pi i \tau H'^2}\,\Theta\left(v^{\zeta, \mu} - \frac{g+h}{2}\right),$$

where the notation is that of § 175, Chap. X., we have

$$\lambda(\zeta, \mu)\,\varpi(\zeta, \mu) = B\Theta\left(v^{\zeta, \mu} - \frac{g+h}{2}\right),$$

wherein B is independent of ζ, and therefore, since the interchange of ζ, μ leaves both sides unaltered, B is also independent of μ. The value of B may be expressed by putting $\zeta = \mu$; thence we obtain, finally,

$$\lambda(\zeta, \mu)\,\varpi(\zeta, \mu) = \Theta(v^{\zeta, \mu} - \tfrac{1}{2}g - \tfrac{1}{2}h)/\Theta(\tfrac{1}{2}g + \tfrac{1}{2}h).$$

This equation may be regarded as equivalent to 2^p equations. For if in one of the p fundamental substitutions $\vartheta_r\zeta = (\alpha_r\zeta + \beta_r)/(\gamma_r\zeta + \delta_r)$, we consider the signs of α_r, β_r, γ_r, δ_r all reversed, the function $\lambda(\zeta, \mu)$, which involves the first powers of these quantities, will take a different value. The function $\varpi(\zeta, \mu)$, the p fundamental circles, and the integrals $v^{\zeta, \mu}$ and their periods $\tau_{n, m}$, and therefore the integers $g_1, \ldots, g_p$, will remain unchanged, if the barriers remain unaltered. But the integer h_r will be increased by unity.

If, on the other hand, the coefficients α, β, γ, δ remaining unaltered, one of the barriers be drawn differently, the left-hand side of the equation remains unaltered; on the right-hand one of $h_1, \ldots, h_p$ will be increased by an integer, say, for example, h_r increased by unity, and therefore each of $\tau_{1, r}, \ldots, \tau_{p, r}$ also increased by unity. Putting u for $v^{\zeta, \mu} - \tfrac{1}{2}g - \tfrac{1}{2}h$, and neglecting integral increments of u, the exponent of the general term of the theta series is increased, save for integral multiples of $2\pi i$, by

$$2\pi i\left(-\tfrac{1}{2}\right) n_r + i\pi n_r^2,$$

which is an even multiple of πi, so that the general term is unchanged.

Ex. i. Prove that the function $\lambda(\zeta, \mu)$ can be written in the form

$$\lambda(\zeta, \mu) = \frac{1}{\zeta-\mu}\left[1 + \sum_i{}' (a_i+\delta_i)\{\zeta, \zeta_i \mid \mu, \mu_i\}\right],$$

where the sign of summation refers to all the substitutions of the group, other than the identical substitution, with the condition that when any substitution occurs its inverse must not occur, and $\{\zeta, \zeta_i \mid \mu, \mu_i\}$ denotes $\dfrac{\zeta-\mu}{\zeta-\mu_i}\Big/\dfrac{\zeta_i-\mu}{\zeta_i-\mu_i}$.

Ex. ii. In case $p=1$, where the fundamental substitution is

$$(\zeta'-B)/(\zeta'-A) = \rho(\zeta-B)/(\zeta-A),$$

putting $e^{2\pi iu} = (\zeta-B)/(\zeta-A)$, $e^{2\pi iv} = (\mu-B)/(\mu-A)$, prove that

$$\zeta-\mu = \frac{B-A}{2i}\frac{\sin\pi(u-v)}{\sin\pi u\sin\pi v}, \quad \{\zeta, \zeta_i \mid \mu, \mu_i\} = 4\rho^i\frac{\sin^2\pi(u-v)}{1-2\rho^i\cos 2\pi(u-v)+\rho^{2i}},$$

and hence

$$\lambda(\zeta, \mu) = \frac{2i\sin\pi u\sin\pi v}{(B-A)\sin\pi(u-v)}\left[1 + \sum_{i=1}^{\infty}\frac{4(-1)^{hi}\rho^{\frac{1}{2}i}(1+\rho^i)\sin^2\pi(u-v)}{1-2\rho^i\cos 2\pi(u-v)+\rho^{2i}}\right].$$

When $h=0$ this becomes *

$$\frac{4i\omega\sin\pi u\sin\pi v}{(B-A)\pi\sigma_3(0)}\frac{\sigma_3[2\omega(u-v)]}{\sigma[2\omega(u-v)]},$$

where the sigma functions are formed with 2ω, $2\omega\tau$ as periods, ω being an arbitrary quantity. Thus (§ 235, Ex.)

$$\varpi(\zeta, \mu)\lambda(\zeta, \mu) = e^{-2\eta\omega(u-v)^2}\frac{\sigma_3[2\omega(u-v)]}{\sigma_3(0)} = \frac{\vartheta_0(u-v)}{\vartheta_0(0)} = \frac{\Theta(u-v-\frac{1}{2})}{\Theta(\frac{1}{2})},$$

where the symbol ϑ_0 is as in Halphen, *Fonct. Ellip.* (Paris, 1886), Vol. I. pp. 260, 252. This agrees with the general result; in putting $\rho^{\frac{1}{2}} = e^{\pi i\tau}$ we have taken $g=1$; and, as stated, h is here taken zero.

When $h=1$ we similarly find

$$\lambda(\zeta, \mu) = \frac{4i\omega\sin\pi u\sin\pi v}{(B-A)\pi\sigma_3(\omega)}\frac{\sigma_3[2\omega(u-v+\frac{1}{2})]}{\sigma[2\omega(u-v)]}e^{-2\eta\omega(u-v)},$$

and hence

$$\varpi(\zeta, \mu)\lambda(\zeta, \mu) = e^{-2\eta\omega(u-v)^2-2\eta\omega(u-v)}\frac{\sigma_3[2\omega(u-v+\frac{1}{2})]}{\sigma_3(\omega)}, = \frac{\Theta(u-v)}{\Theta(0)},$$

also in agreement with the general formula. In these formulae $\Theta(u)$ denotes the series

$$\Sigma e^{2\pi iun+i\pi\tau n^2} = 1 + 2q\cos(2\pi u) + 2q^4\cos(4\pi u) + 2q^9\cos(6\pi u) + \dots\dots,$$

where $q = e^{i\pi\tau}$.

* Cf. Halphen, *Fonct. Ellip.* (Paris, 1886), Vol. I. p. 422.

Ex. iii. Denoting

$$\Sigma'_i \frac{(\gamma_i\mu+\delta_i)^{-m}}{(\mu-\mu_i)^n}, \quad \Sigma'_i (a_i+\delta_i)^2 \frac{(\gamma_i\mu+\delta_i)^{-m}}{(\mu-\mu_i)^n},$$

where the summations include all substitutions of the group except the identical substitution, respectively by $u_{m,n}$, $v_{m,n}$, prove that, when ζ is near to μ,

$$\frac{\varpi(\zeta,\mu)}{\zeta-\mu}=1-\tfrac{1}{2}(\zeta-\mu)^2 u_{2,2}+(\zeta-\mu)^3 u_{2,3}+\tfrac{1}{4}(\zeta-\mu)^4[u_{4,4}-6u_{2,4}+\tfrac{1}{2}u^2_{2,2}+v_{4,4}]+\dots\dots$$

Ex. iv. If z, s be two single-valued functions of ζ, without essential singularities, which are unaltered by the substitutions of the group, the algebraic * relation connecting z and s may be associated with a Riemann surface, whereon ζ is an infinitely valued function; and if z, s be properly chosen, any single-valued function of ζ without essential singularities, which is unaltered by the substitutions of the group, is a rational function on the Riemann surface. But if

$$\{\zeta, z\}=\frac{d^2}{dz^2}\log\frac{d\zeta}{dz}-\tfrac{1}{2}\left(\frac{d}{dz}\log\frac{d\zeta}{dz}\right)^2, \ =\frac{\zeta'''}{\zeta'}-\tfrac{3}{2}\left(\frac{\zeta''}{\zeta'}\right)^2,$$

where $\zeta'=\dfrac{d\zeta}{dz}$, etc., we immediately find that the value $Z=(a\zeta+\beta)/(\gamma\zeta+\delta)$ gives

$$\{Z, z\}=\{\zeta, z\};$$

therefore, as $\{\zeta, z\}$, $=-\{z,\zeta\}\Big/\left(\dfrac{dz}{d\zeta}\right)^2$, is a single-valued function of ζ without essential singularities, and is unaltered by the substitutions of the group, we have

$$\{\zeta, z\}=2I(z, s),$$

where I denotes a rational function. Therefore, if Y denote an arbitrary function, and $P=-\dfrac{d}{dz}\log\left(Y^2\dfrac{d\zeta}{dz}\right)$, Y and ζY are the solutions of the equation

$$\frac{d^2Y}{dz^2}+P\frac{dY}{dz}+\left[I+\tfrac{1}{4}P^2+\tfrac{1}{2}\frac{dP}{dz}\right]Y=0,$$

and if Y be chosen so that $Y^2\Big/\dfrac{dz}{d\zeta}$ is a rational function on the Riemann surface, the coefficients in this equation will also be rational functions. Thus for instance we may take for Y the function $\sqrt{\dfrac{dz}{d\zeta}}$, in which case $P=0$, or we may take for Y the function $\psi(\zeta)$, $=\sqrt{\dfrac{dv}{d\zeta}\dfrac{dv}{d\gamma}}$, considered in § 236, which is uniform on the ζ plane when the barriers are drawn, in which case $P=-\dfrac{d}{dz}\log\dfrac{dv}{dz}$, and the equation takes the form $\dfrac{d^2Y}{dv^2}+R.Y=0$, where R is a rational function, or again we may take for Y the uniform function of ζ, $\lambda(\zeta,\mu)$, considered in § 238†.

* Ex. viii. § 232.

† Cf. Riemann, *Ges. Werke* (Leipzig, 1876), p. 416, p. 415; Schottky, *Crelle*, LXXXIII. (1877), p. 336 ff.

Ex. v. If, as in Ex. iv., we suppose a Riemann surface constructed such that to every point ζ of the ζ plane there corresponds a place (z, s) of the Riemann surface, and in particular to the point $\zeta=\xi$ there corresponds the place (x, y), and if R, S be functions of ξ defined by the expansions

$$\frac{d}{dx}\log \varpi(\zeta, \xi) = -\frac{1}{z-x}+F+(z-x)R+\ldots\ldots, \quad \frac{\varpi(\zeta, \xi)}{\zeta-\xi}=1-\tfrac{1}{2}S(\zeta-\xi)^2+\ldots\ldots,$$

prove that

$$\tfrac{1}{6}\{\xi, x\} = R - \left(\frac{d\xi}{dx}\right)^2 S,$$

and that R, S are rational functions of x and y.

Ex. vi. The last two examples suggest a problem of capital importance—given any Riemann surface, to find a function ζ, which will effect a conformal representation of the surface to such a ζ-region as that here discussed. This problem may be regarded as that of finding a suitable form for the rational function $I(z, s)$. The reader may consult Schottky, *Crelle*, LXXXIII. (1877), p. 336, and *Crelle*, CI. (1887), p. 268, and Poincaré, *Acta Mathematica*, IV. (1884), p. 224, and *Bulletin de la Soc. Math. de France*, t. XI. (18 *May*, 1883), p. 112. In the elliptic case, taking

$$z=\wp\left(\frac{1}{2\pi i}\log\frac{\zeta-B}{\zeta-A}\right), \ =\wp(u),$$

where $\wp$ denotes Weierstrass's function with 1 and τ as periods, it is easy to prove that $\sqrt{\frac{du}{d\zeta}}$ and $\zeta\sqrt{\frac{du}{d\zeta}}$ are the solutions of the equation

$$(4z^3-g_2z-g_3)\frac{d^2Y}{dz^2}+(6z^2-\tfrac{1}{2}g_2)\frac{dY}{dz}+\pi^2 Y=0.$$

239. There is one case of the theory which may be referred to in conclusion. Take p circles $C_1, \ldots, C_p$, exterior to one another, which are all cut at right angles by another circle O; take a further circle C cutting this orthogonal circle O at right angles; invert the circles $C_1, C_2, \ldots$ in regard to C. We shall obtain p circles $C_1', C_2', \ldots, C_p'$ also cutting the orthogonal circle O at right angles. The case referred to is that in which the circles $C_1, C_1', \ldots, C_p, C_p'$ are the fundamental circles and the angles $\kappa_1, \ldots, \kappa_p$ are all zero, so that, if ϑ_n denote one of the p fundamental substitutions, the corresponding points ζ, $\vartheta_n\zeta$ lie on a circle through A_n and B_n. We may suppose that the circles $C_1, \ldots, C_p$ are all interior to the circle C. It can be shewn by elementary geometry that A_n, B_n are inverse points in regard to the circle C as well as in regard to the circle C_n, and further that if ω denote the process of inversion in regard to the circle C and ω_n that of inversion in regard to C_n, the fundamental substitution ϑ_n is $\omega_n\omega$, so that $\omega\vartheta_n\omega=\vartheta_n^{-1}$, or $\omega\vartheta_n=\vartheta_n^{-1}\omega$. Hence if the points of intersection of the circles O, C_n be called a_n', b_n', the points of intersection of O, C_n' be called a_n, b_n, and the points of intersection of O, C be called a, b, it may be shewn without much difficulty that

$$v_n^{a_r, b_r}=P_{n, r}, \ v_n^{a_n, b_n}=\tfrac{1}{2}+Q_n, \ v_n^{a, b}=\tfrac{1}{2}+R, \quad (n, r=1, 2, \ldots, p;\ n\neq r),$$

where $P_{n,r}$, Q_n, R are integers, and the integrations are along the perimeters of the several circles. Hence it follows that the uniform functions of ζ expressed by $e^{2\Pi^{\zeta, c}_{a_r, b_r}}$, $e^{2\Pi^{\zeta, c}_{a, b}}$ are unaltered by the substitutions of the group. Denote them, respectively, by $x_r(\zeta)$ and $x(\zeta)$. Each of them has a single pole of the second order, and a single zero of the second order, and therefore, as in the case of rational functions on a hyperelliptic Riemann surface, we have, absorbing a constant factor in $x_r(\zeta)$, an equation of the form

$$x_r(\zeta) = \frac{x(\zeta) - x(a_r)}{x(\zeta) - x(b_r)}.$$

But it follows also that the function

$$y(\zeta), = e^{\Pi^{\zeta, c}_{a, b} + \Pi^{\zeta, c}_{a_1, b_1} + \dots\dots + \Pi^{\zeta, c}_{a_p, b_p}},$$

is unaltered by the substitutions of the group. Hence we have*, writing y, x for $y(\zeta)$, $x(\zeta)$, etc.,

$$y^2 = xx_1 \dots x_p = x\frac{[x - x(a_1)] \dots\dots [x - x(a_p)]}{[x - x(b_1)] \dots\dots [x - x(b_p)]}.$$

Thus the special case under consideration corresponds to a hyperelliptic Riemann surface; and, for example, the equations $v_n^{a_n, b_n} = \frac{1}{2} + Q_n$, etc., correspond to part of the results obtained in § 200, Chap. XI. It is manifest that the theory is capable of great development. The reader may consult Weber, *Göttinger Nachrichten*, 1886, "Ein Beitrag zu Poincaré's Theorie, u. s. w.," also, Burnside, *Proc. London Math. Soc.* XXIII. (1892), p. 283, and Poincaré, *Acta Math.* III. p. 80 and *Acta Math.* IV. p. 294 (1884); also Schottky, *Crelle*, CVI. (1890), p. 199. For the general theory of automorphic functions references are given by Forsyth, *Theory of Functions* (1893), p. 619. The particular case considered in this chapter is intended only to illustrate general ideas. From the point of view of the theory of this volume, Chapter XIV. may be regarded as an introduction to the theory of automorphic functions (cf. Klein, *Math. Annalen*, XXI. (1883), p. 141, and Ritter, *Math. Annalen*, XLIV. (1894), p. 261).

* The function x here employed is not identical in case $p=1$ with the z of Ex. vi. § 238.

CHAPTER XIII.

ON RADICAL FUNCTIONS.

240. THE reader is already familiar with the fact that if $\operatorname{sn} u$ represent the ordinary Jacobian elliptic function, the square root of $1-\operatorname{sn}^2 u$ may be treated as a single-valued function of u. Such a property is possessed by other square roots. Thus for instance we have*

$$\sqrt{(1-\operatorname{sn} u)(1-k\operatorname{sn} u)}$$

$$= M \sin\frac{\pi}{4K}(K-u)\prod_m \frac{\left[1-2q^m\sin\frac{\pi u}{2K}+q^{2m}\right]\left[1-2q^{m-\frac{1}{2}}\sin\frac{\pi u}{2K}+q^{2m-1}\right]}{1-2q^{2m-1}\cos\frac{\pi u}{K}+q^{4m-2}},$$

where M is a certain constant, and, as usual, $q=e^{-\pi K'/K}$. The single-valuedness of the function $\sqrt{(1-\operatorname{sn} u)(1-k\operatorname{sn} u)}$ can be immediately seen to follow from the fact that *each of the zeros and poles of the function* $(1-\operatorname{sn} u)(1-k\operatorname{sn} u)$ *is of the second order.* It is manifest that we can easily construct other functions having the same property. If now we write $u=u^{x,a}$ and consider the square root on the dissected elliptic Riemann surface, we shall thereby obtain a single-valued function of the place x, whose values on the two sides of either period loop will have a ratio, constant along that loop, which is equal to ± 1.

Ex. Prove that the function

$$\sqrt{(\sqrt{\wp u-e_1}-\sqrt{e_2-e_1})(\sqrt{\wp u-e_1}-\sqrt{e_3-e_1})}$$

is a single-valued function of u.

Further we have, in Chapter XI., in dealing with the hyperelliptic case associated with an equation of the form

$$y^2=(x-a_1)\ldots(x-a_{2p})(x-c),$$

* Cf. Cayley, *Elliptic Functions* (1876), Chap. XI. The function may be regarded as a doubly periodic function, with $8K$, $2iK'$ as its fundamental periods. It is of the fourth order, with K, $5K$, $K+iK'$, $5K+iK'$ as zeros, and iK', $2K+iK'$, $4K+iK'$, $6K+iK'$ as poles.

been led to the consideration of functions of the form $\sqrt{(c-x_1)\dots(c-x_p)}$, which are expressible by theta functions with arguments $u, = u^{x_1, a_1} + \dots\dots + u^{x_p, a_p}$. These functions are not only single-valued functions of the arguments u, but, when the Riemann surface is dissected in the ordinary way, also of every one of the places $x_1, \dots, x_p$. In fact the square root $\sqrt{c-x}$ is a single-valued function of the place x because, c being a branch place, $x-c$ vanishes to the second order at the place, and the point at infinity being a branch place, $x-c$ is there infinite to the second order. The values of the square root $\sqrt{c-x}$ on the two sides of any period loop will have a ratio, constant along that loop, which is equal to ± 1.

241. More generally it may be proved, for any Riemann surface, that if Z be a rational function such that each of its zeros and poles is of the mth order, the mth root, $\sqrt[m]{Z}$, is a single-valued function of position on the dissected surface, with factors at the period loops which are mth roots of unity. And it is easy to prove this in another way by obtaining an expression for such a function. For let $\alpha_1, \dots, \alpha_r$ be the distinct poles of Z, and $\beta_1, \dots, \beta_r$ its distinct zeros, so that the function is of order mr. Let $\Pi_{z, c}^{x, a}$ be the normal elementary integral of the third kind and $v_1^{x, a}, \dots, v_p^{x, a}$ the normal integrals of the first kind. Then when the paths are restricted not to cross the period loops we have* equations

$$m\left(v_i^{\beta_1, \alpha_1} + \dots\dots + v_i^{\beta_r, \alpha_r}\right) = k_i + k_1'\tau_{i, 1} + \dots\dots + k_p'\tau_{i, p}, \qquad (i = 1, 2, \dots, p),$$

wherein $k_1, \dots, k_p, k_1', \dots, k_p'$ are certain integers independent of i. Hence the expression

$$e^{m\left[\Pi_{\beta_1, \alpha_1}^{x, a} + \dots\dots + \Pi_{\beta_r, \alpha_r}^{x, a}\right] - 2\pi i k_1' v_1^{x, a} - \dots\dots - 2\pi i k_p' v_p^{x, a}},$$

wherein a is an arbitrary fixed place, represents the rational function Z, save for an arbitrary constant; and we have

$$\sqrt[m]{Z} = A e^{\Pi_{\beta_1, \alpha_1}^{x, a} + \dots\dots + \Pi_{\beta_r, \alpha_r}^{x, a} - \frac{2\pi i}{m}\left(k_1' v_1^{x, a} + \dots\dots + k_p' v_p^{x, a}\right)},$$

where A is a certain constant. This expression defines $\sqrt[m]{Z}$ on the dissected surface as a single-valued function of position. More accurately it defines one branch of $\sqrt[m]{Z}$, the other $m-1$ branches being obtained by multiplying A by mth roots of unity. So defined, the function $\sqrt[m]{Z}$ is affected, at the period loop α_i, with a factor $e^{-\frac{2\pi i}{m}k_i'}$, and, at the period loop α_i', with the factor $e^{\frac{2\pi i}{m}k_i}$.

242. We have, in chapters X., XI., been concerned with other functions, namely the theta functions which also have the property of being single-

* Chap. VIII. § 155.

valued on the dissected Riemann surface, but affected with a factor for each period loop. They are also simpler than rational functions, in that they do not possess poles. It is therefore of interest to express such functions as $\sqrt[m]{Z}$ by means of theta functions; and the expression has an importance arising from the fact that the theory of the theta functions may be established independently of the theory of the algebraic integrals. To explain this mode of representation consider the quotient

$$\psi(u) = \frac{\vartheta(u-e;\ q)\,\vartheta(u-f;\ r)\ldots\ldots}{\vartheta(u-E;\ Q)\,\vartheta(u-F;\ R)\ldots\ldots},$$

where the numerator and denominator contain the same number of factors, $\vartheta(u, q)$ denotes the function (Chap. X. § 189) given by

$$\Sigma\Sigma\ldots\ldots e^{au^2+2hu(n+q')+b(n+q')^2+2\pi iq(n+q')},$$

$q, r, \ldots, Q, R, \ldots$ denote any characteristics, and $e, f, \ldots, E, F, \ldots$ denote any arguments.

Then by the formula (§ 190)

$$\vartheta(u+\Omega_M;\ q) = e^{\lambda_M(u)+2\pi i(Mq'-M'q)}\,\vartheta(u;\ q),$$

where M, M' denote integers, we have $\psi(u+\Omega_M)/\psi(u) = e^L$, where L is

$$\lambda_M(u-e)+\lambda_M(u-f)+\ldots\ldots-\lambda_M(u-E)-\lambda_M(u-F)-\ldots\ldots$$
$$+\,2\pi iM(q'+r'+\ldots\ldots-Q'-R'-\ldots)-2\pi iM'(q+r+\ldots\ldots-Q-R-\ldots),$$

namely, is

$$-\lambda_M(e+f+\ldots\ldots-E-F-\ldots)+2\pi iM(q'+r'+\ldots\ldots-Q'-R'-\ldots)$$
$$-\,2\pi iM'(q+r+\ldots\ldots-Q-R-\ldots).$$

Thus if

$$e_i+f_i+\ldots\ldots = E_i+F_i+\ldots\ldots,$$

and

$$q_i+r_i+\ldots\ldots-(Q_i+R_i+\ldots) = \frac{1}{m}K_i, \quad (i=1, 2, \ldots, p),$$

$$q_i'+r_i'+\ldots\ldots-(Q_i'+R_i'+\ldots) = \frac{1}{m}K_i',$$

where K_i, K_i' are integers and m is an integer, it follows, for integral values of M, M', that

$$[\psi(u+\Omega_M)/\psi(u)]^m = 1.$$

If now we take $b = i\pi\tau$, as in § 192, and put $u^{x,a}$ for u, $\vartheta(u-e;\ q)$ becomes a single-valued function of x whose zeros are (§§ 190 (L), 179) the places $x_1, \ldots, x_p$, given by

$$e-\Omega_q \equiv u^{x_1, a_1}+\ldots\ldots+u^{x_p, a_p},$$

where $a_1, \dots, a_p$ are p places determined from the place a, just as in § 179 the places $m_1, \dots, m_p$ were determined from the place m; hence, in this case, $\psi(u)$ is the mth root of a rational function, having for zeros places

$$x_1, \dots, x_p, z_1, \dots, z_p, \dots,$$

each m times repeated, and for poles places

$$X_1, \dots, X_p, Z_1, \dots, Z_p, \dots,$$

each m times repeated, these places being subject only to the conditions expressed by the equations

$$u^{x_1, X_1} + \dots\dots + u^{x_p, X_p} + u^{z_1, Z_1} + \dots\dots + u^{z_p, Z_p} + \dots\dots \equiv -\frac{1}{m}\Omega_{K, K'}, \quad \text{(A)}.$$

In this representation we have obtained a function of which the number of m times repeated zeros is a multiple of p, and also the number of m times repeated poles is a multiple of p. It is easy however to remove this restriction by supposing a certain number of the places $x_1, \dots, x_p, z_1, \dots, z_p$ to coincide with places of the set $X_1, \dots, X_p, Z_1, \dots, Z_p, \dots\dots$

243. A rational function on the Riemann surface is characterised by the facts that it is a single-valued function of position, such that itself and its inverse have no infinities but poles, which has, moreover, the same value at the two sides of any period loop. The functions we have described may clearly be regarded as generalisations of the rational functions, the one new property being that the values of the function at the two sides of any period loop have a ratio, constant along that loop, which is a root of unity. For these functions there holds a theorem, expressed by the equations (A) above, which may be regarded as a generalisation of Abel's theorem for integrals of the first kind; and, when the poles of such a function are given, the number of zeros that can be arbitrarily assigned is the same as for a rational function having the same poles, being in general all but p of them; this follows from the theory of the solution of Jacobi's inversion problem (Chap. IX.; cf. also §§ 37, 93). It will be seen in the course of the following chapter that we can also consider functions of a still more general kind, having constant factors at the period loops which are not roots of unity, and possessing, beside poles, also essential singularities; such functions may be called *factorial* functions. The particular functions so far considered may be called *radical* functions; it is proper to consider them first, in some detail, on account of their geometrical interpretation and because they furnish a convenient method of expressing the solution of several problems connected with Jacobi's inversion problem.

244. The most important of the radical functions are those which are square roots of rational functions, and in view of the general theory developed in the next chapter it will be sufficient to confine ourselves to these functions.

In dealing with these we shall adopt the invariant representation by means of ϕ-polynomials, which has already been described*. An integral polynomial of the rth degree in the p fundamental ϕ-polynomials, $\phi_1, \ldots, \phi_p$, will be denoted by $\Phi^{(r)}$, or $\Psi^{(r)}$, when its $2r(p-1)$ zeros are subject to no condition. When all the zeros are of the second order, and fall therefore, in general, at $r(p-1)$ distinct places, the polynomial will be denoted by $X^{(r)}$ or $Y^{(r)}$; we have† already been concerned with such polynomials, $X^{(1)}$, of the first degree in $\phi_1, \ldots, \phi_p$.

It is to be shewn now that the square root $\sqrt{X^{(r)}}$ can properly be associated with a certain characteristic of $2p$ half-integers; and for this purpose it is convenient to utilise the places $m_1, \ldots, m_p$, arising from an arbitrary place m, which have already‡ occurred in the theory of the theta functions. These places are§ such that if a non-adjoint polynomial, Δ, of grade μ, be taken to vanish to the second order at m, there is an adjoint polynomial, $\overline{\psi}$, of grade $(n-1)\sigma+n-3+\mu$, vanishing in the remaining $n\mu-2$ zeros of Δ, whose other zeros consist of the places $m_1, \ldots, m_p$, each repeated. Take now any ϕ-polynomial, ϕ_0, vanishing to the first order at m, and let its other zeros be $A_1, \ldots, A_{2p-3}$; and take a polynomial $\Phi^{(3)}$ vanishing to the second order in each of $A_1, \ldots, A_{2p-3}$; then $\Phi^{(3)}$ will|| contain $5(p-1)-2(2p-3)$, $=p+1$, linearly independent terms, and will have $6(p-1)-2(2p-3)$, $=2p$, further zeros. Let $X^{(1)}$ be any ϕ-polynomial of which all the zeros are of the second order. Consider the most general rational function, of order $2p$, whose poles consist of the place m, this being a pole of the second order, and of the zeros of $X^{(1)}$. This function will contain $2p-p+1$, $=p+1$, linearly independent terms and can be expressed in either of the forms $\Phi^{(3)}/\phi_0^2X^{(1)}$, $\psi/\Delta X^{(1)}$, where ψ is any polynomial of grade $(n-1)\sigma+n-3+\mu$ which vanishes in the $n\mu-2$ zeros of Δ other than m. Since now¶ ψ can be chosen, $=\overline{\psi}$, so that the zeros of this function are the places $m_1, \ldots, m_p$, each repeated, it follows that $\Phi^{(3)}$ can be equally chosen so that this is the case. So chosen it may be denoted by $X^{(3)}$. *Thus the places $m_1, \ldots, m_p$ arise as the remaining zeros of a form $X^{(3)}$ (with $3(p-1)$, $=p+2p-3$, zeros, each of the second order), whose other $2p-3$ separate zeros are zeros of an arbitrary ϕ-polynomial, ϕ_0, which vanishes once at the place m.*

If now $n_1, \ldots, n_{p-1}$ be the places which, repeated, are the zeros of $X^{(1)}$, it follows, since $m, n_1, \ldots, n_{p-1}$, each repeated, are the poles, and $m_1, \ldots, m_p$, each repeated, are the zeros of a rational function, $X^{(3)}/\phi_0^2X^{(1)}$, that, upon the dissected surface, we have

$$v_i^{m_p, m} - v_i^{n_1, m_1} - \ldots\ldots - v_i^{n_{p-1}, m_{p-1}} = -\tfrac{1}{2}(k_i + k_1'\tau_{i,1} + \ldots\ldots + k_p'\tau_{i,p}),$$

* Chap. VI. § 110 ff., and the references there given, and Klein, *Math. Annal.* XXXVI. p. 38.

† Chap. X. § 188, p. 281.

‡ Chap. X. § 179.

§ Chap. X. § 183, Chap. VI. § 92, Ex. ix.

|| Chap. VI. § 111.

¶ Chap. X. § 183.

where $k_1, \ldots, k_p, k_1', \ldots, k_p'$ are certain integers. Hence, as in § 241, it immediately follows that the rational function $X^{(3)}/\phi_0^2 X^{(1)}$, save for a constant factor, is the square of the function

$$e^{\Pi^{x,a}_{m_1,n_1}+\ldots\ldots+\Pi^{x,a}_{m_{p-1},n_{p-1}}+\Pi^{x,a}_{m_p,m}+\pi i(k_1'v_1^{x,a}+\ldots\ldots+k_p'v_p^{x,a})},$$

and therefore that the expression $\sqrt{X^{(3)}}/\phi_0\sqrt{X^{(1)}}$ may be regarded as a single-valued function on the dissected Riemann surface, whose values on the two sides of any period loop have a ratio constant along that loop. These constant ratios are equal to $e^{\pi i k_r'}$ and $e^{-\pi i k_r}$ for the rth loop of the first and second kind respectively. When the places $m_1, \ldots, m_p$ are regarded as given, these equations associate with the form $\sqrt{X^{(1)}}$ a definite characteristic

$$\tfrac{1}{2}k_1, \ldots, \tfrac{1}{2}k_p, \tfrac{1}{2}k_1', \ldots, \tfrac{1}{2}k_p'.$$

Also, if $Y^{(3)}$ be any polynomial which, beside vanishing to the second order in $A_1, \ldots, A_{2p-3}$, vanishes to the second order in places $m_1', \ldots, m_p'$, $Y^{(3)}/X^{(3)}$ is a rational function, and we have equations of the form

$$v_i^{m_1', m_1}+\ldots\ldots+v_i^{m_p', m_p}=\tfrac{1}{2}(\lambda_i+\lambda_1'\tau_{i,1}+\ldots\ldots+\lambda_p'\tau_{i,p}),$$

$$\sqrt{Y^{(3)}}/\sqrt{X^{(3)}}=Ae^{\Pi^{x,a}_{m_1',m_1}+\ldots\ldots+\Pi^{x,a}_{m_p',m_p}-\pi i(\lambda_1'\tau' v_1^{x,a}+\ldots\ldots+\lambda_p' v_p^{x,a})},$$

where $\lambda_1, \ldots, \lambda_p'$ are integers, A is a constant, and the paths of integration are limited to the dissected Riemann surface. These equations associate $\sqrt{Y^{(3)}}$ with the characteristic $\tfrac{1}{2}\lambda_1, \ldots, \tfrac{1}{2}\lambda_p, \tfrac{1}{2}\lambda_1', \ldots, \tfrac{1}{2}\lambda_p'$.

And, as in § 184, Chap. X., we infer that every odd characteristic is associated with a polynomial* $X^{(1)}$, and every even characteristic with a polynomial $Y^{(3)}$, which has $A_1, \ldots, A_{2p-3}$ for zeros of the second order; and it may happen that the polynomial $Y^{(3)}$ corresponding to an even characteristic has the form $\phi_0^2 Y^{(1)}$, in which case the places $m_1', \ldots, m_p'$ consist of the place m and the zeros of a form $Y^{(1)}$.

245. Let now $X^{(2\nu+1)}$ be any polynomial whose zeros consist of $(2\nu+1)(p-1)$ places, $z_1, z_2, \ldots$, each repeated; let ϕ_0 be as before, vanishing in $m, A_1, \ldots, A_{2p-3}$, and $X^{(3)}$ be as before, vanishing to the second order in $A_1, \ldots, A_{2p-3}, m_1, \ldots, m_p$. Then if $\Phi^{(\nu)}$ be any ϕ-polynomial whose zeros are $c_1, c_2, \ldots$, the function

$$\phi_0^2 X^{(2\nu+1)}/[\Phi^{(\nu)}]^2 X^{(3)}$$

* Or in particular cases with a lot of such polynomials, giving rise to coresidual sets of places.

is a rational function of order $2(2\nu+1)(p-1)+2$, whose zeros are $m, z_1, z_2, \ldots$, and whose poles consist of the places $m_1, \ldots, m_p$, and the zeros of $\Phi^{(\nu)}$, each repeated. Hence as before $\phi_0\sqrt{X^{(2\nu+1)}}/\Phi^{(\nu)}\sqrt{X^{(3)}}$ is a single-valued function on the dissected surface, and the form $\sqrt{X^{(2\nu+1)}}$ is associated with a characteristic $\frac{1}{2}q_1, \ldots, \frac{1}{2}q_p, \frac{1}{2}q_1', \ldots, \frac{1}{2}q_p'$, such that, on the dissected surface,

$$v_i^{z_1, m_1} + \ldots\ldots + v_i^{z_p, m_p} + v_i^{z_{p+1}, c_1} + \ldots\ldots = \tfrac{1}{2}(q_i + q_1'\tau_{i,1} + \ldots\ldots + q_p'\tau_{i,p}),$$
$$(i = 1, 2, \ldots, p);$$

and if, instead of $\Phi^{(\nu)}$, we had used any other polynomial $\Psi^{(\nu)}$, the characteristic could, by Abel's theorem, only be affected by the addition of integers.

Suppose now that $Y^{(2\mu+1)}$ is another polynomial, and take a polynomial $\Psi^{(\mu)}$; then if the characteristic of the function $\phi_0\sqrt{Y^{(2\mu+1)}}/\Psi^{(\mu)}\sqrt{X^{(3)}}$ differ from that of $\phi_0\sqrt{X^{(2\nu+1)}}/\Phi^{(\nu)}\sqrt{X^{(3)}}$ only by integers, we have when $x_1, x_2, \ldots$ denote the zeros of $\sqrt{Y^{(2\mu+1)}}$, and $d_1, d_2, \ldots$ denote the zeros of $\Psi^{(\mu)}$, the equation

$$v_i^{x_1, m_1} + \ldots\ldots + v_i^{x_p, m_p} + v_i^{x_{p+1}, d_1} + \ldots\ldots = \tfrac{1}{2}(q_i + q_1'\tau_{i,1} + \ldots\ldots + q_p'\tau_{i,p})$$
$$+ M_i + M_1'\tau_{i,1} + \ldots\ldots + M_p'\tau_{p,i},$$

where $M_1, \ldots, M_p, M_1', \ldots, M_p'$ denote integers; by adding this to the last equation we infer* that $\phi_0^2\sqrt{X^{(2\nu+1)}}\sqrt{Y^{(2\mu+1)}}/\Phi^{(\nu)}\Psi^{(\mu)}X^{(3)}$ is a rational function. Hence†, since there exists a rational function of the form $\phi_0^2X^{(1)}/X^{(3)}$, we infer, *when* $\sqrt{X^{(2\nu+1)}}$, $\sqrt{Y^{(2\mu+1)}}$ *have characteristics differing only by integers, there exists a form* $\Phi^{(\mu+\nu+1)}$ *whose zeros are the separate zeros of* $\sqrt{X^{(2\nu+1)}}$ *and* $\sqrt{Y^{(2\mu+1)}}$, *and we have* $\sqrt{X^{(2\nu+1)}}\sqrt{Y^{(2\mu+1)}} = \Phi^{(\nu+\mu+1)}$.

Hence, all possible forms $\sqrt{Y^{(2\mu+1)}}$, with the same value of μ, whose characteristics, save for integers, are the same, are expressible in the form $\Phi^{(\mu+\nu+1)}/\sqrt{X^{(2\nu+1)}}$, where $\Phi^{(\mu+\nu+1)}$ is a polynomial of the degree indicated, which vanishes once in the zeros of $\sqrt{X^{(2\nu+1)}}$. All such forms $\sqrt{Y^{(2\mu+1)}}$ are therefore expressible by such equations as

$$\sqrt{Y^{(2\mu+1)}} = \lambda_1\sqrt{Y_1^{(2\mu+1)}} + \ldots\ldots + \lambda_{2\mu(p-1)}\sqrt{Y_{2\mu(p-1)}^{(2\mu+1)}},$$

where $\sqrt{Y_1^{(2\mu+1)}}, \ldots, \sqrt{Y_{2\mu(p-1)}^{(2\mu+1)}}$ are special polynomials, and $\lambda_1, \ldots, \lambda_{2\mu(p-1)}$ are constants. The assignation of $2\mu(p-1)-1$, $=(2\mu+1)(p-1)-p$, zeros of $\sqrt{Y^{(2\mu+1)}}$ will determine the constants $\lambda_1, \ldots, \lambda_{2\mu(p-1)}$, and therefore determine the remaining p zeros. When $\mu = 0$ there may be a reduction in the number of zeros determined by the others.

It follows also that the zeros of any form $\sqrt{Y^{(2\mu+1)}}$ are the remaining zeros of a polynomial $\Phi^{(\mu+2)}$ which vanishes in the zeros of a form $\sqrt{X^{(3)}}$ having

* Chap. VIII. § 158. † Chap. VI. § 112.

the same characteristic as $\sqrt{Y^{(2\mu+1)}}$, or a characteristic differing from that of $\sqrt{Y^{(2\mu+1)}}$ only by integers. When the characteristic of $\sqrt{X^{(3)}}$ is odd, and $\sqrt{X^{(3)}} = \Phi^{(1)} \sqrt{X^{(1)}}$, we may take $\Phi^{(\mu+2)}$ to be of the form $\Phi^{(\mu+1)}\,\Phi^{(1)}$.

It can be similarly shewn that if $X^{(2\mu)}$ be a polynomial of even degree, 2μ, in the fundamental ϕ-polynomials, of which all the zeros are of the second order, and $\Phi^{(\mu)}$ be any polynomial of degree μ, the quotient $\sqrt{X^{(2\mu)}}/\Phi^{(\mu)}$ may be interpreted as a single-valued function on the dissected surface, and the form $\sqrt{X^{(2\mu)}}$ may be associated with a certain characteristic of half-integers. Further the zeros of $\sqrt{X^{(2\mu)}}$ are the remaining zeros of a form $\Phi^{(\mu+1)}$ which vanishes in the zeros of a form $\sqrt{X^{(2)}}$ of the same* characteristic as $\sqrt{X^{(2\mu)}}$. Also if $\sqrt{X^{(1)}}$, $\sqrt{Y^{(1)}}$ be two forms whose (odd) characteristics have a sum differing from the characteristic of $\sqrt{X^{(2)}}$ by integers, the ratio $\sqrt{X^{(2)}}/\sqrt{X^{(1)}Y^{(1)}}$ is a rational function; and if we determine $(p-1)$ pairs of odd characteristics, such that the sum of each pair is, save for integers, equal to the characteristic of $\sqrt{X^{(2)}}$, and $\sqrt{X_1^{(1)}}$, $\sqrt{Y_1^{(1)}}$, $\sqrt{X_2^{(1)}}$, $\sqrt{Y_2^{(1)}}$, ..., represent the corresponding forms, there exists an equation of the form

$$\sqrt{X^{(2)}} = \lambda_1 \sqrt{X_1^{(1)} Y_1^{(1)}} + \lambda_2 \sqrt{X_2^{(1)} Y_2^{(1)}} + \ldots\ldots + \lambda_{p-1} \sqrt{X_{p-1}^{(1)}\, Y_{p-1}^{(1)}}.$$

As a matter of fact every characteristic, except the zero characteristic, can, save for integers, be written as the sum of two odd characteristics in $2^{p-2}(2^{p-1}-1)$ ways.

246. In illustration of these principles we consider briefly the geometrical theory of a general plane quartic curve for which $p=3$. We may suppose the equation expressed homogeneously by the coordinates x_1, x_2, x_3 and take the fundamental ϕ-polynomials to be $\phi_1 = x_1$, $\phi_2 = x_2$, $\phi_3 = x_3$. There are then $2^{p-1}(2^p-1) = 28$ double tangents, $X^{(1)}$, of fixed position. There are 2^{2p}, $= 64$, systems of cubic curves, $X^{(3)}$, each touching in six points. Of these six points of contact of a cubic, $X^{(3)}$, of prescribed characteristic, three may be arbitrarily taken; and we have in fact

$$\sqrt{X^{(3)}} = \lambda_1 \sqrt{X_1^{(3)}} + \lambda_2 \sqrt{X_2^{(3)}} + \lambda_3 \sqrt{X_3^{(3)}} + \lambda_4 \sqrt{X_4^{(3)}},$$

where λ_1, λ_2, λ_3, λ_4 are constants, and $\sqrt{X_1^{(3)}}$, $\sqrt{X_2^{(3)}}$, ..., are special forms of the assigned characteristic. The points of contact of all cubics $X^{(3)}$ of given odd characteristic are obtainable by drawing variable conics through the points of contact of the double tangent, D, associated with that odd characteristic. Let Ω_0 be a certain one of these conics and let X_0 denote the corresponding contact-cubic; then the rational function $X_0 D/\Omega_0^2$ has, clearly, no poles, and must be a constant, and therefore, absorbing the constant, we infer that the equation of the fundamental quartic can be written

$$4X_0 D - \Omega_0^2 = 0.$$

* Or a characteristic differing from that of $\sqrt{X^{(2\mu)}}$ by integers.

Three of the conics through the points of contact of D are $x_1D=0$, $x_2D=0$, $x_3D=0$; the corresponding forms of $X^{(3)}$ are x_1^2D, x_2^2D, x_3^2D. Hence all contact cubics of the same characteristic as $\sqrt{D}$ are included in the formula

$$\sqrt{X^{(3)}}=(\lambda_1x_1+\lambda_2x_2+\lambda_3x_3)\sqrt{D}+\sqrt{X_0},$$

or

$$X^{(3)}=X_0+\Omega_0P+DP^2,$$

where $P=\lambda_1x_1+\lambda_2x_2+\lambda_3x_3$, $\lambda_1, \lambda_2, \lambda_3$ being constants; the conic through the points of contact of D which passes through the points of contact of $X^{(3)}$ is given by $\Omega=2\sqrt{DX^{(3)}}$, or $\Omega=2PD+\Omega_0$; and the fundamental quartic can equally be written

$$4X^{(3)}D-\Omega^2=4(X_0+\Omega_0P+DP^2)D-(\Omega_0+2PD)^2=0.$$

If then we introduce space coordinates X, Y, Z, T given by

$$X=x_1,\ Y=x_2,\ Z=x_3,\ T=-\sqrt{X_0/D},$$

so that the general form of $\sqrt{X^{(3)}}$ with the same characteristic as $\sqrt{D}$ is given by

$$\sqrt{X^{(3)}}=\sqrt{D}\,(\lambda_1X+\lambda_2Y+\lambda_3Z-T),$$

we have

$$4X_0(X, Y, Z)\,D(X, Y, Z)=\Omega_0^2(X, Y, Z),$$

$$2TD(X, Y, Z)+\Omega_0(X, Y, Z)=0,$$

where $X_0\,(X, Y, Z)$ is the result of substituting in X_0, for x_1, x_2, x_3, respectively X, Y, Z, etc.; by these equations the fundamental quartic is related to a curve of the sixth order in space of three dimensions, given by the intersection of the quadric surface

$$2TD(X, Y, Z)+\Omega_0(X, Y, Z)=0$$

and the quartic cone

$$4X_0(X, Y, Z)\,D(X, Y, Z)=\Omega_0^2(X, Y, Z);$$

the curve lies also on the cubic surface

$$T^2D(X, Y, Z)+T\Omega_0(X, Y, Z)+X_0(X, Y, Z)=0,$$

which can also be written

$$(T-P)^2\,D(X, Y, Z)+(T-P)\,\Omega(X, Y, Z)+X^{(3)}(X, Y, Z)=0,$$

where P denotes $\lambda_1X+\lambda_2Y+\lambda_3Z$, $\Omega=2PD+\Omega_0$, and $X^{(3)}=DP^2+\Omega_0P+X_0$, as above.

It can be immediately shewn (i) that the enveloping cone of the cubic surface just obtained, whose vertex is the point $X=0=Y=Z$, is the quartic cone whose intersection with the plane $T=0$ gives the fundamental quartic curve, (ii) that the tangent plane of the cubic surface at the point

$X=0=Y=Z$ is the plane $D(X, Y, Z)=0$, (iii) that the planes joining the point $X=0=Y=Z$ to the 27 straight lines of the cubic surface intersect the plane $T=0$ in the 27 double tangents of the fundamental quartic other than D, (iv) that the fundamental quartic curve may be considered as arising by the intersection of an arbitrary plane with the quartic cone of contact which can be drawn to an arbitrary cubic surface from an arbitrary point of the surface.

Thus the theory of the bitangents is reducible to the theory of the right lines lying on a cubic surface. Further development must be sought in geometrical treatises. Cf. Geiser, *Math. Annal.* Bd. I. p. 129, *Crelle* LXXII. (1870); also Frahm, *Math. Annal.* VII. and Toeplitz, *Math. Annal.* XI.; Salmon, *Higher Plane Curves* (1879), p. 231, note; Klein, *Math. Annal.* XXXVI. p. 51.

247. We have shewn that there are 28 double tangents each associated with one of the odd characteristics; the association depends upon the mode of dissection of the fundamental Riemann surface. We have stated moreover (§ 205, Chap. XI.), in anticipation of a result which is to be proved later, that there are $8 . 36 = 288$ ways in which all possible characteristics can be represented by combinations of one, two, or three of seven fundamental odd characteristics. These fundamental characteristics can be denoted by the numbers 1, 2, 3, 4, 5, 6, 7, and in what follows we shall, for the sake of definiteness, suppose them to be either the characteristics so denoted in the table given § 205, or one of the seven sets whose letter notation is given at the conclusion of § 205. Thus the sum of these seven characteristics is the characteristic, which, save for integers, has all its elements zero; or, as we may say, the sum of these characteristics is zero.

A double tangent whose characteristic is denoted by the number i will be represented by the equation $u_i=0$. A combination of two numbers also represents an odd characteristic (§ 205, Chap. XI.), so that there will also be 21 double tangents whose equations are of such forms as $u_{i,j}=0$. The three products $\sqrt{u_1u_{23}}$, $\sqrt{u_2u_{31}}$, $\sqrt{u_3u_{12}}$ will be radical forms, such as have been denoted by $\sqrt{X^{(2)}}$, each with the characteristic 123. Hence if suitable numerical multipliers be absorbed in u_1, u_3, we have (§ 245) an identity of the forms

$$\sqrt{u_1u_{23}}+\sqrt{u_2u_{31}}+\sqrt{u_3u_{12}}=0, \quad (u_2u_{31}+u_3u_{12}-u_1u_{23})^2=4u_2u_3u_{31}u_{12};$$

this must then be a form into which the equation of the fundamental quartic curve can be put. Further, each of the six forms

$$\sqrt{u_2u_{12}},\ \sqrt{u_3u_{13}},\ \sqrt{u_4u_{14}},\ \sqrt{u_5u_{15}},\ \sqrt{u_6u_{16}},\ \sqrt{u_7u_{17}}$$

has the same characteristic, denoted by the symbol 1. Thus, if suitable numerical multipliers be absorbed in u_2, u_4, the equation of the quartic can also be given in the form

$$(u_2u_{12}+u_4u_{14}-u_3u_{13})^2=4u_4u_2u_{12}u_{14}.$$

If therefore

$$f = u_2 u_{31} + u_3 u_{12} - u_1 u_{23}, \quad \phi = u_2 u_{12} + u_4 u_{14} - u_3 u_{13},$$

we have

$$(f - \phi)(f + \phi) = 4u_2 u_{12}(u_3 u_{13} - u_4 u_{14}).$$

Now if $f - \phi$ were divisible by u_2, and $f + \phi$ divisible by u_{12}, the common point of the tangents $u_2 = 0$, $u_{12} = 0$ would make $f = 0$, and therefore be upon the fundamental quartic, $f^2 = 4u_2 u_3 u_{31} u_{12}$; this is impossible when the quartic is perfectly general. Hence, without loss of generality, we may take

$$f - \phi = 2\lambda u_2 u_{12},$$

$$f + \phi = \frac{2}{\lambda}(u_3 u_{13} - u_4 u_{14}),$$

λ being a certain constant, and therefore

$$u_4 u_{14} = u_3 u_{13} - \lambda f + \lambda^2 u_2 u_{12}, \quad = u_3 u_{13} - \lambda(u_2 u_{31} + u_3 u_{12} - u_1 u_{23}) + \lambda^2 u_2 u_{12}.$$

Therefore, when the six tangents u_1, u_2, u_3, u_{23}, u_{31}, u_{12} are given, the tangents u_4, u_{14} can be found by expressing the condition that the right-hand side should be a product of linear factors; as the right-hand is a quadric function of the coordinates this will lead to a sextic equation in λ, having the roots $\lambda = 0$, $\lambda = \infty$; if the other roots be substituted in turn on the right-hand, we shall obtain in turn four pairs of double tangents; these are in fact (u_4, u_{14}), (u_5, u_{15}), (u_6, u_{16}), (u_7, u_{17}). We use the equation obtained however in a different way; by a similar proof we clearly obtain the three equations

$$\begin{aligned} u_4 u_{14} &= u_3 u_{13} - \lambda_1(u_2 u_{31} + u_3 u_{12} - u_1 u_{23}) + \lambda_1^2 u_2 u_{12}, \\ u_4 u_{24} &= u_1 u_{21} - \lambda_2(u_3 u_{12} + u_1 u_{23} - u_2 u_{31}) + \lambda_2^2 u_3 u_{23}, \\ u_4 u_{34} &= u_2 u_{32} - \lambda_3(u_1 u_{23} + u_2 u_{31} - u_3 u_{12}) + \lambda_3^2 u_1 u_{31}, \end{aligned} \qquad \text{(B)}$$

and hence

$$u_4\left(\frac{u_{24}}{\lambda_2} + \frac{u_{34}}{\lambda_3}\right) = u_{23}\left(\lambda_2 u_3 + \frac{u_2}{\lambda_3}\right) + u_1\left(\frac{u_{21}}{\lambda_2} + \lambda_3 u_{31} - 2u_{23}\right);$$

from this we infer that the common point of the tangents u_1, u_4 either lies on u_{23} or on $\lambda_2 u_3 + \frac{u_2}{\lambda_3} = 0$; as the fundamental quartic may be written in the form $\sqrt{Au_4 u_{34}} + \sqrt{Bu_2 u_{23}} + \sqrt{Cu_1 u_{13}} = 0$, it follows that if u_1, u_4, u_{23} intersect, they intersect on the quartic, which is impossible. Hence u_4 must pass through the intersection of u_1 and $\lambda_2 u_3 + \frac{u_2}{\lambda_3} = 0$; now we may assume that the tangents u_1, u_2, u_3 are not concurrent, since else, as follows from the equation $\sqrt{u_1 u_{23}} + \sqrt{u_2 u_{31}} + \sqrt{u_3 u_{12}} = 0$, they would intersect upon the quartic; thus u_4 may be expressed linearly by u_1, u_2, u_3, and we may put

$$u_4 = a_1 u_1 + a_2 u_2 + a_3 u_3 = a_1 u_1 + \frac{1}{h_1}\left(\lambda_2 u_3 + \frac{u_2}{\lambda_3}\right),$$

and so obtain $\lambda_2 = h_1 a_3$, $\lambda_3 = 1/h_1 a_2$, h_1 being a certain constant; then the equation under consideration becomes

$$u_4\left(\frac{u_{24}}{\lambda_2}+\frac{u_{34}}{\lambda_3}\right) = u_{23}h_1(u_4 - a_1u_1) + u_1\left(\frac{u_{21}}{\lambda_2} + \lambda_3 u_{31} - 2u_{23}\right),$$

or

$$u_4\left(\frac{u_{24}}{\lambda_2}+\frac{u_{34}}{\lambda_3} - h_1 u_{23}\right) = u_1\left(\frac{u_{21}}{\lambda_2} + \lambda_3 u_{31} - 2u_{23} - a_1 h_1 u_{23}\right),$$

so that, if k_1 denote a proper constant,

$$\frac{u_{24}}{\lambda_2}+\frac{u_{34}}{\lambda_3} = h_1 u_{23} - \frac{k_1}{h_1}u_1,$$

$$-k_1 u_4 = \frac{u_{12}}{a_3} + \frac{u_{31}}{a_2} - h_1 u_{23}(2 + a_1 h_1).$$

We can similarly obtain the equations

$$-k_2 u_4 = \frac{u_{12}}{a_3} + \frac{u_{23}}{a_1} - h_2 u_{31}(2 + a_2 h_2),$$

$$-k_3 u_4 = \frac{u_{23}}{a_1} + \frac{u_{31}}{a_2} - h_3 u_{12}(2 + a_3 h_3),$$

where h_2, h_3, k_2, k_3 are proper constants; therefore, as u_{23}, u_{31}, u_{12} are not concurrent tangents, since else they would intersect on the fundamental quartic, we infer, by comparing the right-hand sides in these three equations,

$$-\frac{h_1}{k_1}(2 + a_1 h_1) = \frac{1}{k_2 a_1} = \frac{1}{k_3 a_1},\quad -\frac{h_2}{k_2}(2 + a_2 h_2) = \frac{1}{k_3 a_2} = \frac{1}{k_1 a_2},$$

$$-\frac{h_3}{k_3}(2 + a_3 h_3) = \frac{1}{k_1 a_3} = \frac{1}{k_2 a_3},$$

and hence, $k_1 = k_2 = k_3, = k$, say, and $1 + 2h_1 a_1 + a_1^2 h_1^2 = 0$ or $h_1 = -\frac{1}{a_1}$, $h_2 = -\frac{1}{a_2}$, $h_3 = -\frac{1}{a_3}$.

Thus

$$-ku_4 = \frac{u_{23}}{a_1} + \frac{u_{31}}{a_2} + \frac{u_{12}}{a_3},$$

or

$$\frac{u_{23}}{a_1} + \frac{u_{31}}{a_2} + \frac{u_{12}}{a_3} + k(a_1 u_1 + a_2 u_2 + a_3 u_3) = 0. \qquad \text{(C)}$$

Further we obtained the equation

$$\frac{u_{24}}{\lambda_2}+\frac{u_{34}}{\lambda_3} = h_1 u_{23} - \frac{k_1}{h_1}u_1;$$

thus we have

$$\frac{u_{24}}{\lambda_2}+\frac{u_{34}}{\lambda_3}+\frac{u_{23}}{a_1} = ka_1u_1,\quad \frac{u_{34}}{\lambda_3}+\frac{u_{14}}{\lambda_1}+\frac{u_{31}}{a_2} = ka_2u_2,\quad \frac{u_{14}}{\lambda_1}+\frac{u_{24}}{\lambda_2}+\frac{u_{12}}{a_3} = ka_3u_3,$$

and therefore, as $\lambda_2 = -\frac{a_3}{a_1}$, $\lambda_3 = -\frac{a_1}{a_2}$, and similarly $\lambda_1 = -\frac{a_2}{a_3}$, we have, by the equation (C),

$$-\frac{a_3}{a_2}u_{14} = \frac{u_{23}}{a_1} + k(a_2u_2 + a_3u_3),$$

$$-\frac{a_1}{a_3}u_{24} = \frac{u_{31}}{a_2} + k(a_3u_3 + a_1u_1),$$

$$-\frac{a_2}{a_1}u_{34} = \frac{u_{12}}{a_3} + k(a_1u_1 + a_2u_2).$$

But if we put

$$u_5 = b_1u_1 + b_2u_2 + b_3u_3, \quad u_6 = c_1u_1 + c_2u_2 + c_3u_3, \quad u_7 = d_1u_1 + d_2u_2 + d_3u_3,$$

we have also three other equations such as (C), differing from (C) in the substitution respectively of the coefficients b_1, b_2, b_3 c_1, c_2, c_3 and d_1, d_2, d_3 in place of a_1, a_2, a_3, and of three constants, say l, m, n, in place of k. As the tangents u_5, u_6, u_7 are not concurrent (for the fundamental quartic can be written in a form $\sqrt{u_5u_{15}} + \sqrt{u_6u_{16}} + \sqrt{u_7u_{17}} = 0$) we may use these three last equations to determine u_{23}, u_{31}, u_{12} in terms of u_1, u_2, u_3; the expressions obtained must satisfy the equation (C). Thus there exist, with suitable values of the multipliers A, B, C, D, the six equations

$$\frac{A}{a_1} + \frac{B}{b_1} + \frac{C}{c_1} + \frac{D}{d_1} = 0, \qquad Aka_1 + Blb_1 + Cmc_1 + Dnd_1 = 0,$$

$$\frac{A}{a_2} + \frac{B}{b_2} + \frac{C}{c_2} + \frac{D}{d_2} = 0, \qquad Aka_2 + Blb_2 + Cmc_2 + Dnd_2 = 0,$$

$$\frac{A}{a_3} + \frac{B}{b_3} + \frac{C}{c_3} + \frac{D}{d_3} = 0, \qquad Aka_3 + Blb_3 + Cmc_3 + Dnd_3 = 0.$$

From these equations the ratios of the constants k, l, m, n are determinable; suppose the values obtained to be written $\rho k', \rho l', \rho m', \rho n'$, where ρ is undetermined, and k', l', m', n' are definite; then, if we put α_i for $a_i\sqrt{k'}$, β_i for $b_i\sqrt{l'}$, γ_i for $c_i\sqrt{m'}$, δ_i for $d_i\sqrt{n'}$, v_{23} for u_{23}/ρ, v_{31} for u_{31}/ρ, and v_{12} for u_{12}/ρ, the equations obtained consist of

(i) four of the form

$$\frac{v_{23}}{\alpha_1} + \frac{v_{31}}{\alpha_2} + \frac{v_{12}}{\alpha_3} + \alpha_1u_1 + \alpha_2u_2 + \alpha_3u_3 = 0 \qquad \text{(C}'\text{)}$$

in which there occur in turn the sets of coefficients $(\alpha_1, \alpha_2, \alpha_3)$, $(\beta_1, \beta_2, \beta_3)$, $(\gamma_1, \gamma_2, \gamma_3)$, $(\delta_1, \delta_2, \delta_3)$; from any three of these v_{23}, v_{31}, v_{12} may be expressed in terms of u_1, u_2, u_3;

(ii) four sets of the form

$$-\frac{\alpha_3}{\alpha_2}v_{14} = \frac{v_{23}}{\alpha_1} + \alpha_2u_2 + \alpha_3u_3, \quad -\frac{\alpha_1}{\alpha_3}v_{24} = \frac{v_{31}}{\alpha_2} + \alpha_3u_3 + \alpha_1u_1, \quad -\frac{\alpha_2}{\alpha_1}v_{34} = \frac{v_{12}}{\alpha_3} + \alpha_1u_1 + \alpha_2u_2,$$

where $v_{14} = u_{14}/\rho\sqrt{k'}$, $v_{24} = u_{24}/\rho\sqrt{k'}$, $v_{34} = u_{34}/\rho\sqrt{k'}$.

It will be recalled that in the course of the analysis the absolute values, and not merely the ratios of the coefficients in $u_1, u_2, u_3, u_4, u_5, u_6, u_7$, have been definitely fixed. Thus when these seven bitangents are given the values of $a_1, a_2, a_3, b_1, b_2, b_3$, etc. are definite; therefore the equations of the 15 bitangents $v_{23}, v_{31}, v_{12}, v_{14}, v_{24}, v_{34}, \ldots\ldots$ are now determined from the seven given ones in an unique manner, and there is an unique quartic curve expressed by

$$\sqrt{u_1 v_{23}} + \sqrt{u_2 v_{31}} + \sqrt{u_3 v_{12}} = 0,$$

which has the seven given lines as bitangents.

It remains now to determine the remaining six double tangents whose characteristics are denoted by

45, 46, 47, 56, 57, 67.

If the characteristics 1, 2, 3, 4, 5, 6, 7 be taken in the order 1, 4, 5, 2, 3, 6, 7 it is clear that as we have determined the double tangents u_{23}, u_{31}, u_{12} in terms of u_1, u_2, u_3, so we can determine the tangents u_{45}, u_{51}, u_{14} in terms of u_1, u_4, u_5. Thus the tangent u_{45} can be found by substitutions in the foregoing work. For the actual deduction the reader is referred* to the original memoir, Riemann, *Ges. Werke* (Leipzig, 1876), p. 471, or Weber, *Theorie der Abel'schen Functionen vom Geschlecht* 3 (Berlin, 1876), pp. 98—100. Putting $\alpha_1 u_1 = x$, $\alpha_2 u_2 = y$, $\alpha_3 u_3 = z$, $v_{23}/\alpha_1 = \xi$, $v_{31}/\alpha_2 = \eta$, $v_{12}/\alpha_3 = \zeta$, $\beta_i/\alpha_i = A_i$, $\gamma_i/\alpha_i = B_i$, $\delta_i/\alpha_i = C_i$ $(i = 1, 2, 3)$, the quartic has the form

$$\sqrt{x\xi} + \sqrt{y\eta} + \sqrt{z\zeta} = 0,$$

and the 28 double tangents are given by the following scheme, where the number representing the characteristic is prefixed to each

(1) $x = 0$, (2) $y = 0$, (3) $z = 0$, (23) $\xi = 0$, (31) $\eta = 0$, (12) $\zeta = 0$,

(4) $x + y + z = 0$, (5) $A_1 x + A_2 y + A_3 z = 0$, (6) $B_1 x + B_2 y + B_3 z = 0$,

(7) $C_1 x + C_2 y + C_3 z = 0$,

(14) $\xi + y + z = 0$, (24) $\eta + z + x = 0$, (34) $\zeta + x + y = 0$,

(15) $\dfrac{\xi}{A_1} + A_2 y + A_3 z = 0$, (25) $\dfrac{\eta}{A_2} + A_3 z + A_1 x = 0$, (35) $\dfrac{\zeta}{A_3} + A_1 x + A_2 y = 0$,

(16) $\dfrac{\xi}{B_1} + B_2 y + B_3 z = 0$, (26) $\dfrac{\eta}{B_2} + B_3 z + B_1 x = 0$, (36) $\dfrac{\zeta}{B_3} + B_1 x + B_2 y = 0$,

(17) $\dfrac{\xi}{C_1} + C_2 y + C_3 z = 0$, (27) $\dfrac{\eta}{C_2} + C_3 z + C_1 x = 0$, (37) $\dfrac{\zeta}{C_3} + C_1 x + C_2 y = 0$,

* For the theory of the plane quartic curve reference may be made to geometrical treatises; developments in connection with the theta functions are given by Schottky, *Crelle*, CV. (1889), Frobenius, *Crelle*, XCIX. (1885) and *ibid.* CIII. (1887); see also Cayley, *Crelle*, XCIV. and Kohn, *Crelle*, CVII. (1890), where references to the geometrical literature will be found.

$$(67) \quad \frac{x}{1-A_2A_3}+\frac{y}{1-A_3A_1}+\frac{z}{1-A_1A_2}=0,$$

$$(45) \quad \frac{\xi}{A_1(1-A_2A_3)}+\frac{\eta}{A_2(1-A_3A_1)}+\frac{\zeta}{A_3(1-A_1A_2)}=0,$$

$$(75) \quad \frac{x}{1-B_2B_3}+\frac{y}{1-B_3B_1}+\frac{z}{1-B_1B_2}=0,$$

$$(46) \quad \frac{\xi}{B_1(1-B_2B_3)}+\frac{\eta}{B_2(1-B_3B_1)}+\frac{\zeta}{B_3(1-B_1B_2)}=0,$$

$$(56) \quad \frac{x}{1-C_2C_3}+\frac{y}{1-C_3C_1}+\frac{z}{1-C_1C_2}=0,$$

$$(47) \quad \frac{\xi}{C_1(1-C_2C_3)}+\frac{\eta}{C_2(1-C_3C_1)}+\frac{\zeta}{C_3(1-C_1C_2)}=0.$$

Here the six quantities x, y, z, ξ, η, ζ are connected by the equations

$$\left.\begin{aligned} &\xi+\eta+\zeta+x+y+z=0,\\ &\frac{\xi}{A_1}+\frac{\eta}{A_2}+\frac{\zeta}{A_3}+A_1x+A_2y+A_3z=0,\\ &\frac{\xi}{B_1}+\frac{\eta}{B_2}+\frac{\zeta}{B_3}+B_1x+B_2y+B_3z=0,\\ &\frac{\xi}{C_1}+\frac{\eta}{C_2}+\frac{\zeta}{C_3}+C_1x+C_2y+C_3z=0. \end{aligned}\right\} \quad (D)$$

Conversely, if we take arbitrary constants A_1, A_2, A_3, B_1, B_2, B_3, whose number, 6, is, when $p=3$, equal to $3p-3$, namely equal to the number of absolute constants upon which a Riemann surface depends when $p=3$, and, by the first three of the equations (D) determine ξ, η, ζ in terms of the arbitrary lines x, y, z, the last of the equations (D) will determine C_1, C_2, C_3 save for a sign which is the same for all; then it can be directly verified algebraically that the 28 lines here given are double tangents of the quartic curve $\sqrt{x\xi}+\sqrt{y\eta}+\sqrt{z\zeta}=0$.

248. Before leaving this matter we desire to point out further the connection between the two representations of the tangents which have been given. Comparing the two equations of the fundamental quartic curve expressed by the equations (§§ 246, 247)

$$\Omega_0^2=4X_0D, \quad (x\xi+y\eta-z\zeta)^2=4\xi\eta xy,$$

and putting, in accordance therewith,

$$D(x_1, x_2, x_3)=\xi, \quad \Omega_0(x_1, x_2, x_3)=z\zeta-x\xi-y\eta, \quad X_0(x_1, x_2, x_3)=xy\eta$$

and (cf. p. 382) replacing the fourth coordinate T by $T+u$, where

u is an arbitrary linear function of x, y, z or x_1, x_2, x_3, the equation of the cubic surface

$$(T+u)^2 D + (T+u)\,\Omega_0 + X_0 = 0,$$

becomes

$$T^2\xi + T(z\zeta - x\xi - y\eta + 2u\xi) + u^2\xi + u(z\zeta - y\eta - x\xi) + xy\eta = 0,$$

or

$$(T+u)^2\xi + (T+u)(z\zeta - x\xi - y\eta) + xy\eta = 0,$$

which will be found to be the same as

$$(T+u)(T+u-x-z)(T+u-x-\zeta) - (T+u-x)(T+u+y)(T+u+\eta) = 0.$$

Write now

$$v = u - x - z, \quad w = u - x - \zeta, \quad u' = u - x, \quad v' = u + y, \quad w' = u + \eta;$$

then we obtain the result, easy to verify, that if u, v, w, u', v', w' be arbitrary linear functions of the homogeneous space coordinates X, Y, Z, and T be the fourth coordinate, the tangent cone to the cubic surface*

$$(T+u)(T+v)(T+w) - (T+u')(T+v')(T+w') = 0 \qquad \text{(i)}$$

from the vertex $X = 0 = Y = Z$ can be written in the form

$$\sqrt{(P-P')(u-u')} + \sqrt{(u-v')(u-w')} + \sqrt{(u'-v)(u'-w)} = 0,$$

where $P - P' = u + v + w - u' - v' - w'$; we have in fact

$$x = u - u', \quad y = v' - u, \quad z = u' - v, \quad \eta = w' - u, \quad \zeta = u' - w,$$

$$\xi, = -(x + y + z + \eta + \zeta), = P - P'.$$

Now the 27 lines on the cubic surface (i) can be easily obtained†; and thence the forms obtained in § 247, for the bitangents of the quartic, can be otherwise established.

249. *Ex.* i. Prove that when the sum of the characteristics of three bitangents of the quartic is an even characteristic, their points of contact do not lie upon a conic.

By enumerating the constants we infer that it is possible to describe a plane quartic curve having seven arbitrary lines as double tangents. By the investigation of § 247 it follows that only one such quartic can be described when the condition is introduced that no three of the tangents shall have their points of contact upon a conic. By the theory here developed it follows that for a given quartic such a set of seven bitangents can be selected in $8 \cdot 36 = 288$ ways.

Ex. ii. We have given an expression for the general radical form $\sqrt{X^{(3)}}$ of any given odd characteristic. Prove that a radical form $\sqrt{X^{(3)}}$ whose characteristic is even, denoted, suppose, by the index 123, can be written in the form

$$X^{(3)} = \lambda\sqrt{u_1 u_2 u_3} + \lambda_1\sqrt{u_1 u_{12} u_{13}} + \lambda_2\sqrt{u_2 u_{23} u_{21}} + \lambda_3\sqrt{u_3 u_{31} u_{32}},$$

* Any cubic surface can be brought into this form, Salmon, *Solid Geometry* (1882), § 533.

† See Frost, *Solid Geometry* (1886), § 537. The three last equations (D) of § 247 are deducible from the equations occurring in Frost. The three equations correspond to the three roots of the cubic equation used by Frost.

where $\lambda, \lambda_1, \lambda_2, \lambda_3$ are constants, and u_i, u_{ij} denote double tangents of the characteristics denoted by the suffixes, as in § 247.

Ex. iii. If $(\tfrac{1}{2}q, \tfrac{1}{2}q')$, $(\tfrac{1}{2}r, \tfrac{1}{2}r')$ denote any two odd characteristics of half-integers, express the quotient

$$\vartheta(v^{x,z};\ \tfrac{1}{2}q, \tfrac{1}{2}q')/\vartheta(v^{x,z};\ \tfrac{1}{2}r, \tfrac{1}{2}r')$$

algebraically, when $p=3$.

Ex. iv. Obtain an expression of the quotient of any two radical forms $\sqrt{X^{(3)}}$, $\sqrt{Y^{(3)}}$, of assigned characteristics and known zeros, by means of theta functions, p being equal to 3.

250. Noether has given* an expression for the solution of the inversion problem in the general case in terms of radical forms, which is of importance as being capable of great generalization.

Using the places $m_1, \dots, m_p$, associated as in Chap. X. with an arbitrary place m, and supposing them, each repeated, to be the remaining zeros of a form $X^{(3)}$, which vanishes to the second order in each of the places $A_1, \dots, A_{2p-3}$ in which an arbitrary ϕ-polynomial, ϕ_0, which vanishes in m, further vanishes, as in § 244, let $\sqrt{Y^{(3)}}$ be any radical form, and $\Phi^{(1)}$ any ϕ-polynomial whose zeros are $a_1, \dots, a_{2p-2}$. Then (§ 241) the consideration of the rational function $\phi_0^2 Y^{(3)}/[\Phi^{(1)}]^2 X^{(3)}$ leads to the equations

$$[v_i^{x_1, a_1} + v_i^{x_2, a_2} + \dots\dots + v_i^{x_{2p-3}, a_{2p-3}} + v_i^{z, a_{2p-2}}] - [v_i^{z, m} - v_i^{c_1, m_1} - \dots\dots - v_i^{c_p, m_p}]$$
$$= -\tfrac{1}{2}(\sigma_i + \sigma_1'\tau_{i,1} + \dots\dots + \sigma_p'\tau_{i,p}),$$

wherein the places

$$x_1, \dots, x_{2p-3}, c_1, \dots, c_p$$

are the zeros of $\sqrt{Y^{(3)}}$, all of $\sigma_1, \dots, \sigma_p, \sigma_1', \dots, \sigma_p'$ are integers, and z is an arbitrary place; and, as follows from these equations, the places $x_1, \dots, x_{2p-3}$ may be arbitrarily assigned, the places $c_1, \dots, c_p$ and the form $\sqrt{Y^{(3)}}$ being determinate, respectively, from these equations and the equation

$$\log \frac{\phi_0\sqrt{Y^{(3)}}}{\Phi^{(1)}\sqrt{X^{(3)}}} = \text{constant} + \Pi_{x_1, a_1}^{x, a} + \dots\dots + \Pi_{m, a_{2p-2}}^{x, a} + \Pi_{c_1, m_1}^{x, a} + \dots\dots + \Pi_{c_p, m_p}^{x, a}$$
$$+ \pi i\,[\sigma_1' v_1^{x, a} + \dots\dots + \sigma_p' v_p^{x, a}],$$

wherein the place a is arbitrary. Hence if we speak of

$$(\tfrac{1}{2}\sigma_1, \dots, \tfrac{1}{2}\sigma_p, \tfrac{1}{2}\sigma_1', \dots, \tfrac{1}{2}\sigma_p')$$

as the characteristic of $\sqrt{Y^{(3)}}$, it follows, if $\sqrt{Z^{(3)}}$ be another radical form with the characteristic

$$(\tfrac{1}{2}\rho_1, \dots, \tfrac{1}{2}\rho_p, \tfrac{1}{2}\rho_1', \dots, \tfrac{1}{2}\rho_p')$$

and the zeros

$$x_1, \dots, x_{2p-3}, d_1, \dots, d_p,$$

* *Math. Annal.* XXVIII. (1887), p. 354, "Zum Umkehrproblem in der Theorie der Abel'schen Functionen."

that the quotient $\sqrt{Y^{(3)}}/\sqrt{Z^{(3)}}$, which is equal to

$$A e^{\Pi^{x,\,a}_{c_1,\,d_1}+\ldots\ldots+\Pi^{x,\,a}_{c_p,\,d_p}+\pi i[(\sigma_1'-\rho_1')v_1^{x,\,a}+\ldots\ldots+(\sigma_p'-\rho_p')v_p^{x,\,a}]},$$

wherein A is a quantity independent of x, is (§ 187, Chap. X.) also equal to

$$Ce^{\pi i[(\sigma_1'-\rho_1')v_1^{x,\,a}+\ldots\ldots+(\sigma_p'-\rho_p')v_p^{x,\,a}]}\frac{\Theta(v^{x,\,m}-v^{c_1,\,m_1}-\ldots\ldots-v^{c_p,\,m_p})}{\Theta(v^{x,\,m}-v^{d_1,\,m_1}-\ldots\ldots-v^{d_p,\,m_p})},$$

where C is a quantity independent of x; but by the equations here given this is the same as

$$Ce^{\pi i[(\sigma_1'-\rho_1')v_1^{x,\,a}+\ldots\ldots+(\sigma_p'-\rho_p')v_p^{x,\,a}]}$$
$$\frac{\Theta(v^{x,\,a_{2p-2}}+v^{x_1,\,a_1}+\ldots\ldots+v^{x_{2p-3},\,a_{2p-3}}+\tfrac{1}{2}\Omega_\sigma)}{\Theta(v^{x,\,a_{2p-2}}+v^{x_1,\,a_1}+\ldots\ldots+v^{x_{2p-3},\,a_{2p-3}}+\tfrac{1}{2}\Omega_\rho)},$$

where $\tfrac{1}{2}\Omega_\sigma$ denotes p such quantities as $\tfrac{1}{2}(\sigma_i+\sigma_1'\tau_{i,1}+\ldots\ldots+\sigma_p'\tau_{i,p})$; thus, if we put

$$v=v^{x,\,a_{2p-2}}+v^{x_1,\,a_1}+\ldots\ldots+v^{x_{2p-3},\,a_{2p-3}}$$

and recall the formula (§ 175)

$$\Theta(v+\tfrac{1}{2}\Omega_\sigma)=e^{-\pi i\sigma'(v+\frac{1}{2}\sigma+\frac{1}{4}\tau\sigma')}\,\Theta(v;\ \tfrac{1}{2}\sigma,\tfrac{1}{2}\sigma'),$$

we infer that

$$\frac{\sqrt{Y^{(3)}}}{\sqrt{Z^{(3)}}}=E\frac{\Theta(v;\ \tfrac{1}{2}\sigma,\tfrac{1}{2}\sigma')}{\Theta(v;\ \tfrac{1}{2}\rho,\tfrac{1}{2}\rho')},$$

where E is a quantity independent of x.

Now in fact (§ 245) the general radical form $\sqrt{Y^{(3)}}$, of assigned characteristic $(\tfrac{1}{2}\sigma,\tfrac{1}{2}\sigma')$, is given by

$$\lambda_1\sqrt{Y_1^{(3)}}+\ldots\ldots+\lambda_{2p-2}\sqrt{Y_{2p-2}^{(3)}},$$

where $\sqrt{Y_1^{(0)}},\ldots,\sqrt{Y_{2p-2}^{(0)}}$ are special forms of this characteristic, and $\lambda_1,\ldots,\lambda_{2p-2}$ are constants. If we introduce the condition that $\sqrt{Y^{(3)}}$ vanishes at the places $x_1,\ldots,x_{2p-3}$ we infer that $\sqrt{Y^{(3)}}$ is equal to $F\Delta_\sigma^{(3)}(x,x_1,\ldots,x_{2p-3})$, where F is independent of x and $\Delta_\sigma^{(3)}(x,x_1,\ldots,x_{2p-3})$ denotes the determinant

$$\begin{vmatrix}\sqrt{Y_1^{(3)}(x)}, & \ldots\ldots\ldots\ldots, & \sqrt{Y_{2p-2}^{(3)}(x)}\\ \ldots\ldots & \ldots\ldots & \ldots\ldots\\ \sqrt{Y_1^{(3)}(x_i)}, & \ldots\ldots\ldots\ldots, & \sqrt{Y_{2p-2}^{(3)}(x_i)}\\ \ldots\ldots & \ldots\ldots & \ldots\ldots\end{vmatrix}$$

in which i is to be taken in turn equal to $1, 2, \ldots, 2p-3$. Hence we have

$$\frac{\Delta_\sigma^{(3)}(x,x_1,\ldots,x_{2p-3})}{\Delta_\rho^{(3)}(x,x_1,\ldots,x_{2p-3})}=G\frac{\Theta(v;\ \tfrac{1}{2}\sigma,\tfrac{1}{2}\sigma')}{\Theta(v;\ \tfrac{1}{2}\rho,\tfrac{1}{2}\rho')},$$

where, from the symmetry in regard to the places $x, x_1, \ldots, x_{2p-3}$, G is independent* of the position of any of these places, and v is given by

$$v = v^{x, a_{2p-2}} + v^{x_1, a_1} + \ldots\ldots + v^{x_{2p-3}, a_{2p-3}}.$$

To apply this equation to the solution of the inversion problem expressed by p such equations as

$$v^{x_1, \mu_1} + \ldots\ldots + v^{x_p, \mu_p} = u,$$

where $\mu_1, \ldots, \mu_p$ denote p arbitrary given places, we suppose the positions of the places $x_{p+1}, \ldots, x_{2p-3}$ to be given; then instead of $\Delta_\sigma(x, x_1, \ldots, x_{2p-3})$ we have an expression of the form

$$A_1 \sqrt{Y_1^{(3)}(x)} + \ldots\ldots + A_{p+1} \sqrt{Y_{p+1}^{(3)}(x)},$$

where $\sqrt{Y_1^{(3)}(x)}, \ldots, \sqrt{Y_{p+1}^{(3)}(x)}$ denote forms $\sqrt{Y^{(3)}(x)}$ vanishing in the given places $x_{p+1}, \ldots, x_{2p-3}$, and $A_1, \ldots, A_{p+1}$ are unknown constants. Since the arguments u are given, the arguments v are of the form $v^{x, a_{2p-2}} + w$, where w is known. If then in the equation

$$\frac{A_1 \sqrt{Y_1^{(3)}(x)} + \ldots\ldots + A_{p+1} \sqrt{Y_{p+1}^{(3)}(x)}}{B_1 \sqrt{Z_1^{(3)}(x)} + \ldots\ldots + B_{p+1} \sqrt{Z_{p+1}^{(3)}(x)}} = \frac{\Theta(v;\ \tfrac{1}{2}\sigma, \tfrac{1}{2}\sigma')}{\Theta(v;\ \tfrac{1}{2}\rho, \tfrac{1}{2}\rho')}$$

we determine the unknown ratios $A_1 : A_2 : \ldots\ldots : A_{p+1} : B_1 : \ldots\ldots : B_{p+1}$ by the substitution of $2p+1$ different positions for the place x, this equation itself will determine the places $x_1, \ldots, x_p$. They are, in fact, the zeros of either of the forms

$$A_1 \sqrt{Y_1^{(3)}(x)} + \ldots\ldots + A_{p+1} \sqrt{Y_{p+1}^{(3)}(x)},$$

$$B_1 \sqrt{Z_1^{(3)}(x)} + \ldots\ldots + B_{p+1} \sqrt{Z_{p+1}^{(3)}(x)}$$

other than the given zeros $x_{p+1}, \ldots, x_{2p-3}$. If the first of these forms be multiplied by an arbitrary form $\sqrt{Y^{(3)}(x)}$, of characteristic $(\tfrac{1}{2}\sigma, \tfrac{1}{2}\sigma')$, the places $x_1, \ldots, x_p$ are given as the zeros of a rational function of the form

$$A_1 \Phi_1^{(3)}(x) + \ldots\ldots + A_{p+1} \Phi_{p+1}^{(3)}(x),$$

of which $4p-6$ zeros are known, consisting, namely, of the places $x_{p+1}, \ldots, x_{2p-3}$ and the zeros of $\sqrt{Y^{(3)}(x)}$.

In regard to this result the reader may consult Weber, *Theorie der Abel'schen Functionen vom Geschlecht* 3 (Berlin, 1876), p. 157, the paper of Noether (*Math. Annal.* XXVIII.) already referred to, and, for a solution in which the radical forms are mth roots of rational functions, Stahl, *Crelle*, LXXXIX. (1880), p. 179, and *Crelle*, CXI. (1893), p. 104. It will be seen in the following chapter that the results may be deduced from another result of a simpler character (§ 274).

251. The theory of radical functions has far-reaching geometrical applications to problems of the contact of curves. See, for instance, Clebsch, *Crelle*, LXIII. (1864), p. 189. For the theory of the solution of the final algebraic equations see Clebsch and Gordan, *Abel'sche Functnen.* (Leipzig, 1866), Chap. X. Die Theilung; Jordan, *Traité des Substitutions* (Paris, 1870), p. 354, etc.; and now (Aug. 1896), for the bitangents in case $p=3$, Weber, *Lehrbuch der Algebra* (Braunschweig, 1896), II. p. 380.

* For the determination of G see Noether, *Math. Annal.* XXVIII. (1887), p. 368, and Klein, *Math. Annal.* XXXVI. (1890), pp. 73, 74.

CHAPTER XIV.

FACTORIAL FUNCTIONS.

252. THE present chapter is concerned* with a generalisation of the theory of rational functions and their integrals. As in that case, it is convenient to consider the integrals and the functions together from the first. In order, therefore, that the reader may be better able to follow the course of the argument, it is desirable to explain, briefly, at starting, the results obtained. All the functions and integrals considered have certain fixed singularities, at places† denoted by $c_1, \dots, c_k$. A function or integral which has no infinities except at these fixed singularities is described as everywhere finite. The functions of this theory which replace the rational functions of the simpler theory have, beside the fixed singularities, no infinities except poles. But the functions differ from rational functions in that their values are not the same at the two sides of any period loop; these values have a ratio, described as the *factor*, which is constant along the loop; and a system of functions is characterised by the values of its factors. We consider two sets of factors, and, correspondingly, two sets of *factorial functions*, those of the *primary system* and those of the *associated system;* their relations are quite reciprocal. We have then a circumstance to which the theory of rational functions offers no parallel; *there may be everywhere finite factorial functions*‡. The number of such functions of the primary system which are linearly independent is denoted by $\sigma' + 1$; the number of the associated system by $\sigma + 1$. As in the case of algebraical integrals, we may have *everywhere finite factorial integrals.* The number of such integrals of the primary system which are linearly independent is denoted by ϖ, that of the associated system by ϖ'. The factorial integrals of the primary system are not integrals of factorial functions of that system; they are chosen so that the values u, u'

* The subject of the present chapter has been considered by Prym, *Crelle*, LXX. (1869), p. 354; Appell, *Acta Mathematica*, XIII. (1890); Ritter, *Math. Annal.* XLIV. (1894), pp. 261—374. In these papers other references will be found. See also Hurwitz, *Math. Annal.* XLI. (1893), p. 434, and, for a related theory, not considered in the present chapter, Hurwitz, *Math. Annal.* XXXIX. (1891), p. 1. For the latter part of the chapter see the references given in §§ 273, 274, 279.

† In particular the theory includes the case when $k=0$, and no such places enter.

‡ This statement is made in view of the comparison instituted between the development of the theory of rational functions and that of factorial functions. The factorial functions have (unless $k=0$) fixed infinities.

of such an integral on the two sides of a period loop are connected by an equation of the form $u' = Mu + \mu$, where μ is a constant and M is the factor of the primary system of factorial functions which is associated with that period loop. The primary and associated systems are so related that if F be a factorial function of either system, and G' a factorial integral of the other system, FdG'/dx is a rational function without assigned singularities. In the case of the rational functions, the smallest number of arbitrary assigned poles for which a function can always be constructed is $p + 1$. In the present theory, as has been said, it may be possible to construct factorial functions of the primary system without poles; but when that is impossible, or $\sigma' + 1 = 0$, the smallest number of arbitrary poles for which a factorial function of the primary system can always be constructed is $\varpi' + 1$. Similarly when $\sigma + 1 = 0$, the smallest number of arbitrary poles for which a factorial function of the associated system can always be constructed is $\varpi + 1$. Of the two numbers $\sigma + 1$, $\sigma' + 1$, at least one is always zero, except in one case, when they are both unity. When $\sigma' + 1$ is > 0, the everywhere finite factorial functions of the primary system can be expressed linearly in terms of the everywhere finite factorial integrals of the same system. We can also construct factorial integrals of the primary system, which, beside the fixed singularities, have assigned poles; the least number of poles of arbitrary position for which this can be done is $\sigma + 2$. And we can construct factorial integrals of the primary system which have arbitrary logarithmic infinities; the least number of such infinities of arbitrary position is $\sigma + 2$. For the associated system of factors the corresponding numbers are $\sigma' + 2$.

It will be found that all the formulae of the general theory are not immediately applicable to the ordinary theory of rational functions and their integrals. The exceptions, and the reasons for them, are pointed out in footnotes.

The deduction of these results occupies §§ 253—267 of this chapter. The section of the chapter which occupies §§ 271—278, deals, by examples, with the connection of the present theory with the theory of the Riemann theta functions. With a more detailed theory of factorial functions this section would be capable of very great development. The concluding section of the chapter deals very briefly with the identification of the present theory with the theory of automorphic functions.

253. Let $c_1, \dots, c_k$ be arbitrary fixed places of the Riemann surface, which we suppose to be finite places and not branch places. In all the investigations of this chapter these places are to be the same. They may be called the essential singularities of the systems of factorial functions. We require the surface to be dissected so that the places $c_1, \dots, c_k$ are excluded and the surface becomes simply connected. This may be effected in a manner analogous to that adopted in § 180, the places $c_1, \dots, c_k$ occurring instead of

$z_1, \ldots, z_k$. But it is more convenient, in view of one development of the theory, to suppose the loops of § 180 to be deformed until the cuts* between the pairs of period loops become of infinitesimal length. Then the dissection will be such as that represented in figure 9; and this dissection is sufficiently

Fig. 9.

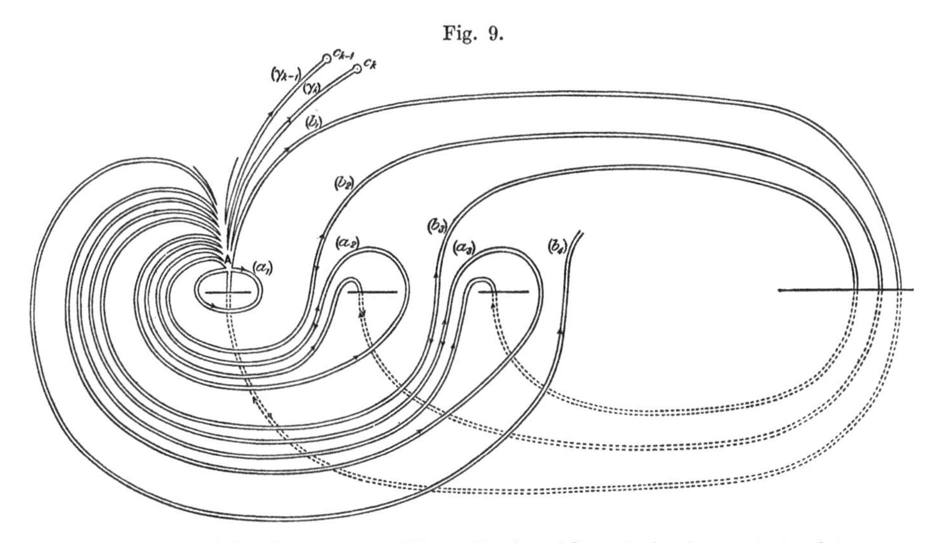

well represented by figure 10. We call the sides of the loops (a_r), (b_r), upon which the letters a_r, b_r are placed, the left-hand sides of these loops, and by the left-hand sides of the cuts $(\gamma_1), \ldots, (\gamma_k)$, to the places $c_1, \ldots, c_k$, we mean

Fig. 10.

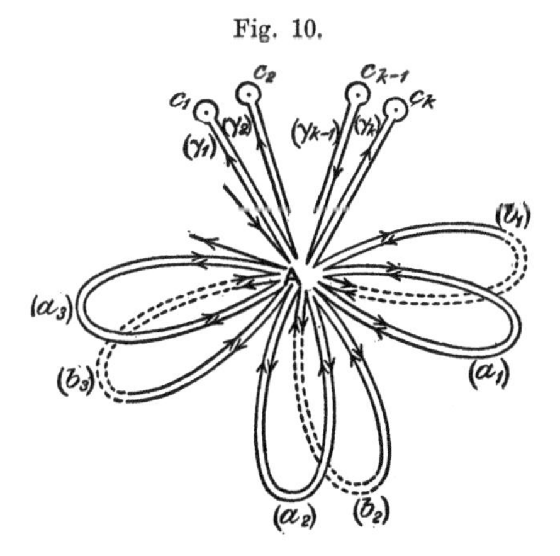

the sides which are on the left when we pass from A to $c_1, \ldots, c_k$ respectively. The consideration of the effect of an alteration in these conventions is postponed till the theory of the transformation of the theta functions has been considered.

* These cuts are those generally denoted by $c_1, \ldots, c_{p-1}$. Cf. Forsyth, *Theory of Functions*, § 181.

254. In connection with the surface thus dissected we take now a series of $2p+k$ quantities

$$\lambda_1, \ldots, \lambda_k, \quad h_1, \ldots, h_p, \quad g_1, \ldots, g_p,$$

which we call the fundamental constants; we suppose no one of $\lambda_1, \ldots, \lambda_k$ to be a positive or negative integer, or zero; but we suppose $\lambda_1 + \ldots + \lambda_k$ to be an integer, or zero; and we consider functions

(1) which are uniform on the surface thus dissected, and have, thereon, no infinities except poles,

(2) whose value on the left-hand side of the period loop (a_i) is $e^{-2\pi i h_i}$ times the value on the right-hand side; whose value on the left-hand side of the period loop (b_i) is $e^{2\pi i g_i}$ times the value on the right-hand side,

(3) which*, in the neighbourhood of the place c_i, are expressible in the form $t^{-\lambda_i}\phi_i$, where t is the infinitesimal at c_i and ϕ_i is uniform, finite, and not zero in the neighbourhood of the place c_i,

(4) which, therefore, have a value on the left-hand side of the cut γ_i which is $e^{-2\pi i\lambda_i}$ times the value on the right-hand side.

Let $\alpha_1, \ldots, \alpha_M, \beta_1, \ldots, \beta_N$ be any places; consider the expression

$$f = Ae^{\Pi_{\beta_1, m}^{x, a} + \ldots + \Pi_{\beta_N, m}^{x, a} - \Pi_{\alpha_1, m}^{x, a} - \ldots - \Pi_{\alpha_M, m}^{x, a} - 2\pi i\left[(h_1+H_1)v_1^{x, a} + \ldots + (h_p+H_p)v_p^{x, a}\right] - \sum_{i=1}^{k}\lambda_i \Pi_{c_i, m}^{x, a}},$$

wherein A is independent of the place x,

$$N - M = \sum_{i=1}^{k} \lambda_i, \qquad \text{(i)},$$

$\Sigma\lambda$ being an integer (or zero), m is an arbitrary place, and $H_1, \ldots, H_p$ are integers. It is clear that this expression represents a function which is uniform on the dissected surface, which has poles at the places $\alpha_1, \ldots, \alpha_M$, and zeros at the places $\beta_1, \ldots, \beta_N$, and that in the neighbourhood of the place c_i this function has the character required. For the period loop (a_i) the function has the factor $e^{-2\pi i(h_i+H_i)} = e^{-2\pi i h_i}$, as desired; for the period loop (b_i) the function has the factor $e^{2\pi i K}$, where

$$K = v_i^{\beta_1, m} + \ldots + v_i^{\beta_N, m} - v_i^{\alpha_1, m} - \ldots - v_i^{\alpha_M, m} - \sum_{r=1}^{p}(h_r+H_r)\tau_{r,i} - \sum_{r=1}^{k}\lambda_i v_i^{c_r, m},$$

and this factor is equal to $e^{2\pi i g_i}$ if only

$$v_i^{\beta_1, m} + \ldots + v_i^{\beta_N, m} - v_i^{\alpha_1, m} - \ldots - v_i^{\alpha_M, m} - \sum_{r=1}^{k}\lambda_i v_i^{c_r, m}$$

$$= g_i + G_i + \sum_{r=1}^{r=p}(h_r+H_r)\tau_{r,i}, \qquad \text{(ii)},$$

G_i being an integer.

* It is intended, as already stated, that the places $c_1, \ldots, c_k$ should be in the finite part of the surface and should not be branch places.

It follows therefore that, subject to the conditions (i) and (ii), such a function as has been described certainly exists.

Conversely it can be immediately proved that any such function must be capable of being expressed in the form here given, and that the conditions (i), (ii) are necessary.

Unless the contrary be expressly stated, we suppose the quantities $\lambda_1, \dots, \lambda_k, h_1, \dots, h_p, g_1, \dots, g_p$ always the same, and express this fact by calling the functions under consideration *factorial functions of the primary system*. The quantities $e^{-2\pi i\lambda_1}, \dots, e^{-2\pi i\lambda_k}, e^{-2\pi ih_1}, \dots, e^{-2\pi ih_p}, e^{2\pi ig_1}, \dots, e^{2\pi ig_p}$ are called *the factors*. It will be convenient to consider with these functions other functions of the same general character but with a different system of fundamental constants,

$$\lambda_1', \dots, \lambda_k', h_1', \dots, h_p', g_1', \dots, g_p',$$

connected with the original constants by the equations

$$\lambda_i + \lambda_i' + 1 = 0, \quad h_i + h_i' = 0, \quad g_i + g_i' = 0;$$

these functions will be said to be *functions of the associated system*. The factors associated therewith are the inverses of the factors of the primary system.

255. As has been remarked, the rational functions on the Riemann surface are a particular case of the factorial functions, arising when the factors are unity and no such places as $c_1, \dots, c_k$ are introduced. From this point of view the condition (i), which can be obtained as the condition that $\int d \log f$, taken round the complete boundary of the dissected surface, is zero, is a generalisation of the fact that the number of zeros and poles of a rational function is the same, and the condition (ii) expresses a theorem generalising Abel's theorem for integrals of the first kind.

Now Riemann's theory of rational functions is subsequent to the theory of the integrals; these arise as functions which are uniform on the dissected Riemann surface, but differ on the sides of a period loop by additive constants. In what follows we consider the theory in the same order, and enquire first of all as to the existence of functions whose differential coefficients are factorial functions. For the sake of clearness such functions will be called *factorial integrals;* and it will appear that just as rational functions are expressible by Riemann integrals of the second kind, so factorial functions are expressible by certain factorial integrals, provided the fundamental constants of these latter are suitably chosen. We define then a factorial integral of the primary system, H, as a function such that dH/dx is a factorial function with the fundamental constants

$$\lambda_1 + 1, \dots, \lambda_k + 1, \; h_1, \dots, h_p, \; g_1, \dots, g_p;$$

thus dH/dx has the same factors as the factorial functions of the primary system, but near the place c_i, dH/dx is of the form $t^{-(\lambda_i+1)}\phi_i$, where ϕ_i is uniform, finite and not zero in the neighbourhood of c_i. Similarly we define a factorial integral of the associated system, H', to be such that dH'/dx is a factorial function with the fundamental constants

$$\lambda_1'+1, \ldots, \lambda_k'+1, \; h_1', \ldots, h_p', \; g_1', \ldots, g_p',$$

or

$$-\lambda_1, \ldots, -\lambda_k, \; -h_1, \ldots, -h_p, \; -g_1, \ldots, -g_p;$$

thus, if f be any factorial function of the primary system, $f\,dH'/dx$ is a rational function on the Riemann surface, for which the places $c_1, \ldots, c_k$ are not in any way special. And similarly, if f' be any factorial function of the associated system, and H any factorial integral of the primary system, $f'\,dH/dx$ is a rational function.

The values of a factorial integral of the primary system, H, at the two sides of any period loop are connected by an equation of the form

$$\overline{H} = \mu H + \Omega,$$

where μ is one of the factors $e^{-2\pi i h_r}$, $e^{2\pi i g_r}$, and Ω is a quantity which is constant along the particular period loop. Near c_i, H is of the form

$$A_i + t^{-\lambda_i}\phi_i + C_i \log t,$$

where A_i is a constant, ϕ_i is uniform, finite, and, in general, not zero in the neighbourhood of c_i, and C_i is a constant, which is zero unless λ_i+1 be a positive integer (other than zero), and may be zero even when λ_i+1 is a positive integer. After a circuit round c_i, H will be changed into

$$\overline{H} = A_i + e^{-2\pi i\lambda_i} t^{-\lambda_i}\phi_i + 2\pi i C_i + C_i \log t;$$

thus, when $C_i = 0$,

$$\overline{H} = He^{-2\pi i\lambda_i} + A_i(1 - e^{-2\pi i\lambda_i}),$$

and when C_i is not zero, and, therefore, λ_i+1 is a positive integer,

$$\overline{H} = H + 2\pi i C_i;$$

in either case we have

$$\overline{H} = \gamma H + \Gamma,$$

where $\gamma = e^{-2\pi i\lambda_i}$, and Γ is constant along the cut (γ_i).

Thus, in addition to the fundamental factors of the system, there arise, for every factorial integral, $2p+k$ new constants, $2p$ of them such as that here denoted by Ω and k of them such as that denoted by Γ. It will be seen subsequently that these are not all independent.

As has been stated we exclude from consideration the case in which any one of $\lambda_1, \ldots, \lambda_k$ is an integer, or zero. Thus the constants C_i will not enter; neither will the corresponding constants for the associated system.

256. Consider now the problem of finding *factorial integrals of the primary system which shall be everywhere finite.* Here, as elsewhere, when we speak of the infinities or zeros of a function, we mean those which are not at the places $c_1, \ldots, c_k$, or which fall at these places in addition to the poles or zeros which are prescribed to fall there.

If V be such a factorial integral, dV/dx is only infinite when dx is zero of the second order, namely $2p-2+2n$ times, at the branch places of the surface. And dV/dx is zero at $x=\infty$, $2n$ times*. Thus, if N denote the number of zeros of dV/dx which are not due to the denominator dx, or, as we may say (cf. § 21) the number of zeros of dV, we have by the condition (i) § 254,

$$N+2n=2p-2+2n+\sum_{i=1}^{k}(\lambda_i+1),$$

so that the number of zeros of dV is $2p-2+\Sigma(\lambda_i+1)$.

Now let f_0 denote a factorial function with the primary system of factors, but with behaviour at c_i like $t^{-(\lambda_i+1)}\phi_i$, where ϕ_i is uniform, finite, and not zero at c_i. Then, if an everywhere finite factorial integral V exists at all, Z, $=f_0^{-1}dV/dx$, will be a rational function on the Riemann surface, infinite at the (say N_0) zeros of f_0, and $2n+2p-2$ times at the branch places of the surface, and zero at the (say M_0) poles of f_0, and $2n$ times at $x=\infty$ (beside being zero at the zeros of dV). Conversely a rational function Z satisfying these conditions will be such that $\int Zf_0dx$ is a function V. Thus *the number of existent functions V which are linearly independent is at least*

$$N_0+2n+2p-2-(2n+M_0)-p+1,\ =p-1+\sum_{i=1}^{k}(\lambda_i+1),$$

provided this be positive. We are therefore sure, when this is the case, that functions V do exist. To find the exact number, let V_0 be one such; then if V be any other, dV/dV_0 is a rational function with poles in the $2p-2+\Sigma(\lambda+1)$ zeros of dV_0; and conversely if R be a rational function whose poles are the zeros of dV_0, the integral $\int RdV_0$ is a function V. Thus† *the number of functions V, when any exist,* is (§ 37, Chap. III.)

$$\varpi,\ =p-1+\Sigma(\lambda+1)+\sigma+1,$$

* These numbers may be modified by the existence of a branch place at infinity. But their difference remains the same.

† For the ordinary case of rational functions $\sigma+1$, as here defined, is equal to unity, and, therefore, omitting the term $\Sigma(\lambda+1)$, we have $\varpi=p$.

where $\sigma+1$ is the number of linearly independent differentials dv, of ordinary integrals of the first kind, which vanish in the $2p-2+\Sigma(\lambda+1)$ zeros of the differential dV_0 of any such function V_0. Since dV/dV_0 is a rational function, the number of differentials dv vanishing in the zeros of dV_0 is the same as the number vanishing in the zeros of dV. Since dv has $2p-2$ zeros, $\sigma+1$ vanishes when $\Sigma(\lambda+1)>0$.

Ex. For the hyperelliptic surface

$$y^2=(x-a)(x-b)(x,1)_{2p},$$

the factorial integrals, V, having the same factors at the period loops as the root function $\sqrt{(x-a)(x-b)}$, and no other factors, are given by

$$\int \sqrt{(x-a)(x-b)}\,(x,1)_{p-2}\frac{dx}{y}$$

and $\varpi=p-1$. Here $k=0$; there are no places $c_1, \dots, c_k$.

257. The number $\sigma+1$ is of great importance; when it is greater than zero, which requires $\Sigma(\lambda+1)$ to be negative or zero, *there are $\sigma+1$ factorial functions of the associated system which are nowhere infinite.*

For if V be an everywhere finite factorial integral of the primary system, and $dv_1, \dots, dv_{\sigma+1}$ represent the linearly independent differentials of integrals of the first kind which vanish in the zeros of dV, the functions

$$\frac{dv_1}{dV}, \dots, \frac{dv_{\sigma+1}}{dV},$$

whose behaviour at a place c_i is like that of $\frac{1}{t^{-(\lambda_i+1)}}\phi_i$, where ϕ_i is uniform, finite and not zero in the neighbourhood of c_i, namely of $t^{-\lambda_i'}\phi_i$, are clearly factorial functions of the associated system, without poles. Conversely if K' denote an everywhere-finite factorial function of the associated system, the integral $\int K'dV$ is the integral of a rational function, and does not anywhere become infinite. Denoting it by v, dv vanishes at the $2p-2+\Sigma(\lambda+1)$ zeros of dV as well as at the $0+\sum_{i=1}^{k}\lambda_i'$, $=-\Sigma(\lambda+1)$, zeros of K' (cf. the condition (i), § 254). Thus, to every factorial integral V we obtain $\sigma+1$ functions K'; and since, when $\sigma+1>0$, the quotient of two differentials dV, dV_0 can* be expressed by the quotient of two differentials dv, dv_0, we cannot thus obtain more than $\sigma+1$ functions K'; while, conversely, to every function K' we obtain a differential dv which vanishes in the zeros of any assigned function V; and, as before, we cannot obtain any others by taking, instead of V, another factorial integral V_0.

* Cf. Chap. VI. § 98.

258. The existence of these everywhere finite factorial functions, K', of the associated system can also be investigated *à priori* from the fundamental equations (i) and (ii) (§ 254). These give, in this case,

$$v_i^{\beta_1, m} + \dots + v_i^{\beta_N, m} = -\sum_{r=1}^{k} (\lambda_r + 1) v_i^{c_r, m} - (g_i + G_i) - \tau_{i,1}(h_1 + H_1) - \dots\dots$$
$$- \tau_{i,p}(h_p + H_p), \qquad \text{(iii)}$$

and $$N = -\sum_{r=1}^{k} (\lambda_r + 1),$$

where $G_1, \dots, G_p, H_1, \dots, H_p$ are integers.

Hence no functions K' exist unless $\Sigma(\lambda+1)$ be a negative integer or be zero; we consider these possibilities separately.

When $\Sigma(\lambda+1)=0$, it is necessary, for the existence of such functions, that the fundamental constants satisfy the conditions

$$\sum_{r=1}^{k} (\lambda_r + 1) v_i^{c_r, m} + g_i + h_1\tau_{i,1} + \dots\dots + h_p\tau_{i,p} \equiv 0, \qquad (i = 1, 2, \dots, p);$$

conversely, when these conditions are fulfilled, taking suitable integers $H_1, \dots, H_p$, it is clear that the function

$$E_0 = Ae^{\sum_{r=1}^{k} (\lambda_r+1)\Pi_{c_r, m}^{x, a} + 2\pi i (h_1+H_1) v_1^{x, a} + \dots\dots + 2\pi i (h_p+H_p) v_p^{x, a}},$$

wherein A is an arbitrary constant, and a, m are arbitrary places, is an everywhere finite factorial function of the associated system, and it can be immediately seen that every such function is a constant multiple of E_0. If then we denote the number of functions K' by $\Sigma+1$ (to be immediately identified with $\sigma+1$), we have, in this case, $\Sigma+1=1$; and there are p functions V, given by $V=\int E_0^{-1} dv$, where dv is in turn the differential of every one of the linearly independent integrals of the first kind; it is easy to see that every function V can be thus expressed. Thus, in the zeros of a differential dV there vanishes one differential dv, so that $\sigma+1=1$. Hence $\sigma+1=\Sigma+1$, and the formula $\varpi = p-1+\Sigma(\lambda+1)+\sigma+1$ is verified.

When $\Sigma(\lambda+1)$ is negative and numerically greater than zero, and the equations (iii) have any solutions, let t denote the number of linearly independent differentials dv which vanish in the places of one and therefore of every set, $\beta_1, \dots, \beta_N$, which satisfies these equations; then* the number of sets which satisfy these equations is ∞^{s-p+t}, where $s = -\Sigma(\lambda+1)$; thus the quotient of two functions K' is a rational function with $\Sigma+1$, $= s-p+t+1$ arbitrary constants, one of these being additive. This is then the number of linearly independent functions K'. If K' be one of these functions, and

* Cf. § 158, Chap. VIII.; § 95, Chap. VI.

$dv_1, \ldots, dv_t$ denote the differentials vanishing in the zeros of K', it is clear that the functions

$$\int \frac{dv_1}{K'}, \ldots, \int \frac{dv_t}{K'}$$

are finite factorial integrals of the primary system, that is, are functions V; conversely if V be any finite factorial integral of the primary system, $\int K' dV$ is an integral, v, of the first kind such that dv vanishes in the zeros of K'. Hence the number t, which expresses the number of differentials dv which vanish in the zeros of K', is equal to the number, ϖ, of functions V. But we have proved that $\varpi = p-1+\Sigma(\lambda+1)+\sigma+1$, and, above, that $t = p-1-s+\Sigma+1$. Hence $\sigma+1 = \Sigma+1$.

Thus we have the results*: *The number, $\sigma+1$, of everywhere finite factorial functions, K', of the associated system is equal to the number of differentials dv which vanish in the $2p-2+\Sigma(\lambda+1)$ zeros of any differential dV;* hence (§ 21, Chap. II.) $\sigma+1$ is less than p, unless $\Sigma(\lambda+1) = -(2p-2)$.

Also, when $\sigma+1>0$, *the number, ϖ, of everywhere finite factorial integrals, V, of the primary system, is equal to the number of differentials dv which vanish in the s, $= -\Sigma(\lambda+1)$, zeros of any function K'.* The argument by which this last result is obtained does not hold when† $\sigma+1=0$. When $\sigma+1>0$, it follows that ϖ is not greater than p.

Similarly when s', $= -\Sigma(\lambda'+1)$, $= \Sigma\lambda$, $= -s-k$, is >0, we can prove, by considering the primary system, that there are $\sigma'+1$ everywhere finite factorial functions K of the primary system, where $\sigma'+1$ is the number of differentials dv vanishing in the $2p-2-\Sigma\lambda$, $= 2p-2+s+k$, zeros of any differential dV; and that, when $\sigma'+1>0$, the number ϖ', of everywhere finite factorial integrals, V', of the associated system is equal to the number of differentials dv vanishing in the s' zeros of any function K. Hence $\sigma'+1=0$ when $s>0$, and, then, no functions K exist. When $s=0$ we have seen that there may or may not be functions K'; but there cannot be functions K unless $k=0$, since otherwise $2p-2+s+k > 2p-2$. And then the existence of functions K depends on the condition whether the fundamental constants be such that

$$\frac{1}{E_0}, = e^{-2\pi i[(h_1+H_1)v_1^{x,a}+\ldots\ldots+(h_p+H_p)v_p^{x,a}]},$$

is a function of the primary system or not, $H_1, \ldots, H_p$ being suitable integers, namely whether there exist relations of the form

$$g_i + G_i + (h_1+H_1)\tau_{i,1} + \ldots\ldots + (h_p+H_p)\tau_{i,p} = 0, \qquad (i=1, 2, \ldots, p),$$

* Which hold for the ordinary case of rational functions, $\sigma+1$ being then unity.

† In the case of the factorial functions which are square roots of rational functions of which all the poles and zeros are of the second order, so that the places $c_1, \ldots, c_k$ are not present, and the numbers g, h are half integers, we have $\varpi = p-1$, $\sigma+1=0$.

where $G_1, \ldots, G_p$ are integers. In such case E_0 is a finite factorial function of the associated system.

On the whole then the theory breaks up into four cases (i) $\sigma+1=0$, $\sigma'+1=0$, (ii) $\sigma+1>0$, $\sigma'+1=0$, (iii) $\sigma+1=0$, $\sigma'+1>0$, (iv) $\sigma+1=1$, $\sigma'+1=1$. Of these the cases (ii) and (iii) are reciprocal.

259. One remark remains to be made in this connection. When $\sigma+1>0$ there are everywhere finite functions, K', of the associated system, given (§ 257) by

$$\frac{dv_1}{dV}, \frac{dv_2}{dV}, \ldots\ldots, \frac{dv_{\sigma+1}}{dV};$$

these have, at any one of the places $c_1, \ldots, c_k$, a behaviour represented by that of $t^{-\lambda'}\phi$; hence the differential coefficients of these functions satisfy all the conditions whereby the differential coefficients, dV'/dx, of the everywhere finite factorial integrals of the associated system, are defined. Therefore* the functions K' are expressible linearly in terms of the functions $V_1', \ldots, V'_{\varpi'}$ by equations of the form

$$K_i', = \frac{dv_i}{dV}, = \lambda_{i,1} V_1' + \ldots\ldots + \lambda_{i,\varpi'} V'_{\varpi'} + \lambda, \qquad (i=1, 2, \ldots, (\sigma+1)),$$

where the coefficients, $\lambda_{i,j}$, λ, are constants.

Hence also the difference $\varpi'-(\sigma+1)$ is not negative. This is also obvious otherwise. For when $\sigma+1>0$, $-\Sigma(\lambda+1)$, $=s$, is zero or positive, and $\sigma+1 \not> p$ (§ 258), and, therefore, $\varpi'-\sigma$, $=p-(\sigma+1)+\sigma'+1+k+s$, can only be as small as zero when $k=0=s$, and $\sigma+1=p$; these are incompatible.

Similarly, when $\sigma'+1>0$, the everywhere finite factorial functions of the original system are linear functions of the factorial integrals $V_1, \ldots, V_\varpi$.

It follows† therefore that of the ϖ periods of the functions $V_1, \ldots, V_\omega$, at any definite period loop, only $\varpi-(\sigma'+1)$ can be regarded as linearly independent; in fact, $\sigma'+1$ of the functions $V_1, \ldots, V_\varpi$ may be replaced by linear functions of the remaining $\varpi-(\sigma'+1)$, and of the functions $K_1, \ldots, K_{\sigma'+1}$.

260. A factorial integral is such that its values at the two sides of a period loop of the first kind are connected by an equation of the form $u'=\mu_i u+\Omega_i$, its values at the two sides of a period loop of the second kind are connected by an equation of the form $u'=\mu'_i u+\Omega'_i$, and its values at the two sides of a loop (γ_i) are connected by an equation of the form $u'=\gamma_i u+\Gamma_i$, where‡ $\Gamma_i=A_i(1-\gamma_i)$. Of the $2p+k$ periods Ω_i, Ω'_i, Γ_i thus

* It is clearly assumed that K'_i is not a constant; thus the reasoning does not apply to the ordinary case of rational functions.

† In the ordinary case of rational functions this number $\varpi-(\sigma+1)$ must be replaced by p. See the preceding note.

‡ § 255. The case where one of $\lambda_1, \ldots, \lambda_k$ is zero or an integer is excluded.

arising, two at least can be immediately excluded. For it is possible, by subtracting one of the constants $A_1, \ldots, A_k$ from the factorial integral, to render one of the periods $\Gamma_1, \ldots, \Gamma_k$ zero; and by following the values of the factorial integral, which is single-valued on the dissected surface, once completely round the sides of the loops, we find, in virtue of $\gamma_1\gamma_2 \ldots \gamma_k = 1$, that

$$\sum_{i=1}^{p} [\Omega_i(1-\mu_i') - \Omega_i'(1-\mu_i)] = \Gamma_1 + \gamma_1\Gamma_2 + \gamma_1\gamma_2\Gamma_3 + \ldots + \gamma_1\gamma_2 \ldots \gamma_{k-1}\Gamma_k.$$

Thus there are certainly not more than $2p-2+k$ linearly independent periods of a factorial integral.

Suppose now that V is any everywhere finite factorial integral of the original system, and V_i' is any one of the corresponding integrals of the associated system. The integral $\int V dV_i'$, taken once completely round the boundary of the surface which is constituted by the sides of the period loops, is equal to zero. By expressing this fact we obtain an equation which is linear in the periods of V and linear in the periods of V_i'. By taking i in turn equal to $1, 2, \ldots, \varpi'$, we thus obtain ϖ' linear equations for the periods of V, wherein the coefficients are the periods of $V_1', \ldots, V'_{\varpi'}$. As remarked above these coefficients are themselves connected by $\sigma+1$ linear equations; so that we thus obtain at most $\varpi'-(\sigma+1)$ linearly independent linear equations for the periods of V. If these are independent of one another and independent of the two reductions mentioned above, it follows that the $2p+k$ periods of V are linearly expressible by only

$$2p-2+k-[\varpi'-(\sigma+1)]$$

periods, at most. Now we have

$$\varpi = p-1+\Sigma(\lambda+1)+\sigma+1,$$
$$\varpi' = p-1-\Sigma(\lambda)+\sigma'+1,$$

and therefore

$$\varpi+\varpi' = 2p-2+k+\sigma+1+\sigma'+1,$$

so that

$$2p-2+k-[\varpi'-(\sigma+1)] = \varpi-(\sigma'+1).$$

Thus $\varpi-(\sigma'+1)$ is the number of periods of a function V which appear to be linearly independent; and, taking account of the existence of the functions $K_1, \ldots, K_{\sigma'+1}$, this is the same as the number of independent linear combinations of the functions $V_1, \ldots, V_\varpi$, which are periodic*. But the conclusions of this article require more careful consideration in particular cases; it is not shewn that the linear equations obtained are always independent, nor that they are the only equations obtainable.

Ex. i. Obtain the lineo-linear relation connecting the periods of the everywhere finite factorial integrals V, V', of the primary and associated system, which is obtained by expressing that the contour integral $\int V dV'$ vanishes.

Ex. ii. In the case of the ordinary Riemann integrals of the first kind, the relation

$$\sum_{i=1}^{p} [\Omega_i(1-\mu_i') - \Omega_i'(1-\mu_i)] = \Gamma_1 + \gamma_1\Gamma_2 + \gamma_1\gamma_2\Gamma_3 + \ldots + \gamma_1\gamma_2 \ldots \gamma_{k-1}\Gamma_k$$

is identically satisfied, and further $k=0$. Thus the reasoning of the text does not hold†.

* We can therefore form linear combinations of the periodic functions V, for which the independent periods shall be $1, 0, \ldots, 0$; $0, 1, \ldots, 0$; etc., as in the ordinary case.

† In that case the numbers $\varpi'-(\sigma+1)$, $2p-2+k$, are to be replaced respectively by p and $2p$. See the note † of § 259.

261. We enquire now how many arbitrary constants enter into the expression of a factorial function of the primary system which has M poles of assigned position.

Supposing one such function to exist, denote it by F_0; then the ratio F/F_0, of any other such function to F, F_0, is a rational function with poles at the zeros of F_0; conversely if R be any rational function with poles at the zeros of F_0, F_0R is a factorial function of the primary system with poles at the assigned poles of F_0. The function R contains

$$N - p + 1 + h + 1$$

arbitrary constants, one of them additive, where N is the number of zeros of F_0, so that $N = M + \sum_{r=1}^{k} \lambda_r$, and $h+1$ is the number of differentials dv vanishing in the zeros of F_0.

But in fact the number of differentials dv vanishing in the zeros of F_0 is the same as the number of differentials dV' vanishing in the poles of F_0, V' being any everywhere finite factorial integral of the associated system.

For if dv vanish in the zeros of F_0, the integral $\int dv/F_0$ is clearly a factorial integral, V', of the associated system without infinities, and such that dV' vanishes in the poles of F_0; conversely if V' be any factorial integral of the associated system such that dV' vanishes in the poles of F_0, the integral $\int F_0 dV'$ is an integral of the first kind, v, such that dv vanishes in the zeros of F_0.

Thus, *the number of arbitrary constants in a factorial function of the primary system, with M given arbitrary poles, is*

$$M + \sum_{r=1}^{k} \lambda_r - p + 1 + h + 1, \; = N - p + 1 + h + 1, \; = M - \varpi' + h + 1 + \sigma' + 1,$$

where N is the number of zeros of the function, and $h+1$ the number of differentials dV' vanishing in the M poles*.

In particular, putting $M = 0$, $h + 1 = \varpi'$ (cf. § 258), we have the formula, already obtained,

$$\sigma' + 1 = \sum_{r=1}^{k} \lambda_r - p + 1 + \varpi'.$$

We can of course also obtain these results by considering the fundamental equations (i) and (ii), § 254.

262. Hence we can determine the smallest value of M for which a factorial function of the primary system with M given poles always exists.

* Counting the additive constant in the expression of a rational function, the last formula holds in the ordinary case.

When $M = \varpi' + 1$ it is not possible to determine a function V', of the form

$$V' = A_1 V_1' + \dots\dots + A_{\varpi'} V'_{\varpi'},$$

wherein $A_1, \dots, A_{\varpi'}$ are constants, to vanish in M arbitrary places; and therefore $h + 1 = 0$. Thus a factorial function of the primary system with $\varpi' + 1$ arbitrary poles will contain, in accordance with the formula of the last Article,

$$\varpi' + 1 + \sum_{r=1}^{k} \lambda_r - p + 1, \ = \sigma' + 2,$$

arbitrary constants.

When $\sigma' + 1 = 0$, this number is 1, and the factorial function is entirely determined save for an arbitrary constant multiplier. Hence we infer that when $\sigma' + 1 = 0$ the smallest value of M is $\varpi' + 1$.

We consider in the next Article how to form the factorial function in question from other functions of the system. Of the existence of such a function we can be sure *à priori* by the formulae (i) (ii) of § 254. Such a function will have $N = \varpi' + 1 + \Sigma\lambda, = p$, zeros. They can be determined to satisfy the equations (ii). Then an expression of the function is given by the general formula of § 254.

When $\sigma' + 1 > 0$, there are $\sigma' + 1$ everywhere finite factorial functions $K_1, \dots, K_{\sigma'+1}$, of the primary system, and the general factorial function with $\varpi' + 1$ poles is of the form

$$F + \lambda_1 K_1 + \dots\dots + \lambda_{\sigma'+1} K_{\sigma'+1},$$

where $\lambda_1, \dots, \lambda_{\sigma'+1}$ are constants, and F is any factorial function with the assigned poles. In this case also there exist no factorial functions with arbitrary poles less than $\varpi' + 1$ in number; the attempt to obtain such functions leads* always to a linear aggregate of $K_1, \dots, K_{\sigma'+1}$.

263. Suppose that $\sigma' + 1 = 0$; we consider the construction of the factorial function of the primary system with $\varpi' + 1$ arbitrary poles.

Firstly let $\sigma + 1 > 0$, so that there are $\sigma + 1$ everywhere finite functions, K', of the associated system, and $\sigma + 1$ differentials dv vanish in the $2p - 2 + \sum_{r=1}^{k} (\lambda_r + 1)$ zeros of any differential dV. Hence $s, = -\sum_{r=1}^{k} (\lambda_r + 1)$, is greater than zero or equal to zero. We take first the case when $s > 0$. Then $\varpi' = p - 1 - \sum_{r=1}^{k} \lambda_r = p - 1 + s + k$, and it is possible to determine a rational function with poles at $\varpi' + 1 = p + s + k$ arbitrary places. This function contains $s + k + 1$ arbitrary constants, one of these being additive. It can therefore be chosen to vanish at the places $c_1, \dots, c_k$, and will then

* For $M = \varpi' - r$, we shall have $h + 1 = r$, and, therefore, $M - \varpi' + h + 1 + \sigma' + 1 = \sigma' + 1$.

contain at least, and in general, $s+1$ arbitrary constants. Taking now any everywhere finite factorial function K' of the associated system, let the rational function be further chosen to vanish in the s zeros of K'; then the rational function is, in general, entirely determined save for an arbitrary constant multiplier. Denote the rational function thus obtained by R. Then R/K' is a factorial function of the primary system with the $\varpi'+1$ assigned poles, and is the function we desired to construct. And since the ratio of two functions K' is a rational function, it is immaterial what function K' is utilised to construct the function required.

This reasoning applies also to the case in which $\sigma+1>0$, $s=0$, unless also $k=0$. Consider then the case in which $\sigma+1>0$, $s=0$ and $k=0$. There is (§ 258) only one function K', of the form

$$E_0 = Ae^{2\pi i[(h_1+H_1)v_1^{x,a}+\dots\dots+(h_p+H_p)v_p^{x,a}]},$$

or $\sigma+1=1$; and E_0^{-1} is a function of the primary system without poles. Thus $\sigma'+1=1$, and the case does not fall under that now being considered, for which $\sigma'+1=0$. The value of ϖ' is p, and the factorial function with $\varpi'+1$ arbitrary poles is of the form $(F+C)E_0$, where $F+C$ is the general rational function with the given poles.

Nextly, let $\sigma+1=0$, as well as $\sigma'+1=0$. Then there exist no functions K' and the previous argument is inapplicable. But, provided $\varpi'+1 \not< 2$, we can apply another method, which could equally have been applied when $\sigma+1>0$. For if P be the factorial function of the primary system with $\varpi'+1$ assigned poles, and V' be one of the ϖ' factorial integrals of the associated system, and v be any integral of the first kind, $P\dfrac{dV'}{dv}$ is a rational function whose poles are at the $\varpi'+1$ poles of P and at the $2p-2$ zeros of dv. Conversely, if R be any rational function with poles at these places (cf. § 37, Ex. ii. Chap. III.), and zeros at the $2p-2-\Sigma\lambda$ zeros of dV', $R\bigg/\dfrac{dV'}{dv}$ is the factorial function required. It contains at least

$$\varpi'+1+2p-2-p+1-(2p-2-\Sigma\lambda), =1,$$

arbitrary constant multiplier.

In case $\varpi'+1<2$, so that $\varpi'=0$, $\Sigma\lambda=p-1$, there are no functions V', and we may fall back upon the fundamental equations of § 254. In this case the least number of poles is 1.

264. Consider now the possibility of forming a factorial integral of the primary system whose only infinities are poles. We shew that it is possible to form such an integral with $\sigma+2$ arbitrary poles, and with no smaller number.

Suppose G to be such a factorial integral, with $\sigma+2$ poles, and, under the hypothesis $\varpi > 0$, let V be an everywhere finite factorial integral, also of the primary system. Then dG/dV is a rational function, with poles at the $2p-2+\Sigma(\lambda+1)$ zeros of dV, and poles at the poles of G; near a pole of G, say c, the form of dG/dV is given by

$$\frac{dG}{dV} = C\left(\frac{1}{t^2} + A + Bt + \dots\right) \div D_c V \cdot \left[1 + t\frac{D_c^2 V}{D_c V} + \dots\right],$$

where t is the infinitesimal for the neighbourhood of the place c, the quantities C, A, B are constants, and $D_c V$ denotes a differentiation in regard to the infinitesimal; this is the same as

$$\frac{dG}{dV} = E\left[-\frac{1}{t^2} + \frac{1}{t}\frac{D_c^2 V}{D_c V} + \text{terms which are finite when } t=0\right],$$

where $E = -C/D_c V$. Thus dG/dV is infinite at a pole of G like a constant multiple of

$$\psi = D_c \Gamma_c^{x,a} - \frac{D_c^2 V}{D_c V}\Gamma_c^{x,a},$$

a being an arbitrary place.

Conversely if R denote a rational function which is infinite to the first order at the zeros of dV, and infinite in the $\sigma+2$ assigned poles of G like functions of the form of ψ, $\int R\,dV$ will be such a factorial integral as desired.

Now R is of the form (§ 20, Chap. II.)

$$A + A_1\Gamma_{e_1}^{x,a} + \dots\dots + A_r\Gamma_{e_r}^{x,a} + B_1\left[D_{x_1}\Gamma_{x_1}^{x,a} - \frac{D_{x_1}^2 V}{D_{x_1} V}\Gamma_{x_1}^{x,a}\right] + \dots\dots$$
$$+ B_{\sigma+2}\left[D_{x_{\sigma+2}}\Gamma_{x_{\sigma+2}}^{x,a} - \frac{D_{x_{\sigma+2}}^2 V}{D_{x_{\sigma+2}} V}\Gamma_{x_{\sigma+2}}^{x,a}\right],$$

wherein a is an arbitrary place, $e_1, \dots, e_r$ denote the zeros of dV, $x_1, \dots, x_{\sigma+2}$ denote the assigned poles of G, and $A, A_1, \dots, A_r, B_1, \dots, B_{\sigma+2}$ are constants; the period of R, in this form, at a general period loop of the second kind, is given by

$$A_1\Omega_i(e_1) + \dots\dots + A_r\Omega_i(e_r) + B_1\left[D_{x_1}\Omega_i(x_1) - \frac{D_{x_1}^2 V}{D_{x_1} V}\Omega_i(x_1)\right] + \dots\dots$$
$$+ B_{\sigma+2}\left[D_{x_{\sigma+2}}\Omega_i(x_{\sigma+2}) - \frac{D_{x_{\sigma+2}}^2 V}{D_{x_{\sigma+2}} V}\Omega_i(x_{\sigma+2})\right],$$

where $\Omega_1(x), \dots, \Omega_p(x)$ are as in § 18, Chap. II., and this must vanish for $i = 1, 2, \dots, p$. Now (§ 258) in the places $e_1, \dots, e_r$ there vanish $\sigma+1$ linear functions of $\Omega_1(x), \dots, \Omega_p(x)$. Thus, from the conditions expressing that the periods of R are zero, we infer $\sigma+1$ linear equations involving only the constants $B_1, \dots, B_{\sigma+2}$, which, since the places $x_1, \dots, x_{\sigma+2}$ are arbitrary, may be assumed to be independent. From these $\sigma+1$ equations we can obtain the ratios $B_1 : B_2 : \ldots\ldots : B_{\sigma+2}$. There remain then, of the p equations expressing that the periods of R are zero, $p-(\sigma+1)$ independent equations containing effectively $r+1$ unknown constants. Thus the number of the constants $A_1, \ldots, A_r, B_1, \ldots, B_{\sigma+2}$ left arbitrary is $r+1-p+\sigma+1$, which is equal to $2p-2+\Sigma(\lambda+1)+1-p+\sigma+1$ or ϖ, and the total number of arbitrary constants in R is $\varpi+1$. Thus we infer that, on the whole, G is of the form*

$$[G]+C_1V_1+\ldots\ldots+C_{\varpi}V_{\varpi}+C,$$

where $[G]$ is a special function with the $\sigma+2$ assigned poles, multiplied by an arbitrary constant, and $C_1, \ldots, C_{\varpi}, C$ are arbitrary constants. And this result shews that $\sigma+2$ is the least number of poles that can be assigned for G. The argument applies to the case when $\sigma+1=0$ provided that $\varpi>0$.

The proof just given supposes $\varpi>0$; but this is not indispensable. Let f_0 be a factorial function with the primary system of multipliers but with a behaviour at the places c_i like $t^{-(\lambda_i+1)}\phi_i$, where ϕ_i is uniform, finite and not zero in the neighbourhood of c_i. Then if, instead of $\int RdV$, we consider an integral $\int Rf_0 dv$, wherein dv is the differential of any Riemann integral of the first kind, and R is a rational function which vanishes in the (say M) poles of f_0, and may become infinite in the zeros of dv and the (say N) zeros of f_0, we shall obtain the same results. It is necessary to take $N>1$ (cf. § 37, Ex. ii. Chap. III.).

265. Another method, holding whether $\varpi=0$ or not, provided $\sigma+1>0$, may be indicated. Let $K'(x)$ be one of the everywhere finite factorial functions of the associated system. Consider the function of x,

$$\psi=\int\frac{1}{K'(x)}\,d\,[\Gamma_c^{x,a}+A\Pi_{c,\gamma}^{x,a}],$$

a, c, γ being any places and A a constant; when x is in the neighbourhood of the place c it is of the form

$$\int\frac{1}{K'(c)}\left[1-t\frac{DK'(c)}{K'(c)}\right]\left[\frac{1}{t^2}+\frac{A}{t}\right]dt,$$

where t is the infinitesimal in the neighbourhood of the place c, and terms which will lead only to positive powers of t under the integral sign are omitted; this is the same as

$$\frac{1}{K'(c)}\int\left\{\frac{1}{t^2}+\left[A-\frac{DK'(c)}{K'(c)}\right]\frac{1}{t}\right\}dt;$$

* In the ordinary case of rational functions, where V is replaced by a Riemann normal integral v, the coefficients of $B_1, \ldots, B_{\sigma+2}$, in the expression for the general period of R, vanish for one value of i, namely when $V=v_i$. Thus $\sigma+1\ (=1)$ pole is sufficient to enable us to construct the factorial integral; it is the ordinary integral of the second kind.

hence if A be $DK'(c)/K'(c)$, the function ψ is infinite at c like $-\frac{1}{t}\frac{1}{K'(c)}$. At the place γ the function ψ is infinite like $-\frac{A}{K'(\gamma)}\log t_\gamma$, where t_γ is the infinitesimal in the neighbourhood of the place γ.

Putting now $M_{c,\gamma}^{x,a} = \Gamma_c^{x,a} + \frac{DK'(c)}{K'(c)}\Pi_{c,\gamma}^{x,a}$, consider the function

$$G(x) = \int \frac{1}{K'(x)} d\left\{A_1 M_{x_1,\gamma}^{x,a} + \ldots + A_{\sigma+2} M_{x_{\sigma+2},\gamma}^{x,a} + B_1 v_1^{x,a} + \ldots + B_p v_p^{x,a}\right\},$$

where a, γ are arbitrary places and $A_1, \ldots, A_{\sigma+2}$, $B_1, \ldots, B_p$ are constants, subject to the conditions

(i) that

$$A_1 D_x M_{x_1,\gamma}^{x,a} + \ldots\ldots + A_{\sigma+2} D_x M_{x_{\sigma+2},\gamma}^{x,a} + B_1 \Omega_1(x) + \ldots\ldots + B_p \Omega_p(x)$$

vanishes at each of the $-\Sigma(\lambda+1)$ zeros of $K'(x)$,

(ii) that

$$A_1 \frac{DK'(x_1)}{K'(x_1)} + \ldots\ldots + A_{\sigma+2}\frac{DK'(x_{\sigma+2})}{K'(x_{\sigma+2})} = 0;$$

the first condition ensures that $G(x)$ is finite at the zeros of $K'(x)$, the second condition ensures that $G(x)$ is finite at the place γ. If we suppose* $v_1^{x,a}, \ldots, v_{\varpi}^{x,a}$ to be those integrals of the first kind whose differentials vanish at the zeros of $K'(x)$ (§ 258), the conditions (i) will involve only the constants $A_1, \ldots, A_{\sigma+2}$, $B_{\varpi+1}, \ldots, B_p$, and if these conditions be independent these $\sigma + 2 + (p - \varpi)$ coefficients will thereby be reduced to

$$\sigma + 2 + p - \varpi + \Sigma(\lambda+1), = 2;$$

thus, if the condition (ii) be independent of the conditions (i), the number of constants finally remaining is $\varpi + 2 - 1 = \varpi + 1$, and the form of $G(x)$ is

$$[G] + C_1 V_1 + \ldots\ldots + C_{\varpi} V_{\varpi} + C$$

as before.

Ex. Prove that, when s, $= -\Sigma(\lambda+1)$, is positive, we have

$$K'(x)\left[D_{x_1} M_{x,\gamma}^{x_1,a} + \ldots + D_{x_s} M_{x,\gamma}^{x_s,a}\right] = D_x\left\{K'(x)\left[\Gamma_{x_1}^{x,\gamma} + \ldots + \Gamma_{x_s}^{x,\gamma}\right]\right\}.$$

266. The factorial integral of the primary system with $\sigma + 2$ arbitrary poles can be simplified. If $x_1, \ldots, x_{\sigma+2}$ be the poles, its most general form may be represented by

$$EG(x_1, \ldots, x_{\sigma+2}) + E_1 V_1 + \ldots\ldots + E_{\varpi} V_{\varpi} + C,$$

* This is to simplify the explanation. In general it is ϖ linear combinations of the normal integrals, whose differentials vanish in the zeros of $K'(x)$. The reduction corresponding to that of the text is then obtained by taking ϖ linear combinations of the conditions (i).

where $E, E_1, \ldots, E_\varpi, C$ are arbitrary constants. Near a place c_1, one of the singular places of the factorial system, the integral will have a form represented by $A_1 + t^{-\lambda_1}\phi$; we may simplify the integral by subtracting from it the constant A_1; the consequence is that the additive period belonging to the loop (γ_1) is zero; further there is one other linear relation connecting the additive periods of the integral, which is obtainable by following the value of the integral once round the boundary of the dissected surface (cf. § 260). Thus the number of periods of the integral is at most $2p-2+k$. We suppose the additive periods of the functions $G(x_1, \ldots, x_{\sigma+2})$, $V_1, \ldots, V_\varpi$, at the loop (γ_1), to be similarly reduced to zero; then the constant C is zero. The linear aggregate $E_1V_1 + \ldots\ldots + E_\varpi V_\varpi$ may be replaced by an aggregate of the non-periodic functions $K_1, \ldots, K_{\sigma'+1}$, and $\varpi - (\sigma'+1)$ of the integrals $V_1, \ldots, V_\varpi$, so that the integral under consideration takes the form

$$EG(x_1, \ldots, x_{\sigma+2}) + C_1V_1 + \ldots + C_{\varpi-(\sigma'+1)}V_{\varpi-(\sigma'+1)} + F_1K_1 + \ldots + F_{\sigma'+1}K_{\sigma'+1},$$

where $C_1, \ldots, C_{\varpi-(\sigma'+1)}, F_1, \ldots, F_{\sigma'+1}$ are constants. We can therefore, presumably, determine the constants $C_1, \ldots, C_{\varpi-(\sigma'+1)}$, so that $\varpi-(\sigma'+1)$ of the additive periods of the integral vanish. The integral will then have $2p-2+k-(\varpi-\sigma'-1), = \varpi'-(\sigma+1)$, periods remaining, together with one period which is a linear function of them. A particular case* is that of Riemann's normal integral of the second kind, for which there are p periods. *As in that case we suppose here that the period loops for which the additive periods of the factorial integral shall be reduced to zero are agreed upon beforehand.* We thus obtain a function

$$F \,.\, G_1(x_1, \ldots, x_{\sigma+2}) + F_1K_1 + \ldots\ldots + F_{\sigma'+1}K_{\sigma'+1},$$

wherein $F, F_1, \ldots, F_{\sigma'+1}$ are arbitrary constants, and $G_1(x_1, \ldots, x_{\sigma+2})$ has additive periods only at $\varpi'-(\sigma+1)$ prescribed period loops, beside a period which is a linear function of these. We may therefore further assign $\sigma'+1$ zeros of the integral and choose F so that the integral is infinite at x_1 like the negative inverse of the infinitesimal. When the integral is so determined we shall denote it by $\Gamma(x_1, x_2, \ldots, x_{\sigma+2})$. The assigned zeros are to be taken once for all, say at $a_1, \ldots, a_{\sigma'+1}$.

267. The factorial function of the primary system with $\varpi'+1$ assigned arbitrary poles can be expressed in terms of the factorial integral of the primary system with $\sigma+2$ assigned poles. Let $x_1, \ldots, x_{\varpi'+1}$ be the assigned poles of the factorial function. Then we may choose the constants $C_1, \ldots, C_{\varpi'-\sigma}$, so that the $\varpi'-(\sigma+1)$ linearly independent periods of the aggregate

$$C_1\Gamma(x_{\sigma+2}, x_1, \ldots, x_{\sigma+1}) + \ldots\ldots + C_{\varpi'-\sigma}\Gamma(x_{\varpi'+1}, x_1, \ldots, x_{\sigma+1})$$

are all zeros. The result is a factorial function with $x_1, \ldots, x_{\varpi'+1}$ as poles,

* Of the result. The reasoning must be amended by the substitution of p, $2p$ for $\varpi'-(\sigma+1)$ and $2p-2+k$ respectively. Cf. the note † of § 260.

which vanishes in the places $a_1, \ldots, a_{\sigma'+1}$. Or, taking arbitrary places $d_1, \ldots, d_{\sigma+1}$ we may choose the constants $E_1, \ldots, E_{\varpi'+1}$ so that the $\varpi' - (\sigma+1)$ linearly independent periods of the aggregate

$$E_1\Gamma(x_1, d_1, \ldots, d_{\sigma+1}) + E_2\Gamma(x_2, d_1, \ldots, d_{\sigma+1}) + \ldots + E_{\varpi'+1}\Gamma(x_{\varpi'+1}, d_1, \ldots, d_{\sigma+1})$$

are all zero, and at the same time the aggregate does not become infinite at $d_1, \ldots, d_{\sigma+1}$. Then the addition, to the result, of an aggregate $F_1K_1 + \ldots\ldots + F_{\sigma'+1}K_{\sigma'+1}$, wherein $F_1, \ldots, F_{\sigma'+1}$ are arbitrary constants, leads to the most general form of the factorial function with $x_1, \ldots, x_{\varpi'+1}$ as poles. For the sake of definiteness we denote by $\psi(x;\ z, t_1, \ldots, t_{\varpi'})$ the factorial function with poles of the first order at $z, t_1, \ldots, t_{\varpi'}$, which is chosen so that it becomes infinite at z like the negative inverse of the infinitesimal, and vanishes at the places $a_1, \ldots, a_{\sigma'+1}$. A more precise notation would be* $\psi(x, a_1, \ldots, a_{\sigma'+1};\ z, t_1, \ldots, t_{\varpi'})$. This function contains no arbitrary constants.

Denoting this function now, temporarily, by ψ, and any everywhere finite factorial integral of the inverse system by V', the value of the integral $\int \psi dV'$, taken round the boundary of the dissected surface formed by the sides of the period loops, is equal to the sum of its values round the poles of ψ. Since $\psi dV'/dx$ is a rational function the value of the integral taken round the boundary is zero. Near a pole of ψ, at which t is the infinitesimal, the integral will have the form

$$\int \left[\frac{A}{t} + B + Ct + \ldots\ldots\right] [(DV') + t(D^2V') + \ldots\ldots]\, dt,$$

where D denotes a differentiation. Thus the value obtained by taking the integral round this pole is $A(DV')$. If then the values of A at the poles $\Gamma_1, \ldots, \Gamma_{\varpi'}$ be denoted by $A_1, \ldots, A_{\varpi'}$, we have, remembering that the value of A at z is -1, the ϖ' equations

$$A_1(DV_1')_1 \ + \ldots\ldots + A_{\varpi'}(DV_1')_{\varpi'} \ = (DV_1')_z\ ,$$

$$\ldots\ldots\ldots\ldots\ldots\ldots\ldots\ldots\ldots\ldots\ldots\ldots\ldots\ldots\ldots\ldots\ldots$$

$$A_1(DV'_{\varpi'})_1 + \ldots\ldots + A_{\varpi'}(DV'_{\varpi'})_{\varpi'} = (DV'_{\varpi'})_z,$$

where $V_1', \ldots, V'_{\varpi'}$ are the ϖ' everywhere finite factorial integrals of the associated system, $(DV_i')_r$ denotes the differential coefficient of V_i' at t_r, and $(DV_i')_z$ denotes the differential coefficient at z. Thus, if $\omega_r(x)$ denote, here, the linear aggregate of the form

$$E_1(DV_1')_x + \ldots\ldots + E_{\varpi'}(DV'_{\varpi'})_x,$$

wherein the constants $E_1, \ldots, E_{\varpi'}$ are chosen so that $\omega_r(t_r) = 1$ and $\omega_r(t_s) = 0$ when t_s is any one of the places $t_1, \ldots, t_{\varpi'}$ other than t_r, we have $A_r = \omega_r(z)$. Hence we infer by the previous article (§ 266) that $\psi(x;\ z, t_1, \ldots, t_{\varpi'})$ is equal to

$$\Gamma(z, d_1, \ldots, d_{\sigma+1}) - \omega_1(z)\,\Gamma(t_1, d_1, \ldots, d_{\sigma+1}) - \ldots\ldots - \omega_{\varpi'}(z)\,\Gamma(t_{\varpi'}, d_1, \ldots, d_{\sigma+1}),$$

* Cf. § 122, Chap. VII. etc.

where $d_1, \ldots, d_{\sigma+1}$ are arbitrary places. For these two functions are infinite at the places $z, t_1, \ldots, t_{\varpi'}$ in the same way and both vanish at the places $a_1, \ldots, a_{\sigma'+1}$.

As in the case of the rational functions, the function $\psi(x; z, t_1, \ldots, t_{\varpi'})$ may be regarded as fundamental, and developments analogous to those given on pages 181, 189 of the present volume may be investigated. We limit ourselves to the expression of any factorial function of the primary system by means of it. The most general factorial function with poles of the first order at the places $z_1, \ldots, z_M$ may be expressed in the form

$$A_1\psi(x; z_1, t_1, \ldots, t_{\varpi'}) + \ldots\ldots + A_M\psi(x; z_M, t_1, \ldots, t_{\varpi'}) + B_1K_1 + \ldots\ldots + B_{\sigma'+1}K_{\sigma'+1},$$

where $A_1, \ldots, A_M, B_1, \ldots, B_{\sigma'+1}$ are constants. The condition that the function represented by this expression may not be infinite at t_r is

$$A_1\omega_r(z_1) + \ldots\ldots + A_M\omega_r(z_M) = 0;$$

in case the ϖ' equations of this form, for $r = 1, 2, \ldots, \varpi'$, be linearly independent, the factorial function contains $M + \sigma' + 1 - \varpi'$ arbitrary constants; but if there be $h+1$ linearly independent aggregates of differentials, of the form

$$C_1dV_1' + \ldots\ldots + C_{\varpi'}dV'_{\varpi'},$$

which vanish in the M assigned poles, then the equations of the form

$$A_1\omega_r(z_1) + \ldots\ldots + A_M\omega_r(z_M) = 0$$

are equivalent to only $\varpi' - (h+1)$ equations, and the number of arbitrary constants in the expression of the factorial function is $M + \sigma' + 1 - \varpi' + h + 1$, in accordance with § 261.

Ex. i. Prove that a factorial integral of the primary system can be constructed with logarithmic infinities only in $\sigma+2$ places, but with no smaller number.

Ex. ii. If the factorial integral $G(x_1, x_2, \ldots, x_{\sigma+2})$ become infinite of the place x_i like $\frac{R_i}{t}$, where t is the infinitesimal at x_i, prove, by considering the contour integral $\int GdK_r'$, where K_r' is one of the $\sigma+1$ everywhere finite factorial functions of the associated system, and G denotes $G(x_1, x_2, \ldots, x_{\sigma+2})$, the $\sigma+1$ equations

$$\sum_{i=1}^{\sigma+2} R_iDK_r'(x_i) = 0,$$

D denoting a differentiation. From these equations the ratio of the residues $R_1, R_2, \ldots, R_{\sigma+2}$ can be expressed.

268. The theory of this chapter covers so many cases that any detailed exhibition of examples of its application would occupy a great space. We limit ourselves to examining the case $p=0$, for which explicit expressions can be given, and, very briefly, two other cases (§§ 268—270).

Consider the case $p=0$, $k=3$, there being three singular places such as have so far in this chapter been denoted by $c_1, c_2, \dots$, but which we shall here denote by α, β, γ, the associated numbers* being $\lambda_1=-3/2$, $\lambda_2=-3/2$, $\lambda_2=-2$. At these places the factorial functions of the associated system behave, respectively, like $t^{-\frac{1}{2}}\phi_1$, $t^{-\frac{1}{2}}\phi_2$, $t^{-1}\phi_3$, and the difference between the number of zeros and poles of such a function is $N'-M'=-\Sigma(\lambda+1)=2$. Thus there exist factorial functions of the associated system with no poles and two zeros. By the general formula of § 254, replacing $\Pi_{c,\gamma}^{x,a}$ by $\log\left(\frac{x-c}{x-\gamma}\Big/\frac{a-c}{a-\gamma}\right)$, the general form of such a function is found to be

$$K'(x)=\frac{Ax^2+Bx+C}{(x-\gamma)(x-\alpha)^{\frac{1}{2}}(x-\beta)^{\frac{1}{2}}}$$

and involves three arbitrary constants, so that $\sigma+1=3$. In what follows $K'(x)$ will be used to denote the special function $1/(x-\gamma)(x-\alpha)^{\frac{1}{2}}(x-\beta)^{\frac{1}{2}}$. The difference between the number of zeros and poles of factorial functions of the primary system is $N-M=-5$; hence $M=0$ is not possible, and $\sigma'+1=0$. Further

$$\varpi, =p-1+\Sigma(\lambda+1)+\sigma+1, =-1-2+3=0,$$
$$\varpi', =p-1-\Sigma\lambda+\sigma'+1 \quad , =-1+5 \quad =4,$$

and the factorial function of the primary system with fewest poles has $\varpi'+1=5$ poles, as also follows from the formula $N-M=-5$. This function is clearly given by

$$P(x)=\frac{(x-\alpha)^{\frac{3}{2}}(x-\beta)^{\frac{3}{2}}(x-\gamma)^2}{(x-x_1)(x-x_2)(x-x_3)(x-x_4)(x-x_5)}.$$

Putting

$$\psi(x)=(x-\alpha)(x-\beta)(x-\gamma),\ f(x)=(x-x_1)(x-x_2)(x-x_3)(x-x_4)(x-x_5),$$
$$\phi(x)=DK'(x)/K'(x)=-\left[(x-\gamma)^{-1}+\tfrac{1}{2}(x-\alpha)^{-1}+\tfrac{1}{2}(x-\beta)^{-1}\right],$$

and putting $\lambda_i=\psi(x_i)/f'(x_i)$, where i is in turn equal to 1, 2, 3, 4, 5 and $f'(x)$ denotes the differential coefficient of $f(x)$, it is immediately clear that $P(x)$ is infinite at x_1 like $\lambda_1/(x-x_1)K'(x_1)$ It can be verified that

$$\sum_1^5\lambda_1=0,\ \sum_1^5 x_1\lambda_1=1,\ \sum_1^5 x_1\lambda_1\phi(x_1)=0,\ \sum_1^5 x_1^2\lambda_1\phi(x_1)=-2,\ \sum_1^5\lambda_1\phi(x_1)=0,$$

and these give

$$\sum_1^5\lambda_1[1+x_1\phi(x_1)]=0,\ \sum_1^5\lambda_1[2x_1+x_1^2\phi(x_1)]=0.$$

The factorial integral G, of the primary system, with $\sigma+2=4$ poles, τ, ξ, η, ζ is (§ 265) given by

$$G(\tau,\xi,\eta,\zeta)=\int\frac{1}{K'(x)}\,d\left\{\sum_1^4 A_1\left[\frac{1}{x-\tau}-\phi(\tau)\log\frac{x-\tau}{x-c}\right]\right\},$$

* It was for convenience of exposition that, in the general theory, the case in which any of the numbers $\lambda_1, \dots, \lambda_k$ are integers, was excluded.

where the sign of summation refers to τ, ξ, η, ζ and the constants A_1, A_2, A_3, A_4 are to be chosen so that (i) the expression

$$A_1\phi(\tau)+A_2\phi(\xi)+A_3\phi(\eta)+A_4\phi(\zeta)$$

is zero, this being necessary in order that $G(\tau, \xi, \eta, \zeta)$ may not become infinite at the place c, and (ii) the expression

$$\sum_1^4 A_1\left[\frac{1}{(x-\tau)^2}+\phi(\tau)\left(\frac{1}{x-\tau}-\frac{1}{x-c}\right)\right]$$

vanishes to the fourth order when x is infinite; the expression always vanishes to the second order when x is infinite; the additional conditions are required because $K'(x)$ is zero to the second order when x is infinite. Taking account of condition (i), we find, by expanding in powers of $\frac{1}{x}$, that the condition (ii) is equivalent to the two

$$\sum_1^4 A_1[1+\tau\phi(\tau)]=0, \quad \sum_1^4 A_1[2\tau+\tau^2\phi(\tau)]=0.$$

Thus, introducing the values of $A_1, \ldots, A_4$ into the expression for $G(\tau, \xi, \eta, \zeta)$, we find, by proper choice of a multiplicative constant,

$$K'(x)\,DG(\tau, \xi, \eta, \zeta)=\begin{vmatrix} \dfrac{1}{(x-\tau)^2}+\dfrac{\phi(\tau)}{x-\tau}, & (\xi), & (\eta), & (\zeta) \\ \phi(\tau), & ., & ., & . \\ 1+\tau\phi(\tau), & ., & ., & . \\ 2\tau+\tau^2\phi(\tau), & ., & ., & . \end{vmatrix} \quad \ldots\ldots(1),$$

in which the second, third and fourth columns differ from the first only in the substitution, respectively, of ξ, η, ζ in place of τ.

The factorial integral $G(\tau, \xi, \eta, \zeta)$ thus determined can in fact be expressed without an integral sign. For we immediately verify that

$$\int dx\,(x-\gamma)\sqrt{(x-\alpha)(x-\beta)}\left[\frac{1}{(x-\tau)^2}+\frac{\phi(\tau)}{x-\tau}\right]$$

is equal, save for an additive constant, to

$$\sqrt{(x-\alpha)(x-\beta)}\left[\frac{\gamma-\tau}{x-\tau}+1+\tau\phi(\tau)+\tfrac{1}{2}\{x-\gamma-\tfrac{1}{2}(\alpha+\beta)\}\phi(\tau)\right]$$

$$+\left[2\tau+\tau^2\phi(\tau)-\{\gamma+\tfrac{1}{2}(\alpha+\beta)\}(1+\tau\phi(\tau))+\left\{\tfrac{1}{2}\gamma(\alpha+\beta)-\tfrac{1}{2}\left(\frac{\alpha-\beta}{2}\right)^2\right\}\phi(\tau)\right]$$

$$\times\log\left\{x-\frac{\alpha+\beta}{2}+\sqrt{(x-\alpha)(x-\beta)}\right\}$$

$$-\frac{2}{\sqrt{(\tau-\alpha)(\tau-\beta)}}\left[(\tau-\gamma)\phi(\tau)+1+\tfrac{1}{2}(\tau-\gamma)\left(\frac{1}{\tau-\alpha}+\frac{1}{\tau-\beta}\right)\right]$$

$$\times\log\frac{\sqrt{(x-\beta)(\tau-\alpha)}+\sqrt{(x-\alpha)(\tau-\beta)}}{\sqrt{x-\tau}},$$

and, by the definition of $\phi(x)$, the coefficient of the logarithm in the last line of this expression is zero; if we substitute these values in the expression found for $G(\tau, \xi, \eta, \zeta)$ we obviously have

$$G(\tau, \xi, \eta, \zeta) = \sqrt{(x-\alpha)(x-\beta)} \begin{vmatrix} \dfrac{\gamma-\tau}{x-\tau}, & (\xi), & (\eta), & (\zeta) \\ \phi(\tau), & ., & ., & . \\ 1+\tau\phi(\tau), & ., & ., & . \\ 2\tau+\tau^2\phi(\tau), & ., & .. & . \end{vmatrix} + \text{constant}, \ldots (2),$$

where the second, third and fourth columns of the determinant differ from the first only in the substitution, in place of τ, respectively of ξ, η, ζ. We proceed now to prove that this determinant is a certain constant multiple of $(x-\alpha)(x-\beta)(x-\mu)/(x-\tau)(x-\xi)(x-\eta)(x-\zeta)$, where μ is determined by

$$\frac{1}{\gamma-\mu} = \frac{1}{\gamma-\tau} + \frac{1}{\gamma-\xi} + \frac{1}{\gamma-\eta} + \frac{1}{\gamma-\zeta} - \frac{3}{2}\left(\frac{1}{\gamma-\alpha} + \frac{1}{\gamma-\beta}\right).$$

If we introduce constants, A, B, C, A', B', C', depending only on α, β, γ, defined by the identities

$$Cx^2 + Bx + A = \frac{2}{\alpha-\beta}(x-\beta)(x-\gamma),$$

$$C'x^2 + B'x + A' = \frac{4}{(\alpha-\beta)^2}(x-\gamma)\left(x - \frac{\alpha+\beta}{2}\right),$$

we can immediately verify that

$$A\phi(x) + B[1 + x\phi(x)] + C[2x + x^2\phi(x)] = -\frac{x-\gamma}{x-\alpha},$$

$$A'\phi(x) + B'[1 + x\phi(x)] + C'[2x + x^2\phi(x)] = -\frac{x-\gamma}{(x-\alpha)(x-\beta)},$$

and hence that

$$\begin{aligned} \frac{\gamma-\tau}{x-\tau} + [A + (x-\alpha)A']\phi(\tau) + [B + (x-\alpha)B'][1+\tau\phi(\tau)] \\ + [C + (x-\alpha)C'][2\tau + \tau^2\phi(\tau)] \\ = (x-\alpha)(x-\beta)\frac{\gamma-\tau}{(\alpha-\tau)(\beta-\tau)}\frac{1}{x-\tau}; \end{aligned}$$

thus

$$G(\tau, \xi, \eta, \zeta) = (x-\alpha)^{\frac{3}{2}}(x-\beta)^{\frac{3}{2}} \begin{vmatrix} \dfrac{\gamma-\tau}{(\alpha-\tau)(\beta-\tau)}\dfrac{1}{x-\tau}, & (\xi), & (\eta), & (\zeta) \\ \phi(\tau), & ., & ., & . \\ 1+\tau\phi(\tau), & ., & ., & . \\ 2\tau+\tau^2\phi(\tau), & ., & ., & . \end{vmatrix} + \text{constant}, \quad \ldots\ldots (3)$$

now it is clear from the equation (2) that $G(\tau, \xi, \eta, \zeta)/\sqrt{(x-\alpha)(x-\beta)}$ is of the form $(x, 1)_3/(x-\tau)(x-\xi)(x-\eta)(x-\zeta)$, where $(x, 1)_3$ denotes an integral cubic polynomial; and since $1/K'(x)$ vanishes when $x=\gamma$, it follows from the equation (1) that the differential coefficient of $G(\tau, \xi, \eta, \zeta)$ vanishes when $x=\gamma$. Hence we have

$$G(\tau, \xi, \eta, \zeta) = L\frac{(x-\alpha)^{\frac{3}{2}}(x-\beta)^{\frac{3}{2}}(x-\mu)}{(x-\tau)(x-\xi)(x-\eta)(x-\zeta)} + M,$$

where μ is such that the differential coefficient of this expression vanishes when $x=\gamma$, and has therefore the value already specified, L is a constant whose value can be obtained from the equation (3) by calculation, and M is a constant which we have not assigned. In the neighbourhood of the place α, $G(\tau, \xi, \eta, \zeta)$ has the form $M + L(x-\alpha)^{\frac{3}{2}}[\lambda + \mu(x-\alpha) + \nu(x-\alpha)^2 + \ldots]$, and similarly in the neighbourhood of the place β. In the neighbourhood of the place γ, $G(\tau, \xi, \eta, \zeta)$ has the form

$$N + (x-\gamma)^2[\lambda' + \mu'(x-\gamma) + \nu'(x-\gamma)^2 + \ldots\ldots],$$

where N is a constant, generally different from M.

In the general case of a factorial integral for $p=0$, $k=3$, the behaviour of the integral at a, β, γ is that of three expressions of the form

$$A+(x-a)^{-\lambda}[P+Q(x-a)+\ldots], \qquad B+(x-\beta)^{-\mu}[P'+Q'(x-\beta)+\ldots],$$
$$C+(x-\gamma)^{-\nu}[P''+Q''(x-\gamma)+\ldots],$$

provided no one of $\lambda+1$, $\mu+1$, $\nu+1$ be a positive integer; herein one of the constants A, B, C may be taken arbitrarily and the others are thereby determined. The factorial integral becomes a factorial function only in the case when all of A, B, C are zero.

We have seen that the factorial function of the primary system with fewest poles has 5 poles; let them be at $\tau, \tau_1, \xi, \eta, \zeta$; then, taking $G(\tau, \xi, \eta, \zeta)$ in the form just found, the factorial function can be expressed in the form

$$P(x) = CG(\tau, \xi, \eta, \zeta) + C_1 G(\tau_1, \xi, \eta, \zeta) + D,$$

when the constants C, C_1, D are suitably chosen.

For clearly D can be chosen so that the function $P(x)$ divides identically by $(x-\alpha)^{\frac{3}{2}}(x-\beta)^{\frac{3}{2}}$. It is then only necessary to choose the ratio $C : C_1$, if possible, so that the function $P(x)$ divides identically by $(x-\gamma)^2$. This requires only that

$$C\frac{x-\mu}{x-\tau} + C_1\frac{x-\mu_1}{x-\tau_1} = \rho\frac{(x-\gamma)^2}{(x-\tau)(x-\tau_1)},$$

where ρ is a constant, or that the expression

$$C(x-\mu)(x-\tau_1) + C_1(x-\tau)(x-\mu_1)$$

divide by $(x-\gamma)^2$. Thus $C : C_1 = -(\gamma-\tau)(\gamma-\mu_1) : (\gamma-\mu)(\gamma-\tau_1)$, and

$$\frac{2\gamma-\mu-\tau_1}{(\gamma-\mu)(\gamma-\tau_1)} = \frac{2\gamma-\mu_1-\tau}{(\gamma-\mu_1)(\gamma-\tau)},$$

or

$$\frac{1}{\gamma-\mu} - \frac{1}{\gamma-\tau} = \frac{1}{\gamma-\mu_1} - \frac{1}{\gamma-\tau_1};$$

this condition is satisfied; both these expressions are by definition equal to

$$\frac{1}{\gamma-\xi} + \frac{1}{\gamma-\eta} + \frac{1}{\gamma-\zeta} - \tfrac{3}{2}\left(\frac{1}{\gamma-\alpha} + \frac{1}{\gamma-\beta}\right).$$

From the theoretical point of view it is however better to proceed as follows—Let the poles of $P(x)$ be at $x_1, \ldots, x_5$. Then $P(x)$ can be expressed in the form

$$P(x) = C_1 G(x_1, \xi, \eta, \zeta) + C_2 G(x_2, \xi, \eta, \zeta) + \ldots\ldots + C_5 G(x_5, \xi, \eta, \zeta) + C,$$

the constants $C, C_1, C_2, \ldots, C_5$ being suitably chosen. This equation requires, by equation (1),

$$K'(x)\, DP = \sum_1^5 C_r \left[\frac{1}{(x-x_r)^2} + \frac{\phi(x_r)}{(x-x_r)}\right] \Delta(\xi, \eta, \zeta)$$

$$+ \begin{vmatrix} 0, & E, & F, & G \\ \sum_1^5 C_r \phi(x_r), & \phi(\xi), & \phi(\eta), & \phi(\zeta) \\ \sum_1^5 C_r [1 + x_r\phi(x_r)], & 1+\xi\phi(\xi), & 1+\eta\phi(\eta), & 1+\zeta\phi(\zeta) \\ \sum_1^5 C_r [2x_r + x_r^2\phi(x_r)], & 2\xi+\xi^2\phi(\xi), & 2\eta+\eta^2\phi(\eta), & 2\zeta+\zeta^2\phi(\zeta) \end{vmatrix}$$

wherein $\Delta(\xi, \eta, \zeta)$ is the minor, in the determinant occurring in equation (1), of the first element of the first row, and $E = (x-\xi)^{-2} + \phi(\xi)(x-\xi)^{-1}$, $F = (x-\eta)^{-2} + \phi(\eta)(x-\eta)^{-1}$, $G = (x-\zeta)^{-2} + \phi(\zeta)(x-\zeta)^{-1}$. If now we take $C_1, \ldots, C_5$ so that

$$\sum_1^5 C_r \phi(x_r) = 0, \quad \sum_1^5 C_r [1 + x_r\phi(x_r)] = 0, \quad \sum_1^5 C_r [2x_r + x_r^2 \phi(x_r)] = 0,$$

this leads to

$$\frac{\Delta(x_1, x_2, x_3)}{\Delta(\xi, \eta, \zeta)} DP = C_4 DG(x_1, x_2, x_3, x_4) + C_5 DG(x_1, x_2, x_3, x_5),$$

and the solution can be completed as before.

There are $\varpi' = 4$ everywhere finite factorial integrals of the associated system; if V' be one of these, then by definition, $\dfrac{dV'}{dx}$ is a factorial function which has at α the form $(x-\alpha)^{-\frac{3}{2}}\phi$, and similarly at β, and has at γ the form $(x-\gamma)^{-2}\phi$. Further dV'/dx is zero to the second order at $x=\infty$. Hence we have

$$V' = \int \frac{(x, 1)_3\, dx}{(x-\alpha)^{\frac{3}{2}}(x-\beta)^{\frac{3}{2}}(x-\gamma)^2},$$

and dV' has $2p-2-\Sigma\lambda = -2+5 = 3$ zeros.

Thus V' can be written in the form

$$V' = R\int \frac{dx}{(x-\alpha)^{\frac{1}{2}}(x-\beta)^{\frac{1}{2}}(x-\gamma)} + \frac{Lx^2+Mx+N}{(x-\gamma)(x-\alpha)^{\frac{1}{2}}(x-\beta)^{\frac{1}{2}}},$$
$$= NK'(x) + MK_1'(x) + LK_2'(x) + RV_0',$$

where N, M, L, R are constants, $K'(x)$, $K_1'(x)$, $K_2'(x)$ are particular, linearly independent, everywhere finite factorial functions of the associated system, and V_0' is a particular everywhere finite factorial integral of the associated system.

Ex. i. In case of a factorial system given by $p=0$, $k=2$, $\lambda_1=-\frac{3}{2}$, $\lambda_2=-\frac{3}{2}$, prove that $\sigma+1=2$, $\sigma'+1=0$, $\varpi=0$, $\varpi'=2$; prove that the factorial function of the primary system with fewest poles is $P(x)=(x-a)^{\frac{3}{2}}(x-\beta)^{\frac{3}{2}}/(x-x_1)(x-x_2)(x-x_3)$; obtain the form of the factorial integral of the second kind of the primary system with fewest poles, and prove that it can be expressed in the form $AP(x)+B$; and shew that the everywhere finite factorial integrals of the associated system are expressible in the form $(Ax+B)/\sqrt{(x-a)(x-\beta)}$, their initial form being

$$V' = \int \frac{(Ax+B)\,dx}{(x-a)^{\frac{3}{2}}(x-\beta)^{\frac{3}{2}}}.$$

Ex. ii. When we take $p=0$ and k, $=2n+2$, places $c_1, \ldots, c_{2n+2}$, and each $\lambda=-\frac{1}{2}$, prove that the original and the associated systems coincide, that $\sigma+1=\sigma'+1=0$, $\varpi=\varpi'=n$, that the everywhere finite factorial integrals, and the integral with one pole are respectively

$$\int \frac{(x, 1)_{n-1}}{\sqrt{f(x)}}\,dx, \qquad \int \left[\frac{f(a)}{(x-a)^2} + \frac{1}{2}\frac{f'(a)}{x-a}\right]\frac{dx}{\sqrt{f(x)}},$$

where $f(x)=(x-c_1)\ldots\ldots(x-c_{2n+2})$. The factorial function with fewest poles is $\sqrt{f(x)}/(x, 1)_{n+1}$; express this in the form

$$\frac{\sqrt{f(x)}}{(x, 1)_{n+1}} = \sum_{i=1}^{n+1}\lambda_i \int \left[\frac{f(a_i)}{(x-a_i)^2} + \frac{1}{2}\frac{f'(a_i)}{x-a_i}\right]\frac{dx}{\sqrt{f(x)}} + \int \frac{(x, 1)_{n-1}}{\sqrt{f(x)}}\,dx + \text{constant},$$

$a_1, \ldots, a_{n+1}$ being the zeros of $(x, 1)_{n+1}$, and determine the $2n+1$ coefficients on the right-hand side.

269. One of the simplest applications of the theory of this chapter is to the case of the root functions already considered in the last chapter; such a function can be expressed in the form e^{ψ}, where

$$\psi = \Pi_{\beta_1, \alpha_1}^{x, \gamma} + \ldots\ldots + \Pi_{\beta_N, \alpha_N}^{x, \gamma} - 2\pi i \sum_1^p (h_i + H_i)\, v_i^{x, \gamma},$$

where $\beta_1, \ldots, \beta_N$ are the zeros, $\alpha_1, \ldots, \alpha_N$ the poles, h_i is a rational numerical fraction, H_i is an integer, and γ is an arbitrary place. The singular places, $c_1, \ldots, c_k$ are entirely absent. The zeros and poles satisfy the equations expressed by

$$v^{\beta_1, \alpha_1} + \ldots\ldots + v^{\beta_N, \alpha_N} = g + G + \tau(h + H),$$

where $G_1, \ldots, G_p$ are integers; and since, if m be the least common denominator of the $2p$ numbers g, h, the mth power of the function is a rational function, there is no function of the system which is everywhere finite, and the same is true of the associated system. Hence $\sigma + 1 = 0 = \sigma' + 1$, $\varpi = \varpi' = p - 1$; thus the function of the system with fewest poles has p poles, and every function of the system can be expressed as a linear aggregate of such functions (§ 267. Cf. § 245, Chap. XIII.).

Ex. i. Prove that when the numbers g, h are any half-integers, the everywhere finite integrals of the system are expressible in the form

$$V = \int \frac{dv}{\phi} \sum_{1}^{p-1} \lambda_i \sqrt{\Phi_i \Psi_i},$$

where v is an arbitrary integral of the first kind, ϕ is the corresponding ϕ-polynomial, and Φ_i, Ψ_i are ϕ-polynomials with $p-1$ zeros each of the second order (cf. § 245, Chap. XIII.). It is in fact possible to represent *any* half-integer characteristic, other than the zero characteristic, as the sum of two odd half-integer characteristics in $2^{p-2}(2^{p-1}-1)$ ways.

Ex. ii. In the hyperelliptic case, when the numbers g, h are any half-integers, prove that the function of the system with $\varpi' + 1 = p$ poles is given by

$$\sqrt{u}\left\{\frac{y}{u\psi(x)} + \sum_{1}^{p} \frac{y_i}{u_i(x - x_i)\psi'(x_i)}\right\},$$

where the places $(x_1, y_1), \ldots$ are the poles in question,

$$\psi(x) = (x - x_1)\ldots(x - x_p), \quad \psi'(x) = d\psi(x)/dx, \quad u = (x - a)(x - b),$$

and a, b are two suitably chosen branch places*, and $u_i = (x_i - a)(x_i - b)$. Shew that in the elliptic case this leads to the function $\dfrac{\sigma(u - v + w)}{\sigma(u - v)} e^{-\eta(u - v)}$.

270. In the case in which the factors at the period loops are any constants, the places $c_1, \ldots, c_k$ being still absent, it remains true that the number of zeros of any function of the system is equal to the number of poles; but here there may be an everywhere finite function of the system, and there will be such a function provided

$$g_i + \tau_{i,1} h_1 + \ldots\ldots + \tau_{i,p} h_p = -[G_i + \tau_{i,1} H_1 + \ldots\ldots + \tau_{i,p} H_p], \quad (i = 1, 2, \ldots, p)$$

in which $G_1, \ldots, H_p$ are integers, the function being, in that case, expressed by

$$E = e^{-2\pi i \sum_{1}^{p} (h_i + H_i) v_i^{x, \gamma}};$$

* For the association of the proper pair of branch places a, b with the given values of the numbers g, h, compare Chap. XI. § 208, Chap. XIII. § 245, and the remark at the conclusion of Ex. i.

then E^{-1} is an everywhere finite function of the associated system, and $\sigma+1=\sigma'+1=1$, $\varpi=\varpi'=p$. It is not necessary to consider this case, for it is clear that every function of the system is of the form ER, R being a rational function.

When $\sigma+1=\sigma'+1=0$ we have $\varpi=p-1=\varpi'$. Then every function of the system can be expressed linearly by means of functions of the system having p poles. If $x_1, \dots, x_p$ be the poles of such a function and $z_1, \dots, z_p$ the zeros, and the relations connecting these be given by

$$v^{z_1, x_1}+\dots\dots+v^{z_p, x_p}=g+G+\tau(h+H).$$

There is beside the expression originally given, a very convenient way of expressing such a function, whose correctness is immediately verifiable, namely

$$\frac{\Theta(u-g-G-\tau h-\tau H)}{\Theta(u)}e^{-2\pi i(h+H)u},$$

wherein

$$u=v^{x, m}-v^{x_1, m_1}-\dots\dots-v^{x_p, m_p},$$

and $m, m_1, \dots, m_p$ are related as in § 179, Chap. X. Omitting a constant factor this is the same as

$$\frac{\Theta(u-g-\tau h)}{\Theta(u)}e^{-2\pi i h u}, =\phi(u), \text{ say};$$

since the difference between the values of the logarithm of $\phi(u)$ at the two sides of any period loop is independent of u, and of x, it follows that $\frac{\partial}{\partial x}\log\phi(u)$ is a rational function of x, and that $\frac{\partial}{\partial u_i}\log\phi(u)$ is a periodic function with $2p$ sets of simultaneous periods; thus the function $\phi(u)$ satisfies linear equations of the form

$$\frac{\partial^2 y}{\partial x^2}=Ry, \quad \frac{\partial^2 y}{\partial u_i\partial u_j}=R_{ij}y, \qquad (i, j=1, 2, \dots, p),$$

where R, R_{ij} are rational functions of x, and $2p$-ply periodic functions of u given* by

$$R=\frac{\partial^2}{\partial x^2}\log\phi(u)+\left[\frac{\partial}{\partial x}\log\phi(u)\right]^2,$$

$$R_{ij}=\frac{\partial^2}{\partial u_i\partial u_j}\log\phi(u)+\left[\frac{\partial}{\partial u_i}\log\phi(u)\right]\left[\frac{\partial}{\partial u_j}\log\phi(u)\right].$$

Ex. The $2p$ constants a, λ can be chosen so that

$$\phi(u)=\frac{\vartheta(u+a)}{\vartheta(u)}e^{\lambda u}$$

satisfies the equations $\phi(u+2\omega)=A\phi(u)$, $\phi(u+2\omega')=A'\phi(u)$, where A, A' each represents p given constants, and the notation is as in § 189, Chap. X.

* Cf. Halphen, *Fonct. Ellipt.*, Prem. Part. (Paris 1886), p. 235, and Forsyth, *Theory of Functions*, pp. 275, 285, for the case $p=1$. By further development of the results given in Chap. XI. of this volume, and in the present chapter, it is clearly possible to formulate the corresponding analytical results for greater values of p.

271. We have seen (§ 261) that the number of arbitrary constants entering into the expression of a factorial function of the primary system with given poles is $N-p+1+h+1$, $=R$ say, where N is the number of zeros of the function, and $h+1$ is the number of linearly independent differentials, dv, of integrals of the first kind, which vanish in the zeros of the function. When $h+1$ vanishes the assigning of the poles of the function, and of $R-1$ of the zeros determines the other $N-R+1$, $=p$, zeros; in any case the assigning of the poles and of $R-1$ of the zeros determines the other $N-R+1$, $=p-(h+1)$, of the zeros. Denote the poles by $\alpha_1, \dots, \alpha_M$ and the assigned zeros by $\beta_1, \dots, \beta_{R-1}$; then the remaining zeros $\beta_R, \dots, \beta_N$ are determined by the congruences

$$v_i^{\beta_1, a} + \dots + v_i^{\beta_{R-1}, a} - v_i^{\alpha_1, a} - \dots - v_i^{\alpha_M, a} - \sum_{r=1}^{k} \lambda_r v_i^{c_r, a} - (g_i + h_1\tau_{i,1} + \dots + h_p\tau_{i,p})$$
$$\equiv -(v_i^{\beta_R, a} + \dots\dots + v_i^{\beta_N, a}),$$

a being an arbitrary place. Now, let the form of the factorial function when the poles are given be

$$C_1F_1(x) + \dots\dots + C_RF_R(x),$$

where $C_1, \dots, C_R$ are arbitrary constants, and $F_1(x), \dots, F_R(x)$ are linearly independent; then, when the zeros $\beta_1, \dots, \beta_{R-1}$ are assigned, the function is a constant multiple of the definite function

$$\Delta(x) = \begin{vmatrix} F_1(x), & \dots\dots, & F_R(x) \\ F_1(\beta_1), & \dots\dots, & F_R(\beta_1) \\ \dots & \dots & \dots \\ F_1(\beta_{R-1}), & \dots\dots, & F_R(\beta_{R-1}) \end{vmatrix};$$

the zeros of this function, other than $\beta_1, \dots, \beta_{R-1}$, are perfectly definite, and are determined by the congruences put down. Let H denote the quantities given by

$$H_i = \sum_{r=1}^{k} \lambda_r v_i^{c_r, a} + g_i + h_1\tau_{i,1} + \dots\dots + h_p\tau_{i,p};$$

take any places $\gamma_1, \dots, \gamma_{h+1}$, of assigned position, and take a place m and p dependent places $m_1, \dots, m_p$ defined as in § 179, Chap. X., and consider the function of x

$$\Theta(v^{x, m} - v^{\gamma_1, m_1} - \dots\dots - v^{\gamma_{h+1}, m_{h+1}} + v^{\beta_1, a} + \dots\dots + v^{\beta_{R-1}, a} - v^{a, m_{h+2}} - \dots\dots$$
$$- v^{a, m_p} - v^{\alpha_1, a} - \dots\dots - v^{\alpha_M, a} - H);$$

if the function does not vanish identically, its zeros, $x_1, \dots, x_p$, are (§ 179, Chap. X.) given by the congruences denoted by

$$-v^{\gamma_1, m_1} - \dots\dots - v^{\gamma_{h+1}, m_{h+1}} + v^{\beta_1, a} + \dots\dots + v^{\beta_{R-1}, a} - v^{a, m_{h+2}} - \dots\dots$$
$$- v^{a, m_p} - v^{\alpha_1, a} - \dots\dots - v^{\alpha_M, a} - H \equiv -v^{x_1, m_1} - \dots\dots - v^{x_p, m_p},$$

or, what is the same thing, by

$$v^{\beta_1, a} + \ldots\ldots + v^{\beta_{R-1}, a} - v^{\alpha_1, a} - \ldots\ldots - v^{\alpha_M, a} - H$$
$$\equiv v^{\gamma_1, x_1} + \ldots\ldots + v^{\gamma_{h+1}, x_{h+1}} - v^{x_{h+2}, a} - \ldots\ldots - v^{x_p, a};$$

now, from what has been said, it follows, comparing these congruences with those connecting the poles and zeros of $\Delta(x)$, that if $x_1, \ldots, x_{h+1}$ be taken at $\gamma_1, \ldots, \gamma_{h+1}$, these congruences determine $x_{h+2}, \ldots, x_p$ uniquely as the places $\beta_R, \ldots, \beta_N$. Thus the zeros of the theta function are the places $\gamma_1, \ldots, \gamma_{h+1}$ together with the zeros, other than $\beta_1, \ldots, \beta_{R-1}$, of the function $\Delta(x)$.

We suppose now M to be as great as $p-1, = r+p-1$, say; as in § 184, p. 269, we take $n_1, \ldots, n_{p-1}$ to be the zeros of a ϕ-polynomial of which all the zeros are of the second order, so that

$$v^{n_p, m} - v^{n_1, m_1} - \ldots\ldots - v^{n_{p-1}, m_{p-1}}$$

is an odd half-period, equal to $\frac{1}{2}\Omega_{s, s'}$ say; and we take the poles $\alpha_{r+1}, \ldots, \alpha_M$ at $n_1, \ldots, n_{p-1}$. Further*, in this article, we denote

$$\Theta(v^{x, z} + \tfrac{1}{2}\Omega_{s, s'})\, e^{\pi i s' v^{x, z}} \text{ by } \lambda(x, z),$$

so that (§ 175, Chap. X.) $\lambda(x, z)$ is also equal to $e^{-\frac{1}{4}\pi i s'(s+\frac{1}{2}\tau s')}\,\Theta(v^{x, z};\ \tfrac{1}{2}s, \tfrac{1}{2}s')$. The function $\lambda(x, z)$ must not be confounded with the function $\lambda(\zeta, \mu)$ of § 238.

Then in fact, denoting the arguments of the theta function by V, we have the following important formula,

$$e^{-2\pi i(h-\frac{1}{2}s')V}\Theta(V) = A\,\frac{\Delta(x)\prod_{j=1}^{r}\lambda(x, \alpha_j)\prod_{j=1}^{h+1}\lambda(x, \gamma_j)\prod_{j=1}^{k}[\lambda(x, c_j)]^{\lambda_j}}{\prod_{j=1}^{R-1}\lambda(x, \beta_j)},$$

where A is a quantity independent of x. In order to prove this it is sufficient to shew (i) that the right-hand side represents a single-valued function of x on the Riemann surface dissected by the $2p$ period loops, (ii) that the right-hand side has no poles and has only the zeros of $\Theta(V)$, and (iii) that the two sides of the equation have the same factor for every one of the $2p$ period loops.

Now the function $\lambda(x, z)$ has no poles; its zeros are the place z, and the places $n_1, \ldots, n_{p-1}$. The places $n_1, \ldots, n_{p-1}$ occur on the right hand

(α) as poles, each once in $\Delta(x)$, each $(R-1)$ times in the product $\prod_{j=1}^{R-1}\lambda(x, \beta_j)$;

(β) as zeros, each r times in $\prod_{j=1}^{r}\lambda(x, \alpha_j)$, $h+1$ times in $\prod_{j=1}^{h+1}\lambda(x, \gamma_j)$, and

* For the introduction of the function $\lambda(x, z)$ see, beside the references given in chapter XIII. (§ 250), also Clebsch u. Gordan, *Abel. Functnen.* pp. 251—256, and Riemann, *Math. Werke* (1876), p. 134.

$\sum_{j=1}^{k} \lambda_j$ times in $\prod_{j=1}^{k} [\lambda(x, c_j)]^{\lambda_j}$; thus these places occur as zeros, on the right hand,

$$M - (p-1) + h + 1 + \Sigma\lambda_j - R, = N - p + 1 + h + 1 - R,$$

times, that is, not at all.

Thus the expression on the right hand may be interpreted as a single-valued function on the Riemann surface dissected by the $2p$ period loops—for we have seen that the places $n_1, \ldots, n_{p-1}$ do not really occur, and the multiplicity, at c_j, in the value of such a factor as $[\lambda(x, c_j)]^{\lambda_j}$ is cancelled by the assigned character of the factorial functions $F(x)$ occurring in $\Delta(x)$. Nextly, the zeros of the denominator of the right-hand side, other than at $n_1, \ldots, n_{p-1}$, are zeros of $\Delta(x)$, and the poles of $\Delta(x)$, other than $n_1, \ldots, n_{p-1}$, are zeros of the product $\prod_{j=1}^{r} \lambda(x, a_j)$, so that the right-hand side remains finite. The only remaining zeros of the right-hand side consist of $\gamma_1, \ldots, \gamma_{h+1}$ and the zeros of $\Delta(x)$ beside $\beta_1, \ldots, \beta_{R-1}$; and we have proved that these are the zeros of $\Theta(V)$. It remains then finally to examine the factors of the two sides of the equation at the period loops. The factors of the left-hand side at the i-th period loops respectively of the first and second kind are (see § 175, Chap. X.)

$$e^{-2\pi i (h_i - \frac{1}{2}s_i')} \text{ and } e^{-2\pi i \sum_{\mu=1}^{p} (h_\mu - \frac{1}{2}s'_\mu)\tau_{\mu, i} - 2\pi i (V_i + \frac{1}{2}\tau_{i, i})};$$

the factor of the right-hand side at the i-th period loop of the first kind is e^ψ, where

$$\psi = -2\pi i h_i + r\pi i s_i' + (h+1)\pi i s_i' + \pi i s_i' \sum_{j=1}^{k} \lambda_j - (R-1)\pi i s_i';$$

now $R = N - p + 1 + h + 1 = r + \sum_{j=1}^{k} \lambda_j + h + 1$; thus $\psi = -2\pi i h_i + \pi i s_i'$, and $e^\psi = e^{-2\pi i (h_i - \frac{1}{2}s_i')}$, or the factors of the two sides of the equation to be proved, at the i-th period loop of the first kind, are the same. Since the factor of $\lambda(x, z)$ at the i-th period loop of the second kind is e^μ where

$$\mu = -2\pi i [v_i^{x, z} + \tfrac{1}{2}s_i + \tfrac{1}{2}s_1'\tau_{i, 1} + \ldots + s'_p\tau_{i, p} + \tfrac{1}{2}\tau_{i, i}] + \pi i (s_1'\tau_{i, 1} + \ldots + s'_p\tau_{i, p}),$$
$$= -2\pi i (v_i^{x, z} + \tfrac{1}{2}s_i + \tfrac{1}{2}\tau_{i, i}),$$

it follows that the factor of the right-hand side at the i-th period loop of the second kind is e^χ where

$$\chi = 2\pi i g_i - 2\pi i \left[\sum_{j=1}^{r} v_i^{x, a_j} + \sum_{j=1}^{h+1} v_i^{x, \gamma_j} + \sum_{j=1}^{k} \lambda_j v_i^{x, c_j} - \sum_{j=1}^{R-1} v_i^{x, \beta_j}\right]$$
$$- \pi i \left[r + h + 1 + \sum_{j=1}^{h} \lambda_j - R + 1\right](\tau_{i, i} + s_i),$$
$$= 2\pi i g_i - \pi i (s_i + \tau_{i, i}) - 2\pi i \left[\sum_{j=1}^{r} v_i^{x, a_j} + \sum_{j=1}^{h+1} v_i^{x, \gamma_j} + \sum_{j=1}^{k} \lambda_j v_i^{x, c_j} - \sum_{j=1}^{R-1} v_i^{x, \beta_j}\right];$$

now we have

$$V_i = v_i^{x,m} - v_i^{\gamma_1, m_1} - \dots - v_i^{\gamma_{h+1}, m_{h+1}} + v_i^{\beta_1, a} + \dots + v_i^{\beta_{R-1}, a} - v_i^{a, m_{h+2}} - \dots - v_i^{a, m_p}$$

$$- v_i^{\alpha_1, a} - \dots - v_i^{\alpha_r, a} - v_i^{n_1, a} - \dots - v_i^{n_{p-1}, a} - \sum_{j=1}^{k} \lambda_j v_i^{c_j, a} - g_i - h_1\tau_{i,1} - \dots - h_p\tau_{i,p},$$

and

$$\tfrac{1}{2}(s_i + s_1'\tau_{i,1} + \dots + s_p'\tau_{i,p}) = v^{m_p, m} - v^{n_1, m_1} - \dots - v^{n_{p-1}, m_{p-1}};$$

thus

$$V_i - \tfrac{1}{2}(s_i + s_1'\tau_{i,1} + \dots + s_p'\tau_{i,p}) = v_i^{x,a} + v_i^{a,\gamma_1} + \dots + v_i^{a,\gamma_{h+1}} + v_i^{\beta_1, a} + \dots + v_i^{\beta_{R-1}, a}$$

$$- v_i^{\alpha_1, a} - \dots - v_i^{\alpha_r, a} - \sum_{j=1}^{k} \lambda_j v_i^{c_j, a} - g_i - h_1\tau_{i,1} - \dots - h_p\tau_{i,p};$$

further

$$0 = -v_i^{x,a} + (h+1)\, v_i^{x,a} - (R-1)\, v_i^{x,a} + r v_i^{x,a} + \sum_{j=1}^{k} \lambda_j v_i^{x,a};$$

hence

$$V_i - \tfrac{1}{2}(s_i + s_1'\tau_{i,1} + \dots + s_p'\tau_{i,p}) = \sum_{j=1}^{h+1} v_i^{x,\gamma_j} - \sum_{j=1}^{R-1} v_i^{x,\beta_j} + \sum_{j=1}^{r} v_i^{x,\alpha_j}$$

$$+ \sum_{j=1}^{k} \lambda_j v_i^{x, c_j} - g_i - h_1\tau_{i,1} - \dots - h_p\tau_{i,p},$$

or

$$\sum_{\mu=1}^{p} (h_\mu - \tfrac{1}{2}s_\mu')\,\tau_{\mu,i} + V_i = -g_i + \tfrac{1}{2}s_i + \sum_{j=1}^{r} v_i^{x,\alpha_j} + \sum_{j=1}^{h+1} v_i^{x,\gamma_j} + \sum_{j=1}^{k} \lambda_j v_i^{x,c_j} - \sum_{j=1}^{R-1} v_i^{x,\beta_j},$$

and thence the identity of the factors taken by the two sides of the equation to be proved, at the i-th period loop of the second kind, is manifest.

And before passing on it is necessary to point out that if the functions $\lambda(x, z)$ be everywhere replaced by $\dfrac{\lambda(x, z)}{\psi}$, and $\Delta(x)$ be replaced by $\psi\Delta(x)$, ψ being any quantity whatever, the value of the right-hand side of the equation is unaltered. For there are R factors $\lambda(x, z)$ occurring in the numerator of the right-hand side of the equation beside $\Delta(x)$, and $R-1$ factors $\lambda(x, z)$ occurring in the denominator of the right-hand side of the equation. In particular ψ may be a function of x.

272. We can now state the following result: Let $a, \alpha_1, \dots, \alpha_r$ be any assigned places; let $n_1, n_2, \dots, n_{p-1}$ be the zeros of a ϕ-polynomial, or of a differential, dv, of the first kind, of which all the zeros are of the second order, and

$$v_i^{m_p, m} - v_i^{n_1, m_1} - \dots - v_i^{n_{p-1}, m_{p-1}} = \tfrac{1}{2}(s_i + s_i'\tau_{i,1} + \dots + s_p'\tau_{i,p}), \quad (i = 1, 2, \dots, p),$$

$m, m_1, \dots, m_p$ being such places as in § 179, Chap. X.; let $h+1$ be the number of linearly independent differentials, dv, which vanish in the zeros of a factorial function of the primary system having $\alpha_1, \dots, \alpha_r, n_1, \dots, n_{p-1}$ as poles, or the number of differentials dV', of everywhere finite factorial integrals of the associated system, which vanish in the places $n_1, \dots, n_{p-1}$, $\alpha_1, \dots, \alpha_r$; let $\gamma_1, \dots, \gamma_{h+1}$ be any assigned places; denote $r + \sum_{j=1}^{k} \lambda_j + h + 1$ by R, and let* $x_1, \dots, x_R$ be any assigned places; let the general factorial function of the primary system having $\alpha_1, \dots, \alpha_r, n_1, \dots, n_{p-1}$ as poles be

$$C_1 F_1(x) + \dots\dots + C_R F_R(x),$$

wherein $C_1, \dots, C_R$ are constants, and let

$$\Delta(x_1, \dots, x_R) = \left| \begin{array}{l} F_1(x_1), \ \dots\dots, F_R(x_1) \\ F_1(x_2), \ \dots\dots, F_R(x_2) \\ \dots\dots\dots\dots\dots\dots\dots\dots \\ F_1(x_R), \ \dots\dots, F_R(x_R) \end{array} \right| \psi(x_1), \dots, \psi(x_R),$$

where $\psi(x)$ denotes any function whatever; let

$$U_i = \sum_{j=1}^{R} v_i^{x_j, a} - \sum_{j=1}^{r} v_i^{\alpha_j, a} - \sum_{j=1}^{h+1} v_i^{\gamma_j, a} - \sum_{j=1}^{k} \lambda_j v_i^{c_j, a},$$

which is independent of a, and let the row of p quantities

$$g_i - \tfrac{1}{2}s_i + (h_1 - \tfrac{1}{2}s_1')\tau_{i,1} + \dots\dots + (h_p - \tfrac{1}{2}s_p')\tau_{i,p}$$

be denoted† by $g - \frac{1}{2}s + \tau(h - \frac{1}{2}s')$; then if, modifying the definition of $\lambda(x, z)$, we put

$$\lambda(x, z) = \frac{\Theta(v^{x,z} + \frac{1}{2}\Omega_{s,s'}) e^{\pi i s' v^{x,z}}}{\psi(x)\psi(z)},$$

we have

$$Ce^{-2\pi i(h-\frac{1}{2}s')[U-(g-\frac{1}{2}s)-\tau(h-\frac{1}{2}s')]} \Theta[U - (g - \tfrac{1}{2}s) - \tau(h - \tfrac{1}{2}s')]$$

$$= \frac{\Delta(x_1, x_2, \dots, x_R)}{\prod\prod_{\substack{i,j=1,2,\dots,R \\ i<j}} \lambda(x_i, x_j)} \prod_{i=1}^{R} \left\{ \prod_{j=1}^{r} \lambda(x_i, \alpha_j) \prod_{j=1}^{h+1} \lambda(x_i, \gamma_j) \prod_{j=1}^{k} [\lambda(x_i, c_j)]^{\lambda_j} \right\},$$

wherein C is a quantity independent of $x_1, \dots, x_R$, which may depend on $c_1, \dots, c_k, \alpha_1, \dots, \alpha_r, \gamma_1, \dots, \gamma_{h+1}$.

273. The formula just obtained is of great generality; before passing to examples of its application it is desirable to explain the origin of a certain function which may be used in place of the unassigned function $\psi(x)$.

We have (§ 187, p. 274), in the notation of § 272,

$$\Pi_{x,z}^{x',z'} = \log \frac{\Theta(v^{x',x} + \frac{1}{2}\Omega_{s,s'})\,\Theta(v^{z',z} + \frac{1}{2}\Omega_{s,s'})}{\Theta(v^{x',z} + \frac{1}{2}\Omega_{s,s'})\,\Theta(v^{z',x} + \frac{1}{2}\Omega_{s,s'})};$$

if the zeros of the rational function of x', $(x' - x)/(x' - z)$, be denoted by

* These replace the $x_1, \beta_1, \dots, \beta_{R-1}$ of § 271.

† So that $V = U - (g - \frac{1}{2}s) - \tau(h - \frac{1}{2}s')$.

$x, x_1, \dots, x_{n-1}$, n being the number of sheets of the fundamental Riemann surface, and the poles of the same function be denoted by $z, z_1, \dots, z_{n-1}$, we have, by Abel's theorem,

$$\Pi_{x_1, z_1}^{x', z'} + \dots\dots + \Pi_{x_{n-1}, z_{n-1}}^{x', z'} = -\Pi_{x, z}^{x', z'} + \log\left(\frac{x'-x}{x'-z}\Big/\frac{z'-x}{z'-z}\right),$$

$$= \log \frac{(x'-x)(z'-z)\,\Theta(v^{x', z} + \tfrac{1}{2}\Omega_{s, s'})\,\Theta(v^{z', x} + \tfrac{1}{2}\Omega_{s, s'})}{(x'-z)(z'-x)\,\Theta(v^{x', x} + \tfrac{1}{2}\Omega_{s, s'})\,\Theta(v^{z', z} + \tfrac{1}{2}\Omega_{s, s'})};$$

now let the places x', z' approach respectively indefinitely near to the places x, z which, firstly, we suppose to be finite places and not branch places; then the right-hand side of the equation just obtained becomes

$$\log\left[\frac{1}{-(x-z)^2}\,\frac{\Theta(v^{x, z} + \tfrac{1}{2}\Omega_{s, s'})\,\Theta(v^{x, z} - \tfrac{1}{2}\Omega_{s, s'})}{X(x)\,X(z)}\right],$$

where

$$X(x) = \sum_{i=1}^{p} \Theta_i(\tfrac{1}{2}\Omega_{s, s'})\,.\,Dv_i^{x, a}, \quad X(z) = \sum_{i=1}^{p} \Theta_i(\tfrac{1}{2}\Omega_{s, s'})\,.\,Dv_i^{z, a},$$

D denoting a differentiation, and a denoting an arbitrary place; but we have (Chap. X. § 175)

$$\Theta(v^{x, z} - \tfrac{1}{2}\Omega_{s, s'}) = e^{\pi i s s'} e^{2\pi i s' v^{x, z}}\,\Theta(v^{x, z} + \tfrac{1}{2}\Omega_{s, s'}) = -e^{2\pi i s' v^{x, z}}\,\Theta(v^{x, z} + \tfrac{1}{2}\Omega_{s, s'});$$

thus, on the whole, when the square roots are properly interpreted, we obtain

$$\lim_{x'=x,\, z'=z} \sqrt{-(x'-x)(z'-z)\,e^{-\Pi_{x, z}^{x', z'}}} = \frac{\Theta(v^{x, z} + \tfrac{1}{2}\Omega_{s, s'})\,.\,e^{\pi i s' v^{x, z}}}{\sqrt{X(x)\,X(z)}}$$

$$= (x-z)\,e^{\frac{1}{2}\Pi_{x_1, z_1}^{x, z} + \dots\dots + \frac{1}{2}\Pi_{x_{n-1}, z_{n-1}}^{x, z}}. \qquad \text{(i)}$$

When the places x, z are finite branch places we obtain a similar result. Denote the infinitesimals at these places by t, t_1, and, when x', z' are near to x, z, respectively, suppose $x' = x + t^{w+1}$, $z' = z + t_1^{w_1+1}$; then from the equation given by Abel's theorem we obtain, if γ denote an arbitrary place,

$$\sum_{r=1}^{w}\left[\Pi_{x_r, \gamma}^{x', z'} - \log t\right] + \sum_{r=w+1}^{n-1} \Pi_{x_r, \gamma}^{x', z'} + \sum_{r=1}^{w_1}\left[\Pi_{\gamma, z_r}^{x', z'} - \log t_1\right] + \sum_{r=w_1+1}^{n-1} \Pi_{\gamma, z_r}^{x', z'}$$

$$= -\log(t^w) - \log(t_1^{w_1}) + \log\left[\frac{t^{w+1} t_1^{w_1+1}}{-(x-z)^2}\,.\,\frac{-\Theta^2(v^{x, z} + \tfrac{1}{2}\Omega_{s, s'})\,.\,e^{2\pi i s' v^{x, z}}}{t t_1 X(x)\,X(z)}\right],$$

where $X(x)$, $X(z)$ are of the same form as before, save that the differentiations $Dv_i^{x, a}$, $Dv_i^{z, a}$, are to be performed in regard to the infinitesimals t, t_1. If the limit of the first member of this equation, as x', z' respectively approach to x, z, be denoted by L, we therefore have

$$\lim_{x'=x,\, z'=z} \sqrt{-t t_1 e^{-\Pi_{x, z}^{x', z'}}} = \frac{\Theta(v^{x, z} + \tfrac{1}{2}\Omega_{s, s'})\,.\,e^{\pi i s' v^{x, z}}}{\sqrt{X(x)\,X(z)}} = (x-z)\,e^{\frac{1}{2}L}. \qquad \text{(ii)}$$

The equations (i), (ii) are very noticeable; *there is no position of x for which the expression $\Theta(v^{x,z}+\tfrac{1}{2}\Omega_{s,s'}) \,.\, e^{\pi i s' v^{x,z}}/\sqrt{X(x)\,X(z)}$ is infinite, and there is only one position of x, namely when x is at z, for which the expression vanishes;* for (§ 188, p. 281) the expression $\sqrt{X(x)}$ vanishes, to the first order, only when x is at one of the places $n_1, \dots, n_{p-1}$, and $\Theta(v^{x,z}+\tfrac{1}{2}\Omega_{s,s'})$ vanishes only when x is at one of the places $z, n_1, \dots, n_{p-1}$; there is no position of x for which $\sqrt{X(x)}$ is infinite. Putting

$$\varpi_1(x, z) = \frac{\Theta(v^{x,z}+\tfrac{1}{2}\Omega_{s,s'})\, e^{\pi i s' v^{x,z}}}{\sqrt{X(x)\,X(z)}},$$

we have further $\varpi_1(x, z) = -\varpi_1(z, x)$, and if t denote the infinitesimal near to z, we have, as x approaches to z, $\text{limit}_{x=z}[\varpi_1(x, z)/t] = 1$. For every position of x and z on the dissected Riemann surface $\varpi_1(x, z)$ has a perfectly determinate value, *save for an ambiguity of sign*, and, as follows from the equations (i), (ii), this value is independent of the characteristic $(\tfrac{1}{2}s, \tfrac{1}{2}s')$.

There are various ways of dealing with the ambiguity in sign of the function $\varpi_1(x, z)$. For instance, let $\phi(x)$ be any ϕ-polynomial vanishing in an arbitrary place m, and in the places $A_1, \dots, A_{2p-3}$ (cf. § 244, Chap. XIII.), and let $Z(x)$ be that polynomial of the third degree in the p fundamental linearly independent ϕ-polynomials which vanishes to the second order in $A_1, \dots, A_{2p-3}$ and in the places $m_1, \dots, m_p$. Further let $\Phi(x)$ be that ϕ-polynomial which vanishes to the second order in the places $n_1, \dots, n_{p-1}$. Then we have shewn (§ 244) that the ratio $\sqrt{Z(x)}/\phi(x)\sqrt{\Phi(x)}$, save for an initial determination of sign for an arbitrary position of x, is single-valued on the dissected Riemann surface; hence instead of the function $\varpi_1(x, z)$ we may use the function

$$E_1(x, z) = \frac{\sqrt{Z(x)\,Z(z)}}{\phi(x)\,\phi(z)} \cdot \frac{\Theta(v^{x,z}+\tfrac{1}{2}\Omega_{s,s'})\, e^{\pi i s' v^{x,z}}}{\sqrt{\Phi(x)\,\Phi(z)}},$$

which has the properties; (i) on the dissected Riemann surface it is a single-valued function of x and of z, (ii) $E_1(x, z) = -E_1(z, x)$, (iii) as a function of x it has, beside the *fixed* zeros $m_1, \dots, m_p$, only the zero given by $x = z$, and it has no infinities beside the *fixed* infinity given by $x = m$, where it is infinite to the first order. At the r-th period loops respectively of the first and second kind it has the factors

$$1,\ e^{-2\pi i\,(v_r^{x,z}+\frac{1}{2}\tau_{r,r})}.$$

But there can be no doubt, in view of the considerations advanced in chapter XII. of the present volume, as to the way in which the ambiguity of the sign of $\varpi_1(x, z)$ ought to be dealt with. Suppose that the Riemann surface now under consideration has arisen from the consideration of the functions there considered (§ 227) which are unaltered by the linear substitutions of the group. Let the places in the region S of the ζ plane which correspond to the places x, z, x', z' of the Riemann surface be denoted by ξ, ζ, ξ', ζ'. Then by comparing the equation obtained in chapter XII. (§ 234),

$$\lim._{\xi'=\xi,\ \zeta'=\zeta} \sqrt{-(\xi'-\xi)(\zeta'-\zeta)e^{-\Pi_{\xi,\zeta}^{\xi',\zeta'}}} = \frac{\Theta(v^{\xi,\zeta}+\tfrac{1}{2}\Omega_{s,s'})\,e^{\pi i s' v^{\xi,\zeta}}}{\sqrt{\dfrac{dv}{d\xi}\dfrac{dv}{d\zeta}}} = \varpi(\xi,\zeta),$$

with the equation here obtained,

$$\lim._{x'=x,\ z'=z} \sqrt{-tt_1 e^{-\Pi_{x,z}^{x',z'}}} = \frac{\Theta(v^{x,z}+\tfrac{1}{2}\Omega_{s,s'})\,e^{\pi i s' v^{x,z}}}{\sqrt{X(x)X(z)}} = \varpi_1(x,z),$$

and noticing that $X(x)$, $\dfrac{dv}{d\xi}$ agree in being differential coefficients of an integral of the first kind, which vanish to the second order at $n_1, \ldots, n_{p-1}$, we deduce the equation

$$\varpi_1(x,z)\Big/\sqrt{\frac{dt}{d\xi}\cdot\frac{dt_1}{d\zeta}} = \varpi(\xi,\zeta);$$

now we have shewn that $\varpi(\xi,\zeta)$ is a single-valued function of ξ and ζ; and any one of the infinite number of values of ξ, which correspond to any value of x, has a continuous and definite variation as x varies in a continuous way; hence it is possible, dividing $\varpi_1(x,z)$ by the factor $\sqrt{\dfrac{dt}{d\xi}\cdot\dfrac{dt_1}{d\zeta}}$, which by itself is of ambiguous sign, to destroy the original ambiguity while retaining the essential character of the function $\varpi_1(x,z)$. The modified function is infinitely many-valued, but each branch is separable from the others by a conformal representation. Thus the question of the ambiguity in the sign of $\varpi_1(x,z)$ is subsequent to the enquiry as to the function ζ which will conformably represent the Riemann surface upon a single ζ plane in a manner analogous to that contemplated in chapter XII. §§ 227, 230*.

In what follows however we do not need to enter into the question of the sign of $\varpi_1(x,z)$. It has been shewn in the preceding article that the final formula obtained is independent of the form taken for the function there denoted by $\psi(x)$. It is therefore permissible, for any position of x, to take for it the expression $\sqrt{X(x)}$, with any assigned sign, without attempting to give a law for the continuous variation of this expression. The advantage is in the greater simplicity of $\varpi_1(x,z)$; for example, when x is at any one

* Klein has proposed to deal with the function $\varpi_1(x,z)$ by means of homogeneous variables. The reader may compare *Math. Annal.* XXXVI. (1890) p. 12, and Ritter, *Math. Annal.* XLIV. (1894) pp. 274—284. In the theory of automorphic functions the necessity for homogeneous variables is well established. Cf. § 279 of the present chapter. For the theory of the function $\varpi_1(x,z)$ in the hyperelliptic case see Klein, and Burkhardt, *Math. Annal.* XXXII. (1888).

of the places $n_1, \ldots, n_{p-1}$, the function $\lambda(x, z)$, as defined in § 271, vanishes independently of z; but this is not the case for $\varpi_1(x, z)$.

Ex. i. Prove that

$$\Pi_{a,\,c}^{x,\,z} = \log \frac{\varpi_1(x, a)\,\varpi_1(z, c)}{\varpi_1(x, c)\,\varpi_1(z, a)}.$$

Ex. ii. Prove that any rational function of which the poles are at $a_1, \ldots, a_M$ and the zeros at $\beta_1, \ldots, \beta_M$, can be put into the form

$$\frac{\varpi_1(x, \beta_1) \ldots\ldots \varpi_1(x, \beta_M)}{\varpi_1(x, a_1) \ldots\ldots \varpi_1(x, a_M)} e^{\lambda_1 v_1^{x,\,a} + \ldots\ldots + \lambda_p v_p^{x,\,a}}$$

where $\lambda_1, \ldots, \lambda_p$ are constants, and a is a fixed place.

In what follows, as no misunderstanding is to be apprehended, we shall omit the suffix in the expression $\varpi_1(x, z)$, and denote it by $\varpi(x, z)$. The function $\varpi(\xi, \zeta)$ of chapter XII. does not recur in this chapter.

274. As an application of the formula of § 272 we take the case of the root form $\sqrt{X^{(3)}(x)}/\Phi(x)\sqrt{X(x)}$, where $X^{(3)}(x)$ is a cubic polynomial of the differential coefficients of the integrals of the first kind, having $3(p-1)$ zeros, each of the second order (cf. § 244, Chap. XIII.). Then the poles $\alpha_1, \ldots, \alpha_r$ are the $2p-2$ zeros of any given polynomial $\Phi(x)$, which is linear in the differential coefficients of integrals of the first kind. Thus $r = 2p-2$, $h+1 = 0$, $R, = r + h + 1 + \sum_1^k \lambda_j = 2p - 2 + 0 + 0 = 2p - 2$; $U = \sum_1^{2p-2} v^{x_j,\,\alpha_j}$, and, taking for the function $\psi(x)$, the expression $\sqrt{X(x)}$, the formula becomes

$$Ce^{-2\pi i[h-\frac{1}{2}s'][U-g+\frac{1}{2}s-\tau(h-\frac{1}{2}s')]}\,\Theta\left[\sum_1^{2p-2} v^{x_j,\,\alpha_j} - (g-\tfrac{1}{2}s) - \tau(h-\tfrac{1}{2}s')\right]$$

$$= \frac{\begin{vmatrix} \sqrt{X_1^{(3)}(x_1)} & \cdot & \sqrt{X_{2p-2}^{(3)}(x_1)} \\ \cdot & \cdot & \cdot \\ \sqrt{X_1^{(3)}(x_{2p-2})} & \cdot & \sqrt{X_{2p-2}^{(3)}(x_{2p-2})} \end{vmatrix} \prod_{i=1}^{2p-2}\prod_{j=1}^{2p-2} \varpi(x_i, \alpha_j)}{\prod\limits_{i<j}^{i,\,j=1,\,2,\,\ldots,\,2p-2}\prod \varpi(x_i, x_j) \prod_{i=1}^{2p-2} \Phi(x_1) \ldots\ldots \Phi(x_{2p-2})}.$$

Herein $\Phi(x)$ is a given polynomial with zeros at $\alpha_1, \ldots, \alpha_{2p-2}$, and the forms $\sqrt{X_1^{(3)}(x)}, \ldots, \sqrt{X_{2p-2}^{(3)}(x)}$ are any set of linearly independent forms, derived as in § 245, Chap. XIII., and having $(-g_1, \ldots, -h_1, \ldots, -h_p)$ for characteristic.

From this formula* that of § 250, Chap. XIII. is immediately obtainable. The result is clearly capable of extension to the case of a function

$$\sqrt{X^{(2r+1)}(x)}/\Phi_r(x)\sqrt{X(x)}.$$

* Cf. Weber, *Theorie der Abel'schen Functionen vom Geschlecht* 3, Berlin, 1876, § 24, p. 156; Noether, *Math. Annal.* XXVIII. (1887), p. 367; Klein, *Math. Annal.* XXXVI. (1890), p. 40. For the introduction of ϕ-polynomials as homogeneous variables cf. §§ 110—114, Chap. VI. of the present volume. See also Stahl, *Crelle*, CXI. (1893), p. 106; Pick, *Math. Annal.* XXIX. "Zur Theorie der Abel'schen Functionen."

275. A general application of the formula of § 272 to the case of rational functions may be made by taking $\alpha_1, \ldots, \alpha_r$ to be any places whatever, r being greater than $p-1$. Then $h+1=0$ and $R=r$; and if the general rational function with poles in $\alpha_1, \ldots, \alpha_r, n_1, \ldots, n_{p-1}$ be

$$A_1F_1(x) + \ldots\ldots + A_{r-1}F_{r-1}(x) + A_r,$$

where $A_1, \ldots, A_r$ are constants, and we take for the function $\psi(x)$ the expression $\sqrt{X(x)}$, and modify the constant C which depends in general upon $\alpha_1, \ldots, \alpha_r$, we obtain the result (cf. § 175, Chap. X.)

$$C\Theta\left[\sum_1^r v^{x_i, a_i};\ \tfrac{1}{2}s, \tfrac{1}{2}s'\right], = Ce^{\pi i s'\left[U+\frac{1}{2}s+\frac{1}{4}\tau s'\right]}\,\Theta\left[\sum_1^r v^{x_i, a_i} + \tfrac{1}{2}s + \tfrac{1}{2}\tau s'\right],$$

$$= \frac{\begin{vmatrix} F_1(x_1), & \ldots, & F_{r-1}(x_1), & 1 \\ \ldots & \ldots & \ldots & \ldots \\ F_1(x_r), & \ldots, & F_{r-1}(x_r), & 1 \end{vmatrix} \prod_{i=1}^{r}\prod_{j=1}^{r} \varpi(x_i, \alpha_j)}{\prod_{\substack{i,j=1,\ldots,r \\ i<j}} \varpi(x_i, x_j) \prod_{\substack{i,j=1,\ldots,r \\ i<j}} \varpi(\alpha_i, \alpha_j)} \sqrt{X(x_1)\ldots X(x_r)\,X(\alpha_1)\ldots X(\alpha_r)}.$$

276. This formula includes many particular cases*. We proceed to obtain a more special formula, deduced directly from the result of § 272. Let $\alpha_1, \ldots, \alpha_r = n_1, \ldots, n_{p-1}$. Then the everywhere finite factorial integrals of the associated system are the ordinary integrals of the first kind, and the number, $h+1$, of dV' which vanish in the places $\alpha_1, \ldots, \alpha_r$, $n_1, \ldots, n_{p-1}$, that is, which vanish to the second order in the places $n_1, \ldots, n_{p-1}$, is 1. The number $R, = r + \sum_j \lambda_j + h + 1, = p\,.\,1 + 0 + 1, = p$. The general function having the poles $n_1^2, \ldots, n_{p-1}^2$ is $F(x) = \Phi(x)/X(x)$, where $X(x)$ is the expression employed in § 273, and $\Phi(x)$ denotes the differential coefficient of the general integral of the first kind. Further

$$U = \sum_1^p v^{x_j, a} - \sum_1^{p-1} v^{n_j, a} - v^{\gamma, a}, = \sum_1^{p-1} v^{x_j, n_j} + v^{x_p, \gamma},$$

γ being an arbitrary place. Hence

$$U - \tfrac{1}{2}s - \tfrac{1}{2}\tau s' = \sum_1^p v^{x_j, m_j} - v^{\gamma, m}, = V \text{ say},$$

and

$$e^{\pi i s'(U+\frac{1}{2}s+\frac{1}{2}\tau s')}\,\Theta(U + \tfrac{1}{2}s + \tfrac{1}{2}\tau s') = e^{\pi i s'(V+\Omega_{s,s'})}\,\Theta(V + \Omega_{s,s'})$$

is equal (§ 175, Chap. X.) to

$$e^{\pi i s'(V+\Omega_{s,s'}) - 2\pi i s'(V+\frac{1}{2}\tau s')}\,\Theta(V), = e^{-\pi i s'(V+s)}\,\Theta(V), = -e^{-\pi i V s'}\,\Theta(V),$$

since ss' is an odd integer. Therefore taking for the function $\psi(x)$ the expression $\sqrt{X(x)}$, $\lambda(x, z)$ is $\varpi(x, z)$, and

$$\Delta(x_1, \ldots, x_R) = \begin{vmatrix} \Phi_1(x_1), & . & , \Phi_p(x_1) \\ . & , & . & , & . \\ \Phi_1(x_p), & . & , \Phi_p(x_p) \end{vmatrix} \div \sqrt{X(x_1)\ldots X(x_p)},$$

* Cf. Klein, *Math. Annal.* XXXVI. p. 38.

where $\Phi(x), \ldots, \Phi_p(x)$ denote $dv_1^{x,a}/dt, \ldots, dv_p^{x,a}/dt$. Thus on the whole

$$Ce^{-\pi i V s'}\Theta(V) = \frac{\Delta(x_1, \ldots, x_p)}{\prod\limits_{i<j}^{i,j=1,\ldots,p}\prod \varpi(x_i, x_j)} \prod_{i=1}^{p} [\varpi(x_i, n_1), \ldots, \varpi(x_i, n_{p-1})\, \varpi(x_i, \gamma)],$$

where C is a quantity which, beside the fixed constants of the surface, depends only on the place γ. Let us denote the expression

$$\frac{\varpi(x_i, n_1), \ldots\ldots, \varpi(x_i, n_{p-1})}{\sqrt{X(x_i)}},$$

which clearly has no zeros or poles, by $\mu(x_i)$; then we proceed to shew that in fact $C = A\mu(\gamma)$, where A is a quantity depending only on the fixed constants of the surface, so that we shall have the formula

$$Ae^{-\pi i s' V}\Theta(V) = \frac{\begin{vmatrix} \Phi_1(x_1) & . & \Phi_p(x_1) \\ . & . & . \\ \Phi_1(x_p) & . & \Phi_p(x_p) \end{vmatrix} \mu(x_1), \ldots, \mu(x_p)\, \varpi(x_1, \gamma), \ldots, \varpi(x_p, \gamma)}{\prod\limits_{i<j}^{i,j=1,\ldots,p}\prod \varpi(x_i, x_j)\, \mu(\gamma)},$$

where

$$V = \sum_1^p v^{x_j, m_j} - v^{\gamma, m}.$$

In this formula γ only occurs in the factors

$$\Psi = \frac{\varpi(x_1, \gamma), \ldots\ldots, \varpi(x_p, \gamma)}{\mu(\gamma)};$$

herein the factor $\sqrt{X(\gamma)}$ occurs once in the denominator of each of $\varpi(x_i, \gamma)$, and p times as a denominator in $\mu(\gamma)$; thus this factor does not occur at all. In determining the factors of Ψ, as a function of γ, it will therefore be sufficient to omit this factor. Thus the factor of Ψ at the i-th period loop of the first kind is $e^{\pi i s'(p - \overline{p-1})}$ or $e^{\pi i s'}$. At the i-th period loop of the second kind the factor of $\Theta(v^{x,z} + \frac{1}{2}\Omega_{s,s'})\, e^{\pi i s' v^{x,z}}$ is $e^{-2\pi i (v_i^{x,z} + \frac{1}{2}\tau_{i,i}) - \pi i s_i}$, and therefore the factor of Ψ is

$$e^{-\pi i s_i - 2\pi i (v^{\gamma, x_p} + v^{n_1, x_1} + \ldots\ldots + v^{n_{p-1}, x_{p-1}} + \frac{1}{2}\tau_{i,i})}.$$

Consider now the expression

$$e^{-\pi i s' V}\Theta(V) = e^{\pi i s'(v^{\gamma, m} - v^{x_1, m_1} - \ldots\ldots - v^{x_p, m_p})}\, \Theta(v^{\gamma, m} - v^{x_1, m_1} - \ldots\ldots - v^{x_p, m_p});$$

at the i-th period loop of the first kind, this function, regarded as depending upon γ, has the factor $e^{\pi i s'}$; at the i-th period loop of the second kind it has the factor

$$e^{\pi i (\tau_{i,1} s'_1 + \ldots\ldots + \tau_{i,p} s'_p) - 2\pi i (v^{\gamma, m} - v^{x_1, m_1} - \ldots\ldots - v^{x_p, m_p} + \frac{1}{2}\tau_{i,i})};$$

but since

$$\pi i\,(s_i + \tau_{i,1}s'_1 + \ldots\ldots + \tau_{i,p}s'_p) = 2\pi i\,(v^{m_p,\,m} - v^{n_1,\,m_1} - \ldots\ldots - v^{n_{p-1},\,m_{p-1}}),$$

it follows that

$$\pi i\,(\tau_{i,1}s'_1 + \ldots\ldots + \tau_{i,p}s'_p) - 2\pi i\,(v^{\gamma,\,m} - v^{x_1,\,m_1} - \ldots\ldots - v^{x_p,\,m_p})$$

is equal to

$$-\pi i s_i - 2\pi i\,(v^{\gamma,\,x_p} + v^{n_1,\,x_1} + \ldots\ldots + v^{n_{p-1},\,x_{p-1}});$$

thus, changing γ into x, we have proved that the function of x

$$e^{\pi i s'\,(v^{x,\,m} - v^{x_1,\,m_1} - \ldots\ldots - v^{x_p,\,m_p})}\,\Theta\,(v^{x,\,m} - v^{x_1,\,m_1} - \ldots\ldots - v^{x_p,\,m_p})$$

has the same factors at the period loop as the function, of x, given by

$$\varpi\,(x, x_1)\,\ldots\ldots\,\varpi\,(x, x_p)/\mu\,(x);$$

it is clear that these functions have the same zeros, and no poles.

Hence the formula set down is completely established*.

277. We pass now to the particular case of the formula of § 272 which arises when the fundamental Riemann surface is hyperelliptic, and associated with the equation

$$y^2 = 4\,(x^{2p+2} + \ldots\ldots).$$

Then the places $n_1, \ldots, n_{p-1}$ are branch places. We suppose also that $\mu+1$ of the places $a_1, \ldots, a_r$ are branch places, say the place for which $x = d_1, \ldots, d_{\mu+1}$, and that $\mu+1$ of the places $x_1, \ldots, x_r$ are branch places, say those at which $x = b_1, \ldots, b_{\mu+1}$. It is assumed that the branch places $n_1, \ldots, n_{p-1}, d_1, \ldots, d_{\mu+1}, b_1, \ldots, b_{\mu+1}$ are different from one another. We put $r-(\mu+1) = \nu$; then the determinant of the functions $F_i(x_j)$, (§ 272), regarded as a function of x_1, is a rational function with poles in $n_1, \ldots, n_{p-1}, a_1, \ldots, a_\nu, d_1, \ldots, d_{\mu+1}$ and zero in $x_2, \ldots, x_\nu, b_1, \ldots, b_{\mu+1}$. Provided ν is not less than μ, such a function is of the form

$$\frac{(x_1-n_1)\ldots(x_1-n_{p-1})(x_1-d_1)\ldots(x_1-d_{\mu+1})(x_1-b_1)\ldots(x_1-b_{\mu+1})(x_1, 1)_{\nu-1-\mu} + y_1(x_1, 1)_{\nu-1+\mu}}{(x_1-n_1)\ldots(x_1-n_{p-1})(x_1-d_1)\ldots(x_1-d_{\mu+1})(x_1-a_1)\ldots(x_1-a_\nu)}$$

where the degrees of $(x_1, 1)_{\nu-1-\mu}$, $(x_1, 1)_{\nu-1+\mu}$ are determined by the condition that the function is not to become infinite when x_1 is infinite. When $\nu=\mu$, the terms $(x_1, 1)_{\nu-1-\mu}$ are to be absent. When $\nu < \mu$, the conditions assigned do not determine the function; we shall suppose $\nu \geqq \mu$. The $2\nu-1$ ratios of the coefficients in the numerator are to be determined by the conditions that the numerator vanishes in $x_2, \ldots, x_\nu$ and in the places conjugate†

* See the references given in the note *, § 274, and in particular Klein, *Math. Annal.* XXXVI. p. 39.

† The place conjugate to (x, y) is $(x, -y)$

to $\alpha_1, \dots, \alpha_\nu$. Hence, save for a factor independent of x_1, the determinant of the functions $F_i(x_j)$ is given by

$$\frac{\sqrt{\psi(x_1)}}{(x_1-n_1)\dots(x_1-d_1)\dots(x_1-\alpha_1)\dots}$$

$$\begin{vmatrix} x_1^{\nu-1-\mu}\sqrt{\psi(x_1)}, & \dots, & \sqrt{\psi(x_1)}, & x_1^{\nu-1+\mu}\sqrt{\phi(x_1)}, & \dots, & \sqrt{\phi(x_1)} \\ x_2^{\nu-1-\mu}\sqrt{\psi(x_2)}, & \dots, & \sqrt{\psi(x_2)}, & x_2^{\nu-1+\mu}\sqrt{\phi(x_2)}, & \dots, & \sqrt{\phi(x_2)} \\ \dots & \dots & \dots & \dots & \dots & \dots \\ -\alpha_1^{\nu-1-\mu}\sqrt{\psi(\alpha_1)}, & \dots, & -\sqrt{\psi(\alpha_1)}, & \alpha_1^{\nu-1+\mu}\sqrt{\phi(\alpha_1)}, & \dots, & \sqrt{\phi(\alpha_1)} \\ -\alpha_2^{\nu-1-\mu}\sqrt{\psi(\alpha_2)}, & \dots, & -\sqrt{\psi(\alpha_2)}, & \alpha_2^{\nu-1+\mu}\sqrt{\phi(\alpha_2)}, & \dots, & \sqrt{\phi(\alpha_2)} \\ \dots & \dots & \dots & \dots & \dots & \dots \end{vmatrix}$$

wherein $\psi(x) = (x-n_1)\dots(x-n_{p-1})(x-d_1)\dots(x-d_{\mu+1})(x-b_1)\dots(x-b_{\mu+1})$, $\phi(x) = y^2/\psi(x)$, and the determinant has 2ν rows and columns; denoting this determinant by $D_{\phi,\psi}$, the determinant of the functions $F_i(x_j)$ (§ 272) is therefore equal to

$$D_{\phi,\psi}\prod_{i=1}^{\nu}\frac{1}{(x_i-\alpha_1)\dots(x_i-\alpha_\nu)\sqrt{(x_i-n_1)\dots(x_i-n_{p-1})}}\sqrt{\frac{(x_i-b_1)\dots(x_i-b_{\mu+1})}{(x_i-d_1)\dots(x_i-d_{\mu+1})}}.$$

Hence, from § 272, taking $\psi(x) = \sqrt{(x-n_1)\dots(x-n_{p-1})}$, so that $\varpi(x, z)$ will denote

$$\frac{\Theta(v^{x,z}+\tfrac{1}{2}\Omega_{s,s'})\,e^{\pi i s' v^{x,z}}}{\sqrt{(x-n_1)\dots(x-n_{p-1})(z-n_1)\dots(z-n_{p-1})}},$$

we have

$$C\Theta\left[\sum_1^{\nu} v^{x_i,\alpha_i}+\sum_1^{\mu+1} v^{b_i,d_i};\ \tfrac{1}{2}s,\ \tfrac{1}{2}s'\right]$$

$$=\prod_{i=1}^{\nu}\prod_{j=1}^{\mu+1}\frac{\varpi(x_i,d_j)}{\varpi(x_i,b_j)}\sqrt{\frac{x_i-b_j}{x_i-d_j}}\ \frac{D_{\phi,\psi}\prod_{i=1}^{\nu}\prod_{j=1}^{\nu}\varpi(x_i,\alpha_j)}{\prod_{i<j}^{i,j=1,\dots,\nu}\varpi(x_i,x_j)\prod_{i=1}^{\nu}\prod_{j=1}^{\nu}(x_i-\alpha_j)},$$

where C is independent of $x_1, \dots, x_\nu$.

Now, if b, d be any two branch places, and a an assigned branch place,

$$\frac{\varpi(x,d)}{\varpi(x,b)} = \frac{\Theta(v^{x,d};\ \tfrac{1}{2}s,\ \tfrac{1}{2}s')}{\Theta(v^{x,b};\ \tfrac{1}{2}s,\ \tfrac{1}{2}s')} = \frac{\Theta(v^{x,a}-v^{d,a};\ \tfrac{1}{2}s,\ \tfrac{1}{2}s')}{\Theta(v^{x,a}-v^{b,a};\ \tfrac{1}{2}s,\ \tfrac{1}{2}s')},$$

and hence, if

$$v_i^{d,a} = \tfrac{1}{2}(\delta_i+\delta_1'\tau_{i,1}+\dots\dots+\delta_p'\tau_{i,p}), \qquad (i=1,2,\dots,p),$$

$$v_i^{b,a} = \tfrac{1}{2}(\beta_i+\beta_1'\tau_{i,1}+\dots\dots+\beta_p'\tau_{i,p}),$$

where $\beta_1, \dots, \beta_p', \delta_1, \dots, \delta_p'$ are integers, we have (§ 175, Chap. X.)

$$\frac{\varpi(x,d)}{\varpi(x,b)} = Ae^{\pi i(\delta'-\beta')v^{x,a}}\frac{\Theta[v^{x,a};\ \tfrac{1}{2}(s-\delta),\ \tfrac{1}{2}(s'-\delta')]}{\Theta[v^{x,a};\ \tfrac{1}{2}(s-\beta),\ \tfrac{1}{2}(s'-\beta')]},$$

where A is independent of x. Thus the expression

$$e^{-\pi i(\delta'-\beta')v^{x,\,a}} \frac{\varpi(x, d)}{\varpi(x, b)} \sqrt{\frac{x-b}{x-d}},$$

which clearly has no poles or zeros, is such that its factors at the period loops are all ± 1. The square of this function is therefore a constant, and the expression itself is a constant.

Therefore if

$$\sum_1^{\mu+1} v_i^{d_i,\,b_i} = \tfrac{1}{2}(\sigma_i + \sigma_1'\tau_{i,1} + \ldots\ldots + \sigma_p'\tau_{i,p}),$$

where $\sigma_1, \ldots, \sigma_p'$ are integers, it follows that the function

$$e^{-\pi i\sigma'(v^{x_1,\,a_1}+\ldots\ldots+v^{x_\nu,\,a_\nu})} \prod_{i=1}^{\nu} \prod_{j=1}^{\mu+1} \frac{\varpi(x_i, d_j)}{\varpi(x_i, b_j)} \sqrt{\frac{x_i-b_j}{x_i-d_j}}$$

is independent of $x_1, \ldots, x_\nu$. Further

$$\Theta(u - \tfrac{1}{2}\sigma - \tfrac{1}{2}\tau\sigma';\ \tfrac{1}{2}s, \tfrac{1}{2}s') = Be^{\pi i\sigma' u}\,\Theta[u;\ \tfrac{1}{2}(s-\sigma), \tfrac{1}{2}(s'-\sigma')]$$

by § 175, Chap. X. Thus on the whole we have

$$C\Theta\left[\sum_1^{\nu} v^{x_i,\,a_i};\ \tfrac{1}{2}(s-\sigma), \tfrac{1}{2}(s'-\sigma')\right]$$

$$= D_{\phi,\psi} \prod_{i=1}^{\nu}\prod_{j=1}^{\nu} \varpi(x_i, a_j) \Big/ \prod_{i<j}^{i,j=1,\ldots,\nu} \varpi(x_i, x_j) \prod_{i=1}^{\nu}\prod_{j=1}^{\nu}(x_i - a_j) \prod_{i<j}^{i,j=1,\ldots,\nu} \varpi(a_i, a_j),$$

where C is independent of $x_1, \ldots, x_\nu$. Hence we can infer that C is in fact independent also of $a_1, \ldots, a_\nu$. For when the sets $x_1, \ldots, x_\nu$, $a_1, \ldots, a_\nu$ are interchanged, $D_{\phi,\psi}$ is multiplied by $(-)^{\nu^2+\nu-\mu} = (-1)^\mu$, and, since $\varpi(x, z) = -\varpi(z, x)$, this is also the factor by which the whole right-hand side is multiplied. The theta function on the left-hand side is also multiplied by ± 1. Thus the square of the ratio of the right-hand side to the theta function on the left is unaltered by the interchange of the set $x_1, \ldots, x_\nu$ with the set $a_1, \ldots, a_\nu$. Thus C^2 is independent of $x_1, \ldots, x_\nu$ and unaltered when $x_1, \ldots, x_\nu$ are changed into $a_1, \ldots, a_\nu$. Hence C is an absolute constant.

It follows that the characteristic $\tfrac{1}{2}(s-\sigma)$, $\tfrac{1}{2}(s'-\sigma')$, and the theta functions, are even or odd according as μ is even or odd.

In the notation of § 200, Chap. XI., the half-periods $\tfrac{1}{2}\Omega_{s,s'}$ are given by

$$\tfrac{1}{2}\Omega_{s,s'} = v^{a_p,\,a} - v^{n_1,\,a_1} - \ldots\ldots - v^{n_{p-1},\,a_{p-1}};$$

hence, if the half-periods given by

$$v^{a_1,\,a} + \ldots\ldots + v^{a_p,\,a}$$

be denoted by $\tfrac{1}{2}\Omega$, the half-periods associated with the characteristic $\tfrac{1}{2}(s-\sigma)$, $\tfrac{1}{2}(s'-\sigma')$ are congruent to expressions given by

$$\tfrac{1}{2}\Omega + v^{n_1,\,a} + \ldots\ldots + v^{n_{p-1},\,a} + v^{b_1,\,a} + \ldots\ldots + v^{b_{\mu+1},\,a} + v^{d_1,\,a} + \ldots\ldots + v^{d_{\mu+1},\,a},$$

while ψ, which is of degree $p+1+2\mu$, is equal to

$$(x-n_1)\dots(x-n_{p-1})(x-b_1)\dots(x-b_{\mu+1})(x-d_1)\dots(x-d_{\mu+1});$$

by means of the formula (§ 201, Chap. XI.)

$$v^{a_1,\,a} + \dots\dots + v^{a_p,\,a} + v^{c_1,\,a} + \dots\dots + v^{c_p,\,a} + v^{c,\,a} \equiv 0,$$

the half-periods associated with the characteristic $\frac{1}{2}(s-\sigma)$, $\frac{1}{2}(s'-\sigma')$ can be reduced to be congruent to expressions denoted by

$$\tfrac{1}{2}\Omega + v^{e_1,\,a} + \dots\dots + v^{e_{p-2\mu},\,a} + v^{e_{p+1-2\mu},\,a},$$

where $e_1, \dots, e_{p-2\mu+1}$ are given by

$$\phi = 4(x-e_1)\dots\dots(x-e_{p+1-2\mu});$$

also, in taking all possible odd half-periods $\frac{1}{2}\Omega_{s,\,s'}$, all possible sets of $p-1$ of the branch places will arise for the set $n_1, \dots, n_{p-1}$. Hence it follows that the formula obtained includes as many results as there are ways of resolving $(x, 1)_{2p+2}$ into two factors $\phi_{p+1-2\mu}$, $\psi_{p+1+2\mu}$, of orders $p+1-2\mu$, $p+1+2\mu$, and (§ 201) that all possible half-integer characteristics arise, each associated with such a resolution. We have in fact, corresponding to $\mu = 0, 1, 2, \dots$, $E\left(\frac{p+1}{2}\right)$, a number of resolutions given by

$$\tfrac{1}{2}\binom{2p+2}{p+1} + \binom{2p+2}{p+3} + \binom{2p+2}{p+5} + \dots\dots, = 2^{2p}.$$

It has been shewn (§ 273) that the expression $\varpi(x, z)$ may be derived, by proceeding to a limit, from the integral $\Pi_{a,\,c}^{x,\,z}$. Hence the formula that has been obtained furnishes a definition of the theta function in terms of the algebraic functions and their integrals, and has been considered from this point of view by Klein, to whom it is due. After the investigation given above it is sufficient to refer* the reader, for further development, to Klein, *Math. Annal.* XXXII. (1888), p. 351, and to the papers there quoted.

Ex. i. Prove that the function $\Theta\left[u;\ \frac{1}{2}(s-\sigma),\ \frac{1}{2}(s'-\sigma')\right]$ vanishes to the μth order for zero values of the arguments.

Ex. ii. In the notation of § 200, Chap. XI., prove, from the result here obtained, that each of the sums

$$\sum_{i=1}^{4r+3} v^{c_i,\,a},\quad \sum_{i=1}^{4r+2} v^{c_i,\,a},\quad v^{a_j,\,a} + \sum_{i=1}^{4r+4} v^{c_i,\,a},\quad v^{a_j,\,a} + \sum_{i=1}^{4r+3} v^{c_i,\,a}$$

represents an odd half-period; here c_i is any one of the places $c, c_1, \dots, c_p$, a_i is any one of the places $a_1, \dots, a_p$, a_j is any one of the places $a_1, \dots, a_p$, and r is an arbitrary integer

* See also Brill, *Crelle*, LXV. (1866), p. 273; and the paper of Bolza, *American Journal*, vol. XVII., referred to § 221, *note*, where Klein's formula is fundamental.

By means of the rule investigated on page 298, of the present volume, the characteristic $\frac{1}{2}(s-\sigma)$, $\frac{1}{2}(s'-\sigma')$ can be immediately calculated from the formula here (p. 436) given for it. Cf., also, Burkhardt, *Math. Annal.* XXXII., p. 426; Thompson, *American Journal*, XV. (1893), p. 91.

whose least value is zero, and whose greatest value is given by the condition that i cannot be greater than $p+1$. Prove also that each of the sums

$$\sum_{i=1}^{4r+1} v^{c_i, a}, \quad \sum_{i=1}^{4r} v^{c_i, a}, \quad v^{a_j, a}+\sum_{i=1}^{4r+2} v^{c_i, a}, \quad v^{a_j, a}+\sum_{i=1}^{4r+1} v^{c_i, a}$$

represents an even half-period. For a more general result cf. the examples of § 303 (Chap. XVII.).

Ex. iii. By taking $\nu=p+1$, $\mu=0$, and the places b, d so that $\frac{1}{2}\Omega_{s,s'}\equiv v^{b,d}$, finally putting $n_1, \dots, n_{p-1}, b, d$ for $a_1, \dots, a_p, a_{p+1}$, obtain, from the formula, the result

$$\frac{\Theta(v^{x,a}+v^{x_1,a_1}+\dots\dots+v^{x_p,a_p})}{\Theta(v^{z,a}+v^{x_1,a_1}+\dots\dots+v^{x_p,a_p})}=\frac{\sqrt{\psi(x)}}{\sqrt{\psi(z)}}\frac{\varpi(x,a)}{\varpi(z,a)}\frac{z-a}{x-a}\prod_{i=1}^{p}\frac{(x-x_i)(z-a_i)}{(x-a_i)(z-x_i)}e^{-\Pi_{x_i,a_i}^{x,z}},$$

where $\Pi_{x_i,a_i}^{x,z}$ replaces $\log\dfrac{\varpi(x,x_i)\,\varpi(z,a_i)}{\varpi(x,a_i)\,\varpi(z,x_i)}$, $\psi(x)=(x-a)\dots(x-a_p)$, and the branch places $a, a_1, \dots, a_p$ are, as in § 203, Chap. XI., such that the theta function in the numerator of the left-hand side vanishes as a function of x at the places $\xi_1, \dots, \xi_p$, conjugate to $x_1, \dots, x_p$; and verify the result *à priori*. By the substitution

$$\frac{(x-x_i)(z-a_i)}{(x-a_i)(z-x_i)}e^{-\Pi_{x_i,a_i}^{x,z}}=e^{\Pi_{\xi_i,a_i}^{x,z}}$$

this formula can be further simplified. Deduce the results

$$\Pi_{x_1,z_1}^{x,z}+\dots\dots+\Pi_{x_p,z_p}^{x,z}=\log\frac{\Theta(v^{x,a}-v^{x_1,a_1}-\dots\dots-v^{x_p,a_p})\,\Theta(v^{z,a}-v^{z_1,a_1}-\dots\dots-v^{z_p,a_p})}{\Theta(v^{z,a}-v^{x_1,a_1}-\dots\dots-v^{z_p,a_p})\,\Theta(v^{x,a}-v^{z_1,a_1}-\dots\dots-v^{z_p,a_p})},$$

$$Z_i(u-v^{x,a})-Z_i(u-v^{z,a})=\Gamma_{x_1}^{x,z}\frac{\partial x_1}{\partial u_i}\Big/\frac{dx_1}{dt}+\dots\dots+\Gamma_{x_p}^{x,z}\frac{\partial x_p}{\partial u_i}\Big/\frac{dx_p}{dt},$$

where $u=v^{x_1,a_1}+\dots\dots+v^{x_p,a_p}$, $Z_i(u)-\dfrac{\partial}{\partial u_i}\log\Theta(u)$, and $\dfrac{dx_1}{dt}, \dots$ are as in § 123, Chap. VII.

These results have already been given (Chap. X.).

278. It is immediately proved, by the formula (§ 187)

$$e^{\Pi_{z,a}^{x,\gamma}}=\frac{\Theta(v^{x,z}+\frac{1}{2}\Omega_{s,s'})\,\Theta(v^{\gamma,a}+\frac{1}{2}\Omega_{s,s'})}{\Theta(v^{x,a}+\frac{1}{2}\Omega_{s,s'})\,\Theta(v^{\gamma,z}+\frac{1}{2}\Omega_{s,s'})}$$

that the general expression of a factorial function given in § 254 can be written in the form

$$\prod_1^N\left[\Theta(v^{x,\beta_i}+\tfrac{1}{2}\Omega_{s,s'})\,e^{\pi i s' v^{x,\beta_i}}\right]$$

$$\div\, e^{2\pi i\sum_1^p(h_i+H_i)v_i^{x,\gamma}}\prod_1^M\left[\Theta(v^{x,a_i}+\tfrac{1}{2}\Omega_{s,s'})\,e^{\pi i s' v^{x,a_i}}\right]\prod_1^k\left[\Theta(v^{x,c_i}+\tfrac{1}{2}\Omega_{s,s'})\,e^{\pi i s' v^{x,c_i}}\right]^{\lambda_i}.$$

And, by the use of the expression $\varpi(x,z)$, this may be put into the form

$$e^{-2\pi i\sum_1^p(h_i+H_i)v_i^{x,\gamma}}\prod_1^N\varpi(x,\beta_i)\prod_1^M\left[\varpi(x,a_i)\right]^{-1}\prod_1^k\left[\varpi(x,c_i)\right]^{-\lambda_i}.$$

Ex. i. In the hyperelliptic case associated with an equation of the form

$$y^2=(x, 1)_{2p+2},$$

if $\bar{x}$ denote the place conjugate to the place x, it follows from the formula of § 273 that

$$\varpi(x, z)=(x-z)\, e^{\frac{1}{2}\Pi^{x, z}_{\bar{x}, \bar{z}}},$$

unless x or z is a branch place.

Ex. ii. In the hyperelliptic case, if $k, k_1, \dots, k_p$ denote branch places, and

$$\phi(x)=(x-k)(x-k_1)\dots(x-k_p)$$

and the equation associated with the surface be $y^2=f(x)$, where $f(x)=\phi(x)\psi(x)$, and if we take places $x, x_1, \dots, x_p, z, z_1, \dots, z_p$, such that

$$v_i^{x_1, k_1}+\dots\dots+v_i^{x_p, k_p}\equiv v_i^{x, k}, \quad v_i^{z_1, k_1}+\dots\dots+v_i^{z_p, k_p}\equiv v_i^{z, k}, \qquad (i=1, 2, \dots, p),$$

then it is easily seen that the rational function having $\bar{x}, x_1, \dots, x_p$ as zeros and $\bar{z}, z_1, \dots, z_p$ as poles, can be put into the form $[y'\phi(x)+y\phi(x')]\div[y'\phi(z)+s\phi(x')]$, where x', y' are the variables and s is the value of y' at the place z. Hence prove, by Abel's theorem, that

$$e^{\frac{1}{2}\Pi^{\bar{x}, \bar{z}}_{x, z}}\frac{\sqrt{\phi(x)\psi(z)}+\sqrt{\phi(z)\psi(x)}}{2\sqrt[4]{f(x)f(z)}}=e^{-\frac{1}{2}(\Pi^{x, z}_{x_1, z_1}+\dots\dots+\Pi^{x, z}_{x_p, z_p})}.$$

Ex. iii. Suppose now that $a, a_1, \dots, a_p$ are the branch places used in chapter XI. (§ 200), so that

$$e^{\Pi^{x, z}_{x_1, z_1}+\dots\dots+\Pi^{x, z}_{x_p, z_p}}=\frac{\Theta(v^{x, a}-v^{x_1, a_1}-\dots\dots-v^{x_p, a_p})\,\Theta(v^{z, a}-v^{z_1, a_1}-\dots\dots-v^{z_p, a_p})}{\Theta(v^{x, a}-v^{z_1, a_1}-\dots\dots-v^{z_p, a_p})\,\Theta(v^{z, a}-v^{x_1, a_1}-\dots\dots-v^{x_p, a_p})},$$

and suppose further that $\frac{1}{2}\Omega, =\frac{1}{2}(s+\tau s')$, is an *even* half-period such that

$$v^{k_1, a_1}+\dots\dots+v^{k_p, a_p}=v^{k, a}+\tfrac{1}{2}\Omega, \quad v^{x_1, a_1}+\dots\dots+v^{x_p, a_p}=v^{x, a}+\tfrac{1}{2}\Omega,$$

and

$$v^{z_1, a_1}+\dots\dots+v^{z_p, a_p}=v^{z, a}+\tfrac{1}{2}\Omega,$$

then deduce that

$$\frac{\Theta(v^{x, z}+\frac{1}{2}\Omega)\, e^{\pi i s' v^{x, z}}}{\Theta(\frac{1}{2}\Omega)}=\varpi(x, z)\frac{\sqrt{\phi(x)\psi(z)}+\sqrt{\phi(z)\psi(x)}}{2(x-z)\sqrt[4]{f(x)f(z)}}.$$

The results of examples i, ii, iii are given by Klein.

Ex. iv. Prove that, if $z, \zeta, c_1, \dots, c_p$ be arbitrary places, and $\gamma_1, \dots, \gamma_p$ be such that the places $\zeta, \gamma_1, \dots, \gamma_p$ are coresidual with the places $z, c_1, \dots, c_p$, then

$$\psi(x, \zeta; z, c_1, \dots, c_p)=\frac{\varpi(x, \zeta)}{\varpi(x, z)\,\varpi(\zeta, z)}\, e^{\Pi^{x, z}_{\gamma_1, c_1}+\dots\dots+\Pi^{x, z}_{\gamma_p, c_p}};$$

hence deduce, by means of the result given in Ex. iv., page 174, that

$$\varpi(z, \zeta)=\frac{e^{\frac{1}{2}(\Pi^{z, \zeta}_{c_1, \gamma_1}+\dots\dots+\Pi^{z, \zeta}_{c_p, \gamma_p})}}{\sqrt{D_\zeta\psi(\zeta, a; z, c_1, \dots, c_p)}},$$

where a is an arbitrary place.

279. The theory of the present chapter may be considered from another point of view. We have already seen, in chapter XII., that the theory of rational functions and their integrals may be derived with a fundamental surface consisting of a portion of a single plane bounded by circles, and the

change of independent variables involved justified itself by suggesting an important function, $\varpi(\zeta, \gamma)$. We explain now*, as briefly as possible, a more general case, in which the singular points, $c_1, \ldots, c_k$, of this chapter, are brought into evidence.

Suppose that a function ζ exists whereby the Riemann surface, dissected as in § 253, can be conformally represented upon the inside of a closed curvilinear polygon, in the plane of ζ, whose sides are arcs of circles†; to the four sides, (a_i), (a_i'), (b_i), (b_i'), of a period-pair-loop are to correspond four sides of the polygon, to the two sides of a cut (γ) are to correspond two sides of the polygon; the polygon will therefore have $2(2p+k)$ sides.

Fig. 11.

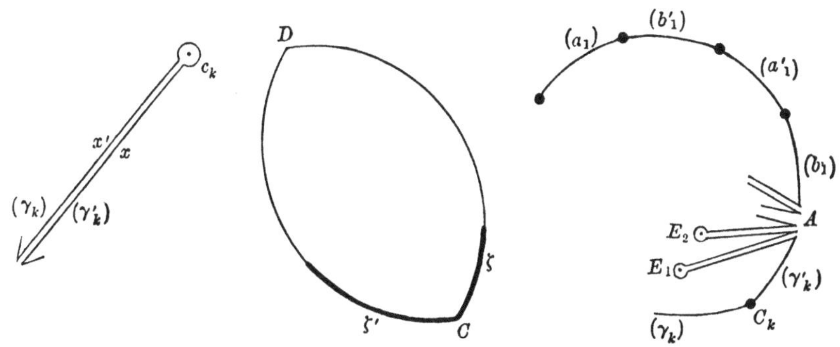

Then it is easily seen that if C be the value of ζ at the angular point C of the polygon, which corresponds to one of the singular points $c_1, \ldots, c_k$ on the Riemann surface, and D be the value of ζ at the other intersection‡ of the circular arcs which contain the sides of the polygon meeting in C, we can pass from one of these sides to the other by a substitution of the form

$$\frac{\zeta'-C}{\zeta'-D} = e^{\frac{2\pi i}{l}} \frac{\zeta-C}{\zeta-D},$$

where $2\pi/l$ is the angle C of the polygon, (l being supposed an integer other than zero); as we pass from a point ζ of one of these sides to the corresponding point of the other side, the argument of the function $[(\zeta-C)/(\zeta-D)]^l$ increases by 2π; if therefore t be the infinitesimal at the corresponding singular point on the Riemann surface, we may write, for small values of t, $(\zeta-C)/(\zeta-D)=t^{\frac{1}{l}}$, so that $\zeta-C=t^{\frac{1}{l}}(C-D)(1-t^{\frac{1}{l}})^{-1}$. Further if ζ, ζ' be corresponding points

* Klein, *Math. Annal.* XXI. (1883), "Neue Beiträge zur Riemann'schen Functionentheorie"; Ritter, *Math. Annal.* XLI. (1893), p. 4; Ritter, *Math. Annal.* XLIV. (1894), p. 342.

† See Forsyth, *Theory of Functions*, chapter XXII., Poincaré, *Acta Math.* vols. I.—V. We may suppose that the polygon is such as gives rise to single-valued automorphic functions.

‡ Supposed to be outside the curvilinear polygon.

on the sides of the polygon which meet in C, we have for small values of t,

$$d\zeta = \frac{1}{l}(C - D)\, t^{\frac{1}{l}-1}\, dt,\ d\zeta' = \frac{1}{l}(C - D)\, t^{\frac{1}{l}-1}\, e^{2\pi i\left(\frac{1}{l}-1\right)}\, dt, \text{ or } d\zeta'/d\zeta = e^{2\pi i\left(\frac{1}{l}-1\right)}$$

ultimately, the factor omitted being a power series in $t^{\frac{1}{l}}$ or $(\zeta - C)/(\zeta - D)$, whose first term is unity.

We shall suppose now that the numbers $\lambda_1, \ldots, \lambda_k$ of this chapter are given by $\lambda_i = -m_i/l_i$, where m_i, l_i are positive integers. Then a function whose behaviour near c_i is that of an expression of the form $t^{-\lambda}\phi$, will, near C_i, behave like $(\zeta - C_i)^{m_i}\phi$, that is, will vanish a certain integral number of times. Further, for a purpose to be afterwards explained, we shall adjoin to the k singular points $c_1, \ldots, c_k$, m others, $e_1, \ldots, e_m$, for each of which the numbers λ are the same and equal to $-\epsilon$, so that, if t be the infinitesimal at any one of the places $e_1, \ldots, e_m$, the factorial functions considered behave like $t^{\epsilon}\phi$ at this place. These additional singular points, like the old, are supposed to be taken out from the surface by means of cuts $(\epsilon_1), \ldots, (\epsilon_m)$; and it is supposed that the corresponding curves in the curvilinear polygon of the ζ-plane are also cuts passing to the interior of the polygon, as in the figure, so that at the point E_1 of the ζ-plane which corresponds to the place e_1 of the Riemann surface, ζ is of the form $\zeta = E_1 + t\phi$, where ϕ is finite and not zero for small values of t, t being the infinitesimal at e_1.

Factorial functions having these new singular points as well as the original singular points will be denoted by a bar placed over the top.

Let dv denote the differential of an ordinary Riemann integral of the first kind which has $p-1$ zeros of the second order, at the places $n_1, \ldots, n_{p-1}$. Consider the function

$$Z_2 = \sqrt{\frac{dv}{d\zeta}}\, e^{-\frac{1}{2}\sum_{i=1}^{k}\left(1-\frac{1}{l_i}\right)\Pi_{c_i,\, c}^{x,\, a} - \sum_{1}^{p-1}\Pi_{n_i,\, c}^{x,\, a} + \frac{\rho}{2m}\sum_{1}^{m}\Pi_{e_i,\, c}^{x,\, a}};$$

where a, c are arbitrary places, and ρ is determined so that Z_2 is not infinite at the place c, or

$$\sum_{i=1}^{k}\left(1 - \frac{1}{l_i}\right) + 2p - 2 = \rho;$$

this function is nowhere infinite on the Riemann surface; it vanishes to the first order only at $\zeta = \infty$; for each of the cuts $(\epsilon_1), \ldots, (\epsilon_m)$ it has a factor $e^{\frac{\pi i\rho}{m}}$; at a singular point c_i it is expressible as a power series in $t^{\frac{1}{l}}$, or $(\zeta - C)/(\zeta - D)$, whose first term is unity. The values of Z_2 at the two sides of a period loop are such that $Z_2'/Z_2 = \sqrt{d\zeta/d\zeta'}$; but since these two sides correspond, on the ζ-plane, to arcs of circles which can be transformed into one another by a substitution of the form $\zeta' = (\alpha\zeta + \beta)/(\gamma\zeta + \delta)$, wherein we suppose $\alpha\delta - \beta\gamma = 1$, it follows that $Z_2'/Z_2 = \gamma\zeta + \delta$. If then we also introduce the function $Z_1, = \zeta Z_2$, we have for the two sides of a period loop, equations of the form

$$Z_2' = \gamma Z_1 + \delta Z_2, \quad Z_1' = \alpha Z_1 + \beta Z_2.$$

Consider now a function

$$f = \overline{K}/Z_2^R,$$

where $\overline{K}$ is a factorial function with the $k+m$ singular points, and $R = 2m\epsilon/\rho$. At a singular point c_i, or C_i, its behaviour is that of a power series in $t^{\frac{1}{l}}$ or $(\zeta - C)/(\zeta - D)$, multiplied by $(\zeta - C_i)^{m_i}$; at a singular point e_i, or E_i, its behaviour is that of a power series in the infinitesimal t multiplied by

$$t^{\epsilon} \bigg/ \left(t^{\frac{\rho}{2m}}\right)^{\frac{2m\epsilon}{\rho}}$$

or unity; at a period loop it is multiplied by a factor of the form $\mu(\gamma\zeta + \delta)^{-R}$, where μ is the factor of $\overline{K}$. The function has therefore the properties of functions expressible by series of the form*

$$\Sigma R(\zeta_i)(\gamma_i \zeta + \delta_i)^R,$$

wherein the notation is, that $\zeta_i = (\alpha_i\zeta + \beta_i)/(\gamma_i\zeta + \delta_i)$ is in turn every combination of substitutions whereby the sides of the curvilinear polygon are related in pairs and $R(\zeta_i)$ is a rational function of ζ_i. The equation connecting the values f', f, of the function f, at the two sides of a period loop, may be put into the form

$$(\gamma Z_1 + \delta Z_2)^R f' = \mu Z_2^R f;$$

and *we may regard* $Z_2^R f$, *or* $\overline{K}$, *as a homogeneous form in the variables* Z_1, Z_2, *of dimension* R.

The difference between the number of zeros and poles of such a factorial function $\overline{K}$ is (§ 254)

$$\Sigma\lambda, = \Sigma\left(-\frac{m_i}{l_i}\right) - \epsilon m, = \Sigma\left(-\frac{m_i}{l_i}\right) - \tfrac{1}{2}\rho R, = \Sigma\left(-\frac{m_i}{l_i}\right) - R(p-1) - \tfrac{1}{2}R\Sigma\left(1 - \frac{1}{l_i}\right),$$

$$= \Sigma\left(-\frac{m_i}{l_i} + \frac{R}{2l_i}\right) - R(p-1) - \tfrac{1}{2}Rk;$$

adding the proper corrections for the zeros of the automorphic form $\overline{K}$ at the angular points $C_1, \ldots, C_k$ (Forsyth, *Theory of Functions*, p. 645) we have, for the excess of the number of zeros of the automorphic form over the number of poles

$$\Sigma\lambda + \Sigma\frac{m_i}{l_i} = -\frac{R}{2}\left[2p - 2 + k + m + 1 - \left(\Sigma\frac{1}{l_i} + m + 1\right)\right]$$

$$= -\frac{R}{2}\left[2p - 2 + q - \Sigma\frac{1}{\mu}\right],$$

where $q = k + m + 1$, $\Sigma\frac{1}{\mu} = \Sigma\frac{1}{l_i} + m + 1$.

We may identify this result with a known formula for automorphic

* Forsyth, *Theory of Functions*, p. 642. The quantity R is, in Forsyth, taken equal to $-2m$.

functions [Forsyth, *Theory of Functions*, p. 648; if in the formula $m\left(n-1-\Sigma\frac{1}{\mu}\right)$, there given, we substitute, by the formula of p. 608, § 293, $n=2N-1+q$, we obtain $m\left(2N-2+q-\Sigma\frac{1}{\mu}\right)$]; for each of the angular points $C_1, \ldots, C_k$ is a cycle by itself, each of the points $E_1, \ldots, E_m$ is a cycle by itself, and the remaining angular points together constitute one cycle (cf. Forsyth, p. 596); the sum of the angles at the first k cycles is $2\pi\Sigma\frac{1}{l_i}$, the sum of the angles at the second m cycles is $2\pi m$, the sum of the angles at the other cycle is 2π*.

There is a way in which the adjoint system of singular points $e_1, \ldots, e_m$ may be eliminated from consideration. Imagine a continuously varying quantity, x_2, which is zero to the first order at $e_1, \ldots, e_m$ and is never infinite, and put $x_1 = xx_2$; the expression $\overline{K}x_2^{-\epsilon}$ may then be regarded as a homogeneous form in x_1, x_2 on the Riemann surface, without singular points at $e_1, \ldots, e_m$; and instead of the function Z_2 we may introduce the form $\zeta_2 = Z_2 x_2^{-\frac{\rho}{2m}}$, which is then without factor for the cuts $(\epsilon_1), \ldots, (\epsilon_m)$, or, as we may say, is *unbranched* at the places $e_1, \ldots, e_m$; and may also put $\zeta_1 = \zeta\zeta_2$.

Thus, (i), a factorial function, considered on the ζ-plane, is a homogeneous automorphic form, (ii), introducing homogeneous variables on the Riemann surface, the consideration of factorial functions may be replaced by the consideration of homogeneous factorial forms.

Ex. Shew that the form

$$P(x, z) = x_2^{\frac{1}{m}} f(z)\, e^{\Pi_{z,c}^{x,a} - \frac{1}{m}\left(\Pi_{e_1,c}^{x,a} + \ldots\ldots + \Pi_{e_m,c}^{x,a}\right) + \sum_{i,j} \lambda_{i,j} v_i^{x,a} v_j^{z,c}},$$

where a, c are arbitrary places and $\lambda_{i,j}$ are constants, is unbranched at $e_1, \ldots, e_m$, that it has no poles, and vanishes only at the place z. Here $f(z)$ is to be chosen so that, when x approaches z, the ratio of $P(x, z)$ to the infinitesimal at z is unity. At the t-th period loop of the second kind the function has a factor $(-)^M$ where

$$M = 2\pi i r + \frac{2\pi i}{m}(q'_2 - q) - \frac{2\pi i}{m}\left(v_t^{e_1,c} + \ldots\ldots + v_t^{e_m,c}\right) + \sum_{i,j} \lambda_{i,j} v_j^{z,c} \tau_{i,t},$$

$q'_2 - q$ denoting the number of circuits, made in passing from one side of the period loop to the other, of x_2 about $x_2=0$ other than those for which x encloses places $e_1, \ldots, e_m$, and r denoting the number of circuits† of x about z.

* The formula is given by Ritter, *Math. Annal.* XLIV. p. 360 (at the top), the quantity there denoted by q being here $-\frac{1}{2}\rho$. We do not enter into the conditions that the automorphic form be single-valued.

† The reader will compare the formula given by Ritter, *Math. Annal.* XLIV. p. 291. It may be desirable to call attention to the fact that the notation $\sigma+1$, $\sigma'+1$, as here used, does not coincide with that used by Ritter. The quantities denoted by him by σ, σ' may, in a sense, be said to correspond respectively to those denoted here, for the factorial system including the singular points $e_1, \ldots, e_m$, by $\sigma'+1$ and ϖ'.

CHAPTER XV.

RELATIONS CONNECTING PRODUCTS OF THETA FUNCTIONS—INTRODUCTORY.

280. As preparatory to the general theory of multiply-periodic functions of several variables, and on account of the intrinsic interest of the subject, the study of the algebraic relations connecting the theta functions is of great importance. The multiplicity and the complexity of these relations render any adequate account of them a matter of difficulty; in this volume the plan adopted is as follows:—In the present chapter are given some preliminary general results frequently used in what follows, with some examples of their application. The following Chapter (XVI.) gives an account of a general method of obtaining theta relations by actual multiplication of the infinite series. In Chapter XVII. a remarkable theory of groups of half-integer characteristics, elaborated by Frobenius, is explained, with some of the theta relations that result; from these the reader will perceive that the theory is of great generality and capable of enormous development. References to the literature, which deals mostly with the case of half-integer characteristics, are given at the beginning of Chapter XVII.

281. Let $\phi(u_1, \dots, u_p)$ be a single-valued function of p independent variables $u_1, \dots, u_p$, such that, if $a_1, \dots, a_p$ be a set of *finite* values for $u_1, \dots, u_p$ respectively, the value of $\phi(u_1, \dots, u_p)$, for any set of finite values of $u_1, \dots, u_p$, is expressible by a converging series of ascending integral positive powers of $u_1 - a_1, u_2 - a_2, \dots, u_p - a_p$. Such a function is an integral analytical function. Suppose further that $\phi(u_1, \dots, u_p)$ has for each of its arguments, independently of the others, the period unity, so that if m be any integer, we have, for $\alpha = 1, 2, \dots, p$, the equation

$$\phi(u_1, \dots, u_\alpha + m, \dots, u_p) = \phi(u_1, \dots, u_p).$$

Then* the function $\phi(u_1, \dots, u_p)$ can be expressed by an infinite series of the form

$$\sum_{n_1=-\infty}^{\infty} \cdots\cdots \sum_{n_p=-\infty}^{\infty} A_{n_1, \dots, n_p} e^{2\pi i(u_1 n_1 + \dots + u_p n_p)},$$

* For the nomenclature and another proof of the theorem, see Weierstrass, *Abhandlungen aus der Functionenlehre* (Berlin, 1886), p. 159, etc.

wherein $n_1, \ldots, n_p$ are integers, each taking, independently of the others, all positive and negative values, and $A_{n_1, \ldots, n_p}$ is independent of $u_1, \ldots, u_p$.

Let the variables $u_1, \ldots, u_p$ be represented, in the ordinary way, each by the real points of an infinite plane. Put $x_1 = e^{2\pi i u_1}, \ldots, x_p = e^{2\pi i u_p}$; then to the finite part of the u_a-plane ($\alpha = 1, \ldots, p$) corresponds the portion of an x_a-plane lying between a circle Γ_a of indefinitely great but finite radius R_a, whose centre is at $x_a = 0$, and a circle γ_a of indefinitely small but not zero radius r_a, whose centre is at $x_a = 0$. The annulus between these circles may be denoted by T_a. Let a_a be a value for x_a represented by a point in the annulus T_a; describe a circle (A_a) with centre at a_a, which does not cut the circle γ_a; then for values of x_a represented by points in the annulus T_a which are within the circle (A_a), u_a may be represented by a series of integral positive powers of $x_a - a_a$; and by the ordinary method of continuation, the values of u_a for all points within the annulus T_a may be successively represented by such series; the most general value of u_a, for any value of x_a, is of the form $x_a + m$, where m is an integer. Thus, in virtue of the definition, $\phi(u_1, \ldots, u_p)$ is a single-valued, and analytical, function of the variables $x_1, \ldots, x_p$, which is finite and continuous for values represented by points within the annuli $T_1, \ldots, T_p$ and upon the boundaries of these. So considered, denote it by $\psi(x_1, \ldots, x_p)$.

Take now the integral

$$\frac{1}{(2\pi i)^p} \iint \cdots \int \frac{\psi(t_1, \ldots, t_p)}{(t_1 - x_1) \ldots (t_p - x_p)} dt_1 \ldots dt_p,$$

wherein $x_1, \ldots, x_p$ are definite values such as are represented by points respectively within the annuli $T_1, \ldots, T_p$; let its value be formed in two ways;

(i) let the variable t_a be taken counter-clockwise round the circumference Γ_a and clockwise round the circumference γ_a ($\alpha = 1, \ldots, p$); when t_a is upon the circumference Γ_a put

$$\frac{1}{t_a - x_a} = \frac{1}{t_a} + \frac{x_a}{t_a^2} + \frac{x_a^2}{t_a^3} + \ldots\ldots = \sum_{h_a=0}^{\infty} \frac{x_a^{h_a}}{t_a^{h_a+1}};$$

when t_a is upon the circumference γ_a put

$$\frac{1}{t_a - x_a} = -\left(\frac{1}{x_a} + \frac{t_a}{x_a^2} + \frac{t_a^2}{x_a^3} + \ldots\right) = -\sum_{k_a=-\infty}^{-1} \frac{x_a^{k_a}}{t_a^{k_a+1}};$$

then the integral is equal to

$$\frac{1}{(2\pi i)^p} \iint \cdots \int \psi(t_1, \ldots, t_p) \prod_{a=1}^{p} \left(dZ_a \sum_{h_a=0}^{\infty} \frac{x_a^{h_a}}{t_a^{h_a+1}} - dz_a \sum_{k_a=-\infty}^{-1} \frac{x_a^{k_a}}{t_a^{k_a+1}} \right),$$

where dZ_a represents an element dt_a taken counter-clockwise along the circumference Γ_a, and dz_a represents an element dt_a taken clockwise along the circumference γ_a; since the component series are uniformly and absolutely convergent, this is the same as

$$\frac{1}{(2\pi i)^p}\sum_{n_1=-\infty}^{\infty}\dots\sum_{n_p=-\infty}^{\infty}\iint\dots\int\psi(t_1, \dots, t_p)\,\frac{x_1^{n_1}\dots x_p^{n_p}}{t_1^{n_1+1}\dots t_p^{n_p+1}}\,dt_1\dots dt_p,$$

where for t_a the course of integration is a single complete circuit coincident with Γ_a when n_a is positive or zero, and a single complete circuit coincident with γ_a when n_a is negative, the directions in both cases being counter-clockwise; thus we obtain, as the value of the integral,

$$\sum_{n_1=-\infty}^{\infty}\dots\sum_{n_p=-\infty}^{\infty}A_{n_1,\dots,n_p}\,x_1^{n_1}\dots x_p^{n_p},$$

where

$$A_{n_1,\dots,n_p}=\frac{1}{(2\pi i)^p}\iint\dots\int\frac{\psi(t_1, \dots, t_p)}{t_1^{n_1+1}\dots t_p^{n_p+1}}\,dt_1\dots dt_p,$$

and the course of integration for t_a may be taken to be any circumference concentric with Γ_a and γ_a, not lying outside the region enclosed by them;

(ii) let the variable t_a be taken round a small circle, of radius ρ_a, whose centre is at the point representing x_a $(\alpha=1, \dots, n)$; putting

$$t_a=x_a+\rho_a e^{i\phi_a},$$

we obtain, as the value of the integral, $\psi(x_1, \dots, x_p)$.

The values of the integral obtained in these two ways are equal*; thus we have

$$\phi(u_1, \dots, u_p)=\sum_{n_1=-\infty}^{\infty}\dots\sum_{n_p=-\infty}^{\infty}A_{n_1,\dots,n_p}\,e^{2\pi i(n_1u_1+\dots+n_pu_p)},$$

where

$$A_{n_1,\dots,n_p}=\int_0^1\dots\int_0^1 e^{-2\pi i(n_1u_1+\dots+n_pu_p)}\,\phi(u_1, \dots, u_p)\,du_1\dots du_p.$$

By the nature of the proof this series is absolutely, and for all finite values of $u_1, \dots, u_p$, uniformly convergent. If $u_a=v_a+iw_a$ $(\alpha=1, \dots, p)$, and M be an upper limit to the value of the modulus of $\phi(u_1, \dots, u_p)$ for assigned finite upper limits of $w_1, \dots, w_p$, given suppose by $|w_a| \ngtr W_a$, we have

$$|A_{n_1,\dots,n_p}| \ngtr Me^{-2\pi(N_1W_1+\dots+N_pW_p)},$$

where $N_a=|n_a|$.

Ex. i. Prove that

$$\frac{\partial}{\partial w_a}\int_0^1\dots\dots\int_0^1 e^{-2\pi i(n_1v_1+\dots+n_pv_p)}\,e^{2\pi(n_1w_1+\dots+n_pw_p)}\,\phi(v_1+iw_1, \dots, v_p+iw_p)\,dv_1\dots dv_p=0.$$

Ex. ii. In the notation of § 174, Chap. X.,

$$e^{i\pi(\tau_{11}n_1^2+\dots+2\tau_{12}n_1n_2+\dots)}=\int_0^1\dots\int_0^1 e^{-2\pi i(n_1u_1+\dots+n_pu_p)}\,\Theta(u_1, \dots, u_p)\,du_1\dots du_p.$$

* Cf., for instance, Forsyth, *Theory of Functions*, p. 47. The reader may also find it of interest to compare Kronecker, *Vorlesungen über ...... Integrale* (Leipzig, 1894), p. 177, and Pringsheim, *Math. Annal.* XLVII. (1896), p. 121, ff.

282. Further it is useful to remark that the series obtained in § 281 is necessarily unique; in other words there can exist no relation of the form

$$\sum_{n_1=-\infty}^{\infty} \dots \sum_{n_p=-\infty}^{\infty} A_{n_1, \dots, n_p} x_1^{n_1} \dots x_p^{n_p} = 0,$$

valid for all values of $x_1, \dots, x_p$ which are given, in the notation of § 281, by $r_a < |x_a| < R_a$, unless each of $A_{n_1, \dots, n_p}$ be zero. For multiplying this equation by $x_1^{-n_1-1} \dots x_p^{-n_p-1} dx_1 \dots dx_p$, and integrating in regard to x_a round a circle, centre at $x_a = 0$, of radius lying between r_a and R_a, $(\alpha = 1, \dots, p)$, we obtain

$$(2\pi i)^p A_{n_1, \dots, n_p} = 0.$$

An important corollary can be deduced. We have remarked (§ 175, Chap. X.) on the existence of 2^{2p} theta functions with half-integer characteristics; it is obvious now that these functions are not connected by any linear equation in which the coefficients are independent of the arguments. For an equation

$$\sum_{s=1}^{2^{2p}} C_{g_s, k_s} \sum_{n_1=-\infty}^{\infty} \dots \sum_{n_p=-\infty}^{\infty} e^{2hu(n+\frac{1}{2}k_s)+b(n+\frac{1}{2}k_s)^2+i\pi g_s(n+\frac{1}{2}k_s)} = 0,$$

where the notation is as in § 174, Chap. X., and k_s, g_s denote rows of p quantities each either 0 or 1, can be put into the form

$$\sum_{N_1=-\infty}^{\infty} \dots \sum_{N_p=-\infty}^{\infty} A_{N_1, \dots, N_p} e^{2\pi i(U_1N_1+\dots+U_pN_p)} = 0,$$

where $2\pi i U_1, \dots, 2\pi i U_p$ are the quantities denoted by hu, $A_{N_1, \dots, N_p}$ is given by

$$A_{N_1, \dots, N_p} = \sum_{g_s} C_{g_s, k_s} e^{b(n+\frac{1}{2}k_s)^2+i\pi g_s(n+\frac{1}{2}k_s)},$$

where the summation includes 2^p terms, and $N_1, \dots, N_p$ take the values arising, by the various values of n and k_s, for the quantities $2n+k_s$; it is clear that the aggregate of the values taken by $2n+k_s$ when n denotes a row of p unrestricted integers, and k_s a row of quantities each restricted to be either 0 or 1, is that of a row of unrestricted integers.

Hence by the result obtained above it follows that $A_{N_1, \dots, N_p} = 0$, for all values of n and k_s. Therefore, if λ denote a row of arbitrarily chosen quantities, each either 0 or 1, we have

$$e^{-b(n+\frac{1}{2}k_s)^2+i\pi\lambda(n+\frac{1}{2}k_s)} A_{N_1, \dots, N_p} = \sum_{g_s} C_{g_s, k_s} e^{i\pi(g_s+\lambda)(n+\frac{1}{2}k_s)} = 0;$$

adding the 2^p equations of this form in which the elements of n are each either 0 or 1, the value of k_s being the same for all, we have

$$\sum_{g_s} C_{g_s, k_s} e^{\frac{1}{2}i\pi k_s(g_s+\lambda)} [1+e^{i\pi\mu_1}] \dots [1+e^{i\pi\mu_p}],$$

where $\mu_1, \dots, \mu_p$ are the elements of the row letter μ given by $\mu = g_s + \lambda$; the product $(1+e^{i\pi\mu_1}) \dots (1+e^{i\pi\mu_p})$ is zero unless all of $\mu_1, \dots, \mu_p$ are even, that is, unless every element of g_s is equal to the corresponding element of λ. Hence we infer that $C_{\lambda, k_s}=0$; and therefore, as λ is arbitrary, that all the 2^{2p} coefficients C_{g_s, k_s} are zero.

Similarly the r^{2p} possible theta functions whose characteristics are rth parts of unity are linearly independent.

283. Another* proof that the 2^{2p} theta functions with half-integer characteristics are linearly independent may conveniently be given here: we have (§ 190), if m and q be integral,

$$\vartheta(u+\Omega_m;\ \tfrac{1}{2}q)=e^{\lambda_m(u)+\pi i(mq'-m'q)}\,\vartheta(u;\ \tfrac{1}{2}q),$$

and therefore if k be integral and $Q'=q'+k'$, $Q=q+k$,

$$e^{-\lambda_m(u)+\pi i(mk'-m'k)}\,\vartheta(u+\Omega_m;\ \tfrac{1}{2}q)=e^{\pi i(mQ'-m'Q)}\,\vartheta(u;\ \tfrac{1}{2}q).$$

Therefore a relation

$$\sum_{s=1}^{2^{2p}} C_s\vartheta(u;\ \tfrac{1}{2}q_s)=0$$

leads to

$$\sum_{s=1}^{2^{2p}} C_s e^{\pi i(mQ_s'-m'Q_s)}\,\vartheta(u;\ \tfrac{1}{2}q_s)=0,$$

where $Q_s=q_s+k$, $Q_s'=q_s'+k'$; in this equation let (m, m') take in turn all the 2^{2p} possible values in which each element of m and m' is either 0 or 1; then as

$$\Sigma e^{\pi i(mQ_s'-m'Q_s)},\ =[1+e^{\pi i(Q_s')_1}]\dots[1+e^{\pi i(Q_s')_p}]\,[1+e^{-\pi i(Q_s)_1}]\dots[1+e^{-\pi i(Q_s)_p}]$$

is zero unless every one of the elements $(Q_s')_1, \dots, (Q_s)_p$ is an even integer, that is, unless $q_s=k$, $q_s'=k'$, we have

$$\sum_m\sum_{s=1}^{2^{2p}} C_s e^{\pi i(mQ_s'-m'Q_s)}\,\vartheta(u;\ \tfrac{1}{2}q_s)=2^{2p}C_k\vartheta(u;\ \tfrac{1}{2}k)=0;$$

thus, for any arbitrary characteristic (k, k'), $C_k=0$. Thus all the coefficients in the assumed relation are zero.

284. We suppose now that we have four matrices ω, ω', η, η', each of p rows and columns, which satisfy the conditions, (i) that the determinant of ω is not zero, (ii) that the matrix $\omega^{-1}\omega'$ is symmetrical, (iii) that, for real values of $n_1, \dots, n_p$, the quadratic form $\omega^{-1}\omega' n^2$ has its imaginary part positive†, (iv) that the matrix $\eta\omega^{-1}$ is symmetrical, (v) that $\eta'=\eta\omega^{-1}\omega'-\tfrac{1}{2}\pi i\bar{\omega}^{-1}$; then the relations (B) of § 140, Chap. VII., are satisfied; we put $a=\tfrac{1}{2}\eta\omega^{-1}$, $h=\tfrac{1}{2}\pi i\omega^{-1}$, $b=\pi i\omega^{-1}\omega'$, so that (cf. Chap. X., § 190)

$$\eta=2a\omega,\quad \eta'=2a\omega'-\bar{h},\quad h\omega=\tfrac{1}{2}\pi i,\quad h\omega'=\tfrac{1}{2}b;$$

* Frobenius, *Crelle*, LXXXIX. (1880), p. 200.

† Which requires that the imaginary part of the matrix $\omega^{-1}\omega'$ has not a vanishing determinant.

as in § 190 we use the abbreviation

$$\lambda_m(u) = H_m(u + \tfrac{1}{2}\Omega_m) - \pi i m m',$$

where

$$H_m = 2\eta m + 2\eta' m', \quad \Omega_m = 2\omega m + 2\omega' m'.$$

We have shewn (§ 190) that a theta function $\vartheta(u, q)$ satisfies the equation

$$\vartheta(u + \Omega_m, q) = e^{\lambda_m(u) + 2\pi i (mq' - m'q)} \vartheta(u, q),$$

m and m' each denoting a row of integers; it follows therefore that, when m, m' each denotes a row of integers, the product of r theta functions,

$$\Pi(u) = \vartheta(u, q^{(1)}) \vartheta(u, q^{(2)}) \ldots\ldots \vartheta(u, q^{(r)}),$$

satisfies the equation

$$\Pi(u + \Omega_m) = e^{r\lambda_m(u) + 2\pi i (mQ' - m'Q)} \Pi(u),$$

wherein Q_i, Q_i' are, for $i = 1, 2, \ldots, p$, the sums of the corresponding components of the characteristics denoted by $q^{(1)}, \ldots, q^{(r)}$.

Conversely*, Q, Q' denoting any assigned rows of p real rational quantities, we proceed to obtain the most general form of single-valued, integral, and analytical function, $\Pi(u)$, which, for all integral values of m and m', satisfies the equation just set down. We suppose r to be an integer, which we afterwards take positive. Under the assigned conditions for the matrices ω, ω', η, η', such a function will be called *a theta function of order* r, *with the associated constants* 2ω, $2\omega'$, 2η, $2\eta'$, *and the characteristic* (Q, Q').

Denoting the function $\vartheta(u; Q)$, of § 189, either by $\vartheta(u; 2\omega, 2\omega', 2\eta, 2\eta'; Q, Q')$ or $\vartheta(u; a, b, h; Q, Q')$, the function $\vartheta(u; 2\omega/r, 2\omega', 2\eta, 2r\eta'; Q, Q'/r)$ is a theta function of the first order with the associated constants $2\omega/r$, $2\omega'$, 2η, $2r\eta'$, and $(Q, Q'/r)$ for characteristic; increasing u by $2\omega m + 2\omega' m'$, where m, m' are integral, the function is multiplied by a factor which characterises it also as a theta function of order r, with the associated constants 2ω, $2\omega'$, 2η, $2\eta'$ and (Q, Q') for characteristic. We have, also,

$$\vartheta(u; ra, rb, rh) = \vartheta\left(u; \frac{2\omega}{r}, 2\omega', 2\eta, 2r\eta'\right) = \vartheta\left(ru; 2\omega, 2r\omega'; \frac{2\eta}{r}, 2\eta'\right) = \vartheta\left(ru; \frac{a}{r}, h, rb\right),$$

where the omitted characteristic is the same for each.

Let k_i be the least positive integer such that $k_i Q_i'$ is an integer, $= f_i$, say; denote the matrix of p rows and columns, of which every element is zero except those in the diagonal, which, in order, are $k_1, k_2, \ldots, k_p$, by k; the inverse matrix k^{-1} is obtained from this by replacing $k_1, \ldots$ respectively by

* Hermite, *Compt. Rend.* t. XL. (1855), and a letter from Brioschi to Hermite, *ibid.* t. XLVII. Schottky, *Abriss einer Theorie der Abel'schen Functionen von drei Variabeln* (Leipzig, 1880), p. 5. The investigation of § 284 is analogous to that of Clebsch and Gordan, *Abel. Funct.*, pp. 190, ff. The investigation of § 285 is analogous to that given by Schottky. Cf. Königsberger, *Crelle*, LXIV. (1865), p. 28.

$1/k_1, \ldots$; in place of the arguments u introduce arguments v determined by the p equations

$$h_{i,1}u_1 + \ldots\ldots + h_{i,p}u_p = k_i v_i, \qquad (i = 1, \ldots, p),$$

which we write $hu = kv$; then, by the equations $h\omega = \frac{1}{2}\pi i$, $h\omega' = \frac{1}{2}b$, it follows that the increments of the arguments v when the arguments u are increased by the quantities constituting the p rows of a period Ω_m, are given by the p rows of U_m defined by

$$kU_m = \pi i m + bm';$$

we shall denote the right-hand side of this equation by Υ_m; thus $U_m = k^{-1}\Upsilon_m = \pi i k^{-1}m + k^{-1}bm'$.

Now we have

$$a(u + \Omega_m)^2 - au^2 = 2au\Omega_m + a\Omega_m^2,$$

and, since* the matrix a is symmetrical, and $H_m = 2a\Omega_m - 2\bar{h}m'$, this is equal to

$$2a\Omega_m u + a\Omega_m^2 = 2a\Omega_m(u + \tfrac{1}{2}\Omega_m) = (H_m + 2\bar{h}m')(u + \tfrac{1}{2}\Omega_m)$$

and therefore equal to

$$\lambda_m(u) + \pi i m m' + 2hum' + h\Omega_m m'$$

or

$$\lambda_m(u) + \pi i m m' + 2kvm' + \Upsilon_m m';$$

thus, by the definition equation for the function $\Pi(u)$, we have

$$e^{-ra(u+\Omega_m)^2}\,\Pi(u + \Omega_m) = e^{-rau^2}\,\Pi(u)\,.\,e^{-r[\pi i m m' + 2(kv + \frac{1}{2}\Upsilon_m)m'] + 2\pi i(mQ' - m'Q)};$$

therefore, if $Q(v)$ denote $e^{-rau^2}\,\Pi(u)$,

$$Q(v + U_m) = Q(v)\,e^{-r[\pi i m m' + 2(kv + \frac{1}{2}\Upsilon_m)m'] + 2\pi i(mQ' - m'Q)};$$

now let $m' = 0$, and $m = ks$, where s denotes a row of integers $s_1, \ldots, s_p$; then $mQ' = ksQ' = k_1 s_1 Q_1' + \ldots\ldots + k_p s_p Q_p' = skQ'$, $= sf$, is also a row of integers; and $U_m = \pi i k^{-1}m + k^{-1}bm' = \pi i s$; thus we have

$$Q(v + \pi i s) = Q(v),$$

or, what is the same thing, the function $Q(v)$ is periodic for each of the arguments $v_1, \ldots, v_n$, separately, the period being πi; it follows then (§ 281) that the function is expressible as an infinite series of terms of the form $C_{n_1, n_2, \ldots, n_p} e^{2(n_1 v_1 + \ldots + n_p v_p)}$, where $n_1, \ldots, n_p$ are summation letters, each of which, independently of the others, takes all integral values from $-\infty$ to $+\infty$, and the coefficients $C_{n_1, \ldots, n_p}$ are independent of $v_1, \ldots, v_n$. This we denote by putting

$$Q(v) = e^{-rau^2}\,\Pi(u) = \Sigma C_n e^{2vn}.$$

To this relation, for the purpose of obtaining the values of the coefficients

* By a fundamental matrix equation, if μ be any matrix of p rows and columns, and u, v be row letters of p elements, $\mu\, u\, v = \bar{\mu}\, v\, u$.

C_n, we apply the equation, obtained above, which expresses the ratio to $Q(v)$ of $Q(v+U_m)$ or $Q(v+k^{-1}\Upsilon_m)$; thence we have

$$\sum_n C_n e^{2(v+k^{-1}\Upsilon_m)n} = \left[\sum_s C_s e^{2vs}\right] e^{-r[\pi imm'+2(kv+\frac{1}{2}\Upsilon_m)m']+2\pi i(mQ'-m'Q)};$$

in this equation, corresponding to a term of the left-hand side given by the summation letter n, consider the term of the right-hand side for which the summation letter s is such that

$$s_i = n_i + rk_i m_i', \qquad (i = 1, 2, \ldots, p);$$

thus $s = n + rkm'$, and $2v_i s_i = 2v_i n_i + 2rk_i v_i m_i'$, or $2vs = 2vn + 2rkvm'$; hence we obtain

$$\sum_n C_n e^{2(v+k^{-1}\Upsilon_m)n} = \left[\sum_n C_{n+rkm'} e^{2vn}\right] e^{-r(\pi imm'+\Upsilon_m m')+2\pi i(mQ'-m'Q)};$$

therefore, equating coefficients of products of the same powers of the quantities $e^{2v_1}, \ldots, e^{2v_p}$, we have

$$C_{n+rkm'} = C_n \,.\, e^{2k^{-1}\Upsilon_m n + r(\pi imm'+\Upsilon_m m') - 2\pi i(mQ'-m'Q)},$$

and this equation holds for all values of the integers denoted by n, m, m'.

By taking the particular case of this equation in which the integers m' are all zero we infer that the quantity

$$\frac{1}{\pi i} k^{-1}\Upsilon_m n - mQ', \; = \frac{1}{\pi i} k^{-1}(\pi im)\, n - mQ', \; = \sum_{s=1}^{p} m_s \left(\frac{1}{k_s} n_s - Q_s'\right)$$

must be an integer for all integral values of the numbers m_s and n_s; therefore the only values of the integers n which occur are those for which the numbers $(n_s - k_s Q_s')/k_s$ are integers; thus, by the definition of k_s, we may put $n = f + kN$, N denoting a row of integers, and $f = kQ'$.

With this value we have

$$k^{-1}\Upsilon_m n - k^{-1}(\pi im)\, n = k^{-1}(bm')\, n = \bar{k}^{-1} n\, (bm') = k^{-1}n \,.\, bm'$$
$$= (k^{-1}f + N) \,.\, bm' = (Q' + N) \,.\, bm' = bm'(Q' + N);$$

hence, as $mQ' = k^{-1}mn - mN$, the equation connecting C_n and $C_{n+rkm'}$ becomes

$$C_{f+rkm'+kN} = C_{f+kN} e^{2bm'(Q'+N)+[r(2\pi im+bm')+2\pi iQ]m'}$$
$$= C_{f+kN} e^{rbm'^2+2bm'N+2\pi iQm'+2bQ'm'},$$

$e^{2\pi irmm'}$ being equal to unity because r is an integer, and $bm'Q' = \bar{b}Q'm' = bQ'm'$; therefore

$$e^{-\frac{1}{r}b(m'r+N)^2} C_{f+k(m'r+N)} = e^{-\frac{1}{r}bN^2} C_{f+kN} \,.\, e^{2\Upsilon_Q m'},$$

Υ_Q being $\pi iQ + bQ'$, or

$$e^{-\frac{1}{r}[b(N+rm')^2+2\Upsilon_Q(N+rm')]} C_{f+k(N+rm')} = e^{-\frac{1}{r}[bN^2+2\Upsilon_Q N]} C_{f+kN};$$

thus, if the right-hand side of this equation be denoted by D_N, we have, for every integral value of m', $D_{N+rm'} = D_N$; therefore every quantity D is equal to a quantity D for which the suffix is a row of positive integers (which may be zero) each less than the numerical value of the integer r. If then ρ be the numerical value of r, the series breaks up into a sum of ρ^p series; let D_μ be the coefficient, in one of these series, in which the integers μ are less than ρ; then the values of the integers N occurring in this series are given by $N = \mu + rM$, M being a row of integers, which, as appears from the work, may be any between $-\infty$ and ∞; and the general term of $Q(v)$ is

$$C_n e^{2vn} = C_n e^{2nv} = C_{f+kN} e^{2(f+kN)v} = D_N e^{\frac{1}{r}(bN^2+2\Upsilon_Q N)+2k(Q'+N)v},$$

$$= D_{\mu+rM} e^{rb\left(M+\frac{\mu}{r}\right)^2+2\Upsilon_Q\left(M+\frac{\mu}{r}\right)+2hu\,(rM+Q'+\mu)},$$

for $k.(Q'+N)v = \bar{k}v(Q'+N) = kv(Q'+N) = hu(Q'+N)$; thus the general term is

$$= D_\mu e^{2rhu\left(M+\frac{Q'+\mu}{r}\right)+rb\left(M+\frac{\mu}{r}\right)^2+2\Upsilon_Q\left(M+\frac{\mu}{r}\right)};$$

now, as $\Upsilon_Q = \pi i Q + bQ'$, and b is a symmetrical matrix, the quantity

$$rb\left(M+\frac{\mu}{r}\right)^2 + 2\Upsilon_Q\left(M+\frac{\mu}{r}\right)$$

is immediately seen to be equal to

$$rb\left(M+\frac{\mu+Q'}{r}\right)^2 + 2\pi i Q\left(M+\frac{\mu+Q'}{r}\right) - 2\pi i\frac{QQ'}{r} - \frac{1}{r}bQ'^2;$$

therefore the general term of $\Pi(u)$, or $e^{rau^2}Q(v)$, with the coefficient D_μ, is $e^{\psi+\chi}$, where

$$\psi = rau^2 + 2rhu\left(M+\frac{Q'+\mu}{r}\right) + rb\left(M+\frac{Q'+\mu}{r}\right)^2 + 2\pi i Q\left(M+\frac{Q'+\mu}{r}\right),$$

$$\chi = -2\pi i\frac{QQ'}{r} - \frac{1}{r}bQ'^2;$$

and this is the general term of the function

$$e^{-2\pi i\frac{QQ'}{r}-\frac{1}{r}bQ'^2}\,\underline{\vartheta}\left(u;\ Q,\ \frac{Q'+\mu}{r}\right),$$

where $\underline{\vartheta}$ denotes a theta function differing only from that before represented (§ 189, Chap. X.) by ϑ, in the change of the matrices a, b, h respectively into ra, rb, rh; the condition for the convergence of the series $\underline{\vartheta}$ requires that r be positive; thus $\rho = r$; recalling the formulae

$$h\Omega_P = \pi i P + bP', \quad \tfrac{1}{2}H_P = a\Omega_P - \bar{h}P',$$

we see, as already remarked on p. 448, that, instead of

$$\omega,\ \omega',\ \eta,\ \eta',$$

the quantities to be associated with the function ϑ are

$$\frac{\omega}{r},\ \omega',\ \eta,\ r\eta';$$

with this notation then we may write, as the necessary form of the function $\Pi(u)$,

$$\Pi(u) = \Sigma K_\mu \vartheta\left(u;\ Q,\ \frac{Q'+\mu}{r}\right),$$

wherein $K_\mu, = D_\mu e^{-2\pi i\frac{QQ'}{r} - \frac{1}{r}bQ'^2}$ is an unspecified constant coefficient, μ denotes a row of p integers each less than the positive integer r, and the summation extends to the r^p terms that arise by giving to μ all its possible values.

From this investigation an important corollary can be drawn; if a single-valued integral analytical function satisfying the definition equation of the function $\Pi(u)$ (p. 448), in which r is a positive integer and the quantities Q, Q' are rational real quantities, be called a theta function of the rth order with characteristic (Q, Q'), then* *any r^p+1 theta functions of the rth order, having the same associated quantities 2ω, $2\omega'$, 2η, $2\eta'$ and the same characteristic, or characteristics differing from one another by integers, are connected by a linear equation or by more than one linear equation, wherein the coefficients are independent of the arguments $u_1, \ldots, u_p$; and therefore any of the functions can be expressed linearly by means of the other r^p functions, provided these latter are not themselves linearly connected.*

For the determining equation satisfied by $\Pi(u)$ is still satisfied if, in place of the characteristic (Q, Q'), we put $(Q+N,\ Q'+N')$, N and N' each denoting a row of p integers; and if

$$\mu + N' \equiv \nu \ (\text{mod. } r),\ \text{say}\ \mu + N' = \nu + rL',$$

we have (§ 190, Chap. X.)

$$\vartheta\left(u;\ Q+N,\ \frac{Q'+N'+\mu}{r}\right) = \vartheta\left(u;\ Q+N,\ \frac{Q'+\nu}{r}+L'\right)$$

$$= e^{2\pi i N\frac{Q'+\nu}{r}}\vartheta\left(u;\ Q,\ \frac{Q'+\nu}{r}\right),$$

and therefore

$$\Sigma K_\mu \vartheta\left(u;\ Q+N,\ \frac{Q'+\mu+N'}{r}\right) = \Sigma H_\nu \vartheta\left(u;\ Q,\ \frac{Q'+\nu}{r}\right),$$

where $H_\nu = K_\mu e^{2\pi i N\frac{Q'+\nu}{r}}$; and the aggregate of the r^p values of $\dfrac{Q'+\nu}{r}$ is the same as that of the values of $\dfrac{Q'+\mu}{r}$.

Thus any r^p+1 theta functions of the rth order, with the same characteristic, or characteristics differing only by integers, and associated with the

* The theorem is attributed to Hermite: cf. *Compt. Rendus*, t. XL. (1855), p. 428.

same quantities 2ω, $2\omega'$, 2η, $2\eta'$, are all expressible as linear functions of the same r^p quantities $\underline{\vartheta}\left(u;\, Q,\, \dfrac{Q'+\mu}{r}\right)$ with coefficients independent of $u_1, \ldots, u_p$. Hence the theorem follows as enunciated.

Ex. i. Prove that the r^p functions $\underline{\vartheta}\left(u;\, Q,\, \dfrac{Q'+\nu}{r}\right)$ are linearly independent (§ 282).

Ex. ii. The function $\vartheta(u+a;\, Q)\,\vartheta(u-a;\, Q)$ is a theta function of order 2 with $(2Q,\, 2Q')$ as characteristic. Hence, if 2^p+1 values for the argument a be taken, the resulting functions are connected by a linear relation.

For example, when $p=1$, we have the equation

$$\sigma^2(a)\,\sigma(u-b)\,\sigma(u+b) - \sigma^2(b)\,\sigma(u-a)\,\sigma(u+a) = \sigma^2(u)\,.\,\sigma(a-b)\,\sigma(a+b).$$

Ex. iii. The function $\vartheta(ru,\, Q)$ is a theta function of order r^2 with $(rQ,\, rQ')$ as characteristic. Prove that if $\underline{\vartheta}$ denote a theta function with the associated constants ω, $r^2\omega'$, $\dfrac{\eta}{r^2}$, η', in place of ω, ω', η, η' respectively, then we have the equations

$$\vartheta\left(u;\, \frac{Q}{r},\, Q'\right) = \sum_{\mu} \underline{\vartheta}\left(ru;\, Q,\, \frac{Q'+\mu}{r}\right),\quad \underline{\vartheta}\left(ru;\, Q,\, \frac{Q'+\mu}{r}\right) = \sum_{\nu} e^{-2\pi i\nu\frac{Q'+\mu}{r}}\,\vartheta\left(u;\, \frac{Q+\nu}{r},\, Q'\right),$$

where the summation letters μ, ν are row letters of p elements all less than r, and each summation contains r^p terms.

Ex. iv. The product of k theta functions, with different characteristics,

$$\vartheta(u+u^{(1)};\, Q^{(1)}) \ldots\ldots \vartheta(u+u^{(k)};\, Q^{(k)})$$

is a theta function of order k for which the quantities

$$\left[\sum_{r=1}^{k} Q^{(r)} - 2\overline{\eta'} \sum_{r=1}^{k} u^{(r)},\quad \sum_{r=1}^{k} Q'^{(r)} + 2\overline{\eta} \sum_{r=1}^{k} u^{(r)}\right],$$

enter as characteristic. Thus a simple case is when $u^{(1)} + \ldots + u^{(k)} = 0$.

For $p=1$ a linear equation connects the five functions

$$\prod_{i=1}^{4} \sigma(u+u_i),\qquad \prod_{i=1}^{4} \sigma(u+u_i+\omega),\qquad \prod_{i=1}^{4} \sigma(u+u_i+\omega'),\qquad \prod_{i=1}^{4} \sigma(u+u_i+\omega+\omega'),$$

$$\sigma\left(2u + \frac{u_1+u_2+u_3+u_4}{4}\right).$$

Ex. v. Any $(p+2)$ theta functions of order r, for which the characteristic and the associated constants ω, ω', η, η' are the same, are connected by an equation of the form $P=0$, where P is an integral homogeneous polynomial in the theta functions. For the number of terms in such a polynomial, of degree N, is greater than $(Nr)^p$, when N is taken great enough. That such an equation does not generally hold for $(p+1)$ theta functions may be proved by the consideration of particular cases.

285. The following, though partly based on the investigation already given, affords an instructive view of the theorem of § 284.

Slightly modifying a notation previously used, we define a quantity, depending on the fundamental matrices ω, ω', η, η', by the equation

$$\begin{aligned}\lambda(u;\, P,\, P') &= H_P(u + \tfrac{1}{2}\Omega_P) - \pi i PP' \\ &= (2\eta P + 2\eta' P')(u + \omega P + \omega' P') - \pi i PP',\end{aligned}$$

where P, P' each denotes a row of p arbitrary quantities. The corresponding quantity arising when, in place of ω, ω', η, η' we take other matrices $\omega^{(1)}$, $\omega'^{(1)}$, $\eta^{(1)}$, $\eta'^{(1)}$ may be denoted by $\lambda^{(1)}(u;\, P, P')$. With this notation, and in case

$$\omega^{(1)},\ \omega'^{(1)},\ \eta^{(1)},\ \eta'^{(1)}$$

are respectively

$$\frac{\omega}{r}\ ,\ \omega'\ ,\ \eta\ ,\ r\eta',$$

where r is an arbitrary positive integer, we have the following identity

$$r\lambda\left[u+\frac{2\omega}{r}s;\ N,\ m'\right]$$
$$=\lambda^{(1)}\left[u+\Omega_m^{(1)};\ k,\ 0\right]+\lambda^{(1)}\left[u;\ m,\ m'\right]-\lambda^{(1)}\left[u;\ s,\ 0\right]-2\pi i m'k,$$

where s, N, m, m', k each denotes a row of p arbitrary quantities subject to the relation

$$s+rN=m+k;$$

this the reader can easily verify; it is a corollary from the result of Ex. ii., § 190.

Let the abbreviation $R(u;\, f)$ be defined by the equation

$$R(u;\, f)=\sum_k e^{-2\eta k\left(u+\omega\frac{k}{r}\right)-2\pi i f\frac{k}{r}}\,\Pi\left(u+2\omega\frac{k}{r}\right),$$
$$=\sum_k e^{-\lambda^{(1)}[u;\, k,\, 0]-2\pi i f\frac{k}{r}}\,\Pi\left(u+2\omega\frac{k}{r}\right),$$

wherein k denotes a row of p positive integers each less than r, and the summation extends to all the r^p values of k thus arising, f is a row of p arbitrary quantities, and $\Pi(u)$ denotes any theta function of order r.

Consider now the value of $R(u+\Omega_m^{(1)};\, f)$; by definition we have

$$\Pi\left[u+2\omega\frac{k}{r}+\Omega_m^{(1)}\right]=\Pi\left(u+2\omega\frac{k+m}{r}+2\omega'm'\right);$$

therefore, if $m+k\equiv s \pmod{r}$, say $m+k=s+rN$, we have, by the definition equation (§ 284) satisfied by $\Pi(u)$,

$$\Pi\left[u+2\omega\frac{k}{r}+\Omega_m^{(1)}\right]=\Pi\left[u+2\omega^{(1)}s+2\omega N+2\omega'm'\right]$$
$$=\Pi\left(u+2\omega^{(1)}s\right)e^{r\lambda[u+2\omega^{(1)}s;\ N,\ m']+2\pi i(NQ'-m'Q)},$$

where (Q, Q') is the characteristic of $\Pi(u)$, and hence

$$R(u+\Omega_m^{(1)};\, f)=\sum_s e^{\psi}\,\Pi\left(u+2\omega^{(1)}s\right),$$

in which

$$\psi=-\lambda^{(1)}\left[u+\Omega_m^{(1)};\ k,\ 0\right]+r\lambda\left[u+2\omega^{(1)}s;\ N,\ m'\right]-2\pi i f\frac{k}{r}+2\pi i(NQ'-m'Q);$$

by the identity quoted at the beginning of this Article, ψ can also be put into the form

$$\psi = \lambda^{(1)}[u;\ m, m'] - \lambda^{(1)}[u;\ s, 0] - 2\pi i m'k - 2\pi i f\frac{k}{r} + 2\pi i (NQ' - m'Q),$$

$$= \lambda^{(1)}[u;\ m, m'] - \lambda^{(1)}[u;\ s, 0] - 2\pi i m'k - 2\pi i m'Q + 2\pi i N(Q' - f) + 2\pi i f\frac{m-s}{r};$$

in the definition equation for $\Pi(u)$, the letters m, m' denote integers; and k has been taken to denote integers; if further f be chosen so that $Q' - f$ is a row of integers, we have, since, by definition, N denotes a row of integers,

$$R(u + \Omega_m^{(1)};\ f) = e^{\lambda^{(1)}[u;\ m, m'] + 2\pi i\left(m\frac{f}{r} - m'Q\right)} \sum_s e^{-\lambda^{(1)}(u;\ s,\ 0) - 2\pi i f\frac{s}{r}}\, \Pi(u + 2\omega^{(1)}s)$$

$$= e^{\lambda^{(1)}[u;\ m, m'] + 2\pi i\left(m\frac{f}{r} - m'Q\right)} R(u;\ f).$$

Hence $R(u;\ f)$ satisfies a determining equation of precisely the same form as that satisfied by $\Pi(u)$, the only change being in the substitution of $\frac{\omega}{r}$, ω', η, $r\eta'$ respectively for ω, ω', η, η'; so* considered $R(u;\ f)$ is a theta function of the first order with $\left(Q, \frac{f}{r}\right)$ as characteristic; putting, in accordance with the definition of f above, $f = Q' + \mu$, where μ is a row of p integers, we therefore have, by § 284,

$$R(u;\ Q' + \mu) = K_{Q'+\mu}\, \vartheta\left(u;\ Q, \frac{Q'+\mu}{r}\right), = K_{Q'+\mu}\, \vartheta\left(ru;\ \frac{a}{r}, h, rb;\ \begin{matrix}\frac{Q'+\mu}{r} \\ Q\end{matrix}\right),$$

(p. 448) where $K_{Q'+\mu}$ is a quantity independent of u, and ϑ is the same theta function as that previously so denoted (§ 284), having, in place of the usual matrices a, b, h, respectively ra, rb, rh.

Remarking now that the series

$$\sum_\mu e^{-2\pi i\frac{\mu k}{r}},$$

wherein μ denotes a row of p integers (including zero), each less than r, and the summation extends to all the r^p terms thus arising, is equal to r^p when the p integers denoted by k are all zero, and is otherwise zero, we infer that the sum

$$\frac{1}{r^p}\sum_\mu R(u;\ Q' + \mu),$$

which, by the definition of $R(u, f)$, putting $f = Q' + \mu$, is equal to

$$\frac{1}{r^p}\sum_k \left[e^{-\lambda^{(1)}(u;\ k,\ 0) - 2\pi i Q'\frac{k}{r}}\, \Pi\left(u + 2\omega\frac{k}{r}\right) \sum_\mu e^{-2\pi i \mu\frac{k}{r}}\right],$$

* $R(u;\ f)$ may also be regarded as a theta function of order r, with the associated constants 2ω, $2\omega'$, 2η, $2\eta'$ and characteristic (Q, f).

is, in fact, equal to $\Pi(u)$. Hence as before we have the equation

$$\Pi(u) = \sum_{\mu} K_{Q+\mu} \vartheta\left(u;\ Q,\ \frac{Q'+\mu}{r}\right).$$

286. *Ex.* i. Suppose that m is an even half-integer characteristic, and that

$$a_1,\ a_2,\ \ldots\ldots,\ a_s$$

are $s, =2^p$, half-integer characteristics such that the characteristic formed by adding the three characteristics m, a_i, a_j is always odd, when i is not equal to j. Thus when m is an integral, or zero, characteristic, the condition is that the characteristic formed by adding two different characteristics a_i, a_j may be odd. The characteristic whose elements are formed by the addition of the elements of two characteristics a, b may be denoted by $a+b$; when the elements of $a+b$ are reduced, by the subtraction of integers, to being less than unity and positive (or zero), the reduced characteristic may be denoted by ab.

For instance when $p=2$, if α, β, γ denote any three odd characteristics, so that* the characteristic $\alpha\beta\gamma$ is even, and if μ be any characteristic whatever, characteristics satisfying the required conditions are given by taking m, a_1, a_2, a_3, a_4 respectively equal to $\alpha\beta\gamma$, μ, $\mu\beta\gamma$, $\mu\gamma\alpha$, $\mu\alpha\beta$; in either case a characteristic ma_ia_j is one of the three α, β, γ and is therefore odd.

When $p=3$, corresponding to any even characteristic m, we can in 8 ways take seven other characteristics α, β, γ, κ, λ, μ, ν, such that the combinations α, β, γ, κ, λ, μ, ν, $m\alpha\beta$, $m\alpha\kappa$, $m\lambda\mu$ constitute all the 28 existent odd characteristics; this is proved in chapter XVII.; examples have already been given, on page 309. Hence characteristics satisfying the conditions here required are given by taking

$$m,\ a_1,\ a_2,\ a_3,\ \ldots,\ a_8$$

respectively equal to

$$m,\ m,\ \alpha,\ \beta,\ \ldots,\ \nu.$$

Now, by § 284, every 2^p+1 theta functions of the second order, with the same periods and the same characteristic, are connected by a linear equation. Hence, if p, q, r denote arbitrary half-integer characteristics, and v, w be arbitrary arguments, there exists an equation of the form

$$A\vartheta(u+w;\ q)\,\vartheta(u-w;\ r) = \sum_{\lambda=1}^{s} A_\lambda \vartheta[u+v;\ (q+r-p-a_\lambda)]\,\vartheta[u-v;\ (p+a_\lambda)],$$

wherein A, A_λ are independent of u; for each of the functions involved is of the second order, as a function of u, and of characteristic $q+r$.

We determine the coefficients A_λ by adding a half period to the argument u; for u put $u+\Omega_{m-a_j-p}$; then by the formula

$$\vartheta(u+\Omega_P,\ q) = e^{\lambda(u;\ P)-2\pi i P'q}\,\vartheta(u;\ P+q),$$

where

$$\lambda(u;\ P) = H_P(u+\tfrac{1}{2}\Omega_P) - \pi i P P',$$

noticing, what is easy to verify, that

$$\lambda(u+v;\ P) + \lambda(u-v;\ P) - \lambda(u+w;\ P) - \lambda(u-w;\ P) = 0$$
$$= -\pi i P'[q+r-p-a_\lambda+p+a_\lambda-q-r],$$

* As the reader may verify from the table of § 204; a proof occurs in Chap. XVII.

we obtain

$$A\vartheta[u+w;\ (m-a_j-p+q)]\,\vartheta[u-w;\ (m-a_j-p+r)]$$
$$=\sum_{\lambda=1}^{s} A_\lambda\vartheta[u+v;\ (m-a_j-a_\lambda+q+r-2p)]\,\vartheta[u-v;\ (m-a_j+a_\lambda)].$$

But since $m-a_j+a_\lambda$ (which, save for integers, is the characteristic ma_ja_λ) is an odd characteristic when j is not the same as λ, we can hence infer, putting $u=v$, that

$$A_\lambda/A=\vartheta[v+w;(m-a_\lambda-p+q)]\vartheta[v-w;(m-a_\lambda-p+r)]/\vartheta[2v;(m-2a_\lambda+q+r-2p)]\vartheta[0;m].$$

Hence the form of the relation is entirely determined. The result can be put into various different shapes according to need. Denoting the characteristic $m+q+r$ momentarily by k, so that k consists of two rows, each of p half-integers, and similarly denoting the characteristic $a_\lambda+p$ momentarily by a_λ, and using the formula for integral M,

$$\vartheta(u;\ q+M)=e^{2\pi iMq'}\vartheta(u;\ q),$$

we have

$$\vartheta[2v;\ (m-2a_\lambda+q+r-2p)]=e^{-4\pi i a_\lambda k'}\vartheta(2v;\ k);$$

we shall denote the right-hand side of this equation by

$$e^{-4\pi i(a_\lambda+p)(m'+q'+r')}\vartheta[2v;\ (m+q+r)];$$

hence the final equation can be put into the form

$$\vartheta[u+w;\ q]\,\vartheta[u-w;\ r]\,\vartheta[2v;\ (m+q+r)]\,\vartheta[0;\ m]$$
$$=\sum_{\lambda=1}^{2^p} e^{4\pi i(a_\lambda+p)(m'+q'+r')}\vartheta[u+v;\ (q+r-p-a_\lambda)]\,\vartheta[u-v;\ (p+a_\lambda)]$$
$$\vartheta[v+w;\ (m-a_\lambda-p+q)]\,\vartheta[v-w;\ (m-a_\lambda-p+r)].$$

It may be remarked that, with the notation of Chap. XI., if $b_1, \ldots, b_p$ be any finite branch places, and A_r denote the characteristic associated with the half-period $u^{b_r, a}$, and we take for the characteristics $a_1, \ldots, a_s$ the 2^p characteristics $A, AA_1 \ldots A_k$, formed by adding an arbitrary half-integer characteristic A to the combinations of not more than p of the characteristics $A_1, \ldots, A_p$, and take for the characteristic m the characteristic associated with the half-period $u^{b_1, a_1}+\ldots+u^{b_p, a_p}$, then each of the hyperelliptic functions $\vartheta(0;\ ma_ia_j)$ vanishes (§ 206), though the characteristic ma_ia_j is not necessarily odd. Hence the formula here obtained holds for any hyperelliptic case when $m, a_1, \ldots, a_s$, have the specified values.

Ex. ii. When $p=2$, denoting three odd characteristics by α, β, γ, we can in Ex. i. take

$$p,\ q,\ r,\ m,\ a_1,\ a_2,\ a_3,\ a_4$$

respectively equal to

$$\alpha\beta\gamma,\ q,\ 0,\ \alpha\beta\gamma,\ 0,\ \beta\gamma,\ \gamma\alpha,\ \alpha\beta,$$

wherein 0 denotes the characteristic of which all the elements are zero, and $\beta\gamma$ denotes the reduced characteristic obtained* by adding the characteristics β and γ. Then the general formula of Ex. i. becomes, putting $v=0$ and retaining the notation m for the characteristic $\alpha\beta\gamma$,

$$\vartheta(u+w;\ q)\,\vartheta(u-w;\ 0)\,\vartheta(0;\ q+m)\,\vartheta(0;\ m)$$
$$=\sum_{\lambda=1}^{4} e^{4\pi i(a_\lambda+m)(m'+q')}\vartheta(u;\ q-m-a_\lambda)\,\vartheta(u;\ m+a_\lambda)\,\vartheta(w;\ q-a_\lambda)\,\vartheta(w;\ a_\lambda).$$

* So that all the elements of $\beta\gamma$ are zero or positive and less than unity.

Ex. iii. As one application of the formula of Ex. ii. we put

$$q=\tfrac{1}{2}\begin{pmatrix}10\\10\end{pmatrix},\quad a=\tfrac{1}{2}\begin{pmatrix}10\\10\end{pmatrix},\quad \beta=\tfrac{1}{2}\begin{pmatrix}01\\11\end{pmatrix},\quad \gamma=\tfrac{1}{2}\begin{pmatrix}01\\01\end{pmatrix},$$

and therefore

$$m=\tfrac{1}{2}\begin{pmatrix}10\\00\end{pmatrix},\quad a_1=\tfrac{1}{2}\begin{pmatrix}00\\00\end{pmatrix},\quad a_2=\tfrac{1}{2}\begin{pmatrix}00\\10\end{pmatrix},\quad a_3=\tfrac{1}{2}\begin{pmatrix}11\\11\end{pmatrix},\quad a_4=\tfrac{1}{2}\begin{pmatrix}11\\01\end{pmatrix};$$

hence we find, comparing the table of § 204, and using the formula

$$\vartheta(u;\, f+M)=e^{2\pi i Mf'}\,\vartheta(u;\, f),$$

where M, $=\begin{pmatrix}M_1'M_2'\\M_1M_2\end{pmatrix}$, consists of integers, $f=\begin{pmatrix}f_1'f_2'\\f_1f_2\end{pmatrix}$, and $Mf'=M_1f_1'+M_2f_2'$, that*

$$\vartheta(u+w;\, q)=-\vartheta_{02}(u+w),\ \vartheta(u-w;\, 0)=\vartheta_5(u-w),\ \vartheta(0;\, q+m)=\vartheta_{12}(0),\ \vartheta(0;\, m)=\vartheta_{01}(0),$$

$$\vartheta(u;q-m-a_1)=\ \vartheta_{12}(u),\vartheta(u;m+a_1)=\ \vartheta_{01}(u),\vartheta(w;q-a_1)=-\vartheta_{02}(w),\vartheta(w;a_1)=\ \vartheta_5(w),$$

$$\vartheta(u;q-m-a_2)=\ \vartheta_5(u),\vartheta(u;m+a_2)=-\vartheta_{02}(u),\vartheta(w;q-a_2)=\ \vartheta_{01}(w),\vartheta(w;a_2)=\ \vartheta_{12}(w),$$

$$\vartheta(u;q-m-a_3)=\ \vartheta_{24}(u),\vartheta(u;m+a_3)=-\vartheta_{04}(u),\vartheta(w;q-a_3)=\ \vartheta_3(w),\vartheta(w;a_3)=\ \vartheta_{14}(w),$$

$$\vartheta(u;q-m-a_4)=-\vartheta_{14}(u),\vartheta(u;m+a_4)=-\vartheta_3(u),\vartheta(w;q-a_4)=\ \vartheta_{04}(w),\vartheta(w;a_4)=-\vartheta_{24}(w),$$

all the factors of the form $e^{4\pi i(a_\lambda+m)(m'+q')}$ being equal to 1; by substitution of these results we therefore obtain

$$\vartheta_{02}(u+w)\,\vartheta_5(u-w)\,\vartheta_{12}(0)\,\vartheta_{01}(0)=\vartheta_{12}\vartheta_{01}\bar\vartheta_{02}\bar\vartheta_5+\vartheta_{02}\vartheta_5\bar\vartheta_{12}\bar\vartheta_{01}+\vartheta_{04}\vartheta_{24}\bar\vartheta_3\bar\vartheta_{14}+\vartheta_3\vartheta_{14}\bar\vartheta_{04}\bar\vartheta_{24},$$

where ϑ_{12} denotes $\vartheta_{12}(u)$, etc., and $\bar\vartheta_{02}$ denotes $\vartheta_{02}(w)$, etc.; this agrees with the formula of §§ 219, 220 (Chap. XI.).

Ex. iv. By putting in the formula of Ex. ii. respectively

$$a=\tfrac{1}{2}\begin{pmatrix}10\\10\end{pmatrix},\quad \beta=\tfrac{1}{2}\begin{pmatrix}11\\01\end{pmatrix},\quad \gamma=\tfrac{1}{2}\begin{pmatrix}01\\11\end{pmatrix},\quad p=q=m=a\beta\gamma=0,$$

obtain the result

$$\vartheta_5(u+w)\,\vartheta_5(u-w)\,\vartheta_5^2=\vartheta_5^2(u)\,\vartheta_5^2(w)+\vartheta_{02}^2(u)\,\vartheta_{02}^2(w)+\vartheta_{24}^2(u)\,\vartheta_{24}^2(w)+\vartheta_{04}^2(u)\,\vartheta_{04}^2(w),$$

which is in agreement with the results of §§ 219, 220.

Dividing the result of Ex. iii. by that of Ex. iv. we obtain an addition formula for the theta quotient $\vartheta_{02}(u)/\vartheta_5(u)$, whereby $\vartheta_{02}(u+w)/\vartheta_5(u+w)$ is expressed by theta quotients with the arguments u and w.

Ex. v. The formula of Ex. ii. may be used in different ways to obtain an expression for the product $\vartheta(u+w;\, q)\,\vartheta(u-w;\, 0)$. It is sufficient that the characteristics m and $q+m$ be even and that the three odd characteristics a, β, γ have the sum m. Thus, starting with a given characteristic q, we express it, save for a characteristic of integers, as the sum of two even characteristics, m and $q+m$, which (unless q be zero) is possible in three ways†, and then express m as the sum of three odd characteristics, a, β, γ, which is possible in two ways‡; then§ we take $a_1=0$, $a_2=\beta\gamma$, $a_3=\gamma a$, $a_4=a\beta$. Taking $q=\tfrac{1}{2}\begin{pmatrix}10\\10\end{pmatrix}$, we have

* In Weierstrass's reduced characteristic symbol the upper row of elements is positive, and the lower row negative; cf. §§ 203, 204, and p. 337, foot-note.

† This is obvious from the table of § 204, or by using the two-letter notation; for instance the symbol $(a_1a_2)\equiv(a_1c)+(a_2c)\equiv(a_1c_1)+(a_2c_1)\equiv(a_1c_2)+(a_2c_2)$.

‡ For example, $(ac)\equiv(a_1a)+(a_2a)+(c_1c_2)\equiv(a_1a_2)+(c_1c)+(cc_2)$. See the final equation of § 201. The six odd characteristics form a set which is a particular case of sets considered in chapter XVII.

§ Moreover we may increase u and w by the same half-period. But the additions of the half-periods P, $P+\Omega_q$ lead to the same result; and, when q is one of a, β, γ, the same result is obtained by the addition of $P+\Omega_m$ and of $P+\Omega_m+\Omega_q$.

$$\tfrac{1}{2}\begin{pmatrix}10\\10\end{pmatrix}\equiv\tfrac{1}{2}\begin{pmatrix}10\\00\end{pmatrix}+\tfrac{1}{2}\begin{pmatrix}00\\10\end{pmatrix}\equiv\tfrac{1}{2}\begin{pmatrix}01\\10\end{pmatrix}+\tfrac{1}{2}\begin{pmatrix}11\\00\end{pmatrix}\equiv\tfrac{1}{2}\begin{pmatrix}00\\11\end{pmatrix}+\tfrac{1}{2}\begin{pmatrix}10\\01\end{pmatrix};$$

putting $m=\tfrac{1}{2}\begin{pmatrix}10\\00\end{pmatrix}$, we may take

$$\alpha=\tfrac{1}{2}\begin{pmatrix}11\\01\end{pmatrix},\quad \beta=\tfrac{1}{2}\begin{pmatrix}11\\10\end{pmatrix},\quad \gamma=\tfrac{1}{2}\begin{pmatrix}10\\11\end{pmatrix}.$$

Hence obtain the result

$$\vartheta_{02}(u+w)\,\vartheta_5(u-w)\,\vartheta_{12}(0)\,\vartheta_{01}(0)=\vartheta_{12}\vartheta_{01}\bar\vartheta_{02}\bar\vartheta_5+\vartheta_{04}\vartheta_{24}\bar\vartheta_{14}\bar\vartheta_3+\vartheta_4\vartheta_{13}\bar\vartheta_{23}\bar\vartheta_{03}+\vartheta_{34}\vartheta_1\bar\vartheta_2\bar\vartheta_0,$$

where, on the right hand, ϑ_{12} denotes $\vartheta_{12}(u)$, etc., and $\bar\vartheta_{02}$ denotes $\vartheta_{02}(w)$, etc. Comparing this result with the result of Ex. iii., namely

$$\vartheta_{02}(u+w)\,\vartheta_5(u-w)\,\vartheta_{12}(0)\,\vartheta_{01}(0)=\vartheta_{12}\vartheta_{01}\bar\vartheta_{02}\bar\vartheta_5+\vartheta_{02}\vartheta_5\bar\vartheta_{12}\bar\vartheta_{01}+\vartheta_{04}\vartheta_{24}\bar\vartheta_3\bar\vartheta_{14}+\vartheta_3\vartheta_{14}\bar\vartheta_{04}\bar\vartheta_{24},$$

we deduce the remarkable identity

$$\vartheta_4(u)\,\vartheta_{13}(u)\,\vartheta_{23}(w)\,\vartheta_{03}(w)+\vartheta_1(u)\,\vartheta_{34}(u)\,\vartheta_0(w)\,\vartheta_2(w)$$
$$=\vartheta_{02}(u)\,\vartheta_5(u)\,\vartheta_{12}(w)\,\vartheta_{01}(w)+\vartheta_3(u)\,\vartheta_{14}(u)\,\vartheta_{04}(w)\,\vartheta_{24}(w),$$

wherein u, w are arbitrary arguments; this is one of a set of formulae obtained by Caspary, to which future reference will be made.

Ex. vi. By taking in Ex. v. the characteristics q, m to be respectively

$$\tfrac{1}{2}\begin{pmatrix}10\\10\end{pmatrix},\quad \tfrac{1}{2}\begin{pmatrix}10\\01\end{pmatrix},$$

and resolving m into the sum $\alpha+\beta+\gamma$ in the two ways

$$\tfrac{1}{2}\begin{pmatrix}10\\10\end{pmatrix}+\tfrac{1}{2}\begin{pmatrix}11\\01\end{pmatrix}+\tfrac{1}{2}\begin{pmatrix}11\\10\end{pmatrix},\quad \tfrac{1}{2}\begin{pmatrix}01\\11\end{pmatrix}+\tfrac{1}{2}\begin{pmatrix}01\\01\end{pmatrix}+\tfrac{1}{2}\begin{pmatrix}10\\11\end{pmatrix},$$

respectively, obtain the formulae

$$\vartheta_{02}(u+w)\,\vartheta_5(u-w)\,\vartheta_0(0)\,\vartheta_2(0)=\vartheta_5\vartheta_{02}\bar\vartheta_2\bar\vartheta_0+\vartheta_0\vartheta_2\bar\vartheta_{02}\bar\vartheta_5\quad-\vartheta_4\vartheta_{13}\bar\vartheta_{24}\bar\vartheta_{04}-\vartheta_{04}\vartheta_{24}\bar\vartheta_{13}\bar\vartheta_4,$$

$$\vartheta_{02}(u+w)\,\vartheta_5(u-w)\,\vartheta_0(0)\,\vartheta_2(0)=\vartheta_0\vartheta_2\bar\vartheta_{02}\bar\vartheta_5-\vartheta_{24}\vartheta_{04}\bar\vartheta_4\bar\vartheta_{13}-\vartheta_{14}\vartheta_3\bar\vartheta_{03}\bar\vartheta_{23}+\vartheta_{34}\vartheta_1\bar\vartheta_{01}\bar\vartheta_{12},$$

and the identity

$$\vartheta_{34}\vartheta_1\bar\vartheta_{01}\bar\vartheta_{12}+\vartheta_4\vartheta_{13}\bar\vartheta_{24}\bar\vartheta_{04}=\vartheta_5\vartheta_{02}\bar\vartheta_0\bar\vartheta_2+\vartheta_{14}\vartheta_3\bar\vartheta_{03}\bar\vartheta_{23}.$$

Putting in this equation $w=0$, we obtain a formula quoted without proof on page 340.

Ex. vii. Obtain the two formulae for $\vartheta_{02}(u+w)\,\vartheta_5(u-w)$ which arise, similarly to those in Exs. v. vi., by taking for m the characteristic $\tfrac{1}{2}\begin{pmatrix}01\\10\end{pmatrix}$, the characteristic q being unaltered.

Ex. viii. Obtain the formulae, for $p=2$,

$$\vartheta_{23}(u+w)\,\vartheta_{23}(u-w)\,\vartheta_5^2(0)=\vartheta_5^2\bar\vartheta_{23}^2+\vartheta_1^2\bar\vartheta_{04}^2-\vartheta_3^2\bar\vartheta_2^2-\vartheta_{13}^2\bar\vartheta_{12}^2,$$

$$\vartheta_{23}(u+w)\,\vartheta_5(u-w)\,\vartheta_5(0)\,\vartheta_{23}(0)=\vartheta_5\vartheta_{23}\bar\vartheta_5\bar\vartheta_{23}+\vartheta_1\vartheta_{04}\bar\vartheta_1\bar\vartheta_{04}-\vartheta_3\vartheta_2\bar\vartheta_3\bar\vartheta_2-\vartheta_{13}\vartheta_{12}\bar\vartheta_{13}\bar\vartheta_{12},$$

where the notation is as in Ex. v.

For tables of such formulae the reader may consult Königsberger, *Crelle*, LXIV. (1865), p. 28, and *ibid.*, LXV. (1866), p. 340. Extensive tables are given by Rosenhain, *Mém. par divers Savants*, (Paris, 1851), t. XI., p. 443; Cayley, *Phil. Trans.* (London, 1881), Vol. 171, pp. 948, 964; Forsyth, *Phil. Trans.* (London, 1883), Vol. 173, p. 834.

Ex. ix. We proceed now to apply the formula of Ex. i. to the case $p=3$; taking the argument $v=0$, the characteristics p, r both zero, and the characteristics $m, a_1, a_2, \ldots\ldots, a_8$ to be respectively $m, m, \alpha, \beta, \ldots\ldots, \nu$, where $\alpha, \beta, \gamma, \kappa, \lambda, \mu, \nu$ are seven characteristics such that the combinations $\alpha, \beta, \gamma, \kappa, \lambda, \mu, \nu, m\alpha\beta, m\alpha\kappa, m\lambda\mu$ are all odd characteristics, m being an even characteristic, and removing the negative signs in the characteristics by such steps* as

$$\begin{aligned}\vartheta(-w;\ m-a_\lambda-p)&=\vartheta(w;\ a_\lambda+p-m)=\vartheta(w;\ a_\lambda+p+m-2m)\\&=e^{-4\pi i m(a'_\lambda+p'+m')}\,\vartheta(w;\ a_\lambda+p+m)\\&=e^{-4\pi i m(p'+a'_\lambda)}\,\vartheta(w;\ p+a_\lambda+m),\end{aligned}$$

the formula becomes†

$$\vartheta(u+w;\ q)\,\vartheta(u-w;\ 0)\,\vartheta(0;\ q+m)\,\vartheta(0;\ m)$$
$$=\sum_{\lambda=1}^{8} e^{-4\pi i(ma'_\lambda+q'a_\lambda)}\,\vartheta(u;\ q+a_\lambda)\,\vartheta(u;\ a_\lambda)\,\vartheta(w;\ q+m+a_\lambda)\,\vartheta(w;\ m+a_\lambda).$$

In order that the left-hand side of this equation may not vanish, the characteristic $q+m$ must be even; now it can be shewn that every characteristic (q), except the zero characteristic, can be resolved into the sum of two even characteristics (m and $q+m$) in ten ways, and that, to every even characteristic (m) there are 8 ways of forming such a set as $\alpha, \beta, \gamma, \kappa, \lambda, \mu, \nu$ (cf. p. 309, Chap. XI.). Hence, for any characteristic q there are various ways of forming such an expression of $\vartheta(u+w;\ q)\,\vartheta(u-w;\ 0)$ in terms of theta functions of u and w; moreover by the addition of the same half-period to u and w, the form of the right-hand side is altered, while the left-hand side remains effectively unaltered. In all cases in which q is even we may obtain a formula by taking $m=0$.

Ex. x. Taking, in Ex. ix., the characteristics q, m both zero, prove in the notation of § 205, when $\alpha, \beta, \ldots\ldots, \nu$ are the characteristics there associated with the suffixes 1, 2, ..., 7, that

$$\vartheta(u+w)\,\vartheta(u-w)\,\vartheta^2=\sum_{i=0}^{7}\vartheta_i^2(u)\,\vartheta_i^2(w).$$

Prove also, taking $m=0$, $q=\frac{1}{2}\begin{pmatrix}111\\000\end{pmatrix}$, that $\vartheta_{456}(u+w)\,\vartheta(u-w)\,\vartheta_{456}\,\vartheta$ is equal to

$$\vartheta(u)\,\vartheta(w)\,\vartheta_{456}(u)\,\vartheta_{456}(w)+\vartheta_4(u)\,\vartheta_4(w)\,\vartheta_{56}(u)\,\vartheta_{56}(w)+\vartheta_5(u)\,\vartheta_5(w)\,\vartheta_{64}(u)\,\vartheta_{64}(w)$$
$$+\vartheta_6(u)\,\vartheta_6(w)\,\vartheta_{45}(u)\,\vartheta_{45}(w)$$
$$-\vartheta_7(u)\,\vartheta_7(w)\,\vartheta_{123}(u)\,\vartheta_{123}(w)-\vartheta_1(u)\,\vartheta_1(w)\,\vartheta_{237}(u)\,\vartheta_{237}(w)-\vartheta_2(u)\,\vartheta_2(w)\,\vartheta_{317}(u)\,\vartheta_{317}(w)$$
$$-\vartheta_3(u)\,\vartheta_3(w)\,\vartheta_{127}(u)\,\vartheta_{127}(w),$$

where ϑ, ϑ_{456} denote respectively $\vartheta(0)$, $\vartheta_{456}(0)$.

Hence we immediately obtain an expression for $\vartheta_{456}(u+w)/\vartheta(u+w)$ in terms of theta quotients $\vartheta_i(u)/\vartheta(u)$, $\vartheta_i(w)/\vartheta(w)$.

Ex. xi. The formula of Ex. i. can by change of notation be put into a more symmetrical form which has theoretical significance. As before let m be any half-integer even characteristic, and let $a_1, \ldots\ldots, a_s$ be s, $=2^p$, half-integer characteristics such that every

* Wherein the notation is that the characteristic p is written $\begin{pmatrix}p_1'\,p_2'\,p_3'\\p_1\,p_2\,p_3\end{pmatrix}$ and p' denotes the row (p_1', p_2', p_3'); and similarly for the characteristics m, a_λ.

† This formula is given by Weber, *Theorie der Abel'schen Functionen vom Geschlecht* 3 (Berlin, 1876), p. 38.

combination ma_ia_j, in which i is not equal to j, is an odd characteristic; let f, g, h be arbitrary half-integer characteristics; let J denote the matrix of substitution given by

$$J=\tfrac{1}{2}\begin{pmatrix} -1 & 1 & 1 & 1 \\ 1 & -1 & 1 & 1 \\ 1 & 1 & -1 & 1 \\ 1 & 1 & 1 & -1 \end{pmatrix},$$

and from the arbitrary arguments u, v, w determine other arguments U, V, W, T by the reciprocal linear equations

$$(U_i,\ V_i,\ W_i,\ T_i)=J\,(u_i,\ v_i,\ w_i,\ 0), \qquad (i=1,\ 2,\ \ldots\ldots,\ p),$$

or, as we may write them,

$$(U,\ V,\ W,\ T)=J\,(u,\ v,\ w,\ 0);$$

further determine the new characteristics F, G, H, K by means of equations of the form

$$(F,\ G,\ H,\ K)=J\,(f,\ g,\ h,\ m),$$

noticing that there are $2p$ such sets of four equations, one for every set of corresponding elements of the characteristics;

then deduce from the equation of Ex. i. that

$$\vartheta(0;\ m)\,\vartheta(u;\ f)\,\vartheta(v;\ g)\,\vartheta(w;\ h)$$
$$=\sum_{\lambda=1}^{2^p} e^{4\pi i a_\lambda a'_\lambda+2\pi i a_\lambda(f'+g'+h'+m')}\,\vartheta(T;\ K+a_\lambda)\,\vartheta(U;\ F+a_\lambda)\,\vartheta(V;\ G+a_\lambda)\,\vartheta(W;\ H+a_\lambda)$$
$$=\sum_{\lambda=1}^{2^p} e^{4\pi i a_\lambda m'}\,\vartheta(U;\ F-a_\lambda)\,\vartheta(V;\ G-a_\lambda)\,\vartheta(W;\ H-a_\lambda)\,\vartheta(T;\ K+a_\lambda).$$

Putting $m=0$, we derive the formula

$$\vartheta(0;\ 0)\,\vartheta(v+w;\ g+h)\,\vartheta(w+u;\ h+f)\,\vartheta(u+v;\ f+g)$$
$$=\sum_{\lambda=1}^{2^p}\vartheta(u+v+w;\ f+g+h+a_\lambda)\,\vartheta(u;\ f-a_\lambda)\,\vartheta(v;\ g-a_\lambda)\,\vartheta(w;\ h-a_\lambda),$$

wherein u, v, w are any arguments and f, g, h are any half-integer characteristics.

Ex. xii. Deduce from Ex. i. that when $p=2$ there are twenty sets of four theta functions, three of them odd and one even, such that the square of any theta function can be expressed linearly by the squares of these four.

287. The number, r^p, of terms in the expansion of $\Pi(u)$ may be expected to reduce in particular cases by the vanishing of some coefficients on the right-hand side. We proceed to shew* that this is the case, for instance, when $\Pi(u)$ is either an odd function, or an even function of the arguments u. We prove first that a necessary condition for this is that the characteristic $(Q,\ Q')$ consist of half-integers.

For, if $\Pi(-u)=\epsilon\Pi(u)$, where ϵ is $+1$ or -1, the equation

$$\Pi(u+\Omega_m)=e^{r\lambda_m(u)+2\pi i(mQ'-m'Q)}\,\Pi(u)$$

gives

$$\epsilon\Pi(-u-\Omega_m)=e^{r\lambda_m(u)+2\pi i(mQ'-m'Q)}\,\epsilon\Pi(-u),$$

* Schottky, *Abriss einer Theorie der Abel'schen Functionen von drei Variabeln* (Leipzig, 1880).

while, the left-hand side of this equation is, by the same fundamental equation, equal to

$$\epsilon e^{r\lambda_{-m}(-u)-2\pi i(mQ'-m'Q)}\,\Pi(-u);$$

hence, for all values of the integers m, m', the expression

$$r[\lambda_m(u)-\lambda_{-m}(-u)]+4\pi i(mQ'-m'Q)$$

must be an integral multiple of $2\pi i$; since, however,

$$\lambda_m(u)=H_m(u+\tfrac{1}{2}\Omega_m)-\pi imm'=\lambda_{-m}(-u),$$

this requires that $2(mQ'-m'Q)$ be an integer; thus $2Q$, $2Q'$ are necessarily integers.

Suppose now that Q, Q' are half-integers; denote them by q, q'; and suppose that $\Pi(u)=\epsilon\Pi(-u)$, where ϵ is $+1$ or -1. Then from the equation

$$\Pi(u)=\sum_\mu K_\mu \underline{\vartheta}\left(u;\ q,\frac{q'+\mu}{r}\right),$$

since, for any characteristic, $\vartheta(u,q)=\vartheta(-u,-q)$, we obtain

$$\Pi(u)=\epsilon\Pi(-u)=\epsilon\sum_\mu K_\mu\underline{\vartheta}\left(-u;\ q,\frac{\mu+q'}{r}\right)=\epsilon\sum_\mu K_\mu\underline{\vartheta}\left(u;\ -q,-\frac{\mu+q'}{r}\right)$$

$$=\epsilon\sum_\mu K_\mu\underline{\vartheta}\left[u;\ q-2q,\frac{\nu+q'}{r}-\frac{\mu+\nu+2q'}{r}\right],$$

where ν is a row of positive integers, each less than r, so chosen that

$$\nu\equiv-(\mu+2q'),\ (\text{mod. } r);$$

thus the aggregate of the values of ν is the same as the aggregate of the values of μ; therefore, by the formula (§ 190), $\vartheta(u;\ q+M,q'+M')=e^{2\pi iMq'}\vartheta(u;\ q,q')$, wherein M, M' are integers, we have

$$\sum K_\nu\underline{\vartheta}\left(u;\ q,\frac{\nu+q'}{r}\right)=\Pi(u)=\epsilon\sum_\mu K_\mu e^{-4\pi iq\frac{\nu+q'}{r}}\underline{\vartheta}\left(u;\ q,\frac{\nu+q'}{r}\right);$$

comparing these two forms for $\Pi(u)$ we see that in the formula

$$\Pi(u)=\sum_\mu K_\mu\underline{\vartheta}\left(u;\ q,\frac{\mu+q'}{r}\right)$$

the values of μ that arise may be divided into two sets; (i) those for which $2\mu+2q'\equiv0$ (mod. r); for such terms the value of ν defined by the previously written congruence is equal to μ, and the transformation effected with the help of the congruence only reproduces the term to which it is applied; thus, *for all such values of* μ *which occur*, $e^{-4\pi iq\frac{\mu+q'}{r}}$ is equal to ϵ; (ii) those terms for which $2\mu + 2q' \not\equiv 0$ (mod. r); for such terms $K_\nu = \epsilon K_\mu e^{-4\pi i q \frac{\nu+q'}{r}}$. Hence on the whole $\Pi(u)$ can be put into the form

$$\sum_\mu K_\mu \underline{\vartheta}\left(u;\ q,\ \frac{\mu+q'}{r}\right) + \sum_\mu K_\mu \left\{\underline{\vartheta}\left(u;\ q,\ \frac{\mu+q'}{r}\right) + \epsilon e^{-4\pi i q\frac{\nu+q'}{r}} \underline{\vartheta}\left(u;\ q,\ \frac{\nu+q'}{r}\right)\right\},$$

where the first summation extends to those values of μ for which $2\mu + 2q' \equiv 0$ (mod. r), and the second summation extends to half those values of μ for which $2\mu + 2q' \not\equiv 0$ (mod. r). The single term

$$\phi(u, \mu) = \underline{\vartheta}\left(u;\ q,\ \frac{\mu+q'}{r}\right) + \epsilon e^{-4\pi i q\frac{\nu+q'}{r}} \underline{\vartheta}\left(u;\ q,\ \frac{\nu+q'}{r}\right),$$

which can also be written in the form

$$\underline{\vartheta}\left(u;\ q,\ \frac{\mu+q'}{r}\right) + \epsilon e^{4\pi i q\frac{\mu+q'}{r}} \underline{\vartheta}\left(u;\ q,\ -\frac{\mu+q'}{r}\right),$$

is even or odd according as $\Pi(u)$ is even or odd; and this is also true for the term $\underline{\vartheta}\left(u;\ q,\ \frac{\mu+q'}{r}\right)$ arising when $2\mu + 2q' \equiv 0$ (mod. r).

Hence if x be the number of values of μ, incongruent for modulus r, which satisfy the congruence $2\mu + 2q' \equiv 0$ (mod. r), and y be the number of these solutions for which also the condition $e^{-4\pi i q\frac{\mu+q'}{r}} = \epsilon$ is satisfied, the number of undetermined coefficients in $\Pi(u)$ is reduced to, at most,

$$y + \tfrac{1}{2}(r^p - x).$$

288. We proceed now to find x and y; we notice that y vanishes when x vanishes, for the terms whose number is y are chosen from among possible terms whose number is x. The result is that *when r is even and the characteristic (q, q') is integer or zero, and $\Pi(-u) = \epsilon\Pi(u)$, the number of terms in $\Pi(u)$ is $\frac{1}{2}r^p + 2^{p-1}\epsilon$; while, when r is odd, or when r is even and the half-integer characteristic (q, q') does not consist wholly of integers, or zeros, the number of terms in $\Pi(u)$ is $\frac{1}{2}r^p + \frac{1}{4}[1-(-)^r]\epsilon e^{4\pi i qq'}$.*

Suppose r is even; then the congruence $2\mu + 2q' \equiv 0$ (mod. r) is satisfied by taking $\mu = M\frac{r}{2} - q'$, and in no other way, M denoting a row of p arbitrary integers. Thus unless q' consists of integers, x is zero, and therefore, as remarked above, y is zero, and the number of terms in $\Pi(u)$ is $\frac{1}{2}r^p$. While, when q' is integral, the incongruent values for μ (modulus r) are obtained by taking the incongruent values for M for modulus 2, in number 2^p; in that case $x = 2^p$; the condition $e^{-4\pi i q\frac{\mu+q'}{r}} = \epsilon$ is the same as $e^{-2\pi i q M} = \epsilon$; when q is integral, this is satisfied by all the 2^p values of M, or by no values of M, according as ϵ is $+1$ or is -1; in both cases $y = 2^{p-1}(1+\epsilon)$; when q is not integral, $p-1$ of the elements of M can be taken arbitrarily and the condition $e^{-2\pi i q M}=\epsilon$ determines the other element, so that $y=2^{p-1}$. Thus, when r is even, we have

(1) when q, q' are both rows of integers (including zero), $x=2^p$, $y=2^{p-1}(1+\epsilon)$, and the number of terms in $\Pi(u)$ is

$$2^{p-1}(1+\epsilon)+\tfrac{1}{2}(r^p-2^p)=\tfrac{1}{2}r^p+2^{p-1}\epsilon,$$

as stated, there being $\frac{1}{2}r^p+2^{p-1}$ terms when $\Pi(u)$ is an even function, and $\frac{1}{2}r^p-2^{p-1}$ terms when $\Pi(u)$ is an odd function;

(2) when q' is integral, and q is not integral, $x=2^p$, $y=2^{p-1}$, and therefore the number of terms in $\Pi(u)$ is

$$2^{p-1}+\tfrac{1}{2}(r^p-2^p)=r^p,$$

in accordance with the result stated;

(3) when q' is not integral, both x and y are zero, and the number of terms is $\frac{1}{2}r^p$, also agreeing with the given formula.

Suppose now that r is odd, then the equation

$$2\mu+2q'=rM,\ \text{or}\ \mu=\frac{rM-2q'}{2},=\text{integer}+\frac{M-2q'}{2},$$

wherein M is a row of integers, requires M to have the form $2q'+2N$, where N is a row of integers, and therefore

$$\frac{\mu}{r}=\frac{rM-2q'}{2r},=N+q'\left(1-\frac{1}{r}\right);$$

this equation, since μ consists of positive integers all less than r, determines the value of N uniquely; hence $x=1$. The condition

$$e^{-4\pi i q\frac{\mu+q'}{r}}=\epsilon,\ \text{or}\ e^{-4\pi i q(q'+N)}=\epsilon,\ \text{or}\ e^{-4\pi i q q'}=\epsilon$$

determines $y=1$ or $y=0$ according as $\epsilon e^{4\pi i q q'}=+1$ or $=-1$; hence the number of terms in $\Pi(u)$ is

$$1+\tfrac{1}{2}(r^p-1),\ \text{or}\ \tfrac{1}{2}(r^p-1),$$

according as $\epsilon e^{4\pi i q q'}=+1$ or -1; this agrees with the given result when r is odd, the number of terms being always one of the numbers $\frac{1}{2}(r^p\pm 1)$.

289. It follows from the investigation just given that if we take products of theta functions, forming odd or even theta functions of order r, with the same half-integer characteristic (q, q'), and associated with the same constants 2ω, $2\omega'$, 2η, $2\eta'$, then when r is even, the number of these which are linearly independent is, at most, $\frac{1}{2}r^p+2^{p-1}\epsilon$ when the characteristic is integral or zero, and is otherwise $\frac{1}{2}r^p$; while, when r is odd, the number which are linearly independent is, at most, $\frac{1}{2}(r^p+\epsilon e^{4\pi i q q'})$, ϵ being ± 1 according as the products are even or odd functions.

Ex. i. In case $p=2$ there are six odd characteristics, and the sum of any three of them is even*, as the reader can easily verify by the table of page 303. Let $\alpha, \beta, \gamma, \delta, \epsilon, \zeta$ denote the odd characteristics, in any order, and let $\alpha\beta\gamma$ denote the characteristic formed by adding the characteristics α, β, γ. Then the product

$$\Pi(u)=\vartheta(u, \alpha)\,\vartheta(u, \beta)\,\vartheta(u, \gamma)\,\vartheta(u, \alpha\beta\gamma)$$

is an odd theta function of the fourth order with integral characteristic. Hence this product can be written in the form

$$\Pi(u)=\Sigma A_\mu \underline{\vartheta}\left(u;\, 0, \frac{\mu}{4}\right),$$

where μ has the 4^2 values arising by giving to each of the two elements of μ, independently of the other, the values 0, 1, 2, 3. Changing the sign of u we have

$$\Pi(u)=-\Sigma A_\mu \underline{\vartheta}\left(-u;\, 0, \frac{\mu}{4}\right), \quad =-\Sigma A_\mu \underline{\vartheta}\left(u;\, 0, -\frac{\mu}{4}\right), \quad =-\Sigma A_\mu \underline{\vartheta}\left(u;\, 0, \frac{\nu}{4}-\frac{\mu+\nu}{4}\right),$$

where ν is chosen so that

$$\mu+\nu\equiv 0 \pmod{4}.$$

This congruence gives 16 values of ν corresponding to the 16 values of μ; of these there are 4 values for which $\mu=\nu$ and $2\mu\equiv 0 \pmod{4}$; these are the values

$$\mu=(0, 0),\ (0, 2),\ (2, 0),\ (2, 2),$$

greater values for the elements of μ being excluded by the condition that these elements must be less than 4. We have by the formula (§ 190) $\vartheta(u;\, q+M)=e^{2\pi i M q'}\,\vartheta(u)$,

$$\Pi(u)=-\Sigma A_\mu \underline{\vartheta}\left(u;\, 0, \frac{\nu}{4}\right);$$

comparing this with the original formula for $\Pi(u)$, we see that

$$A_\nu=-A_\mu,$$

so that the terms in the original formula for $\Pi(u)$ for which $\nu=\mu$ are absent, and the remaining twelve terms may be arranged as six terms in the form

$$\Pi(u)=\Sigma A_\mu\left[\underline{\vartheta}\left(u;\, 0, \frac{\mu}{4}\right)-\underline{\vartheta}\left(u;\, 0, -\frac{\mu}{4}\right)\right]=\Sigma A_\mu\left[\underline{\vartheta}\left(u;\, 0, \frac{\mu}{4}\right)-\underline{\vartheta}\left(-u;\, 0, \frac{\mu}{4}\right)\right],$$

where the summation extends to the following values of μ,

$$\mu=(0, 1),\ (1, 0),\ (1, 1),\ (1, 2),\ (1, 3),\ (2, 3);$$

these values may be interchanged respectively with

$$\mu=(0, 3),\ (3, 0),\ (3, 3),\ (3, 2),\ (3, 1),\ (2, 1),$$

if a proper corresponding change be made in the coefficients A_μ.

The number 6 is that obtained from the formula $\frac{1}{2}r^p+2^{p-1}\epsilon$, by putting $r=4$, $\epsilon=-1$, $p=2$.

Ex. ii. In case $p=2$, denoting the odd characteristics by $\alpha, \beta, \gamma, \delta, \epsilon, \zeta$, and the sum of two of them, say α and β, by $\alpha\beta$, and so on, each of the four products

$$\vartheta(u, \alpha)\,\vartheta(u, \alpha\epsilon\zeta),\quad \vartheta(u, \beta)\,\vartheta(u, \beta\epsilon\zeta),\quad \vartheta(u, \gamma)\,\vartheta(u, \gamma\epsilon\zeta),\quad \vartheta(u, \delta)\,\vartheta(u, \delta\epsilon\zeta),$$

or, in Weierstrass's notation, if $\alpha, \beta, \gamma, \delta, \epsilon, \zeta$ be taken in the order in which they occur in the table of page 303, each of the products

$$\vartheta_{02}(u)\,\vartheta_{34}(u),\quad \vartheta_{24}(u)\,\vartheta_{03}(u),\quad \vartheta_{04}(u)\,\vartheta_{23}(u),\quad \vartheta_{1}(u)\,\vartheta_{5}(u),$$

* This is a particular case of a result obtained in chapter XVII.

is an odd theta function of order 2, and of characteristic differing only by integers from the characteristic denoted by $\epsilon\zeta$, or, in the arrangement here taken, $\frac{1}{2}\begin{pmatrix}10\\11\end{pmatrix}$; thus any three of these products are connected by a linear equation whose coefficients do not depend upon u.

Similarly each of the products

$$\vartheta(u, \alpha\delta\epsilon)\,\vartheta(u, \alpha\delta\zeta),\quad \vartheta(u, \beta\delta\epsilon)\,\vartheta(u, \beta\delta\zeta),\quad \vartheta(u, \gamma\delta\epsilon)\,\vartheta(u, \gamma\delta\zeta),\quad \vartheta(u, \epsilon)\,\vartheta(u, \zeta),$$

or, in Weierstrass's notation, if $\alpha, \beta, \gamma, \delta, \epsilon, \zeta$ be taken in the order in which they occur in the table of p. 303, each of the products

$$\vartheta_{14}(u)\,\vartheta_4(u),\quad \vartheta_{01}(u)\,\vartheta_0(u),\quad \vartheta_{12}(u)\,\vartheta_2(u),\quad \vartheta_{13}(u)\,\vartheta_3(u),$$

is an even theta function of order 2, and of characteristic differing only by integers from the characteristic denoted by $\epsilon\zeta$, or, in the arrangement here taken, $\frac{1}{2}\begin{pmatrix}10\\11\end{pmatrix}$; thus any three of these products are connected by a linear equation whose coefficients do not depend upon u.

Ex. iii. For $p=2$ the number of linearly independent even theta functions of the fourth order and of integral characteristic is $\frac{1}{2}4^2+2=10$. If q, r be any half-integer characteristics, it follows that any eleven functions of the form $\vartheta^2(u, q)\,\vartheta^2(u, r)$ are connected by a linear equation. Taking now, with Weierstrass's notation, the four functions*

$$t=\vartheta_5(u),\quad x=\vartheta_{34}(u),\quad y=\vartheta_{12}(u),\quad z=\vartheta_0(u),$$

it follows that there exists an identical equation

$$A_0t^4+A_1x^4+A_2y^4+A_3z^4+2Ctxyz+F_1y^2z^2+F_2x^2t^2+G_1z^2x^2+G_2y^2t^2+H_1x^2y^2+H_2z^2t^2=0,$$

in which the eleven coefficients $A_0, \ldots\ldots, H_2$ are independent of u.

The characteristics of the theta functions $\vartheta_5(u)$, $\vartheta_{34}(u)$, $\vartheta_{12}(u)$, $\vartheta_0(u)$ may be taken, respectively, to be (cf. § 220, Chap. XI.)

$$\begin{pmatrix}0,&0\\0,&0\end{pmatrix};\ \begin{pmatrix}0,&0\\0,&\frac{1}{2}\end{pmatrix}=\begin{pmatrix}p_1',&p_2'\\p_1,&p_2\end{pmatrix},\text{ say};\ \begin{pmatrix}0,&0\\\frac{1}{2},&0\end{pmatrix}=\begin{pmatrix}q_1',&q_2'\\q_1,&q_2\end{pmatrix},\text{ say};\ \begin{pmatrix}0,&0\\\frac{1}{2},&\frac{1}{2}\end{pmatrix}=\begin{pmatrix}r_1',&r_2'\\r_1,&r_2\end{pmatrix},\text{ say};$$

hence, by the formulae (§ 190)

$$\vartheta(u+\Omega_P;\ q)=e^{\lambda_P(u)-2\pi iP'q}\,\vartheta(u;\ q+P),\ \vartheta(u;\ q+M)=e^{2\pi iMq'}\,\vartheta(u;\ q),$$

wherein M denotes a row of integers, we obtain

$$\vartheta_5(u+\Omega_p)=e^{\lambda_p(u)}\,\vartheta_{34}(u),\ \vartheta_{34}(u+\Omega_p)=e^{\lambda_p(u)}\,\vartheta_5(u),\ \vartheta_{12}(u+\Omega_p)=e^{\lambda_p(u)}\,\vartheta_0(u),$$
$$\vartheta_0(u+\Omega_p)=e^{\lambda_p(u)}\,\vartheta_{12}(u);$$

hence the substitution of $u+\Omega_p$ for u in the identity replaces t, x, y, z respectively by x, t, z, y. Comparing the new form with the original form we infer that

$$A_0=A_1,\quad A_2=A_3,\quad G_1=G_2,\quad H_1=H_2.$$

Similarly the substitution of $u+\Omega_q$ for u replaces t, x, y, z respectively by y, z, t, x; making this change, and then comparing the old form with the derived form, we infer that

$$A_0=A_2,\quad A_1=A_3,\quad F_1=F_2,\quad H_1=H_2.$$

* Which are all even and such that the square of every other theta function is a linear function of the squares of these functions. It can be proved that these functions are not connected by any quadratic relation.

Thus the identity is of the form

$$t^4+x^4+y^4+z^4+2Ctxyz+F(y^2z^2+x^2t^2)+G(z^2x^2+y^2t^2)+H(x^2y^2+z^2t^2)=0.$$

Taking now the three characteristics

$$\begin{pmatrix} f_1', f_2' \\ f_1, f_2 \end{pmatrix}=\begin{pmatrix} 0, \frac{1}{2} \\ 0, 0 \end{pmatrix}, \quad \begin{pmatrix} g_1', g_2' \\ g_1, g_2 \end{pmatrix}=\begin{pmatrix} \frac{1}{2}, 0 \\ 0, 0 \end{pmatrix}, \quad \begin{pmatrix} h_1', h_2' \\ h_1, h_2 \end{pmatrix}=\begin{pmatrix} \frac{1}{2}, \frac{1}{2} \\ 0, 0 \end{pmatrix},$$

and adding to the argument u, in turn, the half-periods Ω_f, Ω_g, Ω_h and then putting $u=0$, we obtain the three equations

$$\vartheta_4^4+\vartheta_{03}^4+G\vartheta_4^2\vartheta_{03}^2=0, \quad \vartheta_{01}^4+\vartheta_2^4+F\vartheta_{01}^2\vartheta_2^2=0, \quad \vartheta_{23}^4+\vartheta_{14}^4+H\vartheta_{23}^2\vartheta_{14}^2=0,$$

where ϑ_4^4 denotes $\vartheta_4^4(0)$, etc., and the notation is Weierstrass's, as in § 220. By these equations the constants F, G, H are determined in terms of zero values of the theta functions. The value of C can then be determined by putting $u=0$ in the identity itself.

Thus we may regard the equation as known; it coincides with that considered in Exx. i. and iv. § 221, Chap. XI., and represents a quartic surface with sixteen nodes. With the assumption of certain relations connecting the zero values of the theta functions, proved by formulae occurring later (Chap. XVII. § 317, Ex. iv.), we can express the coefficients in the equation in terms of the four constants $\vartheta_5(0)$, $\vartheta_{34}(0)$, $\vartheta_{12}(0)$, $\vartheta_0(0)$. We have in fact, if these constants be respectively denoted by d, a, b, c

$$\vartheta_{01}^4+\vartheta_2^4=d^4+a^4-b^4-c^4, \quad \vartheta_4^4+\vartheta_{03}^4=d^4-a^4+b^4-c^4, \quad \vartheta_{23}^4+\vartheta_{14}^4=d^4-a^4-b^4+c^4,$$

$$\vartheta_{01}^2\vartheta_2^2=d^2a^2-b^2c^2, \quad \vartheta_4^2\vartheta_{03}^2=d^2b^2-c^2a^2, \quad \vartheta_{23}^2\vartheta_{14}^2=d^2c^2-a^2b^2;$$

hence the identity under consideration can be put into the form

$$t^4+x^4+y^4+z^4-\frac{d^4+a^4-b^4-c^4}{d^2a^2-b^2c^2}(t^2x^2+y^2z^2)-\frac{d^4+b^4-c^4-a^4}{d^2b^2-c^2a^2}(t^2y^2+z^2x^2)$$

$$-\frac{d^4+c^4-a^4-b^4}{d^2c^2-a^2b^2}(t^2z^2+x^2y^2)+2\frac{dabc\prod^{\epsilon_1,\epsilon_2}[d^2+\epsilon_1a^2+\epsilon_2b^2+\epsilon_1\epsilon_2c^2]}{(d^2a^2-b^2c^2)(d^2b^2-c^2a^2)(d^2c^2-a^2b^2)}txyz=0,$$

where the $\prod^{\epsilon_1,\epsilon_2}$ denotes the product of the four factors obtained by giving to each of ϵ_1, ϵ_2 both the values $+1$ and -1. The quartic surface represented by this equation can be immediately proved to have a node at each of the sixteen points which are obtainable from the four,

$$(d, a, b, c), \ (d, a, -b, -c), \ (d, -a, b, -c), \ (d, -a, -b, c),$$

by writing respectively, in place of d, a, b, c,

(i) (d, a, b, c), (ii) (a, d, c, b), (iii) (b, c, d, a), (iv) (c, b, a, d).

Ex. iv. We have in Ex. iii. obtained a relation connecting the functions

$$\vartheta_5(u), \ \vartheta_{34}(u), \ \vartheta_{12}(u), \ \vartheta_0(u);$$

in Ex. iv. § 221 we have obtained the corresponding relation connecting the functions

$$\vartheta_5(u), \ \vartheta_{01}(u), \ \vartheta_4(u), \ \vartheta_{23}(u);$$

and in Ex. i. § 221 we have explained how to obtain the corresponding relation connecting the functions

$$\vartheta_5(u), \ \vartheta_{23}(u), \ \vartheta_{04}(u), \ \vartheta_1(u).$$

There are* in fact sixty sets of four functions among which such a relation holds; and these sixty sets break up into fifteen lots each consisting of four sets of four functions, such that in every lot all the sixteen theta functions occur, and such that in every lot one of the sets of four consists wholly of even functions while each of the three other sets consists of two odd functions and two even functions. This can be seen as follows: using the letter notation for the sixteen functions, as in § 204, and the derived letter notation for the fifteen ratios of which the denominator is $\vartheta(u)$, as at the top of page 338, it is immediately obvious, as on page 338, that any four ratios of the form

$$1,\ q_{k,l},\ q_{k_1,l_1},\ q_{k_2},$$

in which the letters k, l, k_1, l_1, k_2 *constitute in some order the letters* a_1, a_2, c, c_1, c_2, are connected by a relation of the form in question. Now such a set of four ratios can be formed in fifteen ways; there are firstly six such sets in which all the ratios are even functions of u, obtainable from the set

$$1,\ q_c,\ q_{a_1,c_1},\ q_{a_2,c_2}$$

by permuting the three letters c, c_1, c_2 among themselves in all possible ways; and nextly nine such sets in which two of the ratios are odd functions, obtainable from the set

$$1,\ q_c,\ q_{a_1,a_2},\ q_{c_1,c_2}$$

by taking instead of the pair a_1a_2 each of the three pairs† a_1a_2, aa_1, aa_2, and instead of the pair c_1c_2 each of the three pairs c_1c_2, cc_1, cc_2. Since (§ 204) the letter notation for an odd function consists always of two a-s or two c-s, and for an even function consists of one a and one c, the number of odd and even functions will remain unaltered. Further from each of these fifteen sets we can obtain three other sets of four ratios by the addition of half-periods to the argument u, in such a way that all the sixteen theta functions enter into each lot of sets. The fifteen lots obtained may all be represented by the scheme

$$\begin{array}{cccc} 1, & \alpha, & \beta, & \alpha\beta \\ \alpha_1, & \alpha\alpha_1, & \beta\alpha_1, & \alpha\beta\alpha_1 \\ \beta_1, & \alpha\beta_1, & \beta\beta_1, & \alpha\beta\beta_1 \\ \alpha_1\beta_1, & \alpha\alpha_1\beta_1, & \beta\alpha_1\beta_1, & \alpha\beta\alpha_1\beta_1, \end{array}$$

where 1, α, β, $\alpha\beta$ denote the characteristics of one of the fifteen sets of four theta functions just described, such as $\vartheta(u)$, $\vartheta_c(u)$, $\vartheta_{a_1c_1}(u)$, $\vartheta_{a_2c_2}(u)$, or $\vartheta(u)$, $\vartheta_c(u)$, $\vartheta_{a_1a_2}(u)$, $\vartheta_{c_1c_2}(u)$, $\alpha\beta$ denoting the characteristic formed by the addition of the characteristics α, β; and α_1, β_1 denote any other two characteristics other than α, β, or $\alpha\beta$, and such that $\alpha\beta$ is not the same characteristic as $\alpha_1\beta_1$. This scheme must contain all the sixteen theta functions; for any repetition (such as $\alpha=\beta\alpha_1\beta_1$, for example) would be inconsistent with the hypothesis as to the choice of α_1, β_1 (would be equivalent to $\alpha\beta=\alpha_1\beta_1$). It is easily seen, by writing down a representative of the six schemes in which the first row consists wholly of even functions, and a representative of the nine schemes in which the first row contains two odd functions, that in every scheme there are three rows in which two odd functions occur‡.

Ex. v. There are cases in which the number of linearly connected theta functions, as given by the general theorem, is subject to further reduction. For instance, suppose we

* Borchardt, *Crelle*, LXXXIII. (1877), p. 237. Each of the sixty sets of four functions may be called a Göpel tetrad.

† The letter a, when it occurs in a suffix, is omitted.

‡ A table of the sixty sets of four theta functions is given by Krause, *Hyperelliptische Functionen* (Leipzig, 1886), p. 27.

have $m=2^{p-1}$ odd half-integer characteristics $A_1, \ldots, A_m$, and another half-integer characteristic P, not (integral or) zero, such that the characteristics* $A_1P, \ldots, A_mP$, obtained by adding P to each of $A_1, \ldots, A_m$, are also odd†; suppose further that A is an even half-integer characteristic, and that AP is also an even characteristic, and that the theta functions $\vartheta(u;A)$, $\vartheta(u;AP)$ do not vanish for zero values of the argument. Then, by § 288 the $2^{p-1}+1$ following theta functions of order 2,

$$\vartheta(u;A)\,\vartheta(u;AP),\ \vartheta(u;A_1)\,\vartheta(u;A_1P), \ldots, \vartheta(u;A_m)\,\vartheta(u;A_mP),$$

which are all even functions with a characteristic differing only by integers from the characteristic P, are connected by a linear equation with coefficients independent of u. But in fact, if we put $u=0$, all these functions vanish except the first. Hence we infer that the coefficient of the first function is zero, and that in fact the other 2^{p-1} functions are themselves connected by a linear equation.

Ex. vi. In illustration of the case considered in Ex. v. we take the following :—When $p=3$, it is possible‡, if P be any characteristic whatever, to determine six odd characteristics $A_1, \ldots, A_6$, whose sum is zero, such that the characteristics $A_1P, \ldots, A_6P$ are also odd, and such that all the combinations of three of these, denoted by $A_iA_jA_k$, $A_iA_jA_kP$, are even. By the previous example there exists an equation

$$\lambda\vartheta(u;A_4)\,\vartheta(u;A_4P)$$
$$=\lambda_1\vartheta(u;A_1)\,\vartheta(u;A_1P)+\lambda_2\vartheta(u;A_2)\,\vartheta(u;A_2P)+\lambda_3\vartheta(u;A_3)\,\vartheta(u;A_3P),$$

wherein $\lambda, \lambda_1, \lambda_2, \lambda_3$ are independent of u. Adding to u any half-period Ω_Q, this equation becomes

$$\lambda\vartheta(u;A_4Q)\,\vartheta(u;A_4PQ)$$
$$=\lambda_1\epsilon_1\vartheta(u;A_1Q)\,\vartheta(u;A_1PQ)+\lambda_2\epsilon_2\vartheta(u;A_2Q)\,\vartheta(u;A_2PQ)+\lambda_3\epsilon_3\vartheta(u;A_3Q)\,\vartheta(u;A_3PQ),$$

where $\epsilon_i\,(i=1, 2, 3)$ is a certain square root of unity depending on the characteristics A_4, A_i, P, Q, whose value is determined in the following example. Taking in particular for Ω_Q the half-period associated with the characteristic A_2A_3, so that the characteristics A_2PQ, A_3PQ become respectively the odd characteristics A_3P, A_2P, and putting $u=0$, we infer

$$\lambda\vartheta(0;A_4A_2A_3)\,\vartheta(0;A_4A_2A_3P)=\lambda_1\epsilon_1'\vartheta(0;A_1A_2A_3)\,\vartheta(0;A_1A_2A_3P),$$

where ϵ_1' is the particular value of ϵ_1 when Q is A_2A_3. This equation determines the ratio of λ_1 to λ; similarly the ratios $\lambda_2:\lambda$ and $\lambda_3:\lambda$ are determinable.

Ex. vii. If $\tfrac{1}{2}r$, $\tfrac{1}{2}q$ be half-integer characteristics whose elements are either 0 or $\tfrac{1}{2}$, and $\tfrac{1}{2}k=\tfrac{1}{2}rq$ be their reduced sum, with elements either 0 or $\tfrac{1}{2}$, prove§ that

$$k_a=r_a+q_a-2r_aq_a,\quad k_a'=r_a'+q_a'-2r_a'q_a',\qquad (a=1, 2, \ldots, p),$$

and thence, by the formulae (§ 190)

$$\vartheta(u+\Omega_P;q)=e^{\lambda(u;P)-2\pi iP'q}\,\vartheta(u;P+q),\quad \vartheta(u;q+M)=e^{2\pi iMq'}\,\vartheta(u;q),$$

* A characteristic formed by adding two characteristics A, P is denoted by $A+P$. Its *reduced* value, in which each of its elements is 0 or $\tfrac{1}{2}$, is denoted by AP.

† It is proved in chapter XVII. that, when $p>2$, the characteristic P may be arbitrarily taken, and the characteristics $A_1, \ldots, A_m$ thence determined in a finite number of ways.

‡ This is proved in chapter XVII.

§ Schottky, *Crelle*, CII. (1888), pp. 308, 318.

where M is integral, prove that

$$\vartheta(u+\tfrac{1}{2}\Omega_r\,;\ \tfrac{1}{2}q)=e^{\lambda(u\,;\ \frac{1}{2}r)+\pi i\sum\limits_{a=1}^{p}(r_a q_a q_a'+q_a r_a r_a')-\frac{1}{2}\pi i\sum\limits_{a=1}^{p} r_a' q_a}\,\vartheta(u\,;\ \tfrac{1}{2}rq).$$

If $\frac{1}{2}r$, $\frac{1}{2}a$, $\frac{1}{2}q$ be any reduced characteristics, infer that

$$\frac{\vartheta(u+\frac{1}{2}\Omega_r\,;\ \frac{1}{2}a)\,\vartheta(u+\frac{1}{2}\Omega_r\,;\ \frac{1}{2}aq)}{\vartheta(u+\frac{1}{2}\Omega_r\,;\ \frac{1}{2}q)}=e^{\lambda(u\,;\ \frac{1}{2}r)}\,\epsilon\,\frac{\vartheta(u\,;\ \frac{1}{2}ar)\,\vartheta(u\,;\ \frac{1}{2}aqr)}{\vartheta(u\,;\ \frac{1}{2}qr)},$$

where

$$\epsilon=e^{i\pi\sum\limits_{a=1}^{p}[r_a q_a a_a'+(r_a q_a'+r_a' q_a+r_a')\,a_a]}.$$

Ex. viii. If A_1, A_2, A_3, A_4 denote four odd characteristics, for $p=2$, and B denote an even characteristic, the $\frac{1}{2}2^p+2^{p-1}+1=5$ theta functions, of order 2 and zero (or integral) characteristic, $\vartheta^2(u;\ B)$, $\vartheta^2(u;\ A_1)$, ..., $\vartheta^2(u;\ A_4)$ are, by § 288, connected by a linear equation. As in Ex. v. we hence infer an equation of the form

$$\lambda\vartheta^2(u;\ A_4)=\lambda_1\vartheta^2(u;\ A_1)+\lambda_2\vartheta^2(u;\ A_2)+\lambda_3\vartheta^2(u;\ A_3);$$

adding to u the half-period associated with the characteristic A_2A_3, and putting $u=0$, we deduce by Ex. vii. that

$$\lambda e^{i\pi k_1' a_4}\,\vartheta^2(0;\ A_4A_2A_3)=\lambda_1 e^{i\pi k_1' a_1}\,\vartheta^2(0;\ A_1A_2A_3),$$

where $A_2A_3=\frac{1}{2}k_1$, $A_1=\frac{1}{2}a_1$, $A_4=\frac{1}{2}a_4$. Hence we obtain an equation which we may write in the form

$$\vartheta^2(0;\ A_1A_2A_3)\,\vartheta^2(u;\ A_4)=\binom{A_2A_3}{A_1A_4}\vartheta^2(0;\ A_4A_2A_3)\,\vartheta^2(u;\ A_1)$$
$$+\binom{A_3A_1}{A_2A_4}\vartheta^2(0;\ A_4A_3A_1)\,\vartheta^2(u;\ A_2)+\binom{A_1A_2}{A_3A_4}\vartheta^2(0;\ A_4A_1A_2)\,\vartheta^2(u;\ A_3),$$

where $\binom{A_2A_3}{A_1A_4}$ denotes a certain square root of unity. Such a relation holds between every four of the odd theta functions.

If A_1, ..., A_6 be the odd characteristics, and Q be any other characteristic, the six characteristics A_1Q, ..., A_6Q are said to form a Rosenhain hexad. It follows that the squares of every four theta functions of the same hexad are connected by a linear relation.

CHAPTER XVI.

A DIRECT METHOD OF OBTAINING THE EQUATIONS CONNECTING ϑ-PRODUCTS.

290. THE result given as Ex. xi. of § 286, in the last chapter, is a particular case of certain equations which may be obtained by actually multiplying together the theta series and arranging the product in a different way. We give in this chapter three examples of this method, of which the last includes the most general case possible. The first two furnish an introduction to the method and are useful for comparison with the general theorem. The theorems of this chapter do not require the characteristics to be half-integers.

291. *Lemma.* If b be a symmetrical matrix of p^2 elements, $U, V, u, v, A, B, f, g, q, r, f', g', q', r', M, N, s', t', m, n$ be columns, each of p elements, subject to the equations

$$n+m=2N+s', \quad q'+r'=f', \quad q+r=f, \quad U+V=2u=A,$$
$$-n+m=2M+t', \quad -q'+r'=g', \quad -q+r=g, \quad -U+V=2v=B,$$

then

$$2U(n+q')+b(n+q')^2+2\pi iq(n+q')+2V(m+r')+b(m+r')^2+2\pi ir(m+r')$$
$$=2A\left(N+\frac{s'+f'}{2}\right)+2b\left(N+\frac{s'+f'}{2}\right)^2+2\pi if\left(N+\frac{s'+f'}{2}\right)$$
$$+2B\left(M+\frac{t'+g'}{2}\right)+2b\left(M+\frac{t'+g'}{2}\right)^2+2\pi ig\left(M+\frac{t'+g'}{2}\right).$$

This the reader can easily verify.

Suppose now that the elements of s' and t' are each either 0 or 1, and that n and m take, independently, all possible positive and negative integer values. To any pair of values, the equations $n+m=2N+s'$, $-n+m=2M+t'$ give a corresponding pair of values for integers N and M, and a pair of values for s' and t'. Since $2m=2N+2M+s'+t'$, $s'+t'$ is even, and therefore, since each element of s' and t' is < 2, s' must be equal to t'. Hence by means of the 2^p possible values for s', the pairs (n, m) are divisible into 2^p sets, each characterised by a certain value of s'. Conversely to any assignable integer value for each of the pair (N, M) and any assigned value of s' (< 2) corresponds by the equations $n = N - M$, $m = N + M + s'$ a definite pair of integer columns n, m.

Hence, b being such a matrix that, for real x, bx^2 has its real part negative,

$$\left[\sum_n e^{2U(n+q')+b(n+q')^2+2\pi i q(n+q')}\right]\left[\sum_m e^{2V(m+r')+b(m+r')^2+2\pi i r(m+r')}\right]$$

$$= \sum_{s'} \left[\sum_N e^{2A\left(N+\frac{s'+f'}{2}\right)+2b\left(N+\frac{s'+f'}{2}\right)^2+2\pi i f\left(N+\frac{s'+f'}{2}\right)}\right]$$

$$\left[\sum_M e^{2B\left(M+\frac{s'+g'}{2}\right)+2b\left(M+\frac{s'+g'}{2}\right)^2+2\pi i g\left(M+\frac{s'+g'}{2}\right)}\right];$$

thus, if $\vartheta(u; \lambda)$, or $\vartheta\left(u; \begin{matrix}\lambda'\\ \lambda\end{matrix}\right)$, denote $\sum_n e^{2u(n+\lambda')+b(n+\lambda')^2+2\pi i\lambda(n+\lambda')}$, $\underline{\underline{\vartheta}}(u, \lambda)$ or $\underline{\underline{\vartheta}}\left(u; \begin{matrix}\lambda'\\ \lambda\end{matrix}\right)$ denote $\sum_n e^{4u(n+\lambda')+2b(n+\lambda')^2+2\pi i\lambda(n+\lambda')}$, we have

$$\vartheta(u-v; q)\,\vartheta(u+v; r) = \sum_{s'} \underline{\underline{\vartheta}}\left[u; \begin{matrix}\frac{1}{2}(s'+q'+r')\\ q+r\end{matrix}\right]\underline{\underline{\vartheta}}\left[v; \begin{matrix}\frac{1}{2}(s'-q'+r')\\ -q+r\end{matrix}\right],$$

where the equation on the right contains 2^p terms corresponding to all values of s', which is a column of p integers each either 0 or 1; all other quantities involved are quite unrestricted.

Therefore if a be a symmetrical matrix of p^2 elements and h any matrix of p^2 elements, we deduce, replacing u by hu, and v by hv, and multiplying both sides by $e^{au^2+av^2}$, the result

$$\vartheta(u-v; q)\,\vartheta(u+v; r) = \sum_{\epsilon'} \vartheta_1\left[u; \begin{matrix}\frac{1}{2}(\epsilon'+q'+r')\\ q+r\end{matrix}\right]\vartheta_1\left[v; \begin{matrix}\frac{1}{2}(\epsilon'-q'+r')\\ -q+r\end{matrix}\right],$$

where ϵ' denotes all possible 2^p columns of p elements, each either 0 or 1, and ϑ_1 differs from ϑ only by having $2a$, $2h$, $2b$ instead of a, h, b in the exponent; thus we may write, more fully,

$$\vartheta\left(u-v; \begin{matrix}q'\\ q\end{matrix}\middle|\begin{matrix}2\omega, 2\omega'\\ 2\eta, 2\eta'\end{matrix}\right)\vartheta\left(u+v; \begin{matrix}r'\\ r\end{matrix}\middle|\begin{matrix}2\omega, 2\omega'\\ 2\eta, 2\eta'\end{matrix}\right)$$

$$= \sum_{\epsilon'} \vartheta\left[u; \begin{matrix}\frac{1}{2}(\epsilon'+q'+r')\\ q+r\end{matrix}\middle|\begin{matrix}\omega, 2\omega'\\ 2\eta, 4\eta'\end{matrix}\right]\vartheta\left[v; \begin{matrix}\frac{1}{2}(\epsilon'-q'+r')\\ -q+r\end{matrix}\middle|\begin{matrix}\omega, 2\omega'\\ 2\eta, 4\eta'\end{matrix}\right].$$

Ex. i. When the characteristics q, r are equal half-integer characteristics, say

$$q = r = \tfrac{1}{2}\begin{pmatrix}a'\\ a\end{pmatrix},$$

the equation is

$$e^{\pi i a a'}\,\vartheta\left[u+v; \tfrac{1}{2}\begin{pmatrix}a'\\ a\end{pmatrix}\right]\vartheta\left[u-v; \tfrac{1}{2}\begin{pmatrix}a'\\ a\end{pmatrix}\right] = \sum_{\epsilon'} e^{\pi i a\epsilon'}\,\vartheta_1\left(u; \begin{matrix}\frac{1}{2}(\epsilon'+a')\\ 0\end{matrix}\right)\vartheta_1\left(v; \begin{matrix}\frac{1}{2}\epsilon'\\ 0\end{matrix}\right);$$

multiplying this equation by $e^{\pi i a n}$, when n denotes a definite row of integers, each either

0 or 1, and adding the equations obtained by ascribing to a all the 2^p possible sets of values in which each element of a is either 0 or 1, we obtain

$$2^p \vartheta_1\left(u\,;\,\begin{matrix}\tfrac{1}{2}(n+a')\\0\end{matrix}\right)\vartheta_1\left(v\,;\,\begin{matrix}\tfrac{1}{2}n\\0\end{matrix}\right)=\sum_a e^{\pi i a(n+a')}\,\vartheta\left[u+v\,;\,\tfrac{1}{2}\begin{pmatrix}a'\\a\end{pmatrix}\right]\vartheta\left[u-v\,;\,\tfrac{1}{2}\begin{pmatrix}a'\\a\end{pmatrix}\right];$$

for we have

$$\sum_a e^{\pi i a(\epsilon'+n)}=\prod_{i=1}^{p}\left[1+e^{\pi i(\epsilon'_i+n_i)}\right].$$

Ex. ii. Deduce from Ex. i. that when $p=1$, the ratio of the two functions

$$\vartheta\left[u+a\,;\,\tfrac{1}{2}\begin{pmatrix}0\\0\end{pmatrix}\right]\vartheta\left[u-a\,;\,\tfrac{1}{2}\begin{pmatrix}0\\0\end{pmatrix}\right]+i\vartheta\left[u+a\,;\,\tfrac{1}{2}\begin{pmatrix}1\\1\end{pmatrix}\right]\vartheta\left[u-a\,;\,\tfrac{1}{2}\begin{pmatrix}1\\1\end{pmatrix}\right],$$

$$\vartheta\left[u+b\,;\,\tfrac{1}{2}\begin{pmatrix}1\\0\end{pmatrix}\right]\vartheta\left[u-b\,;\,\tfrac{1}{2}\begin{pmatrix}1\\0\end{pmatrix}\right]+i\vartheta\left[u+b\,;\,\tfrac{1}{2}\begin{pmatrix}0\\1\end{pmatrix}\right]\vartheta\left[u-b\,;\,\tfrac{1}{2}\begin{pmatrix}0\\1\end{pmatrix}\right],$$

is independent of u.

Ex. iii. Prove that the 2^p functions $\vartheta_1\left(u\,;\,\begin{matrix}\tfrac{1}{2}(\epsilon'+a')\\0\end{matrix}\right)$, obtained by varying ϵ', are not connected by any linear equation with coefficients independent of u.

Ex. iv. Prove that if a, a' be integral,

$$\vartheta^2\left[u\,;\,\tfrac{1}{2}\begin{pmatrix}a'\\a\end{pmatrix}\right]=\sum_{\epsilon'} e^{\pi i a\epsilon'}\,\vartheta_1\left[0\,;\,\tfrac{1}{2}\begin{pmatrix}\epsilon'+a'\\0\end{pmatrix}\right]\vartheta_1\left[u\,;\,\tfrac{1}{2}\begin{pmatrix}\epsilon'\\0\end{pmatrix}\right].$$

From this set of equations we can obtain the linear relation connecting the squares of 2^p+1 (or less) assigned theta functions with half-integer coefficients.

Ex. v. Using the notation $|\lambda_{i,j}|$ for the matrix in which the j-th element of the i-th row is $\lambda_{i,j}$, prove that if $u_1, \ldots, u_r, v_1, \ldots, v_r$ be 2.2^p arguments, and $\tfrac{1}{2}\begin{pmatrix}a'\\a\end{pmatrix}$ any half-integer characteristic,

$$\left|\vartheta\left[u_i+v_j\,;\,\tfrac{1}{2}\begin{pmatrix}a'\\a\end{pmatrix}\right]\vartheta\left[u_i-v_j\,;\,\tfrac{1}{2}\begin{pmatrix}a'\\a\end{pmatrix}\right]\right|=\left|\vartheta_1\left[u_i\,;\,\begin{matrix}\tfrac{1}{2}\epsilon_j'a'\\a\end{matrix}\right]\right|\,\left|\vartheta_1\left[v_j\,;\,\begin{matrix}\tfrac{1}{2}\epsilon_i'\\0\end{matrix}\right]\right|,$$

and, denoting the determinant of the matrix on the left hand by $\{u_i, v_j\}$ and the determinant of the second matrix on the right hand by $\{v\}$, deduce that

$$\{v_i, v_j\}=e^{2^{p-1}\pi i A}\{v\}^2, \quad \{u_i, v_j\}=\sqrt{\{u_i, u_j\}\{v_i, v_j\}},$$

where A is the sum of the p elements of the row letter a. When the characteristic $\tfrac{1}{2}\begin{pmatrix}a'\\a\end{pmatrix}$ is odd, $\{u_i, u_j\}$ is a skew symmetrical determinant whose square root is* expressible rationally in terms of the constituents $\vartheta\left[u_i+u_j\,;\,\tfrac{1}{2}\begin{pmatrix}a'\\a\end{pmatrix}\right]\vartheta\left[u_i-u_j\,;\,\tfrac{1}{2}\begin{pmatrix}a'\\a\end{pmatrix}\right]$. For instance when $p=1$, we obtain, with a proper sign for the square root, the equation of three terms†.

Since any 2^p+1 functions of the form $\vartheta\left[u+v_\beta\,;\,\tfrac{1}{2}\begin{pmatrix}a'\\a\end{pmatrix}\right]\vartheta\left[u-v_\beta\,;\,\tfrac{1}{2}\begin{pmatrix}a'\\a\end{pmatrix}\right]$ are connected by a linear equation with coefficients independent of u, it follows that if $u_1, \ldots, u_m$, $v_1, \ldots, v_m$ be any $2m$ arguments, m being greater than 2^p, the determinant of m rows and columns, whose (i, j)th element is $\vartheta\left[u_i+v_j\,;\,\tfrac{1}{2}\begin{pmatrix}a'\\a\end{pmatrix}\right]\vartheta\left[u_i-v_j\,;\,\tfrac{1}{2}\begin{pmatrix}a'\\a\end{pmatrix}\right]$, vanishes identically. When $\tfrac{1}{2}\begin{pmatrix}a'\\a\end{pmatrix}$ is odd and m is even, for example equal to 2^p+2, this determinant is

* Scott, *Theory of determinants* (Cambridge, 1880), p. 71.

† Halphen, *Fonct. Ellip.* (Paris, 1886), t. I. p. 187.

a skew symmetrical determinant whose square root may be expressed rationally in terms of the functions $\vartheta\left[u_i+v_j\,;\ \tfrac{1}{2}\begin{pmatrix}a'\\a\end{pmatrix}\right]\vartheta\left[u_i-v_j\,;\ \tfrac{1}{2}\begin{pmatrix}a'\\a\end{pmatrix}\right]$. The result obtained may be written

$$\{u_i,\ v_j\}^{\frac{1}{2}}=0,$$

wherein* the determinant $\{u_i,\ v_j\}$ has m rows and columns, m being even and greater than 2^p. When m is odd the determinant $\{u_i,\ v_j\}$ itself vanishes.

A proof that for general values of the arguments the corresponding determinant $\{u_i,\ v_j\}$, of 2^p rows and columns, does not identically vanish is given by Frobenius, *Crelle*, XCVI. (1884), p. 102.

A more general formula for the product of two theta functions is given below Ex. ii. § 292.

292. We proceed now to another formula, for the product of four theta functions. Let J denote the substitution

$$\tfrac{1}{2}\begin{pmatrix} -1 & 1 & 1 & 1 \\ 1 & -1 & 1 & 1 \\ 1 & 1 & -1 & 1 \\ 1 & 1 & 1 & -1 \end{pmatrix},$$

and J_{rs} be the element of the matrix which is in the r-th row and the s-th column; then $\sum_{i=1}^{4} J_{ir}\,J_{is}=0$ or 1, according as $r\neq s$, or $r=s\ (r,\ s=1,\ 2,\ 3,\ 4)$.

Let $u_1,\ u_2,\ u_3,\ u_4$ denote four columns, each of p quantities; written down together they will form a matrix of 4 columns and p rows. Let $U_1,\ U_2,\ U_3,\ U_4$ be four other such columns, such that the j-th row of the first matrix $(j=1,\ 2,\ \ldots,\ p)$ is associated with the j-th row of the second by the equation

$$((u_1)_j,\ (u_2)_j,\ (u_3)_j,\ (u_4)_j)=J\,((U_1)_j,\ (U_2)_j,\ (U_3)_j,\ (U_4)_j).$$

Let $v_1,\ v_2,\ v_3,\ v_4$ and $V_1,\ V_2,\ V_3,\ V_4$ be two other similarly associated sets, each of four columns of p elements. Then if h be any matrix whatever, of p rows and columns, we have

$$hu_1v_1+hu_2v_2+hu_3v_3+hu_4v_4=hU_1V_1+hU_2V_2+hU_3V_3+hU_4V_4\,;$$

this is quite easy to prove: an elementary direct verification is obtained by selecting on the left the term $h_{jk}(u_1)_k(v_1)_j+h_{jk}(u_2)_k(v_2)_j+h_{jk}(u_3)_k(v_3)_j+h_{jk}(u_4)_k(v_4)_j$

$$=h_{jk}\sum_{r=1}^{4}\left[J_{r1}(U_1)_k+J_{r2}(U_2)_k+J_{r3}(U_3)_k+J_{r4}(U_4)_k\right]\left[J_{r1}(V_1)_j+J_{r2}(V_2)_j\right.$$
$$\left.+J_{r3}(V_3)_j+J_{r4}(V_4)_j\right]$$

$$=h_{jk}\left\{(\textstyle\sum_r J^2_{r1})\,(U_1)_k\,(V_1)_j+(\sum_r J_{r1}J_{r2})\left[(U_1)_k\,(V_2)_j+(U_2)_k\,(V_1)_j\right]+\ldots\ldots\right\}$$

$$=h_{jk}\left\{(U_1)_k\,(V_1)_j+(U_2)_k\,(V_2)_j+(U_3)_k\,(V_3)_j+(U_4)_k\,(V_4)_j\right\},$$

and this is the corresponding element of $hU_1V_1+hU_2V_2+hU_3V_3+hU_4V_4$.

* The theorem was given by Weierstrass, *Sitzungsber. der Berlin. Ak.* 1882 (I.—XXVI., p. 506), with the suggestion that the theory of the theta functions may be *à priori* deducible therefrom, as is the case when $p=1$ (Halphen, *Fonct. Ellip.* (Paris (1886)), t. I. p. 188). See also Caspary, *Crelle*, XCVI. (1884), and *ibid.* XCVII. (1884), and Frobenius, *Crelle*, XCVI. (1884), pp. 101, 103.

Now we have

$$\vartheta(u_1, q_1)\,\vartheta(u_2, q_2)\,\vartheta(u_3, q_3)\,\vartheta(u_4, q_4)$$
$$= \sum_{n_1,\, n_2,\, n_3,\, n_4} e^{\Sigma a u_r^2 + 2\Sigma h u_r (n_r + q_r') + \Sigma b (n_r + q_r')^2 + 2\pi i \Sigma q_r (n_r + q_r')}.$$

In the exponent here there are four sets each of four columns of p quantities namely the sets

$$u_r,\ n_r,\ q_r,\ q'_r;$$

we suppose each of these transformed by the substitution J. Hence the exponent becomes

$$\sum_{N_1,\, N_2,\, N_3,\, N_4} e^{\Sigma a U_r^2 + 2\Sigma h U_r (N_r + Q_r') + \Sigma b (N_r + Q_r')^2 + 2\pi i \Sigma Q_r (N_r + Q'_r)}$$

wherein the summation extends to all values of N_{rj} given by

$$N_{rj} = \tfrac{1}{2}(n_{1j} + n_{2j} + n_{3j} + n_{4j} - 2n_{rj}),$$

for which all of n_{rj} are integers.

All the values N_{rj} will not be integral. But since $N_{rj} - N_{sj} = n_{sj} - n_{rj}$ the fractional parts of N_{1j}, N_{2j}, N_{3j}, N_{4j} will be the same, $= \tfrac{1}{2}\epsilon_j'$, say, ($\epsilon_j' = 0$ or 1). Let m_{rj} be the integral part of N_{rj}. We arrange the terms of the right hand into 2^p classes according to the 2^p values of ϵ_j'. Then since

$$m_{rj} = \tfrac{1}{2}(n_{1j} + n_{2j} + n_{3j} + n_{4j} - 2n_{rj}) - \tfrac{1}{2}\epsilon_j',$$

every term of the left-hand product, arising from a certain set of values of the $4p$ integers n_{rj}, gives rise to a definite term of the transformed product on the right with a definite value for ϵ_j', while, since

$$n_{rj} = \tfrac{1}{2}(m_{1j} + m_{2j} + m_{3j} + m_{4j} - 2m_{rj}) + \tfrac{1}{2}\epsilon_j',$$

every assignable set of values of the $4p$ integers m_{rj} and value for ϵ_j' (which would correspond to a definite term of the transformed product) will arise, from a certain term on the right, *provided only the values assigned for* m_{rj} *be such that* $\tfrac{1}{2}(m_{1j} + m_{2j} + m_{3j} + m_{4j} + \epsilon_j')$ *is integral.*

Now we can specify an expression involving the quantities

$$\mu_j,\ = \tfrac{1}{2}(m_{1j} + m_{2j} + m_{3j} + m_{4j} + \epsilon_j'),$$

which is 1 or 0 according as $\mu = (\mu_1, \mu_2, \ldots, \mu_p)$ is a column of integers or not. In fact if $\epsilon = (\epsilon_1, \ldots, \epsilon_p)$ be a column of quantities each either 0 or 1—so that ϵ is capable of 2^p values—the expression

$$\frac{1}{2^p}\sum_\epsilon e^{2\pi i \epsilon\mu} = \frac{1}{2^p}(\Sigma e^{2\pi i \epsilon_1 \mu_1})\ldots(\Sigma e^{2\pi i \epsilon_p \mu_p}) = \frac{1}{2^p}(1 + e^{2\pi i \mu_1})(1 + e^{2\pi i \mu_2})\ldots(1 + e^{2\pi i \mu_p})$$

has this property; for when $\mu_1, \ldots, \mu_p$ are not integers they are half-integers.

Hence if the series $\frac{1}{2^p}\sum_\epsilon e^{\pi i\epsilon(m_1+m_2+m_3+m_4+\epsilon')}$ be attached as factor to every term of the transformed product on the right we may suppose the summation to extend to *all* integral values of m_{rj}, *for every value of* ϵ'.

Then the transformed product is

$$\frac{1}{2^p}\sum_{m_1m_2m_3m_4\epsilon} e^{\Sigma a U_r^2+2\Sigma h U_r(m_r+\frac{1}{2}\epsilon'+Q'_r)+\Sigma b(m_r+\frac{1}{2}\epsilon'+Q'_r)^2+2\pi i\Sigma Q_r(m_r+\frac{1}{2}\epsilon'+Q'_r)+\pi i\epsilon(m_1+m_2+m_3+m_4+\epsilon')}$$

$$=\frac{1}{2^p}\Sigma\prod_r e^{aU_r^2+2hU_r(m_r+p'_r)+b(m_r+p'_r)^2+2\pi i p_r(m_r+p'_r)}\cdot e^{-\pi i\epsilon(\Sigma p'_r-\epsilon')},$$

where

$$p_r=\tfrac{1}{2}\epsilon+Q_r,\quad p_r'=\tfrac{1}{2}\epsilon'+Q_r',$$

so that

$$\Sigma p_r'=2\epsilon'+\Sigma Q_r'=2\epsilon'+\Sigma q_r'.$$

Thus we have

$$\vartheta(u_1,q_1)\,\vartheta(u_2,q_2)\,\vartheta(u_3,q_3)\,\vartheta(u_4,q_4)$$

$$=\frac{1}{2^p}\sum_{(\epsilon,\epsilon')} e^{-\pi i\epsilon(\epsilon'+\Sigma q'_r)}\,\vartheta\left[U_1, Q_1+\tfrac{1}{2}\begin{pmatrix}\epsilon'\\ \epsilon\end{pmatrix}\right]\vartheta\left[U_2, Q_2+\tfrac{1}{2}\begin{pmatrix}\epsilon'\\ \epsilon\end{pmatrix}\right]\vartheta\left[U_3, Q_3+\tfrac{1}{2}\begin{pmatrix}\epsilon'\\ \epsilon\end{pmatrix}\right]$$

$$\vartheta\left[U_4, Q_4+\tfrac{1}{2}\begin{pmatrix}\epsilon'\\ \epsilon\end{pmatrix}\right].$$

This very general formula obviously includes the formula of Ex. xi., § 286, Chap. XV. It is clear moreover that a similar investigation can be made for the product of any number, k, of theta-functions, provided only we know of a matrix J, of k rows and columns, which will transform the exponent of the general term of the product into the exponent of the general term of the sum of other products.

It is for this more general case that the next Article is elaborated. It is not necessary for either case that the characteristics $q_1, q_2, \ldots$ should consist of half-integers.

Ex. i. If q be a half-integer characteristic, $=Q$, say, and we use the abbreviation

$$\phi(u, v, w, t;\ Q)=\vartheta(u;\ Q)\,\vartheta(v;\ Q)\,\vartheta(w;\ Q)\,\vartheta(t;\ Q),$$

we have

$$\phi(u+a, u-a, v+b, v-b, Q)=\frac{1}{2^p}\sum_{\epsilon,\epsilon'} e^{-\pi i\epsilon\epsilon'}\phi\left[u+b, u-b, v+a, v-a;\ Q+\tfrac{1}{2}\begin{pmatrix}\epsilon'\\ \epsilon\end{pmatrix}\right],$$

where the summation on the right hand extends to all possible 2^{2p} half-integer characteristics $\frac{1}{2}\begin{pmatrix}\epsilon'\\ \epsilon\end{pmatrix}$; putting $Q+\frac{1}{2}\begin{pmatrix}\epsilon'\\ \epsilon\end{pmatrix}=R$, so that R also becomes all 2^{2p} half-integer characteristics, this is the same as

$$e^{\pi i|Q|}\phi(u+a, u-a, v+b, v-b;\ Q)=\frac{1}{2^p}\sum_R e^{\pi i|Q,R|+\pi i|R|}\phi(u+b, u-b, v+a, v-a;\ R),$$

where,

if $\quad Q=\frac{1}{2}\begin{pmatrix}a'\\ a\end{pmatrix},\quad R=\frac{1}{2}\begin{pmatrix}\beta'\\ \beta\end{pmatrix},\quad$ then $\quad |Q|=aa',\quad |R|=\beta\beta',\quad |Q, R|=a\beta'-a'\beta.$

By adding, or subtracting, to this the formula derived from it by interchange of v and a, we obtain a formula in which only even or odd characteristics R occur on the right hand. Thus, for $p=1$, we derive the equation of three terms.

Ex. ii. If a, β, γ, δ be integers such that $a\gamma$ is positive and $\beta\delta$ is negative, $\rho = a\delta - \beta\gamma$, and r be the absolute value of ρ, prove that

$$\Theta\left(u\,;\ a\gamma\tau\left|\begin{matrix}0\\0\end{matrix}\right.\right)\Theta\left(v\,;\ -\beta\delta\tau\left|\begin{matrix}0\\0\end{matrix}\right.\right) = \sum_{\mu,\nu}\Theta\left(u\delta - v\gamma\,;\ \rho\gamma\delta\tau\left|\begin{matrix}\mu/\rho\\0\end{matrix}\right.\right)\Theta\left(-u\beta + va\,;\ -\rho a\beta\tau\left|\begin{matrix}\nu/\rho\\0\end{matrix}\right.\right)$$

$$= r^{-2p}\sum_{e,f,g,h}\Theta\left(u\delta - v\gamma\,;\ \rho\gamma\delta\tau\left|\begin{matrix}e/\rho\\-\gamma g + \delta h\end{matrix}\right.\right)\Theta\left(-u\beta + va\,;\ -\rho a\beta\tau\left|\begin{matrix}f/\rho\\ag - \beta h\end{matrix}\right.\right),$$

where $\Theta\left(u\,;\ \tau\left|\begin{matrix}\epsilon'\\\epsilon\end{matrix}\right.\right)$ denotes the theta function in which the exponent of the general term is

$$2\pi iu\,(n+\epsilon') + i\pi\tau\,(n+\epsilon')^2 + 2\pi i\epsilon\,(n+\epsilon'),$$

and μ, ν are row letters of p elements, all positive (or zero) and less than r, subject to the condition that $(\delta\mu - \beta\nu)/\rho$, $(a\nu - \gamma\mu)/\rho$ are integral, while e, f, g, h are row letters of p elements which are all positive (or zero) and less than r.

Ex. iii. Taking, in Ex. ii., a, β, γ, δ respectively equal to $1, 1, 1, -k$, we find $\mu = \nu < k+1$, k being positive. Hence, taking $k=3$, prove the formula (Königsberger, *Crelle*, LXIV. (1865), p. 24), of which each side contains 2^p terms,

$$\sum_s \Theta\left(u\,;\ \tau\left|\begin{matrix}\tfrac{1}{2}s'\\\tfrac{1}{2}s\end{matrix}\right.\right)\Theta\left(u\,;\ 3\tau\left|\begin{matrix}\tfrac{1}{2}s'\\\tfrac{1}{2}s\end{matrix}\right.\right) = \sum_s e^{-\pi i s's}\,\Theta\left(0\,;\ \tau\left|\begin{matrix}0\\\tfrac{1}{2}s\end{matrix}\right.\right)\Theta\left(2u\,;\ 3\tau\left|\begin{matrix}0\\\tfrac{1}{2}s\end{matrix}\right.\right),$$

s, s' being rows of p quantities each either 0 or 1.

293. We proceed now to obtain a formula* for the product of any number, k, of theta functions.

We shall be concerned with two matrices X, x, each of p rows and k columns; the original matrix, written with capital letters, is to be transformed into the new matrix by a substitution different for each of the p rows; for the j-th row this substitution is of the form

$$(X_{1,j}, X_{2,j}, \ldots, X_{r,j}, \ldots, X_{k,j}) = \frac{1}{r_j}\,\omega_j\,(x_{1,j}, x_{2,j}, \ldots, x_{r,j}, \ldots, x_{k,j})\,;$$

herein r_j is a positive integer; ω_j is a matrix of k rows and columns, consisting of integers; the determinant formed by the elements of this matrix is supposed other than zero, and denoted by μ_j; bearing in mind that throughout this Article the values of r are $1, 2, \ldots, k$ and the values of j are $1, 2, \ldots, p$, we may write the substitution in the form

$$(X_{r,j}) = \frac{1}{r_j}\,\omega_j\,(x_{r,j}).$$

The substitution formed with the first minors of the determinant of ω_j will be denoted by Ω_j; that formed from Ω_j by a transposition of its rows and columns will be denoted by $\overline{\Omega}_j$. Then the substitution inverse to $\dfrac{1}{r_j}\,\omega_j$ is $\dfrac{r_j}{\mu_j}\,\overline{\Omega}_j$; denoting the former substitution by λ_j, the latter is λ_j^{-1}.

* Prym und Krazer, *Neue Grundlagen...der allgemeinen thetafunctionen*, Leipzig, 1892.

If for any value of j a set of k integers, $P_{r,j}$, be known such that the k quantities

$$\lambda_j^{-1}(P_{r,j}) = \frac{r_j}{\mu_j}\bar{\Omega}_j(P_{r,j})$$

are integers, then it is clear that an infinite number of such sets can be derived; we have only to increase the integers $P_{r,j}$ by integral multiples of μ_j. But the number of such sets in which each of $P_{r,j}$ is positive (including zero) and less than the absolute value of μ_j is clearly finite, since each element has only a finite number of possible values. We shall denote this number by s_j and call it the number of *normal* solutions of the conditions

$$\frac{r_j}{\mu_j}\bar{\Omega}_j(P_{r,j}) = \text{integral};$$

it is the same as the number of sets of k integers, positive (or zero) and less than the absolute value of μ_j, which can be represented in the form $\lambda_j(p_{r,j})$, for integral values of the elements $p_{r,j}$.

The k theta functions to be multiplied together are at first taken to be those given by

$$\Theta_r = \Sigma e^{2V_rN_r + B_rN_r^2}, \qquad (r = 1, \ldots, k),$$

wherein B_r is such a symmetrical matrix that, for real values of the p quantities X, the real part of the quadratic form denoted (§ 174, Chap. X.) by B_rX^2 is negative. The p elements of the row-letters V_r, N_r are denoted by $V_{r,j}$, $N_{r,j}$ $(j = 1, \ldots, p)$. The substitutions λ_j are supposed to be such that the equations $(X_{r,j}) = \lambda_j(x_{r,j})$ transform the sum $\sum_{r=1}^{k} B_rX_r^2$ into a sum $\sum_{r=1}^{k} b_rx_r^2$, in which the matrices b_r are symmetrical and have the property that for real x_r the real part of $b_rx_r^2$ is negative.

Taking now quantities $m_{r,j}$, $v_{r,j}$ determined by

$$(m_{r,j}) = \lambda_j^{-1}(N_{r,j}) = \frac{r_j}{\mu_j}\bar{\Omega}_j(N_{r,j}), \quad (v_{r,j}) = \bar{\lambda}_j(V_{r,j}) = \frac{1}{r_j}\bar{\omega}_j(V_{r,j}),$$

the expressions $\sum_{r=1}^{k} B_rN_r^2$, $\sum_{r=1}^{k} N_rV_r$ are respectively transformed to $\sum_{r=1}^{k} b_rm_r^2$ and

$$\sum_{j=1}^{p} \lambda_j(m_{r,j})(V_{r,j}) = \sum_{j=1}^{p} \bar{\lambda}_j(V_{r,j})(m_{r,j}) = \sum_{r=1}^{k} v_rm_r;$$

hence the product $\prod_{r=1}^{k} \Theta_r$ is transformed into $\sum_{N_1, \ldots, N_k} e^{2\sum_r v_rm_r + \sum_r b_rm_r^2}$, where the quantities $m_{r,j}$ have every set of values such that the quantities $\lambda_j(m_{r,j})$ take all the integral values, $N_{r,j}$, of the original product.

As in the two cases previously considered in this chapter, we seek now to associate *integers* with the quantities $m_{r,j}$. Let $(P_{r,j})$ be any normal solution of the conditions

$$\frac{r_j}{\mu_j}\bar{\Omega}_j(P_{r,j}) = \text{integral}, \ = (p_{r,j}), \text{ say};$$

put, for every value of j,

$$(N_{r,j}) - (P_{r,j}) = \mu_j(M_{r,j}) + (E'_{r,j}), \qquad (r = 1, \ldots, k)$$

wherein $(M_{r,j})$ consists of integers, and $(E'_{r,j})$ consists of positive integers (including zero), of which each is less than the absolute value of μ_j. For an assigned set $(P_{r,j})$ this is possible in one way; then

$$(m_{r,j}) = \frac{r_j}{\mu_j}\bar{\Omega}_j(N_{r,j}) = (p_{r,j}) + r_j\bar{\Omega}_j(M_{r,j}) + \frac{r_j}{\mu_j}\bar{\Omega}_j(E'_{r,j})$$
$$= (n_{r,j}) + \frac{1}{\mu_j}(\epsilon'_{r,j}), \text{ say,}$$

where

$$(n_{r,j}) = (p_{r,j}) + r_j\bar{\Omega}_j(M_{r,j}), \quad (\epsilon'_{r,j}) = r_j\bar{\Omega}_j(E'_{r,j});$$

by this means there is associated with $(N_{r,j})$, corresponding to an assigned set $(P_{r,j})$, a definite set of integers $(n_{r,j})$, and a definite set $(E'_{r,j})$. We do not thus obtain every possible set of integers for $(n_{r,j})$, for we have

$$\frac{1}{r_j}\omega_j(n_{r,j}) = \frac{1}{r_j}\omega_j(p_{r,j}) + \mu_j(M_{r,j}) = (P_{r,j}) + \mu_j(M_{r,j}),$$

so that the values of $n_{r,j}$ which arise are such that $\lambda_j(n_{r,j})$ are integers.

Conversely let $(n_{r,j})$ be any assigned integers such that $\lambda_j(n_{r,j})$ are integers; put

$$\lambda_j(n_{r,j}) = (P_{r,j}) + \mu_j(M_{r,j}),$$

wherein the quantities $M_{r,j}$ are integers, and the quantities $P_{r,j}$ are positive integers (or zero), which are all less than the absolute value of μ_j; this is possible in one way; then taking any set of assigned integers $(E'_{r,j})$, which are all positive (or zero) and less than the absolute value of μ_j, we can define a set of integers $N_{r,j}$ by the equations, wherein $\lambda_j^{-1}(P_{r,j}) = \text{integral}$,

$$(N_{r,j}) = (E'_{r,j}) + (P_{r,j}) + \mu_j(M_{r,j}) = (E'_{r,j}) + \lambda_j(n_{r,j}).$$

Thus, from any set of integers $(N_{r,j})$, arising with a term $e^{\sum_r (2V_rN_r + B_rN_r^2)}$ of the product $\prod_{r=1}^{k}\Theta_r$, we can, by association with a definite normal solution $(P_{r,j})$ of the conditions $\lambda_j^{-1}(P_{r,j}) = \text{integral}$, obtain a definite set $(E'_{r,j})$, and a definite set $(n_{r,j})$ such that $\lambda_j(n_{r,j})$ are integers. And conversely, from any set of integers $(n_{r,j})$ which are such that $\lambda_j(n_{r,j})$ are integral, we can, by association with a definite set $(E'_{r,j})$, obtain a definite normal solution $(P_{r,j})$ and a definite set $(N_{r,j})$.

It follows therefore that if the product $\prod_{r=1}^{k} \Theta_r$ be written down $s_1 \dots s_p$ times, a term $e^{\sum_r (2V_rN_r + B_rN_r^2)}$ being associated in turn with every one of the $s_1 \dots s_p$ normal solutions of the p conditions $\lambda_j^{-1}(P_j) =$ integral, then there will arise, once with every assigned set $(E'_{r,j})$, every possible set $(n_{r,j})$ for which $\lambda_j(n_{r,j})$ are integers.

We introduce now a factor which has the value 1 or 0 according as the integers $(n_{r,j})$ satisfy the conditions $\lambda_j(n_{r,j}) =$ integral, or not. Take k integers $(E_{r,j})$, which are positive (or zero), and less than r_j; put

$$(\epsilon_{r,j}) = \bar{\omega}_j(E_{r,j});$$

then

$$\sum_{j=1}^{p} \sum_{r=1}^{k} \frac{1}{r_j} \epsilon_{r,j}\left(n_{r,j} + \frac{\epsilon'_{r,j}}{\mu_j}\right) = \sum_j \sum_r \frac{1}{r_j} \epsilon_{r,j} m_{r,j} = \sum_j \bar{\lambda}_j(E_{r,j})(m_{r,j}) = \sum_j \lambda_j(m_{r,j})(E_{r,j})$$
$$= \sum_j (N_{r,j})(E_{r,j}) = N_r E_r,$$

and this is integral when N_r is integral, that is, for all the values $(n_{r,j})$ which actually occur; in fact the quantities $N_{r,j}$ defined by

$$(N_{r,j}) = \lambda_j(m_{r,j}) = \frac{1}{r_j}\omega_j\left(n_{r,j} + \frac{\epsilon'_{r,j}}{\mu_j}\right) = \frac{1}{r_j}\omega_j(n_{r,j}) + (E'_{r,j}) = \lambda_j(n_{r,j}) + (E'_{r,j})$$

are integral or not according as $\lambda_j(n_{r,j})$ are integers or not.

Hence, for a given set $n_{r,j}$, and a given set $E'_{r,j}$, the sum

$$\sum_E e^{2\pi i \sum_j \sum_r \frac{1}{r_j} \epsilon_{r,j}\left(n_{r,j} + \frac{\epsilon'_{r,j}}{\mu_j}\right)} = \sum_E e^{2\pi i \sum_r N_r E_r} = \prod_{r,j} \sum_{E_{r,j}} [e^{2\pi i N_{r,j}}]^{E_{r,j}},$$

wherein the summation extends to all positive (and zero) integer values of $(E_{r,j})$ less than r_j, is equal to $r_1^k \dots r_p^k$ when $(N_{r,j})$ are all integral, and otherwise contains a factor of the form

$$(e^{2\pi i r_j N_{r,j}} - 1)/(e^{2\pi i N_{r,j}} - 1),$$

which is zero because $r_j(N_{r,j})$ is certainly integral. Hence if we denote

$$\sum_j \sum_r \frac{1}{r_j} \epsilon_{r,j}\left(n_{r,j} + \frac{\epsilon'_{r,j}}{\mu_j}\right) \text{ by } \sum_r \frac{1}{R} \epsilon_r\left(n_r + \frac{\epsilon'_r}{\mu}\right),$$

R having the values $r_1, \dots, r_p$, then we can write

$$\frac{1}{(r_1 \dots r_p)^k} \sum_E e^{2\pi i \sum_r \frac{1}{R} \epsilon_r \left(n_r + \frac{\epsilon'_r}{\mu}\right)} = 1, \text{ or } 0,$$

according as $\lambda_j(n_{r,j})$ are all integers or not.

If then every term of the transformed series, in which, so far, only those values of $n_{r,j}$ arise for which $\lambda_j(n_{r,j})$ are integers, be multiplied by this factor, and the transformed series be completed by the introduction of terms of the same general form as those which naturally arise in this way, so that now all possible integer values of $(n_{r,j})$ are taken in, the value of the transformed series will be unaltered. In other words we have

$$\prod_r \Theta_r = \prod_r e^{2V_rN_r+B_rN_r^2} = \frac{1}{s_1\ldots s_p(r_1\ldots r_p)^k}\sum_{N_1,\ldots,N_k,E',E}\prod_r e^{2v_rm_r+b_rm_r^2+2\pi i\frac{1}{R}\epsilon_r\left(n_r+\frac{\epsilon'_r}{\mu}\right)},$$

$$=(s_1\ldots s_p)^{-1}(r_1\ldots r_p)^{-k}\sum_{n,E',E}\prod_r e^{2v_r\left(n_r+\frac{\epsilon'_r}{\mu}\right)+b_r\left(n_r+\frac{\epsilon'_r}{\mu}\right)^2+2\pi i\frac{1}{R}\epsilon_r\left(n_r+\frac{\epsilon'_r}{\mu}\right)},$$

wherein all possible integer values of $(n_{r,j})$ arise on the right; thus the right-hand side is equal to

$$(s_1\ldots s_p)^{-1}(r_1\ldots r_p)^{-k}\sum_{E',E}\prod_r \Theta_r\left(v_r;\ \begin{matrix}\epsilon_r'/\mu\\ \epsilon_r/R\end{matrix}\right);$$

and this is the desired form of the transformed product. For convenience we recapitulate the notations; E_r', E_r each denote a column of p integers, positive or zero, such that $E'_{r,j}<|\mu_j|$, $E_{r,j}<r_j$; $(\epsilon'_{r,j})=r_j\overline{\Omega}_j(E'_{r,j})$; $(\epsilon_{r,j})=\overline{\omega}_j(E_{r,j})$; s_j is the number of sets of integral solutions, positive or zero, each less than $|\mu_j|$, of the conditions $\mu_j^{-1}r_j\Omega_j(P_{r,j})=$ integral; $(v_{r,j})=r_j^{-1}\overline{\omega}_j(V_{r,j})$; the function Θ_r is a theta function in which the ordinary matrices a, b, h (§ 189) are respectively 0, b_r, 1; by linear transformation of the variables of the form $V_r=h_rW_r$, and, in case the matrices ω_j be suitable, multiplication by an exponential $e^{\sum_r A_rV_r^2}$, these particularities in the form of the theta functions may be removed.

The number of sets $(E_{r,j})$ is $(r_1\ldots r_p)^k$; the number of sets $(E'_{r,j})$ is $|\mu_1^k\ldots\mu_p^k|$; the product of these numbers is the number of theta-products on the right-hand side of the equation.

Ex. i. We test this formula by applying it to the case already discussed where ω_j is an orthogonal substitution given by

$$\omega_j=\begin{pmatrix}-1 & 1 & 1 & 1\\ 1 & -1 & 1 & 1\\ 1 & 1 & -1 & 1\\ 1 & 1 & 1 & -1\end{pmatrix},\quad =\omega \text{ say},$$

which is independent of j, $r_j=2$, $b_r=b$, $k=4$; then $\mu_j=-16$, $E_{r,j}<2$, $E'_{r,j}<16$, and

$$\frac{r_j}{\mu_j}\overline{\Omega}_j=\left[\frac{1}{r_j}\omega_j\right]^{-1}=\tfrac{1}{2}\omega,\quad \frac{1}{R}(\epsilon_{r,j})=\tfrac{1}{2}\omega(E_{r,j}),\quad \frac{1}{\mu}\epsilon'_{r,j}=\tfrac{1}{2}\omega(E'_{r,j});$$

thence $\frac{1}{R}\epsilon_{1,j}-\frac{1}{R}\epsilon_{2,j}=E_{2,j}-E_{1,j}=$ integral, etc., so that the fractional part of $\frac{1}{R}\epsilon_{r,j}$ is independent of r: similarly the fractional part of $\frac{1}{\mu}(\epsilon'_{r,j})$ is independent of r and we may write $\frac{1}{\mu}(\epsilon'_{r,j})=(\frac{1}{2}\epsilon'_j+L_{1,j},\ \frac{1}{2}\epsilon'_j+L_{2,j},\ \ldots,\ \frac{1}{2}\epsilon'_j+L_{4,j})$ wherein $2L_{r,j}+\epsilon'_j<16$. By the formula

$\vartheta(v, q+N)=e^{2\pi i q' N}\vartheta(v, q)$, when N is integral, we know that $\Theta_r\left(v_r; \begin{matrix}\epsilon'_r/\mu \\ \epsilon_r/R\end{matrix}\right)$ is independent of the integral part of ϵ'_r/μ. Hence the $(16)^{4p}=2^{16p}$ terms on the right-hand side of the general formula, which, for a specified value of $\frac{1}{2}\omega(E_{r,j})$, correspond to all the values of $\frac{1}{2}\omega(E'_{r,j})$, reduce to 2^p terms, in which, since $(E'_{r,j})=\frac{1}{2}\omega(\frac{1}{2}\epsilon'_j+L_{1,j}, \ldots, \frac{1}{2}\epsilon'_j+L_{4,j})$, all values of $\epsilon'(<2)$ arise. Hence there is a factor 2^{15p} and instead of the summation in regard to E, E' we have a summation in regard to E, ϵ', the right hand being in fact

$$C \,.\, 2^{15p}\sum_{E,\,\epsilon'}\prod_r\Theta\left(v_r, \begin{matrix}\frac{1}{2}\epsilon' \\ \frac{1}{2}\omega(E_{r,j})\end{matrix}\right)$$

and containing 2^{4p} terms.

Now put $$\tfrac{1}{2}(E_{1,j}+E_{2,j}+E_{3,j}+E_{4,j})=\tfrac{1}{2}\epsilon_j+M_j,$$

M_j being integral; then the factor of a general term of the expanded right-hand product which contains the quantities $\frac{1}{2}\omega(E_{r,j})$ is

$$\prod_r e^{2\pi i\frac{1}{2}k_r(n_r+\frac{1}{2}\epsilon')},$$

where

$$k_{r,j}=E_{1,j}+E_{2,j}+E_{3,j}+E_{4,j}-2E_{r,j}=\epsilon_j+2(M_j-E_{r,j}),$$

and

$$e^{\frac{1}{2}\pi i\epsilon'\sum_r k_r}=\prod_j e^{\frac{1}{2}\pi i\epsilon_j'(4\epsilon_j+8M_j-2\epsilon_j-4M_j)}=\prod_j e^{\pi i\epsilon_j\epsilon_j'}=e^{\pi i\epsilon\epsilon'},$$

while

$$\sum_j\sum_r \pi i k_{r,j}n_{r,j}\equiv\pi i\sum_j\sum_r\epsilon_j n_{r,j}\ (\text{mod. } 2),\ \equiv\pi i\epsilon\,.\sum_r n_r,$$

so that

$$\prod_r e^{2\pi i\frac{1}{2}k_r(n_r+\frac{1}{2}\epsilon')}=\left[\prod_r e^{2\pi i\frac{1}{2}\epsilon(n_r+\frac{1}{2}\epsilon')}\right]e^{-\pi i\epsilon\epsilon'};$$

therefore the right-hand product consists only of terms of the form $\left[\prod_r\Theta\left(v_r, \begin{matrix}\frac{1}{2}\epsilon' \\ \frac{1}{2}\epsilon\end{matrix}\right)\right]e^{-\pi i\epsilon\epsilon'}$. Hence the 2^{4p} terms arising, for a specified value of ϵ', for all the values of $E_{r,j}$, reduce to 2^p terms, and there is a further factor 2^{3p}—the right hand being

$$C\,.\,2^{18p}\sum_{\epsilon,\,\epsilon'}\left[\prod_r\Theta\left(v_r, \begin{matrix}\frac{1}{2}\epsilon' \\ \frac{1}{2}\epsilon\end{matrix}\right)\right]e^{-\pi i\epsilon\epsilon'},$$

where

$$C=(s_1\ldots s_p)^{-1}(r_1\ldots r_p)^{-k}=(s_1\ldots s_p)^{-1}2^{-4p}=s^{-p}2^{-4p}.$$

To determine the value of C we must know the number (s) of positive integral solutions, each less than 16, of the conditions $\frac{1}{2}\omega(x)=$ integral, $=(y)$ say, namely of the conditions, $x_1+x_2+x_3+x_4=2(x_r+y_r)$. Now of these any positive values of x_1, x_2, x_3, x_4 (<16) are admissible for which $x_1+x_2+x_3+x_4$ is even. They must therefore either be all even, possible in 8^4 ways, or two even, possible in $6\,.\,8^2\,.\,8^2$ ways, or all odd, possible in 8^4 ways. Hence $s=8\,.\,8^4=2^{15}$. Hence $C=1/2^{15p}2^{4p}=1/2^{19p}$ and therefore $C\,.\,2^{18p}=\dfrac{1}{2^p}$.

Making now in the formula thus obtained, which is

$$\prod_r\Theta(V_r, 0)=\frac{1}{2^p}\sum_{\epsilon,\,\epsilon'}e^{-\pi i\epsilon\epsilon'}\prod_r\Theta\left[v_r, \tfrac{1}{2}\begin{pmatrix}\epsilon' \\ \epsilon\end{pmatrix}\right],$$

the substitution $V_r=hU_r$, we have $v_r=\frac{1}{2}(V_1+V_2+V_3+V_4-2V_r)=hu_r$, where $u_r=\frac{1}{2}(U_1+U_2+U_3+U_4-2U_r)$; and if we multiply the left hand by $e^{aU_1^2+aU_2^2+aU_3^2+aU_4^2}$, which is equal to $e^{au_1^2+au_2^2+au_3^2+au_4^2}$, we obtain

$$\prod_r\vartheta(U_r, 0)=\frac{1}{2^p}\sum e^{-\pi i\epsilon\epsilon'}\prod_r\vartheta\left[u_r, \tfrac{1}{2}\begin{pmatrix}\epsilon' \\ \epsilon\end{pmatrix}\right].$$

Therefore if Q_1, Q_2, Q_3, Q_4 denote any characteristics, and, as formerly, Ω_{Q_r} denote the period-part corresponding to Q_r, we have

$$\Pi_r \vartheta(U_r, Q_r) = \Pi_r e^{-\lambda(U_r, Q_r)} \vartheta(U_r + \Omega_{Q_r}, 0) = \Pi_r e^{-\lambda(U_r, Q_r)} \Pi_r \vartheta(U_r + \Omega_{Q_r}, 0),$$

of which the first factor is easily shewn to be $\Pi_r e^{-\lambda(u_r, q_r)}$, if $(q_1, q_2, q_3, q_4) = \frac{1}{2}\omega(Q_1, Q_2, Q_3, Q_4)$; thus

$$\Pi_r \vartheta(U_r, Q_r) = \frac{1}{2^p} \Sigma e^{-\pi i \epsilon\epsilon'} \Pi_r e^{-\lambda(u_r, q_r)} \vartheta\left[u_r + \Omega_{q_r}, \tfrac{1}{2}\binom{\epsilon'}{\epsilon}\right]$$

$$= \frac{1}{2^p} \sum_{\epsilon, \epsilon'} e^{-\pi i \epsilon\epsilon'} \Pi_r e^{-\lambda(u_r, q_r)} e^{\lambda(u_r, q_r) - 2\pi i q'_r(\frac{1}{2}\epsilon)} \vartheta\left[u_r, q_r + \tfrac{1}{2}\binom{\epsilon'}{\epsilon}\right]$$

$$= \frac{1}{2^p} \sum_{\epsilon, \epsilon'} e^{-\pi i \epsilon(\Sigma q'_r + \epsilon')} \vartheta\left[u_1, q_1 + \tfrac{1}{2}\binom{\epsilon'}{\epsilon}\right] \dots \vartheta\left[u_4, q_4 + \tfrac{1}{2}\binom{\epsilon'}{\epsilon}\right],$$

which is exactly the formula previously obtained (§ 292).

Ex. ii. More generally let $\lambda = \frac{1}{r_j}\omega_j$ be any matrix such that the linear equations $(X_r) = \lambda(x_r)$ give

$$X_1^2 + \dots\dots + X_k^2 = m(x_1^2 + \dots\dots + x_k^2),$$

wherein m is independent of $x_1, \dots, x_k$; then, since, by a property of all linear substitutions, the equations $(Y_r) = \lambda(y_r)$ lead to

$$Y_1 \frac{\partial}{\partial X_1} + \dots\dots + Y_k \frac{\partial}{\partial X_k} = y_1 \frac{\partial}{\partial x_1} + \dots\dots + y_k \frac{\partial}{\partial x_k},$$

we have also*

$$Y_1 X_1 + \dots\dots + Y_k X_k = m(y_1 x_1 + \dots\dots + y_k x_k).$$

Hence, if h be any matrix of p rows and columns and

$$(X_{r,j}) = \lambda(x_{r,j}), \qquad (j = 1, \dots, p),$$

we have

$$hX_1Y_1 + hX_2Y_2 + \dots + hX_kY_k = \sum_{i,j} h_{i,j} \sum_r X_{r,j} Y_{r,i} = m \sum_{i,j} h_{i,j} \sum_r x_{r,j} y_{r,i} = m(hx_1y_1 + \dots + hx_ky_k),$$

where X_1, x_1, etc. now denote rows of p quantities.

Thus any orthogonal substitution furnishes a case of our theorem. Taking a case where

$$m = 1, \; r_j = r, \; \omega_j = \omega, \; \mu = \pm r^k, \; E_{r,j} < r, \; E'_{r,j} < |\mu| < r^k,$$

we have

$$\frac{1}{R}(\epsilon_{r,j}) = \frac{1}{r}\omega(E_{r,j}), \; \frac{1}{\mu}(\epsilon'_{r,j}) = \frac{r\bar{\Omega}_j}{\mu}(E'_{r,j}) = \left[\frac{1}{r}\omega\right]^{-1}(E'_{r,j}) = \frac{1}{r}\bar{\omega}(E'_{r,j}),$$

so that the new characteristics will be r-th parts of integers.

Suppose now, in particular, that the substitution is

$$(X_1, \dots, X_r, \dots X_k) = \frac{1}{k}\begin{pmatrix} 2-k & 2 & \dots\dots & 2 \\ 2 & 2-k & \dots\dots & 2 \\ \dots & \dots & \dots & \dots \\ 2 & 2 & \dots\dots & 2-k \end{pmatrix}(x_1, \dots, x_r, \dots, x_k),$$

* Therefore $mxy = XY = \lambda x \cdot \lambda y = \lambda\bar{\lambda}xy$, so that $\lambda\bar{\lambda} = m$; hence the determinant formed with the elements of λ has one of the values $\sqrt{m^k}$.

which gives

$$X_1^2+\ldots+X_k^2=\sum_r\left[\frac{2}{k}(x_1+\ldots+x_k)-x_r\right]^2=k\frac{4}{k^2}(x_1+\ldots+x_k)^2+\Sigma x_r^2-\frac{4}{k}\Sigma x_r(x_1+\ldots+x_k)=\Sigma x_r^2,$$

and

$$X_1+\ldots\ldots+X_k=x_1+\ldots\ldots+x_k,\quad X_1-X_2=x_2-x_1,\text{ etc.}$$

The previous example is a particular case, namely when $k=4$. In what follows we may suppose k odd so that $r_j=k$. When k is even r_j may be taken $=\frac{1}{2}k$. The work is arranged to apply to either case.

The fractional parts of $\frac{1}{\mu}(\epsilon'_{r,j})$ being independent of the suffix r—because

$$\frac{1}{\mu}\epsilon'_{1,j}-\frac{1}{\mu}\epsilon'_{2,j}=E'_{2,j}-E'_{1,j},\text{ etc.},$$

—we may put $\frac{1}{\mu}(\epsilon'_{r,j})=\left(\frac{1}{r}\epsilon_j'+L_{1,j},\ \ldots,\ \frac{1}{r}\epsilon_j'+L_{k,j}\right)$, and may therefore write

$$\prod_r\Theta\begin{pmatrix}v_r, & \epsilon_r'/\mu\\ & \epsilon_r/R\end{pmatrix}\text{ in the form }\prod_r\Theta\begin{pmatrix}v_r, & \epsilon'/r\\ & \epsilon_r/R\end{pmatrix}.$$

The equation

$$\left(\frac{1}{\mu}\epsilon_j'+L_{r,j}\right)=\frac{1}{r}\bar{\omega}(E'_{r,j})=\frac{1}{r}\omega(E'_{r,j})$$

shews that all values of $\frac{1}{r}\epsilon_j'\ (<1)$ do arise. Hence for a given value of $(E_{r,j})$ there are, instead of $|\mu|^{kp}=r^{k^2p}$ terms given by the general formula, only r^p, and the factor $r^{(k^2-1)p}$ divides out.

The values of $\frac{1}{R}(\epsilon_{r,j})$ given by the general formula are in number $|r|^{kp}$, corresponding to all the values of $(E_{r,j})$. As before the fractional part of $\frac{1}{R}(\epsilon_{r,j})$ is independent of r. Let

$$\frac{1}{k}(E_{1,j}+\ldots\ldots+E_{k,j})=\frac{\epsilon_j}{k}+M_j,$$

where $\frac{\epsilon_j}{k}<1$; then

$$\frac{1}{R}(\epsilon_{r,j})=\frac{1}{r}\omega(E_{r,j})=\left(\frac{2}{k}(E_{1,j}+\ldots\ldots+E_{k,j})-E_{r,j}\right)\equiv\left(\frac{2\epsilon_j}{k},\ \frac{2\epsilon_j}{k},\ \ldots\right),\ (\text{mod. }1).$$

The factor in the general term of the expanded product on the right hand which contains $\epsilon_{r,j}$ is

$$K=\prod_j\prod_r e^{2\pi i\frac{1}{r}\epsilon_{r,j}\left(n_{r,j}+\frac{1}{r}\epsilon_j'\right)}.$$

Now

$$\sum_r\frac{1}{r}\epsilon_{r,j}=\sum_r(E_{r,j})=\epsilon_j+kM_j;$$

therefore, as r is k or a factor of k,

$$\prod_r e^{2\pi i\frac{1}{r}\epsilon_{r,j}\frac{1}{r}\epsilon_j'}=e^{2\pi i(\epsilon_j+kM_j)\frac{\epsilon_j'}{r}}=e^{2\pi i\frac{\epsilon_j\epsilon_j'}{r}}$$

and

$$\sum_r\frac{1}{r}\epsilon_{r,j}n_{r,j}=\sum_r\left[\frac{2}{k}(E_{1,j}+\ldots\ldots+E_{k,j})-E_{r,j}\right]n_{r,j}$$

$$=\sum_r\left[\frac{2\epsilon_j}{k}+2M_j-E_{r,j}\right]n_{r,j}\equiv\frac{2}{k}\sum_r\epsilon_j n_{r,j}\ (\text{mod. }1).$$

Hence the factor above is

$$K=\left[\prod_r e^{2\pi i \frac{2\epsilon}{k}\left(n_r+\frac{\epsilon'}{r}\right)}\right] e^{-4\pi i \frac{\epsilon\epsilon'}{r}} \cdot e^{2\pi i \frac{\epsilon\epsilon'}{r}},$$

and the general term of the right hand is

$$\left[\prod_r \Theta\left(v_r, \begin{matrix}\epsilon'/r \\ 2\epsilon/k\end{matrix}\right)\right] e^{-2\pi i \frac{\epsilon\epsilon'}{r}}.$$

Since $\frac{1}{R}(\epsilon_{r,j})=\left(\frac{2\epsilon_j}{k}+2M_j-E_{r,j}\right)$ we may suppose all values of $\epsilon_j<k$ to arise. Hence instead of r^{kp} we have k^p and a factor r^{kp}/k^p divides out.

To evaluate the factor $(r_1 \dots r_p)^{-1}(s_1 \dots s_p)^{-k}$, $=C$, say, we must enquire how many positive solutions exist of the conditions

$$\frac{2}{k}(x_1+\dots\dots+x_k)-x_r=\text{integral},$$

namely, how many solutions of the conditions

$$\frac{2}{k}(x_1+\dots\dots+x_k)=\text{integral},$$

exist, for which each of $x_1, \dots, x_k<r^k$; let s be this number; then $C=s^{-p}r^{-kp}$, and

$$\prod_r \Theta(V_r, 0)=\frac{r^{(k^2-1)p}}{(sk)^p}\sum_{\epsilon, \epsilon'}\left[\prod_r \Theta\left(v_r, \begin{matrix}\epsilon'/r \\ 2\epsilon/k\end{matrix}\right)\right] e^{\frac{-2\pi i\epsilon\epsilon'}{r}},$$

where $\epsilon'<r$, $\epsilon<k$, the number of terms on the right being $(rk)^p$. For values of $\epsilon>\frac{k}{2}$ we may utilise the equation $\vartheta(v, q+N)=e^{2\pi i N q'}\vartheta(v, q)$. For example, when $k=r=3$ there are 3^{2p} terms, corresponding to characteristics $\left(\begin{matrix}\epsilon'/3 \\ 2/3\end{matrix}\right)$. When $k=4$, $r=2$, the characteristics $\frac{2\epsilon}{k}=\frac{\epsilon}{2}$ will, effectively, repeat themselves. We can reduce the number of terms from 8^p or 2^{3p} to 2^{2p}. We shall thus get factors $\left(e^{2\pi i N\cdot\frac{\epsilon'}{2}}\right)^4=1$ and so the formula reduces to that already found.

Ex. iii. Apply the formula of the last example to the orthogonal case given by $\omega_j=\omega$,

$$(X, Y, Z, T, U, V)=\tfrac{1}{2}\omega(x, y, z, t, u, v),$$

$$\omega=\left(\begin{matrix}1 & 1 & 0 & 0 & 1 & -1 \\ 1 & 1 & 0 & 0 & -1 & 1 \\ 1 & -1 & 1 & 1 & 0 & 0 \\ -1 & 1 & 1 & 1 & 0 & 0 \\ 0 & 0 & 1 & -1 & 1 & 1 \\ 0 & 0 & -1 & 1 & 1 & 1\end{matrix}\right), \quad \omega^{-1}=\left(\begin{matrix}1 & 1 & 1 & -1 & 0 & 0 \\ 1 & 1 & -1 & 1 & 0 & 0 \\ 0 & 0 & 1 & 1 & 1 & -1 \\ 0 & 0 & 1 & 1 & -1 & 1 \\ 1 & -1 & 0 & 0 & 1 & 1 \\ -1 & 1 & 0 & 0 & 1 & 1\end{matrix}\right),$$

which lead to $\mu=64$ and

$$X^2+Y^2+Z^2+T^2+U^2+V^2=x^2+y^2+z^2+t^2+u^2+v^2$$
$$X+Y+Z+T+U+V=x+y+z+t+u+v$$
$$Z-T=x-y, \quad U-V=z-t, \quad X-Y=u-v,$$
$$X+Y=x+y, \quad Z+T=z+t, \quad U+V=u+v.$$

CHAPTER XVII.

THETA RELATIONS ASSOCIATED WITH CERTAIN GROUPS OF CHARACTERISTICS.

294. FOR the theta relations now to be considered*, the theory of the groups of characteristics upon which they are founded, is a necessary preliminary. This theory is therefore developed at some length. When the contrary is not expressly stated the characteristics considered in this chapter are half-integer characteristics†; a characteristic

$$\tfrac{1}{2}q = \tfrac{1}{2}\begin{pmatrix} q_1', & q_2', & \dots, & q_p' \\ q_1, & q_2, & \dots, & q_p \end{pmatrix}$$

is denoted by a single capital letter, say Q. The characteristic of which all the elements are zero is denoted simply by 0. If R denote another characteristic of half-integers, the symbol $Q+R$ denotes the characteristic, $S=\tfrac{1}{2}s$,

* The present chapter follows the papers of Frobenius, *Crelle*, LXXXIX. (1880), p. 185, *Crelle*, XCVI. (1884), p. 81. The case of characteristics consisting of n-th parts of integers is considered by Braunmühl, *Math. Annal.* XXXVII. (1890), p. 61 (and *Math. Annal.* XXXII. (1888), where the case $n=3$ is under consideration).

To the literature dealing with theta relations the following references may be given: Prym, *Untersuchungen über die Riemann'sche Thetaformel* (Leipzig, 1882); Prym u. Krazer, *Acta Math.* III. (1883); Krazer, *Math. Annal.* XXII. (1883); Prym u. Krazer, *Neue Grundlagen einer Theorie der allgemeinen Thetafunctionen* (Leipzig, 1892), where the method, explained in the previous chapter, of multiplying together the theta series, is fundamental: Noether, *Math. Annal.* XIV. (1879), XVI. (1880), where groups of half-integer characteristics are considered, the former paper dealing with the case $p=4$, the latter with any value of p; Caspary, *Crelle*, XCIV. (1883), XCVI. (1884), XCVII. (1884); Stahl, *Crelle*, LXXXVIII. (1879); Poincaré, Liouville, 1895; beside the books of Weber and Schottky, for the case $p=3$, already referred to (§§ 247, 199), and the book of Krause for the case $p=2$, referred to § 199, to which a bibliography is appended. References to the literature of the theory of the transformation of theta functions are given in chapter XX. In the papers of Schottky, in *Crelle*, CII. and onwards, and the papers of Frobenius, in *Crelle*, XCVII. and onwards, and in Humbert and Wirtinger (*loc. cit.* Ex. iv. p. 340), will be found many results of interest, directed to much larger generalizations; the reader may consult Weierstrass, *Berlin. Monatsber.*, Dec. 1869, and *Crelle*, LXXXIX. (1880), and subsequent chapters of the present volume.

† References are given throughout, in footnotes, to the case where the characteristics are n-th parts of integers. In these footnotes a capital letter, Q, denotes a characteristic whose elements are of the form q'_i/n, or of the form q_i/n, q_i', q_i being integers, which in the 'reduced' case are positive (or zero) and less than n. The abbreviations of the text are then immediately extended to this case, n replacing 2.

whose elements s_i', s_i are given by $s_i'=q_i'+r_i'$, $s_i=q_i+r_i$. The characteristic, $\frac{1}{2}t$, wherein $t_i'\equiv s_i'$, $t_i\equiv s_i$ (mod. 2) and each of $t_1', \dots, t_p$ is either 0 or 1, is denoted by QR. Unless the contrary is stated it is intended in any characteristic, $\frac{1}{2}q$, that each of the elements q_i', q_i is either 0 or 1. If $\frac{1}{2}q$, $\frac{1}{2}r$, $\frac{1}{2}k$ be any characteristics, we use the following abbreviations

$$|Q| = qq' = q_1q_1' + \dots\dots + q_pq_p', \quad |Q, R| = qr' - q'r = \sum_{i=1}^{p} (q_ir_i' - q_i'r_i),$$

$$|Q, R, K| = |R, K| + |K, Q| + |Q, R|, \quad \binom{Q}{R} = e^{\pi i q' r} = e^{\pi i (q_1' r_1 + \dots + q'_p r_p)};$$

further we say that two characteristics are congruent when their elements differ only by integers, and use for this relation the sign $\equiv$. In this sense the sum of two characteristics is congruent to their difference. And we say that two characteristics Q, R are syzygetic or azygetic according as $|Q, R| \equiv 0$ or $\equiv 1$ (mod. 2), and that three characteristics P, Q, R are syzygetic or azygetic according as $|P, Q, R| \equiv 0$ or $\equiv 1$ (mod. 2).

Ex. Prove that the $2p+1$ characteristics arising in § 202 associated with the half periods u^{a, c_1}, u^{a, a_1}, u^{a, c_2}, ..., u^{a, a_p}, $u^{a, c}$ are azygetic in pairs. Further that if any four of these characteristics, A, B, C, D, be replaced by the four, BCD, CAD, ABD, ABC, the statement remains true; and deduce that every two of the characteristics 1, 2, ..., 7 of § 205 are azygetic.

295. A preliminary lemma of which frequent application will be made may be given at once. Let $a_{1,1}, \dots, a_{1,n}, \dots, a_{r,1}, \dots, a_{r,n}$ be integers, such that the r linear forms

$$U_i = a_{i,1}x_1 + \dots\dots + a_{i,n}x_n, \qquad (i = 1, 2, \dots, r),$$

are linearly independent (mod. 2) for indeterminate values of $x_1, \dots, x_n$; then if $a_1, \dots, a_r$ be arbitrary integers, the r congruences

$$U_1 \equiv a_1, \dots, U_r \equiv a_r, \text{(mod. 2)},$$

have 2^{n-r} sets of solutions* in which each of $x_1, \dots, x_n$ is either 0 or 1. For consider the sum

$$\frac{1}{2^r} \sum_{x_1, \dots, x_n} [1 + e^{\pi i (U_1 - a_1)}] \dots [1 + e^{\pi i (U_r - a_r)}],$$

wherein the 2^n terms are obtained by ascribing to $x_1, \dots, x_n$ every one of the possible sets of values in which each of $x_1, \dots, x_n$ is either 0 or 1. A term in which $x_1, \dots, x_n$ have a set of values which constitutes a solution of the proposed congruences, has the value unity. A term in which $x_1, \dots, x_n$ do not constitute such a solution will vanish; for one at least of its factors will vanish. Hence the sum of this series gives the desired number of sets of

* When the forms $U_1, \dots, U_r$ are linearly independent mod. m, the number of incongruent sets of solutions is m^{n-r}. In working with modulus m we use $\omega = e^{\frac{2i\pi}{m}}$, instead of $e^{i\pi}$; and instead of a factor $1 + e^{\pi i (U_1 - a_1)}$ we use a factor $1 + \mu + \mu^2 + \dots + \mu^{n-1}$, where $\mu = \omega^{U_1 - a_1}$.

solutions of the congruences. Now the general term of the series is typified by such a term as

$$\frac{1}{2^r}\sum_x e^{\pi i(U_1-a_1)+\pi i(U_2-a_2)+\dots+\pi i(U_\mu-a_\mu)},$$

where μ may be 0, or 1, or ..., or p; and this is equal to

$$\frac{1}{2^r}e^{-\pi i(a_1+\dots+a_\mu)}\sum_x e^{\pi i(c_1x_1+\dots+c_nx_n)},$$

where

$$c_i = a_{1,i} + \dots\dots + a_{\mu,i}, \qquad (i=1, 2, \dots, n),$$

and, therefore, equal to

$$\frac{1}{2^r}e^{-\pi i(a_1+\dots+a_\mu)}(1+e^{\pi i c_1})(1+e^{\pi i c_2})\dots(1+e^{\pi i c_n});$$

now, when $\mu > 0$, one at least of the quantities $c_1, \dots, c_n$ must be $\equiv 1$ (mod. 2), since otherwise the sum of the forms $U_1, \dots, U_\mu$ is $\equiv 0$ (mod. 2), contrary to the hypothesis that the r forms $U_1, \dots, U_r$ are independent (mod. 2); hence the only terms of the summations which do not vanish are those arising for $\mu = 0$, and the sum of the series is

$$\frac{1}{2^r}\sum_x . 1,$$

or 2^{n-r}.

Ex. i. If, of all 2^{2p} half-integer characteristics, $\frac{1}{2}q$, the number of even characteristics be denoted by g, and h be the number of odd characteristics, prove by the method here followed that $g-h$, which is equal to $\Sigma e^{\pi i q q'}$, is equal to 2^p. This equation, with $g+h=2^{2p}$, determine the known numbers* $g=2^{p-1}(2^p+1)$, $h=2^{p-1}(2^p-1)$.

Ex. ii. If $\frac{1}{2}a$ denote any half-integer characteristic other than zero, and $\frac{1}{2}q$ become in turn all the 2^{2p} characteristics, the sum $\Sigma e^{\pi i|A,\,Q|}=\Sigma e^{\pi i(aq'-a'q)}$ vanishes. For it is equal to

$$(1+e^{\pi i a_1})(1+e^{\pi i a_2})\dots\dots(1+e^{-\pi i a_1'})\dots\dots(1+e^{-\pi i a'_p}),$$

and if $\frac{1}{2}a$ be other than zero, one at least of these factors vanishes. On the other hand it is obvious that $\Sigma e^{\pi i|0,\,Q|}=2^{2p}$.

We may deduce the result from the lemma of the text. For by what is there proved there are 2^{2p-1} characteristics for which $|A, Q|\equiv 0$ (mod. 2) and an equal number for which $|A, Q|\equiv 1$.

296. We proceed now to obtain a group of characteristics which are such that every two are syzygetic.

Let P_1 be any characteristic other than zero; it can be taken in $2^{2p}-1$ ways.

Let P_2 be any characteristic other than zero and other than P_1, such that

$$|P_1, P_2| \equiv 0 \text{ (mod. 2)};$$

* Among the n^{2p} incongruent characteristics which are n-th parts of integers, there are $n^{p-1}(n^p+n-1)$ for which $|Q|\equiv 0$ (mod. n), and $n^{p-1}(n^p-1)$ for which $|Q|\equiv r$ (mod. n), when r is not divisible by n.

by the previous lemma (§ 295), P_2 can be taken in $2^{2p-1}-2$ ways; also by the definition, if P_1P_2 be the reduced sum* of P_1, P_2,

$$|P_1, P_1P_2| = |P_1, P_1| + |P_1, P_2| \equiv 0 \text{ (mod. 2)}.$$

Let P_3 be any characteristic, other than one of the four $0, P_1, P_2, P_1P_2$, such that the two congruences are satisfied

$$|P_3, P_1| \equiv 0,\ |P_3, P_2| \equiv 0, \text{ (mod. 2)};$$

then P_3 can be chosen in $2^{2p-2}-2^2$ ways; also, by the definition,

$$|P_3, P_1P_2| = |P_3, P_1| + |P_3, P_2| \equiv 0, \text{ (mod. 2)},$$

and

$$|P_3, P_3P_1| \equiv 0, \text{ etc.}$$

Let P_4 be any characteristic, other than the 2^3 characteristics

$$0, P_1, P_2, P_3, P_1P_2, P_2P_3, P_3P_1, P_1P_2P_3,$$

which is such that

$$|P_4, P_1| \equiv 0,\ |P_4, P_2| \equiv 0,\ |P_4, P_3| \equiv 0, \text{ (mod. 2)};$$

then P_4 can be chosen in $2^{2p-3}-2^3$ ways, and we have

$$|P_2P_3, P_4| = |P_2, P_4| + |P_3, P_4| \equiv 0, \text{ (mod. 2), etc.,}$$

and

$$|P_1P_2P_3, P_4| = |P_1, P_4| + |P_2, P_4| + |P_3, P_4| \equiv 0, \text{ (mod. 2)}.$$

Proceeding thus we shall obtain a group of 2^r characteristics,

$$0, P_1, P_2, \ldots, P_1P_2, \ldots, P_1P_2P_3, \ldots,$$

formed by the sums of r fundamental characteristics, and such that every two are syzygetic. The r-th of the fundamental characteristics can be chosen in $2^{2p-r+1}-2^{r-1} = 2^{r-1}(2^{2p-2r+2}-1)$ ways; thus we may suppose r as great as p, but not greater. Such a group will be denoted by a single letter, (P); the r fundamental characteristics, $P_1, P_2, P_3, \ldots$, may be called the *basis* of the group. We have shewn that they can be chosen in

$$(2^{2p}-1)(2^{2p-1}-2)(2^{2p-2}-2^2)\ldots(2^{2p-r+1}-2^{r-1})/\underline{|r},$$

or

$$(2^{2p}-1)(2^{2p-2}-1)(2^{2p-4}-1)\ldots(2^{2p-2r+2}-1)\,2^{\frac{1}{2}r(r-1)}/\underline{|r}$$

ways. But all these ways will not give a different group; any r linearly independent characteristics of the group may be regarded as forming a basis of the group. For instance instead of the basis

$$P_1, P_2, \ldots, P_r$$

we may take, as basis,

$$P_1P_2, P_2, \ldots, P_r,$$

wherein P_1P_2 is taken instead of P_1; then P_1 will arise by the combination

* So that the elements of P_1P_2 are each either 0 or $\frac{1}{2}$.

of P_1P_2 and P_2. Hence, the number of ways in which, for a given group, a basis of r characteristics, $P_1', \ldots, P_r'$, may be selected is

$$(2^r-1)(2^r-2)\ldots(2^r-2^{r-1})/\underline{|r},$$

for the first of them, P_1', may be chosen, other than 0, in 2^r-1 ways; then P_2', other than 0 and P_1', in 2^r-2 ways; then P_3' may be chosen, other than 0, P_1', P_2', $P_1'P_2'$, in 2^r-2^2 ways, and so on, and the order in which they are selected is immaterial.

Hence on the whole the number of different groups, of the form

$$0,\ P_1,\ P_2,\ \ldots,\ P_1P_2,\ \ldots,\ P_1P_2P_3,\ \ldots$$

of 2^r characteristics, in which every two characteristics of the group are syzygetic*, is

$$\frac{(2^{2p}-1)(2^{2p-2}-1)\ldots\ldots(2^{2p-2r+2}-1)}{(2^r-1)(2^{r-1}-1)\ldots\ldots(2-1)}.$$

Such a group may be called a Göpel group of 2^r characteristics. The name is often limited to the case when $r=p$, such groups having been considered by Göpel for the case $p=2$ (cf. § 221, Ex. i.).

297. We now form a set of 2^r characteristics by adding an arbitrary characteristic A to each of the characteristics of the group (P) just obtained; let P, Q, R be three characteristics of the group, and A', A'', A''', the three corresponding characteristics of the resulting set; then

$$|A', A'', A'''| = |AP, AQ, AR| \equiv |P, Q, R| \equiv |Q, R| + |R, P| + |P, Q|, \text{ (mod. 2)},$$

as is immediately verifiable from the definition of the symbols; thus the resulting set is such that every three of its characteristics are syzygetic, that is, satisfy the condition

$$|A', A'', A'''| \equiv 0, \text{ (mod. 2)};$$

this set is not a group, in the sense so far employed; we may choose $r+1$ fundamental characteristics A, A_1, ..., A_r, respectively equal to A, AP_1, AP_2, ..., AP_r, and these will be said to constitute the basis of the system; but the 2^r characteristics of the system are formed from them by taking only combinations which involve an *odd* number of the characteristics of the basis. The characteristics of the basis are not necessarily independent; there may, for instance, exist the relation $A+AP_1\equiv AP_2$, or $A\equiv P_1P_2$. But there can be no relation connecting an *even* number of the characteristics of the basis; for such a relation would involve a relation connecting the set, $P_1, P_2, \ldots, P_r$, of the group before considered, and such a relation was expressly excluded. Hence it follows that there is at most one relation connecting an odd number

* When the characteristics are n-th parts of integers, the number of such syzygetic groups is $(n^{2p}-1)\ldots(n^{2p-2r+2}-1)$ divided by $(n^r-1)\ldots(n-1)$.

of the characteristics of the basis; for two such relations added together would give a relation connecting an even number.

Conversely if $A, A_1, \ldots, A_r$ be any $r+1$ characteristics, whereof no even number are connected by a relation, such that every three of them satisfy the relation

$$|A', A'', A'''| \equiv 0, \text{ (mod. 2)},$$

we can, taking $P_a \equiv A_a A$, obtain r independent characteristics $P_1, \ldots, P_r$, of which every two are syzygetic, and hence, can form such a group (P) of 2^r pairwise syzygetic characteristics as previously discussed. The aggregate of the combinations of an odd number of the characteristics $A, A_1, \ldots, A_r$ may be called a Göpel system* of characteristics. It is such that there exists no relation connecting an even number of the characteristics which compose the system, and every three of the 2^r characteristics of the system satisfy the conditions

$$|A', A'', A'''| \equiv 0, \text{ (mod. 2)}.$$

We shall denote the Göpel system by (AP).

To pass from a definite group, (P), of 2^r pairwise syzygetic characteristics to a Göpel system, the characteristic A may be taken to be any one of the 2^{2p} characteristics. But if it be taken to be any one of the characteristics of the group (P), we shall obtain, for the Göpel system, only the group (P); and more generally, if P denote in turn every one of the characteristics of the group (P), and A be any assigned characteristic, each of the 2^r characteristics AP leads, from the group (P), to the same Göpel system. Hence, from a given group (P) we obtain only 2^{2p-r} Göpel systems. Hence the number of Göpel systems is equal to

$$2^{2p-r}\frac{(2^{2p}-1)(2^{2p-2}-1)\ldots(2^{2p-2r+2}-1)}{(2^r-1)(2^{r-1}-1)\ldots(2-1)}.$$

We shall say that two characteristics, whose difference is a characteristic of the group (P), are congruent, mod. (P). Thus there exist only 2^{2p-r} characteristics which are incongruent to one another, mod. (P).

It is to be noticed that the 2^{2p-r} Göpel systems derived from a given group (P) have no characteristic in common; for if P_1, P_2 denote characteristics of the group, and A_1, A_2 denote two values of the characteristic A, a congruence $A_1P_1 \equiv A_2P_2$ would give $A_2 \equiv A_1P_1P_2$, which is contrary to the hypothesis that A_1 and A_2 are incongruent, mod. (P). Thus the Göpel systems derivable from a given group (P) constitute a division of the 2^{2p} possible characteristics into 2^{2p-r} systems, each of 2^r characteristics. We can however divide the 2^{2p} characteristics into 2^{2p-r} systems based upon any group (Q) of 2^r characteristics; it is not necessary that the characteristics of the group (Q) be syzygetic in pairs.

* By Frobenius, the name Göpel system is limited to the case when $r=p$.

Ex. For $p=2$, $r=2$, the number of groups (P) given by the formula is 15. And the number of Göpel systems derivable from each is 4. We have shewn in Example iv., § 289, Chap. XV., how to form the 15 groups, and shewn how to form the systems belonging to each one. The condition that two characteristics P, Q be syzygetic is equivalent to $|PQ| \equiv |P| + |Q|$ (mod. 2), or in words, two characteristics are syzygetic when their sum is even or odd according as they themselves are of the same or of different character. It is immediately seen that the 15 groups given in § 289, Ex. iv., satisfy this condition. The four systems derivable from any group were stated to consist of one system in which all the characteristics are even and of three systems in which two are even and two odd. We proceed to a generalization of this result.

298. Of the 2^{2p-r} Göpel systems derivable from one group (P), there is a certain definite number of systems consisting wholly of odd characteristics, and a certain number consisting wholly of even characteristics*. We shall prove in fact that when $p > r$ there are $2^{\sigma-1}(2^\sigma + 1)$ of the systems which consist wholly of even characteristics, σ being $p - r$; these may then be described as even systems; and there are $2^{\sigma-1}(2^\sigma - 1)$ systems which may be described as odd systems, consisting wholly of odd characteristics. When $p = r$, there is one even system, and no odd system. In every one of the $2^{2\sigma}(2^r - 1)$ Göpel systems in which all the characteristics are not of the same character, there are as many odd characteristics as even characteristics.

For, if $P_1, \dots, P_r$ be the basis of the group (P), a characteristic A which is such that the characteristics $A, AP_1, \dots, AP_r$ are all either even or odd, must satisfy the congruences

$$|XP_1| \equiv |XP_2| \equiv \dots\dots \equiv |X|, \text{ (mod. 2)}$$

which are equivalent to

$$|X, P_i| \equiv |P_i|, \qquad (i = 1, 2, \dots, r),$$

as is immediately obvious. Since, when $|X, P_1| \equiv |P_1|$, and $|X, P_2| \equiv |P_2|$,

$$\begin{aligned}|X, P_1P_2| \equiv |X, P_1| + |X, P_2| &\equiv |X, P_1| + |X, P_2| + |P_1, P_2| \\ &\equiv |P_1| + |P_2| + |P_1, P_2| \equiv |P_1P_2|,\end{aligned}$$

etc., it follows that these r congruences are sufficient, as well as necessary. These congruences have (§ 295) 2^{2p-r} solutions. If A be any solution, each of the 2^r characteristics forming the Göpel system (AP) is also a solution; for it follows immediately from the definition, if P, Q denote any two characteristics of the group, that

$$\begin{aligned}|APQ| &\equiv |A| + |P| + |Q| + |A, P| + |A, Q| + |P, Q| \\ &\equiv |A| + 2|P| + 2|Q| + |P, Q| \\ &\equiv |A|,\end{aligned}$$

because $|P, Q| \equiv 0$. Hence the 2^{2p-r} solutions of the congruences consist of

* This result holds for characteristics which are n-th parts of integers, provided the group (P) consist of characteristics in which either the upper line, or the lower line, of elements, are zeros.

$2^{2p-r}/2^r = 2^{2p-2r}$ characteristics A, and the characteristics derivable therefrom by addition of the characteristics, other than 0, of the group (P); namely they consist of the characteristics constituting 2^{2p-2r} Göpel systems, these systems being all derived from the group (P). In a notation already introduced, the congruences have 2^{2p-2r} solutions which are incongruent (mod. (P)).

Ex. If S be any characteristic which is syzygetic with every characteristic of the group (P), without necessarily belonging to that group, prove that the 2^{2p-2r} characteristics SA are incongruent (mod. P), and constitute a set precisely like the set formed by the characteristics A.

299. Put now $\sigma = p - r$, and consider, of the $2^{2\sigma}$ Göpel systems just derived, each consisting wholly either of odd or of even characteristics, how many there are which consist wholly of odd characteristics and how many which consist wholly of even characteristics. Let h be the number of odd systems, and g the number of even systems. Then we have, beside the equation

$$g + h = 2^{2\sigma},$$

also

$$g - h = 2^{-2r} \sum_R e^{\pi i |R|} \left[1 + e^{\pi i |R, P_1| - \pi i |P_1|}\right] \dots \left[1 + e^{\pi i |R, P_r| - \pi i |P_r|}\right],$$

wherein $P_1, \dots, P_r$ are the basis of the group (P), and R is in turn every one of the 2^{2p} possible characteristics. For, noticing that the congruence $|RP| \equiv |R|$ is the same as $|R, P| \equiv |P|$, it is evident that the element of the summation on the right-hand side has a zero factor when R is a characteristic for which all of $R, RP_1, \dots, RP_r$ are not of the same character, either even or odd, and that it is equal to $2^{-r} e^{\pi i |R|}$ when these characteristics are all of the same character; while, corresponding to any value of R, say $R = A$, for which all of $R, RP_1, \dots, RP_r$ are of the same character, there arise, on the right hand, 2^r values of R, the elements of the Göpel set (AP), for which the same is true.

Now if we multiply out the right-hand side we obtain

$$2^{2r}(g - h) = \sum_R e^{\pi i |R|} + \sum_{P_1, P_2, \dots} \left[\sum_R e^{\pi i |R| + \pi i |R, P_1| + \dots + |R, P_\mu|}\right] e^{-\pi i |P_1| - \dots - \pi i |P_\mu|},$$

wherein $\sum\limits_{P_1, P_2, \dots}$ denotes a summation extending to every set of μ of the characteristics $P_1, \dots, P_\mu$, and μ is to have every value from 1 to r; but we have, since $P_1, P_2, \dots$, are syzygetic in pairs,

$$|R| + |R, P_1| + \dots\dots + |R, P_\mu| \equiv |RP_1 \dots P_\mu| + |P_1| + \dots\dots + |P_\mu|,$$

and therefore

$$\sum_R e^{\pi i |R| + \pi i |R, P_1| + \dots + \pi i |R, P_\mu| - \pi i |P_1| - \dots - \pi i |P_\mu|} = \sum_R e^{\pi i |RP_1 \dots P_\mu|} = \sum_S e^{\pi i |S|},$$

where $S, = RP_1 \dots P_\mu$, will, as R becomes all 2^{2p} characteristics in turn, also become all characteristics in turn; also $\sum_R e^{\pi i |R|} = \sum_S e^{\pi i |S|}$ is immediately seen to be 2^p; it is in fact the difference between the whole number of even and odd characteristics contained in the 2^{2p} characteristics. Hence

$$2^{2r}(g-h) = 2^p \left[1 + r + \frac{r(r-1)}{2\,!} + \dots\dots + 1\right] = 2^p\,[(1+x)^r]_{x=1} = 2^{p+r},$$

and therefore $g - h = 2^{p-r} = 2^\sigma$.

This equation, with $g + h = 2^{2\sigma}$, when $\sigma > 0$, determines $g = 2^{\sigma-1}(2^\sigma + 1)$ and $h = 2^{\sigma-1}(2^\sigma - 1)$, and when $\sigma = 0$ determines $g = 1$, $h = 0$.

These results will be compared with the numbers $2^{p-1}(2^p+1)$, $2^{p-1}(2^p-1)$, of the even and odd characteristics, which make up the 2^{2p} possible characteristics.

If P_i denote every characteristic of the group (P) in turn, and P_m denote one characteristic of the bases $P_1, \dots, P_r$, and R be such a characteristic that the 2^r characteristics RP_i are not all of the same character, at least one of the r quantities $|R, P_m| + |P_m|$ is $\equiv 1 \pmod{2}$, and therefore the product

$$\prod_{m=1}^{r} \{1 + e^{\pi i |P_m| + \pi i |R,\, P_m|}\}$$

is zero. But, in virtue of the congruences,

$$|P_iP_j| \equiv |P_i| + |P_j|, \quad |R, P_i| + |R, P_j| \equiv |R, P_iP_j|,$$

this product is equal to

$$\sum_{i=1}^{2^r} e^{\pi i |P_i| + \pi i |R,\, P_i|}, \text{ or } e^{-\pi i |R|} \sum_{i=1}^{2^r} e^{\pi i |RP_i|}.$$

Now $e^{\pi i |RP_i|}$ is 1 or -1 according as RP_i is an even or odd characteristic. Hence the system of 2^r characteristics RP_i contains as many odd as even characteristics, and therefore 2^{r-1} of each, unless all its characteristics be of the same character.

300. The $2^{2\sigma}$ Göpel systems thus obtained, each of which consists wholly of characteristics having the same character, either even or odd, have a further analogy with the 2^{2p} single characteristics. We have shewn (§ 202, Chap. XI.) that the 2^{2p} characteristics can all be formed as sums of not more than p of $2p+1$ fundamental characteristics, whose sum is the zero characteristic; we proceed to shew that from the $2^{2\sigma}$ Göpel systems we can choose $2\sigma + 1$ fundamental systems having a similar property for these $2^{2\sigma}$ systems.

Let the $s = 2^{2\sigma}$ Göpel systems be represented by

$$(A_1P), \dots, (A_sP),$$

the first of them, in a previous notation, consisting of A_1 and all characteristics which are congruent to A_1 for the modulus (P), and similarly with the others. Then we prove that it is possible, from $A_1, \dots, A_s$ to choose $2\sigma + 1$ characteristics, which we may denote by $A_1, \ldots, A_{2\sigma+1}$, such that every three of them, say A', A'', A''', satisfy the condition

$$|A', A'', A'''| \equiv 1, \text{ (mod. 2)};$$

but it is necessary to notice that, if P be any characteristic of the group (P),

$$|A'P, A'', A'''|, \equiv |A', A'', A'''| + |P, A''| + |P, A'''|,$$

is $\equiv |A', A'', A'''|$; for $|P, A''|, \equiv |P|$, is also $\equiv |P, A'''|$; hence, if B', B'', B''' be any three characteristics chosen respectively from the systems $(A'P)$, $(A''P)$, $(A'''P)$, the condition $|A', A'', A'''| \equiv 1$ will involve also $|B', B'', B'''| \equiv 1$; hence we may state our theorem by saying that it is possible, from the $2^{2\sigma}$ Göpel systems, to choose $2\sigma + 1$ *systems*, whereof every three are azygetic.

Before proving the theorem it is convenient to prove a lemma; if B be any characteristic not contained in the group (P), in other words not $\equiv 0$ (mod. (P)), and R become in turn all the $2^{2\sigma}$ characteristics $A_1, \ldots, A_s$, then*

$$\sum_R e^{\pi i |R, B|} = 0.$$

For let a characteristic be chosen to satisfy the $r+1$ congruences

$$|X, B| \equiv 1, |X, P_1| \equiv 0, \ldots, |X, P_r| \equiv 0, \text{ (mod. 2)},$$

and, corresponding to any characteristic R which is one of $A_1, \ldots, A_s$, and therefore satisfies the r congruences $|R, P_i| \equiv |P_i|$, take a characteristic $S = RX$; then

$$|S, B| - |R, B| \equiv |X, B| \equiv 1, \text{ and } |S, P_i| = |RX, P_i| \equiv |R, P_i| + |X, P_i| \equiv |P_i|,$$

because $|X, P_i| \equiv 0$; hence the characteristics $A_1, \ldots, A_s$ can be divided into pairs, such as R and S, which satisfy the equation $e^{\pi i |S, B|} = -e^{\pi i |R, B|}$. This proves† that $\sum_R e^{\pi i |R, B|} = 0$.

We now prove the theorem enunciated. Let the characteristic A_1 be chosen arbitrarily from the s characteristics $A_1, \ldots, A_s$; this is possible in $2^{2\sigma}$ ways. Let A_2 be chosen, also from among $A_1, \ldots, A_s$, other than A_1; this is possible in $2^{2\sigma} - 1$ ways. Then A_3 must be one of the characteristics $A_1, \ldots, A_s$, other than A_1, A_2, and‡ must satisfy the congruence $|A_1, A_2, X| \equiv 1$. The number of characteristics satisfying these conditions is equal to

$$\tfrac{1}{2} \sum_R [1 - e^{\pi i |A_1, A_2, R|}],$$

* We have proved an analogous particular proposition, that if B be not the zero characteristic, and R be in turn all the 2^{2p} characteristics, $\sum_R e^{\pi i |R, B|} = 0$ (§ 295, Ex. ii.).

† If R be all the 2^{2p} characteristics in turn, $\sum_R e^{\pi i |0, R|} = 2^{2p}$. If P be one of the group (P), and R be one of $A_1, \ldots, A_s$, so that $|R, P| \equiv |P|$, we have $\sum_R e^{\pi i |P, R|} = e^{\pi i |P|} 2^{2\sigma}$.

‡ We do not exclude the possibility $A_3 \equiv A_1 A_2$. Since $|A_1, A_2, A_1 A_2| \equiv |A_1, A_2|$, it is a possibility only if $|A_1, A_2| \equiv 1$.

wherein R is in turn equal to all the characteristics $A_1, \ldots, A_s$. For a term of this series, in which R satisfies the conditions for A_3, is equal to unity*, while for other values of R the terms vanish. Now, since $|A_1, A_2, R| \equiv |R, A_1A_2| + |A_1, A_2|$, the series is equal to

$$2^{2\sigma-1} - \tfrac{1}{2} e^{\pi i |A_1, A_2|} \sum_R e^{\pi i |R, A_1A_2|};$$

the characteristic A_1A_2 cannot be one of the group (P), for if $A_1A_2 = P$, then $A_2 = A_1P$, which is contrary to the hypothesis that $A_1, \ldots, A_s$ are incongruent for the modulus (P); hence by the lemma just proved the sum of the series is $2^{2\sigma-1}$, and A_3 can be chosen in $2^{2\sigma-1}$ ways.

We consider next in how many ways A_4 can be chosen; it must be one of $A_1, \ldots, A_s$ other than A_1, A_2, A_3 and must satisfy the congruences

$$|A_1, A_2, X| \equiv 1, \quad |A_1, A_3, X| \equiv 1,$$

which, in virtue of the congruence $|A_1, A_2, A_3| \equiv 1$, and the identity

$$|A_2, A_3, X| + |A_3, A_1, X| + |A_1, A_2, X| \equiv |A_1, A_2, A_3|,$$

involve also $|A_2, A_3, X| \equiv 1$. The number of characteristics which satisfy these conditions is equal to

$$2^{-2} \sum_R (1 - e^{\pi i |A_1, A_2, R|})(1 - e^{\pi i |A_1, A_3, R|})$$

or

$$2^{2\sigma-2} - 2^{-2}\sum_R e^{\pi i |A_1, A_2, R|} - 2^{-2}\sum_R e^{\pi i |A_1, A_3, R|} + 2^{-2}\sum_R e^{\pi i |A_1, A_2, R| + \pi i |A_1, A_3, R|},$$

where R is in turn equal to every one of $A_1, \ldots, A_s$; hence, in virtue of the lemma proved, using the equations,

$$|A_1, A_2, R| \equiv |A_1, A_2| + |R, A_1A_2|,$$

$$|A_1, A_2, R| + |A_1, A_3, R| \equiv |A_1, A_2| + |A_1, A_3| + |A_2A_3, R|,$$

the number of solutions obtained is $2^{2\sigma-2}$. But we have

$$|A_1A_2A_3, A_1, A_2| \equiv |A_1, A_2| + |A_1A_2A_3, A_1A_2| \equiv |A_1, A_2| + |A_3, A_1A_2| \equiv |A_1, A_2, A_3| \equiv 1,$$

so that $A_1A_2A_3$ also satisfies the conditions.

Now it is to be noticed that, for an odd number of characteristics $B_1, \ldots, B_{2k+1}$, the condition that every three be azygetic excludes the possibility of the existence of any relation connecting an even number of these characteristics, and for an even number of characteristics $B_1, \ldots, B_{2k}$, the condition that every three be azygetic excludes the possibility of the existence of any relation connecting an even number except the relation $B_1B_2 \ldots B_{2k} \equiv 0$. For, B being any one of $B_1, \ldots, B_{2k+1}$ other than $B_1, \ldots, B_{2m}$, we have, as is easy to verify,

$$|B_1B_2 \ldots B_{2m-1}, B_{2m}, B| \equiv |B_1, B_{2m}, B| + |B_2, B_{2m}, B| + \ldots + |B_{2m-1}, B_{2m}, B|,$$

* It is immediately seen that $|A, B, B| \equiv 0$.

so that the left hand is $\equiv 1$; therefore, as $|B_{2m}, B_{2m}, B| \equiv 0$, we cannot have $B_{2m} = B_1 B_2 \dots B_{2m-1}$. This holds for all values of m not greater than k, and proves the statement.

Hence, $2\sigma + 1$ being greater than 4, we cannot have $A_4 = A_1 A_2 A_3$, for we are determining an odd number, $2\sigma + 1$, of characteristics. On the whole, then, A_4 can be chosen in $2^{2\sigma-2} - 1$ ways.

To find the number of ways in which A_5 can be chosen we consider the congruences

$$|A_1, A_2, X| \equiv 1, \quad |A_1, A_3, X| \equiv 1, \quad |A_1, A_4, X| \equiv 1,$$

which include such congruences as $|A_2, A_3, X| \equiv 1$, $|A_2, A_4, X| \equiv 1$, etc. The characteristic A_5 must be one of $A_1, \dots, A_8$, other than A_1, A_2, A_3, A_4; the condition that A_5 be not the sum of any three of A_1, A_2, A_3, A_4 is included in these conditions. The number of ways in which A_5 can be chosen is therefore

$$2^{-3} \sum_R (1 - e^{\pi i |A_1, A_2, R|})(1 - e^{\pi i |A_1, A_3, R|})(1 - e^{\pi i |A_1, A_4, R|}),$$

where R is in turn equal to every one of $A_1, \dots, A_8$; making use of the fact that $A_1 A_2 A_3 A_4$ is not $\equiv 0$, we find the number of ways to be $2^{2\sigma-3}$.

Proceeding in this way, we find that a characteristic A_{2m+1} can be chosen in a number of ways equal to the sum of a series of the form

$$2^{-(2m-1)} \sum_R [1 - e^{\pi i |A_1, A_2, R|}][1 - e^{\pi i |A_1, A_3, R|}] \dots [1 - e^{\pi i |A_1, A_{2m}, R|}],$$

and therefore in $2^{2\sigma-(2m-1)}$ ways, and that a characteristic A_{2m} can be chosen in $2^{2\sigma-(2m-2)} - 1$ ways, the value $A_{2m} = A_1 A_2 \dots A_{2m-1}$ being excluded. In particular $A_{2\sigma}$ can be chosen in $2^2 - 1$ ways, and $A_{2\sigma+1}$ in 2 ways.

To the $2\sigma + 1$ characteristics thus determined it is convenient* to add the characteristic $A_{2\sigma+2} = A_1 A_2 \dots A_{2\sigma+1}$; if A_i, A_j be any two of $A_1, \dots, A_{2\sigma+1}$ we have

$$|A_{2\sigma+2}, A_i, A_j| \equiv |A_i, A_j, A_1| + \dots\dots + |A_i, A_j, A_{2\sigma+1}| \equiv 1,$$

the expressions $|A_i, A_j, A_i|$, $|A_i, A_j, A_j|$ being both zero. We have then the result: *From the $2^{2\sigma}$ characteristics $A_1, \dots, A_8$ it is possible to choose a set $A_1, \dots, A_{2\sigma+2}$, such that every three of them satisfy the condition*

$$|A', A'', A'''| \equiv 1,$$

in

$$\frac{2^{2\sigma}(2^{2\sigma}-1)\,2^{2\sigma-1}(2^{2\sigma-2}-1)\dots(2^2-1)\,2}{\underline{|2\sigma+2}} = \frac{2^{2\sigma+\sigma^2}(2^{2\sigma}-1)(2^{2\sigma-2}-1)\dots(2^2-1)}{\underline{|2\sigma+2}}$$

ways; there exists no relation connecting an even number of the characteristics $A_1, \dots, A_{2\sigma+2}$ except the prescribed condition that their sum is zero; since the sum of two relations each connecting an odd number is a relation connecting

* In the particular case of § 202, Chap. XI., $A_{2\sigma+2}$ is zero.

an even number, there can be at most* only one independent relation connecting an odd number of the characteristics $A_1, \ldots, A_{2\sigma+2}$. And, as before remarked, to every one of the characteristics $A_1, \ldots, A_{2\sigma+2}$ is associated a Göpel system of 2^r characteristics.

301. The $2^{2\sigma}$ systems $(A_1P), \ldots, (A_sP)$, which have been considered, were obtained by limiting our attention to one group (P) of 2^r pairwise syzygetic characteristics. We are now to limit our attention still further to the sets $A_1, \ldots, A_{2\sigma+2}$ just obtained satisfying the condition that every three are azygetic.

If to any one of the characteristics $A_1, \ldots, A_{2\sigma+2}$, say A_k, we add the characteristic X, the conditions that the resulting characteristic may still be a characteristic of the set $A_1, \ldots, A_s$, are (§ 298) the r congruences $|XA_k, P_i| \equiv |P_i|$, in which $i = 1, \ldots, r$; in virtue of the conditions $|A_k, P_i| \equiv |P_i|$, these are equivalent to the r congruences $|X, P_i| \equiv 0$, which are independent of k; these latter congruences have 2^{2p-r} solutions, but from any solution we can obtain 2^r others by adding to it all the characteristics of the group (P). There are therefore $2^{2p-2r} = 2^{2\sigma}$ congruences X, incongruent with respect to the modulus (P), each of which†, added to the set $A_1, \ldots, A_{2\sigma+2}$, will give rise to a set $A_1', \ldots, A'_{2\sigma+2}$, also belonging to $A_1, \ldots, A_s$. Further $|A_i', A_j', A_k'| \equiv |XA_i, XA_j, XA_k| \equiv |A_i, A_j, A_k| \equiv 1$; and any relation connecting an even number of the characteristics $A_1', \ldots, A'_{2\sigma+2}$ gives a relation connecting the corresponding characteristics of $A_1, \ldots, A_{2\sigma+2}$. Thus the $2^{2\sigma}$ sets derivable from $A_1, \ldots, A_{2\sigma+2}$ have the same properties as the set $A_1, \ldots, A_{2\sigma+2}$.

Hence all the sets $A_1, \ldots, A_{2\sigma+2}$ can be derived from

$$\frac{2^{\sigma^2}(2^{2\sigma}-1)(2^{2\sigma-2}-1)\ldots(2^2-1)}{\underline{|2\sigma+2}}$$

root sets by adding any one of the $2^{2\sigma}$ characteristics X to each characteristic of the *root set*.

302. Fixing attention upon one of these root sets, and selecting arbitrarily $2\sigma+1$ of its characteristics, which shall be those denoted by $A_1, \ldots, A_{2\sigma+1}$, we proceed to shew that of the $2^{2\sigma}$ characteristics X, there is just one such that the characteristics $XA_1, \ldots, XA_{2\sigma+1}$, derived from $A_1, \ldots, A_{2\sigma+1}$, have all the same character, either even or odd. The conditions for this are

$$|XA_1| \equiv |XA_2| \equiv \ldots\ldots \equiv |XA_{2\sigma+1}|,$$

* If the characteristic of which all the elements, except the i-th element of the first line, are zero, be denoted by E_i', and E_i denote the characteristic in which all the elements are zero except the i-th element of the second line, every possible characteristic is clearly a linear aggregate of $E_1', \ldots, E_p', E_1, \ldots, E_p$. Thus when σ has its greatest value, $=p$, there is certainly one relation, at least, connecting any $2\sigma+1$ characteristics.

† It is only in case all the characteristics of the group (P) are even that the values of X can be the characteristics $A_1, \ldots, A_s$.

which are equivalent to the 2σ congruences

$$|X, A_1 A_i| \equiv |A_1| + |A_i|, \quad (i = 2, 3, \ldots, (2\sigma+1));$$

if X be a solution of these congruences, and P be any characteristic of the group (P), we have

$$|XP, A_1 A_i| \equiv |X, A_1 A_i| + |P, A_1| + |P, A_i| \equiv |A_1| + |A_i| + 2|P|,$$

so that XP is also a solution; since the other congruences satisfied by X were in number r, and similarly, associated with any solution, there were 2^r other solutions congruent to one another in regard to the group (P), it follows that the total number of characteristics X satisfying all the conditions is $2^{2p-r-2\sigma-r} = 1$. Thus, as stated, from any $2\sigma+1$ characteristics, $A_1, \ldots, A_{2\sigma+1}$, of a root set, we can derive one set of $2\sigma+1$ characteristics $\bar{A}_1, \ldots, \bar{A}_{2\sigma+1}$, which are all of the same character, their values being of the form $\bar{A}_i = XA_i$.

Starting from the same root set, and selecting, in place of $A_1, \ldots, A_{2\sigma+1}$, another set of $2\sigma+1$ characteristics, say $A_2, \ldots, A_{2\sigma+2}$, we can similarly derive a set of the form

$$X'A_2, \ldots, X'A_{2\sigma+2},$$

consisting of $2\sigma+1$ characteristics of the same character. The question arises whether this can be the same set as $\bar{A}_1, \ldots, \bar{A}_{2\sigma+1}$. The answer is in the negative. For if the set $X'A_2, \ldots, X'A_{2\sigma+2}$ be in some order the same as the set $XA_1, \ldots, XA_{2\sigma+1}$, or the set $XX'A_2, \ldots, XX'A_{2\sigma+2}$ the same as the set $A_1, \ldots, A_{2\sigma+1}$, it follows by addition that $XX'A_1 \equiv A_{2\sigma+2}$ or $XX' \equiv A_1 A_{2\sigma+2}$. Thence the set $A_1 A_2 A_{2\sigma+2}, A_1 A_3 A_{2\sigma+2}, \ldots, A_1 A_{2\sigma+1} A_{2\sigma+2}, A_1$ is the same as $A_1, A_2, \ldots, A_{2\sigma+1}$, or we have 2σ equations of the form $A_1 A_i A_{2\sigma+2} \equiv A_j$, in which $i = 2, \ldots, 2\sigma+1$, $j = 2, \ldots, 2\sigma+1$. Since there is no relation connecting an even number of the characteristics $A_1, \ldots, A_{2\sigma+2}$ except the one expressing that their sum is 0, these equations are impossible*.

Similarly the question may arise whether such a set as $\bar{A}_1, \ldots, \bar{A}_{2\sigma+1}$, of $2\sigma+1$ characteristics of the same character, azygetic in threes, subject to no relation connecting an even number, and incongruent for modulus (P), can arise from two different root sets. The answer is again in the negative. For if $A_1, \ldots, A_{2\sigma+1}$, and $B_1, \ldots, B_{2\sigma+1}$ be two sets taken from different root sets, the $2\sigma+1$ conditions $XA_i \equiv X'B_i$, for $i = 1, \ldots, 2\sigma+1$, to which by addition may be added $XA_{2\sigma+2} \equiv X'B_{2\sigma+2}$, shew that the set $B_1, \ldots, B_{2\sigma+2}$ is derivable from the set $A_1, \ldots, A_{2\sigma+2}$ by addition of the characteristic XX' to every constituent. This is contrary to the definition of root sets. Conversely if $A_1', \ldots, A'_{2\sigma+2}$ be any one of the $2^{2\sigma}$ sets which are derivable from the root set $A_1, \ldots, A_{2\sigma+2}$ by equations of the form $A_i' \equiv ZA_i$, the set of $2\sigma+1$

* To the sets $\bar{A}_1, \ldots, \bar{A}_{2\sigma+1}$ and $X'A_2, \ldots, X'A_{2\sigma+2}$ we may adjoin respectively their respective sums. The two sets of $2\sigma+2$ characteristics thus obtained are not necessarily the same. When σ is odd they cannot be the same, as will appear below (§ 303).

characteristics of the same character, say $\bar{A}_1', \ldots, \bar{A}'_{2\sigma+1}$, which are derivable from $A_1', \ldots, A'_{2\sigma+1}$ by equations of the form $\bar{A}_i' = X'A_i'$, will also be derived from $A_1, \ldots, A_{2\sigma+1}$ by the equations $\bar{A}_i' = XA_i$, in which $X = X'Z$.

On the whole then it follows that there are

$$\frac{2^{\sigma^2}(2^{2\sigma}-1)(2^{2\sigma-2}-1)\ldots(2^2-1)}{\underline{|2\sigma+1}}$$

different sets, $\bar{A}_1, \ldots, \bar{A}_{2\sigma+1}$, of $2\sigma+1$ characteristics of the same character, azygetic in threes, subject to no relation connecting an even number, and incongruent for the modulus (P).

Of the characteristics $\bar{A}_1, \ldots, \bar{A}_{2\sigma+1}$ there can be formed

$$(2\sigma+1, 1) + (2\sigma+1, 3) + \ldots + (2\sigma+1, 2\sigma+1) = 2^{2\sigma}$$

combinations*, each consisting of an odd number; and, since there is no relation connecting an even number of $\bar{A}_1, \ldots, \bar{A}_{2\sigma+1}$, no two of these combinations can be equal. These combinations all belong to the characteristics $A_1, \ldots, A_s$, satisfying the r congruences $|X, P_i| \equiv |P_i|$; for

$$|\bar{A}_1\bar{A}_2 \ldots \bar{A}_{2k-1}, P_i| \equiv |\bar{A}_1, P_i| + \ldots + |\bar{A}_{2k-1}, P_i| \equiv |P_i|.$$

And no two of them are congruent in regard to the modulus (P); for a relation of the form

$$\bar{A}_1 \ldots \bar{A}_{2k-1} \equiv \bar{A}_m\bar{A}_{m+1} \ldots \bar{A}_{m+2\mu}P,$$

wherein P is a characteristic of the group (P), would lead to a relation of the form $\bar{A}_{2\rho} = \bar{A}_1\bar{A}_2 \ldots \bar{A}_{2\rho-1}P$, and thence give $|\bar{A}_1 \ldots \bar{A}_{2\rho-1}P, \bar{A}_{2\rho}, \bar{A}_{2\rho+1}| \equiv 0$, whereas

$$\begin{aligned}|\bar{A}_1 \ldots \bar{A}_{2\rho-1}P, \bar{A}_{2\rho}, \bar{A}_{2\rho+1}| &\equiv |\bar{A}_1 \ldots \bar{A}_{2\rho-1}, \bar{A}_{2\rho}, \bar{A}_{2\rho+1}| + |\bar{A}_{2\rho}, P| + |\bar{A}_{2\rho+1}, P| \\ &\equiv |\bar{A}_1 \ldots \bar{A}_{2\rho-1}, \bar{A}_{2\rho}, \bar{A}_{2\rho+1}| \\ &\equiv |\bar{A}_1, \bar{A}_{2\rho}, \bar{A}_{2\rho+1}| + \ldots + |\bar{A}_{2\rho-1}, \bar{A}_{2\rho}, \bar{A}_{2\rho+1}| \equiv 1.\end{aligned}$$

Thus the $2^{2\sigma}$ combinations, each consisting of an odd number of the characteristics $\bar{A}_1, \ldots, \bar{A}_{2\sigma+1}$, are in fact the characteristics $A_1, \ldots, A_s$. We† call the set $\bar{A}_1, \ldots, \bar{A}_{2\sigma+1}$ a *fundamental set.* We may associate therewith the characteristic $\bar{A}_{2\sigma+2} = \bar{A}_1 \ldots \bar{A}_{2\sigma+1}$, which is azygetic with every two of the set $\bar{A}_1, \ldots, \bar{A}_{2\sigma+1}$; the case in which it has the same character as these will appear in the next article. And it should be remarked that the argument establishes, for the $2^{2\sigma}$ Göpel systems $(A_1P), \ldots, (A_sP)$, the existence of fundamental sets, $(\bar{A}_1P), \ldots, (\bar{A}_{2\sigma+1}P)$, which are Göpel systems, by the odd combinations of the constituents of which, the constituents of the systems $(A_1P), \ldots, (A_sP)$ can be represented.

* Where (n, k) denotes $n(n-1)\ldots(n-k+1)/k\,!$

† By Frobenius the term Fundamental Set is applied to any $2\sigma+2$ characteristics (incongruent mod. (P)) of which every three are azygetic.

303. The characteristics $\bar{A}_1, \ldots, \bar{A}_{2\sigma+1}$ have been derived to have the same character. We proceed to shew now, in conclusion, that this character is the same for every one of the possible fundamental sets, and depends only on σ. Let $\left(\frac{\sigma}{4}\right)$ be the usual sign which is $+1$ or -1 according as σ is a quadratic residue of 4 or not, in other words, $\left(\frac{\sigma}{4}\right)=1$ when σ is $\equiv 1$ or $\equiv 0$ (mod. 4), and $\left(\frac{\sigma}{4}\right)=-1$ when σ is $\equiv 2$ or $\equiv 3$ (mod. 4); then the character of the sets $\bar{A}_1, \ldots, \bar{A}_{2\sigma+1}$ is $\left(\frac{\sigma}{4}\right)$, that is, $\bar{A}_1, \ldots, \bar{A}_{2\sigma+1}$ are even when $\left(\frac{\sigma}{4}\right)=+1$ and are otherwise odd, and the character of the sum $\bar{A}_{2\sigma+2}=\bar{A}_1 \ldots \bar{A}_{2\sigma+1}$ is $e^{\pi i \sigma}\left(\frac{\sigma}{4}\right)$. Or, we may say

when $\sigma \equiv 1$ (mod. 4), $\bar{A}_1, \ldots, \bar{A}_{2\sigma+1}$ are even, $\bar{A}_{2\sigma+2}$ is odd;

when $\sigma \equiv 0$, $\bar{A}_1, \ldots, \bar{A}_{2\sigma+1}$ are even, $\bar{A}_{2\sigma+2}$ is even,

when $\sigma \equiv 2$ (mod. 4), $\bar{A}_1, \ldots, \bar{A}_{2\sigma+1}$ are odd, $\bar{A}_{2\sigma+2}$ is odd;

when $\sigma \equiv 3$, $\bar{A}_1, \ldots, \bar{A}_{2\sigma+1}$ are odd, $\bar{A}_{2\sigma+2}$ is even.

For if $\bar{A}_1, \ldots, \bar{A}_{2\sigma+1}$ be all of character ϵ we have

$$|\bar{A}_1\bar{A}_2 \ldots \bar{A}_{2k+1}| \equiv |\bar{A}_1| + \ldots + |\bar{A}_{2k+1}| + \Sigma |\bar{A}_i, \bar{A}_j|,$$

where $\bar{A}_i, \bar{A}_j$ consist of every pair from $\bar{A}_1, \ldots, \bar{A}_{2k+1}$; also

$$(2k-1)\Sigma|\bar{A}_i, \bar{A}_j| = \Sigma|\bar{A}_i, \bar{A}_j, \bar{A}_h|,$$

where $\bar{A}_i, \bar{A}_j, \bar{A}_h$ consist of every triad from $\bar{A}_1, \ldots, \bar{A}_{2k+1}$; hence, since $|\bar{A}_i, \bar{A}_j, \bar{A}_h| \equiv 1$, and, as is easily seen, $n(n-1)(n-2)/3!$ is even or odd according as n is of the form $4m+1$ or $4m+3$, it follows that $\Sigma|\bar{A}_i, \bar{A}_j|$ is even or odd according as $2k+1$ is of the form $4m+1$ or $4m+3$, therefore $\bar{A}_1\bar{A}_2 \ldots \bar{A}_{2k+1}$ has the character ϵ or $-\epsilon$ according as $2k+1 \equiv 1$ or $\equiv 3$ (mod. 4). Thus the number of combinations of an odd number from $\bar{A}_1, \ldots, \bar{A}_{2\sigma+1}$ which have the character ϵ is

$$\begin{aligned}(2\sigma+1, 1)+(2\sigma+1, 5)+(2\sigma+1, 9)+\ldots \\ &= \tfrac{1}{4}\left\{(1+x)^{2\sigma+1}-(1-x)^{2\sigma+1}+i(1-ix)^{2\sigma+1}-i(1+ix)^{2\sigma+1}\right\}_{x=1} \\ &= 2^{2\sigma-1}+2^{\sigma-\frac{1}{2}}\sin\frac{2\sigma+1}{4}\pi;\end{aligned}$$

this number is $2^{2\sigma-1}+2^{\sigma-1}$ when $\sigma \equiv 0$ or $\sigma \equiv 1$ (mod. 4); otherwise it is $2^{2\sigma-1}-2^{\sigma-1}$; now we have shewn (§ 298) that the characteristics $A_1, \ldots, A_s$ contain respectively $2^{2\sigma-1}+2^{\sigma-1}$, $2^{2\sigma-1}-2^{\sigma-1}$ even and odd characteristics, and (§ 302) that every one of $A_1, \ldots, A_s$ can be formed as an odd combination from $\bar{A}_1, \ldots, \bar{A}_{2\sigma+1}$; hence $\epsilon=+1$ when $\sigma \equiv 0$ or $\sigma \equiv 1$ (mod. 4), and otherwise $\epsilon = -1$; this agrees with the statement made. Further, by the same argument $\bar{A}_1 \bar{A}_2 \dots \bar{A}_{2\sigma+1}$ has the character ϵ or $-\epsilon$ according as $2\sigma + 1 \equiv 1$ or $\equiv 3$ (mod. 4); and this leads to the statement made for $\bar{A}_{2\sigma+2}$.

The reader will find it convenient to remember that the combinations, from the fundamental set $\bar{A}_1, \dots, \bar{A}_{2\sigma+1}$, consisting of 1, 5, 9, 13, ... of them, are all of the same character, and the combinations consisting of 3, 7, 11, ... are all of the opposite character.

Ex. If $A_1, \dots, A_{2p+1}$ be half-integer characteristics azygetic in pairs, and S be the sum of the odd ones of these, prove that a characteristic formed by adding S to a sum of any $p+r$ characteristics of these is even when $r \equiv 0$ or $\equiv 1$ (mod. 4), and odd when $r \equiv 2$ or $\equiv 3$ (mod. 4). (Stahl, *Crelle*, LXXXVIII. (1879), p. 273.)

304. It is desirable now to frame a connected statement of the results thus obtained. It is possible, in

$$(2^{2p}-1)(2^{2p-2}-1)\dots(2^{2p-2r+2}-1)/(2^r-1)(2^{r-1}-1)\dots(2-1)$$

ways, to form a group,

$$0, P_1, P_2, \dots, P_1P_2, \dots, P_1P_2P_3, \dots$$

of 2^r characteristics, consisting of the combinations of r independent characteristics $P_1, \dots, P_r$, such that every two characteristics P, P' of the group are syzygetic, that is, satisfy the congruence $|P, P'| \equiv 0$, (mod. 2). Such a group is denoted by (P), and two characteristics whose difference is a characteristic of the group are said to be congruent for the modulus (P).

From such a group (P), by adding the same characteristic A to each constituent, we form a system, which we call a Göpel system, consisting of the combinations of an odd number of $r+1$ characteristics $A, AP_1, \dots, AP_r$, among an even number of which there exists no relation; this system is such that every three of its constituents, say L, M, N, satisfy the congruence $|L, M, N| \equiv 0$, or, as we say, are syzygetic. Such a Göpel system is represented by (AP).

It is shewn that by taking 2^{2p-r} different values of A and retaining the same group (P), we can thus divide the 2^{2p} possible characteristics into 2^{2p-r} Göpel systems. Among these 2^{2p-r} Göpel systems there are 2^{2p-2r} systems of which all the elements have the same character. Putting $2p - 2r = 2\sigma$ we shew further that $2^{\sigma-1}(2^\sigma+1)$ of these Göpel systems consist wholly of even characteristics, and that $2^{\sigma-1}(2^\sigma-1)$ of them consist wholly of odd characteristics. Putting $s = 2^{2\sigma}$ we denote the $2^{2\sigma}$ Göpel systems which have a distinct character by $(A_1P), \dots, (A_sP)$; and, still retaining the same group (P), we proceed to consider how to represent these $2^{2\sigma}$ systems by means of $2\sigma+1$ fundamental systems.

It appears then that from the characteristics $A_1, \dots, A_s$ we can choose $2\sigma+1$ characteristics $\bar{A}_1, \dots, \bar{A}_{2\sigma+1}$ in

$$2^{\sigma^2}(2^{2\sigma}-1)(2^{2\sigma-2}-1)\dots(2^2-1)/\underline{|2\sigma+1}$$

ways, such that every three of them are azygetic, and all have the same character; this character is not at our disposal but is that of $\left(\frac{\sigma}{4}\right)$; the sum of $\bar{A}_1, \ldots, \bar{A}_{2\sigma+1}$, denoted by $\bar{A}_{2\sigma+2}$, has the character $e^{\pi i \sigma}\left(\frac{\sigma}{4}\right)$. Then all the combinations of 1, 5, 9, ... of $\bar{A}_1, \ldots, \bar{A}_{2\sigma+1}$ have the character $\left(\frac{\sigma}{4}\right)$, and all the combinations of 3, 7, 11, ... have the opposite character. These combinations in their aggregate are the characteristics $A_1, \ldots, A_8$. The characteristics $\bar{A}_1, \ldots, \bar{A}_{2\sigma+1}$ are, like $A_1, \ldots, A_8$, incongruent for the modulus (P). To each of them, say $\bar{A}_i$, corresponds a Göpel system $(\bar{A}_i P)$, to any constituent of which statements may be applied analogous to those made for $\bar{A}_i$ itself.

The characteristic $\bar{A}_{2\sigma+2}$ is such that every three of the set $\bar{A}_1, \ldots, \bar{A}_{2\sigma+2}$ are azygetic. This set is in fact derived, as one of $2\sigma+2$ such, from a set of $2\sigma+2$ characteristics, here called a root set, which satisfies the condition that every three of its constituents are azygetic without satisfying the condition that $2\sigma+1$ of them are of the same character. There are

$$2^{\sigma^2}(2^{2\sigma}-1)\ldots(2^2-1)/\underline{|2\sigma+2}$$

such root sets. It is not possible, from any root set, to obtain another by adding the same characteristic to each constituent of the former set.

The root sets are not the most general possible sets of $2\sigma+2$ characteristics of which every three are azygetic. Of such sets there are

$$2^{\sigma^2+2\sigma}(2^{2\sigma}-1)\ldots(2^2-1)/\underline{|2\sigma+2},$$

but they break up into batches of $2^{2\sigma}$, each derivable from a root set by the addition of a proper characteristic to all the constituents of the root set.

305. As examples of the foregoing theory we consider now the cases $\sigma=0$, $\sigma=1$, $\sigma=2$, $\sigma=p$. When $\sigma=0$, the number of Göpel groups of 2^p pairwise syzygetic characteristics is

$$(2^p+1)(2^{p-1}+1)\ldots\ldots(2+1);$$

from any such group we can, by the addition of the same characteristic to each of its constituents obtain one Göpel system consisting wholly of characteristics of the same *even* character. These results have already been obtained in case $p=2$ (§ 289, Ex. iv.), and, as in that particular case, the 2^p-1 other systems obtainable from the Göpel group by the addition of the same characteristic to each constituent, contain as many odd characteristics as even characteristics.

When $\sigma=1$, we can, from any Göpel group of 2^{p-1} pairwise syzygetic characteristics, obtain 4 Göpel systems, three of them consisting of 2^{p-1} even characteristics and one of 2^{p-1} odd characteristics. The characteristics of the latter (odd) system are obtainable as the sums of three characteristics taken one from each of the three even systems.

When $\sigma=2$, the number of fundamental sets $\bar{A}_1, \ldots, \bar{A}_5$ is

$$\frac{2^4(2^4-1)(2^2-1)}{\underline{|5}}=6;$$

each of them has the character $\left(\frac{\sigma}{4}\right)$, or is odd, and their sum, $\bar{A}_6$, is odd. Among the $2^{2\sigma}=16$ characteristics $A_1, \dots, A_s$ there are $2^{2\sigma-1}-2^{\sigma-1}$ or 6 odd characteristics; these clearly consist of the characteristics $\bar{A}_1, \dots, \bar{A}_6$; the six fundamental sets are obtained by neglecting each of $\bar{A}_1, \dots, \bar{A}_6$ in turn. Among the characteristics $A_1, \dots, A_s$ there are 10 even characteristics, obtainable by combining $\bar{A}_1, \dots, \bar{A}_s$ in threes. And, to each of the characteristics $A_1, \dots, A_s$ corresponds a Göpel system of $2^r=2^{p-\sigma}=2^{p-2}$ characteristics, for the constituents of which similar statements may be made.

Of the cases for which $\sigma=2$, the case $p=2$ is the simplest. After what has been said in Chap. XI., and elsewhere, we can leave that case aside here. For $p=3$ the Göpel systems consist of two characteristics; adopting, for instance, as the group (P), the pair $\frac{1}{2}\begin{pmatrix}000\\000\end{pmatrix}$, $\frac{1}{2}\begin{pmatrix}000\\100\end{pmatrix}$, the condition for the characteristics $A_1, \dots, A_s$, namely $|X, P_1| \equiv |P_1|$, reduces to the condition that the first element of the upper row of the characteristic symbol of X shall be zero; hence the 16 characteristics $A_1, \dots, A_s$ may be taken to be $\frac{1}{2}\begin{pmatrix}0 & a_1' & a_2'\\0 & a_1 & a_2\end{pmatrix}$, where $\frac{1}{2}\begin{pmatrix}a_1' & a_2'\\a_1 & a_2\end{pmatrix}$ represents in turn all the characteristic symbols for $p=2$.

Taking next the case $\sigma=3$, there are $s=2^{2\sigma}=64$ Göpel systems, (AP), each consisting wholly either of odd characteristics or of even characteristics, there being $2^{\sigma-1}(2^\sigma-1)$, $=28$, odd systems, and 36 even systems. From the representatives, $A_1, \dots, A_s$, of these systems, which are incongruent mod. (P), we can choose a fundamental set of 7 characteristics $\bar{A}_1, \dots, \bar{A}_7$ in

$$\frac{2^9(2^6-1)(2^4-1)(2^2-1)}{\lfloor 7}, \quad =288,$$

ways; $\bar{A}_1, \dots, \bar{A}_7$ will be odd, and their sum, $\bar{A}_8$, will be even; for $\left(\frac{\sigma}{4}\right)=(\tfrac{3}{4})=-1$, $e^{\pi i\sigma}\left(\frac{\sigma}{4}\right)=1$. The set $\bar{A}_1, \dots, \bar{A}_7, \bar{A}_8$ is, in accordance with the theory, derived from one of $288/(2\sigma+2)$, $=36$, root sets $A_1, \dots, A_8$ (§ 301), by equations of the form $\bar{A}_i=XA_i$, in which X is so chosen that $\bar{A}_1, \dots, \bar{A}_7$ are of the same character; from this root set we can similarly derive 8 fundamental sets of seven odd characteristics, according as it is A_8 or is one of $A_1, \dots, A_7$ which is left aside. Now the fact is, that, in whichever of the eight ways we pass from the root set to the seven fundamental odd characteristics, the sum of these seven fundamental characteristics is the same. We see this immediately in an indirect way. Let $\bar{A}_1, \dots, \bar{A}_7$ be a fundamental set of odd characteristics derived from the root set $A_1, \dots, A_8$ by the equations $\bar{A}_i=XA_i$; putting $\bar{A}_8=\bar{A}_1\dots\bar{A}_7$, consider the set $\bar{A}_8, \bar{A}_8\bar{A}_1\bar{A}_2, \dots, \bar{A}_8\bar{A}_1\bar{A}_7, \bar{A}_1$, derived from $\bar{A}_1, \dots, \bar{A}_8$ by adding $\bar{A}_8\bar{A}_1$ to each; in the first place it consists of one even characteristic, $\bar{A}_8$, and seven odd characteristics; for

$$|\bar{A}_8\bar{A}_1\bar{A}_i| \equiv |\bar{A}_8|+|\bar{A}_1|+|\bar{A}_i|+|\bar{A}_8, \bar{A}_1, \bar{A}_i| \equiv |\bar{A}_8, \bar{A}_1, \bar{A}_i| \equiv 1, \text{ (mod. 2)},$$

because $\bar{A}_1, \dots, \bar{A}_8$ are azygetic in threes; in the next place

$$|\bar{A}_8, \bar{A}_1, \bar{A}_8\bar{A}_1\bar{A}_i| \equiv |\bar{A}_8, \bar{A}_1, \bar{A}_i| \equiv 1,$$

so that every three of its constituents are azygetic. Hence the characteristics $\bar{A}_8\bar{A}_1\bar{A}_2$, $\dots, \bar{A}_8\bar{A}_1\bar{A}_7, \bar{A}_1$, which, as easy to see, are not congruent to $\bar{A}_1, \dots, \bar{A}_7$ mod. (P), form, equally with $\bar{A}_1, \dots, \bar{A}_7$, a fundamental set, whose sum is likewise $\bar{A}_8$; they are derived from $A_1, \dots, A_8$ by adding $\bar{A}_8\bar{A}_1X$ to each of these. There are clearly six other such fundamental sets, derived from $A_1, \dots, A_8$ by adding respectively $\bar{A}_8\bar{A}_2X, \dots, \bar{A}_8\bar{A}_7X$. Hence to each of the 36 root sets there corresponds a certain even characteristic and to each of these even characteristics there correspond 8 fundamental sets. We can now shew further that the even characteristics, thus associated each with one of the 36 root sets, are in fact the 36 possible* even characteristics of the set $A_1, \dots, A_8$. This again we shew indirectly by shewing how to form the remaining 7.36 fundamental systems from the system $\bar{A}_1, \dots, \bar{A}_7$. The seven characteristics $\bar{A}_8\bar{A}_2\bar{A}_3$, $\bar{A}_8\bar{A}_3\bar{A}_1$, $\bar{A}_8\bar{A}_1\bar{A}_2$, $\bar{A}_4$, $\bar{A}_5$, $\bar{A}_6$, $\bar{A}_7$, are in fact incongruent mod. (P), they are all odd, have for sum $\bar{A}_1\bar{A}_2\bar{A}_3$, which is even, and are azygetic in threes; for $\bar{A}_8\bar{A}_2\bar{A}_3$ is a combination of five of $\bar{A}_1, \dots, \bar{A}_7$, and

$$|\bar{A}_4, \bar{A}_5, \bar{A}_8\bar{A}_2\bar{A}_3| \equiv |\bar{A}_8, \bar{A}_4, \bar{A}_5| + |\bar{A}_2, \bar{A}_4, \bar{A}_5| + |\bar{A}_3, \bar{A}_4, \bar{A}_5| \equiv 1, \quad |\bar{A}_4, \bar{A}_5, \bar{A}_6| \equiv 1,$$

$$|\bar{A}_4, \bar{A}_8\bar{A}_1\bar{A}_2, \bar{A}_8\bar{A}_1\bar{A}_3| \equiv |\bar{A}_8\bar{A}_1\bar{A}_4, \bar{A}_2, \bar{A}_3| \equiv 1, \ |\bar{A}_8\bar{A}_2\bar{A}_3, \bar{A}_8\bar{A}_3\bar{A}_1, \bar{A}_8\bar{A}_1\bar{A}_2| \equiv |\bar{A}_1, \bar{A}_2, \bar{A}_3| \equiv 1,$$

(the modulus in each case being 2); hence these seven characteristics form a fundamental system. There are 35 sets of three characteristics, such as $\bar{A}_1$, $\bar{A}_2$, $\bar{A}_3$, derivable from the seven $\bar{A}_1, \dots, \bar{A}_7$; each of these corresponds to such a fundamental system as that just explained; and each of these fundamental systems is associated with seven other fundamental systems, derived from it by the process whereby the set $\bar{A}_i$, $\bar{A}_i\bar{A}_8\bar{A}_2, \dots, \bar{A}_i\bar{A}_8\bar{A}_7$ is derived from $\bar{A}_1, \dots, \bar{A}_7$.

When $\sigma = p$, a Göpel system consists of one characteristic only; we can, in

$$2^{p^2}(2^{2p}-1)(2^{2p-2}-1)\dots\dots(2^2-1)/\underline{|2p+1}$$

ways, determine a set of $2p+1$ characteristics, all of character $\binom{p}{4}$, of which every three are azygetic; their sum will be of character $e^{\pi i p}\binom{p}{4}$; all the possible 2^{2p} characteristics can be represented as combinations of an odd number of these.

306. We pass now to some applications of the foregoing theory to the theta functions. The results obtained are based upon the consideration of the theta function of the second order defined by

$$\phi(u, a; \tfrac{1}{2}q) = \vartheta(u+a; \tfrac{1}{2}q)\,\vartheta(u-a; \tfrac{1}{2}q),$$

where $\frac{1}{2}q$ is a half-integer characteristic; as theta function of the second order this function has zero characteristic; the addition of any integers to the elements of the characteristic $\frac{1}{2}q$ does not affect the value of the function. By means of the formulae (§ 190, Chap. X.),

$$\vartheta(u+a; \tfrac{1}{2}q+N) = e^{\pi i N q'}\,\vartheta(u+a; \tfrac{1}{2}q),$$

$$\vartheta(u+\tfrac{1}{2}\Omega_k; \tfrac{1}{2}q) = e^{\lambda(u;\, \frac{1}{2}k) - \frac{1}{4}\pi i k' q}\,\vartheta(u; \tfrac{1}{2}k+\tfrac{1}{2}q),$$

wherein N denotes a row of integers and $\lambda(u; s) = H_s(u+\frac{1}{2}\Omega_s) - \pi i s s'$, we immediately find

$$\phi(u+\tfrac{1}{2}\Omega_k, a; \tfrac{1}{2}q) = e^{2\lambda(u;\, \frac{1}{2}k)}\binom{K}{Q}\phi(u, a; \tfrac{1}{2}kq),$$

where $\frac{1}{2}kq$ denotes the sum of the characteristics $\frac{1}{2}k$, $\frac{1}{2}q$; to save the repetition of the $\frac{1}{2}$, this equation will in future be written in the form (cf. § 294)

$$\phi(u+\Omega_K, a; Q) = e^{2\lambda(u;\, K)}\binom{K}{Q}\phi(u, a; KQ);$$

when the contrary is not stated capital letters will denote half-integer characteristics, and KQ will denote the *reduced* sum of the characteristics K, Q, having for each of its elements either 0 or $\frac{1}{2}$.

* Thus, when $p = 3 = \sigma$, the result quoted in § 205, Chap. XI., is justified.

We shall be concerned with groups of 2^r pairwise syzygetic characteristics, such as have been called Göpel groups, and denoted by (P); corresponding to the r characteristics $P_1, \ldots, P_r$ from which such a group is formed, we introduce r fourth roots of unity, denoted by $\epsilon_1, \ldots, \epsilon_r$, which are such that

$$\epsilon_1^2 = e^{\pi i |P_1|}, \ldots, \epsilon_r^2 = e^{\pi i |P_r|};$$

the signs of these symbols are, at starting, arbitrary, but are to be the same throughout unless the contrary be stated. Since the characteristics of the group (P) satisfy the conditions

$$|P_i, P_j| \equiv 0, \text{ (mod. 2)}, \quad \begin{pmatrix} P_i \\ P_j \end{pmatrix} = \begin{pmatrix} P_j \\ P_i \end{pmatrix},$$

we may, without ambiguity, associate with the compound characteristics of the group the $2^r - r$ symbols defined by

$$\epsilon_0 = 1, \quad \epsilon_{i,j} = \epsilon_i \epsilon_j \begin{pmatrix} P_i \\ P_j \end{pmatrix}, \text{ so that } \epsilon_{i,j}^2 = e^{\pi i |P_i| + \pi i |P_j|}, \quad \epsilon_{i,i} = 1,$$

$$\epsilon_{i,j,k} = \epsilon_i \epsilon_{j,k} \begin{pmatrix} P_i \\ P_j P_k \end{pmatrix} = \epsilon_i \epsilon_j \epsilon_k \begin{pmatrix} P_j \\ P_k \end{pmatrix} \begin{pmatrix} P_k \\ P_i \end{pmatrix} \begin{pmatrix} P_i \\ P_j \end{pmatrix} = \epsilon_j \epsilon_{k,i} \begin{pmatrix} P_j \\ P_k P_i \end{pmatrix} = \epsilon_k \epsilon_{i,j} \begin{pmatrix} P_k \\ P_i P_j \end{pmatrix},$$

and

$$\epsilon_j = \epsilon_{i,ij} = \epsilon_i \epsilon_{ij} \begin{pmatrix} P_i P_j \\ P_i \end{pmatrix}, \text{ etc.}$$

Consider now the function* defined by

$$\Phi(u, a; A) = \sum_i \begin{pmatrix} P_i \\ A \end{pmatrix} \epsilon_i \phi(u, a; AP_i),$$

where A is an arbitrary half-integer characteristic, and P_i denotes in turn all the 2^r characteristics of the group (P). Adding to u a half-period Ω_{P_k}, corresponding to a characteristic P_k of the group (P), we obtain

$$\Phi(u + \Omega_{P_k}, a; A) = \sum_i \begin{pmatrix} P_i \\ A \end{pmatrix} \begin{pmatrix} P_k \\ AP_i \end{pmatrix} \epsilon_i e^{2\lambda(u;\, P_k)} \phi(u, a; AP_i P_k);$$

if then $P_h \equiv P_i P_k$, or $P_i \equiv P_h P_k$, we have

$$\begin{pmatrix} P_i \\ A \end{pmatrix} \begin{pmatrix} P_k \\ AP_i \end{pmatrix} \epsilon_i = \begin{pmatrix} P_h \\ A \end{pmatrix} \begin{pmatrix} P_k \\ A \end{pmatrix} \begin{pmatrix} P_k \\ A \end{pmatrix} \begin{pmatrix} P_k \\ P_h \end{pmatrix} \begin{pmatrix} P_k \\ P_k \end{pmatrix} \epsilon_h \epsilon_k \begin{pmatrix} P_h \\ P_k \end{pmatrix} = \epsilon_k e^{\pi i |P_k|} \begin{pmatrix} P_h \\ A \end{pmatrix} \epsilon_h;$$

now, as P_i becomes in turn all the characteristics of the group (P), $P_h, = P_i P_k$, also becomes all the characteristics of the group, in general in a different order; thus we have

$$\Phi(u + \Omega_{P_k}, a; A) = \epsilon_k e^{\pi i |P_k| + 2\lambda(u;\, P_k)} \Phi(u, a; A),$$
$$= \epsilon_k^{-1} e^{2\lambda(u;\, P_k)} \Phi(u, a; A).$$

* If preferred the sign $\begin{pmatrix} P_i \\ A \end{pmatrix}$, whose value is ± 1, may be absorbed in ϵ_i. But there is a certain convenience in writing it explicitly.

If $2\Omega_M$ be any period, we immediately find

$$\Phi(u + 2\Omega_M, a; A) = e^{2\lambda(u; 2M)} \Phi(u, a; A).$$

Thus, $\lambda(u; P_k)$ being a linear function of the arguments $u_1, \dots, u_p$, the function $\Phi(u, a; A)$ is a theta function of the second order with zero characteristic, having the additional property that all the partial differential coefficients of its logarithm, of the second order, have the 2^r sets of simultaneous periods denoted by the symbols Ω_{P_k}.

Ex. i. If S be a half-integer characteristic which is syzygetic with every characteristic of the group (P), prove that

$$\Phi(u+\Omega_S, a; A) = e^{2\lambda(u; S)} \binom{S}{A} \Phi(u, a; AS)$$

$$\Phi(u, a+\Omega_S; A) = e^{2\lambda(a; S)+\pi i|S|+\pi i|S, A|} \binom{S}{A} \Phi(u, a; AS)$$

and

$$\Phi(u+\Omega_S, a+\Omega_S; A) = e^{2\lambda(u; S)+2\lambda(a; S)+\pi i|S, A|} \Phi(u, a; A).$$

Ex. ii. If P_k be any characteristic of the group (P), prove that

$$\Phi(u, a; AP_k) = \binom{P_k}{A} \epsilon_k^{-1} \Phi(u, a; A).$$

Ex. iii. When, as in Ex. i., S is syzygetic with every characteristic of the group (P), shew that

$$e^{\pi i|SP_k|} \Phi(u, a; AP_k) \Phi(v, b; AP_k) = e^{\pi i|S|} \Phi(u, a; A) \Phi(v, b; A).$$

Conversely it can be shewn that if a theta function of the second order with zero characteristic, $\Pi(u)$, which, therefore, satisfies the equation

$$\Pi(u + \Omega_m) = e^{2\lambda_m(u)} \Pi(u),$$

for integral m, be further such that for each of the two half-periods associated with the characteristics $\tfrac{1}{2}m = P$, $\tfrac{1}{2}m = Q$, there exists an equation of the form

$$\Pi(u + \tfrac{1}{2}\Omega_m) = e^{\mu+\nu_1 u_1+\dots+\nu_p u_p} \Pi(u),$$

where $\mu, \nu_1, \dots, \nu_p$ are independent of u, then the characteristics P, Q must be syzygetic. Putting $\nu u = \nu_1 u_1 + \dots\dots + \nu_p u_p$, we infer from the equation just written that

$$\Pi(u + \Omega_m) = e^{\mu+\nu(u+\frac{1}{2}\Omega_m)} \Pi(u + \tfrac{1}{2}\Omega_m) = e^{2\mu+2\nu u+\frac{1}{2}\nu\Omega_m} \Pi(u);$$

comparing this with the equation

$$\Pi(u + \Omega_m) = e^{2\lambda_m(u)} \Pi(u) = e^{2H_m(u+\frac{1}{2}\Omega_m)-2\pi i m m'} \Pi(u)$$

we infer that $\nu = H_m$, $\mu = k\pi i + \tfrac{1}{4}H_m\Omega_m - \pi i m m'$, where k is integral, and hence

$$\Pi(u + \tfrac{1}{2}\Omega_m) = \pm\, e^{-\frac{1}{2}\pi i m m' + 2\lambda(u; \frac{1}{2}m)} \Pi(u).$$

307. In accordance with these indications, let $Q(u)$ denote an analytical integral function of the arguments $u_1, \dots, u_p$ which satisfies the equations

$$Q(u + \Omega_m) = e^{2\lambda(u; m)} Q(u); \quad Q(u + \Omega_{P_k}) = \epsilon_k e^{\pi i|P_k|+2\lambda(u; P_k)} Q(u),$$

for every integral m and every half-integer characteristic P_k of the group (P).

We may regard the group (P) as consisting of part of a group of 2^p pairwise syzygetic characteristics formed by all the combinations of the constituents of the group (P) with the constituents of another pairwise syzygetic group (R) of 2^{p-r} characteristics. Then the 2^p characteristics of the compound group are obtainable in the form P_iR_j, wherein P_i has the 2^r values of the group (P), and R_j has the 2^{p-r} values of the group (R). Since every 2^p+1 theta functions of the second order and the same characteristic are connected by a linear equation, we have

$$CQ(u) = \sum_{i,j} C_{i,j}\,\phi(u, a;\ P_iR_j),$$

where C, $C_{i,j}$ are independent of u and are not all zero*. Hence, adding to u the half-period Ω_{P_k}, we have

$$C\epsilon_k e^{\pi i|P_k|+2\lambda(u;\ P_k)}\,Q(u) = \sum_{i,j} C_{i,j}\,e^{2\lambda(u;\ P_k)}\begin{pmatrix}P_k\\ P_iR_j\end{pmatrix}\phi(u, a;\ P_iP_kR_j),$$

and therefore, as $\epsilon_k e^{\pi i|P_k|} = \epsilon_k^{-1}$,

$$CQ(u) = \sum_{i,j} C_{i,j}\begin{pmatrix}P_k\\ P_iR_j\end{pmatrix}\epsilon_k\phi(u, a;\ P_iP_kR_j);$$

forming this equation for each of the 2^r values of P_k, and adding the results, we have

$$2^rCQ(u) = \sum_{i,j,k} C_{i,j}\begin{pmatrix}P_k\\ P_iR_j\end{pmatrix}\epsilon_k\phi(u, a;\ P_iP_kR_j);$$

herein put $P_h = P_iP_k$, so that as, for any value of i, P_k becomes in turn all the characteristics of the group (P), the characteristic P_h also becomes all the characteristics in turn, in general in a different order; then

$$\epsilon_k\begin{pmatrix}P_k\\ P_iR_j\end{pmatrix} = \epsilon_h\epsilon_i\begin{pmatrix}P_h\\ P_i\end{pmatrix}\begin{pmatrix}P_hP_i\\ P_iR_j\end{pmatrix} = \epsilon_h\epsilon_i\begin{pmatrix}P_h\\ R_j\end{pmatrix}\begin{pmatrix}P_i\\ R_j\end{pmatrix}e^{\pi i|P_i|},$$

and, therefore,

$$2^rCQ(u) = \sum_j\sum_h \epsilon_h\left[\sum_i C_{i,j}\epsilon_i\begin{pmatrix}P_i\\ R_j\end{pmatrix}e^{\pi i|P_i|}\right]\begin{pmatrix}P_h\\ R_j\end{pmatrix}\phi(u, a;\ P_hR_j),$$

$$= \sum_j\sum_h C_j\begin{pmatrix}P_h\\ R_j\end{pmatrix}\epsilon_h\phi(u, a;\ P_hR_j),$$

where

$$C_j = \sum_i C_{i,j}\begin{pmatrix}P_i\\ R_j\end{pmatrix}\epsilon_i e^{\pi i|P_i|},$$

and thus

$$2^rCQ(u) = \sum_j C_j\Phi(u, a;\ R_j).$$

Now the 2^{p-r} functions $\Phi(u, a;\ R_j)$ are not in general connected by any linear relation with coefficients independent of u; for such a relation would be of the form

$$\Sigma H_i\vartheta(u+a;\ AQ_i)\,\vartheta(u-a;\ AQ_i) = 0,$$

* It is proved below (§ 308) that the functions $\phi(u, a;\ P_iR_j)$ are linearly independent, so that, in fact, C is not zero.

wherein H_i is independent of u, and Q_i becomes, in turn, all the constituents of a group (Q) of 2^p pairwise syzygetic characteristics, and we shall prove (in § 308) that such a relation is impossible for general values of the arguments a. Hence, *all theta functions of the second order, with zero characteristic, which satisfy the equation*

$$Q(u+\Omega_{P_k}) = \epsilon_k e^{\pi i |P_k| + 2\lambda(u;\, P_k)} Q(u)$$

for every half-integer characteristic P_k of the group (P), are representable linearly by 2^{p-r}, $=2^\sigma$, of them, with coefficients independent of u. We have shewn that the functions $\Phi(u, a; A)$, defined by the equation

$$\Phi(u, a; A) = \sum_i \binom{P_i}{A} \epsilon_i \vartheta(u+a;\, AP_i)\, \vartheta(u-a;\, AP_i),$$

where the summation includes 2^r terms, are a particular case of such theta functions.

308. Suppose there exists a relation of the form

$$\sum_i H_i \vartheta(u+a;\, AQ_i)\, \vartheta(u+b;\, AQ_i) = 0,$$

where the summation extends to all the 2^p characteristics Q_i of a Göpel group (Q), and H_i is independent of u. Putting for u, $u+\Omega_{Q_a}$, where Q_a is a characteristic of the group (Q), we obtain

$$\sum_i H_i \binom{Q_a}{Q_i} \vartheta(u+a;\, AQ_iQ_a)\, \vartheta(u+b;\, AQ_iQ_a) = 0;$$

hence, if $\epsilon_1, \ldots, \epsilon_p$ are fourth roots of unity associated with a basis $Q_1, \ldots, Q_p$ of the group (Q), as before, and this equation be multiplied by ϵ_a, and the equations of this form obtained by taking Q_a to be, in turn, all the 2^p characteristics of the group (Q), be added together, we have

$$\sum_i \sum_a H_i \binom{Q_a}{Q_i} \epsilon_a \vartheta(u+a;\, AQ_iQ_a)\, \vartheta(u+b;\, AQ_iQ_a) = 0;$$

now let $Q_j \equiv Q_aQ_i$, then for any value of i, as Q_a becomes all the characteristics of the group (Q), Q_j will become all those characteristics; therefore, substituting

$$\binom{Q_a}{Q_i} = \binom{Q_j}{Q_i}\binom{Q_i}{Q_i}, \qquad \epsilon_a = \epsilon_i \epsilon_j \binom{Q_i}{Q_j},$$

we have

$$\sum_i H_i \epsilon_i \binom{Q_i}{Q_i} \sum_j \epsilon_j \vartheta(u+a;\, AQ_j)\, \vartheta(u+b;\, AQ_j) = 0;$$

hence one at least of the expressions

$$\sum_j \epsilon_j \vartheta(u+a;\, AQ_j)\, \vartheta(u+b;\, AQ_j), \qquad \sum_i H_i \epsilon_i^{-1},$$

must vanish.

Here $\epsilon_1, \epsilon_2, \ldots$ have any one of 2^p possible sets of values. The expression $\sum_i H_i \epsilon_i^{-1}$ cannot vanish for every one of these sets; for, multiplying by ϵ_j^{-1}, we have then

$$\sum_i H_i \binom{Q_i}{Q_j} \epsilon_{i,j}^{-1} = 0,$$

where $\epsilon_{i,j}$, like ϵ_i, becomes in turn the symbol associated with every characteristic of the group, and there are 2^p equations of this form; adding these equations we infer $H_j=0$, and, therefore, as j is arbitrary, we infer that all the coefficients are zero.

Hence it follows that there is at least one of the 2^p sets of values for $\epsilon_1, \epsilon_2, \ldots$, for which

$$\sum_j \epsilon_j \vartheta(u+a;\ AQ_j)\, \vartheta(u+b;\ AQ_j)=0.$$

When the arguments $u+a$, $u+b$ are independent, this is impossible; for putting $u+a=U$, $u+b=V$, this is an equation connecting the 2^p functions $\vartheta(U;\ AQ_j)$ in which the coefficients are independent of U (cf. §§ 282, 283, Chap. XV.).

When the arguments $u+a$, $u+b$ are not independent, this equation is not impossible. For instance, if $\epsilon_k=-e^{\frac{1}{2}\pi i |Q_k|}$, it is easy to verify that

$$\epsilon_{h,k}\vartheta(u+\Omega_{Q_k};\ Q_hQ_k)\,\vartheta(u;\ Q_hQ_k)=-\epsilon_h\vartheta(u+\Omega_{Q_k};\ Q_h)\,\vartheta(u;\ Q_h)$$

and hence the equation does hold when $A=0$, $a=\Omega_{Q_k}$, $b=0$, $\epsilon_k=-e^{\frac{1}{2}\pi i|Q_k|}$, for all the values of $\epsilon_1, \ldots, \epsilon_{k-1}, \epsilon_{k+1}, \ldots, \epsilon_p$. For any values of the arguments $u+a$, $u+b$ we infer from the reasoning here given that if the functions $\vartheta(u+a;\ AQ_i)\,\vartheta(u+b;\ AQ_i)$ are connected by a linear equation with coefficients, H_i, independent of u, then (i) they are connected by at least one equation

$$\sum_i \epsilon_i \vartheta(u+a;\ AQ_i)\,\vartheta(u+b;\ AQ_i)=0,$$

for one of the 2^p sets of values of the quantities $\epsilon_1, \epsilon_2, \ldots$, and (ii) similarly, since the 2^p functions $\vartheta(u+a;\ AQ_i)\,\vartheta(u+b;\ AQ_i)$ do not all vanish identically, that the coefficients are connected by at least one equation

$$\sum_i H_i \epsilon_i^{-1}=0.$$

309. The result of § 307 is of great generality; we proceed to give examples of its application (§§ 309—313). The simplest, as well as the most important, case is that in which $\sigma=0$, $r=p$, and to that we give most attention (§§ 309—311).

When $\sigma=0$, any *two* of the functions $\Phi(u, a;\ A)$ are connected by a linear equation, in which the coefficients are independent of u. If v, a, b be any arguments, and A, B any half-integer characteristics, introducing the symbol ϵ to put in evidence the fact that $\Phi(u, a;\ A)$ is formed with one of 2^p possible selections for the symbols $\epsilon_1, \ldots, \epsilon_p$, and so writing $\Phi(u, a;\ A, \epsilon)$ for $\Phi(u, a;\ A)$, we therefore have the fundamental equation

$$\Phi(u, v;\ A, \epsilon)=\frac{\Phi(u, b;\ B, \epsilon)\,\Phi(a, v;\ A, \epsilon)}{\Phi(a, b;\ B, \epsilon)}.$$

By adding the 2^p equations of this form* which arise by giving all the possible sets of values to the fourth roots of unity $\epsilon_1, \ldots, \epsilon_p$, bearing in mind that every symbol ϵ_i, except ϵ_0, $=1$, occurs as often with the positive as with the negative sign, we obtain

$$2^p\vartheta(u+v;\ A)\,\vartheta(u-v;\ A)=\sum_\epsilon\sum_i \binom{P_i}{A}\epsilon_i\vartheta(u+v;\ AP_i)\,\vartheta(u-v;\ AP_i)$$

$$=\sum_\epsilon \frac{\Phi(u, b;\ B, \epsilon)\,\Phi(a, v;\ A, \epsilon)}{\Phi(a, b;\ B, \epsilon)},$$

* Wherein it is assumed that a, b have not such special values that any one of the 2^p quantities $\Phi(a, b;\ B, \epsilon)$ vanishes. Cf. § 308.

whereby the function $\phi(u, v; A)$ is expressed in terms of 2^p functions

$$\Phi(u, b; B, \epsilon).$$

By taking, in the formula

$$\Phi(u, v; A, \epsilon)\,\Phi(a, b; B, \epsilon) = \Phi(u, b; B, \epsilon)\,\Phi(a, v; A, \epsilon),$$

or

$$\sum_i \sum_j \binom{P_i}{A}\binom{P_j}{B} \epsilon_i \epsilon_j \,\phi(u, v; AP_i)\,\phi(a, b; BP_j)$$
$$= \sum_i \sum_j \binom{P_i}{A}\binom{P_j}{B} \epsilon_i \epsilon_j \,\phi(u, b; BP_i)\,\phi(a, v; AP_j),$$

all the 2^p possible sets of values for $\epsilon_1, \dots, \epsilon_p$, and adding the results, we obtain

$$\sum_i \binom{P_i}{AB} e^{\pi i |P_i|}\,\phi(u, v; AP_i)\,\phi(a, b; BP_i)$$
$$= \sum_i \binom{P_i}{AB} e^{\pi i |P_i|}\,\phi(u, b; BP_i)\,\phi(a, v; AP_i);$$

increasing u and b each by the half-period Ω_R, we have

$$\sum_i \binom{RP_i}{AB} e^{\pi i |RP_i|}\,\phi(u, v; ARP_i)\,\phi(a, b; BRP_i)$$
$$= \sum_i \binom{P_i}{AB} e^{\pi i |P_i| + \pi i |R, P_i|}\,\phi(u, b; BP_i)\,\phi(a, v; AP_i);$$

taking R to be all the possible 2^{2p} half-integer characteristics in turn, and adding the resulting equations we deduce*, putting $C = AB$,

$$2^p \phi(u, b; AC)\,\phi(a, v; A)$$
$$= 2^{-p} \sum_i \sum_R \binom{RP_i}{C} e^{\pi i |RP_i|}\,\phi(u, v; RAP_i)\,\phi(a, b; RAP_iC)$$
$$= \sum_S \binom{AS}{C} e^{\pi i |AS|}\,\phi(u, v; S)\,\phi(a, b; SC),$$

where A, C are arbitrary half-integer characteristics, and S becomes all 2^{2p} possible half-integer characteristics in turn; for (Ex. ii. § 295), $\sum_R e^{\pi i |R, P_i|} = 2^{2p}$ when $P_i = 0$, and is otherwise zero, while, for any definite characteristic AP_i, as R becomes all possible characteristics, so does RAP_i. The formula can be simplified by adding the half-period Ω_C to the argument b; the result is obtainable directly by taking $C = 0$ in the formula written.

This agrees with a result previously obtained (§ 292, Chap. XVI.); for a generalisation of it, see below, § 314.

* This equation has been called the Riemann theta formula. Cf. Prym, *Untersuchungen über die Riemann'sche Thetaformel*, Leipzig, 1882.

310. The formula just obtained may be regarded as a particular case of another which is immediately deducible therefrom. Let (K) be a group of 2^μ characteristics formed by taking all the combinations of μ independent characteristics $K_1, \ldots, K_\mu$; if A be any characteristic whatever, we have

$$\sum_K e^{\pi i |A, K|} = (1+e^{\pi i|A, K_1|}) \ldots (1+e^{\pi i |A, K_\mu|}) = 2^\mu, \text{ or } 0,$$

according as $|A, K_i| \equiv 0$ (for $i=1, \ldots, \mu$), or not; hence, putting $C=0$ in the formula of § 309, and replacing the A of that formula by K_i, we deduce

$$2^{p-\mu}\sum_{i=1}^{2^\mu} e^{\pi i|AK_i|}\phi(u, b;\ K_i)\,\phi(a, v;\ K_i) = 2^{-\mu}\sum_{i=1}^{2^\mu} e^{\pi i|AK_i|}\sum_S e^{\pi i|K_i S|}\phi(u, v;\ S)\,\phi(a, b;\ S),$$

where S becomes all 2^{2p} characteristics,

$$= 2^{-\mu}\sum_S e^{\pi i|A|+\pi i|S|}\sum_{i=1}^{2^\mu} e^{\pi i|AS, K_i|}\phi(u, v;\ S)\,\phi(a, b;\ S)$$

$$= 2^{-\mu} e^{\pi i|A|}\sum_R e^{\pi i|AR|}\left(\sum_{i=1}^{2^\mu} e^{\pi i|R, K_i|}\right)\phi(u, v;\ AR)\,\phi(a, b;\ AR),$$

where R becomes all 2^{2p} characteristics,

$$= 2^{-\mu} e^{\pi i|A|}\, 2^\mu \sum_R e^{\pi i|AR|}\phi(u, v;\ AR)\,\phi(a, b;\ AR),$$

where R extends to all the $2^{2p-\mu}$ characteristics for which $|R, K_1| \equiv 0, \ldots, |R, K_\mu| \equiv 0$. Putting $u+\Omega_B$, $a+\Omega_B$ for u, a respectively, and replacing AB by C, we obtain

$$2^{p-\mu}\sum_{i=1}^{2^\mu} e^{\pi i|BCK_i|}\phi(u, b;\ BK_i)\,\phi(a, v;\ BK_i)$$

$$= e^{\pi i|BC|}\sum_{j=1}^{2^{2p-\mu}} e^{\pi i|BCL_j|}\phi(u, v;\ CL_j)\,\phi(a, b;\ CL_j);$$

here (K) is any group of 2^μ characteristics, (L) is an adjoint group of $2^{2p-\mu}$ characteristics defined by the conditions $|L, K| \equiv 0$ (mod. 2), and B, C are arbitrary half-integer characteristics. The formula of the previous Article is obtained by taking $\mu=0$. The formula of the present Article may be regarded as a particular case of that given below in § 315.

311. The function $\phi(u, v;\ A)$ is unaffected by the addition of integers to the half-integer characteristic A; we may therefore suppose that in the functions $\phi(u, v;\ AP_i)$ which have frequently occurred in the preceding Articles, the characteristic AP_i is reduced, all its elements being either 0 or $\frac{1}{2}$. In the applications which now immediately follow (§ 311) it is convenient, to avoid the explicit appearance of certain fourth roots of unity (cf. Ex. vii., p. 469), not to use reduced characteristics. Two, or more, characteristics which are to be added without reduction will be placed with a comma between them; thus A, P_i denotes $A + P_i$. The characteristics P_i are still supposed reduced.

Taking the formula (§ 309)

$$2^p\,\vartheta(u+v;\ A)\,\vartheta(u-v;\ A) = \sum_\epsilon \frac{\Phi(u, b;\ A', \epsilon)\,\Phi(a, v;\ A, \epsilon)}{\Phi(a, b;\ A', \epsilon)},$$

where A' replaces the B of § 309, suppose $a=b$, and put, for

$$u-b, \quad a+v, \quad a-v, \quad u+v, \quad u-v, \quad a+b, \quad a-b, \quad u+b,$$

respectively,

$$U, \quad V, \quad W, \quad U+V, \quad U+W, \quad V+W, \quad 0, \quad U+V+W;$$

then we obtain

$$2^p \vartheta(U+V;\ A)\vartheta(U+W;\ A)$$

$$=\sum_{\epsilon} \frac{\sum_i \sum_j \binom{P_i}{A'}\binom{P_j}{A} \epsilon_i\epsilon_j \vartheta(U+V+W;\ A',P_i)\vartheta(U;\ A',P_i)\vartheta(V;\ A,P_j)\vartheta(W;\ A,P_j)}{\sum_k \binom{P_k}{A'} \epsilon_k \vartheta(V+W;\ A',P_k)\vartheta(0;\ A',P_k)};$$

adding to V and W respectively the half-periods Ω_B, Ω_C, this becomes

$$2^p [U,\ V;\ A,\ B]\,[U,\ W;\ A,\ C]$$

$$=\sum_{\epsilon} \frac{\sum_i \sum_j \nu_i \mu_j t_{i,j} [U,\ V,\ W;\ A',\ B,\ C,\ P_i]\,[U;\ A',\ P_i]\,[V;\ A,\ B,\ P_j]\,[W;\ A,\ C,\ P_j]}{\sum_k \nu_k s_k [V,\ W;\ A',\ B,\ C,\ P_k]\,[0;\ A',\ P_k]}$$

wherein $[U, V;\ A, B]$ denotes $\vartheta[U+V;\ A+B]$, etc., $\mu_i=\binom{P_i}{A}\epsilon_i$, $\nu_i=\binom{P_i}{A'}\epsilon_i$, etc., and, if $B=\frac{1}{2}\binom{\beta'}{\beta}$, $C=\frac{1}{2}\binom{\gamma'}{\gamma}$, $P_i=\frac{1}{2}\binom{q_i'}{q_i}$, then $t_{i,j}$, s_k are fourth roots of unity given by $t_{i,j}=e^{-\frac{1}{2}\pi i(\beta'+\gamma')(q_i+q_j)}$, $s_k=e^{-\frac{1}{2}\pi i(\beta'+\gamma')q_k}$.

In connexion with this formula several results may be deduced.

(α) Putting $W=-V$, $A+B=K$, $A+C=D$, $A'=D$, the formula gives an expression of $\vartheta[U+V;\ K]\,\vartheta[U-V;\ D]$ in terms of the quantities

$$\vartheta[U;\ KP_i],\ \vartheta[V;\ KP_i],\ \vartheta[U;\ DP_i],\ \vartheta[V;\ DP_i],\ \vartheta[0;\ KP_i],\ \vartheta[0;\ DP_i];$$

the expression contains in the denominator only the constants $\vartheta[0;\ KP_i]$, $\vartheta[0;\ DP_i]$; it has been shewn (§ 299) that not all the characteristics KP_i, DP_i can be odd.

Putting further $K=0$, we obtain an expression of $\vartheta[U+V;\ 0]\,\vartheta[U-V;\ D]$ in terms of

$$\vartheta[U;\ P_i],\ \ \vartheta[V;\ P_i],\ \ \vartheta[U;\ DP_i],\ \ \vartheta[V;\ DP_i],\ \ \vartheta[0;\ P_i],\ \ \vartheta[0;\ DP_i].$$

Dividing the former result by the latter we obtain an expression for $\vartheta[U+V;\ K]/\vartheta[U+V;\ 0]$ in terms of theta functions of U and V with the characteristics DP_i, KP_i, P_i, the coefficients being combinations of $\vartheta[0;\ P_i]$, $\vartheta[0;\ DP_i]$, $\vartheta[0;\ KP_i]$ with numerical quantities. In this expression the characteristic D is arbitrary; it may for instance be taken to be zero.

The formulae are very remarkable; replacing, on the right hand, $\epsilon_i e^{\pi i |A, P_i|}$ by ϵ_i, as is clearly allowable, and taking $D=0$, they are both included in the following formula (cf. Ex. viii. § 317)

$$2^p \vartheta[u+v;\ K]\vartheta[u-v;\ 0]$$
$$=\sum_\epsilon \frac{[\sum_a \epsilon_a e^{-\frac{1}{2}\pi i k' q_a}\vartheta(u;K+P_a)\vartheta(u;P_a)][\sum_a \epsilon_a e^{-\frac{1}{2}\pi i k' q_a+\pi i |P_a|}\vartheta(v;K+P_a)\vartheta(v;P_a)]}{\sum_a \epsilon_a e^{-\frac{1}{2}\pi i k' q_a}\vartheta(0;\ K+P_a)\vartheta(0;\ P_a)},$$

where $K=\frac{1}{2}\binom{k'}{k}$, $P_a=\frac{1}{2}\binom{q_a'}{q_a}$, and the summation in regard to a extends to all the 2^p characteristics, P_a, of the group (P).

It is assumed that the characteristic K is such that the denominator on the right hand does not vanish for any one of the 2^p sets of values for the quantities ϵ_a. For instance the case when K is one of the characteristics of the group (P), other than zero, is excluded (cf. § 308).

Ex. i. For $p=1$, if P denote any one of the half-integer characteristics other than zero,

$$\vartheta(u+v)\vartheta(u-v)=\frac{[\vartheta^2(u)\vartheta^2(v)+\vartheta_P^2(u)\vartheta_P^2(v)]\vartheta^2(0)-[\vartheta^2(u)\vartheta_P^2(v)+e^{\pi i|P|}\vartheta_P^2(u)\vartheta^2(v)]\vartheta_P^2(0)}{\vartheta^4(0)-e^{\pi i|P|}\vartheta_P^4(0)},$$

where $\vartheta(u)$, $\vartheta_P(u)$ denote $\vartheta(u;\ 0)$, $\vartheta(u;\ P)$, etc.

Ex. ii. By putting, in case $p=2$,

$$K=\tfrac{1}{2}\binom{10}{10},\quad P_1=\tfrac{1}{2}\binom{01}{01},\quad P_2=\tfrac{1}{2}\binom{01}{11},$$

deduce from the formula of the text that

$$4\vartheta_{12}(0)\vartheta_{01}(0)\vartheta_{02}(u+u')\vartheta_5(u-u')=\sum_{\zeta_1,\zeta_2}[i\zeta_1\zeta_2 A-\zeta_2 B+i\zeta_1 C+D][A'-i\zeta_1 B'-\zeta_2 C'-i\zeta_1\zeta_2 D'],$$

wherein $\zeta_1=\pm 1$, $\zeta_2=\pm 1$, and

$$A=\vartheta_5(u)\vartheta_{02}(u),\quad B=\vartheta_3(u)\vartheta_{14}(u),\quad C=\vartheta_{04}(u)\vartheta_{24}(u),\quad D=\vartheta_{12}(u)\vartheta_{01}(u),$$

A', B', C', D' denoting the same functions of the arguments u'.

Hence obtain the formula given at the bottom of page 457 of this volume.

(β) Putting $B=C$, $V=W=0$, $A'=A$, we obtain

$$2^p\vartheta^2[U;\ A,\ B]=\sum_\epsilon \frac{\sum_i\sum_j \mu_i\mu_j t_{i,j}[U;\ A,\ B,\ B,\ P_i][U;\ AP_i][0;\ A,\ B,\ P_j]^2}{\sum_k \mu_k s_k[0;\ A,\ B,\ B,\ P_k][0;\ A,\ P_k]},$$

which shews that the square of *any* theta function is expressible as a linear function of the squares of the theta functions with the characteristics forming the Göpel system (AP). We omit the proof that these 2^p squares, $\vartheta^2(U;\ AP_i)$, are not in general connected* by any linear relation in which the coefficients are independent of U.

* Cf. the concluding remark of § 308, § 291, Ex. iv. and § 283.

Ex. For $p=2$ obtain the formula

$$(\vartheta_2^4-\vartheta_{01}^4)\,\vartheta_{03}^2(u)=\vartheta_{23}^2\,\vartheta_2^2\,\vartheta_0^2(u)+\vartheta_{14}^2\,\vartheta_{01}^2\,\vartheta_{34}^2(u)-\vartheta_{23}^2\,\vartheta_{01}^2\,\vartheta_{12}^2(u)-\vartheta_{14}^2\,\vartheta_2^2\,\vartheta_5^2(u),$$

where $\vartheta_2=\vartheta_2(0)$, etc.

(γ) There is however a biquadratic relation connecting the functions $\vartheta(u;\,AP_i)$ provided p be greater than 1. In the formula (§ 309)

$$\sum_i e^{\pi i|P_i|}\vartheta(u+v;\,A,P_i)\,\vartheta(u-v;\,A,P_i)\,\vartheta(a+b;\,A,P_i)\,\vartheta(a-b;\,A,P_i)$$
$$=\sum_i e^{\pi i|P_i|}\vartheta(u+b;\,A,P_i)\,\vartheta(u-b;\,A,P_i)\,\vartheta(a+v;\,A,P_i)\,\vartheta(a-v;\,A,P_i),$$

supposing the characteristic A to be chosen so that all the characteristics AP_i are even, as is possible (§ 299) by taking A suitably, substitute for

$$u+v,\quad u-v,\quad a+b,\quad a-b,\quad u+b,\quad u-b,\quad a+v,\quad a-v$$

respectively

$$u+v+w,\quad u-v,\quad a+b+w,\quad a-b,\quad u+b+w,\quad u-b,\quad a+v+w,\quad a-v;$$

then, putting $a=b=0$, we have

$$\sum_i e^{\pi i|P_i|}\vartheta(0;\,A,P_i)\,\vartheta(w;\,A,P_i)\,\vartheta(u-v;\,A,P_i)\,\vartheta(u+v+w;\,A,P_i)$$
$$=\sum_i e^{\pi i|P_i|}\vartheta(u;\,A,P_i)\,\vartheta(v;\,A,P_i)\,\vartheta(u+w;\,A,P_i)\,\vartheta(v+w;\,A,P_i);$$

herein put $w=\Omega_{P_1}$, $v=u+\Omega_{P_2}$, where P_1, P_2 are two of the characteristics belonging to the basis $P_1,\dots,P_p$ of the group (P); then we obtain

$$\sum_i\begin{pmatrix}P_1P_2\\P_i\end{pmatrix}e^{\pi i|P_i|}\vartheta(0;\,A,P_i)\vartheta(0;\,A,P_1,P_i)\vartheta(0;\,A,P_2,P_i)\,\vartheta(2u;\,A,P_1,P_2,P_i)$$
$$=\sum_i\begin{pmatrix}P_1P_2\\P_i\end{pmatrix}e^{\pi i|P_i|}\vartheta(u;\,A,P_i)\vartheta(u;\,A,P_1,P_i)\vartheta(u;\,A,P_2,P_i)\vartheta(u;\,A,P_1,P_2,P_i).$$

Now every characteristic of the group (P) can be given in one of the forms Q_s, Q_sP_1, Q_sP_2, $Q_sP_1P_2$, where Q_s becomes in turn all the characteristics of a group (Q) of 2^{p-2} characteristics; putting

$$\psi(u;\,Q_s)$$
$$=\begin{pmatrix}P_1P_2\\Q_s\end{pmatrix}e^{\pi i|Q_s|}\vartheta(u;\,A,Q_s)\,\vartheta(u;\,A,P_1,Q_s)\,\vartheta(u;\,A,P_2,Q_s)\,\vartheta(u;\,A,P_1,P_2,Q_s),$$

we immediately find

$$\psi(u;\,Q_s)=\psi(u;\,Q_s,P_1)=\psi(u;\,Q_s,P_2)=\psi(u;\,Q_s,P_1,P_2);$$

hence the equation just obtained can be written

$$\sum_{s=1}^{2^{p-2}}\psi(0;\,Q_s)\sum_{m=1}^{4}\frac{\vartheta(2u;\,A,Q_s,R_m)}{\vartheta(0;\,A,Q_s,R_m)}=4\sum_{s=1}^{2^{p-2}}\psi(u;\,Q_s),$$

where R_m has the four values 0, P_1, P_2, P_1+P_2.

Again, if in the formula (§ 309)

$$2^p\,\vartheta(u+v;\,A)\,\vartheta(u-v;\,A)=\sum_\epsilon\frac{\Phi(u,b;\,A,\epsilon)\,\Phi(a,v;\,A,\epsilon)}{\Phi(a,b;\,A,\epsilon)}$$

we add to u the half period Ω_{P_k}, we obtain, after putting $u=v$, $a=b=0$, the result

$$\vartheta(2u;\ A, P_k)\,\vartheta(0;\ A, P_k)=2^{-p}\binom{P_k}{A}\sum_{\epsilon}\frac{1}{\epsilon_k}\frac{\Phi(u, 0;\ A, \epsilon)\,\Phi(0, u;\ A, \epsilon)}{\Phi(0, 0;\ A, \epsilon)}$$

$$=2^{-p}\binom{P_k}{A}\sum_{\epsilon}\frac{1}{\epsilon_k}\frac{\Phi^2(u, 0;\ A, \epsilon)}{\Phi(0, 0;\ A, \epsilon)},$$

where

$$\Phi(u, 0;\ A, \epsilon)=\sum_i\binom{P_i}{A}\epsilon_i\vartheta^2(u;\ AP_i);\quad \Phi(0, 0;\ A, \epsilon)=\sum_i\binom{P_i}{A}\epsilon_i\vartheta^2(0;\ AP_i).$$

By substitution of the value of $\vartheta(2u;\ A, P_k)$ given by this formula, in the formula above, there results the biquadratic relation* connecting the functions $\vartheta(u;\ AP_i)$.

(δ) As an indication of another set of formulae, which are interesting as direct generalizations of the formulae for the elliptic function $\wp(u)$, the following may also be given. Let

$$\delta=\lambda_1\frac{\partial}{\partial v_1}+\dots+\lambda_p\frac{\partial}{\partial v_p},$$

where $\lambda_1, \dots, \lambda_p$ are undetermined quantities, $\delta\vartheta(v)=\vartheta'(v)$, $\delta^2\vartheta(v)=\vartheta''(v)$, and let

$$\wp(v;\ A)=-\delta^2\log\vartheta(v;\ A)=-[\vartheta(v;\ A)\,\vartheta''(v;\ A)-\vartheta'^2(v;\ A)]\div\vartheta^2(v;\ A);$$

then, differentiating the formula

$$2^p\vartheta(u+v;\ A)\,\vartheta(u-v;\ A)=\sum_{\epsilon}\frac{\Phi(u, b;\ A, \epsilon)\,\Phi(a, v;\ A, \epsilon)}{\Phi(a, b;\ A, \epsilon)}$$

twice in regard to v, and afterwards putting $v=0$ and $b=0$, we obtain

$$\wp(u;\ A)=\sum_i C_i\frac{\vartheta^2(u;\ AP_i)}{\vartheta^2(u;\ A)},$$

wherein

$$C_i=\binom{P_i}{A}\sum_{\epsilon}\epsilon_i\frac{\sum_j\binom{P_j}{A}\epsilon_j\vartheta^2(a;\ AP_j)\,\wp(a;\ AP_j)}{\sum_k\binom{P_k}{A}\epsilon_k\vartheta^2(a;\ AP_k)}$$

$$=\sum_{\epsilon}\epsilon_i\frac{\sum_j\epsilon_j\vartheta^2(a;\ AP_j)\,\wp(a;\ AP_j)}{\sum_k\epsilon_k\vartheta^2(a;\ AP_k)},$$

the 2^p quantities C_i being independent of u and of a. By this formula the function $\wp(u;\ A)$ is expressed linearly by the squares of 2^p theta quotients (cf. Chap. XI. § 217).

* Frobenius, *Crelle*, LXXXIX. (1880), p. 204. The general Göpel biquadratic relation has also been obtained algebraically (for Riemann theta functions) by Brioschi, *Annal. d. Mat.*, 2ª Ser., t. x. (1880—1882).

312. These propositions (§§ 309—311) are corollaries from the fact that the functions $\Phi(u, a;\ A, \epsilon)$ are linearly expressible by 2^{p-r} of them; we have considered the case $r=p$ at great length, on account of its importance.

Passing now to the case $r=p-1$, there is a linear relation connecting any three of the functions

$$\Phi(u, a;\ A, \epsilon) = \sum_{i=1}^{2^{p-1}} \binom{P_i}{A} \epsilon_i \vartheta(u+a;\ AP_i)\,\vartheta(u-a;\ AP_i).$$

There is one case in which we can immediately determine the coefficients in this relation; we have $\sigma = p - r = 1$, $2^{2\sigma} = 4$; there are thus four characteristics A, whereof three are even and one odd, which are such that all the 2^{p-1} characteristics (AP) are of the same character. Taking the single case in which these are all odd, we have

$$\Phi(u, a;\ A, \epsilon) = -\Phi(a, u;\ A, \epsilon), \quad \text{and} \quad \Phi(a, a;\ A, \epsilon) = 0;$$

hence, if, in the existing relation

$$\lambda\Phi(u, a;\ A, \epsilon) + \mu\Phi(u, b;\ A, \epsilon) + \nu\Phi(u, c;\ A, \epsilon) = 0,$$

wherein λ, μ, ν are independent of u, we put $u = a$, we infer

$$\mu : \nu = \Phi(c, a;\ A, \epsilon) : \Phi(a, b;\ A, \epsilon);$$

thus the relation is

$$\Phi(b, c;\ A, \epsilon)\,\Phi(u, a;\ A, \epsilon) + \Phi(c, a;\ A, \epsilon)\,\Phi(u, b;\ A, \epsilon) \\ + \Phi(a, b;\ A, \epsilon)\,\Phi(u, c;\ A, \epsilon) = 0,$$

or

$$\sum_{i=1}^{2^{p-1}} \sum_{j=1}^{2^{p-1}} \binom{P_i}{A}\binom{P_j}{A} \epsilon_i \epsilon_j \psi(i, j) = 0,$$

where

$$\psi(i, j) = \vartheta(u+a;\ AP_i)\,\vartheta(u-a;\ AP_i)\,\vartheta(b+c;\ AP_j)\,\vartheta(b-c;\ AP_j) \\ + \vartheta(u+b;\ AP_i)\,\vartheta(u-b;\ AP_i)\,\vartheta(c+a;\ AP_j)\,\vartheta(c-a;\ AP_j) \\ + \vartheta(u+c;\ AP_i)\,\vartheta(u-c;\ AP_i)\,\vartheta(a+b;\ AP_j)\,\vartheta(a-b;\ AP_j).$$

Adding together all the equations thus obtainable, by taking all the 2^{p-1} possible sets of values for the fourth roots of unity $\epsilon_1, \dots, \epsilon_{p-1}$, we obtain

$$\sum_{i=1}^{2^{p-1}} e^{\pi i |P_i|} \psi(i, i) = 0.$$

For instance, when $p=1$, this is the so-called equation of three terms, from which all relations connecting the elliptic functions can be derived. When $p=2$, it is an equation of six terms and there are fifteen such equations, all expressed by

$$\sum_{a, b, c} \vartheta(u+a;\ A)\,\vartheta(u-a;\ A)\,\vartheta(b+c;\ A)\,\vartheta(b-c;\ A) \\ = -e^{\pi i |AB|} \sum_{a, b, c} \vartheta(u+a;\ B)\,\vartheta(u-a;\ B)\,\vartheta(b+c;\ B)\,\vartheta(b-c;\ B),$$

A and B being any two odd characteristics*.

* Cf. Frobenius, *Crelle*, xcvi. (1884), p. 107.

313. Taking next the case $r=p-2$, every 2^2+1, or 5, functions $\Phi(u, a;\, A, \epsilon)$ are connected by a linear relation. In this case there are sixteen characteristics A such that all the 2^{p-2} characteristics (AP) are of the same character, six of them being odd. Denoting the six odd characteristics in any order by $A_1, \ldots, A_6$, and an even characteristic by A, there is an equation of the form

$$\lambda_1\Phi(u, a;\, A_1, \epsilon)+\lambda_2\Phi(u, a;\, A_2, \epsilon)+\lambda_3\Phi(u, a;\, A_3, \epsilon)$$
$$=\Phi(u, a;\, A_4, \epsilon)+\lambda\Phi(u, a;\, A, \epsilon);$$

putting herein $u=a$, this equation reduces to $\lambda\Phi(a, a;\, A, \epsilon)=0$, so that $\lambda=0$. The other coefficients can also be determined; for, if $C=A_2A_3$, we have (§ 306, Ex. i.),

$$\Phi(u+\Omega_C, a;\, A, \epsilon)=e^{2\lambda(u;\, C)}\begin{pmatrix}A_2A_3\\ A\end{pmatrix}\Phi(u, a;\, AA_2A_3, \epsilon);$$

putting therefore for u, in the equation above, the value $a+\Omega_C$, where $C=A_2A_3$, and recalling (§ 303) that $A_1A_2A_3$, $A_4A_2A_3$ are even characteristics, we infer

$$\lambda_1\begin{pmatrix}A_2A_3\\ A_1\end{pmatrix}\Phi(a, a;\, A_1A_2A_3, \epsilon)=\begin{pmatrix}A_2A_3\\ A_4\end{pmatrix}\Phi(a, a;\, A_4A_2A_3, \epsilon).$$

Proceeding similarly with the characteristics A_3A_1, A_1A_2 in turn, instead of A_2A_3, we finally obtain

$$\begin{pmatrix}A_2A_3\\ A_1A_4\end{pmatrix}\Phi(a, a;\, A_4A_2A_3)\,\Phi(u, a;\, A_1)+\begin{pmatrix}A_3A_1\\ A_2A_4\end{pmatrix}\Phi(a, a;\, A_4A_3A_1)\,\Phi(u, a;\, A_2)$$
$$+\begin{pmatrix}A_1A_2\\ A_3A_4\end{pmatrix}\Phi(a, a;\, A_4A_1A_2)\,\Phi(u, a;\, A_3)=\Phi(a, a;\, A_1A_2A_3)\,\Phi(u, a;\, A_4),$$

where, for greater brevity, the ϵ is omitted in the sign of the function Φ (cf. Ex. viii., § 289).

Ex. For $p=2$, deduce the result

$$\vartheta_{34}\vartheta_{34}(2v)\,\vartheta_{02}(u+v)\,\vartheta_{02}(u-v)-\vartheta_{03}\vartheta_{03}(2v)\,\vartheta_{24}(u+v)\,\vartheta_{24}(u-v)+\vartheta_{23}\vartheta_{23}(2v)\,\vartheta_{04}(u+v)\,\vartheta_{04}(u-v)$$
$$=\vartheta_5\vartheta_5(2v)\,\vartheta_1(u+v)\,\vartheta_1(u-v),$$

where $\vartheta_{34}=\vartheta_{34}(0)$, etc. When $v=0$ this is an equation connecting the squares of $\vartheta_{02}(u)$, $\vartheta_{24}(u)$, $\vartheta_{04}(u)$, $\vartheta_1(u)$.

314. The results of §§ 309, 310 are capable of a generalization, obtainable by a repetition of the argument there employed.

A group of 2^k pairwise syzygetic characteristics may be considered as arising by the composition of two such groups. Take $k, =r+s$, characteristics $P_1, \ldots, P_r, Q_1, \ldots, Q_s$, every two of which are syzygetic; form the groups

$$(P)=0,\ P_1, \ldots, P_r,\ P_1P_2, \ldots, P_1P_2P_3, \ldots$$
$$(Q)=0,\ Q_1, \ldots, Q_s,\ Q_1Q_2, \ldots, Q_1Q_2Q_3, \ldots$$

respectively of 2^r and 2^s characteristics; the 2^{r+s} combinations $R_{i,j}=P_iQ_j$ form a group (R) of 2^{r+s} pairwise syzygetic characteristics; for distinctness the fourth roots of unity associated respectively with $P_1, \dots, P_r, Q_1, \dots, Q_s$, may be denoted by $\epsilon_1, \dots, \epsilon_r, \zeta_1, \dots, \zeta_s$; then with $P_{i,i_1}, Q_{j,j_1}, R_{i,j}$ will be associated the respective quantities

$$\epsilon_{i,i_1}=\epsilon_i\epsilon_{i_1}\begin{pmatrix}P_i\\P_{i_1}\end{pmatrix},\quad \zeta_{j,j_1}=\zeta_j\zeta_{j_1}\begin{pmatrix}Q_j\\Q_{j_1}\end{pmatrix},\quad E_{i,j}=\epsilon_i\zeta_j\begin{pmatrix}P_i\\Q_j\end{pmatrix};$$

thus if A be any characteristic

$$\begin{pmatrix}R_{i,j}\\A\end{pmatrix}E_{i,j}=\begin{pmatrix}P_iQ_j\\A\end{pmatrix}\epsilon_i\zeta_j\begin{pmatrix}P_i\\Q_j\end{pmatrix}=\begin{pmatrix}Q_j\\A\end{pmatrix}\zeta_j\cdot\begin{pmatrix}P_i\\AQ_j\end{pmatrix}\epsilon_i=\begin{pmatrix}P_i\\A\end{pmatrix}\epsilon_i\cdot\begin{pmatrix}Q_j\\AP_i\end{pmatrix}\zeta_j.$$

Therefore, using the symbol Ψ for a sum extending to the whole group (PQ),

$$\begin{aligned}\Psi(u, a;\ A, E)&=\sum_{i,j}\begin{pmatrix}R_{i,j}\\A\end{pmatrix}E_{i,j}\,\vartheta(u+a;\ AR_{i,j})\,\vartheta(u-a;\ AR_{i,j})\\&=\sum_j\begin{pmatrix}Q_j\\A\end{pmatrix}\zeta_j\sum_i\begin{pmatrix}P_i\\AQ_j\end{pmatrix}\epsilon_i\,\vartheta(u+a;\ AQ_jP_i)\,\vartheta(u-a;\ AQ_jP_i)\\&=\sum_j\begin{pmatrix}Q_j\\A\end{pmatrix}\zeta_j\,\Phi(u, a;\ AQ_j, \epsilon),\end{aligned}$$

where Φ denotes a sum extending to the 2^r terms corresponding to the characteristics of the group (P).

By the theorem of § 307 the functions obtainable from $\Psi(u, a;\ A, E)$ by taking different values of a and A, and the same group (PQ), are linearly expressible by $2^{p-r-s}=2^{\sigma-s}$ of them, if $\sigma=p-r$, with coefficients independent of u. The 2^s functions $\Phi(u, a;\ AQ_j, \epsilon)$, obtained by varying a and Q_j, are themselves expressible by 2^σ of them.

Thus, taking $r+s=p$, or $s=\sigma$, we have

$$\Psi(u, v;\ A, E)\,\Psi(a, b;\ A, E)=\Psi(u, b;\ A, E)\,\Psi(a, v;\ A, E)$$

or

$$\begin{aligned}&\sum_{j,j_1}\begin{pmatrix}Q_j\\A\end{pmatrix}\begin{pmatrix}Q_{j_1}\\A\end{pmatrix}\zeta_j\zeta_{j_1}\Phi(u, v;\ AQ_j, \epsilon)\,\Phi(a, b;\ AQ_{j_1}, \epsilon)\\&\qquad=\sum_{j,j_1}\begin{pmatrix}Q_j\\A\end{pmatrix}\begin{pmatrix}Q_{j_1}\\A\end{pmatrix}\zeta_j\zeta_{j_1}\Phi(u, b;\ AQ_j, \epsilon)\,\Phi(a, v;\ AQ_{j_1}, \epsilon);\end{aligned}$$

taking for $\zeta_1, \dots, \zeta_s$ all the possible 2^s values, and adding the 2^s equations of this form, we obtain

$$\sum_{j=1}^{2^s}e^{\pi i|Q_j|}\Phi(u, v;\ AQ_j, \epsilon)\,\Phi(a, b;\ AQ_j, \epsilon)=\sum_{j=1}^{2^s}e^{\pi i|Q_j|}\Phi(u, b;\ AQ_j, \epsilon)\,\Phi(a, v;\ AQ_j, \epsilon).$$

Suppose now that $A_1, \dots, A_\lambda$ are the $2^{2\sigma}$ characteristics satisfying the r relations $|X, P_i|\equiv|P_i|$, (mod. 2), and let $C_m=A_1A_m$; then $|C_m, P_i|\equiv 0$; hence, by the formulae of § 306, Ex. i., adding the half period Ω_{C_m} to u and b, and dividing by the factor $e^{\pi i|C_m, A|}$, we have

$$\begin{aligned}&\sum_{j=1}^{2^\sigma}e^{\pi i|C_mQ_j|}\Phi(u, v;\ AC_mQ_j, \epsilon)\,\Phi(a, b;\ AC_mQ_j, \epsilon)\\&\qquad=\sum_{j=1}^{2^\sigma}e^{\pi i|Q_j|+\pi i|C_m, Q_j|}\Phi(u, b;\ AQ_j, \epsilon)\,\Phi(a, v;\ AQ_j, \epsilon);\end{aligned}$$

taking, here, all the $2^{2\sigma}$ values of C_m in turn, and adding the equations, noticing that

$$\sum_{m=1}^{2^{2\sigma}}e^{\pi i|C_m, Q_j|},\ =e^{\pi i|A_1, Q_j|}\sum_{m=1}^{2^{2\sigma}}e^{\pi i|A_m, Q_j|},$$

is zero because Q_j is not a characteristic of the group (P), except for the special value $Q_j=0$, when its value is $2^{2\sigma}$ (§ 300), we derive the formula

$$2^{2\sigma}\Phi(u, b;\ A, \epsilon)\,\Phi(a, v;\ A, \epsilon)=\sum_{j=1}^{2^\sigma}\sum_{m=1}^{2^{2\sigma}}e^{\pi i|C_mQ_j|}\Phi(u, v;\ AC_mQ_j, \epsilon)\,\Phi(a, b;\ AC_mQ_j, \epsilon);$$

now, as already remarked (§ 298, Ex.), if a characteristic S which is syzygetic with every characteristic of the group (P) be added to each of the $2^{2\sigma}$ characteristics $A_1, \ldots, A_\lambda$, the result is another set of $2^{2\sigma}$ characteristics satisfying the same congruences, $|X, P_i| \equiv |P_i|$, as the set $A_1, \ldots, A_\lambda$, and incongruent mod. (P); thus, taking a fixed value of j, we have $C_m Q_j \equiv C_n P_i$, where, as C_m takes its $2^{2\sigma}$ values, C_n also takes the same values in another order, and P_i varies with m. Hence (Ex. iii. § 306) we have

$$e^{\pi i|C_m Q_j|}\Phi(u, v; AC_m Q_j, \epsilon)\Phi(a, b; AC_m Q_j, \epsilon) = e^{\pi i|C_n P_i|}\Phi(u, v; AC_n P_i, \epsilon)\Phi(a, b; AC_n P_i, \epsilon),$$
$$= e^{\pi i|C_n|}\Phi(u, v; AC_n, \epsilon)\Phi(a, b; AC_n, \epsilon),$$

and

$$\sum_{m=1}^{2^{2\sigma}} e^{\pi i|C_m Q_j|}\Phi(u, v; AC_m Q_j, \epsilon)\Phi(a, b; AC_m Q_j, \epsilon)$$
$$= \sum_{m=1}^{2^{2\sigma}} e^{\pi i|C_m|}\Phi(u, v; AC_m, \epsilon)\Phi(a, b; AC_m, \epsilon),$$

and therefore, finally, dividing by a factor 2^σ (there being 2^σ characteristics in (Q)), we have

$$2^\sigma\Phi(u, b; A, \epsilon)\Phi(a, v; A, \epsilon) = \sum_{m=1}^{2^{2\sigma}} e^{\pi i|A_1 A_m|}\Phi(u, v; AA_1 A_m, \epsilon)\Phi(a, b; AA_1 A_m, \epsilon).$$

When $\sigma = p$, this becomes the formula of § 309. We infer that the functions $\Phi(u, a; A, \epsilon)$ are connected by the same relations as the functions of the form $\vartheta(u+a; A)\vartheta(u-a; A)$ when the number of variables (in the latter functions) is σ.

Ex. Prove that, with the notation of the text,

$$2^\sigma\Phi(u, v; A, \epsilon) = \sum_\zeta \frac{\Psi(u, b; A, E)\Psi(a, v; A, E)}{\Psi(a, b; A, E)}.$$

315. The formula of the last Article is capable of a further generalization. Let (R) be a group of 2^μ characteristics, formed with $R_1, \ldots, R_\mu$ as basis, which satisfy the conditions

$$|R, P_1| \equiv 0, \ldots, |R, P_r| \equiv 0.$$

Thus (P) is a sub-group of (R); the group (R) consists of (P), together with groups (RP), whereof the characteristics R form a group of $2^{\mu-r}$ characteristics, whose constituents are incongruent for the modulus (P). The basis of this sub-group of $2^{\mu-r}$ characteristics will be denoted by $R_1, \ldots, R_{\mu-r}$. The total number of characteristics satisfying the prescribed conditions is 2^{2p-r}; thus $\mu \ngtr 2p-r$, and, when $\mu < 2p-r$ the given conditions are not enough to ensure that a characteristic belongs to the group (R).

Then, if F, G be arbitrary characteristics, and R_i become in turn all the characteristics of a group of $2^{\mu-r}$ characteristics of the group (R) which are incongruent mod. (P), we have

$$2^{p-\mu}\sum_{i=1}^{2^{\mu-r}} e^{\pi i|FGR_i|}\Phi(u, b; GR_i, \epsilon)\Phi(a, v; GR_i, \epsilon)$$
$$= 2^{p-\mu-\sigma}\sum_{i=1}^{2^{\mu-r}} e^{\pi i|FGR_i|}\sum_{m=1}^{2^{2\sigma}} e^{\pi i|C_m|}\Phi(u, v; GR_i C_m, \epsilon)\Phi(a, b; GR_i C_m, \epsilon),$$

where $C_m = A_1 A_m$. Since $|R_i, P| \equiv 0$, the constituents of the set $R_i C_m$, where R_i is a fixed characteristic and $m = 1, 2, \ldots, 2^{2\sigma}$, are in some order congruent (mod. (P)) to the constituents of the set C_m; hence (§ 306, Ex. iii.) the series is equal to

$$2^{r-\mu}\sum_{m=1}^{2^{2\sigma}}\sum_{i=1}^{2^{\mu-r}} e^{\pi i|FGR_i| + \pi i|R_i C_m|}\Phi(u, v; GC_m, \epsilon)\Phi(a, b; GC_m, \epsilon),$$

$$= 2^{r-\mu}\sum_{m=1}^{2^{2\sigma}} e^{\pi i|FG| + \pi i|C_m|}\left(\sum_{i=1}^{2^{\mu-r}} e^{\pi i|FGC_m, R_i|}\right)\Phi(u, v; GC_m, \epsilon)\Phi(a, b; GC_m, \epsilon);$$

now $\sum_{i=1}^{2^{\mu-r}} e^{\pi i|L, R_i}$ is zero, unless $|L, R_i| \equiv 0$ (mod. 2) for every characteristic R_i, in which case its value is $2^{\mu-r}$; thus the series is equal to

$$\Sigma e^{\pi i|FG|+\pi i|FGS_m|} \Phi(u, v; FS_m, \epsilon) \Phi(a, b; FS_m, \epsilon),$$

where S_m satisfies the conditions involved in $|S_m, R_i| \equiv 0$, $FGC_m \equiv S_m$, namely the conditions

$$|S_m, R_1| \equiv 0, \ldots, |S_m, R_{\mu-r}| \equiv 0, |FGS_m, P_1| \equiv 0, \ldots, |FGS_m, P_r| \equiv 0;$$

the number of characteristics satisfying these μ conditions is $2^{2p-\mu}$; the number of these which are incongruent for the modulus (P) is $2^{2p-\mu-r} = 2^{2\sigma+r-\mu}$.

Suppose now that $|FG, P_1| \equiv 0, \ldots, |FG, P_r| \equiv 0$; then the characteristics S_m constitute a group satisfying the conditions $|S_m, R| \equiv 0$, where R becomes in turn all the 2^μ characteristics of the group (R). The group (S) of the characteristics S_m may be obtained by combining the characteristics of the group (P) with the characteristics of a group of $2^{2\sigma-\mu+r}$ characteristics which also satisfy these conditions and are incongruent for the modulus (P); putting $\mu = r + \rho$, we have therefore*

$$2^{\sigma-\rho} \sum_{i=1}^{2^\rho} e^{\pi i|FGR_i|} \Phi(u, b; GR_i, \epsilon) \Phi(a, v; GR_i, \epsilon)$$
$$= e^{\pi i|FG|} \sum_{m=1}^{2^{2\sigma-\rho}} e^{\pi i|FGS_m|} \Phi(u, v; FS_m, \epsilon) \Phi(a, b; FS_m, \epsilon).$$

In this equation each of R_i, S_m represents the characteristics, respectively of the groups (R), (S), which are incongruent mod. (P). But it is easy to see (§ 306, Ex. iii.) that we may also regard R_i, S_m as becoming equal to *all* the characteristics, respectively, of the groups (R), (S).

316. We have shewn in Chap. XV. (§ 286, Ex. i.) that a certain addition formula can be obtained for the cases $p = 1, 2, 3$ by the application of one rule. We give now a generalization of that rule, which furnishes results for any value of p.

Suppose that among the $2^{2\sigma}$ characteristics $A_1, A_2, \ldots, A_\lambda$ which, for any Göpel system (P) of 2^r characteristics, satisfy the conditions

$$|X, P_1| \equiv |P_1|, \ldots, |X, P_r| \equiv |P_r|,$$

we have $k + 1 = 2^\sigma + 1$ characteristics $B_1, \ldots, B_k, B$, of which B is even, which are such that, when i is not equal to j, BB_iB_j is an odd characteristic; as follows from § 302 of this chapter, and § 286, Ex. i., Chap. XV., this is certainly possible when $\sigma = 1$, or 2, or 3; and, since

$$|BB_iB_j, P| \equiv |B, P| + |B_i, P| + |B_j, P| \equiv |P|,$$

* The formula is given by Frobenius, *Crelle*, XCVI. p. 95, being there obtained from the formula of § 310, which is a particular case of it. The formula is generalised by Braunmühl to theta functions whose characteristics are n-th parts of integers in *Math. Annal.* XXXVII. (1890), p. 98. The formula includes previous formulae of this chapter.

the characteristics BB_iB_j will be among the set $A_1, \ldots, A_\lambda$, so that all characteristics congruent to BB_iB_j (mod. (P)) are also odd. Then by § 307 there exists an equation of the form*

$$\lambda\Phi(u, c;\ B, \epsilon) = \sum_{m=1}^{k} \lambda_m \Phi(u, a;\ B_m, \epsilon),$$

wherein the coefficients $\lambda, \lambda_1, \ldots, \lambda_k$, are independent of u. Put in this equation $u = a + \Omega_{BB_i}$; then we infer (§ 306, Ex. i.)

$$\lambda\Phi(a, c;\ B_i, \epsilon) = \lambda_i \Phi(a, a;\ B, \epsilon);$$

hence we have

$$\Phi(a, a;\ B, \epsilon)\,\Phi(u, c;\ B, \epsilon) = \sum_{m=1}^{k} e^{\pi i |BB_m|}\,\Phi(a, c;\ B_m, \epsilon)\,\Phi(u, a;\ B_m, \epsilon),$$

which is the formula in question†.

Adding the 2^r equations obtainable from this formula by taking the different sets of values for the fourth roots of unity $\epsilon_1, \ldots, \epsilon_r$, there results

$$\sum_{i=1}^{2^r} e^{\pi i |P_i|}\,\psi_0(BP_i) = \sum_{m=1}^{2^\sigma} \sum_{i=1}^{2^r} e^{\pi i |BB_m| + \pi i |P_i|}\,\psi(B_mP_i),$$

where

$$\psi_0(BP_i) = \vartheta(0;\ BP_i)\,\vartheta(2a;\ BP_i)\,\vartheta(u+c;\ BP_i)\,\vartheta(u-c;\ BP_i),$$
$$\psi(B_mP_i) = \vartheta(a+c;\ B_mP_i)\,\vartheta(a-c;\ B_mP_i)\,\vartheta(u+a;\ B_mP_i)\,\vartheta(u-a;\ B_mP_i).$$

Herein we may replace the arguments

$$2a,\quad u+c,\quad u-c,\quad a+c,\quad a-c,\quad u+a,\quad u-a$$

respectively by

$$U,\ V,\ W,\ \tfrac{1}{2}(U+V-W),\ \tfrac{1}{2}(U-V+W),\ \tfrac{1}{2}(U+V+W),\ \tfrac{1}{2}(-U+V+W),$$

and thence, in case $p = 2$, or $p = 3$, obtain the formula of Ex. xi., § 286, Chap. XV.

Or we may put $a = 0$, and so obtain

$$\sum_{i=1}^{2^r} e^{\pi i |P_i|}\,\vartheta^2(0;\ BP_i)\,\vartheta(u+c;\ BP_i)\,\vartheta(u-c;\ BP_i)$$
$$= \sum_{m=1}^{2^\sigma} \sum_{i=1}^{2^r} e^{\pi i |B_m,\, BP_i|}\,\vartheta^2(u;\ B_mP_i)\,\vartheta^2(c;\ B_mP_i).$$

Other developments are clearly possible, as in § 286, Chap. XV.

Ex. When $\sigma = 1$ there are three even Göpel systems, and one odd; let (BP), (B_1P), (B_2P) be the three even Göpel systems; then we have

$$\Phi(a, a;\ B, \epsilon)\,\Phi(u, c;\ B, \epsilon)$$
$$= e^{\pi i |BB_1|}\,\Phi(a, c;\ B_1, \epsilon)\,\Phi(u, a;\ B_1, \epsilon) + e^{\pi i |BB_2|}\,\Phi(a, c;\ B_2, \epsilon)\,\Phi(u, a;\ B_2, \epsilon),$$

* We may, if we wish, take, instead of the characteristic B on the left hand, any characteristic A such that $|A, P_i| \equiv |P_i|$, $(i = 1, \ldots, 2^r)$.

† For similar results, cf. Frobenius, *Crelle*, LXXXIX. (1880), pp. 219, 220, and Noether, *Math. Annal.* XVI. (1880), p. 327.

where $\Phi(u, a; B, \epsilon)$ consists of 2^{p-1} terms; for instance when $p=1$ we obtain

$$\vartheta(0; B)\vartheta(2a; B)\vartheta(u+c; B)\vartheta(u-c; B)$$
$$=e^{\pi i|BB_1|}\vartheta(a+c; B_1)\vartheta(a-c; B_1)\vartheta(u+a; B_1)\vartheta(u-a; B_1)$$
$$+e^{\pi i|BB_2|}\vartheta(a+c; B_2)\vartheta(a-c; B_2)\vartheta(u+a; B_2)\vartheta(u-a; B_2).$$

317. *Ex.* i. If P be a fixed characteristic and $\Psi(u; A)$ denote the function $\vartheta(u; A)\vartheta(u; A+P)$, prove that

$$\Psi(u+\Omega_P; A)=e^{\frac{1}{2}\pi i|P|+2\lambda(u; P)}\Psi(u; A),$$

and

$$\Psi(u+\Omega_Q; A)/\Psi(u+\Omega_Q; B)=\begin{pmatrix}Q\\AB\end{pmatrix}\Psi(u; A+Q)/\Psi(u; B+Q).$$

Hence, if $B_1, \ldots, B_k, B$ be $k+1=2^{p-1}+1$ characteristics each satisfying the condition $|X, P|\equiv|P|$, such that, when i is not equal to j, BB_iB_j is odd, we have (§ 307) an equation

$$\lambda\Psi(u; A)=\sum_{m=1}^{2^{p-1}}\lambda_m\Psi(u; B_m),$$

where A is any other even characteristic such that $|A, P|\equiv|P|$; putting $u=\Omega_B+\Omega_{B_i}$, we obtain

$$\lambda\begin{pmatrix}BB_i\\AB_i\end{pmatrix}\Psi(0; A+B+B_i)=\lambda_i\Psi(0; B+2B_i)=\lambda_i\begin{pmatrix}P\\B_i\end{pmatrix}\Psi(0; B);$$

therefore

$$\Psi(0; B)\Psi(u; A)=\sum_{m=1}^{2^{p-1}}\begin{pmatrix}BB_m\\AB_m\end{pmatrix}\begin{pmatrix}P\\B_m\end{pmatrix}\Psi(0; A+B+B_m)\Psi(u; B_m).$$

Ex. ii. Obtain applications of the formula of Ex. i. when $p=2, 3, 4$; in these cases $\sigma, =p-1, =1, 2, 3$ respectively, so that we know how to choose the characteristics $B_1, \ldots, B_k, B$ (Ex. i., § 286, Chap. XV., and § 302 of this Chap.).

Ex. iii. From the formula (§ 309)

$$\vartheta(u+b; A)\vartheta(u-b; A)\vartheta(a+v; A)\vartheta(a-v; A)$$
$$=\frac{1}{2^p}\sum_R e^{\pi i|AR|}\vartheta(u+v; R)\vartheta(u-v; R)\vartheta(a+b; R)\vartheta(a-b; R),$$

by putting $a+\Omega_P$ for a, and $b=v=0$, we deduce

$$\vartheta^2(u; A)\vartheta^2(a; AP)=2^{-p}\sum_R e^{\pi i|AR|}\begin{pmatrix}P\\AR\end{pmatrix}\vartheta^2(u; R)\vartheta^2(a; PR),$$

where A, P are any half-integer characteristics and R becomes all the 2^{2p} half-integer characteristics in turn; putting RP for R we also have, from this equation,

$$\vartheta^2(u; A)\vartheta^2(a; AP)=2^{-p}\sum_R e^{\pi i|AR|}\begin{pmatrix}P\\AR\end{pmatrix}e^{\pi i|AR, P|}\vartheta^2(u; RP)\vartheta^2(a; R);$$

therefore

$$[1+e^{\pi i|A, P|+\pi i|P|}]\vartheta^2(0; A)\vartheta^2(0; AP)$$
$$=2^{-p}\sum_R e^{\pi i|AR|}\begin{pmatrix}P\\AR\end{pmatrix}[1+e^{\pi i|P|+\pi i|R, P|}]\vartheta^2(0; R)\vartheta^2(0; PR).$$

The values of R may be divided into two sets, according as $|R, P|+|P|\equiv1$ (mod. 2), or $\equiv0$; for the values of the former set the corresponding terms vanish; the values of R for which $|R, P|+|P|\equiv0$ (mod. 2) may be either odd or even; for the odd values the zero values of the corresponding theta functions are zero; there remain then (§ 299) only $2.2^{p-2}(2^{p-1}+1)$ terms on the right hand corresponding to values of R which satisfy the conditions $|R| \equiv |RP| \equiv 0$ (mod. 2); these values are divisible into pairs denoted by $R=E$, $R=EP$; for such values $1+e^{\pi i |R, P| + \pi i |P|}=2$, and

$$e^{\pi i |AE|}\binom{P}{AE}+e^{\pi i |AEP|}\binom{P}{AEP}$$

$$=e^{\pi i |AE|}\binom{P}{AE}[1+e^{\pi i |AE, P|}]=e^{\pi i |AE|}\binom{P}{AE}[1+e^{\pi i |A, P|+\pi i |P|}];$$

thus, provided $|A, P|+|P| \equiv 0$ (mod. 2),

$$\vartheta^2(;\ A)\,\vartheta^2(;\ AP)=2^{-(p-1)}\sum_E e^{\pi i |AE|}\binom{P}{AE}\vartheta^2(;\ E)\,\vartheta^2(;\ EP), \qquad \text{(i)},$$

wherein $\vartheta^2(;\ A)$ denotes $\vartheta^2(0;\ A)$, etc., and, on the right hand there are $2^{p-2}(2^{p-1}+1)$ terms corresponding to values of E for which $|E| \equiv |EP| \equiv 0$ (mod. 2), only one of the two values, E, EP, satisfying these conditions being taken.

Putting $P=0$, $u=a$, in the second equation of this example, we deduce in order

$$\vartheta^4(u;\ A)=2^{-p}\sum_R e^{\pi i |AR|}\vartheta^4(u;\ R);\ \vartheta^4(u;\ AP)=2^{-p}\sum_R e^{\pi i |APR|}\vartheta^4(u;\ R);$$

so that, by addition,

$$\vartheta^4(u;\ A)+e^{\pi i |A, P|}\vartheta^4(u;\ AP)=2^{-p}\sum_R e^{\pi i |AR|}[1+e^{\pi i |P|+\pi i |R, P|}]\,\vartheta^4(u;\ R);$$

thus, as before,

$$\vartheta^4(;\ A)+e^{\pi i |A, P|}\vartheta^4(;\ AP)=2^{-(p-1)}\sum_E e^{\pi i |AE|}\{\vartheta^4(;\ E)+e^{\pi i |A, P|}\vartheta^4(;\ EP)\}, \quad \text{(ii)}.$$

Ex. iv. Taking $p=2$, let $(P)=0$, P_1, P_2, P_1P_2 be a Göpel group of even characteristics*; let B_1, B_2, B_1B_2 be such characteristics (§ 297) that the Göpel systems (P), (B_1P), (B_2P), (B_1B_2P) constitute all the sixteen characteristics; each of the systems (B_1P), (B_2P), (B_1B_2P) contains two odd characteristics and two even characteristics. Then, in the formulae (i), (ii) of Ex. iii., if P denote any one of the three characteristics P_1, P_2, P_1P_2, the conditions for the characteristics E are $|E, P| \equiv |P| \equiv 0$, $|E| \equiv 0$; the $2 \,.\, 2^{p-2}(2^{p-1}+1)$, $=6$, solutions of these conditions must consist of 0, Q, B and P, QP, BP, where Q is defined by the condition that the characteristics 0, Q, P, QP constitute the group (P), and B is a certain even characteristic chosen from one of the systems (B_1P), (B_2P), (B_1B_2P). Hence, when $P=P_1$, we may, without loss of generality, take for the $2^{p-2}(2^{p-1}+1)=3$ values of E which give rise to different terms in the series (i), (ii), the values 0, P_2, B_1; similarly, when $P=P_2$, we have, for the values of E, $E=0$, P_1, B_2; and when $P=P_1P_2$, $E=0$, P_1, B_1B_2; taking A to be respectively† B_1, B_2, B_1B_2 in these cases, we obtain the six equations

$$\binom{P_1}{B_1}\vartheta^2(;\ 0)\,\vartheta^2(;\ P_1)+e^{\pi i |B_1P_2|}\binom{P_1}{B_1P_2}\vartheta^2(;\ P_2)\,\vartheta^2(;\ P_1P_2)-\vartheta^2(;\ B_1)\,\vartheta^2(;\ B_1P_1)=0,$$

$$\vartheta^4(;\ 0)+\vartheta^4(;\ P_1)+e^{\pi i |B_1P_2|}[\vartheta^4(;\ P_2)+\vartheta^4(;\ P_2P_1)]-[\vartheta^4(;\ B_1)+\vartheta^4(;\ B_1P_1)]=0,$$

$$\binom{P_2}{B_2}\vartheta^2(;\ 0)\,\vartheta^2(;\ P_2)+e^{\pi i |B_2P_1|}\binom{P_2}{B_2P_1}\vartheta^2(;\ P_1)\,\vartheta^2(;\ P_1P_2)-\vartheta^2(;\ B_2)\,\vartheta^2(;\ B_2P_2)=0,$$

$$\vartheta^4(;\ 0)+\vartheta^4(;\ P_2)+e^{\pi i |B_2P_1|}[\vartheta^4(;\ P_1)+\vartheta^4(;\ P_1P_2)]-[\vartheta^4(;\ B_2)+\vartheta^4(;\ B_2P_2)]=0,$$

$$\binom{P_1P_2}{B_1B_2}\vartheta^2(;\ 0)\,\vartheta^2(;\ P_1P_2)+e^{\pi i |B_1B_2P_1|}\binom{P_1P_2}{B_1B_2P_1}\vartheta^2(;\ P_1)\,\vartheta^2(;\ P_2)$$

$$-\vartheta^2(;\ B_1B_2)\,\vartheta^2(;\ B_1B_2P_1P_2)=0,$$

$$\vartheta^4(;0)+\vartheta^4(;P_1P_2)+e^{\pi i |B_1B_2P_1|}[\vartheta^4(;P_1)+\vartheta^4(;P_2)]-[\vartheta^4(;B_1B_2)+\vartheta^4(;B_1B_2P_1P_2)]=0,$$

* There are six such groups (Ex. iv. § 289).

† We easily find $|B_1B_2P_1| \equiv |B_1B_2P_2| \equiv -|B_1B_2|$. Thus the case when B_1B_2 is odd is included by writing B_1P_1 in place of B_1.

wherein $e^{\pi i|B_1P_2|}=e^{\pi i|B_2P_1|}=e^{\pi i|B_1B_2P_1}=-1$. These formulae express the zero values of all the even theta functions in terms of the four $\vartheta(;\,0)$, $\vartheta(;\,P_1)$, $\vartheta(;\,P_2)$, $\vartheta(;\,P_1P_2)$. Thus for instance they can be expressed in terms of ϑ_5, ϑ_{34}, ϑ_{12}, ϑ_0; the equations have been given in Ex. iii., § 289, Chap. XV.

Ex. v. We have in Chap. XVI. (§ 291) obtained the formula

$$\vartheta(u-v;\,q)\,\vartheta(u+v;\,r)=\vartheta\left[u-v;\,\begin{pmatrix}q'\\q\end{pmatrix}\right]\vartheta\left[u+v;\,\begin{pmatrix}r'\\r\end{pmatrix}\right]$$
$$=\sum_{\epsilon'}\vartheta_1\left[u;\,\begin{matrix}\tfrac12(\epsilon'+q'+r')\\q+r\end{matrix}\right]\vartheta_1\left[v;\,\begin{matrix}\tfrac12(\epsilon'-q'+r')\\-q+r\end{matrix}\right],$$

where ϵ' represents a set of p integers, each either 0 or 1, and has therefore 2^p values.

Suppose now that q, r represent the same half-integer characteristic, $=\tfrac12\begin{pmatrix}c'\\c\end{pmatrix}+\tfrac12\begin{pmatrix}0\\k_a\end{pmatrix}$, $=C+K_a$, say; then we immediately find

$$\vartheta_1\left[u;\,\begin{matrix}\tfrac12(\epsilon'+q'+r')\\q+r\end{matrix}\right]\vartheta_1\left[v;\,\begin{matrix}\tfrac12(\epsilon'-q'+r')\\-q+r\end{matrix}\right]=\vartheta_1\left[u;\,\begin{matrix}\tfrac12\epsilon'c'\\k_a\end{matrix}\right].\,e^{\pi icc'}\vartheta_1\left[v;\,\begin{matrix}\tfrac12\epsilon'\\c\end{matrix}\right],$$

where $\epsilon'c'$ denotes the row of p integers, each either 0 or 1, which are given by $(\epsilon'c')_i\equiv\epsilon_i'+c_i'$ (mod. 2); herein the factor $e^{\pi icc'}\vartheta_1\left[v;\,\begin{matrix}\tfrac12\epsilon'\\c\end{matrix}\right]$ is independent of k_a. For K_a we take now, in turn, the constituents

$$0,\ K_1,\ K_2,\ \dots,\ K_p,\ K_1K_2,\ \dots,\ K_1K_2K_3,\ \dots$$

of a Göpel set of 2^p characteristics, in which

$$K_1=\tfrac12\begin{pmatrix}0,\,0,\,0,\,\dots\\1,\,0,\,0,\,\dots\end{pmatrix},\quad K_2=\tfrac12\begin{pmatrix}0,\,0,\,0,\,\dots\\0,\,1,\,0,\,\dots\end{pmatrix},\ \dots,\quad K_p=\tfrac12\begin{pmatrix}0,\,\dots,\,0,\,0\\0,\,\dots,\,0,\,1\end{pmatrix};$$

then denoting $\vartheta[u+v;\,CK_a]\,\vartheta[u-v;\,CK_a]$ by $[CK_a]$, we obtain 2^p equations which are all included in the equation

$$([CK_1],\,\dots,\,[CK_s])=J\left(e^{\pi icc'}\vartheta_1\left[v;\,\begin{matrix}\tfrac12\epsilon_1'\\c\end{matrix}\right],\,\dots,\,e^{\pi icc'}\vartheta_1\left[v;\,\begin{matrix}\tfrac12\epsilon_s'\\c\end{matrix}\right]\right),$$

wherein $s=2^p$, $\epsilon_1',\,\dots,\,\epsilon_s'$ represent the different values of ϵ', and J is a matrix wherein the β-th element of the a-th row is $\vartheta_1\left[u;\,\begin{matrix}\tfrac12\epsilon'_\beta c'\\k_a\end{matrix}\right]$.

The 2^p various values of $\epsilon'_\beta c'$, for an assigned value of c', are, in general in a different order, the same as the various values of ϵ'_β; we may suppose the order of the columns of J to be so altered that the various values of $\epsilon'_\beta c'$ become the values of ϵ'_β *in an assigned order*, the order of the elements $e^{\pi icc'}\vartheta_1\left[v;\,\begin{matrix}\tfrac12\epsilon_1'\\c\end{matrix}\right],\,\dots,\,e^{\pi icc'}\vartheta_1\left[v;\,\begin{matrix}\tfrac12\epsilon_s'\\c\end{matrix}\right]$ being correspondingly altered. When this is done the matrix J is independent of the characteristic C. Now it is possible to choose 2^p characteristics C, say $C_1,\,\dots,\,C_s$ such that the Göpel systems (C_iK) give, together, all the 2^{2p} possible characteristics; then the 2^p equations obtainable from that just written by replacing C in turn by $C_1,\,\dots,\,C_s$, are all included, using the notation of matrices, in the one equation*

$$\left|\,\vartheta[u+v;\,C_aK_\beta]\,\vartheta[u-v;\,C_aK_\beta]\,\right|=\left|\,e^{\pi ic_ac'_a}\vartheta_1\left[v;\,\begin{matrix}\tfrac12\zeta'_\beta c'_a\\c_a\end{matrix}\right]\right|\left|\,\vartheta_1\left[u;\,\begin{matrix}\tfrac12\zeta'_a\\k_\beta\end{matrix}\right]\right|,$$

wherein ζ'_a denotes a row of p integers, each either 0 or 1, and has 2^p values. In each matrix the element written down is the β-th element of the a-th row.

* We can obviously obtain a more general equation by taking 2^{2p} different sets of arguments, the general element of the matrix on the left hand being $\vartheta[u^{(a)}+v^{(\beta)};\,C_aK_\beta]\,\vartheta[u^{(a)}-v^{(\beta)};\,C_aK_\beta]$. Cf. Chap. XV. § 291, Ex. v., and Caspary, *Crelle*, XCVI. (1884), pp. 182, 324; Frobenius, *Crelle*, XCVI. (1884), p. 100. Also Weierstrass, *Sitzungsber. der Ak. d. Wiss. zu Berlin*, 1882, I.—XXVI. p. 506.

Ex. vi. If in Ex. v., $p=2$, and the group (K) consists of the characteristics

$$\tfrac{1}{2}\begin{pmatrix}00\\00\end{pmatrix},\ \tfrac{1}{2}\begin{pmatrix}00\\10\end{pmatrix},\ \tfrac{1}{2}\begin{pmatrix}00\\01\end{pmatrix},\ \tfrac{1}{2}\begin{pmatrix}00\\11\end{pmatrix},$$

while the characteristics C consist of

$$\tfrac{1}{2}\begin{pmatrix}00\\00\end{pmatrix},\ \tfrac{1}{2}\begin{pmatrix}10\\00\end{pmatrix},\ \tfrac{1}{2}\begin{pmatrix}01\\00\end{pmatrix},\ \tfrac{1}{2}\begin{pmatrix}11\\00\end{pmatrix},$$

and the values of ζ' are, in order,

$$(0, 0),\quad (0, 1),\quad (1, 0),\quad (1, 1),$$

shew that the sixteen equations expressed by the final equation of Ex. v. are equivalent to

$$\begin{pmatrix} \begin{bmatrix}00\\11\end{bmatrix}, & \begin{bmatrix}10\\00\end{bmatrix}, & -\begin{bmatrix}01\\10\end{bmatrix}, & \begin{bmatrix}11\\01\end{bmatrix} \\ -\begin{bmatrix}10\\01\end{bmatrix}, & \begin{bmatrix}00\\10\end{bmatrix}, & \begin{bmatrix}11\\00\end{bmatrix}, & \begin{bmatrix}01\\11\end{bmatrix} \\ \begin{bmatrix}01\\00\end{bmatrix}, & -\begin{bmatrix}11\\11\end{bmatrix}, & \begin{bmatrix}00\\01\end{bmatrix}, & \begin{bmatrix}10\\10\end{bmatrix} \\ -\begin{bmatrix}11\\10\end{bmatrix}, & -\begin{bmatrix}01\\01\end{bmatrix}, & -\begin{bmatrix}10\\11\end{bmatrix}, & \begin{bmatrix}00\\00\end{bmatrix} \end{pmatrix} = \begin{pmatrix} a_4, & a_3, & -a_2, & a_1 \\ -a_3, & a_4, & a_1, & a_2 \\ a_1, & a_2, & a_3, & -a_4 \\ a_2, & -a_1, & a_4, & a_3 \end{pmatrix} \begin{pmatrix} \beta_1, & -\beta_4, & \beta_3, & \beta_2 \\ \beta_2, & \beta_3, & \beta_4, & -\beta_1 \\ -\beta_3, & \beta_2, & \beta_1, & \beta_4 \\ \beta_4, & \beta_1, & -\beta_2, & \beta_3 \end{pmatrix}$$

wherein, on the left hand, $\begin{bmatrix}00\\11\end{bmatrix}$ denotes $\vartheta\left[u+v;\ \tfrac{1}{2}\begin{pmatrix}00\\11\end{pmatrix}\right]\vartheta\left[u-v;\ \tfrac{1}{2}\begin{pmatrix}00\\11\end{pmatrix}\right]$, etc., and on the right hand,

$$a_1=\vartheta_1\left[u;\ \tfrac{1}{2}\begin{pmatrix}00\\00\end{pmatrix}\right],\quad a_2=\vartheta_1\left[u;\ \tfrac{1}{2}\begin{pmatrix}10\\00\end{pmatrix}\right],\quad a_3=\vartheta_1\left[u;\ \tfrac{1}{2}\begin{pmatrix}01\\00\end{pmatrix}\right],\quad a_4=\vartheta_1\left[u;\ \tfrac{1}{2}\begin{pmatrix}11\\00\end{pmatrix}\right],$$

$\beta_1, \beta_2, \beta_3, \beta_4$ being respectively the same theta functions with the argument v.

Now if A, B denote respectively the first and second matrices on the right hand, the linear equations

$$(y_1, y_2, y_3, y_4)=A\,(x_1, x_2, x_3, x_4),\quad (x_1, x_2, x_3, x_4)=B\,(z_1, z_2, z_3, z_4)$$

are immediately seen to lead to the results

$$y_1^2+y_2^2+y_3^2+y_4^2=(a_1^2+a_2^2+a_3^2+a_4^2)\,(x_1^2+x_2^2+x_3^2+x_4^2),$$
$$x_1^2+x_2^2+x_3^2+x_4^2=(\beta_1^2+\beta_2^2+\beta_3^2+\beta_4^2)\,(z_1^2+z_2^2+z_3^2+z_4^2)\,;$$

hence if the j-th element of the i-th row of the compound matrix AB, which is the matrix on the left-hand side of the equation, be denoted by $\gamma_{i,j}$, we have

$$\sum_{i=1}^4 \gamma_{i,s}^2=\sum_{i=1}^4 \gamma_{i,r}^2,\ \sum_{i=1}^4 \gamma_{i,r}\gamma_{i,s}=0, \qquad (r\neq s,\ r,\ s=1, 2, 3, 4),$$

and these equations lead to

$$\sum_{j=1}^4 \gamma_{s,j}^2=\sum_{j=1}^4 \gamma_{r,j}^2,\ \sum_{j=1}^4 \gamma_{r,j}\gamma_{s,j}=0.$$

Denoting $\begin{bmatrix}00\\11\end{bmatrix}$, $\begin{bmatrix}10\\00\end{bmatrix}$, by $[a_1c_2]$, $[a_1c_1]$, etc., as in the table of § 204, and interchanging the second and third rows of the matrix on the left-hand side, we may express the result by saying that the matrix

$$\begin{pmatrix} [a_1c_2], & [a_1c_1], & -[a_1c], & [a_2] \\ [a_2c_2], & -[a_2c_1], & [a_2c], & [a_1] \\ -[c_2], & [c_1], & [c], & [a_1a_2] \\ -[cc_1], & -[cc_2], & -[c_1c_2], & [0] \end{pmatrix}$$

gives an *orthogonal* linear substitution of four variables*.

* An algebraic proof may be given ; cf. Brioschi, *Ann. d. Mat.* XIV.

Ex. vii. Deduce from § 309 that

$$2^p \vartheta(u+v\,;\,AP_i)\,\vartheta(u-v\,;\,AP_i)=\sum_{\epsilon}\frac{\epsilon_i^{-1}\left[\sum_{\alpha}\epsilon_\alpha \vartheta^2(u\,;\,AP_\alpha)\right]\left[\sum_{\alpha}\epsilon_\alpha e^{\pi i|AP_\alpha|}\vartheta^2(v\,;\,AP_\alpha)\right]}{\sum_{\alpha}\epsilon_\alpha \vartheta^2(0\,;\,AP_\alpha)},$$

where P_i, P_α are characteristics of a Göpel group (P), of 2^p characteristics. Infer that, *if n be any positive integer, and AP_i be an even characteristic, $\vartheta(nv\,;\,AP_i)$ is expressible as an integral polynomial of order n^2 in the 2^p functions $\vartheta(v\,;\,AP_\alpha)$.*

Ex. viii. If $K=\frac{1}{2}\begin{pmatrix}k'\\k\end{pmatrix}$, $P_\alpha=\frac{1}{2}\begin{pmatrix}q'_\alpha\\q_\alpha\end{pmatrix}$, deduce from § 309, putting

$$a=b=u-U=v-V=\tfrac{1}{2}\Omega_k,$$

that

$$\chi(U+V,\,U-V)\,\chi(0,\,0)=\chi(U,\,U)\,\chi(V,\,-V),$$

where

$$\chi(u,\,v)=\sum_{\alpha}\epsilon_\alpha e^{-\frac{1}{2}\pi ik'q_\alpha}\,\vartheta(u\,;\,K+P_\alpha)\,\vartheta(v\,;\,P_\alpha).$$

CHAPTER XVIII.

TRANSFORMATION OF PERIODS, ESPECIALLY LINEAR TRANSFORMATION.

318. IN the foregoing portion* of the present volume, the fundamental algebraic equation has been studied with the help of a Riemann surface. Much of the definiteness of the theory depends upon the adoption of a specific mode of dissecting the surface by means of period loops; for instance this is the case for the normal integrals, and their periods, and consequently also for the theta functions, which were defined in terms of the periods $\tau_{i,j}$ of the normal integrals of the first kind; it is also the case for the places $m_1, \dots, m_p$ of § 179 (Chap. X.), upon which the theory of the vanishing of the theta functions depends. The question then arises; if we adopt a different set of period loops as fundamental, how is the theory modified, and, in particular, what is the relation between the new theta functions obtained, and the original functions? We have given a geometrical method (§ 183, Chap. X.) of determining the places $m_1, \dots, m_p$ from the place m, from which it appears that they cannot have more than a finite number of positions when m is given, and coresidual places are reckoned equivalent; the enquiry then suggests itself; can they take all these possible positions by a suitable choice of period loops, or is one of these essentially different from the others? The answers to such questions as these are to be sought from the theory of the present chapter.

There is another enquiry, not directly related to the Riemann surface, but arising in connexion with the analytical theory of the theta functions. Taking p independent variables $u_1, \dots, u_p$, and associating with them, in accordance with the suggestion of §§ 138—140 (cf. § 284), the matrices 2ω, $2\omega'$, 2η, $2\eta'$, we are thence able, with the help of the resulting equations

$$2h\omega = \pi i, \quad 2h\omega' = b, \quad \eta = 2a\omega, \quad \eta' = 2a\omega' - \bar{h},$$

to formulate a theta function. But it is manifest that this procedure makes an unsymmetrical use of the columns of periods arising respectively from the matrices ω and ω'; and it becomes a problem to enquire whether this

* References to the literature dealing with transformation are given at the beginning of Chap. XX.

want of symmetry can be removed; and more generally to enquire what general linear functions of the original $2p$ columns of periods, with integral coefficients, can be formed to replace the original columns of periods; and, if theta functions be formed with the new periods, as with the original ones, to investigate the expression of the new theta functions in terms of the original ones.

So far as the theta functions are concerned, it will appear that the theory of the transformation of periods, and of characteristics, includes the consideration of the effect of a modification of the period loops of a Riemann surface; for that reason we give in this chapter the fundamental equations for the transformation of the periods and characteristic of a theta function, when the coefficients of transformation are integers; but the main object of this chapter is to deal with the transformation of the period loops on a Riemann surface. The analytical theory of the expression of the transformed theta functions in terms of the original functions is considered in the two following chapters.

In virtue of the algebraical representation which is possible for quotients of Riemann theta functions (as exemplified in Chap. XI.), the theory of the expression of the transformed theta functions in terms of the original functions, includes a theory of the algebraical transformation of the fundamental algebraical equation associated with a Riemann surface; it is known what success was achieved by Jacobi, from this point of view, in the case of elliptic functions; and some of the earliest contributions to the general theory of transformation of theta functions approach the matter from that side*. We deal briefly with particular results of this algebraical theory in Chap. XXII.

319. Take any undissected Riemann surface associated with a fundamental algebraic equation of deficiency p. The most general set of $2p$ period loops may be constructed as follows:

Draw on the surface any closed curve whatever, not intersecting itself, which is such that if the surface were cut along this curve it would not be divided into two pieces; of the two possible directions in which this curve can be described, choose either, and call it the positive direction; call the side of the curve which is on the left hand when the curve is described positively, the left side; this curve is the period loop (A_1); starting now from any point on the left side of (A_1), a curve can be drawn on the surface, which, without cutting itself, or the curve (A_1), and without dividing the surface, ends at the point of the curve (A_1) at which it began, but on the right side of (A_1); this is the loop (B_1), and the direction in which it has

* See, in particular, Richelot, *Crelle*, XVI. (1837), De transformatione...integralium Abelianorum primi ordinis; in the papers of Königsberger, *Crelle*, LXIV., LXV., LXVII., some of the algebraical results of Richelot are obtained by means of the transformation of theta functions.

been described is its positive direction; its left side is that on the left hand in the positive description of it. The period associated with the loop (A_1), of any Abelian integral, is the constant whereby the value of the integral on the left side of (A_1) exceeds the value on the right side, and is equal to the value obtained by taking the integral along the loop (B_1) in the negative direction, from the end of the loop (B_1) to its beginning. The period associated with the loop (B_1) is similarly the excess of the value of the integral on the left side of the loop (B_1) over its value on the right side, and may be obtained by taking the integral round the loop (A_1) in the positive direction, from the right side of the loop (B_1) to the left side. These periods may be denoted respectively by Ω_1 and Ω_1'.

320. It is useful further to remark that there is no essential reason why what we have called the loops (A_1), (B_1) should not be called respectively the loops $[B_1]$ and $[A_1]$. If this be done, and the positive direction of the (original) loop (B_1) be preserved, the convention as to the relation of the directions of the loops $[A_1]$, $[B_1]$ will necessitate a reversal of the convention as to the positive direction of the (original) loop (A_1). If the periods associated with the (new) loops $[A_1]$, $[B_1]$ be respectively denoted by $[\Omega]$ and $[\Omega']$, we have, therefore, the equations

$$[\Omega]=\Omega', \quad [\Omega']=-\Omega.$$

These equations represent a process—of interchange of the loops (A_1), (B_1), with retention of the direction of (B_1)—which may be repeated. The repetition gives equations which we may denote by

$$\{\Omega\}=[\Omega']=-\Omega, \quad \{\Omega'\}=-[\Omega]=-\Omega',$$

and the two processes are together equivalent to reversing the direction of loop (A_1), and (therefore) of the loop (B_1). The convention that the loop (B_1) shall begin from the left side of the loop (A_1) is not necessary for the purpose of the dissection of the surface into a simply connected surface; but it affords a convenient way of specifying the necessary condition for the convergence of the series defining the theta functions.

321. The pair of loops (A_1), (B_1) being drawn, the successive pairs (A_2), (B_2), ..., (A_p), (B_p) are then to be drawn in accordance with precisely similar conventions—the additional convention being made that neither loop of any pair is to cross any one of the previously drawn loops. If the Riemann surface be cut along these $2p$ loops it will become a p-ply connected surface, with p closed boundary curves. It may be further dissected into a simply connected surface by means of $(p-1)$ further cuts (C_1), ..., (C_{p-1}), taken so as to reduce the boundary to one continuous closed curve.

Upon the p-ply connected surface formed by cutting the original surface along the loops (A_1), (B_1), ..., (A_p), (B_p), the Riemann integrals of the first and second kind are single-valued. In particular if W_1, ..., W_p be a set of linearly independent integrals of the first kind defined by the conditions that the periods of W_r at the loops (A_1), ..., (A_p) are all zero, except that at

(A_r), which is 1, and if $\tau_{r,s}$ be the period of W_r at the loop (B_s), the imaginary part of the quadratic form

$$\tau_{11}n_1^2 + \ldots\ldots + 2\tau_{12}n_1n_2 + \ldots\ldots + \tau_{pp}n_p^2$$

is necessarily positive* for real values of $n_1, \ldots, n_p$. This statement remains true when, for each of the p pairs, the loops (A_r), (B_r) are interchanged, with *e.g.* the retention of the direction of (B_r) and a consequent change in the sign of the period associated with (A_r), as explained above (§ 320); if the loops (A_r), (B_r) be interchanged without the change in the sign of the period associated with (A_r), the imaginary part of the corresponding quadratic form is negative†.

322. In addition now to such a general system of period loops as has been described, imagine another system of loops, which for distinctness we shall call the original system; the loops of the original system may be denoted by (a_r), (b_r) and the periods of any integral, u_i, associated therewith, by $2\omega_{i,r}$, $2\omega'_{i,r}$; the general system of period loops is denoted by (A_r), (B_r), and the periods associated therewith by $[2\omega_{i,r}]$, $[2\omega'_{i,r}]$. For the values of the integral u_i, the circuit of the loop (B_r), in the negative direction, from the right to the left side of the loop (A_r), is equivalent to a certain number, say‡ to $\alpha_{j,r}$, of circuits of the loop (b_j) in the negative direction, together with a certain number, say $\alpha'_{j,r}$, of circuits of the loop (a_j) in the positive direction $(r, j = 1, 2, \ldots, p)$; hence we have

$$[\omega_{i,r}] = \sum_{j=1}^{p} (\omega_{i,j}\alpha_{j,r} + \omega'_{i,j}\alpha'_{j,r}), \qquad (r = 1, 2, \ldots, p);$$

similarly we have equations which we write in the form

$$[\omega'_{i,r}] = \sum_{j=1}^{p} (\omega_{i,j}\beta_{j,r} + \omega'_{i,j}\beta'_{j,r}), \qquad (r = 1, 2, \ldots, p),$$

the interpretation of the integers $\beta_{j,r}$, $\beta'_{j,r}$ being similar to that of the integers $\alpha_{j,r}$, $\alpha'_{j,r}$.

Thus, if $u_1, \ldots, u_p$ denote p linearly independent integrals of the first kind, and the matrices of their periods for the original system of period loops be denoted by 2ω, $2\omega'$, and for the general system of period loops by $[2\omega]$, $[2\omega']$, we have

$$[\omega] = \omega\alpha + \omega'\alpha', \quad [\omega'] = \omega\beta + \omega'\beta',$$

where α, α', β, β' denote matrices whose elements are integers.

* And not zero, since $n_1W_1 + \ldots + n_pW_p$ cannot be a constant. Cf. for instance, Neumann, *Riemann's Theorie der Abel'schen Integrale* (Leipzig, 1884), p. 247, or Forsyth, *Theory of Functions* (1893), p. 447. (Riemann, *Werke*, 1876, p. 124.)

† As previously remarked, p. 247, note.

‡ A circuit of (b_j) in the positive direction furnishing a contribution of -1 to $\alpha_{j,r}$.

If $L_1, \ldots, L_p$ be a set of p integrals of the second kind *associated* with $u_1, \ldots, u_p$, as in § 138, Chap. VII., and satisfying, therefore, the condition

$$\sum_{i=1}^{p} [D_x u_i^{x,a} D_z L_i^{z,c} - D_z u_i^{z,c} D_x L_i^{x,a}] = D_x \left[(z, x) \frac{dz}{dt} \right] - D_z \left[(x, z) \frac{dx}{dt} \right],$$

and the period matrices of $L_1, \ldots, L_p$ at the original and general period loops be denoted respectively by -2η, $-2\eta'$ and $-[2\eta]$, $-[2\eta']$, we have, similarly, for the same values of $\alpha, \alpha', \beta, \beta'$,

$$[\eta] = \eta\alpha + \eta'\alpha', \quad [\eta'] = \eta\beta + \eta'\beta'.$$

We have used the notation Ω_P for the row of P quantities $2\omega P + 2\omega' P'$, where P, P' each denotes a row of p quantities; we extend this notation to the matrix $2\omega\alpha + 2\omega'\alpha'$, where α, α' each denotes a matrix of p rows and columns, and denote this matrix by Ω_α; similarly we denote the matrix $2\eta\alpha + 2\eta'\alpha'$ by H_α; then the four equations just obtained may be written

$$[2\omega] = \Omega_\alpha, \quad [2\omega'] = \Omega_\beta, \quad [2\eta] = H_\alpha, \quad [2\eta'] = H_\beta. \qquad \text{(I.)}$$

Noticing now that the matrices $[2\omega]$, $[2\omega']$, $[2\eta]$, $[2\eta']$ must satisfy the relations obtained in § 140, we have

$$\begin{aligned} \tfrac{1}{2}\pi i &= [\bar{\eta}][\omega'] - [\bar{\omega}][\eta'] = \tfrac{1}{4}(\bar{H}_\alpha \Omega_\beta - \bar{\Omega}_\alpha H_\beta) \\ &= (\bar{\alpha}\bar{\eta} + \bar{\alpha}'\bar{\eta}')(\omega\beta + \omega'\beta') - (\bar{\alpha}\bar{\omega} + \bar{\alpha}'\bar{\omega}')(\eta\beta + \eta'\beta') \\ &= \bar{\alpha}(\bar{\eta}\omega - \bar{\omega}\eta)\beta + \bar{\alpha}'(\bar{\eta}'\omega - \bar{\omega}'\eta)\beta + \bar{\alpha}(\bar{\eta}\omega' - \bar{\omega}\eta')\beta' + \bar{\alpha}'(\bar{\eta}'\omega' - \bar{\omega}'\eta')\beta' \\ &= (\bar{\alpha}\beta' - \bar{\alpha}'\beta)\tfrac{1}{2}\pi i, \end{aligned}$$

in virtue of the relations satisfied by the matrices 2ω, $2\omega'$, 2η, $2\eta'$; and similarly

$$0 = [\bar{\eta}][\omega] - [\bar{\omega}][\eta] = \tfrac{1}{4}(\bar{H}_\alpha \Omega_\alpha - \bar{\Omega}_\alpha H_\alpha) = (\bar{\alpha}\alpha' - \bar{\alpha}'\alpha)\tfrac{1}{2}\pi i,$$

and

$$0 = [\bar{\eta}'][\omega'] - [\bar{\omega}'][\eta'] = \tfrac{1}{4}(\bar{H}_\beta \Omega_\beta - \bar{\Omega}_\beta H_\beta) = (\bar{\beta}\beta' - \bar{\beta}'\beta)\tfrac{1}{2}\pi i;$$

thus we have

$$\bar{\alpha}\beta' - \bar{\alpha}'\beta = 1 = \bar{\beta}'\alpha - \bar{\beta}\alpha', \quad \bar{\alpha}\alpha' - \bar{\alpha}'\alpha = 0, \quad \bar{\beta}\beta' - \bar{\beta}'\beta = 0, \qquad \text{(II.)}$$

namely, the matrices α, β, α', β' satisfy relations precisely similar to those respectively satisfied by the matrices ω, ω', η, η', the $\tfrac{1}{2}\pi i$ which occurs for the latter case being, in the case of the matrices α, β, α', β', replaced by -1; therefore also, as in § 141, the relations satisfied by $\alpha, \beta, \alpha', \beta'$ can be given in the form

$$\alpha\bar{\beta}' - \beta\bar{\alpha}' = 1 = \beta'\bar{\alpha} - \alpha'\bar{\beta}, \quad \alpha\bar{\beta} - \beta\bar{\alpha} = 0, \quad \alpha'\bar{\beta}' - \beta'\bar{\alpha}' = 0. \qquad \text{(III.)}$$

In virtue of these equations, if

$$J = \begin{pmatrix} \alpha, & \beta \\ \alpha', & \beta' \end{pmatrix}$$

denote the matrix of $2p$ rows and columns formed with the elements of the matrices α, β, α', β', we have (cf., for notation, Appendix ii.)

$$\begin{pmatrix} \alpha, & \beta \\ \alpha', & \beta' \end{pmatrix} \begin{pmatrix} \bar{\beta}', & -\bar{\beta} \\ -\bar{\alpha}', & \bar{\alpha} \end{pmatrix} = \begin{pmatrix} \alpha\bar{\beta}' - \beta\bar{\alpha}', & \beta\bar{\alpha} - \alpha\bar{\beta} \\ \alpha'\bar{\beta}' - \beta'\bar{\alpha}', & \beta'\bar{\alpha} - \alpha'\bar{\beta} \end{pmatrix} = \begin{pmatrix} 1 & 0 \\ 0 & 1 \end{pmatrix},$$

and therefore

$$J^{-1}=\begin{pmatrix}\bar{\beta}', & -\bar{\beta}\\ -\bar{a}', & \bar{a}\end{pmatrix},$$

and the original periods can be expressed in terms of the general periods in the form

$$\omega=[\omega]\bar{\beta}'-[\omega']\bar{a}',\quad \omega'=-[\omega]\bar{\beta}+[\omega']\bar{a},$$
$$\eta=[\eta]\bar{\beta}'-[\eta']\bar{a}',\quad \eta'=-[\eta]\bar{\beta}+[\eta']\bar{a}.$$

If 0 denote the matrix of p rows and columns whereof every element is zero, and 1 denote the matrix of p rows and columns whereof every element is zero except those in the diagonal, which are all equal to 1, and if ϵ denote the matrix of $2p$ rows and columns given by

$$\epsilon=\begin{pmatrix}0, & -1\\ 1, & 0\end{pmatrix},\ \text{so that}\ \epsilon^2=\begin{pmatrix}-1, & 0\\ 0, & -1\end{pmatrix}=-1,$$

then it is immediately proved that the relations (II.), (III.) are respectively equivalent to the two equations

$$\bar{J}\epsilon J=\epsilon,\quad J\epsilon\bar{J}=\epsilon,$$

where

$$\bar{J}=\begin{pmatrix}\bar{a}, & \bar{a}'\\ \bar{\beta}, & \bar{\beta}'\end{pmatrix};$$

and it will be noticed that the equations (III.) are obtained from the equations (II.) by changing the elements of J into the corresponding elements of $\bar{J}$.

It follows* from the equation $\bar{J}\epsilon J=\epsilon$ that the determinant of the matrix J is equal to $+1$ or to -1. It will subsequently (§ 333) appear that the determinant is equal to $+1$.

Ex. Verify, for the case $p=2$, that the matrices

$$a=\begin{pmatrix}4, & -20\\ 4, & 1\end{pmatrix},\quad \beta=\begin{pmatrix}-29, & 124\\ -28, & -6\end{pmatrix},$$

$$a'=\begin{pmatrix}-3, & 20\\ -8, & -7\end{pmatrix},\quad \beta'=\begin{pmatrix}22, & -124\\ 56, & 43\end{pmatrix}$$

satisfy the conditions (III.) (Weber, *Crelle*, LXXIV. (1872), p. 72).

323. It is often convenient, simultaneously with the change of period loops which has been described, to make a linear transformation of the fundamental integrals of the first kind, $u_1, \ldots, u_p$. Suppose that we introduce, in place of $u_1, \ldots, u_p$, other p integrals $w_1, \ldots, w_p$, such that

$$u_i=M_{i,1}w_1+\ldots\ldots+M_{i,p}w_p,\qquad (i=1, 2, \ldots, p),$$

or, as we shall write it, $u=Mw$, M being a matrix whose elements are constants and of which the determinant is not zero. We enquire then what are the integrals of the second kind associated with $w_1, \ldots, w_p$. We have (§ 138) denoted $Du_i^{x,a}$ by $\mu_i(x)$, and the matrix of the quantities $\mu_i(c_j)$ by μ;

* For another proof of the relations (II.), (III.) of the text, the reader may compare Thomae, *Crelle*, LXXV. (1873), p. 224. A proof directly on the lines followed here may of course be constructed with the employment only of Riemann's normal elementary integrals of the first and second kind. Cf. § 142.

denote now, also, $Dw_i^{x,a}$ by $\rho_i(x)$, and the matrix of the quantities $\rho_i(c_j)$ by ρ; then we immediately find $\mu = \rho\overline{M}$, and the equation (§ 138)

$$L^{x,a} = \mu^{-1}H^{x,a} - 2au^{x,a}$$

gives

$$\overline{M}L^{x,a} = \rho^{-1}H^{x,a} - 2\overline{M}aMw^{x,a};$$

thus the integrals of the second kind associated with $w_1, \dots, w_p$ are the p integrals given by $\overline{M}L^{x,a}$, and, corresponding to the matrix a for the integrals $L_1^{x,a}, \dots, L_p^{x,a}$, we have, for the integrals $\overline{M}L^{x,a}$, the matrix $\mathrm{a} = \overline{M}aM$. If $2v$, $2v'$ denote the matrices of the periods of the integrals w, and -2ζ, $-2\zeta'$ denote the matrices of the periods of the integrals $\overline{M}L^{x,a}$, so that (§ 139)

$$\zeta = 2\mathrm{a}v, \quad \zeta' = 2\mathrm{a}v' - \tfrac{1}{2}\rho^{-1}\Delta,$$

we therefore have $\omega = Mv$, $\omega' = Mv'$ and

$$\zeta = 2\overline{M}aMv = \overline{M}\eta, \quad \zeta' = 2\overline{M}aMv' - \tfrac{1}{2}\overline{M}\mu^{-1}\Delta = \overline{M}\eta'; \qquad \text{(IV.)}$$

it is immediately apparent from these equations that the matrices v, v', ζ, ζ' satisfy the equations of § 140,

$$v\bar{v}' - v'\bar{v} = 0, \;\; \zeta\bar{\zeta}' - \zeta'\bar{\zeta} = 0, \;\; v'\bar{\zeta} - v\bar{\zeta}' = \tfrac{1}{2}\pi i = \zeta\bar{v}' - \zeta'\bar{v}.$$

324. The preceding Articles have sufficiently shewn how the equations of transformation of the periods arise by the consideration of the Abelian integrals. It is of importance to see that equations of the same character, but of more general significance, arise in connexion with the analytical theory of the theta functions.

Let ω, ω', η, η' be any four matrices of p rows and columns satisfying the conditions (i) that the determinant of ω does not vanish, (ii) that $\omega^{-1}\omega'$ is a symmetrical matrix, (iii) that the quadratic form $\omega^{-1}\omega' n^2$ has its imaginary part positive when $n_1, \dots, n_p$ are real, (iv) that $\eta\omega^{-1}$ is a symmetrical matrix, (v) that $\eta' = \eta\omega^{-1}\omega' - \tfrac{1}{2}\pi i\bar{\omega}^{-1}$. The conditions (i), (ii), (iv), (v) are equivalent to equations of the form of (B) and (C), § 140, and, taking matrices a, b, h such that $a = \tfrac{1}{2}\eta\omega^{-1}$, $h = \tfrac{1}{2}\pi i\omega^{-1}$, $b = \pi i\omega^{-1}\omega'$, or $2h\omega = \pi i$, $2h\omega' = b$, $\eta = 2a\omega$, $\eta' = 2a\omega' - \bar{h}$, the condition (iii) ensures the existence of the function defined by

$$\vartheta\left(u;\, {Q' \atop Q}\right) = \Sigma e^{au^2 + 2hu(n+Q') + b(n+Q')^2 + 2\pi i Q(n+Q')},$$

wherein Q, Q' are any constants (cf. § 174).

Introduce now two other matrices $[\omega]$, $[\omega']$, also of p rows and columns, defined by the equations

$$[\omega] = \omega\alpha + \omega'\alpha', = \tfrac{1}{2}\Omega_\alpha, \text{ say}, \quad [\omega'] = \omega\beta + \omega'\beta', = \tfrac{1}{2}\Omega_\beta, \text{ say},$$

where α, α', β, β', are matrices of p rows and columns whose elements are integers*, it being supposed† that the determinant of the matrix $[\omega]$ does not vanish; and introduce p other variables $w_1, \dots, w_p$ defined by

$$u_i = M_{i,1}w_1 + \dots\dots + M_{i,p}w_p, \qquad (i = 1, 2, \dots, p)$$

or $u = Mw$, where M is a matrix of constants, whose determinant does not vanish; let the simultaneous increments of $w_1, \dots, w_p$ when $u_1, \dots, u_p$ are simultaneously increased by the constituents of the j-th column of $[\omega]$ be denoted by $v_{1,j}, \dots, v_{p,j}$, and the simultaneous increments of $w_1, \dots, w_p$ when $u_1, \dots, u_p$ are simultaneously increased by the elements of the j-th column of $[\omega']$ be denoted by $v'_{1,j}, \dots, v'_{p,j}$; then we have the equations $2Mv = 2[\omega] = \Omega_\alpha$, $2Mv' = 2[\omega'] = \Omega_\beta$, where v, v' denote the matrices of which respectively the (i, j) elements are $v_{i,j}$ and $v'_{i,j}$.

The function $\vartheta\left(u;\, {Q' \atop Q}\right)$ is a function of $w_1, \dots, w_p$; we proceed now to investigate whether it is possible to choose the matrices $\alpha, \alpha', \beta, \beta'$ and the matrix M, so that the function may be regarded as a theta function in $w_1, \dots, w_p$ of order r (cf. Chap. XV. § 284).

Let the arguments $w_1, \dots, w_p$ be simultaneously increased by the constituents of the j-th column of the matrix $2v$; thereby $u_1, \dots, u_p$ will be increased by the constituents of the j-th column of the matrix $[2\omega]$, and, since $\alpha, \alpha', \beta, \beta'$ consist of integers, the function $\vartheta\left(u;\, {Q' \atop Q}\right)$ will (Chap. X. § 190) be multiplied by a factor e^{L_j} where

$$L_j = (H_\alpha)^{(j)}\left[u + \tfrac{1}{2}(\Omega_\alpha)^{(j)}\right] - \pi i\,(\alpha)^{(j)}(\alpha')^{(j)} + 2\pi i\left[(\alpha)^{(j)}Q' - (\alpha')^{(j)}Q\right],$$

$(\alpha)^{(j)}$ denoting the row of p elements forming the j-th column of the matrix α, and $(\Omega_\alpha)^{(j)}$, $(H_\alpha)^{(j)}$ denoting, similarly, the j-th columns of the matrices $2\omega\alpha + 2\omega'\alpha'$, $2\eta\alpha + 2\eta'\alpha'$ respectively; this expression L_j, is linear in $w_1, \dots, w_p$, and can be put into the form

$$L_j = r\,(2\zeta_{1,j}, \dots, 2\zeta_{p,j})\left[(w_1, \dots, w_p) + (v_{1,j}, \dots, v_{p,j})\right] + 2\pi i K_j',$$

where $(w_1, \dots, w_p)$ denotes the row letter whose elements are $w_1, \dots, w_p$, and similarly $(v_{1,j}, \dots, v_{p,j})$ is the row letter formed by the elements of the j-th column of the matrix v, r is a positive integer which is provisionally arbitrary, K_j' and $2\zeta_{1,j}, \dots, 2\zeta_{p,j}$ are properly chosen constants, and $(2\zeta_{1,j}, \dots, 2\zeta_{p,j})$ is the row letter formed of the last of these. Similarly, if the arguments $w_1, \dots, w_p$ be simultaneously increased by $2v'_{1,j}, \dots, 2v'_{p,j}$, the function $\vartheta\left(u;\, {Q' \atop Q}\right)$ takes a factor $e^{L'_j}$, where

$$L_j' = (H_\beta)^{(j)}\left[u + \tfrac{1}{2}(\Omega_\beta)^{(j)}\right] - \pi i\,(\beta)^{(j)}(\beta')^{(j)} + 2\pi i\left[(\beta)^{(j)}Q' - (\beta')^{(j)}Q\right],$$

and, with the same value of r, this can be put into the form

$$L_j' = r\,(2\zeta'_{1,j}, \dots\dots, 2\zeta'_{p,j})\left[(w_1, \dots\dots, w_p) + (v'_{1,j}, \dots\dots, v'_{p,j})\right] - 2\pi i K_j,$$

* The case when $\alpha, \alpha', \beta, \beta'$ are not integers is briefly considered in chapter XX.

† We have $\pi i \omega^{-1}[\omega] = \pi i a + b a'$; we suppose that the determinant of $\pi i a + b a'$ does not vanish.

where $K_j, \zeta'_{1,j}, \ldots, \zeta'_{p,j}$ are properly chosen constants. In these equations we suppose j to be taken in turn equal to 1, 2, ..., p.

Comparing the two forms of L_j we have

$$(H_\alpha)^{(j)}Mw, \text{ or } \bar{M}(H_\alpha)^{(j)}w, = r(2\zeta_{1,j}, \ldots, 2\zeta_{p,j})(w_1, \ldots, w_p),$$

so that the (i, j)th element of the matrix $\bar{M}H_\alpha$ is $2r\zeta_{i,j}$; hence if ζ, ζ' denote respectively the matrices of the quantities $\zeta_{i,j}$ and $\zeta'_{i,j}$, we have

$$\bar{M}H_\alpha = 2r\zeta, \ \bar{M}H_\beta = 2r\zeta'; \qquad \text{(V.)}$$

from these we deduce, in virtue of the equations $2Mv = \Omega_\alpha$, $2Mv' = \Omega_\beta$,

$$\tfrac{1}{2}\bar{H}_\alpha\Omega_\alpha = \tfrac{1}{2}\bar{H}_\alpha . 2Mv = 2r\bar{\zeta}v, \quad \tfrac{1}{2}\bar{H}_\beta\Omega_\beta = \tfrac{1}{2}\bar{H}_\beta . 2Mv' = 2r\bar{\zeta}'v',$$

and therefore, in particular, comparing the (j, j)th elements on the two sides of these equations,

$$\tfrac{1}{2}(H_\alpha)^{(j)}(\Omega_\alpha)^{(j)} = 2r(\zeta)^{(j)}(v)^{(j)}, \quad \tfrac{1}{2}(H_\beta)^{(j)}(\Omega_\beta)^{(j)} = 2r(\zeta')^{(j)}(v')^{(j)},$$

where, as before, $(v)^{(j)}$ is the row letter formed by the elements of the j-th column of the matrix v, etc.; therefore the only remaining conditions necessary for the identification of the two forms of L_j and L_j', are

$$K_j' = (\alpha)^{(j)}Q' - (\alpha')^{(j)}Q - \tfrac{1}{2}(\alpha)^{(j)}(\alpha')^{(j)}, \ -K_j = (\beta)^{(j)}Q' - (\beta')^{(j)}Q - \tfrac{1}{2}(\beta)^{(j)}(\beta')^{(j)},$$

and the p pairs of equations of this form are included in the two

$$K' = \bar{\alpha}Q' - \bar{\alpha}'Q - \tfrac{1}{2}d(\bar{\alpha}\alpha'), \ -K = \bar{\beta}Q' - \bar{\beta}'Q - \tfrac{1}{2}d(\bar{\beta}\beta'), \qquad \text{(VI.)}$$

where K', K are row letters of p elements and $d(\bar{\alpha}\alpha')$, $d(\bar{\beta}\beta')$ are respectively the row letters of p elements constituted by the diagonal elements of the matrices $\bar{\alpha}\alpha'$, $\bar{\beta}\beta'$.

The equations (VI.) arise by identifying the two forms of L_j and L_j'; it is effectively sufficient to identify the two forms of e^{L_j} and $e^{L_j'}$; thus it is sufficient to regard the equations (VI.) as *congruences*, to the modulus 1.

We now impose upon the matrices v, v', ζ, ζ' the conditions

$$\bar{\zeta}v - \bar{v}\zeta = 0 = \bar{\zeta}'v' - \bar{v}'\zeta', \ \bar{\zeta}v' - \bar{v}\zeta' = \tfrac{1}{2}\pi i, \qquad \text{(VII.)}$$

which, as will be proved immediately, are equivalent to certain conditions for the matrices $\alpha, \beta, \alpha', \beta'$; then, denoting $\vartheta\left(u; \begin{smallmatrix} Q' \\ Q \end{smallmatrix}\right)$ by $\phi(w_1, \ldots, w_p)$ or $\phi(w)$, it can be verified* that the $2p$ equations

$$\phi(\ldots, w_r + 2v_{r,j}, \ldots) = e^{L_j}\phi(w), \ \phi(\ldots, w_r + 2v'_{r,j}, \ldots) = e^{L_j'}\phi(w), \ (j = 1, \ldots, p),$$

where L_j, L_j' have the specified forms, lead to the equation

$$\phi(w + 2vm + 2v'm') = e^{r(2\zeta m + 2\zeta' m')(w + vm + v'm') - r\pi i m m' + 2\pi i (mK' - m'K)}\phi(w),$$

wherein m, m' are row letters consisting of any p integers; and this is the

* The verification is included in a more general piece of work which occurs in Chap. XIX.

characteristic equation for a theta function of order r with the associated constants $2v$, $2v'$, 2ζ, $2\zeta'$ (§ 284, p. 448).

The equations (VII.) are equivalent to conditions for the matrices v, v', ζ, ζ', entirely analogous to the conditions (ii), (iv), (v) of § 324 for the matrices ω, ω', η, η'. The condition analogous to (i) of § 324, namely that the determinant of the matrix v do not vanish, is involved in the hypothesis that the determinant of $\pi i\alpha + b\alpha'$ do not vanish. It will be proved below (§ 325) that the remaining condition involved in the definition of a theta function, viz. that the quadratic form $v^{-1}v'n^2$ has its imaginary part positive for real values of $n_1, \dots, n_p$, is a consequence of the corresponding condition for the matrices ω, ω'. We consider first the conditions for the equations (VII.).

In virtue of equations (V.), the equations (VII.) require

$$\bar{H}_\alpha \Omega_\beta - \bar{\Omega}_\alpha H_\beta = 2\bar{H}_\alpha M v' - 2\bar{v}\bar{M}H_\beta = 4r(\bar{\zeta}v' - \bar{v}\zeta') = 2r\pi i,$$

and, similarly,

$$\bar{H}_\alpha \Omega_\alpha - \bar{\Omega}_\alpha H_\alpha = 0, \quad \bar{H}_\beta \Omega_\beta - \bar{\Omega}_\beta H_\beta = 0;$$

but

$$\tfrac{1}{4}(\bar{H}_\alpha \Omega_\beta - \bar{\Omega}_\alpha H_\beta), \; = (\bar{\alpha}\bar{\eta} + \bar{\alpha}'\bar{\eta}')(\omega\beta + \omega'\beta') - (\bar{\alpha}\bar{\omega} + \bar{\alpha}'\bar{\omega}')(\eta\beta + \eta'\beta'),$$
$$= \bar{\alpha}(\bar{\eta}\omega - \bar{\omega}\eta)\beta + \bar{\alpha}(\bar{\eta}\omega' - \bar{\omega}\eta')\beta' + \bar{\alpha}'(\bar{\eta}'\omega - \bar{\omega}'\eta)\beta + \bar{\alpha}'(\bar{\eta}'\omega' - \bar{\omega}'\eta')\beta',$$

and this, by the equations (B), § 140, is equal to

$$\tfrac{1}{2}\pi i(\bar{\alpha}\beta' - \bar{\alpha}'\beta);$$

thus

$$\bar{\alpha}\beta' - \bar{\alpha}'\beta = \bar{\beta}'\alpha - \bar{\beta}\alpha' = r, \qquad \text{(VIII.)}$$

and, similarly,

$$\bar{\alpha}\alpha' - \bar{\alpha}'\alpha = 0, \quad \bar{\beta}\beta' - \bar{\beta}'\beta = 0;$$

and as before (§ 322) these three equations can be replaced by the three

$$\alpha\bar{\beta} = \beta\bar{\alpha}, \quad \alpha'\bar{\beta}' = \beta'\bar{\alpha}', \quad \alpha\bar{\beta}' - \beta\bar{\alpha}' = r = \beta'\bar{\alpha} - \alpha'\bar{\beta}, \qquad \text{(IX.)}$$

the relations satisfied by the matrices α, β, α', β' respectively being *similar to those satisfied by* ω, ω', η, η', *with the change of the* $\frac{1}{2}\pi i$, *which occurs in the latter case, into* $-r$.

The number r which occurs in these equations is called the order of the transformation; when it is equal to 1 the transformation is called a linear transformation.

Ex. i. Prove that, with matrices of $2p$ rows and $2p$ columns,

$$\begin{pmatrix} \alpha & \beta \\ \alpha' & \beta' \end{pmatrix}\begin{pmatrix} \bar{\beta}' & -\bar{\beta} \\ -\bar{\alpha}' & \bar{\alpha} \end{pmatrix} = r\begin{pmatrix} 1 & 0 \\ 0 & 1 \end{pmatrix} = \begin{pmatrix} \bar{\alpha} & \bar{\alpha}' \\ \bar{\beta} & \bar{\beta}' \end{pmatrix}\begin{pmatrix} \beta' & -\alpha' \\ -\beta & \alpha \end{pmatrix},$$

and

$$\begin{pmatrix} \alpha & \beta \\ \alpha' & \beta' \end{pmatrix}\begin{pmatrix} 0 & -1 \\ 1 & 0 \end{pmatrix}\begin{pmatrix} \bar{\alpha} & \bar{\alpha}' \\ \bar{\beta} & \bar{\beta}' \end{pmatrix} = r\begin{pmatrix} 0 & -1 \\ 1 & 0 \end{pmatrix}.$$

The determinant of the matrix will be subsequently proved to be $+r^p$.

Ex. ii. Prove that the equations (V.) of § 324 are equivalent to

$$\begin{pmatrix} M & 0 \\ 0 & r\bar{M}^{-1} \end{pmatrix} \begin{pmatrix} 2\upsilon & 2\upsilon' \\ 2\zeta & 2\zeta' \end{pmatrix} = \begin{pmatrix} 2\omega & 2\omega' \\ 2\eta & 2\eta' \end{pmatrix} \begin{pmatrix} \alpha & \beta \\ \alpha' & \beta' \end{pmatrix}.$$

Ex. iii. If x, y, x_1, y_1 be any row letters of p elements, and X, Y, X_1, Y_1 be other such row letters, such that

$$(X, Y) = \begin{pmatrix} \alpha & \beta \\ \alpha' & \beta' \end{pmatrix} (x, y), \text{ or } \begin{matrix} X = \alpha x + \beta y, & X_1 = \alpha x_1 + \beta y_1, \\ Y = \alpha' x + \beta' y, & Y_1 = \alpha' x_1 + \beta' y_1, \end{matrix}$$

then the equations (VIII.) are the conditions for the self-transformation of the bilinear form $xy_1 - x_1y$, which is expressed by the equation

$$XY_1 - X_1Y = r(xy_1 - x_1y).$$

325. Conversely when the matrices α, α', β, β' satisfy the equations (VIII.), the function $\vartheta\left(u\,;\,{Q' \atop Q}\right)$ satisfies the determining equation for a theta function in $w_1, \dots, w_p$, of order r, with the characteristic (K, K'), and with the associated constants 2υ, $2\upsilon'$, 2ζ, $2\zeta'$; and in virtue of the equations (VII.), the determinant of υ not vanishing, matrices a, b, h, of which the first two are symmetrical, can be taken such that

$$\text{a} = \tfrac{1}{2}\zeta\upsilon^{-1}, \quad \text{h} = \tfrac{1}{2}\pi i\upsilon^{-1}, \quad \text{b} = \pi i\upsilon^{-1}\upsilon';$$

we proceed now to shew* that the real part of the quadratic form $\text{b}n^2$ is negative for real values of $n_1, \dots, n_p$, r being positive, as was supposed.

The quantity, or matrix, obtainable from any complex quantity, or matrix of complex quantities, by changing the sign of the imaginary part of that quantity, or of the imaginary parts of every constituent of that matrix, will be denoted by the suffix 0; and a similar notation will be used for row letters; further the symmetrical matrices $\omega^{-1}\omega'$, $\upsilon^{-1}\upsilon'$ will be denoted respectively by τ and τ', so that $b = \pi i\tau$, $\text{b} = \pi i\tau'$; also τ, τ' will be written, respectively, in the forms $\tau_1 + i\tau_2$, $\tau_1' + i\tau_2'$, where τ_1, τ_2, τ_1', τ_2' are matrices of real quantities. Then, putting

$$x' = \bar{\upsilon}\bar{M}\bar{\omega}^{-1}x, \text{ and therefore } x_0' = \bar{\upsilon}_0\bar{M}_0\bar{\omega}_0^{-1}x_0,$$

where x', x denote rows of p complex quantities, and x_0', x_0 the rows of the corresponding conjugate complex quantities, and recalling that

$$\tau' = \bar{\tau}' = \bar{\upsilon}'\bar{\upsilon}^{-1}, \quad \omega^{-1}M\upsilon = \alpha + \tau\alpha', \quad \omega^{-1}M\upsilon' = \beta + \tau\beta',$$

we have

$$\tau' x' x_0' = \tau'\bar{\upsilon}\bar{M}\bar{\omega}^{-1}x \,.\, \bar{\upsilon}_0\bar{M}_0\bar{\omega}_0^{-1}x_0 = \bar{\upsilon}'\bar{M}\bar{\omega}^{-1}x \,.\, \bar{\upsilon}_0\bar{M}_0\bar{\omega}_0^{-1}x_0$$
$$= (\bar{\beta} + \bar{\beta}'\tau)\,x \,.\, (\bar{\alpha} + \bar{\alpha}'\tau_0)\,x_0;$$

and, if $x = x_1 + ix_2$, $x_0 = x_1 - ix_2$, where x_1, x_2 are real, this is equal to

$$(\bar{\beta} + \bar{\beta}'\tau_1 + i\bar{\beta}'\tau_2)(x_1 + ix_2) \,.\, (\bar{\alpha} + \bar{\alpha}'\tau_1 - i\bar{\alpha}'\tau_2)(x_1 - ix_2)$$

or

$$[\bar{\beta}P + \bar{\beta}'P' + i(\bar{\beta}Q + \bar{\beta}'Q')]\,[\bar{\alpha}P + \bar{\alpha}'P' - i(\bar{\alpha}Q + \bar{\alpha}'Q')],$$

* Hermite, *Compt. Rendus*, XL. (1855), Weber, *Ann. d. Mat.*, Ser. 2, t. ix. (1878—9).

where P, P', Q, Q' are row letters of p real quantities given by

$$P = x_1,\ P' = \tau_1 x_1 - \tau_2 x_2,\ Q = x_2,\ Q' = \tau_1 x_2 + \tau_2 x_1,$$

so that

$$PQ' - P'Q = \tau_2 (x_1^2 + x_2^2);$$

thus the coefficient of i in $\tau' x' x_0'$ is

$$(\bar{\alpha} P + \bar{\alpha}' P')(\bar{\beta} Q + \bar{\beta}' Q') - (\bar{\beta} P + \bar{\beta}' P')(\bar{\alpha} Q + \bar{\alpha}' Q'),$$

which, in virtue of the equations (IX.), is equal to $r(PQ' - P'Q)$ or $r\tau_2 (x_1^2 + x_2^2)$; thus the coefficient of i in $\tau' x' x_0'$ is equal to the coefficient of i in $r\tau x x_0$. Since x' may be regarded as arbitrarily assigned this proves that the imaginary part of $\tau' x' x_0$ is necessarily positive; and this includes the proposition we desired to establish.

Ex. Prove that the equation obtained is equivalent to

$$M_0 v_0 \tau_2' \bar{v} \bar{M} = r \omega_0 \tau_2 \bar{\omega}.$$

326. Of the general formulae thus obtained for the transformation of theta functions, the case of a linear transformation, for which $r = 1$, is of great importance; and we limit ourselves mainly to that case in the following parts of this chapter. We have shewn that a theta function of the first order, with assigned characteristic and associated constants, is unique, save for a factor independent of the argument; we have therefore, for $r = 1$, as a result of the theory here given, the equation

$$\vartheta\left(u;\ 2\omega, 2\omega', 2\eta, 2\eta';\ {Q' \atop Q}\right) = A\vartheta\left(w;\ 2v, 2v', 2\zeta, 2\zeta';\ {K' \atop K}\right).$$

We suppose α, β, α', β' to be any arbitrarily assigned matrices of integers satisfying the equations (VIII.) or (IX.); then there remains a certain redundancy of disposable quantities; we may for instance suppose ω, ω', η, η' and M to be given, and choose v, v', ζ, ζ' in accordance with these equations; or we may suppose ω, ω', v, ζ and ζ' to be prescribed and use these equations to determine M, v', η and η'. It is convenient to specify the results in two cases. We replace u, w respectively by U, W.

(i) $2\omega = 1,\ 2\omega' = \tau,\ \eta = a,\ \eta' = a\tau - \pi i,\ h = \pi i,\ b = \pi i \tau,$
$2v = 1,\ 2v' = \tau',\ \zeta = 0,\ \zeta' = -\pi i\ ,\ \mathrm{a} = 0\ ,\ \mathrm{h} = \pi i\ ,\ \mathrm{b} = \pi i \tau',$
$U = MW,\ M = \alpha + \tau\alpha',\ (\alpha + \tau\alpha')\tau' = \beta + \tau\beta',$

so that, as immediately follows from equations (IX.),

$$(\alpha + \tau\alpha')(\bar{\beta}' - \tau'\bar{\alpha}') = r = (\beta' - \alpha'\tau)(\bar{\alpha} + \bar{\alpha}'\tau),\ U = (\alpha + \tau\alpha')W,\ W = \frac{1}{r}(\bar{\beta}' - \tau'\bar{\alpha}')U,$$

and, because $\eta' = \eta\tau - \pi i$ and $\zeta = 0$,

$$a = \eta = \pi i \alpha' (\alpha + \tau\alpha')^{-1} = \frac{\pi i}{r} \alpha' (\bar{\beta}' - \tau'\bar{\alpha}'),$$

from which we get

$$aU^2 = \frac{\pi i}{r}\,\alpha'(\bar{\beta}' - \tau'\bar{\alpha}')\,U^2 = \pi i\alpha' WU = \pi i\bar{\alpha}'(\alpha + \tau\alpha')\,W^2.$$

These equations satisfy the necessary conditions, and lead, when $r = 1$, to

$$e^{\pi i\bar{\alpha}'(\alpha+\tau\alpha')W^2}\,\Theta\left(U;\tau;\begin{smallmatrix}Q'\\Q\end{smallmatrix}\right) = A\Theta\left(W;\tau';\begin{smallmatrix}K'\\K\end{smallmatrix}\right), \qquad \text{(X.)}$$

where A is independent of $U_1, \ldots, U_p$, and the characteristic (K, K') is determined from (Q, Q') by the equations (§ 324)

$$K' = \bar{\alpha}Q' - \bar{\alpha}'Q - \tfrac{1}{2}d(\bar{\alpha}\alpha'), \quad -K = \bar{\beta}Q' - \bar{\beta}'Q - \tfrac{1}{2}d(\bar{\beta}\beta').$$

The appearance of the exponential factor outside the Θ-function, in equation (X.), would of itself be sufficient reason for using, as we have done, the ϑ-function, in place of the Θ-function, in all general algebraic investigations*.

If in § 324 we put

$$u = 2\omega U, \quad \tau = \omega^{-1}\omega', \quad w = 2v W, \quad \tau' = v^{-1}v'$$

we easily find

$$\pi i\bar{a}'(a + \tau a')\,W^2 = \tfrac{1}{2}\eta\omega^{-1}u^2 - \tfrac{1}{2}r\zeta v^{-1}w^2;$$

thus (§ 189, p. 283) equation (X.) includes the initial equation of this Article.

In general the function occurring on the left side of equation (X.) is a theta function in W of order r with associated constants $2v = 1$, $2v' = \tau'$, $2\zeta = 0$, $2\zeta' = -2\pi i$, and characteristic (K, K').

(ii) A particular case of (i), when the matrix α' consists of zeros, is given by the formulae

$$2\omega = 1,\ 2\omega' = \tau,\ \eta = 0,\ \eta' = -\pi i,\ a = 0,\ h = \pi i,\ b = \pi i\tau,$$
$$2v = 1,\ 2v' = \tau',\ \zeta = 0,\ \zeta' = -\pi i,\ \mathrm{a} = 0,\ \mathrm{h} = \pi i,\ \mathrm{b} = \pi i\tau',$$

$$U = \alpha W,\ \tau' = \alpha^{-1}(\beta + \tau\beta'),\ \tau = \frac{1}{r}(\alpha\tau' - \beta)\,\bar{\alpha},$$

$$\begin{pmatrix}\alpha & \beta\\ \alpha' & \beta'\end{pmatrix} = \begin{pmatrix}\alpha & \beta\\ 0 & r\bar{\alpha}^{-1}\end{pmatrix}, \text{ where } \alpha\bar{\beta} = \beta\bar{\alpha}.$$

Then the function $\Theta\left(U;\tau;\begin{smallmatrix}Q'\\Q\end{smallmatrix}\right)$ or $\Theta\left[\alpha W;\frac{1}{r}(\alpha\tau' - \beta)\,\bar{\alpha};\begin{smallmatrix}Q'\\Q\end{smallmatrix}\right]$ is a theta function in W, of order r, with associated constants $2v = 1$, $2v' = \tau'$, $2\zeta = 0$, $2\zeta' = -2\pi i$, and characteristic (K, K') given by

$$K' = \bar{\alpha}Q', \ -K = \bar{\beta}Q' - r\alpha^{-1}Q - \tfrac{1}{2}d(r\bar{\beta}\bar{\alpha}^{-1}),$$

and, in particular, when $r = 1$ we have

$$\Theta\left(U;\tau;\begin{smallmatrix}Q'\\Q\end{smallmatrix}\right) = A\Theta\left(W;\tau';\begin{smallmatrix}K'\\K\end{smallmatrix}\right), \qquad \text{(XI.)}$$

where A is independent of $U_1, \ldots, U_p$.

* Cf. § 189 (Chap. X.); and for the case $p=1$, Cayley, *Liouville*, x. (1845), or *Collected Works*, Vol. I., p. 156 (1889).

327. It is clear that the results just obtained, for the linear transformation of theta functions, contain the answer to the enquiry as to the changes in the Riemann theta functions which arise in virtue of a change in the fundamental system of period loops. Before considering the results in further detail, it is desirable to be in possession of certain results as to the transformation of the characteristics of the theta function, which we now give; the reader who desires may omit the demonstrations, noticing only the results, and proceed at once to § 332. We retain the general value r for the order of the transformation, though the applications of greatest importance are those for which $r=1$.

As before let $d(\gamma)$ denote the row of p quantities constituted by the diagonal elements of any matrix γ of p rows and columns; in all cases here arising γ is a symmetrical matrix; then we have

$$\left.\begin{array}{ll}\alpha d(\bar{\beta}\beta')+\beta d(\bar{\alpha}\alpha')\equiv rd(\alpha\bar{\beta}), & \bar{\beta}'d(\alpha\bar{\beta})+\bar{\beta}d(\alpha'\bar{\beta}')\equiv rd(\bar{\beta}\beta') \\ \alpha'd(\bar{\beta}\beta')+\beta'd(\bar{\alpha}\alpha')\equiv rd(\alpha'\bar{\beta}'), & \bar{\alpha}'d(\alpha\bar{\beta})+\bar{\alpha}d(\alpha'\bar{\beta}')\equiv rd(\bar{\alpha}\alpha')\end{array}\right\} \pmod{2}$$

and

$$\left.\begin{array}{l}d(\bar{\alpha}\alpha')\,d(\bar{\beta}\beta')\equiv(r+1)\,\Sigma d(\bar{\beta}\alpha')\equiv(r+1)\,\Sigma d(\bar{\beta}'\alpha) \\ d(\alpha\bar{\beta})\,d(\alpha'\bar{\beta}')\equiv(r+1)\,\Sigma d(\alpha\bar{\beta}')\equiv(r+1)\,\Sigma d(\beta\bar{\alpha}')\end{array}\right\} \pmod{2},$$

so that, when $r=1$ or is any odd integer,

$$d(\bar{\alpha}\alpha')\,.\,d(\bar{\beta}\beta')\equiv d(\alpha\bar{\beta})\,.\,d(\alpha'\bar{\beta}')\equiv 0 \pmod{2}.$$

The last result contains the statement that the linear transformation of the zero theta-characteristic is always an even characteristic.

For the equations

$$\beta'\bar{a}-a'\bar{\beta}=r,\quad a\bar{\beta}=\beta\bar{a},$$

give

$$a\bar{\beta}\beta'\bar{a}-\beta\bar{a}a'\bar{\beta}=ra\bar{\beta},$$

and therefore

$$\bar{\beta}\beta'z^2-\bar{a}a'y^2=ra\bar{\beta}x^2,$$

where x is any row letter of p integers, and $z=\bar{a}x$, $y=\bar{\beta}x$; but if γ be a symmetrical matrix of integers and t be any row letter of p integers γt^2, $=\gamma_{11}t_1^2+\dots+2\gamma_{12}t_1t_2+\dots$, is $\equiv\gamma_{11}t_1^2+\dots+\gamma_{pp}t_p^2$, and therefore $\equiv\gamma_{11}t_1+\dots+\gamma_{pp}t_p$, or $\equiv d(\gamma)\,.\,t$, for modulus 2; hence

$$d(\bar{\beta}\beta')\,z-d(\bar{a}a')\,y\equiv rd(a\bar{\beta})\,x \pmod{2}$$

or

$$[ad(\bar{\beta}\beta')+\beta d(\bar{a}a')-rd(a\bar{\beta})]\,x\equiv 0 \pmod{2};$$

and as this is true for any row letter of integers, x, the first of the given equations follows at once. The second of the equations also follows from $\beta'\bar{a}-a'\bar{\beta}=r$, in the same way, and the third and fourth follow similarly from $\bar{\beta}'a-\bar{\beta}a'=r$.

To prove the fifth equation, we have, since $\beta'\bar{a}-a'\bar{\beta}=r$,

$$\bar{\beta}\beta'\bar{a}a'=\bar{\beta}a'\bar{\beta}a'+r\bar{\beta}a'$$

or

$$ba=c^2+rc,$$

where $b=\bar{\beta}\beta'$, $a=\bar{a}a'$, $c=\bar{\beta}a'$; hence, equating the sums of the diagonal elements on the two sides of the equation, we have

$$\sum_{j=1}^{p}\sum_{i=1}^{p} b_{i,j}a_{j,i}=\sum_{j=1}^{p}\sum_{i=1}^{p} c_{i,j}c_{j,i}+r\sum_{i=1}^{p} c_{i,i};$$

therefore, as, unless $i=j$, $b_{i,j}\,a_{j,i}=b_{j,i}\,a_{i,j}$, because a, b are symmetrical matrices, and as

$$c_{i,j}\,c_{j,i}=c_{j,i}\,c_{i,j},$$

we obtain

$$\sum_{i=1}^{p} a_{i,i}\,b_{i,i}\equiv\sum_{i=1}^{p}(c^2_{i,i}+rc_{i,i})\equiv(r+1)\sum_{i=1}^{p} c_{i,i}.$$

The sixth equation is obtained in a similar way, starting from $\bar{\beta}'a-\bar{\beta}a'=r$.

Of the results thus derived we make, now, application to the case when r is odd, limiting ourselves to the case when the characteristic (Q, Q') consists of half-integers; we put then $Q=\frac{1}{2}q$, $Q'=\frac{1}{2}q'$, so that q, q' each consist of p integers; then K, K' are also half-integers, respectively equal to $\frac{1}{2}k$, $\frac{1}{2}k'$, say, where

$$k'=\bar{a}q'-\bar{a}'q-d(\bar{a}a'),\quad -k=\bar{\beta}q'-\bar{\beta}'q-d(\bar{\beta}\beta').$$

In most cases of these formulae, it is convenient to regard them as congruences, to modulus 2. This is equivalent to neglecting additive *integral* characteristics.

From these equations we derive immediately, in virtue of the equations of the present Article

$$q\equiv ak+\beta k'+d(a\bar{\beta}),\quad q'\equiv a'k+\beta'k'+d(a'\bar{\beta}')\quad (\text{mod. } 2)$$

and

$$qq'\equiv kk'\quad (\text{mod. } 2).$$

Further if μ, μ' be row letters of p integers, and

$$\nu'=\bar{a}\mu'-\bar{a}'\mu-d(\bar{a}a'),\quad -\nu=\bar{\beta}\mu'-\bar{\beta}'\mu-d(\bar{\beta}\beta'),$$

we find, also in virtue of the equations of the present Article,

$$k\nu'-k'\nu\equiv q\mu'-q'\mu+(\mu'+q')\,d(a\bar{\beta})+(\mu+q)\,d(a'\bar{\beta}'),\quad (\text{mod. } 2);$$

therefore, if also

$$\sigma'=\bar{a}\rho'-\bar{a}'\rho-d(\bar{a}a'),\quad -\sigma=\bar{\beta}\rho'-\bar{\beta}'\rho-d(\bar{\beta}\beta'),$$

we have

$$k\nu'-k'\nu+\nu\sigma'-\nu'\sigma+\sigma k'-\sigma'k\equiv q\mu'-q'\mu+\mu\rho'-\mu'\rho+\rho q'-\rho'q\quad (\text{mod. } 2).$$

Denoting the half-integer characteristics $\frac{1}{2}\begin{pmatrix}q'\\q\end{pmatrix}$, $\frac{1}{2}\begin{pmatrix}\mu'\\\mu\end{pmatrix}$, $\frac{1}{2}\begin{pmatrix}\rho'\\\rho\end{pmatrix}$ by A, B, C, and the characteristics $\frac{1}{2}\begin{pmatrix}k'\\k\end{pmatrix}$, $\frac{1}{2}\begin{pmatrix}\nu'\\\nu\end{pmatrix}$, $\frac{1}{2}\begin{pmatrix}\sigma'\\\sigma\end{pmatrix}$, which we call *the transformed characteristics*, by A', B', C', we have therefore the results (§ 294)

$$|A|\equiv|A'|,\quad |A, B, C|\equiv|A', B', C'|,\quad (\text{mod. } 2)$$

or, in words, *in a linear transformation of a theta function with half-integer characteristic, and in any transformation of odd order, an odd (or even) characteristic transforms into an odd (or even) characteristic, and three syzygetic (or azygetic) characteristics transform into three syzygetic (or azygetic) characteristics.*

Of these the first result is immediately obvious when $r=1$ from the equation of transformation (§ 326), by changing w into $-w$.

Hence also it is obvious that if A be an even characteristic for which $\vartheta(0; A)$ vanishes, then the transformed characteristic A' is also an even characteristic for which the transformed function $\underline{\vartheta}(0; A')$ vanishes.

328. If in the formula of linear transformation of theta functions with half-integer characteristic, which we may write

$$\vartheta\left[u;\ \tfrac{1}{2}\begin{pmatrix} q' \\ q \end{pmatrix}\right] = A\underline{\vartheta}\left[w;\ \tfrac{1}{2}\begin{pmatrix} k' \\ k \end{pmatrix}\right],$$

we replace u by $u + \frac{1}{2}\Omega_m = u + \omega m + \omega' m'$, where m, m' denote rows of integers, and, therefore, since $\omega = M(v\bar{\beta}' - v'\bar{\alpha}')$, $\omega' = M(-v\bar{\beta} + v'\bar{\alpha})$, (cf. Ex. i., § 324), replace w by $w + vn + v'n'$, where

$$n' = \bar{\alpha}m' - \bar{\alpha}'m, \quad -n = \bar{\beta}m' - \bar{\beta}'m,$$

we obtain (§ 189, formula (L))

$$\vartheta\left[u;\ \tfrac{1}{2}\begin{pmatrix} q'+m' \\ q+m \end{pmatrix}\right] = A'\underline{\vartheta}\left[w;\ \tfrac{1}{2}\begin{pmatrix} k'+n' \\ k+n \end{pmatrix}\right],$$

where A' is independent of $u_1, \dots, u_p$, and $k'+n'$, $k+n$ are obtainable from $q'+m'$, $q+m$ by the same formulae whereby k', k are obtained from q', q, namely

$$k' + m' = \bar{\alpha}(q'+m') - \bar{\alpha}'(q+m) - d(\bar{\alpha}\alpha'),$$
$$-(k+m) = \bar{\beta}(q'+m') - \bar{\beta}'(q+m) - d(\bar{\beta}\beta');$$

these formulae are different from those whereby n', n are obtained from m', m; for this reason it is sometimes convenient to speak of $\frac{1}{2}\begin{pmatrix} q' \\ q \end{pmatrix}$ as a *theta characteristic*, and of $\frac{1}{2}\begin{pmatrix} m' \\ m \end{pmatrix}$ as a *period characteristic;* as it arises here the difference lies in the formulae of transformation; but other differences will appear subsequently; these differences are mainly consequences of the obvious fact that, when half-integer characteristics which differ by integer characteristics are regarded as identical, the sum of any odd number of theta characteristics is transformed as a theta characteristic, while the sum of any even number of theta characteristics is transformed as a period characteristic. In other words, a period characteristic is to be regarded as the (sum or) difference of two theta characteristics.

It will appear for instance that the characteristics associated in §§ 244, 245, Chap. XIII. with radical functions of the form $\sqrt{X^{(2\nu+1)}}$ are to be regarded as theta characteristics—and the characteristics associated in § 245 with radical functions of the form $\sqrt{X^{(2\mu)}}$, which are defined as sums of characteristics associated with functions $\sqrt{X^{(2\nu+1)}}$, are to be regarded as period characteristics.

We may regard the distinction* thus explained somewhat differently, by taking as the fundamental formula of linear transformation that which expresses $\vartheta\left[u\,;\ \tfrac{1}{2}\begin{pmatrix}q'\\q\end{pmatrix}\right]$ in terms of $\vartheta\left[w+\tfrac{1}{2}\Omega_r\,;\ \tfrac{1}{2}\begin{pmatrix}l'\\l\end{pmatrix}\right]$, where

$$r'=d\,(\bar{a}a'),\quad r=d\,(\bar{\beta}\beta'),$$

and

$$l'=k'+d\,(\bar{a}a')=\bar{a}q'-\bar{a}'q,\quad -l=-k+d\,(\bar{\beta}\beta')=\bar{\beta}q'-\bar{\beta}'q.$$

In the following pages we shall always understand by 'characteristic,' a theta characteristic; when it is necessary to call attention to the fact that a characteristic is a period characteristic this will be done.

329. It is clear that the formula of linear transformation of a theta function with any half-integer characteristic is obtainable from the particular case

$$\vartheta\,(u)=A\underline{\vartheta}\left[w\,;\ \tfrac{1}{2}\begin{pmatrix}r'\\r\end{pmatrix}\right],$$

where $r'=d\,(\bar{\alpha}\alpha')$, $r=d\,(\bar{\beta}\beta')$, by the addition of half periods to the arguments. It is therefore of interest to shew that matrices α, β, α', β' can be chosen, satisfying the equations

$$\alpha\bar{\beta}=\beta\bar{\alpha},\quad \alpha'\bar{\beta}'=\beta'\bar{\alpha}',\quad \alpha\bar{\beta}'-\beta\bar{\alpha}'=1,$$

which will make the characteristic $\tfrac{1}{2}\begin{pmatrix}r'\\r\end{pmatrix}$ equal to any even half-integer characteristic.

Any even half-integer characteristic, being denoted by

$$\tfrac{1}{2}\begin{pmatrix}k_1'\ldots k_p'\\k_1\ldots k_p\end{pmatrix},$$

we may, momentarily, call $\begin{pmatrix}k_i'\\k_i\end{pmatrix}$ the i-th column of the characteristic; then the columns may be of four sorts,

$$\begin{pmatrix}0\\0\end{pmatrix},\ \begin{pmatrix}1\\0\end{pmatrix},\ \begin{pmatrix}0\\1\end{pmatrix},\ \begin{pmatrix}1\\1\end{pmatrix},$$

but the number of columns of the last sort must be even; we build now a matrix

$$\begin{pmatrix}\alpha & \beta\\ \alpha' & \beta'\end{pmatrix}$$

* Theta characteristics have also been named eigentliche Charakteristiken and Primcharakteristiken; they consist of $2^{p-1}(2^p-1)$ odd and $2^{p-1}(2^p+1)$ even characteristics. The period characteristics have been called Gruppencharakteristiken and Elementarcharakteristiken or sometimes relative Charakteristiken. For them the distinction of odd and even is unimportant—while the distinction between the zero characteristic—which cannot be written as the sum of two different theta characteristics—and the remaining $2^{2p}-1$ characteristics, is of great importance. The distinction between theta characteristics and period characteristics has been insisted on by Noether, in connection with the theory of radical forms—Cf. Noether, *Math. Annal.* XXVIII. (1887), p. 373, Klein, *Math. Annal.* XXXVI. (1890), p. 36, Schottky, *Crelle*, CII. (1888), p. 308. The distinction is in fact observed in the *Abel'sche Functionen* of Clebsch and Gordan, in the manner indicated in the text.

of $2p$ rows and columns by the following rule*—Corresponding to a column of the characteristic of the first sort, say the i-th column, we take $\alpha_{i,i}=\beta'_{i,i}=1$, but take every other element of the i-th row and i-th column of α and β', and every element of the i-th row and i-th column of β and α' to be zero; corresponding to a column of the characteristic of the second sort, say the j-th column, we take $\alpha_{j,j}=\beta'_{j,j}=\alpha'_{j,j}=1$, but take every other element of the j-th row and j-th column of α, β', α', and every element of the j-th row and column of β, to be zero; corresponding to a column of the characteristic of the third sort, say the m-th column, we take $\alpha_{m,m}=\beta_{m,m}=\beta'_{m,m}=1$, but take every other element of the m-th row and column of α, β, β' and every element of the m-th row and column of α' to be zero; corresponding to a pair of columns of the characteristic of the fourth sort, say the ρ-th and σ-th, we take $\alpha_{\rho,\rho}=\beta_{\rho,\rho}=\beta'_{\rho,\rho}=1$, $\alpha_{\sigma,\sigma}=\alpha'_{\sigma,\sigma}=\beta'_{\sigma,\sigma}=1$, $\alpha_{\sigma,\rho}=1$, $\beta_{\rho,\sigma}=-1$, $\alpha'_{\sigma,\rho}=1$, $\beta'_{\rho,\sigma}=-1$, and take every other element of the ρ-th row and column and of the σ-th row and column, of each of the four matrices α, α', β, β', to be zero. Then it can be shewn that the matrix thus obtained satisfies all the necessary conditions and gives $k'=d(\bar{\alpha}\alpha')$, $k=d(\bar{\beta}\beta')$.

Consider for instance the case $p=5$, and the characteristic

$$\tfrac{1}{2}\begin{pmatrix} 0 & 1 & 0 & 1 & 1 \\ 0 & 0 & 1 & 1 & 1 \end{pmatrix};$$

the matrix formed by the rules from this characteristic is

$$\begin{array}{ccccc|ccccc}
1 & 0 & 0 & 0 & 0 & 0 & 0 & 0 & 0 & 0 \\
0 & 1 & 0 & 0 & 0 & 0 & 0 & 0 & 0 & 0 \\
0 & 0 & 1 & 0 & 0 & 0 & 0 & 1 & 0 & 0 \\
0 & 0 & 0 & 1 & 0 & 0 & 0 & 0 & 1 & -1 \\
0 & 0 & 0 & 1 & 1 & 0 & 0 & 0 & 0 & 0 \\
\hline
0 & 0 & 0 & 0 & 0 & 1 & 0 & 0 & 0 & 0 \\
0 & 1 & 0 & 0 & 0 & 0 & 1 & 0 & 0 & 0 \\
0 & 0 & 0 & 0 & 0 & 0 & 0 & 1 & 0 & 0 \\
0 & 0 & 0 & 0 & 0 & 0 & 0 & 0 & 1 & -1 \\
0 & 0 & 0 & 1 & 1 & 0 & 0 & 0 & 0 & 1
\end{array}$$

and it is immediately verified that this satisfies the equations for a linear transformation (§ 324 (IX.), for $r=1$), and gives, for the diagonal elements of $\bar{a}a'$, $\bar{\beta}\beta'$, respectively, the elements 01011 and 00111.

Since we can transform the zero characteristic into any even characteristic, we can of course transform any even characteristic into the zero characteristic; for instance, when there is an even theta function which vanishes for zero values of the arguments, we can, by making a linear transformation, take for this function the theta function with zero characteristic.

* Clebsch and Gordan, *Abel. Fctnen* (Leipzig, 1866), p. 318.

Ex. For the hyperelliptic case, when $p=3$, the period loops being taken as in § 200, the theta-function whose characteristic is $\frac{1}{2}\begin{pmatrix}1 & 0 & 1\\ 1 & 1 & 1\end{pmatrix}$ vanishes for zero arguments (§ 203); prove that the transformation given by

$$a=\begin{vmatrix} 1 & 0 & 0 \\ 0 & 1 & 0 \\ -1 & 0 & 1 \end{vmatrix}, \quad \beta=\begin{vmatrix} -1 & 0 & 0 \\ 0 & -1 & 0 \\ 1 & 0 & 0 \end{vmatrix}, \quad a'=\begin{vmatrix} 0 & 0 & -1 \\ 0 & 0 & 0 \\ 0 & 0 & -1 \end{vmatrix}, \quad \beta'=\begin{vmatrix} 1 & 0 & 1 \\ 0 & 1 & 0 \\ 0 & 0 & 1 \end{vmatrix},$$

is a linear transformation and gives an equation of the form

$$\vartheta\left[u;\ \tfrac{1}{2}\begin{pmatrix}1 & 0 & 1\\ 1 & 1 & 1\end{pmatrix}\right]=A\vartheta[w;\ 0],$$

where A is independent of $u_1, \ldots, u_p$.

330. We have proved (§ 327) that if three half-integer theta characteristics be syzygetic (or azygetic) the characteristics arising from them by any linear transformation are also syzygetic (or azygetic). It follows therefore that a Göpel system of 2^r characteristics, syzygetic in threes (§ 297, Chap. XVII.), transforms into such a Göpel system. Also the $2^{2\sigma}$ Göpel systems of § 298, having a definite character, that of being all odd or all even, transform into systems having the same character. And the $2\sigma+1$ fundamental Göpel systems (§ 300), which satisfy the condition that any three characteristics chosen from different systems of these are azygetic, transform into such systems; moreover since the linear transformation of a characteristic which is the sum of an odd number of other characteristics is the sum of the transformations of these characteristics, the transformations of these $2\sigma+1$ systems possess the property belonging to the original systems, that all the $2^{2\sigma}$ Göpel systems having a definite character are representable by the combinations of an odd number of them. It follows therefore that the theta relations obtained in Chap. XVII., based on the properties of the Göpel systems, persist after any linear transformation.

331. But questions are then immediately suggested, such as these: What are the simplest Göpel systems from which all others are obtainable* by linear transformation? Is it possible to derive the $2^{2\sigma}$ Göpel systems of § 298, having a definite character, by linear transformation, from systems based upon the $2^{2\sigma}$ characteristics obtainable by taking all possible half-integer characteristics in which $p-\sigma$ columns consist of zeros? Are the fundamental sets of $2p+1$ three-wise azygetic characteristics, by the odd combinations of which all the 2^{2p} half-integer characteristics can be represented (§ 300), all derivable by linear transformation from one such set?

We deal here only with the answer to the last question—and prove the following result: *Let $D, D_1, \ldots, D_{2p+1}$ be any $2p+2$ half-integer characteristics, such that, for $i<j$,*

* An obvious Göpel group of 2^p characteristics is formed by all the characteristics in which the upper row of elements are all zeros, and the lower row of elements each $=0$ or $\frac{1}{2}$.

$i=1, \ldots, 2p$, $j=2, \ldots, 2p+1$, we have $|D, D_i, D_j|=1$; then it is possible to choose a half-integer characteristic E, and a linear transformation, such that the characteristics

$$ED,\ ED_1,\ \ldots,\ ED_{2p+1}$$

transform into

$$0,\ \lambda_1,\ \ldots,\ \lambda_{2p+1},$$

where $\lambda_1, \ldots, \lambda_{2p+1}$ are certain characteristics to be specified, of which (by § 327) every two are azygetic. It will follow that if $D', D_1', \ldots, D'_{2p+1}$ be any other set of $2p+2$ characteristics of which every three are azygetic, a characteristic E', and a linear transformation, can be found such that, with a proper characteristic E, the set $ED, ED_1, \ldots, ED_{2p+1}$ transforms into $E'D', E'D_1', \ldots, E'D'_{2p+1}$. It will be shewn that the characteristics $\lambda_1, \ldots, \lambda_{2p+1}$ can be written down by means of the hyperelliptic half-periods denoted (§ 200) by u^{a, c_1}, u^{a, a_1}, $u^{a, c_2}, \ldots, u^{a, a_p}$, $u^{a, c}$; it has already been remarked (§ 294, Ex.) that the characteristics associated with these half-periods are azygetic in pairs. The proof which is to be given establishes an interesting connexion between the conditions for a linear transformation and the investigation of § 300, Chap. XVII.

Taking an Abelian matrix,

$$\begin{pmatrix} \alpha & \beta \\ \alpha' & \beta' \end{pmatrix},$$

for which

$$\bar{\alpha}\alpha' - \bar{\alpha}'\alpha = 0, \quad \bar{\beta}\beta' - \bar{\beta}'\beta = 0, \quad \bar{\alpha}\beta' - \bar{\alpha}'\beta = 1,$$

define characteristics of integers by means of the equations

$$a_r = \begin{pmatrix} \alpha'_{1,r}, & \alpha'_{2,r}, & \ldots, & \alpha'_{p,r} \\ \alpha_{1,r}, & \alpha_{2,r}, & \ldots, & \alpha_{p,r} \end{pmatrix}, \quad b_r = \begin{pmatrix} \beta'_{1,r}, & \beta'_{2,r}, & \ldots, & \beta'_{p,r} \\ \beta_{1,r}, & \beta_{2,r}, & \ldots, & \beta_{p,r} \end{pmatrix}, \quad a_r' = -a_r,$$

where $\alpha'_{s,r}$ is the r-th element of the s-th row of the matrix α', etc. and $r=1, 2, \ldots, p$; then the symbol which, in accordance with the notation of § 294, Chap. XVII., we define by the equation

$$|A_r, B_s| = \alpha_{1,r}\beta'_{1,s} + \ldots + \alpha_{p,r}\beta'_{p,s} - \alpha'_{1,r}\beta_{1,s} - \ldots - \alpha'_{p,r}\beta_{p,s},$$

is the (r, s)-th element of the matrix $\bar{\alpha}\beta' - \bar{\alpha}'\beta$, and may be denoted by $(\bar{\alpha}\beta' - \bar{\alpha}'\beta)_{r,s}$; thus the conditions for the matrices $\alpha, \alpha', \beta, \beta'$ are equivalent to the $p(2p-1)$ equations

$$|A_r, B_r| = 1, \quad |A_r, B_s| = 0, \quad |A_r, A_s| = 0, \quad |B_r, B_s| = 0, \quad (r \neq s,\ r, s = 1, 2, \ldots, p),$$

whereof the first gives p conditions, the second $p(p-1)$ conditions, and the third and fourth each $\frac{1}{2}p(p-1)$ conditions. It is convenient also to notice, what are corollaries from these, the equations

$$|B_s, A_r| = -|A_r, B_s| = 0, \quad |B_r, A_r| = -|A_r, B_r| = -1, \quad |B_r, A_r'| = -|A_r', B_r| = |A_r, B_r| = 1.$$

Consider now the $2p+1$ characteristics, of integers, given by

$$a_1,\ b_1,\ a_1'b_1a_2,\ a_1'b_1b_2,\ a_1'b_1a_2'b_2a_3,\ a_1'b_1a_2'b_2b_3,\ \ldots,\ a_1'b_1\ldots b_{p-1}b_p,\ a_1'b_1\ldots a_p'b_p,$$

whereof the first $2p$ are pairs of the type

$$a_1'b_1\ldots a'_{r-1}b_{r-1}a_r, \quad a_1'b_1\ldots a'_{r-1}b_{r-1}b_r,$$

for $r=1, 2, \ldots, p$, and $a_1'b_1a_2$ means the sum, without reduction, of the characteristics a_1', b_1, a_2, and so in general. The sum of these characteristics is a characteristic consisting wholly of even integers. If these characteristics be denoted, in order, by $c_1, c_2, \ldots, c_{2p+1}$, it immediately follows, from the fundamental equations connecting $a_1, \ldots, b_p$, that

$$c_{i,1}c'_{j,1} + \ldots - c'_{i,1}c_{j,1} - \ldots = 1, \quad \left(i < j,\ \begin{matrix} i = 1, 2, \ldots, 2p \\ j = 2, 3, \ldots, 2p+1 \end{matrix}\right).$$

Thus the $(2p+1)$ half-integer characteristics derivable from $c_1, c_2, \ldots, c_{2p+1}$, namely $C_1=\frac{1}{2}c_1, \ldots, C_{2p+1}=\frac{1}{2}c_{2p+1}$, are azygetic in pairs.

Conversely let $D, D_1, \ldots, D_{2p+1}$ be any half-integer characteristics such that, for $i<j$, $i=1, \ldots, 2p$, $j=2, \ldots, 2p+1$, we have $|D, D_i, D_j|=1$, so that (§ 300, p. 496) there exist connecting them only two relations (i) that their sum is a characteristic of integers, and (ii) a relation connecting an odd number of them; putting $C_i=D'D_i$ $(i=1, \ldots, 2p)$, where $D'=-D$, we obtain a set of independent characteristics $C_1, \ldots, C_{2p}$, such that for $i<j$,

$$|C_i, C_j|=1, \quad \begin{pmatrix} i=1, 2, \ldots, 2p-1 \\ j=2, 3, \ldots, 2p \end{pmatrix};$$

taking $C_{2p+1}=C_1'C_2C_3'C_4\ldots C'_{2p-1}C_{2p}$, where $C'_{2r-1}=-C_{2r-1}$, we have also the $2p$ equations

$$|C_m, C_{2p+1}|=1, \quad (m=1, 2, \ldots, 2p).$$

Thus putting $C_1=\frac{1}{2}c_1, \ldots, C_{2p+1}=\frac{1}{2}c_{2p+1}$, we can obtain an Abelian matrix by means of the equations, previously given,

$$c_{2r-1}=a_1'b_1\ldots a'_{r-1}b_{r-1}a_r, \quad c_{2r}=a_1'b_1\ldots a'_{r-1}b_{r-1}b_r, \quad c_{2p+1}=a_1'b_1\ldots a_p'b_p,$$

the i-th column of this matrix consisting of the elements of the lower and upper rows of the integer characteristic a_i or b_i, according as $i<p+1$ or $i>p$. We proceed now to find the result of applying the linear transformation, given by this Abelian matrix, to the half-integer characteristics $C_1, \ldots, C_{2p+1}$.

The equations for the transformation of the characteristic $\frac{1}{2}\begin{pmatrix} q' \\ q \end{pmatrix}$ to the characteristic $\frac{1}{2}\begin{pmatrix} k' \\ k \end{pmatrix}$, which are (§ 324, VI.),

$$k'=\bar{a}q'-\bar{a}'q-d(\bar{a}a'), \quad -k=\bar{\beta}q'-\bar{\beta}'q-d(\bar{\beta}\beta'),$$

are equivalent, in the notation here employed, to

$$k_i'=|A_i, Q|-[d(\bar{a}a')]_i, \quad -k_i=|B_i, Q|-[d(\bar{\beta}\beta')]_i, \quad (i=1, 2, \ldots, p),$$

where $A_i=\frac{1}{2}a_i$, $Q=\frac{1}{2}q$; taking

$$Q=\tfrac{1}{2}a_1'b_1\ldots a'_{r-1}b_{r-1}a_r, \; =\tfrac{1}{2}a_1'b_1\ldots a'_{r-1}b_{r-1}b_r, \text{ and } =\tfrac{1}{2}a_1'b_1\ldots a_p'b_p,$$

in turn, we immediately find that the transformations of the characteristics $C_{2r-1}, C_{2r}, C_{2p+1}$, are given, omitting integer characteristics, by

$$\tfrac{1}{2}\begin{pmatrix} d(\bar{a}a') \\ d(\bar{\beta}\beta') \end{pmatrix}+\tfrac{1}{2}\begin{pmatrix} 1\,1\ldots1\,0\,0\ldots0 \\ 1\,1\ldots1\,1\,0\ldots0 \end{pmatrix}, \; \tfrac{1}{2}\begin{pmatrix} d(\bar{a}a') \\ d(\bar{\beta}\beta') \end{pmatrix}+\tfrac{1}{2}\begin{pmatrix} 1\,1\ldots1\,1\,0\ldots0 \\ 1\,1\ldots1\,0\,0\ldots0 \end{pmatrix}, \; \tfrac{1}{2}\begin{pmatrix} d(\bar{a}a') \\ d(\bar{\beta}\beta') \end{pmatrix}+\tfrac{1}{2}\begin{pmatrix} 1\,1\ldots1 \\ 1\,1\ldots1 \end{pmatrix},$$

or, say, by

$$\tfrac{1}{2}\begin{pmatrix} d(\bar{a}a') \\ d(\bar{\beta}\beta') \end{pmatrix}+\tfrac{1}{2}\begin{pmatrix} 1 \\ 1 \end{pmatrix}^{r-1}\begin{pmatrix} 0 \\ 1 \end{pmatrix}\begin{pmatrix} 0 \\ 0 \end{pmatrix}^{p-r}, \; \tfrac{1}{2}\begin{pmatrix} d(\bar{a}a') \\ d(\bar{\beta}\beta') \end{pmatrix}+\tfrac{1}{2}\begin{pmatrix} 1 \\ 1 \end{pmatrix}^{r-1}\begin{pmatrix} 1 \\ 0 \end{pmatrix}\begin{pmatrix} 0 \\ 0 \end{pmatrix}^{p-r}, \; \tfrac{1}{2}\begin{pmatrix} d(\bar{a}a') \\ d(\bar{\beta}\beta') \end{pmatrix}+\tfrac{1}{2}\begin{pmatrix} 1 \\ 1 \end{pmatrix}^{p},$$

respectively.

Now let the characteristics

$$\tfrac{1}{2}\begin{pmatrix} 0 \\ 1 \end{pmatrix}\begin{pmatrix} 0 \\ 0 \end{pmatrix}^{p-1}, \; \tfrac{1}{2}\begin{pmatrix} 1 \\ 0 \end{pmatrix}\begin{pmatrix} 0 \\ 0 \end{pmatrix}^{p-1}, \ldots, \tfrac{1}{2}\begin{pmatrix} 1 \\ 1 \end{pmatrix}^{r-1}\begin{pmatrix} 0 \\ 1 \end{pmatrix}\begin{pmatrix} 0 \\ 0 \end{pmatrix}^{p-r}, \; \tfrac{1}{2}\begin{pmatrix} 1 \\ 1 \end{pmatrix}^{r-1}\begin{pmatrix} 1 \\ 0 \end{pmatrix}\begin{pmatrix} 0 \\ 0 \end{pmatrix}^{p-r}, \ldots, \tfrac{1}{2}\begin{pmatrix} 1 \\ 1 \end{pmatrix}^{p},$$

be respectively denoted by

$$\lambda_1, \lambda_2, \ldots, \lambda_{2r-1}, \lambda_{2r}, \ldots, \lambda_{2p+1};$$

then we have proved that the half-integer characteristic DD_i transforms, save for an integer characteristic, into $\lambda_i+\frac{1}{2}\begin{pmatrix} r' \\ r \end{pmatrix}$, where $r=d(\bar{\beta}\beta')$, $r'=d(\bar{a}a')$; since the transformation of the sum of two characteristics is the sum of their transformations added to $\frac{1}{2}\begin{pmatrix} r' \\ r \end{pmatrix}$, and since the characteristic $\frac{1}{2}\begin{pmatrix} s' \\ s \end{pmatrix}$, where $s'=d(\alpha'\bar{\beta}')$, $s=d(\alpha\bar{\beta})$, transforms into the zero characteristic (§ 327), it follows that the transformation of the characteristic $\frac{1}{2}\begin{pmatrix} s' \\ s \end{pmatrix}+DD_i$ is the characteristic λ_i; hence, putting $E=\frac{1}{2}\begin{pmatrix} s' \\ s \end{pmatrix}+D$, and omitting integer characteristics, the characteristics

$$ED,\ ED_1,\ \ldots,\ ED_{2p+1}$$

transform, respectively, into

$$0,\ \lambda_1,\ \ldots,\ \lambda_{2p+1};$$

and this is the result we desired to prove.

The number of matrices of integers, of the form

$$\begin{pmatrix} \alpha & \beta \\ \alpha' & \beta' \end{pmatrix},$$

in which $\bar{\alpha}\alpha'-\bar{\alpha}'\alpha=0$, $\bar{\beta}\beta'-\bar{\beta}'\beta=0$, $\bar{\alpha}\beta'-\bar{\alpha}'\beta=1$, is infinite; but it follows from the investigation just given that if all the elements of these matrices be replaced by their smallest positive residues for modulus 2, the number of different matrices then arising is finite, being equal to the number of sets of $2p+1$ half-integer characteristics, with integral sum, of which every two characteristics are azygetic. As in § 300, Chap. XVII., this number is

$$(2^{2p}-1)\,2^{2p-1}\,(2^{2p-2}-1)\,2^{2p-3}\ldots\ldots(2^2-1)\,2\,;$$

we may call this the number of incongruent Abelian matrices, for modulus 2. Similarly the number* of incongruent Abelian matrices for modulus n is

$$(n^{2p}-1)\,n^{2p-1}\,(n^{2p-2}-1)\,n^{2p-3}\ldots\ldots(n^2-1)\,n.$$

Ex. By adding suitable integers to the characteristics denoted by 1, 2, 3, 4, 5, 6, 7 in the table of § 205, for $p=3$, we obtain respectively

$$\tfrac{1}{2}\begin{pmatrix} -1 & 0 & 0 \\ -1 & 0 & 0 \end{pmatrix},\ \tfrac{1}{2}\begin{pmatrix} -1 & -1 & 0 \\ 0 & -1 & 0 \end{pmatrix},\ \tfrac{1}{2}\begin{pmatrix} -1 & -1 & 1 \\ 0 & 0 & 1 \end{pmatrix},\ \tfrac{1}{2}\begin{pmatrix} -1 & 0 & 1 \\ 0 & 1 & 1 \end{pmatrix},$$

$$\tfrac{1}{2}\begin{pmatrix} 0 & 0 & -1 \\ 1 & 0 & -1 \end{pmatrix},\ \tfrac{1}{2}\begin{pmatrix} 0 & 1 & -1 \\ 1 & 1 & 0 \end{pmatrix},\ \tfrac{1}{2}\begin{pmatrix} 0 & 1 & 0 \\ 1 & 1 & 1 \end{pmatrix};$$

denoting these respectively by $C_1, C_2, \ldots, C_7$, we find, for $i<j$, that

$$|C_i,\ C_j|=1, \qquad (i=1,\ \ldots,\ 6\,;\ j=2,\ \ldots,\ 7).$$

The equations of the text

$$c_{2r-1}=a_1'b_1\ldots\ldots a'_{r-1}b_{r-1}a_r, \quad c_{2r}=a_1'b_1\ldots\ldots a'_{r-1}b_{r-1}b_r,$$

give

$$a_r=c_1c_2'\ldots\ldots c_{2r-3}c'_{2r-2}c_{2r-1}, \quad b_r=c_1c_2'\ldots\ldots c_{2r-3}c'_{2r-2}c_{2r},$$

and therefore, in this case, we find

$$a_1=\begin{pmatrix} -1 & 0 & 0 \\ -1 & 0 & 0 \end{pmatrix}, \quad a_2=\begin{pmatrix} -1 & 0 & 1 \\ -1 & 1 & 1 \end{pmatrix}, \quad a_3=\begin{pmatrix} 0 & 0 & -1 \\ 0 & 0 & -1 \end{pmatrix},$$

$$b_1=\begin{pmatrix} -1 & -1 & 0 \\ 0 & -1 & 0 \end{pmatrix}, \quad b_2=\begin{pmatrix} -1 & 1 & 1 \\ -1 & 2 & 1 \end{pmatrix}, \quad b_3=\begin{pmatrix} 0 & 1 & -1 \\ 0 & 1 & 0 \end{pmatrix};$$

* Another proof is given by Jordan, *Traité des Substitutions* (Paris, 1870), p. 176.

hence the linear substitution, of the text, for transforming the fundamental set of characteristics $C_1, \ldots, C_7$ is

$$\begin{pmatrix} -1 & -1 & 0 & 0 & -1 & 0 \\ 0 & 1 & 0 & -1 & 2 & 1 \\ 0 & 1 & -1 & 0 & 1 & 0 \\ \hline -1 & -1 & 0 & -1 & -1 & 0 \\ 0 & 0 & 0 & -1 & 1 & 1 \\ 0 & 1 & -1 & 0 & 1 & -1 \end{pmatrix}$$

From this we find $\frac{1}{2}\begin{pmatrix} s' \\ s \end{pmatrix} = \frac{1}{2}\begin{pmatrix} d(a'\bar{\beta}') \\ d(a\bar{\beta}) \end{pmatrix} = \frac{1}{2}\begin{pmatrix} 2 & 0 & 2 \\ 1 & 2 & 1 \end{pmatrix}$; since the sum of $C_1, \ldots, C_7$ is an integral characteristic, it follows by the general theorem, that *if the characteristic* $\frac{1}{2}\begin{pmatrix} 0 & 0 & 0 \\ 1 & 0 & 1 \end{pmatrix}$ *be added to each of* $C_1, \ldots, C_7$, *and then the linear transformation given by the matrix be applied, they will be transformed respectively into the characteristics* $\lambda_1, \ldots, \lambda_7$.

A further result should be mentioned. On the hyperelliptic Riemann surface suppose the period loops drawn as in the figure (12);

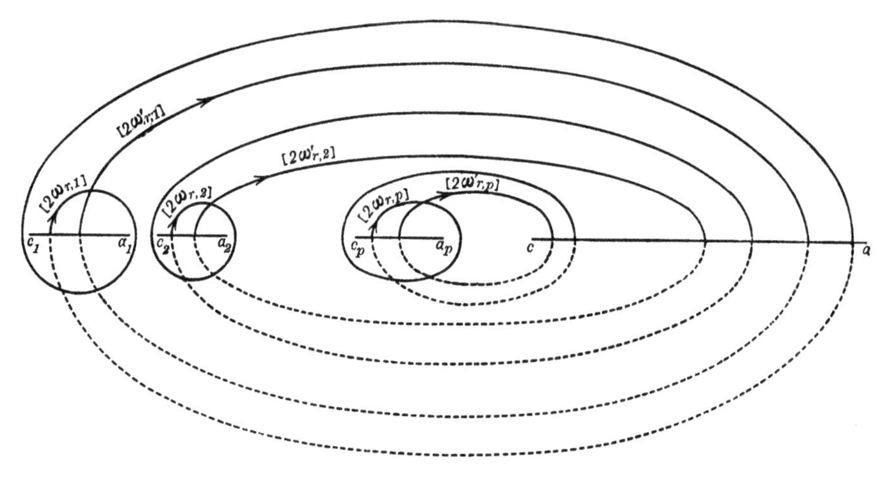

Fig. 12.

then the characteristics associated with the half-periods $u^{a, c_1}, u^{a, a_1}, \ldots, u^{a, c_p}, u^{a, a_p}$, $u^{a, c}$ will be, save for integer characteristics, respectively $\lambda_1, \lambda_2, \ldots, \lambda_{2p}, \lambda_{2p+1}$; this the reader can immediately verify by means of the rule given at the bottom of page 297 of the present volume.

Ex. Prove that if the characteristics $0, \lambda_1, \ldots, \lambda_{2p+1}$ be subjected to the transformation given by the Abelian matrix of $2p$ rows and columns which is denoted by $\begin{pmatrix} 1, & -1 \\ 0, & 1 \end{pmatrix}$, then, save for integer characteristics, λ_i is changed to $\Sigma_i + \frac{1}{2}\begin{pmatrix} 0 \\ 1 \end{pmatrix}^p$, where

$$\Sigma_{2r-1} = \frac{1}{2}\begin{pmatrix} 1 \\ 0 \end{pmatrix}^{r-1}\begin{pmatrix} 0 \\ 1 \end{pmatrix}\begin{pmatrix} 0 \\ 0 \end{pmatrix}^{p-r}, \quad \Sigma_{2r} = \frac{1}{2}\begin{pmatrix} 1 \\ 0 \end{pmatrix}^{r-1}\begin{pmatrix} 1 \\ 1 \end{pmatrix}\begin{pmatrix} 0 \\ 0 \end{pmatrix}^{p-r}, \quad \Sigma_{2p+1} = \frac{1}{2}\begin{pmatrix} 1 \\ 0 \end{pmatrix}^{p}, \quad (r=1, 2, \ldots, p),$$

are the characteristics which arise in § 200, Chap. XI. as associated with the half-periods $u^{a,\,c_r}$, $u^{a,\,a_r}$, $u^{a,\,c}$ respectively. The characteristics $\Sigma_1, \dots, \Sigma_{2p+1}$ satisfy the $p(2p-1)$ conditions $|\Sigma_i, \Sigma_j|=1$, for $i<j$.

332. We proceed now to shew how any linear transformation may be regarded as the result of certain very simple linear transformations performed in succession. As a corollary from the investigation we shall be able to infer that every linear transformation may be associated with a change in the method of taking the period loops on a Riemann surface; we have already proved the converse result, that every change in the period loops is associated with matrices, α, α', β, β', belonging to a linear substitution (§ 322).

It is convenient to give first the fundamental equations for a composition of two transformations of any order. It has been shewn (§ 324) that the equations for the transformation of a theta function of the first order, in the arguments u, with characteristic (Q, Q') and associated constants 2ω, $2\omega'$, 2η, $2\eta'$, to a theta function of order r, in the arguments w, where $u=Mw$, with characteristic (K, K') and associated constants 2υ, $2\upsilon'$, 2ζ, $2\zeta'$, are

$$K'=\bar{\alpha}Q'-\bar{\alpha}'Q-\tfrac{1}{2}d(\bar{\alpha}\alpha'),\quad -K=\bar{\beta}Q'-\bar{\beta}'Q-\tfrac{1}{2}d(\bar{\beta}\beta'),$$

$$\begin{pmatrix} M, & 0 \\ 0, & r\bar{M}^{-1}\end{pmatrix}\begin{pmatrix} 2\upsilon, & 2\upsilon' \\ 2\zeta, & 2\zeta'\end{pmatrix}=\begin{pmatrix} 2\omega, & 2\omega' \\ 2\eta, & 2\eta'\end{pmatrix}\begin{pmatrix} \alpha, & \beta \\ \alpha', & \beta'\end{pmatrix};$$

and from the last equation, writing it in the form $\mu\mathsf{U}=\Omega\Delta$, it follows, in virtue of the equations $\Omega\epsilon\bar{\Omega}=-\tfrac{1}{2}\pi i\epsilon$, $\mathsf{U}\epsilon\bar{\mathsf{U}}'=-\tfrac{1}{2}\pi i\epsilon$ (§ 140, Chap. VII.), and the easily verifiable equation $\bar{\mu}\epsilon\mu=r\epsilon$, where the matrix ϵ is given by

$$\epsilon=\begin{pmatrix} 0 & -1 \\ 1 & 0\end{pmatrix},$$

that also $\bar{\Delta}\epsilon\Delta=r\epsilon$, as in Ex. i., § 324. And, just as in § 324, it can be proved that equations for the transformation of a theta function of order r in the arguments w, with characteristic (K, K'), and associated constants 2υ, $2\upsilon'$, 2ζ, $2\zeta'$, to a theta-function of order rs, in the arguments u_1, given by $w=Nu_1$, with characteristic (Q_1, Q_1'), and associated constants $2\omega_1$, $2\omega_1'$, $2\eta_1$, $2\eta_1'$, are

$$Q_1'=\bar{\gamma}K'-\bar{\gamma}'K-\tfrac{1}{2}rd(\bar{\gamma}\gamma'),\quad -Q_1=\bar{\delta}K'-\bar{\delta}'K-\tfrac{1}{2}rd(\bar{\delta}\delta'),$$

$$\begin{pmatrix} N, & 0 \\ 0, & s\bar{N}^{-1}\end{pmatrix}\begin{pmatrix} 2\omega_1, & 2\omega_1' \\ 2\eta_1, & 2\eta_1'\end{pmatrix}=\begin{pmatrix} 2\upsilon, & 2\upsilon' \\ 2\zeta, & 2\zeta'\end{pmatrix}\begin{pmatrix} \gamma, & \delta \\ \gamma', & \delta'\end{pmatrix};$$

and writing the last equation in the form $\nu\Omega_1=\mathsf{U}\nabla$, we infer as before that $\bar{\nabla}\epsilon\nabla=s\epsilon$.

Now from the equations $\mu\mathsf{U}=\Omega\Delta$, $\nu\Omega_1=\mathsf{U}\nabla$, we obtain $\mu\nu\Omega_1=\mu\mathsf{U}\nabla=\Omega\Delta\nabla$, or, if $\Delta_1=\Delta\nabla$,

$$\begin{pmatrix} MN, & 0 \\ 0, & rs\overline{MN}^{-1}\end{pmatrix}\begin{pmatrix} 2\omega_1, & 2\omega_1' \\ 2\eta_1, & 2\eta_1'\end{pmatrix}=\begin{pmatrix} 2\omega, & 2\omega' \\ 2\eta, & 2\eta'\end{pmatrix}\Delta_1;$$

from this equation we find as before that the matrix Δ_1, given by

$$\Delta_1 = \Delta\nabla = \begin{pmatrix} \alpha\gamma+\beta\gamma', & \alpha\delta+\beta\delta' \\ \alpha'\gamma+\beta'\gamma', & \alpha'\delta+\beta'\delta' \end{pmatrix} = \begin{pmatrix} \alpha_1, & \beta_1 \\ \alpha_1', & \beta_1' \end{pmatrix}, \text{ say,}$$

satisfies the equation $\bar{\Delta}_1 \epsilon \Delta_1 = rs\epsilon$. Similarly from the two sets of equations transforming the characteristics, by making use of the equations

$$d(\bar{\alpha}_1\alpha_1') \equiv \bar{\gamma} d(\bar{\alpha}\alpha') + \bar{\gamma}' d(\bar{\beta}\beta') + rd(\bar{\gamma}\gamma'),$$
$$d(\bar{\beta}_1\beta_1') \equiv \bar{\delta} d(\bar{\alpha}\alpha') + \bar{\delta}' d(\bar{\beta}\beta') + rd(\bar{\delta}\delta'), \quad (\text{mod. } 2),$$

which can be proved by the methods of § 327, we immediately find

$$Q_1' \equiv \bar{\alpha}_1 Q' - \bar{\alpha}_1' Q - \tfrac{1}{2} d(\bar{\alpha}_1\alpha_1'), \quad -Q_1 \equiv \bar{\beta}_1 Q' - \bar{\beta}_1' Q - \tfrac{1}{2} d(\bar{\beta}_1\beta_1'), \quad (\text{mod. } 2).$$

Hence any transformation of order rs may be regarded as compounded of two transformations, of which the first transforms a theta-function of the first order into a theta function of the r-th order, and the second transforms it further into a theta function of order rs.

It follows therefore that the most general transformation may be considered as the result of successive transformations of prime order. It is convenient to remember that the matrix of integers, Δ_1, associated with the compound transformation, is equal to $\Delta\nabla$, the matrix Δ, associated with the transformation which is first carried out, being the left-hand factor.

One important case should be referred to. The matrix

$$r\Delta^{-1} = \begin{pmatrix} \bar{\beta}' & -\bar{\beta} \\ -\bar{\alpha}' & \bar{\alpha} \end{pmatrix}$$

is easily seen to be that of a transformation of order r; putting it in place of ∇, the final equations for the compound transformation ∇_1 may be taken to be

$$u_1 = ru, \quad 2\omega_1 = 2\omega, \quad 2\omega_1' = 2\omega', \quad 2\eta_1 = 2\eta, \quad 2\eta_1' = 2\eta'.$$

The transformation $r\Delta^{-1}$ is called *supplementary* to Δ (cf. Chap. XVII., § 317, Ex. vii.).

333. Limiting ourselves now to the case of linear transformation, let A_k $(k = 2, 3, \ldots, p)$ denote the matrix of $2p$ rows and columns indicated by

$$A_k = \begin{pmatrix} \mu_k, & 0 \\ 0, & \mu_k \end{pmatrix},$$

where μ_k has unities in the diagonal except in the first and k-th places, in which there are zeros, and has elsewhere zeros, except in the k-th place of the first row, and the k-th place of the first column, where there are unities; let B denote the matrix of $2p$ rows and columns indicated by

$$B = \left(\begin{array}{cccccccc} 0 & & & & -1 & & & \\ & 1 & & & & 0 & & \\ & & 1 & & & & 0 & \\ & & & 1 & & & & 0 \\ 1 & & & & 0 & & & \\ & 0 & & & & 1 & & \\ & & 0 & & & & 1 & \\ & & & 0 & & & & 1 \end{array}\right),$$

which has unities in the diagonal, except in the first and $(p+1)$-th places, where there are zeros, and has elsewhere zeros except in the $(p+1)$-th place of the first row, where there is -1, and the $(p+1)$-th place of the first column, where there is $+1$; let C denote the matrix of $2p$ rows and columns indicated by

$$C = \left(\begin{array}{cccccccc} 1 & & & & -1 & & & \\ & 1 & & & & 0 & & \\ & & 1 & & & & 0 & \\ & & & 1 & & & & 0 \\ 0 & & & & 1 & & & \\ & 0 & & & & 1 & & \\ & & 0 & & & & 1 & \\ & & & 0 & & & & 1 \end{array}\right),$$

which has unities everywhere in the diagonal and has elsewhere zeros, except in the $(p+1)$-th place of the first row, where it has -1; let D denote the matrix of $2p$ rows and columns indicated by

$$D = \left(\begin{array}{cccccccc} 1 & & & & 0 & -1 & & \\ & 1 & & & -1 & 0 & & \\ & & 1 & & & & 0 & \\ & & & 1 & & & & 0 \\ 0 & & & & 1 & & & \\ & 0 & & & & 1 & & \\ & & 0 & & & & 1 & \\ & & & 0 & & & & 1 \end{array}\right),$$

which has unities everywhere in the diagonal and has elsewhere zeros, except in the $(p+2)$-th place of the first row and the $(p+1)$-th place of the second row, in each of which there is -1. It is easy to see that each of these matrices satisfies the conditions (IX.) of § 324, for $r=1$.

Then it can be proved that every matrix of $2p$ rows and columns of integers,

$$\begin{pmatrix} \alpha, & \beta \\ \alpha', & \beta' \end{pmatrix},$$

for which $\alpha\bar{\beta} = \beta\bar{\alpha}$, $\alpha'\bar{\beta}' = \beta'\bar{\alpha}'$, $\alpha\bar{\beta}' - \beta\bar{\alpha}' = 1$, can be written* as a product of positive integral powers of the $(p+2)$ matrices $A_2, \dots, A_p, B, C, D$. The proof of this statement is given in the Appendix (II) to this volume.

We shall therefore obtain a better understanding of the changes effected by a linear transformation by considering these transformations in turn. We have seen that any linear transformation may be considered as made up of two processes, (i) the change of the fundamental system of periods, effected by the equations

$$[\omega] = \omega\alpha + \omega'\alpha', \quad [\omega'] = \omega\beta + \omega'\beta',$$
$$[\eta] = \eta\alpha + \eta'\alpha', \quad [\eta'] = \eta\beta + \eta'\beta',$$

(ii) the change of the arguments, effected by the equation $u = Mw$, and leading to

$$[\omega] = Mv, \quad [\omega'] = Mv', \quad \zeta = \bar{M}[\eta], \quad \zeta' = \bar{M}[\eta'];$$

of these we consider here the first process. Applying the equations†

$$[\omega] = \omega\alpha + \omega'\alpha', \quad [\omega'] = \omega\beta + \omega'\beta',$$

respectively for the transformations A_k, B, C, D, we obtain the following results:

For the matrix (A_k) we have

$$[\omega_{r,1}] = \omega_{r,k}, \quad [\omega_{r,k}] = \omega_{r,1}, \quad [\omega'_{r,1}] = \omega'_{r,k}, \quad [\omega'_{r,k}] = \omega'_{r,1}, \quad (r = 1, 2, \dots, p);$$

or, in words, if $2\omega_{r,i}$, $2\omega'_{r,i}$ be called the i-th pair of periods for the argument u_r, the change effected by the substitution A_k is an interchange of the first and k-th pairs of periods—no other change whatever being made.

When we are dealing with p quantities, the interchange of the first and k-th of these quantities can be effected by a composition of the two processes (i) an interchange of the first and second, (ii) a cyclical change whereby the second becomes the first, the third becomes the second, ..., the p-th becomes the $(p-1)$-th, and the first becomes the p-th. Such a cyclical change is easily seen to be effected by the matrix

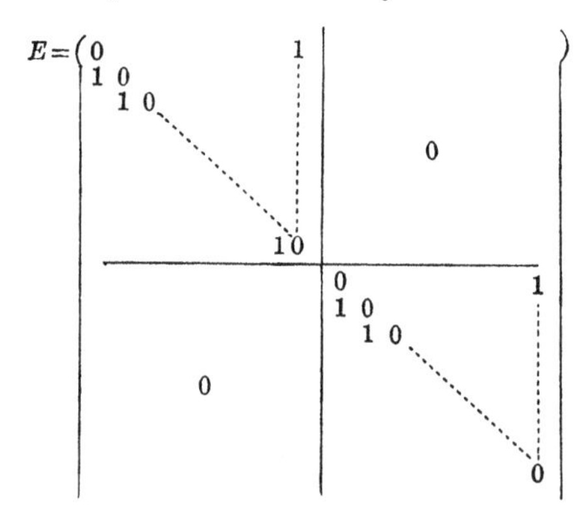

* Other sets of elementary matrices, by the multiplication of which any Abelian matrix can be formed, can easily be chosen. One other obvious set consists of the matrices obtained by interchanging the rows and columns of the matrices A_k, B, C, D.

† We may state the meaning of the matrices A_k, B, C, D somewhat differently in accordance with the property remarked in Ex. iii., § 324.

which verifies the equations (IX.) § 324, for $r=1$. Hence the matrices $A_3, \dots, A_p$ can each be represented by a product of positive powers of the matrices E and A_2. Thereby the $(p+2)$ elementary matrices $A_2, \dots, A_p, B, C, D$ can be replaced by only 5 matrices E, A_2, B, C, D*.

Considering next the matrix B we obtain

$$[\omega_{r,1}] = \omega'_{r,1}, \quad [\omega'_{r,1}] = -\omega_{r,1}, \quad [\omega_{r,i}] = \omega_{r,i}, \quad [\omega'_{r,i}] = \omega'_{r,i}, \quad \begin{pmatrix} r = 1, 2, \dots, p \\ i = \quad 2, \dots, p \end{pmatrix},$$

so that this transformation has the effect of interchanging $\omega_{r,1}$ and $\omega'_{r,1}$, changing the sign of one of them; no other change is introduced.

The matrix C gives the equation

$$[\omega'_{r,1}] = \omega'_{r,1} - \omega_{r,1}, \quad (r = 1, 2, \dots, p),$$

but makes no other change.

The matrix D makes only the changes expressed by the equations

$$[\omega'_{r,1}] = \omega'_{r,1} - \omega_{r,2}, \quad [\omega'_{r,2}] = \omega'_{r,2} - \omega_{r,1}.$$

In applying these transformations to the case of the theta functions we notice immediately that A_k, C and D all belong to the case considered in § 326 (ii), in which the matrix $\alpha' = 0$.

Thus in the case of the transformation A_k we have

$$\Theta\left(u;\ \tau \,\middle|\, {}^{Q'}_{Q}\right) = A\Theta\left(w;\ \tau' \,\middle|\, {}^{K'}_{K}\right),$$

where w differs from u only in the interchange of u_1 and u_k, τ' differs from τ only in the interchange of the suffixes 1 and k in the constituents $\tau_{r,s}$ of the matrix τ, and K, K' differ from Q, Q' only in the interchange of the first and k-th elements both in Q and Q'. Thus in this case the constant A is equal to 1.

In the case of the matrix (C), the equations of § 326 (2) give

$$\Theta\left(u;\ \tau \,\middle|\, {}^{Q'}_{Q}\right) - A\Theta\left(w;\ \tau' \,\middle|\, {}^{K'}_{K}\right),$$

where

$u=w$, $\tau'=\tau$ save that $\tau'_{1,1}=\tau_{1,1}-1$, and $K'=Q'$, $K=Q$ save that $K_1=Q_1+Q_1'-\frac{1}{2}$;

now the general term of the left-hand side, or

$$e^{2\pi i u(n+Q') + i\pi\tau(n+Q')^2 + 2\pi i Q(n+Q')}$$

is equal to

$$e^{2\pi i w(n+K') + i\pi\tau'(n+K')^2 + i\pi(n_1+Q_1')^2 + 2\pi i K(n+K') - 2i\pi(Q_1'-\frac{1}{2})(n_1+Q_1')}$$

$$= e^{-i\pi(Q_1'^2 - Q_1')}\, e^{2\pi i w(n+K') + i\pi\tau'(n+K')^2 + 2\pi i K(n+K')};$$

thus in the case of the transformation (C) the constant A is equal to $e^{-i\pi(Q_1'^2 - Q_1')}$; when Q_1' is a half-integer, this is an eighth root of unity.

* See Krazer, *Ann. d. Mat.*, Ser. II., t. xii. (1884). The number of elementary matrices is stated by Burkhardt to be further reducible to 3, or, in case $p=2$, to 2; *Götting. Nachrichten*, 1890, p. 381.

In the case of the matrix (D), the equations of § 326 (ii) lead to

$$\Theta\left(u;\ \tau \,\Big|\, {}^{Q'}_{Q}\right) = A\Theta\left(w;\ \tau' \,\Big|\, {}^{K'}_{K}\right),$$

where $u=w$, $\tau'=\tau$ save that $\tau'_{1,2}=\tau_{1,2}-1$, $\tau'_{2,1}=\tau_{2,1}-1$, and $K'=Q'$, $K=Q$ save that $K_1=Q_1+Q_2'$, $K_2=Q_2+Q_1'$; now we have

$$e^{2\pi i u(n+Q')+i\pi\tau(n+Q')^2+2\pi i Q(n+Q')} = e^{2i\pi(n_1n_2-Q_1'Q_2')}\, e^{2\pi i w(n+K')+i\pi\tau'(n+K')^2+2\pi i K(n+K')};$$

thus, in the case of the matrix (D) the constant A is equal to $e^{-2\pi i Q_1'Q_2'}$.

We consider now the transformation (B)—which falls under that considered in (i) § 326. In this case $\pi i\bar{\alpha}'(\alpha+\tau\alpha')w^2$ is equal to $\pi i\tau_{1,1}w_1^2$, and the equation $(\alpha+\tau\alpha')\tau'=\beta+\tau\beta'$ leads to the equations

$$\tau'_{1,1}=-1/\tau_{1,1},\quad \tau'_{1,r}=\tau_{1,r}/\tau_{1,1},\quad \tau'_{r,s}=\tau_{r,s}-\tau_{1,r}\tau_{1,s}/\tau_{1,1},$$

or, the equivalent equations $(r, s=2, 3, \dots, p)$,

$$\tau_{1,1}=-1/\tau'_{1,1},\quad \tau_{1,r}=-\tau'_{1,r}/\tau'_{1,1},\quad \tau_{r,s}=\tau'_{r,s}-\tau'_{1,r}\tau'_{1,s}/\tau'_{1,1};$$

also $u_1=\tau_{1,1}w_1$, $u_r=\tau_{1,r}w_1+w_r$, so that $w_1=-\tau'_{1,1}u_1$, $w_r=u_r-\tau'_{1,r}u_1$, and $\tau_{1,1}w_1^2=-\tau'_{1,1}u_1^2$; further we find

$$K'=Q' \text{ save that } K_1'=-Q_1, \text{ and } K=Q \text{ save that } K_1=Q_1';$$

with these values we have the equation

$$e^{\pi i\tau_{1,1}w_1^2}\,\Theta\left(u;\ \tau \,\Big|\, {}^{Q'}_{Q}\right) = A\Theta\left(w;\ \tau' \,\Big|\, {}^{K'}_{K}\right).$$

334. To determine the constant A in the final equation of the last Article we proceed as follows*:—We have

(i) $$\int_0^1 e^{2\pi i m w}\,dw = 0 \text{ or } 1,$$

according as m is an integer other than zero, or is zero;

(ii) if α be a positive real quantity other than zero, and β, γ, δ be real quantities,

$$\int_{-\infty}^{\infty} e^{(-\alpha+i\beta)(x+\gamma+i\delta)^2}\,dx = \sqrt{\frac{\pi}{\alpha-i\beta}},$$

where for the square root is to be taken that value of which the real part is positive†;

* For indications of another method consult Clebsch u. Gordan, *Abel. Funct.*, § 90; Thomae, *Crelle*, LXXV. (1873), p. 224.

† By the symbol $\sqrt{\mu}$, where μ is any constant quantity, is to be understood that square root whose real part is positive, or, if the real part be zero, that square root whose imaginary part is positive.

(iii) with the relations connecting u, w and τ, τ' given in the previous Article,

$$un = (wn)_1 + (\tau_{1,1} n_1 + \ldots\ldots + \tau_{1,p} n_p) w_1,$$

where $(wn)_1$ denotes $w_2 n_2 + \ldots\ldots + w_p n_p$;

(iv) the series representing the function $\Theta(w, \tau')$ is uniformly convergent for all finite values of $w_1, \ldots, w_p$, and therefore, between finite limits, the integral of the function is the sum of the integrals of its terms.

Therefore, taking the case when $\binom{Q'}{Q}$ and therefore $\binom{K'}{K}$ are $\binom{0}{0}$, and integrating the equation

$$e^{\pi i \tau_{1,1} w_1^2} \Theta(u; \tau) = A_0 \Theta(w; \tau'),$$

in regard to $w_1, \ldots, w_p$, each from 0 to 1, we have

$$A_0 = \sum_{n_1=-\infty}^{\infty} \sum_{n_2, \ldots, n_p}^{-\infty, \infty} \int_0^1 \ldots \int_0^1 e^{\pi i \tau_{1,1} w_1^2 + 2\pi i (wn)_1 + 2\pi i (\tau_{1,1} n_1 + \ldots + \tau_{1,p} n_p) w_1 + i\pi \tau n^2} dw_1 \ldots dw_p,$$

where, on the right hand, the integral is zero except for $n_2 = 0, \ldots, n_p = 0$; thus

$$A_0 = \sum_{n_1=-\infty}^{\infty} \int_0^1 e^{\pi i \tau_{1,1} w_1^2 + 2\pi i \tau_{1,1} n_1 w_1 + i\pi \tau_{1,1} n_1^2} dw_1$$

$$= \sum_{n_1=-\infty}^{\infty} \int_0^1 e^{\pi i \tau_{1,1} (w_1 + n_1)^2} dw_1$$

$$= \int_{-\infty}^{\infty} e^{\pi i \tau_{1,1} x^2} dx;$$

hence since the real part of $\pi i \tau_{1,1}$ is negative (§ 174), we have

$$A_0 = \sqrt{\frac{\pi}{-\pi i \tau_{1,1}}} = \sqrt{\frac{i}{\tau_{1,1}}},$$

where the square root is to be taken of which the real part is positive.

Hence

$$e^{\pi i \tau_{1,1} w_1^2} \Theta(u; \tau) = \sqrt{\frac{i}{\tau_{1,1}}} \Theta(w; \tau'),$$

and from this equation, by increasing w by $K + \tau' K'$, we deduce that

$$e^{\pi i \tau_{1,1} w_1^2} \Theta\left(u; \tau \,\middle|\, {Q' \atop Q}\right) = \sqrt{\frac{i}{\tau_{1,1}}} \cdot e^{2\pi i Q_1 Q_1'} \Theta\left(w; \tau' \,\middle|\, {K' \atop K}\right).$$

Hence, when the decomposition of any linear transformation into transformations of the form A_k, B, C, D is known, the value of the constant factor, A, can be determined.

335. But, save for an eighth root of unity, we can immediately specify the value in the general case; for when Q, Q' are zero, the value of the constant A has been found to be unity for each of the transformations A_k, C, D, and for the transformation B to have a value which is in fact equal to $\sqrt{i/|M|}$, $|M|$ denoting the determinant of the matrix M. Hence for a transformation which can be put into the form

$$\begin{pmatrix} a & \beta \\ a' & \beta' \end{pmatrix} = \dots B^{r_2} \dots A_k^{\rho} \dots D^{\nu} \dots B^{r_1} \dots C^{\mu} \dots A_k^{\lambda} \dots,$$

if the values of the matrix M for these component transformations be respectively

$$\dots M_2^{r_2} \dots 1 \dots 1 \dots M_1^{r_1} \dots 1 \dots 1 \dots,$$

the value of the constant A, when Q, Q' are zero, for the complete transformation, will be

$$\dots \left(\sqrt{\frac{i}{|M_2|}}\right)^{r_2} \dots\dots \left(\sqrt{\frac{i}{|M_1|}}\right)^{r_1} \dots\dots;$$

but if the complete transformation give $u = Mw$, we have $M = \dots M_2 M_1 \dots$; thus, for any transformation we have the formula

$$e^{\pi i \bar{a}'(a+\tau a')w^2} \Theta(u, \tau) = \frac{\epsilon}{\sqrt{|M|}} \Theta\left[w, \tau' \middle| \tfrac{1}{2}\begin{pmatrix} r' \\ r \end{pmatrix}\right],$$

where $M = a + \tau a'$, $u = Mw$, and ϵ is an eighth root of unity, r, r' being as in § 328, p. 544.

Putting $2\omega u$, $2vw$ for u, w, as in § 326, this equation is the same as

$$\frac{1}{\sqrt{|\omega|}} \vartheta(u;\ 2\omega, 2\omega', 2\eta, 2\eta') = \frac{\epsilon}{\sqrt{|M|\,|v|}} \vartheta\left[w;\ 2v, 2v', 2\zeta, 2\zeta' \middle| \tfrac{1}{2}\begin{pmatrix} r' \\ r \end{pmatrix}\right]$$

where $|\omega|$ is the determinant of the matrix ω, etc.

Of such composite transformations there is one which is of some importance, that, namely, for which

$$\begin{pmatrix} a & \beta \\ a' & \beta' \end{pmatrix} = \begin{pmatrix} 0 & -1 \\ 1 & 0 \end{pmatrix},$$

so that

$$[\omega_{r,i}] = \omega'_{r,i}, \quad [\omega'_{r,i}] = -\omega_{r,i}; \qquad (r, i = 1, 2, \dots, p).$$

Then

$$M = \tau, \quad \tau\tau' = -1, \quad u = \tau w, \quad \pi i \bar{a}'(a + \tau a') w^2 = \pi i \tau w^2 = \pi i u w = -\pi i \tau' u^2.$$

We may suppose this transformation obtained from the formula given above for the simple transformation B—thus—Apply first the transformation B which interchanges $\omega_{r,1}$, $\omega'_{r,1}$ with a certain change of sign of one of them; then apply the transformation $A_2 B A_2$ which effects a similar change for the pair $\omega_{r,2}$, $\omega'_{r,2}$; then the transformation $A_3 B A_3$, and so on. Thence we eventually obtain the formula

$$e^{\pi i \tau w^2} \Theta\left(u;\ \tau \middle| \begin{smallmatrix} Q' \\ Q \end{smallmatrix}\right) = \sqrt{\frac{i}{\tau_{1,1}}} \sqrt{\frac{i}{\tau'_{2,2}}} \sqrt{\frac{i}{\tau''_{3,3}}} \dots\dots e^{2\pi i (Q_1 Q_1' + \dots + Q_p Q_p')} \Theta\left(w;\ \tau' \middle| \begin{smallmatrix} -Q \\ Q' \end{smallmatrix}\right),$$

where

$$\tau'_{2,2} = \tau_{2,2} - \frac{\tau_{1,2}^2}{\tau_{1,1}}, \quad \tau''_{3,3} = \tau'_{3,3} - \frac{\tau'^2_{2,3}}{\tau'_{2,2}}, \dots,$$

and, save for an eighth root of unity,

$$\sqrt{\frac{i}{\tau_{1,1}}} \sqrt{\frac{i}{\tau'_{2,2}}} \sqrt{\frac{i}{\tau''_{3,3}}} \dots = \frac{1}{\sqrt{|\tau|}},$$

where $|\tau|$ is the determinant of the matrix τ.

The result can also be obtained immediately, and the constant obtained by an integration as in the simple case of the transformation B; we thus find, for the value of the constant here denoted by $\sqrt{\frac{i}{\tau_{1,1}}}\sqrt{\frac{i}{\tau'_{2,2}}}\dots$, the integral*

$$\int_{-\infty}^{\infty}\dots\int_{-\infty}^{\infty} e^{\pi i\tau x^2}\, dx_1\dots dx_p.$$

Ex. i. Prove that another way of expressing the value of this integral is

$$e^{\frac{1}{2}i\sum_{r=1}^{p}\tan^{-1}\lambda_r}/\sqrt[4]{|\tau\tau_0|},$$

where, if the matrix τ be written $\rho+i\sigma$, $|\tau\tau_0|$ is the determinant of the matrix $\rho^2+\sigma^2$, which is equal to the square of the modulus of the determinant of the matrix τ, also $\lambda_1, \dots, \lambda_p$ are the (real) roots of the determinantal equation $|\rho-\lambda\sigma|=0$, and $\tan^{-1}\lambda_r$ lies between $-\pi/2$ and $\pi/2$. Of the fourth root the positive real value is to be taken.

Ex. ii. For the case $p=1$, the constant for any linear transformation is given by

$$e^{\pi i a'(a+\tau a')w^2}\Theta\left[u;\ \tau\,\middle|\,\tfrac{1}{2}\binom{s'}{s}\right]\div\Theta(w;\ \tau')=L\sum_{\mu=0}^{a'-1}e^{-\frac{\pi i a}{a'}(\mu+\frac{1}{2}a')^2}$$

$$=L\sqrt{a'}\left(\frac{a'}{a}\right)e^{-\frac{\pi i a}{4}}\ \text{or}\ L\sqrt{a'}\left(\frac{a}{a'}\right)e^{-\frac{\pi i}{4}[a+(a-1)(a'-1)]}$$

according as a or a' is odd; where a' is positive, and

$$\begin{aligned}as'-a's&=aa',\\ \beta s'-\beta's&=\beta\beta',\end{aligned}\qquad L=e^{\frac{\pi i a'}{4a}s^2}\sqrt{\frac{i}{a'(a+\tau a')}}.$$

336. Returning now to consider the theory more particularly in connexion with the Riemann surface, we prove first that every linear transformation of periods such as

$$[\omega]=\omega\alpha+\omega'\alpha',\quad [\omega']=\omega\beta+\omega'\beta',$$

where

$$\alpha\bar{\beta}-\beta\bar{\alpha}=0,\quad \alpha'\bar{\beta}'-\beta'\bar{\alpha}'=0,\quad \alpha\bar{\beta}'-\beta\bar{\alpha}'=1,$$

can be effected by a change in the manner in which the period loops are taken. For this it is sufficient to prove that each of the four elementary types of transformation, A_k, B, C, D, from which, as we have seen, every such transformation can be constructed, can itself be effected by a change in the period loops.

The change of periods due to substitutions A_k can clearly be effected without drawing the period loops differently, by merely numbering them

* Weber has given a determination of the constant A for a general linear transformation by means of such an integral, and thence, by means of multiple-Gaussian series. See *Crelle*, LXXIV. (1872), pp. 57 and 69.

differently—attaching the numbers 1, k to the period-loop-pairs which were formerly numbered k and 1. No further remark is therefore necessary in regard to this case.

The substitution B, which makes only the change given by

$$[\omega_{r,1}] = \omega'_{r,1}, \quad [\omega'_{r,1}] = -\omega_{r,1},$$

can be effected, as in § 320, by regarding the loop (b_1) as an $[a_1]$ loop, with retention of its positive direction; thus the direction of the (old) loop (a_1), which now becomes the $[b_1]$ loop, will be altered; the change is shewn by comparing the figure of § 18 (p. 21) with the annexed figure (13).

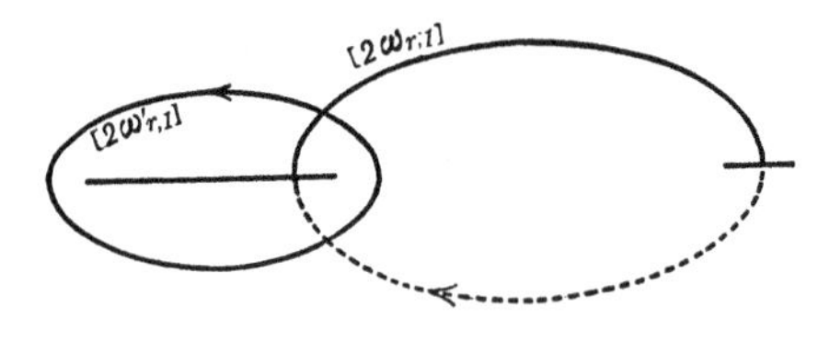

FIG. 13.

The change, due to the substitution C, which is given by

$$[\omega'_{r,1}] = \omega'_{r,1} - \omega_{r,1},$$

is to be effected by drawing the loop $[a_1]$ in such a way that a circuit of it (which gives rise to the value $[2\omega'_{r,1}]$ for the integral u_r) is equivalent to a circuit of the original loop (a_1) taken with a circuit of the loop (b_1) from the positive to the negative side of the original loop (a_1).

This may be effected by taking the loop $[a_1]$ as in the annexed figure (14) (cf. § 331).

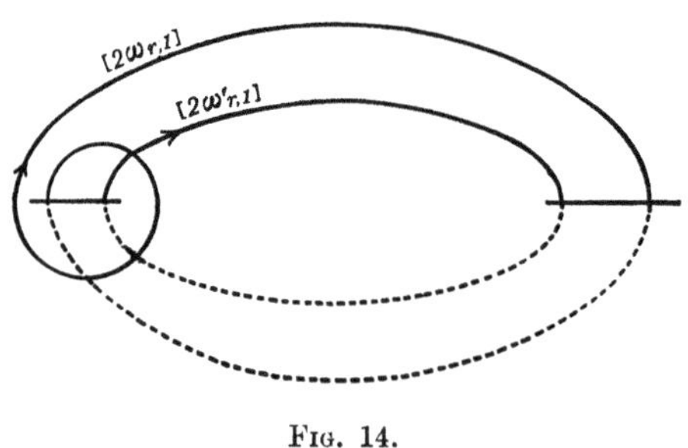

FIG. 14.

For the transformation D the only change introduced is that given by

$$[\omega'_{r,1}] = \omega'_{r,1} - \omega_{r,2}, \quad [\omega'_{r,2}] = \omega'_{r,2} - \omega_{r,1},$$

and this is effected by drawing the loops $[a_1]$, $[a_2]$, so that a circuit of

$[a_1]$ is equivalent to a circuit of the (original) loop (a_1) together with a circuit of (b_2), in a certain direction, and similarly for $[a_2]$. This may be done as in the annexed diagram (Fig. 15).

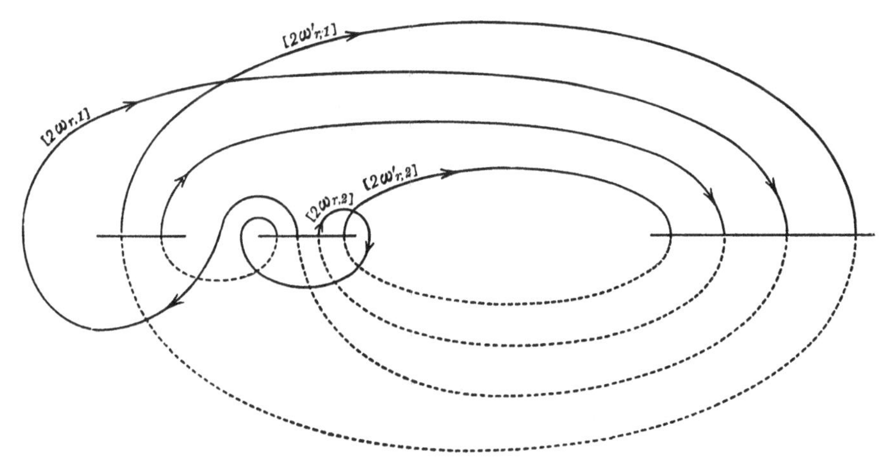

FIG. 15.

For instance the new loop $[a_2]$ in this diagram (Fig. 15) is a deformation of a loop which may be drawn as here (Fig. 16);

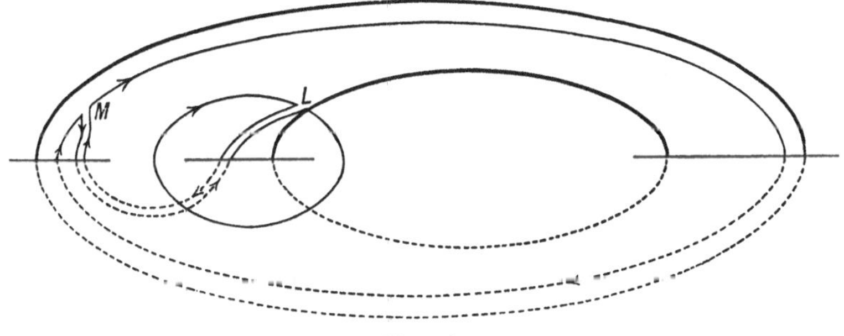

FIG. 16.

since the integrand of the Abelian integral u_r is single-valued on the Riemann surface, independently of the loops, the doubled portion from L to M is self-destructive; and a circuit of this new loop $[a_2]$ gives $\omega'_{r,2} - \omega_{r,1}$, as desired.

Hence the general transformation can be effected by a composition of the changes here given. It is immediately seen, for any of the linear transformations of § 326, that if the arguments there denoted by $U_1, \ldots, U_p$ be a set of normal integrals of the first kind for the original system of period loops, then $W_1, \ldots, W_p$ are a normal set for the new loops associated with the transformation.

337. Coming next to the question of how the theory of the vanishing of the Riemann theta function, which has been given in Chap. X., is modified

by the adoption of a different series of period loops, we prove first that when a change is made equivalent to the linear transformation

$$[\omega] = \omega\alpha + \omega'\alpha', \quad [\omega'] = \omega\beta + \omega'\beta',$$

the places $m_1, \dots, m_p$ of § 179, Chap. X., derived from any place m, upon which the theory of the vanishing of the theta function depends, become changed into places $m_1', \dots, m_p'$ which satisfy the p equations

$$u_i^{m_1', m_1} + \dots + u_i^{m_p', m_p} \equiv \tfrac{1}{2}[d(\alpha\bar{\beta})]_i + \tfrac{1}{2}\tau_{i,1}[d(\alpha'\bar{\beta}')]_1 + \dots + \tfrac{1}{2}\tau_{i,p}[d(\alpha'\bar{\beta}')]_p, \quad (i = 1, \dots, p),$$

wherein $u_1, \dots, u_p$ denote the normal integrals of the first kind for the original system of period loops.

For let $w_1, \dots, w_p$ be the normal integrals of the first kind for the new period loops, and let $m_1', \dots, m_p'$ be the places derived from the place m, in connexion with the new system of period loops, just as $m_1, \dots, m_p$ were derived from the original system. In the equation of transformation

$$e^{\pi i \bar{a}'(a+\tau a')w^2}\,\Theta\left[u;\ \tau \,\middle|\, \begin{matrix}\tfrac{1}{2}d(\alpha'\bar{\beta}') \\ \tfrac{1}{2}d(\alpha\bar{\beta})\end{matrix}\right] = A_1\Theta(w;\ \tau'),$$

put

$$w = w^{x,m} - w^{x_1, m_1'} - \dots - w^{x_p, m'_p},$$

so that the right-hand side of the equation vanishes when x is at any one of the places $m_1', \dots, m_p'$; then we also have

$$u = u^{x,m} - u^{x_1, m_1'} - \dots - u^{x_p, m_p'};$$

hence the function

$$\Theta\left[u^{x,m} - u^{x_1,m_1'} - \dots - u^{x_p, m_p'};\ \tau \,\middle|\, \begin{matrix}\tfrac{1}{2}d(\alpha'\bar{\beta}') \\ \tfrac{1}{2}d(\alpha\bar{\beta})\end{matrix}\right]$$

vanishes when x is at any one of the places $x_1, \dots, x_p$; therefore, by a proposition previously given (Chap. X., § 184 (X.)), the places $m_1', \dots, m_p'$ satisfy the equivalence stated above.

It is easy to see that this equivalence may be stated in the form

$$w_i^{m_1', m_1} + \dots + w_i^{m'_p, m_p} \equiv \tfrac{1}{2}[d(\bar{\beta}\beta')]_i + \tfrac{1}{2}\tau'_{i,1}[d(\bar{\alpha}\alpha')]_1 + \dots + \tfrac{1}{2}\tau'_{i,p}[d(\bar{\alpha}\alpha')]_p, \quad (i = 1, 2, \dots, p).$$

It may be noticed also that, of the elementary transformations associated with the matrices A_k, B, C, D, of § 333, only the transformation associated with the matrix C gives rise to a change in the places $m_1, \dots, m_p$; for each of the others the characteristic $[\tfrac{1}{2}d(\alpha\bar{\beta}), \tfrac{1}{2}d(\alpha'\bar{\beta}')]$ vanishes.

338. From the investigation of § 329 it follows, by interchanging the rows and columns of the matrix of transformation, that a linear trans-

formation can be taken for which the characteristic $[\frac{1}{2}d(\alpha\bar{\beta}), \frac{1}{2}d(\alpha'\bar{\beta}')]$ represents any specified even characteristic; thus all the $2^{p-1}(2^p+1)$ sets*, $m_1', \ldots, m_p'$, which arise by taking the characteristic $\frac{1}{2}\binom{\mu'}{\mu}$ in the equivalence

$$u^{m_1', m_1} + \ldots\ldots + u^{m'_p, m_p} \equiv \tfrac{1}{2}\Omega_{\mu, \mu'}$$

to be in turn all the even characteristics, can arise for the places $m_1', \ldots, m_p'$. In particular, if $\frac{1}{2}\Omega_{\mu, \mu'}$ be an even half-period for which $\Theta(\frac{1}{2}\Omega_{\mu, \mu'})$ vanishes, we may obtain for $m_1', \ldots, m_p'$ a set consisting of the place m and $p-1$ places $n_1', \ldots, n'_{p-1}$, in which $n_1', \ldots, n'_{p-1}$ are one set of a co-residual lot of sets of places in each of which a ϕ-polynomial vanishes to the second order (cf. Chap. X., § 185).

Ex. If in the hyperelliptic case, with $p=3$, the period loops be altered from those adopted in Chap. XI., in a manner equivalent to the linear transformation given in the Example of § 329, the function $\Theta(w; \tau')$, defined by means of the new loops, will vanish for $w=0$; and the places m_1', m_2', m_3', arising from the place a (§ 203, Chap. XI.), as $m_1, \ldots, m_p$ arise from m in § 179, Chap. X., will consist† of the place a itself and two arbitrary conjugate places, z and $\bar{z}$.

339. We have, on page 379 of the present volume, explained a method of attaching characteristics to root forms $\sqrt{X^{(1)}}, \sqrt{Y^{(3)}}$; we enquire now how these characteristics are modified when the period loops are changed. It will be sufficient to consider the case of $\sqrt{Y^{(3)}}$; the case of $\sqrt{X^{(1)}}$ arises (§ 244) by taking $\phi_0\sqrt{X^{(1)}}$ in place of $\sqrt{Y^{(3)}}$. Altering the notation of § 244, slightly, to make it uniform with that of this chapter, the results there obtained are as follows; the form $X^{(3)}$ is a polynomial of the third degree in the fundamental ϕ-polynomials, which vanishes to the second order in each of the places $A_1, \ldots, A_{2p-3}, m_1, \ldots, m_p$, where $A_1, \ldots, A_{2p-3}$ are, with the place m, the zeros of a ϕ-polynomial ϕ_0; the form $Y^{(3)}$ is a polynomial, also of the third degree in the fundamental ϕ-polynomials, which vanishes to the second order in each of the places $A_1, \ldots, A_{2p-3}, \mu_1, \ldots, \mu_p$; if

$$u_i^{\mu_1, m_1} + \ldots + u_i^{\mu_p, m_p} = \tfrac{1}{2}(q_i + q_i'\tau_{i,1} + \ldots + q_p'\tau_{i,p}), \quad (i = 1, 2, \ldots, p),$$

where $u_1, \ldots, u_p$ are the Riemann normal integrals of the first kind, the characteristic associated with the form $Y^{(3)}$ is that denoted by $\frac{1}{2}\binom{q'}{q}$; and‡ it may be defined by the fact that the function $\sqrt{Y^{(3)}}/\sqrt{X^{(3)}}$, which is single-valued on the dissected Riemann surface, takes the factors $(-1)^{q_i'}, (-1)^{q_i}$ respectively at the i-th period loops of the first and second kind.

Take now another set of period loops; let $m_1', \ldots, m_p'$ be the places

* Or lot of sets, when the equivalence has not an unique solution.

† Cf. the concluding remark of § 185.

‡ Integer characteristics being omitted.

which, for these loops, arise as $m_1, \ldots, m_p$ arise for the original set of period loops; let $Z^{(3)}$ be the form which, for the new loops, has the same character as has the form $X^{(3)}$ for the original loops, so that $Z^{(3)}$ vanishes to the second order in each of $A_1, \ldots, A_{2p-3}, m_1', \ldots, m_p'$; then from the equivalences (§ 337)

$$w_i^{m_1', m_1} + \ldots + w_i^{m'_p, m_p} \equiv \tfrac{1}{2}[d(\bar{\beta}\beta')]_i + \tfrac{1}{2}\tau'_{i,1}[d(\bar{\alpha}\alpha')]_1 + \ldots + \tfrac{1}{2}\tau'_{i,p}[d(\bar{\alpha}\alpha')]_p, \quad (i = 1, \ldots, p),$$

where $w_1, \ldots, w_p$ are the normal integrals of the first kind, it follows, as in § 244, that the function $\sqrt{Z^{(3)}}/\sqrt{X^{(3)}}$ is single-valued on the Riemann surface dissected by the new system of period loops, and at the r-th new loops, respectively of the first and second kind, has the factors

$$e^{-\pi i[d(\bar{\alpha}\alpha')]_r}, \quad e^{\pi i[d(\bar{\beta}\beta')]_r}.$$

The equations of transformation,

$$[\omega] = \omega\alpha + \omega'\alpha', \quad [\omega'] = \omega\beta + \omega'\beta',$$

of which one particular equation is that given by

$$[\omega_{n,r}] = \omega_{n,1}\alpha_{1,r} + \ldots\ldots + \omega_{n,p}\alpha_{p,r} + \omega'_{n,1}\alpha'_{1,r} + \ldots + \omega'_{n,p}\alpha'_{p,r}, \quad (n, r = 1, \ldots, p),$$

express the fact (cf. § 322) that a negative circuit of the new loop $[b_r]$ is equivalent to $\alpha_{i,r}$ negative circuits of the original loop (b_i) and $\alpha'_{i,r}$ positive circuits of the original loop (a_i); thus a function which has the factors $e^{-\pi i q_i'}$, $e^{\pi i q_i}$ at the i-th original loops, will at the r-th new loop $[a_r]$ have the factor $e^{-\pi i l_r'}$, where l_r' is an integer which is given by

$$-l_r' \equiv \sum_{i=1}^{p} [-q_i'\alpha_{i,r} + q_i\alpha'_{i,r}], \qquad (\text{mod. } 2);$$

thus the factors of $\sqrt{Y^{(3)}}/\sqrt{X^{(3)}}$ at the new period loops are given by $e^{-\pi i l'}$, $e^{\pi i l}$, where l, l' are rows of integers such that

$$l' \equiv \bar{\alpha}q' - \bar{\alpha}'q, \quad -l \equiv \bar{\beta}q' - \bar{\beta}'q, \qquad (\text{mod. } 2).$$

Therefore the factors of $\sqrt{Y^{(3)}}/\sqrt{Z^{(3)}} = (\sqrt{Y^{(3)}}/\sqrt{X^{(3)}})/(\sqrt{Z^{(3)}}/\sqrt{X^{(3)}})$, at the new period loops, are given by $e^{-\pi i k'}$, $e^{\pi i k}$, where

$$k' \equiv \bar{\alpha}q' - \bar{\alpha}'q - d(\bar{\alpha}\alpha'), \quad -k \equiv \bar{\beta}q' - \bar{\beta}'q - d(\bar{\beta}\beta'), \qquad (\text{mod. } 2);$$

now the characteristic associated with $\sqrt{Y^{(3)}}$ corresponding to the original system of period loops may be defined by the factors of $\sqrt{Y^{(3)}}/\sqrt{X^{(3)}}$ at those loops; similarly the characteristic which belongs to $\sqrt{Y^{(3)}}$ for the new system of loops is defined by the factors of $\sqrt{Y^{(3)}}/\sqrt{Z^{(3)}}$, and is therefore $\frac{1}{2}\binom{k'}{k}$; the equations just obtained prove therefore that *the characteristic associated with $\sqrt{Y^{(3)}}$ is transformed precisely as a theta characteristic.*

The same result may be obtained thus; the p equations of the form

$$u_i^{\mu_1, m_1}+\dots+u_i^{\mu_p, m_p}=\tfrac{1}{2}(q_i+q_1\tau_{i,1}+\dots+q_p\tau_{i,p}), \quad (i=1, \dots, p),$$

are immediately seen, by means of the equation $(a+\tau a')(\bar{\beta}'-\tau'\bar{a}')=1$ to lead to p equations expressible by

$$w^{\mu_1, m_1}+\dots+w^{\mu_p, m_p}=\tfrac{1}{2}(\bar{\beta}'q-\bar{\beta}q')+\tfrac{1}{2}\tau'(\bar{a}q'-\bar{a}'q);$$

subtracting from these the equations

$$w_i^{m_1', m_1}+\dots+w_i^{m_p', m_p}\equiv\tfrac{1}{2}[d(\bar{\beta}\beta')]_i+\tfrac{1}{2}\tau'_{i,1}[d(\bar{a}a')]_1+\dots+\tfrac{1}{2}\tau'_{i,p}[d(\bar{a}a')]_p, \quad (i=1, \dots, p),$$

we obtain equations from which (as in § 244) the characteristic of $\sqrt{Y^{(3)}}$, for the new loops, is immediately deducible.

Similar reasoning applies obviously to the characteristics of the forms $\sqrt{X^{(2\nu+1)}}$ considered on page 380 (§ 245). But the characteristic for a form $\sqrt{X^{(2\mu)}}$ (p. 381), which is obtained by consideration of the single-valued function $\sqrt{X^{(2\mu)}}/\Phi^{(\mu)}$—into which the form $\sqrt{X^{(3)}}$, depending on the places $m_1, \dots, m_p$, does not enter—is transformed in accordance with the equations

$$k'\equiv\bar{\alpha}q'-\bar{\alpha}'q, \quad -k\equiv\bar{\beta}q'-\bar{\beta}'q, \quad (\text{mod. } 2),$$

and may be described as a *period-characteristic*, as in § 328.

340. Having thus investigated the dependence of the characteristics assigned to radical forms upon the method of dissection of the Riemann surface, it is proper to explain, somewhat further, how these characteristics may be actually specified for a given radical form. The case of a form $\sqrt{X^{(2\mu)}}$ differs essentially from that of a form $\sqrt{X^{(2\nu+1)}}$. When the zeros of a form $\sqrt{X^{(2\mu)}}$ are known, and the Riemann surface is given with a specified system of period loops, the factors of a function $\sqrt{X^{(2\mu)}}/\Phi^{(\mu)}$ at these loops may be determined by following the value of the function over the surface, noticing the places at which the values of the function branch—which places are in general only the fixed branch places of the Riemann surface; the process is analogous to that whereby, in the case of elliptic functions, the values of $\sqrt{\wp(u+2\omega_1)-e_1}/\sqrt{\wp(u)-e_1}$, $\sqrt{\wp(u+2\omega_2)-e_1}/\sqrt{\wp(u)-e_1}$ may be determined, by following the values of $\sqrt{\wp(u)-e_1}$ over the parallelogram of periods. But it is a different problem to ascertain the factors of the function $\sqrt{Y^{(3)}}/\sqrt{X^{(3)}}$ at the period loops, because the form $\sqrt{X^{(3)}}$ depends upon the places $m_1, \dots, m_p$, and we have given no elementary method of determining these places; the geometrical interpretation of these places which is given in § 183 (Chap. X.), and the algebraic process resulting therefrom, does not distinguish them from other sets of places satisfying the same conditions; the distinction in fact, as follows from § 338, cannot be made algebraically unless the period loops are given by algebraical equations. Nevertheless we

may determine the characteristic of a form $Y^{(3)}$, and the places $m_1, \dots, m_p$, by the following considerations*:—It is easily proved, by an argument like that of § 245 (Chap. XIII.), that if there be a form $\sqrt{X^{(1)}}$ having the same characteristic as $\sqrt{Y^{(3)}}$, there exists an equation of the form $\sqrt{X^{(1)}}\sqrt{Y^{(3)}} = \Phi^{(2)}$; and conversely, if $q+1$ linearly independent polynomials, of the second degree in the p fundamental ϕ-polynomials, vanish in the zeros of $\sqrt{Y^{(3)}}$, and $\Psi^{(2)}$ denote the sum of these $q+1$ polynomials, each multiplied by an arbitrary constant, that we have an equation $\sqrt{Y^{(1)}}\sqrt{Y^{(3)}} = \Psi^{(2)}$, where $\sqrt{Y^{(1)}}$ is a linear aggregate of $q+1$ radical forms like $\sqrt{X^{(1)}}$, all having the same characteristic as $\sqrt{Y^{(3)}}$; in general, since a form $\Psi^{(2)}$ can contain at most $3(p-1)$ linearly independent terms (§ 111, Chap. VI.), and the number of zeros of $\sqrt{Y^{(3)}}$ is $3(p-1)$, we have $q+1=0$; in any case the value of $q+1$ is capable of an algebraic determination, being the number of forms $\Phi^{(2)}$ which vanish in assigned places. Now the number of linearly independent forms $\sqrt{X^{(1)}}$ with the same characteristic is even or odd according as the characteristic is even or odd (§§ 185, 186, Chap. X.); hence, without determining the characteristic of $\sqrt{Y^{(3)}}$ we can beforehand ascertain whether it is even or odd by finding whether $q+1$ is even or odd. Suppose now that $\mu_1, \dots, \mu_p$ and $\mu_1', \dots, \mu_p'$ are two sets of places such that

$$(m^3, A_1, \dots, A_{2p-3}) \equiv (\mu_1^2, \dots, \mu_p^2) \equiv (\mu_1'^2, \dots, \mu_p'^2),$$

m being an arbitrary place, and $m, A_1, \dots, A_{2p-3}$ being the zeros of any ϕ-polynomial ϕ_0; so that $\mu_1, \dots, \mu_p$ and $\mu_1', \dots, \mu_p'$ are two sets arbitrarily selected from 2^{2p} sets which can be determined geometrically as in § 183, Chap. X. (cf. § 244, Chap. XIII.); let $Y^{(3)}$ vanish to the second order in each of $\mu_1, \dots, \mu_p, A_1, \dots, A_{2p-3}$ and $Y_1^{(3)}$ vanish to the second order in each of $\mu_1', \dots, \mu_p', A_1, \dots, A_{2p-3}$; by following the values of the single-valued function $\sqrt{Y_1^{(3)}}/\sqrt{Y^{(3)}}$ on the Riemann surface, we can determine its factors at the period loops; at the r-th period loops of the first and second kind let these factors be $(-1)^{k_r'}$, $(-1)^{k_r}$ respectively; then if $\frac{1}{2}(q_1, \dots, q_p')$ and $\frac{1}{2}(Q_1, \dots, Q_p')$ be respectively the characteristics of $\sqrt{Y^{(3)}}$ and $\sqrt{Y_1^{(3)}}$, which we wish to determine, we have (§ 244)

$$k_r' \equiv Q_r' - q_r', \quad k_r \equiv Q_r - q_r, \quad (\text{mod. } 2).$$

Take now, in turn, for $\mu_1', \dots, \mu_p'$, all the possible 2^{2p} sets which, as in § 183, are geometrically determinable from the place m; and, for the same form $\sqrt{Y^{(3)}}$, determine the 2^{2p} characteristics of all the functions $\sqrt{Y_1^{(3)}}/\sqrt{Y^{(3)}}$ arising

* Noether, *Jahresbericht der Deutschen Mathematiker Vereinigung*, Bd. iii. (1894), p. 494, where the reference is to Fuchs, *Crelle*, LXXIII. (1871); cf. Prym, *Zur Theorie der Functionen in einer zweiblättrigen Fläche* (Zürich, 1866).

by the change of the forms $\sqrt{Y_1^{(3)}}$; then there exists one, and only one, characteristic, $\frac{1}{2}\binom{s'}{s}$, satisfying the condition that the characteristic

$$\tfrac{1}{2}\binom{s'}{s}+\tfrac{1}{2}\binom{k'}{k}$$

is even when $\sqrt{Y_1^{(3)}}$ has an even characteristic and odd when $\sqrt{Y_1^{(3)}}$ has an odd characteristic; for, clearly, the characteristic $\frac{1}{2}\binom{q'}{q}$ is a value for $\frac{1}{2}\binom{s'}{s}$ which satisfies the condition, and if $\frac{1}{2}\binom{\sigma'}{\sigma}$ were another possible value for $\frac{1}{2}\binom{s'}{s}$ we should have

$$(k+\sigma)(k'+\sigma') \equiv (k+q)(k+q') \qquad (\text{mod. } 2),$$

or

$$k(\sigma'-q')+k'(\sigma-q) \equiv qq'-\sigma\sigma'$$

for all the 2^{2p} possible values of $\frac{1}{2}\binom{k'}{k}$; and this is impossible (Chap. XVII., § 295).

Hence we have the following rule:—*Investigate the factors of $\sqrt{Y_1^{(3)}}/\sqrt{Y^{(3)}}$ for an arbitrary form $\sqrt{Y^{(3)}}$ and all 2^{2p} forms $\sqrt{Y_1^{(3)}}$; corresponding to each form $\sqrt{Y_1^{(3)}}$ determine, by the method explained in the earlier part of this Article, whether its characteristic is even or odd; then, denoting the factors of any function $\sqrt{Y_1^{(3)}}/\sqrt{Y^{(3)}}$ respectively at the first and second kinds of period loops by quantities of the form $(-1)^{k'}, (-1)^{k}$, determine the characteristic $\frac{1}{2}\binom{q'}{q}$, satisfying the condition that the characteristic $\frac{1}{2}\binom{q'+k'}{q+k}$ is, for every form $\sqrt{Y_1^{(3)}}$, even or odd according as the characteristic of that form, $\sqrt{Y_1^{(3)}}$, is even or odd; then $\frac{1}{2}\binom{q'}{q}$ is the characteristic of the form $\sqrt{Y^{(3)}}$; this being determined the characteristic of every form $\sqrt{Y_1^{(3)}}$ is known; the particular form $\sqrt{Y_1^{(3)}}$ for which the characteristic, thus arising, is actually zero, is the form previously denoted by $\sqrt{X^{(3)}}$—namely the form vanishing in the places $m_1, \dots, m_p$ which are to be associated (as in § 179, Chap. X.) with the particular system of period loops of the Riemann surface which has been adopted.*

Thus the method determines the places $m_1, \dots, m_p$ and determines the characteristic of every form $\sqrt{Y^{(3)}}$; the characteristic of any other form $\sqrt{Y^{(2\nu+1)}}$ is then algebraically determinable by the theorems of § 245 (p. 380).

341. For the hyperelliptic case we have shewn, in Chap. XI., how to express the ratios of the 2^{2p} Riemann theta functions with half-integer characteristics by means of algebraic functions; the necessary modification

of these formulae when the period loops are taken otherwise than in Chap. XI., follows immediately from the results of this chapter. If the change in the period loops be that leading to the linear transformation which is associated with the Abelian matrix formed with the integer matrices α, β, α', β', we have (§ 324)

$$\vartheta\left[u;\ \tfrac{1}{2}\binom{q'}{q}\right] = A\vartheta_1\left[w;\ \tfrac{1}{2}\binom{k'}{k}\right],$$

where

$$k' = \bar{\alpha}q' - \bar{\alpha}'q - d(\bar{\alpha}\alpha'), \quad -k = \bar{\beta}q' - \bar{\beta}'q - d(\bar{\beta}\beta').$$

If now, considering as sufficient example the formula of § 208 (Chap. XI.), we have

$$u_r^{b,\,a} \equiv q_1\omega_{r,1} + \dots + q_p\omega_{r,\,p} + q_1'\omega'_{r,1} + \dots + q_p'\omega'_{r,p},$$

then we have

$$w_r^{b,\,a} \equiv l_1 v_{r,1} + \dots + l_p v_{r,\,p} + l_1' v'_{r,1} + \dots + l_p' v_{r,\,p},$$

where

$$l' = \bar{\alpha}q' - \bar{\alpha}'q = k' + d(\bar{\alpha}\alpha'), \quad -l = \bar{\beta}q' - \bar{\beta}'q = -k + d(\bar{\beta}\beta');$$

therefore, if the characteristic $\frac{1}{2}(d(\bar{\beta}\beta'), d(\bar{\alpha}\alpha'))$ be denoted by μ, the function $\vartheta_1\left[w;\ \frac{1}{2}\binom{k'}{k}\right]$ is a constant multiple of $\vartheta_1\left[w;\ \frac{1}{2}\binom{l'}{l} + \mu\right]$; and we may denote the latter function by $\vartheta_1[w|w^{b,a} + \mu]$. Thus the formula of § 208 is equivalent to

$$\sqrt{(b - x_1)\dots(b - x_p)} = C\,\frac{\vartheta_1(w|w^{b,a} + \mu)}{\vartheta_1(w|\mu)},$$

where C is independent of the arguments $w_1, \dots, w_p$, and, as in § 206,

$$w_r = w_r^{x_1,\,a_1} + \dots + w_r^{x_p,\,a_p}, \qquad (r = 1, 2, \dots, p).$$

Similar remarks apply to the formula of §§ 209, 210. It follows from § 337 that the characteristic μ is that associated with the half-periods

$$w^{m_1',\,a_1} + \dots + w^{m_p',\,a_p},$$

where $m_1', \dots, m_p'$ are the places which, for the new system of period loops, play the part of the places $m_1, \dots, m_p$ of § 179, Chap. X. It has already (§ 337) been noticed that for the elementary linear substitutions A_k, B, D the characteristic μ is zero.

342. In case the roots $c_1, a_1, c_2, a_2, \dots, c$, in the equation associated with the hyperelliptic case

$$y^2 = 4(x - c_1)(x - a_1)(x - c_2)(x - a_2)\dots(x - c_p)(x - a_p)(x - c),$$

be *real* and in *ascending* order of magnitude, we may usefully modify the notation of § 200, Chap. XI. Denote these roots, in order, by $b_{2p}, b_{2p-1}, \dots, b_0$, so that b_{2i}, b_{2i-1} are respectively c_{p-i+1}, a_{p-i+1} and b_0 is c, and interchange the period loops (a_i), (b_i), with retention of the direction of (b_i), as in the figure annexed (Fig. 17).

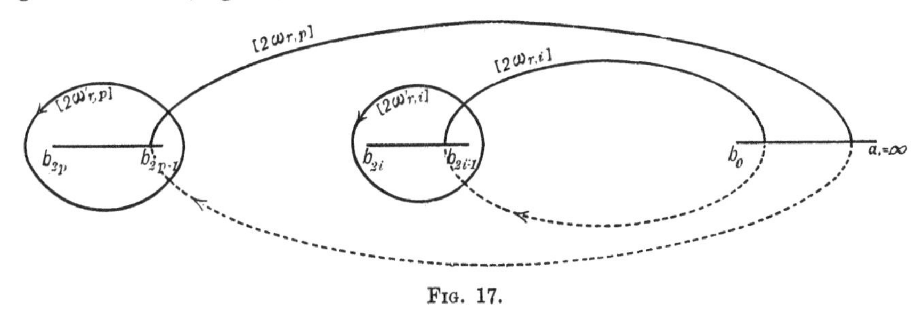

FIG. 17.

Then if $U_1^{x,a}, \ldots, U_p^{x,a}$ are linearly independent integrals of the first kind, such that $dU_r^{x,a}/dx = \psi_r/y$, where ψ_r is an integral polynomial in x, of degree $p-1$ at most, with only *real* coefficients, the half-periods

$$U_r^{b_{2i},\, b_{2i-1}} = [\omega_{r,i}], \quad U_r^{b_{2i-1},\, b_{2i-2}} = [\omega'_{r,i}] - [\omega'_{r,i-1}], \quad (i = 1, 2, \ldots, p;\ [\omega']_{r,0} = 0),$$

are respectively real and purely imaginary, so that $[\omega'_{r,i}]$ is also purely imaginary; if now $w_1^{x,a}, \ldots, w_p^{x,a}$ be the normal integrals, so that

$$U_r = [2\omega_{r,1}]\, w_1 + \ldots + [2\omega_{r,p}]\, w_p, \quad w_r = L_{r,1} U_1 + \ldots + L_{r,p} U_p,$$

then the second set of periods of $w_1^{x,a}, \ldots, w_p^{x,a}$, which are given by

$$\tau'_{r,i} = L_{r,1}[2\omega'_{1,i}] + \ldots + L_{r,p}[2\omega'_{p,i}], \qquad (r, s = 1, 2, \ldots, p),$$

are also purely imaginary*; forming with these the theta function $\Theta(w;\ \tau')$, the theta function of Chap. XI. is given (§ 335) by

$$e^{\pi i \tau w^2}\, \Theta\left(u;\ \tau \,\middle|\, {}^{Q'}_{Q}\right) = A e^{2\pi i QQ'}\, \Theta\left(w;\ \tau' \,\middle|\, {}^{-K}_{K'}\right),$$

where K, K' are obtainable from Q, Q' respectively by reversing the order of the p elements, and A is the constant $\sqrt{i/\Delta_1}\,\sqrt{i\Delta_1/\Delta_2}\,\sqrt{i\Delta_2/\Delta_3}\ldots$, in which $\Delta = \tau_{1,1}$, $\Delta_2 = \tau_{1,1}\tau_{2,2} - \tau^2_{1,2}$, etc. We find immediately that

$$U_r^{b_{2i-1},\, a} = -[\omega_{r,i}] - \ldots - [\omega_{r,p}] + [\omega'_{r,i}], \quad U_r^{b_{2i},a} = -[\omega_{r,i+1}] - \ldots - [\omega_{r,p}] + [\omega'_{r,i}],$$

$(i = 0, 1, \ldots, p)$, and may hence associate with b_{2i-1}, b_{2i} the respective odd and even characteristics

$$\{2i-1\} = \tfrac{1}{2}\begin{pmatrix} 0\ldots0 & 1 & 0\ldots & 0 \\ 0\ldots0 & -1 & -1\ldots & -1 \end{pmatrix}, \quad = \tfrac{1}{2}\begin{pmatrix}0\\0\end{pmatrix}^{i-1}\begin{pmatrix}1\\-1\end{pmatrix}\begin{pmatrix}0\\-1\end{pmatrix}^{p-i},$$

$$\{2i\} = \tfrac{1}{2}\begin{pmatrix} 0\ldots0 & 1 & 0\ldots & 0 \\ 0\ldots0 & 0 & -1\ldots & -1 \end{pmatrix} = \tfrac{1}{2}\begin{pmatrix}0\\0\end{pmatrix}^{i-1}\begin{pmatrix}1\\0\end{pmatrix}\begin{pmatrix}0\\-1\end{pmatrix}^{p-i},$$

* The quantities $\tau_{i,j}$ of Chap. XI. (of which the matrix is given in terms of the $\tau'_{i,j}$ of § 342 by $\tau\tau' = -1$) are also purely imaginary when $c_1, a_1, \ldots, c_p, a_p, c$ are real and in ascending order of magnitude.

and may denote the theta functions with these characteristics respectively by $\Theta_{2i-1}(w;\tau')$, $\Theta_{2i}(w;\tau')$; if $b_k, b_l, b_m, \dots,$ be any of the places $b_{2p}, \dots, b_0$, not more than p in number, and if, with $0 \ngtr q_i < 2$, $0 \ngtr q_i' < 2$, we have

$$U_r^{b_k, a} + U_r^{b_l, a} + \dots \equiv -q_1[\omega_{r,1}] - \dots - q_p[\omega_{r,p}] + q_1'[\omega'_{r,1}] + \dots + q_p'[\omega'_{r,p}],$$

then the function whose characteristic is $\frac{1}{2}\binom{q'}{-q}$ may be denoted by

$$\Theta_{k,l,m\dots}(w;\tau').$$

This function is equal to, or equal to the negative of, the function with characteristic $\frac{1}{2}\binom{q'}{q}$, according as the characteristic is even or odd.

We have thus a number notation for the 2^{2p} half-integer characteristics*, equally whether the surface be hyperelliptic or not; this notation is understood to be that of Weierstrass (Königsberger, *Crelle*, LXIV. (1865), p. 20). For the numerical definition of the half-periods, which are given by the rule at the bottom of p. 297, precise conventions are necessary as to the allocation of the signs of the single valued functions $\sqrt{x-b_r}$ on the Riemann surface (cf. Chap. XXII.).

In the hyperelliptic case $p=2$, the characteristics of the theta functions given in the table of § 204 are supposed to consist of positive elements less than unity; when Q_1, Q_2, Q_1', Q_2' are each either 0 or $\frac{1}{2}$, the formula of the present article gives

$$e^{\pi i\tau w^2}\,\Theta\left[u;\tau \,\middle|\, \begin{matrix} Q_1' & Q_2' \\ Q_1 & Q_2 \end{matrix}\right] = Ae^{-2\pi i QQ'}\,\Theta\left[w;\tau' \,\middle|\, \begin{matrix} Q_2 & Q_1 \\ -Q_2' & -Q_1' \end{matrix}\right];$$

the number notations for the transformed characteristics are then immediately given by the table of § 204. The result is that the numbers

02, 24, 04, 1, 13, 3, 5, 23, 12, 2, 01, 0, 14, 4, 34, 03

are respectively replaced by

3, 1, 13, 24, 04, 02, 5, 0, 4, 2, 34, 23, 14, 12, 01, 03.

* For convenience in the comparison of results in the analytical theory of theta functions, it appears better to regard it as a notation for the characteristics rather than for the functions.

CHAPTER XIX.

ON SYSTEMS OF PERIODS AND ON GENERAL JACOBIAN FUNCTIONS.

343. THE present chapter contains a brief account of some general ideas which it is desirable to have in mind in dealing with theta functions in general and more especially in dealing with the theory of transformation.

Starting with the theta functions it is possible to build up functions of p variables which have $2p$ sets of simultaneous periods—as for instance by forming quotients of integral polynomials of theta functions (Chap. XI., § 207), or by taking the second differential coefficients of the logarithm of a single theta function (Chap. XI., § 216, Chap. XVII., § 311 (δ)). Thereby is suggested, as a matter for enquiry, along with other questions belonging to the general theory of functions of several independent variables, the question whether every such multiply-periodic function can be expressed by means of theta functions*. Leaving aside this general theory, we consider in this chapter, in the barest outline, (i) the theory of the periods of an analytical multiply-periodic function, (ii) the expression of the most general single valued analytical integral function of which the second logarithmic differential coefficients are periodic functions.

344. If an uniform analytical function of p independent complex variables $u_1, \dots, u_p$ be such that, for every set of values of $u_1, \dots, u_p$, it is unaltered by the addition, respectively to $u_1, \dots, u_p$, of the constants $P_1, \dots, P_p$, then $P_1, \dots, P_p$ are said to constitute a period column for the function. Such a column will be denoted by a single letter, P, and P_k will denote any one of $P_1, \dots, P_p$. It is clear that if each of $P, Q, R, \dots$ be period columns for the function, and $\lambda, \mu, \nu, \dots$ be any definite integers, independent of k, then the column of quantities $\lambda P_k + \mu Q_k + \nu R_k + \dots$ is also a period column for the function; we shall denote this column by $\lambda P + \mu Q + \nu R + \dots$, and say that it is a linear function of the columns $P, Q, R, \dots$, the coefficients $\lambda, \mu, \nu, \dots$, in this case, but not necessarily

* Cf. Weierstrass, *Crelle*, LXXXIX. (1880), p. 8.

always, being integers. The real parts of the new column are the same linear functions of the real parts of the component columns, as also are the imaginary parts. More generally, when the p quantities $\lambda P_k + \mu Q_k + \nu R_k + \ldots$ are zero for the same values of $\lambda, \mu, \nu, \ldots$, we say that the columns $P, Q, R, \ldots$ are connected by a linear equation; it must be noticed, for the sake of definiteness, that it does not thence follow that, for instance, P is a linear function of the other columns, unless it is known that λ is not zero.

It is clear moreover that any $2p+1$, or more, columns of periods are connected by at least one linear equation with real coefficients (that is, an equation for each of the p positions in the column—p equations in all, with the same coefficients); for, in order to such an equation, the separation of real and imaginary gives $2p$ linear equations to be satisfied by the $2p+1$ real coefficients; allowing possible zero values for coefficients these equations can always be satisfied.

For instance the periods $\Omega = \Omega_1 + i\Omega_2$, $\omega = \omega_1 + i\omega_2$, $\omega' = \omega_1' + i\omega_2'$, are connected by an equation

$$t\Omega + x\omega + y\omega' = 0,$$

in which however, if $\omega_1\omega_2' - \omega_2\omega_1' = 0$, also $t = 0$.

Thus, for any periodic function, there exists a least number, r, of period columns, with r lying between 1 and $2p+1$, which are themselves not connected by any linear equation with real coefficients, but are such that every other period column is a linear function of these columns with real finite coefficients. Denoting such a set* of r period columns by $P^{(1)}, \ldots, P^{(r)}$, and denoting any other period column by Q, we have therefore the p equations

$$Q_k^{(r)} = \lambda_1 P_k^{(1)} + \ldots\ldots + \lambda_r P_k^{(r)}, \qquad (k = 1, 2, \ldots, p),$$

wherein $\lambda_1, \ldots, \lambda_r$ are independent of k, and are real and not infinite. *It is the purpose of what† follows to shew, in the case of an uniform analytical function of the independent complex variables $u_1, \ldots, u_p$,* (I.) *that unless the function can be expressed in terms of less than p variables which are linear functions of the arguments $u_1, \ldots, u_p$, the coefficients $\lambda_1, \ldots, \lambda_r$ are rational numbers,* (II.) *that, $\lambda_1, \ldots, \lambda_r$ being rational numbers, sets of r columns of periods exist in terms of which every existing period column can be linearly expressed with integral coefficients.*

Two lemmas are employed which may be enunciated thus:—

(α) If an uniform analytical function of the variables $u_1, \ldots, u_p$ have a column of infinitesimal periods, it is expressible as a function of less than p variables which are linear functions of $u_1, \ldots, u_p$. And conversely, if such

* It will appear that the number of such sets is infinite; it is the number r which is unique.

† These propositions are given by Weierstrass. *Abhandlungen aus der Functionenlehre* (Berlin, 1886), p. 165 (or *Berlin. Monatsber.* 1876).

uniform analytical function of $u_1, \ldots, u_p$ be expressible as a function of less than p variables which are linear functions of $u_1, \ldots, u_p$, it has columns of infinitesimal periods.

(β) Of periods of an uniform analytical function of the variables $u_1, \ldots, u_p$, which does not possess any columns of infinitesimal periods, there is only a finite number of columns of which every period is finite.

345. To prove the first part of lemma (α) it is sufficient to prove that when the function $f(u_1, \ldots, u_p)$ is *not* expressible as a function of less than p linear functions of $u_1, \ldots, u_p$, then it has *not* any columns of infinitesimal periods.

We define as an *ordinary* set of values of the variables $u_1, \ldots, u_p$ a set $u_1', \ldots, u_p'$, such that, for absolute values of the differences $u_1 - u_1', \ldots, u_p - u_p'$ which are within sufficient (not vanishing) nearness to zero, the function, $f(u_1, \ldots, u_p)$, can be represented by a converging series of positive integral powers of these differences—the possibility of such representation being the distinguishing mark of an analytical function; other sets of values of the variables are distinguished as *singular* sets of values*.

Then if the function be not expressible by less than p linear functions of $u_1, \ldots, u_p$, there can exist no set of constants $c_1, \ldots, c_p$ such that the function

$$c_1 \frac{\partial f}{\partial u_1} + \ldots + c_p \frac{\partial f}{\partial u_p}$$

vanishes for all ordinary sets of values of the variables; for this would require f to be a function of the $p-1$ variables $c_i u_1 - c_1 u_i \; (i = 2, \ldots, p)$. Hence there exist sets of ordinary values such that not all the differential coefficients $\partial f/\partial u_1, \ldots, \partial f/\partial u_p$ are zero; let $u_1^{(1)}, \ldots, u_p^{(1)}$ be such an ordinary set of values; for all values of $u_1, \ldots, u_p$ in the immediate neighbourhoods respectively of $u_1^{(1)}, \ldots, u_p^{(1)}$, the statement remains true that not all the partial differential coefficients are zero.

Then, similarly, the determinants of two rows and columns formed from the array

$$\left| \begin{matrix} \dfrac{\partial f}{\partial u_1^{(1)}}, & \dfrac{\partial f}{\partial u_2^{(1)}}, & \ldots, & \dfrac{\partial f}{\partial u_p^{(1)}} \\ \dfrac{\partial f}{\partial u_1}, & \dfrac{\partial f}{\partial u_2}, & \ldots, & \dfrac{\partial f}{\partial u_p} \end{matrix} \right|$$

do not all vanish for every ordinary set of values of the variables; let $u_1^{(2)}, \ldots, u_p^{(2)}$ be an ordinary set for which they do not vanish; for all values of

* The ordinary sets of values constitute a continuum of $2p$ dimensions, which is necessarily limited; the limiting sets of values are the singular sets. Cf. Weierstrass, *Crelle*, LXXXIX. (1880), p. 3.

$u_1, \dots, u_p$ in the immediate neighbourhoods respectively of $u_1^{(2)}, \dots, u_p^{(2)}$, the statement remains true that not all these determinants are zero.

Proceeding step by step in the way thus indicated we infer that there exist sets of ordinary values of the variables, $(u_1^{(1)}, \dots, u_p^{(1)}), \dots, (u_1^{(p)}, \dots, u_p^{(p)})$, such that the determinant, Δ, of p rows and columns in which the k-th element of the r-th row is $\partial f(u_1^{(r)}, \dots, u_p^{(r)})/\partial u_k^{(r)}$, does not vanish; and since these are ordinary sets of values of the arguments, this determinant will remain different from zero if (for $r = 1, \dots, p$) the set $u_1^{(r)}, \dots, u_p^{(r)}$ be replaced by $v_1^{(r)}, \dots, v_p^{(r)}$, where $v_k^{(r)}$ is a value in the immediate neighbourhood of $u_k^{(r)}$.

This fact is however inconsistent with the existence of a column of infinitesimal periods. For if $H_1, \dots, H_p$ be such a column, of which the constituents are not all zero, we have

$$0 = f(u_1^{(r)} + H_1, \dots, u_p^{(r)} + H_p) - f(u_1^{(r)}, \dots, u_p^{(r)}), \qquad (r = 1, \dots, p),$$

$$= \sum_{k=1}^{p} H_k \frac{\partial f}{\partial u_k} [u_1^{(r)} + \theta_1 H_1, \dots, u_p^{(r)} + \theta_p H_p],$$

where $\theta_1, \dots, \theta_p$ are quantities whose absolute values are $\not> 1$, and the bracket indicates that, after forming $\partial f/\partial u_k$, we are (for $m = 1, \dots, p$) to substitute $u_m^{(r)} + \theta_m H_m$ for $u_m^{(r)}$; these p equations, by elimination of $H_1, \dots, H_p$ give zero as the value of a determinant which is obtainable from Δ by slight changes of the sets $u_1^{(r)}, \dots, u_p^{(r)}$; we have seen above that such a determinant is not zero.

To prove the converse part of lemma (α) we may proceed as follows. Suppose that the function is expressible by m arguments $v_1, \dots, v_m$ given by

$$v_k = a_{k,1} u_1 + \dots + a_{k,p} u_p, \qquad (k = 1, \dots, m),$$

wherein $m < p$. The conditions that $v_1, \dots, v_m$ remain unaltered when $u_1, \dots, u_p$ are replaced respectively by $u_1 + tQ_1, \dots, u_p + tQ_p$ are satisfied by taking $Q_1, \dots, Q_p$ so that

$$a_{k,1} Q_1 + \dots\dots + a_{k,p} Q_p = 0, \qquad (k = 1, \dots, m),$$

and since $m < p$ these conditions can be satisfied by finite values of $Q_1, \dots, Q_p$ which are not all zero. The additions of the quantities $tQ_1, \dots, tQ_p$ to $u_1, \dots, u_p$, not altering $v_1, \dots, v_m$, will not alter the value of the function f. Hence by supposing t taken infinitesimally small, the function has a column of infinitesimal periods.

346. As to lemma (β), let $P_k = \rho_k + i\sigma_k$ be one period of any column of periods, $(k = 1, \dots, p)$, wherein ρ_k, σ_k are real, so that, in accordance with the condition that the function has no column of infinitesimal periods, there

is an assignable real positive quantity ϵ such that not all the $2p$ quantities ρ_k, σ_k are less than ϵ. Then if μ_k, ν_k be $2p$ specified positive integers, there is at most one column of periods satisfying the conditions

$$\mu_k\epsilon \ngtr |\rho_k| < (\mu_k+1)\,\epsilon, \qquad \nu_k\epsilon \ngtr |\sigma_k| < (\nu_k+1)\,\epsilon, \qquad (k=1, \ldots, p);$$

wherein $|\rho_k|$, $|\sigma_k|$ are the numerical values of ρ_k, σ_k; for if $\rho_k' + i\sigma_k'$ were one period of another column also satisfying these conditions, the quantities $\rho_k' - \rho_k + i(\sigma_k' - \sigma_k)$ would form a period column wherein every one of the $2p$ quantities $\rho_k' - \rho_k$, $\sigma_k' - \sigma_k$ was numerically less than ϵ.

Hence, since, if g be any assigned real positive quantity, there is only a finite number of sets of $2p$ positive integers μ_k, ν_k such that each of the $2p$ quantities $\mu_k\epsilon$, $\nu_k\epsilon$ is within the limits $(-g, g)$, it follows that there is only a finite number of columns of periods $P_k = \rho_k + i\sigma_k$, such that each of ρ_k, σ_k is numerically less than g. And this is the meaning of the lemma.

347. We return now to the expression (§ 344) of the most general period column of the function f by real linear functions of r period columns, of finite periods, in the form

$$Q = \lambda_1 P^{(1)} + \ldots\ldots + \lambda_r P^{(r)};$$

here the suffix is omitted, and we make the hypothesis that the function is not expressible by fewer than p linear combinations of $u_1, \ldots, u_p$.

Consider, first, the period columns Q for which $\lambda_2 = \lambda_3 = \ldots = \lambda_r = 0$ and $0 < \lambda_1 \ngtr 1$. Since there are no columns of infinitesimal periods, there is a lower limit to the values of λ_1 corresponding to existing period columns Q satisfying these conditions; and since there is only a finite number of period columns of wholly finite periods, there is an existing period for which λ_1 is equal to this lower limit. Let $\lambda_{1,1}$ be this least value of λ_1, and $Q^{(1)}$ be the corresponding period column Q.

Consider, next, the period columns Q for which $\lambda_3 = \lambda_4 = \ldots = \lambda_r = 0$, and $0 \ngtr \lambda_1 \ngtr 1$, $0 < \lambda_2 \ngtr 1$. As before there are period columns of this character in which λ_2 has a least value, which we denote by $\lambda_{2,2}$. If there exist several corresponding values of λ_1, let $\lambda_{1,2}$ denote one of these, and denote $\lambda_{1,2}P^{(1)} + \lambda_{2,2}P^{(2)}$ by $Q^{(2)}$.

In general consider the period columns of the form

$$\lambda_1 P^{(1)} + \ldots\ldots + \lambda_m P^{(m)}, \qquad (m \ngtr r),$$

wherein

$$0 \ngtr \lambda_1 \ngtr 1, \ \ldots\ldots, \ 0 \ngtr \lambda_{m-1} \ngtr 1, \ 0 < \lambda_m \ngtr 1.$$

Since there are no infinitesimal periods, there is a lower limit to the values of λ_m corresponding to existing period columns satisfying these conditions; since there is only a finite number of period columns of wholly finite periods, there is at least one existing column Q for which λ_m is equal to this lower limit; denote this value of λ_m by $\lambda_{m,m}$, and denote by $\lambda_{1,m}, \ldots, \lambda_{m-1,m}$ values arising in an actual period column $Q^{(m)}$ given by

$$Q^{(m)} = \lambda_{1,m}P^{(1)} + \lambda_{2,m}P^{(2)} + \ldots + \lambda_{m,m}P^{(m)};$$

there may exist more than one period column in which the coefficient of $P^{(m)}$ is $\lambda_{m,m}$.

Thus, taking $m = 1, 2, \ldots, r$, we obtain r period columns $Q^{(1)}, \ldots, Q^{(r)}$. In terms of these any period column Q, $= \lambda_1 P^{(1)} + \ldots + \lambda_r P^{(r)}$, in which $\lambda_1, \ldots, \lambda_r$ are real, can be uniquely written in the form

$$N_1 Q^{(1)} + \ldots + N_r Q^{(r)} + \mu_1 P^{(1)} + \ldots + \mu_r P^{(r)},$$

wherein $N_1, \ldots, N_r$ are integers, and $\mu_1, \ldots, \mu_r$ are real quantities which are zero or positive and respectively less than $\lambda_{1,1}, \ldots, \lambda_{r,r}$. For, putting λ_r into the form $N_r\lambda_{r,r} + \mu_r$, where N_r is an integer and μ_r, if not zero, is positive and less than $\lambda_{r,r}$, we have

$$\begin{aligned} Q &= \lambda_1 P^{(1)} + \ldots + \lambda_r P^{(r)} \\ &= \lambda_1' P^{(1)} + \ldots + \lambda'_{r-1} P^{(r-1)} + N_r Q^{(r)} + \mu_r P^{(r)}, \end{aligned}$$

where

$$\lambda_1' = \lambda_1 - N_r\lambda_{1,r}, \; \ldots, \; \lambda'_{r-1} = \lambda_{r-1} - N_r\lambda_{r-1,r};$$

and herein the column $Q' = \lambda_1' P^{(1)} + \ldots + \lambda'_{r-1} P^{(r-1)}$ can quite similarly be expressed in the form

$$Q' = \lambda_1'' P^{(1)} + \ldots + \lambda''_{r-2} P^{(r-2)} + N_{r-1} Q^{(r-1)} + \mu_{r-1} P^{(r-1)},$$

and so on.

But now, if $N_1 Q^{(1)} + \ldots + N_r Q^{(r)} + \mu_1 P^{(1)} + \ldots + \mu_r P^{(r)}$ be a period column, it follows, as $N_1, \ldots, N_r$ are integers, that also $\mu_1 P^{(1)} + \ldots + \mu_r P^{(r)}$ is a period column; and this in fact is only possible when each of $\mu_1, \ldots, \mu_r$ is zero. For, by our definition of $Q^{(r)}$, the coefficient μ_r is zero; then, by the definition of $Q^{(r-1)}$, the coefficient μ_{r-1} is zero; and so on.

On the whole we have the proposition (II., § 344)—*if*

$$Q^{(m)} = \lambda_{1,m}P^{(1)} + \ldots + \lambda_{m,m}P^{(m)}, \qquad (m = 1, \ldots, r),$$

be that real linear combination of the first m columns from $P^{(1)}, \ldots, P^{(r)}$ in which the m-th coefficient $\lambda_{m,m}$ has the least existing value greater than zero and not greater than unity, or be one such combination for which $\lambda_{m,m}$ satisfies the same condition, then every period column is expressible as a linear combination of the columns $Q^{(1)}, \ldots, Q^{(r)}$ with integral coefficients.

It should be noticed that $Q^{(1)}, \ldots, Q^{(r)}$ are not connected by any linear equation with real coefficients, or the same would be true of $P^{(1)}, \ldots, P^{(r)}$. And it should be borne in mind that the expression of any period column by means of *integral* coefficients, in terms of $Q^{(1)}, \ldots, Q^{(r)}$, is a consequence of the fact that the function $f(u_1, \ldots, u_p)$ has only a limited number of period columns which consist wholly of finite periods. Conversely the period columns, of finite periods, obtainable with such integral coefficients, are limited in number.

Another result (I., § 344) is thence obvious—*The coefficients in the linear expression of any period column in terms of $P^{(1)}, \ldots, P^{(r)}$ are rational numbers.*

For by the demonstration of the last result it follows that the period columns $P^{(1)}, \ldots, P^{(r)}$ can be expressed with integral coefficients in terms of $Q^{(1)}, \ldots, Q^{(r)}$ in the form

$$P^{(m)} = N_1^{(m)} Q^{(1)} + \ldots + N_r^{(m)} Q^{(r)}, \qquad (m = 1, \ldots, r);$$

from these equations, since the columns $P^{(1)}, \ldots, P^{(r)}$ are not connected by any linear relation with real coefficients, the columns $Q^{(1)}, \ldots, Q^{(r)}$ can be expressed as linear combinations of $P^{(1)}, \ldots, P^{(r)}$ with only rational numbers as coefficients; hence any linear combinations of $Q^{(1)}, \ldots, Q^{(r)}$ with integral coefficients is a linear combination of $P^{(1)}, \ldots, P^{(r)}$ with rational-number coefficients.

It needs scarcely* to be remarked that the set of period columns $Q^{(1)}, \ldots, Q^{(r)}$, in terms of which any other column can be expressed with integral coefficients, is not the only set having this property.

348. We consider briefly the application of the foregoing theory to the case of uniform analytical functions of a single variable which do not possess any infinitesimal periods. It will be sufficient to take the case when the function has two periods which have not a real ratio; this is equivalent to excluding singly periodic functions.

If $2\omega_1$, $2\omega_2$ be two periods of the function, whose ratio is not real, and 2Ω be any other period, it is possible to find two real quantities λ_1, λ_2 such that

$$\Omega = \lambda_1\omega_1 + \lambda_2\omega_2;$$

then of periods of the form $2\lambda_1\omega_1$, in which $0 < \lambda_1 \not> 1$, of which form periods do exist, $2\omega_1$ itself being one, there is one in which λ_1 has a least value, other than zero—as follows because the function has only a finite number of finite periods. Denote this least value by μ_1, and put $\Omega_1 = \mu_1\omega_1$. Of periods of the form $2\lambda_1\omega_1 + 2\lambda_2\omega_2$ in which $0 \not> \lambda_1 \not> 1$, $0 < \lambda_2 \not> 1$, there is a finite number, and therefore one, in which λ_2 has the least value arising, say μ_2; let one of the corresponding values of λ_1 be λ; put $\Omega_2 = \lambda\omega_1 + \mu_2\omega_2$. Then any period $2\Omega = 2\lambda_1\omega_1 + 2\lambda_2\omega_2$ is of the form $2N_1\Omega_1 + 2N_2\Omega_2 + 2\nu_1\omega_1 + 2\nu_2\omega_2$, where ν_1, ν_2 are (zero or) positive and respectively less than μ_1 and μ_2, and N_1, N_2 are integers, such that $\lambda_2 = N_2\mu_2 + \nu_2$, $\lambda_1 - N_2\lambda = N_1\mu_1 + \nu_1$. But the existence of a period $\Omega - 2N_1\Omega_1 - 2N_2\Omega_2 = 2\nu_1\omega_1 + 2\nu_2\omega_2$ with $\nu_1 < \mu_1$, $\nu_2 < \mu_2$ is contrary to the definition of μ_1 and μ_2, unless ν_1 and ν_2 be both zero. Hence every period is expressible in the form

$$\Omega = 2N_1\Omega_1 + 2N_2\Omega_2,$$

where N_1, N_2 are integers.

In other words, *a uniform analytical function of a single variable without infinitesimal periods cannot be more than doubly periodic*†.

* For the argument compare Weierstrass (l. c., § 344), Jacobi, *Ges. Werke*, t. ii., p. 27, Hermite, *Crelle*, XL. (1850), p. 310, Riemann, *Crelle*, LXXI. (1859) or *Werke* (1876), p. 276. See also Kronecker, "Die Periodensysteme von Functionen reeller Variabeln," *Sitzungsber. der Berl. Akad.*, 1884, (Jun. bis Dec.), p. 1071.

† Cf. Forsyth, *Theory of Functions* (1893), §§ 108, 107. It follows from these Articles, in this order, that any three periods of a uniform function of one variable can be expressed, with

It follows also that every period is expressible by $2\omega_1$, $2\omega_2$ with only rational-number coefficients.

349. *Ex.* i. If r quantities be connected by k homogeneous linear equations with integral coefficients ($r>k$), it is possible to find $r-k$ other quantities, themselves expressible as linear functions of the r quantities with integral coefficients, in terms of which the r quantities can be linearly expressed with integral coefficients.

Ex. ii. If $P^{(1)}, \dots, P^{(r)}$ be r columns of real quantities, each containing $r-1$ constituents, a column $N_1P^{(1)}+\dots+N_rP^{(r)}$ can be formed, in which $N_1, \dots, N_r$ are integers, whose $r-1$ constituents are within assigned nearness of any $r-1$ assigned real quantities (cf. Chap. IX., § 166, and Clebsch u. Gordan, Abels. Funct., p. 135).

Ex. iii. An uniform analytical function of p variables, having r period columns $P^{(1)}, \dots, P^{(r)}$, each of p constituents, and having a further period column expressible in the form $\lambda_1P^{(1)}+\dots+\lambda_rP^{(r)}$, wherein $\lambda_1, \dots, \lambda_r$ are real, will necessarily have a column of infinitesimal periods if even one of the coefficients $\lambda_1, \dots, \lambda_r$ be irrational.

From this result, taken with Ex. i., another demonstration of the proposition of the text (§ 347) can be obtained. The result is itself a corollary from the reasoning of the text.

Ex. iv. If $u_1^{x,a}, \dots, u_p^{x,a}$ be linearly independent integrals of the first kind, on a Riemann surface, and the periods, $2\omega_{r,s}$, $2\omega'_{r,s}$, of the integral $u_r^{x,a}$ be written $\rho_{r,s}+i\sigma_{r,s}$, $\rho'_{r,s}+i\sigma'_{r,s}$, shew that the vanishing of the determinant of $2p$ rows and columns which is denoted by

$$\begin{vmatrix} \rho_{r,1}, \dots, \rho_{r,p}, & \rho'_{r,1}, \dots, \rho'_{r,p} \\ \sigma_{r,1}, \dots, \sigma_{r,p}, & \sigma'_{r,1}, \dots, \sigma'_{r,p} \end{vmatrix},$$

would involve* the equation

$$(M_1-iN_1)\,u_1^{x,a}+\dots+(M_p-iN_p)\,u_p^{x,a}=\text{constant},$$

where $M_1, N_1, \dots, M_p, N_p$ are the minors of the elements of the first column of this determinant and are supposed not all zero.

The vanishing of this determinant is the condition that the period columns of the integrals should be connected by a homogeneous linear relation with real coefficients.

350. The argument of the text has important bearings on the theory of the Inversion Problem discussed in Chap. IX. The functions by which the solution of that problem is expressed have $2p$ columns of periods in terms of which all other period columns can be expressed linearly with integral coefficients; these $2p$ columns are not connected by any linear equation with integral coefficients (§ 165), and, therefore, are not connected by any linear equation with real coefficients.

It has been remarked (§ 174, Chap. X.) that the Riemann theta functions whereby the $2p$-fold periodic functions expressing the solution of the Inversion Problem can be built up, are not the most general theta functions possible. The same is therefore presumably true of the $2p$-fold periodic functions themselves. Weierstrass has stated a theorem†

integral coefficients, in terms of two periods. These two periods, and any fourth period of the function, can, in their turn, be expressed integrally by two other periods—and so on. The reasoning of the text shews that when the function has no infinitesimal periods, the successive processes are finite in number, and every period can be expressed, with integral coefficients, in terms of two periods.

* Forsyth, *Theory of Functions* (1893), p. 440, Cor. ii.

† *Berlin, Monatsber.* Dec. 2, 1869, *Crelle*, LXXXIX. (1880). For an application to integrals of radical functions, Cf. Wirtinger, *Untersuchungen über Thetafunctionen* (Leipzig, 1895), p. 77.

whereby it appears that the most general $2p$-fold periodic functions that are possible can be supposed to arise in the solution of a generalised Inversion Problem; this Inversion Problem differs from that of Jacobi in that the solution involves multiform periodic functions*; the theorems of the text possess therefore an interest, so far as they hold, in the case of such multiform functions. The reader is referred to Weierstrass, *Abhandlungen aus der Functionenlehre* (Berlin, 1886), p. 177, and to Casorati, *Acta Mathematica*, t. viii. (1886).

351. We pass now to a brief account of a different theory which is necessary to make clear the position occupied by the theory of theta functions. Considering, *à priori*, uniform integral analytical functions which, like the theta functions, are such that their partial logarithmic differential coefficients of the second order are periodic functions, we investigate certain relations which must necessarily hold among the periods, and we prove that all such functions can be expressed by means of theta functions.

Suppose that to the p variables $u_1, \ldots, u_p$ there correspond σ columns of quantities $a_i^{(j)}$ $(i=1, \ldots, p, j=1, \ldots, \sigma)$ and σ columns of quantities $b_i^{(j)}$—according to the scheme

$$\begin{array}{c|c|c|}
u_1 & a_1^{(1)}, \ldots, a_1^{(\sigma)} & b_1^{(1)}, \ldots, b_1^{(\sigma)} \\
u_2 & a_2^{(1)}, \ldots, a_2^{(\sigma)} & b_2^{(1)}, \ldots, b_2^{(\sigma)} \\
\cdot & \cdot \quad \cdot\cdot \quad \cdot & \cdot \quad \cdot\cdot \quad \cdot \\
\cdot & \cdot \quad \cdot\cdot \quad \cdot & \cdot \quad \cdot\cdot \quad \cdot \\
u_p & a_p^{(1)}, \ldots, a_p^{(\sigma)} & b_p^{(1)}, \ldots, b_p^{(\sigma)}
\end{array};$$

and suppose $\phi(u)$ to be an uniform, analytical function of $u_1, \ldots, u_p$ which for finite values of $u_1, \ldots, u_p$ is finite and continuous—which further has the property expressed by the equations

$$\phi(u + a^{(j)}) = e^{2\pi i b^{(j)}[u+\frac{1}{2}a^{(j)}]+2\pi i c^{(j)}} \phi(u), \qquad (j=1, \ldots, \sigma), \qquad \text{(I.)}$$

wherein $b^{(j)}$ is a symbol for a column $b_1^{(j)}, \ldots, b_p^{(j)}$ and $c^{(j)}$ is a single quantity depending only on j. The aggregate of $c^{(1)}, \ldots, c^{(\sigma)}$ will be called the characteristic or the parameter of $\phi(u)$; $a_i^{(j)}$ will finally be denoted by $a_{i,j}$. We suppose that the columns $a^{(j)}$ are *independent*, in the sense that there exists no linear equation connecting them of which the coefficients are rational numbers; but it is not assumed that the columns $a^{(j)}$ constitute all the independent columns for which the function ϕ satisfies an equation of the form (I.). Also we suppose that the equation (I.) is not satisfied for any column of wholly infinitesimal quantities put in place of $a^{(j)}$. The reason for this last supposition is that in such case it is possible to express ϕ as the product of an exponential of a quadric function of $u_1, \ldots, u_p$, multiplied into a function of less than p variables, these fewer variables being linear functions of $u_1, \ldots, u_p$. The function $\phi(u)$ in the most general

* With a finite number of values.

case is a generalisation of a theta function; it will be distinguished by the name of a *Jacobian function*; but, for example, it may be a theta function, for which, when $\sigma < 2p$, the columns $a^{(j)}$ are σ of the $2p$ columns of quasi-periods, $2\omega^{(j)}$.

A consequence of the two suppositions is that in the matrix of σ columns and $2p$ rows, of which the $(2i-1)$th and $2i$-th rows are formed respectively by the real and imaginary parts of the row $a_i^{(1)}, \ldots, a_i^{(\sigma)}$, not every determinant of σ rows and columns can vanish. For if with σ arbitrary real variables $x_1, \ldots, x_\sigma$ we form $2p$ linear functions, the $(2i-1)$th and $2i$-th of these having for coefficients the $(2i-1)$th and $2i$-th rows of the matrix of σ columns and $2p$ rows just described, the condition that every determinant from this matrix with σ rows and columns should vanish, is that all these $2p$ linear functions should be expressible as linear functions of at most $\sigma - 1$ of them. Now it is possible to choose *rational integer* values of $x_1, \ldots, x_\sigma$ to make all of these $\sigma - 1$ linear functions infinitesimally small*; they cannot be made simultaneously zero since the σ columns of periods are independent. Therefore every one of the $2p$ linear functions would be infinitesimally small for the same integer values of $x_1, \ldots, x_\sigma$. Thus there would exist a column of infinitesimal quantities expressible in the form $x_1 a^{(1)} + \ldots + x_\sigma a^{(\sigma)}$. Now it will be shewn to be a consequence of the coexistence of equations (I.) that also an equation of the form (I.) exists when $a^{(j)}$ is replaced by an expression $x_1 a^{(1)} + \ldots + x_\sigma a^{(\sigma)}$, wherein $x_1, \ldots, x_\sigma$ are integers. This however is contrary to our second supposition above.

Hence also the matrix of σ columns and $2p$ rows, wherein the $(2i-1)$th and $2i$-th rows consist of $a_i^{(1)}, \ldots, a_i^{(\sigma)}$ and the quantities which are the conjugate complexes of these respectively, is such that not every determinant of σ rows and columns formed therefrom is zero.

And also, by the slightest modification of the argument, σ cannot be $> 2p$. The case when σ is equal to $2p$ is of especial importance; in fact the case $\sigma < 2p$ can be reduced to this† case.

352. Consider now the equations (I.). We proceed to shew that in order that they should be consistent with the condition that $\phi(u)$ is an uniform function, it is necessary, if a, b denote the matrices of p rows and σ columns which occur in the scheme of § 351, that the matrix of σ rows and columns‡, expressed by

$$\bar{a}b - \bar{b}a, \qquad (A),$$

should be a skew symmetrical one of which each element is a rational

* Chap. IX., § 166.

† When $\sigma = 2p$, the hypothesis of no infinitesimal periods is a consequence of the other conditions (cf. § 345).

‡ The notation already used for square matrices can be extended to rectangular matrices. See, for example, Appendix II., at the end of this volume (§ 406).

integer. Denote it by k, so that $k_{\alpha\alpha}=0$, $k_{\alpha\beta}=-k_{\beta\alpha}$. But further also we shew that it is necessary, if x denote a column of σ quantities and x_1 denote the column whose elements are the conjugate complexes of those of x, that for all values, other than zero, satisfying the p equations

$$ax=0, \qquad \text{(B)},$$

the expression $ikxx_1$ should be positive. We shew that $ikxx_1$ cannot be zero unless, beside ax, also ax_1 be zero: a condition only fulfilled by putting each of the elements of $x=0$ (as follows because the σ columns of periods are independent and there are no infinitesimal periods). The condition (B) is in general inoperative when $\sigma < p+1$.

353. Before giving the proof it may be well to illustrate these results by shewing that they hold for the particular case of the theta functions for which (cf. § 284, Chap. XV.)

$$\sigma=2p, \quad a=|2\omega, 2\omega'|, \quad 2\pi ib=|2\eta, 2\eta'|,$$

and therefore

$$ax=2\omega X+2\omega' X'=\Omega_X, \quad bx=\frac{1}{2\pi i}H_X,$$

where X is a column of p quantities, X' a column of p quantities, and $x=\begin{vmatrix} X \\ X' \end{vmatrix}$. Let $y=\begin{vmatrix} Y \\ Y' \end{vmatrix}$, where, similarly, each of Y and Y' is a column of p quantities; then*

$$XY'-X'Y=\frac{1}{2\pi i}(H_X\Omega_Y-H_Y\Omega_X)=ay\,.\,bx-ax\,.\,by=(\bar{a}b-\bar{b}a)\,xy=kxy,$$

but

$$XY'-X'Y=\sum_{i,j}^{1\ldots p}[X_iY_i'-X_j'Y_j]=\sum_{i,j}^{1\ldots p}(x_iy_{i+p}-x_{j+p}y_j)=\sum_{i,j}^{1\ldots p}[\epsilon_{i+p,i}x_iy_{i+p}+\epsilon_{j,j+p}x_{j+p}y_j],$$

where $\epsilon_{i+p,i}=+1=-\epsilon_{i,i+p}$ and $\epsilon_{i,j}=0$ when $i\sim j$ is not equal to p; thus we may write

$$kxy=XY'-X'Y=\epsilon xy,$$

namely, the matrix k is in the case of the theta functions the matrix ϵ, of $2p$ rows and columns, which has already been employed (Chap. XVIII., § 322).

It can be similarly shewn that in the case of theta functions of order r, $k=r\epsilon$.

Next if a, b, h denote the matrices occurring in the exponents of the exponential in the theta series, we have†

$$\mathrm{h}\Omega_X=\pi iX+\mathrm{b}X',$$

namely $\mathrm{h}\,.\,ax=\pi iX+\mathrm{b}X'$. Hence the equations $ax=0$ give $X=-\frac{1}{\pi i}\mathrm{b}X'$. If X_1, X_1' denote the conjugate complexes of X, X' we have therefore $X_1=\frac{1}{\pi i}\mathrm{b}_1X_1'$.

Hence $ikxx_1=i\epsilon xx_1=i(XX_1'-X'X_1)=-\frac{1}{\pi}[\mathrm{b}X'X_1'+\mathrm{b}_1X_1'X']=-\frac{1}{\pi}(\mathrm{b}+\mathrm{b}_1)X'X_1'$, since $\mathrm{b}=\bar{\mathrm{b}}$ and $\mathrm{b}_1=\bar{\mathrm{b}}_1$. Thus if $\mathrm{b}=\mathrm{c}+i\mathrm{d}$, $\mathrm{b}_1=\mathrm{c}-i\mathrm{d}$, the quantity $-\mathrm{c}X'X_1'$ is positive unless each element of X' is zero, namely, the real part of $\mathrm{b}X'X_1'$ is negative for *all* values of X' (except zero). If $X'=m+in$, $\mathrm{b}\,(m^2+n^2)$ is equal to $\mathrm{b}m^2+\mathrm{b}n^2$; and the condition that this be negative is just the condition that the theta series converge.

* For the notation see Appendix II.

† Chap. X. § 190, Chap. VII. § 140.

354. Passing from this case to the proof of equations (A), (B) of § 352, we have, from equation (I.),

$$\phi\left[u+a^{(1)}+a^{(2)}\right]=e^{2\pi i b^{(1)}\left[u+a^{(2)}+\frac{1}{2}a^{(1)}\right]+2\pi i c^{(1)}}\phi\left(u+a^{(2)}\right)$$

$$=e^{2\pi i b^{(1)}\left[u+a^{(2)}+\frac{1}{2}a^{(1)}\right]+2\pi i c^{(1)}+2\pi i b^{(2)}\left[u+\frac{1}{2}a^{(2)}\right]+2\pi i c^{(2)}}\phi(u)$$

$$=e^{2\pi i\left[b^{(1)}+b^{(2)}\right]\left[u+\frac{1}{2}a^{(1)}+\frac{1}{2}a^{(2)}\right]+2\pi i\left[c^{(1)}+c^{(2)}\right]}e^{L_{12}}\phi(u),$$

where $L_{12}=\pi i\left[b^{(1)}a^{(2)}-b^{(2)}a^{(1)}\right]$, $=-L_{21}$. Since the left-hand side of the equation is symmetrical in regard to a_1 and a_2, $e^{L_{12}}$ must be $=e^{L_{21}}$, and hence $L_{12}/\pi i$ is a rational integer, $=k_{21}$ say, such that $k_{12}=-k_{21}$.

Obviously, in $k_{12}=a^{(1)}b^{(2)}-a^{(2)}b^{(1)}$, the part $a^{(1)}b^{(2)}$ is formed by compounding the first column of the matrix a (of σ columns and p rows) with the second column of the matrix b. Similarly with $a^{(2)}b^{(1)}$. Namely k_{12} is the (1, 2)th element of $k=\bar{a}b-\bar{b}a$. Since similar reasoning holds for every element, it follows that the matrix k is a skew symmetrical matrix of integers. Conversely, if this be so, it is easy to prove by successive steps the equation

$$\phi\left(u+a^{(1)}m_1+a^{(2)}m_2+\ldots+a^{(\sigma)}m_\sigma\right)/\phi(u)$$

$$=e^{2\pi i\left[b^{(1)}m_1+\ldots+b^{(\sigma)}m_\sigma\right]\left[u+\frac{a^{(1)}m_1+\ldots+a^{(\sigma)}m_\sigma}{2}\right]+2\pi i\left(c^{(1)}m_1+\ldots+c^{(\sigma)}m_\sigma\right)+\pi i L}, \qquad \text{(II.)}$$

where

$$L=\sum_{\substack{\alpha=1,\,\ldots,\,\sigma\\ \beta=2,\,\ldots,\,\sigma}}^{\alpha<\beta} k_{\alpha\beta}m_\alpha m_\beta,$$

and $m_1, \ldots, m_\sigma$ are integers; this equation may be represented* by

$$\phi(u+am)=\phi(u)\,e^{2\pi i bm\left[u+\frac{am}{2}\right]+2\pi i cm+\pi i\sum^{\alpha<\beta}k_{\alpha\beta}m_\alpha m_\beta}$$

In fact, assuming the equation (II.) to be true for one set $m_1, \ldots, m_\sigma$, we have, by the equations (I.),

$$\phi\left[u+am+a^{(1)}\right]=e^{2\pi i b^{(1)}\left[u+am+\frac{1}{2}a^{(1)}\right]+2\pi i c^{(1)}}\phi(u+am),$$

$$=e^{2\pi i bm\left[u+\frac{1}{2}am\right]+2\pi i b^{(1)}\left[u+am+\frac{1}{2}a^{(1)}\right]+2\pi i cm+2\pi i c^{(1)}+\pi i\sum^{\alpha<\beta}k_{\alpha\beta}m_\alpha m_\beta}\phi(u),$$

$$=e^{2\pi i\left[bm+b^{(1)}\right]\left[u+\frac{1}{2}am+\frac{1}{2}a^{(1)}\right]+2\pi i\left[cm+c^{(1)}\right]+\pi i\sum^{\alpha<\beta}k_{\alpha\beta}m_\alpha m_\beta+\pi i R}\phi(u),$$

* For the notation see Appendix II.—or thus—

$$bm\,.\,u=\sum_i\left[b_{i1}m_1+\ldots\ldots+b_{i\sigma}m_\sigma\right]u_i$$

$$=\left(\sum_i b_{i1}u_i\right)m_1+\ldots\ldots+\left(\sum_i b_{i\sigma}u_i\right)m_\sigma$$

$$=\left(\sum_i b_i^{(1)}u_i\right)m_1+\ldots\ldots+\left(\sum_i b_i^{(\sigma)}u_i\right)m_\sigma$$

$$=b^{(1)}u\,.\,m_1+\ldots\ldots+b^{(\sigma)}u\,.\,m_\sigma$$

$$=b^{(1)}m_1\,.\,u+\ldots\ldots+b^{(\sigma)}m_\sigma\,.\,u.$$

where R is equal to $b^{(1)} . am - bm . a^{(1)}$, namely equal to

$$\sum_i b_i^{(1)} [a_i^{(1)} m_1 + \dots + a_i^{(\sigma)} m_\sigma] - \sum_j [b_j^{(1)} m_1 + \dots + b_j^{(\sigma)} m_\sigma] a_j^{(1)} = k_{21} m_2 + \dots + k_{\sigma 1} m_\sigma,$$

so that

$$R + \sum^{\alpha<\beta} k_{\alpha\beta} m_\alpha m_\beta$$

$$= k_{21} m_2 + \dots + k_{\sigma 1} m_\sigma + k_{12} m_1 m_2 + \dots + k_{1\sigma} m_1 m_\sigma + k_{23} m_2 m_3 + \dots + k_{2\sigma} m_2 m_\sigma + \dots,$$

$$= 2 (k_{21} m_2 + \dots + k_{\sigma 1} m_\sigma) + k_{12} (m_1 + 1) m_2 + \dots + k_{1\sigma} (m_1 + 1) m_\sigma + k_{23} m_2 m_3 + \dots;$$

hence

$$e^{\pi i R + \pi i \sum\limits_{\alpha<\beta} k_{\alpha\beta} m_\alpha m_\beta} = e^{\pi i \sum\limits_{\alpha<\beta} k_{\alpha\beta} m_\alpha' m_\beta'},$$

where

$$[m_1', \dots, m_\sigma'] = [m_1 + 1, m_2, \dots, m_\sigma];$$

therefore

$$\phi [u + am'] = e^{2\pi i bm' [u + \frac{1}{2} am'] + 2\pi i cm' + \pi i \sum\limits_{\alpha<\beta} k_{\alpha\beta} m_\alpha' m_\beta'} \phi (u).$$

Similarly we can take the case $\phi (u + am - a^{(1)})$, noticing that equation (I.) can be written

$$\phi (v - a^{(j)}) = \phi (u) e^{-2\pi i b^{(j)} [v - \frac{1}{2} a^{(j)}] - 2\pi i c^{(j)}},$$

where $v = u + a^{(j)}$.

355. The theorem (A) is thus proved. The theorem (B) is of a different character, and may be made to depend on the fact that a one-valued function of a single complex variable cannot remain finite for all values of the variable.

Consider the expression

$$L (\xi) = e^{-2\pi i b\xi (v + \frac{1}{2} a\xi) - 2\pi i c\xi} \phi (v + a\xi),$$

wherein $\xi_1, \dots, \xi_\sigma$ are real quantities.

Then $L (\xi + m)/L (\xi)$, wherein $m_1, \dots, m_\sigma$ are rational integers, is equal to $e^{\pi i k m\xi + \pi i \sum^{\alpha<\beta} k_{\alpha\beta} m_\alpha m_\beta}$, as immediately follows from equation (I.), and is therefore a quantity whose modulus is unity. Now when $\xi_1, \dots, \xi_\sigma$ are each between 0 and 1 and v is finite, $L (\xi)$ is finite. Its modulus is therefore finite for *all* real values of ξ; let G be an upper limit to the modulus of $L (\xi)$; G can be determined by considering values of ξ between 0 and 1. Let now $x_1, \dots, x_\sigma$ be such that $ax = 0$, and let x_1 denote the column of quantities which are the conjugate complexes of the elements of the column x. Put $\xi = x + x_1$, so that $a\xi = ax_1$.

Then

$$\phi (v + ax_1) = \phi (v + a\xi) = e^{\pi i b\xi . a\xi + 2\pi i (c + bv) \xi} L (\xi),$$

wherein an upper limit of the modulus of $L (\xi)$ is a positive quantity G whose value may be taken large enough to be unaffected by replacing x by any other solution of $ax=0$; it is necessary in fact only to consider the modulus of $L(\xi)$ when ξ is between 0 and 1.

We have

$$b\xi \,.\, a\xi = b(x+x_1)\,.\,a(x+x_1) = bx\,.\,ax_1 + bx_1\,.\,ax_1$$
$$= bx\,.\,ax_1 - bx_1\,.\,ax + bx_1\,.\,ax_1 = kxx_1 + \bar{a}bx_1^2,$$

$(c+\bar{b}v)\,\xi,\ = w(x+x_1)$, say, $= wx + w_1x_1 + (w-w_1)\,x_1$,

where $w = c + \bar{b}v$; therefore

$$e^{\pi i b\xi\,.\,a\xi + 2\pi i(c+\bar{b}v)\xi}\,L(\xi) = e^{i\pi kxx_1 + i\pi\bar{a}bx_1^2 + 2\pi i(w-w_1)x_1}\,e^{2\pi i(wx+w_1x_1)}\,L(\xi);$$

this equation is the same as

$$e^{-i\pi\bar{a}bx_1^2 - 2\pi i(w-w_1)x_1}\,\phi(v+ax_1) = e^{\rho}K,$$

where

$$K,\ = L(\xi)\,e^{2\pi i(wx+w_1x_1)},$$

has the same modulus as $L(\xi)$, less than G, and where

$$\rho = i\pi kxx_1$$
$$= i\pi\Sigma k_{ij}\,[x_j(x_1)_i - x_i(x_1)_j] = i\pi\Sigma k_{ij}\begin{vmatrix} y_j+iz_j, & y_j-iz_j \\ y_i+iz_i, & y_i-iz_i \end{vmatrix} = 2i\pi\Sigma k_{ij}i\begin{vmatrix} y_j, & -z_j \\ y_i, & -z_i \end{vmatrix}$$

$= 2\pi\Sigma k_{ij}(y_jz_i - y_iz_j) = 2\pi kyz$, is a real quantity ($x$ being equal to $y+iz$).

Now if x be any solution of the equations $ax=0$, then μ_1x is also a solution, μ being any arbitrary complex quantity and μ_1 its conjugate complex. Replace x throughout by μ_1x, and therefore ξ by $\mu_1x+\mu x_1$. Then the equation just written becomes

$$e^{-i\pi\bar{a}b\mu^2x_1^2 - 2\pi i(w-w_1)\mu x_1}\,\phi(v+\mu ax_1) = e^{\rho\mu\mu_1}\,.\,K,$$

K having also its modulus $< G$.

Herein the left side, if not independent of μ, is, for definite constant values of v and x, a one-valued continuous (analytical) function of μ which is finite for all finite values of μ. Hence it must be infinite for infinite values of μ. Hence ρ must be positive, *viz., values of x such that $ax=0$ are such that the real quantity $ikxx_1$ is necessarily positive* provided only the expression

$$e^{-\mu^2 i\pi\bar{a}bx_1^2 - 2\pi i\mu(w-w_1)x_1}\,\phi(v+\mu ax_1)$$

is not independent of μ.

Now if this expression be independent of μ, it is equal to $\phi(v)$, the value obtained when $\mu=0$, and therefore

$$e^{-i\pi\mu^2\bar{a}bx_1^2}\,\frac{\phi(v+\mu ax_1)}{\phi(v)} = e^{2\pi i\mu(w-w_1)x_1};$$

here the left side is a function of v provided ax_1 be not zero; when ax_1 is zero its value is unity; we take these possibilities in turn:

(i) Suppose first ax_1 is not zero,

then

$$(w - w_1) x_1 = (\bar{b}v - \bar{b}_1 v_1) x_1 = bx_1 . v - b_1 x_1 . v_1$$

must, like the left side, be a function of v and therefore a linear function, say $\frac{1}{2\pi i}(Bv + C)$, so that

$$\phi(v + \mu a x_1) = \phi(v) e^{A\mu^2 + Bv\mu + C\mu}, \text{ where } A = i\pi \bar{a} b x_1^2;$$

hence $\mu a x_1$ represents a column of periods* for the function $\phi(v)$—and this for arbitrary values of μ.

In this case however $\phi(v)$ would be capable of a column of infinitesimal periods, contrary to our hypothesis.

Hence ρ must be positive for values of x such that $ax = 0$, $ax_1 \neq 0$.

(ii) But in fact as there are σ columns of independent periods we cannot simultaneously have $ax = 0$, $ax_1 = 0$. For the last is equivalent to $a_1 x = 0$; and $ax = 0$, $a_1 x = 0$, together, involve that every determinant of σ rows and columns in the matrix $\begin{vmatrix} a \\ a_1 \end{vmatrix}$ is zero—and thence involve the existence of infinitesimal periods (§ 351).

Hence $ikxx_1$ is necessarily positive for values of x, other than zero, satisfying $ax = 0$; and this is the theorem (B).

Remark i. From the existence of two matrices a, b of p rows and σ columns, for which $\bar{a}b - \bar{b}a$ is a skew symmetrical matrix of integers k such that $ikxx_1$ is positive for values of x other than zero satisfying $ax=0$, can be inferred that in the matrix of σ columns and $2p$ rows, $\begin{vmatrix} a \\ a_1 \end{vmatrix}$, not every determinant of σ rows and columns can vanish—and also that the σ columns of quantities which form the matrix a are independent, namely that we cannot have the p equations $a_{i1} x^{(1)} + \ldots + a_{i\sigma} x^{(\sigma)} = 0$ satisfied by rational integers $x^{(1)}, \ldots, x^{(\sigma)}$. For then, also, $a_1 x = 0$, since $x = x_1$.

Remark ii. In the matrix k, if σ be not less than p, all determinants of $2(\sigma - p)$ rows and columns cannot be zero. In the matrix a, not all determinants of $\frac{1}{2}\sigma$ or $\frac{1}{2}(\sigma+1)$ rows and columns can be zero. In particular when $\sigma = 2p$, for the matrix k, the determinant is not zero; for the matrix a, not all determinants of p rows and columns can be zero.

Let ξ, η be columns each of σ quantities. Then the coexistence of the 3 sets of equations

$$a\xi = 0, \quad a_1\eta = 0, \quad \bar{k}(\xi + \eta) = 0$$

is inconsistent with the conditions (A) and (B) (§ 352), except for zero values of ξ and η. The second of them obviously gives also $a\eta_1 = 0$.

For from these equations we infer that $k\eta_1\xi = a\xi . b\eta_1 - b\xi . a\eta_1$ is zero, and also

$$\bar{k}(\xi + \eta) . \eta_1 = k\eta_1(\xi + \eta) = k\eta_1\xi + k\eta_1\eta,$$

and therefore also $k\eta_1\eta$ is zero. But by condition (B) the vanishing of $k\eta_1\eta$ when, as here, $a\eta_1 = 0$, enables us to infer $\eta = 0$.

* We use the word period for the quantities $a^{(j)}$ occurring in our original equation (I.).

Similarly

$$k\xi\xi_1 = \bar{k}\xi_1\xi = \bar{k}(\xi_1+\eta_1)\,.\,\xi - \bar{k}\eta_1\xi = \bar{k}(\xi_1+\eta_1)\,.\,\xi - k\xi\eta_1 = \bar{k}(\xi_1+\eta_1)\,.\,\xi - (\bar{a}b - \bar{b}a)\,\xi\eta_1$$
$$= \bar{k}(\xi_1+\eta_1)\,.\,\xi - (a\eta_1\,.\,b\xi - b\eta_1\,.\,a\xi)$$

is zero when $\bar{k}(\xi+\eta_1)=0$, $a\eta_1=0$, $a\xi=0$. Thence by condition (B), since $a\xi=0$, ξ is zero.

Suppose now that the number of the p linear functions $a\xi$ which are linearly independent is ν, so that all determinants of $(\nu+1)$ rows and columns of the matrix a are zero, but not all determinants of ν rows and columns; and that the number of the σ linear functions $k\xi$ which are linearly independent is 2κ*, so that in the matrix k all determinants of $2\kappa+1$ rows and columns vanish, but not all of 2κ rows and columns. Then we can choose $2\nu+2\kappa$ linearly independent linear functions from the $2p+\sigma$ functions $a\xi$, $a_1\eta$, $\bar{k}(\xi+\eta)$. If this number, $2\nu+2\kappa$, of independent functions, were less than the number 2σ of variables ξ, η, the chosen independent functions could be made to vanish simultaneously for other than zero values of the variables, and then all the linear functions dependent on these must also vanish.

Hence

$$2\nu+2\kappa \overset{=}{>} 2\sigma \text{ or } \nu+\kappa \overset{=}{>} \sigma.$$

Now

$$\nu \overset{=}{<} p,\quad 2\kappa \overset{=}{<} \sigma\,;\quad \text{hence } \nu \overset{=}{>} \tfrac{1}{2}\sigma,\quad 2\kappa \overset{=}{>} 2(\sigma-p).$$

Remark iii. It follows from (ii) that if $k=0$, then $\nu=\sigma$ and $\sigma \leqq p$. Also that a function of p variables which is everywhere finite, continuous and one-valued for finite values of the variables and has no infinitesimal periods cannot be *properly* periodic (without exponential factors) for more than p columns of independent periods; in every set of σ independent periods of such a function the determinants of σ rows and columns are not all zero. The proof is left to the reader.

Remark iv. When $\sigma=2p$ we can put $a=|2\omega, 2\omega'|$, wherein the square matrix 2ω is chosen so that its determinant is not zero. When we write $a=|2\omega, 2\omega'|$ we shall always suppose this done.

356. *Ex.* i. Prove that the exponential of any quadric function of $u_1, \ldots, u_p$ is a Jacobian function of the kind here considered, for which the matrix k is zero.

Ex. ii. Prove that the product of any two or more Jacobian functions, ϕ, with the same number of variables and the same value for σ, is a function of the same character, and that the matrix k of the product is the sum of the matrices k of the separate factors.

Ex. iii. If ϕ be considered as a function of other variables v than u, obtained from them by linear equations of the form $u=\mu+cv$ (μ being any column of p quantities, and c any matrix of p rows and columns), prove that the matrix k of the function ϕ, regarded as a function of v, is unaltered.

Obtain the transformed values of a, b, c and $bm(u+\frac{1}{2}am)+cm$. (Cf. Ex. i., § 190, Chap. X.)

Ex. iv. If instead of the periods a we use $a'=ag$, where g is a matrix of integers with σ rows and columns, prove that $\phi(u+a'm)$ is of the form $e^{2\pi i b'm(u+\frac{1}{2}a'm)+2\pi i c m}\phi(u)$, and that $k'=\bar{g}kg$; and also that kxy becomes changed to $k'x'y'$ by the linear equations $x=gx'$, $y=gy'$. In such case the form $k'x'y'$ is said to be *contained* in kxy. When the relation is reciprocal, or $g^2=1$, the forms are said to be equivalent. Thus to any function ϕ there corresponds a class of equivalent forms k. (Cf. Chap. XVIII., § 324, Ex. i.)

Examples iii. and iv. contain an important result which may briefly be summarised by

* That the number must be even is a known proposition, Frobenius, *Crelle*, LXXXII. (1877), p. 242.

saying that for Jacobian functions, *qua* Jacobian functions, there is no theory of transformation of periods such as arises for the theta functions. A transformed theta function is a Jacobian function; the equations of Chap. XVIII. (§ 324) are those which are necessary in order that, for this Jacobian function, the matrix k should be the matrix ϵ, or $r\epsilon$ (cf. § 353).

Ex. v. If A be a matrix of $2p$ rows and σ columns of which the first p rows are the rows of a and the second p rows those of b, prove that

$$\bar{A}\bar{\epsilon}A = k.$$

In fact if $\xi = Ax$, $\xi' = Ax'$, then

$$kx'x = ax \,.\, bx' - ax' \,.\, bx = \Sigma\,[\xi_i\xi'_{i+p} - \xi_j'\xi_{j+p}] = \epsilon\xi\xi'$$
$$= \epsilon Ax \,.\, Ax' = \bar{A}\epsilon A \,.\, x'x.$$

Hence also when $\sigma = 2p$ the determinant of A is the square root of the determinant of k, which in that case, being a skew symmetrical determinant of even order, is a perfect square.

Ex. vi. Shew that when $\sigma = 2p$ and with the notation $a = |2\omega, 2\omega'|$, $2\pi i b = |2\eta, 2\eta'|$, that

$$\bar{A}\bar{\epsilon}A = \frac{2}{\pi i}\begin{vmatrix} \bar{\omega}\eta - \bar{\eta}\omega, & \bar{\omega}\eta' - \bar{\eta}\omega' \\ \bar{\omega}'\eta - \bar{\eta}'\omega, & \bar{\omega}'\eta' - \bar{\eta}'\omega' \end{vmatrix},$$

the notation being an abbreviated one for a matrix of $2p$ rows and columns. Thus in the case when $k = \epsilon$, the equation of Ex. v. expresses the Weierstrass equations for the periods (Chap. VII., § 140).

Ex. vii. In the case of the theta functions we shewed (§ 140, and p. 533) that the relations connecting the periods could be written in two different ways, one of which was associated with the name of Weierstrass, the other with that of Riemann. We can give a corresponding transformation of the equations (A), (B) (§ 352) in this case, provided $\sigma = 2p$, the determinant of the matrix k not being zero.

As to the equation (A), writing it in the equivalent form given in Ex. v., we immediately deduce

$$Ak^{-1}\bar{A} = \epsilon, \qquad \text{(A}'\text{)},$$

which is the transformation of equation (A).

As to the equation (B), let x be a column of $\sigma = 2p$ arbitrary quantities, and determine the column z, of $\sigma = 2p$ elements, so that the $2p$ equations expressed by $az = 0$, $bz = x$, are satisfied. Then

$$\bar{a}x = \bar{a}bz = (\bar{a}b - \bar{b}a)\,z = kz, \ = \mu, \text{ say; so that } k^{-1}\mu = z,\ k^{-1}\mu_1 = z_1;$$

thus

$$ikzz_1 = i\,(\bar{a}b - \bar{b}a)\,zz_1 = i\,(az_1 \,.\, bz - az \,.\, bz_1) = iaz_1 \,.\, bz = iaz_1x = i\bar{a}xz_1 = i\mu z_1$$
$$= ik^{-1}\mu_1\mu = ik^{-1}\bar{a}_1x_1 \,.\, \bar{a}x = iak^{-1}\bar{a}_1x_1x\,;$$

therefore, the form

$$iak^{-1}\bar{a}_1x_1x \qquad \text{(B}'\text{)},$$

is positive for all values of the column x, other than zero. This is the transformed form of equations (B).

Ex. viii. When $a = |2\omega, 2\omega'|$, $b = \dfrac{1}{2\pi i}|2\eta, 2\eta'|$, $\sigma = 2p$, we have

$$A\epsilon\bar{A} = \begin{vmatrix} 2\omega, & 2\omega' \\ \dfrac{\eta}{\pi i}, & \dfrac{\eta'}{\pi i} \end{vmatrix}\begin{vmatrix} 0 & -1 \\ 1 & 0 \end{vmatrix}\begin{vmatrix} 2\bar{\omega}, & \dfrac{\bar{\eta}}{\pi i} \\ 2\bar{\omega}', & \dfrac{\bar{\eta}'}{\pi i} \end{vmatrix} = \begin{vmatrix} -4\,(\omega\bar{\omega}' - \omega'\bar{\omega}), & -\dfrac{2}{\pi i}(\omega\bar{\eta}' - \omega'\bar{\eta}) \\ \dfrac{2}{\pi i}(\eta'\bar{\omega} - \eta\bar{\omega}'), & -\dfrac{1}{(\pi i)^2}(\eta\bar{\eta}' - \eta'\bar{\eta}) \end{vmatrix}$$

Hence when $k=\epsilon$, the equation (A′) of Ex. vii., equivalent to $A\epsilon\bar{A}=-\epsilon$, expresses the Riemann equations for the periods (Chap. VII., § 140). In the same case the equation (B′), of Ex. vii., expresses that

$$i a\epsilon\bar{a}_1 x_1 x = \sum_{\nu=1}^{p} \sum_{\kappa, \lambda=1}^{p} \left[(a_1)_{\kappa,\nu} a_{\lambda,\nu+p} - a_{\lambda,\nu} (a_1)_{\kappa,\nu+p}\right] x_\lambda (x_1)_\kappa$$

is negative for all values of x other than zero.

Ex. ix. When $p=1$, the two conditions (B), (B′), or

$$i\epsilon x x_1 = \text{positive for } ax=0, \quad i a\epsilon\bar{a}_1 x_1 x = \text{negative for arbitrary } x,$$

become, for $a=|2\omega, 2\omega'|$, if the elements of x be denoted by x and x', and the conjugate imaginaries by x_1, x_1', respectively,

$$i(\omega\omega_1)^{-1}(\omega\omega_1' - \omega'\omega_1) x' x_1' = \text{positive}, \quad i(\omega_1\omega' - \omega\omega_1') x x_1 = \text{negative},$$

and if $\omega=\rho+i\sigma$, $\omega_1=\rho-i\sigma$, $\omega'=\rho'+i\sigma'$, $\omega_1'=\rho'-i\sigma'$, these conditions are equivalent to

$$\rho\sigma' - \rho'\sigma > 0,$$

and express that the real part of $i\omega'/\omega$ is negative.

357. Suppose now that $\sigma=2p$; we proceed (§ 359) to consider how to express the Jacobian function. Two arithmetical results, (i) and (ii), will be utilised, and these may be stated at once: (i) *if k be a skew symmetrical matrix whose elements are integers, with $2p$ rows and columns, and ϵ have the signification previously attached to it, it is possible to find a matrix g, of $2p$ rows and columns, whose elements are integers, such that** $k=\bar{g}\epsilon g$. For instance when $p=1$, we can find a matrix such that

$$\begin{vmatrix} 0 & k_{12} \\ -k_{12} & 0 \end{vmatrix} = \begin{vmatrix} g_{11} & g_{21} \\ g_{12} & g_{22} \end{vmatrix} \begin{vmatrix} 0 & -1 \\ 1 & 0 \end{vmatrix} \begin{vmatrix} g_{11} & g_{12} \\ g_{21} & g_{22} \end{vmatrix} = \begin{vmatrix} g_{21}g_{11} - g_{11}g_{21} & g_{21}g_{12} - g_{11}g_{22} \\ g_{22}g_{11} - g_{12}g_{21} & g_{22}g_{12} - g_{12}g_{22} \end{vmatrix},$$

namely, such that $k_{12}=g_{21}g_{12}-g_{11}g_{22}$; for this we can in fact take g_{11}, g_{12} arbitrarily. In general the $4p^2$ integers contained in g are to satisfy $p(2p-1)$ conditions.

Ex. i. If a be a matrix of integers, of p rows and columns, and λ be an integer, and

$$k = \begin{vmatrix} 0, & -\lambda\bar{a} \\ \lambda a, & 0 \end{vmatrix},$$

g may have either of the two following forms

$$g_1 = \begin{vmatrix} \lambda, & 0 \\ 0, & \bar{a} \end{vmatrix}, \quad g_2 = \begin{vmatrix} \lambda a, & 0 \\ 0, & 1 \end{vmatrix} = \begin{vmatrix} \lambda, & 0 \\ 0, & \bar{a} \end{vmatrix} \begin{vmatrix} a, & 0 \\ 0, & \bar{a}^{-1} \end{vmatrix}, \; = g_1\mu, \text{ say},$$

for we immediately find $\bar{\mu}k\mu=k$.

* For a proof see Frobenius, *Crelle*, LXXXVI. (1879), p. 165, *Crelle*, LXXXVIII. (1880), p. 114.

Ex. ii. If μ be any matrix of integers, with $2p$ rows and columns, such that $\bar{\mu}\epsilon\mu=\epsilon$ (cf. § 322, Chap. XVIII.), we have, if $k=\bar{g}\epsilon g$, also $k=\bar{g}\bar{\mu}^{-1}\epsilon\mu^{-1}g$, and instead of g we may take the matrix $\mu^{-1}g$.

(ii) If g be a given matrix of integers, of $2p$ rows and columns, and x be a column of $2p$ elements, the conditions, for x, that the $2p$ elements gx should be prescribed integers cannot always be satisfied, however the elements of x (which are necessarily rational numerical fractions) are chosen. If for any rational values of x, integral or not, gx be a row of integers, and we put $x=y+L$, where y has all its elements positive (or zero) and less than unity, and L is a row of integers (including zero), then $gx=gy+gL=gy+M$, where M is a row of integers; in this case the row gx will be said to be congruent to gy for modulus g. The result to be utilised* is, that *the number of incongruent rows gx, namely, the number of integers which can be represented in the form gx while each element of x is zero or positive and less than unity, is finite. It is in fact equal to the absolute value of the determinant of g.* For instance when g is $\begin{vmatrix} g_{11} & g_{12} \\ g_{21} & g_{22} \end{vmatrix}$ there are $g_{11}g_{22}-g_{12}g_{21}$ integer pairs which can be written $g_{11}x_1+g_{12}x_2$, $g_{21}x_1+g_{22}x_2$, for (rational) values of x_1, x_2 less than unity. The reader may verify, for instance, that when $g=\begin{vmatrix} 6 & 3 \\ 1 & 2 \end{vmatrix}$, the 9 ways are given (cf. p. 637, *Footnote*) by

	1	2	3	4	5	6	7	8	9
x_1, x_2	0, 0	$\frac{1}{9}, \frac{4}{9}$	$\frac{5}{9}, \frac{2}{9}$	$\frac{1}{3}, \frac{1}{3}$	$\frac{2}{9}, \frac{8}{9}$	$\frac{7}{9}, \frac{1}{9}$	$\frac{4}{9}, \frac{7}{9}$	$\frac{2}{3}, \frac{2}{3}$	$\frac{8}{9}, \frac{5}{9}$
$6x_1+3x_2, x_1+2x_2$	0, 0	2, 1	4, 1	3, 1	4, 2	5, 1	5, 2	6, 2	7, 2

To prove the statement in general let t be the number required, of integers representable in the form gx, when $x<1$. Consider how many integers could be obtained in the form gX when X is restricted only to have all its elements less than (a positive number) N. Corresponding to any one of the t integers obtained in the former case we can now obtain $N-1$ others by increasing only one of the elements of x in turn by 1, 2, ..., $N-1$. This can be done independently for each element of x. Hence the number of integers gX is tN^σ where σ, here to be taken $=2p$, is the number of elements in x. Let one of these integers be called M. Then $g\frac{X}{N}=\frac{M}{N}$ or say $gx=\frac{M}{N}$, wherein x is less than unity. Now when N is very great, the

* Cf. Appendix ii, § 418, and the references there given, and Frobenius, *Crelle*, xcvii. (1884), p. 189.

variation of $z=\frac{M}{N}$, as M changes, approaches to that of a continuous quantity, and the number of its values, being the same as the number of values of M, is

$$\iint \ldots (N dz_1) \ldots (N dz_\sigma),$$

where $z_1, \ldots, z_\sigma$ vary from zero to all values which give to x, in the equations $gx = z$, a value less than unity. Now this integral is

$$N^\sigma \iint \ldots \frac{\partial (z_1, \ldots, z_\sigma)}{\partial (x_1, \ldots, x_\sigma)} dx_1 \ldots dx_\sigma = N^\sigma |g| . \iint \ldots dx_1 \ldots dx_\sigma = N^\sigma |g|.$$

Since this is equal to tN^σ, it follows that t is equal to $|g|$, as was stated.

358. Supposing then that the matrix g, with $2p$ rows and columns each consisting of integers, has been determined so that $k = \bar{a}b - \bar{b}a = \bar{g}\epsilon g$, we consider the expression of the Jacobian function when $\sigma = 2p$. The determinant of k not being zero, the determinant of g is not zero.

Put $K = ag^{-1}$, so that K is a matrix of p rows and $2p$ columns, and $a = Kg$; put similarly $b = Lg$; also, take a row of $2p$ quantities denoted by C, such that $c = \bar{g}C + \frac{1}{2}[g]$, where c is the parameter (§ 351) of the Jacobian function, and $[g]$ is a row of $2p$ quantities of which one element is

$$[g]_\alpha = \sum_{\kappa=1}^{\kappa=p} g_{\kappa, \alpha}\, g_{p+\kappa, \alpha}, \qquad (\alpha = 1, \ldots, 2p);$$

take x, x', X, X', rows of $2p$ quantities such that

$X = gx$, $X' = gx'$, so that $ax = Kgx = KX$, $bx = LX$, $ax' = KX'$, $bx = LX'$;

then as

$$kx'x, \; = ax \,.\, bx' - ax' \,.\, bx, = (\bar{K}L - \bar{L}K)\, X'X,$$

is also equal to

$$\bar{g}\epsilon g x' x = \epsilon g x' \,.\, g x = \epsilon X' X,$$

we have

$$\bar{K}L - \bar{L}K = \epsilon, \qquad \text{(C)},$$

so that

$$KxLx' - Kx'Lx = (\bar{K}L - \bar{L}K)\, x'x = \epsilon x'x = \sum_{i,j}^{1, \ldots, p} (x_i x'_{i+p} - x_j' x_{j+p});$$

further, as $ikxx_1$ is positive for $ax = 0$, we have

$$i\epsilon XX_1 = \text{positive when } KX = 0, \qquad \text{(D)};$$

thus, if A denote the matrix $\begin{vmatrix} K \\ L \end{vmatrix}$, we have, from the equation (C),

$$\bar{A}\bar{\epsilon}A = -A\epsilon\bar{A} = \epsilon, \qquad \text{(E)},$$

and, if z be a row of p arbitrary quantities, and X be a row of $2p$ quantities such that $KX = 0$, $LX = z$, so that $\bar{K}z = \bar{K}LX = (\bar{K}L - \bar{L}K)X = \epsilon X$, and therefore $\epsilon\bar{K}z = -X$, $\bar{K}_1 z_1 = \epsilon X_1$, we have

$$iK_1\epsilon\bar{K}zz_1 = \text{positive, for arbitrary } z \text{ other than zero,} \qquad \text{(F)};$$

for

$$iK_1\epsilon\bar{K}zz_1 = -iK_1Xz_1 = -i\bar{K}_1z_1X = -i\epsilon X_1X = i\epsilon XX_1.$$

If we now change the notation by writing $K = |2\omega, 2\omega'|$, $2\pi iL = |2\eta, 2\eta'|$, and introduce the matrices a, b, h of p rows and columns defined by

$$\text{a} = \tfrac{1}{2}\eta\omega^{-1}, \quad \text{h} = \tfrac{1}{2}\pi i\omega^{-1}, \quad \text{b} = \pi i\omega^{-1}\omega',$$

it being assumed, in accordance with Remark iv. (§ 355) *that the determinant of the matrix* ω *is not zero*, then the equation (E) shews (cf. Ex. viii., § 356) that the matrices a, b are symmetrical, and that $\eta' = \eta\omega^{-1}\omega' - \frac{1}{2}\pi i\bar{\omega}^{-1}$, so that we can also write

$$\eta = 2\text{a}\omega, \quad \eta' = 2\text{a}\omega' - \text{h}', \quad 2\text{h}\omega = \pi i, \quad 2\text{h}\omega' = \text{b};$$

also, by actual expansion,

$$iK_1\epsilon\bar{K} = 4i\omega_1[\omega_1^{-1}\omega_1' - \bar{\omega}'\bar{\omega}^{-1}]\bar{\omega} = -\frac{1}{\pi}\omega_1[\text{b}_1 + \bar{\text{b}}]\bar{\omega} = -\frac{1}{\pi}\omega_1[\text{b}_1 + \text{b}]\bar{\omega}$$

$$= -\frac{2}{\pi}\omega_1\text{c}\bar{\omega}, \text{ if } \text{b} = \text{c} + i\text{d};$$

thus

$$iK_1\epsilon\bar{K}zz_1 = -\frac{2}{\pi}\text{c}t_1t, \text{ where } t = \bar{\omega}z,\ z \text{ and } t \text{ being rows of } p \text{ arbitrary quantities};$$

and therefore, by the equation (F), for real values of $n_1, \ldots, n_p$ other than zero, the quadratic form $\text{b}n^2$ has its real part essentially negative.

Hence we can define a theta function by the equation

$$\vartheta\left(u;\ \begin{matrix}\gamma' \\ -\gamma\end{matrix}\right) = \sum_n e^{\text{a}u^2 + 2\text{h}u(n+\gamma') + \text{b}(n+\gamma')^2 - 2\pi i\gamma(n+\gamma')},$$

wherein γ, γ' are rows of p quantities given by $C = (\gamma', \gamma)$, that is, $C_r = \gamma_r'$, $C_{p+r} = \gamma_r$, for $r < p + 1$. Denoting this function by $\vartheta(u;\ C)$ and taking μ for a row of $2p$ integers, the function is immediately seen (§ 190, Chap. X.) to satisfy the equation

$$\vartheta(u + K\mu;\ C) = e^{2\pi iL\mu(u + \frac{1}{2}K\mu) + 2\pi iC\mu + \pi i \sum\limits_{\alpha,\beta}^{\alpha<\beta} \epsilon_{\alpha,\beta}\mu_\alpha\mu_\beta}\, \vartheta(u;\ C),$$

which is the definition equation for a Jacobian function of periods K, L and parameter C, for which the matrix k is ϵ.

Further, if μ be a matrix of integers with $2p$ rows and columns, such that $\bar{\mu}\epsilon\mu = \epsilon$, and (Ex. ii., § 357) we replace g by $\mu^{-1}g$, the matrices K, L are replaced by $K\mu$ and $L\mu$. Thus instead of the theta function $\vartheta(u;\ C)$ we obtain a linear transformation of this theta function (cf. § 322, Chap. XVIII.).

359. Proceeding further to obtain the expression for the general value of the Jacobian function ϕ, let $\phi(u; \nu)$ denote

$$\phi(u+K\nu)\, e^{-2\pi i L\nu\,(u+\frac{1}{2}K\nu)-2\pi i C\nu+2\pi i nn'},$$

where $\nu_i = n_i$, $\nu_{i+p} = n_i'$, for $i < p+1$. Then, since $a = Kg$, and therefore $aN = KgN$, we have

$$\phi(u+aN, \nu) = \phi(u+KgN, \nu) = \phi(u+K\mu, \nu), \qquad (1),$$

where μ denotes the row gN, so that $aN = K\mu$, N being a column of $2p$ integers and therefore μ a column of integers; thus $\phi(u+aN, \nu)$ is equal to

$$\phi(u+aN+K\nu)\, e^{-2\pi i L\nu\,(u+K\mu+\frac{1}{2}K\nu)-2\pi i C\nu+\pi i nn'} = \phi(u+K\nu)\, e^{R},$$

where

$$R = 2\pi i bN(u+K\nu+\tfrac{1}{2}aN) + 2\pi i cN + \pi i \sum^{\alpha<\beta} k_{\alpha\beta} N_\alpha N_\beta$$
$$- 2\pi i L\nu(u+K\mu+\tfrac{1}{2}K\nu) - 2\pi i C\nu + \pi i nn',$$

by the properties of ϕ, N being a column of integers; thus $\phi(u+aN, \nu)$ is equal to

$$\phi(u, \nu)\, e^{2\pi i bN(u+\frac{1}{2}aN)+2\pi i cN+\pi i \sum^{\alpha<\beta} k_{\alpha\beta}N_\alpha N_\beta+2\pi i(bN\,.\,K\nu - L\nu\,.\,K\mu)}.$$

Now $bN = LgN = L\mu$, therefore

$$bN\,.\,K\nu - L\nu\,.\,K\mu = (\bar{K}L - \bar{L}K)\,\mu\nu = \epsilon\mu\nu = mn' - m'n,$$

where $\mu_i = m_i$, $\mu_{i+p} = m_i'$, etc. for $i < p+1$. If then we take ν, as well as μ, to consist of integers, it will follow that

$$\phi(u+aN, \nu) = \phi(u, \nu)\,.\,e^{2\pi i bN(u+\frac{1}{2}aN)+2\pi i cN+\pi i \sum^{\alpha<\beta} k_{\alpha\beta}N_\alpha N_\beta},$$

and therefore that

$$\frac{\phi(u+aN)}{\phi(u)} = \frac{\phi(u+aN, \nu)}{\phi(u, \nu)} = e^{2\pi i bN(u+\frac{1}{2}aN)+2\pi i cN+\pi i \sum^{\alpha<\beta} k_{\alpha\beta}N_\alpha N_\beta}.$$

Next

$$\phi(u, \mu+\nu) = \phi(u+K\mu+K\nu)\, e^{-2\pi i(L\mu+L\nu)(u+\frac{1}{2}K\mu+\frac{1}{2}K\nu)-2\pi i(C\mu+C\nu)+\pi i(m+m')(n+n')} \qquad (2),$$

and this

$$= \phi(u+K\mu, \nu)\, e^{M},$$

where

$$M = 2\pi i L\nu(u+K\mu+\tfrac{1}{2}K\nu) + 2\pi i C\nu - \pi i nn' - 2\pi i(L\mu+L\nu)(u+\tfrac{1}{2}K\mu+\tfrac{1}{2}K\nu)$$
$$- 2\pi i(C\mu+C\nu) + \pi i(m+m')(n+n');$$

therefore

$$\frac{\phi(u+K\mu, \nu)}{\phi(u, \mu+\nu)}\, e^{-[2\pi i L\mu\,(u+\frac{1}{2}K\mu)+2\pi i C\mu-\pi i mm']}$$
$$= e^{2\pi i L\mu\,(\frac{1}{2}K\nu)-2\pi i L\nu\,(\frac{1}{2}K\mu)+\pi i mm'+\pi i nn'-\pi i(m+m')(n+n')},$$

of which the exponent of the right side is

$$\pi i\,[(\bar{K}L-\bar{L}K)\,\mu\nu - mn' - m'n] = \pi i\,[mn' - m'n - (mn'+m'n)] = -\,2\pi i m'n,$$

so that, since μ, ν consist of integers, the right side is unity.

Hence we have

$$\frac{\phi\,(u+K\mu,\,\nu)}{\phi\,(u,\,\mu+\nu)} = e^{2\pi i L\mu\,(u+\frac{1}{2}K\mu)+2\pi i C\mu-\pi i m m'}.$$

It is to be carefully noticed that this equation does not require $\mu\equiv 0\,(\text{mod. } g)$.

We suppose now that $\mu\equiv 0\,(\text{mod. } g)$. Then $cN+\frac{1}{2}\overset{\alpha<\beta}{\Sigma} k_{\alpha\beta}N_\alpha N_\beta \equiv C\mu - \frac{1}{2}mm'$ (mod. unity) and $L\mu = bN$, $K\mu = aN$, as will be proved immediately (§ 360); thus

$$\frac{\phi(u+aN)}{\phi\,(u)} = \frac{\phi(u+aN,\nu)}{\phi\,(u,\,\nu)} = \frac{\phi(u+aN,\nu)}{\phi\,(u,\,\mu+\nu)} = e^{2\pi i bN\,(u+\frac{1}{2}aN)+2\pi i cN+\pi i\overset{\alpha<\beta}{\Sigma} k_{\alpha\beta}N_\alpha N_\beta},$$

and therefore $\phi\,(u,\,\mu+\nu) = \phi\,(u,\,\nu)$ for integer values ν and any integer values μ that can be written in the form gN, for integer N; namely $\phi\,(u,\,\nu)$ is unaltered by adding to ν any set of integers congruent to zero for the matrix modulus g.

The set of $|g|$ integers gr, wherein r has all rational fractional values less than unity will now be denoted by ν, each value of ν denoting a column of $2p$ integers—in particular $r=0$ corresponds to a set of integers $\equiv\mu\,(\text{mod. } g)$. And ν' shall denote a special one of the sets of integers which are similarly a representative incongruent system for the transposed matrix modulus $\bar{g}$, such that $\nu'=gr'$, the quantities r' being a set of fractions less than 1. With the assigned values for ν, let

$$\psi\,(u) = \sum_\nu e^{-2\pi i r'\nu}\,\phi\,(u,\,\nu);$$

then

$$\psi\,(u+K\lambda) = \sum_\nu e^{-2\pi i r'\nu}\,\phi\,(u+K\lambda,\,\nu) = \sum_\nu e^{2\pi i r'\nu}\,e^{2\pi i L\lambda\,(u+\frac{1}{2}K\lambda)+2\pi i C\lambda-\pi i l l'}\,\phi\,(u,\,\lambda+\nu)$$

for any set of integers λ, as has been shewn (λ being such that, for $i<p+1$, $\lambda_i = l_i$, $\lambda_{i+p} = l_i'$).

If now $\nu+\lambda=\rho$, so that ρ also describes, with ν, a set of integers incongruent in regard to modulus g, those for which the necessary fractions s, in $\rho = gs$, are >1 being replaced, by the theorem proved*, by others for which the necessary fractions are <1, so that the range of values for ρ is precisely that for ν, then we have

$$\begin{aligned}\psi\,(u+K\lambda) &= \sum_\nu e^{-2\pi i r'\rho+2\pi i r'\lambda}\,e^{2\pi i L\lambda\,(u+\frac{1}{2}K\lambda)+2\pi i C\lambda-\pi i l l'}\,\phi\,(u,\,\rho),\\ &= e^{2\pi i r'\lambda+2\pi i L\lambda\,(u+\frac{1}{2}K\lambda)+2\pi i C\lambda-\pi i l l'}\sum_\nu e^{-2\pi i r'\nu}\,\phi\,(u,\,\nu),\\ &= e^{2\pi i r'\lambda+2\pi i L\lambda\,(u+\frac{1}{2}K\lambda)+2\pi i C\lambda-\pi i l l'}\,\psi\,(u).\end{aligned}$$

* That $\phi\,(u,\,\nu)$ is unaltered when to ν is added a column $\equiv 0\,(\text{mod. } g)$.

Hence, by the result of § 284, Chap. XV., we have

$$\psi(u) = A_{\nu'} \vartheta(u, C + r'),$$

the theta function depending on the a, b, h derived in this chapter (§ 358).

Now let ν' describe a set of incongruent values for the modulus $\bar{g}$; then

$$\sum_{\nu'} A_{\nu'} \vartheta(u, C + r') = \Sigma \psi(u) = \sum_{\nu} \sum_{\nu'} e^{-2\pi i r' \nu} \phi(u, \nu);$$

and since $\nu' = \bar{g} r'$, we have $\nu' r = \bar{g} r' r = g r r' = \nu r'$; thus

$$\sum_{\nu'} e^{-2\pi i r' \nu} = \sum_{\nu'} (e^{-2\pi i r \nu'}) = \sum_{\nu'} (e^{-2\pi i r_1})^{\nu'_1} (e^{-2\pi i r_2})^{\nu'_2} \dots (e^{-2\pi i r_{2p}})^{\nu'_{2p}};$$

this sum can be evaluated:

when $\nu \equiv 0$ (mod. g), or the numbers r are zero, its value is equal to the number of incongruent columns for modulus $\bar{g}$, $= |g|$. Since $k = \bar{g} \epsilon g$, we have $|k| = (|g|)^2$, so that $|g| = \sqrt{|k|}$.

when $\nu \not\equiv 0$ (mod. g), so that some of $r_1, \dots, r_{2p}$ are fractional, its value is zero, as is easy to prove (see below, § 360).

Hence we have the following fundamental equation:

$$\sqrt{|k|}\, \phi(u) = \sum_{\nu'} A_{\nu'} \vartheta(u, C + \nu'),$$

which was the expression sought.

Thus between $\sqrt{|k|} + 1$ functions ϕ with the same periods and parameters there exists a homogeneous linear relation with constant coefficients.*

Ex. i. Prove that a product of n functions ϕ is a function ϕ for which $\sqrt{|k|}$ is changed into $n^p \sqrt{|k|}$. In fact the periods are na, nb.

Ex. ii. Prove that the number of homogeneous products of n factors selected from $p+2$ functions ϕ of the same periods and parameters is greater than $n^p \sqrt{|k|}$ when n is large enough. And infer that there exists a homogeneous polynomial relation connecting any $p+2$ functions ϕ of the same periods and parameters. (Cf. Chap. XV., § 284, Ex. v.)

360. We now prove the two results assumed.

(*a*) If $\mu \equiv 0$ (mod. g) or $\mu = gN$, where N are integers, then

$$cN + \tfrac{1}{2} \sum^{\alpha<\beta} k_{\alpha\beta} N_\alpha N_\beta \equiv C\mu - \tfrac{1}{2} m m' \quad \text{(mod. unity).}$$

For

$$k_{\alpha\beta} = (\bar{g}\epsilon g)_{\alpha\beta} = \sum_{\gamma} (\bar{g})_{\alpha\gamma} (\epsilon g)_{\gamma\beta} = \sum_{\gamma=1}^{2p} (\bar{g})_{\alpha\gamma} \sum_{\lambda=1}^{p} [\epsilon_{\gamma,\lambda} g_{\lambda,\beta} + \epsilon_{\gamma,\lambda+p} g_{\lambda+p,\beta}]$$

$$= \sum_{\gamma=1}^{p} g_{\gamma\alpha} \sum_{\lambda=1}^{p} [\epsilon_{\gamma\lambda} g_{\lambda\beta} + \epsilon_{\gamma,\lambda+p} g_{\lambda+p,\beta}] + \sum_{\gamma=1}^{p} g_{\gamma+p,\alpha} \sum_{\lambda=1}^{p} [\epsilon_{\gamma+p,\lambda} g_{\lambda,\beta} + \epsilon_{\gamma+p,\lambda+p} g_{\lambda+p,\beta}]$$

$$= -\sum_{\gamma=1}^{p} g_{\gamma,\alpha} g_{\gamma+p,\beta} + \sum_{\gamma=1}^{p} g_{\gamma+p,\alpha} g_{\gamma,\beta} = \sum_{\gamma=1}^{p} [g_{\gamma+p,\alpha} g_{\gamma,\beta} - g_{\gamma,\alpha} g_{\gamma+p,\beta}]$$

$$= \sum_{\gamma=1}^{p} [g_{\gamma+p,\alpha} g_{\gamma,\beta} - g_{\gamma,\alpha} g_{\gamma+p,\beta}];$$

* Weierstrass, *Berl. Monatsber.*, 1869; Frobenius, *Crelle*, xcvii. (1884); Picard, Poincaré, *Compt. Rendus*, xcvii. (1883), p. 1284.

therefore

$$\overset{\alpha<\beta}{\Sigma} k_{\alpha,\beta} N_\alpha N_\beta = \sum_{\gamma=1}^{p} \overset{\alpha<\beta}{\Sigma} [g_{\gamma+p,\alpha} N_\alpha \,.\, g_{\gamma,\beta} N_\beta - g_{\gamma,\alpha} N_\alpha \,.\, g_{\gamma+p,\beta} N_\beta]$$

$$\equiv \sum_{\gamma=1}^{p} \overset{\alpha<\beta}{\Sigma} [g_{\gamma+p,\alpha} N_\alpha \,.\, g_{\gamma,\beta} N_\beta + g_{\gamma,\alpha} N_\alpha \,.\, g_{\gamma+p,\beta} N_\beta], \quad (\text{mod. } 2),$$

$$\equiv \sum_{\gamma=1}^{p} \left\{ \overset{\alpha<\beta}{\Sigma} g_{\gamma+p,\alpha} N_\alpha \,.\, g_{\gamma,\beta} N_\beta + \overset{\alpha<\beta}{\Sigma} g_{\gamma,\beta} N_\beta \,.\, g_{\gamma+p,\alpha} N_\alpha \right\}$$

$$\equiv \sum_{\gamma=1}^{p} \Sigma\Sigma g_{\gamma+p,\alpha} N_\alpha \,.\, g_{\gamma,\beta} N_\beta, \qquad (\text{mod. } 2),$$

where the $\Sigma\Sigma$ indicates that the summation extends to every pair α, β except those for which $\alpha = \beta$; thus

$$\overset{\alpha<\beta}{\Sigma} k_{\alpha\beta} N_\alpha N_\beta + \sum_{\gamma=1}^{p} \sum_{\alpha=1}^{2p} g_{\gamma+p,\alpha} N_\alpha \,.\, g_{\gamma,\alpha} N_\alpha$$

$$\equiv \sum_{\gamma=1}^{p} [g_{\gamma,1} N_1 + \ldots\ldots + g_{\gamma,2p} N_{2p}] [g_{\gamma+p,1} N_1 + \ldots\ldots + g_{\gamma+p,2p} N_{2p}]$$

$$\equiv \sum_{\gamma=1}^{p} \mu_\gamma \,.\, \mu_{\gamma+p} \equiv mm', \qquad (\text{mod. } 2);$$

therefore, since $\tfrac{1}{2} N_\alpha^2 \equiv \tfrac{1}{2} N_\alpha$ (mod. unity), and therefore

$$\tfrac{1}{2} \sum_{\gamma=1}^{p} g_{\gamma+p,\alpha} N_\alpha \,.\, g_{\gamma,\alpha} N_\alpha \equiv \tfrac{1}{2} [g] N,$$

we have

$$cN + \tfrac{1}{2} \overset{\alpha<\beta}{\Sigma} k_{\alpha\beta} N_\alpha N_\beta \equiv cN + \tfrac{1}{2} mm' - \tfrac{1}{2}[g] N \equiv \{\bar{g}C + \tfrac{1}{2}[g]\} N + \tfrac{1}{2} mm' - \tfrac{1}{2}[g] N, \quad (\text{mod. } 1),$$

$$\equiv gN \,.\, C + \tfrac{1}{2} mm' \equiv \mu C + \tfrac{1}{2} mm' \equiv C\mu - \tfrac{1}{2} mm', \text{ as required.}$$

(*b*) If $r_1, \ldots\ldots, r_{2p}$ be any set of rational fractions all less than unity and not all zero and such that the row $gr = \nu$ consists of integers, and $(\nu'_1, \ldots\ldots, \nu'_{2p}), = \nu'$, be every integer row in turn which can be represented in the form $\bar{g}r'$ for values of r' less than unity, then

$$\sum_{\nu'} (e^{-2\pi i r_1})^{\nu'_1} \,.\, (e^{-2\pi i r_2})^{\nu'_2} \ldots\ldots (e^{-2\pi i r_{2p}})^{\nu'_{2p}}$$

is zero. Since, as remarked (§ 359), the sum can also be written

$$\sum_{r'} (e^{-2\pi i \nu_1})^{r'_1} \ldots\ldots (e^{-2\pi i \nu_{2p}})^{r'_{2p}},$$

wherein $\nu_1, \ldots, \nu_{2p}$ are integers, the sum is unaffected by the addition of any integers to any one or more of the representants $r'_1, \ldots, r'_{2p}$, namely it has the same value for all sets, ν', of incongruent columns (for the modulus $\bar{g}$). If to each of any set of incongruent columns ν' we add the column $(0, \ldots, 0, \lambda_i, 0, \ldots, 0)$, all of whose elements are zero except that occupying the i-th place, which is an integer, we shall obtain another set of incongruent columns.

Suppose then in the above sum r_i is fractional. Add to every one of the incongruent sets ν' the column (0, 0, ..., 1, 0, ..., 0), of which every element except the i-th is zero. In the summation everything is unaffected except the powers of $e^{-2\pi i r_i}$, which are multiplied by $e^{-2\pi i r_i}$. Hence the sum is unaffected when multiplied by $e^{-2\pi i r_i}$, and must therefore be zero.

We put down the figures for a simple case given by

$$p=1, \quad g=\begin{vmatrix} 4 & 5 \\ 1 & 2 \end{vmatrix};$$

then $gr=(4r_1+5r_2,\ r_1+2r_2)$ and the equations $gr=\nu$ give

$$\left.\begin{array}{r} 4r_1+5r_2=\nu_1 \\ r_1+2r_2=\nu_2 \end{array}\right\} \therefore \left\{\begin{array}{l} 3r_1=2\nu_1-5\nu_2 \\ 3r_2=4\nu_2-\ \nu_1\,; \end{array}\right.$$

thus the values of r_1, r_2 and ν_1, ν_2 are given by the table

$r_1,\ r_2$	0, 0	$\frac{1}{3},\ \frac{1}{3}$	$\frac{2}{3},\ \frac{2}{3}$
$\nu_1,\ \nu_2$	0, 0	3, 1	6, 2

Similarly $\bar{g}r'=(4r'_1+r'_2,\ 5r'_1+2r'_2)$, and the equations $\bar{g}r'=\nu'$ give

$$\left.\begin{array}{r} 4r'_1+\ r'_2=\nu'_1 \\ 5r'_1+2r'_2=\nu'_2 \end{array}\right\} \therefore \left\{\begin{array}{l} 3r'_1=2\nu'_1-\ \nu'_2 \\ 3r'_2=4\nu'_2-5\nu'_1\,; \end{array}\right.$$

thus the values of r'_1, r'_2 and ν'_1, ν'_2 are given by the table

$r'_1,\ r'_2$	0, 0	$\frac{1}{3},\ \frac{2}{3}$	$\frac{2}{3},\ \frac{1}{3}$
$\nu'_1,\ \nu'_2$	0, 0	2, 3	3, 4

Thus the sum in question is

$$\begin{aligned} &(e^{-2\pi i r_1})^0\,(e^{-2\pi i r_2})^0+(e^{-2\pi i r_1})^2\,(e^{-2\pi i r_2})^3+(e^{-2\pi i r_1})^3\,(e^{-2\pi i r_2})^4 \\ &=(e^{-2\pi i\nu_1})^0\,(e^{-2\pi i\nu_2})^0+(e^{-2\pi i\nu_1})^{\frac{1}{3}}\,(e^{-2\pi i\nu_2})^{\frac{2}{3}}+(e^{-2\pi i\nu_1})^{\frac{2}{3}}\,(e^{-2\pi i\nu_2})^{\frac{1}{3}} \\ &=1+e^{-2\pi i\,(2r_1+3r_2)}+e^{-2\pi i\,(3r_1+4r_2)}=1+e^{-\frac{2\pi i}{3}(\nu_1+2\nu_2)}+e^{-\frac{2\pi i}{3}(2\nu_1+\nu_2)}. \end{aligned}$$

For $r_1=r_2=\nu_1=\nu_2=0$, these terms are each unity; for

$$(r_1,\ r_2)=(\tfrac{1}{3},\ \tfrac{1}{3}),\quad (\nu_1,\ \nu_2)=(3,\ 1)$$

these terms are

$$1+e^{-2\pi i\,(\frac{2}{3})}+e^{-2\pi i\,(\frac{1}{3})}=1+e^{-\frac{2\pi i}{3}(2)}+e^{-\frac{2\pi i}{3}(1)}$$

or zero.

For $(r_1,\ r_2)=(\frac{2}{3},\ \frac{2}{3})$, $(\nu_1,\ \nu_2)=(6,\ 2)$, these terms are

$$1+e^{-2\pi i\,(\frac{1}{3})}+e^{-2\pi i\,(\frac{2}{3})}=1+e^{-\frac{2\pi i}{3}(1)}+e^{-\frac{2\pi i}{3}(2)}$$

or zero.

361. We give now an example of the expression of ϕ functions. Take the case in which $p=1$, and

$$k=\begin{vmatrix} 0 & -3 \\ 3 & 0 \end{vmatrix};$$

the conditions $\bar{a}b - \bar{b}a = k$, and $\bar{g}\epsilon g = k$, if $a = (a, a')$, $b = (b, b')$, become

$$ab' - a'b = -3, \qquad g_{12}g_{21} - g_{11}g_{22} = -3;$$

taking for instance

$$g = \begin{vmatrix} 4 & 5 \\ 1 & 2 \end{vmatrix},$$

we have, if $x = (x, x')$, $x_1 = (x_1, x_1')$, and $ax + a'x' = 0$, the equation

$$ikxx_1 = 3i(xx_1' - x'x_1) = -\frac{3ix'x_1'}{aa_1}(a'a_1 - aa_1') = \frac{6x'x_1'}{aa_1}(\alpha\beta' - \alpha'\beta),$$

where $a = \alpha + i\beta$, $a' = \alpha' + i\beta'$. Thus, beside $ab' - a'b = -3$, we must have $\alpha\beta' > \alpha'\beta$. The quantities a, b, a', b' are otherwise arbitrary.

The equations $a = Kg$, $b = Lg$ give $(a, a') = (4K + K', 5K + 2K')$; therefore

$$3K = 2a - a', \quad 3L = 2b - b',$$
$$3K' = 4a' - 5a, \quad 3L' = 4b' - 5b;$$

further the equation $c = gC + \frac{1}{2}[g]$ gives

$$(c, c') = \begin{vmatrix} 4 & 1 \\ 5 & 2 \end{vmatrix} (C, C') + \tfrac{1}{2}(4, 10) = (4C + C' + 2, 5C + 2C' + 5),$$

so that

$$3C = 2c - c' + 1, \quad 3C' = 4c' - 5c - 10.$$

Also, from $K = |2\omega, 2\omega'|$, $2\pi iL = |2\eta, 2\eta'|$, with

$$\mathrm{a} = \frac{\eta}{2\omega}, \quad \mathrm{h} = \frac{\pi i}{2\omega}, \quad \mathrm{b} = 2\mathrm{h}\omega',$$

we obtain

$$\mathrm{a} = \pi i(2b - b')/(2a - a'), \quad \mathrm{b} = \pi i(4a' - 5a)/(2a - a'), \quad \mathrm{h} = 3\pi i/(2a - a').$$

If then $\vartheta(u; C)$ denote the theta function, with characteristic $\begin{pmatrix} C \\ -C' \end{pmatrix}$, given by

$$\vartheta(u; C) = \Sigma e^{\mathrm{a}u^2 + 2\mathrm{h}u(n+C) + \mathrm{b}(n+C)^2 - 2\pi iC'(n+C)},$$

then the Jacobian function, with a, b as periods, and c as parameter, is given by

$$3\phi(u) = \sum_{\nu'} A_{\nu'}\vartheta(u; C + r'),$$

where, in the three terms of the right hand, r' is in turn equal to $\begin{pmatrix} 0 \\ 0 \end{pmatrix}$, $\begin{pmatrix} 1/3 \\ 2/3 \end{pmatrix}$, $\begin{pmatrix} 2/3 \\ 1/3 \end{pmatrix}$.

The function $\phi(u)$ may in fact be considered as a theta function of the third order; its various expressions, obtainable by taking different forms for the matrix g, are transformations of one another, in the sense of Chap. XVIII. and XX.

362. The theory of the expression of a Jacobian function which has been given is for the case when $\sigma = 2p$. Suppose $\sigma < 2p$, and that we have two matrices a, b, each of p rows and σ columns, such that $\bar{a}b - \bar{b}a$, $= k$, is a skew symmetrical matrix of integers, for which $ikxx_1$ is a positive form for all values satisfying $ax = 0$, other than those for which also $a_1x = 0$, or $x = 0$; then it is possible* to determine other $2p - \sigma$ columns of quantities, and thence to construct matrices, A, B, of $2p$ columns (whereof the first σ columns are those of a, b), such that $\bar{A}B - \bar{B}A = K$ is a skew symmetrical matrix of integers for which $iKxx_1$ is positive when $Ax = 0$, except when $x = 0$ or $A_1x = 0$.

There will then correspond to the set A, B a function Φ, involving $\sqrt{|K|}$ arbitrary coefficients, such that, for integral n,

$$\Phi(u + An) = e^{2\pi iBn(u + \frac{1}{2}An) + 2\pi iCn + \sum_{\alpha<\beta} K_{\alpha,\beta} n_\alpha n_\beta} \Phi(u).$$

The function $\phi(u)$, which is subject only to the condition that

$$\phi(u + an) = e^{2\pi ibn(u + \frac{1}{2}an) + 2\pi icn + \sum_{\alpha<\beta} k_{\alpha,\beta} n_\alpha n_\beta} \phi(u),$$

is then obtained by regarding $\phi(u)$ as a particular case of $\Phi(u)$, in which the added columns in A, B are arbitrary except that they must be such that the necessary conditions for A, B are satisfied.

For further development the reader should consult Frobenius, *Crelle*, XCVII. (1884), pp. 16, 188, and *Crelle*, CV. (1889), p. 35.

* Frobenius, *Crelle*, XCVII. (1884), p. 24.

CHAPTER XX.

TRANSFORMATION OF THETA FUNCTIONS.

363. IT has been shewn in Chapter XVIII. that a theta function of the first order, in the arguments u, with characteristic (Q, Q'), say $\vartheta(u, Q)$, may be regarded as a theta function of the r-th order in the arguments w, with characteristic (K, K'), provided certain relations, (I), (II), of § 322, p. 532, are satisfied. Let this theta function in w be denoted by $\Pi(w, K)$. We confine ourselves in this chapter, unless the contrary be stated, to the case when (Q, Q') is a half-integer characteristic. Then the function $\vartheta(u, Q)$ is odd or even; therefore, since $u = Mw$, the function $\Pi(w, K)$ is an odd or even function of the arguments w. Now we have shewn, in Chap. XV. (§ 287), that every such odd, or even, theta function of order r, is expressible as a linear function of functions of the form

$$\psi_r(w;\ K, K'+\mu) = \vartheta\left[rw;\ 2v, 2rv', 2\zeta/r, 2\zeta' \,\middle|\, \begin{matrix}(K'+\mu)/r\\ K\end{matrix}\right] + \epsilon\vartheta\left[-rw;\ 2v, 2rv', 2\zeta/r, 2\zeta' \,\middle|\, \begin{matrix}(K'+\mu)/r\\ K\end{matrix}\right],$$

where ϵ is ± 1, according as the function is even or odd. The most important result of the present chapter is that the functions $\psi_r(w;\ K, K'+\mu)$ which occur can be expressed as integral polynomials of the r-th degree in 2^p theta functions $\vartheta\left(w;\ 2v, 2v', 2\zeta, 2\zeta' \,\middle|\, \begin{matrix}R'\\ R\end{matrix}\right)$, whose characteristics are those of a Göpel system of half-integer characteristics (Chap. XVII., § 297); the earlier part (§§ 364—370) of the chapter is devoted to proving this theorem.

The theory is different according as r is odd or even. When r is odd, ϵ is $e^{\pi i |Q|}$, and we have shewn (§ 327 Chap. XVIII.) that, for odd values of r, $|Q| \equiv |K|$, (mod. 2); the theory deals then only with functions

$$\psi_r(w;\ K, K'+\mu)$$

in which $\epsilon = e^{\pi i |K|}$. When r is even, ϵ, though still equal to $e^{\pi i |Q|}$, may or may not be equal to $e^{\pi i |K|}$, according to the integer matrix which determines the transformation; but in this case, also, the value of ϵ in the functions $\psi_r(w; K, K'+\mu)$ which occur is determinate.

The proof of the theorem is furnished by obtaining actual expressions for the functions $\psi_r(w; K, K'+\mu)$ as integral polynomials of the r-th degree in the 2^p functions $\vartheta\left(w; 2v, 2v', 2\zeta, 2\zeta' \middle| \begin{matrix} R' \\ R \end{matrix}\right)$; the coefficients arising in these polynomials are theta functions whose arguments are r-th parts of periods, of the form $(2vm + 2v'm')/r$. The completion of the theory of the transformation requires that these coefficients should be expressed in terms of constants depending on theta functions with half integer characteristics (§ 373).

Further the theory requires that the coefficients in the expression of the function $\Pi(w; K)$ by the functions $\psi_r(w; K, K'+\mu)$ should be assigned in general. In simple cases this is often an easy matter. The general case is reduced to simpler cases by regarding the general transformation of the r-th order as arising from certain standard transformations for which there is no difficulty as to the coefficients, by the juxtaposition of linear transformations (§§ 371—2)*.

364. It follows from § 332, Chap. XVIII. that any transformation may be obtained by composition of transformations for which the order r is a prime number. It is therefore sufficient theoretically to consider the two cases when $r = 2$, and when r is an odd prime number. We begin with the former case, and shew that the transformed theta function can be expressed as a quadric polynomial in 2^p theta functions belonging to a special Göpel system. A more general expression is given later (§ 370).

* For the transformation of theta functions, and of Abelian functions, the following may be consulted. Jacobi, *Crelle*, VIII. (1832), p. 416; Richelot, *Crelle*, XII. (1834), p. 181, and *Crelle*, XVI. (1837), p. 221; Rosenhain, *Crelle*, XL. (1850), p. 338, and *Mém. par divers Savants*, t. XI. (1851), pp. 396, 402; Hermite, *Liouville*, Ser. 2, t. III. (1858), p. 26, and *Comptes Rendus*, t. XL. (1855); Königsberger, *Crelle*, LXIV. (1865), p. 17, *Crelle*, LXV. (1866), p. 335, *Crelle*, LXVII. (1867), p. 58; Weber, *Crelle*, LXXIV. (1872), p. 69, and *Annali di Mat.* Ser. 2, t. IX. (1878); Thomae, *Ztschr. f. Math. u. Phys.*, t. XII. (1867), and *Crelle*, LXXV. (1872), p. 224; Kronecker, *Berlin. Monatsber.*, 1880, pp. 686, 854; H. J. S. Smith, *Report on the Theory of Numbers*, *British Association Reports*, 1865, Part VI., § 125 (cf. Weber, *Acta Math.*, VI. (1885), p. 342; Weber, *Elliptische Functionen* (1891), p. 103; Dirichlet, in *Riemann's Werke* (1876), p. 438; Cauchy, *Liouville*, V. (1841), and *Exer. de Math.*, II., p. 118; Gauss, *Werke* (1863), t. II., p. 11 (1808), etc.; Kronecker, *Berlin. Sitzungsber.* 1883; Frobenius, *Crelle*, LXXXIX. (1880), p. 40, *Crelle*, XCVII. (1884), pp. 16, 188, *Crelle*, CV. (1889), p. 35; Wiltheiss, *Crelle*, XCVI. (1884), p. 21; the books of Krause, *Die Transformation der Hyperelliptischen Functionen* (1886), (and the bibliography there given), *Theorie der Doppeltperiodischen Functionen* (1895); Prym u. Krazer, *Neue Grundlagen einer Theorie der allgemeinen Thetafunctionen* (1892), Zweiter Teil. See also references given in Chap. XXI., of the present volume, and in Appendix II.

By means of the equations $u = Mw$, a function $\vartheta\left(u;\ 2\omega, 2\omega', 2\eta, 2\eta' \middle| \begin{matrix} Q' \\ Q \end{matrix}\right)$, with half-integer characteristic $\begin{pmatrix} Q' \\ Q \end{pmatrix}$, becomes a theta function in w, $\Pi(w;\ K, K')$, of order 2, with the associated constants $2\upsilon, 2\upsilon', 2\zeta, 2\zeta'$ and the characteristic (K, K'), where (§ 324, Chap. XVIII.)

$$2M\upsilon = 2\omega\alpha + 2\omega'\alpha', \quad 2M\upsilon' = 2\omega\beta + 2\omega'\beta', \quad 2\bar{M}(\eta\alpha + \eta'\alpha') = 4\zeta,$$

$$2\bar{M}(\eta\beta + \eta'\beta') = 4\zeta', \quad K' = \bar{\alpha}Q' - \bar{\alpha}'Q - \tfrac{1}{2}d(\bar{\alpha}\alpha'), \quad -K = \bar{\beta}Q' - \bar{\beta}'Q - \tfrac{1}{2}d(\bar{\beta}\beta'),$$

and

$$\bar{\alpha}\alpha' = \bar{\alpha}'\alpha, \quad \bar{\beta}\beta' = \bar{\beta}'\beta, \quad \bar{\alpha}\beta' - \bar{\alpha}'\beta = \bar{\beta}'\alpha - \bar{\beta}\alpha' = 2;$$

this theta function in w, $\Pi(w;\ K, K')$, can by § 287, p. 463, be expressed as a linear aggregate of terms of the form

$$\psi_r(w;\ K, K'+\mu) = \vartheta\left[rw;\ 2\upsilon, 2r\upsilon', 2\zeta/r, 2\zeta' \middle| \begin{matrix} (K'+\mu)/r \\ K \end{matrix}\right]$$

$$+ \epsilon\vartheta\left[-rw;\ 2\upsilon, 2r\upsilon', 2\zeta/r, 2\zeta' \middle| \begin{matrix} (K'+\mu)/r \\ K \end{matrix}\right],$$

r being equal to 2; here ϵ, $= e^{4\pi i QQ'}$, is ± 1, according as the original function, that is, according as the function $\Pi(w;\ K, K')$, is even or odd. For brevity we put $w = 2\upsilon W$, $\upsilon\tau' = \upsilon'$, and denoting by $\Theta(W, \tau')$ the series $\Sigma e^{2\pi i Wn + i\pi\tau' n^2}$, we consider the function

$$\Psi_r(W;\ K, K'+\mu) = \Theta\left[rW;\ r\tau' \middle| \begin{matrix} (K'+\mu)/r \\ K \end{matrix}\right] + \epsilon\Theta\left[-rW;\ r\tau' \middle| \begin{matrix} (K'+\mu)/r \\ K \end{matrix}\right],$$

which is equal to $e^{-\frac{1}{2}r\zeta\upsilon^{-1}w^2}\psi_r(w;\ K, K'+\mu)$. Throughout the chapter the symbols $\vartheta\left(w \middle| \begin{matrix} K' \\ K \end{matrix}\right)$, $\Theta\left(W \middle| \begin{matrix} K' \\ K \end{matrix}\right)$ denote respectively

$$\vartheta\left[w;\ 2\upsilon, 2\upsilon', 2\zeta, 2\zeta' \middle| \begin{matrix} K' \\ K \end{matrix}\right], \quad \Theta\left(W;\ \tau' \middle| \begin{matrix} K' \\ K \end{matrix}\right).$$

Taking the final formula of § 291, p. 472, replacing $\omega, \omega', \eta, \eta', \begin{pmatrix} q' \\ q \end{pmatrix}, \begin{pmatrix} r' \\ r \end{pmatrix}$ respectively by $\upsilon, \upsilon', \zeta, \zeta', \tfrac{1}{2}\begin{pmatrix} \alpha' \\ \alpha \end{pmatrix}, \tfrac{1}{2}\begin{pmatrix} \alpha' \\ \alpha \end{pmatrix} + \begin{pmatrix} K' \\ K \end{pmatrix}$, multiplying both sides of the equation by $e^{\pi i \alpha(\mu - K' - \alpha')}$, where μ is a row of integers each either 0 or 1, and adding the 2^p equations obtainable by giving α all values in which each of its elements is 0 or 1, we obtain

$$\sum_\alpha e^{\pi i\alpha(\mu-K')}\,\Theta\left[V-U;\ \tau'\middle|\tfrac{1}{2}\binom{\alpha'}{\alpha}\right]\Theta\left[V+U;\ \tau'\middle|\tfrac{1}{2}\binom{\alpha'}{\alpha}+\binom{K'}{K}\right]$$

$$=\sum_\alpha e^{\pi i\alpha(\epsilon'+\mu)}\sum_{\epsilon'}\Theta\left[2V;\ 2\tau'\middle|\begin{matrix}\tfrac{1}{2}(\epsilon'+K')\\ K\end{matrix}\right]\Theta\left[2U;\ 2\tau'\middle|\begin{matrix}\tfrac{1}{2}(\epsilon'+K'+\alpha')\\ K\end{matrix}\right],$$

and hence, replacing V, U respectively by W, 0,

$$2^p\Theta\left[2W;\ 2\tau'\middle|\begin{matrix}\tfrac{1}{2}(\mu+K')\\ K\end{matrix}\right]\Theta\left[0;\ 2\tau'\middle|\begin{matrix}\tfrac{1}{2}(\mu+K'+\alpha')\\ K\end{matrix}\right]$$

$$=\sum_\alpha e^{\pi i\alpha(\mu-K')}\,\Theta\left[W;\ \tau'\middle|\tfrac{1}{2}\binom{\alpha'}{\alpha}\right]\Theta\left[W;\ \tau'\middle|\tfrac{1}{2}\binom{\alpha'}{\alpha}+\binom{K'}{K}\right].$$

This may be regarded as the fundamental equation for quadric transformation; we consider various cases of it.

(i) When (K, K') is the zero characteristic we obtain

$$\Theta\left[2W;\ 2\tau'\middle|\begin{matrix}\tfrac{1}{2}\mu\\ 0\end{matrix}\right]=2^{-p}\sum_\alpha e^{\pi i\alpha\mu}\,\Theta^2\left[W;\ \tau'\middle|\tfrac{1}{2}\binom{\alpha'}{\alpha}\right]\Big/\Theta\left[0;\ 2\tau'\middle|\begin{matrix}\tfrac{1}{2}(\mu+\alpha')\\ 0\end{matrix}\right],$$

the right-hand side being independent of α', which for simplicity may be put $=0$.

We can infer that in any quadric transformation, *when the transformed function has zero characteristic, it can be expressed as a linear aggregate of the* 2^p *squares* $\vartheta^2\left(w\middle|\tfrac{1}{2}\binom{\alpha'}{\alpha}\right)$, *in which* α' *is an arbitrary row of integers (each* 0 *or* 1) *and* α *has all possible values in which its elements are either* 0 *or* 1.

(ii) When $K'=0$, $K=\tfrac{1}{2}n$ is not zero, we obtain

$$\Theta\left[2W;\ 2\tau'\middle|\tfrac{1}{2}\binom{\mu}{n}\right]\Theta\left[0;\ 2\tau'\middle|\tfrac{1}{2}\binom{\mu+\alpha'}{n}\right]$$

$$=2^{-p}\sum_\alpha e^{\pi i\alpha\mu}(1+e^{\pi i n(\mu+\alpha')})\,\Theta\left[W;\ \tau'\middle|\tfrac{1}{2}\binom{\alpha'}{\alpha}\right]\Theta\left[W;\ \tau'\middle|\tfrac{1}{2}\binom{\alpha'}{\alpha+n}\right],$$

where on the right side only 2^{p-1} terms are to be taken in the summation in regard to α, two values of α whose difference is a row of elements congruent (mod. 2) to the elements of n not being both admitted. When $\tfrac{1}{2}\binom{\mu}{n}$ is an even characteristic we may put $\alpha'=0$; when $\tfrac{1}{2}\binom{\mu}{n}$ is an odd characteristic we may put $\alpha'=\mu$.

In this case, as before, only 2^p theta functions enter on the right hand, and their characteristics form a special Göpel system.

The cases (i) and (ii) give the transformation of any theta function when the matrix, of 2^p rows and columns, associated with the transformation* is

* For the notation, cf. Chap. XVIII., §§ 322, 324.

$\begin{pmatrix}20\\01\end{pmatrix}$. It can be shewn that by adjunction of linear transformations every quadric transformation is reducible to this case (cf. § 415 below); so that theoretically no further formulae are required. As it may often be a matter of difficulty to obtain the linear transformations necessary to reduce any given quadric transformation to this one, it is proper to give the formulae for the functions

$$\Psi_2(W;\ K, K'+\mu) = \Theta\left[2W;\ 2\tau' \left| \begin{matrix}\tfrac{1}{2}(\mu+K')\\ K\end{matrix}\right.\right] + \epsilon\Theta\left[-2W;\ 2\tau' \left| \begin{matrix}\tfrac{1}{2}(\mu+K')\\ K\end{matrix}\right.\right];$$

by this means the problem is reduced to finding the coefficients in the expression of any theta function in w, of the second order, in terms of functions $\Psi_2(W;\ K, K'+\mu)$ (see § 372 below). Hence we add the following case.

(iii) When K' is not zero, we deduce, by changing the sign of W in the fundamental formula, the equation

$$2^p\Theta\left[0;\ 2\tau' \left| \begin{matrix}\tfrac{1}{2}(\mu+K'+\alpha')\\ K\end{matrix}\right.\right]\Psi_2(W;\ K, K'+\mu)$$
$$= \sum_{\alpha} e^{\pi i\alpha(\mu-K')}\, C_\alpha \Theta\left[W;\ \tau' \left| \tfrac{1}{2}\begin{pmatrix}\alpha'\\ \alpha\end{pmatrix}\right.\right]\Theta\left[W;\ \tau' \left| \tfrac{1}{2}\begin{pmatrix}\alpha'\\ \alpha\end{pmatrix} + \begin{pmatrix}K'\\ K\end{pmatrix}\right.\right],$$

where, putting $K=\tfrac{1}{2}k$, $K'=\tfrac{1}{2}k'$, we have $C_\alpha = 1+\epsilon e^{\pi i k(k'+\alpha')+\pi i\alpha k'}$. When ϵ is $+1$, there are 2^{p-1} values of α for which $\alpha k' \equiv k(k'+\alpha')+1$ (§ 295, Chap. XVII.); for these values $C_\alpha = 0$; when $\epsilon = -1$, there are 2^{p-1} values of α for which $\alpha k' \equiv k(k'+\alpha')$; for these values $C_\alpha = 0$. In either case it follows that the right side of the equation contains only 2^{p-1} terms, and contains only 2^p theta functions whose characteristics are a special Göpel system.

It is easy to see that the results of cases (ii) and (iii) can be summarised as follows: *when the characteristic* (K, K') *is not zero the transformed function is a linear aggregate of* 2^{p-1} *products of the form* $\vartheta[w;\ A, P_i]\,\vartheta[w;\ A, K, P_i]$ *wherein the* 2^{p-1} *characteristics* P_i *are of the form* $\tfrac{1}{2}\begin{pmatrix}0\\ \alpha\end{pmatrix}$, $K=\begin{pmatrix}K'\\ -K\end{pmatrix}$, *and* A, K *are such that** $e^{\pi i|K|+\pi i|A,K|}=\epsilon$.

These results are in accordance with § 288, Chap. XV.; there being $2^{p-1}(1+\epsilon)$ linearly independent theta functions of the second order with zero characteristic and of character ϵ, namely 2^p such *even* functions and no odd functions, and there being 2^{p-1} linearly independent theta functions of the second order with characteristic other than zero.

365. *Ex.* i. When $p=1$, the results of case (i), if we put $\Theta_{gh}(W;\ \tau')$ for $\Theta\left[W;\ \tau' \left| \tfrac{1}{2}\begin{pmatrix}g\\ -h\end{pmatrix}\right.\right]$, as is usual, are

$$\Theta_{00}(2W;\ 2\tau') = \frac{\Theta_{00}^2(W;\ \tau')+\Theta_{01}^2(W;\ \tau')}{2\Theta_{00}(2\tau')} = \frac{\Theta_{10}^2(W;\ \tau')+\Theta_{11}^2(W;\ \tau')}{2\Theta_{10}(2\tau')},$$

* For the notation, see Chap. XVII., § 294.

and

$$\Theta_{10}(2W;\ 2\tau')=\frac{\Theta_{00}^2(W;\ \tau')-\Theta_{01}^2(W;\ \tau')}{2\Theta_{10}(2\tau')}=\frac{\Theta_{10}^2(W;\ \tau')-\Theta_{11}^2(W;\ \tau')}{2\Theta_{00}(2\tau')},$$

where $\Theta(2\tau')$ denotes $\Theta(0;\ 2\tau')$. If then we introduce the notations

$$\sqrt{k}=\frac{\Theta_{10}(2\tau')}{\Theta_{00}(2\tau')},\quad \sqrt{k'}=\frac{\Theta_{01}(2\tau')}{\Theta_{00}(2\tau')},\quad \sqrt{\lambda}=\frac{\Theta_{10}(\tau')}{\Theta_{00}(\tau')},\quad \sqrt{\lambda'}=\frac{\Theta_{01}(\tau')}{\Theta_{00}(\tau')},$$

$$\sqrt{x}=\frac{1}{\sqrt{k}}\frac{\Theta_{11}(2W;\ 2\tau')}{\Theta_{01}(2W;\ 2\tau')},\quad \sqrt{y}=\sqrt{\frac{k'}{k}}\frac{\Theta_{10}(2W;\ 2\tau')}{\Theta_{01}(2W;\ 2\tau')},\quad \sqrt{z}=\sqrt{k'}\frac{\Theta_{00}(2W;\ 2\tau')}{\Theta_{01}(2W;\ 2\tau')},$$

$$\sqrt{\xi}=\frac{1}{\sqrt{\lambda}}\frac{\Theta_{11}(W;\ \tau')}{\Theta_{01}(W;\ \tau')},\quad \sqrt{\eta}=\sqrt{\frac{\lambda'}{\lambda}}\frac{\Theta_{10}(W;\ \tau')}{\Theta_{01}(W;\ \tau')},\quad \sqrt{\zeta}=\sqrt{\lambda'}\frac{\Theta_{00}(W;\ \tau')}{\Theta_{01}(W;\ \tau')},$$

we find by multiplying the equations above that

$$\Theta_{00}^4(W;\ \tau')-\Theta_{01}^4(W;\ \tau')=\Theta_{10}^4(W;\ \tau')-\Theta_{11}^4(W;\ \tau'),$$

and therefore that

$$\lambda^2+\lambda'^2=1,$$

so that also

$$k^2+k'^2=1;$$

while, comparing the two forms for $\Theta_{00}(2W;\ 2\tau')$, putting $W=0$, we obtain

$$\sqrt{k}=\frac{\lambda}{1+\lambda'},\ \text{ or }\ k=\frac{1-\lambda'}{1+\lambda'},\ \text{ giving }\ \lambda=\frac{2\sqrt{k}}{1+k};$$

further the equations for $\Theta_{00}(2W;\ 2\tau')$ and $\Theta_{10}(2W;\ 2\tau')$ give the results

$$\frac{\zeta+\lambda'}{\eta+\lambda'\xi}=1+\lambda',\quad \frac{\zeta-\lambda'}{\eta-\lambda'\xi}=1-\lambda',$$

from which we find

$$\eta=1-\xi,\ \zeta=1-\lambda^2\xi;\ \text{thus also}\ y=1-x,\ z=1-k^2x.$$

Ex. ii. The equations of case (ii), also for $p=1$, give

$$\Theta_{01}(2W;\ 2\tau')=\frac{\Theta_{00}(W;\ \tau')\Theta_{01}(W;\ \tau')}{\Theta_{01}(2\tau')},\quad \Theta_{11}(2W;\ 2\tau')=\frac{\Theta_{10}(W;\ \tau')\Theta_{11}(W;\ \tau')}{\Theta_{01}(2\tau')}.$$

From these we have by division

$$\sqrt{x}=(1+\lambda')\frac{\sqrt{\xi(1-\xi)}}{\sqrt{1-\lambda^2\xi}},$$

while from these and the results of Ex. 1, we find

$$\sqrt{y}=[1-(1+\lambda')\xi]/\sqrt{1-\lambda^2\xi},\quad \sqrt{z}=[1-(1-\lambda')\xi]/\sqrt{1-\lambda^2\xi}.$$

Ex. iii. When $p=1$, by considering the change in the value of the function

$$\vartheta_{01}^2(w)\frac{d}{dw}\left[\frac{\vartheta_{11}(w)}{\vartheta_{01}(w)}\right]$$

when w is increased by a period, we immediately find that it is a theta function in w of the second order with characteristic $\frac{1}{2}\binom{1}{0}$; hence by the result of case (iii) above, the function is a constant multiple of $\vartheta_{10}(w)\vartheta_{00}(w)$; determining the constant by putting $w=0$, we obtain the equation

$$\Theta_{00}(\tau')\Theta_{10}(\tau')[\Theta'_{11}(W;\ \tau')\Theta_{01}(W;\ \tau')-\Theta'_{01}(W;\ \tau')\Theta_{11}(W;\ \tau')]$$
$$=\Theta'_{11}(\tau')\Theta_{01}(\tau')\Theta_{10}(W;\ \tau')\Theta_{00}(W;\ \tau'),$$

which is immediately seen to be equivalent to

$$\frac{\Theta'_{11}(\tau')\,\Theta_{00}(\tau')}{\Theta_{01}(\tau')\,\Theta_{10}(\tau')}\,W=\int_0^\xi \frac{d\xi}{\sqrt{4\xi(1-\xi)(1-\lambda^2\xi)}}.$$

[We may obtain the theta relation, here deduced, from the addition formula of Ex. i., § 286, p. 457; taking therein $m=\frac{1}{2}\begin{pmatrix}0\\-1\end{pmatrix}$, $a_1=\frac{1}{2}\begin{pmatrix}1\\-1\end{pmatrix}$, $a_2=\frac{1}{2}\begin{pmatrix}0\\-1\end{pmatrix}$, $w=0$, $q=\frac{1}{2}\begin{pmatrix}1\\0\end{pmatrix}$, $r=p=\frac{1}{2}\begin{pmatrix}0\\0\end{pmatrix}$, we immediately derive

$$\vartheta_{10}(u)\,\vartheta_{00}(u)\,\vartheta_{11}(2v)\,\vartheta_{01}(0)=\vartheta_{00}(v)\,\vartheta_{10}(v)\,[\vartheta_{01}(u-v)\,\vartheta_{11}(u+v)-\vartheta_{01}(u+v)\,\vartheta_{11}(u-v)];$$

if, for small values of v, this equation be expanded in powers of v, and the coefficients of v on the two sides be put equal, there results the equation in question.]

Ex. iv. By differentiating the second result of Ex. ii., putting $W=0$, and putting $W=0$ in the first result of the same example and in the second value for $\Theta_{00}(2W;\,2\tau')$ in Ex. i., we obtain

$$\frac{\Theta'_{11}(2\tau')}{\Theta_{00}(2\tau')\,\Theta_{01}(2\tau')\,\Theta_{10}(2\tau')}=\frac{\Theta'_{11}(\tau')}{\Theta_{00}(\tau')\,\Theta_{01}(\tau')\,\Theta_{10}(\tau')},$$

so that the second of these functions is unaltered by replacing τ' by $2^n\tau'$, n being as large as we please. Hence we immediately find from the series for the functions, by putting $\tau'=\infty$, that each of these fractions is equal to π. Hence if the integral occurring in the last example be denoted by J we have $J=\pi\Theta_{00}^2(\tau')\,W$. In precisely the same way we find $I=2\pi\Theta_{00}^2(2\tau')\,W$, where I is an integral differing only from J by the substitution of x for ξ and k for λ. Hence

$$I/J=2\Theta_{00}^2(2\tau')/\Theta_{00}^2(\tau'),\ =1+\lambda',$$

as follows from the first result of Ex. 1.

From these results we are justified in writing the formula of Ex. ii. in the form

$$\mathrm{sn}\left[(1+\lambda')J;\ \frac{1-\lambda'}{1+\lambda'}\right]=\frac{(1+\lambda')\,\mathrm{sn}(J,\lambda)\,\mathrm{cn}(J,\lambda)}{\mathrm{dn}(J,\lambda)};$$

and this is Landen's first transformation for Elliptic functions.

Ex. v. The preceding examples deal, in the case $p=1$, with the quadric transformation associated with the matrix $\begin{pmatrix}2&0\\0&1\end{pmatrix}$. Prove when $p=1$ that for any matrix of quadric transformation the transformed theta function is expressible linearly in terms of one or more of the eight functions

$$\Theta=\Theta_{00}(2W;\,2\tau'),\quad \Theta_2=\Theta_{10}(2W;\,2\tau'),\quad \Theta_0=\Theta_{01}(2W;\,2\tau'),\quad \Theta_1=\Theta_{11}(2W;\,2\tau'),$$

$$\Theta_4=\Theta\left(2W;\,2\tau'\left|\begin{matrix}1/4\\0\end{matrix}\right.\right)+\Theta\left(2W;\,2\tau'\left|\begin{matrix}-1/4\\0\end{matrix}\right.\right),\quad \Theta_5=\Theta\left(2W;\,2\tau'\left|\begin{matrix}1/4\\0\end{matrix}\right.\right)-\Theta\left(2W;\,2\tau'\left|\begin{matrix}-1/4\\0\end{matrix}\right.\right),$$

$$\Theta_6=\Theta\left(2W;\,2\tau'\left|\begin{matrix}1/4\\1/2\end{matrix}\right.\right)+i\Theta\left(2W;\,2\tau'\left|\begin{matrix}-1/4\\1/2\end{matrix}\right.\right),\quad \Theta_7=\Theta\left(2W;\,2\tau'\left|\begin{matrix}1/4\\1/2\end{matrix}\right.\right)-i\Theta\left(2W;\,2\tau'\left|\begin{matrix}-1/4\\1/2\end{matrix}\right.\right).$$

Prove in particular that the functions arising for the transformation associated with the matrix $\begin{pmatrix}1&0\\0&2\end{pmatrix}$ are expressed as follows:

$$\Theta_{00}(W;\,\tfrac{1}{2}\tau')=\Theta+\Theta_2,\quad \Theta_{01}(W;\,\tfrac{1}{2}\tau')=\Theta-\Theta_2,\quad \Theta_{10}(W;\,\tfrac{1}{2}\tau')=\Theta_4,\quad \Theta_{11}(W;\,\tfrac{1}{2}\tau')=-i\Theta_5;$$

and that the functions arising for the transformation associated with the matrix $\begin{pmatrix}1 & 1\\ 0 & 2\end{pmatrix}$ are expressed as follows:

$$\Theta_{00}(W;\ \tfrac{1}{2}\tau'-\tfrac{1}{2})=\Theta-i\Theta_2,\quad \Theta_{01}(W;\ \tfrac{1}{2}\tau'-\tfrac{1}{2})=\Theta+i\Theta_2,$$

$$\Theta_{10}(W;\ \tfrac{1}{2}\tau'-\tfrac{1}{2})=e^{-\frac{3\pi i}{8}}\Theta_6,\quad \Theta_{11}(W;\ \tfrac{1}{2}\tau'-\tfrac{1}{2})=e^{\frac{\pi i}{8}}\Theta_7.$$

Obtain from the formulae of the text the expressions of the functions Θ_4, Θ_5, Θ_6, Θ_7 of the form

$$\Theta_4=C_4\Theta_{00}(W)\Theta_{10}(W),\ \Theta_5=C_5\Theta_{01}(W)\Theta_{11}(W),\ \Theta_6=C_6\Theta_{01}(W)\Theta_{10}(W),\ \Theta_7=C_7\Theta_{00}(W)\Theta_{11}(W),$$

where C_4, C_5, C_6, C_7 are constants.

Ex. vi. The reason why the matrices $\begin{pmatrix}2 & 0\\ 0 & 1\end{pmatrix}$, $\begin{pmatrix}1 & 0\\ 0 & 2\end{pmatrix}$, $\begin{pmatrix}1 & 1\\ 0 & 2\end{pmatrix}$ are selected in Ex. v. will appear subsequently (§ 415); the matrix $\begin{pmatrix}1 & 0\\ 0 & 2\end{pmatrix}$ gives the transformation which is *supplementary* to that given by $\begin{pmatrix}2 & 0\\ 0 & 1\end{pmatrix}$; it gives results leading to the equation

$$\operatorname{sn}[(1+k)u,\ 2\sqrt{k}/(1+k)]=(1+k)\operatorname{sn}(u,k)/[1+k\operatorname{sn}^2(u,k)];$$

by combination of these results with those for the matrix $\begin{pmatrix}2 & 0\\ 0 & 1\end{pmatrix}$ we obtain the multiplication formula

$$\Theta_{11}(2W;\ \tau')=A\Theta_{11}(W;\ \tau')\Theta_{01}(W;\ \tau')\Theta_{10}(W;\ \tau')\Theta_{00}(W;\ \tau'),$$

where A is a constant (cf. Ex. vii., § 317, Chap. XVII. and § 332, Chap. XVIII.).

The matrix associated with any quadric transformation can be put into the form

$$\Omega\begin{pmatrix}2 & 0\\ 0 & 1\end{pmatrix}\Omega',$$

where Ω, Ω' are matrices of linear transformations; for instance we have

$$\begin{pmatrix}0 & -1\\ 1 & 0\end{pmatrix}\begin{pmatrix}2 & 0\\ 0 & 1\end{pmatrix}\begin{pmatrix}0 & 1\\ -1 & 0\end{pmatrix}=\begin{pmatrix}1 & 0\\ 0 & 2\end{pmatrix}$$

with the corresponding equations

$$U=\tau W_1,\quad W_1=2W_2,\quad W_2=-\tau_2 W_3;\qquad \tau_1=-1/\tau,\quad \tau_2=\tau_1/2,\quad \tau_3=-1/\tau_2,$$

from which we have, for instance,

$$\Theta_{10}(W_3;\ \tfrac{1}{2}\tau_3)=\Theta_{10}(U;\ \tau)=e^{\frac{-\pi i U^2}{\tau}}\sqrt{\frac{i}{\tau}}\,\Theta_{01}(W_1;\ \tau_1)=e^{\frac{-\pi i U^2}{\tau}}\sqrt{\frac{i}{\tau}}\,\Theta_{01}(2W_2;\ 2\tau_2)$$

$$=e^{\frac{-\pi i U^2}{\tau}}E\Theta_{00}(W_2;\ \tau_2)\Theta_{01}(W_2;\ \tau_2)=F\Theta_{00}(W_3;\ \tau_3)\Theta_{10}(W_3;\ \tau_3),$$

(E, F being constants) whereby the transformation formula for $\Theta_{10}(W_3;\ \tfrac{1}{2}\tau_3)$ is obtained from those for $\Theta_{10}(2W;\ 2\tau')$, with the help of those arising for linear transformation.

366. We pass now to the case when the order of transformation is any odd number, dealing with the matter in a general way. Simplifications that can theoretically be always introduced by means of linear transformations are considered later (§ 372).

We first investigate a general formula* whereby the function

$$\vartheta\left[rw;\ 2\upsilon,\ 2r\upsilon',\ 2\zeta/r,\ 2\zeta'\,\middle|\,\begin{matrix}(K'+\mu)/r\\ K\end{matrix}\right]$$

can be expressed in terms of products of functions with associated constants 2υ, $2\upsilon'$, 2ζ, $2\zeta'$. We shall then afterwards employ the formulae developed in Chap. XVII., to express these products in the required form.

Let σ, σ' be two matrices each of p rows and m columns, whose constituents are any constants; let the j-th columns of these be denoted respectively by $\sigma^{(j)}$ and $\sigma'^{(j)}$, so that the values of j are $1, 2, \ldots, m$; let Υ_σ denote the matrix $2\upsilon\sigma + 2\upsilon'\sigma'$, which has p rows and m columns, and let the j-th column of this matrix, which is given by $2\upsilon\sigma^{(j)} + 2\upsilon'\sigma'^{(j)}$, be denoted by $\Upsilon_\sigma^{(j)}$; also, K, K' being rows of any p real rational elements, let Υ_K, Z_K denote the rows $2\upsilon K + 2\upsilon' K'$, $2\zeta K + 2\zeta' K'$; and use the abbreviation

$$\varpi(w;\ K, K') = Z_K(w + \tfrac{1}{2}\Upsilon_K) - \pi i K K';$$

finally, let $s = (s^{(1)}, \ldots, s^{(m)})$ be a column of m integers whose squares have the sum r, so that

$$\sum_j [s^{(j)}]^2 = r;$$

then, using always $\vartheta(w)$ for $\vartheta(w;\ 2\upsilon, 2\upsilon', 2\zeta, 2\zeta')$, *the function*

$$\Pi(w) = e^{-r\varpi[w;\ K/r,\ K'/r]} \prod_{j=1}^{m} \vartheta\left[s^{(j)}\left(w + \frac{\Upsilon_K - \Upsilon_\sigma s}{r}\right) + \Upsilon_\sigma^{(j)}\right]$$

is, in w, a theta function of order r with associated constants 2υ, $2\upsilon'$, 2ζ, $2\zeta'$ and characteristic (K, K').

For when the arguments w are increased by the elements of the row Υ_N, where N, N' are rows of p integers, the function

$$\vartheta\left[s^{(j)}\left(w + \frac{\Upsilon_K - \Upsilon_\sigma s}{r}\right) + \Upsilon_\sigma^{(j)}\right]$$

is multiplied by a factor e^{ψ_j}, where ψ_j is equal to

$$[2\zeta N s^{(j)} + 2\zeta' N' s^{(j)}]\left[s^{(j)}\left(w + \frac{\Upsilon_K - \Upsilon_\sigma s}{r}\right) + \Upsilon_\sigma^{(j)} + \upsilon N s^{(j)} + \upsilon' N' s^{(j)}\right] - \pi i\,[N s^{(j)}][N' s^{(j)}],$$

that is

$$[s^{(j)}]^2\left\{Z_N\left(w + \frac{\Upsilon_K - \Upsilon_\sigma s}{r} + \tfrac{1}{2}\Upsilon_N\right) - \pi i N N'\right\} + Z_N \Upsilon_\sigma^{(j)} s^{(j)};$$

the sum of the m values of ψ_j is given by

$$\begin{aligned}\sum_{j=1}^{m} \psi_j &= r\,\{Z_N(w + \tfrac{1}{2}\Upsilon_N) - \pi i N N'\} + Z_N\Upsilon_K - Z_N\Upsilon_\sigma s + Z_N\Upsilon_\sigma s\\ &= r\varpi(w;\ N, N') + Z_N \Upsilon_K;\end{aligned}$$

* Königsberger, *Crelle*, LXIV. (1865), p. 28. See Rosenhain, *Crelle*, XL. (1850), p. 338, and *Mém. par divers Savants*, t. XI. (1851), p. 402.

also, when w is increased by Υ_N, the function $-r\varpi\,[w\,;\ K/r,\ K'/r)$ is increased by $-Z_K\Upsilon_N$; thus the complete resulting factor of $\Pi\,(w)$ is

$$e^{r\varpi\,(w\,;\ N,\ N')+Z_N\Upsilon_K-Z_K\Upsilon_N},$$

of which (§ 190, p. 285) the exponent is equal to

$$r\varpi\,(w\,;\ N,\ N')+2\pi i\,(NK'-N'K)\,;$$

thus (§ 284, p. 448) $\Pi\,(w)$ is a theta function in w, of the r-th order with $(K,\ K')$ as characteristic.

Therefore (§ 284, p. 452) we have an equation

$$\Pi\,(w)=\sum_{\mu}A_\mu\,\vartheta\left[rw\,;\ 2v,\ 2rv',\ 2\zeta/r,\ 2\zeta'\ \Big|\ \begin{matrix}(K'+\mu)/r\\ K\end{matrix}\right],$$

where μ is a row of p integers each positive (including zero) and less than r, and the coefficients A_μ are independent of w. *The coefficients A_μ are independent of $K,\ K'$*, as we see immediately by first proving the equation which arises from this equation by putting K and K' zero, and then, in that equation, replacing w by $w+2vK/r+2v'K'/r$.

In this equation, replace K by $K+h$, where h is a row of p integers, each positive (including zero) and less than r; then, using the equation previously written (§ 190, p. 286), for integral M, in the form

$$\vartheta\,(u\,;\ q+M)=e^{2\pi iMq'}\,\vartheta\,(u\,;\ q),$$

we find

$$e^{-r\varpi\,[w\,;\ (K+h)/r,\ K'/r]-2\pi i\,(K'+\epsilon)\,h/r}\prod_{j=1}^{m}\vartheta\left[s^{(j)}\left(w+\frac{2vh+\Upsilon_K-\Upsilon_\sigma s}{r}\right)+\Upsilon_\sigma^{(j)}\right]$$

$$=\sum_{\mu}A_\mu e^{2\pi i(\mu-\epsilon)h/r}\,\vartheta\left[rw\,;\ 2v,\ 2rv',\ 2\zeta/r,\ 2\zeta'\ \Big|\ \begin{matrix}(K'+\mu)/r\\ K\end{matrix}\right],$$

where ϵ is taken to be any row of p integers each positive (or zero) and less than r; ascribing now to h all the possible r^p values, and using the fact that

$$r^{-p}\sum_{h}e^{2\pi i(\mu-\epsilon)h/r}=1,\ \text{ or }\ 0,$$

according as $\mu-\epsilon\equiv 0$ or $\not\equiv 0$, (mod. r), we infer, by addition, the equation

$$C_\mu\vartheta\left[rw\,;\ 2v,\ 2rv',\ 2\zeta/r,\ 2\zeta'\ \Big|\ \begin{matrix}(K'+\mu)/r)\\ K\end{matrix}\right]$$

$$=\sum_{h}e^{\psi}\prod_{j=1}^{m}\vartheta\left[s^{(j)}\left(w+\frac{2vh+\Upsilon_K-\Upsilon_\sigma s}{r}\right)+\Upsilon_\sigma^{(j)}\right],$$

where

$$\psi=-r\varpi\,[w\,;\ (K+h)/r,\ K'/r]-2\pi i\,(K'+\mu)\,h/r,$$

and $C_\mu,\ =r^pA_\mu$, is independent of w and of the characteristic $(K,\ K')$.

367. We put down now two cases of this very general formula:—

(α) if each of the matrices σ, σ' consist of zeros, and each of the m integers $s^{(1)}, \dots, s^{(m)}$ be unity, so that $m=r$, we obtain

$$C_\mu \vartheta\left[rw;\ 2v,\ 2rv',\ 2\zeta/r,\ 2\zeta' \,\middle|\, \begin{matrix}(K'+\mu)/r\\ K\end{matrix}\right]$$
$$= \sum_h e^{-r\varpi[w;\ (K+h)/r,\ K'/r] - 2\pi i (K'+\mu) h/r}\, \vartheta^r\left[w + \frac{2vh+\Upsilon_K}{r}\right].$$

In using this equation we shall make the simplification which arises by putting $w = 2vW$, $v^{-1}v' = \tau'$, and

$$\Theta(W, \tau') = e^{-\frac{1}{2}\zeta v^{-1} w^2}\, \vartheta(w) = \sum_n e^{2\pi i W n + i\pi\tau' n^2};$$

then the equation can be transformed without loss of generality, by means of the relations connecting the matrices v, v', ζ, ζ' (cf. § 284, p. 447), to the form

$$C_\mu e^{-2\pi i K'[W+\frac{1}{2}\tau' K'/r] - 2\pi i K K'/r}\, \Theta\left(rW;\ r\tau' \,\middle|\, \begin{matrix}(K'+\mu)/r\\ K\end{matrix}\right)$$
$$= \sum_h e^{-2\pi i \mu h/r}\, \Theta^r\left[W + \frac{h+K+\tau' K'}{r};\ \tau'\right], \qquad \text{(I)}$$

where C_μ is independent of W and of K and K'.

This equation is of frequent application in this chapter; it is of a different character from the multiplication formula given Chap. XVII., § 317, Ex. vii., whereby the function $\Theta(rW, \tau')$ was expressed by functions $\Theta(W, \tau')$ with different characteristics but the same *period*, τ'.

Ex. i. When $r=2$, $p=2$, we have

$$C_0\Theta(2W, 2\tau') = \Theta^2(W_1, W_2;\ \tau') + \Theta^2(W_1+\tfrac{1}{2}, W_2;\ \tau') + \Theta^2(W_1, W_2+\tfrac{1}{2};\ \tau') + \Theta^2(W_1+\tfrac{1}{2}, W_2+\tfrac{1}{2};\ \tau').$$

Ex. ii. If λ, μ, h be rows of p integers each less than r, prove that the ratio

$$\sum_h e^{-2\pi i \mu h/r}\,\Theta^r\left[W+\frac{h}{r}\,\middle|\,\begin{matrix}\lambda/r\\ 0\end{matrix}\right] \div \sum_\mu e^{-2\pi i \mu h/r}\,\Theta^r\left[W+\frac{h}{r}\right]$$

is independent of W.

(β) if the matrix σ' consist of zeros, and if each of the m integers $s^{(1)}, \dots, s^{(m)}$ be unity, so that $m=r$, and if the matrix σ, of p rows and r columns, have, for the constituents of every one of its rows, the elements

$$0,\ \frac{1}{r},\ \frac{2}{r},\ \dots\dots,\ \frac{r-1}{r},$$

then the matrix Υ_σ will have, for the constituents of its i-th row, the elements

$$0,\ \frac{\Omega_i}{r},\ \frac{2\Omega_i}{r},\ \dots\dots,\ \frac{(r-1)\,\Omega_i}{r},$$

where Ω_i is the sum of the elements of the i-th row of the matrix $2v$, so that

$$\Omega_i = 2 \sum_{h=1}^{p} v_{i,h};$$

also the i-th of the p elements denoted by $\frac{1}{r}\Upsilon_\sigma s$ will be

$$\frac{1}{r}\left[\frac{\Omega_i}{r} + \dots\dots + \frac{(r-1)\Omega_i}{r}\right] = \frac{r-1}{2r}\Omega_i,$$

and therefore the i-th of the elements of $\Upsilon_\sigma^{(j)} - \frac{1}{r}\Upsilon_\sigma s$ will be

$$\frac{j-1}{r}\Omega_i - \frac{r-1}{2r}\Omega_i.$$

Thus, denoting the row $(\Omega_1, \dots, \Omega_p)$ by Ω, the theorem is

$$C_\mu \vartheta\left[rw,\ 2v,\ 2rv',\ 2\zeta/r,\ 2\zeta' \,\middle|\, \begin{matrix}(K'+\mu)/r \\ K\end{matrix}\right]$$
$$= \sum_h e^{\psi} \prod_{j=1}^{r} \vartheta\left[w + \frac{2vh + \Upsilon_K}{r} + \left(\frac{j-1}{r} - \frac{r-1}{2r}\right)\Omega\right],$$

where ψ has the same value as in § 366. And as before this result can be written without loss of generality in the form

$$C_\mu e^{-2\pi i K'[W+\frac{1}{2}\tau' K'/r] - 2\pi i K K'/r}\, \Theta\left[rW,\ r\tau' \,\middle|\, \begin{matrix}(K'+\mu)/r \\ K\end{matrix}\right]$$
$$= \sum_h e^{-2\pi i \mu h/r} \phi\left(U + \frac{h + K + \tau' K'}{r}\right), \tag{II}$$

where $U = W - (r-1)/2r$ and, for any value of u,

$$\phi(u) = \Theta(u;\ \tau')\,\Theta\left(u + \frac{1}{r};\ \tau'\right) \dots\dots \Theta\left(u + \frac{r-1}{r};\ \tau'\right);$$

the number of different terms on the right side of this equation is r^{p-1}; for if m be a positive integer less than r, the two values of h expressed by $h = (h_1, \dots, h_p)$ and $h = (h_1', \dots, h_p')$, in which $h_1' \equiv h_1 + m, \dots, h_p' \equiv h_p + m$, (mod. r), give the same value for $\phi\left(U + \frac{h + K + \tau' K'}{r}\right)$.

Ex. i. For $p=2$, $r=2$, we obtain

$$\tfrac{1}{2}C_0\Theta(2W,\ 2\tau') = \Theta(W_1 - \tfrac{1}{4},\ W_2 - \tfrac{1}{4};\ \tau')\,\Theta(W_1 + \tfrac{1}{4},\ W_2 + \tfrac{1}{4};\ \tau')$$
$$+ \Theta(W_1 + \tfrac{1}{4},\ W_2 - \tfrac{1}{4};\ \tau')\,\Theta(W_1 - \tfrac{1}{4},\ W_2 + \tfrac{1}{4};\ \tau').$$

Ex. ii. For $p=2$, $r=3$, we obtain, omitting the period τ' on the right side,

$$\tfrac{1}{3}C_0\Theta(3W;\ 3\tau') = \Theta(W_1,\ W_2)\,\Theta(W_1 - \tfrac{1}{3},\ W_2 - \tfrac{1}{3})\,\Theta(W_1 + \tfrac{1}{3},\ W_2 + \tfrac{1}{3})$$
$$+ \Theta(W_1,\ W_2 - \tfrac{1}{3})\,\Theta(W_1 + \tfrac{1}{3},\ W_2)\,\Theta(W_1 - \tfrac{1}{3},\ W_2 + \tfrac{1}{3})$$
$$+ \Theta(W_1 + \tfrac{1}{3},\ W_2 - \tfrac{1}{3})\,\Theta(W_1 - \tfrac{1}{3},\ W_2)\,\Theta(W_1,\ W_2 + \tfrac{1}{3}).$$

368. We consider now the expression of the function

$$\Psi_r(W;\, K,\, K'+\mu) = \Theta\left[rW;\ r\tau' \left| \begin{matrix} (K'+\mu)/r \\ K \end{matrix} \right.\right] + \epsilon\Theta\left[-rW;\ r\tau' \left| \begin{matrix} (K'+\mu)/r \\ K \end{matrix} \right.\right],$$

in terms of functions $\Theta\left[W;\ \tau' \left| \begin{matrix} R' \\ R \end{matrix} \right.\right]$, in the case when r is odd. We suppose as before (K, K') to be a half-integer characteristic, and we suppose $\epsilon = e^{\pi i |K|}$, so that ϵ is ± 1 according as the characteristic (K, K') is even or odd*. It follows from § 327, Chap. XVIII., if (K, K') has arisen by transformation of order r from a characteristic (Q, Q'), that ϵ is also equal to $e^{\pi i |Q|}$ and is ± 1 according as the function is even or odd.

It is immediately seen that equation (I) (§ 367) can be put into the form

$$C_\mu e^{2\pi i (r-1) K' \left[W + \frac{K}{r} - \frac{r-1}{2r}\tau' K'\right]} \Theta\left[rW;\ r\tau' \left| \begin{matrix} (K'+\mu)/r \\ K \end{matrix} \right.\right]$$

$$= \sum_h e^{-2\pi i \left(K' + \frac{\mu}{r}\right) h} \Theta^r \left[W + \frac{h - (r-1)(K + \tau' K')}{r} \left| \begin{matrix} K' \\ K \end{matrix} \right.\right];$$

from this equation by changing the sign of W, we deduce the result

$$C_\mu e^{2\pi i \frac{r-1}{r} K' [K - (r-1)\tau' K'/2]} \Psi_r [W;\ K,\ K'+\mu]$$

$$= \sum_h e^{-2\pi i \left(K' + \frac{\mu}{r}\right) h} \left\{ e^{-2\pi i (r-1) K' W} \Theta^r \left[W + a \left| \begin{matrix} K' \\ K \end{matrix} \right.\right] + e^{2\pi i (r-1) K' W} \Theta^r \left[W - a \left| \begin{matrix} K' \\ K \end{matrix} \right.\right] \right\},$$

where we have replaced $\epsilon e^{-4\pi i r K K'}, = \epsilon e^{-\pi i r |K|}$, by unity, and a denotes the expression $[h - (r-1)(K + \tau' K')]/r$, which is an r-th part of a period. We proceed to shew that *the function*

$$e^{-2\pi i (r-1) K' W} \Theta^r \left[W + a \left| \begin{matrix} K' \\ K \end{matrix} \right.\right] + e^{2\pi i (r-1) K' W} \Theta^r \left[W - a \left| \begin{matrix} K' \\ K \end{matrix} \right.\right]$$

can be expressed as an integral polynomial of the r-th degree in 2^p *functions* $\Theta^r [W;\ \tau' \,|\, AP_i]$, *where* AP_i *are the characteristics of any Göpel system of half-integer characteristics whereof* (K, K') *is one characteristic.*

From the formula of § 311, p. 513, putting $C = 0$, $A' = A$, $B = P = \frac{1}{2}\begin{pmatrix} q' \\ q \end{pmatrix}$, and replacing $U, V, W, \begin{pmatrix} P_i \\ A \end{pmatrix} \epsilon_i, \begin{pmatrix} P_j \\ A \end{pmatrix} \epsilon_j$ respectively by $W, a, b, \epsilon_i, \epsilon_j$ we obtain, if $P_a = \frac{1}{2}\begin{pmatrix} q_a' \\ q_a \end{pmatrix}$,

* Thus, when $2(K'+\mu) = rm$, m being integral,

$$\epsilon = e^{2\pi i K (rm - 2\mu)} = e^{2\pi i K m} = e^{4\pi i K \frac{K'+\mu}{r}},$$

as in § 287, Chap. XV., and

$$\Psi_r(W;\ K,\ K'+\mu) \text{ reduces to } 2\Theta\left[rW,\ r\tau' \left| \begin{matrix} m/2 \\ K \end{matrix} \right.\right].$$

$$2^p\Theta(W+a;\ A+P)\Theta(W+b;\ A)$$

$$=\sum_{\epsilon}\frac{\chi(a,b;\ P,\epsilon)}{\chi(a+b,0;\ P,\epsilon)}\sum_{\alpha}\epsilon_\alpha e^{-\frac{1}{2}\pi i q' q_\alpha}\Theta(W+a+b;\ A+P+P_\alpha)\Theta(W;\ A+P_\alpha),$$

where

$$\chi(u,v;\ P,\epsilon)=\sum_{\alpha}\epsilon_\alpha e^{-\frac{1}{2}\pi i q' q_\alpha}\Theta(u;\ A+P+P_\alpha)\Theta[v;\ A+P_\alpha];$$

the function $\chi(u,v;\ A,P,\epsilon)$ may be immediately shewn to be unaltered by the addition of an integral characteristic to the characteristic P_α of one of its terms; we may therefore suppose all these characteristics to be reduced characteristics, each element being 0 or $\frac{1}{2}$.

Hence we get

$$2^p\Theta^2(W+a;\ A)=\sum_{\epsilon}\frac{\chi(a,a;0,\epsilon)}{\chi(2a,0;0,\epsilon)}\sum_{\alpha}\epsilon_\alpha\Theta(W+2a;\ A+P_\alpha)\Theta(W;\ A+P_\alpha),$$

and hence $2^{2p}\Theta^3(W+a;\ A)$ is equal to

$$\sum_{\epsilon}H_1\sum_{\alpha}\epsilon_\alpha\Theta(W;\ A+P_\alpha)\sum_{\epsilon'}H_2\sum_{\beta}\epsilon_\beta' e^{-\frac{1}{2}\pi i q_\alpha' q_\beta}\Theta(W+3a;\ A+P_\alpha+P_\beta)\Theta(W;A+P_\beta),$$

where

$$H_1=\frac{\chi(a,a;\ 0,\epsilon)}{\chi(2a,0;\ 0,\epsilon)},\quad H_2=\frac{\chi(2a,a;\ P_\alpha,\epsilon')}{\chi(3a,0;\ P_\alpha,\epsilon')};$$

proceeding in this way we obtain $2^{(r-1)p}\Theta^r(W+a;\ A)$

$$=\sum_{\epsilon_1}H_1\sum_{\alpha_1}\Theta_1\sum_{\epsilon_2}H_2\sum_{\alpha_2}\Theta_2\ldots\sum_{\epsilon_{r-1}}H_{r-1}\chi(W+ra,\ W;\ P_{\alpha_1}+\ldots+P_{\alpha_{r-2}};\ \epsilon_{r-1}),\qquad\text{(III)}$$

where each of P_{α_1}, P_{α_2}, ... becomes in turn all the characteristics of the group (P), and ϵ_1, ϵ_2, ... relate respectively to the groups described by P_{α_1}, P_{α_2}, ..., and further

$$H_m=\chi[ma,a;\ P_{\alpha_1}+\ldots+P_{\alpha_{m-1}},\epsilon_m]\div\chi[(m+1)a,0;\ P_{\alpha_1}+\ldots+P_{\alpha_{m-1}},\epsilon_m],$$
$$(m=1,\ldots,r-1),$$

$$\Theta_m=\epsilon_{\alpha_m}e^{\lambda_m}\Theta(W;\ A+P_{\alpha_m}),\quad \lambda_m=-\tfrac{1}{2}\pi i(q'_{\alpha_1}+\ldots+q'_{\alpha_{m-1}})q_{\alpha_m},$$
$$(m=1,\ldots,r-2).$$

The equation (III) expresses $\Theta^r(W+a;\ A)$ as an integral polynomial which is of the $(r-1)$th degree in functions $\Theta(W;\ A+P_\alpha)$, whose characteristics belong to the Göpel system (AP), and is of the first degree in functions $\Theta[W+ra;\ A+P_\alpha]$. But it does not thence follow when a is an r-th part of a period, that $\Theta^r(W+a;\ A)$ can be expressed as an integral polynomial of the r-th degree in functions $\Theta[W;\ A+P_\alpha]$; for instance if the Göpel system be taken to be one of which all the characteristics are even (§ 299, Chap. XVII.), it is not the case that the function $\Theta^3(W+\frac{1}{3})$, which is neither odd nor even, or the function $\Theta^3(W+\frac{1}{3})-\Theta^3(W-\frac{1}{3})$, which is odd, can be expressed as an integral polynomial of the third degree in the functions of this Göpel system; differential coefficients of these functions will enter into the expression. The reason is found in the fact noticed in § 308, p. 510; *the denominator of H_{r-1} may vanish.*

Noticing however, when P is any characteristic of the Göpel group (P), that $\chi(-u,-v;\ P,\epsilon)=e^{\pi i|P|+\pi i|A,P|}\chi(u,v;\ P,\epsilon)$, so that the coefficients H_m are unaltered by change of the sign of a, and putting the characteristic $A=\binom{K'}{K}$, we infer, from the equation (III), that

$$2^{(r-1)p}\left[e^{-2\pi i(r-1)K'W}\Theta^r(W+a;\ A)+e^{2\pi i(r-1)K'W}\Theta^r(W-a;\ A)\right]$$

is equal to

$$\sum_{\epsilon_1}H_1\sum_{a_1}\dots\dots\sum_{\epsilon_{r-1}}H_{r-1}\left[e^{-2\pi i(r-1)K'W}\chi(W+ra,\ W;\ P,\epsilon_{r-1})\right.$$
$$\left.+e^{2\pi i(r-1)K'W}\chi(W-ra,\ W;\ P,\epsilon_{r-1})\right],$$

where P denotes $P_{a_1}+\dots+P_{a_{r-2}}$; and it can be shewn that when a becomes equal to $[h-(r-1)(K+\tau'K')]/r$, the limit of the expression

$$U=H_{r-1}\left[e^{-2\pi i(r-1)K'W}\chi(W+ra,W;P,\epsilon_{r-1})+e^{2\pi i(r-1)K'W}\chi(W-ra,a;P,\epsilon_{r-1})\right],$$

if it is not a quadratic polynomial in functions $\Theta(W;\ AP_a)$, *is zero.* The consequence of this will be that $\Psi_r[W;\ K,K'+\mu]$ is expressible as a polynomial involving only the functions $\Theta(W;\ AP_a)$.

For the fundamental formula of § 309, p. 510, immediately gives*, for any values of a, b,

$$\chi(W+a,\ W+b;\ P,\epsilon)\chi(a+b,0;\ P,\epsilon)=\chi(a,b;\ P,\epsilon)\chi(W+a+b,\ W;\ P,\epsilon),$$

and hence, replacing ϵ_{r-1} simply by ϵ, the expression U is equal to

$$\sum_a\epsilon_a e^{-\frac{1}{2}\pi i q'q_a}\left\{e^{-2\pi i(r-1)K'W}\Theta(W+a;\ A+P_a)\Theta[W+(r-1)a;\ A+P+P_a]\right.$$
$$\left.+e^{2\pi i(r-1)K'W}\Theta(W-a;\ A+P_a)\Theta[W-(r-1)a;\ A+P+P_a]\right\},$$

where $P,=\frac{1}{2}\binom{q'}{q}$, is used for $P_{a_1}+\dots\dots+P_{a_{r-2}}$ and $\epsilon_1,\epsilon_2,\dots$ for $(\epsilon_{r-1})_1$, $(\epsilon_{r-1})_2,\dots$. Replacing ra in this expression by the period $h-(r-1)(K+\tau'K')$, and omitting an exponential factor depending only on r, h, K, K' and P, it becomes

$$\sum_a\zeta_a e^{-\frac{1}{2}\pi i q'q_a}\left\{\Theta[W+a;\ A+P_a]\Theta[W-a;\ A+P+P_a]\right.$$
$$\left.+\Theta[W-a;\ A+P_a]\Theta[W+a;\ A+P+P_a]\right\},$$

* We take the case when the characteristics B, A of § 309 are equal. It is immediately obvious from the equation here given that in the expressions here denoted by H_m, the value of the half-integer characteristic A is immaterial.

A being as before taken $=\binom{K'}{K}$ and $\zeta_a = \epsilon_a e^{\pi i[h-(r-1)K]q'_a+\pi i(r-1)K'q_a}$; and this is immediately shewn to be the same as

$$\left(1+\zeta_P\binom{A}{P}e^{-\frac{1}{2}\pi i|P|}\right)\sum_a \zeta_a e^{-\frac{1}{2}\pi i q' q_a}\Theta(W+a;\ A+P_a)\,\Theta(W-a;\ A+P+P_a),$$

where ϵ_P is the fourth root of unity associated with the characteristic P of the Göpel group (P), which is to be taken equal to 1 in case $P=0$. Thus the expression vanishes when $\zeta_P = -e^{\frac{1}{2}\pi i|P|}\binom{A}{P}$. Hence, in order to prove that when the expression U is not a quadratic polynomial in functions $\Theta(W;\ AP_a)$, it is zero, it is sufficient to prove that the only case in which U is not such a quadratic polynomial is when $\zeta_P = -e^{\frac{1}{2}\pi i|P|}\binom{A}{P}$.

Now the denominator of H_{r-1} is

$$\sum_a \epsilon_a e^{-\frac{1}{2}\pi i q' q_a}\Theta[ra;\ A+P+P_a]\,\Theta[0;\ A+P_a],$$

where P still denotes $P_{a_1}+\dots+P_{a_{r-2}}$ and ϵ_a has the set of values of ϵ_{r-2}; save for a non-vanishing exponential factor this is equal to

$$\sum_a \zeta_a \Theta^2(0;\ AP_a),$$

or
$$\left(1+\zeta_P\binom{A}{P}e^{-\frac{1}{2}\pi i|P|}\right)\sum_\beta \zeta_\beta e^{-\frac{1}{2}\pi i q' q_\beta}\Theta[0;\ A+P+P_\beta]\,\Theta[0;\ A+P_\beta],$$

according as $P=0$ or not, where, in the second form, P_β is to describe a group of 2^{p-1} characteristics such that the combination of this group with the group $(0, P)$ gives the Göpel group (P). We shall assume that, when ζ_P is not equal to $-e^{\frac{1}{2}\pi i|P|}\binom{A}{P}$, neither of these expressions vanishes for general values of the periods τ'.

Since the function $\Psi_r(W;\ K, K'+\mu)$ is certainly finite, we do not examine the finiteness of the coefficients H_m when m is less than $r-1$, these coefficients being independent of W; further, in a Göpel system (AP), any one of the characteristics AP_a may be taken as the characteristic A; the change being only equivalent to adding the characteristic P_a to each characteristic of the group (P); hence (§ 327, Chap. XVIII.), our investigation gives the following result:—*Let any 2^p functions $\vartheta\left(u;\ 2\omega,\ 2\omega',\ 2\eta,\ 2\eta'\ \middle|\ \begin{matrix}Q'\\Q\end{matrix}\right)$, whose (half-integer) characteristics form a Göpel system, syzygetic in threes, be transformed by any transformation of odd order; let (AP) be the Göpel system formed by the transformed characteristics $\binom{K'}{K}$; then every one of the* *original functions is an integral polynomial of order r in the 2^p functions** $\vartheta(w;\ 2v,\ 2v',\ 2\zeta,\ 2\zeta' \mid AP)$: as follows from § 288, Chap. XV., the number of terms in the polynomial is at most, and in general, $\frac{1}{2}(r^p+1)$.

For the cases $p=1, 2, 3$, and for any hyperelliptic case, it is not necessary to use the addition formula developed in Chap. XVIII. We may use instead the addition formula of § 286, Chap. XV. It is however then further to be shewn that only 2^p theta functions enter in the final formula. For the case $p=3$ the reader may consult Weber, *Ann. d. Mat.* 2ª Ser., t. IX. (1878), p. 126.

369. We give an example of the application of the method here followed.

Suppose $p=1$, $r=3$, and that the transformation is that associated with the matrix $\begin{pmatrix}3 & 0\\ 0 & 1\end{pmatrix}$; then (§ 324, Chap. XVIII.) taking $M=3$, the function

$$\vartheta\left[u;\ 2\omega,\ 2\omega',\ 2\eta,\ 2\eta' \,\middle|\, \tfrac{1}{2}\begin{pmatrix}0\\ -1\end{pmatrix}\right],$$

or $\vartheta_{01}(u)$, is equal to $\vartheta_{01}(3w;\ 2v,\ 6v',\ 2\zeta/3,\ 2\zeta')$ or $\frac{1}{2}e^{\frac{1}{2}\eta\omega^{-1}u^2}\Psi_3(W;\ -\frac{1}{2},\ 0)$. Now we have, with $a=(h+1)/3$,

$$C_0\Psi_3(W;\ -\tfrac{1}{2},\ 0)=\sum_h\left[\Theta_{01}^3(W+a)+\Theta_{01}^3(W-a)\right];$$

also $\Theta_{01}^3(W+a)$ is equal to

$$\frac{\chi(a,\ a;\ 0,\ \epsilon)}{\chi(2a,\ 0;\ 0,\ \epsilon)}\sum_\alpha \epsilon_\alpha\Theta(W;\ A+P_\alpha)\sum_{\epsilon'}\frac{\chi(2a,\ a;\ P_\alpha,\ \epsilon')}{\chi(3a,\ 0;\ P_\alpha,\ \epsilon')}\sum_\beta \epsilon'_\beta e^{-\frac{1}{2}\pi i q'_\alpha q_\beta}\Theta(W+3a;\ A+P_\alpha+P_\beta)\Theta(W;\ A+$$

if we take the Göpel system to be $\frac{1}{2}\begin{pmatrix}0\\ -1\end{pmatrix}$, $\frac{1}{2}\begin{pmatrix}1\\ 0\end{pmatrix}$, so that $P_1=\frac{1}{2}\begin{pmatrix}1\\ 1\end{pmatrix}$, this is equal to

$$\tfrac{1}{4}\sum_\epsilon \frac{\Theta_{01}^2(a)+\epsilon_1\Theta_{10}^2(a)}{\Theta_{01}(2a)\Theta_{01}+\epsilon_1\Theta_{10}(2a)\Theta_{10}}\Theta_{01}(W)\sum_{\epsilon'}\frac{\Theta_{01}(2a)\Theta_{01}(a)+\epsilon_1'\Theta_{10}(2a)\Theta_{10}(a)}{\Theta_{01}(3a)\Theta_{01}+\epsilon_1'\Theta_{10}(3a)\Theta_{10}}E_0$$

$$+\tfrac{1}{4}\sum_\epsilon \frac{\Theta_{01}^2(a)+\epsilon_1\Theta_{10}^2(a)}{\Theta_{01}(2a)\Theta_{01}+\epsilon_1\Theta_{10}(2a)\Theta_{10}}\epsilon_1\Theta_{10}(W)\sum_{\epsilon'}\frac{\Theta_{10}(2a)\Theta_{01}(a)-i\epsilon_1'\Theta_{01}(2a)\Theta_{10}(a)}{\Theta_{10}(3a)\Theta_{01}-i\epsilon_1'\Theta_{01}(3a)\Theta_{10}}E_1,$$

where Θ_{01} denotes $\Theta_{01}(0)$, etc., and

$$E_0=\Theta_{01}(W+3a)\Theta_{01}(W)+\epsilon_1'\Theta_{10}(W+3a)\Theta_{10}(W),$$
$$E_1=\Theta_{10}(W+3a)\Theta_{01}(W)-i\epsilon_1'\Theta_{01}(W+3a)\Theta_{10}(W).$$

Now, in accordance with the general rules, the denominator of the fraction

$$\frac{\Theta_{10}(2a)\Theta_{01}(a)-i\epsilon_1'\Theta_{01}(2a)\Theta_{10}(a)}{\Theta_{10}(3a)\Theta_{01}-i\epsilon_1'\Theta_{01}(3a)\Theta_{10}}$$

vanishes when $\epsilon_1'=-e^{\frac{1}{2}\pi i\begin{pmatrix}A\\ P\end{pmatrix}}e^{\pi i(h-2K)q_1'+\pi i 2K'q_1}$, namely, as $\begin{pmatrix}K'\\ K\end{pmatrix}=\frac{1}{2}\begin{pmatrix}0\\ -1\end{pmatrix}=A$, when $\epsilon_1'=-ie^{\pi i(h+1)}$, and $a=(h+1)/3$; in fact, putting $a=(h+1+x)/3$,

$$\Theta_{10}(3a)\Theta_{01}-i\epsilon_1'\Theta_{01}(3a)\Theta_{10}=e^{\pi i(h+1)}\Theta_{10}(x)\Theta_{01}-i\epsilon_1'\Theta_{01}(x)\Theta_{10},$$
$$=\tfrac{1}{2}e^{\pi i(h+1)}\left[\Theta_{10}''\Theta_{01}-\Theta_{01}''\Theta_{10}\right]x^2,$$

* The expression of the transformed theta function in terms of $2^p=4$ theta functions is given by Hermite, *Compt. Rendus*, t. XL. (1855), for the case $p=2$. For the general hyperelliptic case cf. Königsberger, *Crelle*, LXIV. (1865), p. 32.

for small values of x, when $i\epsilon_1'=e^{\pi i(h+1)}$, because the differential coefficients of the even functions, being odd functions, vanish for zero argument; thus the denominator of the fraction vanishes to the second order. We find similarly, for $i\epsilon_1'=e^{\pi i(h+1)}$, $a=\frac{1}{3}(h+1+x)$, that the numerator of this fraction is equal to

$$e^{\pi i(h+1)}\left[\Theta'_{01}\left(\frac{h+1}{3}\right)\Theta_{10}\left(\frac{h+1}{3}\right)-\Theta'_{10}\left(\frac{h+1}{3}\right)\Theta_{01}\left(\frac{h+1}{3}\right)\right]x\,;$$

in the same case also we find that the expression E_1 is equal to

$$e^{\pi i\,(h+1)}\left[\Theta'_{10}(W)\,\Theta_{01}(W)-\Theta'_{01}(W)\,\Theta_{10}(W)\right]x,$$

while the expression $\Theta_{10}(W-3a)\,\Theta_{01}(W)-i\epsilon_1'\Theta_{01}(W-3a)\,\Theta_{10}(W)$ is equal to the negative of this. Thus the function $\Theta_{01}^3(W+a)$ can be expressed by the functions $\Theta_{10}(W)$, $\Theta_{01}(W)$, and their differential coefficients of the first order; but the function $\Theta_{01}^3(W+a)+\Theta_{01}^3(W-a)$ can be expressed by the functions $\Theta_{10}(W)$, $\Theta_{01}(W)$ only.

In the function $\Theta_{01}^3(W+a)+\Theta_{01}^3(W-a)$ the part

$$\sum_{\epsilon'}\frac{\Theta_{10}(2a)\,\Theta_{01}(a)-i\epsilon_1'\Theta_{01}(2a)\,\Theta_{10}(a)}{\Theta_{10}(3a)\,\Theta_{01}-i\epsilon_1'\Theta_{01}(3a)\,\Theta_{10}}E_1$$

furnishes only the single term for which $i\epsilon_1'=-e^{\pi i\,(h+1)}$, namely,

$$4e^{\pi i\,(h+1)}\frac{\Theta_{01}\left(\frac{h+1}{3}\right)\Theta_{10}\left(\frac{h+1}{3}\right)}{\Theta_{01}\,\Theta_{10}}\Theta_{01}(W)\,\Theta_{10}(W).$$

Ex. i. Prove that the final result is that $\frac{1}{2}C_0\vartheta_{01}(u)$ is equal to

$$\frac{\Theta_{01}^3(\frac{1}{3})\,\Theta_{01}-\Theta_{10}^3(\frac{1}{3})\,\Theta_{10}}{[\Theta_{01}^2(\frac{1}{3})\,\Theta_{01}^2+\Theta_{10}^2(\frac{1}{3})\,\Theta_{10}^2][\Theta_{01}^4+\Theta_{10}^4]}\{[\Theta_{01}^2(\tfrac{1}{3})\,\Theta_{01}^2+\Theta_{10}^2(\tfrac{1}{3})\,\Theta_{10}^2]\,\vartheta_{01}^3(w)$$
$$-[\Theta_{10}^2(\tfrac{1}{3})\,\Theta_{01}^2-\Theta_{01}^2(\tfrac{1}{3})\,\Theta_{10}^2]\,\vartheta_{01}(w)\,\vartheta_{10}^2(w)\}$$
$$+\frac{\Theta_{10}^2(\frac{1}{3})\,\Theta_{01}^2(\frac{1}{3})\,[\Theta_{10}(\frac{1}{3})\,\Theta_{01}+\Theta_{01}(\frac{1}{3})\,\Theta_{10}]}{[\Theta_{01}^2(\frac{1}{3})\,\Theta_{01}^2+\Theta_{10}^2(\frac{1}{3})\,\Theta_{10}^2]\,\Theta_{01}\,\Theta_{10}}\vartheta_{01}(w)\,\vartheta_{10}^2(w)+\tfrac{1}{2}\vartheta_{01}^3(w),$$

where Θ_{01}, Θ_{10} denote $\Theta_{01}(0)$ and $\Theta_{10}(0)$ respectively.

Ex. ii. Prove that

$$\Theta_{01}(W-\tfrac{1}{3})\,\Theta_{10}(W+\tfrac{1}{3})-\Theta_{10}(W-\tfrac{1}{3})\,\Theta_{01}(W+\tfrac{1}{3})$$
$$=2\,\frac{\Theta'_{10}(\frac{1}{3})\,\Theta_{01}(\frac{1}{3})-\Theta'_{01}(\frac{1}{3})\,\Theta_{10}(\frac{1}{3})}{\Theta_{10}\Theta''_{01}-\Theta_{01}\Theta''_{10}}[\Theta'_{01}(W)\,\Theta_{10}(W)-\Theta'_{10}(W)\,\Theta_{01}(W)].$$

370. General formulae for the quadric transformation are also obtainable. The results are different, as has been seen, according as the characteristic (K, K') of the transformed function is zero (including integral) or not. The results are as follows:—

When (K, K') is zero, the transformed function can be expressed as a linear aggregate of the 2^p functions $\vartheta^2(w\,|\,A, P_i)$, whose characteristics are those of any Göpel system.

When (K, K') is not zero, the transformed function can be expressed as a linear aggregate of the 2^{p-1} products $\vartheta(w \mid A, P_i)\, \vartheta(w \mid A, K, P_i)$, in which the characteristics P_i are those of any Göpel group whereof the characteristic K, $=(K, K')$, is one constituent, and A is a characteristic such that $|A, K| \equiv |K|$, or $|A, K| \equiv |K| + 1$ (mod. 2), according as the function to be expressed is even or odd*.

When (K, K') is zero, the equation (I), § 367, putting $K = K' = \mu = 0$, and then increasing W by $\frac{1}{2}\mu\tau'$, where μ is a row of quantities each either 0 or 1, gives

$$C\Theta\left(2W;\ 2\tau' \,\middle|\, \begin{matrix}\mu/2\\ 0\end{matrix}\right) = \sum_h e^{-\pi i \mu h}\, \Theta^2\left(W + \tfrac{1}{2}h;\ \tau' \,\middle|\, \begin{matrix}\mu/2\\ 0\end{matrix}\right);$$

hence, from the fundamental formula of § 309 (p. 510), writing therein $v = 0$, $u = W + a$, $b = a = h/2$, $A = \frac{1}{2}\begin{pmatrix}\mu\\ 0\end{pmatrix}$, $P_i = \frac{1}{2}\begin{pmatrix}q_i'\\ q_i\end{pmatrix}$, and $\begin{pmatrix}P_i\\ A\end{pmatrix} e^{\pi i h q_i'} \epsilon_i = \zeta_i$, we obtain

$$2^p C\Theta\left(2W;\ 2\tau' \,\middle|\, \begin{matrix}\mu/2\\ 0\end{matrix}\right)$$

$$= \sum_h e^{-\pi i \mu h} \sum_\zeta \frac{\sum_i \begin{pmatrix}P_i\\ A\end{pmatrix} e^{\pi i h q_i'} \zeta_i \Theta^2\left(0;\ \tau' \,\middle|\, \tfrac{1}{2}\begin{pmatrix}\mu\\ h\end{pmatrix} + P_i\right)}{\sum_i \zeta_i \Theta^2(0;\ \tau' \mid AP_i)} \sum_i \zeta_i \Theta^2(W;\ \tau' \mid AP_i),$$

where C is independent of μ. It is assumed that the sum $\sum_i \zeta_i \Theta^2(0;\ \tau' \mid AP_i)$ is different from zero for each of the 2^p sets of values of the fourth roots ζ_i. This formula suffices to express any theta function of the second order with zero characteristic.

When (K, K') is other than zero, by putting in the equation (I), § 367, $r = 2$, $\mu = 0$, adding $\frac{1}{2}\tau' h'$ to W, where h' is a row of quantities each either 0 or 1, and then changing the sign of W, we obtain

$$Ce^{-\pi i \lambda'(K + \frac{1}{2}\tau'\lambda')}\, \Psi_2(W;\ K, K' + h') = \sum_h \left[e^{2\pi i \lambda' W} \Theta^2(W + a) + \epsilon e^{-2\pi i \lambda' W} \Theta^2(W - a)\right],$$

where $\lambda = K + h$, $\lambda' = K' + h'$, and C is the same constant as before, independent of W, K, K', h', and $a = \frac{1}{2}\lambda + \frac{1}{2}\tau'\lambda'$, the period τ' being omitted on the right side. Hence, taking the fundamental formula of § 309 (p. 510), putting therein $v = 0$, $u = W + a$, $b = a$, $A = 0$, $B = A$, and then writing $a = \frac{1}{2}\lambda + \frac{1}{2}\tau'\lambda' + \frac{1}{2}x$, where x is a row of p equal quantities, we find, provided $|K, P_i| \equiv 0$, (mod. 2),

* When (K, K') is zero, the function is necessarily even (§ 288, p. 463), and therefore $|K| \equiv |Q|$. We have seen (§ 327, Chap. XVIII.) that this is always true when r is odd. When r is 2, it is not always so, as is obvious by considering the transformation, for $p = 1$, in which $\alpha = 2$, $\beta = 0$, $\alpha' = 0$, $\beta' = 1$, and $(Q, Q') = (\frac{1}{2}, \frac{1}{2})$; then we find $(K, K') = (\frac{1}{2}, 1)$; thus $|Q| = 1$, $|K| = 2$.

and $\epsilon = e^{\pi i|K| + \pi i|A, K|}$, that $2^p C\Psi_2(W;\ K, K'+h')$ is equal to the limit, when x vanishes, of the expression

$$e^{\pi i K'(K+\frac{1}{2}\tau' K')} \sum_h e^{-\pi i h h'} \sum_\zeta E_\zeta \sum_i \zeta_i e^{-\pi i K' q_i} \Theta(W;\ A, P_i)\{\Theta(W+x \mid A, K, P_i) + \Theta(W-x \mid A, K, P_i)\},$$

where $\zeta_i = \begin{pmatrix} P_i \\ A \end{pmatrix} e^{\pi i (hq'_i - h'q_i)} \epsilon_i$, and

$$E_\zeta = \frac{\sum_i \zeta_i e^{-\pi i (h-2A) q_i' - \pi i h' x}\, \Theta^2\left(\tfrac{1}{2}x + \tfrac{1}{2}K + \tfrac{1}{2}\tau' K' \;\middle|\; \tfrac{1}{2}\begin{pmatrix} h' \\ h \end{pmatrix} + P_i\right)}{\sum_i \zeta_i e^{-\pi i K' q_i} \Theta(x;\ A, K, P_i)\, \Theta(0;\ A, P_i)}.$$

It can easily be proved (cf. § 308, p. 510) that the denominator of E_ζ vanishes, for $x=0$, for the 2^{p-1} sets of values of the fourth roots ζ_i in which the fourth root corresponding to the characteristic K of the group (P) has the value $-\begin{pmatrix} A \\ K \end{pmatrix} e^{\frac{1}{2}\pi i|K|}$, and that the corresponding expressions

$$U_\zeta = E_\zeta \sum_i \zeta_i e^{-\pi i K' q_i} \Theta(W;\ A, P_i)\{\Theta(W+x \mid A, K, P_i) + \Theta(W-x \mid A, K, P_i)\}$$

have the limit zero; the summation $\sum_\zeta$ is therefore to be taken only to extend to the 2^{p-1} sets of values in which this fourth root $= +\begin{pmatrix} A \\ K \end{pmatrix} e^{\frac{1}{2}\pi i|K|}$. It may however happen that the denominator of E_ζ vanishes for other sets of values of the fourth roots ζ_i, when $x=0$. We assume that for such sets of values the sum multiplying E_ζ in the expression U_ζ does not vanish for $x=0$; by recurring to the proof of the formula of § 308, it is immediately seen that this is equivalent to assuming that the expression

$$\sum_i \epsilon_i \Theta^2(U;\ P_i)$$

is not zero for general values of the arguments U for any set of values of the fourth roots ϵ_i (cf. (β), p. 514). That being so, the value of E_ζ when its denominator vanishes for $x=0$, can always be obtained from the limiting expression given, by expanding its numerator and denominator in powers of x.

Ex. Applying the formula of this page for the case $p=1$ to the function

$$\Theta_{11}(2W;\ 2\tau') = \tfrac{1}{2}\Psi_2(W;\ -\tfrac{1}{2}, 1),$$

for which $(K, K') = (-\frac{1}{2}, 0)$ and $h'=1$, we immediately find that the Göpel system in terms of which the function can be expressed is (A, AP_1), where $A = \frac{1}{2}\begin{pmatrix} 1 \\ 0 \end{pmatrix}$, $P_1 = K = \frac{1}{2}\begin{pmatrix} 0 \\ -1 \end{pmatrix}$; we are to exclude the value of the expression U_ζ in which $\zeta_1 = -\begin{pmatrix} A \\ K \end{pmatrix} = 1$; the value of E_ζ for $\zeta_1 = -1$ is easily found to be

$$E_\zeta = e^{-\pi i(x-h)}\left[\Theta_{10}^2(\tfrac{1}{2}x-\tfrac{1}{4})-\Theta_{10}^2(\tfrac{1}{2}x+\tfrac{1}{4})\right]\div\Theta_{11}(x)\,\Theta_{10}(0)$$

of which both numerator and denominator vanish for $x=0$. The final result of the formula is

$$C\Theta_{11}(2W;\ 2\tau')=-4\Theta_{10}(\tfrac{1}{4};\ \tau')\,\Theta_{10}'(\tfrac{1}{4};\ \tau')\,\Theta_{11}(W;\ \tau')\,\Theta_{10}(W;\ \tau')/\Theta_{11}'(0;\tau')\,\Theta_{10}(0;\ \tau').$$

Prove this result, and also

$$C\Theta_{01}(2W;\ 2\tau')=2\Theta_{00}^2(\tfrac{1}{4};\ \tau')\,\Theta_{00}(W;\ \tau')\,\Theta_{01}(W;\ \tau')/\Theta_{00}(0;\ \tau')\,\Theta_{01}(0;\ \tau'),$$

and (cf. § 365) obtain the formulae

$$\Theta_{10}(\tfrac{1}{4};\ \tau')\,\Theta_{10}'(\tfrac{1}{4};\ \tau')=-\frac{\pi}{2}\Theta_{10}^2(0;\ \tau')\,\Theta_{00}^2(\tfrac{1}{4};\ \tau'),$$

$$\Theta_{00}^4(\tfrac{1}{4};\ \tau')=\tfrac{1}{2}\Theta_{00}(0;\ \tau')\,\Theta_{01}(0;\ \tau')\left[\Theta_{00}^2(0;\ \tau')+\Theta_{01}^2(0;\ \tau')\right],$$

$$\Theta_{00}^2(0;\ 2\tau')=\tfrac{1}{2}\left[\Theta_{00}^2(0;\ \tau')+\Theta_{01}^2(0:\ \tau')\right],$$

$$C=\sqrt{2\left[\Theta_{00}^2(0;\ \tau')+\Theta_{01}^2(0;\ \tau')\right]}.$$

371. The preceding investigations of this chapter enable us to specify in all cases the form of the function $\vartheta\left(u;\ 2\omega,2\omega',2\eta,2\eta'\,\middle|\,\begin{matrix}Q'\\Q\end{matrix}\right)$ or $\vartheta\left(u\,\middle|\,\begin{matrix}Q'\\Q\end{matrix}\right)$ when expressed in terms of functions $\vartheta\left(w;\ 2v,2v',2\zeta,2\zeta'\,\middle|\,\begin{matrix}K'\\K\end{matrix}\right)$ or $\vartheta\left(w\,\middle|\,\begin{matrix}K'\\K\end{matrix}\right)$. In many particular cases it is convenient to start from this form and determine the coefficients in the expression by particular methods. But it is proper to give a general method. For this purpose we should consider two stages, (i) the determination of the coefficients in the expression of the function $\vartheta\left(u\,\middle|\,\begin{matrix}Q'\\Q\end{matrix}\right)$ by means of functions $\psi_r(w;\ K,K'+\mu)$, (ii) the determination of the coefficients in the expression of the functions $\psi_r(w;\ K,K'+\mu)$ by means of functions $\vartheta\left(w\,\middle|\,\begin{matrix}K'\\K\end{matrix}\right)$. The preceding formulae of this chapter enable us to give a complete determination of the latter coefficients in a particular form, namely, in terms of theta functions whose arguments are fractional parts of the periods $2v, 2v'$; but this is by no means to be regarded as the final form.

372. Dealing first with the coefficients in the expression of the function $\vartheta\left(u\,\middle|\,\begin{matrix}Q'\\Q\end{matrix}\right)$ by functions $\psi_r(w;\ K,K'+\mu)$, there is one case in which no difficulty arises, namely, when the transformation is that associated with the matrix $\begin{pmatrix}r&0\\0&1\end{pmatrix}$; then $\vartheta\left(u\,\middle|\,\begin{matrix}Q'\\Q\end{matrix}\right)$ is equal to $\vartheta\left(rw;\ 2v,2rv',2\zeta/r,2\zeta'\,\middle|\,\begin{matrix}K'/r\\K\end{matrix}\right)$, the row K' being in fact equal to rQ', namely $\vartheta\left(u\,\middle|\,\begin{matrix}Q'\\Q\end{matrix}\right)$ is $\tfrac{1}{2}\psi_r(w;\ K,K')$.

Now it can be shewn*, that if Ω_r be the matrix associated with any transformation of order r, and r be a prime number, or a number without square factors, then linear transformations, Ω, Ω', can be determined such that $\Omega_r = \Omega \begin{pmatrix} r & 0 \\ 0 & 1 \end{pmatrix} \Omega'$. Hence, in cases in which the matrices Ω, Ω' have been calculated, it is sufficient, first to carry out the transformation Ω upon the given function $\vartheta\left(u \,\middle|\, \begin{matrix} Q' \\ Q \end{matrix}\right)$; then to use the formulae for the transformation $\begin{pmatrix} r & 0 \\ 0 & 1 \end{pmatrix}$, whereby the original function appears as an integral polynomial of order r in 2^p theta functions; and finally to apply the transformation Ω' to these 2^p theta functions. All cases in which the order of transformation is not a prime number may be reduced to successive transformations of prime order (§ 332, Chap. XVIII.).

We can however make a statement of greater practical use, as follows. It is shewn in the Appendix II. (§§ 415, 416) that the matrix associated with any transformation of order r can be put into the form $\Omega \begin{pmatrix} A & B \\ 0 & B' \end{pmatrix}$, where Ω is the matrix of a linear transformation, and that, in whichever of the possible ways this is done, the determinant of the matrix B' is the same for all. In all cases in which this has been done the required coefficients are given by the equation

$$\frac{1}{\sqrt{|\omega|}} \vartheta\left(u;\ 2\omega, 2\omega', 2\eta, 2\eta' \,\middle|\, \begin{matrix} Q' \\ Q \end{matrix}\right)$$

$$= \frac{r^{\frac{1}{2}p}\epsilon}{\sqrt{|M|\,|v|\,|B'|}} e^{-\frac{\pi i}{r} dK' + \frac{\pi i}{r^2} \gamma K'^2} \sum_{\mu} e^{-\frac{\pi i}{r} d\mu - \frac{\pi i}{r^2}\gamma\mu^2} \vartheta\left[rw;\ 2v, 2rv', 2\zeta/r, 2\zeta' \,\middle|\, \begin{matrix} (K'+\mu)/r \\ K \end{matrix}\right],$$

wherein, (Q, Q') being a half-integer characteristic, ϵ is an eighth root of unity, $u = Mw$, $|M|$ is the determinant of the matrix M, etc., μ is in turn every row of integers each positive (or zero) and less than r, which satisfies the condition that the p quantities $\frac{1}{r} B'\mu$ are integral, and, finally, γ denotes the symmetrical matrix $\bar{B}B'$, while d denotes the row of integers formed by the diagonal elements of γ. It is shewn in the Appendix II., that the resulting range of values for μ is independent of how the original matrix is resolved into the form in question. For any specified form of the linear transformation Ω the value of ϵ can be calculated (as in Chap. XVIII., §§ 333–4); if ϵ_0

* Cf. Appendix II.; and for details in regard to the case $p=3$, Weber, *Ann. d. Mat.*, Ser. 2ª, t. IX. (1878—9). We have shewn (Chap. XVIII., § 324, Ex. i.) that the determinant of the matrix of transformation is $\pm r^p$. From the result quoted here it follows that that determinant is $+r^p$.

denote its value when the characteristic (Q, Q') is zero, its value for any other characteristic is given by

$$\epsilon/\epsilon_0 = e^{-\pi i[Q_1'+\frac{1}{2}d(\bar{\rho}\rho')][Q_1+\frac{1}{2}d(\bar{\sigma}\sigma')]+\pi i Q Q'},$$

where $\Omega = \begin{pmatrix}\rho & \sigma\\ \rho' & \sigma'\end{pmatrix}$, and $Q_1' = \bar{\rho}Q' - \bar{\rho}'Q - \frac{1}{2}d(\bar{\rho}\rho')$, $-Q_1 = \bar{\sigma}Q' - \bar{\sigma}'Q - \frac{1}{2}d(\sigma\sigma')$.

To prove this formula, we have first (§ 335, Chap. XVIII.), if $\Omega = \begin{pmatrix}\rho & \sigma\\ \rho' & \sigma'\end{pmatrix}$, the equation

$$\frac{1}{\sqrt{|\omega|}}\vartheta\left(u;\ 2\omega,\ 2\omega',\ 2\eta,\ 2\eta'\ \Big|\ \begin{matrix}Q'\\ Q\end{matrix}\right) = \frac{\epsilon}{\sqrt{|M_1|\,|\omega_1|}}\vartheta\left(u_1;\ 2\omega_1,\ 2\omega_1',\ 2\eta_1,\ 2\eta_1'\ \Big|\ \begin{matrix}Q_1'\\ Q_1\end{matrix}\right),$$

where $u = M_1 u_1$, $M_1\omega_1 = \omega\rho + \omega'\rho'$, etc. Writing $u_1 = 2\omega_1 U_1$, $\omega_1' = \omega_1\tau_1$, we have

$$\vartheta\left(u_1;\ 2\omega_1,\ 2\omega_1',\ 2\eta_1,\ 2\eta_1'\ \Big|\ \begin{matrix}Q_1'\\ Q_1\end{matrix}\right) = e^{\frac{1}{2}\eta_1\omega_1^{-1}u_1^2}\,\Theta\left(U_1;\ \tau_1\ \Big|\ \begin{matrix}Q_1'\\ Q_1\end{matrix}\right),$$

and the equations $u_1 = M_2 w$, $M_2 v = \omega_1 A$, $M_2 v' = \omega_1 B + \omega_1' B'$, give, if $w = 2vW$, $v' = v\tau'$, and in virtue of $A\bar{B}' = r$, the equations $U_1 = AW$, $r\tau_1 = A\tau'\bar{A} - B\bar{A}$, while, by the equation $r\zeta = \bar{M}_2\eta_1 A$, we find $\eta_1\omega_1^{-1}u_1^2 = r\zeta v^{-1}w^2$. Now it is immediately seen that the exponent of the general term of $\Theta\left(U_1;\ \tau_1\ \Big|\ \begin{matrix}Q_1'\\ Q_1\end{matrix}\right)$ gives

$$2\pi i U_1 n + i\pi\tau_1 n^2 = 2\pi i r W\left(m + \frac{\mu}{r}\right) + \pi i r\tau'\left(m + \frac{\mu}{r}\right)^2 + \pi i d\left(m + \frac{\mu}{r}\right)$$
$$- i\pi(\gamma m^2 + dm) - 2\pi i\bar{B}\frac{B'\mu}{r}m - \frac{\pi i d\mu}{r} - \frac{\pi i}{r^2}\gamma\mu^2,$$

wherein $\gamma = \bar{B}B'$, and d denotes the row of diagonal elements of γ, and m, μ are obtained by putting $\bar{A}n = rm + \mu$, m being a row of integers, and μ a row of integers each less than r and positive (including zero); this equation is equivalent to $n - B'm = \frac{1}{r}B'\mu$; corresponding to every n it determines an unique m and an unique μ for which $\frac{B'\mu}{r}$ is integral; corresponding to any assigned μ for which $\frac{B'\mu}{r}$ is integral, and an assigned m, the equation determines an unique n. Since then $\gamma m^2 + dm$ is an even integer, and, for the terms which occur, $\bar{B}\frac{B'\mu}{r}m$ is an integer, we have

$$\Theta(U_1, \tau_1) = \sum_{\mu} e^{-\frac{\pi i d\mu}{r} - \frac{\pi i}{r^2}\gamma\mu^2}\,\Theta\left[rW;\ r\tau'\ \Big|\ \begin{matrix}\mu/r\\ d/2\end{matrix}\right].$$

Increasing, in this equation, U_1 by $Q_1 + \tau_1 Q_1'$, we hence deduce

$$\Theta\left(U_1;\ \tau_1\ \Big|\ \begin{matrix}Q_1'\\ Q_1\end{matrix}\right) = e^{-\frac{\pi i}{r}dK' + \frac{\pi i}{r^2}\gamma K'^2}\sum_{\mu} e^{-\frac{\pi i d\mu}{r} - \frac{\pi i}{r^2}\gamma\mu^2}\,\Theta\left[rW;\ r\tau'\ \Big|\ \begin{matrix}(K'+\mu)/r\\ K\end{matrix}\right],$$

where $K' = \bar{A}Q_1'$, $-K = \bar{B}Q_1' - \bar{B}'Q_1 - \frac{1}{2}d(\bar{B}B')$, so that (K, K') is the characteristic of the final theta function of w. Since now the matrix $Mv\bar{B}' = M_1M_2v\bar{B}' = M_1\omega_1A\bar{B}' = rM_1\omega_1$, and therefore $|M||v||B'| = r^p|M_1||\omega_1|$, we have, by multiplying the last obtained equation by $e^{\frac{1}{2}\eta_1\omega_1^{-1}u_1^2} = e^{\frac{1}{2}r\zeta v^{-1}w^2}$, the formula which was given above.

Ex. i. When $p=1$, the transformation associated with the matrix $\begin{pmatrix}1 & 0\\ 0 & 3\end{pmatrix}$ gives rise to the function $\Theta(W; \frac{1}{3}\tau')$; we have

$$\Theta(W; \tfrac{1}{3}\tau') = \Theta(3W; 3\tau') + \Theta\left(3W; 3\tau' \left| \begin{matrix} 1/3 \\ 0 \end{matrix}\right.\right) + \Theta\left(3W; 3\tau' \left| \begin{matrix} -1/3 \\ 0 \end{matrix}\right.\right).$$

Other simple examples have already occurred for the quadric transformations (§ 365).

Ex. ii. Prove when $p=2$, by considering the transformation of order r (r odd) for which

$$a = \begin{pmatrix}1, & -\mu\\ 0, & r\end{pmatrix},\quad \beta = \begin{pmatrix}2\lambda, & 0\\ 0, & 0\end{pmatrix},\quad a' = \begin{pmatrix}0 & 0\\ 0 & 0\end{pmatrix},\quad \beta' = \begin{pmatrix}r & 0\\ \mu & 1\end{pmatrix},$$

that

$$\Theta\left[u_1 - \mu u_2, ru_2; \frac{1}{r}(\tau_{11} - 2\mu\tau_{12} + \mu^2\tau_{22} - 2\lambda), 2\tau_{12} - 2\mu\tau_{22}, r\tau_{22}\right]$$
$$= \tfrac{1}{2}\psi(0, 0) + \sum_{n=1}^{\frac{1}{2}(r-1)} e^{-\frac{2\pi i}{r}n^2\lambda}\,\psi(n, -n\mu),$$

where $\psi(n_1, n_2)$ denotes $\Theta\left(ru; r\tau \left| \begin{matrix} n_1/r, & n_2/r \\ 0 \end{matrix}\right.\right) + \Theta\left(ru; r\tau \left| \begin{matrix} -n_1/r, & -n_2/r \\ 0 \end{matrix}\right.\right)$. (Wiltheiss, *Crelle*, XCVI. (1884), pp. 21, 22.)

373. In regard now to the question of the coefficients which enter in the expression of the functions $\psi_r(w; K, K'+\mu)$ by means of functions $\vartheta\left(w \left| \begin{matrix} K' \\ K \end{matrix}\right.\right)$, the problem that arises is that of the determination of these coefficients in terms of given constants, as for instance the zero values of the original theta functions. The theory of this determination must be omitted from the present volume. In the case when the order of the transformation is odd these coefficients arise in this chapter expressed in terms of theta functions, $\vartheta\left(\frac{2vm + 2v'm'}{r}; 2v, 2v', 2\zeta, 2\zeta'\right)$, whose arguments are r-th parts of the periods $2v$, $2v'$. By means of two supplementary transformations, Δ, $r\Delta^{-1}$, (as indicated § 332, Chap. XVIII.), or by means of the formulae of Chap. XVII. (as indicated in Ex. vii., § 317, Chap. XVII.), we can obtain equations for functions $\vartheta(rw; 2v, 2v', 2\zeta, 2\zeta')$ as integral polynomials of degree r^2 in functions $\vartheta(w; 2v, 2v', 2\zeta, 2\zeta')$. By means of these equations the functions $\vartheta\left(\frac{2vm + 2v'm'}{r}; 2v, 2v', 2\zeta, 2\zeta'\right)$ are determined in terms of functions $\vartheta(0; 2v, 2v', 2\zeta, 2\zeta')$; or this determination may arise by elimination from the original equations of transformation, without use of the multiplication equations. There remains then further the theory of the relations connecting the functions $\vartheta(0; 2v, 2v', 2\zeta, 2\zeta')$ and the functions $\vartheta(0; 2\omega, 2\omega', 2\eta, 2\eta')$, which is itself a matter of complexity.

For the case $p=1$, the reader may consult, for instance, Weber, *Elliptische Functionen* (Braunschweig, 1891), Krause, *Theorie der doppeltperiodischen Functionen* (Erster Band, Leipzig, 1895). For the case $p=2$, Krause, *Hyperelliptische Functionen* (Leipzig, 1886), Königsberger, *Crelle*, LXIV., LXV., LXVII. For the form of the general results, the chapter, *Die Theilung*, of Clebsch u. Gordan, *Abel'sche Functionen* (Leipzig, 1866), which deals with the theta functions arising on a Riemann surface, may be consulted. For the hyper-elliptic case, see also Jordan, *Traité des Substitutions* (Paris, 1870), p. 365, and Burkhardt, *Math. Annal.* XXXV., XXXVI., XXXVIII. (1890—1).

In particular cases, knowing the form of the expression of the functions

$$\vartheta(u;\ 2\omega,\ 2\omega',\ 2\eta,\ 2\eta')$$

in terms of functions $\vartheta(w;\ 2v,\ 2v',\ 2\zeta,\ 2\zeta')$, we are able to determine the coefficients by the substitution of half-periods coupled with expansion of the functions in powers of the arguments. See, for instance, the book of Krause (*Hyperelliptische Functionen*) and Königsberger, *as above*.

Ex. i. In case $p=2$, $r=3$, the function $\Theta_5(3W,\ 3\tau')$ is a cubic polynomial of the functions $\Theta_5(W,\ \tau')$, $\Theta_{34}(W,\ \tau')$, $\Theta_1(W,\ \tau')$, $\Theta_{02}(W,\ \tau')$, of which the characteristics are respectively $\frac{1}{2}\begin{pmatrix}0, & 0\\ 0, & 0\end{pmatrix}$, $\frac{1}{2}\begin{pmatrix}0, & 0\\ 0, & -1\end{pmatrix}$, $\frac{1}{2}\begin{pmatrix}1, & 0\\ -1, & -1\end{pmatrix}$, $\frac{1}{2}\begin{pmatrix}1, & 0\\ -1, & 0\end{pmatrix}$; these form a Göpel system. The only products of these functions which are theta functions of the third order and of zero characteristic are those contained in the equation

$$\Theta_5(3W,\ 3\tau')=A\phi_5^3+B\phi_5\phi_{34}^2+C\phi_5\phi_1^2+D\phi_5\phi_{02}^2+E\phi_{34}\phi_1\phi_{02},$$

where $\phi_5=\Theta_5(W,\ \tau')$, etc.; this equation contains the right number $\frac{1}{2}(r^p+1)=5$ of terms on the right side. Putting instead of the arguments W_1, W_2 respectively

$$W_1,\ W_2-\tfrac{1}{2};\quad W_1-\tfrac{1}{2}+\tfrac{1}{2}\tau_{11},\ W_2-\tfrac{1}{2}+\tfrac{1}{2}\tau_{21};\quad W_1-\tfrac{1}{2}+\tfrac{1}{2}\tau_{11},\ W_2+\tfrac{1}{2}\tau_{21},$$

we obtain in turn

$$\Theta_{34}(3W,\ 3\tau')=\ \ A\phi_{34}^3+B\phi_{34}\phi_5^2+C\phi_{34}\phi_{02}^2+D\phi_{34}\phi_1^2+E\phi_5\phi_1\phi_{02},$$
$$\Theta_1\ (3W,\ 3\tau')=-A\phi_1^3\ -B\phi_1\phi_{02}^2+C\phi_1\ \phi_5^2\ +D\phi_1\phi_{34}^2+E\phi_5\phi_{02}\phi_{34},$$
$$\Theta_{02}(3W,\ 3\tau')=-A\phi_{02}^3-B\phi_{02}\phi_1^2+C\phi_{02}\phi_{34}^2+D\phi_{02}\phi_5^2+E\phi_5\phi_1\phi_{34},$$

whereby the Göpel system of functions $\Theta_5(3W,\ 3\tau')$, $\Theta_{34}(3W,\ 3\tau')$, $\Theta_1(3W,\ 3\tau')$, $\Theta_{02}(3W,\ 3\tau')$ is expressed by means of the Göpel system ϕ_5, ϕ_{34}, ϕ_1, ϕ_{02}.

From the first two equations, by putting the arguments zero, we obtain

$$A=\frac{\overline{\Theta}_5\Theta_5-\overline{\Theta}_{34}\Theta_{34}}{\Theta_5^4-\Theta_{34}^4},\quad B=\frac{\overline{\Theta}_{34}\Theta_5^3-\overline{\Theta}_5\Theta_{34}^3}{\Theta_5\Theta_{34}(\Theta_5^4-\Theta_{34}^4)},$$

where $\overline{\Theta}_5=\Theta_5(0;\ 3\tau')$, etc., and $\Theta_5=\Theta_5(0;\ \tau')$, etc.; by the addition of other even half-periods to the arguments, for instance, those associated with the characteristics

$$\tfrac{1}{2}\begin{pmatrix}1, & 1\\ 0, & 0\end{pmatrix},\ \tfrac{1}{2}\begin{pmatrix}0, & 1\\ 0, & 0\end{pmatrix},\ \tfrac{1}{2}\begin{pmatrix}0, & 0\\ -1, & 0\end{pmatrix},$$

we can obtain expressions for C, D, E; these substitutions give respectively

$$\Theta_{23}(3W;\ 3\tau')=A\phi_{23}^3-B\phi_{23}\phi_{24}^2+C\phi_{23}\phi_{04}^2-D\phi_{23}\phi_{03}^2+E\phi_{24}\phi_{04}\phi_{03},$$
$$\Theta_4\ (3W;\ 3\tau')=A\phi_4^3\ -B\phi_4\ \phi_3^2\ -C\phi_4\ \phi_{14}^2+D\phi_4\ \phi_{13}^2-E\phi_3\ \phi_{14}\phi_{13},$$
$$\Theta_{12}(3W;\ 3\tau')=A\phi_{12}^3+B\phi_{12}\phi_0^2\ +C\phi_{12}\phi_2^2\ +D\phi_{12}\phi_{01}^2+E\phi_0\ \phi_2\ \phi_{01};$$

putting herein $W=0$ we obtain in succession the values of D, C and E, expressed in terms of the constants previously used, $\overline{\Theta}_5$, $\overline{\Theta}_{34}$, Θ_5, Θ_{34} and the constants $\overline{\Theta}_{23}$, $\overline{\Theta}_4$, $\overline{\Theta}_{12}$, Θ_{23}, Θ_{03}, Θ_4, Θ_{14}, Θ_{12}, Θ_0, Θ_2, Θ_{01}. Thus the zero values of each of the ten even functions $\Theta(W;\tau')$ enter in the expression of the coefficients A, B, C, D, E; there remains then the question of the expression of the zero values of the ten even functions in terms of four independent quantities (cf. Ex. iv., § 317, Chap. XVII.), and the question of the relations connecting the constants $\overline{\Theta}_5$, $\overline{\Theta}_{34}$, etc., and the constants Θ_5, Θ_{34}, etc. (cf. the following example).

Ex. ii. Denoting $\Theta_{01}(0;3\tau')\,\Theta_{01}(0;\tau')$ by C_{01}, etc., shew that when $p=2$ the result of Ex. iii., § 292 (p. 477) gives the equations

$$C_{01}+C_2=C_5+C_{34}-C_{12}-C_0,$$
$$C_4+C_{03}=C_5-C_{34}+C_{12}-C_0,$$
$$C_{23}+C_{14}=C_5-C_{34}-C_{12}+C_0,$$

these being the only equations derivable from that result. By these equations, in virtue of the relations connecting the ten constants $\Theta(0;\tau')$, and the relations connecting the ten constants $\Theta(0;3\tau')$, (for the various even characteristics), the three ratios

$$\Theta_{34}(0;3\tau')/\Theta_5(0,3\tau'),\quad \Theta_{12}(0;3\tau')/\Theta_5(0;3\tau'),\quad \Theta_0(0;3\tau')/\Theta_5(0;3\tau')$$

are determinable in terms of the three

$$\Theta_{34}(0;\tau')/\Theta_5(0;\tau'),\quad \Theta_{12}(0,\tau')/\Theta_5(0;\tau'),\quad \Theta_0(0;\tau')/\Theta_5(0;\tau').$$

By addition of these equations we obtain

$$C_{01}+C_2+C_4+C_{03}+C_{23}+C_{14}+C_{34}+C_{12}+C_0=3C_5.$$

Obtain similarly from the result of Ex. iii., § 292, for any value of p, the equation

$$\Sigma\Theta\left[0;\,3\tau'\,\middle|\,\tfrac{1}{2}\binom{s'}{s}\right]\Theta\left[0;\,\tau'\,\middle|\,\tfrac{1}{2}\binom{s'}{s}\right]=(2^p-1)\,\Theta(0;3\tau')\,\Theta(0;\tau'),$$

where the summation on the left extends to all even characteristics except the zero characteristic; for instance, when $p=1$, this is the equation

$$\Theta_{01}(0;3\tau')\,\Theta_{01}(0;\tau')+\Theta_{10}(0;3\tau')\,\Theta_{10}(0;\tau')=\Theta_{00}(0;3\tau')\,\Theta_{00}(0;\tau'),$$

namely (cf. Ex. i., § 365 of this chapter) it is the modular equation for transformation of the third order which is generally written in the form (Cayley, *Elliptic Functions*, 1876, p. 188),

$$\sqrt{k\lambda}+\sqrt{k'\lambda'}=1.$$

As here in the case $p=2$, so for any value of p, we obtain, from the result of Ex. iii., § 292, 2^p-1 modular equations for the cubic transformation.

Ex. iii. From the formula of § 364 we obtain modular equations for the quadric transformation, in the form

$$2^p\,\Theta\left[0;\,2\tau'\,\middle|\,\tfrac{1}{2}\binom{k'}{k}\right]\Theta\left[0;\,2\tau'\,\middle|\,\tfrac{1}{2}\binom{k'+s'}{k}\right]=\sum_s e^{\pi isk'}\,\Theta\left[0;\,\tau'\,\middle|\,\tfrac{1}{2}\binom{s'}{s}\right]\Theta\left[0;\,\tau'\,\middle|\,\tfrac{1}{2}\binom{s'}{k+s}\right],$$

where s is a row of p quantities each either 0 or 1, so that the right side contains 2^p terms, and k, k', s' are any rows of p quantities each either 0 or 1.

374. In the fundamental equations of transformation we have considered only the case when the matrices α, α', β, β' are matrices of integers; the analytical theory can be formulated in a more general way, as follows; the argument is an application of the results of Chap. XIX.

Suppose we have the relations expressed (cf. Ex. ii., § 324, Chap. XVIII.) by

$$\begin{pmatrix} M, & 0 \\ 0, & r\bar{M}^{-1}\end{pmatrix}\begin{pmatrix} 2\upsilon, & 2\upsilon' \\ 2\zeta, & 2\zeta'\end{pmatrix} = \begin{pmatrix} 2\omega, & 2\omega' \\ 2\eta, & 2\eta'\end{pmatrix}\begin{pmatrix}\alpha, & \beta \\ \alpha', & \beta'\end{pmatrix},$$

where r is a positive rational number, M is any matrix of p rows and columns, whose determinant does not vanish, α, β, α', β' are matrices of p rows and columns whose elements are *rational numbers not necessarily integers*, ω, ω', η, η' are matrices of p rows and columns satisfying the equations (B), § 140 (Chap. VII.), and υ, υ', ζ, ζ' are similar matrices satisfying similar conditions; then, as necessarily follows, the matrices α, β, α', β' satisfy the relation (viii) of § 324 (Chap. XVIII.).

If now x, y be any matrices of p rows and columns, the relations supposed are immediately seen to be equivalent to

$$\begin{pmatrix} M, & 0 \\ 0, & r\bar{M}^{-1}\end{pmatrix}\begin{pmatrix} 2\upsilon x, & 2\upsilon' y \\ 2\zeta x, & 2\zeta' y\end{pmatrix} = \begin{pmatrix} 2\omega, & 2\omega' \\ 2\eta, & 2\eta'\end{pmatrix}\begin{pmatrix}\alpha x, & \beta y \\ \alpha' x, & \beta' y\end{pmatrix};$$

we suppose that x, y *are such matrices of integers that* αx, βy, $\alpha' x$, $\beta' y$ *are matrices of integers, and, at the same time, such that* rx *is a matrix of integers*; such matrices x, y can be determined in an infinite number of ways.

Let u, w be two rows of p arguments connected by the equations $u = Mw$; when the arguments w are simultaneously increased by the elements of the row of quantities denoted by $2\upsilon xm + 2\upsilon' ym'$, in which m, m' are rows of p integers, the arguments u are increased by the elements of the row $2\omega n + 2\omega' n'$, where $n = \alpha xm + \beta ym'$, $n' = \alpha' xm + \beta' ym'$ are rows of integers. The resulting factor of the function $\vartheta(u;\, 2\omega, 2\omega', 2\eta, 2\eta')$ is e^R, where, if $H_\alpha = 2\eta\alpha + 2\eta'\alpha'$, etc., (cf. (v), § 324, Chap. XVIII.), R is given by

$$\begin{aligned} R &= H_n(u + \tfrac{1}{2}\Omega_n) - \pi i n n' \\ &= (H_\alpha xm + H_\beta ym')(Mw + M\upsilon xm + M\upsilon' ym') - \pi i n n' \\ &= (\bar{M}H_\alpha xm + \bar{M}H_\beta ym')(w + \upsilon xm + \upsilon' ym') - \pi i n n' \\ &= r(2\zeta xm + 2\zeta' ym')(w + \upsilon xm + \upsilon' ym') - \pi i n n'; \end{aligned}$$

now, since $\bar{\beta}'\alpha = r + \bar{\beta}\alpha'$, and because αx, βy, $\alpha' x$, $\beta' y$, rx are matrices of integers, we have

$$\begin{aligned} nn' &= \bar{x}\bar{\alpha}'\alpha x m^2 + (\bar{y}\bar{\beta}\alpha' x + \bar{y}\bar{\beta}'\alpha x)\, mm' + \bar{y}\bar{\beta}'\beta y m'^2 \\ &\equiv fm + f'm' + r\bar{y}x mm' \pmod{2}, \end{aligned}$$

where f, f' denote respectively the rows of integers formed by the diagonal elements of the symmetrical matrices $\bar{x}\bar{\alpha}'\alpha x$, $\bar{y}\bar{\beta}'\beta y$ (cf. § 327, Chap. XVIII.).

Thus, if we denote $\vartheta(u;\, 2\omega, 2\omega', 2\eta, 2\eta')$ by $\phi(w)$, we have

$$\phi(w + 2\upsilon xm + 2\upsilon' ym') = e^{r(2\zeta xm + 2\zeta' ym')(w + \upsilon xm + \upsilon' ym') + \pi i(fm + f'm') + \pi i(-r\bar{y}x)mm'}\, \phi(w).$$

Further if a, b denote the matrices of $2p$ columns and p rows, given respectively by

$$a=(2vx,\ 2v'y),\quad 2\pi i b=(2r\zeta x,\ 2r\zeta' y),$$

we have

$$\begin{aligned}\tfrac{1}{2}\frac{\pi i}{r}(\bar{a}b-\bar{b}a)&=\begin{pmatrix}\bar{x}\bar{v}\\ \bar{y}\bar{v}'\end{pmatrix}(\zeta x,\ \zeta' y)-\begin{pmatrix}\bar{x}\bar{\zeta}\\ \bar{y}\bar{\zeta}'\end{pmatrix}(vx,\ v'y)\\ &=\begin{pmatrix}\bar{x}(\bar{v}\zeta-\bar{\zeta}v)x, & \bar{x}(\bar{v}\zeta'-\bar{\zeta}v')y\\ \bar{y}(\bar{v}'\zeta-\bar{\zeta}'v)x, & \bar{y}(\bar{v}'\zeta'-\bar{\zeta}'v')y\end{pmatrix}\\ &=\tfrac{1}{2}\pi i\begin{pmatrix}0, & -\bar{x}y\\ \bar{y}x, & 0\end{pmatrix};\end{aligned}$$

so that $\bar{a}b-\bar{b}a=k$, say, is a skew symmetrical matrix of integers given by

$$\bar{a}b-\bar{b}a=k=\begin{pmatrix}0, & -r\bar{x}y\\ r\bar{y}x, & 0\end{pmatrix},$$

and we have

$$\overset{\alpha<\beta}{\Sigma}\, k_{\alpha,\beta}m_\alpha m_\beta'=\sum_{\alpha,\beta}(-r\bar{x}y)_{\alpha,\beta}m_\alpha m_\beta'=-r\bar{y}xmm',\qquad (\alpha,\beta=1,\ldots,p).$$

Finally, let λ, μ be rows of p quantities, the rows of conjugate complex quantities being denoted by λ_1, μ_1, and let λ, μ be taken so that the row of quantities $a(\lambda,\mu)$ consists of zeros, or

$$a(\lambda,\mu)=2vx\lambda+2v'y\mu=0,$$

so that $x\lambda=-\tau' y\mu$, where* $\tau'=v^{-1}v'$, is a symmetrical matrix, $=\rho'+i\sigma'$, say, ρ' and σ' being matrices of real quantities; then by

$$x\lambda_1=-\tau_1' y\mu_1=-(\rho'-i\sigma')\,y\mu_1,$$

we have

$$\begin{aligned}ik(\lambda,\mu)(\lambda_1,\mu_1)&=-ir(\bar{x}y\mu,\ -\bar{y}x\lambda)(\lambda_1,\mu_1)=-ir(\bar{y}x\lambda_1\mu-\bar{y}x\lambda\mu_1)\\ &=ir\bar{y}(\tau_1' y\mu_1\mu-\tau' y\mu\mu_1)=ir\bar{y}[(\rho'-i\sigma')-(\rho'+i\sigma')]\,y\mu\mu_1\\ &=2r\bar{y}\sigma' y\mu\mu_1=2r\sigma'\nu\nu_1,\end{aligned}$$

in which $\nu=y\mu$, $\nu_1=y\mu_1$; as in § 325, Chap. XVIII., since r is positive, the form $r\sigma'\nu\nu_1$ is necessarily positive except for zero values of μ.

On the whole, comparing formula (II), § 354, Chap. XIX., the function $\phi(w)$ satisfies the conditions of §§ 351—2, Chap. XIX., necessary for a Jacobian function of w in which the periods and characteristic are given † by

$$a=(2vx,\ 2v'y),\quad 2\pi i b=(2r\zeta x,\ 2r\zeta' y),\quad c=(\tfrac{1}{2}f,\ \tfrac{1}{2}f').$$

* The determinant of the matrix v is supposed other than zero, as in Chap. XVIII., § 324.

† In § 351, Chap. XIX., the row letters have σ elements; in the present case σ is equal to $2p$, and it is convenient to represent the corresponding row letters by two constituents, each of p elements; and similarly for the matrices of $2p$ columns and p rows.

To this function we now apply the result of § 359, Chap. XIX., in order to express it by theta functions of w. The condition for the matrix of integers there denoted by g, namely $\bar{g}\epsilon g = k$, is satisfied by $g = \begin{pmatrix} rx, & 0 \\ 0, & y \end{pmatrix}$, for

$$\begin{pmatrix} r\bar{x}, & 0 \\ 0, & \bar{y} \end{pmatrix}\begin{pmatrix} 0, & -1 \\ 1, & 0 \end{pmatrix}\begin{pmatrix} rx, & 0 \\ 0, & y \end{pmatrix} = \begin{pmatrix} r\bar{x}, & 0 \\ 0, & \bar{y} \end{pmatrix}\begin{pmatrix} 0, & -y \\ rx, & 0 \end{pmatrix} = \begin{pmatrix} 0, & -r\bar{x}y \\ r\bar{y}x, & 0 \end{pmatrix};$$

hence, with the notation of § 358, Chap. XIX.,

$$K = ag^{-1} = (2vx,\ 2v'y)\begin{pmatrix} \frac{1}{r}x^{-1}, & 0 \\ 0, & y^{-1} \end{pmatrix} = (2v/r,\ 2v'),$$

$$2\pi i L = 2\pi i b g^{-1} = (2r\zeta x,\ 2r\zeta' y)\begin{pmatrix} \frac{1}{r}x^{-1}, & 0 \\ 0, & y^{-1} \end{pmatrix} = (2\zeta,\ 2r\zeta').$$

Hence, as our final result, by § 359, Chap. XIX., *the function* $\phi(w)$, *or* $\vartheta(u;\ 2\omega,\ 2\omega',\ 2\eta,\ 2\eta')$, *can be expressed as a sum of constant multiples of functions** $\vartheta(w;\ 2v/r,\ 2v',\ 2\zeta,\ 2\zeta')$ *with different characteristics, the number of such terms being at most* $\sqrt{|K|} = r^p |x|\,|y|$, *where* $|x|$, $|y|$ *denote the determinants of the matrices* x, y. This is an extension of the result obtained when the matrices $\alpha, \beta, \alpha', \beta'$ are formed with integers; as in that case there will be a reduction in the number of terms, from $r^p |x|\,|y|$, owing to the fact that the function $\phi(w)$ is even. A similar result holds whatever be the characteristic of the function $\vartheta(u;\ 2\omega,\ 2\omega',\ 2\eta,\ 2\eta')$. The generalisation is obtained quite differently by Prym and Krazer, *Neue Grundlagen einer Theorie der allgemeinen Thetafunctionen* (Leipzig, 1892), *Zweiter Theil*, which should be consulted.

Ex. Denoting by E the matrix of p rows and columns of which the elements are zero, other than those in the diagonal, which are each unity, and taking for the matrices a, β, a', β' respectively $\frac{m}{n}E$, 0, 0, $\frac{n}{m}E$, where m, n are integers without common factor, we have the formula

$$m^p\,\Theta(u;\ \tau) = \sum_r \sum_s \Theta\left(\frac{n}{m}u;\ \frac{n^2}{m^2}\tau \,\middle|\, \begin{matrix} ms/n \\ nr/m \end{matrix}\right),$$

wherein r, s are rows of p positive integers, in which every element of r is 0 or numerically less than m, and every element of s is 0 or numerically less than n. This formula includes that of § 284, Ex. iii. (Chap. XV.); it is a particular case of a formula given by Prym and Krazer (*loc. cit.*, p. 77).

To obtain a verification—the general term of the right side is e^{ψ}, where

$$\psi = 2\pi i u\left(\frac{n}{m}N + s\right) + i\pi\tau\left(\frac{n}{m}N + s\right)^2 + 2\pi i r\left(\frac{n}{m}N + s\right);$$

* That is, functions $\vartheta(rw,\ 2v,\ 2rv',\ 2\zeta/r,\ 2\zeta')$; cf. § 284, p. 448.

hence $\sum_r e^{\psi}=0$ unless N/m is integral; when N/m is integral, $=M$, say, then $\sum_r e^{\psi}=m^p e^{\phi}$, where

$$\phi=2\pi iuK+i\pi\tau K^2,$$

K, $=nM+s$, obtaining all integral values when M takes all integral values and s takes all integral values (including zero) which are numerically less than n.

375. The theory of the transformation of theta functions may be said to have arisen in the problem of the algebraical transformation of the hyperelliptic theta quotients considered in Chap. XI. of this volume. To practically utilise the results of this chapter for that problem it is necessary to adopt conventions sufficient to determine the constant factors occurring in the algebraic expression of these theta quotients (cf. §§ 212, 213), and to define the arguments of the theta functions in an algebraical way. The reader is referred* to the forthcoming volumes of Weierstrass's lectures.

It has already (§ 174, p. 248) been remarked that when $p>3$ the most general theta function cannot be regarded as arising from a Riemann surface; for the algebraical problems then arising the reader is referred to the recent papers of Schottky and Frobenius (*Crelle*, CII. (1888), and following) and to the book of Wirtinger, *Untersuchungen über Thetafunctionen* (Leipzig, 1895).

* Cf. Rosenhain, *Mém. p. divers Savants*, XI. (1851), p. 416 ff.; Königsberger, *Crelle*, LXIV. (1865), etc.

CHAPTER XXI.

COMPLEX MULTIPLICATION OF THETA FUNCTIONS. CORRESPONDENCE OF POINTS ON A RIEMANN SURFACE.

376. IN the present chapter some account is given of two theories; the former is a particular case of the theory of transformation of theta functions; the latter is intimately related with the theory of transformation of Riemann theta functions. Not much more of the results of these theories is given than is necessary to classify the references which are given to the literature.

377. In the transformation of the function $\Theta(u;\ \tau)$, to a function of the arguments w, with period τ' (§ 324, Chap. XVIII.), the following equations have arisen

$$u = Mw, \quad M = \alpha + \tau\alpha', \quad M\tau' = \beta + \tau\beta';$$

there* are cases, for special values of τ, in which τ' is equal to τ. We investigate necessary conditions for this to be so; and we prove, under a certain hypothesis, that they are sufficient. The results are stated in terms of the matrix of integers associated with the transformation; we do not enter into the investigation of the values of τ to which the results lead. We limit ourselves throughout to the function $\Theta(u;\ \tau)$; the change to the function $\vartheta(u;\ 2\omega, 2\omega', 2\eta, 2\eta')$ can easily be made.

Suppose that, corresponding to a matrix $\Delta = \begin{pmatrix} \alpha & \beta \\ \alpha' & \beta' \end{pmatrix}$, of $2p$ rows and columns, for which

$$\alpha\bar{\beta} = \beta\bar{\alpha}, \quad \alpha'\bar{\beta}' = \beta'\bar{\alpha}', \quad \alpha\bar{\beta}' - \beta\bar{\alpha}' = r = \beta'\bar{\alpha} - \alpha'\bar{\beta},$$

where r is a positive integer, there exists a matrix τ satisfying the equation

$$(\alpha + \tau\alpha')\,\tau = \beta + \tau\beta',$$

which is such that, for real values of $n_1, \ldots, n_p$, the imaginary part of the quadratic form τn^2 is positive.

* References to the literature for the case $p=1$ are given below (§ 383). For higher values of p, see Kronecker, *Berlin. Monatsber.* 1866, p. 597, or *Werke*, Bd. I. (Leipzig, 1895), p. 146; Weber, *Ann. d. Mat.*, Ser. 2, t. IX. (1878—9), p. 140; Frobenius, *Crelle*, XCV. (1883), p. 281, where other references are given; Wiltheiss, *Bestimmung Abelscher Funktionen mit zwei Argumenten* u. s. w. Habilitationsschrift, Halle, 1881 (E. Karras), and *Math. Annal.* XXVI. (1886), p. 130.

In that case, as follows from Chap. XX., the function $\Theta[(\alpha+\tau\alpha')w;\ \tau]$, when multiplied by a certain exponential of the form $e^{\gamma w^2}$, is expressible as an integral polynomial of the r-th degree in 2^p functions $\Theta[w;\ \tau]$; on this account we say that there exists a *complex multiplication**, or a *special transformation*, belonging to the matrix Δ. The equation $(\alpha+\tau\alpha')\tau=\beta+\tau\beta'$ is equivalent to $(\bar{\beta}'-\tau\bar{\alpha}')\tau=-\bar{\beta}+\tau\bar{\alpha}$; this arises from the supplementary matrix

$$r\Delta^{-1}=\begin{pmatrix}\bar{\beta}' & -\bar{\beta}\\ -\bar{\alpha}' & \bar{\alpha}\end{pmatrix},$$

just as the former equation arises from Δ; we put $M=\alpha+\tau\alpha'$, $N=\bar{\beta}'-\tau\bar{\alpha}'$; we denote by $|\Delta-\lambda|$ the determinant of the matrix $\Delta-\lambda E$, where E is the matrix unity of $2p$ rows and columns, and λ is a single quantity; similarly we denote by $|M-\lambda|$ the determinant of the matrix $M-\lambda E'$, where E' is the matrix unity of p rows and columns.

Then we prove first, that *when there exists such a complex multiplication, to every root of the equation in λ of order p given by $|M-\lambda|=0$, there corresponds a conjugate complex root of the equation $|N-\lambda|=0$; that the $2p$ roots of the equation $|\Delta-\lambda|=0$ are constituted by the roots of the two equations $|M-\lambda|=0$, $|N-\lambda|=0$, or $|\Delta-\lambda|=|M-\lambda|\,|N-\lambda|$; and that all these roots are of modulus $\sqrt{r}$. Hence when $r=1$, they can be shewn to be all roots of unity.*

378. The equations of the general transformation, of order r, and its supplementary transformation, namely

$$M=a+\tau a',\quad M\tau'=\beta+\tau\beta',\quad N=\bar{\beta}'-\tau'\bar{a}',\quad N\tau=-\bar{\beta}+\tau'\bar{a},$$

give

$$(a+\tau a')\tau'=\beta+\tau\beta';$$

hence, if $\tau=\tau_1+i\tau_2$, where τ_1 and τ_2 are matrices of real quantities, and similarly $\tau'=\tau_1'+i\tau_2'$, we have by equating imaginary parts

$$(a+\tau_1 a')\tau_2'=\tau_2(\beta'-a'\tau_1');$$

therefore the two matrices

$$M\tau_2'=(a+\tau_1 a')\tau_2'+i\tau_2 a'\tau_2',\quad \tau_2\bar{N}=\tau_2(\beta'-a'\tau_1')-i\tau_2 a'\tau_2'$$

are conjugate imaginaries, $=f+ig$ and $f-ig$, say.

Now suppose $\tau'=\tau$; then from

$$M\tau_2=f+ig,\quad \tau_2\bar{N}=f-ig,$$

we have, if λ be any single quantity, and M_0 be the matrix whose elements are the conjugate complexes of the elements of M,

$$(M_0-\lambda)\tau_2=f-ig-\lambda\tau_2=\tau_2(\bar{N}-\lambda),$$

and hence, as $|\tau_2|$ is not zero,

$$|M_0-\lambda|=|\bar{N}-\lambda|,$$

* The name *principale Transformation* has been used (Frobenius, *Crelle*, xcv.).

which shews that to any root of the equation $|M-\lambda|=0$ there corresponds a conjugate complex root of the equation $|N-\lambda|=0$. Further we have, if $\tau_0=\tau_1-i\tau_2$,

$$\begin{pmatrix}1 & \tau\\ 1 & \tau_0\end{pmatrix}\begin{pmatrix}\alpha & \beta\\ \alpha' & \beta'\end{pmatrix}=\begin{pmatrix}M & M\tau\\ M_0 & M_0\tau_0\end{pmatrix}=\begin{pmatrix}M & 0\\ 0 & M_0\end{pmatrix}\begin{pmatrix}1 & \tau\\ 1 & \tau_0\end{pmatrix},$$

and writing this equation in the form

$$t\Delta=\mu t,$$

where

$$t=\begin{pmatrix}1 & \tau\\ 1 & \tau_0\end{pmatrix},\quad \mu=\begin{pmatrix}M & 0\\ 0 & M_0\end{pmatrix},$$

it easily follows that the determinant of the matrix t is not zero, and that, if λ be any single quantity, we have

$$t(\Delta-\lambda)=(\mu-\lambda)t,$$

so that

$$|\Delta-\lambda|=|\mu-\lambda|=|M-\lambda|\,|M_0-\lambda|=|M-\lambda|\,|N-\lambda|.$$

Thus the roots of the equation $|\Delta-\lambda|=0$ are constituted by the roots of the equations

$$|M-\lambda|=0,\quad |N-\lambda|=0.$$

Further, from a result previously obtained (Chap. XVIII., § 325, Ex.), when, as here, $\tau'=\tau$ and $2\omega=1$, $2\upsilon=1$, we have

$$M_0\tau_2\overline{M}=r\tau_2 \text{ or } M\tau_2\overline{M}_0=r\tau_2\,;$$

also as, for real values of $n_1, \ldots, n_p$, the form $\tau_2 n^2$ is a positive form, it can be put into the shape $m_1^2+\ldots\ldots+m_p^2$, $=Em^2$, say, E being the matrix unity of p rows and columns, and m being a row of quantities given by $m=Sn$, where S is a matrix of real elements; the equation $\tau_2 n^2=E\,.\,Sn\,.\,Sn$ gives $\tau_2=\overline{S}ES=\overline{S}S$; for distinctness we shall write

$$\tau_2=\overline{K}K_0,$$

$K=K_0=S$ being conjugate complex matrices. Take now a matrix $R=K\overline{M}K^{-1}$; then

$$\overline{R}R_0=\overline{K}^{-1}M\overline{K}K_0\overline{M}_0K_0^{-1}=\overline{K}^{-1}M\tau_2\overline{M}_0K_0^{-1}=r\overline{K}^{-1}\tau_2K_0^{-1}=r\,;$$

thus if λ be a root of $|M-\lambda|=0$, and therefore, as $R-\lambda=K(\overline{M}-\lambda)K^{-1}$, also a root of $|R-\lambda|=0$, and if z, $=x+iy$, be a row of p quantities such that $Rz=\lambda z=E\lambda z$, where E is the matrix unity of p rows and columns, we have

$$\overline{R}R_0z_0z=R_0z_0\,.\,Rz=E\lambda_0z_0\,.\,E\lambda z=\lambda\lambda_0\,.\,Ez_0z$$

or

$$(\lambda\lambda_0-r)\,Ez_0z=0.$$

Therefore as Ez_0z, which is equal to $\sum_{m=1}^{p}(x_m^2+y_m^2)$, is not zero, it follows that $\lambda\lambda_0=r$; in other words, all the roots of the equations $|M-\lambda|=0$, $|\Delta-\lambda|=0$, are of modulus $\sqrt{r}$.

Suppose now that $r=1$, so that the roots of the equation $|\Delta-\lambda|=0$ are all of modulus unity; then we prove for an equation

$$x^n+Ax^{n-1}+Bx^{n-2}+\ldots\ldots+N=0,$$

of any order, wherein the coefficients $A, B, \ldots, N$ are rational integers, and the coefficient of the highest power of x is unity, that if all the roots be of modulus unity, they are also roots of unity*; so that, for instance, there is no root of the form $e^{i\sqrt{2}}$.

* Kronecker, *Crelle*, LIII. (1857), p. 173; *Werke*, Bd. I. (1895), p. 103.

Let the roots be $e^{i\alpha}$, $e^{i\beta}$, ..., so that

$$A=-(\cos\alpha+\cos\beta+\dots),\quad B=\cos(\alpha+\beta)+\cos(\alpha+\gamma)+\dots,\ \dots;$$

then A lies between $-n$ and n, and B lies between $\pm\tfrac{1}{2}n(n-1)$, etc.; hence there can only be a finite number, say k, of equations of the above form, whereof all the roots are roots of unity. Thus, if $x_1, \dots, x_n$ be the roots of our equation, so that, for any positive integer μ, the roots of the equation

$$F_\mu(x)=(x-x_1^\mu)(x-x_2^\mu)\dots(x-x_n^\mu)=0,$$

are also roots of unity, it follows that, of the equations

$$F_1(x)=0,\quad F_2(x)=0,\ \dots,\quad F_{k+1}(x)=0,$$

there must be two at least which are identical. Hence, supposing $F_\mu(x)=0$, $F_\nu(x)=0$ to be identical, we have n equations of the form

$$x_1^\mu=x_{r_1}^\nu,\quad x_2^\mu=x_{r_2}^\nu,\ \dots.$$

Choosing from these equations the cycle given by

$$x_1^\mu=x_{r_1}^\nu,\quad x_{r_1}^\mu=x_{s_1}^\nu,\ \dots,\quad x_{m_1}^\mu=x_1^\nu,$$

consisting, suppose, of σ equations, we infer that

$$x_1^{\mu^\sigma}=x_1^{\nu^\sigma},$$

and, hence, that x_1 is a $(\mu^\sigma-\nu^\sigma)$-th root of unity.

Ex. Prove that, when $M=\alpha+\tau\alpha'$, $M\tau'=\beta+\tau\beta'$,

$$\begin{pmatrix}1&\tau\\1&\tau_0\end{pmatrix}\begin{pmatrix}\alpha&\beta\\\alpha'&\beta'\end{pmatrix}=\begin{pmatrix}M&0\\0&M_0\end{pmatrix}\begin{pmatrix}1&\tau'\\1&\tau_0'\end{pmatrix};$$

and deduce*, if $\Delta=\begin{pmatrix}\alpha&\beta\\\alpha'&\beta'\end{pmatrix}$ and

$$H=\tfrac{1}{2}\begin{pmatrix}1&1\\\tau_0&\tau\end{pmatrix}\begin{pmatrix}\tau_2^{-1}&0\\0&\tau_2^{-1}\end{pmatrix}\begin{pmatrix}1&\tau\\1&\tau_0\end{pmatrix}=\begin{pmatrix}\tau_2^{-1}&\tau_2^{-1}\tau_1\\\tau_1\tau_2^{-1}&\tau_1\tau_2^{-1}\tau_1+\tau_2\end{pmatrix},$$

that

$$\bar{\Delta}H\Delta=rH'.$$

Hence, when $\tau'=\tau$, if z be a row of $2p$ elements, and $x=\Delta z$, we have

$$Hx^2=rHz^2,$$

which expresses a self-transformation of the quadratic form Hz^2, which has real coefficients. Cf. Hermite, *Compt. Rendus*, XL. (1855), p. 785; Laguerre, *Journ. de l'éc. pol.*, t. XXV., cah. XLII. (1867), p. 215; Frobenius, *Crelle*, XCV. (1883), p. 285.

379. Conversely, let

$$\Delta=\begin{pmatrix}\alpha&\beta\\\alpha'&\beta'\end{pmatrix}$$

be a matrix of integers of $2p$ rows and columns, such that

$$\bar{\alpha}\alpha'=\bar{\alpha}'\alpha,\quad \bar{\beta}\beta'=\bar{\beta}'\beta,\quad \bar{\alpha}\beta'-\bar{\alpha}'\beta=r=\bar{\beta}'\alpha-\bar{\beta}\alpha',$$

* Cf. Chap. XVIII. § 325, Ex.

where r is a positive integer; and suppose that the roots of the equation $|\Delta-\lambda|=0$ are all complex and of modulus $\sqrt{r}$. *Under the special hypothesis* that the roots of* $|\Delta-\lambda|=0$ *are all different, we prove now that a matrix* τ *can be determined such that* (i) τ *is a symmetrical matrix,* (ii) *for real values of* $n_1, \ldots, n_p$ *the imaginary part of the quadratic form* τn^2 *is positive,* (iii) *the equation*

$$(\alpha+\tau\alpha')\,\tau=\beta+\tau\beta'$$

is satisfied. Thus every such matrix Δ gives rise to a complex multiplication.

380. We utilise the following lemma, of which we give the proof at once.—If h be a matrix of n rows and columns, such that the determinant $|h+\lambda|$, wherein λ is a single quantity, vanishes to the first order when λ vanishes, and if x, y be rows of n quantities other than zero, such that

$$hx=0, \quad \bar{h}y=0,$$

then the quantity xy, $=x_1y_1+\ldots\ldots+x_ny_n$, is not zero.

Denoting the row x by ξ_1, its elements being $\xi_{11}, \ldots, \xi_{1n}$, determine other $n(n-1)$ quantities $\xi_{i,j}$ $(i=2, \ldots, n;\ j=1, \ldots, n)$ such that the determinant $|\xi|$ does not vanish; similarly, denoting y by η_1, determine $n(n-1)$ further quantities $\eta_{i,j}$ such that the determinant $|\eta|$ does not vanish. Then consider the determinant of the matrix $\eta(h+\lambda)\bar{\xi}$; the (r, s)-th element of this matrix is

$$\sum_i \eta_{r,i}\sum_j h_{i,j}\xi_{s,j}+\lambda\sum_i \eta_{r,i}\xi_{s,i}=\sum_j \xi_{s,j}\sum_i h_{i,j}\eta_{r,i}+\lambda\sum_i \eta_{r,i}\xi_{r,i},$$

$(i=1, \ldots, n;\ j=1, \ldots, n)$, and when $r=1$ we have

$$\sum_i h_{i,j}\eta_{r,i}=h_{1,j}\eta_{1,1}+\ldots\ldots+h_{n,j}\eta_{1,n}=(\bar{h}y)_j=0,$$

while when $s=1$, we have

$$\sum_j h_{i,j}\xi_{s,j}=h_{i,1}\xi_{1,1}+\ldots\ldots+h_{i,n}\xi_{1,n}=(hx)_i=0;$$

thus the $(1, 1)$-th element of this matrix is λxy, and every other element in the first row and column has the factor λ; thus the determinant of the matrix is of the form $\lambda[Axy+\lambda B]$. But the determinant of the matrix is equal to $|h+\lambda|\,|\xi|\,|\eta|$, and therefore by hypothesis vanishes only to the first order when λ vanishes. Thus xy is not zero.

381. Suppose now that $\lambda, \lambda_0, \mu, \mu_0, \ldots$ are the roots of the equation $|\Delta-\lambda|=0$, where λ and λ_0, and μ and μ_0, etc. are conjugate complexes. It is possible to find two rows x, x', each of p quantities, to satisfy the equations

$$\alpha x+\beta x'=\lambda x, \quad \alpha'x+\beta'x'=\lambda x', \text{ or, say, } (\Delta-\lambda)(x, x')=0, \qquad \text{(i)},$$

and similarly two rows z, z', each of p quantities, to satisfy the equations

$$\alpha z+\beta z'=\mu z, \quad \alpha'z+\beta'z'=\mu z', \qquad \text{(ii)};$$

from equations (i), if x_0 be the conjugate imaginary to x, etc., it follows, since $\lambda\lambda_0=r$, that

$$\alpha x_0+\beta x_0'=\frac{r}{\lambda}x_0, \quad \alpha'x_0+\beta'x_0'=\frac{r}{\lambda}x_0',$$

and hence, in virtue of the relations satisfied by the matrices $\alpha, \beta, \alpha', \beta'$, we have

$$\bar{\beta}'x_0-\bar{\beta}x_0'=\lambda x_0, \quad -\bar{\alpha}'x_0+\bar{\alpha}x_0'=\lambda x_0',$$

* For the general case, see Frobenius, *Crelle*, xcv. (1883).

which belong to the supplementary matrix $r\Delta^{-1}$ just as the equations (i) belong to the matrix Δ; for our purpose however they are more conveniently stated by saying that $t=x_0'$, $t'=-x_0$, satisfy the equations

$$(\bar{\Delta}-\lambda)(t, t')=0;$$

hence as x, x' satisfy the equations

$$(\Delta-\lambda)(x, x')=0,$$

it follows from the lemma just proved, putting $n=2p$, that $tx+t'x'$ is not zero; in other words the quantity

$$xx_0'-x'x_0$$

is not zero. Further from the equations (i), (ii) we infer

$$\lambda\mu(xz'-x'z)=(\alpha x+\beta x')(\alpha' z+\beta' z')-(\alpha' x+\beta' x')(\alpha z+\beta z');$$

and by the equations satisfied by the matrices α, β, α', β' this is easily found to be the same as

$$(\lambda\mu-r)(xz'-x'z)=0;$$

thus, as the equation $\lambda\mu=r$ would be the same as $\lambda=\lambda_0$, we have

$$xz'-x'z=0.$$

Also we have

$$\alpha z_0+\beta z_0'=\mu_0 z_0, \quad \alpha' z_0+\beta' z_0'=\mu_0 z_0';$$

thus we deduce, as in the case just taken, that

$$(\lambda\mu_0-r)(xz_0'-x'z_0)=0;$$

and hence as $\lambda\mu_0-r$, $=r(\lambda/\mu-1)$, is not zero, we have

$$xz_0'-x'z_0=0.$$

If we put $x=x_1+ix_2$, $x_0=x_1-ix_2$, $x'=x_1'+ix_2'$, $x_0'=x_1'-ix_2'$, the quantity

$$xx_0'-x'x_0=-2i(x_1x_2'-x_1'x_2)$$

is seen to be a pure imaginary; if in equations (i) λ be replaced by λ_0, the sign of $xx_0'-x'x_0$ is changed, but the quantity is otherwise unaltered; since then the equations (i) determine only the ratios of the constituents of the rows x, x', *we may suppose the sign of the imaginary part of* λ *in equations* (i), *and the resulting values of the constituents of* x *and* x', *to be so taken that*

$$xx_0'-x'x_0=-2i;$$

this we shall suppose to be done; and we shall suppose that the conditions for the $(p-1)$ similar equations, such as

$$zz_0'-z'z_0=-2i,$$

are also satisfied. With this convention, let the constituents of x and x' be denoted by

$$\xi_{1,1}, \ldots, \xi_{1,p}, \xi'_{1,1}, \ldots, \xi'_{1,p};$$

similarly let the constituents of the rows z, z', which are taken corresponding to the root μ, be denoted by

$$\xi_{2,1}, \ldots, \xi_{2,p}, \xi'_{2,1}, \ldots, \xi'_{2,p},$$

and so on for all the p roots λ, μ, ... Then the equations $xx_0'-x'x_0=-2i$, $zz_0'-z'z_0=-2i$, etc., are all expressed by the statement that the diagonal elements of the matrix

$$\xi\bar{\xi}_0'-\xi'\bar{\xi}_0$$

are each equal to $-2i$. When r is not equal to s $(r, s<p+1)$, the (1, 2)-th element of this matrix is

$$xz_0'-x'z_0,$$

which we have shewn to be zero; similarly every element of the matrix, other than a diagonal element, is zero; we may therefore write

$$\xi\bar{\xi}_0' - \xi'\bar{\xi}_0 = -2i.$$

Take now a row of p quantities, t, and define the rows X, X' by the equations

$$X = \bar{\xi}t, \quad X' = \bar{\xi}'t,$$

so that

$$X_0 = \bar{\xi}_0 t_0, \quad X_0' = \bar{\xi}_0' t_0;$$

then

$$-2it_0 t = \xi\bar{\xi}_0' t_0 t - \xi'\bar{\xi}_0 t_0 t = XX_0' - X'X_0;$$

hence it follows that the determinant of the matrix ξ' is not zero, since otherwise it would be possible to determine t, with constituents other than zero, so that $X' = 0$, and therefore also $X_0' = 0$; as this would involve $-2it_0 t = 0$, it is impossible.

382. If now the matrix τ be determined from the equations

$$x + \tau x' = 0, \quad z + \tau z' = 0, \ldots,$$

where x, x' are determined, as explained, from a proper value of λ, etc., or, what is the same thing, if τ be defined by

$$\xi + \xi'\bar{\tau} = 0,$$

then

$$\xi\bar{\xi}' - \xi'\bar{\xi} = \xi'\tau\bar{\xi}' - \xi'\bar{\tau}\bar{\xi}' = \xi'(\tau - \bar{\tau})\bar{\xi}';$$

but the equations of the form $xz' - x'z = 0$ are equivalent to

$$\xi\bar{\xi}' - \xi'\bar{\xi} = 0;$$

now, since the determinant $|\xi'|$ does not vanish, a row of quantities t can be determined so that $X' = \bar{\xi}'t$, for an arbitrary value of X'; thus for this arbitrary value we have

$$(\tau - \bar{\tau}) X'^2 = 0,$$

and therefore

$$\tau = \bar{\tau},$$

or the matrix τ is symmetrical.

Further, from the equation $\xi + \xi'\bar{\tau} = 0$, we have

$$\xi\bar{\xi}_0' - \xi'\bar{\xi}_0 = \xi'\tau_0\bar{\xi}_0' - \xi'\tau\bar{\xi}_0' = \xi'(\tau_0 - \tau)\bar{\xi}_0',$$

and hence, if $\tau = \rho + i\sigma$, since $\xi\bar{\xi}_0' - \xi'\bar{\xi}_0 = -2i$, we have

$$1 = \xi'\sigma\bar{\xi}_0', \text{ or } t_0 t = \sigma X_0' X',$$

where t is a row of any p quantities and $X' = \bar{\xi}'t$; hence, since the determinant $|\xi'|$ does not vanish, it follows, if X' be any row of p quantities, that $\sigma X_0' X'$ is positive; in particular when $n_1, \ldots, n_p$ are real, the imaginary part of the quadratic form τn^2 is positive.

Finally from the equations

$$\alpha x + \beta x' = \lambda x, \quad \alpha' x + \beta' x' = \lambda x',$$

putting $x = -\tau x'$, we infer

$$(\beta - \alpha\tau)\, x' = -\lambda\tau x', \quad (\beta' - \alpha'\tau)\, x' = \lambda x',$$

and therefore

$$\tau(\beta' - \alpha'\tau)\, x' + (\beta - \alpha\tau)\, x' = 0,$$

or

$$[\beta + \tau\beta' - (\alpha + \tau\alpha')\,\tau]\, x' = 0,$$

and hence

$$[\beta + \tau\beta' - (\alpha + \tau\alpha')\,\tau]\, \bar{\xi}' = 0,$$

from which, as $|\xi'|$ is not zero, we obtain

$$\beta + \tau\beta' - (\alpha + \tau\alpha')\,\tau = 0.$$

We have therefore completely proved the theorem stated.

It may be noticed, as follows from the equation $\xi + \xi'\tau = 0$, that we may form a theta function with associated constants given by

$$2\omega = 2\xi', \quad 2\omega' = -2\xi;$$

these will then satisfy the equations

$$\omega'\bar{\omega} - \omega\bar{\omega}' = 0, \quad \omega\bar{\omega}_0' - \omega'\bar{\omega}_0 = -2i;$$

the former equation always holds; the matrix ω can be determined so that the latter holds, as is easy to see.

Ex. Prove that by cogredient linear substitutions of the form

$$u' = cu, \quad w' = cw,$$

we can reduce the equations $u = Mw$ to the form

$$u_1' = \mu_1 w_1', \ \ldots, \ u_p' = \mu_p w_p',$$

where $\mu_1, \ldots, \mu_p$ are the roots of $|M - \lambda| = 0$.

383. For an example we may take the case $p = 1$; suppose that a, β, a', β' are such integers that $a\beta' - a'\beta = r$, a positive integer, and that the roots of the equation

$$(a + \tau a')\,\tau = \beta + \tau\beta'$$

are imaginary; if $a' = 0$, the condition that τ should not be a rational fraction requires that

$$\begin{pmatrix} a & \beta \\ a' & \beta' \end{pmatrix} = \begin{pmatrix} a & 0 \\ 0 & a \end{pmatrix},$$

where $a^2 = r$, and then the equation for τ is satisfied by all values of τ; this case is that of a multiplication by the rational number a, and we may omit it here; when a' is not zero we have

$$2a'\tau = -(a - \beta') \pm \sqrt{(a + \beta')^2 - 4r},$$

and therefore $(a + \beta')^2 < 4r$; this of itself is sufficient to ensure that the roots of the equation

$$\begin{pmatrix} a - \lambda & \beta \\ a' & \beta' - \lambda \end{pmatrix} = \lambda^2 - \lambda(a + \beta') + r = 0$$

are unequal, conjugate imaginaries, of modulus $\sqrt{r}$.

If then r be any given positive integer and h be a positive or negative integer numerically less than $2\sqrt{r}$, and a, a' be any integers such that $(a^2-ha+r)/a'$ is integral, $=-\beta$, we obtain a special transformation corresponding to the matrix

$$\Delta=\begin{pmatrix} a & \beta \\ a' & h-a \end{pmatrix}$$

for a value of τ given by

$$\tau=\frac{h-2a}{2a'}+i\frac{\sqrt{4r-h^2}}{2\,|\,a'\,|},$$

where $|\,a'\,|$ is the absolute value of a', and the square root is to be taken positively; the corresponding value of M is $a+\tau a'$. Hence by the results of Chap. XX., the function

$$\Theta\left[\tfrac{1}{2}(h\pm i\sqrt{4r-h^2})\,w\,;\ \frac{h-2a\pm i\sqrt{4r-h^2}}{2a'}\right],$$

when multiplied by a certain exponential of the form $e^{\lambda w^2}$, is expressible as an integral polynomial of order r in two functions $\Theta\left[w\,;\ \dfrac{h-2a\pm i\sqrt{4r-h^2}}{2a'}\right]$ with different characteristics.

The expression for the elliptic functions is obtainable independently as in the general case of transformation. When

$$M\upsilon=\omega a+\omega' a',\quad M\upsilon'=\omega\beta+\omega'\beta',\quad a\beta'-a'\beta=r,\quad u=Mw,$$

if to any two integers m, m' we make correspond two integers n, n' and two integers k, k', each positive (or zero) and less than r, by means of the equations

$$rn+k=m\beta'-m'\beta,\quad rn'+k'=-ma'+m'a,$$

or the equivalent equations

$$m=na+n'\beta+\frac{1}{r}(ak+\beta k'),\quad m'=na'+n'\beta'+\frac{1}{r}(a'k+\beta'k'),$$

then we immediately infer from the equation

$$\wp(u)=u^{-2}+\sum_{m}\sum_{m'}{}'\,[(u+2m\omega+2m'\omega')^{-2}-(2m\omega+2m'\omega')^{-2}],$$

by using n, n', instead of m, m', as summation letters, that

$$M^2\wp(Mw\,|\,2\omega,\,2\omega')=\wp(w\,|\,2\upsilon,\,2\upsilon')+\sum_{k}\sum_{k'}{}'\left[\wp\left(w+\frac{2\upsilon k+2\upsilon' k'}{r}\,\Big|\,2\upsilon,\,2\upsilon'\right)-\wp\left(\frac{2\upsilon k+2\upsilon' k'}{r}\right)\right],$$

wherein the summation refers to the $r-1$ sets k, k' other than $k=k'=0$, for which (§ 357, p. 589) the congruences

$$ak+\beta k'\equiv 0,\quad a'k+\beta' k'\equiv 0\pmod{r}$$

are satisfied*.

This formula is immediately applicable to the case when there is a complex multiplication; we may then put

$$2\omega=2\upsilon=1,\quad 2\omega'=2\upsilon'=\tau,\quad \beta'=h-a,\quad -\beta=(a^2-ha+r)/a',\quad \tau=(h-2a\pm i\sqrt{4r-h^2})/2a',$$

* When these congruences have a solution $(k_0,\,k_0')$, in which k_0, k_0' have no common factor, i.e. (Appendix II., § 418) when a, a', β, β' have no common factor, the remaining solutions are of the form $(\lambda k_0,\,\lambda k_0')$, where $\lambda<r$; in that case taking integers x, x' such that $k_0x'-k_0'x=1$, it is convenient to take $2\upsilon k_0+2\upsilon' k_0'$ and $2\upsilon x+2\upsilon' x'$ as the periods of the functions $\wp$ on the right side.

and $M=(h\pm i\sqrt{4r-h^2})/2$, as above, where $h^2<4r$. The application of the resulting equation is sufficiently exemplified by the case of $r=2$ given below (Exx. ii., iii.).

In the particular case where $r=1$, the condition $h^2<4r$ shews that h can have only the values 0 or $+1$ or -1; in this case the values n, n' given by

$$m=na-n'\frac{a^2-ha+r}{a'},\quad m'=na'+n'(h-a)$$

are integral when m and m' are integral; hence as $-\frac{a^2-ha+r}{a'}+(h-a)\tau=M\tau$, we immediately find

$$g_2=60\sum_m\sum_n{}'\frac{1}{(m+m'\tau)^4}=\left(\frac{2}{h+i\sqrt{4-h^2}}\right)^4 g_2\,;\ g_3=140\sum_m\sum_n{}'\frac{1}{(m+m'\tau)^6}=\left(\frac{2}{h+i\sqrt{4-h^2}}\right)^6 g_3.$$

Thus when $h=0$ we have $g_3=0$, and if a, a' be any integers such that $(a^2+1)/a'$ is integral, we have $\tau=(\pm i-a)/a'$, the upper or lower sign being taken according as a' is positive or negative. In this case the function $\wp(u)$ satisfies the equation

$$(\wp' u)^2=4(\wp u)^3-g_2\wp(u),$$

where

$$g_2=60\sum_{m\,n}\sum{}'\frac{1}{[m+n(\pm i-a)/a']^4}.$$

When $h=1$ we have $g_2=0$, and if a, a' be any integers such that $(a^2-a+1)/a'$ is integral, we have $\tau=(1-2a\pm i\sqrt{3})/a'$; in this case

$$(\wp' u)^2=4(\wp u)^3-g_3.$$

When $h=-1$, we have $g_2=0$, and, if $(a^2+a+1)/a'$ be integral, then $\tau=(-1-2a\pm i\sqrt{3})/a$.

Ex. i. Denoting the general function $\wp u$ by $\wp(u;\,g_2,\,g_3)$, it is easy to prove that the arc of the lemniscate $r^2=a^2\cos 2\theta$ is given by $a^2/r^2=\wp(s/a;\,4,\,0)$; when n is any prime number of the form $4k+1$ the problem of dividing the perimeter of the curve into n equal parts is reducible to the solution of an equation of order k—when n is a prime number of the form $2^\lambda+1$, the problem can be solved by the ruler and compass only. (Fagnano, *Produzioni Matematiche*, (1716), Vol. II.; Abel, *Œuvres*, 1881, t. I., p. 362, etc.) It is also easy to prove that the arc of the curve $r^3=a^3\cos 3\theta$ is given by $a^2/r^2=\wp(s/a;\,0,\,4)$; when n is a prime number of the form $6k+1$, the problem of dividing the perimeter of this curve into n equal parts is reducible to the solution of an equation of order k (Kiepert, *Crelle*, LXXIV. (1872), etc.). These facts are consequences of the linear special transformations of the theta functions connected with the curves.

Ex. ii. In case $r=2$, taking $a=4$, $a'=9$, $h=0$, we have $\tau=(-4+i\sqrt{2})/9$, and

$$-2\wp(i\sqrt{2}\,.\,w)=\wp(w)+\wp\left(w+\frac{\tau}{2}\right)-\wp(\tau/2).$$

By expanding this equation in powers of w, and equating the coefficients of w^2, we find easily that, if $\wp(\tau/2)=e$, then $g_2=\frac{15}{2}e^2$, and $g_3=-\frac{7}{2}e^3$; hence we infer that by means of the transformation

$$-2\xi=x+\frac{9}{8(x-1)}$$

we obtain

$$\int_\xi^\infty\frac{d\xi}{\sqrt{8\xi^3-15\xi+7}}=i\sqrt{2}\int_x^\infty\frac{dx}{\sqrt{8x^3-15x+7}},$$

which can be directly verified. *It is manifest that when $r=2$, $h=0$, we are led to this equation for all values of a and a'.*

Ex. iii. Prove that if $m=\frac{1}{2}(h+i\sqrt{8-h^2})$, the substitution

$$m^2\xi=x+\frac{3m^4-3}{m^4+4}\frac{1}{x-1}$$

gives the equation

$$\int_\xi^\infty \frac{d\xi}{\sqrt{(m^4+4)\xi^3-15\xi-(m^4-11)}}=m\int_x^\infty \frac{dx}{\sqrt{(m^4+4)x^3-15x-(m^4-11)}}.$$

This includes all such equations obtainable when $r=2$. Complex multiplication arises for the five cases $h=0$, $h=\pm 1$, $h=\pm 2$.

Ex. iv. When $r=3$ and $p=1$, we see by considering the matrix

$$\begin{pmatrix}\alpha & \beta\\ \alpha' & \beta'\end{pmatrix}=\begin{pmatrix}1 & -2\\ 1 & 1\end{pmatrix}=\begin{pmatrix}1 & 1\\ 0 & 1\end{pmatrix}\begin{pmatrix}0 & -1\\ 1 & 0\end{pmatrix}\begin{pmatrix}1 & 1\\ 0 & 3\end{pmatrix}$$

that the function $\Theta_{1,1}[(1+i\sqrt{2})w;\ i\sqrt{2}]$ is expressible as a cubic polynomial in the functions $\Theta_{0,1}(w;\ i\sqrt{2})$, $\Theta_{1,1}(w;\ i\sqrt{2})$. The actual form of this polynomial is calculable by the formulae of Chap. XXI. (§§ 366, 372), by applying in order the linear substitutions $\begin{pmatrix}1 & 1\\ 0 & 1\end{pmatrix}$, $\begin{pmatrix}0 & -1\\ 1 & 0\end{pmatrix}$ and then the cubic transformation $\begin{pmatrix}1 & 1\\ 0 & 3\end{pmatrix}$. Hence deduce that $k=\sqrt{2}-1$ and

$$\text{sn}[(1+i\sqrt{2})W]=(1+i\sqrt{2})\,\text{sn}\,W[1-\text{sn}^2 W/\text{sn}^2\gamma]/[1-k^2\,\text{sn}^2 W.\,\text{sn}^2\gamma],$$

where $\gamma=2(K-iK')/3$, K being (§ 365, Chap. XXI.) $=\pi\Theta_{00}^2$, and $iK'=\tau K$.

For the complex multiplication of elliptic functions the following may be consulted: Abel, *Œuvres*, t. I. (1881), p. 379; Jacobi, *Werke*, Bd. I., p. 491; Sohnke, *Crelle*, XVI. (1837), p. 97; Jordan, *Cours d'Analyse*, t. II. (1894), p. 531; Weber, *Elliptische Functionen* (1891), Dritter Theil; Smith, *Report on the Theory of Numbers*, British Assoc. Reports, 1865, Part VI.; Hermite, *Théorie des équations modulaires* (1859); Kronecker, *Berlin. Sitzungsber.* (1857, 1862, 1863, 1883, etc.), *Crelle*, LVII. (1860); Joubert, *Compt. Rendus*, t. L. (1860), p. 774; Pick, *Math. Annal.* XXV., XXVI.; Kiepert, *Math. Annal.* XXVI. (1886), XXXII. (1888), XXXVII., XXXIX.; Greenhill, *Proc. Camb. Phil. Soc.* IV., V. (1882—3), *Quart. Journal*, XXII. (1887), *Proc. Lond. Math. Soc.* XIX. (1888), XXI. (1890); Halphen, *Liouville*, (1888); Weber, *Acta Math.* XI. (1887), *Math. Annal.* XXIII., XXXIII. (1889), XLIII. (1893); Etc.

384. We come now to a different theory*, leading however in one phase of it, to the fundamental equations which arise for the transformation of theta functions, that namely of the correspondence of places on a Riemann surface. The theory has a geometrical origin; we shall therefore speak either of a Riemann surface, or of the plane curve which may be supposed to be represented by the equation associated with the Riemann surface, according to convenience. The nature of the points under consideration may be illustrated by a simple example. If at a point x of a curve the tangent be drawn, intersecting the curve again in $z_1, z_2, \ldots, z_{n-2}$, we may say that to the point x, regarded as a variable point, there correspond the $n-2$ points

* For references to the literature of the geometrical theory, see below, § 387, Ex. iv., p. 647. The theory is considered from the point of view of the theory of functions by Hurwitz, *Math. Annal.* XXVIII. (1887), p. 561; *Math. Annal.* XXXII. (1888), p. 290; *Math. Annal.* XLI. (1893), p. 403. See also, Klein-Fricke, *Modulfunctionen*, Bd. II. (Leipzig, 1892), p. 518, and Klein, *Ueber Riemann's Theorie* (Leipzig, 1882), p. 67. For (1, 1) correspondence in particular see the references given in § 393, p. 654.

$z_1, \dots, z_{n-2}$. To any point z of the curve, regarded as arising as one of a set $z_1, \dots, z_{n-2}$, there will reciprocally correspond all the points, $x_1, x_2, \dots, x_{m-2}$, which are points of contact of tangents drawn to the curve from z. Such a correspondence is described as an $(n-2, m-2)$ correspondence. A point of the curve for which x coincides with one of the points $z_1, \dots, z_{n-2}$ corresponding to it, is called a coincidence; such points are for instance the inflexions of the curve.

In general an (r, s) correspondence on a Riemann surface involves that any place x determines uniquely r places $z_1, \dots, z_r$, while any place z, regarded as arising as one of a set $z_1, \dots, z_r$, determines uniquely s places $x_1, \dots, x_s$. The investigation of the possible methods of this determination is part of the problem.

385. Suppose such an (r, s) correspondence to exist; let the positions of z that correspond to any variable position of x be denoted by $z_1, \dots, z_r$, and the positions of x that correspond to any variable position of z be denoted by $x_1, \dots, x_s$; and denote by $c_1, \dots, c_r$ the positions of $z_1, \dots, z_r$ when x is at the particular place a, and by $a_1, \dots, a_s$ the positions of $x_1, \dots, x_s$ when z is at the particular place c; denoting linearly independent Riemann normal integrals of the first kind by $v_1, \dots, v_p$, consider the sum

$$v_i^{z_1, c_1} + \dots\dots + v_i^{z_r, c_r}$$

as a function of x; since it is necessarily finite we clearly have equations of the form

$$M_{i,1} v_1^{x, a} + \dots\dots + M_{i,p} v_p^{x, a} \equiv v_i^{z_1, c_1} + \dots\dots + v_i^{z_r, c_r}, \qquad (i = 1, \dots, p),$$

where $M_{i,1}, \dots, M_{i,p}$ are constants. On the dissected surface the omitted aggregate of periods of the integral v_i indicated by the sign $\equiv$ is self-determinative; if the paths of integration be not restricted from crossing the period loops the sign $\equiv$ can be replaced by the sign of equality (cf. Chap. VIII. §§ 153, 158).

If now x describe the kth period loop of the second kind, from the right to the left side of the kth period loop of the first kind, the places $z_1, \dots, z_r$ will describe corresponding curves and eventually resume, in some order, the places they originally occupied; since, on the dissected Riemann surface $v_i^{z_1, c_1} + v_i^{z_2, c_2} = v_i^{z_2, c_1} + v_i^{z_1, c_2}$, we may suppose each of them actually to resume its original position; hence we have an equation

$$M_{i,k} = \alpha_{i,k} + \tau_{i,1} \alpha'_{1,k} + \dots\dots + \tau_{i,p} \alpha'_{p,k},$$

wherein $\alpha_{i,k}, \alpha'_{i,k}, \dots$ are integers; similarly by taking x round the kth period loop of the first kind we obtain

$$M_{i,1} \tau_{1,k} + \dots\dots + M_{i,p} \tau_{p,k} = \beta_{i,k} + \tau_{i,1} \beta'_{1,k} + \dots\dots + \tau_{i,p} \beta'_{p,k};$$

we have therefore $2p^2$ equations expressible in the form

$$M = \alpha + \tau\alpha', \qquad M\tau = \beta + \tau\beta',$$

wherein α, α', β, β' are matrices of integers, of p rows and columns.

Consider next, as a function of x, the integral

$$\int v_m^{x, a}\, d\left[\Pi_{z_1, c_1}^{z, c} + \ldots\ldots + \Pi_{z_r, c_r}^{z, c}\right],$$

wherein z, c are, primarily, arbitrary positions, independent of x, and $\Pi_{z_1, c_1}^{z, c}$ is the Riemann normal integral of the third kind. The subject of integration becomes infinite when any one of the places $z_1, \ldots, z_r$ coincides with z, or, in other words, when z is among the places corresponding to x, and this happens when x is at any one of the places $x_1, \ldots, x_s$, which correspond to z; the subject of integration similarly becomes infinite when x is at any one of the places $a_1, \ldots, a_s$, which correspond to the particular position of z denoted by c; it is assumed that c does not coincide with any one of the places $c_1, \ldots, c_r$. The sum of the values obtained when the integral is taken, in regard to x round the infinities $x_1, \ldots, x_s, a_1, \ldots, a_s$, is, save for an additive aggregate* of periods of the integral v_m, equal to

$$2\pi i\,(v_m^{x_1, a_1} + \ldots\ldots + v_m^{x_s, a_s}).$$

This quantity is then equal to the value obtained when x is taken round the period loops on the Riemann surface. Consider first, for the sake of clearness, the contribution arising as x describes the kth period loop of the second kind; if x described the left side of this period loop in the negative direction, from the right to the left side of the kth period loop of the first kind, the aggregates of the paths described by $z_1, \ldots, z_r$ would, in the notation just previously adopted, be equivalent to $\alpha_{\lambda, k}$ negative circuits of the λth period loop of the second kind, and $\alpha'_{\lambda, k}$ positive circuits of the λth period loop of the first kind ($\lambda = 1, \ldots, p$). In the actual contour integration under consideration the description by x of the left side of the kth period loop of the second kind is to be in the positive direction; hence the contribution arising for the integral as x describes both sides of the kth period loop of the second kind is

$$-2\pi i \tau_{m, k} \sum_{\lambda=1}^{p} \alpha'_{\lambda, k}\, v_\lambda^{z, c};$$

similarly the contribution as x describes the sides of the kth period loop of the first kind is

$$2\pi i E_{m, k} \sum_{\lambda=1}^{p} \beta'_{\lambda, k}\, v_\lambda^{z, c},$$

* Which vanishes when paths can be drawn on the dissected surface connecting $a_1, \ldots, a_s$ respectively to $x_1, \ldots, x_s$, so that simultaneous positions on these paths are simultaneous positions of $x_1, \ldots, x_s$. Cf. Chap. VIII. § 153; Chap. IX. § 165.

where $E_{m,k}=0$ unless $m=k$, and $E_{m,m}=1$. Taking therefore all the period loops into consideration, that is, $k=1, \ldots, p$, we obtain

$$v_m^{x_1, a_1} + \ldots\ldots + v_m^{x_s, a_s} \equiv \sum_{\lambda=1}^{p} \beta'_{\lambda, m}\, v_\lambda^{z, c} - \sum_{k=1}^{p}\sum_{\lambda=1}^{p} \tau_{m,k}\, \alpha'_{\lambda,k}\, v_\lambda^{z, c} \equiv \sum_{\lambda=1}^{p} N_{m,\lambda}\, v_\lambda^{z, c},$$

where
$$N_{m,\lambda} = \beta'_{\lambda, m} - \sum_{k=1}^{p} \tau_{m,k}\, \alpha'_{\lambda,k},$$

so that $N_{m,\lambda}$ is the (m, λ)th element of the matrix

$$N = \bar{\beta}' - \tau\bar{\alpha}';$$

since the equations $M = \alpha + \tau\alpha'$, $M\tau = \beta + \tau\beta'$ give

$$-\bar{\beta} + \tau\bar{\alpha} = (\bar{\beta}' - \tau\bar{\alpha}')\,\tau,$$

we have also

$$N\tau = -\bar{\beta} + \tau\bar{\alpha}.$$

These equations express the sum $v_m^{x_1, a_1} + \ldots + v_m^{x_s, a_s}$ in terms of integrals $v_\lambda^{z, c}$ in a manner analogous to the expression originally taken for $v_i^{z_1, c_1} + \ldots + v_i^{z_r, c_r}$ in terms of integrals $v_\lambda^{x, a}$, the difference being the substitution, for the matrix $\begin{pmatrix}\alpha & \beta\\ \alpha' & \beta'\end{pmatrix}$, of the matrix $\begin{pmatrix}\bar{\beta}' & -\bar{\beta}\\ -\bar{\alpha}' & \bar{\alpha}\end{pmatrix}$.

386. The theory of correspondence of points of a Riemann surface now divides into two parts according as the equation, which arises by elimination, either of the matrix M or the matrix N, namely,

$$\tau\alpha'\tau + \alpha\tau - \tau\beta' - \beta = 0,$$

is true independently of the matrix τ, in virtue of special values for the matrices α, β, α', β', or, on the other hand, is true for more general values of these matrices, in virtue of a special value for the matrix τ.

We take the first possibility first; it is manifest that for any value of τ the equation is satisfied if

$$\alpha = -\gamma E,\ \beta = 0,\ \alpha' = 0,\ \beta' = -\gamma E,$$

where γ is any single integer, and E is the matrix unity of p rows and columns; conversely, if the equations are to hold independently of the value of τ, we must have the n^2 equations

$$\sum_{i,j}^{1\ldots p} \alpha'_{i,j}\tau_{m,i}\tau_{\lambda,j} = 0, \quad \sum_{i=1}^{p} \alpha_{m,i}\tau_{i,\lambda} = \sum_{j=1}^{p} \tau_{m,j}\beta'_{j,\lambda}, \quad \beta_{m,\lambda} = 0, \quad (m, \lambda = 1, \ldots, p),$$

and, for general values of τ, these give

$$\alpha'_{i,j} = 0, \quad \alpha_{m,m} = \beta'_{\lambda,\lambda}, \quad \alpha_{m,m'} = \beta'_{\lambda,\lambda'} = 0, \quad \beta_{m,\lambda} = 0, \quad (m \neq m',\ \lambda \neq \lambda'),$$

which are equivalent to the results taken above.

With these values we have, as the particular forms of the general equations of § 385,

$$v_i^{z_1, c_1} + \ldots\ldots + v_i^{z_r, c_r} + \gamma v_i^{x, a} \equiv 0,$$

$$v_m^{x_1, a_1} + \ldots\ldots + v_m^{x_s, a_s} + \gamma v_m^{z, c} \equiv 0. \qquad (i, m = 1, \ldots, p).$$

Let the value on the dissected surface of the left side of the first of these equivalences be

$$g_i + g_1' \tau_{i,1} + \ldots\ldots + g'_p \tau_{i,p},$$

where $g_1, \ldots, g_p, g_1', \ldots, g'_p$ are integers. Consider now the function

$$\phi(x, z; a, c) = e^{\Pi_{z_1, c_1}^{z, c} + \ldots\ldots + \Pi_{z_r, c_r}^{z, c} + \gamma \Pi_{x, a}^{z, c} - 2\pi i (g_1' v_1^{z, c} + \ldots\ldots + g'_p v_p^{z, c})},$$

wherein $z_1, \ldots, z_r$ are the places corresponding to x, and $c_1, \ldots, c_r$ their positions when x is at a, and z, c are arbitrary places. In virtue of the equations just obtained it is a rational function of z, and rational in the place c (cf. Chap. VIII., § 158). Regarded as a function of x it is also rational; for the quotient of its values at the two sides of a period loop of the second kind, which, by what has just been shewn, must be rational in z, is, by the properties of the integral of the third kind, necessarily of the form

$$e^{2\pi i (K_1 v_1^{z, c} + \ldots\ldots + K_p v_p^{z, c})},$$

where $K_1, \ldots, K_p$ are integers; this quotient, as a function of z, has no infinities; being a rational function of z, it is therefore a constant, and therefore unity, since it reduces to unity when z is at c; hence $\phi(x, z; a, c)$, as a function of x, has no factors at the period loops; as it can have no infinities but poles it is therefore a rational function of x; it is similarly rational in a. As a function of x it vanishes when one of $z_1, \ldots, z_r$ coincides with z, that is, when x coincides with one of $x_1, \ldots, x_s$.

We have therefore the result. *Associated with any (r, s) correspondence which can exist upon a perfectly general Riemann surface, it is possible to construct a function $\phi(x, z; a, c)$, rational in the variable places x, z and the fixed places a, c, which, regarded as a function of x vanishes to the first order at the places $x_1, \ldots, x_s$, which correspond to z, and vanishes to order γ (if γ be positive), at the place z; which, as a function of x, is infinite to the first order when x coincides with any one of the places $a_1, \ldots, a_s$ which correspond to c, and is infinite to order γ (γ being positive) when x is at c; which, as a function of z, has similarly (for γ positive) the zeros $z_1, \ldots, z_r$, x^γ and the poles $c_1, \ldots, c_r$, a^γ.* An analogous statement can be made when γ is negative.

Ex. i. Some examples may be given to illustrate the form of this rational function. On a plane cubic curve we do in fact obtain a (1, 4) correspondence, for which $\gamma = 2$, by taking for the point z_1 which corresponds to x, the point in which the tangent at

x meets the curve again, and therefore, for the points x_1, x_2, x_3, x_4 which correspond to z, the points of contact of tangents to the curve drawn from z. The value $\gamma=2$ is obtained from Abel's theorem, which clearly gives the equation

$$v^{z_1, c_1}+2v^{x, a}\equiv 0$$

as representative of the fact that a straight line meets the curve twice at x and once at z_1. Denote the equation of the curve in the ordinary symbolical way by $A_x^3=0$; then the equation $A_x^2A_z=0$, for a fixed position of x, represents the tangent at x; and for a fixed position of z, represents the polar conic of the point z, which vanishes once in the points of contact, x_1, x_2, x_3, x_4, of tangents drawn from z and vanishes also twice at z, where it touches the curve; then consider the function

$$\frac{A_x^2A_z}{A_x^2A_c \cdot A_a^2A_z};$$

when z, a, c are fixed, this function of x vanishes to the first order at x_1, x_2, x_3, x_4 and to the second order at z, and is infinite to the first order at the places a_1, a_2, a_3, a_4 which correspond to c, and infinite to the second order at c; when z, a, c are fixed, this function of z vanishes to the first order at z_1, and to the second order at x, and is infinite to the first order at the place c_1, which corresponds to a, and infinite to the second order at a.

Ex. ii. More generally for any plane curve of order n, and deficiency p, if to a point x we make correspond the $r=n-2$ points $z_1, \dots, z_{n-2}$, in which the tangent at x meets the curve again, and to a point z the $s=2n+2p-4$ points of contact $x_1, \dots, x_s$ of tangents drawn to the curve from z (so that, for instance, when the curve has κ cusps, κ of the points $x_1, \dots, x_s$ will be the same for all positions of z), we shall have an (r, s) correspondence for which $\gamma=2$. If $A_x^n=0$ be the equation of the curve, the function

$$\frac{A_x^{n-1}A_z}{A_x^{n-1}A_c \cdot A_a^{n-1}A_z},$$

regarded as a function of x, for fixed positions of z, a, c (of which a and c are not to be multiple points), has for zeros the places $x_1, \dots, x_s, z^2$, for poles the places $a_1, \dots, a_s, c^2$, and regarded as a function of z, has for zeros the places $z_1, \dots, z_r, x^2$, and for poles the places $c_1, \dots, c_r, a^2$.

Ex. iii. If from a point x a tangent be drawn to a plane curve, and the corresponding points be the points other than the point of contact, in which the tangent meets the curve again, we have

$$v^{z_1, c_1}+\dots\dots+v^{z_{n-3}, c_{n-3}}+2v^{z', c'}+v^{x, a}\equiv 0,$$

where z' is the point of contact of one of the tangents drawn from x, there being as many such equations as tangents to the curve from x; since the $2n+2p-4$ points z' lie on the first polar of x, it follows by Abel's theorem that

$$\Sigma v^{z', c'}+2v^{x, a}\equiv 0;$$

therefore

$$v^{z_1, c_1}+\dots\dots+v^{z_r, c_r}+(2n+2p-8)\, v^{x, a}\equiv 0,$$

so that $\gamma=2n+2p-8$. As a function of z the function $\phi(x, z;\, a, c)$ has therefore the $(n-3)(2n+2p-4)$ zeros $z_1, \dots, z_r$, which correspond to x, as well as the zero x, of the $(2n+2p-8)$th order, and has as poles the places $c_1, \dots, c_r$, which correspond to a, as well as the zero a, of the $(2n+2p-8)$th order.

For instance for a plane quartic, there are 10 places corresponding to x, one for each of the tangents that can be drawn from x to the curve; the function $\phi(x, z;\, a, c)$, as a function of z, vanishes to the first order at each of these ten places, and vanishes to the sixth order at x; its infinities are the places similarly derived from the fixed position, a, of x. We can build up this function in the manner suggested by the use already made of Abel's theorem in the determination of the value of γ; for a fixed position of x, let $T(z)=0$ be the equation, in the variable z, for the ten tangents to the quartic drawn from z; let $P(z)=0$ be the first polar of x; the quotient

$$T(z)/P^2(z)$$

vanishes when z is at the places $z_1, \ldots, z_{10}$, and vanishes when z is at x to order $10-2(2)=6$; let $T_a(z)$, $P_a(z)$ represent what $T(z)$, $P(z)$ become when x is at a; then the function of z

$$\frac{T(z)}{P^2(z)} \bigg/ \frac{T_a(z)}{P_a^2(z)}$$

has the same behaviour as has the function $\phi(x, z; a, c)$ as a function of z. From this function, by multiplication by a factor involving x but independent of z, we can form a symmetrical expression in x and z; this will be the function $\phi(x, z; a, c)$. In fact, denoting the equation of the quartic curve by $A_x^4=0$, and expressing the fact that the line joining the point x of the curve to the point ζ not on the curve should touch the curve, *viz.*, by equating to zero the discriminant in λ of $(A_x+\lambda A_\zeta)^4-A_x^4$, we obtain an equation of the form

$$A_\zeta^4[\zeta^6, x^6]=(A_xA_\zeta^3)^2[9(A_x^2A_\zeta^2)^2-16A_xA_\zeta^3 \,.\, A_x^3A_\zeta],$$

which represents the tangents to the curve drawn from x. Replacing ζ by z, a point on the curve, so that $A_z^4=0$, we have, since $A_xA_\zeta^3=0$ is the first polar of x,

$$T(z)/P^2(z)=9(A_x^2A_z^2)^2-16A_xA_z^3 \,.\, A_x^3A_z;$$

hence

$$\phi(x, z; a, c)=\frac{9(A_x^2A_z^2)^2-16A_xA_z^3 \,.\, A_x^3A_z}{[9(A_a^2A_z^2)^2-16A_aA_z^3 \,.\, A_a^3A_z][9(A_x^2A_c^2)^2-16A_xA_c^3 \,.\, A_x^3A_c]}.$$

Ex. iv. If a (1, 1) correspondence exists, the rational function of x, denoted by $\phi(x, z; a, c)$, is of order $\gamma+1$.

387. A problem of great geometrical interest is to determine the number of positions of x, in which x coincides with one of the places $z_1, \ldots, z_r$, which correspond to it. This is called the number of *coincidences*.

A simple way to determine this number is to consider the rational function of x obtained as the limit when $z=x$, of the ratio $\phi(x, z; a, c)/(x-z)^2$; putting

$$\phi(x; a, c)=\lim_{z=x}[\phi(x, z; a, c)/(x-z)^2],$$

and bearing in mind that if t be the infinitesimal on the Riemann surface, dx/dt vanishes to the first order at every finite branch place, and is infinite to the second order at every infinite place of the surface, we immediately find from the properties of the function $\phi(x, z; a, c)$, on the hypothesis that none of the branch places of the surface are at infinity, the following result; the rational function of x denoted by $\phi(x; a, c)$ vanishes to the first order at every place x of the surface at which x coincides with one of the places

$z_1, \ldots, z_r$ which correspond to it, vanishes also to order 2γ at each of the n infinite places of the surface, and is infinite to order γ at each of the branch places of the surface and at each of the places a, c, while it is infinite to the first order at each of the places $c_1, \ldots, c_r$ which correspond to a, and at each of the places $a_1, \ldots, a_s$ which correspond to c; hence, denoting the number of coincidences by C we have

$$C + 2n\gamma = (2n + 2p - 2)\gamma + 2\gamma + r + s,$$

so that*

$$C = r + s + 2p\gamma.$$

The same result is obtained when there are branch places at infinity. The argument has assumed γ to be positive; a similar argument, when γ is negative, leads to the same result.

Ex. i. The number, i, of inflexions of a plane curve of order n and deficiency p is given (Ex. ii. § 386) by

$$i + h = n - 2 + 2n + 2p - 4 + 4p = 3(n + 2p - 2),$$

where h is the number of coincidences arising other than inflexions, as for instance at the multiple points of the curve. In determining h it must be remembered that we have not excluded the possibility of there being fixed positions of x which correspond to z for all positions of z; for instance in the case of a curve with cusps all these cusps have been reckoned among the places $x_1, \ldots, x_s$ which correspond to z. Therefore for a curve with κ cusps, h will contain a term 2κ; for a curve with only δ double points and κ cusps, the formula is the well-known one

$$i - \kappa = 3(m - n),$$

where m is the class of the curve, equal to $n(n-1) - 2\delta - 3\kappa$.

Ex. ii. Obtain the expression of the function $\phi(x;\ a, c)$ determined by the limit

$$\{A_x^{n-1} A_z / (x-z)^2 \,.\, A_x^{n-1} \dot{A}_c \,.\, A_a^{n-1} \dot{A}_z\}_{z=x},$$

where $A_x^n = 0 = A_z^n = A_a^n = A_c^n$. (Cf. Ex. ii. § 386.)

Ex. iii. The number of double tangents of a curve of order n and deficiency p may be obtained from Ex. iii. § 386, if we notice that a double tangent, touching at P and Q, will arise both when P is a coincidence, and when Q is a coincidence; hence if τ be the number of double tangents, and h the number of coincidences not giving rise to double tangents, we have

$$2\tau + h = 2(n-3)(2n + 2p - 4) + 2p(2n + 2p - 8) = 4\sigma(\sigma + 1) - 8p,$$

where $\sigma = n + p - 3$. For instance for a curve with no singular points other than δ double points and κ cusps, there will be a contribution to h equal to twice the number of those improper double tangents which are constituted by the tangents to the curve from the cusps and the lines joining the cusps in pairs. The number of tangents, t, from a cusp is given (cf. § 9, Chap. I., Ex.) by

$$t + \kappa - 1 = 2(n-2) + 2p - 2, \text{ or } t = 2n - 5 - \kappa + 2p = n^2 - n - 3 - 2\delta - 3\kappa.$$

There will not arise any such contribution corresponding to a double point, since the two

* This result was first given by Cayley; see, for references, Ex. iv. below.

points of the curve that there correspond are different places (cf. § 2, Chap. I.); hence we have

$$h = 2\kappa t + \kappa^2 - \kappa\,;$$

and therefore
$$\tau = 2\sigma(\sigma+1) - 4p - \kappa t - \tfrac{1}{2}(\kappa^2 - \kappa)\,;$$

substituting the values for σ, p and t, we find the ordinary formula equivalent to

$$\tau = \delta + \tfrac{1}{2}(m-n)(m+n-9),$$

where m is the class of the curve.

Ex. iv. The points of contact of the double tangents of a quartic curve $A_x^4 = 0$ lie upon a curve whose equation is obtainable by determining the limit, when $z = x$, of the expression

$$[9(A_x^2 A_z^2)^2 - 16 A_x A_z^3 \,.\, A_x^3 A_z]/(x-z)^2.$$

For the result, cf. Dersch, *Math. Annal.* VII. (1874), p. 497.

For the general geometrical theory the reader will consult geometrical treatises; the following references may be given here; Clebsch-Lindemann-Benoist, *Leçons sur la Géométrie* (Paris, 1879—1883), t. I. p. 261, t. II. p. 146, t. III. p. 76; Chasles, *Compt. Rendus*, t. LVIII. (1864); Chasles, *Compt. Rendus*, t. LXII. (1866), p. 584; Cayley, *Compt. Rendus*, t. LXII. (1866), p. 586, and *London Math. Soc. Proc.* t. I. (1865—6), and *Phil. Trans.* CLVIII. (1868) (or *Coll. Works*, V. 542; VI. 9; VI. 263); Brill, *Math. Annal.* t. VI. (1873), and t. VII. (1874). See also Lindemann, *Crelle*, LXXXIV. (1878); Bobek, *Sitzber. d. Wiener Akad.*, XCIII. (ii. Abth.), (1886), p. 899; Brill, *Math. Annal.* XXXI. (1887), XXXVI. (1890); Castelnuovo, *Rend. Acc. d. Lincei*, 1889; Zeuthen, *Math. Annal.* XL. (1892), and the references there given.

Ex. v. If we use the equation (Chap. X. § 187)

$$e^{\Pi_{x,a}^{z,c}} = \frac{\Theta(v^{x,z} + \tfrac{1}{2}\Omega)\,\Theta(v^{a,c} + \tfrac{1}{2}\Omega)}{\Theta(v^{x,c} + \tfrac{1}{2}\Omega)\,\Theta(v^{a,z} + \tfrac{1}{2}\Omega)},$$

where Ω is an odd half-period, equal to $\lambda + \tau\lambda'$ say, λ, λ' being each rows of p integers, and form the rational function of x and a,

$$R(x, a) = lim_{\substack{z=x \\ c=a}} (-1)^\gamma \frac{\Theta^\gamma(v^{x,z} + \tfrac{1}{2}\Omega)\,\Theta^\gamma(v^{a,c} + \tfrac{1}{2}\Omega)}{\phi(x, z\,;\, a, c)}$$

$$= (-1)^\gamma \frac{[\sum_m \Theta'_m(\tfrac{1}{2}\Omega)\,.\,Dv_m^{x,}]^\gamma\,[\sum_m \Theta'_m(\tfrac{1}{2}\Omega)\,.\,Dv_m^{a,}]^\gamma}{[\phi(x, z\,;\, a, c)/(x-z)^\gamma (a-c)^\gamma]_{x=z,\, a=c}},$$

we have

$$\Theta(v^{x,a} + \tfrac{1}{2}\Omega)\,.\,e^{\pi i\lambda' v^{x,a}} = [R(x, a)]^{\frac{1}{2\gamma}}\, e^{\frac{1}{2\gamma}(\Pi_{z_1, c_1}^{x,a} + \dots\dots + \Pi_{z_r, c_r}^{x,a} - 2\pi i g' v^{x,a})},$$

which is a generalisation of the equation (i), p. 427.

The function $R(x, a)$ vanishes when x is at any one of the places $c_1, \dots, c_r$, which correspond to a, and when x is at any one of the places $a_1, \dots, a_s$ which correspond to the position a of the place c; it vanishes also 2γ times at each of the zeros of the function $\Theta(v^{x,a} + \tfrac{1}{2}\Omega)$. It is infinite C times, namely when x has any of the positions in which it coincides with one of the places $z_1, \dots, z_r$ which correspond to it. In the particular case of Ex. i. p. 427, the function $R(x, a)$ is $(x-a)^2 X(x)$, and the equation $C = r + s + 2p\gamma$ expresses that the number of branch places (where two places for which x is the same coincide) is $2(n-1) + 2p$.

Ex. vi. Determine the periods of the function of x expressed by

$$\Pi_{z_1, c_1}^{x,a} + \dots\dots + \Pi_{z_r, c_r}^{x,a},$$

where $z_1, \ldots, z_r$ are the places corresponding to x, and $c_1, \ldots, c_r$ are the places corresponding to a.

Ex. vii. If there be upon the same Riemann surface two correspondences, an (r, s) correspondence and an (r', s') correspondence, then to any place z will correspond, in virtue of the first correspondence, the places $x_1, \ldots, x_s$, and to any one of these latter, say x_i, will correspond, in virtue of the second correspondence, say $z'_{i,1}, \ldots, z'_{i,r'}$; conversely to any place z' will correspond, in virtue of the second correspondence, the places $x_1, \ldots, x_{s'}$, and to any one of these latter, say x_i, will correspond, in virtue of the first correspondence, say $z_{i,1}, \ldots, z_{i,r}$; we have therefore an $(r's, rs')$ correspondence of the points (z, z'). In virtue of the equations

$$v^{x_1, a_1'} + \ldots\ldots + v^{x_{s'}, a'_{s'}} + \gamma' v^{z', c'} \equiv 0,$$

$$v^{z_{i,1}, c_{i,1}} + \ldots\ldots + v^{z_{i,r}, c_{i,r}} + \gamma v^{x_i, a'_i} \equiv 0, \qquad (i = 1, \ldots, s'),$$

we have

$$\sum_{i=1}^{s'} \sum_{j=1}^{r} v^{z_{i,j}, c_{i,j}} - \gamma\gamma' v^{z', c'} \equiv 0.$$

Hence* we can make the inference. *If upon the same Riemann surface there be two correspondences, an (r, s) correspondence of places x, z, and an (r', s') correspondence of places x', z', then the number of common corresponding pairs of these two correspondences, for which both x, x' coincide, and also z and z', is*

$$r's + rs' - 2\gamma\gamma' p.$$

388. We have so far considered only those correspondences† which can exist on any Riemann surface. We give now some results‡ relating to correspondences which can only exist on Riemann surfaces of special character, more particularly (1, 1) correspondences.

We prove first that any $(1, s)$ correspondence is associated with equations which are identical in form with those which have arisen in considering the special transformation of theta functions. For any such correspondence, in which to any place x corresponds the single place z, and to any position of z the places $x_1, \ldots, x_s$, we have shewn that we have the equations $(i = 1, \ldots, p)$

$$v_i^{z, c} \equiv M_{i,1} v_1^{x, a} + \ldots\ldots + M_{i,p} v_p^{x, a}, \quad M = \alpha + \tau\alpha', \quad M\tau = \beta + \tau\beta',$$

$$v_i^{x_1, a_1} + \ldots\ldots + v_i^{x_s, a_s} \equiv N_{i,1} v^{z, c} + \ldots\ldots + N_{i,p} v_p^{z, c}, \quad N = \bar{\beta}' - \tau\bar{\alpha}', \quad N\tau = -\bar{\beta} + \tau\bar{\alpha};$$

hence

$$s v_i^{z, c} \equiv M_{i,1} \sum_{m=1}^{s} v_1^{x_s, a_s} + \ldots\ldots + M_{i,p} \sum_{m=1}^{s} v_p^{x_s, a_s}$$

$$\equiv \sum_{k=1}^{p} M_{i,k} (N_{k,1} v_1^{z, c} + \ldots\ldots + N_{k,p} v_p^{z, c})$$

$$\equiv L_{i,1} v_1^{z, c} + \ldots\ldots + L_{i,p} v_p^{z, c},$$

* Provided the $(r's, rs')$ correspondence is not an identity.

† Called by Hurwitz, Werthigkeit-correspondenzen, γ being the Werthigkeit.

‡ For other results, see Klein-Fricke, *Modulfunctionen*, Bd. II. (Leipzig, 1892), pp. 540 ff.

where $L_{i,m}$ is the (i, m)th element of the matrix $L, = MN$. This matrix is therefore equal to s. Now

$$MN = M(\ \bar{\beta}' - \tau\bar{\alpha}') = (\alpha + \tau\alpha')\bar{\beta}' - (\beta + \tau\beta')\bar{\alpha}' = \alpha\bar{\beta}' - \beta\bar{\alpha}' + \tau(\alpha'\bar{\beta}' - \beta'\bar{\alpha}'),$$

$$MN\tau = M(-\bar{\beta} + \tau\bar{\alpha}) = -(\alpha + \tau\alpha')\bar{\beta} + (\beta + \tau\beta')\bar{\alpha} = -(\alpha\bar{\beta} - \beta\bar{\alpha}) + \tau(\beta'\bar{\alpha} - \alpha'\bar{\beta}),$$

which we may write in the form

$$MN = H + \tau B, \quad MN\tau = -A + \tau\bar{H};$$

if now $\tau = \tau_1 + i\tau_2$, where τ_1, τ_2 are matrices of real quantities, it follows by equating to zero the imaginary part in the equation

$$MN - s = H - s + \tau B = 0$$

that $\tau_2 B = 0$; since for real values of $n_1, \ldots, n_p$ the quadratic form $\tau_2 n^2$ is necessarily positive, the determinant of the matrix τ_2 is not zero; hence we must have $B = 0$; hence also $H = s$ and $A = 0$; or

$$\alpha\bar{\beta} = \beta\bar{\alpha}, \quad \alpha'\bar{\beta}' = \beta'\bar{\alpha}', \quad \alpha\bar{\beta}' - \beta\bar{\alpha}' = \beta'\bar{\alpha} - \alpha'\bar{\beta} = s;$$

and these equations, with the equation $(\alpha + \tau\alpha')\tau = \beta + \tau\beta'$, are identical in form with those already discussed in this chapter (§§ 377, ff.).

We are able then as in the former case to deduce certain conditions for the matrices $\alpha, \beta, \alpha', \beta'$, which in their general form necessarily involve special values for the matrix τ.

389. In particular, in order that a (1, 1) correspondence* may exist, the roots of the equation $|M - \lambda| = 0$ must be conjugate imaginaries of the roots of the equation $|N - \lambda| = 0$, must be all of modulus unity, and must be roots of the equation $|\Delta - \lambda| = 0$, where $\Delta = \begin{pmatrix} \alpha & \beta \\ \alpha' & \beta' \end{pmatrix}$. They must therefore be roots of unity. For the sake of definiteness we shall suppose $p > 1$ and that Δ and τ are such that the roots of $|M - \lambda| = 0$ are all different; this excludes the case already considered when $\Delta = \begin{pmatrix} -\gamma & 0 \\ 0 & -\gamma \end{pmatrix}$. Supposing a (1, 1) correspondence to exist, for which this condition is satisfied, if in the fundamental equations $(i = 1, \ldots, p)$

$$v_i^{z, c} \equiv M_{i,1} v_1^{x, a} + \ldots\ldots + M_{i,p} v_p^{x, a},$$

we introduce other integrals of the first kind, say $V_1^{x, a}, \ldots, V_p^{x, a}$, where

$$V_i^{x, a} = c_{i,1} v_1^{x, a} + \ldots\ldots + c_{i,p} v_p^{x, a},$$

and therefore also

$$V_i^{z, c} = c_{i,1} v_1^{z, c} + \ldots\ldots + c_{i,p} v_p^{z, c},$$

* The (1, 1) correspondence for the case $p=1$ is considered in an elementary way in § 394. The reader may prefer to consult that Article before reading the general investigation.

then we can put the fundamental equations into the form

$$V_i^{z,c} = \lambda_i V_i^{x,a};$$

for this it is necessary that λ_i should be a root of the equation $|M-\lambda|=0$, and that the p quantities $c_{i,1}, \ldots, c_{i,p}$ should be determined from the equations

$$c_{i,1} M_{1,r} + \ldots\ldots + c_{i,p} M_{p,r} = \lambda_i c_{i,r}, \qquad (r = 1, \ldots, p);$$

under the prescribed conditions the determinant of the matrix c will be different from zero.

Hence as λ_i is a root of unity, it can be shewn, when $p > 1$, that *every such* (1, 1) *correspondence is periodic, with a finite period*; that is, if the place corresponding to x be z_1, the place corresponding to the position z_1, of x, be z_2, the place corresponding to the position z_2, of x, be z_3, and so on, then after a finite number of stages one of the places $z_1, z_2, z_3, \ldots$ coincides with x. In order to prove this, suppose that all the roots of the equation $|M-\lambda|=0$ are k-th roots of unity; then denoting the place x by z_0 and the place a by c_0, the equations of the correspondence may be written

$$dV_i^{z_1, c_1} = \lambda_i dV_i^{z_0, c_0}, \; dV_i^{z_2, c_2} = \lambda_i dV_i^{z_1, c_1}, \ldots\ldots, dV_i^{z_k, c_k} = \lambda_i dV_i^{z_{k-1} c_{k-1}};$$

these give

$$dV_i^{z_k, c_k} = \lambda_i^k dV_i^{z_0, c_0} = dV_i^{z_0, c_0},$$

and therefore

$$c_{i,1} [dv_1^{z_k, c_k} - dv_1^{z_0, c_0}] + \ldots\ldots + c_{i,p} [dv_p^{z_k, c_k} - dv_p^{z_0, c_0}] = 0;$$

hence on the dissected Riemann surface we have equations of the form

$$v_r^{z_k, c_k} - v_r^{z_0, c_0} = \lambda_r + \lambda_1' \tau_{r,1} + \ldots\ldots + \lambda_p' \tau_{r,p}, \qquad (r = 1, \ldots, p),$$

where $\lambda_1, \ldots, \lambda_p'$ are integers. Thus either $z_k = z_0$ and $c_k = c_0$, which is the result we wish to obtain, or else there is a rational function expressed by

$$e^{\Pi_{z_k, c_k}^{x, a} - \Pi_{z_0, c_0}^{x, a} - 2\pi i (\lambda_1' v_1^{x, a} + \ldots\ldots + \lambda_p' v_p^{x, a})},$$

which is of the second order, having z_k, c_0 as zeros and z_0, c_k as poles; now a surface on which there is a rational function of the second order is necessarily hyperelliptic (Chap. V. § 55)—but, on a hyperelliptic surface, for which $p > 1$, of the two poles of such a function either determines the other, and of the two zeros either determines the other; it is not possible to construct such a function whereof, as here, one pole c_k is fixed, and the other arbitrary and variable (§ 52).

Hence we must have $z_k = z_0$, and $c_k = c_0$, which proves the result enunciated.

There is no need to introduce the integrals V in order to establish this result. It is known (Cayley, *Coll. Works*, Vol. II. p. 486) that if $\lambda_1, \lambda_2, \ldots$ be the roots of the equation $|M-\lambda|=0$, the matrix M satisfies the equation $(M-\lambda_1)(M-\lambda_2)\ldots\ldots=0$; when the roots

$\lambda_1, \lambda_2, \ldots$ are different k-th roots of unity it can thence be inferred that the matrix M satisfies the equation $M^k=1$; then from the successive equations $dv^{z_1, c_1}=Mdv^{z_0, c_0}$, $dv^{z_2, c_2}=Mdv^{z_1, c_1}$, etc., we can infer $dv^{z_k, c_k}=dv^{z_0, c_0}$, and hence as before that $z_k=z_0$, $c_k=c_0$.

A proof of the periodicity of the (1, 1) correspondence, following different lines, and not assuming that the roots of the equation $|M-\lambda|=0$ are different, is given by Hurwitz, *Math. Annal.* XXXII. (1888), p. 295, for the cases when $p>1$. It will be seen below that the cases $p=0$, $p=1$ possess characteristics not arising for higher values of p (§ 394).

390. Assuming the periodicity of the (1, 1) correspondence, we can shew that all Riemann surfaces upon which a (1, 1) correspondence exists, can be associated with an algebraic equation of particular form. As before let k be the index of the periodicity, and let $\omega=e^{2\pi i/k}$; let S, T be any two rational functions on the surface, and let the values of S at the successive places $x, z_1, z_2, \ldots, z_{k-1}, x$ which arise by the correspondence be denoted by $S, S_1, \ldots, S_{k-1}, S$, and similarly for T; then the values of the functions

$$s = S + \omega^{-1}S_1 + \ldots\ldots + \omega^{-(k-1)}S_{k-1}$$
$$t = T + \quad T_1 + \ldots\ldots + \quad T_{k-1}$$

at the place z_r are respectively

$$s_r = S_r + \omega^{-1}S_{r+1} + \ldots\ldots + \omega^{-(k-1)}S_{r+k-1} = \omega^r s, \text{ and } t;$$

hence it can be inferred (cf. Chap. I., § 4) that there exists a rational relation connecting s^k and t. Conversely S and T can be chosen of such generality that any given values of s and t arise only at one place of the original Riemann surface. Thus the surface can be associated with an equation of the form

$$(s^k, t) = 0,$$

wherein every power of s which enters is a multiple of k.

Such a surface is clearly capable of the periodic (1, 1) transformation expressed by the equations

$$s' = \omega s, \; t' = t.$$

The following further remarkable results may be mentioned*:

(α) The index of periodicity k cannot be greater than $10(p-1)$.

(β) When $k > 2p-2$ the Riemann surface can be associated with an equation of the form

$$s^k = t^{k_1}(t-1)^{k_2}(t-c)^{k_3}.$$

(γ) When $k > 4p-4$, the Riemann surface can be associated with an equation of the form

$$s^k = t^{k_1}(t-1)^{k_2}.$$

Herein k_1, k_2, k_3 are positive integers less than k.

* Hurwitz, *Math. Annal.* XXXII. (1888), p. 294.

391. We can deduce from § 389 that in the case of a (1, 1) correspondence the number of coincidences is not greater than $2p+2$. In the case of a hyperelliptic surface, when the correspondence is that in which conjugate places—of the canonical surface of two sheets—are the corresponding pairs, the coincidences are clearly the branch places, and their number is $2p+2$; for all other (1, 1) correspondences on a hyperelliptic surface, the number of coincidences cannot be greater than 4.

For, when the surface is not hyperelliptic, let g denote a rational function which is infinite only at one place z_0 of the surface, to an order $p+1$; and let g' be the value of the same function at the place z_1, which corresponds to z_0; then the function $g'-g$ is of order $2p+2$, being infinite to order $p+1$ at z_0 and to order $p+1$ at the place z_{-1} to which z_0 corresponds; now every coincidence of the correspondence is clearly a zero of $g'-g$; thus the number of coincidences is not greater than $2p+2$. In the case of a hyperelliptic surface

$$y^2=(x,\ 1)_{2p+1},$$

we may similarly consider the function $x'-x$, of order 4;—unless the correspondence be that given by $y'=-y$, $x'=x$, for which $x'-x$ is identically zero. We thus obtain the result that the number of coincidences cannot be greater than 4, except for the (1, 1) correspondence $y'=-y$, $x'=x$.

It can be shewn for the most general possible (r, s) correspondence, associated with the equations

$$v_i^{z_1,\, c_1}+\dots\dots+v_i^{z_r,\, c_r}\equiv M_{i,1}v_1^{x,\, a}+\dots\dots+M_{i,p}v_p^{x,\, a},\quad M=a+\tau a',\quad M\tau=\beta+\tau\beta',$$

by equating the value obtained for the following integral, taken round the period loops,

$$\int d\,(\Pi_{z_1,\, c_1}^{x,\, c}+\dots\dots+\Pi_{z_r,\, c_r}^{x,\, c}),$$

to the value obtained for the integral taken round the infinities of the subject of integration, that the number of coincidences is

$$C=r+s-(a_{11}+\dots\dots+a_{pp}+\beta'_{11}+\dots\dots+\beta'_{pp}).$$

Since $a_{11}+\dots\dots+\beta'_{pp}$ is the sum of the roots of the equation $|\Delta-\lambda|=0$, it follows for a (1, 1) correspondence, in which all the $2p$ roots of $|\Delta-\lambda|=0$ are roots of unity, that $C \not> 2p+2$. For any (r, s) correspondence belonging to a matrix $\Delta=\begin{pmatrix}-\gamma & 0\\ 0 & -\gamma\end{pmatrix}$, the same formula gives $C=r+s+2p\gamma$, as already found.

We have remarked (§ 386, Ex. iv.) for the case of a (1, 1) correspondence associated with a matrix Δ of the form $\begin{pmatrix}-\gamma & 0\\ 0 & -\gamma\end{pmatrix}$, the existence of a rational function of order $1+\gamma$. For any such (1, 1) correspondence, if p be >1, γ must be equal to $+1$ in order that the number $1+1+2p\gamma$ of coincidences may be $\not> 2p+2$. Thus such a correspondence involves the existence of a rational function of order 2, and involves therefore that the surface be hyperelliptic. This is also obvious from the fact that such a correspondence is associated with equations of the form

$$v_i^{z,\, c}+\gamma v_i^{x,\, a}\equiv 0,\qquad (i=1,\ \dots,\ p);$$

conversely, for $\gamma=1$, equations of this form are known to hold for any hyperelliptic surface, associated with the correspondence of the conjugate places of the surface. From the considerations here given, it follows for $p>1$ that for a (1, 1) correspondence the number of coincidences can in no case be $>2p+2$.

392. In conclusion it is to be remarked that on any Riemann surface for which $p > 1$, there cannot be an infinite number of (1, 1) correspondences. For consider the places of the Riemann surface that can be the poles of rational functions of order $<(p+1)$ which have no other poles (§§ 28, 31, 34—36, Chap. III.). Denote these places momentarily as g-places. As such a (1, 1) correspondence is associated with a linear transformation of integrals of the first kind, which does not affect the zeros of the determinant Δ, of § 31, it follows that the place corresponding to a g-place must also be a g-place. Now, when the surface is not hyperelliptic, every g-place cannot be a coincidence of the correspondence; for we have shewn (Chap. III., § 36) that then the number of distinct g-places is greater than $2p+2$; and we have shewn in this chapter (§ 391) that the number of coincidences in a (1, 1) correspondence, when $p > 1$, can in no case be $> 2p+2$. Therefore, when the surface is not hyperelliptic, a (1, 1) correspondence must give rise to a permutation among the g-places; since the number of such permutations is finite, the number of (1, 1) correspondences must equally be finite. But the result is equally true for a hyperelliptic surface; for we have shewn (§ 391) that for such a surface the number of coincidences of a (1, 1) correspondence cannot be greater than 4, except in the case of a particular one such correspondence; since the number of distinct g-places is $2p+2$, every (1, 1) correspondence other than this particular one must give rise to a permutation of these g-places. As the number of such permutations is finite, the number of (1, 1) correspondences must equally be finite.

It is proved by Hurwitz* that the number of (1, 1) correspondences, when $p > 1$, cannot be greater than $84(p-1)$. In case $p=3$, a surface is known to exist having this number of (1, 1) correspondences†.

393. The preceding proof§ (§ 392) is retained on account of its ingenuity. It can however be replaced by a more elementary proof‡ by means of the remark that a (1, 1) correspondence upon a Riemann surface can be represented by a rational, reversible transformation of the equation of the surface into itself. Let the equation of the surface be $f(x, y)=0$; let (z, s) be the place corresponding to (x, y); then z, s are each rational functions of x and y such that $f(z, s)=0$; conversely x, y are each

* *Math. Annal.* XLI. (1893), p. 424.

† Klein, *Math. Annal.* XIV. (1879), p. 428; *Modulfunctionen*, t. I., 1890, p. 701.

§ Hurwitz, *Math. Annal.* XLI. (1893), p. 406.

‡ Weierstrass, *Math. Werke*, Bd. II. (Berlin, 1895), p. 241.

rational functions of z, s. To give a formal demonstration we may proceed as follows; supposing the number of sheets of the Riemann surface to be n, let $z_1, \ldots, z_n$ denote the places corresponding to the n places $x_1^{(0)}, \ldots, x_n^{(0)}$ for which $x=0$, and let $z_1', \ldots, z'_n$ denote the n places corresponding to the places $x_1^{(\infty)}, \ldots, x_n^{(\infty)}$ for which x is infinite; as x is a rational function on the surface we have, for suitable paths of integration (cf. Chap. VIII. § 154)

$$v_i^{x_1^{(0)}, x_1^{(\infty)}} + \ldots\ldots + v_i^{x_n^{(0)}, x_n^{(\infty)}} = 0, \qquad (i=1, \ldots, p);$$

hence from the equations

$$v_i^{z, c} \equiv M_{i,1} v_1^{x, a} + \ldots\ldots + M_{i,p} v_p^{x, a},$$

we have

$$v_i^{z_1', z_1} + \ldots\ldots + v_i^{z'_n, z_n} \equiv 0, \qquad (i=1, \ldots, p);$$

there exists therefore (Chap. VIII., § 158) a rational function having the places $z_1, \ldots, z_n$ as zeros, and the places $z_1', \ldots, z_n'$ as poles; regarding this as a function of z, s and denoting it by $\phi(z, s)$, it is clear therefore that $x/\phi(z, s)$ is a constant, which may be taken to be 1. Hence $x=\phi(z, s)$, etc.

For the theorem that for $p>1$ the number of (1, 1) correspondences is limited the reader may consult, Schwarz, *Crelle*, LXXXVII. (1879), p. 139, or *Gesamm. Math. Abhand.*, Bd. II. (Berlin, 1890), p. 285; Hettner, *Götting. Nachr.* (1880), p. 386; Noether, *Math. Annal.*, XX. (1882), p. 59; Poincaré, after Klein, *Acta Math.*, VII. (1885); Klein, *Ueber Riemann's Theorie u. s. w.* (Leipzig, 1882), p. 70 etc.; Noether, *Math. Annal.*, XXI. (1883), p. 138; Weierstrass, *Math. Werke*, Bd. II. (Berlin, 1895), p. 241; Hurwitz, *Math. Annal.*, XLI. (1893), p. 406.

394. In regard to the (1, 1) correspondence for the case $p=1$, some remarks may be made. The case $p=0$ needs no consideration here; any (1, 1) correspondence is expressible by an equation of the form

$$Att' + Bt + Ct' + D = 0;$$

thus there exists a triply infinite number of (1, 1) correspondences.

In case $p=1$, if there be a (1, 1) correspondence, whereby the variable place x corresponds to x', and a, a' be simultaneous positions of x and x', it is immediately shewn, if $v^{x, a}$ denote the normal integral of the first kind, that there exists an equation of the form

$$v^{x', a'} \equiv \mu v^{x, a},$$

wherein μ is a constant independent both of a and x. From this equation, by supposing x to describe the period loops, we deduce equations of the form

$$\mu = \alpha + \tau\alpha', \quad \mu\tau = \beta + \tau\beta', \qquad \text{(i)},$$

where α, α', β, β' are integers. By supposing x' to describe the period loops we deduce equations of the form

$$1 = \mu(\gamma + \tau\gamma'), \quad \tau = \mu(\delta + \tau\delta'), \qquad \text{(ii)},$$

where γ, γ', δ, δ' are integers. The expression of these integers in terms of α, α', β, β' is known from the general considerations of this chapter; it is however interesting to consider the equations independently. From the equations (ii) we deduce

$$\delta' - \tau\gamma' = \mu(\gamma\delta' - \gamma'\delta), \quad \delta - \tau\gamma = -\tau\mu(\gamma\delta' - \gamma'\delta);$$

if now $\gamma\delta' - \gamma'\delta = 0$, either γ' and γ are zero, which is inconsistent with $1 = \mu(\gamma + \tau\gamma')$, or else τ is a rational fraction; it is known that in that case the deficiency of the surface is not 1 but 0; we may therefore exclude that case; if $\gamma\delta' - \gamma'\delta$ be not zero, we have

$$\mu = \frac{\delta' - \tau\gamma'}{\gamma\delta' - \gamma'\delta} = a + \tau a', \quad \tau\mu = \frac{\tau\gamma - \delta}{\gamma\delta' - \gamma'\delta} = \beta + \tau\beta';$$

hence, unless τ be a rational fraction, we have

$$\frac{\delta'}{\gamma\delta' - \gamma'\delta} = a, \quad \frac{-\gamma'}{\gamma\delta' - \gamma'\delta} = a', \quad \frac{\gamma}{\gamma\delta' - \gamma'\delta} = \beta', \quad \frac{-\delta}{\gamma\delta' - \gamma'\delta} = \beta,$$

and therefore

$$1 = (a\beta' - a'\beta)(\gamma\delta' - \gamma'\delta);$$

thus $a\beta' - a'\beta = \gamma\delta' - \gamma'\delta = +1$ or -1; let ϵ denote their common value; then we deduce

$$\delta' = \epsilon a, \quad \gamma' = -a'\epsilon, \quad \gamma = \beta'\epsilon, \quad \delta = -\beta\epsilon;$$

by these the equations (ii) lead to

$$a + \tau a' = \mu, \quad \beta + \tau\beta' = \mu\tau,$$

that is, to the equations (i).

Further, from the equations (i) we deduce in turn

$$\tau^2 a' + \tau(a - \beta') - \beta = 0, \quad \mu^2 - \mu(a + \beta') + \epsilon = 0,$$

so that μ is a root of the equation

$$\begin{vmatrix} a - \mu & \beta \\ a' & \beta' - \mu \end{vmatrix} = 0;$$

now if a' be zero, the first of equations (i) gives $\mu = a$, and, therefore, as τ cannot be the rational fraction $\beta/(a - \beta')$, the second of equations (i) gives $a = \beta'$, $\beta = 0$; the equations

$$\mu = a = \beta', \quad a' = \beta = 0, \quad a\beta' - a'\beta = \epsilon$$

give $\mu^2 = \epsilon$, or, since μ, $= a$, is an integer, they require $\epsilon = +1$ and $\mu = +1$ or $\mu = -1$; the equations corresponding to $\mu = +1$ and $\mu = -1$ are

$$v^{x', a'} \equiv v^{x, a} \text{ and } v^{x', a'} + v^{x, a} \equiv 0;$$

these do belong to existing correspondences—of the kind considered in §§ 386, 387, *the coefficient* γ *being* ± 1*. *But they differ from the* (1, 1) *correspondences which are possible when* $p > 1$, *in each containing an arbitrary parameter;*

if next, a' be not zero, the equation for τ gives

$$2\tau a' = -(a - \beta') \pm \sqrt{(a + \beta')^2 - 4\epsilon},$$

so that, as τ cannot be real, we must have

$$(a + \beta')^2 - 4\epsilon < 0,$$

* For instance, on a plane cubic curve, the former equation is that in which to a point of argument u we make correspond the point of argument $u +$ constant; the line joining these two points envelopes a curve of the sixth class, which in case the difference of arguments be a half-period becomes the Cayleyan, doubled; while the latter equation is that in which we make correspond the two variable intersections of a variable straight line passing through a fixed point of the cubic.

and this shews that, in this case also, $\epsilon=1$. Hence the equations are reduced to precisely the same form as those already considered for the special transformation of theta functions (§ 383); and the result is that the only special surfaces, having $p=1$, for which there exists a (1, 1) correspondence are those which may be associated with one of the two equations

$$y^2=4x^3-g_2x, \quad y^2=4x^3-g_3;$$

the former has the obvious (1, 1) correspondence given by $x'=-x$, $y'=iy$; the latter has the obvious correspondence given by $x'=e^{\frac{2i\pi}{3}}x$, $y'=y$; the index of periodicity is 2 in the former case and 3 in the latter case.

Ex. Consider the (1, 2) correspondence on a surface for which $p=1$ in a similar way. For the equation

$$y^2=8x^3-15x+7$$

shew that a (1, 2) correspondence is given (cf. Ex. ii. § 383) by

$$-2\xi=x+\frac{9}{8(x-1)}, \quad \eta=y\,\frac{i\sqrt{2}}{4}\,\frac{x^2-2x-\frac{1}{8}}{(x-1)^2}.$$

CHAPTER XXII.

DEGENERATE ABELIAN INTEGRALS.

395. THE present chapter contains references to parts of the existing literature dealing with an interesting application of the theory of transformation of theta functions.

It was remarked by Jacobi* for the case $p=2$, that if the fundamental algebraic equation be of the form

$$y^2 = x(x-1)(x-\kappa)(x-\lambda)(x-\kappa\lambda),$$

an hyperelliptic integral of the first kind is reducible to elliptic integrals; in fact, putting $\xi = x + \kappa\lambda/x$, we immediately verify that

$$\frac{(x \pm \sqrt{\kappa\lambda})\,dx}{\sqrt{x(x-1)(x-\kappa)(x-\lambda)(x-\kappa\lambda)}} = \frac{d\xi}{\sqrt{(\xi \mp 2\sqrt{\kappa\lambda})(\xi - 1 - \kappa\lambda)(\xi - \kappa - \lambda)}}.$$

396. Suppose more generally that for any value of p there exists an integral of the first kind

$$U = \lambda_1 u_1 + \ldots\ldots + \lambda_p u_p,$$

wherein $u_1, \ldots, u_p$ denote the normal integrals of the first kind, which is reducible to the form

$$\int \frac{d\xi}{\sqrt{R(\xi)}},$$

$R(\xi)$ being a cubic polynomial in ξ, such that ξ and $\sqrt{R(\xi)}$ are rational functions on the original Riemann surface; then there exist p pairs of equations of the form

$$\lambda_i = b_i'\,\Omega - a_i'\,\Omega', \quad \lambda_1\tau_{i,1} + \ldots\ldots + \lambda_p\tau_{i,p} = -b_i\Omega + a_i\Omega', \qquad (i = 1, \ldots, p),$$

wherein a_i, b_i, a_i', b_i' are integers; we may suppose Ω' to be chosen so that the $2p$ integers

$$a_1, \ldots, a_p, a_1', \ldots, a_p'$$

have no common factor and so that

$$a_1b_1' + a_2b_2' + \ldots\ldots + a_pb_p' - a_1'b_1 - a_2'b_2 - \ldots\ldots - a_p'b_p = r,$$

* *Crelle*, VIII. (1832), p. 416.

where r is a *positive* integer; we assume that r is not zero. Eliminating the quantities $\lambda_1, \ldots, \lambda_p$, and putting $\omega = \Omega'/\Omega$, we have the p equations

$$b_i + b_1'\tau_{i,1} + \ldots + b_p'\tau_{i,p} = \omega(a_i + a_1'\tau_{i,1} + \ldots + a_p'\tau_{i,p}), \qquad (i = 1, \ldots, p);$$

if therefore the matrix of integers, $\Delta = \begin{pmatrix} \alpha & \beta \\ \alpha' & \beta' \end{pmatrix}$, of $2p$ rows and columns, wherein the first column consists of the integers $a_1, \ldots, a_p'$ in order, and the $(p+1)$th column consists of the integers $b_1, \ldots, b_p'$ in order, be determined to satisfy the conditions for a transformation of order r,

$$\bar{\alpha}\alpha' = \bar{\alpha}'\alpha, \quad \bar{\beta}\beta' = \bar{\beta}'\beta, \quad \bar{\alpha}\beta' - \bar{\alpha}'\beta = r,$$

(§ 420, Appendix II.), then it immediately follows from the equation for the transformed period matrix τ', namely

$$(\alpha + \tau\alpha')\tau' = \beta + \tau\beta',$$

that $\tau'_{11} = \omega$, $\tau'_{12} = 0, \ldots, \tau'_{1p} = 0$; to see this it is sufficient to compare the elements of the first columns of the two matrices $\beta + \tau\beta'$, $(\alpha + \tau\alpha')\tau'$. In other words, when there exists such a degenerate integral of the first kind as here supposed, it is possible*, by a transformation of order r, to arrive at periods τ' for which the theta function $\vartheta(w, \tau' \mid q)$ is a product of an elliptic theta function, in the variable w_1, and a theta function of $(p-1)$ variables, $w_2, \ldots, w_p$.

397. It can however be shewn that in the same case it is possible by a *linear* transformation to arrive at a period matrix τ'' for which

$$\tau''_{13} = 0, \; \tau''_{14} = 0, \; \ldots, \; \tau''_{1p} = 0,$$

while τ''_{12}, $= 1/r$, is a rational number. We shall suppose† two rows x, x', each of p integers, to be determined satisfying the equations

$$ax' - a'x = 1, \quad bx' - b'x = 0,$$

such that the $2p$ elements of $rx - b$, $rx' - b'$ have unity as their greatest common factor, a denoting the row $a_1, \ldots, a_p$, etc., and suppose (§ 420) a matrix of integers, of $2p$ rows and columns,

$$\Delta = \begin{pmatrix} \gamma & \delta \\ \gamma' & \delta' \end{pmatrix} = \left(\begin{array}{ll|l} a, & rx - b, \ldots & x, \ldots \\ a', & rx' - b', \ldots & x', \ldots \end{array}\right)$$

to be determined, satisfying the conditions for a linear transformation,

$$\bar{\gamma}\gamma' = \bar{\gamma}'\gamma, \quad \bar{\delta}\delta' = \bar{\delta}'\delta, \quad \bar{\gamma}\delta' - \bar{\gamma}'\delta = 1,$$

wherein the first column consists of the elements of a and a', the second column consists of the elements of $rx - b$ and $rx' - b'$, and the $(p+1)$th

* This theorem is due to Weierstrass, see Königsberger, *Crelle*, LXVII. (1867), p. 73; Kowalevski, *Acta Math.* IV. (1884), p. 395. See also Abel, *Œuvres*, t. I. (1881), p. 519.

† The proof that this is possible is given in Appendix II., § 419. It may be necessary, beforehand, to make a linear transformation of the periods Ω, Ω'.

column consists of the elements of x and x'; the conditions for a linear transformation, so far as they affect these three columns only, are

$$a(rx'-b')-a'(rx-b)=0,\ ax'-a'x=1,\ (rx-b)x'-(rx'-b')x=0,$$

and these are satisfied in virtue of the equation $ab'-a'b=r$. Then the equation for the transformed period matrix τ'', namely

$$(\gamma+\tau\gamma')\tau''=\delta+\tau\delta',$$

leads to $\tau''_{3,1}=0, \ldots, \tau''_{p,1}=0$ if only the p equations

$$[\gamma_{i,1}+(\tau\gamma')_{i,1}]\,\tau''_{1,1}+[\gamma_{i,2}+(\tau\gamma')_{i,2}]\,\tau''_{2,1}=\delta_{i,1}+(\tau\delta')_{i,1}, \qquad (i=1, \ldots, p),$$

which are obtained by equating corresponding elements of the first columns of the matrices $\delta+\tau\delta'$, $(\gamma+\tau\gamma')\tau''$, are satisfied; these p equations are included in the single equation

$$\tau''_{1,1}[a+\tau a']+\tau''_{2,1}[rx-b+\tau(rx'-b')]=x+\tau x',$$

and are satisfied* by $\tau''_{1,1}=\omega/r$, $\tau''_{2,1}=1/r$; for we have, as the fundamental condition, the equation

$$\omega(a+\tau a')=b+\tau b'.$$

398. It follows therefore in case $p=2$ that the matrix τ'' has the form

$$\begin{pmatrix}\tau''_{11}, & 1/r\\ 1/r, & \tau''_{22}\end{pmatrix};$$

hence it immediately follows that beside the integral of the first kind already considered, which is expressible as an elliptic integral, there is another having the same property. In virtue of the equations here obtained the first integral having this property can be represented, after division by Ω, in the form

$$U=(b'-r\tau''_{1,1}a')\,u,$$

where u denotes the row of 2 integrals u_1, u_2; consider now the integral

$$V=[rt'-a'-r\tau''_{2,2}(rx'-b')]\,u,$$

where t' is a row of two elements, these being the constituents of the first column of the matrix δ'; the periods of V at the first set of period loops are given by the row of quantities

$$rt'-a'-r\tau''_{2,2}(rx'-b'),$$

* See Kowalevski, *Acta Math.* IV. (1884), p. 400; Picard, *Bulletin de la Soc. Math. de France*, t. XI. (1882—3), p. 25, and *Compt. Rendus*, XCII. XCIII. (1881); Poincaré, *Bulletin de la Soc. Math. de France*, t. XII. (1883—4), p. 124; Poincaré, *American Journal*, vol. VIII. (1886), p. 289.

and are linear functions of the two quantities $1, r\tau''_{2,2}$; the periods of V at the second set of period loops are given by

$$[\tau(rt'-a')]_i - r\tau''_{2,2}[\tau(rx'-b')]_i, \qquad (i=1, 2);$$

now the equation $(\gamma+\tau\gamma')\tau''=\delta+\tau\delta'$ gives

$$(\gamma+\tau\gamma')_{i,1}\tau''_{1,2}+(\gamma+\tau\gamma')_{i,2}\tau''_{2,2}=(\delta+\tau\delta')_{i,2}, \qquad (i=1, 2),$$

and hence we have

$$\tau''_{1,2}[a+\tau a']+\tau''_{2,2}[rx-b+\tau(rx'-b')]=t+\tau t',$$

where t is the row formed by the constituents of the first column of the matrix δ; therefore, as $\tau''_{1,2}=1/r$, the periods of V at the second set of period loops are expressible in the form

$$-(rt-a)_i+r\tau''_{2,2}(rx-b)_i, \qquad (i=1, 2),$$

and these are also linear functions of the two quantities $1, r\tau''_{2,2}$. Hence it may be inferred that the integral V is reducible to an elliptic integral.

399. It has been shewn in the last chapter that for special values of the periods τ there exist transformations of the theta functions into theta functions for which the transformed periods are equal to the original periods. It can be shewn* that for the special case now under consideration such a transformation holds. Suppose that a theta function ϑ, with period τ, is transformed, as described above, into a theta function ϕ, with period τ', for which $\tau'_{1,2}=0=\ldots=\tau'_{1,p}$, by a transformation associated with the matrix $\Delta=\begin{pmatrix}\alpha & \beta\\ \alpha' & \beta'\end{pmatrix}$; suppose further that there exists, associated with a matrix $H=\begin{pmatrix}\lambda & \mu\\ \lambda' & \mu'\end{pmatrix}$, a transformation whereby the theta function ϕ is transformed into another theta function with the same period τ'; then it is easy to prove that there exists a corresponding transformation of the theta function ϑ whereby it becomes changed into a theta function with the same period τ, namely the transformation is that associated with the matrix

$$\begin{pmatrix}f & g\\ f' & g'\end{pmatrix}=\begin{pmatrix}\alpha & \beta\\ \alpha' & \beta'\end{pmatrix}\begin{pmatrix}\lambda & \mu\\ \lambda' & \mu'\end{pmatrix}\begin{pmatrix}\bar{\beta}' & -\bar{\beta}\\ -\bar{\alpha}' & \bar{\alpha}\end{pmatrix};$$

to prove this it is only necessary to shew that the equations

$$(\lambda+\tau'\lambda')\tau'=\mu+\tau'\mu', \quad (\alpha+\tau\alpha')\tau'=\beta+\tau\beta'$$

give the equation

$$(f+\tau f')\tau=g+\tau g'.$$

* Wiltheiss, *Math. Annal.* XXVI. (1886), p. 127.

Hence it follows that in order to determine a transformation of the function ϑ which leaves the period τ unaltered, it is sufficient to determine a transformation of the function ϕ which leaves the period τ' unaltered; this determination is facilitated by the special values of $\tau'_{1,2}, \dots, \tau'_{1,p}$; and in fact we immediately verify that the equation $(\lambda + \tau'\lambda')\tau' = \mu + \tau'\mu'$ is satisfied by taking $\lambda' = \mu = 0$ and by taking each of λ and μ' to be the matrix in which every element is zero except the elements in the diagonal, each of these elements being 1 except the first, which is -1.

400. Thus for the case $p = 2$, supposing $r = 2$, the original function ϑ is transformed into a theta function with unaltered period τ, by means of the transformation of order 4 associated with the matrix,

$$\begin{pmatrix} \alpha & \beta \\ \alpha' & \beta' \end{pmatrix} \begin{pmatrix} m & 0 \\ 0 & m \end{pmatrix} \begin{pmatrix} \bar{\beta}', & -\bar{\beta} \\ -\bar{\alpha}', & \bar{\alpha} \end{pmatrix}, \quad = \Delta M \nabla \text{ say,}$$

where m denotes the matrix $\begin{pmatrix} -1 & 0 \\ 0 & 1 \end{pmatrix}$; the matrix ∇ is equal to $2\Delta^{-1}$, and it is easy to see that this transformation of order 4 is equivalent to a multiplication, with multiplier 2, together with a *linear* transformation associated with the matrix

$$\Delta M (\tfrac{1}{2}\nabla).$$

We have therefore the result; when, in case $p = 2$, there exists a transformation of the second order whereby the periods τ are changed into periods τ' for which $\tau'_{1,2} = 0$, then there exists a linear transformation whereby the periods τ are changed into the same periods τ, or what we have called in the last chapter a complex multiplication.

401. The transcendental results thus obtained enable us to specify the algebraic conditions for the existence of an integral of the first kind which is reducible to an elliptic integral.

Thus for instance when $p = 2$, to determine all the cases in which an integral of the first kind can be reduced to an elliptic integral by means of a transformation of the second order, $\Delta = \begin{pmatrix} \alpha & \beta \\ \alpha' & \beta' \end{pmatrix}$, it is sufficient to consider the conditions that the transformed even theta function $\vartheta\left(w;\ \tau' \middle| \tfrac{1}{2}\begin{pmatrix} 1 & 1 \\ 1 & 1 \end{pmatrix}\right)$ may vanish for zero values of w; for when $\tau'_{1,2} = 0$ this function breaks up into the product of two odd elliptic theta functions. By means of the formulae* for transformation of the second order, it can be shewn† that this condition leads to the equation

$$-\vartheta_2^2 \vartheta_{01}^2 \vartheta_{34}^2 + \vartheta_{14}^2 \vartheta_0^2 \vartheta_{23}^2 = 0,$$

* Chap. XX. § 364.

† Königsberger, *Crelle*, LXVII. (1867), p. 77.

and by means of the relations expressing the constants of the fundamental algebraic equation in terms of the zero values of the even theta functions* it can be shewn that this is equivalent to the condition that the fundamental algebraic equation may be taken to be of the form

$$y^2 = x(x-1)(x-\kappa)(x-\lambda)(x-\kappa\lambda),$$

so that the case obtained by Jacobi is the only one possible for transformations of the second order.

In the same case of $p=2$, $r=2$, the same result follows more easily from the existence, deduced above, of a complex multiplication belonging to a transformation of the first order. For it follows from this fact that the algebraic equation can be taken in a form in which it can be transformed into itself by a transformation in which the independent variable is transformed by an equation of the form

$$x = \frac{A\xi + B}{C\xi - A},$$

and this leads† to the form, for the fundamental algebraical equation,

$$s^2 = (z^2 - a^2)(z^2 - b^2)(z^2 - c^2),$$

which is immediately identified with the form above by putting

$$x = \sqrt{\kappa\lambda}\,(z+1)/(z-1),$$

the quantities a, b, c being respectively

$$1, \quad (\sqrt{\kappa\lambda}+1)^2/(\sqrt{\kappa\lambda}-1)^2, \quad (\sqrt{\kappa}+\sqrt{\lambda})^2/(\sqrt{\kappa}-\sqrt{\lambda})^2.$$

Similarly for $p=3$, when the surface is not hyperelliptic, it can be shewn‡ from the relations connecting the theta functions when a theta function is the product of an elliptic theta function and a theta function of two variables, that the only cases in which an integral of the first kind can be reduced to an elliptic integral are those in which the fundamental algebraic equation can be taken to be of the form

$$\sqrt{x(Ax+By)} + \sqrt{y(Cx+Dy)} + \sqrt{1+Fx+Gy} = 0.$$

The Riemann surface associated with this equation possesses a (1, 1) correspondence given by the equations

$$\xi = -x/(1+Fx+Gy), \quad \eta = -y/(1+Fx+Gy).$$

* Cf. Ex. v. p. 341. By means of the substitution $x = c_1 + (a_1 - c_1)\xi$, the branch places can be taken at $\xi = 0, 1, \kappa, \lambda, \mu$, wherein, if c_1, a_1, c_2, a_2, c be real and in ascending order, $0, 1, \kappa, \lambda, \mu$ are in ascending order of magnitude. For complete formulae, when the theta functions are regarded as primary, and the algebraic equation as derived, see Rosenhain, *Mém. p. divers Savants*, XI. (1851), p. 416 ff.

† Wiltheiss, *Math. Annal.* XXVI. (1886), p. 134.

‡ Kowalevski, *Acta Math.* IV. (1884), p. 403.

402. But the problem of determining the algebraic equations for which an associated integral of the first kind reduces to an elliptic integral may be considered algebraically, by beginning with an elliptic integral and transforming it into an Abelian integral. The reader may consult Richelot, *Crelle*, XVI. (1837); Malet, *Crelle*, LXXVI. (1873), p. 97; Brioschi, *Compt. Rendus*, LXXXV. (1877), p. 708; Goursat, *Bulletin de la Soc. Math. de France*, t. XIII. (1885), p. 143, and *Compt. Rendus*, C. (1885), p. 622; Burnside, *Proc. Lond. Math. Soc.* vol. XXIII. (1892), p. 173.

403. The paper of Königsberger already referred to (*Crelle*, LXVII.) deals with the case of a transformation of the second order, for $p=2$. For the case of a transformation of the third order, when $p=2$, consult, beside the papers of Goursat (*loc. cit.* § 402), also Hermite, *Ann. de la Soc. Scient. de Bruxelles*, 1876, and Burkhardt, *Math. Annal.* XXXVI. (1890), p. 410. For the case $p=2$, and a transformation of the fourth order, see Bolza, *Ueber die Reduction hyperelliptischer Integrale u. s. w.*, Götting. Dissertation (Berlin, Schade, 1885), or *Sitzungsber. der Naturforsch. Ges. zu Freiburg* (1885). The paper of Kowalevski (*Acta Math.* IV.) deals with the case of a transformation of the second order for $p=3$. See further the references given in this chapter, and Poincaré, *Compt. Rendus*, t. XCIX. (1884), p. 853; Biermann, *Sitzungsber. der Wiener Akad.* Bd. LXXXVII. (ii. Abth.) (1883), p. 983.

APPENDIX I.

On Algebraic Curves in Space.

404. Given an algebraic curve (C) in space, let a point O be found, not on the curve, such that the number of chords of the curve that pass through O is finite; let the curve be projected from O on to any arbitrary plane, into the plane curve (f), and referred to homogeneous coordinates ξ, η, τ in that plane, whose triangle of reference has such a position that the curve does not pass through the angular point η, and has no multiple points on the line $\tau=0$; let the curve (C) be referred to homogeneous coordinates ξ, η, ζ, τ of which the vertex ζ of the tetrahedron of reference is at O. Putting $x=\xi/\tau$, $y=\eta/\tau$, $z=\zeta/\tau$, it is sufficient to think of x, y, z as Cartesian coordinates, the point O being at infinity. Thus the plane curve (f) is such that y is not infinite for any finite value of x, and its equation is of the form $f(y, x)=y^m+A_1y^{m-1}+\ldots\ldots+A_m=0$, where $A_1, \ldots, A_m$ are integral polynomials in x; the curve (C) is then of order m; we define its deficiency to be the deficiency of (f); to any point (x, y) of (f) corresponds in general only one point (x, y, z) of (C), and, on the curve (C), z is not infinite for any finite values of x, y.

Now let $f'(y)=\partial f(y, x)/\partial y$; let ϕ be an integral polynomial in x and y, so chosen that at every finite point of (f) at which $f'(y)=0$, say at $x=a$, $y=b$, the ratio $(x-a)\phi/f'(y)$ vanishes to the first order at least; let $\alpha=\Pi(x-a)$ contain a simple factor corresponding to every finite value of x for which $f'(y)=0$; let $y_1, \ldots, y_m$ be the values of y which, on the curve (f), belong to a general value of x, so that to each pair (x, y_i) there belongs, on the curve (C), only one value of z; considering the summation

$$\sum_{i=1}^{m} \frac{(c-y_1)\ldots\ldots(c-y_m)}{c-y_i}\,\alpha\left[\frac{z\phi}{f'(y)}\right]_{y=y_i},$$

where c is an arbitrary quantity, we immediately prove, as in § 89, Chap. VI., that it has a value of the form

$$\alpha\,(c^{m-1}u_1+c^{m-2}u_2+\ldots\ldots+u_m),$$

where $u_1,\ldots, u_m$ are integral polynomials in x; putting y_i for c, after division by α, we therefore infer that z can be represented in the form

$$z=\psi/\phi,$$

where ϕ, ψ are integral polynomials in x and y, whereof ϕ is arbitrary, save for the conditions for the fractions $(x-a)\phi/f'(y)$. This is Cayley's monoidal expression of a curve in space with the adjunction of the theorem, described by Cayley as the capital theorem of Halphen, relating to the arbitrariness of ϕ (Cayley, *Collect. Works*, Vol. v. 1892, p. 614).

It appears therefore that a curve in space may be regarded as arising as an interpretation of the relations connecting three rational functions on a Riemann surface; and, within a finite neighbourhood of any point of the curve in space, the coordinates of the points of the curve may be given by series of integral powers of a single quantity t, this being the quantity we have called the infinitesimal for a Riemann surface; to represent the whole curve only a finite number of different infinitesimals is necessary. More generally the representation by means of automorphic functions holds equally well for curves in space. And the theory of Abelian integrals can be developed for a curve in space precisely as for a plane curve, or can be deduced from the latter case; the identity of the deficiency for the curve in space and the plane curve may be regarded as a corollary. Also we can deduce the theorem that, of the intersections with a curve in space of a variable surface, not all can be arbitrarily assigned, the number of those whose positions are determined by the others being, for a surface of sufficiently high order, equal to the deficiency of the curve.

Ex. If through $p-1$ of the generators of a quadric surface, of the same system, a surface of order $p+1$ be drawn, the remaining curve of intersection is representable by two equations of the form

$$y^2 = (x, 1)_{2p+2}, \quad zu_1 = u_2,$$

where $(x, 1)_{2p+2}$ is an integral polynomial in x of order $2p+2$, and u_1, u_2 are respectively linear and quadric polynomials in x and y.

For the development of the theory consult, especially, Noether, *Abh. der Akad. zu Berlin vom Jahre* 1882, pp. 1 to 120; Halphen, *Journ. École Polyt.*, Cah. LII. (1882), pp. 1—200; Valentiner, *Acta Math.*, t. II. (1883), pp. 136—230. See also, Schubert, *Math. Annal.* XXVI. (1885); Castelnuovo, *Rendiconti della R. Accad. dei Lincei*, 1889; Hilbert, *Math. Annal.*, XXXVI. (1890).

APPENDIX II.

ON MATRICES*.

405. A SET of n quantities

$$(x_1, \ldots, x_n)$$

is often denoted by a single letter x, which is then called a *row letter*, or a column letter. By the sum (or difference) of two such rows, of the same number of elements, is then meant the row whose elements are the sums (or differences) of the corresponding elements of the constituent rows. If m be a single quantity, the row letter mx denotes the row whose elements are $mx_1, \ldots, mx_n$. If x, y be rows, each of n quantities, the symbol xy denotes the quantity $x_1y_1 + \ldots\ldots + x_ny_n$.

406. The set of n equations denoted by

$$x_i = a_{i,1}\xi_1 + \ldots\ldots + a_{i,p}\xi_p, \qquad (i = 1, \ldots\ldots, n)$$

where n may be greater or less than p, can be represented in the form $x = a\xi$, where a denotes a rectangular block of np quantities, consisting of n rows each of p quantities, the r-th quantity of the i-th row being $a_{i,r}$. Such a block of quantities is called a *matrix*; we call $a_{i,r}$ the (i, r)th element of the matrix. The sum (or difference) of two matrices, of the same number of rows and columns, is the matrix formed by adding (or subtracting) the corresponding elements of the component matrices. Two matrices are equal only when all their elements are equal; a matrix vanishes only when all its elements are zero. If $\xi_1, \ldots, \xi_p$ be expressible by m quantities $X_1, \ldots, X_m$ by the equations

$$\xi_r = b_{r,1} X_1 + \ldots\ldots + b_{r,m} X_m, \qquad (r = 1, 2, \ldots\ldots, p),$$

so that $\xi = bX$, where b is a matrix of p rows and m columns, then we have

$$x_i = c_{i,1} X_1 + \ldots\ldots + c_{i,m} X_m, \qquad (i = 1, \ldots\ldots, n),$$

or $x = cX$, where

$$c_{i,s} = a_{i,1} b_{1,s} + \ldots\ldots + a_{i,p} b_{p,s}, \qquad \begin{pmatrix} i = 1, \ldots\ldots, n \\ s = 1, \ldots\ldots, m \end{pmatrix},$$

* The literature of the theory of matrices, or, under a slightly different aspect, the theory of bilinear forms, is very wide. The following references may be given: Cayley, *Phil. Trans.* 1858, or *Collected Works*, vol. II. (1889), p. 475; Cayley, *Crelle*, L. (1855); Hermite, *Crelle*, XLVII. (1854); Christoffel, *Crelle*, LXIII. (1864) and LXVIII. (1868); Kronecker, *Crelle*, LXVIII. (1868) or *Gesam. Werke*, Bd. I. (1895), p. 143; Schläfli, *Crelle*, LXV. (1866); Hermite, *Crelle*, LXXVIII. (1874); Rosanes, *Crelle*, LXXX. (1875); Bachmann, *Crelle*, LXXVI. (1873); Kronecker, *Berl. Monatsber.*, 1874; Stickelberger, *Crelle*, LXXXVI. (1879); Frobenius, *Crelle*, LXXXIV. (1878), LXXXVI. (1879), LXXXVIII. (1880); H. J. S. Smith, *Phil. Trans.*, CLI. (1861), also, *Proc. Lond. Math. Soc.*, 1873, pp. 236, 241; Laguerre, *J. d. l'éc. Poly.*, t. XXV., cah. XLII. (1867), p. 215; Stickelberger, *Progr. poly. Schule, Zürich*, 1877; Weierstrass, *Berl. Monats.* 1858, 1868; Brioschi, *Liouville*, XIX. (1854); Jordan, *Compt. Rendus*, 1871, p. 787, and *Liouville*, 1874, p. 35; Darboux, *Liouville*, 1874, p. 347.

$c_{i,s}$ being the (i, s)th element of a matrix of n rows and m columns; it arises from the equations $x=a\xi$, $\xi=bX$, whereof the result may be written $x=abX$; hence we may formulate the rule: *A matrix* a *may be multiplied into another matrix* b *provided the number of columns of* a *be the same as the number of rows of* b; *the* (i, s)*th element of the resulting matrix is the result of multiplying*, in accordance with the rule given above, *the* i-*th row of* a *by the* s-*th column of* b. Thus, for multiplication, matrices are not generally commutative, but, as is easy to see, they are associative.

The matrix whose (i, s)th element is $c_{s,i}$, where $c_{s,i}$ is the (s, i)th element of any matrix c of n rows and m columns, is called the transposed matrix of c, and may be denoted by $\bar{c}$; it has m rows and n columns, and, briefly, is obtained by interchanging the rows and columns of c. The matrix which is the transposed of a product of matrices is obtained by taking the factor matrices in the reverse order, each transposed; for example, if a, b, c be matrices,

$$\overline{abc}=\bar{c}\bar{b}\bar{a}.$$

407. The matrices which most commonly occur are square matrices, having an equal number of rows and columns. With such a matrix is associated a determinant, whose elements are the elements of the matrix. When the determinant of a matrix, a, of p rows and columns, does not vanish, the p linear equations expressed by $x=a\xi$ enable us to represent the quantities $\xi_1, \dots, \xi_p$ in terms of $x_1, \dots, x_p$; the result is written $\xi=a^{-1}x$, and a^{-1} is called the inverse matrix of a; the (i, r)th element of a^{-1} is the minor of $a_{r,i}$ in the determinant of the matrix a, divided by this determinant itself. The inverse of a product of square matrices is obtained by taking the inverses of the factor matrices in reverse order; for example, if a, b, c be square matrices, of the same number of rows and columns, for each of which the determinant is not zero, we have

$$(abc)^{-1}=c^{-1}b^{-1}a^{-1}.$$

The inverse of the transposed of a matrix is the transposed of its inverse; thus

$$(\bar{a})^{-1}=(\overline{a^{-1}}).$$

The determinant of a matrix a being represented by $|a|$, we clearly have $|ab|=|a||b|$.

408. Finally, the following results are of frequent application in this volume: (i) If a be a matrix of n rows and p columns, and ξ a row of p quantities, the symbol $a\xi$ denotes a row of n quantities; if η be a row of n quantities, the product of these two rows, or $(a\xi)(\eta)$, is denoted by $a\xi\eta$. When $n=p$ this must be distinguished from the matrix which would be denoted by $a\,.\,\xi\eta$—this latter never occurs. We have then

$$a\xi\eta=\sum_{i=1}^{n}\sum_{r=1}^{p} a_{i,r}\,\xi_r\,\eta_i,$$

and this is called a bilinear form; we also clearly have the noticeable equation

$$a\xi\eta=\bar{a}\eta\xi\,;$$

(ii) if b be a matrix of n rows and q columns, the product of the two rows $a\xi$, $b\eta$, wherein η is now a row of q quantities, is given by either $(\bar{b}a)\,\xi\eta$ or $(\bar{a}b)\,\eta\xi$, so that we have

$$a\xi\,.\,b\eta=\bar{b}a\xi\eta=\bar{a}b\eta\xi.$$

The result of multiplying any square matrix, of p rows and columns, by the matrix E, of p rows and columns, wherein all the elements are zero except the diagonal elements, which are each unity, is to leave the multiplied matrix unaltered. For this reason the matrix E is often denoted simply by 1, and called the matrix unity of p rows and columns.

409. *Ex.* i. If a bilinear form axy, wherein x, y are rows of p quantities, and a is a square matrix of p rows and columns, be transformed into itself by the linear substitution $x=R\xi$, $y=S\eta$, where R, S are matrices of p rows and columns, then $aR\xi\,.\,S\eta=a\xi\eta$; hence

$$\bar{S}aR=a.$$

Ex. ii. If h be an arbitrary matrix of p rows and columns, such that the determinants of the matrices $a\pm h$ do not vanish, and the determinant of the matrix a do not vanish, prove that

$$(a+h)\,a^{-1}\,(a-h)=a-ha^{-1}\,h=(a-h)\,a^{-1}\,(a+h)\,;$$

hence shew that if

$$R=a^{-1}\,(a-h)\,(a+h)^{-1}\,a, \quad \bar{S}=a\,(a-h)^{-1}\,(a+h)\,a^{-1},$$

the substitutions $x=R\xi$, $y=S\eta$ transform axy into $a\xi\eta$.

For a substitution in which $R=S$ see Cayley, *Collected Works*, vol. II. p. 505. Cf. also Taber, *Amer. Journ.*, vol. XVI. (1894) and *Proc. Lond. Math. Soc.*, vol. XXVI. (1895).

Ex. iii. The matrices, of two rows and columns,

$$E=\begin{pmatrix}1 & 0\\ 0 & 1\end{pmatrix}, \quad J=\begin{pmatrix}0 & -1\\ 1 & 0\end{pmatrix},$$

give $E^2=E$, $J^2=-E$; and the determinant of the matrix

$$xE+yJ=\begin{pmatrix}x, & -y\\ y, & x\end{pmatrix}$$

vanishes, for real values of x, y, only when $x=0$, $y=0$.

Ex. iv. The matrices, of four rows and columns,

$$e=\begin{pmatrix}1&0&0&0\\0&1&0&0\\0&0&1&0\\0&0&0&1\end{pmatrix}, \quad j_1=\begin{pmatrix}0&1&0&0\\-1&0&0&0\\0&0&0&-1\\0&0&1&0\end{pmatrix}, \quad j_2=\begin{pmatrix}0&0&1&0\\0&0&0&1\\-1&0&0&0\\0&-1&0&0\end{pmatrix}, \quad j_3=\begin{pmatrix}0&0&0&1\\0&0&-1&0\\0&1&0&0\\-1&0&0&0\end{pmatrix},$$

give $j_1^2=j_2^2=j_3^2=-e$, $j_2j_3=-j_3j_2=j_1$, $j_3j_1=-j_1j_3=j_2$, $j_1j_2=-j_2j_1=j_3$, $j_1j_2j_3=-e$.

Hence these matrices obey the laws of the fundamental unities of the quaternion analysis. Further the determinant of the matrix

$$ex+j_1x_1+j_2x_2+j_3x_3=\begin{pmatrix}x & x_1 & x_2 & x_3\\ -x_1 & x & -x_3 & x_2\\ -x_2 & x_3 & x & -x_1\\ -x_3 & -x_2 & x_1 & x\end{pmatrix}$$

which is equal to $(x^2+x_1^2+x_2^2+x_3^2)^2$, vanishes, for real values of x, x_1, x_2, x_3, only when each of x, x_1, x_2, x_3 is zero. (Frobenius, *Crelle*, LXXXIV. (1878), p. 62.)

410. In the course of this volume we are often concerned with matrices of $2p$ rows and $2p$ columns. Such a matrix may be represented in the form

$$\mu=\begin{pmatrix}a & b\\ c & d\end{pmatrix},$$

wherein a, b, c, d are square matrices with p rows and columns; if μ' be another such matrix given by

$$\mu'=\begin{pmatrix}a' & b'\\ c' & d'\end{pmatrix}$$

the (i, r)th element of the product $\mu'\mu$, when i and r are both less than $p+1$ is

$$a'_{i,1}\, a_{1,r} + \ldots\ldots + a'_{i,p}\, a_{p,r} + b'_{i,1}\, c_{1,r} + \ldots\ldots + b'_{i,p}\, c_{p,r},$$

and this is the sum of the (i, r)th elements of the matrices $a'a$, $b'c$; similarly when i and r are not both less than $p+1$; hence we may write

$$\begin{pmatrix} a' & b' \\ c' & d' \end{pmatrix} \begin{pmatrix} a & b \\ c & d \end{pmatrix} = \begin{pmatrix} a'a+b'c, & a'b+b'd \\ c'a+d'c, & c'b+d'd \end{pmatrix},$$

the law of formation for the product matrix being the same as if $a, b, c, d, a', b', c', d'$ were single quantities.

Ex. Denoting the matrices $\begin{pmatrix} 1 & 0 \\ 0 & 1 \end{pmatrix}$, $\begin{pmatrix} 0 & -1 \\ 1 & 0 \end{pmatrix}$ respectively by 1 and j, the matrices of Ex. iv. can be denoted by

$$e = \begin{pmatrix} 1 & 0 \\ 0 & 1 \end{pmatrix},\quad j_1 = \begin{pmatrix} -j & 0 \\ 0 & j \end{pmatrix},\quad j_2 = \begin{pmatrix} 0 & 1 \\ -1 & 0 \end{pmatrix},\quad j_3 = \begin{pmatrix} 0 & -j \\ -j & 0 \end{pmatrix}.$$

411. We proceed now to prove the proposition* assumed in § 333, Chap. XVIII. Retaining the definitions of the matrices A_k, B, C, D there given, and denoting A_k^{-1}, B^{-1}, C^{-1}, D^{-1} respectively by a_k, b, c, d, we find

$$a_k = A_k, \text{ so that } A_k^2 = 1,$$

and

$$b = \left(\begin{array}{cccccccc} 0 & & & & 1 & & & \\ & 1 & & & & 0 & & \\ & & 1 & & & & 0 & \\ & & & 1 & & & & 0 \\ -1 & & & & 0 & & & \\ & 0 & & & & 1 & & \\ & & 0 & & & & 1 & \\ & & & 0 & & & & 1 \end{array}\right),\quad c = \left(\begin{array}{cccccccc} 1 & & & & 1 & & & \\ & 1 & & & & 0 & & \\ & & 1 & & & & 0 & \\ & & & 1 & & & & 0 \\ 0 & & & & 1 & & & \\ & 0 & & & & 1 & & \\ & & 0 & & & & 1 & \\ & & & 0 & & & & 1 \end{array}\right),\quad d = \left(\begin{array}{cccccccc} 1 & & & & 01 & & & \\ & 1 & & & 10 & & & \\ & & 1 & & & & 0 & \\ & & & 1 & & & & 0 \\ 0 & & & & 1 & & & \\ & 0 & & & & 1 & & \\ & & 0 & & & & 1 & \\ & & & 0 & & & & 1 \end{array}\right)$$

so that b, c, d differ respectively from B, C, D only in the change of the sign of the elements which are not in the diagonal. It is easy moreover to verify such facts as the following

$$B^4 = 1,\quad (BC)^3 = 1,\quad DA_2 = A_2D,\quad A_kBA_kB = BA_kBA_k,\quad B^2DB^2A_2 = A_2B^2DB^2,$$

which are equivalent respectively with

$$b^4 = 1,\ (cb)^3 = 1,\ a_2d = da_2,\ ba_kba_k = a_kba_kb,\ a_2b^2db^2 = b^2db^2a_2\,;$$

but such results are immediately obvious from the interpretations of the matrices a_k, b, c, d which are now to be given.

Let Δ denote any matrix of $2p$ rows and columns, and let the four products

$$\Delta a_k,\ \Delta b,\ \Delta c,\ \Delta d$$

* For a shorter proof of an equivalent result the reader may consult C. Jordan, *Traité des Substitutions* (Paris, 1870), p. 174. The theorem was first given by Kronecker, "Ueber bilineare Formen," *Monatsber. Berl. Akad.* 1866, *Crelle*, LXVIII. or in *Werke* (Leipzig, 1895), Bd. I. p. 160; the proof here given follows the lines there indicated.

be formed; the resulting matrices will differ from Δ in respects which are specified in the following statements:

(i) a_k interchanges the first and k-th columns (of Δ), and, at the same time, the $(p+1)$th and $(p+k)$th columns $(1<k<p+1)$. For the sake of uniformity we introduce also a_1, $=1$.

(ii) b interchanges the first and $(p+1)$th columns, at the same time changing the signs of the elements of the new first column.

(iii) c adds the first column to the $(p+1)$th.

(iv) d adds the first and second columns respectively to the $(p+2)$th and the $(p+1)$th.

Hence we have these results: if the matrices denoted by the following symbols be placed at the right side of any matrix Δ, of $2p$ rows and columns, so that the matrix Δ acts upon them, the results mentioned will accrue:—

$l_k=a_kb^2a_k$, changes the signs of the k-th and $(p+k)$th columns (of Δ),

$t_k=a_kb\,a_k$, interchanges the k-th and $(p+k)$th columns (of Δ), giving the new k-th column an opposite sign to that it had before its change of place,

$t'_k=a_kb^3a_k$, interchanges the k-th and $(p+k)$th columns, giving the new $(p+k)$th column a changed sign.

$m_k=a_kb^2cb^2a_k$, adds the k-th column to the $(p+k)$th.

$m'_k=a_kb^3cbcb^3a_k=a_kb^2c^{-1}b^2a_k$, subtracts the k-th column from the $(p+k)$th.

$n_k=a_kb^2cbca_k=a_kbc^{-1}b^3a_k$, adds the $(p+k)$th column to the k-th.

$n'_k=a_kb^3cba_k$, subtracts the $(p+k)$th column from the k-th.

$g_{r,s}=a_ra_2a_sa_2b^3dba_2a_sa_2a_r$, subtracts the s-th column from the r-th, and, at the same time, adds the $(p+r)$th column to the $(p+s)$th.

$g'_{r,s}=a_ra_2a_sa_2bdb^3a_2a_sa_2a_r$, adds the s-th column to the r-th, and, at the same time, subtracts the $(p+r)$th from the $(p+s)$th column.

$f_{r,s}=t_sg_{r,s}t'_s$, adds the $(p+r)$th and $(p+s)$th columns respectively to the s-th and r-th columns.

$f'_{r,s}=t_sg'_{r,s}t'_s$, subtracts the $(p+r)$th and $(p+s)$th columns respectively from the s-th and r-th columns.

To this list we add the matrix a_k, whose effect has been described, and the matrix b^2, which changes the sign both of the first and of the $(p+1)$th columns; then it is to be shewn that a product, P, of positive integral powers of these matrices, can be chosen such that, if Δ be any Abelian matrix of integers, given by

$$\Delta=\begin{pmatrix}\alpha & \beta\\ \alpha' & \beta'\end{pmatrix},\ \text{where } \alpha\bar{\beta}=\beta\bar{\alpha},\ \alpha'\bar{\beta}'=\beta'\bar{\alpha}',\ \alpha\bar{\beta}'-\beta\bar{\alpha}'=1,$$

the product ΔP is the matrix unity—of which every element is zero except those in the diagonal, each of which is 1. Hence it will follow that $\mu=P^{-1}$; namely that every such Abelian matrix can be written as a product of positive integral powers of the matrices A_k, B, C, D. Up to a certain point of the proof we shall suppose the matrix Δ to be that for a transformation of any order, r.

In the matrices a_k, a_r, a_s, each of k, r, s is to be $<p+1$; and in general each of k, r, s is >1; but for the sake of uniformity it is convenient, as already stated, to introduce a matrix $a_1=1$; then each of k, r, s may have any positive value less than $p+1$.

412. Of the matrix Δ we consider first the first row, and of this row we begin with the p-th and $2p$-th elements, $a_{1,p}$, $\beta_{1,p}$; if the numerically greater of these elements be not a positive integer, use the matrix l_p to make it positive*—form, that is, the product Δl_p. Then, let γ be the greater, and δ the less of these two elements; if δ is positive, use the matrix m'_p or the matrix n'_p, as many times as possible, to subtract from γ the greatest possible multiple† of δ (i.e. if ν be the matrix upon which we are operating, $=\Delta$ or $=\Delta l_p$, form one of the products $\nu(m'_p)^r$, $\nu(n'_p)^s$); if δ is negative, use m_p or n_p to add to γ the greatest possible multiple of δ; so that, in either case, the remainder, γ', from γ, is numerically less than δ and positive. Now, by the matrix l_p, take the element δ to be positive‡; then again, by application of m_p or n_p or m'_p or n'_p replace δ by a positive quantity numerically less than γ'. Let this process alternately acting on the remainder from γ and δ, be continued until either γ or δ is replaced by zero. Then use the matrix t_p or t'_p to put this zero element at the $2p$-th place of the first row of the matrix, Δ', which, after all these changes, replaces Δ.

Let a similar process of alternate reduction and transposition be applied to Δ', until the $(1, 2p-1)$th element of the resulting matrix is zero. And so on. Eventually we arrive, in continuing the operation, at a matrix instead of Δ, in which there is a zero in each of the places formerly occupied by $\beta_{1,1}, \dots\dots, \beta_{1,p}$.

Now apply the processes given by b^2, l_p, $g_{1,p}$, $g_{p,1}$, and eventually a_p, if necessary, to reduce the $(1, p)$th element to zero. Then the processes b^2, l_{p-1}, $g_{1,p-1}$, $g_{p-1,1}$, a_{p-1}, as far as necessary, to reduce the $(1, p-1)$th element to zero; and so on, till the places, which in the original matrix were occupied by $a_{1,2}, \dots, a_{1,p}$, are all filled by zeros.

Consider now the second row of the modified matrix. Beginning with the $(2, p)$th and $(2, 2p)$th elements, use the specified processes to replace the latter by a zero. Next replace, similarly, the $(2, 2p-1)$th element by a zero; and so on, finally replacing the $(2, p+2)$th element by a zero. The necessary processes will not affect the fact that all the elements in the first row, except the $(1, 1)$th element, are zero. Next reduce the elements occupying the $(2, p)$th, ..., $(2, 3)$th places to zero.

Proceeding thus we eventually have (i) the $(r, s+p)$th element zero, for every $r<p$ and every $s<p$, in which $s \geqq r$, (ii) the (r, s)th element zero, for every $r<p$ and every $s<p$, in which $s>r$. In other words the matrix has a form which may be represented, taking $p=4$, by the matrix ρ,

$$\rho = \left(\begin{array}{cccccccc} a_{11} & 0 & 0 & 0 & 0 & 0 & 0 & 0 \\ a_{21} & a_{22} & 0 & 0 & \beta_{21} & 0 & 0 & 0 \\ a_{31} & a_{32} & a_{33} & 0 & \beta_{31} & \beta_{32} & 0 & 0 \\ a_{41} & a_{42} & a_{43} & a_{44} & \beta_{41} & \beta_{42} & \beta_{43} & 0 \\ a'_{11} & a'_{12} & a'_{13} & a'_{14} & \beta'_{11} & \beta'_{12} & \beta'_{13} & \beta'_{14} \\ \cdot & \cdot & \cdot & \cdot & \cdot & \cdot & \cdot & \cdot \\ \cdot & \cdot & \cdot & \cdot & \cdot & \cdot & \cdot & \cdot \\ a'_{41} & a'_{42} & a'_{43} & a'_{44} & \beta'_{41} & \beta'_{42} & \beta'_{43} & \beta'_{44} \end{array} \right);$$

since now the original matrix is an Abelian matrix, and each of the matrices a_k, b, c, d is an Abelian matrix, it follows (Chap. XVIII., § 324) that $a\bar{\beta} = \beta\bar{a}$; if the original matrix be

* The changes of sign of the other elements of the same column which enter therewith do not concern us.

† The simultaneous subtractions, effected by the matrix m'_p, of the other elements of the column, do not concern us. Similar remarks apply to following cases.

‡ It is not absolutely necessary to use the matrix l_p in this or in the former case; but it conduces to clearness.

for greater generality supposed primarily to be associated with a transformation of order r, the value $r=1$ being introduced later, the determinant of the matrix is $\pm r^p$ (§ 324, Ex. i.) and is not zero; hence comparing in turn the 1st, 2nd, ..., rows of the matrices $a\bar{\beta}$ and $\beta\bar{a}$ we deduce that in the matrix ρ the elements $\beta_{21}, \beta_{31}, \beta_{32}, \ldots$ of the matrix β which are on the left side of the diagonal are also zero; thus, in ρ, every element of the matrix β is zero. Apply now to the matrix ρ the relation

$$a\bar{\beta}' - \beta\bar{a}' = r,$$

which in this case reduces to $a\bar{\beta}'=r$. Then it is immediately found that the elements of the matrix β' which are on the left side of the diagonal are also zero—and also that

$$a_{11}\beta'_{11} = \ldots\ldots = a_{pp}\beta'_{pp} = r.$$

The resulting form of the matrix ρ may then be shortly represented by

$$\sigma = \begin{pmatrix} \triangle & \\ \square & \nabla \end{pmatrix}.$$

If now to the matrix σ we apply the processes given by the matrices $g_{1,2}$ or $g'_{1,2}$ and l_2, we may suppose a_{21} numerically less than a_{22}, and a_{22} positive; if then we apply the processes given by the matrices $g_{1,3}$ or $g'_{1,3}$ and l_3, and the processes given by the matrices $g_{2,3}$ or $g'_{2,3}$ and l_3, we may suppose a_{31}, a_{32} numerically less than a_{33}, and may suppose a_{33} to be positive. Proceeding thus we may eventually suppose all the elements of any row of the matrix a which are to the left of its diagonal to be less than the diagonal elements of that row—and may suppose that all the elements of the diagonal of the matrix a are positive; this involves that the diagonal elements of β' are positive, and in particular when r is a prime number involves that these elements are each 1 or r.

Further we may reduce the elements of the matrix a' which are in the diagonal of a', and those which are to the left of this diagonal, by means of the diagonal elements of the matrix β'. We begin with the elements of the last row of a'; by means of the processes given by the matrices n_p or n'_p we may suppose a'_{pp} to be numerically less than β'_{pp}; by means of the processes given by the matrices $f_{p,p-1}$ or $f'_{p,p-1}$ we may suppose $a'_{p,p-1}$ to be numerically less than $\beta'_{p,p}$; in general by means of the processes given by $f_{p,s}$ or $f'_{p,s}$ we may suppose $a'_{p,s}$ to be numerically less than $\beta'_{p,p}$. Similarly by the processes given by n_{p-1} or n'_{p-1} we may suppose $a'_{p-1,p-1}$ numerically less than $\beta'_{p-1,p-1}$, and by the processes $f_{p-1,s}$ or $f'_{p-1,s}$, where $s<p-1$, we may suppose $a'_{p-1,s}$ numerically less than $\beta'_{p-1,p-1}$. The general result is that in every row of the matrix a' we may suppose the diagonal element, and the elements to the left of the diagonal, to be all numerically less than the diagonal element of the same row of the matrix β'.

413. If then we take the case when $r=1$ we have the result that it is possible to form a product Ω of the $p+2$ matrices a_k, b, c, d, such that the product $\Delta\Omega$ has a form which may be represented, taking $p=3$, by

$$\Delta\Omega = \begin{pmatrix} 1 & 0 & 0 & 0 & 0 & 0 \\ 0 & 1 & 0 & 0 & 0 & 0 \\ 0 & 0 & 1 & 0 & 0 & 0 \\ 0 & a'_{12} & a'_{13} & 1 & \beta'_{12} & \beta'_{13} \\ 0 & 0 & a'_{23} & 0 & 1 & \beta'_{23} \\ 0 & 0 & 0 & 0 & 0 & 1 \end{pmatrix},$$

wherein all the elements of each of the matrices a and β' to the left of the diagonals are zero, and all the elements of the matrix a' both in the diagonal, and to the left of the diagonal, are zero. Applying then the condition $a\bar{\beta}'=1$, we find that the elements of the matrix β' to the right of its diagonal are also zero, so that $\beta'=a=1$. Then finally, applying the condition $a'\bar{\beta}'=\beta'\bar{a}'$, equivalent to $a'=\bar{a}'$, we have $a'=0$. Thus the reduced matrix is the matrix unity of $2p$ rows and columns, *and* Δ, $=\Omega^{-1}$, *is expressed as a product of positive integral powers of the* $p+2$ *matrices* A_k, B, C, D, as desired. Since the determinant of each of the matrices A_k, B, C, D is $+1$, the determinant of the linear matrix Δ is also $+1$.

414. In the particular case $p=1$ the only matrices of the $p+2$ matrices A_k, B, C, D which are not nugatory are the two matrices B and C; we denote these here by U and V and put further

$$u=U^{-1}=\begin{pmatrix}0&1\\-1&0\end{pmatrix},\quad v=V^{-1}=\begin{pmatrix}1&1\\0&1\end{pmatrix},\quad v_1=uvu^3vu^3,\quad w=uvu^3,\quad w_1=u^2vu^3vu^2\,;$$

then we immediately verify the facts denoted by the following table

	u	u^2	u^3	v	v_1	w	w_1
(ξ,η)	$(-\eta,\xi)$	$(-\xi,-\eta)$	$(\eta,-\xi)$	$(\xi,\eta+\xi)$	$(\xi,\eta-\xi)$	$(\xi-\eta,\eta)$	$(\xi+\eta,\eta)$

of which, for example, the first entry means that if $\Delta=\begin{pmatrix}a&\beta\\a'&\beta'\end{pmatrix}$ be any matrix of 2 rows and columns, and we form the product Δu, then the columns ξ, η of the matrix Δ are interchanged, and at the same time the sign of the new first column is changed; we have in fact

$$\begin{pmatrix}a&\beta\\a'&\beta'\end{pmatrix}\begin{pmatrix}0&1\\-1&0\end{pmatrix}=\begin{pmatrix}-\beta&a\\-\beta'&a'\end{pmatrix};$$

hence it is immediately shewn, as in the more general case, that every matrix $\Delta=\begin{pmatrix}a&\beta\\a'&\beta'\end{pmatrix}$, for which the integers a, β, a', β' satisfy the relation $a\beta'-a'\beta=1$, can be expressed as a product of positive integral powers of the two matrices

$$U=\begin{pmatrix}0&-1\\1&0\end{pmatrix},\quad V=\begin{pmatrix}1&-1\\0&1\end{pmatrix}.$$

415. Combining the final result for the decomposition of a linear Abelian matrix with the results obtained for any Abelian matrix of order r we arrive at the following statement, whereof the parts other than the one which has been formally proved may be deduced from that one, or established independently: let $\Delta=\begin{pmatrix}a&\beta\\a'&\beta'\end{pmatrix}$ be any Abelian matrix of order r; then it is possible to find a linear matrix Ω expressible as a product of positive integral powers of the $(p+2)$ matrices A_k, B, C, D, which will enable us to write $\Delta=\Delta_i\Omega$, where Δ_i is an Abelian matrix of order r having any one, arbitrarily chosen, of the four forms representable by

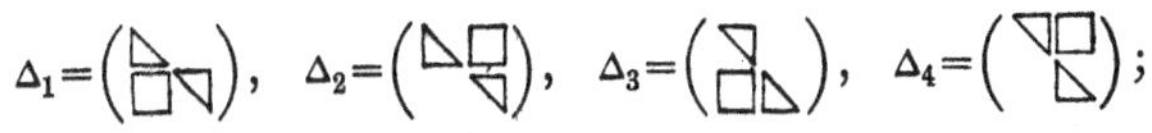

$$\Delta_1=\begin{pmatrix}\text{◺}&\\\square&\text{◹}\end{pmatrix},\quad \Delta_2=\begin{pmatrix}\text{◺}&\square\\&\text{◹}\end{pmatrix},\quad \Delta_3=\begin{pmatrix}\text{◹}&\\\square&\text{◺}\end{pmatrix},\quad \Delta_4=\begin{pmatrix}\text{◹}&\square\\&\text{◺}\end{pmatrix};$$

and it is also possible to choose the linear matrix Ω to put Δ into the form $\Delta=\Omega\Delta_i$, where Δ_i is also any one, arbitrarily chosen, of these same four forms. It follows that the determinant of the matrix Δ is $+r^p$. In virtue of the equations $a_{ii}\beta'_{ii}=r\ (i=1, \ldots, p)$, which hold for any one of the matrices Δ_1, Δ_2, Δ_3, Δ_4, and the inequalities which may also be supposed to hold among the other elements, as exemplified, § 412, for the case of Δ_1, it is easy to find the number of different existing reduced matrices of any one of these forms. For instance when $p=2$, the number when r is a prime number is $1+r+r^2+r^3$; for $p=3$, and r

a prime number, it is $1+r+r^2+2r^3+r^4+r^5+r^6$; for details the reader may consult Hermite, *Compt. Rendus*, t. XL. (1855), p. 253, Wiltheiss, *Crelle*, XCVI. (1884), pp. 21, 22, and the book of Krause, *Die Transformation der Hyperelliptischen Functionen* (Leipzig, 1886), which deal with the case $p=2$; for the case $p=3$, see Weber, *Annali di Mat.* Ser. 2ª, t. IX. (1878), p. 139, where also the reduction to the form $\Delta=\Omega\begin{pmatrix} r & 0 \\ 0 & 1 \end{pmatrix}\Omega'$, in which Ω, Ω' are linear matrices, is considered. Cf. also Gauss, *Disq. Arith.*, § 213; Eisenstein, *Crelle*, XXVIII. (1844), p. 327; Hermite, *Crelle*, XL., p. 264, XLI. (1851), p. 192; Smith, *Phil. Trans.* CLI. (1861), *Arts.* 13, 14.

416. Considering (cf. § 372) any reduction, of the form

$$\begin{pmatrix} \alpha & \beta \\ \alpha' & \beta' \end{pmatrix} = \begin{pmatrix} \rho & \sigma \\ \rho' & \sigma' \end{pmatrix} \begin{pmatrix} A & B \\ 0 & B' \end{pmatrix}, \text{ or say } \Delta=\Omega\Delta_0,$$

where $\begin{pmatrix} \rho & \sigma \\ \rho' & \sigma' \end{pmatrix}$ is a linear matrix, we prove that however this reduction be effected, (i) the determinant of the matrix B' is the same, save for sign, (ii) if μ be a row of p positive integers each less than r (including zero), the rows determined by the condition, $\frac{1}{r}B'\mu=$integral, are the same. For any other reduction of this kind, say $\Delta=\Omega'\Delta'_0$, must be such that

$$\Omega'=\begin{pmatrix} \rho & \sigma \\ \rho' & \sigma' \end{pmatrix} \begin{pmatrix} \bar{q}' & -\bar{q} \\ -\bar{p}' & \bar{p} \end{pmatrix}, \quad \Delta'_0=\begin{pmatrix} p & q \\ p' & q' \end{pmatrix} \begin{pmatrix} A & B \\ 0 & B' \end{pmatrix},$$

where $\begin{pmatrix} p & q \\ p' & q' \end{pmatrix}$ is a linear matrix; the condition that the matrix α' of the matrix Δ'_0 should vanish, namely $p'A=0$, requires (since $|A|\,|B'|=r^p$ and therefore $|A|$, the determinant of A, is not zero) that $p'=0$; thus the reduction $\Delta=\Omega'\Delta'_0$ can be written

$$\begin{pmatrix} \alpha & \beta \\ \alpha' & \beta' \end{pmatrix} = \begin{pmatrix} \rho\bar{q}', & -\rho\bar{q}+\sigma\bar{p} \\ \rho'\bar{q}', & -\rho'\bar{q}+\sigma'\bar{p} \end{pmatrix} \cdot \begin{pmatrix} pA, & pB+qB' \\ 0, & q'B' \end{pmatrix}.$$

Now $pq'=1$; therefore $|q'|=\pm 1$; thus $|q'B'|=\pm|B'|$, which proves the first result. Also, if μ be a row of integers such that $\frac{1}{r}B'\mu$ is a row of integers, $=m$ say, then $\frac{1}{r}q'B'\mu$, $=q'm$, is also a row of integers; while if $\frac{1}{r}q'B'\mu$ be a row of integers, $=n$ say, then $\frac{1}{r}\bar{p}q'B'\mu$, which is equal to $\frac{1}{r}B'\mu$, is equal to $\bar{p}n$, and is also a row of integers; since $q'B'$ is the matrix which, for the reduction $\Delta=\Omega'\Delta'_0$, occupies the same place as that occupied, for the reduction $\Delta=\Omega\Delta_0$, by the matrix B', the second result is also proved.

417. Considering any rectangular matrix whose constituents are integers, if all the determinants of $(l+1)$ rows and columns formed from this matrix are zero, but not all determinants of l rows and columns, the matrix is said to be of rank l. The following theorem is often of use, and is referred to § 397, Chap. XXII.; *In order that a system of simultaneous not-homogeneous linear equations, with integer coefficients, should be capable of being satisfied by integer values of the variables, it is necessary and sufficient that the rank l of, and the greatest common divisor of all determinants of order l which can be formed from, the matrix of the coefficients of the variables in these equations, should be unaltered when to this matrix is added the column formed by the constant terms in these equations.* For the proof the reader may be referred to H. J. S. Smith, *Phil. Trans.* CLI. (1861), *Art.* 11, and to Frobenius, *Crelle*, LXXXVI. (1879), pp. 171—2.

418. Consider a matrix of $n+1$ columns and $n+1$ or more rows, whose constituents are integers, of which the general row is denoted by

$$a_i\, b_i \ldots\ldots k_i,\, l_i,\, e_i;$$

let Δ be the greatest common divisor of the determinants formed from this matrix with $n+1$ rows and columns; let Δ' be the greatest common divisor of the determinants formed from this matrix with n rows and columns; then, since every determinant of the $(n+1)$th order may be written as a linear aggregate of determinants of the n-th order, the quotient Δ/Δ' is integral, $=M$, say. Then *the* $n+1$ *or more simultaneous linear congruences*

$$U_i = a_i x + b_i y + \ldots\ldots + k_i z + l_i t + e_i u \equiv 0 \pmod{M}$$

have just Δ *incongruent sets of solutions, and have a solution whose constituents have unity as their highest common divisor.* Frobenius, *Crelle*, LXXXVI. (1879), p. 193.

Also, *if in the m linear forms* ($m <=$ or $> n+1$)

$$U_i = a_i x + b_i y + \ldots\ldots + k_i z + l_i t + e_i u, \qquad (i=1, \ldots, m),$$

the greatest common divisor of the $m(n+1)$ *coefficients be unity, it is possible to determine integer values of* $x, y, \ldots, t, u$, *such that the m forms have unity as their greatest common divisor;* in particular, when $n=1$, *if the* $2m$ *numbers* a_i, b_i *have unity as their greatest common divisor, and the* $\frac{1}{2}m(m-1)$ *determinants* $a_i b_j - a_j b_i$ *be not all zero, it is possible to find an integer x so that the m forms* $a_i x + b_i$ *have unity as their greatest common divisor.* Frobenius, *loc. cit.*, p. 156.

419. The theorem of § 418 includes the theorem of § 357, p. 589; it also includes the simple result stated § 383, p. 637, note. It also justifies the assumption made in § 397, that the periods Ω, Ω' may be taken so that the simultaneous equations $ax' - a'x = 1$, $bx' - b'x = 0$ can be solved in integers in such a way that the $2p$ elements $rx - b$, $rx' - b'$ have unity as their greatest common divisor; assuming that r is not zero so that the $p(2p-1)$ determinants $a_i b_j - a_j b_i$, $a_i b_j' - a_j' b_i$, $a_i' b_j' - a_j' b_i'$ are not all zero, and that Ω' has been taken so that the $2p$ integers $a_1, \ldots, a_p, a_1', \ldots, a_p'$ have no common divisor other than unity, the necessary and sufficient condition for the solution of the equations $ax' - a'x = 1$, $bx' - b'x = 0$ is (§ 417) that the greatest common divisor, say M, of the $p(2p-1)$ binary determinants spoken of should divide each of the $2p$ integers $b_1, \ldots, b_p'$; if this condition is not already satisfied we may proceed as follows: find two coprime integers (§ 418) which satisfy the $2p$ congruences

$$\lambda b_i' + \mu a_i' \equiv 0, \quad \lambda b_i + \mu a_i \equiv 0 \pmod{M}, \qquad (i=1, \ldots, p),$$

and thence two integers ρ, σ such that $\lambda\sigma - \mu\rho = 1$; put $\Omega_1' = \lambda\Omega' + \mu\Omega$, $\Omega_1 = \rho\Omega' + \sigma\Omega$, $B_i = b_i\lambda + a_i\mu$, $A_i = b_i\rho + a_i\sigma$, $B_i' = b_i'\lambda + a_i'\mu$, $A_i' = b_i'\rho + a_i'\sigma$; then

$$b_i\Omega - a_i\Omega' = B_i\Omega_1 - A_i\Omega_1', \quad b_i'\Omega - a_i'\Omega' = B_i'\Omega_1 - A_i'\Omega_1',$$

and the greatest common divisor of the $p(2p-1)$ binary determinants $A_i B_j - A_j B_i$, $A_i B_j' - A_j' B_i$, $A_i' B_j' - A_j' B_i'$, which is equal to M, divides the $2p$ integers $B_1, \ldots, B_p'$; thus M is the greatest common divisor of these $2p$ integers; next put $\Omega_2 = M\Omega_1$, $\Omega_2' = \Omega_1'$, $\mathrm{b}_i = B_i/M$, $\mathrm{b}_i' = B_i'/M$, $\mathrm{a}_i = A_i$, $\mathrm{a}_i' = A_i'$; then the greatest common divisor of the $p(2p-1)$ binary determinants $\mathrm{a}_i\mathrm{b}_j - \mathrm{a}_j\mathrm{b}_i$, etc., is unity, and this is also the greatest common divisor of the $2p$ integers $\mathrm{b}_1, \ldots, \mathrm{b}_p'$. Now let (x, x') be any solution of the equations $\mathrm{a}x' - \mathrm{a}'x = 1$, $\mathrm{b}x' - \mathrm{b}'x = 0$, so that $(rx - \mathrm{b}, rx' - \mathrm{b}')$ is a solution of the equations $\mathrm{a}\xi' - \mathrm{a}'\xi = 0$, $\mathrm{b}\xi' - \mathrm{b}'\xi = 0$; let (ξ, ξ') be an independent solution of these latter equations (Smith, *Phil. Trans.*, CLI. (1861), *Art.* 4) so that the $p(2p-1)$ binary determinants $x_i\xi_j - x_j\xi_i$, etc., are not all zero, so chosen that the $2p$ elements ξ_i, ξ_i' have unity as their highest common divisor; then if h be any integer, the $2p$ elements $x_i + \mathrm{h}\xi_i$, $x_i' + \mathrm{h}\xi_i'$ form a solution of the equations $\mathrm{a}x' - \mathrm{a}'x = 1$, $\mathrm{b}x' - \mathrm{b}'x = 0$; let h be chosen so that the $2p$ elements $rx_i - \mathrm{b}_i + \mathrm{h}r\xi_i$, $rx_i' - \mathrm{b}_i' + \mathrm{h}r\xi_i'$ have no common factor greater than unity (§ 418). Putting $X = x + \mathrm{h}\xi$,

$X' = x' + \mathrm{h}\xi'$, the first column of the matrix in § 397 will consist of the elements of (a, a′), the $(p+1)$th column will consist of the elements of (b, b′), the second column will consist of the elements of $rX - \mathrm{b}$, $rX' - \mathrm{b}'$; and since these latter have unity as their greatest common factor, it is possible to construct the $(p+2)$th and all other columns of this matrix (§ 420).

420. A theorem is assumed in § 396, which has an interest of its own—*If of an Abelian matrix of order r there be given the constituents of the first r columns, and also the constituents of the* $(p+1)$*th, ...,* $(p+r)$*th columns* $(r<p)$, *it is always possible to determine the remaining* $2(p-r)$ *columns.* For a general enunciation the reader may refer to Frobenius, *Crelle*, LXXXIX. (1880), p. 40. We explain the method here by a particular case; suppose that of an Abelian matrix of order r, for $p=3$, there be given the first and $(p+1)$th columns; denote the matrix by

$$\begin{pmatrix} a & x & t & b & y & u \\ a' & x' & t' & b' & y' & u' \end{pmatrix};$$

the elements of the given columns will satisfy the relation $ab' - a'b = r$; it is required to determine in order the second, the fifth, the third and the sixth columns; the relations arising from the equations

$$\bar{a}a' - \bar{a}'a = 0, \quad \bar{\beta}\beta' - \bar{\beta}'\beta = 0, \quad \bar{a}\beta' - \bar{a}'\beta = r$$

so far as they affect these columns respectively, are as follows:

$$\left.\begin{aligned} ax' - a'x &= 0 \\ bx' - b'x &= 0 \end{aligned}\right\} \text{(i)}, \quad \left.\begin{aligned} ay' - a'y &= 0 \\ by' - b'y &= 0 \\ xy' - x'y &= r \end{aligned}\right\} \text{(ii)}, \quad \left.\begin{aligned} at' - a't &= 0 \\ bt' - b't &= 0 \\ xt' - x't &= 0 \\ yt' - y't &= 0 \end{aligned}\right\} \text{(iii)}, \quad \left.\begin{aligned} au' - a'u &= 0 \\ bu' - b'u &= 0 \\ xu' - x'u &= 0 \\ yu' - y'u &= 0 \\ tu' - t'u &= r \end{aligned}\right\} \text{(iv)};$$

now let (x, x') be a solution of equations (i) in which the $2p$ constituents have no common factor other than unity; determine 2 rows of p elements ξ, ξ' such that $x\xi' - x'\xi = 1$, and denote $a\xi' - a'\xi$ by A and $b\xi' - b'\xi$ by B; then it is immediately verified that the values

$$y = r\xi - (Ab - Ba), \quad y' = r\xi' - (Ab' - Ba'),$$

satisfy equations (ii); next let (t, t') be a solution of equations (iii) in which the $2p$ constituents have no common factor other than unity; determine 2 rows of p elements, v, v', such that $tv' - t'v = 1$, and denote $av' - a'v$, $bv' - b'v$, $xv' - x'v$, $yv' - y'v$ respectively by A, B, X, Y; then it is immediately verified that the values

$$u = rv - (Ab - Ba) - (Xy - Yx), \quad u' = rv' - (Ab' - Ba') - (Xy' - Yx')$$

satisfy the equations (iv).

INDEX OF AUTHORS QUOTED. THE NUMBERS REFER TO THE PAGES.

TABLE OF SOME FUNCTIONAL SYMBOLS.

Riemann's normal elementary integrals

of first kind, generally, $v_1^{x,a}, \ldots, v_p^{x,a}$, p. 15. For periods, p. 16,

of second kind, $\Gamma_z^{x,a}$; periods of, $\Omega_1, \ldots, \Omega_p$, or $\Omega_1(z), \ldots, \Omega_p(z)$, pp. 15, 21,

of third kind, $\Pi_{z,c}^{x,a}$, p. 15.

Integral, rational, functions, g_i, or $g_i(x, y)$, or $g_i(y, x)$, pp. 55, 61.

ϕ-polynomials, special functions, numerators of differential coefficients of integrals of the first kind, $\phi_1, \ldots, \phi_{n-1}$, p. 61. Also $\phi_1, \ldots, \phi_p$, p. 146.

(x, ξ), $=\dfrac{\phi_0(x, y) + \phi_1(x, y)\, g_1(\xi, \eta) + \ldots + \phi_{n-1}(x, y)\, g_{n-1}(\xi, \eta)}{(x-\xi)\, f'(y)}$, p. 68.

Elementary integral of third kind, $P_{z,c}^{x,a}$, p. 68. (Canonical integral), $Q_{z,c}^{x,a}$, p. 185. (Canonical integral), $R_{z,c}^{x,a}$, p. 194.

Integrals of second kind, associated with given system of integrals of first kind, $L_i^{x,a}$, p. 193; periods of, 196. Also $H_z^{x,a}$, p. 182, and $F_z^{x,a}$, p. 291, are used for integrals of second kind.

$\psi(x, a;\ z, c_1, \ldots, c_p)$, pp. 77, 171, 177. This is called Weierstrass's fundamental rational function.

$\psi(x, a;\ z, c)$, pp. 174, 175, 178, 200.

$\bar{E}(x, z)$, pp. 171, 178 (Prime function).

$E(x, z)$, pp. 176, 178, 205 (Prime function).

Matrices, see Appendix II., p. 666.

$\Theta(u, \tau;\ Q, Q')$ or $\Theta\left(u, \tau \,\middle|\, \begin{matrix} Q' \\ Q \end{matrix}\right)$ or $\Theta\left(u \,\middle|\, \begin{matrix} Q' \\ Q \end{matrix}\right)$ or $\Theta(u;\ Q, Q')$

$$=\Sigma e^{2\pi iu(n+Q')+i\pi\tau(n+Q')^2+2\pi iQ(n+Q')}, \text{ p. } 248.$$

$\vartheta(u;\ Q, Q')$ or $\vartheta\left(u \,\middle|\, \begin{matrix} Q' \\ Q \end{matrix}\right) = \Sigma e^{au^2+2hu(n+Q')+b(n+Q')^2+2\pi iQ(n+Q')}$, p. 283.

$\zeta_i(u) = \dfrac{\partial}{\partial u_i} \log \vartheta(u)$, p. 287.

$\wp_{i,j}(u) = -\dfrac{\partial^2}{\partial u_i \partial u_j} \log \vartheta(u)$, p. 292. See also p. 516.

$\omega_i(x)$ (Differential coefficient of integral of first kind), p. 169. Also $\mu_i(x)$, p. 192.

$\nu_{i,j}$, p. 192. $\bar{\nu}_{i,j}$, p. 288.

$W(x, z;\ c_1, \ldots, c_p)$, p. 174.

$\varpi(\zeta, \gamma)$, p. 360 (Prime function). But for $\varpi(x, z)$, see pp. 430, 428.

$\lambda(\zeta, \mu)$, p. 367.

$|Q|$, $|Q, R|$, $\begin{pmatrix} Q \\ R \end{pmatrix}$, p. 487.

$\Phi(u, a;\ A)$, p. 509.

$\phi(u)$, a Jacobian function, p. 579, ff.

$\psi_r(w;\ K, K'+\mu)$, $\Psi_r(W;\ K, K'+\mu)$, p. 601.

SUBJECT INDEX TO THE PAGES OF THIS VOLUME.